The Science and Technology of Unconventional Oils

The Science and Technology of Unconventional Oils

Finding Refining Opportunities

Maria Magdalena Ramirez-Corredores

ELSEVIER

ACADEMIC PRESS
An imprint of Elsevier

Academic Press is an imprint of Elsevier
125 London Wall, London EC2Y 5AS, United Kingdom
525 B Street, Suite 1800, San Diego, CA 92101-4495, United States
50 Hampshire Street, 5th Floor, Cambridge, MA 02139, United States
The Boulevard, Langford Lane, Kidlington, Oxford OX5 1GB, United Kingdom

Library of Congress Cataloging-in-Publication Data
A catalog record for this book is available from the Library of Congress

British Library Cataloguing-in-Publication Data
A catalogue record for this book is available from the British Library

ISBN: 978-0-12-801225-3

For information on all Academic Press publications visit our website at https://www.elsevier.com/books-and-journals

Publisher: John Fedor
Acquisition Editor: Kostas KI Marinakis
Editorial Project Manager: Sarah Jane Watson
Production Project Manager: Maria Bernard
Cover Designer: Alan Studholme

Typeset by TNQ Books and Journals

To Adrian and Miguel, the newest members of my family and who keep my life enlightened.

Contents

* Chapter contributed by Dr. Lucia M. Petkovic

Biography

Dr. Magdalena Ramirez-Corredores is a distinguished scientific researcher at Idaho National Laboratory in the Chemistry and Radiation Measurements Department, working on biomass science and process development. She gained her previous industrial research experience at KiOR LLC, BP International Ltd., and PDVSA Intevep. She is a fellow researcher with a successful record of producing innovative solutions and breakthrough technologies in hydrocarbon and alternative feeds processing and in catalysis: biomass conversion to fuels and chemicals; unconventional oils processing; fossil and bio-oil hydrotreatment; light ends; and pretreatment, separation, and purification processes. She has 60 inventions described in 49 US patents (11 cases still pending in the United States) and more than 180 worldwide awarded patents. She has authored 115 publications (one book, seven monographs, 62 articles, and 45 conference papers). Her total number of citations (excluding all authors' self-citations) is above 1300, for an *h*-index of 19. She earned her doctorate in chemistry at England's University of Bath in 1980 and her B.Sc. in Chemistry at Universidad Central de Venezuela, Caracas, in 1974. She is also a recipient of the highest industrial research awards from her country of origin, Venezuela: the Exceptional Contributions Award, PDVSA 2001, and National Award on Technological Research, CONICIT 1991.

Preface

The declining offer of conventional oils, together with the increasing availability of unconventional oils has created new needs and problems for the refining industry. Additionally, unconventional oils might represent the major component of the future market, since the largest existing reserves are of this type of crude oil. Recent records of oil price volatility, of both high and low prices have been pushing a renewed interest in these crude oils. Initially, the problems were associated with the adjustment required to meet product quality while the needs were related to adaptation to this market diversification. The refiners were prompted to search and develop technologies for the processing of these newly introduced unconventional oils.

The first unconventional oils to flood the market were the heavy and extraheavy oils, and bitumen. The poor quality of these oil grades moved the producer to sell them under discounted prices, offering the refiner the opportunity of challenging their capabilities, trying to gain a price differential benefit. Thus they became known as *opportunity crudes*. Two different broad classes of opportunity crudes can be distinguished: the highly viscous and contaminated crude oils (heavy, extraheavy, and bitumen) and the highly acid oils of more recent incorporation in the market plate. Opportunity crudes have the potential to provide significant economic benefits for refiners, if they can overcome the many technical difficulties for their processing. Only those refiners with technology to support flexibility will enjoy a competitive advantage. The so-called *bottom of the barrel* (BotB) refers to the heavier fraction of the oil that contains molecules whose conversion or removal is mandatory if a smooth refining operation is desired. The BotB fraction represents a large proportion in heavy and extraheavy oils, and bitumen.

This book, *The Science and Technology of Unconventional Oils: Finding Refining Opportunities* intends to report the collective physical and chemical knowledge of unconventional oils (heavy, extraheavy, sour/acid, and shale oil) and the issues associated with their refining for the production of transportation fuels. It will focus on the discussion of the scientific results and technology activities of the refining of unconventional oils. The presence of reactive and refractory compounds and components that negatively impact refining processing (the "bad actors") are discussed and analyzed. The commercially available technologies, with their reported improvements and emerging ideas, concepts, and technologies, are described. This comprehensive overview constitutes the basis for establishing technology gaps, and in return sets the science and technology needs to be addressed in the future. In summary, this book incorporates the relevant knowledge of processing unconventional crude oils and the BotB fraction, describing the related commercially available and emerging technologies to contribute to the identification of existing gaps.

The presence of compounds, whose behavior harmfully affects the refining processes, gives rise to the main challenge that the refiner has to overcome. Asphaltenes, metal-bearing compounds, and naphthenic acids constitute those bad actors that jeopardize the smooth operation of the refinery. Their physical properties and reactivity have severe consequences not only in refining, but also in production and transportation. A lack of comprehensive knowledge of these molecules still subsists and may be the reason for the inexistence of cost-effective technologies.

In recent times, the oil industry has journeyed through scenarios of high price volatility, characterized by increasing supplies from emerging markets, and emerging crude qualities, including the shale oil boom in the United States. Although these crudes can be purchased at discounted prices, refiners are struggling to keep margins in good shape. The increase in the proportion of unconventional oils into the refinery diet, together with increasingly stringent environmental requirements, makes the fuel manufacturing process one of enormous complexity.

Hence the opportunity for cost-effective technologies for abating the effect of these compounds remains open, even after more than five decades of their incorporation in the market. Most likely, the inclusion of new units dedicated to change the properties of the new blend diet into the specifications of the currently operating units might have to be considered. More drastic solutions may call for reinventing the oil refinery, based on an "out-of-the-box" brand-new way of refining.

The book consists of eight chapters. Chapter 1 contains an introductory motivation for providing a status of the collective knowledge of the compounds that make refining (and production) of unconventional oils more difficult and costly; together with a general overview of the impact these compounds may have on the refinery processes. Then, Chapters 2–4 individually present the information associated with asphaltenes, metal compounds, and acidity. Next, in Chapters 5 and 6, respectively, commercially available conventional technologies and emerging ideas and technologies are described. In Chapter 7, shale oils are briefly considered. The conclusions, future trends, and final remarks are collected in Chapter 8. At this point, the author would like to thank Dr. Lucia Petkovic from Idaho National Laboratory for her valuable contribution to the writing of the chapter on shale oil (Chapter 7). This chapter complements the others dealing with opportunity crude oils and allows the book to cover the whole spectrum of unconventional oils.

Fundamental knowledge of problematic compounds arises from their separation, identification, and molecular characterization. In recent years, significant advances have been achieved: on the one side, the economic incentives driven by the oil industry, and on the other, the remarkable improvements made in analytical techniques. Nonetheless, further advances are still needed, specifically in structure–reactivity relationships.

A broad range of process areas have been examined for abatement of these compounds, including physical and chemical treatments, as well as improvement to and integration/combination of existing processes. With the exception of asphaltenes removal (deasphalting), the lack of commercial application of standalone processes for the abatement of the other compounds indicates that a suitable cost/benefit balance has not been achieved yet.

Although the R&D history is long, it seems that opportunities still exist for developing new processes and technologies, which need to be more economic, effective, and efficient than the current mitigation strategies in place by the use of conventional technologies. It seems then that unconventional situations are calling for unconventional solutions.

Maria Magdalena Ramirez-Corredores
June 2016

CHAPTER 1

Unconventional Oils

1. Crude Oil

Petroleum is a naturally occurring mixture of hydrocarbons, from which transportation fuels and a great variety of chemicals are derived by different physical and chemical processes. Most of the imaginable organic compounds form part of the petroleum matrix, namely, alkanes (paraffins), alkenes (olefins), alkynes, cyclic hydrocarbons, aromatics, etc.

1.1 Composition

Hydrocarbon compounds include saturated acyclic alkanes (normal and branched paraffins), cycloalkanes (naphthenes), olefins, and aromatic hydrocarbons. Naphthene molecules are formed by saturated rings and side chains. According to the number of naphthenic rings, they might be monocyclic, bicyclic, tricyclic, etc.; likewise, aromatics also present normal or branched chains and naphthenic rings. They are categorized by the number of aromatic rings in the molecule, i.e., monoaromatic, diaromatic, triaromatic, etc., regardless of any other substituent.

Besides carbon and hydrogen, other elements are also present in the molecular formula of part of the oil components: the heteroatomic compounds. Nonmetallic elements such as sulfur, nitrogen, and oxygen might be present in the low percentage range, while metals such as vanadium, nickel, and iron (when present) can be found in ppm levels. The presence of all these heteroatoms in the crude oil is considered highly inconvenient, either for environmental reasons or for their contaminating and poisoning effects during refining processes. For on-road transportation fuels, S levels have to be kept below 50 ppm in most countries since 2006. Among the oxygen compounds, naphthenic acids (NAs) are regarded as one of the most troublesome compounds. In summary, the ranges of elemental composition of crude oil are presented in Table 1.1 and examples of representative compounds in Figs. 1.1 and 1.2.

Thus, crude oil refining separates this mixture of compounds into the products required by the market. Obviously, quality adjustments have to be included to satisfy performance specifications and governmental regulations. The higher and broader the contamination in the crude oil, the higher the operational complexity of the involved refining processes.

The Science and Technology of Unconventional Oils. http://dx.doi.org/10.1016/B978-0-12-801225-3.00001-2

Table 1.1: Elemental Composition of Crude Oil

Element	Composition Range (wt%)
Carbon	83–87
Hydrogen	10–14
Sulfur	0.05–8.0
Nitrogen	0.05–3.0
Oxygen	0.05–1.5
Metals	0–0.2

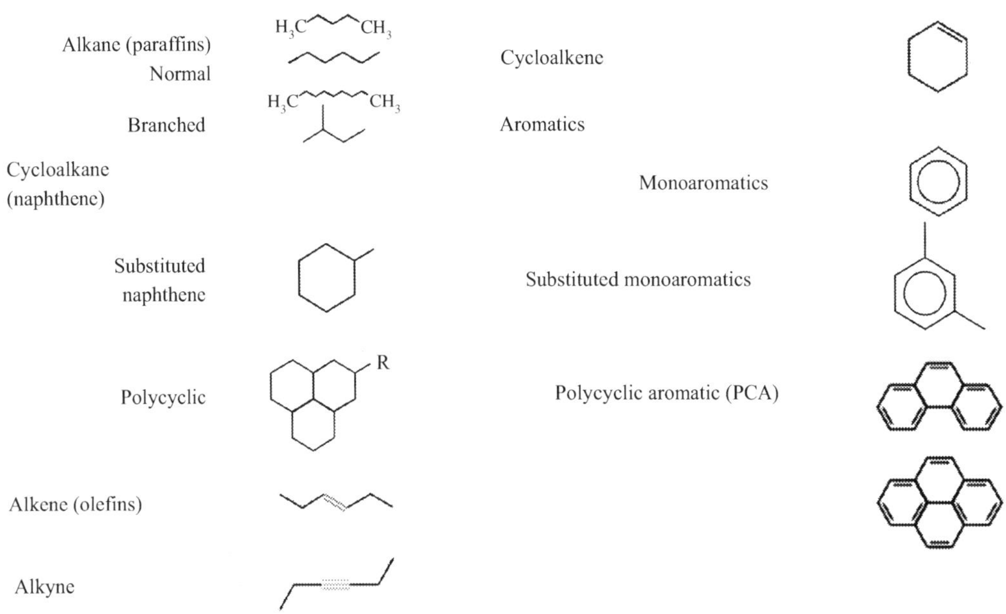

Figure 1.1
Representative hydrocarbons.

Crude oil can be contaminated either with naturally occurring materials or by the addition of other compounds (nonnaturally occurring materials added during production or transportation). The latter additives will not be considered here; however, it is convenient at this point to mention the concentration ranges of those naturally occurring contaminants, as shown in Table 1.2 (adapted from Ref. [1]). The removal of the contaminating (water-soluble) salts, water, and sediments is the first step prior to any other refining process. This unit is known as the desalter. In simple terms, the process consists of a water wash, in which the aqueous phase is separated by the action of an electric field. The process might also involve solids filtration if needed and the use of additives for scavenging H_2S, chemical demulsifiers, dispersants, and metal removal improvers; a basic solution is sometimes added to favor NAs removal.

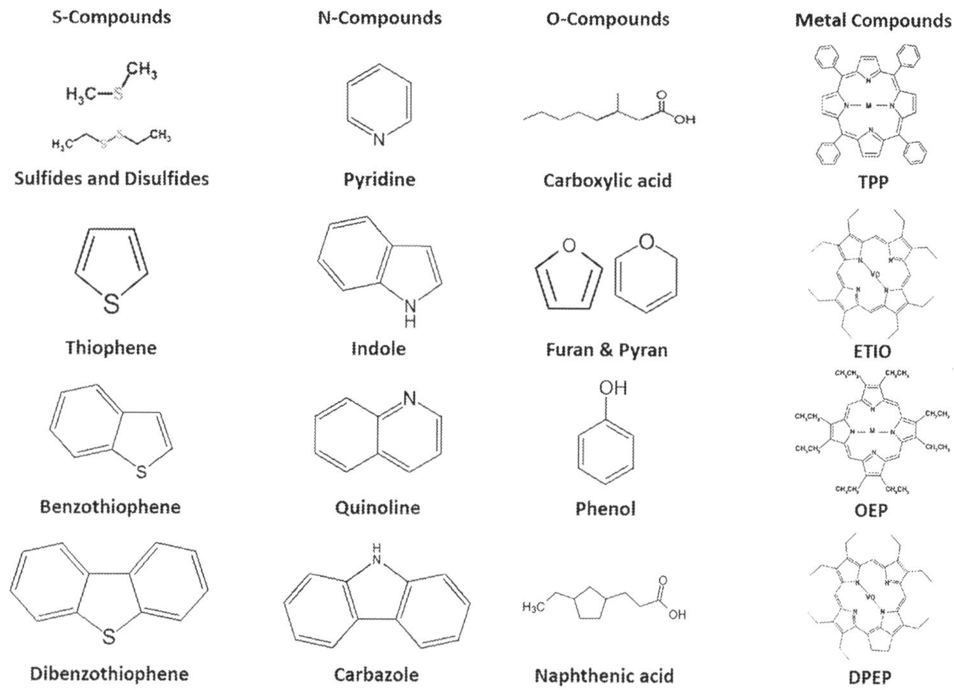

Figure 1.2
Representative heteroatomic compounds present in crude oil.

Table 1.2: Concentration Ranges of Naturally Occurring Impurities

Impurity	General Range
Salts	10–1000 ptb
Water	0.1–2 vol%
Sediment	1–500 ptb
S-compounds	0.1–5 wt% as S
Organometallic compounds	Ni, V, Fe, As: 5–400 ppm as Me
Naphthenic acids	0.03–0.4 vol%
N-compounds	0.05–15 vol%
Nonacidic O-compounds (resins, cresols, etc.)	0–2 wt%

The comprehensive study of heavy oil composition and characterization carried out by Boduszynski et al. [2–9] in the different boiling point fractions, from the lightest compounds down to the "bottom of the barrel" (BotB), led them to conclude a *gradual and continuous increase* of aromaticity, molecular weight (MW), and heteroatom content with increasing boiling point. Thus Boduszynski's model of crude oil is also known as the "***continuous model***". Therefore, the BotB fraction is composed by heavy macro-molecules, which cannot be separated by distillation, e.g., resins and asphaltenes. Although these compounds are not a boiling point type of fraction, the continuity of the continuous model has been extrapolated to extend even to this nondistillable residue [3].

In Boduszynski's approach two characterization views were considered: the bottom up and the top down. The former is based on "average molecular parameters." The data illustrate the changes in relative concentrations of the "molecular building blocks" with increasing atmospheric equivalent boiling point and decreasing solubility of heavy crude components. The latter uses high-pressure liquid chromatography for further separation of the heavy ends by compound classes prior to characterization [7].

Another feature that characterizes the heavier crude oils is a higher concentration of heteroatomic compounds (S and metals were exemplified in Table 1.2). The concentration of heteroatomic compounds increases with increasing boiling point of the fraction where they are present. Thus one may say that they are concentrated in the BotB. For this reason, subsequent to distillation, other processes, typically referred as downstream processes, purify these streams, further fractionate, and finish/polish them to provide the required performance specifications and to meet environmental and governmental regulations of the final products. Environmental regulations began to strengthen in the 1970s and became harsher over the years. These regulations affect refinery emissions, but the pressure has been strongest for clean transportation fuels. See, for instance, how the European specifications of gasoline and diesel have changed through time (Table 1.3).

Table 1.3: Evolution of European Specifications for Gasoline and Diesel

	Euro I	Euro II	Euro III	Euro IV	Euro V
	1993	1995–96	2000	2005	2009
Gasoline (95/85)					
Sulfur, ppm (max)	1000	500	150	50/10	10
Benzene, wt% (max)	5			1	
Aromatics, wt% (max)	—	—	42	35	
Olefins, wt% (max)	—	—		18	
Oxygen, wt% (max)	2.5			2.7	
Reid vapor pressure (RVP), kPa (max)	<80			<60	
Final boiling point (FBP), °C (max)	215			210	
Diesel					
Cetane number (min)	49			51	
Cetane index (min)			46		
Sulfur, ppm (max)	2000	500	350	50/10	10
Density, g/cm^3 (max)	0.860			0.845	
Distillation T-95, °C	370		360	340	

While complexity of the molecular components increases, their reactivity decreases. The increase in density is associated with the degree of molecular condensation of the constituting compounds. The feature that characterizes the changes in the degree of condensation is the C/H ratio: the higher the degree of condensation, the higher the C/H ratio [and as mentioned, the higher the density, the lower the American Petroleum Institute (API) gravity]. As shown in Fig. 1.3 and in Table 1.3, the number of carbon atoms (carbon number) in the molecules composing each fraction increases as the boiling point increases, and so do the MW and the complexity of the molecular configurations.

To exemplify this complexity increase, Boduszynski [7] summarized the atmospheric equivalent boiling point of the corresponding *n*-alkane and the number of isomers as a function of the increasing carbon number in the molecule (Table 1.4).

The immense number of compounds and their isomers makes the analysis of crude oil and the interpretation of the data an extremely tedious and difficult activity. In fact, the heavier the cut, the more difficult the identification task for individual compounds. Separation techniques and methodologies have been developed to characterize the "heavy ends" of crude oil into families or types of compounds. One of these separation methods, based on fractionation and chromatography, is the so-called SARA, which stands for the separated compound class, i.e., S for "saturates," A for "aromatics," R for "resins," and A

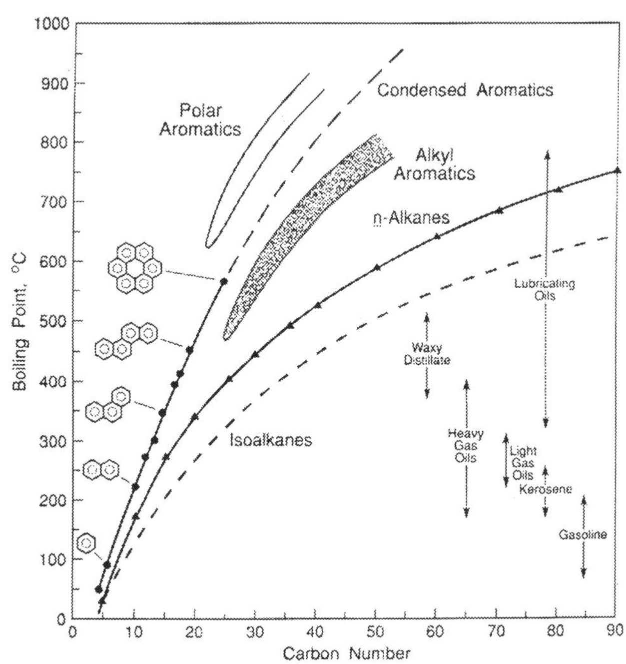

Figure 1.3
Variation of boiling point with increasing molecular weight.

Table 1.4: Molecular Complexity Example

Carbon Number	AEBP (°C)	Number of Isomers
5	36	3
8	125	18
10	173	75
12	216	355
15	270	4347
20	343	36.6×10^{4}
25	402	36.7×10^{6}
30	450	41.1×10^{8}
35	490	49.3×10^{10}
40	522	62.4×10^{12}
45	550	82.2×10^{14}
60	560	221.5×10^{20}
80	672	1056.4×10^{28}
100	708	5920×10^{36}

AEBP, atmospheric equivalent boiling point.

for "asphaltenes" (the reader is directed to Ref. [10] for more details on the SARA method). Resins and asphaltenes are the heavier compounds present in the BotB, and will be revisited in following chapters. Partial and sequential separation stages of the complex mixture are intended to obtain simpler fractions, which will not reveal their composition but instead the groups of isolate components of the more complex matrix. Those fractions are more amenable for pursuing their molecular characterization. A great number of preparative techniques have been used for the staged separation, and an example is given in Fig. 1.4, reported in Ref. [7]. Further characterization of each of these compound groups might be undertaken by GC, GCxGC, GC-MS, GC-FIMS, etc.

As can be derived from this discussion, crude oil might be composed of more than 50,000 molecular entities, of which 20,000 contain heteroatoms (N, O, S, metals) with distinct elemental compositions (C, H, N, O, S, Me). Previously, one could think of the titanic task of a full molecular characterization of crude oil as an utopic wish. However, Prof. Alan Marshall from Florida State University developed a methodology, tools, and protocols to attain results in a very time- and cost-effective way. Petroleomics, derived from genomics and coined by Marshall [11], is used to describe the study of all components of crude oil and their effect on oil properties and reactivity. Ultrahigh-resolution Fourier transform ion cyclotron resonance mass spectrometry (FT-ICR MS) was the selected technique. This technique is now widely employed and became directly associated with petroleomics. The characterization of vacuum gas oil, vacuum resid (VR), extracted asphaltenes, and polycyclic aromatic hydrocarbons is the subject of several papers from Prof. Marshall's group and will be considered in the following chapters. Some of them correspond to the PhD dissertation of G.C. Klein [12]. One of the first confirmed findings is the lack of reliability of the traditional

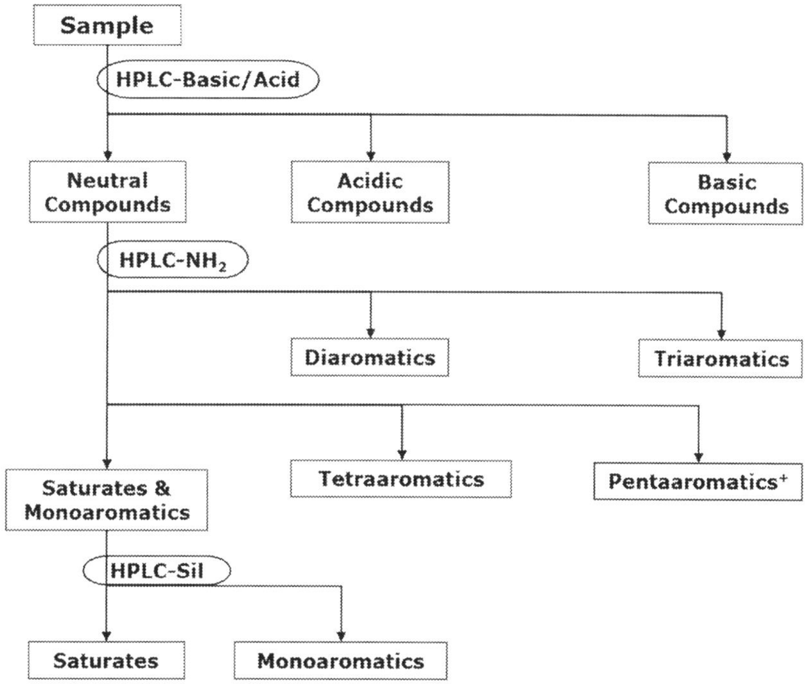

Figure 1.4
Preparative high-pressure liquid chromatography (HPLC) separation scheme.

acidity measurement known as total acid number (TAN). Further details will also be given in the corresponding chapter. Some of the work is also the result of collaboration with the oil industrial sector and provides information on the nature of less reactive compounds [13], the characterization of asphaltenes with regard to their agglomeration, coking, and deposition tendencies [14], and on the features associated with fouling [15]. The Department of Chemistry of Warwick University (UK) under Prof. Professor P.J. Derrick has similar equipment, and the area of petroleomics is coordinated by Dr. Mark Barrow. They have worked on the characterization of NAs and their corrosion abilities [16].

More examples and specific details will be given in the following chapters. However, the complexity of these mixtures and of their constituting molecules is emphasized once more. As it has been confirmed, new and more advanced techniques are needed for a comprehensive molecular characterization of crude oil and for the understanding of the behavior of its components.

1.2 Classes of Oil and Quality Description

In principle, crude oil quality is defined by its composition. However, since price and quality are associated, easy-to-measure physical properties are typically used to define quality. Thus boiling point, viscosity, and density [specific gravity (SG) relative to water is most commonly used in the oil industry] are considered as markers for oil quality (Fig. 1.3). The values of these properties are known to increase with increasing MW.

API gravity (in degrees API) is one of the units used as a measure of SG.[1] The average MW and the mean boiling point reflect the increasing proportion of heavier compounds with the decrease in API gravity. During refining, the mixture of compounds is separated, in the first steps by boiling point, through distillation (the residue obtained at atmospheric pressure is then subjected to vacuum distillation to deepen the first fractionation). In Table 1.5 the boiling point range of the different streams produced by distillation is summarized. In the same way that compounds with low boiling points (low MW) are called "light ends," those with high boiling points (high MW) are called "heavy ends." The heavy ends are also known to constitute the BotB and a detailed description of their nature, behavior, and processing represents the core of this book.

Although as a typically accepted generality MW is associated with boiling point, one might keep in mind that the boiling point also depends on the interaction forces that keep the molecules together, and this in turn is determined by the composition and the molecular structure. Thus, for instance, lighter polar compounds would have higher boiling points than some of the heavier nonpolar compounds. A good comparison was provided for the incorrectness of using the terms "heavy" and "high boiling" as synonyms for "high MW" (see Fig. 2 reported in Ref. [8]).

Crude oil might be classified as extralight, light, median, heavy, and extraheavy depending on the SG range and/or boiling point profile. The ranges of values of the API gravity for

Table 1.5: Boiling Point Ranges of Crude Oil Streams Produced by Distillation

Stream	Boiling Point Range (°C)	Carbon Number Range
Light ends (refinery gas and liquefied petroleum gas)	<40	C1–C4
Naphtha, gasoline	40–215	C5–C17
Light gas oil (LGO), kerosene, and diesel	215–342	C8–C18
Heavy gas oil (HGO) and fuel oil (FO)	342–370	C12–> C20
Light vacuum gas oil (LVGO), lube oil, and waxes	370–509	>C20
Heavy vacuum gas oil (HVGO)	509–550	>C35
Vacuum resid (VR), asphalt, and coke	>550	>C50

[1] API gravity = (141.5 ÷ SG) − 131.5.

Table 1.6: Selected Properties of Different Types of Crude Oils

Property	Extralight	Light	Median	Heavy	Extraheavy
API gravity (°)	>39	31.1–39	22.3–31.1	10.0–22.3	>10
H/C ratio	2.2	1.9	1.8	1.5	0.8
Density (g/cc)	<0.83	0.87–0.83	0.92–0.87	1.0–0.92	>1.0
Viscosity (cSt)	<2	2–10	10–100	100–5000	>5000
Mean boiling point (MBP, °C)	<350	250–450	300–500	400–600	>500
Molecular weight	<200	200–275	275–400	400–600	>600
Sulfur (wt%)	<0.1	0.1–1	0.5–1	1–3	>3
Metals (ppm)	ND	<10	10–100	100–300	>300
Conradson carbon	<1	1–4	2–12	10–16	>12

ND, no data.

each type of crude are >40 for extralight, 30–40 for light, 20–30 for median, 10–20 for heavy, and >10 for extraheavy. To emphasize the differences among these ranges of density-based grades, a selected set of properties of crude oils is presented in Table 1.6.

As expected, the distillation yields of the streams mentioned in Table 1.5 change with the type of crude oil distilled. As exemplified in Fig. 1.5 (some streams have been regrouped to emphasize differences), the proportion of heavy ends increases with decreasing API gravity of the crude class/grade.

Viscosity is another property that distinguishes among the crude oil classes and brings into the picture another class of oil, namely, *oil sands*. The term sand refers, in a general sense,

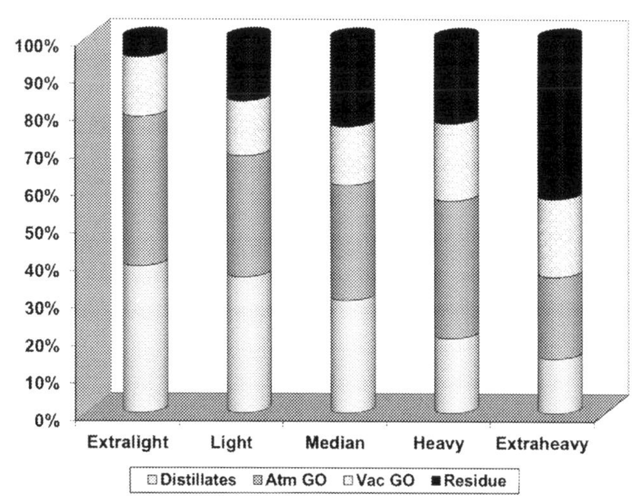

Figure 1.5
Exemplifying distillation yields for API-categorized crude oils. Distillates (gasoline + kero + diesel); Atm GO (atmospheric gas oil or HGO); Vac GO (VGO).

to any noncemented sandstone with high permeability containing immobile viscous oil. Oil sands result in a mixture of sand with bitumen, water, clay, and other substances. Bitumen gives the name *tar sands* to the oil sands since it is also known as tar. Oil sands contain high-S oil with viscosities ranging up to $2*10^6$ cP. The density of raw bitumen ranges from about 960 to 1020 kg/m^3. This density value makes the API gravity of bitumen to fall into the extraheavy class range, but it is excessively more viscous. This distinction between bitumen and extraheavy crude oil is not very precise. In terms of API gravity, they both fall in the class of less than 10 degrees, but bitumen is much more viscous than extraheavy oil, which could flow more easily than bitumen. Bitumen is occasionally associated with an oil degradation product, comprising the refractory heavy molecules left behind in the sands, after lighter molecules are released. In the cases of bitumen, heavy, or extraheavy oils, viscosity reduction is required prior to any attempts at pumping and transportation. In other words, viscosity reduction is mandatory for enabling market accessibility. Typical pipelines designs require viscosity below 350 cP and density below 940 kg/m^3, at temperatures of 7.5–18.5°C. Those differences in viscosity and density for different classes are illustrated in Fig. 1.6. In drawing Fig. 1.6, we have allowed a certain degree of overlapping between classes considering that quality from the same well changes frequently.

Although there is not a general consensus within the oil industry to set the property value ranges that categorize each class of oil, the United Nations Institute for Training and Research (UNITAR) and the API have established formal quality ranges for light, medium, and heavy oil in terms mainly of API gravity and S-content. These property value ranges are collected in Table 1.7.

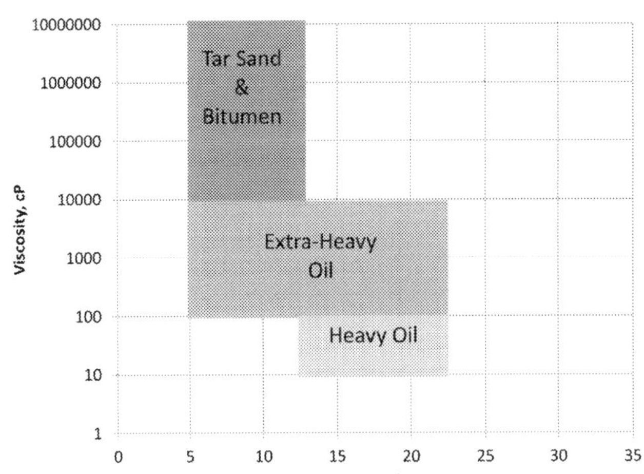

Figure 1.6
Viscosity and API gravity distinction among crude oil classes and bitumen.

Table 1.7: Value Ranges of Quality Attributes for United Nations Institute for Training and Research (UNITAR) Definition of Oil Classes

	Light Sweet	Light Sour	Medium Sour	Heavy Sweet	Heavy Sour
API gravity (°)	30–40	30–40	26–35	10–25	10–25
Sulfur (wt%)	<0.5	0.5–1.5	~1.1	<0.5	1.0–3.0
Metals (ppm)	Low	Low	Low–medium	Low–high	High
Acidity	Low	Low–medium	Low–medium	Low	Medium–high
Nitrogen (ppm)	Varies	Varies	Varies	Varies	Varies

As can be seen, the quality ranges are the basis for defining crude oil grades or classes as:

- *Light sweet*: Crude oils of very high quality, ideal for refining light products (light gases, gasoline components). Light sweet crudes contain small amounts of sulfur and have API gravity greater than 30 degrees. Important crude oils in the world oil trade that belong to this class are Brent and West Texas Intermediate (WTI).
- *Light sour*: Crude oils with low density and sulfur level greater than 0.5%. Sulfur content has to be reduced by oil refining, which results in additional costs. Arabian Extra Light Export crude oil belongs to this class.
- *Medium/medium sweet*: Crude oils with API gravity from 26 to 35 degrees and sulfur level below 1.1%. This class includes Daqing and Forcados export crude oils.
- *Medium sour*: Unlike the previous class of crudes medium sour crude oils contain a high amount of sulfur (over 1.1%). Arabian Light Export and Kuwait Export Blend crude oils belong to this class.
- *Heavy sweet*: 10 to 25 API gravity degree oils containing little or no sulfur. To this class belong Marlim Export and Cano Limon crude oils.
- *Heavy sour*: Low-quality crude oils with high density and high sulfur content. Examples for this class are Oriente Export and Maya Heavy Export crude oils.

Commercial grades of oils derived from tar sands vary because bitumen has to be modified to meet pipeline specifications for viscosity and density, and/or to meet refinery desired yields or qualities. Thus the viscosity-reduced crude oil, enabled for pumping, transporting, and distribution, is known generally as synthetic crude or *syncrude*. This need for bitumen viscosity reduction imposes the upstream sector to install and operate upgraders. Since the level of required upgrading would depend on the original viscosity of the crude or bitumen, the chosen technology for the *primary upgraders* varies and also the severity at which these are operated. In consequence, the original term syncrude had to be broadened to account for the different ways in which viscosity reduction was achieved, thus the following quality grades are derived:

- *Diluted bitumen* or *dilbit*: A controlled blend of (~70%) bitumen with a (~30%) diluent. Diluents also vary among naphtha, natural gas liquids, ultralight sweet crudes, condensates, and light refinery components. Density of dilbits falls in the range of 650–750 kg/m^3.

- *Synthetic crude oil* (*SCO*) or *syncrude*: A partially upgraded bitumen or heavy oil, whose viscosity has been reduced to enable pumping, transportation, and distribution. Density of SCOs falls in the range of 840–870 kg/m^3. SCO has a higher refinery value than dilbit since it is nominally residue-free or at least minimized residue (improved yield/value).
- *Synthetic bitumen oil* or *synbit*: A blend of (~50%) SCO and (~50%) bitumen having similar properties to a medium sour crude. In principle, it has a higher refinery value than dilbit but lower than SCO.
- *Syndilbits* are blends of diluent, hydrotreated light SCO, and bitumen. Their properties and behavior fall between dilbits and synbits.
- *Coker bitumen* (*cokerbit*): A diluted bitumen for which coker distillates are used as diluent. It may have higher refinery value than dilbit, depending on the quality and level of diluent.
- *Medium SCO*: A partially upgraded bitumen to meet the quality grade of a medium sour crude.
- *Sweet SCO*: A partially upgraded bitumen to meet the quality grade of a medium sweet crude.

The quality differences among some of these grades produced in Canada (bitumen) and Venezuela (Orinoco Belt) are exemplified in Fig. 1.7 from data reported for the API gravity and sulfur content in Ref. [17]. These data showed an outliner point for synbit that

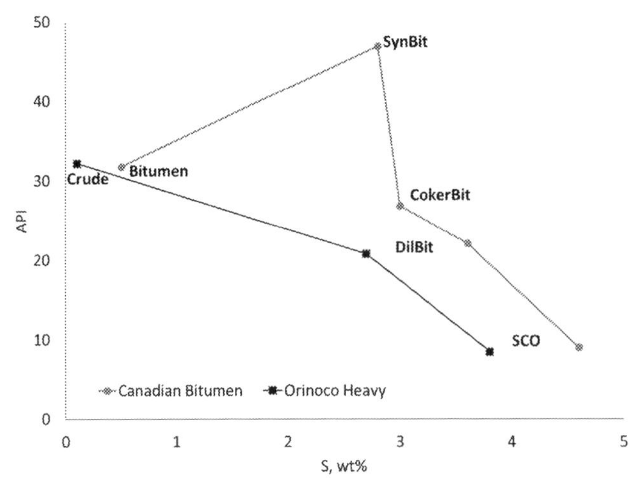

Figure 1.7
Quality of commercial grades of Canadian bitumen and Venezuelan Orinoco belt oil. *SCO*, synthetic crude oil.

surprisingly had an API gravity higher than SCO, though it is supposedly a 50/50 mixture of SCO with viscous bitumen. Although the value might be mistaken, it was included as such because in the article [17] it was reported twice, in a table and in a drawing. Grade production choice depends on many variables, some technical but mostly economical such as: market price, availability or logistic accessibility of diluent, production technologies, and producer and refiners agreements. The basic properties of commercially available crude oils of all different grades can be accessed through a web tool [18].

2. Oil Availability

As an energy source, petroleum constitutes the second share of the market after coal. Actually, the transport sector consumes most of the oil supply, with a minor proportion going to a primary source (heat and power) and the rest to the petrochemical industry. In this section, the supply trends are summarized, with special emphasis on the impact of opportunity crude (OCs) oils.

The total existing amount of petroleum in a given reservoir, which is economically (currently or potentially) recoverable, is called resource. The commercially recoverable quantity of petroleum under defined conditions with existing technologies is known as reserves. Since there is uncertainty involved in reserves assessment, they are classified as proved or unproved, and the latter as probable and possible, in decreasing order of certainty.

The definition of reserves is more financial than physical. Of course the oil has to be in the reservoir to count as reserve; hence one way of increasing reserves is by discovering new fields and new resources but it is understood that less than 13% of the oil in the planet is waiting to be discovered [19]. Oil price increases is a way of making resources of costly production to become commercially feasible reserves. Technology improvements could decrease production cost and enable resources to become reserves, as well. Finally, new technologies could give access to resources otherwise unavailable and eventually bring operation costs down to commercially exploitable levels.

2.1 Crude Oil

Oil production and reserves estimations are reviewed and updated periodically, as long as information becomes available. A number of agencies and institutions monitor these assessments and evaluations locally, regionally, and globally. Information is available through the websites of the Energy Information Administration [20], International Energy Agency [21], World Energy Council [22], World Petroleum Council [23], the Organization of the Petroleum Exporting Countries (OPEC) [24], and other consultancy companies. BP annually publishes statistical information, which is available from its website [25].

In Fig. 1.8 the production (P), reserves (R), and remaining exploitation time (R/P), as per 2007 (A) and 2013 (B) are shown by region (data were taken from Ref. [25] for 2007 and from Ref. [26] for 2013). As can be seen, in general, oil is unevenly distributed geographically throughout the planet in limited oil reservoirs. Furthermore, in less than a decade the relative proportion of the pie slices have changed significantly. The R/P of the Middle East was the largest (82 years) by 2007, but this moved to South and Central America (almost 124 years) in 2013.

To examine these changes in more detail, the three parameters (R, P, and R/P) reported in Fig. 1.8A by region have been plotted in Fig. 1.9 for the top 20 countries in decreasing order of R/P.

It can be seen that some of the countries with the largest reserves will run out of oil faster than others with more modest reserves. Although for Iraq and Kuwait the published data indicate R/P > 100 years, the growth in their production during the examined period of time has caused a number of changes in the market map. A comparison of the production

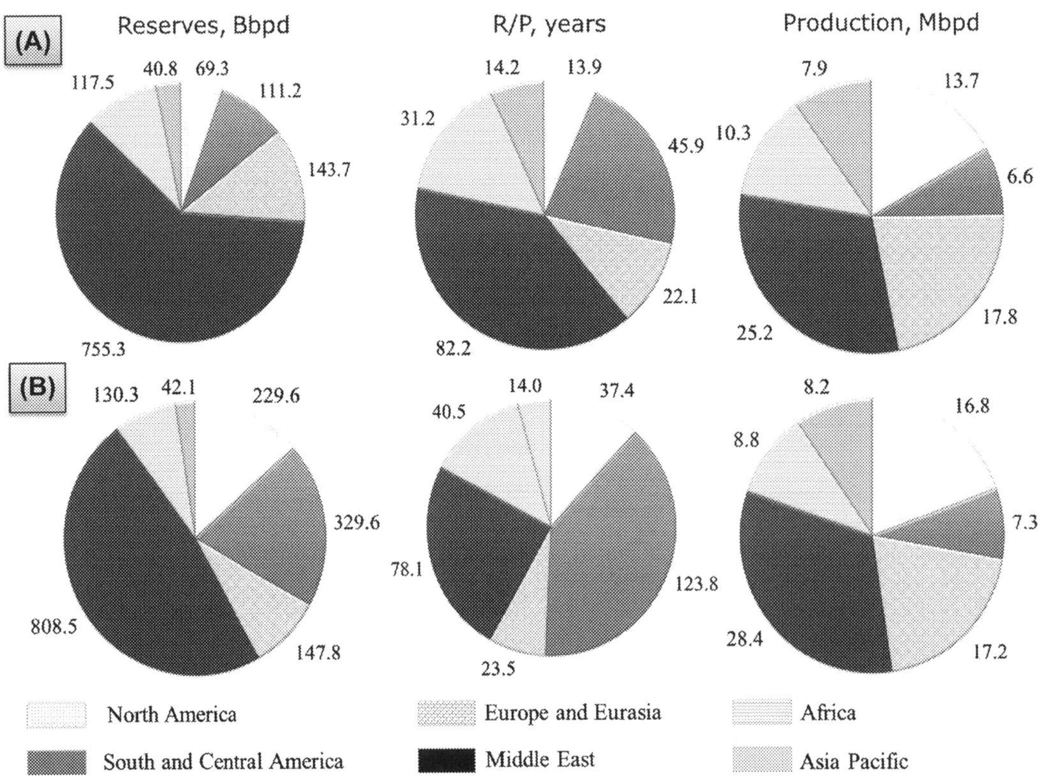

Figure 1.8
Regional distribution of crude oil: reserves, production, and remaining exploitation time (R/P). (A) 2007; (B) 2013.

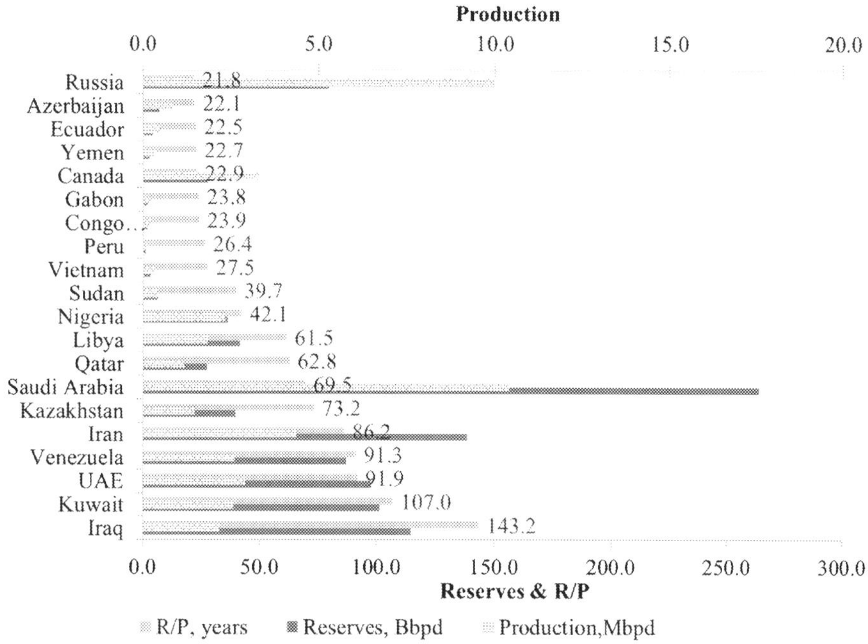

Figure 1.9
Top 20 countries with largest remaining exploitation time (R/P) (years) of remaining reserves.

for 2007 and 2013 for these countries is presented in Fig. 1.10; in this case and to be consistent with Fig. 1.8B, the top 30 countries (with largest R/P) have to be included.

The observed market map changes are also caused by producing countries with new emerging growth, such as Algeria, Angola, Brazil, Colombia, China, and the United States. These changes in the map are accompanied by changes in the quality of the crude slate. Although API gravity and S-content (heaviness and sourness) remain the most widely accepted quality parameters (as will be considered in the next subsection), by the end of the last century, a new quality parameter entered the crude oil market: acidity. High-acid crudes (HACs) were first introduced in the market about that time and their production growth has not stopped since. In fact, HACs capture the fastest-growing segment of global oil production. California, Brazil, the North Sea, Russia, China, India, and West Africa are the largest producers. Canadian oil sands-derived commercially available crudes are deemed highly acidic. A large proportion of the world's oil supply is heavy and sour [27].

2.2 Opportunity Crude Oils

Although oil price is affected by market conditions and geopolitics, quality is the differentiating feature among different crude oils. The quality difference creates a price

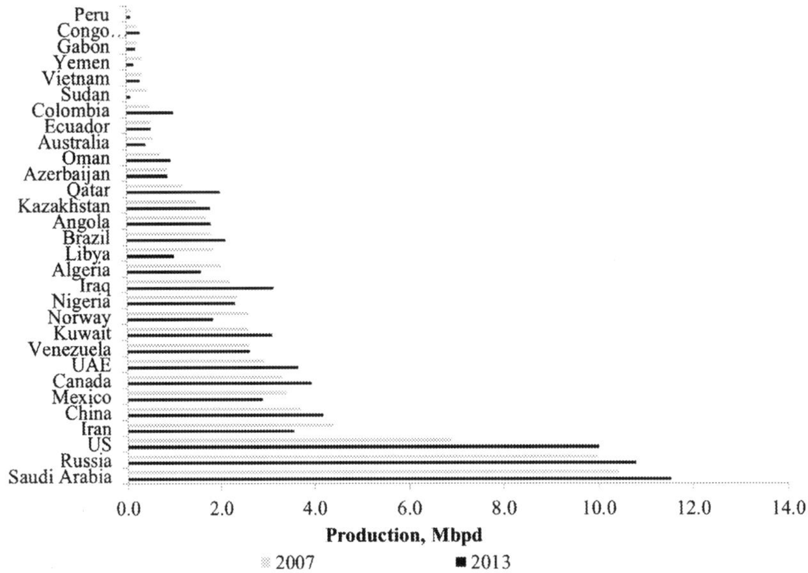

Figure 1.10
Oil production in the top 30 countries with largest remaining exploitation time (R/P) (years) of remaining reserves.

differential between the benchmark high-quality crude (light and sweet) and the low-quality crude (heavy and sour/acid). The so-called "opportunity crudes" (OCs) or "advantageous crudes" associated with "unconventional oils" correspond to crude oils that are sold at discounted prices in view of their lower quality. Notwithstanding, the opportunity regards their prices and not an intrinsic profitability. In fact, they represent extraordinary technical challenges not only to the refiner, but also to the producer and to the transporters. Besides, heavy/extraheavy oils and bitumen, acid and sour crude oils are the other class of OCs.

The economic opportunity of these types of crude, i.e., their price discount, is exemplified in Fig. 1.11 (for a heavy oil in A [28] and for an HAC in B [29]). It has been noticed that in a time of ample supply, the discounts contract.

Although price differentials are affected by several factors, Tahmassebi [30] has listed as the most important: oil prices, heavy fuel oil demand, world average refinery complexity and utilization rate, and the quality of crude supply. Refinery complexity and utilization rate are associated with the volume of OCs that could be possibly refined. There are physical, economic, and logistical constraints on this volume. First, OCs require a certain level of complexity in the refinery; a simple refinery cannot handle this type of crude. Logistical constraints refer to the accessibility of the crude to the market (pipelines, transportation,

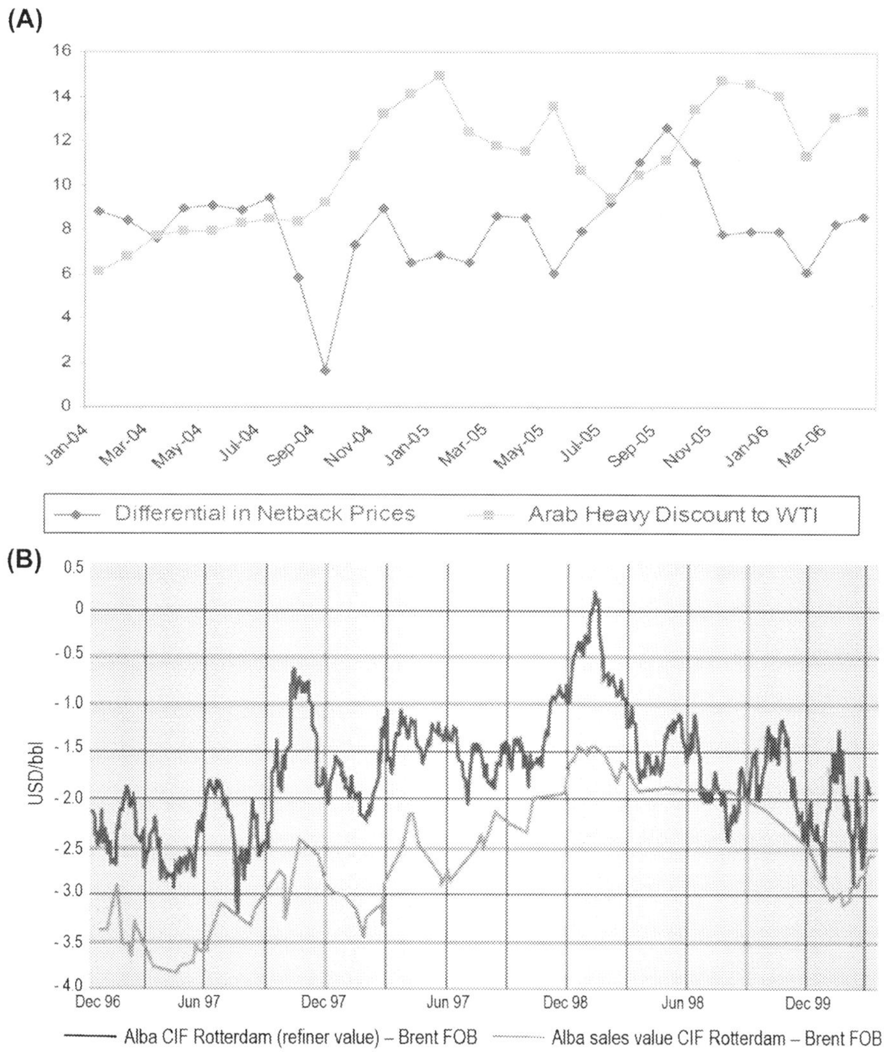

Figure 1.11

Crude oil price differential and discount ($/bbl). (A) Heavy oil. *(Reproduced from Fattouh B. Opec's discounts on heavy crude oil: is a new policy instrument taking shape? Oxford Institute for Energy Studies. Oxford Energy Comment; June 2006. 9 pp., with permission from the Oxford Institute of Energy.)*; (B) High-acid crude (HAC). *(Reproduced from Skippins J, Johnson D, Davies R. Corrosion-mitigation program improves economics for processing naphthenic crudes. Oil Gas J. 2000;98(37):64–68, with permission from Pennwell.)* *CIL*, cost, insurance and freight; *FOB*, free on board; *WTI*, West Texas Intermediate.

distribution, etc.). Finally, the economy is associated with the processing of the crude, whether the refinery needs new investments, operating costs, etc. A good development example that had to face these three constraints was the exploitation of Canadian crude oils and bitumen [31].

Regarding quality and the benchmark crude Brent, it seems that each degree of API values 18.8¢ at a price of US$40/bbl, while at a price of US$60 it would be 27.2¢ (indicating a crude price dependence of the differentials). Sulfur dependence was illustrated by ExxonMobil [32], Hartman [33], and Wang [34]. Each percentage point of S-content lowered the price by US$1.46 a barrel, and each degree increase in API raised the price by 13.2¢ per barrel, for 1992 crude oil average price (using a set of 55 samples) [34]. Bacon and Tordo [35] have correlated the discount as a function of API gravity, sulfur content, and acidity in a work for the World Bank. This model predicts that each degree of API raises the price US$0.007 per dollar of Brent and each additional one percentage point of sulfur lowers the price US$0.056 per dollar of Brent. In terms of acidity, the calculated discounted price for crudes with TAN < 0.5 showed no discount, but above this TAN value, each point of excess TAN lowers the price of crude by $0.051 per dollar of Brent. This excess is defined as TAN of crude oil minus TAN of Brent (taken as 0.07) [36]. A larger set of properties was considered in an attempt to find correlations between quality, yields, and price. Although no correlation was found, it was clear that larger refining benefits could be found during high price periods [37].

The price differential is also said to become aggravated when producers have difficulties in accessing the market. In the case of Western Canadian oil the spread (price differential) with WTI was expected to hover around the $23/bbl range through December 2015 [38]. A referential discount for Canadian oil can be found back in December 2012 when it was traded at a discount of more than $42/bbl, while by November 2013 the discount showed an optimistic move to $41.79/bbl [39], and later to $23.37/bbl [40]. Since relative quality cannot be responsible for all this variation, only market volatility can be blamed. Thus price differential and discount are highly influenced by oil prices/market [41] (Fig. 1.12; [42]). For instance, in the 1970s, supply from the Middle East suffered the disruption of the Arab embargo and the Iranian revolution, creating a fast growth of quality price differential. The "lions" crude oils in price definition (the referential markers) have traditionally been Brent (from the North Sea) and WTI (from the United States). The spread between these two was typically $5, but recent times have seen this spread widen up to $20/bbl. The reason is because of the dramatic change that the US crude oil landscape experimented through the past few years. Changes are caused by a combination of a fast growth of supply, as well as a mismatch between pipeline infrastructure and available refinery capacity. This variability in WTI pricing has affected the discounting of Canadian prices relative to this North American benchmark [42].

The end of 2014 and the first quarter of 2015 confronted deep changes in the formation process of crude oil price. These recent changes were induced by a revolutionary move in oil market dynamics caused by the accelerated increase in US unconventional production. The previous 3 years faced remarkably stable price levels, since the US supply growth was offset by disruptions in other parts of the world. However, behind this price stability,

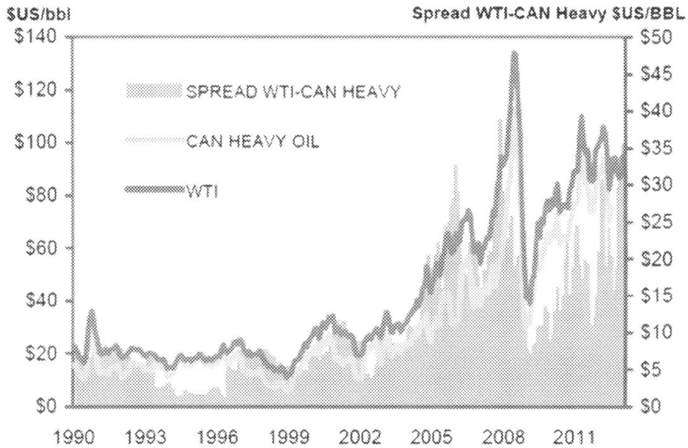

Figure 1.12

Price and price differential for Western Canadian and West Texas Intermediate (WTI). *Reproduced from Bendiner J, Burleton D, Preston L. Drilling down on crude oil price differentials. TD Econ March 2013: 6 pp. Available from: http://www.td.com/document/PDF/economics/special/DrillingDownOnCrudeOilPriceDifferentials.pdf, with permission from TD Economics.*

adjustments in price differentials and changes in the trading flow map were the leveraging parts [43].

However, modeled or calculated discount cannot be taken as fixed rules. In fact, the variability of the discount is also related to market trends. ChevronTexaco considers that the variability of the discount is caused by changes in OPEC quotas, changes in the HAC supply/demand balance, increased Marlim production, and the fuel oil demand for HACs [44]. A corrected discount of −$0.75/TAN was introduced as a consequence of the latter analysis. Regarding the OPEC quotas, this organization has clarified that discounting was not a policy instrument for cutting supply, although they did not rule out the possibility of considering it in the future [28]. Nevertheless, OPEC's producers in situations of retarded demand or oversupply cut the discount of the heavy oils to decrease their attractiveness and affect refining margins to induce a reduction in their demand. The effectiveness of this practice is good enough, particularly in the current situation where not many refineries are equipped to handle these crudes.

In summary, prior to considering the incorporation of a given crude to a refinery diet, any refiner would use their own valuation models that include all factors affecting the price (as discussed earlier) and the refinery configuration and its available technology [34,45–53].

2.2.1 Heavy, Extraheavy, Bitumen, and Tar Sands

UNITAR in 1982 decided to rationalize presentations and discussion points to set the definitions for heavy crude oil and tar sands bitumen. In the opinion of the expert panel, these crude oils and bitumen may be characterized first by viscosity and then by density. In determining the international resource base, viscosity should be used first to differentiate between crude oils, on one hand, and bitumen, on the other. Subsequently, density should be used to differentiate among extraheavy crude oils, heavy crude oils, and other crude oils.

Bitumen has viscosities greater than 10,000 cP. Crude oils have viscosities less than or equal to 10,000 cP. These viscosities are gas free as measured and referenced to original reservoir temperature. Among crude oils, extraheavy crude oils have densities greater than 1000 kg/m^3 (API gravities @ 15.60°C of less than 10 degrees). Heavy crude oils have densities from 934 to 1000 kg/m^3 (API gravities from 20 to 10 degrees) inclusive. UNITAR did not provide classification for any other crude oil with a density of less than 934 kg/m^3 (API gravities > 20 degrees) [54].

Besides the higher viscosity and lower API gravity, heavy crudes are characterized also by their difficulty to vaporize, high conradson carbon (CCR) or microcarbon residue (MCR), high asphaltenes content, and high levels of heteroatom contaminants (S, N, O, and metals). These contaminants concentrate in the BotB fractions. The challenges and opportunities existing for the processing of heavy oils are associated with a disproportion of larger molecules and/or high-boiling point compounds (the BotB), with the characteristics just mentioned.

Heavy oil resources are largely concentrated in Canada and Venezuela, which hold respectively some 1.5 and 2.5 trillion barrels. In Canada, the greatest accumulation of oil sands occurs in vast deposits exceeding 60 m thick of sands with 30% porosity at relatively shallow depths (0–600 m). Smaller deposits of oil sands can be found in China, Russia, India, Indonesia, and Ecuador [55]. If reserves can be proved with a 20% recovery rate, these two countries alone would account for more proven reserves than the conventional reserves of the Middle East. In fact, with more than 175 billion barrels accepted as proven in 2003, Canada had at that time the second largest proven reserves in the world, after Saudi Arabia. Alberta province holds more than 97% of Canada's reserves, and in Alberta most of the deposits are located in Saskatchewan. Thus Alberta oil sands are considered the world's largest formation of bitumen. The estimated total potential oil reserve value of these deposits is 315 billion barrels. The bitumen in Athabasca, Cold Lake, and Peace River regions remains a substantial opportunity for long-term development, particularly considering their location within a politically stable country with existing infrastructure linked to US markets. Worldwide, Venezuela follows Canada as the second largest commercial developer. In this country, most of the OCs' reserves are concentrated in the southeast of the country, north of the Orinoco River, known as the *Orinoco oil belt*. Since technically these are not bitumen reserves, they have been reclassified as extraheavy oil deposits. They exhibit a very low API gravity but have not been degraded to the same extent as oil sands. Venezuelan climate in the Orinoco belt region (40–50°C) contributes to the lower viscosity of the oil compared to that of Canadian bitumen, facilitating also their extraction through known horizontal drilling techniques.

At the moment this book was prepared, the growth in heavy oil production was well represented by the growing situation of Canada resources. Under moderate growth assumptions, the Canadian Association of Petroleum Producers reported an expected growth of more than 75% in the period 2008 to 2020 (Fig. 1.13 from Ref. [56]).

Most technological developments have taken place in Canada, where an attractive tax and royalty regime for heavy oils and oil sands, introduced in 1996, has prompted major new investment from private industry. Russia also has significant heavy oil deposits [57].

This growing trend in heavy oil production becomes more pronounced by the decline of the light crude oil supply, leading to an increasing proportion of heavy crude oils in the available market pool. Consequently, the refiner is forced to increase the flexibility of the process units or to see shrinkage in its margins. The US trend can be seen in Fig. 1.14 (data from the Energy Information Agency, EIA [58]), i.e., the figure shows the distribution of imported crude oil by type of crude. In this plot, the change in time of the percentages of total imported crude oil is categorized in terms of three types of crudes: heavy, HO (API < 20.0), median, MO (20.0–40.0), and light, LO (API > 40.0). It is observed that the HO component started to increase at the end of the 1970s, and more steeply by 1999. The MO component abruptly began also to increase by 1983, but the whole effect was aggravated by the declining supply of LO.

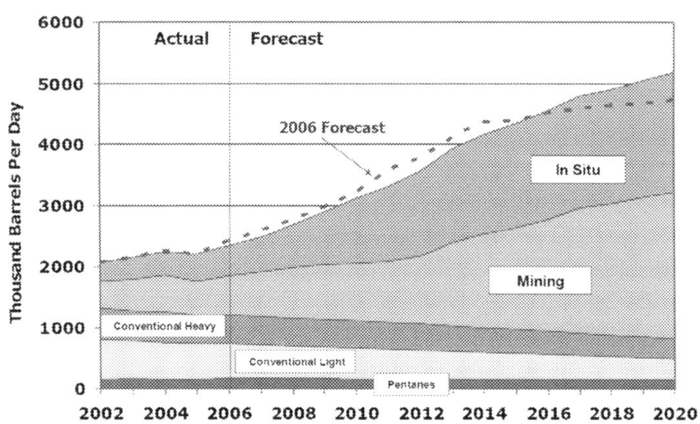

Figure 1.13

Growth of crude oil production (including heavy oils and tar sands) in Canada. *Reproduced from Canadian Association of Petroleum Producers (CAPP). Crude oil forecast, markets and pipeline expansions. Canada; 2014. 52 pp. Available from: http://www.capp.ca/getdoc.aspx?DocId=247759&DT=NTV, with permission from the Canadian Association of Petroleum Producers.*

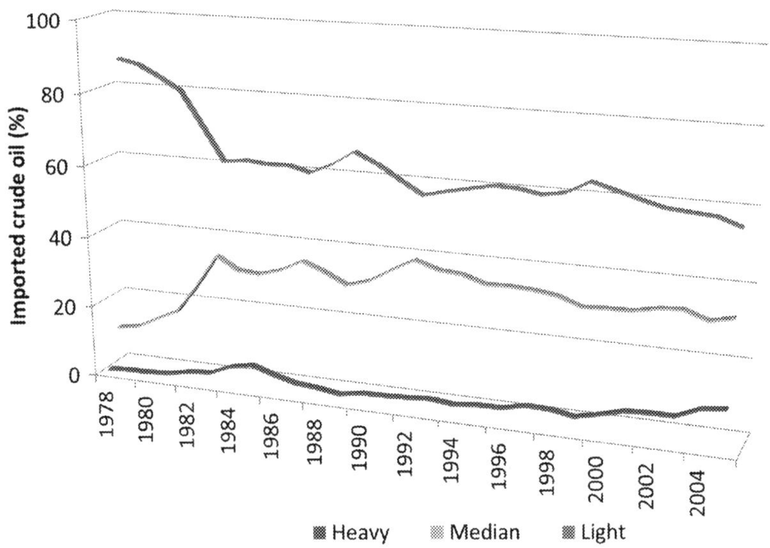

Figure 1.14
Percentage of US imported crude for different ranges of API gravity.

Market analysis and forecasting reported by the EIA in 2006 predicted the average API gravity of the crude oil fed to refining processes in the United States of about 29 by 2010, and this value was supposed to keep decreasing continuously thereafter. In fact, the trend of the average API gravity of the imported crude oils to the United States during the last 25 years (data from the EIA site [59]) shown in Fig. 1.15 confirmed that trend was not new.

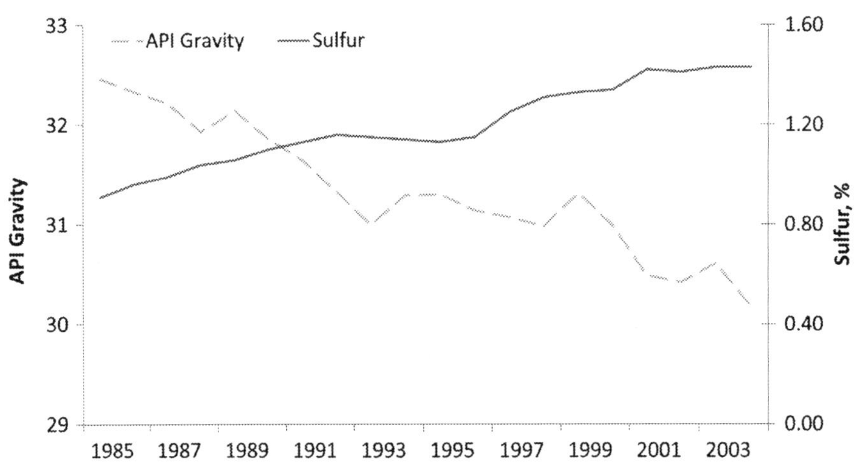

Figure 1.15
API gravity and sulfur content of the US annual refinery diet.

The decline in the API gravity indicates the increasing use of heavier components in the refinery diet during the last three decades. The direct consequence is a decrease in the quality of the feedstock slate, as exemplified by the increase of the average sulfur content (also seen in Fig. 1.15). Unfortunately, a decrease in quality might be typically accompanied by an increase in all other contaminants (nitrogen, metals, and asphaltenes). Although the chosen examples correspond to the United States, the global market [60] shows a steady growth in the low-quality crude oil (OCs) supply (Fig. 1.16A, heavy and HAC supply were included in this figure), leading to a continuous increase in the average

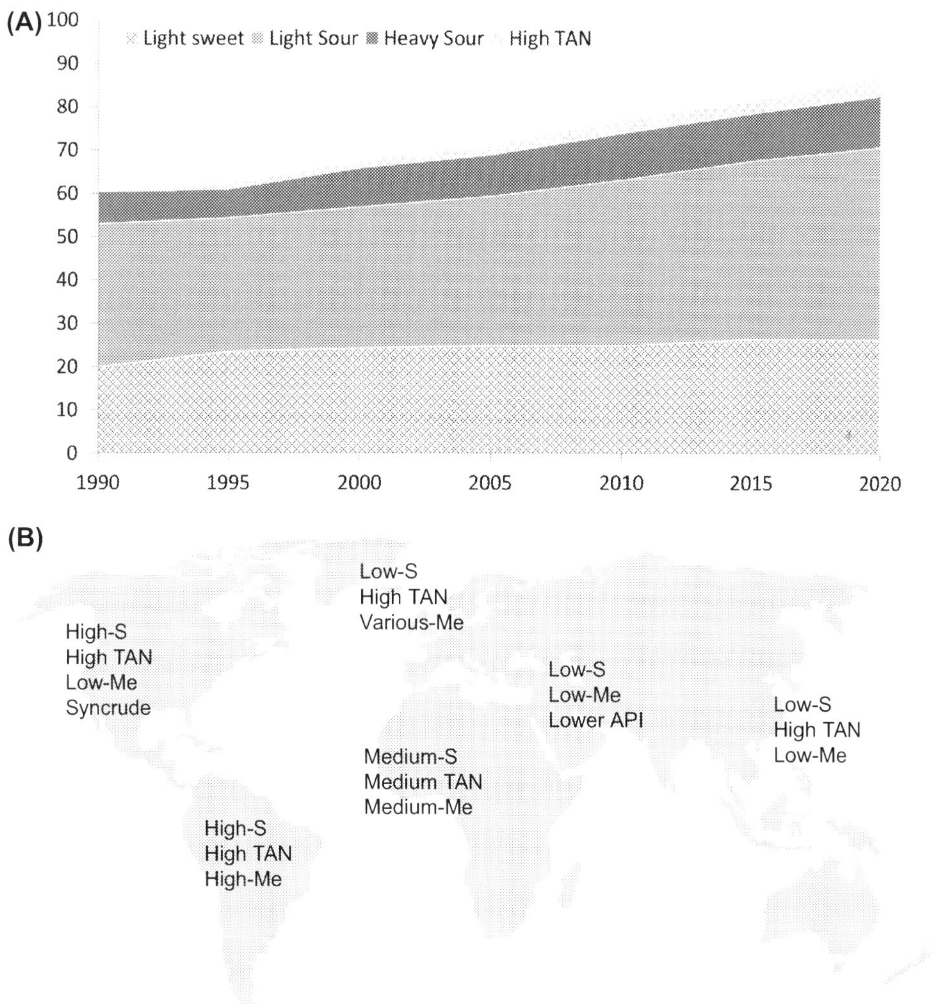

Figure 1.16

World production of opportunity crudes by crude type. *TAN*, total acid number. *Reproduced from Energy Information Administration Annual energy outlook 2006 with projections to 2030. Report No. DOE/EIA-0383; 2006. 242 pp., with permission from the International Energy Agency.*

concentration of unwanted contaminants (S, N, and metals). The concentration of the polluting contaminants increases with the weight fraction and so the heavier the oil, the larger the concentration of the contaminants. A summary of the regional distribution of OC oil quality throughout the world is presented in Fig. 1.16B.

2.2.2 Acid/Sour Crude Oils

Acid compounds are present in both heavy and light crude oils. In Table 1.8 we have collected a list of representative HACs and their acidity value (TAN in mg_{KOH}/g). African, Chinese, and Venezuelan crudes are among the most acid. In fact, Sudan has started to report exceedingly high-TAN values for some of its crudes [61].

In recent years, HAC has accounted for an increasing share of the global oil supply. According to an Oct. 2004 note from the World Bank Group [52], crude with a TAN greater than 1.0 grew to 9.5% of global supply in 2004 from 7.5% in 1998. Indeed, HACs represent the fastest-growing segment of global oil production, as can be seen in Fig. 1.16A. The regional supply of HCAs is represented by crudes exploited in North and South America, Africa, and the Far East (Fig. 1.17, elaborated with data taken from Refs. [44,62]). As can be seen production has grown rapidly in all regions.

Regional HACs supply has been reported to steadily increase since 1998 [44,62]. In the Americas a plateau was forecasted to have been reached by 2010 and indeed this might be the case, because of the decreasing Venezuelan production. In Brazil's long-term strategy, the needed investment in refinery capacity and conversion units is the key factor for channeling larger volumes of heavy acid oils to the international market. In this case, the future of acid oil producers will depend on investments in refineries close to oil product consumption centers. For Brazil, this means investments in modifying its refineries and setting up partnerships in the downstream segment for consumer centers absorbing all products of high added value, such as the United States and even Southeast Asia and Western Europe [63].

North Sea showed a similar trend, peaking at the end of the previous decade and declining since then. West African supply, where crude is coming on line from first-time producers such as Equatorial Guinea and Sudan as well as Chad, shows an even greater increase in TAN. In 2001, 5% of West African crude had a TAN greater than 1.0, but by 2006 this share was expected to rise to 13%, as forecast by the World Bank [52]. It was estimated that as much as an additional 600 kbpd of high-TAN crudes would come from West Africa by 2007−2008 [44,62]. In the Far East, there is also a growing output of high-TAN crudes with significant additional volumes coming from China. Many of the new Chinese offshore crudes (Penglai, QHD, CFD, and others) are high in acid content and as much as 400 kbpd of additional volumes of high-TAN materials were expected to be produced by

Table 1.8: Examples of High-Acid Crudes From Different Countries

Source Country	Crude	TAN	Source Country	Crude	TAN	Source Country	Crude	TAN
Angola	B31N	0.84	Congo	Emeraude	3.72	Trinidad	Trintopec	3.1
	Benguela heavy	1.1	Egypt	El Morgan	1.43	United States	Charenton	2
	CLOV	0.86	Equatorial Guinea	Ceiba	0.77		Edgerly field	1.79
	Dalia/Camelia	1.5		Zafiro	0.83		Friburg	1.59
	Kuito	2.4	Gabon	Gombe Marin	2		Louisiana main pass	4
	Mondo	0.98		Oguendjo	1.69		LT WC SR blend	1.3
	Negage	1	Indonesia	Bima	2.68		Poseidon	1.86
	Pazflor	1.33		Duri	1.46		TTTI grade A	1.31
	Saxi/Batuque	0.9	Iran	Ardechir	1.36		WC sour heavy	2.83
Australia	Wandoo	1.49	Ivory Coast	Baobap	2.06	UK	Alba	1.59
Brazil	Albacora	2.4	Malaysia	Miri light	3.8		Bressay	2.6
	Marlim	1.21	Norway	Balder	1.5		Captain	2.1
	Marlim Leste	1		Grane	2.04		Clair	1.2
	Marlim Sul	1		Heidrun	2.87		Gryphon/Maclure	2.99
	Roncador	1.5		Hermod	2.15		Harding	2.89
Cameroon	Lokele	2.35		Troll	0.76		Leadon	3.41
Chad	Doba	4.77	Sudan	Ambar-1	1.0–2.0		Mariner	2.1
China	Bozhong	1.57		Bong W1	0.7		SGS	2.3
	Duli	1.4		Fal-1	3.2–10.4	Venezuela	Bachaquero	3.13
	Liaohe	2.12		Fal-2	3.1–5.3		BCF-15; 17; 22	1.8–3.0
	Liuhua	1.12		Fula	1.0–16.2		Boscan	1.1
	Penglai	3.57		Gassab-1	1.42		Caripito	1.7
	Qinghuangdao	2.49		Nile blend	1.06		Laguna	2.82
	Shengli-GIC	1.16		Moleeta-1	2.04		Merey	0.8
	Xinjiang	4.51		Palogue-1	2.6–3.64		Morichal	2.83
Colombia	Barranca	2.07		Palogue South-2	3.27		Pilon	1.6
	Permian mixed	1.73		Wengi-2	2.53		Tia Juana	3.61

TAN, Total acid number.

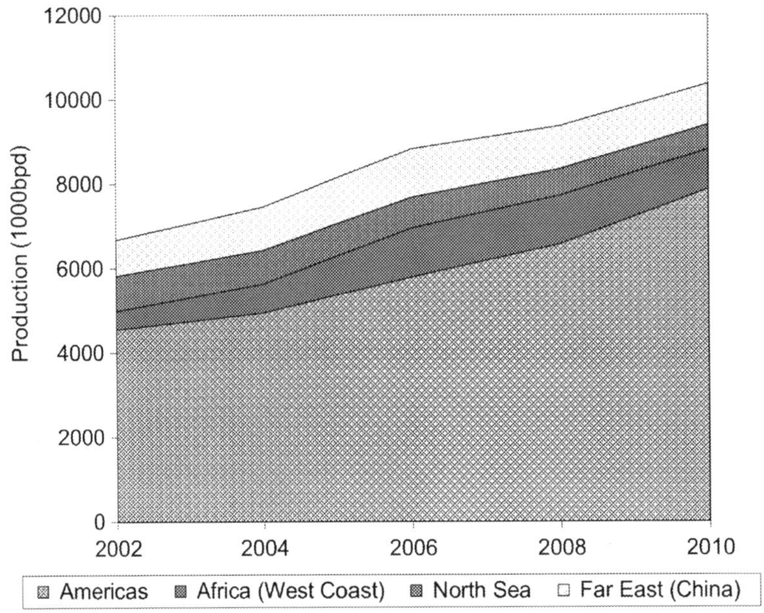

Figure 1.17
Regional production of high-acid crudes (HACs).

2007–2008 [44,62]. Although some producing wells of HACs are known to be declining (e.g., North Sea), the emerging production more than offsets that decline [64].

The production growth of HACs is larger and faster in countries east of Suez than it is in those situated at the west of Suez, as reported in Ref. [47], when the assessment only involves HAC-producing countries (Fig. 1.18 elaborated with the data reported in [47,65]). A 2008 article from *Oil and Gas Journal* [47] summarized and collected the data on the market for HACs, included in a report from Asia Pacific Energy Consulting (APEC) [65].

The study stated that incremental high-TAN crude output should rise by 1.8 million bpd during 2005–2010, with 61% of that gain expected in west-of-Suez production regions. In fact, HACs made up 25.4% of west-of-Suez crude supply in high-TAN-exporting countries in 2006; in contrast, it made up 11.4% of crude output in high-TAN-exporting countries worldwide (see Fig. 1A and B reported in Ref. [47]). The figures of Ref. [47] show consistently that west-of-Suez output is far greater than east-of-Suez output and continues throughout the timeframe of the study to dominate the world's total of internationally traded acidic crude. In summary, HACs' production, trade, and use will increase globally.

Trade has been sketched by Skippins et al. [62] and is shown in Fig. 1.19. Asia–Pacific is considered the focus of interregional acidic crude trade through the last and next decades, according to Quing [66]. China is considered the market leader for HACs and would

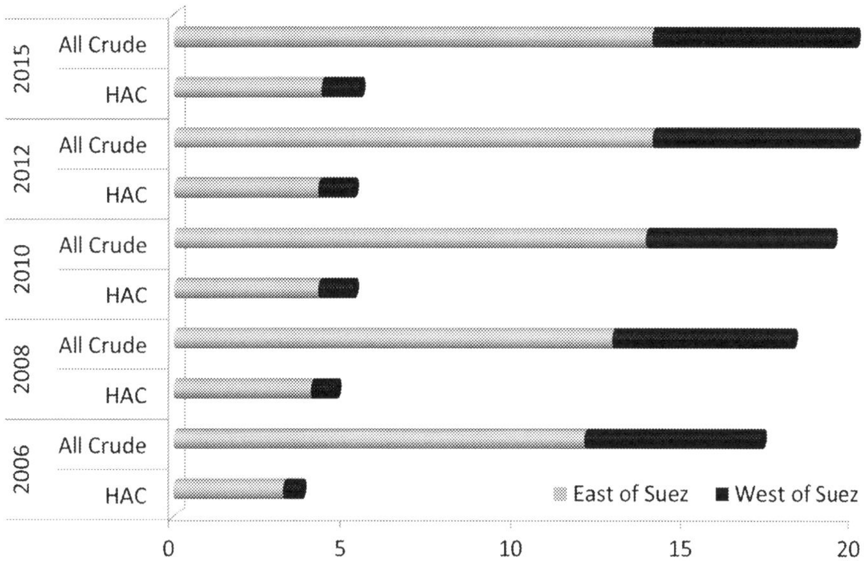

Figure 1.18
Regional production growth: high-acid crudes (HACs) producer countries only.

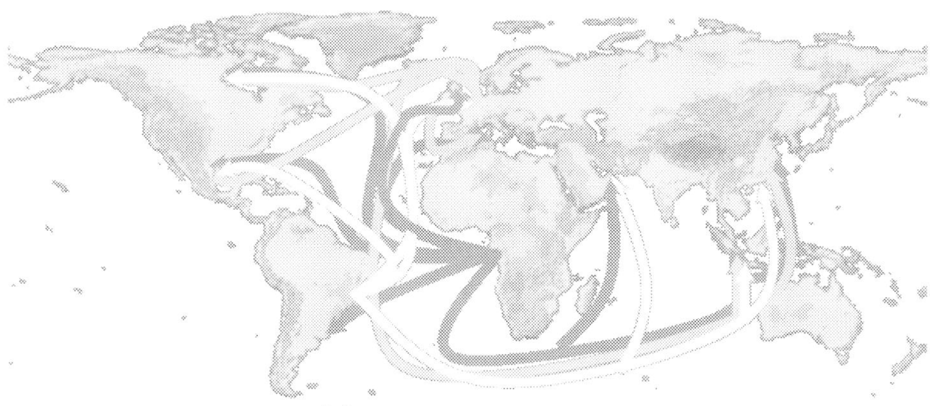

Figure 1.19
High-acid crudes (HACs) trade flow. *Reproduced from Skippins J, Bell K, Kronk J, Bagdasarian A, Johnson D. High acid crude. Proc. Crude Oil Quality Association Meeting. New Orleans, LA, USA; May 2002. ChevronTexaco, 180 pp., with permission from the Crude Oil Quality Association.*

probably continue to dominate high-TAN production, consumption, and imports for Asia–Pacific for at least the medium term.

APEC expects [65], however, a significant increase in high-TAN use by South Korea, Taiwan, Malaysia, Singapore, and India. Chinese exports to the US West Coast may also rise, depending on system needs of Chevron and ConocoPhillips.

2.3 Shale Oil

Shale oil is extracted from the oil shale by thermal extracting methods; the raw shale oil is typically light and sweet (high API gravity and low S-content). Regardless of these quality benefits, shale oil tends to be discounted in the market, because it offers the poor characteristics of OCs. This is because of a high organic O-content, which may cause high SG, particularly in cases where a larger presence of heavy paraffins (waxes) is found. Additionally, H/C ratio and diesel cetane values tend to be lower than conventional crude oils. Although there is a heavy end cut in these crudes, the proportion of the $360°C^{-}$ is anomalously high in comparison with crude oils with similar API.

Characterization results of shale oils are scarce and show large variability [67–71]. The content variability of heteroatomic moieties in shale oil is the consequence of the several methods employed in its recovery from the source rock. Regardless of extraction method, O-content is typically higher than that of N- or S-content, but in some instances N- or S-content may be the greatest. In general, S-content in pyrolysis-recovered shale oils is comparable to that present in conventional crude oils, but N-content is higher [72–77]. The identification of 30,000 compounds by FT-ICR MS coupled with electrospray ionization and atmospheric pressure photoionization allowed the definition of chemical features that distinguished shale oils from conventional oils. The double bond equivalent (DBE) distributions of the N1 and aromatic hydrocarbon classes were lower in the shale oils than in conventional oil. O2 compounds with DBE = 1 were abundant in the shale oils, whereas O2 compounds with DBE = 3–4 predominated in conventional oil [78].

The presence of waxes is partially responsible for the incompatibility behavior exhibited by this type of crude oil. Furthermore, the high contamination with minerals leads to scaling and fouling observed in pipelines and heat exchangers. It is believed that deposited waxes serve as an organic matrix where minerals are incrusted. These minerals are mainly Ca based (e.g., calcite, dolomite, etc.).

Estonia has been one of the world leaders in the exploitation of its oil shale resources, currently followed by the United States, where recently a boom of utilization of shale resources took place. In fact, the United States is expecting to use these resources to win an energy independence status. By 2015, the imported/exported volumes were balanced.

3. Impact in the Oil Industry

The oil industry is a molecular trader whose business is based on the price of such molecules and on their properties. In fact, refining economics is highly impacted by crude cost, type and quality of products, and disposition of low-value stranded streams and by-products. Economically, refining can be considered a necessary evil, historically known as a low profitability industry, requiring high capital investments. This level of investment can be understood by the number of processing units that might be involved in the refining process. In Table 1.9, a list of these different processes and a set of brief descripting terms are shown. Part of the processes listed concern direct processing of hydrocarbon feeds, while others regard the processing of by-products generated by the former.

The poor margins experienced in the last decades of the 20th century were probably the concerted effect of oil prices and quality that led to the closure of many refineries around the globe. Furthermore, since the crude price is the most significant element contributing to refinery margins, refiners turn toward processing OCs, whenever possible. However, the economic opportunity brings enormous challenges to the refiner based on the fact that these lowest-quality crude oils (highly contaminated) have to be refined into high-quality products (contaminant free or minimized).

The crude price and its inherent molecular properties act together to determine the economic value of a product. Therefore the composition of crude oil is the cornerstone of any attempt for molecular management of refining processes. Not only in the petroleum industry, but also in the chemical industry in general, the understanding of the molecular structure and of the compositional knowledge would resolve one side of the equation. The other half is behavior, performance, and/or properties. The role of refining is to move these molecules around in the oil matrix and convert some of them to yield the products under the specifications fixed by the market and governmental regulations. Since refining products are hydrocarbons, any heteroatom is considered an impurity and has to be removed. For instance, the composition of Canadian oil sands results in feedstocks that are very difficult to handle in the desalter, hard to vaporize, thermally unstable, corrosive, and produce unstable and reactive coker product (high diolefin content) [79]. In general, OCs are known to cause problems from the refinery entrance all the way down to the waste water treatment units; examples are desalter upsets, heat exchangers fouling and scaling, fire heaters, corrosion in atmospheric and vacuum towers, plugging of catalytic process units, catalyst deactivation, etc.

The heavier crude oils are considered to be either immature and/or degraded. It is believed that residue processability depends on the maturation degree of the original crude oil [80], though there is no systematic study explaining such dependence. The characteristics and properties of a crude oil seem to be correlated not only to its geological origin (kerogen), but also to the degree of maturation that it has received through aging. Upon maturation, a negative effect is observed on the stability and coking tendency of residual feedstocks. The main reason for this appears to be that while the asphaltenes become more aromatic upon aging, the resins become less aromatic, thus causing an increasing gap in aromaticity between the two. The increasing gap is responsible for a behavior that causes the toughest processing problems. Additionally, metals in crude oils vary with the depth and age of the reservoir rock [81].

Table 1.9: Summary of Refining Processes

Process	Action	Method	Purpose	Feedstock(s)	Product(s)
Fractionation Processes					
Atm. distillation	Separation	Thermal	Separate fractions	Desalted crude oil	Gas, gas oil, distillate, residual
Vacuum distillation	Separation	Thermal	Separate w/o cracking	Atmospheric residue	Gas oil, lube stock, residual
Conversion Processes—Decomposition					
Catalytic cracking	Alteration	Catalytic	Upgrade gasoline	Gas oil, coke distillate	Gasoline, petrochemical feedstock
Coking	Polymerize	Thermal	Convert vacuum residuals	Gas oil, coke distillate	Gasoline, petrochemical feedstock
Hydro-cracking	Hydrogenate	Catalytic	Convert to lighter HCs	Gas oil, cracked oil, residue	Lighter, higher-quality products
Hydrogen steam reforming	Decompose	Thermal/catalytic	Produce hydrogen	Desulfurized gas, O_2, steam	Hydrogen, CO, CO_2
Steam cracking	Decompose	Thermal	Crack large molecules	Atm HFO/distillate	Cracked naphtha, coke, residual
Visbreaking	Decompose	Thermal	Reduce viscosity	Atmospheric residue	Distillate, tar
Conversion Processes—Unification					
Alkylation	Combining	Catalytic	Reaction of olefins and isoparaffins (C_5^-)	Tower isobutane/cracker olefin	Iso-octane (alkylate)
Grease compounding	Combining	Thermal	Combine soaps and oils	Lube oil, fatty acid, alky metal	Lubricating grease
Polymerizing	Polymerize	Catalytic	Reaction of two or more olefins	Cracker olefins	High-octane naphtha, petrochemical stocks

Conversion Processes—Alteration or Rearrangement					
Catalytic reforming	Alteration dehydration	Catalytic	Upgrade low-octane naphtha	Coker/hydro-cracker naphtha	High oct. reformate/ aromatic
Isomerization	Rearrange	Catalytic	Convert straight chain to branch	Butane, pentane, hexane	Isobutane/pentane/hexane
Treatment Processes					
Amine treating	Treatment	Absorption	Remove acidic contaminants	Sour gas, HCs w/CO_2 and H_2S	Acid free gases and liquid HCs
Desalting	Dehydration	Absorption	Remove contaminants	Crude oil	Desalted crude oil
Drying and sweetening	Treatment	Absorption/ thermal	Remove H_2O and S-compounds	Liq HCs, LPG, alky feedstock	Sweet and dry hydrocarbons
Furfural extraction	Solvent extr.	Absorption	Upgrade mid distillate and lubes	Cycle oils and lube feed-stocks	High quality diesel and lube oil
Hydrotreating	Hydrogenation	Catalytic	Remove S, N, and metals, saturate HCs	Naphtha, distillates, residues, cracked HCs	Olefins, cracker feed, gasoline, distillate, lube
Phenol extraction	Solvent extr.	Absorption/ thermal	Improve visc. index, color	Lube oil base stocks	High quality lube oils
Solvent deasphalting	Treatment	Absorption	Remove asphalt	Vac. tower residual, propane	Heavy lube oil, asphalt
Solvent dewaxing	Treatment	Cool/filter	Remove wax from lube stocks	Vac. tower lube oils	Dewaxed lube basestock
Solvent extraction	Solvent extr.	Abspt precip.	Separate unsat. oils	Gas oil, reformate, distillate	High-octane gasoline
Sweetening	Treatment	Catalytic	Remove H_2S, convert mercaptan	Untreated distillate/ gasoline	High-quality distillate/ gasoline

A larger proportion of heteroatomic compounds in unconventional crude oils bring about a series of processing issues, which has been associated with the presence of asphaltenes, metal compounds, and/or acid compounds. The particular impact of the different compounds will be discussed in the corresponding chapters.

Higher severity processes are required to convert the heavier, more recalcitrant compounds. These complex/recalcitrant compounds are more abundant the heavier the oil is. Refinery economics dictates the need to increase severity at or near to conditions at which asphaltene precipitation starts. Refinery operation at maximum conversion for OCs jeopardizes reliability, e.g., for HACs corrosion would increase and for heavy oils asphaltene-induced fouling would show up. In fact, these major components of unconventional oils are partially or totally responsible for problems associated with solid deposition/fouling, emulsion formation, flocculation, catalyst deactivation, storage instability, and corrosion.

On the other hand, the environmental requirements continue to become increasingly stringent, imposing arduous pressures on the load for the refining units, regardless of the conversion option the refiner chooses. The regulations impose tougher quality specifications on the products side of the oil market, particularly stringent for the transportation fuels. The demand for fuel oil (the heaviest of the fuels) is decreasing, and consequently more complex residues have to be converted to higher-quality fuels. Therefore the increase in the unconventional oils proportion into the refinery diet, together with the increasingly stringent environmental requirements, make the fuel manufacturing process one of enormous complexity. For these reasons, API gravity and S-content were traditionally taken as quality indicators. The lower the API, the larger the proportion of VR in the crude oil. Consequently, fuel oil yield would increase. Similarly, increasing the S-content imposes increments in hydroprocessing needs. Then, the impact of crude oil quality on refining processing costs (RPCs) can be easily visualized. Refiners use their own models for the estimation and prediction of processing costs. An example has been reported in Ref. [46] in which a comparison of the estimated RPC is presented for light low S-containing crude oils with heavier and higher S-containing examples. The lowest calculated RPC of about $4.1 was obtained for Seria export light (Fig. 3 in Ref. [46]). The impact of API gravity and S-content on heavier and sourer crude oils on RPC is reproduced in Fig. 1.20.

As discussed, decreasing API aggravates processing and negatively affects RPCs. Processing OC oils may result unattractive, based purely on their RPCs. Some refineries with technological capabilities may consider processing under circumstances in which crude oil price differentials are larger than their RPC.

HACs are not free from difficulties and bring new challenges to the refiner. The increasing HAC supply volume call on refinery capacity increases [82,83], providing the crudes are

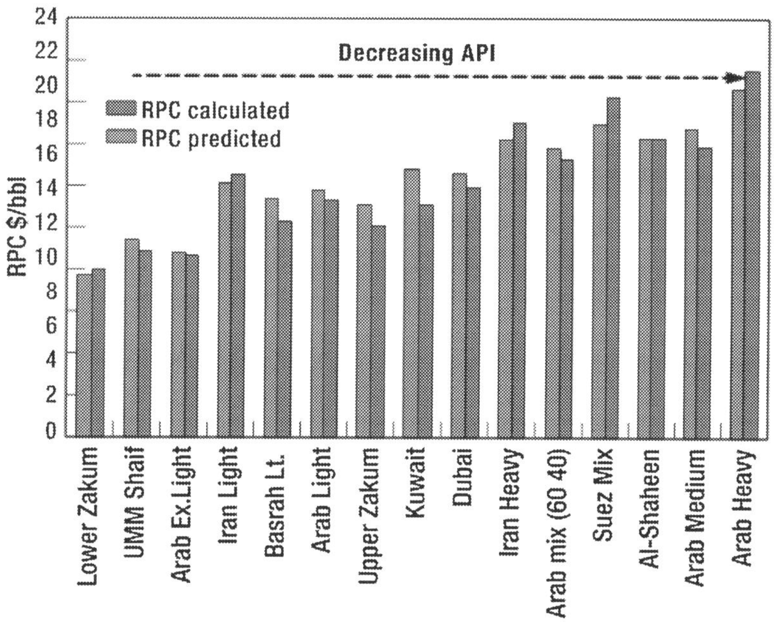

Figure 1.20

Refining processing costs for various high S-containing crude oils. *RPCs*, refining processing costs. *Reproduced from Voolapalli RK, Parihar P, Kumar R. Improve feedstock selection for your refinery. Hydrocarb Proc. 2012;(6):65–71, with permission from Gulf Pub.*

blended down to a feed limit of 0.5 with a 0.2 TAN crude. Under this situation, dilution could help the refiner to take advantage of the discount opportunity [84]. Nevertheless, new refining strategies have to be defined for processing this type of crude (see, for instance, [66,85]) and even other process technologies could improve margins by facilitating processing. The Chinese Huizou refinery is an example; designed for 240 kbpd it constitutes the first large-scale refinery for the direct processing of HACs [66]. In general, high discounts for HACs have important implications for developing countries. Depending on the relative rate of construction of adequate refineries and of the share growth of HACs, the discount could be expected to increase.

Simple refineries are not equipped for the processing of this type of crude oil. In some instances and depending on the actual refinery diet, probably the inclusion of new units dedicated to changing the properties of the new crude blend into the specifications of the currently operating units might have to be considered [86]. The ideal processing technology would be one that converts a heavy or sour feedstock into another that meets current specifications, yielding economically higher benefits. The use of existing conventional technologies would lead to an increase in the complexity of the refinery, probably to a level of that exemplified in Fig. 1.21.

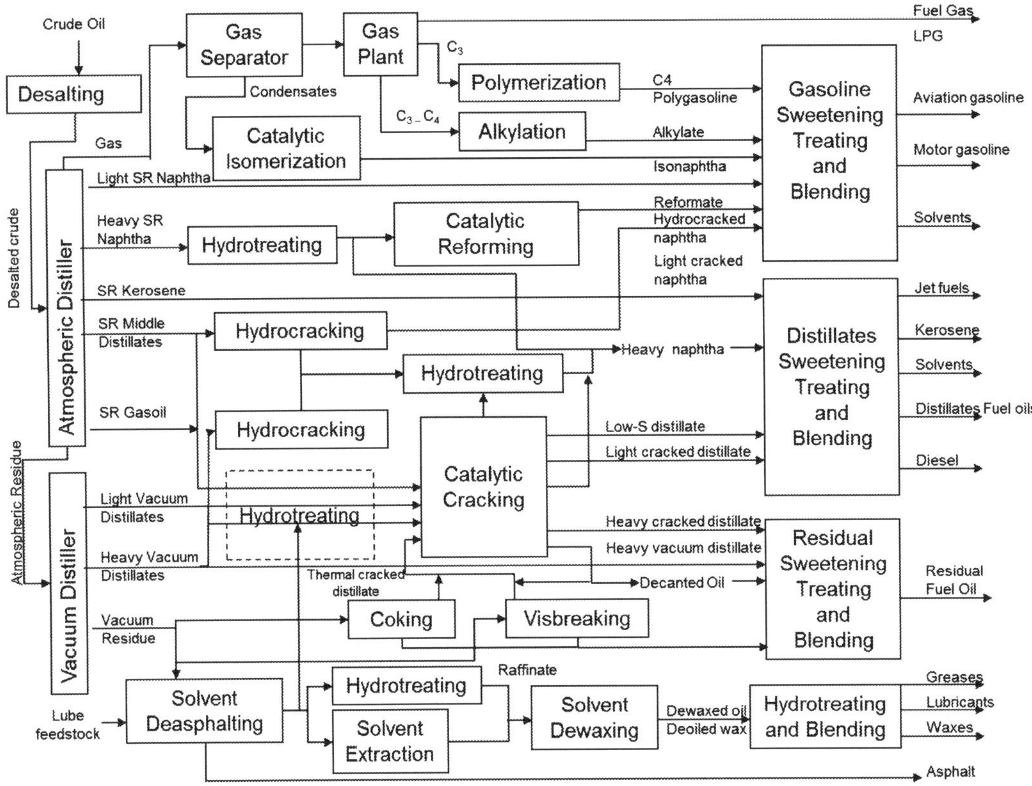

Figure 1.21

Example scheme of a complex refinery. *SR*, straight run. *Reproduced from Ramirez-Corredores MM, Borole AP. Biocatalysis in oils refining. Studies Surf. Sci. Catal., vol. 164. Amsterdam Elsevier; 2007. 406 pp., with permission from Elsevier.*

Additionally, operating conditions have to be flexible enough to yield similar benefits on feedstock changes. The incorporation of unconventional oils into the refinery diet represents technical challenges that have to be overcome prior to purchase. These challenges vary among the geographical regions because of market, product specifications, and environmental regulations [87]. For instance, the evolution into more stringent environmental regulations in Europe has taken place faster than anywhere else in the world. The availability of conventional (light and sweet) oil also changes within the regions. Therefore the pressures on the refiners for the potential incorporation of unconventional oils and the capabilities of existing refineries for handling it differ geographically and will drive their strategic decisions. The procedure for the selection of the unconventional oil to purchase differs from that of conventional crude oils. Besides economic evaluation, a careful consideration of operational concerns has to be included in the decision-making process. A new crude, particularly an unconventional oil, would bring both operational and capital expenditures, whose accurate evaluation is required and are particular for any given refinery's configuration [88].

Nevertheless, one remaining question prevails: is it the time to reinvent refining? This question is the inspiration of the present book, since at this point it is clear that conventional technologies cannot provide a comprehensive solution for unconventional troublesome crude oils.

4. Summary

During the last decade, the oil industry faced a scenario with records of high oil prices, price volatility, and increasing supplies of emerging unconventional crude oils. The poor quality of these oil grades moves the producer to sell them under discounted prices, offering refining the opportunity to challenge their capabilities. Thus they were called OCs. Two different broad classes of OCs can be distinguished: the highly viscous and contaminated crude oils (heavy, extraheavy, and bitumen) and the acid/sour oils.

OCs have the potential to provide significant economic benefits for refiners since these crude oils can be purchased at discounted prices. To take advantage of the potential improved profitability, the refinery has to be upgraded for increasing feedstock flexibility. Refiners are struggling to keep margins in good shape by trying to overcome many technical challenges, by mitigating effects for maintaining unit integrity and reliability, by monitoring, controlling, and adjusting product quality, and by making sure environmental compliance is in place.

Nevertheless, the opportunity for cost-effective technologies for abating the effect of these compounds remains open. Most likely, the inclusion of new units dedicated to change the properties of the new blend diet into the specifications of the currently operating units might have to be considered. More drastic solutions may call for reinventing refining. Only those refiners with technology to support flexibility will enjoy a competitive advantage.

The so-called BotB refers to the heavier molecules whose conversion or removal is mandatory, if a smooth refining operation is desired. Asphaltenes, metal-bearing compounds, and NAs constitute the "bad" actors present in OCs. Their physical properties and reactivity have severe consequences not only in refining, but also in production and transportation. A lack of comprehensive knowledge of the bad actors molecules may be the reason for the inexistence of cost-effective technologies. This book will focus on the discussion of the existing relevant knowledge on processing unconventional oils, and will identify the technology gaps to be closed in the future.

Different needs and problems have prompted the refining industry into the search and development of technologies for processing unconventional oils. Originally, they would have filled the role of product quality improvements and market diversification. Nowadays, their increasing availability, based on the existing huge reserves, together with the declining offer of light oils is pushing a renewed interest in this area. The increase in their proportion into the refinery diet, together with the increasingly stringent environmental requirements and the speciation trend, as a whole makes the fuel manufacturing process one of enormous complexity. Probably the inclusion of new units dedicated to change the properties of the new crude blend diet into the specifications of the currently operating units might have to be considered in the short term.

The objective of the present work is to set a baseline for the processing of unconventional oils by discussing relevant knowledge and identifying technology gaps. This book consists of eight chapters. This first chapter contains the motivation for a collective deployment of the existing knowledge on the compounds that make refining (and production as well) of unconventional oils more difficult and costly and the current refining situation in terms of the impact these compounds may cause. Then, the second through fourth chapters, respectively present the information associated with each particular class of compounds (asphaltenes, metals, and NAs), their characterization, and behavior. In the fifth and sixth chapters the processes for the abatement of these contaminants are included. The former deals with the commercially available technologies, and the emerging concepts and processes are considered in the latter. Then, a new actor competing in the nonconventional scene is presented in Chapter 7, shale oil. The conclusions, future possibilities, and final remarks are collected in Chapter 8.

References

[1] Kremer LN. Crude oil and quality variations. PTQ 2004:87–93. Autumm.

[2] Boduszynski MM, Rechsteiner CE. Molecular weight distributions in heavy crudes. In: Proc. 7th UNITAR Intern. Conf. 'Heavy Crude Tar Sands'. Beijing, China; October 27–30, 1998. 9 pp.

[3] Boduszynski MM, Rechsteine r CE, Shafizadeh ASG, Carlson RMK. Composition and properties of heavy crudes. In: Proc. 7th UNITAR Intern. Conf. 'Heavy Crude Tar Sands'. Beijing, China; October 27–30, 1998. 12 pp.

[4] Boduszynski MM. Characterization of 'heavy' crude components. In: Proc. ACS Div. Pet. Chem. Preprints. Chicago, IL, USA, 30. ACS; 1985. p. 626–35.

[5] Boduszynski MM. Limitations of average structure determination for heavy ends in fossil fuels. Pet Sci Technol 1984;2(3):211–32.

[6] Boduszynski MM, Altgelt KH. Composition of heavy petroleums. 4. Significance of the extended Atmospheric Equivalent Boiling Point (AEBP) scale. Energy Fuels 1992;6(1):72–6.

[7] Boduszynski MM. Composition of heavy petroleums. 2. Molecular characterization. Energy Fuels 1988;2(5):597–613.

[8] Boduszynski MM. Composition of heavy petroleums. 1. Molecular weight, hydrogen deficiency, and heteroatom concentration as a function of atmospheric equivalent boiling point up to 1400. Degree.F (760c). Energy Fuels 1987;1(1):2–11.

[9] Altgelt KH, Boduszynski MM. Composition of heavy petroleums. 3. An improved boiling point-molecular weight relation. Energy Fuels 1992;6(1):68–72.

[10] Jewell DM, Albaugh EW, Davis BE, Ruberto RG. Integration of chromatographic and spectroscopic techniques for the characterization of residual oils. Ind Eng Chem Fund 1974;13(3):278–82.

[11] Marshall AG, Rodgers RP. Petroleomics: the next grand challenge for chemical analysis. Acc Chem Res 2004;37(1):53–9.
[12] Klein GC. Petroleomics: Applications in the fingerprinting of the acidic and basic crude oil components detected by electrospray ionization fourier transform ion cyclotron resonance mass spectrometry. Doctor of Philosophy Thesis from Florida State University, Department of Chemistry and Biochemistry; October 12, 2005. 145 pp.
[13] Fu J, Kim S, Rodgers RP, Hendrickson CL, Marshall AG, Qian K. Nonpolar compositional analysis of vacuum gas oil distillation fractions by electron ionization fourier transform ion cyclotron resonance mass spectrometry. Energy Fuels 2006;20(2):661–7.
[14] Mullins OC, Sheu EY, Hammami A, Marshall AG. Asphaltenes, heavy oils, and petroleomics. New York: Springer-Verlag Inc.; 2006. 800 pp.
[15] Klein GC, Kim S, Rodgers RP, Marshall AG, Yen A, Asomaning S. Mass spectral analysis of asphaltenes. I. Compositional differences between pressure-drop and solvent-drop asphaltenes determined by electrospray ionization fourier transform ion cyclotron resonance mass spectrometry. Energy Fuels 2006;20(5):1965–72.
[16] Barrow MP, Headley JV, Peru KM, Derrick PJ. Fourier transform ion cyclotron resonance mass spectrometry of principal components in oilsands naphthenic acids. J Chromatogr A 2004;1058(1–2):51–9.
[17] Sayles S, Routt DM. Unconventional crude oil selection and compatibility. San Antonio, TX: Proc. NPRA Annual Meeting; 2011. p. 380–400.
[18] Crude Quality Inc. Crudemonitor. Ca. 2015. Available from: http://www.crudemonitor.ca/home.php.
[19] Haaland O, Klovning R, Sem T. The future of the world's extra heavy oil resources - competition and potential. In: Proc. 7th UNITAR Intern. Conf. 'Heavy Crude Tar Sands'. Beijing, China, 1; October 27–30, 1998. 18 pp.
[20] US Department of Energy. Energy information administration petroleum & other liquids. 2015. Available from: http://www.eia.gov/petroleum/.
[21] International Energy Agency. Online data services. IEA; 2015.
[22] World Energy Council (WEC). 2015. Available from: www.worldenergy.org.
[23] World Petroleum Council Reserves & resources. 2015. Available from: http://www.world-petroleum.org/index.php?/Reserves-Resources/reserves-a-resources.html.
[24] Organization of the Petroleum Exporting Countries (OPEC). Datagraphs. 2015. Available from: http://www.opec.org/opec_web/en/data_graphs/335.htm.
[25] BP International Plc. BP statistical review of world energy. 2008. Available from: http://www.bp.com/subsection.do?categoryId=9023761&contentId=7044545.
[26] BP International Plc. BP statistical review of world energy. 2014. Available from: http://www.bp.com/subsection.do?categoryId=9023761&contentId=7044545.
[27] Yeung TW. Evaluating opportunity crude processing. PTQ 2006;11(Q4):93–6.
[28] Fattouh B. Opec's discounts on heavy crude oil: is a new policy instrument taking shape?. Oxford Institute for Energy Studies. Oxford Energy Comment; June 2006. 9 pp.
[29] Skippins J, Johnson D, Davies R. Corrosion-mitigation program improves economics for processing naphthenic crudes. Oil Gas J 2000;98(37):64–8.
[30] Tahmassebi H. Crude oil and product value differentials: historical perspective and outlook. Energy 1986;11(4–5):343–59.
[31] Alberta Energy. Enbridge and Alberta energy and utilities board appendix a: markets and pricing for Alberta bitumen production. In: Technical report os#1: oil sands economics royalty. ed. Edmonton, Canada: Alberta Department of Energy; 2007. 21 pp.
[32] Rabinow RA. ExxonMobil pipeline company. In: Quality bank policy applying on crude petroleum from points offshore Texas to points onshore Texas102. F.E.R.C.; 2001. 6 pp.
[33] Hartmann JCM. Crude valuation for crude selection. PTQ 2003:123–7. Winter.
[34] Wang Z. Hedonic prices for crude oil. Appl Econ Lett 2003;10:857–61.

[35] Bacon R, Tordo S. Crude oil price differentials and differences in oil qualities: a statistical analysis. Report No. WB.34906. Technical paper No. 81. Washington, DC, USA: ESMAP; October 01, 2005. 40 pp.
[36] Kittrell N. Removing acid from crude oil. In: Proc. Crude Oil Quality Association Meeting. New Orleans, LO; February 2006. 19 pp.
[37] Stratiev D, R.Dinkov, Petkov K, Stanulov K. Evaluation of crude oil quality. Pet Coal 2010;52(1):35−43.
[38] Switzer M. Explaining oil price differentials. In: Best of BNN; November 21, 2013. Calgary, Canada, 1 pp.
[39] Cryderman K. Oil price discount roars back amid glut. In: The globe mail; November 06, 2013. Calgary, Canada, 2 pp.
[40] Katusa M. The tricky calculus of oil price differentials. In: The world of energy; January 29, 2014. Calgary, Canada, 1 pp.
[41] Hume N. Differential between oil prices widens as US crude stocks rise. Financ Times 2014. 1 pp. Available from: http://www.ft.com/cms/s/0/c9de3e62-8911-11e3-bb5f-00144feab7de.html#ixzz2zYZg5FEW.
[42] Bendiner J, Burleton D, Preston L. Drilling down on crude oil price differentials. TD Econ March 2013. 6 pp. Available from: http://www.td.com/document/PDF/economics/special/DrillingDownOnCrudeOilPrice Differentials.pdf.
[43] Fattouh B, Sen A. New swings for West African crudes. The Oxford Institute for Energy Studies: Oxford Energy Comment; August 2014. 7 pp.
[44] Shafizadeh A, Mcateer G, Sigmon J. High acid crudes. In: Proc. Crude Oil Quality Association Meeting. New Orleans; January 30, 2003. Chevron Texaco, 50 pp. Available from: http://www.coqg.org/20030130High%20Acid%20Crudes.pdf.
[45] Lordo S. Impact and management of opportunity crudes on the desalting and wastewater process. In: Proc. Opportunity Crudes Conf.: challenges, benefits. Houston, TX. Hydrocarbon Publishing Co; April 28−May 2, 2008 [Unpublished presentation].
[46] Voolapalli RK, Parihar P, Kumar R. Improve feedstock selection for your refinery. Hydrocarb Proc 2012;(6):65−71.
[47] Processing News. Refiner demand for discounted feeds will increase trade of high-acid crudes. Oil Gas J 2008;106(11):52−5.
[48] Arora A, Mukherjee U. Refinery configurations for maximising middle distillates. PTQ 2011;16(3). 9 pp.
[49] Arora A, Mukherjee U. Refinery configurations for maximum conversion to middle distillates. In: Proc. NPRA Annual Meeting. San Antonio, TX, USA; March 20−22, 2011. Paper AM-11−19, 18 pp.
[50] Birch C. Achieving maximum crude oil value depends on accurate evaluation. Oil Gas J 2002;100(2). 54−56, 58.
[51] Petroleum Resources Branch. A review of selected issues affecting the price of crude oil. Report No. Ottawa, Ontario: Natural Resources Canada; October 2010.
[52] Bacon R, Tordo S. Crude oil prices: predicting price differentials based on quality. View Point, The World Bank; 2004 [Note #275], 4 pp.
[53] Cao X. Optimum selection of vacuum residue processing scheme in oil price boosting area. Pet Proc Petrochem 2009;40(1):1−9.
[54] Martinez AR. Report of working group on definitions. In: Proc. 1st Intern. Conf. The future of Heavy Crude and Tar Sands. Edmonton, Alberta, Canada; 1982. p. 97−8. Available from: http://www.oildrop.org/Lib/Conf/2nddef.html.
[55] Dusseault MB. Hydrocarbons from non-conventional and alternative fossil resources. In: Treccani P, editor. Encyclopaedia of hydrocarbons. Roma: Istituto Della Enciclopedia Italiana, Fondata Da Giovanni Treccani S.P.A. Marchesi Grafiche Editoriali S.P.A.; 2007. p. 21−55.
[56] Canadian Association of Petroleum Producers (CAPP). Crude oil forecast, markets and pipeline expansions. Canada. 2014. 52 pp. Available from:, http://www.capp.ca/getdoc.aspx?DocId=247759&DT=NTV.
[57] IEA. Resources to reserves: oil & gas technologies for the energy markets of the future. Report No. 2005. 130 pp.
[58] Energy Information Administration. Annual energy outlook 2006 with projections to 2030. Report No. DOE/EIA-0383. 2006. 242 pp.

[59] Energy Information Administration. Percentages of total imported crude oil by API gravity. 2006. Available from: http://tonto.eia.doe.gov/dnav/pet/pet_move_ipct_k_a.htm.
[60] International Energy Agency. World energy outlook. Paris, France: OECD/IEA; 2012. 690 pp. Available from: http://www.iea.org/publications/freepublications/publication/name. 49561,en.html.
[61] Dou L, Cheng D, Li M, Xiao K, Shi B, Li Z. Unusual high acidity oils from the great palogue field, melut basin, Sudan. Org Geochem 2008;39(2):210–31.
[62] Skippins J, Bell K, Kronk J, Bagdasarian A, Johnson D. High acid crude. In: Proc. Crude Oil Quality Association Meeting. New Orleans, LA, USA; May 2002. ChevronTexaco, 180 pp.
[63] Szklo AS, Machado G, Schaeffer R, Felipe Simoes A, Barboza Mariano J. Placing Brazil's heavy acid oils on international markets. Energy Policy 2006;34(6):692–705.
[64] Fedorak PM. Overview of naphthenic acids analyses at the University of Alberta. In: Proc. CONRAD/OSERN Symp. Coast Terrace Inn, Edmonton, Alberta; May 12–13, 2003. 50 pp.
[65] Troner A. Buyer of first resort: Asia pacific & bitumen feedstock. In: Proc. 7th Asian bitumen Conf. Bangkok, Thailand; October 19–21, 2011. Unpublished Presentation, 34 pp. Available from: http://www.apecconsulting.com/PDF/WebBitumenSpeechOct2011Sidebar.pdf.
[66] Qing W. Processing high TAN crude: Part I. PTQ 2011;16(Q1):35–43.
[67] Qin KZ, Wang RA, Jia SS, Lao YX, Guo SH. Characterization and geochemistry of devonian oil shale north Alabama-south central Tennessee. In: Proc. ACS National Meeting, symposium on characterization and chemistry of oil shales petroleum chemistry. St. Louis, Missouri; 1984. p. 104–12.
[68] Rheams KF, Neathery TL. Characterization and geochemistry of devonian oil shale north Alabama-south central Tennessee. In: Proc. ACS National Meeting, symposium on characterization and chemistry of oil shales petroleum chemistry. St. Louis, Missouri; 1984. p. 113–25.
[69] Dalai NS, Suryan MS, Shen MS. Electron Spin Resonance and thermogravimetric investigations of the pyrolysis of eastern shales. Preprints of the ACS symposium on characterization and chemistry of oil shales. In: Proc. ACS National Meeting, symposium on characterization and chemistry of oil shales petroleum chemistry. St. Louis, Missouri; 1984. p. 34–40.
[70] Jong KM, Petrakis L, Takayasu M, Friedlaender FJ. Physicochemical properties of magnetically separated oil shale solids. In: Proc. ACS National Meeting, symposium on characterization and chemistry of oil shales petroleum chemistry. St. Louis, Missouri; 1985. p. 159–65.
[71] Levy JH, Stuart WI. Thermal properties of australian oil shales: characterization by thermal analysis and infrared spectrophotometry. In: Proc. ACS National Meeting, symposium on characterization and chemistry of oil shales petroleum chemistry. St. Louis, Missouri; 1984. p. 61–70.
[72] Chen X, Shen B, Sun J, Wang C, Shan H, Yang C, et al. Characterization and comparison of nitrogen compounds in hydrotreated and untreated shale oil by electrospray ionization (ESI) fourier transform ion cyclotron resonance mass spectrometry (FT-ICR MS). Energy Fuels 2012;26(3):1707–14.
[73] Kumar R, Bansal V, Badhe RM, Madhira ISS, Sugumaran V, Ahmed S, et al. Characterization of Indian origin oil shale using advanced analytical techniques. Fuel 2013;113:610–6.
[74] Tong J, Liu J, Han X, Wang S, Jiang X. Characterization of nitrogen-containing species in huadian shale oil by electrospray ionization fourier transform ion cyclotron resonance mass spectrometry. Fuel 2013;104:365–71.
[75] Savest N, Hruljova J, Oja V. Characterization of thermally pretreated kukersite oil shale using the solvent-swelling technique. Energy Fuels 2009;23(12):5972–7.
[76] Hsu CS, Lobodin VV, Rodgers RP, Mckenna AM, Marshall AG. Compositional boundaries for fossil hydrocarbons. Energy Fuels 2011;25(5):2174–8.
[77] Ohmes RA, Sayles S. Crude oil property and contaminant characterization and tracking in the modern heavy oils refinery. In: Proc. 242nd ACS Nat. Meeting. Denver, CO; August 28–September 1, 2011. 1 pp.
[78] Bae E, Na JG, Chung SH, Kim HS, Kim S. Identification of about 30 000 chemical components in shale oils by electrospray ionization (ESI) and atmospheric pressure photoionization (appi) coupled with 15 t fourier transform ion cyclotron resonance mass spectrometry (FT-ICR MS) and a comparison to conventional oil. Energy Fuels 2010;24(4):2563–9.

[79] Process Notes. Oil sands crude – profits and problems? PTQ 2012. 1 pp.
[80] Van Den Berg FGA, Heijnis RMA, Stamps PA, Kramer PA. A geochemical framework for understanding residue properties. In: Proc. AIChE Spring Nat. Meeting. New Orleans, LA; March 11–14, 2002. p. 1418–23.
[81] Abu-Elgheit M, Khalil SO, Barakat AO. Characterization of crude oils by new trace-metal indexes. ACS Div Pet Chem Prepr 1979;24(3):793–7.
[82] Houlton G. Global crude oil supply and quality trends. In: Proc. Crude Oil Quality Association Meeting. San Antonio, TX, USA. Purvin & Gertz; February 24, 2011. 25 pp.
[83] Purvin & Gertz Inc. Study on oil refining and oil markets. Buenos Aires, Calgari, Dubai, Houston, London, Los Angeles, Moscow, Singapore: Prepared for the European Commission; January 2008. Report No. L2293/mg, 377 pp. Available from: http://ec.europa.eu/dgs/energy_transport/security/studies/doc/energy/2008_01_study_oil_refining_oil_markets_en.pdf.
[84] Skippins J, Johnson D, Davies R. Evaluation of the economics for the processing of naphthenic crudes. In: Proc. Intern. Conf. corrosion in refinery petrochemical power generation plants. Venice, Italy; May 18–19, 2000. p. 275–86.
[85] Qing W. Processing high TAN crude: Part II. PTQ 2011;16(1). 7 pp.
[86] Process Notes. Opportunity knocks. PTQ 2011. 1 pp.
[87] Carruthers G, Sayles S, Mark Routt D. Bottom of the barrel: unconventional crudes (part one). Hydrocarb Eng 2011;16(10):45–50.
[88] Carruthers G, Sayles S, Mark Routt D. Bottom of the barrel: unconventional crudes (part two). Hydrocarb Eng 2012;17(3):33–6.
[89] Ramirez-Corredores MM, Borole AP. Biocatalysis in oils refining. In: Studies Surf. Sci. Catal, vol. 164. Amsterdam Elsevier; 2007. 406 pp.

CHAPTER 2

Asphaltenes

1. Introduction

The major constituents of crude oils are fractionated according to their boiling point, as described in Chapter 1. However, the heaviest fraction, the residue of distillation, can be separated into two solubility classes, based on their solubility in light alkanes: a soluble fraction, the maltenes, and an insoluble fraction, the asphaltenes. The nondistillable constituents are also referred as the bottom of the barrel (BotB).

Maltenes contain saturates (S), aromatics (A) and the heavier compounds are named resins (R). Asphaltenes are usually present in most viscous and heavy crude oils. They are considered to be the highest molecular weight (MW) component of crude oil. Consequently, being asphaltenes, the heaviest compounds of the residue fraction, they can be regarded as the bottom of the BotB. Although in some respects (e.g., MW), the borderline between asphaltenes and resins is a matter of definition, solvent classification remains a useful tool to categorize the heavy components in petroleum according to their solubility.

Since their identification in 1837 [1], they have been the subject of numerous studies, as will become evident later. Nevertheless, a full understanding of their molecular constituents and the phenomenology of their behavior has not yet been achieved.

These heavy molecules cause problems not only to refining, but also to production and transportation. Studies carried out on the "live oil" (the crude oil present downhole the oil well) and on deposits collected from pipelines have brought about controversies and uncertainties regarding the characterization results obtained from asphaltenes isolated from produced crude oils (dead oils). From a production/transportation standpoint, understanding the asphaltene behavior and preventing their deposition require practices for sampling and characterization [2] that might differ from those followed from a refining viewpoint. The differences among asphaltenes are brought about not only from where in the oil value chain they are being sampled, but also from which crude oil they originate and how they are isolated. Although in some instances live oil asphaltenes will be discussed, the scope of this book will focus on dead oils.

There was strong evidence to support the prevailing sentiment that the fraction, which is coined "asphaltenes," is greatly influenced and defined by the very method utilized to

The Science and Technology of Unconventional Oils. http://dx.doi.org/10.1016/B978-0-12-801225-3.00002-4

isolate it. Molecular characterization and fundamental knowledge are underpinned by the separation and identification of the BotB compounds, and their completion is far from accomplished. The lack of capabilities for the isolation of individual molecules has precluded the success of most of the molecular characterization work. The complex nature of the examined mixture of compounds creates severe complications for explaining the experimental results. In recent years, significant advances have been achieved because of the remarkable improvements in analytical techniques. Nonetheless, further advances are still needed, specifically in structure—behavior relationships. This chapter will discuss their characteristics and behavior, and will try, when possible, to create bridges between the two.

2. Definition

The term *asphaltenes* was coined by Boussingault [1] in 1837 for the distillation residue of bitumen, since it was observed to be an asphalt-like material. At that time, asphaltenes were characterized as being insoluble in alcohol and soluble in turpentine. Marcusson [3], in the early part of this century, introduced a solvent precipitation method to fractionate the crude oil in two separated fractions, using petroleum naphtha as solvent. These two fractions were named *petrolenes*, later referred as *maltenes* (the soluble part), and asphaltenes (the insoluble part). In operational terms, Nellensteyn, in the 1930s, suggested that the nuclei of asphaltene micelles consist of microcrystalline graphite particles [4]. A more elaborated colloidal definition was then introduced by Pfeiffer and Saal in 1940, stating that asphaltenes might consist of highly aromatic molecular hydrocarbons, which were the centers of the micelles formed by sorption of resin molecules on the surface or even at the interior of the asphaltene "particles" [5]. Later on, the use of a single compound as the antisolvent (AS) was introduced, suggesting *n*-pentane at first and later *n*-heptane. *Asphaltenes were then defined as that portion of the crude oil that was soluble in toluene and insoluble in n-heptane*. In consequence, asphaltenes are not a specific family of chemicals with common functionality that differ in their MW.

These operational-based definitions result vague and broad, and leave a conceptual gap from a molecular compound concept. Regardless of its simplification and operational nature, it became the most accepted and used definition for asphaltenes. Regrettably, this definition does not include their physical state in the oil, nor their chemical or molecular nature. Asphaltene yield increases with decreasing carbon number (CN) of the alkane AS used for precipitation [6,7]. There is a school of thought that considers that the higher the yield, the higher the contamination, typically by resins of the collected asphaltene fraction. Another school of thought considers that yields affected with the method and conditions employed involve precipitation of asphaltenes of different structures. Hence there are pentane, hexane, or heptane asphaltenes, for instance. Properties affected by the choice of AS include MW, polarity, and aromaticity [8] (Fig. 2.1).

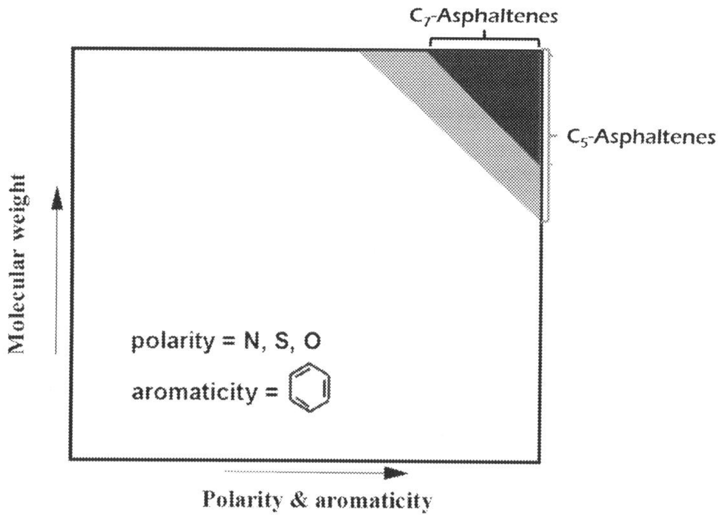

Figure 2.1

Asphaltene properties variation with antisolvent employed. *Reproduced from Long RB. The concept of asphaltenes. In: Bunger JW, Li N, editor. Adv. Chem. Series 195: Chemistry of asphaltenes. Washington, DC: ACS; 1981. pp. 17–27, with permission from ACS Publications.*

Asphaltenes precipitate as a solid powder. Therefore as such asphaltenes may not be present in petroleum, but rather be formed in the process of action of an AS on the system, resulting as an associative combination of molecules having a higher density than the solution, and hence becoming separated from the system by precipitation. Unger attributed the formation of asphaltenes to the presence of paramagnetic molecules that have positive potential energies of interaction with respect to saturated hydrocarbon molecules (repulsion) or other types of molecules with σ-bound atoms [9]. Meanwhile, other authors consider the interaction between the aromatic systems to be responsible for aggregation (stacking via π-bonding, see Section 4.3). Thus during aggregation, molecules are distinguished as those bearing aromatic systems and those with steric inhibition that would disrupt stacking [10].

From an operational level definition to a molecular level of understanding, a big gap of uncertainty still exists. Those first approaches to reach a molecular definition of asphaltenes were the foundations for a long-lasting colloidal model of the asphaltene structural organization that is still considered for explaining some of their behavioral features. In fact, from the very beginning and even today, asphaltenes are believed to exist as discrete or dispersed colloidal particles in crude oil, while the resin molecules act as the transition between the rest of the oil [11]. The larger/heavier molecules would occupy the core, and aromaticity and size would decrease toward the outer particle. Still, whether a single asphaltene molecule or more than one occupies the core is still unclear [12].

This ill-defined nature of the asphaltenes disregards their molecular complexity and has led to the assumption that they can be considered as a single simple component of crude oil. An improvement to this view is to ponder these compounds as a polydispersed material in terms of MW, shape, and size (exhibiting size, shape, and MW within a broad distribution range), keeping in mind that this fraction is not uniform in chemical composition. Meanwhile, their physical nature in the crude oil is explained in terms of isolated entities, chemically behaving as macromolecules. This physical nature is highly affected by asphaltenes' main feature: aggregation. While aggregation might lead to the precipitation of the solid asphaltenes and could define their physical appearance, their chemical performance seems to be better explained based on micellar and colloidal behavior [13].

Chemical characterization of this family of compounds indicates that they are highly aromatic molecules bearing heteroatomic moieties. Aliphatic alkyl structures are also present, but the existence of naphthenic cycles is still subject to controversy, as is the degree of condensation of the aromatic rings. The typical approach of asphaltene molecular representation is based on a molecular reconstruction, which translates all available characterization information of the precipitated mixture into an average model molecule. The nonunique solutions of this "top-down" approach suffer a significant drawback, as will be illustrated in the next sections. A "bottom-up" approach has also been explored by many groups. It involves well-defined model molecules and systematic structure variations. A molecular definition of asphaltenes is yet to be established, though significant advances have been gained in the understanding of the molecular characteristics of these compounds.

The physical definition of asphaltenes considered a hierarchical system in which molecules tend to associate into molecular nanoaggregates. These nanoaggregates may be dispersed in the oil matrix or undertake further association into a micellar architecture. However, larger colloidal particles of asphaltene are unstable and undergo phase separation forming a solid mass [14].

As well as asphaltenes, resins are also defined operationally. Thus resins are the material that precipitates with addition of propane, but are soluble in *n*-heptane. Since the propane/*n*-heptane pair is not universally accepted, the general concept is that *resins are insoluble in lower MW n-alkanes and soluble in higher MW n-alkanes*. A more specific definition of resins was proposed by Demirba as: *a polar fraction of petroleum that is soluble in n-alkanes and aromatic solvents and insoluble in ethyl acetate* [15].

3. Isolation

Asphaltenes are present in crude oil at higher concentrations when the API gravity is lower or the specific gravity/density of the oil is higher. Preferentially, the residue from distillation is used for their separation. Based on the definition, asphaltenes are first precipitated by using a normal alkane AS. The use of a single-component precipitant was suggested by Strieter in 1941, and the use of *n*-pentane for the isolation of Athabasca asphaltene was first reported by Pasternak and Clark in 1951. The asphaltene yield decreases with the CN of the AS, for instance, the precipitated fraction from Athabasca bitumen was 50% with liquid propane, while it was 17, 11, and 9% with *n*-pentane, *n*-heptane, and *n*-decane, respectively [16,17]. Yield increases in the order: terminal olefin > *n*-paraffin > isoparaffin [18].

Besides the nature of the AS, other factors affecting the yield and quality of the precipitated material include (1) the use of an asphaltene solvent to dilute the crude oil or bitumen, (2) the AS-to-oil ratio, (3) the ratio of solvent to AS, (4) the contact time of the asphaltenes to supernatant liquid, and (5) the operating p and T conditions. The use of a good solvent for asphaltenes might involve either starting with a liquid–liquid extraction of the vacuum resid (VR) with an aromatic solvent or using an aromatic dilution of the VR. Mixtures of paraffinic ASs have been tested, e.g., normal isoparaffin mixtures [19]. At least 40 volumes of the liquid AS should be used to ensure complete asphaltene precipitation [6]. Multistage methods have been considered in which, for instance, the AS-to-oil ratio is changed for each stage [20,21]. A contact time of at least 8 h has been recommended [22]. Although the evidence is scarce, a too long contact time might allow the resins to be adsorbed onto the asphaltenes [6].

Another issue to pay attention to is the possible retention of extraneous materials by the precipitated asphaltene, e.g., the paraffinic AS, other light compounds, inorganic salts and minerals, etc. AS retention has been demonstrated by a sequential precipitation paraffin removal with increasing CN of the AS. For this demonstration, the C_5-precipitated material was mixed with *n*-hexane. Then, the insoluble fraction was washed with *n*-heptane, and the C_7-insoluble material was mixed with *n*-octane. Rough characterization of the retained paraffinic material indicated an increase in the CN of this material, with the increase of the AS MW [23]. Light paraffins were reported to be trapped by asphaltenes during precipitation [24]. Heavy paraffins can be coprecipitated with the asphaltenes when waxy crude oils are treated. A method for the removal of waxes from the coprecipitate has been proposed, in which waxes and asphaltenes are first precipitated together on a ground polytetrafluoroethylene (PTFE)-packed column using methyl ethyl ketone at −20°C and then the precipitate is sequentially eluted with solvents of increasing polarity at different temperatures. Eluting solvents included heptane at −20°C (for low polarity oils and moderately branched alkanes, possibly containing naphthenic components), heptane at 60°C (for *n*-alkanes with carbon atoms higher than C_{20} and slightly branched alkanes), toluene at ~25°C (for asphaltenes), and then methylene chloride at ~25°C (for higher polarity asphaltene components). Tested crude oils included Minnelusa (Powder River Basin Field, Wyoming), Tensleep (Oregon Basin Field, Wyoming), LC (Alaska), Dakota (Fourteen Mile Field, Wyoming), and Gullfaks (Norway) [25]. Retention of minerals was detected by measuring the ash content of the precipitated asphaltenes directly from solutions of oil sand bitumen. This inorganic contamination could be reduced from 3.2 wt % to less than 0.5 wt% by centrifugation [26]. Other heavy compounds (450–650 amu), including paraffins and resins, have been reported to be trapped as well [24].

In current practice, when high-quality asphaltenes and maximized deasphalted oil (DAO) yield are targeted, the choice of precipitant is usually *n*-heptane. Nevertheless, even the *n*-heptane precipitate always contains high melting point paraffins and chemisorbed-*n*-heptane-insoluble resinous and other maltene materials (carboxylic acids, fluorenones, fluorenols, polycyclic terpenoids, thiolane- and thiane-derived and acyclic sulfoxides, carbazoles, quinolines, vanadyl porphyrins, etc.) as well as low-MW asphaltene fragments. Further washing with AS is typically recommended as a first attempt to remove part of these contaminants [27].

Asphaltenes quantification is also based on precipitation, requiring that standard analytical procedures have to be followed and specified (e.g., ASTM D893, D2006, D2007, D3279, D4124, D7061, D7827). The four broad components of the residue—saturates, aromatics, resins, and asphaltenes (SARA)—are separated and quantified using an ASTM method (D2007 or D4124, shown as a general scheme in Fig. 2.2A) or IP method (Fig. 2.2B), based on extraction (*n*-alkane) and adsorption (on alumina and/or silica gel). A complete fractionation of VR's solubility classes is included in Fig. 2.2C; examples of this type of fractionation applied to bitumen samples from Athabasca oil sands can be found in Refs. [28–30]. Two additional fractions, carboids and carbenes, are included in this last figure, which corresponds to toluene-insoluble compounds. These fractions can be found in few crude oils, but these are mainly present in thermally treated VRs or asphaltene-containing fractions or feeds. Carboids and carbenes are products from the early stages of asphaltene decomposition, characterized by a higher degree of aromaticity than the original asphaltenes and a slightly reduced MW. Carbenes, which are insoluble in benzene or toluene but soluble in carbon disulfide, are the first product of asphaltene decomposition. Carboids, which are insoluble in any organic solvent, are the true precursors to coke. Carboids are condensed, crosslinked polymers in which the greatest part of the carbon atoms is aromatic [31].

The issues associated with the solubility/adsorption–elution protocols have been discussed by Kharrat et al. [32] and are mainly associated with the physical and chemical properties of asphaltenes that are considered in this chapter. Modifications to this protocol have been suggested by researchers from the Alberta Research Council [33–36]. These modifications addressed the resin–asphaltene interactions and more particularly polar–polar interactions.

Depending on the desired quality, different sequences of solvent extractions are employed for isolation. Further purification of asphaltenes can be achieved by two different fractionation methods, and might be referred to as the "coarse method" (entire subset of

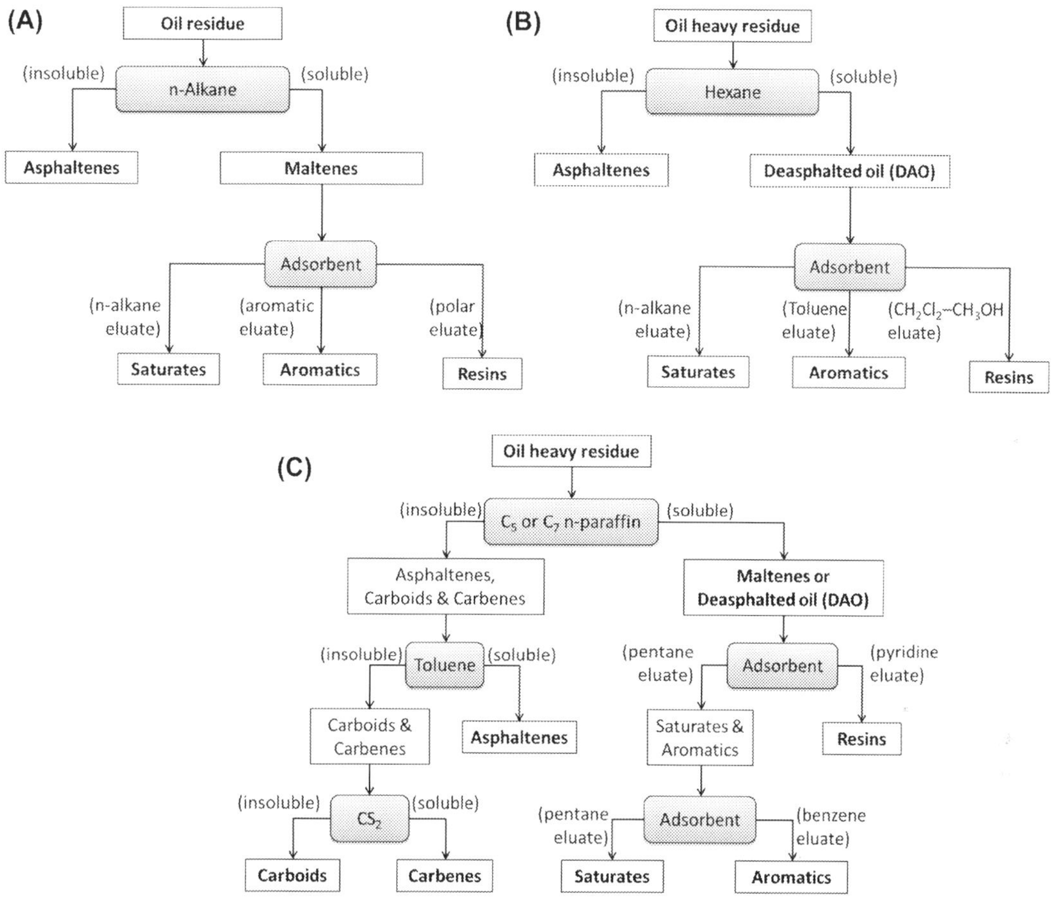

Figure 2.2
Residue separation in four components (A) ASTM D2007 or D4124, (B) IP 469, (C) complete fractionation of VR's solubility classes.

chemical species) and the "fine method" (discrete subset of asphaltenes that precipitate within the range of two solvent conditions). Several solutions and reprecipitations are needed to increase asphaltene quality. A solvent/AS combination is used and in both methods; the total asphaltene fraction is typically dispersed in a "good" solvent (e.g., toluene) at a fixed solute concentration and a flocculating solvent, the AS (e.g., *n*-heptane), is added to induce partial precipitation. During a coarse fractionation, two asphaltene fractions (i.e., insoluble and soluble) are isolated by precipitation at a given solvent condition. Typically, the ratio of flocculants/solvent is varied, so that several pairs of more and less soluble fractions are isolated [37]; an example scheme is shown in Fig. 2.3 that uses mixtures of toluene/*n*-heptane (heptol) at two different ratios (40/60 and 15/85) [38].

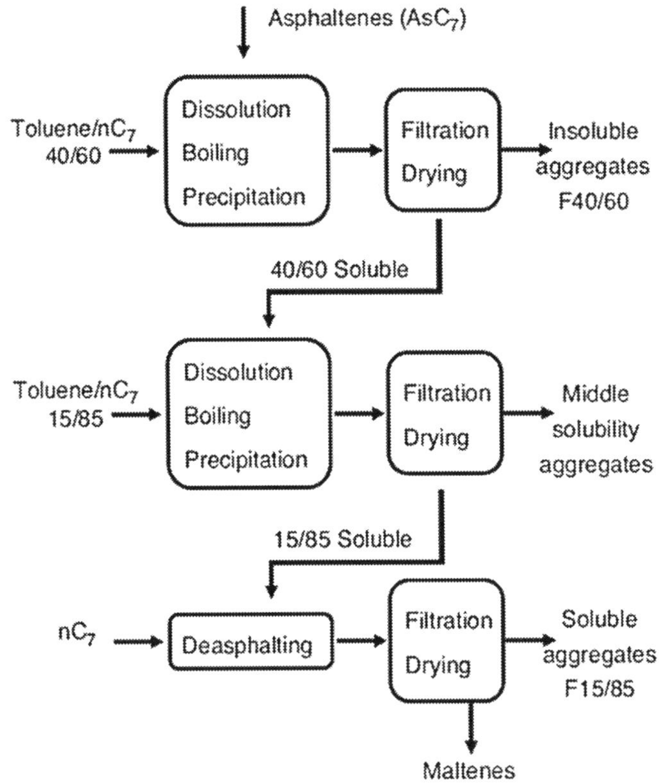

Figure 2.3

Asphaltene subfractionation by sequential flocculation using toluene/*n*-heptane mixtures. *Reproduced from Marques J, Merdrignac I, Baudot A, Barré L, Guillaume D, Espinat D, Brunet S. Asphaltenes size polydispersity reduction by nano- and ultrafiltration separation methods – comparison with the flocculation method. Oil Gas Sci Technol Rev IFP 2008;63(1):139–49, with permission from IFP.*

Supercritical fluid extraction (SFE) can be used for asphaltene extraction from crude oil. Unfortunately, carbon dioxide, one of the SFE choices, does not have sufficient solvent strength to extract polar compounds. In this case, a polar modifier has to be added [39]. One of the typical ASs, *n*-pentane, has also been employed at SFE conditions for both extraction and fractionation [40]. SFE was carried out on tar sands with mixtures of alkane-aromatic solvents at different temperatures. The extracts were further fractionated into oils, asphaltenes, preasphaltenes, and gases. The highest yield of SFE (24.3 wt% dry basis) was obtained with an *n*-pentane/benzene (1/1, v/v) mixture at about 380°C. Evolution of light alkanes (C_4^-) from the asphaltenes at higher temperatures indicated the presence of labile moieties as molecular substituents [41]. However, these light molecules could have been also trapped during asphaltene separation.

Filtration (micro-, nano-, and ultrafiltration) has been employed as a means of separation because it least disturbs the physical state of the asphaltenes. Both inorganic and organic membranes have been used for the separation of asphaltenes from the crude oil. Ceramic monolith membranes with average pore size in the range 0.02–1.4 μm were used to separate asphaltenes from Cold Lake heavy oil at temperatures below 190°C. Although fouling restricted the flow rate through the membrane, up to 80% asphaltene recovery was achieved. Fouling gave rise to a gel layer on the membrane that increased its resistance. An increase in the asphaltene content of the feed increased membrane fouling and so resistance [42]. Single-tube ceramic membranes (average pore diameter 0.02–0.1 μm) were used for the ultrafiltration of Cold Lake heavy oil. Filtration conditions of temperatures in the range of 80–160°C, transmembrane pressure of about 600 kPa, and cross-flow velocities between 2 and 10 m/s were tested. Similarly to the previously described work (Ref. [42]), ultrafiltration resulted in rapid fouling of the membranes with a reduction of permeate flux from an initial value of 660 kg/m^2 day to ~60 kg/m^2 day after 6 h on stream, when a 0.1 μm membrane was operated at $\Delta P = 600$ kPa, 120°C, and a velocity of 7 m/s. Asphaltene retention was 80% over this period of time (6 h) [43]. CeraMem Corp. tested ceramic membranes for the removal of asphaltenes from vacuum gas oil (VGO). In this case, vulcanization of the membrane with sulfur and addition of fluidized catalytic cracking (FCC) catalyst fines improved asphaltene separation [44]. Physical separation by nanofiltration using a zirconia membrane with a 20-nm average pore size proved to be somewhat effective [45]. Membranes of Ultracel YM and Anopore with pore sizes between 3 and 20 nm were tested for evaluation of diffusion and cluster size of asphaltenes indicating that pore sizes below that range need to be used for an effective separation of asphaltene nanoaggregates [46,47]. Polyethersulfone porous membranes were used for the physical fractionation of C_7-asphaltenes by cross-flow membrane ultrafiltration. This type of membrane (UP020 membrane) presented a 20-kDa MW cut-off and was produced by NADIR and provided by Alting, Hoerdt, France [48]. A two-stage filtration of C_7-asphaltenes was carried out using UP020 in the first step at 10 bar and then a dense polyimide nanofiltration membrane of 400-Da MW cut-off (Starmem 240 membrane, provided by MET, Imperial College, London, UK), at 30 bar [38,49].

Asymmetric ceramic monolith membranes with pore sizes of 0.2 μm and 50 nm were investigated for the separation of asphaltenes (and heavy metals) from three Iranian crude oils (1–10 wt% asphaltene content), at a pressure gradient of 200 kPa and temperature range of 75–190°C. The original nanometer particles of asphaltene aggregated to micrometer size during heating of a crude oil sample to testing temperatures. These micrometer particles could be separated using those membranes and achieved 60–87 wt% of asphaltene recovery [50,51]. Ultrafiltration has to recognize the magnitude of the smallest asphaltene cluster size and the fact that a certain degree of association would be implied. Safaniya (also known as Arab heavy) VR, containing 23 and 13.4 wt% of C_5- and C_7-asphaltenes, respectively, was separated by solvent-free nanofiltration using 5, 10, 20,

50, and 100 nm ceramic membranes at 200°C. While the entire VR passed through the 100-nm membrane, it was all retained by the 5-nm membrane [52,53]. Practical applications of filtration for the separation of asphaltenes from crude oil (or VRs) have been claimed in several awarded patents; examples are given in Refs. [54–57].

A Russian group reported a method for the isolation of asphaltenes consisting of dissolving the sample in benzene under normal conditions and subjecting it to the effect of a direct current voltage electric field [58].

In an attempt to simplify characterization and to further understand physical and chemical properties, the isolated asphaltenes have been subfractionated ("fine" fractionation, as defined earlier). Various methods and techniques have been explored, such as chromatographic techniques [59–63], sequential elution solvent chromatography [64,65], extraction with solvent mixtures [21,66–73], dialysis fractionation [74], ultracentrifugation [75], as well as chemical fractionation (e.g., acid–base separation [76,77], reactivity/composition [67,78,79]) and atypical precipitations by the addition of flocculants [78,80–83]. In cases where mass balances were reported, the total recovery of asphaltenes was always lower than 92%. The characterization results obtained with such subfractions will be discussed in the following sections.

Some of the methodologies developed for the separation and purification of asphaltene fractions have also been adapted to resin fractions. An example can be found in Ref. [84], where reversed phase chromatography and adsorption chromatography normal phase chromatography were considered and demonstrated the polydispersity of these fractions, though this polydispersity degree was much lower than that observed in asphaltenes. A review article includes details of isolation methods for the recovery of resins [12].

4. Properties

The physical, chemical, and physicochemical characterization of macromolecules is in general a big challenge. Asphaltenes, being a solubility class and as such a mixture of different molecules, elevates the level of complications. Furthermore, the composition of this mixture depends not only on the source (original crude oil), but also on the isolation method employed for the separation of the asphaltenic fraction. Even though there have been established experimental standardized protocols for the analytical determination of asphaltenes, researchers have been employing and exploring other recovery methods for their characterization studies. Some suggestions have been made for the definition of consistent methods, methodologies, and techniques for sample preparation, particularly when the target is a comparative exercise among several crude oils (see, for instance, Ref. [85]).

4.1 Physical Properties

The solvent-separated asphaltenes have a dark brown color, but in a single pass, chemisorbed resins and low-MW asphaltene fragments can be retained. After solvent evaporation, the residual asphaltene is jet black and the chemisorbed material is deep reddish brown. Although asphaltenes have a dark color, some components exhibit specific colors in solution [e.g., fluorenones (bright red), polycondensed aromatics and condensed thiophenes (yellow/orange), and various vanadyl porphyrins (violet, green, yellow, etc.)].

The physical state of asphaltenes in crude oil is not yet quite understood and a subject of long-lasting controversies. Representations and interpretations that consider asphaltenes as a liquid, colloidal system, and/or dispersed solid will be discussed in the next sections. Probably, this issue arises because asphaltenes are a solubility class and as such contain a broad mixture of compounds, which may coexist in any of these physical states. It has been recognized that asphaltenes either in alkane solvents or native oils can exhibit polymorphism and may also comprise two phases over a broad range of conditions. Clearly, asphaltenes intraact and interact with one another and with solvent media on a range of length scales [86].

The statistical analysis of the physical properties as a function of asphaltene concentration of more than 400 asphaltene-containing crude oils showed multiple peaking of oil viscosity (up to two orders of magnitude) [87], peaking of oil density, and a frequency of oil appearance [88] at specific asphaltene contents close to structural phase boundaries observed in asphaltene and oil solutions. This statistical analysis also revealed that the formation and evolution of asphaltenes in solutions, as well as in native petroleum, proceeds via multiple structurally distinct phases. Thus phases are delimited by well-defined boundaries on the temperature–composition (T–C) phase diagram of asphaltenes at ambient pressure [89]. This type of phase diagram will be discussed in conjunction with the incompatibility behavior of crude oil blends later (see Section 5.1 and Fig. 2.36).

Similarly, phase transition studies were interpreted as the result of the existence of a multiphasic system, rather than a biphasic one [90]. Furthermore, reasoning for the colloidal nature of asphaltenes was presented for the case of Chinese crude oils [91–94]. Two regimes separated by a critical point have been distinguished in the phase diagram: a colloidal regime and a "swollen asphaltene" regime, the latter being much larger [95].

The precipitated solid material is amorphous [96], though crystallinity has been observed in certain circumstances [97,98]. The presence of liquid crystals in both fractions of Athabasca VR, asphaltenes, and maltenes has been evidenced by using polarized light microscopy, differential scanning calorimetry, and mid- and near-infrared photoacoustic spectroscopy with depth profiling [99]. Involvement of liquid-crystalline phases was postulated during phase transition in precipitation processes. Considering phase splitting as entropy driven led to the thought that the asphaltene amorphous phase may become an ordered mesophase precursor for the formation of the liquid-crystalline phase [90].

The solid asphaltenes have a density of approximately 1.2 g/cm^3 and are thermally infusible, i.e., have no melting point, thus they would decompose upon heating, leaving a carbonaceous deposit. Density has been estimated using molecular dynamics (MD) simulations of average molecular structures representative of various asphaltenes (such as those presented in Figs. 2.19B and 2.31). Although calculated values were considerably lower than experimental values, this study could be employed for predicting the possible effects of density on the various types of structural models. Density increases were found when the H/C ratio decreased. Density decreases were larger for amphoteric molecules when the degree of aromatic polycondensation increased and with the presence of heteroatoms (the magnitude of the increase depended on functionality) [100]. The polydispersity of MW and molecular size is also reflected in the density. The specific gravity evaluated in the SARA fractions showed a wide range of values for the asphaltenes (Athabasca and Cold Lake) that was not exhibited by the other fractions [101].

A summary of the trends in the properties of C_7- and C_5-asphaltenes, resins, and maltenes has been provided. The increasing trend of viscosity, density, polarity, and MW follows the ranking as purified C_7-asphaltenes > raw C_7-asphaltenes > raw C_5-asphaltenes > resins > maltenes [102].

The colloidal structure of bitumen is the basis of a three-category classification [103–105]:

- "Sol" bitumen behaves as Newtonian fluids and exhibits a noninteracting micelles structure;
- "Sol–gel" bitumen shows elastic behavior in the initial deformation stages and super-micelles and giant supermicelles are present;
- "Gel" bitumen with a 3D structure derived from extensive intermolecular interaction shows high resilience.

This latter structure was assigned to be responsible for the rheological improvement of bitumen characteristics, observed upon addition of phosphorous compounds. Addition of polyphosphoric acid $(HPO_3)_n$ to Safaniya VR increased its stiffness, improved its elasticity, and lowered its thermal susceptibility [106]. Meanwhile, addition of ethylene glycol or diglycidyl ether improved the rheological characteristics, with important decreases in viscosity [107].

Viscosity behavior has enormous impact on crude oil production and transportation. The understanding of the effect of asphaltenes on the viscosity of crude oil is of vital importance. Subfractionation into narrow fractions and their characterization have been one of the approaches followed. Asphaltenes were removed from the VR by stepwise extraction or precipitation to recover subfractions that monotonically differ in MW, molecular size, or polarity. In all cases, increases in any of these asphaltene properties lead to enormous decreases in viscosity, of up to two orders of magnitude [40,108–115]. Another approach has been the reconstruction of VR by mixing its maltenes and asphaltenes fractions. In this way reconstructed VRs were obtained containing different volume fractions of asphaltenes and their viscosity was evaluated. The measured viscosity of these reconstructed VRs at different temperatures [116] is shown in Fig. 2.4 as a function of the volume fraction of asphaltenes.

The viscosity trend was explained based on other experimental results on MW and molecular and particle sizes (discussed in corresponding sections later). Asphaltenes were observed to behave similarly in maltenes as they do in toluene, forming molecular clusters (nanoaggregates). Size and mass of clusters decrease with temperature increases (see discussion later referring to Fig. 2.28). The viscosity trend with temperature was explained in terms of changes in solvation. Thus aggregates dissociate into molecular clusters with temperature, yielding a decrease of solvation with decreasing cluster size. The non-Newtonian behavior found at low temperatures was understood as a consequence of decreases in the repulsive interactions between clusters upon lowering temperature [116]. Changes in viscosity with asphaltene aggregation were the basis for the definition of a viscometric method for detecting precipitation onset [117], the principles of which were reported in Ref. [118].

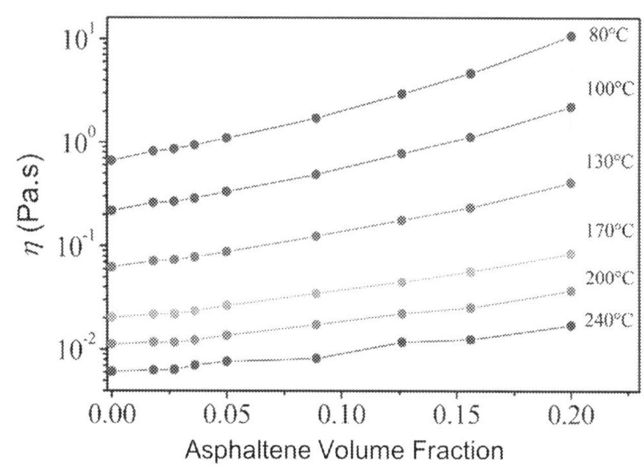

Figure 2.4

Viscosity of reconstructed vacuum resids (VRs) as a function of asphaltene volume fractions at different temperatures. *Reproduced from Eyssautier J, Hénaut I, Levitz P, Espinat D, Barré L. Organization of asphaltenes in a vacuum residue: a small-angle X-ray scattering (SAXS)—viscosity approach at high temperatures. Energy Fuels 2012;26(5):2696–704, with permission from ACS Publications.*

A rheological study of asphaltenes solutions indicated that behavior in xylene can hardly be compared to that in crude oil, or in VR (in maltenes). The resins present in the maltenes fraction were thought to have a strong dissociation effect that prevents asphaltenes association and makes aggregation a reversible process. This transient network of asphaltenes was responsible for behavior as a shear thinning fluid, observed at low temperature. Hence maltenes addition was recommended for the reduction of crude oil viscosity. Moreover, because of the sensitivity of asphaltenes to polar compounds when (naphtha) dilution is employed as viscosity reducer, inclusion of a polar additive could be beneficial. The addition of a small amount of alcohol would improve asphaltenes stability [119]. The viscoelastic properties are also affected by temperature. A temperature increase facilitates asphaltene aggregate diffusion reducing the associated relaxation time. The Brownian effects are high enough above 60°C to prevent formation of aggregate suprastructures [120].

The physicochemical properties of VRs from nine crude oils were reported by Gawel et al. [121]. The SARA fractions were obtained and characterized. Their API and viscosity results are presented in Fig. 2.5A and B in plots that visualize any possible correlation with the resins and asphaltenes contents. These plots seem to indicate that both of these properties, API and viscosity, are affected by the total resins + asphaltenes content within the range of composition examined. Nevertheless, at asphaltenes contents below 5%, viscosity seems to be highly affected by asphaltenes content since viscosity increased steeply within three orders of magnitude along a narrow increase in asphaltenes content. Regarding API, the plot shown in Fig. 2.5C includes data from Mansoori [122,123] and tries to verify, if any, a correlation with the R/A ratio, as proposed by several researchers including Mansoori, Creek [124], and others [40,100,125–128]. Although that plot showed a lack of correlation between API and R/A ratio, it did show a monotonous exponential decrease with the increase in the total content of resins and asphaltenes. A comparable study was carried out with Iranian heavy crude oil, resulting in consistent similarities [129].

The effect of pressure (0.01–34 MPa) on viscosity was found to be negligible when viscosity was measured at temperatures ranging from 25 to 180°C [130].

The surface tension of C_7-asphaltenes from a variety of heavy oils and bitumen was found to vary in the range of 15–19 mN/m at 358°C. A decreasing trend with temperature was observed, reaching a value of ~5 mN/m at 400°C, explained as being caused by the cracking and coking that start to take place at those high temperatures [131].

The dipole moment of resins and asphaltenes was found to be in the ranges of about 2–3D and 3–7D, respectively, for a variety of crudes considered and ASs employed. The dipole moments of asphaltenes were higher than those of resins for a given petroleum fluid. However, the resins from a different crude oil may have higher dipole moment than

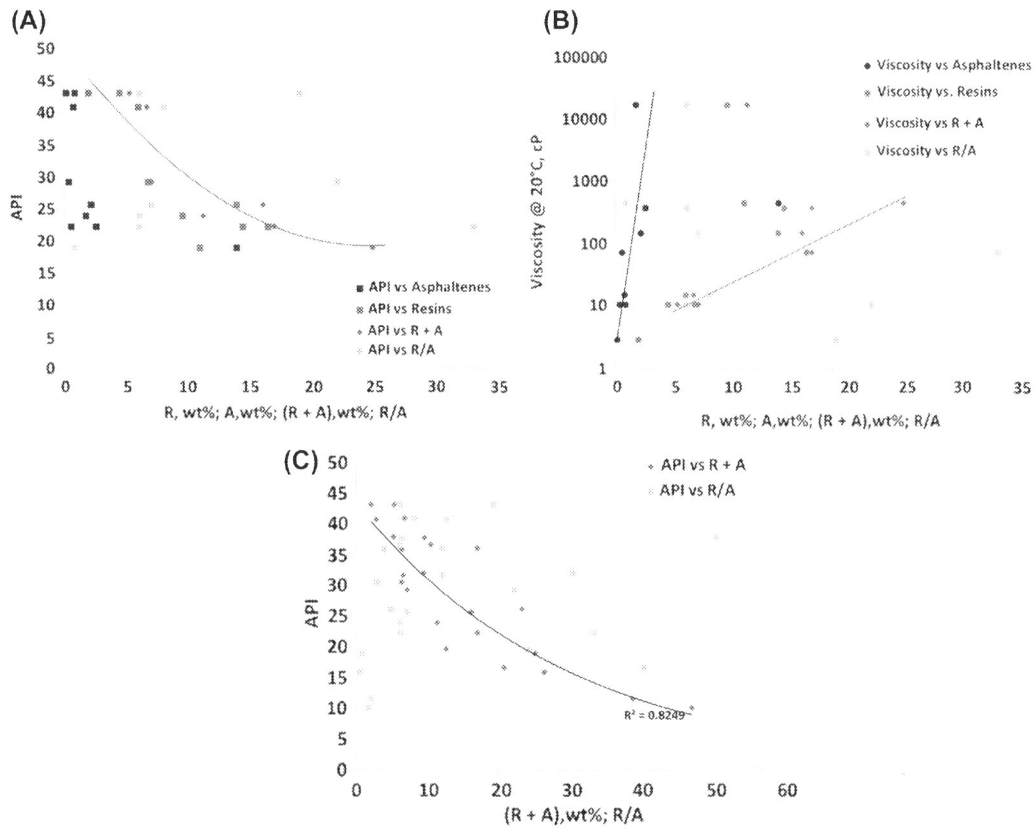

Figure 2.5
Effect of resin and asphaltene contents in API gravity and viscosity.

asphaltenes from another petroleum fluid [132]. Then, in principle the effect that resins have on asphaltenes from their own crude oil will not necessarily be the same on asphaltenes in a different crude oil.

Dielectric properties and conductivity affect operations of certain refining units, such as the desalter. Formation and breaking of emulsions may be also affected by these electrical properties. Conductivity decreases with increases in aggregation. An effect of mobility seems to be controlling conductivity and thus decreasing aggregate size would increase conductivity by increasing mobility of the charge carrier [133]. This decrease in conductivity upon aggregation supports an explanation of electron transfer between asphaltene molecules as the main mechanism in forming aggregates [134]. Consistently, a drastic change in dielectric response was observed near the critical point when flocculation occurred [135] and decrease continues through precipitation [136]. Conductivity is also higher for subfractions of higher polarity [68]. For all considered cases, surface conductivity was found to be predominant over bulk conductivity in solid asphaltenes.

Unidirectional electron hopping between spatially close, shallow localized traps was suggested as the major transport mechanism of charge carriers at asphaltene surfaces. The dielectric constant of solid asphaltenes below 35 and 40°C was found to be both frequency and temperature independent and was evaluated as $\varepsilon = 4.3-5.4$ [137].

Asphaltenes have paramagnetic centers with concentration of free radicals in the range of $10^{19}-10^{20}$ spin/g, as evaluated by electron paramagnetic resonance (EPR). Characterization results and quantum mechanics calculations of the forbidden gap size showed that asphaltenes with structural units of 6−13 aromatic rings varied from 4.92 to 6.49 eV for the molecular fragments and from 2.84 to 3.20 eV for the free radical form. This paramagnetic fraction of asphaltenes might be considered as an amorphous compensatory organic broadband semiconductor [138].

4.2 Solubility

Asphaltenes are a solubility class of compounds, characterized by their solubility in aromatics and immiscibility in light *n*-paraffins. Asphaltene solubility has been associated with its aromaticity and polarity rather than with the molecular size or dimensions of the asphaltene constituents [8,139−141]. In a more general context, asphaltenes as polar compounds are soluble in polar solvents (pyridine, methane chloride, carbon tetrachloride, carbon disulfide, and light aromatics) and insoluble in nonpolar solvents [142]. Nonetheless, insolubility in *n*-paraffins varies with the CN of the alkane and can be associated with the volume needed for flocculation of the asphaltenes. In fact, the volume of *n*-paraffin at the flocculation point increases as the *n*-paraffin CN increases, reaching a maximum at a CN of 9 or 10, and then decreases. This fact introduces a paradox to the solvency quality since a smaller volume of a larger *n*-paraffin (CN $>$ 11) is needed to begin precipitating in comparison to a smaller *n*-paraffin (CN $<$ 7), even though larger volumes of the large *n*-paraffin would precipitate much less (and more aromatic) asphaltenes than a smaller volume of the smaller paraffin. Several models have been proposed to explain solubility/precipitation for a large variety of *n*-paraffins (from pentane to hexadecane) [143−146]. Cycloalkanes are better solvents for asphaltenes than *n*-alkanes are. In fact, cyclohexane has been reported to dissolve as much as 40% of C_7-asphaltenes [147].

DAO represents the fraction of the oil in which asphaltenes were originally dissolved. Consequently, DAOs appear to be the ideal natural solvent for asphaltenes. It is believed that this is because of the high content of aromatics and resins ($\sim$30 wt%). In fact, nuclear magnetic resonance (NMR) results indicated that $\pi-\pi$ interactions are the most prominent interactions between maltenes and asphaltenes and might be responsible for the high solvent power of aromatics and resins [148]. Resins are known to be a good solvent for asphaltenes [149].

Lindbergh oil has been found to be a good solvent for asphaltenes [150]. Although the Lindbergh oil already contains about 14% of its own asphaltenes, it is still undersaturated and can take up more asphaltenes. According to these authors, straight-run gas oils with low saturate concentration below 35 wt% are desirable solvents for asphaltenes. For any stream with low resin contents (<20 wt%), saturate concentration above about 30 wt% would determine asphaltene insolubility.

Solubility could be inferred from the effect of additives on asphaltene precipitation onset. It was found that aromatics and hydrogen donor compounds increased the onset ratio by a slight to moderate degree, heteroatom compounds increased the onset ratio by a moderate degree, whereas surfactants increased the onset ratio by the highest degree [77,151–153]. Thus additives increasing the onset ratio can be considered solubility enhancers.

Asphaltenes are insoluble in CO_2 as evidenced from the induced precipitation with the incorporation of this compound into the oil system (e.g., CO_2 flooding during oil recovery [154]). The kinetics of precipitation by CO_2 was studied under isothermal (in the range of 25–65°C) and isobaric (at 17.2 MPa) conditions. The reaction order (m) for asphaltenes and the reaction order (n) for CO_2 were greater than one ($m + n > 4$), indicating that precipitation is a complex process and not an elementary step. Factors affect CO_2-induced precipitation in an opposite way to the same factor in the heptane-induced case. CO_2-induced precipitation increases with API and decreases paraffinicity, oil aromaticity, and asphaltenes content [155]. Meanwhile at supercritical conditions, CO_2 is a good solvent for asphaltenes, but as mentioned, in other conditions it would promote flocculation and precipitation. In fact, the proposed mechanism by which CO_2 destabilizes water-in-crude oil emulsions involves flocculation and precipitation steps [156]. At increasing pressures and in the presence of compressed (AS or aliphatic) gas, crystallization would occur. Processes involving high pressures of aliphatic gas phase should be subjected to an asphaltene crystallization risk [157].

Although in general, solubility increases with temperature, reported studies concluded the contrary for asphaltenes. An explanation was given in terms of the decrease in surface tension of the AS and the consequent increase in its solvating power [158–160]. However, at temperatures below room temperature (RT ~ 25°C), other researchers have found increases in solubility with temperature increase. Changes in the viscosity of the medium and in the disaggregation of the asphaltene clusters were provided as explanations [161–164]. The effect of pressure has been studied in conditions of interest for oil production. Thus solubility has been found to decrease from well pressure to bubble point[1] and then increases again with further increases in pressure. Volatilization of light compounds with the decrease in pressure leads to increase in MW of the remaining fluid [165,166].

[1] The bubble point is the condition at which the oil exhibits its minimum density and minimum solubility.

A more detailed study of solubility has been carried out on subfractions separated from the precipitated asphaltene fraction. The solubility of acid, basic and neutral (ABN) asphaltene subfractions differs by virtue of the differences in chemical behavior. The differences were more marked for the neutral fraction, while the acid and basic asphaltenes exhibited only subtle differences. The neutral fraction showed a higher H/C ratio (lower aromaticity), a much lower nitrogen content (by a factor of 3–4), and a lower carbonyl content. Although the acid and basic fractions were chemically similar, their solubility differed; compared to the original whole fraction of asphaltenes, the acid subfraction was remarkably less soluble in mixtures of heptane and toluene, while the basic asphaltenes were significantly more soluble [80,167]. It could be expected that the least soluble subfraction would be the most likely to (aggregate and) precipitate.

Asphaltenes that have been subfractionated by polarity with mixtures of pentane–methylene chloride showed a decrease in solubility when polarity increased. The most polar fraction also contained the higher metals concentration. In the whole VR, asphaltenes with lower polarity seem to interact and solvate well with the higher polarity asphaltenes and inhibit their aggregation. Dissolution of asphaltenes was found to obey first-order kinetics and fit into a shrinking core model. Dissolution rate constants decreased with the polarity of the fraction [168]. The fact that the most polar fraction was the least soluble seems to be in agreement with the findings of the least soluble asphaltene fraction having the most complex structure [169,170].

Similar results have been found by Kaminski et al. [78] and Acevedo et al. [74]. Kaminski et al. used the same AS/S, pentane in CH_2Cl_2 mixtures with increasing amounts of pentane, for fractionating a solution of C_7-asphaltenes in CH_2Cl_2 [78]. In this work, the polarity of the recovered asphaltene fraction was expected to decrease with the increasing amount of pentane used for its precipitation. Analysis of these recovered fractions indicated a concentrating effect of metals in the higher polarity fractions. Furthermore, a direct effect of the metal content on asphaltene dissolution with amphiphile/alkane solvents was also found. Neither inhibition nor synergy was found to exist among the asphaltene molecules, since the unfractionated samples appear to behave as a sum of their fractions. These resulting fractions of increased polarity exhibited a decreased kinetics of redissolution in heptane solutions of surfactants (resin-like solvents) [153]. The decrease in dissolution rate constants with an increase in polarity might introduce difficulties for remediation when high polar asphaltenes are present [78].

Acevedo et al. employed a dialysis method, with tetrahydrofuran (THF) and acetone as solvent [74]. In this work, seven fractions of soluble and insoluble asphaltenes were collected by increasing the amount of THF in acetone (from 40% to 100% THF). Solubility in toluene changed upon fractionation. The soluble asphaltenes contained in the first three fractions showed similar but decreased solubility in toluene that decreases even further in the fourth and fifth fractions and becomes insoluble for fractions six and seven. Similarly, the seven fractions containing the insoluble asphaltenes in THF/acetone were totally insoluble in toluene. The first insoluble fraction could recover its solubility in toluene when mixed with the corresponding soluble fraction. However, this behavior was not observed for the others. The first three fractions were only soluble in pure THF. Furthermore, the other three fractions of insoluble (in THF/acetone mixture) asphaltenes could not be dissolved in typical asphaltene solvents, such as pyridine, nitrobenzene, chloroform, or carbon tetrachloride. Any of the tested solvents could not dissolve the seventh fraction of insoluble asphaltenes. Consistent with Acevedo's work was the work of Yang et al. showing that the solubility in toluene (or aromatic solvents) changed along the subfractions precipitated stepwise with heptane from bitumen [83]. The solubility of these six subfractions of Athabasca bitumen asphaltenes depended on their composition, which varied in aromaticity and in metalloporphyrin (vanadyl) content. While the first three subfractions (with lower H/C ratios as well as higher metalloporphyrin content) were more sensitive to the aromaticity of the solvent, the other three with higher H/C were somehow less sensitive to the aromaticity of the solvent, i.e., solubility in heptol mixtures was higher.

Changes in the toluene solubility behavior of asphaltenes fractionated using the PNP (*p*-nitrophenol) method [171,172] were interpreted as asphaltenes being composed using a colloidal phase, formed by a lower soluble fraction (A1) dispersed by a soluble asphaltene fraction (A2). In this method, 48–70% of the total asphaltenes in a cumene solution are precipitated by PNP. The precipitated solids were extracted with chloroform and an aqueous solution of sodium hydroxide. The recovered PNP complex rendered a soluble and insoluble fraction upon dissolution in toluene. Structurally, the insoluble fraction was thought to have a rigid and flat core of polycondensed aromatic and naphthenic rings (continental model, see Section 4.4.3), while the soluble fraction was more flexible with the presence of a more open structure connected by aliphatic chains (archipelago model, see Section 4.4.3). Besides all the chemical modifications that the proposed method might introduce to the nature of the asphaltenes, it was only one type of asphaltene (that derived from a Furrial crude oil) that yielded a detectable amount of the less-soluble fraction. A wider variety of asphaltenes were needed to understand better the structure–property relationships and validate the models more broadly. Five other crude oils, stable and unstable, were considered, namely, Boscan, Carabobo, Ceuta, Hamaca, and Monagas. The fraction A1 was always larger than the fraction A2. The solubility behavior of these two fractions was associated with their chemical composition and molecular structure. Regarding chemical composition, the most important feature was a higher metal concentration of the A1 fraction, compared to that of the A2 [173]. Further details on the molecular structures proposed for these fractions and their aggregation behavior and surface properties will be discussed in the corresponding sections. The solubility of PNP-fractionated Hamaca asphaltenes in a variety (57) of solvents was evaluated and the solubility parameters (SPs) were calculated. Good and bad solvents were identified for each fraction. These good and bad solvents were xylene and diethyl ether for A1, respectively, and *n*-butyl acetate and oleic acid for A2, respectively. It was concluded that the calculated SPs explained well the observed solubility changes [682].

Definition of SPs for evaluating the relative solvency behavior of a specific solvent (or solute) would assign similar values to similar chemical species since in principle two chemically similar species are miscible. In turn, two compounds will dissolve one another when their SPs are similar. SPs of asphaltenes are experimentally measured by the titration method, therefore precipitation conditions such as time, solvent, and solute (asphaltene or oil) concentration, temperature, and detection means will affect the determination [174]. Furthermore, the existence of a toluene-insoluble subfraction in a toluene-soluble fraction is indicative of the complexity of the asphaltene aggregation during precipitation.

Hildebrand SP [175,176] is derived from the cohesive energy density[2] of the solvent (intermolecular interaction energy per unit volume) and is obtained from the heat of vaporization per unit molar volume. The heat of mixing two materials (1 and 2) is dependent on the difference between their SPs squared $(SP_1-SP_2)^2$. If the SPs are identical, the heat of mixing is zero and the dissolution/mixing process is driven by the entropy term $T\Delta S$ alone ($\Delta G = \Delta H - T\Delta S \leq 0$), and mixing will occur. If the SPs are not identical, the term $(SP_1-SP_2)^2$ will have a net positive value, which will cause the energy term ΔH to oppose the entropy term. If the entropy term is less than the energy term, mixing or dissolution will not occur. The combination of the entropy of mixing of different size molecules with the heat of mixing from SPs, as expressed by the polymer solution Flory–Huggins (F–H) theory, expressed asphaltene solubility well [177,178].

Based on Hildebrand SPs, asphaltenes and resins appeared to be a continuum or family of complex molecules with a variation in MW and polarity, rather than two fractions containing chemically different compounds [179].

Meanwhile, the Hansen SP comprises three components[3]: dispersion forces (van der Waals interaction), polarity (related to dipole moment, dipole–dipole interactions), and hydrogen bonding. These three values can be considered as coordinates for a point in three dimensions also known as the Hansen space, which describes a sphere where solute–solvent interactions of miscible compounds take place. These three dimensions describe solubility spaces either in 2D (zones or maps) or in 3D (envelopes).
A compilation of both Hildebrand and Hansen SPs can be found in the literature in

---

[2] The term solubility parameter was introduced by Hildebrand and Scott from whom the Hildebrand SPs were defined as the square root of the cohesive energy density.

[3] Proposed by Hansen, C.M. in his graduate dissertation: The three-dimensional SP and solvent diffusion coefficient. PhD thesis from Danish Technical Press, 1967.

Refs. [180,181], respectively. Hansen SPs have been successfully used for the estimation of flocculation offsets [175,176,182–184].

Interaction forces that control solubility are different whether the solvent is aromatic or polar (e.g., acetone as in Ref. [82]). Thus polar interactions dominate solubility behavior in highly polar solvents, while for solvents with low polarity, the solubility behavior is dominated by dispersion interactions. In fact, neither MW nor size seems to affect asphaltene solubility more than aromaticity, aliphaticity, and polarity of both asphaltene molecules and those of the solvent itself [82]. Assuming that dispersion forces govern asphaltene precipitation, a model based on refractive index (RI) was defined [185]. Any other forces were considered of secondary importance in the description of asphaltene–resin interaction. The London dispersion properties of a material are characterized by the dependence of its RI on wavelength. Buckley et al. derived an expression for the SP of a given oil from this model as a function of RI (Eq. 2.1).

$$SP_{\text{Buckley et al.}} = \left(\frac{\sqrt{3}}{384}\frac{\pi h\nu}{\sigma^3}\right)^{\frac{1}{2}} \frac{\sigma^3}{V/N} \frac{RI^2 - 1}{(RI^2 + 2)^{3/4}} \tag{2.1}$$

In this equation, h is the Planck constant, ν is the wavenumber, σ is the diameter of the molecular hard sphere, V is the molar volume, and N is Avogadro's number. The trends found for alkanes, both normal and branched, can be used to predict flocculation onset. Similarly, the molar volume of AS was key in the determination of flocculation onset in a wide range of conditions [186]. A parallelism between Buckley's ideas and a unique crude oil parameter was found to exist when that parameter was the crude oil screening factor (COSF), defined according to the model described in Ref. [187]. This model considered asphaltene molecules that were discotic seven-center Lennard–Jones (LJ) molecules, the resins were single spheres, and the surrounding crude oil was a continuum characterized by the COSF. The COSF describes a crude oil in terms of its Hamaker constant and the dielectric constants. The LJ potential or, in general, any simple intermolecular potential may consist of two major contributions: a short-range repulsion force (electrostatic and exchange forces) and a longer-range dispersion force (London or van der Waals forces). COSF was found to measure the deviation from LJ potential. It was also found that the behavior of the system was influenced primarily by the change in the Hamaker constant of the media because of the negligible change of the dielectric constant within similar chemical families [187]. The determination of Hamaker constant of asphaltenes from calculated surface energies, based on contact angle measurements, indicated a prevalence of dispersion forces for asphaltene interactions in oil systems [188], which explains the parallelism between RI-based SP and aggregation behavior derived from the model based on COSF [187].

According to the Linear Solvation Energy Relationship (LSER), the SP_{LSER} accounts for four types of effective interactions: dispersion interactions, dipolarity/polarizability, hydrogen bond basicity, and hydrogen bond acidity [189,190]. The evaluation of this type of SP for crude oils and asphaltenes using flocculation threshold data and inversed chromatography has been reported. The factorization of Hildebrand's SPs into the coefficients of SP_{LSER} components indicated that oil and asphaltenes SPs are determined mainly by dispersion interactions [191].

In a Heithaus titration, an asphaltene solution containing a known amount of asphaltenes (W_a), in a known volume of solvent (V_s), is titrated with an AS. The volume of AS (V_{AS}) is recorded at the flocculation point [192]. The flocculation ratio (FR, Eq. 2.2) and dilution concentration (C_a, Eq. 2.3) are calculated as:

$$\mathrm{FR} = \frac{V_s}{(V_s + V_{\mathrm{AS}})} \tag{2.2}$$

$$C_a = \frac{W_a}{(V_s + V_{\mathrm{AS}})} \tag{2.3}$$

A plot of FR versus C_a is made and the intercepts are determined as the Heithaus parameters, $\mathrm{FR}_{\max}$ and $C_{\min}$. These values are used to calculate SPs for asphaltenes (peptizability of asphaltenes, Eq. 2.4), maltenes (solvent power, Eq. 2.5), and the overall compatibility of residuum, C_R (Eq. 2.6, see Section 5 for further discussion on compatibility).

$$\mathrm{SP}_a = 1 - \mathrm{FR}_{\max} \quad \text{Peptizability of asphaltenes} \tag{2.4}$$

$$\mathrm{SP}_m = \mathrm{FR}_{\max}\frac{1}{(C_{\min} + 1)} \quad \text{Solvent power of maltenes} \tag{2.5}$$

$$C_R = \frac{\mathrm{SP}_m}{(1 - \mathrm{SP}_a)} = \frac{1}{C_{\min} + 1} \quad \text{Compatibility of residuum} \tag{2.6}$$

Wiehe proposed a two-dimensional SP consisting of a complexing component and a force field component. The complexing component measures the interaction energy that requires a specific orientation between an atom of one molecule and a second atom of a different molecule, such as H-bonding and electron donor–electron acceptor interactions. The field force component accounts for the interaction energy of the liquid that is not destroyed by changes in the orientation of the molecules, including van der Waals and dipole interactions [193]. The analysis of the SPs evaluated indicate that:

- Asphaltenes are insoluble in liquids of low field force SP component and in liquids of moderate and high complexing SP component;
- Fifty percent aromatic carbons in asphaltenes provide them with a high field force SP component and a preference for liquids with a low complexing SP component;
- Fifty percent aliphatic carbons cause asphaltenes to be insoluble in even moderately complexing liquids;
- Every solvent for asphaltenes is a solvent for resins and every solvent for resins is a solvent for aromatics [194].

In petroleum macromolecules, the main interaction is by van der Waals forces that are strongest for aromatics. Hence SPs are primarily determined by the field force component and so the complexing component can be neglected. Since aromaticity increases with decreases in the H/C ratio, in compounds with increasing MW, SP will follow the inversed trend of the H/C ratio [195]. The alkyl substituents (with the higher H/C ratio) surrounding the aromatic cores would be preferentially lost during thermal cracking with a consequent increase in SP. Hence changes in solubility are expected to reflect the thermal history of the asphaltenes [194].

The evaluation of SPs of narrow subfractions of VR indicated a decrease in solubility with the increase in MW of the subfraction [196], in agreement with the results reported in Refs. [169,170]. A correlation was developed to define the SP in terms of oil properties: density (ρ), MW, and H/C ratio (Eqs. 2.7 and 2.8). In these equations, the SP are defined as 25°C and the average value of SP is that at which solvent–oil molecule interaction is negligible, while the maximum value refers to the case in which the solvent effect on oil molecule interaction is significant enough that the oil molecules appear to have little affinity for each other [196]. Theoretically calculated SPs indicated a direct proportionality to the H/C ratio of the fractions of VRs [197] and visbroken VR [198].

$$\mathrm{SP}_{av} = 16.14(\mathrm{MW})^{0.0166}\left(\frac{\rho}{\mathrm{H/C}}\right)^{0.4293} \tag{2.7}$$

$$\mathrm{SP}_{\max} = 19.92(\mathrm{MW})^{0.0166}\left(\frac{\rho}{\mathrm{H/C}}\right)^{0.3788} \tag{2.8}$$

According to the work of Painter et al. [199], asphaltenes with H/C ratios above 1.0 would exhibit SPs in excess of 22 $\mathrm{MPa}^{0.5}$ and some of their components would undergo a microphase separation to form clusters in toluene solutions. A question remains as to whether or not steric or kinetic factors are responsible for the apparent stabilization of fractal cluster dispersions in toluene against further aggregation. The polydispersed nature of asphaltenes gives rise to polydispersed SPs, which have been used by Painter et al. [199] for the creation of miscibility maps (an example is shown in Fig. 2.6) and of miscibility factors that govern asphaltene solubility. The miscibility maps are based on SPs of asphaltenes and solvents, and on the molar volume of solvents. The solubility factors include: asphaltene self-association, degree of self-association of strongly polar and H-bonding solvents; and free volume differences between asphaltene components and solvents.

Asphaltene solubility has such an impact on crude blending and other refining operating units (see Sections 5.1 and 5.2) that SPs and models have been defined for assessment or measurement of these phenomena by the oil industry. For instance, SPs calculated theoretically could explain the observed changes in toluene solubility [200]. Experimentally evaluated SPs of dead heavy oil and bitumen have limited practical

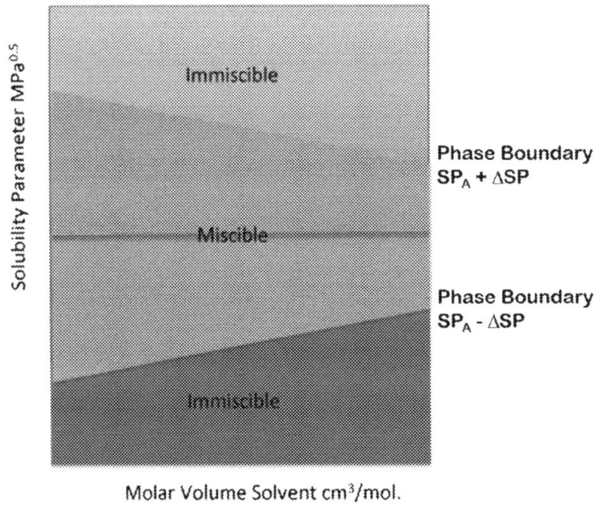

Figure 2.6

Miscibility maps showing the boundaries of solubility in terms of solubility parameters (SPs) and molar volume of solvent. *Reproduced from Painter P, Veytsman B, Youtcheff J. Guide to asphaltene solubility. Energy Fuels 2015;29(5):2951–61, with permission from ACS Publications.*

applications for live oils and less- to low-asphaltene-containing crude oils. The p and T conditions for commercial production processes largely differ from those regimes applied at laboratory conditions. The composition of the separated phases in a real oil system are mixtures of hydrocarbons with varying molecular sizes and complexities [201].

Experimentally derived fitting models have been defined for predicting the precipitation onset in the presence of different ASs. Rassamdana et al. [202] developed a scaling equation based on features of the aggregation processes under the effects of pure alkane ASs. In this equation (Eq. 2.9), the critical ratio R_c at the onset of precipitation depends only on the MW of the AS ($\mathrm{MW_{AS}}$) employed. The critical ratio is the AS-to-crude ratio at which onset of asphaltene precipitation for a given AS starts for the first time. In this equation, c is a constant on the order of 10^{-2} and T is the temperature in °C [203]. In agreement with experimental measurements, this equation predicted R_c ~0.63 at the onset of precipitation for propane, at T of 60°C and 2000 psig.

$$R_c = c(\mathrm{MW_{AS}}T)^{\frac{1}{4}} \tag{2.9}$$

A similar equation was proposed for the prediction of the weight percentage of precipitated asphaltenes (W). The equations defining the three variables (R_c, W, and $\mathrm{MW_{AS}}$) were combined in two (Eqs. 2.10 and 2.11) to simplify the general scale equation for asphaltenes precipitation onset (Eq. 2.12) [202,203].

$$X = R_c/(\mathrm{MW_{AS}})^z \tag{2.10}$$

$$Y = W / R_c^{z'} \tag{2.11}$$

$$Y = A_1 + A_2X + A_3X^2 + A_3X^3 \tag{2.12}$$

An experimental evaluation of these equations was carried out on Cold Lake VR and Athabasca atmospheric tower bottoms as asphaltene sources, diluted with pure *n*-alkanes, a lube oil base-stock-Paraflex (PFX), a heavy vacuum gas oil (HVGO) and a resin-enriched fraction (REF) recovered from the Cold Lake VR by SFE and fractionation [204]. Although a good match was found with the pure alkanes, in the presence of the multicomponent diluents (PFX, HVGO, REF, and their blends), two additional variables had to be combined with the variable X in the scaling equation. The saturates content (Sa) and the density in g/cm^3 at 20°C ($\rho_{D,20}$) were included as shown in Eq. (2.13). Several variables were tested to account for asphaltene precipitation from the different feeds in the scaling equation. A colloidal instability index (CII) was defined, as per Eq. (2.14), and incorporated in the scaling equation through Y' (Eq. 2.15). The CII is not as good an indicator as the SPs might be, since it involves only fractions quantity, but no elements of chemical interactions are considered.

$$X' = X\left[\frac{Sa}{\rho_{D,20}}\right]^{1.5} \tag{2.13}$$

$$\text{CII} = \frac{saturates + asphaltenes}{aromatics + resins} \tag{2.14}$$

$$Y' = Y(\text{CII})^2 \tag{2.15}$$

The suitability of RI for the determination of asphaltene precipitation onset might be because of the convenience of determining and incorporating the effect of temperature or pressure. For instance, inducing asphaltene self-precipitation with temperature increase has shown that onset occurs at a characteristic RI of 1.42 for considered crude oils [205]. Therefore the onset temperature of self-precipitation could be determined using a linear correlation between RI and temperature, which is specific of each crude oil. Dilute solutions of crude oil and other correlations were needed for the assessment of RI [205].

Verdier et al. [206] defined an RI-based SP (e.g., Eq. 2.1) to account for the effect of pressure on solubility. This dependence was introduced by incorporating an equation of state (EOS) for the evaluation of the density (ρ) (see Eq. 9 in Ref. [206]). The $\text{SP}_{\text{Verdier et al.}}$ is shown in Eq. 2.16, in which R_m is the molar refraction and MW_w is the average number MW.

$$\text{SP}_{\text{Verdier et al.}} = 53.827\frac{R_m}{\text{MW}_w}\rho(T,P) + 2.418 \tag{2.16}$$

The comparison of calculated $SP_{\text{Verdier et al.}}$ with RI-based SPs showed discrepancies that increased with the pressure. At 1 bar, the difference was about 1 $MPa^{0.5}$, while it increased to 6.5 $MPa^{0.5}$ at higher pressures.

Another property associated with asphaltene solubility is surface tension [207] and Hildebrand's initial work defined an SP (SP_{old}) from a correlation found with surface tension (γ) and the inverse of the cubic root of the molar volume (Eq. 2.17) [208]. According to Speight, asphaltenes would be insoluble in nonpolar solvents with surface tension lower than 25 dynes/cm and soluble in polar solvents with a surface tension higher than 25 dynes/cm [209].

$$SP_{old} = \frac{\gamma}{\sqrt[3]{V}} \tag{2.17}$$

Mitchell and Speight evaluated SP_{old} [17] and found a threshold SP_{old} value of 4.2, below which asphaltenes will precipitate. Speight [210] reported the solubility of asphaltenes (appeared to be C_5-asphaltenes) in different hydrocarbon and hydrocarbonaceous solvents. All these data are collected in Table 2.1. The high solubility observed in certain solvents may represent resin solubility, rather than asphaltene solubility.

Table 2.1: Surface Tension-Based Solubility Parameter (SP) and Asphaltene Solubility in Different Solvents [17]

Solvent	SP_{Old}	Solubility (wt%)	Solvent	SP_{Old}	Solubility (wt%)
n-Paraffins			Ethers		
Pentane	3.2	0	Ethyl ether	3.5	15.3
Hexane	3.5	18.3	*n*-Propyl ether	3.8	22.9
Heptane	3.8	30.1	*n*-Butyl ether	4.0	30.0
Octane	3.9	38.9	*n*-Amy1 ether	4.2	36.5
Nonane	4.0	40.3	Ethyl-1-butyl ether	3.5	19.5
Decane	4.1	43.2	Ketones		
2-Methyl paraffins			Acetone	5.3	4.1
Isohexane	3.4	7.6	3-Pentanone	5.2	17.8
Isoheptane	3.7	22.1	Acetophenone	1.8	100
Isooctane	3.8	29.9	2-Pentamone	5.1	27.0
Isononane	3.9	38.3	Methylethyl ketone	5.4	18.2
Isodecane	3.9	39.8	Methyl isobutyl ketone	4.7	11.5
Terminal olefins			Esters		
Pentene	3.4	2.8	Methyl acetate	5.6	37.7
Hexene	3.6	21.3	Ethyl acetate	4.9	15.3
Heptene	3.8	32.9	Propyl acetate	4.8	12.4
Octene	4.0	47.2	Ethyl propionate	4.8	13.2
Nonene	4.1	47.1	Alcohols		
Decene	4.1	43.9	*n*-Amyl alcohol	5.2	3.3
			Ethylene glycol	12.2	4.2

The oil SP proposed by EniTecnologie (Eq. 2.18) is based on density and viscosity (Eq. 2.19), properties readily available in the PVT report. A model (OCCAM Model) was developed taking into account the main experimental evidence, evaluating the SPs of oil and asphaltenes independently, from onset flocculation experiments. This thermodynamic model based on regular solution theory defined an EOS, which included a parameter to account for variable critical interactions. The model allowed a better representation of the experimental data, especially the presence of a maximum in the onset volume for a series of n-alkane precipitants [211]. An experimental correlation is used to estimate oil SPs and no additional experiments are required. The linear fitting parameters k_1 and k_2 in Eq. (2.18) are determined from this correlation, and μ is the viscosity. In Eq. (2.19), the parameters k_3 and k_4 are size related, N is Avogadro number, h is Planck constant, and ϑ is the specific volume in cm^3/g [212].

$$\mathrm{SP}_{Eni} = k_1 + k_2\left[Ln\left(Ln\left(\frac{\mu}{\mu_0}\right)\right)\right] \tag{2.18}$$

$$\mu_0 = \frac{Nh}{\vartheta}\left(\frac{k_3}{k_4}\right)^2 \tag{2.19}$$

Most of these equations indicate a nonlinear correlation between the solubility (and therefore the insoluble/precipitated asphaltene amount) and the volume ratio of AS/S, as found experimentally for instance by Tojima et al. [169] but opposite to the report of Trejo et al. [170], who found a linear correlation between %precipitated asphaltenes versus %V-heptane (in the heptol mixture).

As solubility varies with conditions, precipitation also does. The effect of temperature (40–100°C), pressure (1.5–4.5 MPa), contact time (0.5–6 h), and solvent-to-oil ratio (2:1 to 5:1 mL/g) was studied for the precipitation process. The chemical composition of precipitated solids was highly impacted by temperature. The demicellation of asphaltenes retaining resins or lower molecular components may occur with the increase of temperature. Precipitation increased with pressure increases [213].

The role of resins in the solubilization, precipitation, and stabilization of asphaltenes remains a subject for discussion. Partition of resins and low-MW asphaltenes is one of the complicating factors in the assessment of solubility. Both resins and low-MW asphaltenes have been found to play a solvating effect for the higher-MW asphaltenes that favor their solubility [168,169]. The structure of the resin molecules consists of a polar head group and a long hydrocarbon tail that acts as a surfactant molecule. The head groups of these molecules orient themselves toward the polar surface of asphaltenes, with their hydrocarbon tails extending into the oil phase, thus forming a micellar structure. Since resins are soluble in light alkanes, the micellar structures break apart by the action of light paraffins [201].

These molecules, resins, and low-MW asphaltenes undergo a reversible adsorption equilibrium on large asphaltene clusters, leading to their equipartitioning between the solution phase and the insoluble solid. The partition coefficients were estimated from the concentration profiles between the solution and solid phases. Results indicated that: (1) equipartitioning represents a complicating factor in the recovery of the whole asphaltene fraction, by either the solvent extraction or the precipitation methods, (2) resins and asphaltenes mainly associate via aromatic stacking, (3) the AS-soluble asphaltene coprecipitated material may be viewed as a fifth compound class fraction of crude oils, and (4) the presence of this fraction can profoundly affect the chemical and physical properties of asphaltene and the colloidal stability of its solutions [214].

However, increasing the resins content of a crude oil (or VR) with additional sources of resins from other crudes may not have the same effect. Various resins and dodecyl benzene sulfonic acid (DBSA) amphiphiles were added to three different petroleum fluids to measure precipitation with *n*-pentane. Results show that resins with a high dipole moment were more effective than resins with a low dipole moment, though the addition of resins increased the amount of precipitated asphaltenes. Asphaltenes with higher dipole moment were more prone to precipitation. Addition of DBSA amphiphiles showed a retrograde phenomenon, initial increases of precipitated asphaltene amount, and beyond a certain concentration precipitation decreases. The occurring interactions might be more complex than those of purely dispersion forces [215]. DBSA has been proven to be more effective than several commercial additives in keeping the asphaltene particles dispersed in solution and preventing them from settling at lower concentrations [216].

As may be evident at this point, precipitation of asphaltenes is not an exclusively solubility-derived issue. The conventional understanding of a solute precipitation from its solution does not apply to asphaltenes. In fact, there are some other factors that complicate the phase behavior of asphaltenes. Precipitation from toluene solutions is an example, since asphaltene concentration affects precipitation kinetics in an unexpected way. Two distinct regions for the effect of asphaltene concentration were identified. It has been found that for asphaltene concentration below 1%, concentration accelerated precipitation kinetics and aggregation. However, above that threshold the opposite trend was observed. Meanwhile, the total amount of precipitated asphaltenes increased monotonically with asphaltene concentration. These observations were explained to be caused by (1) monotonic increases in unstable asphaltenes with increases in total concentration, and (2) increase in the solvency power of the solution and/or of the stable asphaltenes (increases in solvency power) [217]. Spiecker et al. [80] have shown that the abundance of the soluble fraction of asphaltenes increases the stability of the insoluble fraction, indicating a sort of self-stabilizing effect of asphaltenes. A strong cooperative effect of the soluble fraction to the less soluble described as polar and with H-bonding characteristics was suggested. They also proposed solubility mechanisms for the considered studied cases.

In some instances (B6 and Hondo asphaltenes), solubility was governed by polar interactions, while in others (Arab heavy or Safaniya and Canadon Seco) there was an additional contribution of a π-interaction mechanism.

The reversibility [218] or irreversibility of the dissolution−precipitation−redissolution processes [219−224] is other evidence of the complexity of the factors affecting these processes. The observed hysteresis [221−224] indicates a lower solubility of the aggregated asphaltenes than that of the originally dispersed asphaltenes in the crude oil. These differences are accentuated when observed in the oil well (with live oils) or in produced oils (dead oils). While precipitation of asphaltenes in model systems resembled that of the live crude oil, it was almost completely reversible in the live system with repressurization, but only a partial redissolution was observed with the model systems [225,226]. Reversibility has been a controversial subject and in some instances is stated in definite terms such as *flocculation of asphaltene in paraffinic crude oils is known to be irreversible* [15] or *the process of flocculation is reversible* (in regard to precipitation with heptol solutions) [227].

Mass balances of asphaltene precipitation have provided more evidence of the complexity of the involved mechanisms and the potentiality of several coexisting equilibria. The heavier asphaltenes seem to precipitate first and somehow inhibit the precipitation of the lighter asphaltenes [228].

Measurements of the onsets of asphaltene precipitation above and redissolution below the bubble-point pressures on live oils at specified temperatures indicated that the onset pressure is often significantly higher than the bubble-point pressure. As expected from purely solubility principles, the onset pressure of asphaltene [229,230] increases at low temperature and the bulk density of the oil increases with redissolution. Partial reversibility was observed on pressurization−depressurization scans. The kinetics of precipitation and redissolution may be quite different from each other and hence may introduce a time dependence that should be considered during testing. In the timeframe of a minute, reversibility may become significant [229,230]. In fact, the slower kinetics of redissolution has been considered responsible of the lack of complete reversibility of the precipitation−dissolution process [231].

Redissolution of asphaltenic deposits has been the subject of investigations that indicate the lack of total reversibility. The conditions and additives (e.g., surfactants, such as ethoxylated nonylphenol, Renex; hexadecyltrimethylammonium bromide; hexadecylpyridinium chloride; sodium dodecylsulfate; sodium dodecylbenzenesulfonate; etc.) required to increase solubility of precipitated asphaltenes is clear evidence of the lower solubility of agglomerated asphaltenes. The presence of (natural or added) surfactants in the oil was found to inhibit asphaltene precipitation [152,232].

As mentioned, heavy paraffins (waxes) could coprecipitate during asphaltene precipitation [233–243]. Since asphaltenes are insoluble in paraffins, the possibility of paraffin content synergy on asphaltene precipitation was investigated by analyzing wax deposits as well as asphaltenic deposits. It was found that asphaltenes and waxes do not interact synergistically for coprecipitating in solid organic deposits [233]. A different approach was followed and opposite conclusions were reached by Oh and Deo [244] by adding organic solid (heavy) waxes to the considered crude oil (Rangely, Colorado, USA). The addition of heavy *n*-alkanes brought about a quicker onset of asphaltene precipitation. If the heavy waxes are precipitated or removed from the crude oil, the remaining oil becomes more polar and a better solvent for the asphaltenes, making their precipitation unlikely [244].

Propane injection in the oil well has been considered as a method for heavy oil recovery. Then, the propensity of asphaltene deposition upon propane injection needed to be considered. A range of pressures ($P = 300-850$ kPa) were scanned at $T = 20.8°C$ through the oil–propane system in a see-through windowed high-pressure saturation cell. No observable precipitation was reported at pressures below 780 kPa. At 850 kPa, asphaltenes precipitated, with a significant increase in solubility and oil-swelling factor. Density, viscosity, and aromaticity of the flashed-off heavy oil upon deasphalting were lower than those of the original heavy crude oil [245].

Clearly at this point, it appears that asphaltene solubility is a very complex phenomenon that depends on many factors. The diversity of parameters that has been considered in the definition of SPs and the lack of universality of any of these for a reliable and predictable trend of asphaltene solubility indicate that comparison among different solvents has to be done exclusively on a case-by-case basis. However, within a particular type of solvent, specific interactions and dependence may apply and a particular trend from a given type of SP may hold valid. Factors like H-bonding and electronic interactions (electron–donor-withdrawing character) that appreciably affect solvent power and considered in Hansen's SPs have succeeded in predicting precipitation onset in many situations. For all these reasons, isolation of asphaltenes by precipitation depend on many more factors than the relative proportion of AS/asphaltenes, such as the chemical nature of AS and on the possible change to it introduced by the presence of other compounds in the oil matrix.

The quantitative description of asphaltene solubility is taken as the basis for estimating asphaltene precipitation. Similarly, the evaluation of the flocculation point is vital for their processing. As asphaltene concentration increases, the solvent–asphaltene interaction forces that drive solubility start to compete with asphaltene–asphaltene association forces, which drive agglomeration. Rearrangements of these molecular interactions should take place during precipitation. Discrepancies between methods (equilibrium precipitation and flocculation titration) employed to determine either solubility or the incipient precipitation of asphaltenes have been found. Although it has been assigned to the slow kinetics of precipitation [22], other factors that drive aggregation might be affecting the results as well. While solubility is typically determined visually, the use of more sensitive techniques may reveal the existence of nanoaggregates in systems previously defined as solutions. This is the case with toluene, the default asphaltene solvent, even though it may be present also in other good solvents. Somehow these nanoaggregates do not coalesce further into larger clusters or flocs probably by steric and/or kinetic factors. The results of Eyssautier et al. on scattering techniques to toluene solutions of an asphaltene indicated the presence of fractal clusters [52,246,247]. Formation of fractal clusters in toluene solutions of asphaltenes has been postulated by Hoepfner et al., also based on the use of scattering experiments [14,248–250]. The formation of fractal clusters has also been reported to occur in porous media under diffusion-limited processes [251]. According to Evdokimov et al., at least two main asphaltene fractions that differ in their toluene solubility may be forming the asphaltenic phase. The most solvophobic fraction is on the edge of instability at concentrations as low as 10–15 mg/L. The other one would become unstable above 100–150 mg/L. Presumably, one of the key factors affecting instability is the solvent-mediated interaction between asphaltenes, determined by the solvophobic effect [252]. Clearly, macroscopic determinations of solubility, visual or otherwise, are not conclusive enough for the presence of a thermodynamic single-phase solution.

Similarly, the evaluation of the flocculation point is vital for the processing of heavy oils or asphaltenic streams. Methodologies for the study and detection of flocculation and of precipitations have been defined using a variety of techniques: electrical conductivity [133,135,137,253–257], optical techniques (transmission, scattering, microscopy, etc.) [248,252,258–270], gravimetric [147,271–274], ultrasound and acoustic methods [275,276], NMR [82,86,170,239,277–292], attenuated total reflectance-Fourier transform infrared imaging [293], etc.

4.3 Aggregation

The chemical nature of the asphaltene molecules and the complex nature of the corresponding environment may result in their agglomeration. At this point, the reader is encouraged to keep in mind that asphaltenes are typically isolated by precipitation. The starting point is the physical state of asphaltenes in the oil and then how association begins and gives rise to agglomeration. Aggregation might occur by a complicated mechanism and be limited by kinetic [249,294–296] and steric factors [14,297], since asphaltenes are bulky and heavy molecules diffusing in a highly viscous medium.

4.3.1 Primary Association

Molecules may associate with nanoscale clusters. The level of clustering depends not only on the nature of the asphaltene, but also on concentration and conditions. From here to aggregation, one line of thought considers these nanoscale aggregates to be colloidal structures (e.g., [298,299]), but a second view considers that colloid formation is not needed. According to the latter, aggregation may occur by a liquid–liquid-phase separation that results in the formation of a solvent-rich phase in equilibrium with asphaltene-rich clusters [13,273,300–302]. Aliphatic hydrocarbons and more particularly *n*-alkanes are considered flocculating agents. As mentioned, this flocculating action depends on the CN of the aliphatic chain. No aggregates were observed in the presence of alkanes longer than 28 C-atoms [303].

The understanding of the association phenomena is indispensable not only for the determination of true molecular parameters (weight and size), but also for the development of solutions for some of the problems asphaltenes cause. A broad range of techniques, both experimental and theoretical, have been applied for such complex study. Typically, asphaltene solutions are considered in these studies, and association/agglomeration is induced by addition of solvent–AS systems. Nevertheless, one has to keep in mind that the observed behaviors in model solutions do not necessarily have to resemble the current situation in crude oil and even less that of live oils. In oil, asphaltene particles might be present partly dissolved, partially in micellar forms, and/or partially in (steric) colloidal form, dictated by the polarity of the local environment.

As mentioned, a fraction of the asphaltenes would tend to associate and aggregate even in the best of the solvents—toluene. In dilute solutions, the observed threshold concentration for self-assembly of asphaltene monomers was reported to be below 10 mg/L [252]. Stacking as an initial mode of association might occur via interaction of π-orbitals by means of charge transfer complexes of the aromatic rings, as appears from X-ray diffraction results [304]. Stacking involves five to six molecular layers but only two or three according to, respectively, Refs. [304,305]. Asphaltene molecules have one or perhaps two fused ring systems per molecule [306]; a strong π-association might only occur among large fused aromatic-ring systems [307]. However, stacking via aromatic core interaction was postulated to occur also between colloidal particles [252]. In terms of size, the stack diameter was found not to change with subfractionation [71], but upon agglomeration of the primary particles; a size increase even above 200 nm can be expected. Molecular association leads to the formation of nanoclusters (or nanoaggregates), which have been detected in toluene solutions, dead oils, live oils, and bitumen [38,53,247,307–316]. Nevertheless, these nanostructures, which represent only a small fraction of the total asphaltenes, may remain stable and according to Refs. [53,317] these nanoaggregates would not grow by association with other molecules present.

The current understanding of the asphaltene molecular structure (see Section 4.4 below) does not match exclusively with any of these stacking theories. Furthermore, the presence of alkyl substituents in the aromatic rings forming the asphaltene molecule might represent a steric hindrance for stacking [286]. Additionally, steric effects might also inhibit aggregate growth and in consequence intervene in their size, though for those hindered molecules other forces might prevail and a different association mechanism would take place. Although steric effects might hinder stacking [318], the interaction of π-orbitals is still a valid means for association.

Murgich has listed the intermolecular forces that govern the aggregation phenomenon in asphaltenes as:

1. The intermolecular charge transfer;
2. The short-range exchange repulsion energy; and
3. The weak inductive interaction;
4. The electrostatic (coulombic) interaction between the molecular charges;
5. The van der Waals interaction.

In his opinion, this general and better-defined terminology should be employed rather than less precise terms such as H-bonding or the so-called π−π interaction. Furthermore, one should also keep in mind that the presence of these latter interactions in a given aggregate does not prevent the action of any other force on other fragments of the molecules from occurring [319].

Six subfractions precipitated from bitumen by a stepwise procedure with increasing heptane/bitumen ratio showed consistent results in terms of aromaticity, since the most aromatic subfraction was the first to be collected [83]. (It is worth pointing out here that the proportion of vanadyl porphyrins in this first fraction was also the largest. Metal contaminants will be the subject of the next chapter and their characteristics will be discussed then). These findings seem to agree with the earlier-discussed hypothesis of π−π aromatic as a first mode of interaction for asphaltene molecules to associate. On the other hand, the fact that composition changes are observed throughout the subfractions is an indication that molecular structure varies and interactions driving the aggregation most likely vary as well.

An approach for the understanding of the π−π stacking association of asphaltene molecules postulated the use of pyrene and dipyrenyl decane as model compounds. However, these molecules did not show significant association in *o*-dichlorobenzene solution. Incorporation of polar functional groups, such as ketones and hydroxyls, gave stronger association of pyrene derivatives but only up to the formation of dimers [320]. Although it is hard to imagine that a single molecule could present enough features to simulate asphaltene behavior, it could be possible to propose model compounds for a single type of behavior. Another molecule, hexa-tert-butyl-hexa-perihexabenzocoronene (Fig. 2.7), has been proposed as a model compound and showed heptane-induced association, a growth rate dependent on toluene/heptane ratio, and similar diffusion-limited aggregation kinetics of real asphaltenes [321]. Other model compounds that have been used to study π−π interactions and H-bonding, as well as emulsion stabilization, were

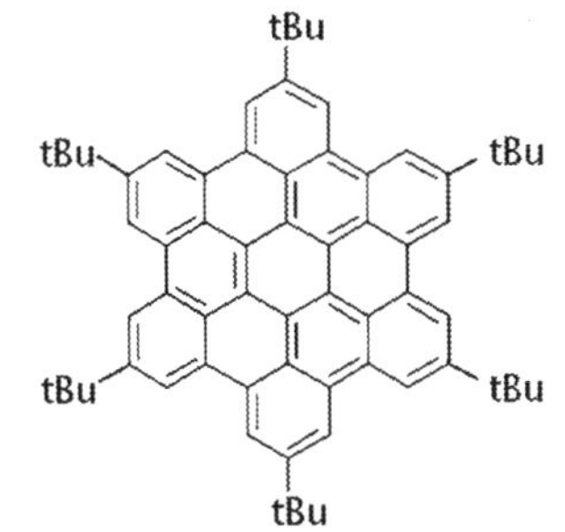

Figure 2.7

Asphaltene model compound: hexa-tert-butylhexa-perihexabenzocoronene (tBu). *Reproduced from Breure B, Subramanian D, Leys J, Peters CJ, Anisimov MA. Modeling asphaltene aggregation with a single compound. Energy Fuels 2012;27(1):172–6, with permission from ACS Publications.*

4,4′-bis(2-pyren-1-yl-ethyl)-2,2′-bipyridine and 4,4′-bis[2-(9-anthryl)ethyl]-2,2′-bipyridine [322], substituted acridine [323], and polyaromatic, acidic model compounds derived from amino acids β-alanine, phenylalanine, and tryptophan, which were used additionally to model interfacial behavior and polar interactions [324].

Primary association toward micellization was studied by Fourier transform infrared spectroscopy on asphaltene solutions from three Chinese VRs (Liaohe, Gudao, and Shengli). Liu el al. concluded that the H-bond was one of the main association forces for the asphaltene molecules; additional to H-bonding, contributions from the π–π interaction, polarity induction, as well as electrostatic forces were also identified [325].

The effect of the heteroatomic moieties on association could be studied by specific subfractionation of the VR. More drastic changes in composition upon subfractionation can be obtained when polar solvents were used [82], in which case the subfractions differ in the distribution of heteroatomic compounds. These physicochemical studies and the characterization of asphaltene agglomerates have enlightened the understanding of the chemistry behind agglomeration. For instance, sulfur is mainly present as slightly polar thiophenic groups, which improbably can contribute to intermolecular associations. Unlike sulfur, oxygen and nitrogen functional groups introduce a greater polarity to the molecules, enabling participation in strong intermolecular associations. Carboxylic acids, carbonyls, phenols, pyrroles, and pyridines that are capable of participating in proton or donor–acceptor interactions have also been identified in asphaltenes [82,326]. The heavier aggregates showed the highest aromaticity, carboxylic acid, and perhaps alcohol content [61]. The content of O-containing compounds in isolated or precipitated asphaltenic fractions has been found to be far below 4% [327,328], which might indicate that the heavy aggregates are only a small proportion of the agglomerated population. A high oxygen content was found in the most polar fraction of asphaltenes, characterized by being soluble in highly polar solvents, such as *N*-methyl pyrrolidone and insoluble in mildly polar or nonpolar solvents. This highly polar fraction plays an important role in aggregation though this role is not clear yet [329]. In summary, the intermolecular bonds are mainly alkenyl bridges [319,330–333]. The attaching molecules are 50% aromatic and 50% polar. The 60% of sulfur bonds (in sulfide linkages) are intramolecular holding together core segments. Meanwhile, 40% are intermolecular linking low-MW compounds to the core and represent 6.4% of the aggregated asphaltene fraction [334].

The changes observed in the near-UV–visible absorption spectra of toluene solutions with the changes in asphaltene concentration reflect the association and aggregation phenomena. Meanwhile, the presence of other crude oil constituents (e.g., resins) were not a determining factor as the concentration was for association. Molecular asphaltene structures could be observed at concentrations below 1–2 mg/L. Dimers predominated in the range of 5–15 mg/L and dimer pairs (tetramers) were effectively formed at concentrations close to 90 mg/L, which can be considered a quasispherical nanocluster with diameters of about 2 nm. This gradual increase of average complexity assemblies of coexistent molecular aggregates was thought to be inconsistent with the notions of "critical micelle concentration" (CMC) described in the next paragraphs [259].

Measurements of surface tension and vapor pressure osmometry (VPO) of asphaltene solutions of increasing concentration showed an increase in number average MW that was associated with an increased degree of association. However, the observed linear decrease in interfacial tension with concentration indicated that no micelles were formed and the aggregation observed with VPO does *not* appear to be caused by micellization [335].

In solution, asphaltene micellization was postulated very early [5] as a structural model of crude oils. More recently, H-bonding was assigned as the type of interaction responsible for micelle formation. More particularly, this H-bonding was postulated to involve the heteroatomic moieties [336]. A CMC was thought to exist as a prelude to agglomeration. In this view, the resins added to the asphaltene–solvent medium participate in the formation of micelles and are not involved as cosolvent. CMC ranges from 2 to 18 g/L for different crude oils [302,337–340]. Results of the aggregation kinetics indicated the existence of two types of aggregation: diffusion-limited aggregation and reaction-limited aggregation. At asphaltene concentrations in toluene below the critical micelle concentration of about 3 g/L, the aggregation kinetics appears to be solely limited by diffusion. However, above the CMC, reaction-limited aggregation takes place at least in the initial stage of particle growth [341].

For the VR of Arabian medium/heavy crude oil, CMC was shown to be between 3.5 wt% and 4.5 wt%. Micellization rather than aggregation was proposed based on the spontaneity of the phenomenon. Above CMC, the size of the micellar particles at the CMC did not change with increasing asphaltene concentration [342]. Instead, micelle association via coacervation has been proposed by Priyanto et al. [343] that led to sizes speculatively thought larger than 25 nm. The formation of micellar particles has been reported by other authors as well [344–346]. Measurements of CMC in a Brazilian crude showed larger values of CMC for C_5-asphaltenes in toluene, nitrobenzene, and pyridine than those of the corresponding C_7-asphaltenes [347]. The CMC values were shown to depend on the type of asphaltenes and the solvent used. Examination of the onset data for carefully fractionated asphaltenes established that two asphaltenes with different characteristics could have the same CMC values. It is observed that a heavier fraction has the same CMC value as a lighter parent asphaltene. Consequently, CMC does not seem to be the only determinant factor in asphaltene aggregation [348].

A plot of the onset values (determined by heptane titration followed with near-infrared spectroscopy) versus asphaltene concentrations gave distinct break points, named critical aggregation concentrations (CACs) by Oh et al. [271]. CMC values are typically determined from surface tension measurements. The reported CAC values were 3.0, 3.7, 5.0, and 8.2 g/L for toluene, trichloroethylene, THF, and pyridine, respectively. CAC and CMC values were similar [271]. However, other evidence indicated that asphaltenes could begin associating below the CMC range. This other evidence prioritized a stepwise mechanism rather than the formation of finite-size micelles and the existence of CMC [66,302,349,350].

As mentioned, according to Acevedo et al. and based on the results of their fractionation studies [74], micelle structure shows free radicals trapped in the core of an aromatic structure, surrounded by the more soluble asphaltene molecules (Fig. 2.8). The presence of an unpaired electron in asphaltenes was indicated by EPR characterization [351]; whether free radicals are responsible for that signal is still uncertain. The polycyclic aromatic

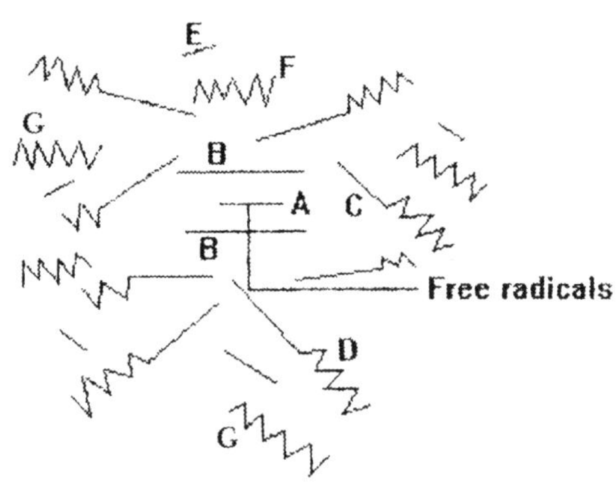

Figure 2.8
Micelle model proposed by Acevedo et al. (A) free radicals, (B) even aromatic structures, (C) soluble asphaltene molecules, (D), resins, (E) aromatics, and (G) saturates. *Reproduced from Acevedo S, Escobar G, Ranaudo MA, Pinate J, Amorın A, Diaz M, et al. Observations about the structure and dispersion of petroleum asphaltenes aggregates obtained from dialysis fractionation and characterization. Energy Fuels 1997;11:774–8, with permission from ACS Publications.*

hydrocarbons caging the free radicals would avoid intermolecular reactions. The presence of an unpaired electron in asphaltene aggregates would increase intermolecular associations only if transannular electron delocalization is possible.

This micelle structure is the core of the colloidal model, in which a layer of resin molecules covered the micelle and is enclosed in another layer of aromatic molecules, followed by another one of saturates. This structure is also the basis of the physical model of the oil residue, which has been used to explain solubility and compatibility [195,304,352]. The model proposed by Yen [304] tried to explain the association, agglomeration, up to crystallite structures (see Fig. 2.9) within a hierarchical type of organization.

Details of the micelle structure of Yen's model [352] are shown in Fig. 2.10; in this structure, bonds of sulfide, ether, aliphatic chain, and/or naphthenic rings constitute the bridge among the other building rings. The micelle size was reported to be of about 16–20 Å high and 8–16 Å wide [353].

The hierarchical view provided by Yen envisions the asphaltene phenomenology at different levels, particularly association toward aggregation. However, the uncertainties surrounding asphaltene molecular structure at the time this model was proposed introduced a high degree of inaccuracy.

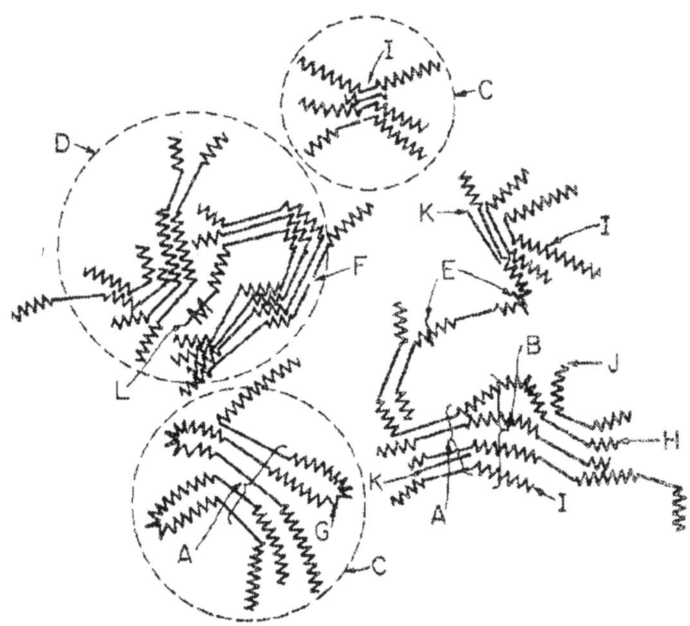

Figure 2.9

Yen model (A) crystallite, (B) chain bundle, (C) particle, (D) micelle, (E) weak link, (F) gap and hole, (G) intracluster, (H) intercluster, (I) resins, (J) single layer, (K) porphyrin, (L) metal. *Reproduced from Dickie JP, Yen TF. Macrostructures of the asphaltic fractions by various instrumental methods. Anal Chem 1967;39(14):1847–52, with permission from ACS Publications.*

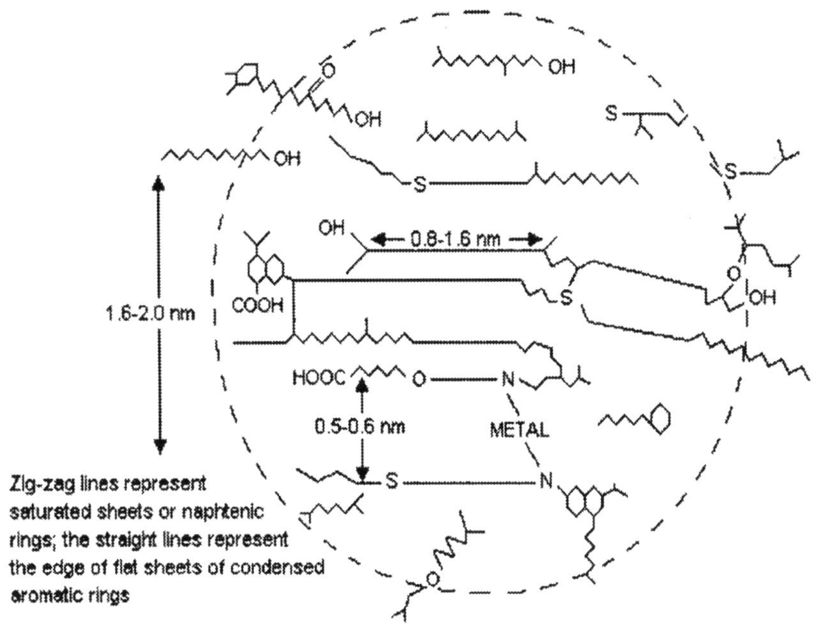

Figure 2.10

Details of the micelle structure according to Yen model. *Reproduced from Yen TF. Structure of petroleum asphaltene and its significance. Energy Sources Part A Recovery Util Environ Eff 1974;1(4):447–63, with permission from Taylor & Francis.*

More recently, a supramolecular assembly of molecules has been proposed, which in addition to aromatic π–π stacking combined cooperative binding by H-bonding, acid–base interactions, metal coordination complexes, and interactions between cycloalkyl and alkyl groups to form hydrophobic pockets. An example of this type of assembly is given in Fig. 2.11 [354].

This supramolecular model of Gray (University of Alberta) includes a range of architectures and molecular structural types that the authors considered may coinhabit in the crude oil. The arrangement creates porous networks and host–guest complexes. The latter may include organic clathrates, in which occluded guest molecules stabilize the assembly of a cage.

4.3.2 Particle Growth

As well as the various proposals of asphaltene molecular structure that have given rise to different modes of association, particle growth hypotheses would result from the view of associated molecules. The asphaltene colloidal particles are thought to have core–shell (core–corona) structures. Presumably, the cores would form stacked aggregates of "insoluble" asphaltenes, while the stabilizing shells/coronas may be composed of asphaltene molecules with higher solubility [252]. This model agrees with the basis

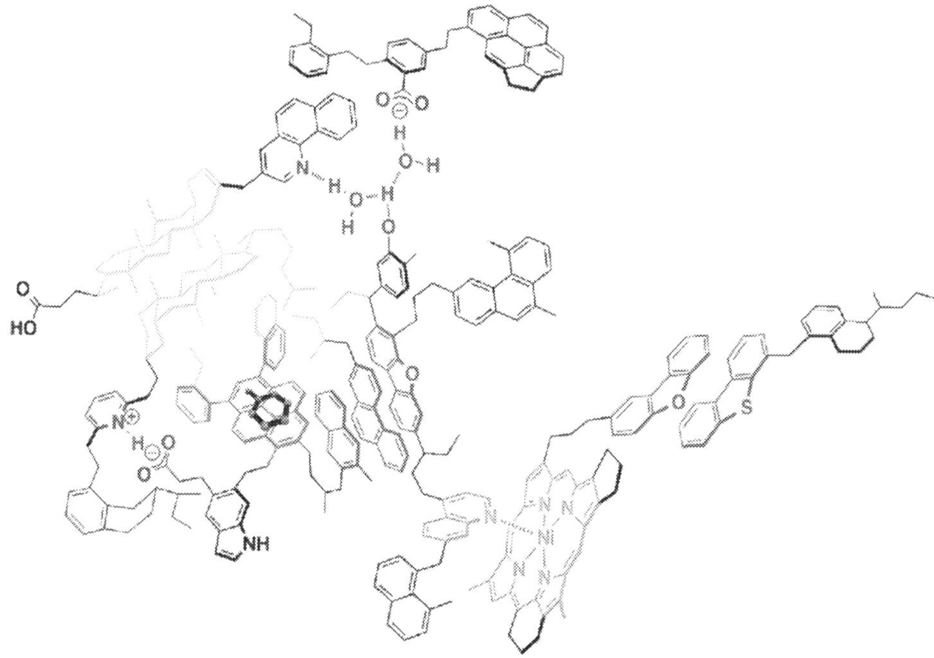

Figure 2.11

Supramolecular assembly representative of asphaltene aggregate including acid−base interactions and H-bonding (*blue*), metal coordination complex (*red*), a hydrophobic pocket (*orange*), π−π stacking (face to face, *dark green*; within a clathrate containing toluene, *light green*). *Reproduced from Gray MR, Tykwinski RR, Stryker JM, Tan X. Supramolecular assembly model for aggregation of petroleum asphaltenes. Energy Fuels 2011;25(7):3125−34, with permission from ACS Publications.*

of Acevedo's aggregation explanation. In Acevedo's aggregate structure (rosary type) the particle core is made up of the least soluble fraction A1, surrounded by the more soluble A2 [171,172], which act as solvating/dispersing agent [355] (see proposed molecular structures in the corresponding section later; Fig. 2.20A). Studies of aggregation and dissociation of solutions of asphaltenes in resins (solvent) indicated that the colloidal aggregates grow by subsequent incorporation of A1 asphaltenes, while the A2-type asphaltenes remained dissolved in the resins [356]. In this rosary type of structure, the colloidal A1−A2 aggregates are connected by the alkyl chains, rather than being stacked through the π−π system. Although the structure of A1 monomer is rather rigid, the alkyl connection makes the rosary-type structures very flexible and a large number of folded and unfolded conformers could be envisioned [355].

Yen's model was later modified by Mullins [357] focused on aggregation (Fig. 2.12). In this Yen−Mullins model, a molecule of asphaltenes is in the range of ~1.5 nm; a nanoaggregate consists of about six (stacked) molecules with a size of about 2 nm, and the cluster contains about eight nanoaggregates starting at about 5 nm [358]. Previous work

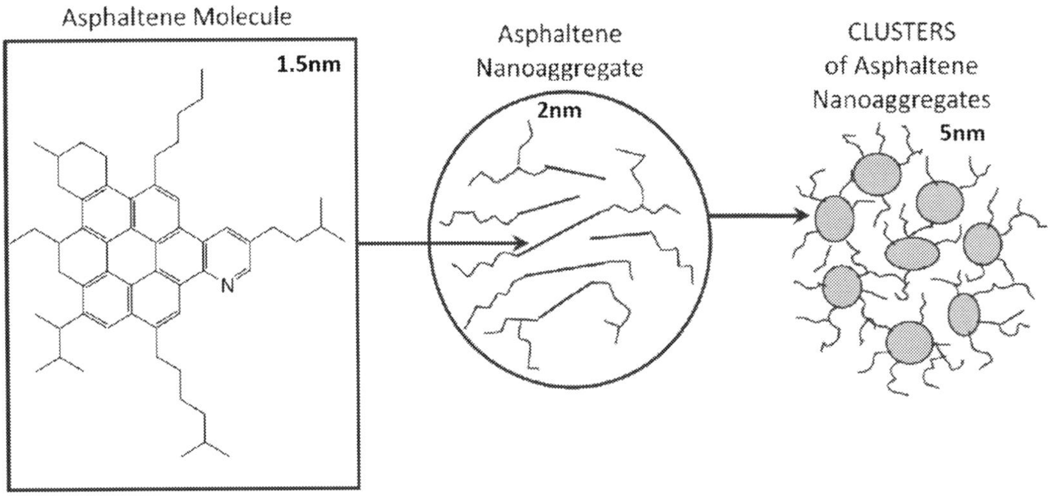

Figure 2.12
Modified Yen–Mullins model (I) asphaltene molecule, (II) asphaltene nanoaggregate, (III) clusters of asphaltene nanoaggregates. *Reproduced from Mullins OC. The modified Yen model. Energy Fuels 2010;24(4):2179–207, with permission from ACS Publications.*

from Mullins' group concluded that resins did not interact with these nanoaggregates; neither do they interact with the asphaltene molecules [317]. Hence according to Mullins, resins should not be considered surfactants of asphaltene molecules, and neither are asphaltene nanoaggregates a standard micelle system.

The work of Eyssautier et el. evidenced this type of hierarchical aggregation up to the mesoscale organization [52]. Thus molecules form nanoaggregates that assemble into fractal clusters in a second aggregation step, which precedes the clusters' aggregation step. The persistence of this multiscale organization, from free monomers to high aggregation number fractal clusters, from RT to at least 300°C, was also observed. An additional coherent scattering domain, attributed by others to the crystallite size of aromatic stacking, also persists at 300°C. Evidence of stacking was also found by characterizing the coke formed in the microcarbon residue test made from a sponge coke-forming sweet VR feed, for which local nucleation was observed at very small selected areas, with diameters of 0.25–1.25 μm. Electron diffraction of these areas, for VRs with the lowest heteroatom content, showed (002) stacking layers, with stacking distances of ~3.4 Å [264]. Other NMR results are also consistent with the hierarchal Yen–Mullins model [359].

Meanwhile, the oil residue model proposed by Wiehe [195] serves as a basis for understanding compatibility (Fig. 2.13), which is believed to be an aggregation consequence of changes in solubility.

a s a
a s a s s
a a R R R a s
s s R A A R a s
a s R A A R s a
s a a R R a s
a s a s s
s s a

A = Asphaltenes (Solute)
R = Resins (Dispersant)
a = Aromatics (Solvent)
s = Saturates (Nonsolvent)

Figure 2.13
Model of an oil residue: *a*, aromatic; *A*, asphaltene; *R*, resin; *s*, saturate. *Reproduced from Wiehe IA, Kennedy RJ. The oil compatibility model and crude oil incompatibility. Energy Fuels 2000;14:56–9, with permission from ACS Publications.*

Two questions have always been asked: what forces drive the association and at what point does aggregation lead to precipitation? The forces that drive association somehow have to be stronger than those that keep the asphaltenes in solution. Asphaltene flocculation could be induced by changes in solubility (adding a paraffinic AS). When flocculation by *n*-heptane addition of a toluene solution of asphaltenes was followed by cryomicroscopy it appears as a phase separation process of two liquids with different asphaltene concentrations. The aggregates grew as the amount of *n*-heptane increased, but solvent was trapped inside these large particles. A pure asphaltene phase does not precipitate as a classic case of precipitation. Concentrated asphaltene suspensions in *n*-heptane exhibit aggregates with a fractal-like structure [360].

In the absence of shaking or stirring, flocculation has been found to proceed continuously forming fractal-like flocs of several microns in size with a very loose structure. Above a size of 1.5–2 μm sedimentation starts and growth continues. Flocs reach sizes in the range of 4–5 μm when ultimate sedimentation takes place [227], in agreement with fractal-like diffusion-limited aggregation [361,362]. Under shaking, the unstable fractal aggregates dissociate and basic aggregated particles of an order of 1 μm are formed. The kinetics of aggregation–flocculation of asphaltenes was proposed to be divided into three stages. The first one was a nucleation stage corresponding to the formation of asphaltene clusters of a critical size. The second stage was the growth of these clusters into basic aggregates by absorbing either asphaltene molecules or small micelles from the solution. The coalescence of these basic aggregates into fractal structures constitutes the third stage [227].

Although flocculation has been considered to be reversible [227], the reversibility of precipitation is a subject of discussion. The redissolution of the precipitate has shown a hysteresis loop when the conditions are returned to preflocculation point [221,222,224,363]. Only a fraction of the total asphaltenes would precipitate and the magnitude of the fraction left in solution will depend on the AS employed. In fact, aggregation and precipitation are controlled by different intermolecular forces: while strong forces induce association, precipitation is caused by dispersive forces among the aggregates.

4.3.3 Factors Affecting Aggregation

Asphaltene aggregation behavior is likely controlled by the polydispersity, chemical composition, and steric arrangement or interconnectivity of functional groups in the asphaltene monomers [364]. Regardless of physical properties and of geographical origin, calorimetric data indicated that asphaltene aggregates form a polydispersed system with low cohesion energy [365]. Filtration has proven the polydispersity of size, since different amounts could be recovered using different pore size membranes [366].

A particle size distribution from 33 to 252 Å (obtained by small-angle X-ray scattering, SAXS) of Safaniya asphaltenes was observed upon fractionation by ultracentrifugation [75]. However, a combination of three techniques is needed to cover the wide range of aggregate sizes. Thus asphaltene aggregate sizes in toluene solutions were studied by SAXS, small-angle neutron scattering (SANS), and dynamic light scattering (DLS). Pressure showed only a minor effect within the considered temperatures. A huge modification of asphaltene macrostructure was observed over a wide temperature range. At high temperatures, reversible aggregation of asphaltene leads to stable small entities. Irreversible aggregation of asphaltene and a large increase of the aggregate size occur upon decreasing the temperature [367]. Instead, reversibility of the aggregation process, detected by RI changes, was observed for temperature- (RT—120°C) and pressure- (0.65–2.75 MPa) induced aggregation [368].

Concentration, temperature, and time affect both the equilibrium of aggregate dissociation and aggregation rate. In concentrated solutions van der Waals forces predominate; in diluted solutions coulombic forces contribute most to aggregation [334]. The high MW molecules are covalently surrounded by a varying number of smaller ones, which are held together by intermolecular bonds. Bond formation might have been evidenced by the fact that the heat of solution is exothermic, while the heat of dilution of a concentrated asphaltene solution was endothermic (attributed to bond dissociation in aggregates or micelles in solution) [369] and by the observation of a hysteresis between precipitation and redissolution [222]. Once the concentration of the aggregates becomes large, cohesion forces such as van der Waals move the aggregates to interact with each other, leading to precipitation [370].

The effects of concentration were also explained in terms of the colloidal model. At concentrations below CMC, mainly molecular association takes place. Meanwhile, at concentrations above CMC in aromatic solvents the aggregates would be formed from elementary asphaltene particles and the character of the aggregation is solely determined by the diffusion of the aggregates. Above the CMC, micelle aggregation occurs. As the asphaltene concentration increases, the micelles should become larger and the potential barrier increases [341,371,372]. Finally, at higher concentrations the self-association of asphaltene micelles, because of the increase in micelle concentration in an aromatic medium, would be favored [373]. The effect of concentration has been explained based on classic Derjaguin, Landau, Verwey, and Overbeek theory, according to which during aggregation of colloidal particles two factors affect the kinetics: diffusion and reaction. Fast aggregation is typically governed by diffusion. In this case, one is in the presence of diffusion-limited aggregation (DLA). Another situation occurs when there is chemical interaction between the particles. In reaction-limited aggregation (RLA) the particles interact in a special manner, e.g., chemical, steric, etc., and not every contact between two particles results in their sticking. In other words, particles sticking may be an activated process. Thus at concentrations below CMC a DLA process takes place that may be a micellization stage. Above CMC the initial stages of aggregation become RLA and then change back to DLA [371]. However, the view of a colloidal nature of asphaltene-containing mixture derives from the fact that asphaltene molecules are in a medium that presents a much smaller average MW. In this picture, asphaltenes and oil constitute a true solution and the colloidal behavior is caused by the big difference in dimensions between asphaltenes and oil (lyophilic colloid). This means that system behavior could be described by means of the traditional thermodynamic equations for liquid—liquid equilibrium, avoiding preconception of a colloid [374].

The effects of temperature, heat, dispersant additives (ionic liquids), solvents, and ultrasound on disaggregation of nanoaggregates in toluene solutions has been assessed by considering Rayleigh scattering on the apparent absorption of visible radiation. While solvents and ultrasound did not change the particle size, the other considered factors did increase it [375,376].

The effect to the chemical nature of the asphaltenes on aggregation behavior was study by evaluating the flocculation and SPs of subfractions of C_7-asphaltenes that have been subfractionated using mixtures of methylene chloride as polar solvent and *n*-pentane as flocculant. Although there was not a clear trend among the subfractions, it was found that the subfraction 40/60 was more prone to flocculate, exhibiting also the largest diffusion coefficient. The subfraction with the least flocculating behavior (10/90) was also the one with the lowest diffusion coefficient. Hence it was suggested that asphaltenes with larger sizes, MW, and aromaticity had the greater flocculation capability [110].

The crystalline organization of precipitated asphaltene particles was derived from the X-ray diffraction pattern, which was interpreted as the result of a lamellar arrangement, schematically represented in Fig. 2.14. Some of the lattice parameters, such as the interlayer distance (d_M), the interchain distance (d_γ), the diameter of the aromatic clusters

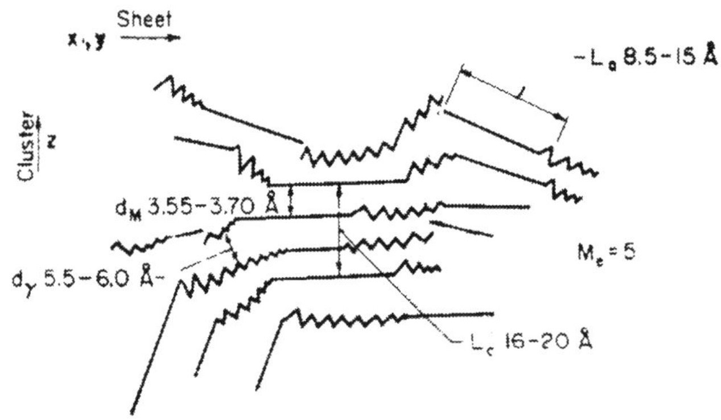

Figure 2.14

Dimensions of the proposed lamellar structure of asphaltene. *Reproduced from Yen TF, Erdman JG, Pollack SS. Investigation of the structure of petroleum asphaltenes by X-Ray diffraction. Anal Chem 1961;33(11):1587–94, with permission from ACS Publications.*

perpendicular to the plane of the sheets (L_α), and the diameter of the aromatic sheets (L_c) were determined from the X-ray diffraction pattern [97].

These parameters were also evaluated for asphaltenes from four Turkish crude oils [377]. Asphaltenes from a Kuwaiti AR was subfractionated in 13 narrow fractions, some of which were then characterized by NMR and XRD [378]. A comparison of the values obtained for those Turkish and Kuwaiti asphaltenes and those of Yen [97] are collected in Table 2.2, showing good agreement.

Although asphaltenes separated with *n*-heptane typically are shiny black solids and have been assumed to be crystalline from this appearance, they are in fact amorphous. Crystallinity has been observed in the presence of contaminating wax crystals [96]. Regardless, asphaltene crystals, from colorless to brown, with different shape and crystallographic structures were reported to be observed in a petrographic study in samples containing inorganic contaminants [98].

The lamellar arrangement can be the consequence of the stacking array derived from the $\pi-\pi$ interactions mentioned earlier. A descriptive model for asphaltene association and

Table 2.2: Comparison of Crystallite Parameters of Turkish [377] and Kuwaiti [378] Asphaltenes, and Those of Yen's Results [97]

	Turkish Asphaltenes				Kuwaiti Asphaltenes		
					Atmospheric Residue	Fraction 4	Yen's
d_M (Å)	3.54	3.46	3.54	3.64	3.5	3.6	3.55–3.70
d_γ (Å)	5.96	5.89	6.27	5.99	4.4	4.5	5.5–6.0
L_c (Å)	3.95	6.1	10.81	10.55	8.9	9.9	8.5–15
M	2.12	2.77	4.07	3.9	6	5	5

precipitation based on these intermolecular forces was proposed. This 2D stacking leads to highly flexible monomolecular sheets in which the spontaneous bend out of the aromatic plane would cause the formation of hollow spherical vesicles [370].

Acidity and basicity have been considered not to be very important in aggregation [74], instead petroleum bases are thought to stabilize asphaltenes by an acid–base mechanistic interaction [379]. Characterization of asphaltene deposits indicated a high content of high MW polar compounds [68]. Therefore, some possible explanations include: (1) acid–base interactions modify the local environment of the asphaltenes making them more sensitive to physical conditions; (2) the acid, and/or base polar moieties are sterically inaccessible; or (3) the polar moieties in the high-MW asphaltenes do not bear acid or basic functionalities.

The more polar asphaltenes are also the less soluble class and prone to aggregate, forming sparingly soluble agglomerates. Therefore low solubility favors agglomeration. The aggregates derived from this less soluble subfraction are the result of all sorts of interaction forces: H-bonding, $\pi-\pi$ interaction, and electron donor–acceptor interactions between the polar and aromatic moieties. The formed aggregates had fractal dimensions in the range of 1.7–2.1, in the absence of solvating resins. In the presence of resins, these become more compact with fractal dimension in the range of ~3. From a molecular point of view, these less-soluble asphaltenes were thought to be of archipelago type [380].

Aggregation energies of asphaltenes have been modeled and evaluated by IFP in single-component solvents (pyridine, toluene, and *n*-heptane). The combined knowledge of the systematic studies of the influence of molecular structure on solvation and aggregation with analytical information is expected to provide better predictions on heavy oil processability and probably will launch new process ideas [38,381–384].

4.4 Characteristics of Molecules and Clusters

The elucidation of the molecular structure of asphaltenes has been a research challenge for more than 70 years. Characterization results indicated that asphaltenes are condensed aromatic cores containing branches and bridges of hydrocarbonaceous and heteroatomic moieties. The presence of high concentrations of heteroatoms, such as nitrogen, oxygen, and sulfur as noncyclic and heterocyclic groups, has been proved. Metals contamination, mainly from nickel and vanadium compounds, is another characterization target, for which the chemical nature is not yet fully understood.

A clear difference with the other boiling point-defined components of crude oils is introduced by defining asphaltenes as a solubility class. The molecular compounds existing in this class of component have been difficult to isolate and consequently to identify. Characterization and modeling are probably the most popular study areas of asphaltene physical chemistry. The complexity of the molecules has given rise to long-standing controversies. Subfractionation of the asphaltene fraction has also been carried out for simplifying the complicated objective of characterization (see, for instance, Ref. [27]).

A diversity of characterization techniques has been employed for the characterization of subfractions in the help for defining speciation, i.e., the chemical nature of these moieties and functional groups. A great deal of methodologies for characterizing heavy oils and modeling asphaltene molecules and agglomeration have been proposed (e.g., Ref. [385]).

4.4.1 Characterization Concerns

Since the main concerns for asphaltenes analyses regards their agglomeration and precipitation, ideal methods for separation, isolation, and purification are required prior to any characterization attempt. Characterization has to overcome complicating limitations, for instance, some of the major issues that characterization work has to face include:

1. The isolated asphaltenes contain aggregates of molecules of different and high MW;
2. The isolated fraction (and subfractions) is a mixture of compounds widely different;
3. Condensed polynuclear aromatic ring systems tends to form graphite-like stacks, even those containing aliphatic chains and heteroatoms, such as those supposedly present in the asphaltenes;
4. Macromolecular structures might enclose microporous units that could entrain smaller molecules; and/or
5. Asphaltenes cannot be represented as a solid phase intrinsically insoluble in hydrocarbon media.

UV spectra for toluene solutions of petroleum and coal asphaltenes showed prominent peaks at 288–310 nm (4.3–4.0 eV), near the absorption edge of the solvent (toluene) at 285 nm. These peaks have been demonstrated to originate from experimental artifacts. Below the absorption edge the "zero line" corresponds to negligible transmittance, which cannot be further affected by the solute (asphaltenes) [386].

According to Strauss et al. [387], data derived from fluorescence decay and depolarization kinetic times were wrong because of inappropriate instrumentation, misleading results, and misinterpretations. In this regard, the conclusion regarding the absence of bichromophoric-type molecules was mistaken and consequently a unique continental type of molecule was also a mistake. In their opinion, asphaltene fraction is a mixture of a plethora of different, unknown components, with unknown concentrations along with innumerable different, unknown, and some known chromophores portraying widely different absorption coefficients, fluorescence quantum yields, and kinetic decay times. The asphaltene core was supposedly a single polycondensed ring system, and then the MW was obtained by the fluorescence depolarization with the rotational correlation time method. This is definitely not so, and the results reported in Refs. [10,81,306,388–390] were wrong. These potential uncertainties from results and interpretations of fluorescence data were somehow anticipated also by Ascanius et al. They found that the asphaltene fraction insoluble in *N*-methyl-2-pyrrolidone (NMP, up to 53%) hardly exhibited any UV−visible light absorption or fluorescence. This result implies that a substantial fraction of asphaltenes will not be represented in a fluorescence spectrum and will derive severe implications in the capacity of fluorescence spectroscopy to analyze asphaltenes [391]. These findings and the presence of NMP-insoluble fraction might explain the enormous value of MW evaluated by laser desorption mass spectrometry (LDMS) of the NMP-insoluble residue of asphaltenes of up to *m/z* 200,000. This group also found agreement between size-exclusion chromatography (SEC) results (using NMP as mobile phase) and LDMS, both showing very wide mass ranges for the whole asphaltene fraction [392]. On the other hand, the fact that a multivariate calibration technique was needed to elaborate a prediction model to quantify asphaltenes and resins from their fluorescence spectroscopy intensity data [393] provides further evidence of the need for cautious interpretation of the results from this technique. Furthermore, caution has also to be exercised when using NMP as solvent, extractant, or eluent in any other application or handling of asphaltenes. For instance, Karaca et al. [394] fractionated an AR into seven fractions by silica column chromatography, sequentially eluting with pentane, toluene, acetonitrile, pyridine, NMP, and water. SEC in NMP as eluent and synchronous UV-fluorescence were used for characterization of the fractions. SEC showed a bimodal distribution, in which the area of the band corresponding to the larger sizes increased through the fractionation sequence. Although this could reflect a correlation between size and polarity, it could also reflect the NMP asphaltene solubility issue.

Besides the discussed issues, directly derived from the use of fluorescence spectroscopy in asphaltenes characterization, the variation of the fluorescence intensity from asphaltene brought about new implications for the determination of asphaltene MW by laser ionization mass spectrometry (LIMS). The fluorescence intensity in the spectral range of 360−620 nm exhibits a progressively decreasing trend with an increasing MW of the fractions. The photophysical analysis of Strauss et al. [395] translated this into a decrease in the ionization capabilities of the laser on asphaltene molecules. Then, it is evident that the LIMS method for MW and MW distribution measurements would suffer from a progressively severe bias against higher MW species, ultimately leading to the complete loss of detection beyond a critical MW [395].

Initially, molecular mass distribution was the main contribution from the low-resolution MS methods [143,396−428], while later high resolution provided a more detailed speciation [429−470]. In recent times, detailed characterization came from mass spectroscopic techniques and/or its combination and integration with other methods and methodologies. The main questioning on the use of MS concerns sampling: preparation, volatilization representativeness, ionizability, etc. but also on mass analysis. Regarding ionizability, ionization sources and methodologies have been compared (e.g., [423,426,471–473]) and confirmed differing results from different sources. Two laser-based techniques have been compared in terms of volatilization, ionization, and mass analysis: laser desorption laser ionization MS (L^2MS) and surface-assisted laser desorption/ionization (SALDI) MS. In L^2MS, laser desorption, being a thermal process, induces a nonselective volatilization; the occurring single-photon ionization is a soft and universal method applicable to virtually any organic molecule; and mass analysis occurs by time-of-flight mass spectrometry, which has a nearly constant sensitivity across a broad mass range. Moreover, fragmentation and multiple charging are minimized in L^2MS, hence nearly all components of asphaltenes are detected as disaggregated molecules. Meanwhile, SALDI detects asphaltenes in the form of nanoaggregates [473].

Concerns extend to behavioral characteristics, such as stability, typically assessed through onset determination. Many different techniques have been employed: optical microscopy, spectrophotometry, rheometry, electrical measurements, and other techniques have been proposed in the literature to monitor the mixture status. These many techniques are found as well in many more methodologies. Therefore inconsistent results and lack of comparability is the status quo. The need for standardization is emphasized or at least the definition of common practices should be established. Correra et al. [266] have made suggestions in regard to common practices and remarks about titration:

- To obtain the true onset, continuous titration is to be avoided;
- In stepwise titrations, the time between successive precipitant additions must be at least 15 min;
- A minimum or maximum in the monitored variable is not necessarily associated with the onset condition;
- A better estimate of the onset can be obtained with the "slope" criterion.

4.4.2 Chemical Features

According to Boduszynski's continuous model of crude oil, the average molecular parameters are clear evidence of the gradual and continuous increase of aromaticity, MW, and heteroatom content with increasing boiling point [398,474–481]. He reported the average molecular parameters for the atmospheric residue (AR) of two different heavy oils: Kern River and San Joaquin Valley (Offshore California, California). The comparative data for the heaviest fractions are presented in Table 2.3 [479]. These data have been arranged in increasing MW order to facilitate the comparison. (The average MW values were obtained by VPO and the results were interpreted as being caused by intermolecular associations).

Asphaltenes are heteroatomically substituted; besides carbon and hydrogen, these compounds contain sulfur, nitrogen, and oxygen. Comparison of the asphaltenes recovered from deposits collected in pipelines with those isolated from the same crude oil indicated

Table 2.3: Average Molecular Parameters[a] of Atmospheric Residue (AR) Fractions [479]

	KR1	Cal1	KR2	Cal2	KR3	KR4	Cal3	Cal4
Molecular weight	876	955	1464	2254	3064	5625	5697	16,348
Average Number of Atoms/Molecule								
C	62.4	66.2	103.8	154.6	218.1	395.8	387.5	1112.1
C_{arom}	17.7	16.4	38.6	41.4	97.3	214.1	151.9	487.1
H	93.3	101.6	149.7	229.9	265.7	457	496.8	1315.3
S	356	1611	608	4120	1388	2514	11,358	34,739
N	745	532	1464	1481	4815	9643	7121	23,938
O	515	310	873	873	2298	4641	2991	12,261
V	8	8	1	7	10	30	104	452
Ni	1	2	3	2	23	38	24	103
Fe			1	1	15	42	16	88

[a]Heteroatom concentrations have been multiplied by 10^3.

a higher concentration of the heteroatoms in the former [446]. A clear trend of increasing heteroatomic concentration, as well as aromaticity, with average MW of the fraction can be immediately inferred from Table 2.3. A mass percentage atomic composition is collected in Table 2.4 [479].

In Table 2.5, the elemental composition of some examples of C_5- and C_7-asphaltenes isolated from three Mexican crude oils [482] and Canadian, Iranian, Iraqi, and Kuwaiti crude oils [29,210] are compared.

In a larger number of samples, the atomic H/C ratios have been found to vary between 1.0 and 1.3 and N, S, and O contents could be found above 1% and as high as about 10% [27,30,80,476,479]. The composition of these elements varies within wider ranges, among the larger set of samples from 0.3% to 4.9% for O, 0.3–10.3% for S, and 0.6–3.3% for N. A summary of compositional ranges of C_7-asphaltenes has been given by Cimino et al. [141] and is reproduced in Table 2.6.

Table 2.4: Atomic Composition of Average Molecule Studied in Ref. [479]

	KR1	Cal1	KR3	Cal3	KR5	KR6	Cal5	Cal7
C	39.7	38.9	40.5	39.5	44.3	45.5	42.8	44.5
H	59.3	59.7	58.4	58.8	54.0	52.5	54.8	52.6
S	0.23	0.95	0.24	1.05	0.28	0.29	1.25	1.39
N	0.47	0.31	0.57	0.38	0.98	1.11	0.79	0.96
O	0.33	0.18	0.34	0.22	0.47	0.53	0.33	0.49
V	0.005	0.005	0.000	0.002	0.002	0.003	0.011	0.018
Ni	0.0006	0.0012	0.0012	0.0005	0.0047	0.0044	0.0026	0.0041
Fe	0.0000	0.0000	0.0004	0.0003	0.0030	0.0048	0.0018	0.0035

Table 2.5: Elemental Analysis and Metal Content of (C_5 and C_7) Asphaltenes From Certain Opportunity Crudes

	Maya		Isthmus		Olmeca		Canada		Iran		Iraq		Kuwait	
Solvent	n-C_5	n-C_7	n-C_5	n-C_7	n-C_5	n-C_7	n-C_5	n-C_7	n-C_5	n-C_7	n-C_5	n-C_7	n-C_5	n-C_7
Elemental Analysis (wt%)														
Carbon	81.23	81.62	83.90	83.99	86.94	87.16	79.5	78.4	83.8	84.2	81.7	80.7	82.4	82
Hydrogen	8.11	7.26	8.00	7.30	7.91	7.38	8	7.6	7.5	7	7.9	7.1	7.9	7.3
Oxygen	0.97	1.02	0.71	0.79	0.62	0.64	3.8	4.6	2.3	1.4	1.1	1.5	1.4	1.9
Nitrogen	1.32	1.46	1.33	1.35	1.33	1.34	1.2	1.4	1.4	1.6	0.8	0.9	0.9	1
Sulfur	8.25	8.46	6.06	6.48	3.20	3.48	7.5	8	5	5.8	8.5	9.8	7.4	7.8
Atomic Ratios														
H/C	1.198	1.067	1.144	1.043	1.092	1.016	1.21	1.16	1.07	1	1.16	1.06	1.14	1.07
O/C	0.009	0.009	0.006	0.007	0.005	0.006	0.036	0.044	0.021	0.012	0.01	0.014	0.014	0.017
N/C	0.014	0.015	0.013	0.014	0.013	0.013	0.013	0.015	0.014	0.016	0.008	0.01	0.009	0.01
S/C	0.038	0.039	0.027	0.029	0.014	0.015	0.035	0.038	0.022	0.026	0.039	0.016	0.034	0.036
Metals (wppm)														
Nickel	268.7	320.2	155.4	180.4	81.9	157.7	–	–	–	–	–	–	–	–
Vanadium	1216.6	1509.2	710.3	746.6	501.0	703.8	–	–	–	–	–	–	–	–

Table 2.6: Compositional Ranges of C_7 Asphaltenes [141]

Yield of asphaltenes (oil-based wt%)	<30
H/C ratio	0.8–1.4
Sulfur (wt%)	0.5–10.0
Nitrogen (wt%)	0.6–2.6
Oxygen (wt%)	0.3–4.8
Aromaticity factor	0.45–0.70
n (average number of atoms of C per alkyl substituent)	4–7

In summary, the chemical composition of asphaltenes appears to be polydispersed. The elemental profiles of C/H, S, N, and O for the SARA fractions were reported by Gawel et al. [121] and are reproduced here in Fig. 2.15A–D, respectively. As can be seen, the largest difference in composition was found in the asphaltene fractions of the oils, followed by the resin fractions. These fractions had significantly higher abundance of oxygen and nitrogen heteroatoms than the saturated and aromatic fractions. Furthermore, the highest oxygen content in the asphaltene fraction corresponded to the lightest considered crude oils, while it was the highest in the resin fraction for the heaviest oils.

The elemental profiles of (20–30) narrow fractions from B6, Canadon Seco (CS), and Hondo (HO) asphaltenes were statistically analyzed assuming that the data were "normally" distributed. While H/C, N/C, and S/C ratios in B6 and HO asphaltenes

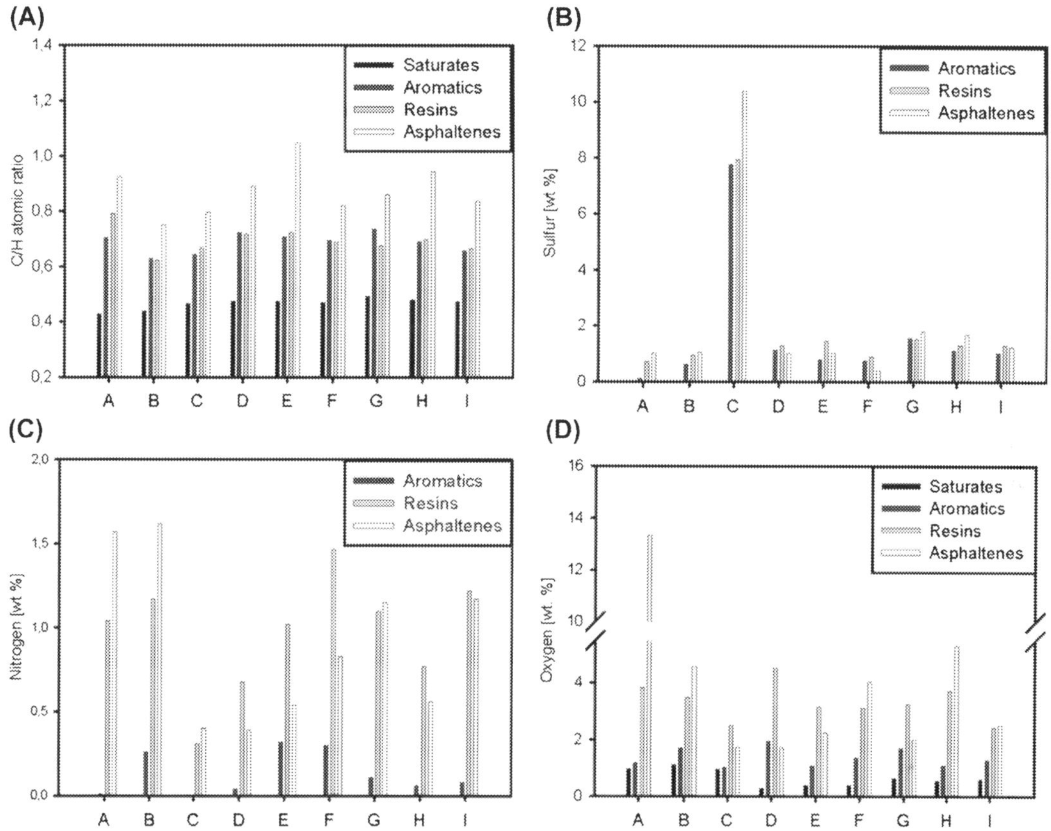

Figure 2.15

Elemental profiles for the saturates, aromatics, resins, and asphaltenes (SARA) fractions of nine different crude oils: (A) C/H, (B) S, (C) N, and (D) O. *Reproduced from Gawel B, Eftekhardadkhah M, Øye G. Elemental composition and fourier transform infrared spectroscopy analysis of crude oils and their fractions. Energy Fuels 2014;28(2):997–1003, with permission from ACS Publications.*

appeared to obey a Gaussian distribution, metals in the three asphaltene samples concentrated in the first precipitated subfractions. The interactions of metal compounds seemed to be important in the asphaltene aggregation mechanism, as indicated by the relatively high correlation between the N/C ratio and vanadium and nickel contents. On the basis of these results, polar and H-bonding interactions appear to be more important than dispersion interactions in the initial precipitation of B6 and HO asphaltenes from heptol. Meanwhile, the low concentrations of nitrogen, vanadium, and nickel in CS asphaltenes relative to the B6 and HO fractions make those polar interactions less important than the dispersion interactions [37].

Characterization of subfractionated asphaltenes also lead to differentiating chemical features of the molecular structure. The low-MW subfractions have very different chemical characteristics from heavy fractions. In general, when MW increased, the size of aromatic moieties decreased, and the degree of substitution on available aromatic carbons increased, implying the presence of fewer aliphatic carbons and higher aromaticity in low-MW fractions. As expected, metals were not uniformly distributed throughout the subfractions [108].

Chemical characterization of VR fractions and subfractions for the determination of structural features of the molecules has been widely reported [63,68,113,115,128, 327,328,368,396,483–526]. Other review articles deal with the chemical composition of asphaltenes, either with a variety of source crude oils or with the analytical methods employed [1,27,480,482,495,527,528].

Great effort has been centered on the speciation and functionality of the heteroatomic moieties. In terms of chemical nature, it can be expected that acids will contain more O, and the bases somewhat more N and S. Nevertheless, S-compounds [460,491,529–534] fall in the boundaries of ABN, since sulfur is found to be homogeneously distributed within these three types of subfractions.

Sulfur was reported to be present in asphaltenes as thiophenic, sulfide, and sulfoxides [535]. High-resolution MS (Fourier transform ion cyclotron resonance mass spectrometry, FT-ICR-MS) has confirmed the presence of aromatic S-species [446,447,460,463]. Thiophenic and sulfidic moieties [533,536], as well as occasional high sulfoxide concentrations [537], were identified by X-ray absorption near edge structure (XANES). Other techniques and studies confirmed speciation assignments [327,416,455,461,469,518,523,528,538–542]. Quantitative results indicated that sulfur exists as mainly thiophenic heterocycles (65–85%) but also as sulfidic groups [533], confirmed by potentiometric titrations, which also demonstrated that the thiophenic-S-content grows with MW increase [352]. Geochemical studies indicated that similarly to carbon, the aromaticity of the sulfur compounds increased with maturity [536,543,544].

The sulfide-S-content in asphaltenes of different origins varied considerably, for instance, it was found to range from 13% to 65% of the total-S in Alberta crude oil asphaltenes [27]. The unambiguous evidence for the presence of aliphatic sulfides in asphaltene was provided by the presence of the aliphatic sulfones in the oxidation reaction product. The combination of ^{33}S NMR with reactivity (oxidation) studies permitted the observation of many types of sulfone molecules; upon reaction, however, no assignment to any particular sulfone types was possible [506]. Ignasiak et al. [496] experimentally showed that at least 65% of the sulfur was present in sulfide bonds, while other considerations suggested a true figure of at least 90%. The experimental results of the reaction between asphaltenes and potassium naphthalide were erroneously interpreted as being caused by the exclusive cleavage of C–S–C bonds, and no C–C bond was broken. This erroneous interpretation was combined with elemental analysis and NMR data to conclude that Athabasca asphaltene molecules consisted of a sulfur polymeric framework embedded in an alicyclic diaromatic structure with some alkyl substituents, and held together by sulfide linkages [496,545]. The chemical features of the suggested monomeric structure are reproduced in Fig. 2.16.

Nitrogen species include cyclic N-species [326,449,537,546]. The chemical nature of N-compounds in bitumen and heavy oils from Utah was profiled as from about 54% to 62% nonbasic and very weakly basic (mostly amide and indole-type nitrogen) and from about 30% to 35% weakly basic (mostly of pyridine-type nitrogen) [547]. Early studies on potentiometric quantitative determination of basic nitrogen of asphaltene fractions showed the presence of NH-groups, as well as five-membered ring pyrrolic nitrogen and six-membered ring pyridinic nitrogen groups [548]. Meanwhile, aliphatic-N has not been detected either by X-ray photoelectron spectroscopy (XPS) [549] or by XANES [537], which demonstrated the aromatic nature of all asphaltene nitrogen-containing moieties. Other potentiometric titrations agreed with this assignment of N-species being mainly pyrrolic and to a minor extent pyridinic structures [550]. Initially, fractionation was needed to simplify the complexity of identifying such a complex mixture [72]. Then, techniques like low-resolution MS [416], high-performance liquid chromatography-SEC in combination with FTIR and elemental analysis [541], and NMR [551,552] were used to provide insights and guidelines for further characterization studies. In addition, high-resolution MS techniques (FT-ICR) provided more detailed information on the local environment of N, particularly the association with other heteroatoms [430,434,463,468,469,489,553–556]. Evaluation of the double bond equivalent values for the atmospheric pressure photoionization (APPI) positive-ion species MS spectra could distinguished clearly between pyrrolic and pyridinic species in molecules having between 33 and 37 C-atoms [462]. Characterization of the acid–base subfractionated asphaltenes showed that the acidic and neutral fraction contained both longer carbon chain–lower aromaticity and less polar–shorter carbon chain–higher aromaticity N-compounds [468]. Comparison of the FT-ICR-MS characterization results on N and O speciation from SARA separation and chemical fractionation indicated that neutral

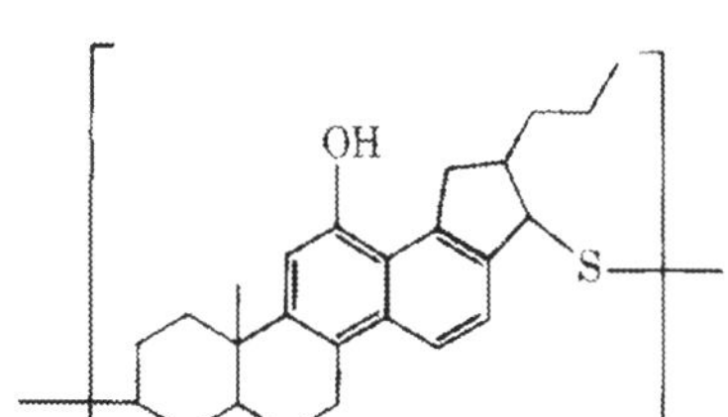

Figure 2.16

Hypothetical S-containing monomer of suggested polymer asphaltene structure. *Reproduced from Ignasiak T, Kemp-Jones AV, Strausz OP. The molecular structure of Athabasca asphaltene. Cleavage of the carbon-sulfur bonds by radical ion electron transfer reactions. J Org Chem 1977;42(2):312–20, with permission from ACS Publications.*

nitrogen species are typically of N_1 and N_1O_1 types and these species are entrained in the asphaltene fraction during deasphalting [465].

Oxygen species include carboxylic acids, carbonyls, and phenols. Oxygen as hydroxyl can be found in phenols, alcohols, and carboxylic acids; as carbonyl in esters, ketones, and amides; and bridging (ethers) functional groups as well as aromatic heterocycles [328,352,496,497]. Yen's work identified quinoid groups and ketones in an Athabasca asphaltene [545], though it did not detect quinone or semiquinone groups [352]. Intra- and intermolecular hydrogen bonding of hydroxo-groups was found to be close to completion by infrared and derivatization studies. According to several authors [487,497,557,558], a considerable portion of these OH-phenol groups coexist with two or more functions on the same aromatic ring or on adjacent peripheral sites, or sites adjacent to a carbonyl function, in a condensed aromatic system. Subfractionation by MW indicated an increase in carbonyl groups with a decrease in MW [63]. Characterization (FT-ICR-MS) of subfractionated asphaltenes located most of the (O_2, O_2S, O_2S_2, O_3, and O_4) oxygen-containing compounds in the more polar fractions [468], in agreement with the presence of carboxo-groups. The presence of a carboxylic functional group as a free acid in the asphaltene fraction can have an enormous effect on interfacial activity [466]. According to Shi et al. [465], SARA-fractionated asphaltenes were concentrated with highly aromatic and acidic O-containing molecules. This work reported a lower MW of the O-containing molecules in the asphaltene fraction than those present in the aromatic fraction. The presence of other heteroatoms in the same O-containing molecule was reported by Pantoja et al., from the characterization [matrix-assisted laser desorption/ionization (MALDI)-time-of-flight] of asphaltenes from 30 different crude oil samples. Sulfur (and O) was found in 12% of the molecules and nitrogen (and O) in 17% [416].

A solvent extraction scheme selectively fractionated the O-species from the whole asphaltene fraction. The soluble fraction (in a 15% alcohol + acetone mixed solvent) contained 40% of the oxygen present in the asphaltene fraction [496].

Subfractionation of the VR using supercritical fluids (*n*-pentane) provides asphaltene fractions that might be a closer representation of industrially produced asphaltenes. The characterization of 10 narrow subfractions revealed unexpected results. The contaminant heteroatoms (S, N, Ni, and V), together with the Conradson carbon residue (CCR), showed uneven distributions [559]. In Fig. 2.17 these results are compiled, including the MW obtained by VPO measurements and the SARA proportions.

One notorious aspect of asphaltene composition is that the precipitated fraction concentrates the heavy metals (Ni, V, and Fe) present in the crude oil (see also Ref. [560]). Whether the metal heteroatoms are part of the asphaltene molecules or they play any role in the precipitation/aggregation process remains uncertain. Less than 20% of the bottom of the BotB concentrated out of the whole residue: 30% of the S, 44% of the

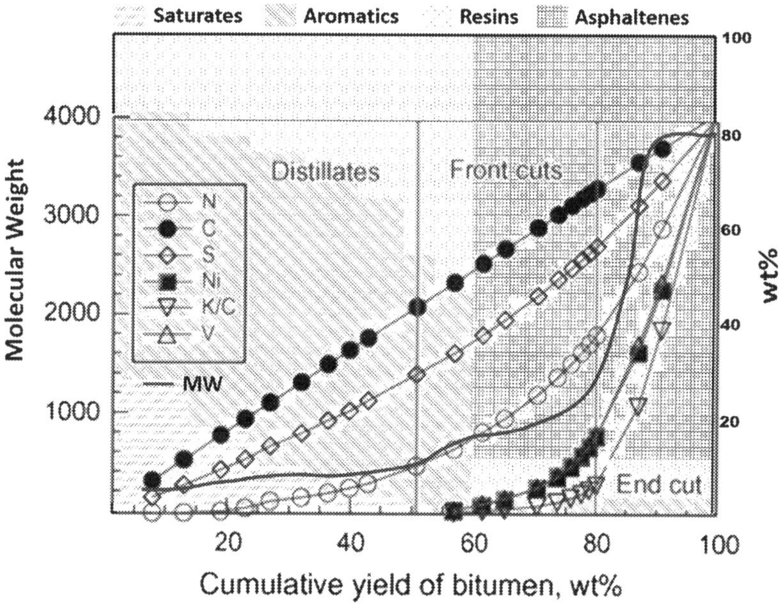

Figure 2.17

Compilation of characterization results from subfractions obtained by supercritical extraction. *Reproduced from Zhao S, Sparks BD, Kotlyar LS, Chung KH. Correlation of processability and reactivity data for residua from bitumen, heavy oils and conventional crudes: characterization of fractions from super-critical pentane separation as a guide to process selection. Catal Today 2007;125(3–4):122–36, with permission from Elsevier.*

N, 62% of CCR, and 80% of Ni and/or V. The profiles for N, Ni, V, and CCR showed similar trends, indicating that these species might be somehow associated. VPO MW and SARA fractions were quantified and indicated that separation is basically dependent on MW and that the end cut concentrates the asphaltene fraction with a larger contamination with resins and fewer aromatics.

Characterization of asphaltene subfractions showed differentiating chemical features among them. Typically, metals exhibit a nonuniform type of distribution, as it was the case for subfractions of Athabasca VR when subfractionation was carried out in terms of polarity using mixtures of THF/n-hexane. Eight narrow subfractions were recovered, starting with a 30/70 and adding 5% increments in n-C_6 up to 100% n-C6 and the remaining maltenes were also separated in S, A, and R. Another VR from a different Canadian heavy oil, Morgan, showed a decreasing metal content with decreasing polarity of the subfraction. Meanwhile, the other heteroatoms, N, S, and O, showed the same decreasing trend on the elemental content with decreasing polarity in both Canadian crude oils.

The removal of asphaltenes by an incremental change in polarity caused a monotonous decrease in SP of the deasphalted residue and its viscosity [115]. SFE was employed to separate four Chinese VRs into 15–17 narrow fractions using *n*-pentane as fluid. These fractions were obtained at constant temperature (~240°C, slightly above the pentane critical temperature of 197°C) by scanning the pressure from 4.0 to 12.0 MPa over 8 h (critical pressure of pentane is 3.4 MPa). Total liquid product recovery varied from 71.7 to 87.8 wt%, depending on the crude. Under such conditions, asphaltenes and resins are expected to be extracted. The reported composition of the VRs showed a total asphaltene and resin content ranging from ~25 to ~54 wt%, indicating that not all the heavy compounds were extracted and part was retained by the raffinate. The trends observed in the evaluated physicochemical properties agree well with the previous discussion. The incremental removal of the heavy components lead to the following monotonous changes of the properties of the raffinate: decreases in density, average MW, kinematic viscosity, heteroatomic contaminants content, and CCR, meanwhile, the H/C ratio increased [40].

The definition of the C-skeleton has been the subject of a long-standing controversy. Spectroscopic information of asphaltenes isolated using different methodologies and various crude oil sources have been reported in the literature and are the basis of proposed molecular structures. Carbon is estimated to be 50% aromatic and 50% aliphatic in asphaltenes [63]. Spectroscopic data support the hypothesis that asphaltene molecular structure contains condensed polynuclear aromatic rings bearing alkyl side chains. In this section the reported data will be shown and further discussion on the incorporation of these data in building a molecular model will be further discussed in the next section. However, at this point it is worth saying that the greatest controversy has been on the degree of condensation of the aromatic rings. The number of rings in the condensed system varies from as low as 3 up to 15 to 20 in the larger systems [209,210,352,399,496,519–521,561–565]. Although the population of asphaltenes with an average of 1–3 rings in the condensed system was small, they were observed in subfractions with high polarity by fluorescence emission spectroscopy [389]. Analytical pyrolysis [pyrolysis gas chromatography mass spectrometry (Pyr-GC/MS)] confirmed the presence of single rings (alkylbenzenes and cycloparaffins). The composition of the pyrolysate was roughly 14% paraffins, 44% cycloparaffins, 24% alkylbenzenes, 17% other aromatics, and 1% benzothiophenes, giving a proportion of 58% nonaromatics and 42% aromatics. The major presence of monocyclic unsaturated compounds in the pyrolysate was explained as a consequence of the high tendency of forming coke of the polycyclic compounds [566]. Low aromaticity and low degree of condensation were also confirmed in fractions of different polarity [394].

In the preliminary work of Strausz's group, the observation of monocyclic to pentacyclic aromatics seemed to indicate a maximum degree of condensation of 5, which nowadays is known to be larger. The identification of these families of naphthalenes, phenantrenes, chrysenes, and pyrenes was based on thermolysis results [496,508,511,516,567]. Reactivity studies indicated the existence of a higher degree of condensation and allowed the postulation of possible building units present in the asphaltene molecules in a proportion close to 60% of the molecule: cyclic and thiophenic sulfur, aliphatic and cyclic alkyl groups, fluorenes, and polycyclic aromatics. The remaining 40% was believed to be formed by larger polycyclic aromatic hydrocarbons (PAHs) and heteroatomic moieties [519]; the next step was connecting the building units to a model that explained the reactivity results. That hypothetical molecule became the first proposal of an asphaltene molecule (the derived model will be discussed in the next section); the C/H ratio of this molecule was 1.18 and had an MW of 6191 Da. Several characteristics of this model molecule have been observed in other studies (characterization, reactivity, and their combination), i.e., acyclic and cyclic saturated hydrocarbons, aromatic hydrocarbons and thiophenic aromatics [417], alkyl-aromatics [568], the existence of terminal O and OH groups for intermolecular H-bridges [497], the presence of intramolecular bridging S [545], H-bondings through N [569], polycyclic aromatics [570], etc. Thus special attention has been given to the characterization of individual moieties and functional groups, for instance, the length of the alkyl chains surrounding or bridging the aromatic rings.

NMR results indicated a preference of condensed arrangement over linear fusion of the aromatic rings (e.g., ovalene vs decacene) [400,571], confirming the presence of pericyclic aromatic condensation, as observed also by X-ray Raman spectra [572,573]. A decrease in aliphatic-to-aromatic hydrogen ratios with increasing MW was observed by H-NMR, which indicated an increase in the degree of condensation of the ring system [27]. Scanning tunneling microscopy, high-resolution transmission electron microscopy, and Raman spectroscopy provided evidence that the most probable asphaltene polycyclic is composed of 6–7 fused aromatic rings [574] within a distribution that goes from 3 to 11 rings [285,305,575–578]. A model built with 7 condensed rings and containing heteroatoms as per the experimentally determined distribution could render a good match of the measured optical absorption and fluorescence emission spectra of asphaltene fraction. The MW of the model also agrees well with the centroid of the MW distribution [579]. A smaller level of condensation was reflected by the FT-ICR-MS method that identified aromatic hydrocarbons appearing as 1–5 ring aromatics and naphthenoaromatics. Mass spectrometry-Z distribution pattern peaked at condensed aromatic core structures, such as benzene, fluorene, and chrysene [463]. The presence of naphthenic rings was found to increase with MW increases, being almost absent in the low-MW end [63,503]. Meanwhile, a larger degree of condensation (11–13 rings) was reported to be exhibited by the heaviest fraction isolated from toluene solutions, by precipitation with toluene/pentane mixtures of increasing pentane content [580]. Slightly larger numbers (<15) have also been reported from a Pyr-GC/MS study [400]. Instead, absorption, fluorescence, and excitation spectra of Athabasca asphaltene and five of its GPC fractions indicated PAH units with an average of three aromatic rings [505].

Consistent with this idea of a small number for the degree of condensation were the results from a combination of magnetic susceptibility and ESR [581], and thermolysis studies [582–585].

Cyr et al. [63] identified methyl, methylene, or methine carbon; but no quaternary aliphatic carbons were observed. They also concluded that the aromatic core of the asphaltene was surrounded by alkyl chains with a mean length of 7 Cs for the lowest MW of the different gel permeation chromatography (GPC)-separated asphaltene fractions (MW 1200) and increasing to 12 Cs for the highest MW fraction (MW 16,900) [63]. Bestougeff and Byramjee concluded that the aromatic rings are extensively substituted (70–80%) with long alkyl chains [27]. Meanwhile, infrared indicated a mean size of the aliphatic carbon chain of about 4 Cs [400,571].

Reactivity studies were performed based on ruthenium ion-catalyzed oxidation (RICO) of asphaltenes by sodium meta-per-iodate in water–tetrachloromethane–acetonitrile solvent systems at RT. RICO has provided evidence for the presence of structural features on an immature asphaltene that were not reported previously and were not observed in mature molecules. These are: *n*-alkyl side chains attached to aromatic rings (showing a slight even-to-odd CN preference) along with small amounts of R–C1–C3-branched *n*-alkyls; *n*-alkenyl bridges connecting aromatic rings; an increased concentration of unsubstituted cyclohexane rings condensed to aromatic rings compared to other asphaltenes; a relatively low level of condensed aromatic structures; and an apparent low level of unsubstituted terminal aromatic rings compared to other asphaltenes from mature resources [500,515,518,520,522–525]. The results of structural characterization studies were thought to be moving away from the older ideas that asphaltene molecules contained large polynuclear aromatic systems. Furthermore, the possibility of the variety of functional groups present playing important roles in the observed asphaltene behavior was also suggested [30].

The subfractionation of C_7-asphaltenes by sequential elution solvent chromatography revealed an increase in the polycondensation of the aromatic rings with the increase in MW obtained by SEC [65]. Although an increase in aromaticity is expected to occur with increasing asphaltene MW, this does not necessarily imply an increase in the degree of condensation. Furthermore, this conclusion was based in questioned MW (SEC) values and moreover when subfractionation was not performed on an asphaltene solution, but on an acetonitrile slurry that might have contained agglomerates that could not be dissociated through the elution process.

The distinguishing chemical features between coal and petroleum asphaltenes have been summarized by Yen [586] and are collected in Table 2.7. In general, petroleum asphaltenes are larger, heavier, more condensed, and further substituted with longer C-chains.

Table 2.7: Distinguishing Chemical Features Between Coal and Petroleum Asphaltenes [586]

Feature	Petroleum	Coal
Aromaticity	0.2–0.5	0.6–0.7
Degree of condensation $(H/C)_{arom}$	0.3–0.5	
Length of side alkyl chain (C#)	4–6	42,371
Degree of substitution (%)	50–70	30–50
Molecular weight (Da)	800–2500	400–600
Reactivity	Lower	Higher
Degree of association (#mol/cluster)	5–7	2–4
Cluster size (Å)	10–15	7–14
Polarity (heteroatom/C)	0.05	0.08
Hydroxyl and pyrrolic substitution	Lower	Higher

4.4.3 Molecular Structures and Models

Molecular structures and models have been derived from the characterization results and have been used for understanding not only behavior, but also process performance. Interest has been mainly focused on their real size and weight, since those will be probed on monitoring precipitation. Molecular models have to be taken as an attempt to understand behavior and to explain characterization results from asphaltenes, but not as a representation of an average molecule.

A variety of molecular models have been proposed, probably because of the nature of a solubility class fraction, which contains thousands of different compounds. These compounds might have precipitated from the titration of a VR with a light *n*-paraffin by virtue of their MW, size, polarity, aromaticity, or a combination of factors.

Initially, the top-down approach was followed and more recently the bottom-up approach has received much more attention. Any of these approaches proposes an "average" asphaltene model molecule, which has to be regarded as a mere representation, rather than a proposal for the physical existence of a single molecular topology and/or morphology. The previous section detailed the chemical features determined by the characterization of asphaltenes isolated by different methods from different sources. Bestougeff and Byramjee [27] listed the 10 basic elements that any model has to consider and include:

1. Total number of cycles;
2. Number of internal C—C bonds connecting C-aromatics;
3. Number of aromatic cycles;
4. Number of peripheral C—C bonds on aromatics;
5. Number of naphthenic cycles;
6. Number of C-atoms in methyl and methylene groups;

7. Degree of ring condensation;
8. Nature and probable location of functional groups;
9. Number and average chain length;
10. Substituting CH, CH_2, and CH_3 groups in aromatics.

Additionally, any model of a core of condensed aromatic rings should include small amounts of sulfur, nitrogen, and oxygen as well. These models have been focused on the explanation of the two main features of asphaltenes: their association/aggregation behavior and their MW. The low H/C ratio suggests that asphaltenes might have significant aromatic arrangements and the high heteroatom content found in the asphaltene fraction is responsible for their high polarity. The existing discrepancy and long-lasting controversy in asphaltene molecular structure concerns the degree of condensation of these aromatic rings, since that will imply significant differences in the properties of the resulting molecule. Still, these molecular species remain poorly understood, constituting the most chemically complex molecules in oil.

Speight and Moschopedis [209] made an attempt to include their NMR findings in one of the first proposals of asphaltene molecular structure: "…it is difficult to visualize these postulated structures as part of the asphaltene molecule. The fact is that all methods employed for structural analysis involve, at some stage or another, assumptions that, although based on data concerning the more volatile fractions of petroleum, are of questionable validity when applied to asphaltenes…" The proposed structure responding to the formula $C_{79}H_{92}N_2S_2O_3$ with an MW of 3449, whose chemical features were described in the previous section, is shown in Fig. 2.18.

Another one of the first structural models of an asphaltene molecule, known as the Strausz model, was based on thermolysis [496,508,511,516,567], RICO reactivity [500,515,518,520,522–525], and analytical results [328,514–516,587] and corresponds to the formula $C_{420}H_{496}N_6S_{14}O_4V$ with an MW of 6191 Da and H/C of 1.18 (Fig. 2.19 [519]).

The chemical features of this molecule include:

- A 2,5-*n*-alkyl thiolane and an oxide of it;
- A 2,6-*n*-alkyl thiane;
- 2,5-*n*-alkyl thiophenes;
- 2,4-*n*-alkyl benzothiophenes;
- 1,9-*n*-alkyl dibenzothiophenes;
- A 4,9-*n*-alkyl fluorene;
- *o*-di-*n*-alkyl benzenes;
- An *n*-alkanoic acid ester;
- A pentacyclic naphthenic ring system;

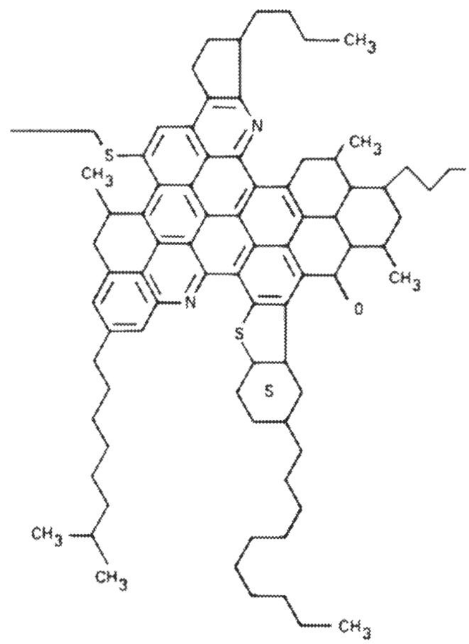

Figure 2.18
Speight hypothetical structure of a petroleum asphaltene. *Reproduced from Speight JG, Moschopedis SE. On the molecular nature of petroleum asahaltenes. In: Bunger JW, Li NC, editors. Adv. chem. series 195: chemistry of asphaltenes: ACS; 1981. p. 1–15, with permission from ACS Publications.*

- An *n*-alkyl pyridine;
- An *n*-alkyl quinoline;
- An *o*-di-*n*-alkyl naphthalene;
- A decalin;
- Octahydrophenanthrenes, etc.;
- Condensation products of some of the aforementioned aromatic structural units;
- A deoxophylloerythroetioporphyrin; and
- A biphenyl linkage and linkages based on RICO results.

This model is characterized by a low level of polycondensation of the aromatic rings and by the presence of branches and connecting bridges. The molecular structure of this type of model is known as "archipelago structure."

Several molecular models have been suggested for explaining asphaltene behavior and characterization results [171,200,285,306,355,483,588–594], Fig. 2.20 collects some of the model structures proposed for asphaltene molecules. Fig. 2.20A shows the proposed molecular structures for asphaltenes in subfractions A1 and A2, from the PNP method (see more details in Sections 4.2 and 4.3) [200,355], among other characteristics these two structures differ in the degree of condensation of the polyaromatic system. Theoretical calculations (MM) were used to estimate the SPs (Hildebrand type) of these

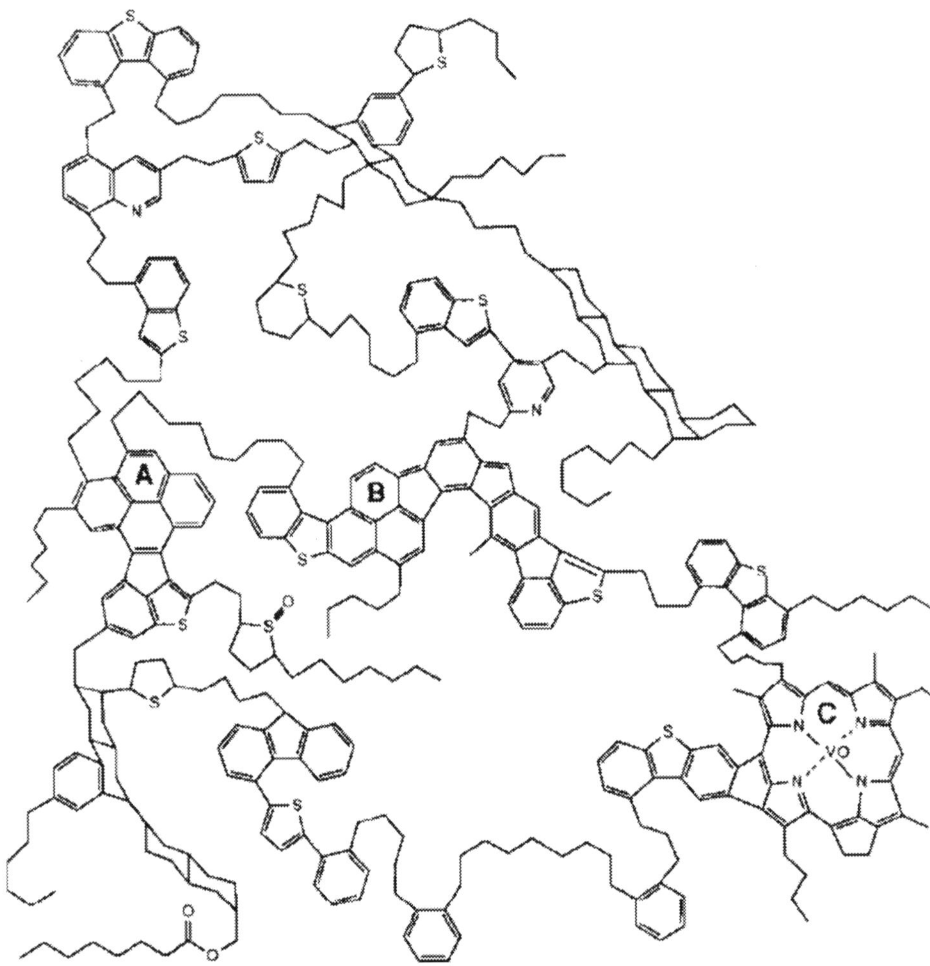

Figure 2.19

Strausz hypothetical representation of an asphaltene molecule. *Reproduced from Strausz OP, Mojelsky TW, Lown EM. The molecular structure of asphaltene: an unfolding story fuel 1992;71:1355–63, with permission from Elsevier.*

two models, which consistently agree with the observed solubility of the respective fractions [200]. Fig. 2.20B showed an archipelago structure proposed by Sheremata et al. [588] for an Athabasca asphaltene. Fig. 2.20C [285,306,589,590] and D [483] include structures proposed by various authors, generalized as the "continental model," "pericondensed structure," or "island model," which are characterized as mainly formed by polycondensed cycles of substituted aromatic moieties.

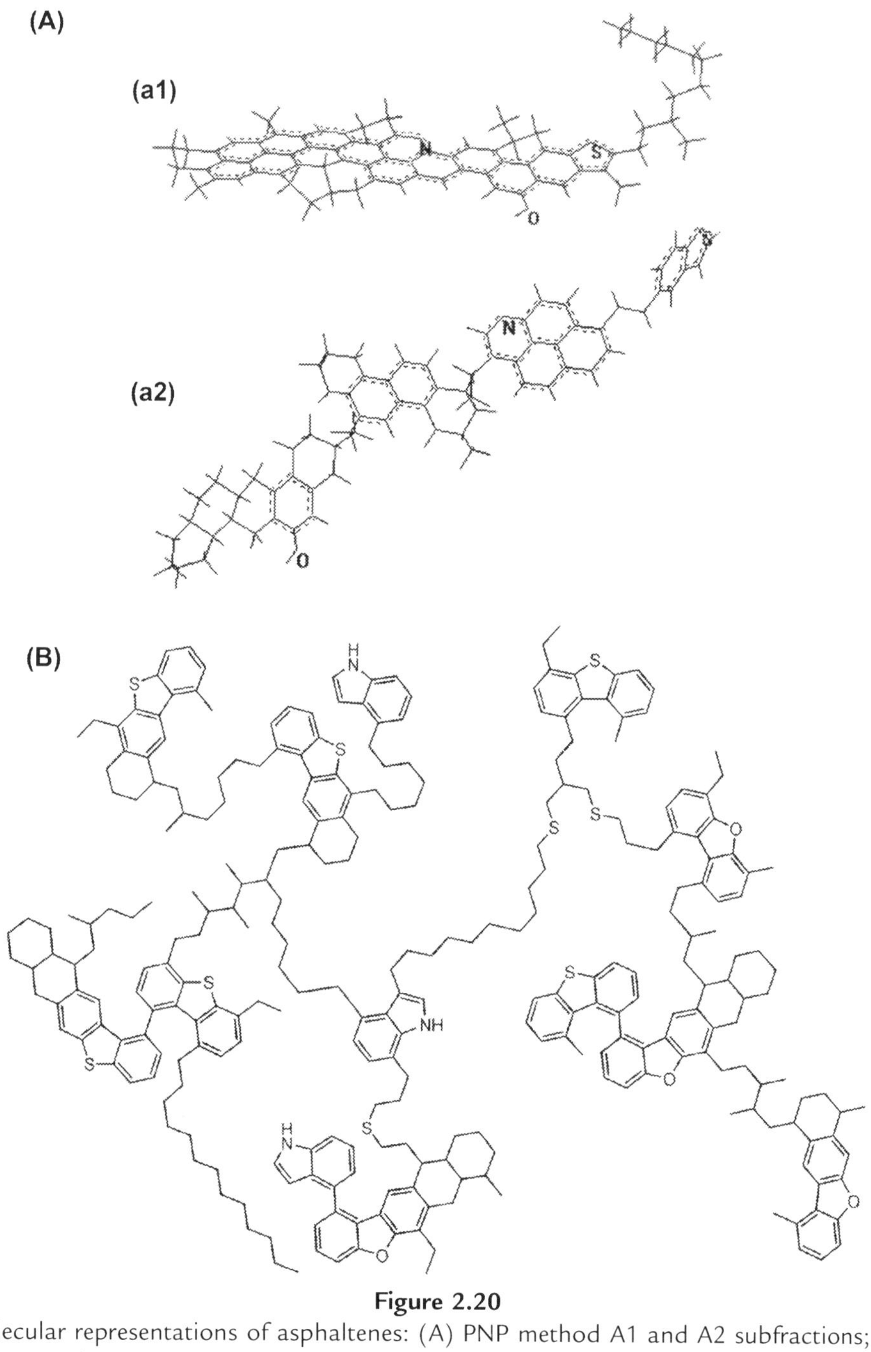

Figure 2.20

Molecular representations of asphaltenes: (A) PNP method A1 and A2 subfractions; (B) Athabasca asphaltene. *Reproduced (A) from Acevedo S, Castro A, Negrin JG, Fernandez A, Escobar G, Piscitelli V, et al. Relations between asphaltene structures and their physical and chemical properties: the rosary-type structure. Energy Fuels 2007;21(4):2165-75, with permission from ACS Publications; (B) from Sheremata JM, Gray MR, Dettman HD, Mccaffrey WC. Quantitative molecular representation and sequential optimization of Athabasca asphaltenes. Energy Fuels 2004;18(5):1377-84, with permission from ACS Publications.*

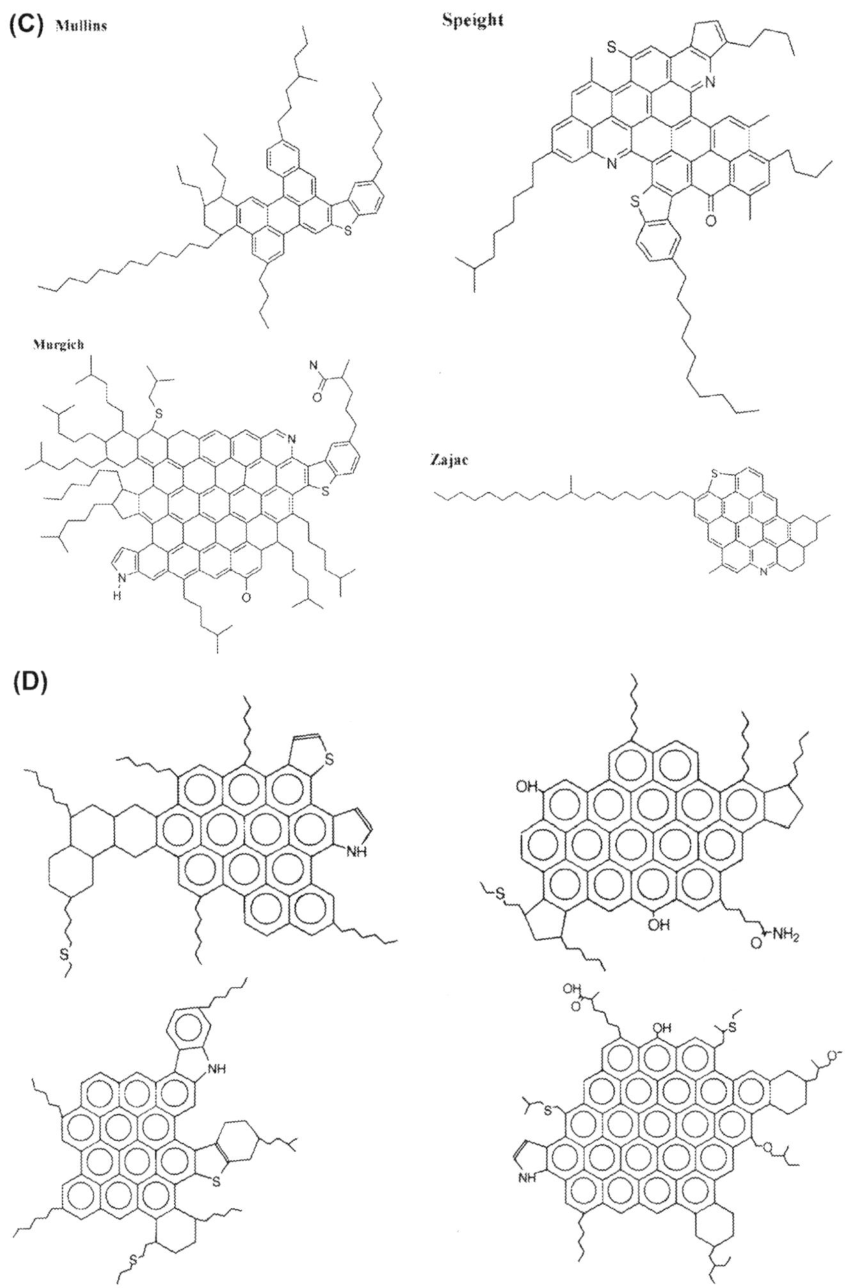

Figure 2.20 cont'd

Molecular representations of asphaltenes: (C) various authors' proposals [285]; (D) based on structural parameters. *Reproduced (C) from Groenzin H, Mullins OC. Molecular size and structure of asphaltenes from various sources. Energy Fuels 2000;14(3):677-84; Murgich J, Rodriguez MJ, Aray Y. Molecular recognition and molecular mechanics of micelles of some model asphaltenes and resins. Energy Fuels 1996;10(1):68-76, with permission from ACS Publications; Parkash S, Moschopedis S, Speight J. Physical properties and surface characteristics of asphaltenes. Fuel 1979;58(12):877-82, with permission from Elsevier; (D) from Leon O, Rogel E, Espidel J, Torres G. Asphaltenes: structural characterization, self-association, and stability behavior. Energy Fuels 2000;14(1):6-10, with permission from ACS. Publications.*

Similarly, a combination of solid-state ^{13}C NMR, XPS, and elemental abundance was used to characterize the average chemical structure of several C_7-asphaltenes from VRs of different origins (Campana, Mid-Continent US, San Joaquin Valley, Lloydminster Wainwright, Maya, and Heavy Canadian) [595]. The proposed molecular structures are shown in Fig. 2.21; in these models core links represent either inter- or intramolecular bonds. Among the considered asphaltene aromaticity, heteroatomic moieties, and naphthenic cycles were the main differentiating features.

Evidence for the continental-type structure seems to have been found by the examination of 23 model compounds and two petroleum asphaltene samples by L^2MS. Comparison of the model compounds spectra with those of the asphaltenes showed similar fragmentation patterns (little to no fragmentation) when the model compound had one aromatic core with and without various pendant alkyl groups, whereas all model compounds having more than one aromatic core show energy-dependent fragmentation [473,596].

Obviously, such complexity gives rise to many possible configurations and a single unique description can only be used for modeling and/or understanding purposes. In the continental model, the rings are surrounded by alkyl side chains of an average length of 5–6 carbons. The continental models explain well the $\pi-\pi$-association hypothesis of their aromatic flat surfaces, allowing aggregate formation by stacking. In the "archipelago" model aliphatic chains interconnect groups of aromatic regions of smaller degree of polycondensation as those considered in the continental model. The molecules proposed for this model would also have larger overall area and would give rise to many more molecular conformations. Regarding aggregation, any $\pi-\pi$ interaction of these molecules would exhibit steric limitations.

The traces of heavy metals such as Ni and V sometimes are not included in the molecular model because these are present in ppm level and will be present in only a few of the asphaltene molecules. The prevalence of one structure over the other has been another subject of controversy, while it seems to be fair to assume the existence of both. The various modes of aggregation, the differences in solubility, and as seen later the variety of surface properties and reactivity, as well as the wide range of stability, indicate a plausible existence of both types (continental and archipelago) of molecules in real asphaltenic fractions and even including other types of unknown structures. Transmission electron microscopy has shown an amorphous matrix (probably formed by archipelago-type molecules) with regions of highly regular aromatic stacking (formed by continental-type molecules) [597].

The fragmentation, decomposition, or crackability of the continental structure is supposed to be lower than that of the archipelago type. Laser desorption–laser ionization mass spectroscopy [426,596], laser-induced acoustic-desorption studies, and laser-induced, acoustic-desorption, electron-impact mass spectrometry [419] showed

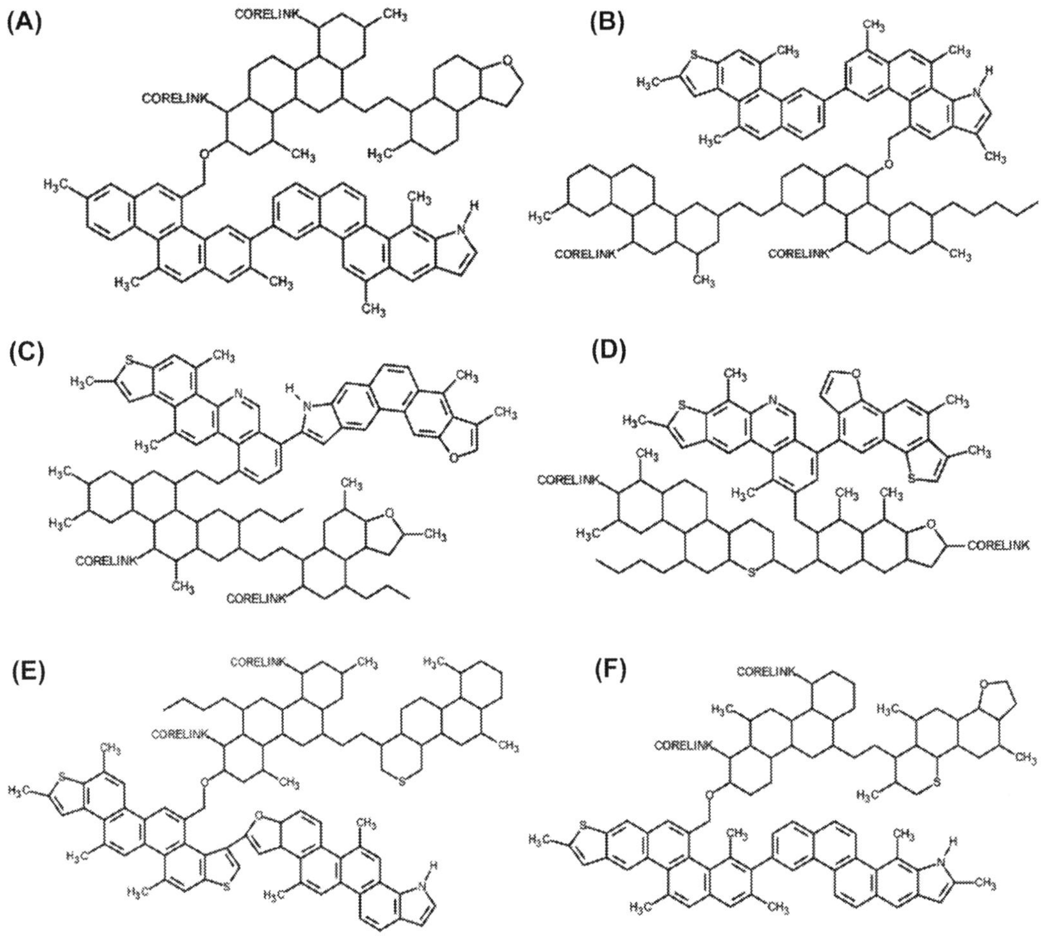

Figure 2.21

Proposed molecular structures of asphaltenes of different origin: (A) Campana, (B) Mid-Continent US, (C) San Joaquin Valley, (D) Lloydminster Wainwright, (E) Maya and Heavy Canadian. *Reproduced from Siskin M, Kelemen SR, Eppig CP, Brown LD, Afeworki M. Asphaltene molecular structure and chemical influences on the morphology of coke produced in delayed coking. Energy Fuels 2006;20(3):1227–34, with permission from ACS Publications.*

that the expected fragmentation was consistent with the experimentally observed results that were indicative of a predominance of the continental structure. This predominance seems to have been confirmed by the observed reactivity toward catalytic hydrogenation [485]. On the other hand, Borton et al. concluded that the MS fragmentation patterns cannot unambiguously differentiate between continental- and archipelago-type structures [399]. Nevertheless, their findings seemed to agree better with the continental model.

Based on experimental results from X-ray diffraction, SAXS, and SANS, association and aggregation models of molecular structures of resins and asphaltenes have been proposed for explaining interactions that drive association and micellization. The model employed for the asphaltene molecular structure is one of the many reported in Ref. [564] and responds to an archipelago-type model (Fig. 2.22A). An aggregate of three asphaltene layers was manually combined (Fig. 2.22B) to reproduce shape and data reported by various authors [598]. The asphaltene layer was constructed with four of those molecules, by matching the irregularities of the brick-shaped molecules through van der Waals and $\pi-\pi$ interactions (Fig. 2.22C). Cavities were left behind in forming the aggregates, which served as host for inorganic contaminants. The molecular, layer, and aggregate dimensions were also included in Fig. 2.22A and B [598].

Similarly, a resin molecule (Fig. 2.23A) was proposed [598] by manually building reported chemical features, such as long paraffinic chains, naphthenic rings, condensed aromatic–naphthenic systems, indene rings, pyrrole, N–H-group, carbonyl-group, and S-containing groups [500–502,515,599]. Some of these resin molecules of quasirectangular shape were connected manually to the asphaltene layer shown in Fig. 2.22C for assembling a layer of the micelle structure shown in Fig. 2.23B. A computer-built three-layer structure of the micelle-reproduced dimensions ($\sim 10 \times 13$ nm), which can be typically found in the literature.

A more rational way of building molecular structural models consisted of a two-step reconstruction algorithm and used partial analytical data [600]. A diagram of this building algorithm is presented in Fig. 2.24. The first step creates an initial mixture of molecules by a Monte Carlo sampling method (stochastic reconstruction). The second step modified the molar fractions of the molecules for fitting the analytical data or properties and improved the representativeness of the generated mixture (reconstruction by entropy maximization; Fig. 2.25A). The ultimate objective of this work was to provide a molecular mixture resembling a VR that could be used for kinetic modeling.

The diagram used to classify the molecules into one of the SARA fractions is shown in Fig. 2.25B, and it was based on the solubility characteristics reported in Refs. [601,602].

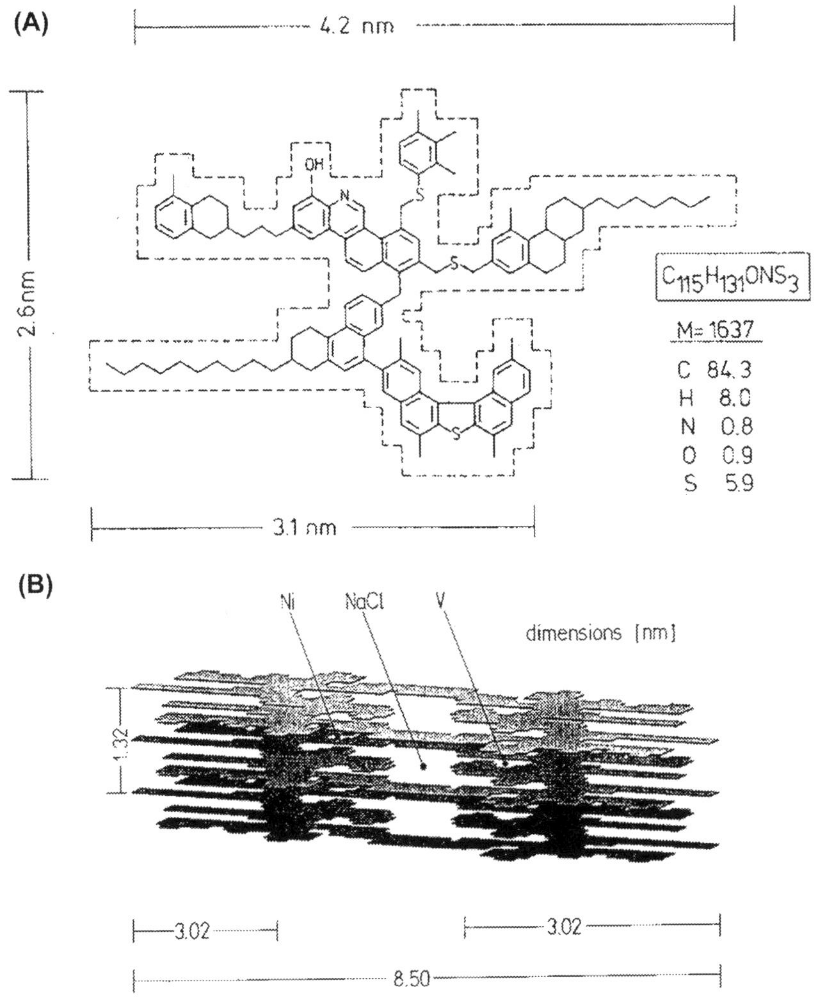

Figure 2.22

Asphaltene model structures: (A) molecule, (B) aggregate, and (C) layer. *Reproduced from Jovanovic JA. Models of an asphaltene aggregate and a micelle of the petroleum colloid system. Hem Ind 2000;54(6):270–75, with permission from the Association of Chemical Engineers of Serbia.*

Similarly, a compositional model based on structural attributes was developed to predict BotB properties. The defined structural attributes are illustrated in Fig. 2.26. A library was built with more than 400 cores and 200 side chains/intercore linkages to generate around 400,000 molecules that can be used to predict properties and behavior [603].

A similar approach of molecular construction based on structural attributes resulted in models that were subjected to molecular mechanics (MM) and MD calculations for predicting thermodynamic properties [604].

(C)

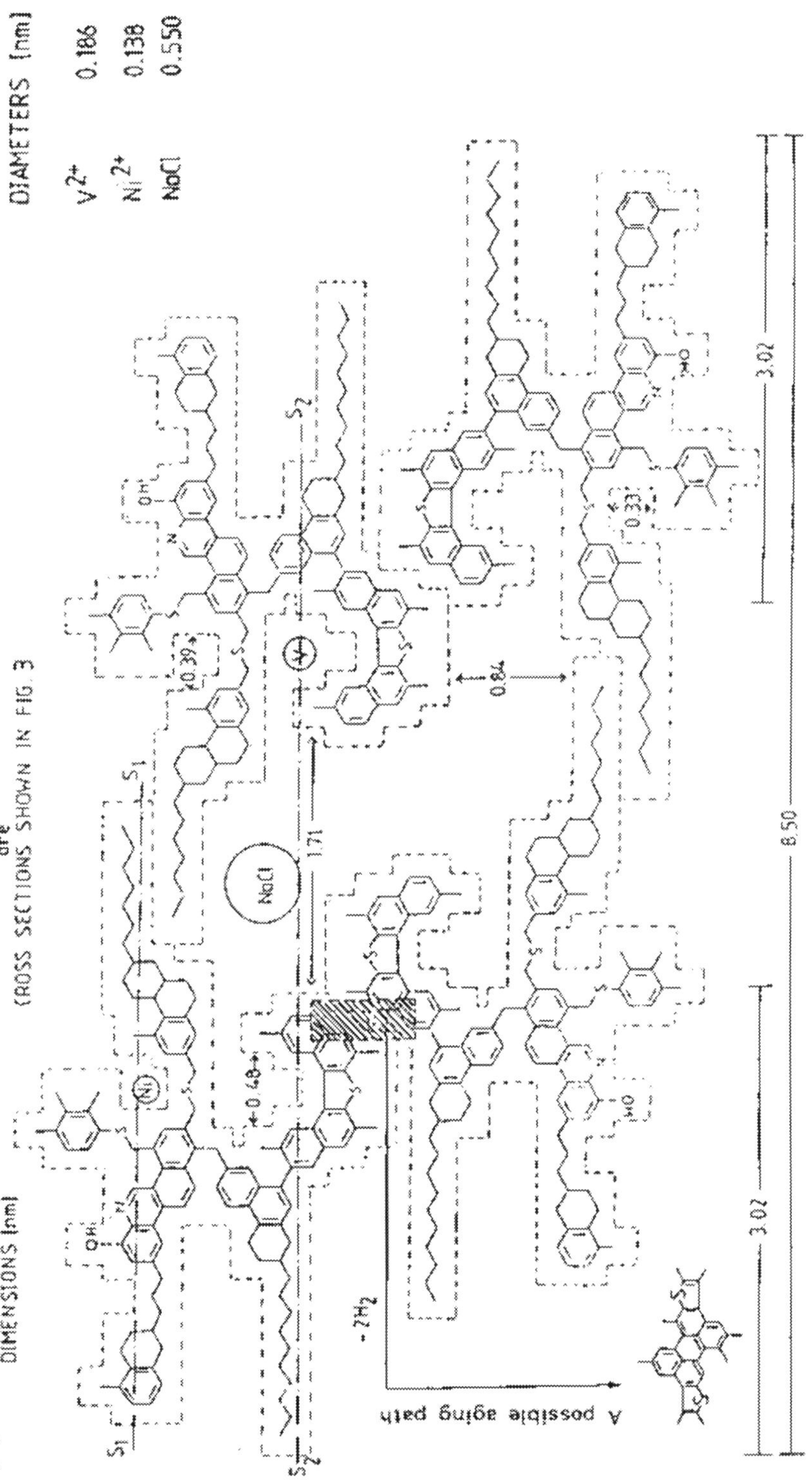

Figure 2.22 cont'd

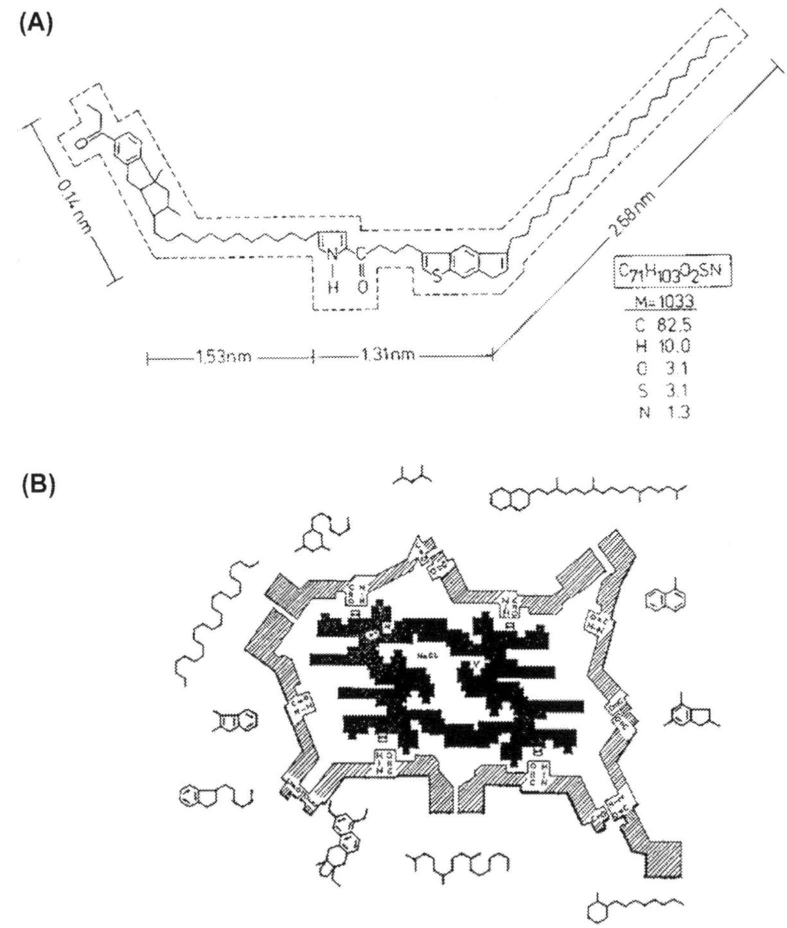

Figure 2.23

Asphaltene micelle structure: (A) resin molecule, (B) micelle. *Reproduced Jovanovic JA. Models of an asphaltene aggregate and a micelle of the petroleum colloid system. Hem Ind. 2000;54(6):270–75, with permission from the Association of Chemical Engineers of Serbia.*

4.4.4 Molecular Weight

A number of techniques have been used through the years for the determination of MW of asphaltenes and a couple of critical reviews on their pros and cons have also been published [346,605]. Isolated asphaltenes have shown to present an MW distribution that depends not only on the isolation method and source, but also on the technique used for its assessment. Early reported MW determinations include MW values from 900 to 2000 by viscosity measurements [11,606], 2000–3000 by equal vapor pressure method [607], 1000–4000 from light absorption coefficients [608], 2500–4000 by ebulloscopic method [609], 600–6000 with cryogenia, 1000–5000 with VPO [101,607,610], up to 80,000 with osmotic pressure method [611] or SEC [60,65,340,388,394,403,507,612–617],

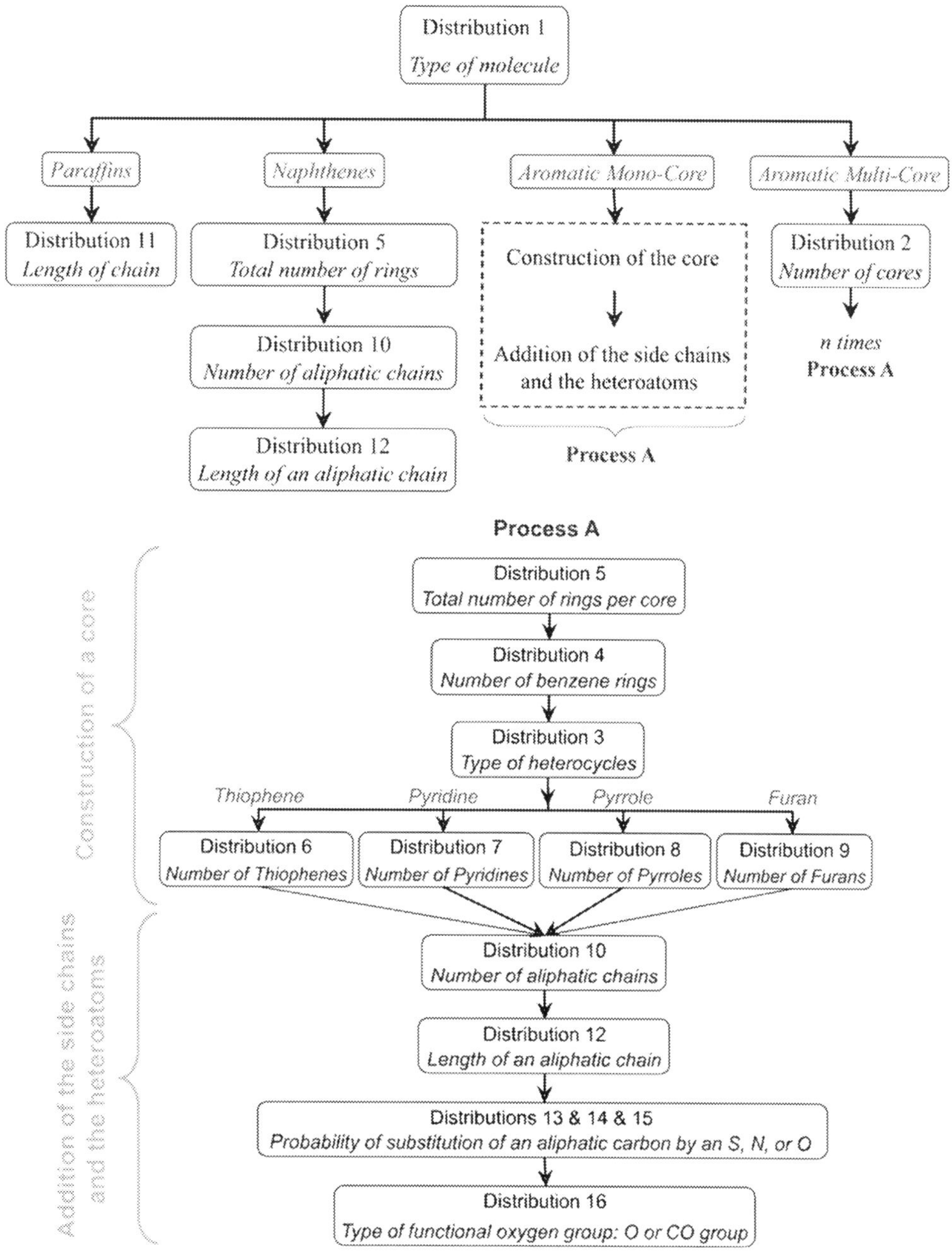

Figure 2.24

Building diagram example of the stochastic reconstruction step of reconstruction algorithm. *Reproduced from De Oliveira LP, Vazquez AT, Verstraete JJ, Kolb M. Molecular reconstruction of petroleum fractions: application to vacuum residues from different origins. Energy Fuels 2013;27(7):3622–41, with permission from ACS Publications.*

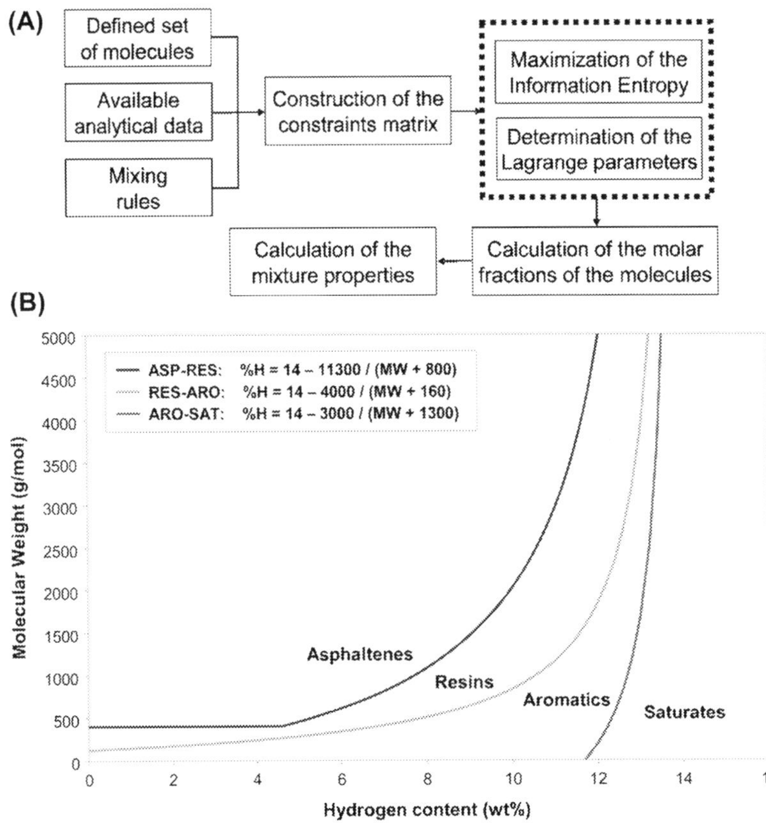

Figure 2.25

Reconstruction algorithm: (A) flow diagram of the entropy maximization step, (B) molecular classification diagram. *Reproduced from De Oliveira LP, Vazquez AT, Verstraete JJ, Kolb M. Molecular reconstruction of petroleum fractions: application to vacuum residues from different origins. Energy Fuels 2013;27(7):3622–41, with permission from ACS Publications.*

80,000–140,000 with a monomolecular film method [5], and up to 300,000 Da with ultracentrifugation [11].

The determination of the MW of asphaltenes is strongly hampered by the tendency of these molecules to associate with each other and with other petroleum constituents. Even in much diluted solutions, asphaltene association has been detected. MW and its distribution, along with the dynamic association of individual covalent molecules, not only represent attributes consistent with the assumed colloidal behavior of asphaltene solutions, but also create complicating factors for the definition of methodologies for analysis. The MW of 14 asphaltene samples was determined by VPO in different solvents (benzene, Bz; methylene bromide, CH_2Br_2; pyridine, Pyr; and nitrobenzene, NO_2–Bz), and at various temperatures (37, 100, 115, and 130°C) [607]. The variation in the measured MW values

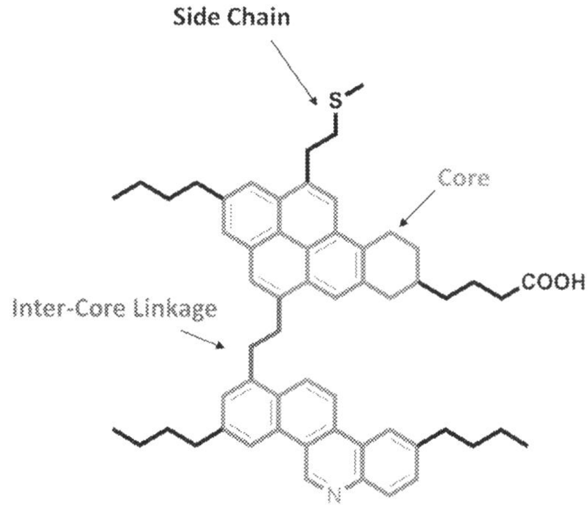

Figure 2.26
Example of defined attributes in an archipelago-type molecule used for representing molecular structures. *Reproduced from Zhang L, Hou Z, Horton SR, Klein MT, Shi Q, Zhao S, et al. Molecular representation of petroleum vacuum resid. Energy Fuels 2014;28(3):1736–49, with permission from ACS Publications.*

provide clear evidence of the different degrees of association achieved in each employed solvent at each of the applied temperatures (Fig. 2.27).

Obviously, the experimental measurement of both MW and its distribution will depend not only on the aggregation state, but also on how the asphaltene cut was obtained. It starts by using agglomeration/precipitation as the separation technique. Thus, n-C_7 yields smaller amounts (by 15–98%) of asphaltenes than that obtained with n-C_5, the difference being resins and low-MW asphaltene components. Thus the MW of the asphaltene fraction is found to increase with increasing alkane (precipitant) CN. Although the higher MW can be interpreted as being caused by a decrease in the proportion of coprecipitated (lower MW) maltenes, other explanations are the increased ability to aggregate by their increased polarity or the increased proportion of heavier heteroatoms. This latter explanation is in agreement with the observed trend of compounds containing N, S, and/or O, which are also increasingly present in the heavier fractions. Hence one could imagine that many methods historically employed for the evaluation of MW and its distribution have been measuring asphaltene clusters, "micelles," or aggregates rather than asphaltene molecules.

The first technique used for MW determination was GPC on crosslinked polystyrene gels (SEC). Broad MW distributions were found spanning from 700 to more than 40,000 g-mol when fractionating several different asphalt samples. Each sample consisted of two portions of similar quantities, one peaking around 1000 MW, the other around 14,000 [60]. Strausz fractionated resins I, resins II, and asphaltene using n-pentane as precipitant and separated the asphaltenes into five finer fractions, whose MW ranged from 1200 to 17,000

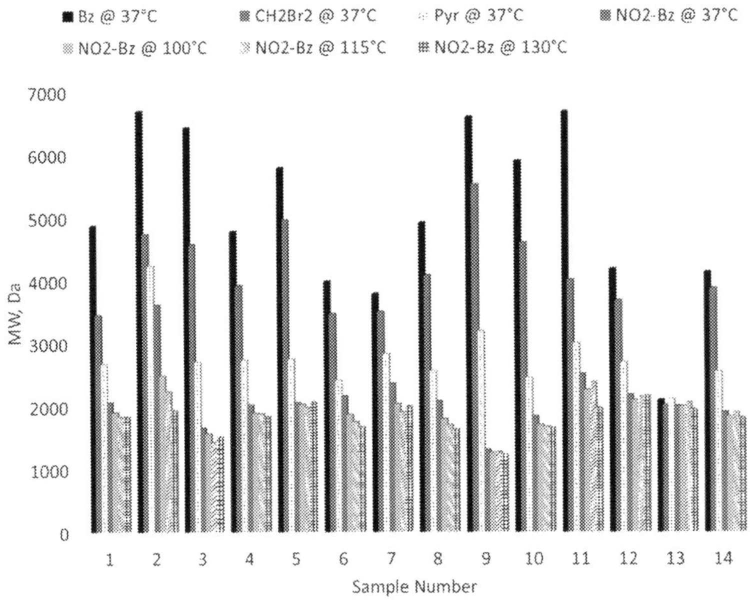

Figure 2.27
Molecular weight (MW) of 14 asphaltenes of different origins in various solvents at several temperatures.

(by SEC). A comparison of SEC and VPO showed consistent MW results with Colombian asphaltenes, when SEC was calibrated with phthalocyanines [616].

The detection of very high MWs via SEC has been associated with nanoaggregation and induced flocculation by the eluting solvent [618]. In fact, flocculation induced by one of the elution solvents, NMP [406], has been confirmed and argued against high-MW values [614]. Moreover, strong interaction between polystyrene particles and asphaltenes was showed by atomic force microscopy (AFM), which may translate into aggregation of asphaltenes in the SEC columns. Furthermore, the existence of asphaltenes (nano)-aggregates in solvents, such as toluene, NMP (in which case an insoluble phase has been detected [391]), and THF, would either damage the column packing or change its porosity creating major limitations of SEC for the analysis of asphaltene MWs [615].

The introduction of VPO for measuring the MW of asphaltenes led to considerable variations and inconsistent data with that derived by means of SEC and viscosity-based methods. Results were found to depend upon the nature of the solvent as well as the temperature. The recognition of these inconsistencies suggested revision of the existing models for the physical structure of asphaltenes [607]. In fact, the number average MW of asphaltenes in benzene and THF as solvents was studied using this technique at three different temperatures (37, 45, and 60°C). Both variables, the nature of solvent and the temperature were found to affect the degree of association, which explained the large spread of discrepancy [619]. Association was confirmed by Yarranton et al. who also evaluated the effect of asphaltene concentration and compared C_5- with C_7-asphaltenes. For diluted samples (0.5 kg/m^3), the number average MW was ~1800 Da and increased up to ~10,000 Da for concentrations of 20 kg/m^3 [335]. The MW of nanoaggregate clusters dissolved in a 80:20 THF:acetonitrile mixture was evaluated by capillary electrophoresis to be in the range of 3000–4000 Da [620].

As mentioned, acidic asphaltenes are the least soluble type of asphaltenes and the most prone to associate. Yarranton et al. determined the VPO number average MW of subfractions precipitated with heptol mixture, with increasing proportion of heptane [143,621]. Their results indicate an apparent MW increase with decreasing solubility, with the least soluble fraction showing MWs in the range of 7000–8000 Da. This MW range could be explained to be caused by the formation of stacks of four to six asphaltene molecules, with real MW of 1200–1400 Da, which will be also consistent with an aggregate size of 5–10 nm.

Plasma desorption mass spectrometry [622] has produced reasonably good results comparable to those obtained with VPO [607]. However, in the latter case, the error is sufficiently high that it is difficult to obtain an accurate distribution from a set of asphaltene subfractions. On the other hand, VPO compared well with laser desorption ionization-time-of-flight in providing number average MW of asphaltenes and derivatized compounds [623]. Another comparison was attained with laser desorption high-resolution MS that yielded an MW distribution in the range 100–1500 Da for both Cold Lake and Athabasca asphaltenes. The average MW from these distributions was 523 and 463 for the Cold Lake and Athabasca asphaltenes, respectively. Meanwhile, VPO showed an opposite comparison of 1020 and 2468, respectively, for these asphaltenes [624].

The two extreme solubility fractions (soluble fraction-F15/85 and insoluble fraction-F40/60) obtained by the flocculation method using S/AS (toluene/n-C_7) mixtures in different proportions (shown in Fig. 2.3) mainly differ in their H/C ratio, nickel and vanadium content, and average chemical structure. The F15/85 fraction concentrated smaller aggregates than those in the F40/60 fraction, but with no significant size polydispersity difference. The larger aggregates also contained a larger proportion of the metal contaminants [38].

Average MW increases with polarity for asphaltenes with an average MW of 850–1100 Da; the higher polarity fractions showed an average MW in the range of 900–1500 Da, while the lowest polar fraction fell in the range of 300–700 Da [168].

Nanofiltration rendered sized fractions of less than 10 nm (P10), between 10 and 20 nm (P10–P20), and from 20 to 50 nm (P20–P50). The mass-to-radius of gyration dependence for these fractions is shown in Fig. 2.28. Two distinct generic behaviors were observed: (1)

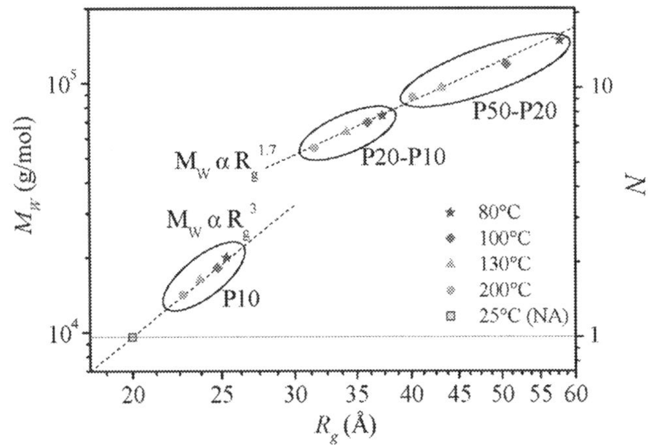

Figure 2.28

Mass-to-radius of gyration dependence for nanosized fractions. *Reproduced from Eyssautier J, Espinat D, Gummel J, Levitz P, Becerra M, Shaw J, et al. Mesoscale organization in a physically separated vacuum residue: comparison to asphaltenes in a simple solvent. Energy Fuels 2012;26(5):2680–87, with permission from ACS Publications.*

mass increased rapidly with the radius for the smallest clusters, which might correspond to a compact growth of particles; and (2) for the larger aggregate fractions, a power law relation between mass and radius of gyration indicates loose aggregates, most likely mass fractals [52].

A review of the VPO, GPC, ultrafiltration, ultracentrifugation viscosity, SAXS, infrared spectroscopy, solubilization, and interfacial tension methods indicated a range of 1200–2700 amu for the number average MW and an MW range of 1000–10,000 amu or higher [605].

In the case of Athabasca C_5-asphaltene fraction, a feature of a trimodal MW distribution has been observed, indicating three different MW domains [498]. However, redistribution on time has also been observed to occur (primarily toward lower MW and size).

More recently, the use of high-resolution mass spectroscopic techniques (e.g., FT-ICR-MS; field ionization mass spectrometry) coupled with the different preparation and ionization methods has shown ranges of 400–1500 Da with a maximum around 750–850 Da, with much more consistency than ever before. Ionization methods included: electrospray ionization; APPI; LDI and MALDI; atmospheric pressure chemical ionization [408,423, 444,449,585,625–627], as well as time-resolved fluorescence depolarization and fluorescence correlation spectroscopy (FCS) [388,628]. The average MW of asphaltenes measured with L^2MS and SALDI were in the range of 600–700 Da with an upper mass limit near 1500 Da [473]. These authors were inclined to a continental geometry of the asphaltenes and showed evidences for the formation of stable nanoaggregates of about seven molecules [473]. They also thought their results provided confirmation of many components of the Yen–Mullins model (Fig. 2.12 [357]).

These techniques and the deep characterization studies carried out on fractionated asphaltenes support the idea that the solubility class captures a variety of molecules and not a unique molecular typology (polydispersed chemistry) and explain the phenomenology undergone by asphaltenes. In fact, these techniques have shown the existence of more than 17,000 different molecular species [439,446,447,451,458,629] in the BotB fraction. Thus it is fair to assume that some molecules would resemble the continental type of model with up to 7–8 fused rings, while others would look like the archipelago model with a variety of linking chains.

Subfractionation with S/AS mixtures and evaluation of the MW distribution by LDI-MS resulted in the data exemplified in Fig. 2.29 along the details of the extractant mixtures. As can be seen, the width of the distribution, as well as the MW, increased with the decrease in the proportion of the AS in the mixture [73].

4.4.5 Size and Shape of Molecules and Particles

Asphaltene-associated sizes are reported in terms of the measuring technique and the structural view considered. Thus sizes of molecules, micelles, colloidal particles, nanoaggregates, and microaggregates have been reported.

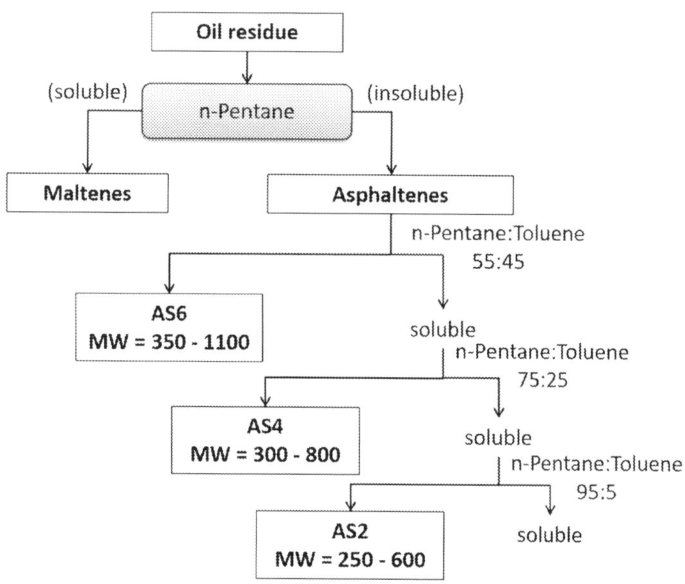

Figure 2.29

Subfractionation scheme using antisolvent/solvent mixtures and molecular weight (MW) range of subfractions. *Reproduced from Yang MG, Eser S. Fractionation and molecular analysis of a vacuum residue asphaltenes. ACS Div Fuel Chem. Preprints 1999;44(4):768–71, with permission from ACS Publications.*

The largest dimension of asphaltene structures observed by scanning tunneling microscopy (STM) averaged 10.4 Å. This value seems to correspond to a condensed ring portion containing 6–7 fused aromatic rings. Three representative models of such structure could be used to derive by NMR an average dimension of 11.1 Å in very good agreement with the STM-evaluated average [285]. A similar size of about 1 nm was obtained for dilute toluene solutions by fluorescence correlation spectroscopy [630]. A maximum identified molecular diameter has been reported to be of ~13 Å [631].

Subfractionation of a Canadian VR by SFE into 13 narrow subfractions was carried out, from which five cuts were characterized. Their VPO MW and their molecular size were determined. Molecular size was derived from measurements of diffusion coefficients. A monotonous increase of size with MW was found for the first four fractions (average diameter varied from 1.08 to 1.75 nm), but an abrupt jump (4.68 nm) was observed with the end cut. Size distribution showed a near monodispersity for the four narrow fractions while the end cut was polydispersed (~3–7 nm). All these indications suggested a preferential enrichment in asphaltenes for the end cut and the presence of aggregates in such subfraction [632].

Maya-derived asphaltenes were studied by SANS in d_{10}-L-methylnaphthalene from 20–400°C; at lower temperatures the colloidal materials were rod-shaped particles with a fairly uniform radius (10 Å) and polydispersed in length (20–500 Å). At high temperatures (400°C), spherical particles formed with diameters of ~12 Å [578].

The hydrodynamic radius of asphaltenes in deuterated chloroform solution has been determined from the values of diffusion coefficients using the Stokes–Einstein equation. Asphaltenes from Khafji, Iranian light, and Maya VR and a resin from Maya VR were measured by pulsed-field gradient spin-echo 1H NMR. The largest radius was observed for 30 g/L solutions, asphaltenes showed a value of 3.8 nm, and the resin was 1.5 nm. For the 0.1 g/L solution, the radii varied between 3.5 and 4.2 nm (for the slow diffusion coefficient value component) and 0.6–0.7 nm (for the fast diffusion coefficient value component) [282]. Spherical particles with average radius and degree of polydispersity 33.8 Å and 15.4%, respectively, were found to fit the SAXS data of C_7-asphaltenes from Ratawi crude oil. There were particles with radii as small as 20 Å and as large as 50 Å within the size distribution of these asphaltenes [633].

For the case of the bituminous coal asphaltenes (which appear to be smaller than petroleum asphaltenes [306,408,410,421,586], e.g., Jobo's asphaltene ca. 25 Å [634]), it appeared that the colloidal size of the lower MW molecules was smaller than 10 Å [388], falling below the resolution of the unit employed. The hydrodynamic radius derived from diffusion coefficient showed values in the range of about 6 Å for coal asphaltenes and in the range of 10–12 Å for petroleum asphaltenes [635]. These values are slightly lower than those found later for asphaltenes from four vacuum residues of Daqing, Liaohe,

Shengli, and Gudao crude oils. Size calculation (28–50 Å) was based on MWs measured using benzene, nitrobenzene, and chlorobenzene as solvents [636]. The micelle diameters for these Chinese VR asphaltenes ranged between 5 and 10 nm. Micelle size of asphaltenes from Arabian heavy were measured by dielectric loss. The radius ranged from 8.4 to 10.4 nm for asphaltene concentrations of 5–20% [637]. Diameters within this range were evaluated by transmission electron microscopy for solutions of the PNP-fractionated asphaltenes A1 and A2 fractions. The resins fraction from the same crude oil was employed as solvent for the asphaltenes. The (number, weight, and *z*-) average diameters of A1 asphaltene were always larger (5.1–8.4 nm) than those of the A2 asphaltenes (3.2–6.5 nm), in agreement with the observed differences in solubility. Diameters were evaluated at 150, 200, and 250°C; for the A1 asphaltenes diameter passes through a minimum at 200°C, while it decreases with temperature for A2 asphaltenes [311]. The VPO MW of the A2 asphaltenes was ~1000 amu [638].

The measured molecular size of C_5-asphaltenes depended on the wavelength of the excitation source used for the measurement. Subfractions precipitated with increasing pentane in pentane–toluene mixtures did not show significant changes in size, indicating that the molecular structure did not vary among the different subfractions; rather their population did. These population differences of the same set of molecules between the subfractions were responsible for the measured size observations [81]. Application of fluorescence depolarization measurements for estimating the size of asphaltene molecules (MW in the range of 500–1000 amu) and of model compounds (average MW of 750 amu) indicated that a strong correlation existed between the size of an individual fused ring system in an asphaltene molecule and the overall size of this corresponding molecule. Using simple theoretical models and model-independent comparisons with known chromophores, a range of asphaltene molecular diameters of 10–20 Å was obtained [390]. From all this, it was concluded that the asphaltene molecules have one or at most two fused ring systems per molecule, which support the archipelago-type model. The asphaltene molecules from coal were confirmed to be much smaller than petroleum asphaltenes. A continuous MW/size distribution was found from the resins through the asphaltenes [306].

Most recent studies reflect a high level of consistency among the MS techniques, confirming smaller sizes than originally ascribed. FCS was used to determine translational diffusion coefficients of asphaltene and model compounds under a broad range of concentrations including ultralow concentrations. The FCS results were found to agree well with those of time-resolved fluorescence depolarization. Again, a correlation between molecular size and asphaltene chromophore size was found, supporting a molecular archipelago-type architecture, with one or two PAHs per molecule [628]. Fractionation of asphaltenes with *n*-methyl pyrrolidinone to render soluble and insoluble fractions showed that both fractions of asphaltenes were monomeric [388].

The estimation of the solvent effect on colloid size determined by SAXS from asphaltene dispersed in heavy petroleum residuum and in liquefied coal was reported. Regardless of the solvent used, asphaltenes from Jobo exhibited the same size in all the solvents. The results also indicated that the radius of gyration of asphaltenes from Jobo AR was of about 25 Å. Since the size was independent of concentration, it was concluded that the basic unit of the Jobo asphaltenes was a macromolecule of this size. Meanwhile, Tia Juana asphaltenes showed smaller particles in the better solvents; sizes were 38–50 Å in benzene, 32–37 Å in chlorobenzene, and 28–32 Å in nitrobenzene [634]. Regarding particle size, the examination of asphaltenes from different crudes indicated the presence of spherical primary colloidal particles, with average diameters in the range of 7–9 nm, and with an apparent Gaussian distribution [306,631].

Using SANS and SAXS, two distinct ranges of sizes were distinguished in an asphaltene fraction. Dilute solutions and high temperatures showed the smaller sizes of ~40 Å, while the other group of ~1000 Å was seen at high concentrations and low temperatures [639]. Clearly, the first group falls within molecular, micellar, and nanocluster sizes, while the latter corresponds to aggregated particles.

Similarly, AFM results of the asphaltenes' supramolecular structure [640] with a resolution of about 1 μm showed objects of a quasi-gel-like structure, which might comprise micelle-associated asphaltene molecules. The dispersed phase consists of about 144 particles per 1 μm^2, as average concentration, with a distance of about 20–50 nm between the micelles. The thickness for a particle of 100 nm diameter was less than 3 nm, indicating that these structures were formed by stacking.

The micelle diameters were also evaluated and found to be 90–110 Å in benzene and 50–90 Å in chlorobenzene [636]. Pollack and Yen proposed diameters for spherical particles of asphaltenes in the range of 30–50 Å [641].

XRD experiments at 30, 150, and 300°C showed that the values of the layer distance between aromatic sheets of asphaltenes did not change with temperature from a value of about 3.6 Å, but the number of aromatic sheets in a stacked cluster decreased from 8 to 5, from the lowest temperature to 300°C [642]. SANS measurements showed a fractal network for asphaltenes of Maya when diluted in decalin stable up to 350°C. In other studied solvents (1-methylnaphthalene and quinoline), asphaltenes aggregated in the form of a prolate ellipsoid with a high aspect ratio at 25°C that became smaller with temperature increases. These ellipsoids transformed into a compact sphere with a size of around 50 Å in diameter at 350°C [643].

Particle size and shape of precipitated asphaltenes depend not only on the nature of asphaltenes and their source, but also on precipitation and recovery conditions. Both properties showed broad polydispersity. The complex nature of the molecules and the probable existence of different structural types complicate morphological matters. Consequently, there would be a wide spectrum of various sizes, forms, and shapes. Monodispersed models of spherical [644], rod [645], and round disk [646] shapes have been proposed but they fail to fit the data. Toluene solutions of asphaltenes from several sources showed particles with an average radius in the range of 0.63–1.8 nm when analyzed by SANS [360]. This size range was broader to 0.8–20 nm for the thin disk particles ($\sim$0.8–1 nm thick) of toluene solutions of asphaltenes and resins, analyzed by SAXS and SANS. The radius of the particles decreased with the decrease in resins content [149], which may agree with the findings of a narrow radius range for asphaltene solutions. Polydispersed oblate cylinders could fit the SANS data from C_7-asphaltenes of Hondo, Canadon Seco, and Arab heavy crude oils [647], consistent with the reported flat disks shape. Particle sizes on the order of 10 nm have been detected in VR with SAXS [648]. Flat-like particles (disks?) were observed by Loh et al. [347]. Near-infrared spectral data coupled to Rayleigh theory of light scattering produced a measure of 30 nm in radius for the size of asphaltene aggregates [226,649]. However, nonspherical particles of solid asphaltene were precipitated and recovered from AS-diluted Cold Lake bitumen, with sizes that varied from 3.5 to 7.0 μm when the dilution ratio went from 100:1 to 40:1 [270].

Ultrafiltration using Gore-Tex membranes with nominal pore sizes as small as 30 nm and a transmembrane differential pressure no greater than 60 psi demonstrated that asphaltenes originally present in conventional oils and in water-in-oil emulsions of seven considered VRs (two North American, two Middle East, one South American, one Russian and one South Asian), containing up to 7.6% asphaltenes, were no larger than 30 nm [650].

Light scattering study of asphaltene aggregation in heptol showed an incremental aggregate size during the initial stages, then decreased and reached a steady size by 6 h. This steady size was shown to increase with increasing heptane concentration. Aggregate size ranged from 110 to 133 nm [651].

A variety of techniques and measurements (density, VPO, elemental analysis, FT-ICR-MS, time-resolved fluorescence emission spectra, SAXS, DLS, membrane diffusion, Rayleigh scattering, nanofiltration, interfacial tension, interfacial adsorption, and surface force) have been used to assess asphaltene self-association and the physical dimensions of molecules and nanoaggregates [47]. A single source of asphaltenes was considered and narrow fractions were precipitated using heptol mixtures for this multitechnique effort of multiple groups from several organizations. Results indicated that about 90% of the asphaltenes self-associated. Asphaltene nanoaggregates were found to be smaller and more aromatic than bulk asphaltenes. In the case of the densest, highest MW, and most polar subfraction, the asphaltene molecules that associate were larger and less aromatic. The average VPO MW was about 850 Da for the monomer and up to 30,000 Da for the nanoaggregates. The measured size of the nanoaggregates depended on the used technique. It was 20 nm from nanofiltration, 14 nm from SAXS and DL, and 5–9 nm from diffusion and Rayleigh scattering. Film studies showed a swelling factor of 4 in the presence of a solvent. The hierarchical scale of association and aggregation was defined as ~1 nm for the molecules, incrementing up to one order of magnitude for the nanoaggregates to ~5–10 nm (i.e., 5–10 molecules) and massively increasing in two to three orders of magnitude for bulk aggregation ~1000 nm (4×10^8 molecules per particle). The formed aggregates exhibited discoid shape and high porosity. The discrepancies in measured values from the different techniques was explained in view of the presence of nanoaggregates and flocs, and the relative sensitivity of the techniques toward these types of particles [47].

4.5 Surface Properties

Asphaltenes and their flocs and particles are known to be surface-active materials. Surface activity gives rice to a broad range of interactions leading to dispersion, emulsion formation, films, adsorption, etc. This surface activity might be related to the large proportion of heteroatomic moieties distributed among a great variety of functional groups. In general, surface activity can be caused by the heteroatomic substitutions that provide polarity to the molecule, i.e., moieties containing sulfur, nitrogen, oxygen, and metal that form polar groups.

4.5.1 Emulsions

Asphaltenes may act as a surfactant having a nonpolar part with affinity to oil or to neutral/apolar compounds and a polar part prone to associate with water or to polar compounds [652]. In an oil–water system, the polar extreme will be immersed in the aqueous phase, and the nonpolar (bulky and heavy) will extend into the oil, sterically keeping the droplets apart and inhibiting coalescence [37,653,654].

The effect of the naphthenic acid content of DAO on the interfacial properties of the asphaltenes at the crude oil–water interface led to the hypothesis of naphthenic acid–asphaltene acid–base complexes formation. Therefore basic nitrogen-containing asphaltenes would interact with naphthenic acids to form such complexes. These complexes would be responsible for the high propensity for aggregation at the hydrocarbon–water interface and for the observed high interfacial activity of asphaltenes in crude oils [655]. During crude oil production, crude oil–water mixtures may be recovered and the first step is their separation, i.e., dewatering step. An intermediate layer (the *rag layer*) at the crude oil–water interface may be formed and represents a source of problems. Characterization of one of these rag layers showed a high concentration of naphthenic acids compared to the bulk upper oil. This preferential partitioning of the most surface-active naphthenic acids into the complex fluid rag layer [656], and their interaction with asphaltenes, would lead to a more complex rag layer in the case of heavy acid crude oils. The amphiphilic nature of asphaltenes and naphthenic acids synergistically stabilizes the water and solids in the crude oil continuous phase of the complex fluid rag layer. Water-in-oil emulsions, solids-in-oil dispersions, oil-in-oil dispersions, and oil-in-water-in-oil multiple emulsions would coexist in the continuous oil phase of the complex fluid rag layer [657].

Indeed, the role of asphaltenes (and resins as well) in the formation of water-in-oil and oil-in-water emulsions has been widely documented. Reviews have been published as book chapters by Kilpatrick et al. [658,659], from which the graphical description of the three types of oil–water emulsions are reproduced in Fig. 2.30.

Historically, the paramount issues associated with the upstream formation of water–oil emulsions and their stability drove huge R&D efforts that continue as a topic under study, even today. These emulsions are considered to be a sort of "special" colloidal dispersion. In fact, very early asphaltenes were realized as natural surfactants [660,661]. Characterization of subfractions has indicated that as little as 2% of the total asphaltenes may have interfacial activity [662,663].

The emulsion-stabilizing capabilities of flocculates or asphaltene particles have been shown to be superior to those of molecular asphaltenes [661]. The repulsive interaction

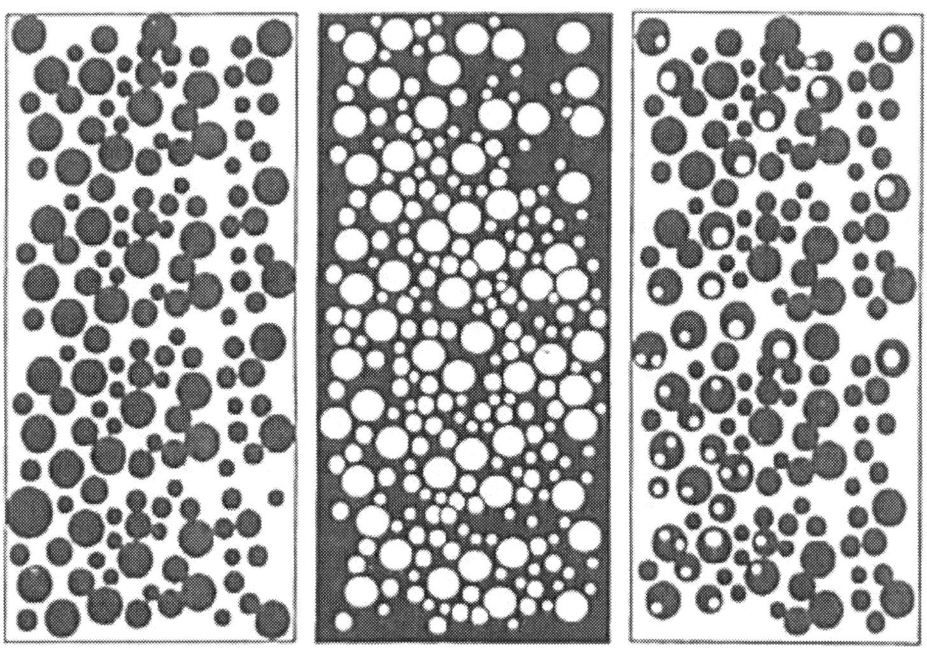

Figure 2.30
Graphical description of oil–water emulsions. *Reproduced from Kilpatrick PK, Spiecker PM. Asphaltene emulsions. In: Sjoblom J, editor. Encyclopedia of emulsion technology. New York: Dekker; 2001. pp. 707–30, with permission from Dekker.*

forces between asphaltene surfaces in toluene had a steric nature and could be well described by the scaling theory of macromolecules. In toluene, the water content and temperature were found to have only a marginal effect on the interaction forces. These colloidal forces measured between asphaltenes in toluene indicate that the water droplets in naphtha-diluted bitumen are stable because of the steric repulsive forces between the adsorbed asphaltenes on water droplets [664].

Studies of model emulsions indicated that the primary factors governing their stability were the aromaticity of the crude medium, the concentration of asphaltenes, and the availability of solvating resins in the oil (i.e., the resin/asphaltene, R/A ratio). The model emulsions were the most stable when the crude medium was 30–40% aromatic and in many cases at small R/A ratios (R/A $\leq$1), implicating that asphaltenes were the most effective stabilizing agents. The resultant emulsion stability depended on the types of resins and asphaltenes in the medium, indicating the presence of specific resin–asphaltene interactions. The most effective interfacially active components were the most polar and/or condensed portions of the resin and asphaltene fractions [665].

Stability of water–oil emulsions has been found to decrease with increasing R/A ratio [666]. The lower stability at high R/A ratios was explained as corresponding to asphaltene monomers or small oligomers strongly interacting with resin molecules. These interactions rendered the aggregates less interfacially active and thus reduced emulsion stability. In the absence of resins, the more polar and/or higher MW asphaltenes were insoluble in heptol mixtures. The addition of resins dissolved this insoluble material and aggregate size increased up to the solubility limit; and at this point, emulsion stability maximized. It is clear that asphaltene chemistry dictates emulsion stability. The most polar species typically required significantly higher resin concentrations to disrupt asphaltene interactions and completely destabilize emulsions. Aggregation and film formation are likely driven by polar heteroatom interactions, such as H-bonding, which allow asphaltenes to absorb, consolidate, and form cohesive films at the oil–water interface [667].

A characterization study of real emulsions from Kuwaiti oils emphasized the results on sulfur speciation. It was indicated that sulfur present in asphaltenes stabilizing the water–oil emulsions was probably in the form of fused ring type or as –C–S–C–, and a small amount of sulfur is found as SH groups. Whether these functional groups play any role in emulsion stabilization was not clarified [668,669].

Kilpatrick and Spiecker [658] have pointed out that asphaltenes form a strong, viscoelastic network by physically self-crosslinking at the oil–water interfaces. Based on this, they have summarized the factors affecting emulsions stability as those variables that affect the kinetics of the process and the magnitude of the film strength: (1) aromaticity of the medium; (2) concentration and chemistry of the resin fraction and other solvating species; and (3) polarity and specific chemistry of asphaltenes.

A positive effect of emulsion formation is a viscosity reduction that enhances flowability. In these cases, stability is needed for the transporting period of time. However, emulsion might break and four processes are the known causes: (1) creaming [11,670,671] (caused by the difference in density of the constituting phases), (2) flocculation or aggregation (discussed in the previous section), (3) coalescence (irreversible larger drops created by fusion of the smaller droplets) [672–676], and (4) phase inversion [677]. Creaming, sedimentation, and flocculation are reversible phenomena caused by the fact that there is no rupture of the stabilizing layer/film at the interface.

4.5.2 Films

Asphaltenes play a most important role in the formation of the interfacial film that provides at times a long-lasting kinetic stability. The asphaltenes' tendency to form mechanically strong interfacial films when mixed with water [678] was reassured by the fact that the presence of an alcohol in the oil phase was not enough to break the film layer [679], and the presence of acid compounds neither prevented nor inhibited its formation [678]. An increase of the mechanical strength of the film prevents coalescence of the emulsion droplets by effectively separating oil from water. Accordingly, in this interfacial film, asphaltenes reduce hydrophobic effects; meanwhile acid groups are believed to reduce interfacial tension [678].

Dilation rheological studies have led to the proposal of a four-step pathway for the formation of asphaltene/resin films in the oil–water interface. These four steps are: (1) asphaltenes and resins coaggregate into polydispersed oblate cylindrical aggregates and diffuse to the oil–water interface; (2) these aggregates adsorb on the interface and begin to consolidate into a film; (3) resins in the inner area of the film crawl through the film to the interface; and (4) most asphaltene contacts are displaced at the interface [680]. The observed changes in the viscoelastic properties anticipated a combination of diffusion-exchange and surface-rearrangement mechanisms [681], which agree well with the proposed pathway in Ref. [680]. Furthermore, AFM and Langmuir technique led to a rather different interpretation of the characteristics of asphaltene and asphaltene/resin films. AFM was employed to examine the topography of the film layers formed by asphaltenes, resins, and their mixtures. Pure asphaltene monolayers consisted of rigid film with a close-packed structure, while those of resins were continuous open networks. Mixed films of these two fractions show that a gradual increase in resin concentration leads to an opening of the rigid asphaltene structure toward a more resin-like configuration [682]. Meanwhile, Langmuir technique showed similar results to those from AFM and additionally indicated that the size of asphaltene aggregates increased when the aliphatic proportion in the solvent and the asphaltene concentration increased. The resin films showed high compressibility, indicating a collapse of the monomolecular film. Resins aggregation did not progress to the same extent as asphaltenes aggregation. Resins predominate in the asphaltene/resin film when their concentration exceeded 40 wt% [683].

The results of interfacial shear viscosity measurements showed that the structure of the asphaltene film at the water–oil interface changed from a 2D to a 3D network upon asphaltene concentration increase. Three types of film were distinguished, namely, expanded liquid film, condensed liquid film, and solid-like 3D network film. Moreover, the film structure formed by asphaltene molecules differed in morphology and strength from that formed by asphaltene particles, probably because of the differences in adsorption and migration processes of asphaltene molecules compared to those of asphaltene particles at the interface [684].

The compositional characterization of the interfacial films showed that the rigid acidic and basic asphaltene multilayered films near or at the critical bitumen concentration played a stabilizing role. As the diluent volume increased beyond critical dilution, the asphaltene basic and nonpolar species were displaced from the oil–water interface either by precipitation or by the lack of effective dispersion by the resin molecules [466]. It is important to point out that in the case of toluene–water interfaces, the film consisted of a monolayer and became more flexible when propanol was added to the oil phase but retained its structure [679]. The acidic subfraction of asphaltenes (ion-exchanged subfractionation, according to [80,167]) formed much stronger films as evidenced by interface rheology experiments. This was explained in terms of the accessibility of the functional group. Since the basic asphaltenes did not form such a strong film, it was supposed that their functional groups might be simply inaccessible. Meanwhile, for the acid asphaltenes, the functional groups should lie on the periphery of the molecular structure, prone for binding other asphaltene molecules [658]. The collaborative role of the naphthenic acids in filmmaking will be further considered in Chapter 4. The interaction between naphthenic acids and asphaltenes depends on the asphaltene type. Dramatic decreases in asphaltene diffusion coefficients upon increases in their concentration were observed [685].

Mingyuan et al. [686] associated the interfacial activity of asphaltenes to the presence of acidic and open chain carbonyl groups. The elastic nature of the open chain carbonyls (as opposed to carbonyls in ring systems) was suggested as a way of introducing a strong H-bonding in emulsified systems. The multiple functional groups present in asphaltenes change with their composition. Consequently, the rheological properties would be different from one to another crude oil. As an example, surface tension values ranged between 15 and 19 mN/m at 358°C and viscosities from 9 to 16 Pa*s for asphaltenes from a variety of heavy oils and bitumen [131].

4.5.3 Adsorption

Surface activity of asphaltenes is also reflected in adsorption [687]. Thus asphaltenes can act in both roles: adsorbent and adsorbate.

Asphaltene structure has been postulated to hold a substantial microporosity capable of either adsorption or occlusion of other hydrocarbons present in the crude oil. While adsorption might occur at molecular level, occlusion could only take place in larger aggregates [688]. These sorption capabilities of asphaltenes (and resins) have been applied in their direct use of adsorbents [689] and in the preparation of carbon adsorbents [690–692].

In an experiment in which a small aliquot of a dilute asphaltene–toluene solution was added to a large volume of resin–heptane solution, the resins were adsorbed on the asphaltene particles but they could not prevent asphaltene flocculation. The asphaltenes–resins interaction enthalpy was determined to be around 3–5 kJ/mol, indicating a weak interaction [693]. Although this could be seen as a contrast with the idea that resins are able to peptize and stabilize asphaltenes, these experimental conditions are far from representing the crude oil environment of asphaltenes.

As discussed in Section 4.2 and based on PNP-fractionated asphaltenes from Furrial crude oil [171,172,200], a modified colloidal model was proposed in which insoluble asphaltenes (A1 fraction) are dispersed by the action of more soluble ones (A2 fraction). The MM calculated SPs could be correlated with trends in other surface-related properties, such as flocculation, adsorption, vaporization, and affinity [694].

The asphaltene surface area was proven to affect wax crystallization and the strength of the wax network. The effect of asphaltenes was rationalized associating the strong dependence of wax crystallization with the degree of asphaltene dispersion [235].

Asphaltenes in toluene solutions were observed to bind water. The amount of bound water increased with increasing asphaltene concentration, though the number of water molecules per asphaltene molecule decreased, indicating that this ratio depended on the aggregation state of the asphaltenes [695].

The covalent nature of the molecular structure of asphaltenes was used to explain their adsorption capabilities. In the same way, the understanding of the interactions with squalane revealed insights into the asphaltenes–resins interactions. Thus the squalane adsorption on asphaltenes was viewed as an adduction of the squalane molecule into an appropriate-size cavity in the asphaltene structure by the interaction of the olefinic π-bond of the former with the aromatic ring systems of the latter. This cavity was suggested to be present in the covalent molecular structure, rather than in the micelle-like aggregates, since in the micelle-like aggregates the dynamic equilibrium would continuously rearrange the structure, rendering the adduction (clathration) process ineffective. The number of those cavities might be evaluated by determining squalane uptake. This information could then be used to discriminate whether resins unimolecularly saturate the asphaltene surface as an adsorbate or reach a steady-state equipartition distribution of the adsorbate between the oil and the adsorbed phase. Experimentally, precipitated solid asphaltenes were capable of adsorbing significantly more resinous material than the crude oil originally contained [526].

Adsorption of asphaltenes at the water/toluene interface was found to be a diffusion-limited process. The changes observed as a function of pH at asphaltene concentrations above CMC were speculatively interpreted as being caused by a chemical reaction that may occur at pH below 4 [696].

Adsorption of asphaltenes and resins onto a hydrophilic surface from solutions with varying aromatic/aliphatic character showed that asphaltenes adsorbed to a larger degree than resins. Preadsorbed asphaltenes were not displaced from the surface by the resins, and neither did resins adsorb onto the asphaltene-coated surface. Resins and asphaltenes associated when both are present in the same solution and in this case the two constituents in the bulk liquid adsorbed to the surface in the form of mixed aggregates [697].

These two surface-derived effects (interfacial tension and adsorption) together make the asphaltenes a vivid interface that will connect solid surfaces to the oil matrix. Studies of asphaltenes adsorption on silica surfaces showed decreases of the adsorbed amount of asphaltenes with the polarity of the used solvent or with increases of the amount of resins present. This observation was associated with the evolution of the asphaltenes' aggregate size and masses. Hence there was a direct correlation between the amount of asphaltenes adsorbed at the interface and the mass of the aggregates in the bulk. Furthermore, asphaltene aggregate conformation changes from fractal in solution to constant density at the surface contact [384]. These relations between adsorption and interfacial activity also explain the role of solids in water–oil emulsion stability. In fact, oil field emulsions retain solids that are reported to provide a stabilizing effect [698–700]. The size of these stabilizing solids is in the range of submicron to micron [701–707]. Since, typically, the solids originate from the formation, aluminosilicate clays have been identified in emulsions formed during production [708,709]. The stability of these emulsions increases with a decrease in particle size and an increase in particle concentration [710]. The synergetic effect between solids and asphaltenes in emulsion stabilization is reflected by the statement given by Lordo [698]: "...In general, it is easier to obtain solids that are asphaltene-free than asphaltenes that are solids free..."

Among other properties, the presence of transition metal ions or groups within the asphaltene molecules underpins the possibility for electronic interactions, providing d-orbitals that might lodge electrons from other adsorbates or create a vehicle for a charge transfer mechanism to adsorb onto a given adsorbent surface. Highly stable porphyrin-like structures host vanadium, cobalt, copper, and nickel ionic groups. Iron and aluminum can be linked to other heteroatoms (S, O, etc.) through weaker interactions than those found on porphyrin N-type bonds. When Cu is present the electrical surface charge might change from negative to a positive charge. The reported uptake of heavy metal ions by asphaltenes from solution was higher for V, moderate for Ni, and negligible for Cu; nevertheless, the nature of such interaction remains unknown [711]. Asphaltene interaction with metals [712] and metal surfaces [713–716] has been documented, since it is seen as the initial step toward deposition in pipelines, fouling in heat exchangers, etc. Asphaltene aggregation and adsorption are slow processes; a stepwise adsorption mechanism has been proposed for the creation of adsorbed multilayers that favors a flocculation environment and eventually leads to plugging problems [717]. Adsorption on metallic nanoparticles showed a decrease in surface affinity in the order of Ni > ZnO > CoS > Co [718].

Additionally to adsorption on metals [716,719], asphaltenes also adsorb onto oxides, clays, and other minerals [333,670,708,717,719–741]. Furthermore, asphaltenes are also known to adsorb other molecules, such as the resin molecules [742–746]. On silica, a multilayer adsorption was observed when high concentration solutions were employed [747–749]. On alumina, surface acidity showed a better capacity for asphaltene adsorption than basic or neutral surfaces [750]; besides the smaller the particle, the larger the adsorption [751]. The adsorption on minerals was diminished by the presence of an aqueous phase [752], which represents an advantageous finding for production purposes since one of the most employed heavy oil and bitumen recovery technologies uses water vapor for enhancing recovery (steam-assisted gravity drainage, SAGD).

A variety of nanoparticles has been tested for asphaltene adsorption. Different types of silica and alumina nanoparticles, including metal impregnated (Ni and Pd), were considered and, in general, adsorption equilibria was rapidly attained. Best results were achieved with alumina-supported hygroscopic salts of the metals. Thus for asphaltene solutions, adsorption seemed to preclude flocculation and precipitation. A porous media was incorporated to test the effect of the injection of nanofluids in the inhibition of agglomeration, precipitation, and deposition of asphaltenes on the rock surfaces. The results indicated that nanoparticles were capable of restoring production after damage caused by asphaltene and improved recovery during the displacement tests [753]. However, it is worth pointing out that the preparation of the nanofluid incorporated a surfactant and blank experiments showing the effect of surfactant-containing fluids were not reported in that work.

4.6 Stability

Typically, low stability is associated with high reactivity. However, the instability of asphaltenes is accompanied by low reactivity, as will be discussed in the next section. In both cases, stability and reactivity are a partial consequence of the high aromaticity of the molecular array. Regarding asphaltenes, stability is mostly associated with their phase changes and as such it is a consequence of their solubility, aggregation, and surface properties all together and within the context created by the surrounding medium. The relative contribution of each of these drivers and the interdependence among them make the assessment of stability and the identification of its causes a rather complicated challenge.

According to Speight, the stability of petroleum is based on a delicately balanced harmony within a three-phase system consisting of the asphaltene constituents, the aromatic fraction (including the resin constituents), and the saturate fraction [30]. This harmony is broken upon flocculation and the system evolution is a consequence of its dynamic behavior and continuous change of properties. For instance, Christiansen et al. [261] indicated that SPs at the onset are constantly offset between measurements from batch and from continuous systems. The batch method typically indicates greater instability than the continuous method.

In terms of stability, live oils are the study case for production issues, while dead oils are for transportation and refining. However, the knowledge derived in each case might be of interest to the other. As long and as much as possible, asphaltenes should be studied in their natural form. The observed trend in asphaltene stability has been associated with the R/A ratio. Thus effect of light paraffins for breaking the harmony of the system has been explained as being caused by the solubility of resins in the paraffin. The stability of the asphaltene micelle was postulated to depend on: (1) the "peptizability" of the asphaltene core; (2) the peptizing power of the resins; (3) the relative amounts of asphaltenes and resins; and (4) the aromaticity of the oil phase [113,608]. Electrodeposition experiments led to the belief that a positively charged asphaltene nucleus is surrounded by a cloud of negatively charged resins, which effectively screen the nuclear charge. The electrical properties of asphaltenes could be rationalized by this model, in which the peptizing effect of the resins serve as an intermediate continuous stage between asphaltenes and the other maltene components, keeping a harmonic phase equilibrium in stable systems [754].

Nevertheless, the role of resins on the stabilization of asphaltenes in crude oil has been explained in many different ways. The surfactant/peptizing effect of the resins on contributing to the solvent power (polarity and aromaticity) of the medium is one explanation. However, this role of resins in asphaltene dispersion has been considered an article of faith that has never been scientifically validated. Whether asphaltenes are dispersed in resin-coated inverse micelles, with resins "peptizing" asphaltenes, has long been a subject of debate. Many researchers have assigned a stabilizing role to the resin–asphaltene interaction [755]. Thus Schabron and Speight [756] have pointed out that the potential for graphite-type stacking by the asphaltene molecules in the center of a micelle might not be as great as the potential for the micelles forming by asphaltene–resin interactions rather than by asphaltene–asphaltene interactions. Resins contribute to the solvent power (polarity and aromaticity) of the medium. Destabilization of the asphaltenes takes place when the solvating power of the medium is reduced to the point when asphaltenes and micelles are no longer compatible with the medium [756]. Although a tricritical phenomenon has not been observed experimentally, a near-tricritical coexistence of a monomer phase and a micellar phase has been theoretically anticipated because of the coupling between the micellization and phase separation [757]. As has been discussed, asphaltenes in crude oil assume various forms depending on the oil aromaticity/paraffinicity, and on the other compounds present in the oil. Molecules or monomers, molecular clusters (nanoaggregates), micelles, and aggregates can all coexist in the oil matrix.

A molecular thermodynamic approach was used to envision the influence of resins on asphaltene aggregation, and thus in stabilization. The model predicted the formation of mixed resin–asphaltene aggregates described as aromatic cores composed of stacked aromatic sheets surrounded by aliphatic chains. The developed analytical expression for the free energy of mixed aggregation incorporated solubility, mixing, and interfacial and steric effects. A decrease of aggregate size with the presence of resins and an increase on the CMC were predicted, in agreement with the experimentally observed behavior. The effectiveness of different resins to split the asphaltene aggregates was found to depend on the solubility of their polyaromatic ring [758].

Destabilization of the asphaltenes leading to phase separation would occur when the solvating power of the medium toward asphaltene molecules or micelles is reduced to a point at which they are no longer fully compatible with the oil medium [756]. A high-resolution microscope and image processing software were used to detect and determine the amount of deposited asphaltene as well as its size distribution at different conditions in a pressure–volume–temperature (PVT) visual cell. To address wettability changes, surface properties were studied by contact angle measurement. The results verify that the amount of asphaltene deposition increases when the pressure increases and the quantity of asphaltene deposition decreases as the R/A ratio in these samples increases. At high R/A ratios, asphaltenes were found to be more stable, but stability depended on asphaltene source (original crude oil). The results showed that, as the pressure increased, the stability of asphaltene decreased more than expected. The observed changes in surface properties indicate that in the presence of resins, the surfaces become more water-wet and their roughness decreases [759].

The interactions between asphaltene and resin molecules [224,273,621,709,710,760] have been suggested as the controlling driver for either keeping asphaltenes in solution [37,648] or the onset of flocculation [761]. In the former, resins adsorbed on asphaltenes when the surface tension of the medium decreased [648] and in this case polar and H-bonding interactions appeared minimized compared to dispersion interactions [37].

Thus during dilution with an AS, such as *n*-heptane, desorption of resins would be the cause of flocculation.

The stabilizing effect of type II resins has been experimentally demonstrated. In fact, regardless of the crude source, asphaltene stabilization can be achieved by adding type II resins from one crude to another. In the reported example, resins were isolated from Boscan, Hamaca, El Furrial, and Cerro Negro crude oil samples and the stabilization effect of resins II was confirmed [762].

According to Creek [124], the differences between precipitation and aggregation are the forces that drive each of these processes. He also distinguished aggregation from precipitation as distinct steps in a completely reversible process. Strong interaction sites located at the periphery of the asphaltene molecules drive aggregation, starting with the reversible association in 2D sheets (lamellar array). Meanwhile, precipitation eventually occurs, as determined by van der Waals attractions between aggregates [124]. In this proposal, precipitation is only a consequence of aggregation, which may be considered part of the real situation by other authors.

The unstable asphaltenes are characterized by a relatively small molecular size, higher aromaticity, and higher heteroatom content, compared to those of the more stable asphaltenes [763]. As seen in Fig. 2.31 (from Ref. [764]), unstable crude oils are characterized by low hydrogen-to-carbon ratios (1.1–1.2), high values of aromaticity (fraction of aromatic carbon = 0.4–0.5), and high degree of condensation of the aromatic rings. A lamellar arrangement (as conceived in Yen's model, see Fig. 2.9) agrees well with these findings on particular chemical features exhibited by unstable and troublesome asphaltenes (aromaticity, polycondensation, and heteroatoms).

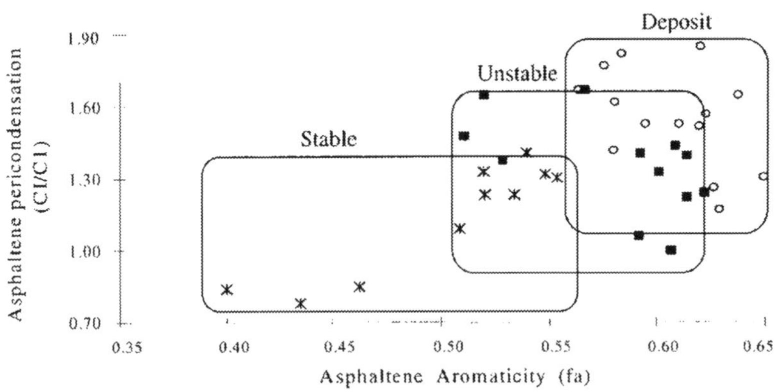

Figure 2.31

Molecular parameters of asphaltenes from (*) stable crudes; (■) unstable crudes; (o) deposits. *Reproduced from Carbognani L, Orea M, Fonseca M. Complex nature of separated solid phases from crude oils. Energy Fuels 1999;13:351–58, with permission from ACS Publications.*

Based on these features, obtained experimentally by elemental analysis and H-NMR of C_7-asphaltenes from Venezuelan crude oils, molecular structures were proposed to represent an average molecule of the asphaltenes present in those crude oils. Thus the proposed molecular structures of asphaltenes from stable crude oils BC5, BC6, MG27, MO21, MO29, and CN are shown in Fig. 2.32A, while those of unstable crude oils CO2, BO7, VG3, and FU1 are shown in Fig. 2.32B [100]. The similarity among the chemical features shown in the molecules of asphaltenes from stable crude oils BC5, BC6, and CN and those of unstable crudes VG3 and FU1 indicates that stability is not merely derived from those cited features.

Those features discussed in a previous work by Leon et al. [483] (see average molecular structures in Fig. 2.20B; some also shown in the work of Ref. [100]) included a lower degree of aromatic condensation in stable asphaltenes and a higher degree for unstable asphaltenes. However, the lowest degree of condensation is observed in the structure FU1 that comes from an unstable crude oil. Apparently, the degree of condensation is not one of the relevant chemical features affecting stability. The chemical features described by Rogel and Carbognani [100] were also present in the asphaltenes recovered from a deposit of precipitated solids. Notwithstanding, it is worth pointing out here that the solids typically contained inorganic materials, probably minerals from the oil well. In fact, other chemical features can enlighten the mechanisms or interactions giving rise to the minerals' retention by the organic matter. For instance, when less than 10 wt% of minerals was present in the deposits, the organics were totally soluble in aromatic solvents (containing more than 80% aromatics), indicating its asphaltenic nature. However, when the mineral content was greater than 20 wt%, insoluble organics could not be removed from the mineral matrix, either by aromatic solvents or by the addition of alcohols. This seems to constitute evidence of a strong mineral–asphaltene interaction that favors agglomeration and precipitation. Further discussion of these interactions were given in Section 4.5.

Alkylbenzene-derived amphiphile molecules can stabilize asphaltenes in alkane-based solutions [153,248,765]. In the case of alkylbenzene sulfonic acid, it has been found that a sufficiently long alkyl tail could effectively reduce asphaltene deposition [766]. As mentioned, acid–base interactions have been found possible in asphaltenes, hence the feasibility of stabilization and dissolution of asphaltenes using amphiphile solutions. The variety of particle shapes found in precipitated asphaltenes indicates that asphaltene association does not require any specific direction or orientation. The evaluation of the effects of acidic alkylbenzene-derived and basic alkylpyridine amphiphiles on asphaltene stability demonstrated a positive action of the former. The amphiphile molecules adsorb on the asphaltene surfaces by their head group, while the tail groups create a stable steric alkyl layer around asphaltene molecules. The chemical structure of the amphiphiles as well as the nature of the remaining oil acting as solvent influence the stabilization mechanism by the following processes:

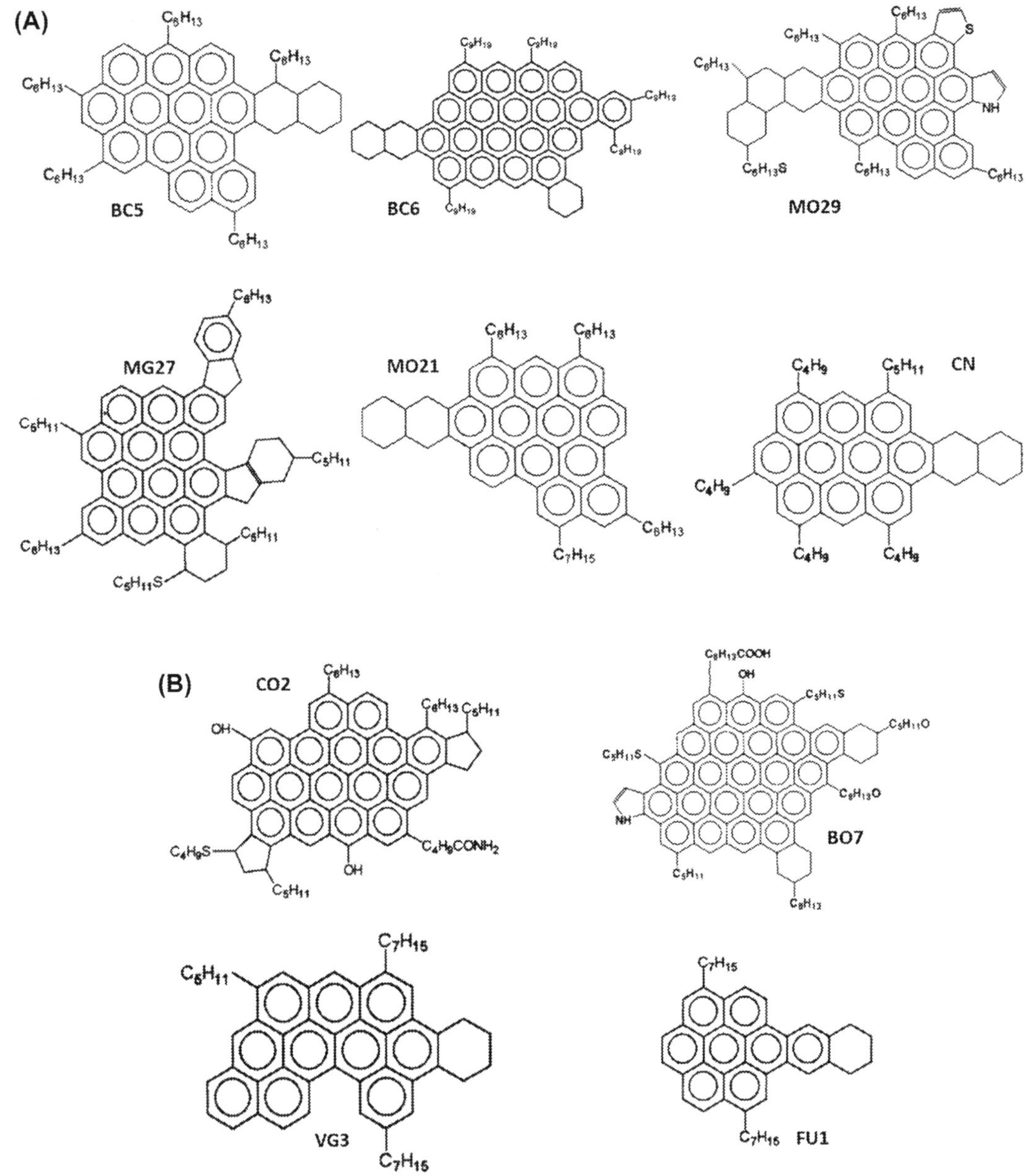

Figure 2.32

Proposed average molecular structures of asphaltenes from: (A) stable and (B) unstable crudes. *Reproduced from Rogel E, Carbognani L. Density estimation of asphaltenese using molecular dynamics simulations. Energy Fuels 2003;17(2):378–86, with permission from ACS Publications.*

- Increasing the polarity (or the acidity) of the head group strengthens the interaction with asphaltenes, and consequently amphiphile's effectiveness on stabilization increases;
- Increasing the tail length improves effectiveness to sterically stabilize asphaltenes even though it might reduce asphaltenes–amphiphiles interaction; and
- The influence of alkane varies with the types of amphiphiles [201].

In the presence of light gases, such as methane, ethane, propane, and N_2, asphaltenes are usually more stable as pressure and temperature increase; however, experimental measurements indicate that, in the presence of CO_2, asphaltenes become more stable as the temperature decreases [767]. Additionally, pressure increases were found to favor stability [768]. A model based on perturbed chain-statistical associating fluid theory (PC-SAFT) considered for an EOS indicated that CO_2 can act as an inhibitor or a promoter of asphaltene precipitation depending upon the range of temperature, pressure, and composition studied. At fixed pressure and live oil composition, CO_2 addition increases the asphaltene stability below the crossover temperature, whereas above this point, the asphaltene becomes less stable when the CO_2 concentration is increased [767]. In general, temperature has a significant effect on stability and the dependence is crude specific. No correlation has been found yet [769].

The flocculation onset measurements of recombined long residua showed differences with that of the original VR. These differences were larger for those crude oils with higher stability [770]. Instability of asphaltenes in crude oils has been associated with the most polar asphaltenes, characterized by high metal content and higher MW [168]. Comparison of the kinetics of asphaltene aggregation for an unstable (Furrial) and a stable (Boscan) crude oil showed two different behaviors. In the case of Boscan, the aggregate size increased rapidly initially, while size increased gradually for Furrial asphaltene aggregates. The aggregation process is initially determined by the diffusion of the aggregates of Boscan crude oil. In this stable crude, the attractive interactions between flocs and aggregate asphaltenes control the kinetics. In unstable crudes, it seems that the aggregation kinetics is initiated by increments of the number of particles and not by their growth, up to a threshold where the aggregation process becomes governed by the diffusion of colloid particles [771,772].

The CII defined according to Eq. (2.14) is based on SARA composition of a VR. The Gaestel Stability Index (GSI) is defined similarly but instead uses a SAPA (saturates, aromatics, polar resins and asphaltenes) composition in which the polar resins are recovered in a slightly different method [773]. Thus the higher the index value, the less stable the overall system. Therefore CII and GSI provide indications of the propensity for coke formation for a particular residuum [774]. The propensity to form coke was also found to correlate with the trend for aggregation, in good agreement with that relation found with CII and GSI [775].

In production, crude oils containing low amounts of asphaltenes cause big and complicated problems. Deposition of asphaltenes from low-asphaltene-containing oils is in fact one of the bigger problems for the production side of the oil industry. This can probably be explained from results that indicate that the stability of crude oil does not depend on the amount of asphaltenes present but on the characteristic nature of the asphaltenes [168]. A study on deposits found in pipelines indicated that most of the deposits considered were constituted for more than 80% asphaltenes. However, the same study also indicated that only less than 10% of the crude oil asphaltenes took part in the deposition process [124].

4.7 Reactivity

The chemical complexity of these compounds mentioned earlier, which tangles the assessment of molecular identification indirectly, implicates difficulties for the understanding of reactivity. We have already discussed the fact that similar MW is not a consequence of similar boiling points and vice versa. The BotB compounds not only are the residue of distillation, but also cover a wide range of MW compounds. According to the basis of the continuous Boduszynski's model (see Section 4.4.1), asphaltenes should be an extension of maltene compositional space to higher and higher CN; instead they overlap the MW space and are an extension to higher degrees of aromaticity [454].

Regardless of the broad distribution of MW in the asphaltene fraction (400 to ~5000+ amu), it has been found that the functional groups present in the molecules do not vary significantly [63,162], indicating that chemical properties would not differ broadly, though will vary from one source or isolation method to another. There is also some evidence that the heteroatomic functional groups tend to concentrate slightly at the upper end of the molar mass distribution [776].

The chemical features that determine the reactivity of asphaltene molecules include, but are not limited to, aromaticity, degree of ring condensation, functional groups present, number and length of branches, and acid–base–neutral chemical nature. Additionally, it can be expected that physical properties will affect also the reactivity, e.g., aggregation would affect accessibility, creating steric limitations, and together with solubility would give rise to mass transfer limitations.

Attempts have been made to define a characterization index for assessing reactivity and/or processability. The most widely accepted index is the UOP K, also known as Watson K (K_w), defined by Watson and Nelson for hydrocarbon refinery streams [777] almost 100 years ago (Eq. 2.20); in this equation, BP is the boiling point and SG the specific gravity at 15.6°C. K_w value is typically 13 for paraffins, 12 for hydrocarbons whose chain and ring weights are equivalent, 11 for pure naphthenes, and 10 for pure aromatics. This index categorizes crude oils as paraffinic ($K_w > 12.2$), naphthenic ($12.2 > K_w > 10.5$), or aromatic ($K_w < 10.5$). Since paraffins are the most reactive toward cracking, a $K_w > 12$ indicates good conversion performance and in general "crackability" increases when K_w increases.

$$K_w = \frac{\sqrt[3]{1.8\ \text{BP}}}{\text{SG}} \tag{2.20}$$

However, in the case of the asphaltenic fraction, determination or evaluation of BP becomes challenging. Other indexes have been postulated for the assessment of a characteristic reactivity and/or processability. This is the case for the K_H of Shi et al. [778] shown in (Eq. 2.21).

$$K_H = 10\frac{\text{H/C}}{\rho \text{MW}^{0.1236}} \tag{2.21}$$

The propensity of coke formation is typically associated with the CCR value of the hydrocarbon stream. A correlation between CCR and K_H was adjusted to fit the equation shown in Eq. (2.22).

$$\text{CCR} = 2451K_H^2 + 44.1K_H + 200 \tag{2.22}$$

Similarly to the previous categorization of crude oils in terms of K_w, K_H has also been used to assess the processability of VRs as $K_H > 7.5$ is a VR adaptable to secondary processing, $6.5 < K_H < 7.5$ corresponds to intermediate processability that would anticipate some issues, and $K_H < 6.5$ represents a difficult-to-process VR.

Zhao et al. [67] preferred to use viscosity (η at 70°C) instead of density for defining their K_R reactivity index (Eq. 2.23); their correlation equation with CCR is included in Eq. (2.24).

$$K_R = 10\frac{[\text{H/C}]^2}{\eta^{0.1305}\text{MW}^{0.1236}} \tag{2.23}$$

$$\text{CCR} = \frac{65.685}{K_R} - 5.81 \tag{2.24}$$

These previous equations found good agreement with CCR prediction; however, it has been demonstrated that CCR by itself is not sufficient for coke yield prediction from VR subfractions [779]. Additional work included the characterization of eight VRs, their SARA fractions, and five narrow asphaltene fractions. The results were used to establish new correlations for the evaluation of physical properties, such as BP (Eq. 2.25), and critical temperature and pressure [780].

$$\text{BP} = 79.23M_n^{0.3709}\rho^{0.1326} \tag{2.25}$$

Ovalles et al. [781] defined a reactivity index (R_{feed}) incorporating structural parameters of the asphaltene molecules; it is shown in Eq. (2.26). In this equation f_a is the aromaticity and γ is the degree of polyaromatic condensation, defined as the ratio of (total number of bridged aromatic carbons)/(total number of aromatic carbons). The equation could fit well the data from hydrodenitrogenation (HDN) and microcarbon residue (MCR) reduction reactions.

$$R_{feed} = \frac{5 \times 10^4}{[Asph]^{2.9} \times f_a^{4.4} \times \gamma^{0.05}} \quad (2.26)$$

4.7.1 Thermal Conversion

The thermal conversion of BotB fractions gives indications of the chemical reactivity of these compounds. Thermal conversion occurs via free radicals (as confirmed by the promotion observed by the addition of radical initiators [782]). Hence stability and nature of products depends on the mechanism of saturation of these radicals. Thus increase in severity increases the probability of radicals condensation, in particular arene radicals that would lead to coke formation. The reactivity of the radicals formed by aromatic heteroatomic moieties is such that these can be considered fast-reacting coke precursors. Additionally, alkyl radicals may transfer hydrogen and end in the formation of olefins that would also be a potential source of coke.

Thermal conversion will break the most labile C–C bonds, thus alkyl-substituted branches will be the first to be broken. An obvious increase in the saturate/aromatic (S/A) ratio of the total product would be observed, which is typically reported as an increase in aromaticity of the heavier products, such as is reported in Refs. [783,784]. About 60 wt% of Athabasca asphaltenes were converted to light hydrocarbons at temperatures below 525°C [785]. A linear correlation was found between the aliphatic-C content and the yield of volatile compounds and between S-evolution and the yields of pyrolysis products for the thermolysis of C_7-asphaltenes from different crude oils [786].

The heaviest fractions exhibited larger compositional change than the lighter fractions upon thermal conversion. The asphaltene fraction was the most affected, with a decrease in their stability mainly caused by becoming more insoluble in an increasingly aliphatic oil. The change in the asphaltenes' SP distribution was larger than a concomitant increase in the SPs' of the lighter fractions [128].

The trends observed during asphaltene thermal conversion has been associated with the transformation of the macrostructure induced by microstructural changes caused by conversion. More particularly, the low reactivity of metal compounds compared to that of the alkyl substituents has been proposed as largely responsible for the macrostructural changes observed for the asphaltene fraction [787]. The observed macrostructural changes observed by Hauser et al. included reductions of the average distance between the aromatic sheets and the stack height of the aromatic sheets, and insignificant changes in the average diameter of the aromatic sheets [788].

The lower S- and N-content and the higher H/C ratio of the liquid product of asphaltene thermal conversion was interpreted as being caused by the breaking of C—heteroatom bonds and to H-transfer from the heavier fragments, which would convert into preasphaltene molecules and/or evolve to coke [789]. However, both coke and asphaltenes in the residue product are known to concentrate the heteroatomic compounds and become more aromatic. As mentioned in previous paragraphs, alkyl substituents of asphaltenes break apart forming saturate compounds that will increase the H/C ratio of the liquid product. Furthermore, the heteroatomic asphaltenes are known to be the least reactive and to be categorized as coke precursors. Hence C-rejection and the preferential coking of heteroatomic compounds are responsible for the lower S- and N-content and the higher H/C ratio of the liquid product.

The reactivity of maltenes and asphaltenes from Arabian light, Arabian heavy, Hondo, and Maya VRs was compared [790]. Asphaltenes could be converted selectively to maltenes at temperatures of 400—425°C (as also reported by Savage et al. [791]), while above 450°C, conversion led predominantly to coke. Maya asphaltenes were the least reactive. Thermal conversion of maltenes sequentially yielded asphaltene and then coke. Finally, the comparison of the reactivity of the whole residue with that of the separate components indicated that certain maltenes could enhance the reactivity of the asphaltenes [790]. Savage et al. did not test temperatures above 450°C, but they evaluated the kinetic rate constants. Additional findings in their previous work reported the formation of more H-deficient asphaltenes. Secondary thermal degradation produced coke. The coke and maltenes fractions cracked to lighter products via secondary reactions. The incorporation of H-donor to the reacting system made asphaltene conversion slower but more selective to maltenes [791].

The thermal cracking of Cold Lake vacuum bottoms showed that asphaltene converted faster when reacted alone, and lower when reacted within the whole vacuum bottoms. MW-based subfractions of asphaltenes indicated an increasing trend in coke yield with increasing MW [792]. In agreement with these findings, the coking tendency was observed to increase by increasing asphaltene/maltene ratios in the residue [111]. The size of the aromatic core [413] of the asphaltene molecules has also been associated with the reactive tendency toward coking [793]. The molecular changes observed by Michael et al. on their thermally cracked studied feed were represented by average molecular structures as illustrated in Fig. 2.33 [592].

Asphaltenes from three different crude oils were subjected to thermal conversion in the presence of hydrogen at 4 MPa and temperatures in the range of 350—430°C. The

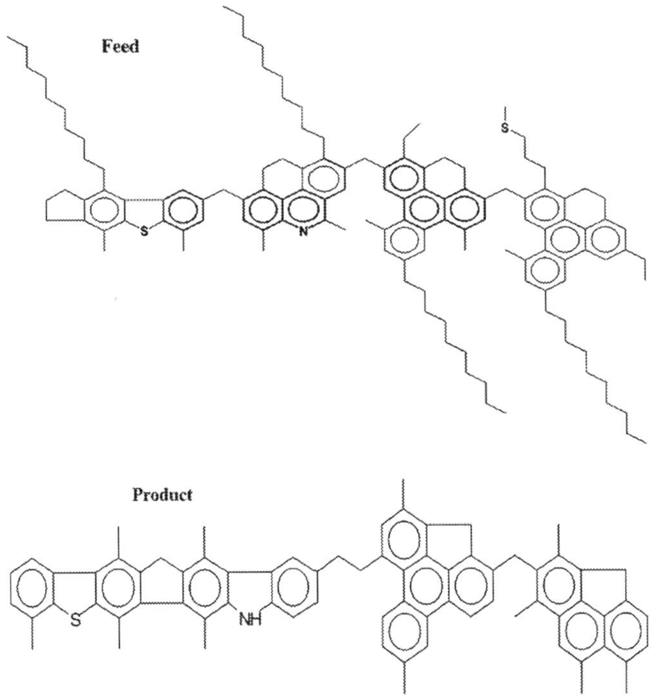

Figure 2.33

Representation of an average chemical structure of an asphaltene and its thermally cracked product. *Reproduced from Michael G, Al-Siri M, Khan ZH, Ali FA. Differences in average chemical structures of asphaltene fractions separated from feed and product oils of a mild thermal processing reaction. Energy Fuels 2005;19:1598–605, with permission from ACS Publications.*

apparent activation energies fell in the range between 170 and 255 kJ/mol, with no detected hydrogen incorporation into the products [794]. The apparent activation energy for the thermolysis of asphaltenes from other crude oils was found to range between 122 and 220 kJ/mol [795].

In thermal conversion, an empirically derived correlation allowed the prediction of the time (t_{max}) at which the maximum oil and gas yield could be reached (Eq. 2.27). In this equation, y_0 is the initial weight fraction of asphaltenes, k the overall rate constant, k_1 the rate constant for oil + gas formation, k_2 the rate constant for coke formation, and C_0 is the initial asphaltene concentration [796].

$$t_{max} = \frac{\sqrt{\frac{y_0}{k}} - 1}{C_0[k_1 + k_2]} \tag{2.27}$$

The kinetics of the cracking reactions can be derived from the kinetics of coke formation by incorporating phase separation and hydrogen transfer rates [797]. The presence of heteroatoms modifies reactivity, even some of the N and S heterocyclic aromatics were found to be unreactive and when the aromatic-C/unreactive-C ratio was <16, shot coke was found as the major solid [595]. A defined severity index SI [798] was based on kinetic parameters of bitumen coking (Eq. 2.28). SI could be considered a thermal reactivity index for asphaltenes [799]. The time for coke onset (t^*) and the kinetics were experimentally determined; see, for instance, Ref. [798].

$$SI = t^* e^{\left[\left(-\frac{Ea}{R}\right)\left(\frac{1}{T}-\frac{1}{700}\right)\right]} \tag{2.28}$$

Asphaltenes precipitated from Athabasca bitumen using different AS/bitumen ratios showed a minimum effect of this ratio on the reactivity of the resulting precipitated asphaltenes at temperatures of about 525°C. At this temperature, 60 wt% of the hydrocarbons were volatilized (probably cracked molecules) and the remainder was transformed into coke [785]. Under mild thermal cracking conditions a wide range of boiling point products were observed. Meanwhile, at more severe conditions, aromaticity correlates with solid yield (35–75%) [800].

The propensity to form coke of the various fractions constituting VRs was examined by evaluating their CCR and MCR. The purpose was to find a correlation between CCR or MCR and chemical properties of the VR and its fraction. Among the reported correlations, the best fitting was found between MCR and the H/C atomic ratio, as can be seen in Fig. 2.34 [801].

Mitigating coke formation or maximizing liquid yield has driven the study of the effect of solvents and of the incorporation of hydrogen and H-donors. In the case of hydrogen, asphaltene cracking produced higher liquid yield, since most of the saturated radicals remained in the oil and less frequently combined to coke. Additionally, radical saturation decreased the oil reactivity toward cracking [802]. The thermal conversion of asphaltenes at 430°C in the presence of solvents such as maltenes, 1-methynaphthalene, naphthalene, and tetralin showed that only solvents with H-donating ability could suppress coke formation [677,797].

It seems then that cracking and hydrogen transfer are intriguingly correlated to the yields and quality of the product slate. Thus aromaticity increased while the liquid yield decreased when pyrolysis was carried out in the presence of water [803]. Thermal conversion in the presence of steam is of interest because of the possible changes that hydrocarbon molecules may undergo under steam stimulation conditions during heavy oil production. Three types of Chinese heavy oils were tested and the molecular changes were monitored by SARA fraction. Resins and asphaltenes were reduced in the range 20–30 wt %, while aromatics and saturates fractions increased. This conversion resulted in a substantial viscosity reduction of heavy oils (as high as 42%) [804].

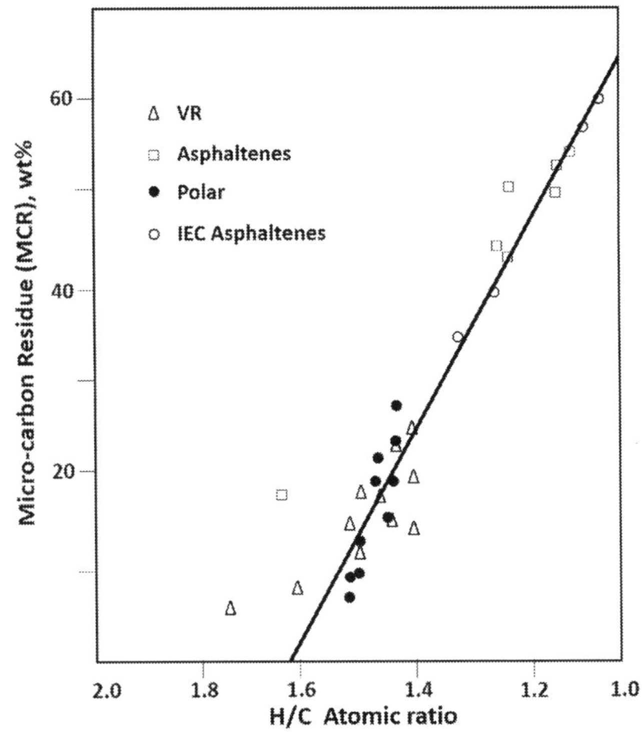

Figure 2.34
Correlation between microcarbon residue (MCR) and H/C ratio of vacuum resid (VR) and its fractions. *IEC*, ion exchanged chromatography. *Reproduced from Schabron JF, Speight JG. Correlation between carbon residue and molecular weight. ACS Div Fuel Chem. Preprints 1997;42(2):386–89, with permission from ACS Publications.*

An on-column precipitation method has been developed for an assessment of the thermal history of the asphaltenes, which indirectly could provide insights into reactivity. It has been found that the fraction of C_7-asphaltenes soluble in cyclohexane decreases upon thermal conversion; supposedly these asphaltenes are thermally converted into coke precursors. The method consists in the precipitation of a portion of residua sample in heptane on a PTFE-packed column and then eluting with a sequence of increasing polarity solvents, namely, cyclohexane, toluene, and methylene chloride [147].

4.7.2 Hydroprocessing

Asphaltenes may react under hydroprocessing (HDP) conditions though the reaction pathway would depend on their composition and molecular characteristics. In general, hydrogenation of the polycondensed aromatic rings has been observed at lower temperatures [805–807] and some asphaltenes were converted to resins [807]. Below a certain temperature threshold hydrogenation dominates the process chemistry and above that temperature the prevalent reaction becomes cracking. This temperature threshold varies among the scientific researchers, e.g., 370°C [806], 390°C [808], or 420°C [805].

The reactivity toward HDP of asphaltenes with a high ratio of internal aromatic carbon to peripheral aromatic carbon (above 3) was low whereas those with a ratio between 1.5 and 2 were more reactive [809]. Asphaltenes with higher polarity have been reported to be less reactive than nonpolar asphaltenes [810].

The characterization of asphaltenes as a function of time on-stream indicated that the following reactions took place under HDP [811]:

- Aromatization of cycloaliphatic groups present in asphaltene molecules;
- Cyclodehydrogenation of alkyl chains to form other naphthene ring systems;
- Dehydrogenation of the new naphthene ring systems to form more aromatic rings;
- Dealkylation of aromatic ring systems.

These reactions lead to shortening of the alkyl chains and an increase in the aromaticity of reacted asphaltenes. The number of aromatic rings diminishes when pressure and temperature are increased and space velocity is reduced. The percentage of substituted aromatic rings also diminishes with severity [812]. These results on aromaticity differed from the HDP of DAO doped with increasing amounts of asphaltenes. In this case, the aromaticity and the condensation degree increased [813], as also reported for hydrodemetallization (HDM) and hydrodesulfurization (HDS) of Kuwait AR [814]. Additionally, typical decreases in the MW, H/C ratio of asphaltene, S-content, and increase in N-content were also reported [813].

Under hydrotreating (HDT) conditions, the reactivity of asphaltenes for the removal of heteroatoms is not as high as desirable, indicated by the fact that hydrotreated bottom products resulted in more aromaticity and contained more nitrogen, sulfur, and metals [805]. Thus, in general, heteroatom removal from asphaltenes is typically more difficult compared to other hydrocarbon fractions (e.g., [815]). Part of the sulfur can be removed and favors asphaltene decomposition. However, aromatically bound sulfur is more difficult to remove and mostly remains in the asphaltene molecule [816]. The S-containing species with ring sizes smaller than 5 rings can be effectively processed, while larger ring systems are more resistant to react [817]. The reactivity of the S-species toward hydrocracking (HCK) differed also and S was mostly removed from the compounds present at the top subfractions [818].

Under HDT, the N-content in asphaltenes either does not change [806,816] or even increases [805,813]. Nitrogen is aromatically bound in the asphaltene molecules [488,537] and their hydrogenation is thermodynamically unfavorable [819]. The refractory or partially converted asphaltenes are present in the hydrotreated product as extended open ring systems as high as 11 rings. HDP tends to remove or reduce the substituting alkyl chains, but the core ring system remains with and/or without heteroatoms [817].

The propensity to coke formation represents a drawback for catalytic processes. The observation that reduction in CCR parallels heteroatom removal might be caused by the presence of these contaminants preferentially in the coke precursor species [815].

While CCR is often associated with the coking tendency of the feed, Chung et al. have found that coking was significant for the bottom subfractions and negligible for the top subfractions, regardless of a similar CCR content for all subfractions. Within this bottom subfraction, coking increased with the degree of aromatic condensation or the asphaltene molecules [580]. The factors affecting coke deposition on an NiMo/γ-Al_2O_3 catalyst during HDP of Athabasca bitumen and Maya crude oil were visualized at three different scales:

- Macroscopic scale: number, nature, and composition of phases present;
- Nanoscale: asphaltene nanoaggregation within phases; and
- Molecular scale: hydrogen solubility by phase.

Two sets of experiments were considered: (1) feedstock, permeates, and retentates (from nanofiltration with ceramic membrane), and (2) bitumen diluted with *n*-dodecane, *n*-decane, or 1-methylnaphthalene. Coke deposition showed to be diffusion limited, while vanadium deposition, arising primarily from the maltene fraction, was not diffusion limited [316]. Although coking and coke deposition are the main drawbacks of treating asphaltenic fractions, the high content of heteroatomic species and particularly metal compounds are the source of problems that jeopardize the processing of these fractions. Some of the trends observed in HDP of metal-containing VRs and/or asphaltenes aggravate with increasing metal content. The issues directly associated with metal compounds will be discussed in the next chapter. Indirectly, metals have been associated with the most refractory and unstable asphaltenes. However, the profile of the major metals, V, Ni, and to a lesser extent Fe, within the maltenes–asphaltenes fractions seems to depend on the source crude oil. When VRs from several crude oils and crude blends were separated by ultracentrifugation, the proportion of precipitate was larger than the total amount of asphaltenes originally present, even though some asphaltenes remained in the supernatant. In this study, metals also concentrated in the precipitate though the partition between the raffinate and the precipitate depended on the crude source and the type of metal [820].

At low severity HDP, the stability of the product was higher than that when the severity was increased, in which case asphaltene stability was reduced [806]. During HDP, deposits of asphaltenes or coke on the catalyst are indistinguishable. Additionally, light paraffins are produced. Hence more refractory asphaltenes may deposit on the catalyst, while the more reactive asphaltenes would convert to lighter compounds and coke. It could be expected that at higher severity, formation of more paraffins and coke, together with a higher proportion of more refractory unreacted asphaltenes, would lead to larger instability. In fact, Christiansen et al. [261] concluded that the instability of the hydrotreated product was mainly caused by changes on the solvency power of the oil. The instability of the hydrotreated product not only affects product specifications, but also gives rise to sedimentation problems [821]. The preservation of the polycondensed aromatic rings of the refractory and of the unreacted asphaltenes favors the stacking mechanism of aggregation, as it has been observed by electron microscopy [822].

HDP of Maya crude oil improved the quality of the oil by decreasing API gravity by about 8°, 78 wt% S, 52.5 wt% nickel and vanadium, and 62 wt% asphaltene content [805,823]. Fractionation and characterization of the hydrotreated asphaltenes indicated that the more refractory asphaltenes toward demetallization were also more aromatic [114].

Subfractionation using supercritical pentane [559] showed characteristics in the obtained narrow cuts that could be associated with their reactivity and provide directions on the processing options [818,824]. Characterization results indicated an asymmetry between SARA components and refractory/problematic compounds. As pointed out earlier, the characteristic features that determined reactivity are common among asphaltenes from different sources. These authors associated processability problems with asphaltene content, rather than with the nature and/or reactivity of asphaltene molecules [824]. Hence removing these problematic bottoms prior to HCK was suggested for enabling processing [580].

Supercritical fractionation was employed for the residues of hydrocracked VGO and of thermally cracked VR [818,825]. Although the technique was proven to be useful in fractionating the residues from the liquid product of the corresponding reacted/converted feed, in both cases the results were interpreted as if the residue was the solely result of nonconversion of refractory compounds and not the possible recombination of reactive formed fragments.

Additionally to the reactions just described, other structural changes of asphaltenes have been reported to occur during HDP. On one side, the aromatic sheets (d_m), the distance between the aliphatic chains and naphthenic sheets (d_c), and the average diameter of the aromatic sheet (L_a) remained substantially unchanged with severity. On the other side, the cluster diameter (L_c) decreased in size as the severity increased (see comparative data in Table 2.2 and illustrative example in Fig. 2.14) [826].

All these findings are another confirmation of the magnitude of the complexity of the BotB compounds and of their behavior and reactivity, both individually and as a dynamic system, as they should be considered.

4.7.3 Oxidation

Reactivity toward oxidation reactions is high [827], although very slow at temperatures lower than 380°C and maximizing in the range of 480–580°C [828]. The kinetic parameters change over reaction time; the order of reaction initially was between 1 and 2, but at longer reaction times the reaction becomes first order [829].

A Ru ion-catalyzed oxidation reaction was the basis of the RICO method developed by Strausz for the structural characterization of asphaltenes. Under the conditions of the RICO method, aromatic carbons selectively convert into CO_2, carboxylic acids and esters groups while leaving aliphatic and naphthenic structures of asphaltenes virtually unaffected [500,515,518,520,522–525]. The application of the RICO method with characterization purposes indirectly has confirmed the reactivity of asphaltenes toward oxidation under the conditions employed [359,411,493,830–833].

Sato et al. oxidized asphalt at temperatures in the range 340–400°C and under supercritical water, at water densities in the range 0–0.5 g/cm^3, under an argon and air atmosphere. Inert atmospheres favor cracking, while 38 wt% desulfurization was achieved under oxidative conditions [834]. Formic acid under similar conditions converted asphaltenes to lighter products at the lower temperature window since higher temperatures favored coke yield [835]. Oxidation with permanganate converted the CH_2/CH_3 groups to carboxylic acids and other carbonyl functional groups [836].

The interaction of crude oil with iron from the pipelines was studied by characterizing the asphaltene deposits on the tube walls. Magnetic and nonmagnetic iron phases were identified as oxidized magnetite (magnetic) and iron oxyhydroxides (nonmagnetic). The crude oil also showed an increase in the content of C=C and C–O bonds, and an increase of the thermal stability, indicating possible iron-catalyzed reactions [837].

The adsorptive capabilities of nanoparticles of Co_3O_4, Fe_3O_4, and NiO have been found to parallel their catalytic activity for oxidation reactions. In fact, the adsorption affinity toward asphaltenes and the oxidation activity follow the order NiO $>$ Co_3O_4 $>$ Fe_3O_4 [838].

Under photooxidation conditions, the observed changes included an increase in carbonyl, phenolic, sulfoxide, and carboxylic groups, without significant alteration of the hydrocarbon skeletons of polyaromatic and alkyl compounds associated with the asphaltene matrix [839]. Photoirradiation of asphaltenes induced their degradation into gaseous products (hydrogen, methane, ethane, propane, formaldehyde, carbon monoxide, and dioxide) and condensation products, which consisted of highly condensed aromatic structures and oxidation products insoluble in organic solvents [840].

4.7.4 Other Reactions

Mechanochemical treatment led to the identification of the most labile bonds in the molecular structure of asphaltenes as bridges of the types: $-(CH_2)-$, –CO–, –S–, –CS–, and $-COCH_2-$ [841].

Supercritical water in the presence of H_2 and CO_2 promoted the hydrogenation of asphaltene through a reverse water–gas-shift reaction and probably suppressed coke formation [842].

Reaction with alkali metal seemed to activate C–S bonds, probably by generating thiols. This reaction applied as a pretreatment prior to thermal conversion was observed to enhance desulfurization, validating the probable thiols formation hypothesis. The mixture of lithium – liquid ammonia in ethanol induced asphaltene reduction. In subsequent pyrolysis, this added hydrogen served as a means for saturating and stabilizing the free radicals produced during thermolysis, thus increasing the yield of pentane-soluble material [843].

Under conditions for Friedel–Crafts alkylation reactions, the nature of chemical reactive sites present in asphaltenes was studied. Clear indications were given on the aromatic C–H group undergoing alkyl substitution reaction in the presence of strong Lewis acid $AlCl_3$ [836,844] or $ZnCl_2$ [844].

Biodegradation and bioconversion of asphaltenes have been reviewed [845–849]. The interactions between microbes and the high MW components of crude oils include oxidation of aliphatic and aromatic carbon groups, oxidation of naphthenic acids, and oxidation and desulfurization of aromatic and aliphatic sulfur groups [213,850,851].

Dunn and Yen studied asphaltene conversion with ultrasound at RT and pressures. The main occurring reactions were cracking and dehydrogenation. In the absence of a hydrogen source, dehydrogenation reactions were favored over cracking reactions (4:1 ratio), while in the presence of a hydrogen source selectivity shifted toward cracking (about 1:1) [852].

Chemolysis reactions (Ni_2B reduction, BBr_3 hydrolysis, and RICO) lead to bond cleavage, in the order of C–S > C–O > C_{arom}–C [327,522,523]. Reactivity has to be examined throughout a wider scope, for instance, Athabasca oil sand asphaltenes, which degraded relatively easily to lower MW species. Under reductive conditions, asphaltenes did not exhibit any MW changes on reacting with amides in liquid ammonia [504].

Preliminary results for the molecular changes induced on asphaltenes by the interaction with an H-plasma have been reported. Observations included the presence of ionic fragments C^+, CH^+, C_2^+, NH^+, and V^+, and disaggregation of the original asphaltene aggregates [853].

In general and summarizing, the low reactivity of asphaltenes has been related to the extent of intrinsic polarity [349] and to the high aromaticity of the molecular system.

4.8 Models and Theoretical Studies

The intent of this section is not to discuss the validity of any given model, nor to evaluate them or to establish any judgment on any approach or result. Instead and similarly to our view on the structural models, the benefit of the doubt will be given to any approach of the theoretical, thermodynamic, or phenomenological models. The main purpose of this book is to establish bridges between science and technology; in consequence, our attention is directed to the findings and predictions made by the models. Hopefully, new ideas could arise for the abatement of the issues originated by this family of compounds.

Typically, the objective of any modeling approach has been the understanding of properties and behaviors of asphaltenes within a broad range of environments and conditions. Probably, aggregation and precipitation/deposition might be the most studied phenomena and the increasing number of proposed models and modeling techniques still continues. Nevertheless, the interests and objectives of modeling differ depending of the oil sector involved, namely, upstream or downstream. Larger efforts have been pursued in the production side, with objectives such as:

- Instability and interactions, solubility, association, aggregation, precipitation, deposition;
- Thermodynamic models: prediction of phase equilibria, properties of compounds like, e.g., critical (or pseudocritical) properties;
- Correlations: for the estimation of situations on real process/system conditions.

Of course, some of the findings and estimations will be valid for downstream, and certainly any derived knowledge will be valuable.

In modeling, two main interpretations on the physical state of asphaltenes in crude oil have been defined; basically:

1. Asphaltenes are a solubility class of compounds that are dissolved in the oil (solution model). Accordingly, precipitation occurs upon passing a threshold defined by solute/solvent ratio [165,220]; or
2. According to the colloidal model (e.g., Refs. [5,319,334,854]), asphaltenes are insoluble colloidal solids dispersed in the crude oil by peptizing/surfactant effects of other molecules (e.g., resins). Precipitation will occur when partition of the peptizing/surfactant molecules favors the supernatant media over the asphaltene colloid.

In a lesser extent, asphaltenes have been thought of as solids for describing precipitation just as a general solid−liquid phase split [855−860]. Theoretical, thermodynamic, empirical, and phenomenological modeling studies have been carried out for these models of the physical state of asphaltenes. However, it is fair to assume that the high associating propensity of asphaltenes with each other, their surface activity, and their polydispersity make them partly in solution and partly in suspension [861].

A variety of approaches using different theories and methodologies have been used to explain the properties, phenomenology, and phase behavior from the aforementioned models of asphaltene physical state. These include quantum chemistry theory and methods: first principle, semiempirical, and density functional theory [492,862−869];

stochastic, MM, and MD methods [314,330,331,350,589,870–875]; thermodynamics and solution theories: F–H-type [144,184,202,221,855,876–879]; regular solution theory [143–146,165,177,184,224,880–889]; integral equation theory [890]; perturbed chain and other statistical associating fluid theory (SAFT) [891–898]; standard cubic, Peng–Robinson, noncubic, and other EOSs [126,891,899–909]; and polymer solutions [910]. Reviews on modeling have been also published, e.g., [122,897,911,912].

4.8.1 Molecular and Association Modeling

The modeling of association and aggregation has been pursued by different approaches, some of them using a priori proposed molecular models. Semiempirical calculations were used to assess the length of the alkyl chains in asphaltenes and results were verified by their correspondence with experimental DRIFT data [913]. Quantum-chemical calculations of MM-optimized molecular and radical fragments (naphthenoaromatic type) showed a nonplanar, convex structure. The structure became planar (plate shape) when aromatization corresponds to a high degree of condensed rings (continental model) [914]. These calculations seem to agree well with AFM results of the asphaltenes' supramolecular structure [640] and with the associative stacking structure proposed by Unger, for the association of molecular fragments into complex structural units to form the observed particles of petroleum asphaltenes [915]. A Monte Carlo algorithm was used to build molecular structures according to an average MW of 750 Da; elemental analysis and chemical features were deduced from NMR results. Molecular representations could be obtained that fitted the continental and the archipelago type of model structures [916]. The generated optimized structures were used for modeling association with MD calculations. MD predicted asphaltene association for both types of structural models [314,917–920].

Models explaining the associating behavior of asphaltene molecules may be based on Wertheim's association theory (as it has been used in some of the SAFT-based modeling methods) and/or on the energetics of interactions. An example is an analytical function that contains a first term associated with bond interaction energies and a second term associated with nonbonding interactions (coulombic and van der Waals forces) [591]. As discussed earlier, the perturbation of the balance of forces that keep the asphaltene in solution or dispersed and those that drive their association and aggregation shift the equilibrium to either side (Fig. 2.12). A theoretical model (based on MM and MD calculations) explaining the resins' peptizing behavior [radial distribution functions and associated potentials of mean force (PMF)] has been proposed [874,890]. According to this model, the repulsive barriers characteristic of aggregate systems showed by PMF demonstrated: (1) a strong aggregation effect in precipitant media (e.g., alkanes); (2) formation of stable asphaltene cores peptized by resins in intermediate precipitants (e.g., aromatics); and (3) formation of typical solutions in highly dispersive media (e.g., polar solvents) [874].

Simple models and regular solution theory have been used to explain asphaltene association and its consequences. Estimation of the flocculation onset using Hildebrand-type SPs and assuming a dispersed-colloid model predicted values in good agreement with those experimentally determined, only if solvation of particles and immobilization of dispersion medium components were considered. The SP distribution (the only polydispersity considered) was assumed to be Gaussian and the flocculation points were based on Scatchard–Hildebrand solubility theory within a continuous thermodynamics framework [921,922]. The estimated flocculation point agreed well with experimental values obtained for systems of crude oil + solvent + precipitant, when the precipitant was an *n*-paraffin. The colloidal disperse phase was found to exhibit a similar behavior to that of polymer solutions [883,923]. This simple model was improved by considering a modified F–H interaction parameter to determine a criterion for metastability. The bitumen was composed of a distribution of components defined by their Hildebrand SPs, though for polar ASs the SPs were rather of the Hansen type [184]. The flocculation onset was estimated for various precipitating agents [182]. Such a model was tested for predicting: (1) a stabilizing effect by a dispersion interaction with amphiphilic compounds through H-bonds of about $10[\mathrm{MJ\ m^{-3}}]^{0.5}$ [924]; (2) coke formation during the cracking of residues [925,926]; and (3) the cloud and pour point [927]. A spherical region was used to describe the solubility space and the use of Hansen-type SPs indicated that the precipitation mechanisms were different for polar and nonpolar solvents [183].

Regarding the efficacy of flocculation inhibitors, a Monte Carlo calculation was used for trying to understand the reasons for failures in the inhibitory action. The simulation results showed that the inhibitor molecules with more polar head were effective in nonpolar solvents, by exhibiting high adsorption on the asphaltene surface. Meanwhile, at high concentration the self-association of the inhibitors caused a decline in their adsorption preference on the asphaltene surfaces [928].

Experimental data of metal distribution among the SARA fractions under solvent deasphalting conditions within a range of temperatures and pressures was the basis for the modeling of the association between metal-containing molecules and molecules of resins or asphaltenes. Thermodynamic equilibria of proposed reactions that may occur during precipitation of asphaltenes were employed [929].

4.8.2 Aggregation, Micellization, Precipitation, and Deposition

The polymorphism and the possible multiphasic state of asphaltenes calls for a coordinated use of multiple experimental techniques and theoretical approaches for the study of phase behavior and precipitation. Models must capture the relevant physics and chemistry at the molecular, nano-, and macrolength scales to be predictive [86].

Either theoretical (e.g., [184]) or empirical (e.g., [122,930,931]) models have been proposed for the understanding and prediction of precipitation and deposition of both live and dead oils. For the former theoretical model, polarity and the formation of H-bonds play an important role in solvation that can be used to model the overlapping of thermodynamic equilibria and the kinetics of coagulation and aggregation in the process of precipitation [184].

One of the theoretical approaches has been the construction of molecular structures that represent the characterization results and use these as input into a simulation algorithm. An example of this approach was given by Yang et al. [662,663] who represented the molecular structures of the interfacial active asphaltenes and the remaining asphaltenes as the structures shown in Fig. 2.35A and B, respectively. As mentioned earlier, characterization results indicated that the former only comprised less than 2% of the total [662]. MD modeling confirmed the interfacial activity of the structure shown in Fig. 2.35A and the lack of activity for the structure in Fig. 2.35B [663]. These structures not only differ in composition, their functionality is also different.

Meanwhile, the basis for precipitation is considered to be the association thermodynamic that leads to micellization [930]. The micellization process as a prelude to aggregation and precipitation was modeled within a thermodynamic framework based on direct minimization of the Gibbs free energy. The predictions made from the proposed model reproduced well the experimental observations [932].

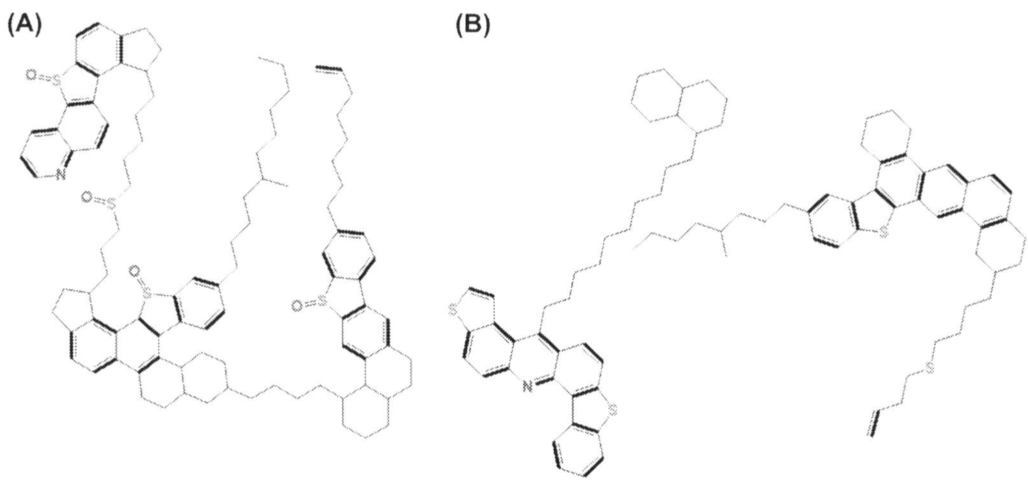

Figure 2.35
Molecular representations of (A) interfacially active asphaltene and (B) nonactive asphaltene. *Reproduced from Yang F, Tchoukov P, Dettman HD, Teklebrhan RB, Liu L, Dabros T, et al. Asphaltene subfractions responsible for stabilizing water-in-crude oil emulsions: Part 2. Molecular representations and molecular dynamic simulations. Energy Fuels 2015;29(8):4783–94, with permission from ACS Publications.*

The role and action mechanism of precipitation inhibitors have been predicted using a thermodynamic model of micellization. Stabilization of the asphaltene micelles in crude was thought to be because of strong interactions between an asphaltene and an amphiphiles inhibitor molecule. The adsorption enthalpy was suggested as the most important criterion of inhibiting action. Meanwhile, the lower effectiveness of resins was justified as being caused by a weaker interaction between resin and asphaltene molecules. Thus the adsorption energy is proposed to dominate micellar formation, while other parameters would have little effect on micellar stability [905].

A theoretical study on *n*-alkane titration of asphaltene solutions was performed using the statistical association fluid theory for potentials of variable range EOS in the framework of the McMillan–Mayer theory. The phase equilibria of asphaltene precipitation under composition changes (titration) could be predicted at ambient pressure and temperature [280].

Two basic mechanisms of organic deposition considered asphaltenes either dissolved (monomers) or suspended (polymers) in the oil, and are based on statistical mechanics of particles. These models are the continuous thermodynamic (CT) model and the steric colloidal (SC) model. According to the CT model, the dissolved heavy organics may or may not form a solid phase depending on the thermodynamic conditions of temperature, pressure, and composition. Meanwhile, in the SC model the heavy organic solid colloidal particles suspended in the oil are stabilized by the adsorbed resins on their surface. Utilization of kinetic theory of aggregation enables one to develop a fractal aggregation (FA) model, which combines the ideas of the two proposed CT and SC models. The FA model is capable of describing several situations, such as the phenomena of organic deposition, growing mechanism of heavy organic aggregates, the geometrical aspects of aggregates, the size distributions of precipitated organics, and the solubility of heavy organics in an oil under the influence of miscible solvents [933]. An empirical model served to validate a colloidal approach in formulating a general scaling law for representing colloid deposition [934].

Prediction of precipitation and deposition has been attempted by introducing a generalized corresponding states principle. The effects of variations of pressure, injection of miscible fluids, and temperature were considered in the model [935]. In this regard, MD results obtained with an asphaltene molecular model were implemented into an EOS-based model for prediction of precipitation and deposition [218]. Flocculation, precipitation, and deposition have been modeled using an EOS, improved by allowing the precipitated asphaltene to react and convert into larger particles. Reaction rates could be modulated to assign a categorization as fully irreversible, fully reversible, or partially reversible. This model was validated through a wide range of pressure, temperature, and composition conditions [936]. A model using Hildebrand SPs of dead oil tried to estimate the SPs of the live oil under reservoir conditions by incorporating the dissolved gas composition along with the PVT properties of the live oil. Then, EOS was used to predict instability of live oils [937].

Extension of the PC-SAFT EOS model [767] to mixtures containing dissolved gases, such as methane, CO_2, and ethane, incorporated a derivation of a new mixing rule for SPs of mixtures containing liquids and dissolved gases [891]. The model accurately predicted the crude oil bubble point and density as well as asphaltene precipitation conditions [892]. A named "one-third" rule was defined for estimating crude oil properties such as SP, viscosity, thermal conductivity, diffusivity, and heat capacity as a function of mass density. The one-third rule is based on the observation that the molar refractivity is approximately proportional to the MW of a hydrocarbon molecule. The proportionality constant is approximately equal to one-third for hydrocarbons and crude oil systems [938]. The predicted properties are in good agreement with experimental results. The one-third rule and the evaluated properties were included as part of the method for assessment of asphaltene instability trend and was used to predict the asphaltene precipitation onset at reservoir conditions [939]. The thermodynamic model was then proven to predict phase behavior of the polydispersed asphaltene systems in a wide range of temperatures, pressures, and compositions [940,941]. PC-SAFT EOS showed superior prediction capabilities than cubic-plus-association [942].

Deposition has been approached by using statistical mechanics of polydispersed polymer solutions together with kinetic theory of aggregation. The developed model was able to predict both reversible and irreversible heavy organic depositions. The possibility of extrapolating data from the model to those of field conditions was claimed by the authors [870]. However, the lack of real data precludes the confirmation of such statement, for now.

Deposition in the oil well was approached by modeling adsorption of asphaltenes on mineral materials by using a Polanyi's modified theory. Additionally, the Dubinin–Astakhov model was used to correlate the predicted adsorption characteristic curves [943]. The effect of adsorption on deposition was empirically modeled by fitting the experimental data to a Henry adsorption isotherm type. Initially, before asphaltene precipitation onset, the adsorption process was controlled by adsorption kinetics for a crude oil–heptol system. Then, the adsorption process became governed by diffusion and convective transfer. Finally, after asphaltene precipitation, transport of asphaltene was suggested to follow a multistep process that included precipitation, aggregation, diffusion, advection, and deposition [944,945].

Empirical phenomenological and mathematical modeling has also been applied [907,930,946–948] for the prediction of asphaltene behavior. Although a phenomenological model can predict behavior, the polydispersity of asphaltenes limits the applicability of these models. For instance, the model of Sabbagh et al. predicted the onset and precipitation amount over a broad range of compositions (bitumen with asphaltene content between about 5 and about 22%), temperatures from 0 to 100°C, and pressures up to 7 MPa successfully, but failed in more diluted systems [907].

A particular case of phenomenological models is the compatibility model. Since compatibility is one of the most impacting features of asphaltenes on the oil industry operations, it will be considered in an individual section. These models are based either on the thermodynamics or on the phenomenology associated with solubility and aggregation. Theoretical models have been used for the evaluation of SPs (see Section 4.2). One of these models built structural molecular models and estimated the SPs considering the interactions that define solubility [949], based on the Hansen sphere method[4] (Hansen parameters [181]) or Hildebrand parameters [175,176]. Other researchers calculated and compared single-component (Hildebrand SPs [950]) and three-component SPs (Hansen type) [144].

In refining, the ultimate objective of most of the SPs is to design a compatibility model to minimize the need for characterization data (see, for instance, Refs. [195,951]). These SPs could explain the observed solubility changes in nonpolar and slightly polar organic solvents (e.g., toluene [200]), but failed to predict behavior in polar solvents. Hansen SPs have also been used to estimate a sort of solubility space, which indicates that bitumen stability depends on the mutual solubility of all its components, somehow contradictory to a bitumen model consisting of a dispersion of asphaltene micelles [949,952]. SPs have also been estimated for pure liquids [177] and to discriminate among different *n*-paraffins [953]. Computation of RI-based SPs (Eq. 2.16) and its application to field conditions through modeling was considered by Verdier for predicting precipitation and deposition [954]. MD calculations showed that the asphaltene SPs decreased with aggregation increases. Thus the aggregation of the asphaltenes helps the solubilization of the lower SP fraction in solvents with low SPs. On the other hand, aggregation could generate larger particles in these solvents for which the buoyancy forces overcome the Brownian forces and, in consequence, the aggregates would settle out from the solution [955].

A mechanical model was based on particle population theories. A population balance is achieved when aggregation and fragmentation reaches an equilibrium. In this model for application under dynamic flow conditions, aggregation rate depends upon the rate at which collisions occur and on the probability of cohesion of particles during a collision. Meanwhile, fragmentation was mainly caused by shear forces. Furthermore, fragmentation and restructuring processes produce smaller and more compact aggregates after a sufficiently

[4] The basis of the Hansen sphere method is that the total energy of vaporization of a liquid consists of several individual parts, arising from (atomic) dispersion forces, (molecular) permanent dipole–permanent dipole forces, and (molecular) H-bonding (electron exchange). Thus the Hansen SP is formed by the contribution of the dispersive, polar, and H-bonding forces.

long time within the same fractal dimension [956–958]. A phenomenological model, based on kinetic results, has also been proposed for the estimation of particle growth and it was claimed to fit well the particle size range for asphaltene aggregates in heptol [959].

4.8.3 Properties

Prediction of properties, particularly rheological properties under conditions of asphaltene association and agglomeration, has been another modeling target. Combined methodologies have been employed involving theoretical calculations, empirical calculations, and phenomenological models, as well as comparisons and verification through experimental studies [112,127,182,184,883,921,924,927,960–962]. The asphaltene nanoaggregates were found to be nonspherical disk-shaped particles with low thickness-to-diameter ratio, when viscosity data for asphaltene suspensions were reinterpreted in terms of the Krieger–Dougherty model. This ratio depended on the nature of the asphaltene/oil system and increased with temperature increase [963].

The viscosity behavior of reconstructed VRs was modeled using two typical viscosity equations for colloidal systems, namely, the generalized Pal–Rhodes equation and the Mooney equation. Eleven reconstituted heavy oil samples with different asphaltene contents at six different constant temperatures were considered. The asphaltene particles were found to be significantly solvated in the medium and to be nonspherical as indicated by the solvation constant and by the shape factor derived from the Pal–Rhodes equation. The changes in packing volume fraction with temperature derived from the Mooney equation were explained as being caused by resins desorption [964].

A thermodynamic model considered that most of the asphaltene material was located inside the aggregates, rather than in its monomeric form. A common bulk-phase EOS could be employed for estimation of asphaltene fugacity. Incorporation of the resin molecules into the model and the definition of the resin structure and mode of interaction with asphaltenes are examples of the overcome challenges. The shape of resins became a parameter to be considered in the modeling. The model predicted a deformation of the resin shell of the asphaltene aggregates and played an important role in defining particle size and particle size distribution [965].

Viscosity has been evaluate with empirical equations based on compositional data and other properties such as boiling point, API gravity, and MW [966], on API and temperature [967] and on a mixing rule [968].

A methodology following a structure–function approach was aimed to predict macroscopic properties of asphaltenic fluids from nanoscale aggregate description using parameters derived from scattering measurements (SAXS and SANS). A single parameter, the mass fractal dimension D_f of aggregates, was enough for estimations of macroscopic properties (e.g., viscosity) and prediction of certain behaviors. D_f showed interdependence with the gyration radius, weight-average MW, and the second virial constant. Prediction of aggregation, interfacial activity, and adsorption as well as stability of emulsions and of hydroconversion effluents matched well with experimental observations [383].

The Langmuir EOS was applied to the modeling of the interfacial tension of asphaltenes at the oil–water interface as a function of interfacial coverage, as well as to the modeling of elasticity versus interfacial tension. An estimation of the size of the polyaromatic cores for Norwegian asphaltenes resulted in a molecular area of 0.32 nm^2, which corresponds to an average size of ~6.2 rings/molecule-core [969].

A review of characterization and modeling published in 2006 concluded that there was still room for development of experimental techniques and methodologies to provide robust data to the models. The complexity of acquiring data under reservoir deposition was pointed out [970]. Probably, the study of behavior and characterization of dead oils under refining process conditions are easier to undertake; nevertheless, results are not going to be less complicated.

5. Impact on the Oil Industry

The poor definition of asphaltenes has been affecting the activities and objectives of the R&D efforts. Determination of physical and chemical properties is so complex that new methodologies and many techniques were and are required. Interpretation of results is similarly complicated; sometimes modeling is used to support such intricate interpretations. Thus long-standing controversies are still in place regarding mainly molecular structure and physical state. However, whatever these molecules are, they are real and their properties and behavior give rise to immeasurable problems in production, separation, transportation, and refining.

The physical state of asphaltenes in crude oil has been proposed, based on two main interpretations. One considers asphaltenes as a class of compounds that are dissolved in the surrounding medium (i.e., the rest of the oil) and precipitate after the oil solubility falls below a certain threshold [220]. The second hypothesis on the state of asphaltenes in oil fluids considers a specific stabilizing effect of resin molecules, i.e., the asphaltenes are considered to be insoluble colloidal solids that are peptized by adsorbed resin molecules on their surface [319]. According to this latter model, the resin fraction favors asphaltene solubility in the crude matrix [971], and the phenomenon of asphaltene separation is likely governed by the partitioning of the resins between the surfaces of the asphaltene colloids and the supernatant solution. It has also been hypothesized that at process temperatures typical of distillation and reaction and based on the melting behavior of asphaltenes, the majority of asphaltenes likely exist either in a molten phase or as a solution [972].

Asphaltenes are large molecules that would aggregate in crude oil over a wide range of concentrations and conditions, precipitate under nonpolar environments, exhibit strong adsorption on and adhesion to a wide range of surfaces, are elastic under tension and porous to smaller compounds, and occlude other compounds present in the oil, which otherwise would remain soluble.

In general, the issues arising from asphaltenes for upstream operations could be distinguished from those downstream in terms of the scale of asphaltene structure. The meso-/macroscale impacts more on upstream operations, while the micro-/mesoscale affects more downstream. In these terms, agglomeration and stability are concerned more with upstream, while molecular interactions, reactivity, and surface activity affect largely refining units, processes, and catalysts.

Speight [30] has summarized the problems caused by asphaltenes in the oil industry as:

- During recovery and transportation operations: well bore plugging and pipeline deposition;
- During field storage and pipeline transportation: emulsions formation with contaminating water;
- During crude oil and product storage: sedimentation and plugging induced by oxidation;
- Upon heating: thermal degradation makes asphaltenes more aromatic (loss of aliphatic chains) and less soluble leading to sedimentation and coke formation.

The insolubility of asphaltenes in light paraffin liquids as well as in other fluids (e.g., CO_2) is a source of problems during crude oil production operations. Asphaltene precipitation can cause formation damage and well bore plugging, requiring expensive treatment and cleanup procedures. A tool for predicting asphaltene deposition (ADEPT) has been developed. The occurrence, magnitude, and profile of asphaltene deposition in a well bore could be calculated. The simulator consists of a thermodynamic module and a deposition module. The thermodynamic module uses a PC-SAFT EOS (see discussion in Section 4.8) to describe the phase behavior of oil [973,974].

More details of the problems and reported situations caused by asphaltenes will be given in the next paragraphs and sections. In oil recovery processes, the flowability of the medium is affected by the presence of asphaltenes. Observations on asphaltene gradients in crude oil, heavy oil gradients, viscosity gradients, tar mat formation, bitumen deposition, and asphaltene flow assurance are typical. Perturbations on asphaltene stability causes phase separation, which might plug the oil-bearing rock formation near a well. Similar problems can be initiated by asphaltene agglomeration, which may accumulate and plug the porous structure of the reservoir matrix, well bores, and flow lines. Asphaltenes also aggregate at oil—water interfaces and stabilize water-in-oil emulsions or at oil—solid interfaces where they can alter surface-wetting properties [96]. The deposition problem is typically addressed by defining at lab scale the conditions that determined the asphaltene deposition envelope and either considering injection of flocculation inhibitor [975] or planning the frequency of cleaning. Different methodologies for the determination of the onset of the asphaltene precipitation at different conditions have been proposed; see, for instance, Ref. [976] and reviews of laboratory techniques (gravimetry, acoustic resonance, light scattering, high pressure microscopy, particle size analysis, and filtration) have also been published [977,978]. The selection of the suitable inhibitor (or combination of inhibitors), the dosage, the frequency, and the injection point require comprehensive testing and evaluation [975].

Asphaltene precipitation from live crude oils has been observed to occur during pressure reduction [229]. Enhanced oil recovery (EOR) is a well-known process in the petroleum industry for increasing oil production in declining oil wells by stimulation. In a miscible displacement process, carbon dioxide and natural gas are considered to be two of the most effective agents for such stimulation technique. However, injection of any of these two agents (CO_2 or natural gas) into an oil reservoir might cause asphaltene deposition, which changes the flow behavior and the equilibrium properties of the fluids. This deposits formation is a function of (1) the composition of the crude oil, (2) the displacement agents, and (3) the reservoir conditions (pressure and temperature). Consequently, plugging (of the formation, wellbores, and production facilities) and isolation of oil from the flowing fluid change the wettability and permeability properties of the reservoir, and eventually reduce the efficiency of the EOR process [202,203,979–983].

Recent results and comparisons within several oil reservoirs indicated that asphaltene may accumulate by gravitation and create a sort of tar mat on the formation rock. These mats are not equilibrated and do not match any simple model. Possibly, this carbonaceous coat on rock surfaces seals off porosity, thereby preventing equilibration in the mat [358]. The asphaltenes concentrated in these mats exhibit similar chemical composition and MW to those in the crude oil [984].

The deposits found in production and transportation lines are complex materials, consisting of organic and inorganic compounds. The organic part is typically highly asphaltenic, though waxes and resins may be present as well. The solubility of these deposits is as complex as the nature of their constitution. It has been found that the aromatic content of a dissolving media was key for increasing solvent power toward highly insoluble components. Similarly, pressure, which is one of the major players in deposition, was found to be a minor driver for the solubilization of solid deposits. Contrarily, temperature was assessed to exert the major role on solubilization. On the other hand, chemical additives effective for avoiding asphaltene precipitation failed to show any improvements on redissolution of the solids already formed [985]. These chemical additives may inhibit asphaltene deposition; however, the required amount per unit of asphaltene present in the oil is typically higher, for field operations than that estimated from laboratory tests. The reason given was that the asphaltene content at the production field was much smaller than that tested in the oils samples in the lab [263].

The formation of asphaltene-based sludge during transportation is another problem for discharging operations. To comply with specifications for pipeline operation, several processes have been developed for onsite modifications of viscosity, which are needed prior to any attempts for pumping and transporting of virgin heavy crude oils from the oil fields to the refining infrastructure [986].

Some of the conventional solutions for production and transportation problems (drag-reducing additives, viscosity reduction, asphaltene dispersants, etc.) are the source of operational problems during refining. In refinery processes, the thermally induced changes in crude oil composition can result in phase separation of the asphaltenes (considered the main coke precursors). Consequently, asphaltenes are considered the main causes of fouling in heated equipment, i.e., furnace tubes, heat exchangers, etc. Furthermore, their recalcitrant behavior and propensity to form coke or coke precursors are the basis for catalyst deactivation during residue processing. Thus the asphaltene-induced or caused catalyst deactivation occurs either by physicochemical interactions, simply fouling, and/or by coke deposition. While chemical interactions might result in irreversible deactivation, most fouling and coke making are reversible. Last but not least is the phase separation caused during blending of high-asphaltene-containing crude oil with another more paraffinic or, in general, with a crude oil with lower solvent power. In this case, a compatibility issue may appear to make operations run into difficulties. A more detailed description and discussion on the impact and troubles created by asphaltenes during refining follows.

5.1 Compatibility

If heavier, predominantly aromatic oil-containing asphaltenes is blended into lighter, predominantly aliphatic crude oil containing fewer asphaltenes, some of the asphaltenes can aggregate into micron scale or larger structures because of a reduction in the effective solvent quality of the resulting blend. If this occurs, even over a small range of mixing compositions, the blend is referred as "incompatible"; otherwise it is "compatible."

Blending and dilution are considered means for viscosity reduction with pipeline transportation purposes of heavy oils and bitumen. Then, compatibility might become an issue that may result in asphaltene precipitation, fouling, deposition, accumulation, and even plugging in the pipelines. Thus the terms "compatible" and "incompatible" refer to blends or mixtures of different crude oils; however, a given crude oil under physical or chemical treatments may undergo changes that affect "self-compatibility" or "self-incompatibility." Soluble Solutions, Inc. have identified more than 20 self-incompatible crude oils (containing insoluble asphaltenes) and more than 400 pairs of incompatible crude oils [987]. Compatibility then becomes one of the main concerns when incorporating heavy crude oils into the refinery diet. It requires careful planning and scheduling, besides determination of compatible–incompatible ranges.

The colloidal model was initially used to describe compatibility and incompatibility as the phase changes from sol to gel. According to the colloidal model, asphaltenes are micelle dispersions in the viscous oils. The micelle consists of an asphaltene core surrounded by polar, aromatic molecules. Compatibility is then determined by the degree to which the micelles form extended gel structures. For compatible cases, the dispersed materials are well peptized by the solvent in view of: (1) their small size or amount, (2) the limited or inexistence of strong associations, and/or (3) an effective solvation. In an incompatible system, associations of dispersed materials progress presumably for the lack of peptizing efficiency of the solvent. Rogel et al. measured the surface tension of several asphaltene fractions in nitrobenzene, THF, cyclohexane, and toluene and based on the colloidal model found a linear relationship between CMC and SPs of the solvents for all the solvents. An empirical correlation was derived and suggested that higher compatibility between solvent and asphaltene fractions led to significantly large CMCs in the same solvent [988]. Now that the CMC concept has been overturned, this correlation becomes invalidated.

Despite all the criticisms of the colloidal model, it has supported the development of compatibility models and of automated equipment for the determination of compatibility ranges. One of these is the Heithaus method, which is based on a titration with heptane of toluene-diluted asphaltene solutions (see Section 4.2). The method is described by a set of equations (Eqs. 2.2–2.6), from which three parameters are used to define the compatibility behavior of a system. High values of SP_A, SP_M, and C_R correspond to peptizable asphaltenes, good solvency power of maltenes, and a compatible system, respectively. A stable system may be composed of asphaltenes that are not readily peptizable, but which are dispersed in maltenes that have good solvent characteristics, or the reverse [192].

Traditional methods of detecting blend incompatibility range from spot tests on filter paper to direct optical microscopy observations of the aggregates. For these methods, the high absorbance of the oil matrix obliges to redisperse the asphaltene sample at very dilute concentrations in nonabsorbing solvents [562]. Optical spectroscopy has been used also for detecting incompatibility in blends [989], but this method tells little about the relative concentrations and structures of asphaltenes in nanoparticles and aggregates. SANS helps in examining the recovered asphaltenes [990]. Original crude oil mixtures can be examined directly by SANS to probe asphaltene aggregation. Significant neutron scattering length density difference between the hydrogen-poor asphaltenes and the surrounding oil, together with the small SANS length scales, makes it ideally suited for these purposes and thus average size, nanoscale particles concentration, and volume fraction of microscale aggregates can be assessed simultaneously [309]. Three different onset indicators have been identified by analyzing both compatible and incompatible crude oil blends. The most obvious indicator of incompatibility is the presence of a large surface-scattering intensity from asphaltene aggregates at low wave number. A second indicator is the relative reduction in the strength of the scattering from micelle-like nanoparticles that have been incorporated into the aggregates. A third indicator is a reduction in the effective size of the nanoparticles that remain in solution after the aggregates have formed [309,990]. It was also observed that larger asphaltene nanoparticles are more susceptible to aggregation than smaller asphaltene nanoparticles. The systematic decrease in asphaltene nanoparticles size that remain in the mixtures after aggregation is caused by the reduction in the effective solvent quality, i.e., the surrounding oil becomes more aliphatic. It seems then that there exist smaller nanoparticles (asphaltenes or maltenes), which do not aggregate even in an incompatible regime. These smaller asphaltene nanoparticles do not interact with other asphaltenes and remain dispersed, even in a more paraffinic environment.

Furthermore, not only can blending lead to incompatibility, but during handling and processing, and if under certain conditions the chemical bitumen environment is changed drastically, the bitumen might become unstable. Then, a risk for precipitation of the least soluble component is created. Examples are aging, oxidation, visbreaking, or fluxing bitumens. If incompatibility-induced precipitation occurs during processing at temperatures in the range of 350°C, coke flakes can be formed from the asphaltenes [951]. In situations of paraffin-induced incompatibility, the paraffin–asphaltene formed complexes are easier to redissolve than the original asphaltenes are [991]. Incompatibility is also observed among intermediate streams, e.g., VR is incompatible with catalytic cracking bottom (CCB) oil when the blending ratio of CCB oil is increased [992].

Most integrated oil companies have developed compatibility models for predicting compatibility ranges. These models differ mainly in the definition and determination of SPs (see Section 4.2). For instance, in the Exxon compatibility model [195], the SP of a mixture is the volumetric average SP. The physical definition of the oil residue for this model was shown in Fig. 2.12. This compatibility model considers two key parameters, namely, the insolubility number I_{N} (Eq. 2.28) and the solubility blending number S_{BN} (Eq. 2.29); in these equations, the subscripts are f for flocculation, Hep for heptane, and Tol for toluene, besides the oil itself. These parameters measure the asphaltene insolubility I_{N} and the solvent power of the oil for dissolving asphaltenes S_{BN}. Therefore compatibility[5] is determined based on $S_{\mathrm{BN}} > I_{\mathrm{N}}$. However, occasionally the presence of resins may result in compatibility of crude oil blends predicted to be incompatible [987].

$$I_{\mathrm{N}} = 100\frac{(\mathrm{SP}_f - \mathrm{SP}_{Hep})}{(\mathrm{SP}_{Tol} - \mathrm{SP}_{Hep})} \tag{2.29}$$

$$S_{\mathrm{BN}} = 100\frac{(\mathrm{SP}_{oil} - \mathrm{SP}_{Hep})}{(\mathrm{SP}_{Tol} - \mathrm{SP}_{Hep})} \tag{2.30}$$

[5] A blending compatibility calculator is available from Crude Monitor on the web (http://www.crudemonitor.ca/tools/compatibility_calculator/compatcalc.php).

A test for the determination of I_N and S_{BN} consists in the evaluation of two other parameters: the heptane dilution HD and the toluene equivalence TE. The former is the maximum volume (in mL) of n-heptane that can be blended with 5 mL of testing oil without precipitating asphaltenes; the latter is the minimum % toluene required in 10 mL of heptol (the test liquid) for keeping the asphaltenes soluble from 2 g of oil [993,994]. Then, I_N and S_{BN} can be calculated using Eqs. (2.31) and (2.32), respectively [995].

$$I_N = \frac{\text{TE}}{\left(1 - \frac{V_H}{25\rho}\right)} \tag{2.31}$$

$$S_{BN} = I_n\left(1 + \frac{V_H}{5}\right) \tag{2.32}$$

The slower kinetics of redissolution requires that asphaltenes never precipitate during blending [195,951,987,996]. Hence a convenient mixing order has to be established when blending two crude oils with predicted range of incompatibility. In this regard, blending should be made such that the blend compositions result in S_{BN} decreases [997].

A comparison of the predicted compatibility by CII and I_N/S_{BN} among five considered crude oils indicated that the latter was more accurate in predicting the incompatibility range in blends between pairs of crude oils [997]. GE has also announced the development of indexes based on asphaltene stability for the assessment of crude oil compatibility, though no details were given on the methodology [998].

A criteria for anticipating incompatibility has been defined based on a phase–concentration–temperature diagram (Fig. 2.36) proposed by Evdokimov [87–89]. This

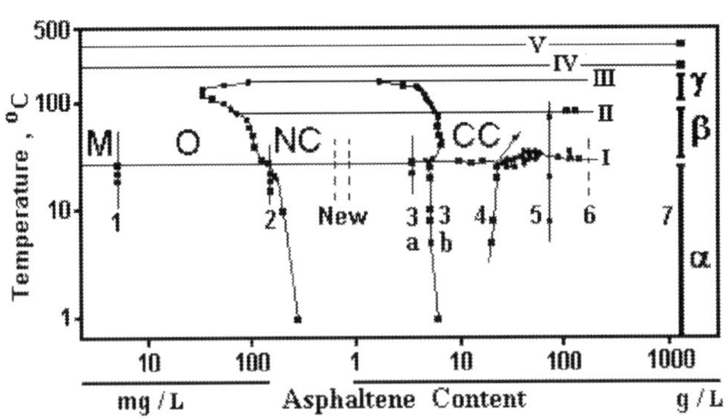

Figure 2.36

Multiple structural phases of asphaltenes. *Reproduced from Evdokimov IN. The importance of asphaltene content in petroleum III — new criteria for prediction of incompatibility in crude oil blends. Pet Sci Technol 2010;28(13):1351–57, with permission from Taylor & Francis.*

type of diagram resulted from the evaluation of more than 400 crude oil samples. The boundaries delimit seven phase regions in terms of asphaltene concentration as:

1. Transition from a solution of monomers (M) to a solution of oligomers (O): ~5–7 mg/L;
2. Emergence of nanocolloids (NC) with particles 2–4 nm in diameter: ~100–150 mg/L;
3. Appearance of colloidal clusters (CC):
   a. 1.7–3.1 g/L; and
   b. 6–8 g/L;
4. Structural transformations forming fractal flocs ≥0.1 μm in size: ~28 g/L;
5. Structural transformations forming small aggregates: ~55–65 g/L;
6. Structural transformations forming aggregates: ~140–160 g/L;
7. A new boundary: 0.4–0.8 g/L (revealed by statistical analysis not in laboratory experiments).

In this diagram, the temperature-defined phase boundaries were: (I) at ~25–35°C; (II) at ~100°C, and (III) at ~180°C, which apparently arise from competitive contributions of various types of intermolecular interactions. According to these criteria, incompatibility would exist when the crude blend reaches asphaltene concentrations (in g/L) close to the boundaries, while compatibility arises for crude blends with asphaltene concentrations far from the boundaries [89].

The negative impact of oil incompatibility has pushed beyond model development toward the fast assessment of incompatibility ranges. Thus several methods have been proposed and automated equipment is currently available commercially [232,995,999–1005]. Therefore careful selection of blending light crude oil and/or diluent must be exercised. Evaluation of a light crude oil and its distillates, such as kerosene or diesel for blending and diluting Middle East heavy crude oil, indicated that kerosene performed well and the addition of 0.5–2 wt% of a solvent mixture of hexanol and toluene to the crude oil and/or the diluent improved compatibility [1006].

5.2 Fouling and Deposition

Fouling is the formation of an unanticipated phase that hinders the processing of a different phase. Typically, the fouling phase is a solid but in gas processing it could be a liquid or it could be an emulsion when distinct liquid phases are being processed. Continuous fouling gives rise to unwanted accumulation of the deposited material that requires periodic shutdown of the involved unit for its cleaning.

In general, both organic and inorganic materials would foul and most of the organic fouling is caused by asphaltenes. One of the reasons is derived from the surface-active nature of asphaltenes, namely, adsorption on solid surfaces, which represents an omnipresent problem throughout the entire value chain of the oil industry. At production level [1007], it may cause reservoir damage,while in transportation, it would foul pipelines and equipment. Fouling will extend to the upgrading and refining units.

Adsorption on metal surfaces may lead to plugging of refining units. The tendency for asphaltenes to adsorb on heated metal surfaces increases as the oil mixture approaches compositions at which asphaltenes precipitate, which is aggravated in the case of nearly incompatible blends [1008−1011]. It has been reported that the loading capacity of metals decreases in the order stainless steel < iron < aluminum [714]. Thus stainless steel should be recommended for asphaltene-containing feeds to minimize plugging. Regarding iron in particular, asphaltenes showed a tendency to form sludge in the presence of Fe(II) and Fe(III) during acid stimulation of the oil well. This sludging tendency is greater for Fe(III) than it is for Fe(II). Reducing agents and/or oxygen scavengers can reduce sludging, but it has been found that cannot stop it [1012].

Additionally to adsorption, asphaltene fouling is intrinsically associated with asphaltene stability. Thus during production, asphaltenes could separate from the oil upon depressurization, but in other cases the addition of lift, injection gas, or solvents could also cause destabilization. Instability can be induced by variations of temperature, pressure, composition, flow regime, and wall and electrokinetic effects. Asphaltene deposition is an enormous problem around the world. The economic implications of this problem are no less serious, considering that well workover cost could be as high as a quarter of a million dollars. In some instances the problem can cause formation damage and the well has to be shut down, luckily temporarily. A survey of field cases and experiences occurred by 1988, covered a broad range of cause−effect situations that may have been repeated in other places in the world in more recent times [1013].

Laboratory testing of four different crude oils indicated that the fouling rates of individual oils decreased with decreasing asphaltene and suspended solids content, decreased CII, and increasing S_{BN}. The effect of blending on fouling rate appeared to be nonlinear and showed a better correlation with CII than it did with S_{BN} [1014]. It has also been postulated that during deposition, the least soluble asphaltenes were the first to precipitate forming cluster-like deposits. These cluster-like deposits collected more asphaltenes from the passing flow, forming island-like deposits. The solubility in the micellar fluids of cluster-like deposits appears lower than that of island-like deposits [201].

Asphaltenes with a high degree of aromatic condensation are difficult to hydrogenate [793]. The difficulties in hydrogenation cause an increase in aromaticity of the asphaltenes in the hydroprocessed product. These hydroprocessed asphaltenes are more unstable and tend to agglomerate and precipitate [169]. The instability of the processed liquid product leads to the formation of sediments and deposits downstream the HDT reactor [1015,1016]. Analysis of sediments and deposits indicated that these materials consist of insoluble continental-type asphaltenes [1017,1018].

The polynuclear aromatic systems present in asphaltenes are highly stable. Under thermal decomposition conditions, these polynuclear aromatic systems would yield substantial amounts of coke and only the limited number of alkyl substituents would convert into lower boiling point products. At temperatures of 430–550°C, for which coke formation was an inevitable part, fouling will occur regardless of asphaltene stability, and far from flocculation onset conditions. Nevertheless, within the whole operating range of process furnaces in refineries, the stability of asphaltene in the blend is very important, not just the onset of precipitation [376]. Coke precursors entrained in the gas phase of the delayed coker fractionator are responsible for coke deposition in the tower. The shape of the mesophase coke particles was attributed to the aromatic nature of the precursors [1019,1020]. Solid precipitation and fouling takes place when heavy oils are subjected to heat, blended with incompatible solvent or cracked. The heat-induced deposition from VRs of Redwater BC, CA Coastal, Boscan, MAXCL, and Vistar crude oils was found to be insignificant at low temperatures (below 100°C) and started above 175°C [1021]. If the blend of crude oils needed to pass through furnace tubes, or any other refinery unit at high temperature, the blend stability should be maximized and/or somehow ensured that blend composition would not fall within an incompatible range.

Thus in refineries with coking units, one of the units most affected with asphaltenes deposition and fouling is probably the coker furnace. A coking stability index (CSI) test, which determines the relative stability of coker furnace feedstocks, has been defined based on asphaltene precipitation onset. The feed is titrated with *n*-heptane and the inflection point of the titration curve, corresponding to the point of asphaltene precipitation, is taken as the CSI.The lower the CSI value, the larger the propensity of asphaltenes to foul would be [1022–1024]. Other indices that could anticipate the fouling capabilities of a crude oil, its fractions, or blends have been investigated and defined. Asphaltene fouling capacity has been associated with the CII, S_{BN}, and/or the R/A ratio of the crude oil or the feedstock [1025]. It has been estimated that hydrocarbon feeds with CII values greater than 2 were potential fouling candidates [1026,1027]. The threshold value for S_{BN} of about 100 has been somehow arbitrarily fixed to indicate that asphaltenes in crude oils with an $S_{BN} < 100$ would have a high tendency to foul [1014]. Regarding the R/A ratio, a value of 2.5 has been found to maximize fouling rate in heat exchangers with a surface temperature of 230°C, while a value above 5.8 zeroed the rate out. This study also found that feeds with high values of CII would not foul if the R/A ratio would be high enough [1028].

Inorganic compounds associated with the BotB fractions also give rise to deposits and fouling materials, forming scales on different process units. Heavy oils or bitumen contain residual salty water and inorganic solids when the hot water extraction or the SAGD processes are employed. Downstream, during solvent extraction or deasphalting processes, a mixed precipitate of those inorganics and the asphaltene fraction will interfere with the normal operation of the units [708]. Iron sulfide product of any unit corrosion is one of the most frequently found material in the deposits. During catalytic hydrotreatment of asphaltene-containing feeds, asphaltene-coated iron sulfide deposited rapidly [1029].

An eight-step method has been proposed for the prevention and mitigation of deposition and for the remediation of its effects on production wells, pipelines, and processing plants: (1) predictive modeling and analysis; (2) dual completion of oil wells; (3) compatibility tests of injection fluids before applications; (4) consideration of the compositional gradient of heavy organics in reservoirs in production scheme design; (5) application of mechanical removal technologies for deposits; (6) application of solvent for dissolution of deposits; (7) hot oil treatment of the in situ deposits; and (8) use of dispersant to stabilize the heavy organics [123]. Common practices upstream include: pressure-drop minimization in the production facility, removal of incompatible materials from crude oil streams, reduction of shear, and limitation of blending of stock liquids. Additionally, chemical treatments comprise the use of several types of additives such as antifoulants, dispersants, and aromatic solvents ([122] and references therein). Although asphaltene dispersants would increase asphaltene stability and inhibit deposition, these will not disaggregate nanoaggregates and will not prevent their fouling [375,376]. A method for the field evaluation of antifouling additives has been developed [1030]. Dispersing additives has also been recommended to inhibit deposition and fouling on catalytic beds (pipes, heat exchangers, etc.). An asphaltene additive (polypropylene oxide-phosphide) proven to inhibit flocculation of small micelles (length <0.004 m) into particles (length $= 0.02-0.03$ m), at low temperatures has been tested under HCK conditions. Addition of 1 wt% of such additive resulted in a 10 percentage point improvement of the heavy residue conversion [793].

In summary, asphaltene fouling is caused directly by adsorption, precipitation, compatibility, or stability, and/or indirectly by thermal decomposition or coking, and by the interaction with inorganic particles. The distinction between organic and inorganic fouling and the tracing of the fouling precursors to their source may provide ideas on alternative mitigation actions. Moreover, during processing, physical and chemical changes might occur to the asphaltenes but also to the surrounding media disturbing the harmonic balance that keeps asphaltenes in solution. These changes are exemplified in Fig. 2.37 by showing the variation in SP as a function of the H/C ratio [29]. Hence decreases in solvency power of the media and increasing insolubility of asphaltene products obviously lead to fouling and deposition.

Economically, fouling increases operating and maintenance costs, as well as it decreases margins. Sources of economic impact include: production loss during frequent shutdowns because of fouling, increases in energy consumption, increases in environmental compliance costs, capital expenditures (including excess surface area, stronger foundations, provisions for extra space, increased transport and installation costs, costs of

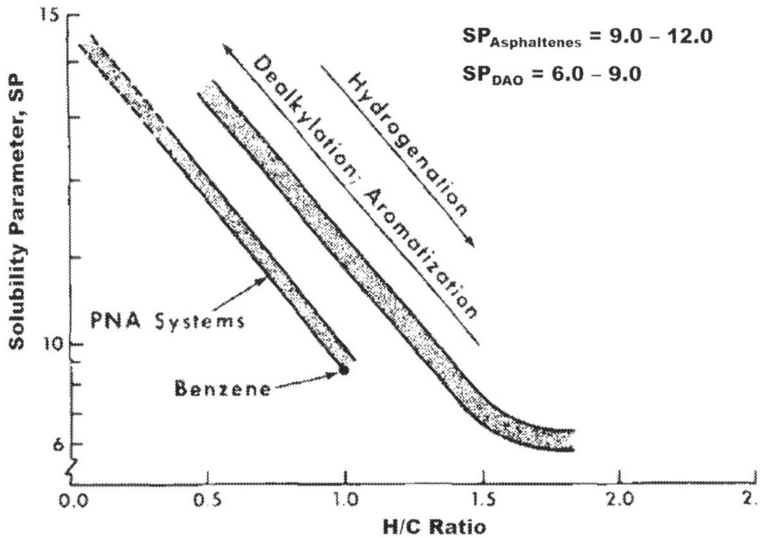

Figure 2.37

Relationship of the solubility parameter and the H/C ratio. *PNA*, Polynuclear aromatic; *SP*, solubility parameter. *Reproduced from Speight JG. Chemical and physical studies of petroleum asphaltenes. In: Yen TF, Chilingarian GV, editors. Asphaltene asphalts. I. Developments in pet. Science. Amsterdam, The Netherlands: Elsevier; 1994. pp. 7–66, with permission from Elsevier.*

antifouling equipment and installation of online cleaning devices and treatment plants, and increased cost of disposal of replacing parts and a larger heat exchanger in this particular case of fouling), increases in staff and other costs for removing fouling deposits, antifouling chemicals and related devices operation, disposal of cleaning chemicals after cleaning, etc. [1031]. The most fouled refinery unit (66% of the total fouling) is the crude distillation unit, and in this unit energy cost increases represent 75% of the total economic impact of fouling [1032]. Inasmuch as coke is an insulator, the increasing layers of coke would demand an increasing heat input, leading to even higher degrees of overheating and, consequently, faster coke formation [1033].

5.3 Catalyst Deactivation

During catalytic processing, unreacted asphaltenes could flocculate, precipitate, or deposit on the surface or pore mouth of the catalyst, fouling or otherwise deactivating the catalyst. On the other hand, thermally converted asphaltenes will produce coke that deactivates the catalyst in the same way that unconverted asphaltenes do (fouling and pore plugging). Moreover, heteroatomic moieties present or trapped by the asphaltenes may irreversibly poison the catalyst-active sites, as is the case of metal compounds (the particular case of metals deactivation will be discussed in more detail in Chapter 3). Thus C-deposition is a reversible type of deactivation; nitrogen and metal deactivation may not be.

Refining catalytic units exposed to the asphaltene-containing streams concern the FCC and the HDT processes. The highly active zeolite catalyst employed in FCC is poisoned by heteroatomic compounds. Cracking activity relies on surface acidity, both Lewis and Bronsted. Thus in FCC, basic N-compounds might neutralize the active acid sites, while carbonaceous deposits on the catalyst during reaction will be burnt off in the regeneration stage. Additionally, and although the catalyst and reacting system are prepared for a certain degree of coke formation, excessive coke laydown would alter the heat balance of the unit and create operational problems. Furthermore, the restricted size of the pore system of the zeolite may be easily plugged with large asphaltene molecules, creating a different activity profile along the riser (the reactor). The VGO or any other blend used as feed for FCC is preferentially hydrotreated prior to being fed to the FCC unit.

HDP catalysts are multifunctional, exhibiting activity for a number of reactions for the removal of contaminating heteroatoms. Major reactions include HDS, HDN, HDM, and hydrodeoxygenation. These reactions occur under high hydrogen pressure and involve the hydrogenolysis of C-heteroatom bonds. Additionally, hydrogenation of double bonds and of aromatics would also take place.

Catalyst deactivation during hydrotreatment of asphaltene-containing feeds results from the accumulation of carbonaceous deposits. The low reactivity and high aromaticity of asphaltenes make them coke precursors. Coke accumulates rapidly during the first hours of reaction, then its concentration remains roughly constant. Nevertheless, a significant loss in catalyst activity results from this rapid accumulation [1034–1037]. The continuous loss in activity over time of stream leaves more unreacted asphaltenes that may remain adsorbed on the catalyst surface obstructing and eventually plugging the pore mouth of the catalyst [1034–1037].

In fixed bed HDP, C-deposits start to build up at the top of the bed accelerating an increase in reactor pressure drop and reducing the life cycle of the run. A catalyst deactivation study showed that fouling rate was directly correlated with asphaltene aromaticity and process temperature [1038]. For treating asphaltene-containing feeds a guard bed and graded catalyst beds are typically used, with the purpose of preventing pressure drops and protecting high active catalysts from severe fouling and poisoning [1039].

At the top of the reactor, the presence of ferric ions dissolved from pipes and instruments as corrosion product causes asphaltene association and clustering, as mentioned earlier [201]. Therefore in corroded units extreme care should be exercised to avoid accelerating pressure-drop problems.

Fouling and deposition of asphaltenes imposed frequent cleaning of units. Fluids and additives that have been proven to work in the upstream sector might work downstream.

Asphaltenes can be dissolved by alkane-based fluids containing amphiphilic compounds. Two types of compounds have been proven to work, namely, alkylbenzene-derived amphiphiles, nonylphenol, and DBSA. Several factors affect the rate of asphaltene dissolution by these micellar fluids:

- Type and concentration of amphiphile: the rate of asphaltene dissolution appears to follow Langmuir—Hinshelwood kinetics with respect to the concentration of amphiphiles, i.e., rate increases steadily at low concentrations and reaches a plateau at higher concentrations, where all asphaltenes may have been dissolved;
- Type of solvent: rate of asphaltene dissolution increases with a decrease in the CN of the alkane in the micellar fluid;
- Temperature: rate of dissolution increases with increasing micellar fluid temperature and follows an Arrhenius temperature dependence with an activation energy between 4 and 7 kcal/mol; and
- Flow rate: higher flow rates of micellar fluids increase rate of dissolution [201].

Solubility of asphaltenes is one of the properties that set the basis of the known asphaltene removal technologies (solvent deasphalting, discussed in Chapter 5). Consequently, the understanding of solubility and identifying the causes of changes in solubility would help in process control and operability. On the other hand, the understanding and identification of causes that induced changes in solubility under other process conditions at which an asphaltene-containing stream needs to be processed would also be needed for the smooth operation of such process. The flocculation rate increased with increased asphaltene concentration and since the flocculation process is reversible [1040], solubility could be used as a way of controlling flocculation.

Asphaltene solubility decreases with the increase in *n*-paraffin proportion of paraffin—aromatic mixtures and is more sensitive to the aromaticity of the solvent. In other words, solubility is improved by decreases in H/C ratio of the medium [83]. Hence during HDT (some unsaturated hydrocarbons will be saturated, H/C ratio increased) some refractory compounds in these subfractions can be expected to become less soluble and more prone to precipitate and foul.

Several review articles on HDP catalyst deactivation have been published through the years and the reader is invited to see more details in Refs. [1041—1046].

5.4 Other Problems

Another consequence of asphaltene surface activity is the formation of water—oil and oil—water emulsions. Emulsion stability (both the stability of the emulsions and that of asphaltene dispersion in the crude oil matrix through the resin molecules) becomes an issue that leads to unit upset and failure to meet product specifications.

The formation of solid-stabilized emulsions at the production site has been discussed in previous sections (e.g., Section 4.5). Furthermore (and also mentioned earlier), asphaltene aggregates tend to form a viscoelastic film, which prevents the coalescence of droplets from water-in-oil emulsions. The oil phase has to be separated or recovered before feeding it to the refining process units. The desalting unit is the most seriously impacted with emulsion formation, emulsion breaking, sedimentation, and films (rag layer stability), in addition to tank farm operation and wastewater management [700]. Moreover, the direct impact of asphaltene-associated issues on the desalter includes: increase in slop oil generation, reduction in throughput, increase of rag layer by the solids or asphaltene-stabilized emulsion, frequent stops for cleaning, contaminated desalted crude with water and/or solids carryover, increasing decay of desalting and dehydration performance, and contamination of the desalter effluent with oil undercarry and with an increased proportion of oil-coated solids. These latter problems not only jeopardize the desalter operation but also cascade a series of problems downstream, among which one could cite: increased corrosion potential in crude units, increased fouling and scaling on heat exchanger network, downstream catalyst deactivation and unit operability (particularly those units dealing with the BotB fractions, i.e., VGO, VR), and increased energy consumption. It is clear then that demulsifying additives have to be used, with the purpose of breaking these emulsions [700,704]. Additionally, asphaltene-stabilizing additives are needed when dispersion stability is jeopardized or aggregation requires mitigation [1047], for instance, amphiphile additives [765,1048].

One of the modes of action of an antifoulant/dispersant additive is by steric inhibition of aggregation. These dispersants may be polymers or polyfunctional molecules that adsorb or anchor to the particle on several sites, blocking self-interaction. Polymer dispersants may be applied ahead of the desalting operation, since these are hydrophobic and oil-soluble materials, with high enough MW to remain in the bottoms of any subsequent distillation, minimizing issues in downstream desalter units [1049]. On the other hand, dispersing additives have been recommended. Some of these additives incorporate inorganic elements in their formulation (e.g., Si [1050]) that will interact and poison refining catalysts. Notwithstanding, the stabilizing effect of the resins in keeping the asphaltenes in solution has to be accounted for. Any handling or processing condition that moves the R/A ratio away from optimum will lead potentially to asphaltene deposition and in consequence operability issues. A nonintrusive means of breaking the emulsions is the use of microwaves. Microwave radiation can enhance the demulsification rate by an order of magnitude compared with conventional heating [1051].

A combination of an asphaltene stabilizer and a demulsifier was observed to introduce synergies that facilitate demulsification of asphaltene-stabilized emulsions compared to the use on any single additive. However, this type of pair is crude specific and careful selection of additives should be exercised, apart from the refining benefits that mitigation of the emulsion-derived problems would bring to the refiner [1052].

The need for using more than one additive in the units affected by asphaltenes presence obliges to check compatibility and stability under process conditions. In the case of desalting, it has been found that phenolic resin additives were compatible with asphaltenic crude in desalting efficiency, whereas the polyhydric alcohol series were more efficient in dewatering efficiency and a combination effectively behaved during desalting [1053]. Paraffinic crude oils introduce additional challenges in the proper selection of demulsifiers. Diamine series polymerized with propylene oxide (PO) were found to be the most compatible with this type of crude oil for dewatering, whereas polyamine series polymerized with PO and ethylene oxide (EO) were the most compatible for desalting. Meanwhile, for high-paraffin crude oil, a triblock copolymer structure of PO—EO—PO for polyhydric alcohol was more effective than a diblock PO—EO [1054].

6. General Remarks

This chapter has tried to review and describe what is scientifically known about asphaltenes. These compounds, the heaviest molecules present in crude oil (the bottom of the BotB), have attracted R&D attention for almost a century. Still, accurate identification of their molecular structure is lacking. Efforts continue and more sophisticated techniques have contributed with more details toward that ultimate objective.

Their properties and physicochemical behavior have been assessed, particularly the most troublesome. Even though crude oil can be considered a continuous system (the Boduszynski's model [481]), in terms of MW and boiling point (among other properties) of the hydrocarbon classes, the fact that asphaltenes are a solubility class not only changes the picture, but also complicates it beyond imagination. Rather, asphaltenes may be regarded as a mixture of compounds representative of various chemical classes. Thus asphaltenes are a polydispersed system, exhibiting polydispersion of MW, boiling point, molecular size, and many more properties.

Molecular models have been proposed, some of them based on molecular details provided by characterization techniques and methodologies. Some of these models have been used for the theoretical calculation of physical and physicochemical properties. Refined first principle methods as well as semiempirical methods have been applied. Additionally, empirical, statistic, stochastic, and phenomenological models have been applied for predicting physicochemical behavior, particularly stability and precipitation, as well as for the calculation of physicochemical properties. However, in Correra's opinion [1055], the acritical use of models is another source of confusion to an already complex problem. An impressive number of researchers had the tendency to set up and employ models founded on not-proven hypotheses or pictures, with an astonishing number of fitting parameters, instead of trying to understand a priori the physical nature of the phenomenon.

Although some researchers have tried to average molecular features into a single molecular structure for different purposes, the author of the present book sympathizes with the idea of trying to understand asphaltenes' behavior, keeping in mind the presence of a variety of molecules. These molecules will differ not only in their composition, but also in their structural features. Accordingly, molecules with a continental (pericondensed) structure most likely will aggregate via stacking by interactions through the aromatic system. Meanwhile, those molecules with archipelago structures would have more flexibility and a greater number of possible interaction forces and interacting sites (see Section 4.4.3).

Fortunately, nowadays characterization techniques and computing methods and hardware have advanced to a point where a higher level of complexity could be approached. Isolation and chemical subfractionation have evolved, contributed to simplifying the complexity of characterization, and provided complementary details on solubility and reactivity features. These details have been fed back into the modeling work providing further insights into the precipitation mechanisms and paving the way for formulating new ideas for alternative processes to solvent deasphalting. Nevertheless, some of the features of the original Pfeiffer and Saal-devised colloidal model [5] prevail, particularly those that explain rheological behavior. The present version of this model may be considered a hybrid solution–colloidal model, in which asphaltene nanoaggregates are dispersed by amphiphilic resins. The resins intermediate between asphaltenes and the other components of the maltenes fraction (aromatics and saturate), forming a continuum. Some of the experimental results that support this rationale where presented and discussed earlier. However, micellization, CMC, and colloidal properties have not been needed to interpret further additional results.

The current consensus for the macrostructural organization of asphaltenes is a hierarchical model. Gray's supramolecular assembly of molecules incorporates all possible interactions and detects experimental arrangements, namely, aromatic $\pi-\pi$ stacking, H-bonding, acid–base interactions, metal coordination complexes, interactions between cycloalkyl and alkyl groups, and the formation of hydrophobic pockets. Additionally, it includes a range of architectures and molecular structural types creating porous networks and host–guest complexes. The latter may include organic clathrates, in which occluded guest molecules stabilize the assembly of a cage [354].

Asphaltenes' physical (mainly density and viscosity), physicochemical (solubility, aggregation, surface activity, and phase stability), and chemical (low reactivity) properties give rise to numerous and severe problems in the oil industry. This negative impact caused by the BotB compounds throughout the entire value chain of the oil industry has imposed pressure on R&D for finding technological solutions. Although fundamental research has never stopped, parallel applied research was carried out in search for methods and processes that cope cost-effectively with industry operations.

Upstream problems are mainly caused by deposition and emulsion formation, such as well bore plugging, fouling and plugging in pipelines, sludge formation and sedimentation during storage and transportation, emulsification, etc. Meanwhile, downstream incompatibility, precipitation, emulsion stabilization, deposition, and low reactivity lead to fouling in refinery units, desalter upsets, catalyst deactivation, failure in meeting product specifications, poor quality of fuel oil, etc.

As a whole, the cumulative knowledge underpins solutions for crude oil incompatibility, deposits formation, fouling, plugging, emulsification, etc. New findings on the temperature dependence of properties, on the behavior under supercritical conditions, and on adsorption (on metals, oxides, clays, and, in general, asphaltene surface properties) can be synergistically added together for structuring a brand new concept for deasphalting. These three areas of knowledge might be used for selective and controlled removal/conversion. However, the exploration of additives and supercritical extraction may lead to another alternative, for which not only its technical viability has to be demonstrated, but also its economics has to be determined.

The increasing proportion of heavy oils in the refinery diet is part of the present situation and that is not likely to change in the future. Although these oils are sold at discounted prices, the refiner has to face increases in refining costs and tight margins to satisfy the market requirements.

Current practices typically concern mitigation and include: discontinuing purchase of the asphaltenic crude oil; limiting blends to compatible mixtures; and/or treating the crude oil with required additives, such as asphaltene stabilizers, demulsifiers, dispersants, etc.

Development of a cost-effective alternative for current solvent deasphalting (discussed in Chapter 5) would increase the flexibility of the refinery toward the processing of complex diets. Moreover, the current and future specifications of fuels will demand the application of those cost-effective solutions for the removal of contaminants and for facilitating downstream processes. A complete package of proprietary technologies for the treatment and processing of heavy oils and bitumen represents a competitive advantage for the possessor.

In summary, asphaltene precipitation and deposition mechanisms, and their chemical or molecular structures, are less well characterized and consequently less understood than the mitigation actions of the factual situations they cause, such as emulsion breaking, corrosion protection, and dispersion/stability control. However, these mitigating methods currently available are extremely limited and are highly specific to the crude source. Each problem has to be defined on a case-by-case basis for finding a possible effective solution that then needs to be tested and proven to be suitable and acceptable. Furthermore, the complex nature of asphaltenes typically leads to a poor definition of the problems, which in turn derives into an unclear or misleading selection of the chemical handling additives or potential solution.

On the other hand, numerous ideas have been proposed and tested at R&D level over the last 50 years (see Chapter 6) that have failed in reaching the refinery level. Nevertheless, new ideas are still needed to feed the technology development pipeline. These ideas would probably derive from the careful and holistic analysis of the cumulative knowledge of the characteristics, features, and behavior of the molecules that have to be treated. The many ways in which asphaltenes behave and react, depolymerize or disaggregate, and/or decompose could be the basis or inspiration for new processes that contribute to their abatement and render the most value for the refiner.

References

[1] Boussingault JB. Memoire sur la composition des bitumens. Ann Chim Phys 1837;64:141–51.

[2] Nagarajan NR, Dong C, Mullins OC, Honarpour MM. Challenges of heavy oil fluid sampling and characterization. In: Proc. SPE Annual Technical Conf. Exhibition. San Antonio, TX, vol. 2; 2012. p. 924–35.

[3] Marcusson JZ. Chemical composition of natural asphalts. Angew Chem 1916;29:346–9.

[4] Nellensteyn FJ. Theoretical aspect of the relation of bitumen to solid matter. In: Proc. 11th World Pet. Congress. London, UK, vol. 2; August 28–September 2, 1983. p. 616–8.

[5] Pfeiffer JP, Saal RNJ. Asphaltic bitumen as colloid system. J Phys Chem 1940;44(2):139–49.

[6] Speight JG, Long RB, Trowbridge TD. On the definition of asphaltenes. In: Proc. ACS Div. Fuel Chem. Preprints. Kansas City, MO, USA, vol. 27. ACS; 1982. p. 268–75.

[7] Calles JA, Dufour J, Marugán J, Peña JL, Giménez-Aguirre R, Merino-García D. Properties of asphaltenes precipitated with different *n*-alkanes. A study to assess the most representative species for modeling. Energy Fuels 2008;22(2):763–9.

[8] Long RB. The concept of asphaltenes. In: Bunger JW, Li N, editors. Adv. chem. series 195: chemistry of asphaltenes. Washington, DC: ACS; 1981. p. 17–27.

[9] Unger F. The nature of resins and asphaltenes. ACS Div Fuel Chem 1997;42(2):445–8. Preprints.

[10] Buenrostro-Gonzalez E, Groenzin H, Lira-Galeana C, Mullins OC. The overriding chemical principles that define asphaltenes. Energy Fuels 2001;15(4):972–8.

[11] Ray BR, Witherspoon PA, Grim RE. A study of the colloidal characteristics of petroleum using the ultracentrifuge. J Phys Chem 1957;61(10):1296–302.

[12] Andersen SI, Speight JG. Petroleum resins: separation, character, and role in petroleum. Pet Sci Technol 2001;19(1–2):1–34.

[13] Sirota EB. Physical structure of asphaltenes. Energy Fuels 2005;19:1290–6.

[14] Hoepfner MP, Fogler HS. Multiscale scattering investigations of asphaltene cluster breakup, nanoaggregate dissociation, and molecular ordering. Langmuir 2013;29(49):15423–32.

[15] Demirba A. Asphaltene yields from five types of fuels via different methods. Energy Conv Manag 2002;43(8):1091–7.

[16] Corbett LW, Petrossi U. Differences in distillation and solvent separated asphalt residua. Ind Eng Chem Prod Res Dev 1978;17(4):342–6.

[17] Mitchell DL, Speight JG. The solubility of asphaltenes in hydrocarbon solvents. Fuel 1973;52(2):149–52.

[18] Centeno G, Trejo F, Ancheyta J, Carlos A. Precipitación de asfaltenos del crudo maya en un sistema a presión. J Mex Chem Soc 2004;48(3):179–88.

[19] Navarro L, Álvarez M, Grosso JL, Navarro U. Separation and characterization of resins and asphaltenes from castilla crude oil. Evaluation of its molecular interaction. CT Cienc Tecnol Futuro 2004;2(5):53–67.

[20] Fossen M, Kallevik H, Knudsen KD, Sjöblom J. Asphaltenes precipitated by a two-step precipitation procedure. 1. Interfacial tension and solvent properties. Energy Fuels 2007;21(2): 1030–7.

[21] Fossen M, Sjöblom J, Kallevik H, Jakobsson J. A new procedure for direct precipitation and fractionation of asphaltenes from crude oil. J Dispers Sci Technol 2007;28(1):193–7.

[22] Speight JG, Long RB, Trowbridge TD. Factors influencing the separation of asphaltenes from heavy petroleum feedstocks. Fuel 1984;63(5):616–20.

[23] Teixeira MAG, Gonçalves MLA. Evaluation of paraffinic material in asphaltenes isolated by precipitation with light alkanes. Pet Sci Technol 2000;18(3):273–86.

[24] Acevedo S, M. CTJ, Carrier H, Bouyssiere B, Lobinski R. Trapping of paraffin and other compounds by asphaltenes detected by laser desorption ionization-time of flight mass spectrometry (LDI-TOF MS): role of A1 and A2 asphaltene fractions in this trapping. Energy Fuels 2009;23(2):842–8.

[25] Goual L, Schabron JF, Turner TF, Towler BF. On-column separation of wax and asphaltenes in petroleum fluids. Energy Fuels 2008;22(6):4019–28.

[26] Ignasiak TM, Kotlyar L, Longstaffe FJ, Strausz OP, Montgomery DS. Separation and characterization of clay from Athabasca asphaltene. Fuel 1983;62(3):353–62.

[27] Bestougeff MA, Byramjee RJ. Chemical constitution of asphaltenes. In: Yen TF, Chilingarian GV, editors. Asphaltenes asphalts. 1. Developments in pet. science. Elsevier; 1994. p. 67–94.

[28] Yoon S, Bhatt SD, Lee W, Lee HY, Jeong SY, Baeg JO, et al. Separation and characterization of bitumen from Athabasca oil sand. Korean J Chem Eng 2009;26(1):64–71.

[29] Speight JG. Chemical and physical studies of petroleum asphaltenes. In: Yen TF, Chilingarian GV, editors. Asphaltene asphalts. I. Developments in pet. science. Amsterdam, The Netherlands: Elsevier; 1994. p. 7–66.

[30] Speight JG. Petroleum asphaltenes – part 1: asphaltenes, resins and the structure of petroleum. Oil Gas Sci Technol Rev IFP 2004;59(5):467–77.

[31] Speight JG. Initial reactions in the coking of residua. ACS Div Pet Chem 1987;32(2):413–8. Preprints.

[32] Kharrat AM, Zacharia J, Cherian VJ, Anyatonwu A. Issues with comparing SARA methodologies. Energy Fuels 2007;21(6):3618–21.

[33] Fuhr BJ, Holloway LR, Hammami A. Analytical considerations related to asphaltenes and waxes in the same crudes. Energy Fuels 1999;13(2):336–9.

[34] Fuhr BJ, Hawrelechko C, Holloway LR, Huang H. Comparison of bitumen fractionation methods. Energy Fuels 2005;19(4):1327–9.

[35] Fuhr B, Scott K, Dettman H, Salmon S. Fractionation of bitumen by distillation, SARA analysis and gel permeation chromatography. In: Proc. 229th ACS Nat. Meeting, Div. Pet. Chem. Preprints. San Diego, CA, vol. 50. PETR-131; March 13–17, 2005. p. 264–5.

[36] Fuhr BJ, Holloway LR, Reichert C. Rapid analytical characterization of residues from heavy oil and bitumen upgrading processes. J Can Pet Technol 1986;25(5):28–32.

[37] Gawrys KL, Blankenship GA, Kilpatrick PK. On the distribution of chemical properties and aggregation of solubility fractions in asphaltenes. Energy Fuels 2006;20:705–14.

[38] Marques J, Merdrignac I, Baudot A, Barré L, Guillaume D, Espinat D, Brunet S. Asphaltenes size polydispersity reduction by nano- and ultrafiltration separation methods – comparison with the flocculation method. Oil Gas Sci Technol Rev IFP 2008;63(1):139–49.

[39] Rudzinski WE, Aminabhavi TM. A review on extraction and identification of crude oil and related products using supercritical fluid technology. Energy Fuels 2000;14(2):464–75.

[40] Shi TP, Hu YX, Xu ZM, Su T, Wang RA. Characterizing petroleum vacuum residue by supercritical fluid extraction and fractionation. Ind Eng Chem Res 1997;36(9):3988–92.

[41] Demirba A. Recovery of asphaltenes from tar sand by supercritical fluid extraction. Pet Sci Technol 2000;18(7):771–81.
[42] Lai WC, Smith KJ. Heavy oil microfiltration using ceramic monolith membranes. Fuel 2001;80(8):1121–30.
[43] Duong A, Chattopadhyaya G, Kwok W, Smith K. An experimental study of heavy oil ultrafiltration using ceramic membranes. Fuel 1997;76(9):821–8.
[44] Bishop B, Goldsmith R, Schucker R, Rawlins K. Ceramic membrane process for upgrading vacuum residual oil. In: Proc. AIChE Spring Nat. Meeting. New Orleans, LA; 2004. 1 pp.
[45] Amundaraín Hurtado JL, Chodakowski M, Long B, Shaw JM. Characterization of physically and chemically separated Athabasca asphaltenes using small-angle x-ray scattering. Energy Fuels 2011;25(11):5100–12.
[46] Dechaine GP. Solubility and diffusion of vanadium compounds and asphaltene aggregates (Ph.D. thesis from University of Alberta, Department of Chem. Mat. Eng.). January 26, 2010. 246 pp.
[47] Yarranton HW, Ortiz DP, Barrera DM, Baydak EN, Barré L, Frot D, et al. On the size distribution of self-associated asphaltenes. Energy Fuels 2013;27(9):5083–106.
[48] Barbier J, Marques J, Caumette G, Merdrignac I, Bouyssiere B, Lobinski R, et al. Monitoring the behaviour and fate of nickel and vanadium complexes during vacuum residue hydrotreatment and fraction separation. Fuel Proc Technol 2014;119:185–9.
[49] Barre L, Baudot A, Brunet S, Espinat D, Guillaume DG, Marques J, et al. Asphaltenes separation by membrane filtration - comparison with the flocculation method. In: Proc. Molecular Structure of Heavy Oils Coal Liquefaction Products. Lyon, France; April 12–13, 2007. 12 pp.
[50] Ashtari M, Bayat M, Sattarin M. Investigation on asphaltene and heavy metal removal from crude oil using a thermal effect. Energy Fuels 2011;25(1):300–6.
[51] Ashtari M, Ashrafizadeh SN, Bayat M. Asphaltene removal from crude oil by means of ceramic membranes. J Pet Sci Eng 2012;82–83:44–9.
[52] Eyssautier J, Espinat D, Gummel J, Levitz P, Becerra M, Shaw J, et al. Mesoscale organization in a physically separated vacuum residue: comparison to asphaltenes in a simple solvent. Energy Fuels 2012;26(5):2680–7.
[53] Zhao B, Shaw JM. Composition and size distribution of coherent nanostructures in Athabasca bitumen and Maya crude oil. Energy Fuels 2007;21(5):2795–804.
[54] Arod J, Bartoli B, Bergez P, Biedermann J, Caminade P, Martinet J-M, et al. Process for the treatment of a hydrocarbon charge by high temperature ultrafiltration, Patent No. US4411790 (also published as EP0041013, FR2482975, ES8300841). Commissariat Energie Atomique; Raffinage Cie Francaise; October 25, 1983.
[55] Osterhuber EJ. Upgrading of heavy oils by solvent dissolution and ultrafiltration, Patent No. US4797200. Exxon Research and Engineering Co; January 10, 1989.
[56] Feimer JL, Desjardine LT. Multi-stage ultrafiltration process (OP-3711), Patent No. US5256297 (also published as WO9413756). Exxon Research and Engineering Co.; October 26, 1993.
[57] Smith KJ. Upgrading heavy oil by ultrafiltration using ceramic membrane, Patent No. US5785860. Univ British Columbia; July 28, 1998.
[58] Samedova FI, Mir-Babaev MF. Separation of asphaltenes by physical method. Khimiya Tekhnologiya Topliv Masel. Chem Technol Fuels Oils 1995;31(5):251–2.
[59] Dettman H, Inman A, Salmon S, Scott K, Fuhr B. Chemical characterization of GPC fractions of Athabasca bitumen asphaltenes isolated before and after thermal treatment. Energy Fuels 2005;19(4):1399–404.
[60] Altgelt KH. Fractionation of asphaltenes by gel permeation chromatography. J Appl Polym Sci 1965;9(10):3389–93.
[61] Woods JR, Kung J, Kingston D, Kotlyar L, Sparks B, Mccracken T. Canadian crudes: a comparative study of SARA fractions from a modified HPLC separation technique. Oil Gas Sci Technol Rev IFP 2008;63(1 special issue):151–63.

[62] Nabiullina ER, Kudasheva FK. Petroleum asphaltene fractions. Chem Technol Fuels Oils (Translation of Khimiya i Tekhnologiya Topliv i Masel) 1989;24(11–12):510–2.
[63] Cyr N, Mcintyre DD, Toth G, Strausz OP. Hydrocarbon structural group analysis of athabasca asphaltene and its GPC fractions by 13C NMR. Fuel 1987;66(12):1709–14.
[64] Jacobs FS, Filby RH. Liquid chromatographic fractionation of oil-sand and crude oil asphaltenes. Fuel 1983;62(10):1186–92.
[65] Suelves I, Islas CA, Millan M, Galmes C, Carter JF, Herod AA, et al. Chromatographic separations enabling the structural characterisation of heavy petroleum residues. Fuel 2003;82(1):1–14.
[66] Merino-Garcia D, Murgich J, Andersen SI. Asphaltene self-association: modeling and effect of fractionation with a polar solvent. Pet Sci Technol 2004;22(7–8):735–58.
[67] Zhao S, Xu Z, Xu C, Chung KH, Wang R. Systematic characterization of petroleum residua based on SFEF. Fuel 2005;84(6):635–45.
[68] Wattana P, Fogler HS, Yen A, Del Carmen Garcia M, Carbognani L. Characterization of polarity-based asphaltene subfractions. Energy Fuels 2005;19(1):101–10.
[69] Quintero LCN, Seid PR, Tasic L, Leal KZ, Menezes SMC. Fractionation of an asphaltene from a Brazilian vacuum residue. In: Proc. AIChE Spring Nat. Meeting. New Orleans, LA; 2008. 1 pp.
[70] Navarro Quintero LC, Seidl PR, Tasic L, Leal K, Menezes SMC. Subfractionation of asphaltenes from Brazilian vacuum residues. In: Proc. 235th ACS Nat. Meeting. New Orleans, LA; April 6–10, 2008. 1 pp.
[71] Andersen SI, Jensen JO, Speight JG. X-ray diffraction of subfractions of petroleum asphaltenes. Energy Fuels 2005;19(6):2371–7.
[72] Andersen SI, Keul A, Stenby E. Variation in composition of subfractions of petroleum asphaltenes. Pet Sci Technol 1997;15(7–8):611–45.
[73] Yang MG, Eser S. Fractionation and molecular analysis of a vacuum residue asphaltenes. ACS Div Fuel Chem 1999;44(4):768–71. Preprints.
[74] Acevedo S, Escobar G, Ranaudo MA, Pinate J, Amorın A, Diaz M, et al. Observations about the structure and dispersion of petroleum asphaltenes aggregates obtained from dialysis fractionation and characterization. Energy Fuels 1997;11:774–8.
[75] Fenistein D, Barré L. Experimental measurement of the mass distribution of petroleum asphaltene aggregates using ultracentrifugation and small-angle X-ray scattering. Fuel 2001;80(2):283–7.
[76] Gould KA, Long RB. A new technique for the acid/base separation of petroleum and coal-derived fractions. Fuel 1986;65(4):572–6.
[77] Nalwaya V, Tangtayakom V, Piumsomboon P, Fogler S. Studies on asphaltenes through analysis of polar fractions. Ind Eng Chem Res 1999;38(3):964–72.
[78] Kaminski TJ, Fogler HS, Wolf N, Wattana P, Mairal A. Classification of asphaltenes via fractionation and the effect of heteroatom content on dissolution kinetics. Energy Fuels 2000;14(1):25–30.
[79] Boukir A, Aries E, Guiliano M, Asia L, Doumenq P, Mille G. Subfractionation, characterization and photooxidation of crude oil resins. Chemosphere 2001;43(3):279–86.
[80] Spiecker PM, Gawrys KL, Kilpatrick PK. Aggregation and solubility behavior of asphaltenes and their subfractions. J Colloid Interface Sci 2003;267(1):178–93.
[81] Groenzin H, Mullins OC, Eser S, Mathews J, Yang MG, Jones D. Molecular size of asphaltene solubility fractions. Energy Fuels 2003;17(2):498–503.
[82] Buenrostro-Gonzalez E, Andersen SI, Garcia-Martinez JA, Lira-Galeana C. Solubility/molecular structure relationships of asphaltenes in polar and nonpolar media. Energy Fuels 2002;16(3):732–41.
[83] Yang X, Hamza H, Czarnecki J. Investigation of subfractionstof Athabasca asphaltenes and their role in emulsion stability. Energy Fuels 2004;18(3):770–7.
[84] Islas-Flores CA, Buenrostro-Gonzalez E, Lira-Galeana C. Fractionation of petroleum resins by normal and reverse phase liquid chromatography. Fuel 2006;85(12–13):1842–50.
[85] Seidl PR, Chrisman ECAN, Silva RC, De Menezes SMC, Teixeira MAG. Critical variables for the characterization of asphaltenes extracted from vacuum residues. Pet Sci Technol 2004;22(7–8):961–71.

[86] Maham Y, Chodakowski MG, Zhang X, Shaw JM. Asphaltene phase behavior: prediction at a crossroads. Fluid Phase Equilib 2005;227(2):177–82.
[87] Evdokimov IN. The importance of asphaltene content in petroleum II-multi-peak viscosity correlations. Pet Sci Technol 2010;28(9):920–4.
[88] Evdokimov IN. The importance of asphaltene content in petroleum-the revision of some persistent stereotypes. Pet Sci Technol 2010;28(7):756–63.
[89] Evdokimov IN. The importance of asphaltene content in petroleum III – new criteria for prediction of incompatibility in crude oil blends. Pet Sci Technol 2010;28(13):1351–7.
[90] Marczak W, Dafri D, Modaressi A, Zhou H, Rogalski M. Physical state andmaging of flocculated asphaltenes. Energy Fuels 2007;21(3):1256–62.
[91] Li S, Fu H, Liu C, Liang W. Colloidal natures of two typical chinese vacuum residua I: colloidal structures in terms of SFEF fractions. ACS Div Fuel Chem 1999;44(4):772–7. Preprints.
[92] Li S, Fu H, Liu C, Liang W, Zhang Z. Colloidal natures of two typical chinese vacuum residua II: revisit and characterization of colloidal structures. ACS Div Fuel Chem 1999;44(4):778–82. Preprints.
[93] Li S, Liu C, Que G, Liang W, Yajie Z. Colloidal structures of three chinese petroleum vacuum residues. Fuel 1996;75(8):1025–9.
[94] Li S, Liu C, Que G, Liang W. Colloidal structures of vacuum residua and their thermal stability in terms of saturate, aromatic, resin and asphaltene composition. J Pet Sci Eng 1999;22(1–3):37–45.
[95] Sirota EB. Swelling of asphaltenes. Pet Sci Technol 1998;16(3–4):415–31.
[96] Buckley JS. What are asphaltenes?. 2010. Available from: http://baervan.nmt.edu/Petrophysics/group/intro-2-asphaltenes.pdf.
[97] Yen TF, Erdman JG, Pollack SS. Investigation of the structure of petroleum asphaltenes by X-Ray diffraction. Anal Chem 1961;33(11):1587–94.
[98] Barakat MAK, Shimy TM, Mostafa YM. Mineralogy of asphaltenes separated from crude oils and tar pollutants by X-ray, polarizing microscope, IR and NMR techniques. Pet Sci Technol 1999;17(7):677–91.
[99] Bagheri SR, Masik B, Arboleda P, Wen Q, Michaelian KH, Shaw JM. Physical properties of liquid crystals in Athabasca bitumen fractions. Energy Fuels 2012;26(8):4978–87.
[100] Rogel E, Carbognani L. Density estimation of asphaltenese using molecular dynamics simulations. Energy Fuels 2003;17(2):378–86.
[101] Peramanu S, Pruden BB, Rahimi P. Molecular weight and specific gravity distributions for Athabasca and Cold Lake bitumens and their saturate, aromatic, resin, and asphaltene fractions. Ind Eng Chem Res 1999;38(8):3121–30.
[102] Carbognani Ortega LA, Rogel E, Vien J, Ovalles C, Guzman H, Lopez-Linares FA, et al. Effect of precipitating conditions on asphaltene properties and aggregation. Energy Fuels 2015;29(6):3664–74.
[103] Yen TF. Macrostructure of petroleum asphalt. In: Proc. Symp. on Chem. Characterization of Asphalts. Washington, DC. USA, vol. 35. ACS; 1990. p. 314–9.
[104] Yen TF. Colloidal aspect of a macrostructure of petroleum asphalt. ACS Symp Chem Charact Asphalts 1992;10(4–6):723–33.
[105] Yen TF. Multiple structural orders of asphaltenes. In: Yen TF, Chilingarian GV, editors. Asphaltenes asphalts. 1. Developments in pet. science. Amsterdam: Elsevier; 1994. p. 111–23.
[106] Giavarini C, Mastrofini D, Scarsella M, Barré L, Espinat D. Macroseructure and rheological properties of chemically modified residues and bitumens. Energy Fuels 2000;14(2):495–502.
[107] Shiryaeva RN, Kudasheva FK, Gimaev RN, Sagitova CK. Rheological properties of crude oils with a high resin and asphaltene content. Effect of an electromagnetic field and modifiers. Chem Technol Fuels Oils (Translation of Khimiya i Tekhnologiya Topliv i Masel) 2006;42(3):202–5.
[108] Semple KM, Cyr N, Fedorak PM, Westlake DWS. Characterization of asphaltenes from Cold Lake heavy oil: variations in chemical structure and composition with molecular size. Can J Chem 1990;68:1092–9.

[109] Liu D, Wang Z, Zhou J, Deng W, Liang S, Que G. Characterization of vacuum residua and their substitution by using FT-IR technique. Proc ACS Div Pet Chem 2001;46:256–8. Preprints.
[110] Östlund JA, Wattana P, Nydén M, Fogler HS. Characterization of fractionated asphaltenes by UV-Vis and NMR self-diffusion spectroscopy. J Colloid Interface Sci 2004;271(2):372–80.
[111] Avid B, Sato S, Takanohashi T, Saito I. Characterization of asphaltenes from Brazilian vacuum residue using heptane-toluene mixtures. Energy Fuels 2004;18(2):1792–7.
[112] Angle CW, Lue L, Dabros T, Hamza HA. Viscosities of heavy oils in toluene and partially deasphalted heavy oils in heptol in a study of asphaltenes self-interactions. Energy Fuels 2005;19(5): 2014–20.
[113] Carbognani L, Gonzalez MF, Pereira-Almao P. Characterization of Athabasca vacuum residue and its visbroken products. Stability and fast hydrocarbon group-type distributions. Energy Fuels 2007;21(3):1631–9.
[114] Trejo F, Ancheyta J. Characterization of asphaltene fractions from hydrotreated Maya crude oil. Ind Eng Chem Res 2007;46(23):7571–9.
[115] Kharrat AM. Characterization of Canadian heavy oils using sequential extraction approach. Energy Fuels 2009;23(2):828–34.
[116] Eyssautier J, Hénaut I, Levitz P, Espinat D, Barré L. Organization of asphaltenes in a vacuum residue: a small-angle X-ray scattering (SAXS)–viscosity approach at high temperatures. Energy Fuels 2012;26(5):2696–704.
[117] Escobedo J, Mansoori GA. Viscometric determination of the onset of asphaltene flocculation: a novel method. SPE Prod Facil 1995;10(2):115–8.
[118] Escobedo J, Mansoori GA. Viscometric principles of onsets of colloidal asphaltene flocculation in paraffinic oils and asphaltene micellization in aromatics. SPE Prod Facil 1997;12(2): 116–22.
[119] Barre L, Coustet PC, Moan M, Pina A. Composition and heavy oil rheology. Oil Gas Sci Technol Rev IFP 2004;59(5):489–501.
[120] Meyer V, Pilliez J, Habas JP, Montel F, Creux P. Rheological evidence of the diffusionnal aggregation of asphaltenes in extra-heavy crude oils. Energy Fuels 2008;22(5):3154–9.
[121] Gaweł B, Eftekhardadkhah M, Øye G. Elemental composition and fourier transform infrared spectroscopy analysis of crude oils and their fractions. Energy Fuels 2014;28(2):997–1003.
[122] Mansoori GA. Asphaltene deposition and its control. 1995. Available from: http://tigger.uic.edu/~mansoori/Asphaltene.Deposition.and.Its.Control_html.
[123] Mansoori GA. Remediation of asphaltene and other heavy organic deposits in oil wells and in pipelines. SOCAR Proc 2010;2010(4):12–23.
[124] Creek JL. Freedom of action in the state of asphaltenes: escape from conventional wisdom. Energy Fuels 2005;19(4):1212–24.
[125] Molina VD, Uribe UN, Murgich J. Correlations between SARA fractions and physicochemical properties with 1H NMR spectra of vacuum residues from Colombian crude oils. Fuel 2010;89(1):185–92.
[126] Quiñones-Cisneros SE, Andersen SI, Creek J. Density and viscosity modeling and characterization of heavy oils. Energy Fuels 2005;19(4):1314–8.
[127] Rahimian I, Laux H. Density investigation of asphaltenes, coke precursors and solid deposits [dichtebestimmung von asphaltenen, koksartigen ruckstanden und festen ablagerungen]. Erdol Erdgas Kohle 2000;116(1):15–7.
[128] Xu Z, Van Den Berg FGA, Sun X, Xu C, Zhao S. Detailed characterization of virgin heavy oil resid and its thermally cracked resid. Energy Fuels 2013;28(3):1664–73.
[129] Davarpanah L, Vahabzadeh F, Dermanaki A. Structural study of asphaltenes from iranian heavy crude oil. Oil Gas Sci Technol Rev IFP 2013:15. http://dx.doi.org/10.2516/ogst/2012066. Available from:.
[130] Barrufet MA, Setiadarma A. Experimental viscosities of heavy oil mixtures up to 450 K and high pressures using a mercury capillary viscometer. J Pet Sci Eng 2003;40(1–2):17–26.

[131] Asprino OJ, Elliott JAW, Mccaffrey WC, Gray MR. Fluid properties of asphaltenes at 310–530 °C. Energy Fuels 2005;19:2026–33.
[132] Goual L, Firoozabadi A. Measuring asphaltenes and resins, and dipole moment in petroleum fluids. AIChE J 2002;48(11):2646–63.
[133] Sheu EY, De Tar MM, Storm DA. Dielectric properties of asphaltene solutions. Fuel 1994;73(1): 45–50.
[134] Sheu EY, Storm DA, De Tar MM. Asphaltenes in polar solvents. J Noncryst Solids 1991;131–133(Part 1):341–7.
[135] Sheu EY, Acevedo S. A dielectric relaxation study of precipitation and curing of Furrial crude oil. Fuel 2006;85(14–15):1953–9.
[136] Vrålstad H, Spets Ø, Lesaint C, Lundgaard L, Sjöblom J. Dielectric properties of crude oil components. Energy Fuels 2009;23(11):5596–602.
[137] Evdokimov IN, Losev AP. Electrical conductivity and dielectric properties of solid asphaltenes. Energy Fuels 2010;24(7):3959–69.
[138] Dolomatov MY, Dezortsev SV, Shutkova SA. Asphaltenes of oil and hydrocarbon distillates as nanoscale semiconductors. J Mat Sci Eng B 2012;2(B):151–7.
[139] Long RB. The concept of asphaltenes. ACS Div Pet Chem 1979;24(4):891–900. Preprints.
[140] Long RB, Speight JG. Studies in petroleum composition. 2. Scale-up studies for separating heavy feedstocks by adsorption. Ind Eng Chem Res 1989;28(10):1503–7.
[141] Cimimo R, Correra S, Bianco AD. Solubility and phase behavior of asphaltenes in hydrocarbon media. In: Sheu EY, Mullins OC, editors. Asphaltenes: fundamentals applications. New York: Plenum Press; 1993. p. 97–126.
[142] Bestougeff M, Burollet PF, Byramjee RJ. Heavy crude oils and their classification. In: Proc. 2nd UNITAR Intern. Conf. 'Heavy Crude Tar Sands': Future of Heavy Crude and Tar Sands. New York, NY, USA; Caracas, Venezuela. McGraw-Hill Inc., Coal Age Mining Information Services; 1984. p. 12–6.
[143] Yarranton HW, Masliyah JH. Molar mass distribution and solubility modeling of asphaltenes. AIChE J 1996;42(12):3533–43.
[144] Mannistu KD, Yarranton HW, Masliyah JH. Solubility modeling of asphaltenes in organic solvents. Energy Fuels 1997;11(3):615–22.
[145] Yarranton HW. Asphaltene self-association. J Dispers Sci Technol 2005;26(1):5–8.
[146] Yarranton HW, Fox WA, Svrcek WY. Effect of resins on asphaltene self-association and solubility. Can J Chem Eng 2007;85(5):635–42.
[147] Schabron JF, Rovani Jr JF. On-column precipitation and re-dissolution of asphaltenes in petroleum residua. Fuel 2008;87(2):165–76.
[148] Stapf S, Ordikhani-Seyedlar A, Ryan N, Mattea C, Kausik R, Freed DE, et al. Probing maltene–asphaltene interaction in crude oil by means of NMR relaxation. Energy Fuels 2014;28(4):2395–401.
[149] Bardon C, Barre L, Espinat D, Guille V, Li MH, Lambard J, et al. Colloidal structure of crude oils and suspensions of asphaltenes and resins. Fuel Sci Technol Intern 1996;14(1–2):203–42.
[150] Jamaluddin AKM, Nazarko TW, Sills S, Fuhr BJ. Deasphalted oil: a natural asphaltene solvent. SPE Prod Facil 1996;11(3):161–5.
[151] Peramanu S, Clarke PF, Pruden BB. Flow loop apparatus to study the effect of solvent, temperature and additives on asphaltene precipitation. J Pet Sci Eng 1999;23(2):133–43.
[152] Mohamed RS, Loh W, Ramos ACS, Delgado CC, Almeida VR. Reversibility and inhibition of asphaltene precipitation in Brazilian crude oils. Pet Sci Technol 1999;17(7):877–96.
[153] Permsukarome P, Chang C, Fogler HS. Kinetic study of asphaltene dissolution in amphiphile/alkane solutions. Ind Eng Chem Res 1997;36(9):3960–7.
[154] Hamouda AA, Chukwudeme EA, Mirza D. Investigating the effect of CO_2 flooding on asphaltenic oil recovery and reservoir wettability. Energy Fuels 2009;23(2):1118–27.

[155] Idem RO, Ibrahim HH. Kinetics of CO_2-induced asphaltene precipitation from various saskatchewan crude oils during CO_2 miscible flooding. J Pet Sci Eng 2002;35(3–4):233–46.
[156] Zaki NN, Carbonell RG, Kilpatrick PK. A novel process for demulsification of water-in-crude oil emulsions by dense carbon dioxide. Ind Eng Chem Res 2003;42(25):6661–72.
[157] Bungert B, Sadowski G, Arlt W. Separations and material processing in solutions with dense gases. Ind Eng Chem Res 1998;37:3208–20.
[158] Hotier G, Robin M. Effects of different thinners on heavy petroleum products: measuring, interpreting and forecasting asphaltene flocculation. Rev Inst Fr Pet 1983;38(1):101–20.
[159] Lhioreau C, Briant J, Tindy R. Influence de la flocculation des asphaltenes. Rev Inst Fr Pét 1967;22:797–806.
[160] Speight JG. The chemical and physical structure of petroleum: effects on recovery operations. Pet Sci Eng 1999;22(1–3):3–15.
[161] Ali LH, Al-Ghannam KA. Investigations into asphaltenes in heavy crude oils. I. Effect of temperature on precipitation by alkane solvents. Fuel 1981;60(11):1043–6.
[162] Andersen SI. Effect of precipitation temperature on the composition of n-heptane asphaltenes. Fuel Sci Technol Intern 1994;12(1):51–74.
[163] Andersen SI, Birdi KS. Influence of temperature and solvent on the precipitation of asphaltenes. Fuel Sci Technol Intern 1990;8(6):593–615.
[164] Rogacheva OV, Gimaev RN, Gubaidullin VZ, Danil'yan TD. Aggregate stability of petroleum dispersed systems. Colloid J USSR 1984;46(5):828–32.
[165] Hirschberg A, Dejong LNJ, Schipper BA, Meijer JG. Influence of temperature and pressure on asphaltene flocculation. Soc Pet Eng J 1984;24(3):283–93.
[166] Pasadakis N, Varotsis N, Kallithrakas N. The influence of pressure on the asphaltenes content and composition in oils. Pet Sci Technol 2001;19(9–10):1219–27.
[167] Spiecker PM. The impact of asphaltene chemistry and solvation on emulsion and interfacial film formation (Ph.D. thesis from North Carolina State University, Chem. Eng.). 2001. 302 pp.
[168] Wattana P. Precipitation and characterization of petroleum asphaltenes (Ph.D. thesis from University of Michigan, Chem. Eng.). 2004. 185 pp.
[169] Tojima M, Suhara S, Imamura M, Furuta A. Effect of heavy asphaltene on stability of residual oil. Catal Today 1998;43(3–4):347–51.
[170] Trejo F, Centeno G, Ancheyta J. Precipitation, fractionation and characterization of asphaltenes from heavy and light crude oils. Fuel 2004;83(16):2169–75.
[171] Gutierrez LB, Ranaudo MA, Mendez B, Acevedo S. Fractionation of asphaltene by complex formation with p-nitrophenol. A method for structural studies and stability of asphaltene colloids. Energy Fuels 2001;15:624–8.
[172] Castillo J, Fernandez A, Ranaudo MA, Acevedo S. New techniques and methods for the study of aggregation, adsorption, and solubility of asphaltenes. Impact of these properties on colloidal structure and flocculation. Pet Sci Technol 2001;19(1–2):75–106.
[173] Marcano F, Flores R, Chirinos J, Ranaudo MA. Distribution of Ni and V in A1 and A2 asphaltene fractions in stable and unstable Venezuelan crude oils. Energy Fuels 2011;25(5):2137–41.
[174] Andersen SI, Speight JG. Asphaltene precipitation and incipient flocculation in mixed solvents. ACS Div Fuel Chem 1989;37(3):1335–41. Preprints.
[175] Hildebrand J, Scott RL. Regular solutions. Englewood Cliffs, NJ, USA: Prentice-Hall Inc; 1962. 180 pp.
[176] Hildebrand J, Scott RL. The solubility of nonelectrolytes. 3rd ed. New York: Reinhold; 1950. 488 pp.
[177] Wiehe IA, Yarranton HW, Akbarzadeh K, Rahimi PM, Teclemariam A. The paradox of asphaltene precipitation with normal paraffins. Energy Fuels 2005;19:1261–7.
[178] Mullins OC, Pomerantz AE, Ballard Andrews A, Zuo JY. Asphaltenes explained for the nonchemist. PETROPHYSICS 2015;56(3):266–75.
[179] Gonzalez G, Sousa MA, Lucas EF. Asphaltenes precipitation from crude oil and hydrocarbon media. Energy Fuels 2006;20(6):2544–51.

[180] Hildebrand JH, Prausnitz JM, Scott RL. Regular and related solutions. New York, NY, USA: Van Nostrand Reinhold; 1970. 228 pp.
[181] Hansen CM. Hansen solubility parameters: a user's handbook. 2nd ed. Boca Raton, FL: CRC Press; 2007. 546 pp.
[182] Laux H, Rahimian I. Colloid-disperse crude oil systems. Erdol Erdgas Kohle 1994;47(11):430–5.
[183] Laux H, Rahimian I. Boundaries of colloid stability of crude oils [grenzen der kolloidstabilita?T von erdo?Len]. Erdol Erdgas Kohle 1998;114(1):25–9.
[184] Laux H, Rahimian I, Butz T. Thermodynamics and mechanism of stabilization and precipitation of petroleum colloids. Fuel Proc Technol 1997;53(1–2):69–79.
[185] Buckley JS, Hirasaki GJ, Liu Y, Von Drasek S, Wang JX, Gill BS. Asphaltene precipitation and solvent properties of crude oils. Pet Sci Technol 1998;16(3–4):251–85.
[186] Wang JX, Buckley JS. An experimental approach to prediction of asphaltene flocculation. In: Proc. SPE Intern. Symp. Oilfield Chem. Houston, TX; 2001. p. 179–87.
[187] Aguilera-Mercado B, Herdes C, Murgich J, Müller EA. Mesoscopic simulation of aggregation of asphaltene and resin molecules in crude oils. Energy Fuels 2006;20(1):327–38.
[188] Fotland P, Askvik KM. Determination of hamaker constants for asphaltenes in a mixture of pentane and benzene. Colloids Surf A Physicochem Eng Asp 2008;324(1–3):22–7.
[189] Taft RW, Abboud JLM, Kamlet MJ, Abraham MH. Linear solvation energy relations. J Solut Chem 1985;14(3):153–86.
[190] Abraham MH. Scales of solute hydrogen-bonding: their construction and application to physicochemical and biochemical processes. Chem Soc Rev 1993;22(2):73–83.
[191] Mutelet F, Ekulu G, Solimando R, Rogalski M. Solubility parameters of crude oils and asphaltenes. Energy Fuels 2004;18(3):667–73.
[192] Heithaus JJ. Measurement and significance of asphaltene peptization. J Inst Pet 1962;48(458):45–53.
[193] Wiehe IA. Two-dimensional solubility parameter mapping of heavy oils. In: Proceedings of the 1995 International Symposium on the Thermodynamics of Heavy Oils and Asphaltenesvol. 14 (1–2); 1996. p. 284–312.
[194] Wiehe IA, Liang KS. Asphaltenes, resins, and other petroleum macromolecules. Fluid Phase Equilib 1996;117(1–2 pt 2):201–10.
[195] Wiehe IA, Kennedy RJ. The oil compatibility model and crude oil incompatibility. Energy Fuels 2000;14:56–9.
[196] Xingyi W, Zhiming X, Suoqi Z, Chunming X, Keng H,C. Solubility parameters of bitumen-derived narrow vacuum resid fractions. Energy Fuels 2009;23(1):386–91.
[197] Rogel E. Theoretical estimation of the solubility parameter distributions of asphaltenes, resins, and oils from crude oils and related materials. Energy Fuels 1997;11(4):920–5.
[198] Rogel E. Theoretical approach to the stability of visbroken residues. Energy Fuels 1998;12(5):875–80.
[199] Painter P, Veytsman B, Youtcheff J. Guide to asphaltene solubility. Energy Fuels 2015;29(5):2951–61.
[200] Acevedo S, Escobar O, Echevarria L, Gutierrez LB, Mendez B. Structural analysis of soluble and insoluble fractions of asphaltenes isolated using the PNP method. Relation between asphaltene structure and solubility. Energy Fuels 2004;18(2):305–11.
[201] Chang CL, Nalwaya V, Singh P, Fogler HS. Removal of asphaltene and paraffin deposits using micellar solutions and fused reactions. Report No. DOE/PC/91008-6. Ann Arbor, MI: The University of Michigan; May 1998 (149 + 29) 176 pp.
[202] Rassamdana H, Dabir B, Nematy M, Farhani M, Sahimi M. Asphalt flocculation and deposition: I. The onset of precipitation. AIChE J 1996;42(1):10–22.
[203] Rassamdana H, Farhani M, Dabir B, Mozaffarian M, Sahimi M. Asphalt flocculation and deposition. V. Phase behavior in miscible and immiscible injections. Energy Fuels 1999;13(1):176–87.
[204] Hong E, Watkinson P. A study of asphaltene solubility and precipitation. Fuel 2004;83(14–15):1881–7.
[205] Bayat M, Sattarin M, Teymouri M. Prediction of asphaltene self-precipitation in dead crude oil. Energy Fuels 2008;22(1):583–6.

[206] Verdier S, Duong D, Andersen SI. Experimental determination of solubility parameters of oils as a function of pressure. Energy Fuels 2005;19(4):1225–9.

[207] Nellensteyn FJ. The colloidal structure of bitumens. Sci Pet 1938;4:2760–3.

[208] Hildebrand JH. Solubility and internal pressure. J ACS 1917;39(11):2297–301.

[209] Speight JG, Moschopedis SE. On the molecular nature of petroleum asphaltenes. In: Bunger JW, Li NC, editors. Adv. chem. series 195: chemistry of asphaltenes. ACS; 1981. p. 1–15.

[210] Speight JG. Structure of petroleum asphaltenes – current concepts. Alta Res Counc Inf Ser 1978;(81). 32 pp.

[211] Correra S. Asphaltene, or: how and why I became a fan of the regular solutions theory. In: Proc. 12th Joint European Thermodynamics Conf. Brescia, Italy; July 1–5, 2013. 6 pp.

[212] Correra S, Merino-Garcia D. Asphaltene deposition: determination of the solvent power of crude oil. In: Proc. Offshore Mediterranean Conf. Exhibition. Ravenna, Italy; March 16–18, 2005. 10 pp.

[213] Pineda LA, Trejo F, Ancheyta J. Correlation between properties of asphaltenes and precipitation conditions. Pet Sci Technol 2007;25(1 & 2):105–19.

[214] Strausz OP, Torres M, Lown EM, Safarik I, Murgich J. Equipartitioning of precipitant solubles between the solution phase and precipitated asphaltene in the precipitation of asphaltene. Energy Fuels 2006;20(5):2013–21.

[215] Goual L, Firoozabadi A. Effect of resins and DBSA on asphaltene precipitation from petroleum fluids. AIChE J 2004;50(2):470–9.

[216] Marcano F, Moura LGM, Cardoso FMR, Rosa PTV. Evaluation of the chemical additive effect on asphaltene aggregation in dead oils: a comparative study between ultraviolet–visible and near-infrared-laser light scattering techniques. Energy Fuels 2015;29(5):2813–22.

[217] Haji-Akbari N, Teeraphapkul P, Fogler HS. Effect of asphaltene concentration on the aggregation and precipitation tendency of asphaltenes. Energy Fuels 2014;28(2):909–19.

[218] Lopez-Chavez E, Pacheco-Sanchez JH, Martinez Magadan JM, Castillo-Alvarado FDL, Soto-Figueroa C, Garcia-Cruz I. Methodology for predicting the phase envelope of a heavy crude oil and its asphaltene deposition onset. Pet Sci Technol 2007;25(1 & 2):19–39.

[219] Andersen SI. Hysteresis in precipitation and dissolution of petroleum asphaltenes. Fuel Sci Technol Intern 1992;10(10):1743–9.

[220] Andersen SI. Dissolution of solid Boscan asphaltenes in mixed solvents. Fuel Sci Technol Intern 1994;12(11–12):1551–77.

[221] Andersen SI, Stenby EH. Thermodynamics of asphaltene precipitation and dissolution investigation of temperature and solvent effects. Fuel Sci Technol Intern 1996;14(1–2):261–87.

[222] Peramanu S, Singh C, Agrawala M, Yarranton HW. Investigation on the reversibility of asphaltene precipitation. Energy Fuels 2001;15(4):910–7.

[223] Roux J-N, Broseta D, Demé B. SANS study of asphaltene aggregation: concentration and solvent quality effects. Langmuir 2001;17(16):5085–92.

[224] Yarranton HW, Beck J, Alboudwarej H, Svrcek WY. Asphaltene self-association and precipitation modeling. In: Proc. ACS Div. Pet. Chem. Preprints. Boston, MA, vol. 47; 2002. p. 336–7.

[225] Aske N, Kallevik H, Johnsen EE, Sjöblom J. Asphaltene aggregation from crude oils and model systems studied by high-pressure NIR spectroscopy. Energy Fuels 2002;16(5):1287–95.

[226] Aske N. Characterisation of crude oil components, asphaltene aggregation and emulsion stability by means of near infrared spectroscopy and multivariate analysis (DOKTOR INGENIØR thesis from Norwegian University of Science and Technology, Department of Chem. Eng.). June 2002. 51 pp.

[227] Anisimov MA, Yudin IK, Nikitin V, Nikolaenko G, Chernoutsan A, Toulhoat H, et al. Asphaltene aggregation in hydrocarbon solutions studied by photon correlation spectroscopy. J Phys Chem 1995;99(23):9576–80.

[228] Andersen SI, Lira-Galeana C, Stenby EH. On the mass balance of asphaltene precipitation. Pet Sci Technol 2001;19(3–4):457–67.

[229] Joshi NB, Mullins OC, Jamaluddin A, Creek J, Mcfadden J. Asphaltene precipitation from live crude oil. Energy Fuels 2001;15(4):979–86.
[230] Hammami A, Phelps CH, Monger-Mcclure T, Little TM. Asphaltene precipitation from live oils: an experimental investigation of onset conditions and reversibility. Energy Fuels 2000;14(1):14–8.
[231] Wiehe IA, Kennedy RJ. Process for blending potentially incompatible petroleum oils to avoid being nearly incompatible, Patent No. US5997723 (Also published as WO0031006, EP1133461, CA2347798, TW499422B). Exxon Research and Engineering Co.; February 16, 2002.
[232] Ramos ACDS, Rolemberg MP, Moura LGMD, Zilio EL, Santos MDFPD, González G. Determination of solubility parameters of oils and prediction of oil compatibility. J Pet Sci Eng 2013;102:36–40.
[233] Yang X, Kilpatrick PK. Asphaltenes and waxes do not interact synergistically and coprecipitate in solid organic deposits. Energy Fuels 2005;19(4):1360–75.
[234] Garcıa MDC. Crude oil wax crystallization. The effect of heavy n-paraffins and flocculated asphaltenes. Energy Fuels 2000;14:1043–8.
[235] Kriz P, Andersen SI. Effect of asphaltenes on crude oil wax crystallization. Energy Fuels 2005;19(3):948–53.
[236] Shiryaeva RN, Kudasheva FK, Gimaev RN, Makeeva NS. Inhibiting the deposition of tars, asphaltenes, and waxes in oil pipelines with chemical reagents. Chem Technol Fuels Oils (Translation of Khimiya i Tekhnologiya Topliv i Masel) 2009;45(3):214–6.
[237] Merino-Garcia D, Margarone M, Correra S. Kinetics of waxy gel formation from batch experiments. Energy Fuels 2007;21(3):1287–95.
[238] Fuhr BJ, Cathrea C, Coates L, Kalra H, Majeed AI. Properties of asphaltenes from a waxy crude. Fuel 1991;70(11):1293–7.
[239] Venkatesan R, Östlund JA, Chawla H, Wattana P, Nydén M, Fogler HS. The effect of asphaltenes on the gelation of waxy oils. Energy Fuels 2003;17(6):1630–40.
[240] Thanh NX, Hsieh M, Philp RP. Waxes and asphaltenes in crude oils. Org Geochem 1999;30(2–3):119–32.
[241] Tinsley JF, Jahnke JP, Dettman HD, Prud'home RK. Waxy gels with asphaltenes 1: characterization of precipitation, gelation, yield stress, and morphology. Energy Fuels 2009;23(4):2056–64.
[242] Jahnke JP, Tinsley JF, Prud'homme RK, Dettman HD. Waxy gels with asphaltenes 2-reduction of network strength with polymer additives. In: Proc. AIChE Annual Meeting. Salt Lake City, UT; 2007. 1 pp.
[243] Tinsley JF, Jahnke JP, Adamson DH, Guo X, Amin D, Kriegel R, et al. Waxy gels with asphaltenes 2: use of wax control polymers. Energy Fuels 2009;23(4):2065–74.
[244] Oh K, Deo MD. Effect of organic additives on the onset of asphaltene precipitation. Energy Fuels 2002;16(3):694–9.
[245] Luo P, Gu Y. Characterization of a heavy oil-propane system in the presence or absence of asphaltene precipitation. Fluid Phase Equilib 2009;277(1):1–8.
[246] Eyssautier J, Frot D, Barré L. Structure and dynamic properties of colloidal asphaltene aggregates. Langmuir 2012;28(33):11997–2004.
[247] Eyssautier J, Levitz P, Espinat D, Jestin J, Gummel J, Grillo I, et al. Insight into asphaltene nanoaggregate structure inferred by small angle neutron and X-ray scattering. J Phys Chem B 2011;115(21):6827–37.
[248] Hoepfner MP, Boas Favero CV, Fogler HS. Asphaltene instability measured by small-angle neutron scattering. In: Proc. AIChE Annual Meeting. Minneapolis, MN; October 16–21, 2011. 1 pp.
[249] Hoepfner MP, Limsakoune V, Chuenmeechao V, Maqbool T, Scott Fogler H. A fundamental study of asphaltene deposition. Energy Fuels 2013;27(2):725–35.
[250] Hoepfner MP, Vilas Boîas Fávero C, Haji-Akbari N, Fogler HS. The fractal aggregation of asphaltenes. Langmuir 2013;29(28):8799–808.
[251] Sahimi M, Rassamdana H. Formation, growth and precipitation of fractal molecular aggregates in porous media. Phys A Stat Mech Appl 1997;240(3–4):419–31.

[252] Evdokimov IN, Eliseev NY, Akhmetov BR. Asphaltene dispersions in dilute oil solutions. Fuel 2006;85(10–11):1465–72.
[253] Fotland P, Anfindsen H, Fadnes FH. Detection of asphaltene precipitation and amounts precipitated by measurement of electrical conductivity. Fluid Phase Equilib 1993;82:157–64.
[254] Penzes S, Speight JG. Electrical conductivities of bitumen fractions in non-aqueous solvents. Fuel 1974;53(3):192–7.
[255] Anfindsen H, Fadnes FH, Fotland P. Method for the measurement of precipitation of asphaltene in oil, Patent No. US5420040 (Also published as EP0544354, CA2083218). Norsk Hydro AS; May 30, 1995.
[256] Li M, Liu C, Liang W. Study on deposition of asphaltene by electrical conductivity method. Acta Pet Sin Pet Proc Sect 1998;14(4):78–9.
[257] Zhang LL, Yang GH, Sun ZC, Que GH. Study on tile occurrence of asphaltene precipitation in middle East AR through mass fraction normalized conductivity. Acta Pet Sin Pet Proc Sect 2002;18(6):56–60.
[258] Wang JX, Brower KR, Buckley JS. Advances in observation of asphaltene destabilization. In: Proc. SPE Intern. Symp. Oilfield Chem. Houston, TX, USA; 1999. p. 441–4.
[259] Evdokimov IN, Eliseev NY, Akhmetov BR. Assembly of asphaltene molecular aggregates as studied by near-UV/visible spectroscopy: I. Structure of the absorbance spectrum. J Pet Sci Eng 2003;37(3–4):135–43.
[260] Eser S, Wang G, Wincek RT. Chemistry of carbonaceous mesophase development and petroleum coke texture. In: Proc. 240th ACS Nat. Meeting. Boston, MA; August 22–26, 2010. 17 pp.
[261] Christiansen S, Andersen SI, Buckley JS. Detecting and characterizing asphaltene instability. Pet Sci Technol 2004;22(7–8):719–33.
[262] Anisimov MA, Yudin IK. Dynamic light scattering monitoring of asphaltene aggregation. In: Proc. 238th ACS Nat. Meeting. Washington, DC; August 16–10, 2009. 1 pp.
[263] Aquino-Olivos MA, Buenrostro-Gonzalez E, Andersen SI, Lira-Galeana C. Investigations of inhibition of asphaltene precipitation at high pressure using bottomhole samples. Energy Fuels 2001;15(1):236–40.
[264] Dorset DL, Siskin M, Brown LD. Molecular assemblies in asphaltenes and their high-temperature coke products. 2. Levels of molecular organization. Energy Fuels 2011;25(10):4580–5.
[265] Wang JX, Brower KR, Buckley JS. Observation of asphaltene destabilization at elevated temperature and pressure. SPE J 2000;5(4):420–5.
[266] Correra S, Capuano F, Panariti N. Onset of asphaltene precipitation: analysis of measurement methods. In: Proc. 5th Intern. Conf. on Pet. Phase Behaviour Fouling. Banff, Canada; June 13–17, 2004. 5 pp.
[267] Castillo J, Canelo NC, Acevedo S, Carrier H, Daridon JL. Optical fiber extrinsic refractometer to measure RI of samples in a high pressure and temperature systems: application to wax and asphaltene precipitation measurements. Fuel 2006;85(14–15):2220–8.
[268] Arciniegas LM, Babadagli T. Quantitative and visual characterization of asphaltenic components of heavy-oil after solvent interaction at different temperatures and pressures. Fluid Phase Equilib 2014;366:74–87.
[269] Acevedo S, Ranaudo MA, Pereira JC, Castillo J, Fernández A, Pérez P, et al. Thermo-optical studies of asphaltene solutions: evidence for solvent-solute aggregate formation. Fuel 1999;78(9):997–1003.
[270] Li H, Wan WK. Investigation of the asphaltene precipitation process from Cold Lake bitumen by confocal scanning laser microscopy. In: Proc. SPE Intern. Heavy Oil Symp. Calgary, Canada; June 19–21, 1995. p. 709–40.
[271] Oh K, Ring TA, Deo MD. Asphaltene aggregation in organic solvents. J Colloid Interface Sci 2004;271(1):212–9.
[272] Schabron JF, Rovani Jr JF. Asphaltene determinator: on-column precipitation and redissolution of asphaltenes. In: Proc. 233rd ACS Nat. Meeting, Div. Pet. Chem. Chicago, IL, vol. 52; March 25–29, 2007. p. 54–7.
[273] Sztukowski DM, Jafari M, Alboudwarej H, Yarranton HW. Asphaltene self-association and water-in-hydrocarbon emulsions. J Colloid Interface Sci 2003;265(1):179–86.

[274] Tavakkoli M, Grimes MR, Liu X, Garcia CK, Correa SC, Cox QJ, et al. Indirect method: a novel technique for experimental determination of asphaltene precipitation. Energy Fuels 2015;29(5):2890–900.

[275] Jones GM, Povey MJW. Measurement and control of asphaltene agglomeration in hydrocarbon liquids, Patent No. US7360403. Baker Hughes; April 22, 2008.

[276] Han W, Birchak JR, Storm BH, Ritter TE. Acoustic sensor for fluid characterization, Patent No. US6672163. Halliburton Energy Services, Inc.; January 06, 2004.

[277] Pedersen KS, Rønningsen HP. Effect of precipitated wax on viscosity – a model for predicting non-newtonian viscosity of crude oils. Energy Fuels 2000;14(1):43–51.

[278] Li SP, Tang F, Shen BX, Zhang BL, Yang JY, Xu XR. A study of the characteristics in blended crudes at the optimal activation state: Part I. Pet Sci Technol 2008;26(13):1499–509.

[279] Durand E, Clemancey M, Lancelin JM, Verstraete J, Espinat D, Quoineaud AA. Aggregation states of asphaltenes: evidence of two chemical behaviors by 1H diffusion-ordered spectroscopy nuclear magnetic resonance. J Phys Chem C 2009;113(36):16266–76.

[280] Buenrostro-Gonzalez E, Lira-Galeana C, Gil-Villegas A, Wu J. Asphaltene precipitation in crude oils: theory and experiments. AIChE J 2004;50(10):2552–70.

[281] Fossen M, Kallevik H, Knudsen KD, Sjöblom J. Asphaltenes precipitated by a two-step precipitation procedure. 2. Physical and chemical characteristics. Energy Fuels 2011;25(8):3552–67.

[282] Kawashima H, Takanohashi T, Iino M, Matsukawa S. Determining asphaltene aggregation in solution from diffusion coefficients as determined by pulsed-field gradient spin-echo 1H NMR. Energy Fuels 2008;22(6):3989–93.

[283] Östlund JA, Löfroth JE, Holmberg K, Nyden M. Flocculation behavior of asphaltenes in solvent/nonsolvent systems. J Colloid Interface Sci 2002;253(1):150–8.

[284] Chen Y, Wang Y, Wu C, Xia F. Laboratory experiments and field tests of an amphiphilic metallic chelate for catalytic aquathermolysis of heavy oil. Energy Fuels 2008;22(3):1502–8.

[285] Zajac GW, Sethi NK, Joseph JT. Molecular imaging of petroleum asphaltenes by scanning-tunneling-microscopy – verification of structure from C-13 and proton nuclear-magnetic-resonance data. Scanning Microsc 1994;8(3):463–70.

[286] Seidl PR, Chrisman ECa N, Carvalho CCV, Leal KZ, De Menezes SMC. NMR analysis of asphaltenes separated from vacuum residues by selected solvents. J Dispers Sci Technol 2004;25(3):349–53.

[287] Opedal NVDT. NMR as a tool to follow destabilization of water-in-oil emulsions (Ph.D. thesis from Norwegian University of Science and Technology, Department of Chem. Eng.). September 2011. 177 pp.

[288] Miknis FP, Pauli AT, Michon LC, Netzel DA. NMR imaging studies of asphaltene precipitation in asphalts. Fuel 1998;77(5):399–405.

[289] Sanchez-Minero F, Ancheyta J, Silva-Oliver G, Flores-Valle S. Predicting SARA composition of crude oil by means of NMR. Fuel 2013;110:318–21.

[290] Nielsen KE, Dittmer J, Malmendal A, Nielsen NC. Quantitative analysis of constituents in heavy fuel oil by 1H nuclear magnetic resonance (NMR) spectroscopy and multivariate data analysis. Energy Fuels 2008;22(6):4070–6.

[291] Östlund JA, Andersson SI, Nydén M. Studies of asphaltenes by the use of pulsed-field gradient spin echo NMR. Fuel 2001;80(11):1529–33.

[292] Miknis FP, Pauli AT. Nuclear magnetic resonance imaging studies of asphaltene precipitation in aged and unaged asphalts. ACS Div Fuel Chem 1997;42(2):374–7. Preprints.

[293] Gabrienko AA, Lai CH, Kazarian SG. In situ chemical imaging of asphaltene precipitation from crude oil induced by *n*-heptane. Energy Fuels 2014;28(2):964–71.

[294] Hoepfner MP, Maqbool T, Scott Fogler H. Effects of precipitant concentration on asphaltene deposition. In: Proc. AIChE Annual Meeting. Nashville, TN; 2009. 1 pp.

[295] Maqbool T, Raha S, Hoepfner MP, Fogler HS. Modeling the aggregation of asphaltene nanoaggregates in crude oil-precipitant systems. Energy Fuels 2011;25(4):1585–96.

[296] Maqbool T, Srikiratiwong P, Fogler HS. Effect of temperature on the precipitation kinetics of asphaltenes. Energy Fuels 2011;25(2):694–700.
[297] Wang S, Liu J, Zhang L, Masliyah J, Xu Z. Interaction forces between asphaltene surfaces in organic solvents. Langmuir 2010;26(1):183–90.
[298] Espinat D, Ravey JC, Guille V, Lambard J, Zemb T, Cotton JP. Colloidal macrostructure of crude oil studied by neutron and X-ray small angle scattering techniques. J Phys IV JP 1993;3(8):181–4.
[299] Espinat D, Ravey JC. Colloidal structure of asphaltene solutions and heavy-oil fractions studied by small-angle neutron and X-ray scattering. In: Proc. SPE Intern. Symp. Oilfield Chem. New Orleans, LA, USA; March 2–5, 1993. p. 365–73.
[300] Agrawala M, Yarranton HW. An asphaltene association model analogous to linear polymerization. Ind Eng Chem Res 2001;40(21):4664–72.
[301] Sirota EB, Lin MY. Physical behavior of asphaltenes. Energy Fuels 2007;21(5):2809–15.
[302] Merino-Garcia D, Andersen SI. Calorimetric evidence about the application of the concept of CMC to asphaltene self-association. J Dispers Sci Technol 2005;26(2):217–25.
[303] Wang J, Buckley JS. Asphaltene stability in crude oil and aromatic solvents-the influence of oil composition. Energy Fuels 2003;17(6):1445–51.
[304] Dickie JP, Yen TF. Macrostructures of the asphaltic fractions by various instrumental methods. Anal Chem 1967;39(14):1847–52.
[305] Sharma A, Groenzin H, Tomita A, Mullins OC. Probing order in asphaltenes and aromatic ring systems by HRTEM. Energy Fuels 2002;16(2):490–6.
[306] Groenzin H, Mullins OC. Molecular size and structure of asphaltenes from various sources. Energy Fuels 2000;14(3):677–84.
[307] Andreatta G, Goncalves CC, Buffin G, Bostrom N, Quintella CM, Arteaga-Larios F, et al. Nanoaggregates and structure-function relations in asphaltenes. Energy Fuels 2005;19:1282–9.
[308] Lisitza NV, Freed DE, Sen PN, Song YQ. Study of asphaltene nanoaggregation by nuclear magnetic resonance (NMR). Energy Fuels 2009;23(3):1189–93.
[309] Mason TG, Lin MY. Asphaltene nanoparticle aggregation in mixtures of incompatible crude oils. Phys Rev E 2003;67(5). Paper #050401, 4 pp.
[310] Priyanto S, Mansoori GA, Suwono A. Measurement of property relationships of nano-structure micelles and coacervates of asphaltene in a pure solvent. Chem Eng Sci 2001;56:6933–9.
[311] Acevedo S, García LA, Rodríguez P. Changes of diameter distribution with temperature measured for asphaltenes and their fractions A1 and A2. Impact of these measurements in colloidal and solubility issues of asphaltenes. Energy Fuels 2012;26(3):1814–9.
[312] Mostowfi F, Indo K, Mullins OC, Mcfarlane R. Asphaltene nanoaggregates studied by centrifugation. Energy Fuels 2009;23(3):1194–200.
[313] Zeng H, Song YQ, Johnson DL, Mullins OC. Critical nanoaggregate concentration of asphaltenes by direct-current (dc) electrical conductivity. Energy Fuels 2009;23(3):1201–8.
[314] Headen TF, Boek ES, Skipper NT. Evidence for asphaltene nanoaggregation in toluene and heptane from molecular dynamics simulations. Energy Fuels 2009;23(3):1220–9.
[315] Mansoori GA. Nanoscale structures of asphaltene molecules. Asphaltene steric-colloid and asphaltene micelles and vesicles. 2005. Available from: http://tigger.uic.edu/~mansoori/Asphaltene.Molecule_html.
[316] Zhao B. High temperature behaviours of asphaltene aggregates in heavy feedstocks and in mixtures with diluents (Ph.D. thesis from University of Alberta, Department of Chem. Mat. Eng.). 2008. 197 pp.
[317] Mullins OC, Betancourt SS, Cribbs ME, Dubost FX, Creek JL, Andrews AB, et al. The colloidal structure of crude oil and the structure of oil reservoirs. Energy Fuels 2007;21(5):2785–94.
[318] Carbognani L, Rogel E. Solid petroleum asphaltenes seem surrounded by alkyl layers. Pet Sci Technol 2003;21(3–4):537–56.
[319] Murgich J. Intermolecular forces in aggregates of asphaltenes and resins. Pet Sci Technol 2002;20(9–10):983–97.

[320] Akbarzadeh K, Bressler DC, Wang J, Gawrys KL, Gray MR, Kilpatrick PK, et al. Association behavior of pyrene compounds as models for asphaltenes. Energy Fuels 2005;19(4):1268–71.
[321] Breure B, Subramanian D, Leys J, Peters CJ, Anisimov MA. Modeling asphaltene aggregation with a single compound. Energy Fuels 2012;27(1):172–6.
[322] Tan X, Fenniri H, Gray MR. Water enhances the aggregation of model asphaltenes in solution via hydrogen bonding. Energy Fuels 2009;23(7):3687–93.
[323] Fu Q, He T, Guo W, Zhao L, Chai Y, Zhou T, et al. Theoretical study of substituent effect on acridine as one model molecule of asphaltene. J Mol Struct Theochem 2009;906(1–3):6–10.
[324] Nordgård EL, Sjoblom J. Model compounds for asphaltenes and c80 isoprenoid tetraacids. Part I: synthesis and interfacial activities. J Dispers Sci Technol 2008;29(8):1114–22.
[325] Liu D, Kong X, Li M, Wang Z. Study on the aggregation of residue-derived asphaltene molecules. Energy Fuels 2010;24(6):3624–7.
[326] Mullins OC. Molecular structure and aggregation of asphaltenes and petroleomics. In: Proc. SPE Annual Technical Conf. Exhibition. Dallas, TX; 2005. p. 1527–36.
[327] Peng P, Morales-Izquierdo A, Hogg A, Strausz OP. Molecular structure of Athabasca asphaltene: sulfide, ether, and ester linkages. Energy Fuels 1997;11(6):1171–87.
[328] Frakman Z, Ignasiak TM, Lown EM, Strausz OP. Oxygen compounds in Athabasca asphaltene. Energy Fuels 1990;4(3):263–70.
[329] Daaou M, Bendedouch D, Modarressi A, Rogalski M. Properties of the polar fraction of hassi-messaoud asphaltenes. Energy Fuels 2012;26(9):5672–8.
[330] Murgich J. Molecular simulation and the aggregation of the heavy fractions in crude oils. Mol Simul 2003;29(6):451–61.
[331] Murgich J, Abanero JA, Strausz OP. Molecular recognition in aggregates formed by asphaltene and resin molecules from the Athabasca oil sand. Energy Fuels 1999;13(2):278–86.
[332] Murgich J, Isea R, Abanero J, Strausz OP. The stability of the asphaltene and resin aggregates and their chemical reactivity. ACS Div Fuel Chem 1999;44(4):746–9. Preprints.
[333] Murgich J, Jesus Rodríguez M, Izquierdo A, Carbognani L, Rogel E. Interatomic interactions in the adsorption of asphaltenes and resins on kaolinite calculated by molecular dynamics. Energy Fuels 1998;12(2):339–43.
[334] Strausz OP, Peng P, Murgich J. About the colloidal nature of asphaltenes and the MW of covalent monomeric units. Energy Fuels 2002;16(4):809–22.
[335] Yarranton HW, Alboudwarej H, Jakher R. Investigation of asphaltene association with vapor pressure osmometry and interfacial tension measurements. Ind Eng Chem Res 2000;39(8):2916–24.
[336] Li S, Liu C, Que G, Liang W, Zhu Y. A study of the interactions responsible for colloidal structures in petroleum residua. Fuel 1997;76(14–15):1459–63.
[337] Andersen SI, Birdi KS. Aggregation of asphaltenes as determined by calorimetry. J Colloid Interface Sci 1991;142(2):497–502.
[338] Andersen SI, Christensen SD. The critical micelle concentration of asphaltenes as measured by calorimetry. Energy Fuels 2000;14(1):38–42.
[339] Andersen SI, Del Rio JM, Khvostitchenko D, Shakir S, Lira-Galeana C. Interaction and solubilization of water by petroleum asphaltenes in organic solution. Langmuir 2001;17(2):307–13.
[340] Andersen SI. Concentration effects in HPLC-SEC analysis of petroleum asphaltenes. J Liq Chromatogr 1994;17(19):4065–79.
[341] Yudin IK, Nikolaenko GL, Gorodetskii EE, Markhashov EL, Agayan VA, Anisimov MA, et al. Crossover kinetics of asphaltene aggregation in hydrocarbon solutions. Phys A Stat Theor Phys 1998;251(1–2):235–44.
[342] Storm DA, Barresi RJ, Sheu EY. Evidence for the micellization of asphaltenic molecules in vacuum residue. In: Proc. 210th ACS Nat. Meeting. Chicago, IL, USA, vol. 40. ACS; August 20–24, 1995. p. 776–9.

[343] Priyanto S, Mansoori GA, Suwono A. Structure & properties of micelles and micelle coacervates of asphaltene macromolecule. In: Proc. AIChE Annual Meeting. Reno, NV. USA; November 4–9, 2001. Session 90: Nanotools, 1 pp.
[344] Galtsev VE, Ametov IM, Grinberg OY. Asphaltene association in crude oil as studied by ENDOR. Fuel 1995;74(5):670–3.
[345] Mansoori GA. Modeling of asphaltene and other heavy organic depositions. J Pet Sci Eng 1997;17(1–2):101–11.
[346] Rogacheva OV, Rimaev RN, Gubaidullin VZ, Khakimov DK, Speight JG, Wernick DL, et al. Molecular weight and association of asphaltenes: a critical review. Colloid J USSR 1980;42(1): 490–9.
[347] Loh W, Mohamed RS, Ramos ACS. Aggregation of asphaltenes obtained from a Brazilian crude oil in aromatic solvents. Pet Sci Technol 1999;17(1–2):147–63.
[348] Oh K, Oblad SC, Hanson FV, Deo MD. Examination of asphaltenes precipitation and self-aggregation. Energy Fuels 2003;17(2):508–9.
[349] Juyal P, Merino-Garcia D, Andersen SI. Effect on molecular interactions of chemical alteration of petroleum asphaltenes. I. Energy Fuels 2005;19(4):1272–81.
[350] Murgich J, Merino-Garcia D, Andersen SI, Del Rio JM, Galeana CL. Molecular mechanics and microcalorimetric investigations of the effects of molecular water on the aggregation of asphaltenes in solutions. Langmuir 2002;18(23):9080–6.
[351] Yen TF, Erdman JG, Saraceno AJ. Investigation of the nature of free radicals in petroleum asphaltenes and related substances by electron spin resonance. Anal Chem 1962;34(6): 694–700.
[352] Yen TF. Structure of petroleum asphaltene and its significance. Energy Sources Part A Recover Util Environ Eff 1974;1(4):447–63.
[353] Faber J. Towards small scale use of asphalt as a fuel. Report No. Chemiewinkel Rapport C102. The Netherlands: University of Groningen; January 2002. 30 pp.
[354] Gray MR, Tykwinski RR, Stryker JM, Tan X. Supramolecular assembly model for aggregation of petroleum asphaltenes. Energy Fuels 2011;25(7):3125–34.
[355] Acevedo S, Castro A, Negrin JG, Fernandez A, Escobar G, Piscitelli V, et al. Relations between asphaltene structures and their physical and chemical properties: the rosary-type structure. Energy Fuels 2007;21(4):2165–75.
[356] Acevedo S, Zuloaga C, Rodríguez P. Aggregation-dissociation studies of asphaltene solutions in resins performed using the combined freeze fracture-transmission electron microscopy technique. Energy Fuels 2008;22(4):2332–40.
[357] Mullins OC. The modified Yen model. Energy Fuels 2010;24(4):2179–207.
[358] Mullins OC, Seifert DJ, Zuo JY, Zeybek M. Clusters of asphaltene nanoaggregates observed in oilfield reservoirs. Energy Fuels 2013;27(4):1752–61.
[359] Dutta Majumdar R, Gerken M, Mikula R, Hazendonk P. Validation of the Yen–Mullins model of Athabasca oil-sands asphaltenes using solution-state 1H NMR relaxation and 2d HSQC spectroscopy. Energy Fuels 2013;27(11):6528–37.
[360] Barre L, Espinat D, Rosenberg E, Scarsella M. Colloidal structure of heavy crudes and asphaltene solutions. Oil Gas Sci Technol Rev IFP 1997;52(2):161–75.
[361] Gmachowski L, Paczuski M. Fractal dimension of asphaltene aggregates determined by turbidity. Colloids Surf A Physicochem Eng Asp 2011;384(1–3):461–5.
[362] Liu YC, Sheu EY, Chen SH, Storm DA. Fractal structure of asphaltenes in toluene. Fuel 1995;74(9):1352–6.
[363] Acevedo S, Ranaudo MA, Escobar G, Gutierrez LB, Gutierrez X. A unified view of the colloidal nature of asphaltenes. In: Sheu EY, Mullins OC, editors. Asphaltene constituents: fundamentals applications. New York: Plenum Press; 1995. p. 131–54.

[364] Gawrys KL. How asphaltenes aggregate: role of chemistry and solvent (Ph.D. thesis from North Caroline State University, Chem. Eng.). 2005. 437 pp.

[365] Ekulu G, Nicolas C, Achard C, Rogalski M. Characterization of aggregation processes in crude oils using differential scanning calorimetry. Energy Fuels 2005;19:1297–302.

[366] Savvidis TG, Fenistein D, Barré L, Béhar E. Aggregated structure of flocculated asphaltenes. AIChE J 2001;47(1):206–11.

[367] Espinat D, Fenistein D, Barré L, Frot D, Briolant Y. Effects of temperature and pressure on asphaltenes agglomeration in toluene. A light, X-ray, and neutron scattering investigation. Energy Fuels 2004;18(5):1243–9.

[368] Nomura M, Rahimi Parviz M, Koseoglu Omer R. Advances in characterization and utilization of asphaltenes. In: Heavy hydrocarbon resources. Washington, DC: ACS; 2005. p. 1–17.

[369] Zhang Y, Takanohashi T, Sato S, Kondo T, Saito I, Tanaka R. Dissolution and dilution of asphaltenes in organic solvents. Energy Fuels 2003;17(1):101–6.

[370] Porte G, Zhou H, Lazzeri V. Reversible description of asphaltene colloidal association and precipitation. Langmuir 2002;19(1):40–7.

[371] Yudin IK, Nikolaenko GL, Gorodetskii EE, Kosov VI, Melikyan VR, Markhashov EL, et al. Mechanisms of asphaltene aggregation in toluene-heptane mixtures. J Pet Sci Eng 1998;20(3–4):297–301.

[372] Yudin IK, Nikolaenko GL, Gorodetskii EE, Markhashov EL, Frot D, Briolant Y, et al. Universal behavior of asphaltene aggregation in hydrocarbon solutions. Pet Sci Technol 1998;16(3–4):395–414.

[373] Monte MBM, Coelho RR, Middea A. Investigation of molecular weight and aggregation of asphaltenes in organic solvents using surface tension measurements. Pet Sci Technol 2004;22(7–8):991–1001.

[374] Wang JX, Buckley JS. A two-component solubility model of the onset of asphaltene flocculation in crude oils. Energy Fuels 2001;15(5):1004–12.

[375] Derakhshesh M, Gray MR, Dechaine GP. Dispersion of asphaltene nanoaggregates and the role of rayleigh scattering in the absorption of visible electromagnetic radiation by these nanoaggregates. Energy Fuels 2013;27(2):680–93.

[376] Derakhshesh M. Asphaltene aggregation and fouling behavior (Ph.D. thesis from University of Alberta, Chemical and Materials Engineering). 2012. 194 pp.

[377] Yasar M, Akmaz S, Gurkaynak MA. Investigation of the molecular structure of Turkish asphaltenes. Pet Sci Technol 2009;27(10):1044–61.

[378] Ali FA, Ghaloum N, Hauser A. Structure representation of asphaltene GPC fractions derived from Kuwaiti residual oils. Energy Fuels 2006;20(1):231–8.

[379] Barth T, Høiland S, Fotland P, Askvik KM, Myklebust R, Erstad K. Relationship between the content of asphaltenes and bases in some crude oils. Energy Fuels 2005;19:1624–30.

[380] Gawrys KL, Spiecker PM, Kilpatrick PK. The role of asphaltene solubility and chemical composition on asphaltene aggregation. Pet Sci Technol 2003;21(3–4):461–89.

[381] Barre L, Palermo T, Simon S. Relationships between solution and interfacial properties of asphaltene aggregates. In: Proc. Molecular Structure of Heavy Oils Coal Liquefaction Products. Lyon, France; April 12–13, 2007. 8 pp.

[382] Barré L, Simon S, Palermo T. Solution properties of asphaltenes. Langmuir 2008;24(8):3709–17.

[383] Barré L, Jestin J, Morisset A, Palermo T, Simon S. Relation between nanoscale structure of asphaltene aggregates and their macroscopic solution properties. Oil Gas Sci Technol Rev IFP 2009;64(5):617–28.

[384] Simon S, Jestin J, Palermo T, Barre L. Relation between solution and interfacial properties of asphaltene aggregates. Energy Fuels 2009;23(1):306–13.

[385] Akbarzadeh K, Dhillon A, Svrcek WY, Yarranton HW. Methodology for the characterization and modeling of asphaltene precipitation from heavy oils diluted with n-alkanes. Energy Fuels 2004;18(5):1434–41.

[386] Evdokimov IN, Losev AP. On the nature of UV/Vis absorption spectra of asphaltenes. Pet Sci Technol 2007;25(1 & 2):55–66.

[387] Strausz OP, Safarik I, Lown EM, Morales-Izquierdo A. A critique of asphaltene fluorescence decay and depolarization-based claims about molecular weight and molecular architecture. Energy Fuels 2008;22(2):1156−66.

[388] Badre S, Carla Goncalves C, Norinaga K, Gustavson G, Mullins OC. Molecular size and weight of asphaltene and asphaltene solubility fractions from coals, crude oils and bitumen. Fuel 2006;85(1):1−11.

[389] Ralston CY, Mitra-Kirtley S, Mullins OC. Small population of one to three fused-aromatic ring moieties in asphaltenes. Energy Fuels 1996;10(3):623−30.

[390] Groenzin H, Mullins OC. Asphaltene molecular size and structure. J Phys Chem A 1999;103(50):11237−45.

[391] Ascanius BE, Garcia DM, Andersen SI. Analysis of asphaltenes subfractionated by *N*-methyl-2-pyrrolidone. Energy Fuels 2004;18(6):1827−31.

[392] Al-Muhareb E, Morgan TJ, Herod AA, Kandiyoti R. Characterization of petroleum asphaltenes by size exclusion chromatography, UV-fluorescence and mass spectrometry. Pet Sci Technol 2007;25(1 & 2):81−91.

[393] Riveros L, Jaimes B, Ranaudo MA, Castillo J, Chirinos J. Determination of asphaltene and resin content in Venezuelan crude oils by using fluorescence spectroscopy and partial least squares regression. Energy Fuels 2006;20(1):227−30.

[394] Karaka F, Millan-Agorio M, Morgan TJ, Bull ID, Herod A, Kandiyoti R. The pentane- and toluene-soluble fractions of a petroleum residue and three coal tars by size exclusion chromatography and UV-fluorescence spectroscopy. Oil Gas Sci Technol Rev IFP 2008;63(1):129−37.

[395] Strausz OP, Safarik I, Lown EM. Cause of asphaltene fluorescence intensity variation with molecular weight and its ramifications for laser ionization mass spectrometry. Energy Fuels 2009;23(3):1555−62.

[396] Akmaz S, Iscan O, Gurkaynak MA, Yasar M. The structural characterization of saturate, aromatic, resin, and asphaltene fractions of Batiraman crude oil. Pet Sci Technol 2011;29(2):160−71.

[397] Ávila BMF, Pereira VB, Gomes AO, Azevedo DA. Speciation of organic sulfur compounds using comprehensive two-dimensional gas chromatography coupled to time-of-flight mass spectrometry: a powerful tool for petroleum refining. Fuel 2014;126:188−93.

[398] Boduszynski MM. Composition of heavy petroleums. 1. Molecular weight, hydrogen deficiency, and heteroatom concentration as a function of atmospheric equivalent boiling point up to 1400°F (760°c). Energy Fuels 1987;1(1):2−11.

[399] Borton D, Pinkston DS, Hurt MR, Tan X, Azyat K, Scherer A, et al. Molecular structures of asphaltenes based on the dissociation reactions of their ions in mass spectrometry. Energy Fuels 2010;24(10):5548−59.

[400] Calemma V, Rausa R, D'anton P, Montanari L. Characterization of asphaltenes molecular structure. Energy Fuels 1998;12(2):422−8.

[401] Caumette G, Lienemann C-P, Merdrignac I, Bouyssiere B, Lobinski R. Element speciation analysis of petroleum and related materials. J Anal At Spectrom 2009;24(3):263−76.

[402] Cunico RL, Sheu EY, Mullins OC. Molecular weight measurement of ug8 asphaltene using APCI mass spectroscopy. Pet Sci Technol 2004;22(7−8):787−98.

[403] Domin M, Herod A, Kandiyoti R, Larsen JW, Lazaro MJ, Li S, et al. A comparative study of bitumen molecular-weight distributions. Energy Fuels 1999;13(3):552−7.

[404] Green JA, Green JB, Grigsby RD, Pearson CD, Reynolds JW, Shay JY, et al. Analysis of heavy oils: method development and application to Cerro Negro heavy petroleum. Report No. NIPER-459. Bartlesville, OK: National Institute for Petroleum and Energy Research; 1989. 533 pp.

[405] Herod AA. Limitations of mass spectrometric methods for the characterization of polydisperse materials. Rapid Commun Mass Spectrom 2010;24(17):2507−19.

[406] Herod AA, Bartle KD, Kandiyoti R. Characterization of heavy hydrocarbons by chromatographic and mass spectrometric methods: an overview. Energy Fuels 2007;21(4):2176−203.

[407] Herod AA, Bartle KD, Morgan TJ, Kandiyoti R. Analytical methods for characterizing high-mass complex polydisperse hydrocarbon mixtures: an overview. Chem Rev 2012;112(7):3892–923.

[408] Hortal AR, Hurtado P, Martinez-Haya B, Mullins OC. Molecular-weight distributions of coal and petroleum asphaltenes from laser desorption/ionization experiments. Energy Fuels 2007;21(5):2863–8.

[409] Hortal AR, Martínez-Haya B, Lobato MD, Pedrosa JM, Lago S. On the determination of molecular weight distributions of asphaltenes and their aggregates in laser desorption ionization experiments. J Mass Spectrom 2006;41(7):960–8.

[410] Hurt MR, Borton DJ, Choi HJ, Kenttämaa HI. Comparison of the structures of molecules in coal and petroleum asphaltenes by using mass spectrometry. Energy Fuels 2013;27(7):3653–8.

[411] Martínez-Haya B, Hortal AR, Hurtado P, Lobato MD, Pedrosa JM. Laser desorption/ionization determination of molecular weight distributions of polyaromatic carbonaceous compounds and their aggregates. J Mass Spectrom 2007;42(6):701–13.

[412] Mendez A, Meneghini R, Lubkowitz J. Physical and chemical characterization of petroleum products by GC-MS. J Chromatogr Sci 2007;45(10):683–9.

[413] Miller JT, Fisher RB, Thiyagarajan P, Winans RE, Hunt JE. Subfractionation and characterization of Mayan asphaltene. Energy Fuels 1998;12(6):1290–8.

[414] Morgan TJ, George A, Alvarez-Rodriguez P, Millan M, Herod AA, Kandiyoti R. Estimating molecular masses of petroleum-derived fractions: high mass (>2000 u) materials in maltenes and asphaltenes from Maya crude oil. J Chromatogr A 2010;1217(24):3804–18.

[415] Panda SK, Andersson JT, Schrader W. Mass-spectrometric analysis of complex volatile and nonvolatile crude oil components: a challenge. Anal Bioanal Chem 2007;389(5):1329–39.

[416] Pantoja PA, Mendes MA, Nascimento CAO. Contribution of mass spectrometry in assessing quality of petroleum fractions. The use of mass spectrometry for assessing asphaltenes. J Pet Sci Eng 2013;109:198–205.

[417] Payzant JD, Rubinstein I, Hogg AM, Strausz OP. Field-ionization mass spectrometry: application to geochemical analysis. Geochim Cosmochim Acta 1979;43(8):1187–93.

[418] Petrova LM, Abbakumova NA, Zaidullin IM, Borisov DN. Polar-solvent fractionation of asphaltenes from heavy oil and their characterization. Pet Chem 2013;53(2):81–6.

[419] Pinkston DS, Duan P, Gallardo VA, Habicht SC, Tan X, Qian K, et al. Analysis of asphaltenes and asphaltene model compounds by laser-induced acoustic desorption/fourier transform ion cyclotron resonance mass spectrometry. Energy Fuels 2009;23(11):5564–70.

[420] Pomerantz AE, Hammond MR, Morrow AL, Mullins OC, Zare RN. Two-step laser mass spectrometry of asphaltenes. JACS 2008;130(23):7216–7.

[421] Pomerantz AE, Hammond MR, Morrow AL, Mullins OC, Zare RN. Asphaltene molecular-mass distribution determined by two-step laser mass spectrometry. Energy Fuels 2009;23(3):1162–8.

[422] Pomerantz AE, Mullins OC, Paul G, Ruzicka J, Sanders M. Orbitrap mass spectrometry: a proposal for routine analysis of nonvolatile components of petroleum. Energy Fuels 2011;25(7):3077–82.

[423] Rizzi A, Cosmina P, Flego C, Montanari L, Seraglia R, Traldi P. Laser desorption/ionization techniques in the characterization of high molecular weight oil fractions. Part 1: asphaltenes. J Mass Spectrom 2006;41(9):1232–41.

[424] Roussis SG, Proulx R. Probing the molecular weight distributions of non-boiling petroleum fractions by Ag^+ electrospray ionization mass spectrometry. Rapid Commun Mass Spectrom 2004;18(15):1761–75.

[425] Sabbah H, Morrow AL, Pomerantz AE, Mullins OC, Tan X, Gray MR, et al. Comparing laser desorption/laser ionization mass spectra of asphaltenes and model compounds. Energy Fuels 2010;24(6):3589–94.

[426] Sabbah H, Pomerantz AE, Wagner M, Müllen K, Zare RN. Laser desorption single-photon ionization of asphaltenes: mass range, compound sensitivity, and matrix effects. Energy Fuels 2012;26(6):3521–6.

[427] Wu Q, Pomerantz AE, Mullins OC, Zare RN. Laser-based mass spectrometric determination of aggregation numbers for petroleum- and coal-derived asphaltenes. Energy Fuels 2013;28(1):475–82.

[428] Abdallah WA, Taylor SD. Study of asphaltenes adsorption on metallic surface using XPS and TOF-SIMS. J Phys Chem C 2008;112(48):18963–72.
[429] Arnaud CH. Digging into asphaltenes. Chem Eng News 2009;87(38):12–7.
[430] Bae E, Na JG, Chung SH, Kim HS, Kim S. Identification of about 30 000 chemical components in shale oils by electrospray ionization (ESI) and atmospheric pressure photoionization (APPI) coupled with 15 T fourier transform ion cyclotron resonance mass spectrometry (FT-ICR MS) and a comparison to conventional oil. Energy Fuels 2010;24(4):2563–9.
[431] Barrow MP, Headley JV, Peru KM, Derrick PJ. Fourier transform ion cyclotron resonance mass spectrometry of principal components in oilsands naphthenic acids. J Chromatogr A 2004;1058(1–2):51–9.
[432] Barrow MP, Headley JV, Peru KM, Derrick PJ. Data visualization for the characterization of naphthenic acids within petroleum samples. Energy Fuels 2009;23(5):2592–9.
[433] Barrow MP, Mcdonnell LA, Feng X, Walker J, Derrick PJ. Determination of the nature of naphthenic acids present in crude oils using nanospray fourier transform ion cyclotron resonance mass spectrometry: the continued battle against corrosion. Anal Chem 2003;75(4):860–6.
[434] Fu J, Klein GC, Smith DF, Kim S, Rodgers RP, Hendrickson CL, et al. Comprehensive compositional analysis of hydrotreated and untreated nitrogen-concentrated fractions syncrude oil by electron ionization, field desorption ionization, and electrospray ionization ultrahigh-resolution FT-ICR mass spectrometry. Energy Fuels 2006;20(3):1235–41.
[435] Gaspar A, Zellermann E, Lababidi S, Reece J, Schrader W. Characterization of saturates, aromatics, resins, and asphaltenes heavy crude oil fractions by atmospheric pressure laser ionization Fourier transform ion cyclotron resonance mass spectrometry. Energy Fuels 2012;26(6):3481–7.
[436] Headley JV, Peru KM, Mohamed MH, Wilson L, Mcmartin DW, Mapolelo MM, et al. Electrospray ionization Fourier transform ion cyclotron resonance mass spectrometry characterization of tunable carbohydrate-based materials for sorption of oil sands naphthenic acids. Energy Fuels 2013;27(4):1772–8.
[437] Hsu CS, Hendrickson CL, Rodgers RP, Mckenna AM, Marshall AG. Petroleomics: advanced molecular probe for petroleum heavy ends. J Mass Spectrom 2011;46(4):337–43.
[438] Hughey CA, Galasso SA, Zumberge JE. Detailed compositional comparison of acidic NSO compounds in biodegraded reservoir and surface crude oils by negative ion electrospray Fourier transform ion cyclotron resonance mass spectrometry. Fuel 2007;86(5–6):758–68.
[439] Hughey CA, Rodgers RP, Marshall AG. Resolution of 11 000 compositionally distinct components in a single electrospray ionization Fourier transform ion cyclotron resonance mass spectrum of crude oil. Anal Chem 2002;74(16):4145–9.
[440] Hughey CA, Rodgers RP, Marshall AG, Qian K, Robbins WK. Identification of acidic NSO compounds in crude oils of different geochemical origins by negative ion electrospray Fourier transform ion cyclotron resonance mass spectrometry. Org Geochem 2002;33(7):743–59.
[441] Hughey CA, Rodgers RP, Marshall AG, Walters CC, Qian K, Mankiewicz P. Acidic and neutral polar NSO compounds in smackover oils of different thermal maturity revealed by electrospray high field Fourier transform ion cyclotron resonance mass spectrometry. Org Geochem 2004;35(7):863–80.
[442] Juyal P, Yen AT, Rodgers RP, Allenson S, Wang J, Creek J. Compositional variations between precipitated and organic solid deposition control (OSDC) asphaltenes and the effect of inhibitors on deposition by electrospray ionization Fourier transform ion cyclotron Resonance (FT-ICR) mass spectrometry. Energy Fuels 2009;24(4):2320–6.
[443] Kekäläinen T, Pakarinen JMH, Wickström K, Lobodin VV, Mckenna AM, Jänis J. Compositional analysis of oil residues by ultrahigh-resolution Fourier transform ion cyclotron resonance mass spectrometry. Energy Fuels 2013;27(4):2002–9.
[444] Klein GC. Petroleomics: applications in the fingerprinting of the acidic and basic crude oil components detected by electrospray ionization Fourier transform ion cyclotron resonance mass spectrometry (Doctor of Philosophy thesis from Florida State University, Department of Chemistry and Biochemistry). October 12, 2005. 145 pp.

[445] Klein GC, Angström A, Rodgers RP, Marshall AG. Use of saturates/aromatics/resins/asphaltenes (SARA) fractionation to determine matrix effects in crude oil analysis by electrospray ionization Fourier transform ion cyclotron resonance mass spectrometry. Energy Fuels 2006;20(2):668–72.

[446] Klein GC, Kim S, Rodgers RP, Marshall AG, Yen A. Mass spectral analysis of asphaltenes. II. Detailed compositional comparison of asphaltenes deposit to its crude oil counterpart for two geographically different crude oils by ESI FT-ICR MS. Energy Fuels 2006;20(5):1973–9.

[447] Klein GC, Kim S, Rodgers RP, Marshall AG, Yen A, Asomaning S. Mass spectral analysis of asphaltenes. I. Compositional differences between pressure-drop and solvent-drop asphaltenes determined by electrospray ionization Fourier transform ion cyclotron resonance mass spectrometry. Energy Fuels 2006;20(5):1965–72.

[448] Klitzke CF, Corilo YE, Siek K, Binkley J, Patrick J, Eberlin MN. Petroleomics by ultrahigh-resolution time-of-flight mass spectrometry. Energy Fuels 2012;26(9):5787–94.

[449] Marshall AG, Kim DG, Klein GC, Stanford LA, Purcell JM, Schaub TM, et al. Compositional characterization of petroleum by ultrahigh-resolution Fourier transform ion cyclotron resonance mass spectrometry with multiple ionization sources. In: Proc. ACS Nat. Meeting. San Francisco, CA, vol. 232; 2006. 1 pp.

[450] Marshall AG, Rodgers RP. Petroleomics: the next grand challenge for chemical analysis. Acc Chem Res 2004;37(1):53–9.

[451] Marshall AG, Rodgers RP. Petroleomics: chemistry of the underworld. Proc Nat Acad Sci U.S.A 2008;105(47):18090–5.

[452] Mckenna AM, Blakney GT, Xian F, Glaser PB, Rodgers RP, Marshall AG. Heavy petroleum composition. 2. Progression of the Boduszynski model to the limit of distillation by ultrahigh-resolution FT-ICR mass spectrometry. Energy Fuels 2010;24(5):2939–46.

[453] Mckenna AM, Donald LJ, Fitzsimmons JE, Juyal P, Spicer V, Standing KG, et al. Heavy petroleum composition. 3. Asphaltene aggregation. Energy Fuels 2013;27(3):1246–56.

[454] Mckenna AM, Marshall AG, Rodgers RP. Heavy petroleum composition. 4. Asphaltene compositional space. Energy Fuels 2013;27(3):1257–67.

[455] Mckenna AM, Purcell JM, Rodgers RP, Marshall AG. Heavy petroleum composition. 1. Exhaustive compositional analysis of Athabasca bitumen HVGO distillates by Fourier transform ion cyclotron resonance mass spectrometry: a definitive test of the Boduszynski model. Energy Fuels 2010;24(5):2929–38.

[456] Mckenna AM, Williams JT, Putman JC, Aeppli C, Reddy CM, Valentine DL, et al. Unprecedented ultrahigh resolution FT-ICR mass spectrometry and parts-per-billion mass accuracy enable direct characterization of nickel and vanadyl porphyrins in petroleum from natural seeps. Energy Fuels 2014;28(4):2454–64.

[457] Pakarinen JMH, Teravainen MJ, Pirskanen A, Wickstrom K, Vainiotalo P. A positive-ion electrospray ionization Fourier transform ion cyclotron resonance mass spectrometry study of Russian and North Sea crude oils and their six distillation fractions. Energy Fuels 2007;21(6): 3369–74.

[458] Podgorski DC, Corilo YE, Nyadong L, Lobodin VV, Bythell BJ, Robbins WK, et al. Heavy petroleum composition. 5. Compositional and structural continuum of petroleum revealed. Energy Fuels 2013;27(3):1268–76.

[459] Purcell JM, Hendrickson CL, Rodgers RP, Marshall AG. Atmospheric pressure photoionization proton transfer for complex organic mixtures investigated by Fourier transform ion cyclotron resonance mass spectrometry. J Am Soc Mass Spectrom 2007;18(9):1682–9.

[460] Purcell JM, Juyal P, Kim DG, Rodgers RP, Hendrickson CL, Marshall AG. Sulfur speciation in petroleum: atmospheric pressure photoionization or chemical derivatization and electrospray ionization Fourier transform ion cyclotron resonance mass spectrometry. Energy Fuels 2007;21(5):2869–74.

[461] Purcell JM, Merdrignac I, Rodgers RP, Marshall AG, Gauthier T, Guibard I. Stepwise structural characterization of asphaltenes during deep hydroconversion processes determined by atmospheric pressure photoionization (APPI) Fourier transform ion cyclotron resonance (FT-ICR) mass spectrometry. Energy Fuels 2010;24(4):2257−65.

[462] Purcell JM, Rodgers RP, Hendrickson CL, Marshall AG. Speciation of nitrogen containing aromatics by atmospheric pressure photoionization or electrospray ionization Fourier transform ion cyclotron resonance mass spectrometry. J Am Soc Mass Spectrom 2007;18(7):1265−73.

[463] Schaub TM, Rodgers RP, Marshall AG, Qian K, Green LA, Olmstead WN. Speciation of aromatic compounds in petroleum refinery streams by continuous flow field desorption ionization FT-ICR mass spectrometry. Energy Fuels 2005;19(4):1566−73.

[464] Senko MW, Hendrickson CL, Pasa-Tolic L, Marto JA, White FM, Guant S, et al. Electrospray ionization Fourier transform ion cyclotron resonance at 9.4 t. Rapid Commun Mass Spectrom 1996;10(14):1824−8.

[465] Shi Q, Hou D, Chung KH, Xu C, Zhao S, Zhang Y. Characterization of heteroatom compounds in a crude oil and its saturates, aromatics, resins, and asphaltenes (SARA) and non-basic nitrogen fractions analyzed by negative-ion electrospray ionization Fourier transform ion cyclotron resonance mass spectrometry. Energy Fuels 2010;24(4):2545−53.

[466] Stanford LA, Rodgers RP, Marshall AG, Czarnecki J, Wu XA. Compositional characterization of bitumen/water emulsion films by negative- and positive-ion electrospray ionization and field desorption/ionization Fourier transform ion cyclotron resonance mass spectrometry. Energy Fuels 2007;21(2):963−72.

[467] Tong J, Liu J, Han X, Wang S, Jiang X. Characterization of nitrogen-containing species in Huadian shale oil by electrospray ionization Fourier transform ion cyclotron resonance mass spectrometry. Fuel 2013;104:365−71.

[468] Wang S, Yang C, Xu C, Zhao S, Shi Q. Separation and characterization of petroleum asphaltene fractions by ESI FT-ICR MS and UV-vis spectrometer. Sci China Chem 2013;56(7):856−62.

[469] Xu CM, Liu Y, Zhao SQ, Shi Q. Compositional analysis of petroleum asphaltenes by negative ion electrospray high resolution FT-ICR mass spectrometry. Zhongguo Shiyou Daxue Xuebao (Ziran Kexue Ban) J China Univ Pet Ed Nat Sci 2013;37(5):190−5.

[470] Zhang L, Xu Z, Shi Q, Sun X, Zhang N, Zhang Y, et al. Molecular characterization of polar heteroatom species in Venezuela Orinoco petroleum vacuum residue and its supercritical fluid extraction subfractions. Energy Fuels 2012;26(9):5795−803.

[471] Qian K, Edwards KE, Siskin M, Olmstead WN, Mennito AS, Dechert GJ, et al. Desorption and ionization of heavy petroleum molecules and measurement of molecular weight distributions. Energy Fuels 2007;21(2):1042−7.

[472] Liu M, Zhao SQ, Yang YY. Development of mass spectrometry on structure and composition of heavy oil. Xiandai Huagong/Modern Chem Ind 2013;33(6):140−4.

[473] Pomerantz AE, Wu Q, Mullins OC, Zare RN. Laser-based mass spectrometric assessment of asphaltene molecular weight, molecular architecture, and nanoaggregate number. Energy Fuels 2015;29(5):2833−42.

[474] Boduszynski MM, Rechsteiner CE. Molecular weight distributions in heavy crudes. In: Proc. 7th UNITAR Intern. Conf. 'Heavy Crude Tar Sands'. Beijing, China; October 27−30, 1998. 9 pp.

[475] Boduszynski MM, Rechsteiner CE, Shafizadeh ASG, Carlson RMK. Composition and properties of heavy crudes. In: Proc. 7th UNITAR Intern. Conf. 'Heavy Crude Tar Sands'. Beijing, China; October 27−30, 1998. 12 pp.

[476] Boduszynski MM. Characterization of 'heavy' crude components. In: Proc. ACS Div. Pet. Chem. Preprints. Chicago, IL, USA, vol. 30. ACS; 1985. p. 626−35.

[477] Boduszynski MM. Limitations of average structure determination for heavy ends in fossil fuels. Pet Sci Technol 1984;2(3):211−32.

[478] Boduszynski MM, Altgelt KH. Composition of heavy petroleums. 4. Significance of the extended atmospheric equivalent boiling point (AEBP) scale. Energy Fuels 1992;6(1):72–6.
[479] Boduszynski MM. Composition of heavy petroleums. 2. Molecular characterization. Energy Fuels 1988;2(5):597–613.
[480] Altgelt KH, Boduszynski MM. Composition of heavy petroleums. 3. An improved boiling point-molecular weight relation. Energy Fuels 1992;6(1):68–72.
[481] Altgelt K, Boduszynski M. Composition and analysis of heavy petroleum fractions. New York: Marcel Dekker, CRC Press; 1993. 512 pp.
[482] Ancheyta J, Centeno G, Trejo F, Marroquin G, Garcia JA, Tenorio E, et al. Extraction and characterization of asphaltenes from different crude oils and solvents. Energy Fuels 2002;16(5):1121–7.
[483] Leon O, Rogel E, Espidel J, Torres G. Asphaltenes: structural characterization, self-association, and stability behavior. Energy Fuels 2000;14(1):6–10.
[484] Oyekunle LO. Certain relationships between chemical composition and properties of petroleum asphalts from different origin. Oil Gas Sci Technol Rev IFP 2006;61(3):433–41.
[485] Rueda-Velásquez RI, Freund H, Qian K, Olmstead WN, Gray MR. Characterization of asphaltene building blocks by cracking under favorable hydrogenation conditions. Energy Fuels 2013;27(4):1817–29.
[486] Buenrostro-Gonzalez E, Espinosa-Pena M, Andersen SI, Lira-Galeana C. Characterization of asphaltenes and resins from problematic Mexican crude oils. Pet Sci Technol 2001;19(3–4):299–316.
[487] Navarro Frometa AE, Boffill Perez K, Padron Bueno L, Doblado Rodriguez E. Characterization of Cuban oil asphaltenes. Rev Cuba Quim 1999;11(1):43–50.
[488] Al-Hajji AA, Muller H, Koseoglu OR. Characterization of nitrogen and sulfur compounds in hydrocracking feedstocks by Fourier transform ion cyclotron mass spectrometry. Oil Gas Sci Technol Rev IFP 2008;63(1):115–28.
[489] Dalmaschio GP, Malacarne MM, De Almeida VMDL, Pereira TMC, Gomes AO, De Castro EVR, et al. Characterization of polar compounds in a true boiling point distillation system using electrospray ionization FT-ICR mass spectrometry. Fuel 2014;115:190–202.
[490] Yang C, Wang ZD, Yang ZY, Hollebone B, Brown CE, Landriault M, et al. Chemical fingerprints of Alberta oil sands bitumen and related petroleum products. Environ Forensics 2011;12:173–88.
[491] Mckay JF, Harnsberger PM, Erickson RB, Cogswell TE, Latham DR. Composition of petroleum heavy ends 2. Characterization of compound types in petroleum >675 C residues. Fuel 1981;60(1):17–26.
[492] Coelho RR, Hovell I, Rajagopal K. Elucidation of the functional sulphur chemical structure in asphaltenes using first principles and deconvolution of mid-infrared vibrational spectra. Fuel Proc Technol 2012;97:85–92.
[493] Wang Z, Que G, Liang W, Qian J. Investigation on chemical structure of resins and pentane asphaltenes in vacuum residua. Acta Pet Sin Pet Proc Sect 1999;15(6):39–46.
[494] Borges B, Del Carpio E, De Navarro CU, Ortega N. Morphological and spectroscopic characterization of asphaltenes from different heavy oils. Rev Latinoam Metal Mat 2014;34(2):218–29.
[495] Merdrignac I, Espinat D. Physicochemical characterization of petroleum fractions: the state of the art. Oil Gas Sci Technol Rev IFP 2007;62(1):7–32.
[496] Ignasiak T, Kemp-Jones AV, Strausz OP. The molecular structure of Athabasca asphaltene. Cleavage of the carbon-sulfur bonds by radical ion electron transfer reactions. J Org Chem 1977;42(2):312–20.
[497] Ignasiak T, Strausz OP, Montgomery DS. Oxygen distribution and hydrogen bonding in Athabasca asphaltene. Fuel 1977;56(4):359–65.
[498] Selucky M, Kim SS, Skinner F, Strausz OP. Structure-related properties of Athabasca asphaltenes and resins as indicated by chromatographic separation. In: Bunger JW, Li NC, editors. Adv. chem. series 195: chemistry of asphaltenes. Washington, DC: ACS; 1981. p. 83–118.
[499] Strausz OP. Some recent advances in the chemistry of oil sand bitumen. In: Proc. 1st Intern. Conf. Future of Heavy Crude Oils and Tar Sands. Edmonton, Alberta, Canada. Unitar by McGraw Hill; 1981. p. 187–94.

[500] Mojelsky TW, Montgomery DS, Strausz OP. Ruthenium (viii) catalyzed oxidation of high molecular weight components of Athabasca oil sand bitumen. AOSTRA J Res 1985;2:131–7.
[501] Mojelsky TW, Montgomery DS, Strausz OP. Pyrolytic probes into the aliphatic core of oxidized Athabasca asphaltene. AOSTRA J Res 1986;3:43–51.
[502] Mojelsky TW, Montgomery DS, Strausz OP. The basic nitrogen compounds in Athabasca maltene. AOSTRA J Res 1986;38(1):19–28.
[503] Payzant JD, Montgomery DS, Strausz OP. Sulfides in petroleum. Org Geochem 1986;9(6):357–69.
[504] Wachowska H, Ignasiak T, Strausz OP, Carson D, Ignasiak B. Application of non-reductive alkylation in liquid ammonia to studies on macromolecular structure of coals and bitumen-derived asphaltene. Fuel 1986;65(8):1081–4.
[505] Yokota T, Scriven F, Montgomery DS, Strausz OP. Absorption and emission spectra of Athabasca asphaltene in the visible and near ultraviolet regions. Fuel 1986;65(8):1142–9.
[506] Mcintyre DD, Strausz OP. Sulphur-33 nuclear magnetic resonance of oxidized petroleum asphaltene. Magnetic Reson Chem 1987;25(1):36–8.
[507] Strausz OP. Structural studies on residua. In: Proc. ACS Div. Pet. Chem. Preprints. Denver, CO, USA, vol. 32. ACS; 1987. p. 396–7.
[508] Payzant JD, Montgomery DS, Strausz OP. The identification of homologous series of benzo[b] thiophenes, thiophenes, thiolanes, and thianes assessing linear carbon framework in the pyrolysis oil of Athabasca asphaltene. AOSTRA J Res 1988;4:117–31.
[509] Strausz OP. Structural investigations of Alberta oil sand bitumens. In: Proc. ACS Nat. Meeting, Div. Pet. Chem. Preprints. Toronto, ON, Canada, vol. 33. ACS; 1988. p. 264–8.
[510] Payzant JD, Mcintyre DD, Mojelsky TW, Torres M, Montgomery DS, Strausz OP. The identification of homologous series of thiolanes and thianes possessing a linear carbon framework from petroleums and their interconversion under simulated geological conditions. Org Geochem 1989;14(4):461–73.
[511] Payzant JD, Mojelsky TW, Strausz OP. Improved methods for the selective isolation of the sulfide and thiophenic classes of compounds from petroleum. Energy Fuels 1989;3(4):449–54.
[512] Strausz OP. Structural studies on resids: correlation between structure and activity. In: Heck RH, Degnan TF, editors. Fundamentals of resid upgrading. New York: AIChE; 1989. p. 1–6.
[513] Strausz OP. Recent advances in the chemistry of alberta bitumens and heavy oils. In: Proc. 4th UNITAR Intern. Conf. 'Heavy Crude Tar Sands'. Edmonton, Alberta, Canada; August 7–12, 1989. p. 605–28.
[514] Strausz OP. Structural features of Athabasca bitumen related to upgrading performance. In: Proc. ACS Nat. Meeting, Div. Pet. Chem. Preprints. Dallas, TX, USA, vol. 34. ACS; 1989. p. 395–400.
[515] Strausz OP. Bitumen and heavy oil chemistry. AOSTRA-Technical Handbook on Oilsands, Bitumens Heavy Oils. 1989. p. 34–73.
[516] Payzant JD, Lown EM, Strausz OP. Structural units of Athabasca asphaltene: the aromatics with a linear carbon framework. Energy Fuels 1991;5(3):441–4.
[517] Strausz OP, Lown EM. Structural features of Athabasca bitumen related to upgrading performance. Fuel Sci Technol Intern 1991;9(3):269–81.
[518] Mojelsky TW, Ignasiak TM, Frakman Z, Mcintyre DD, Lown EM, Montgomery DS, et al. Structural features of Alberta oil sand bitumen and heavy oil asphaltenes. Energy Fuels 1992;6(1):83–96.
[519] Strausz OP, Mojelsky TW, Lown EM. The molecular structure of asphaltene: an unfolding story. Fuel 1992;71:1355–63.
[520] Strausz OP, Lown EM, Mojelsky TW. Molecular structure of asphaltene. In: Proc. ACS Nat. Meeting, Div. Pet. Chem. Preprints. San Diego, CA, USA, vol. 39. ACS; March 13–18, 1994. p. 199–207.
[521] Strausz OP, Lown EM, Mojelsky TW, Peng P. About the structure of petroleum asphaltenes. ACS Div Fuel Chem 1998;43(4):917–23. Preprints.
[522] Peng P, Fu J, Sheng G, Morales-Izquierdo A, Lown EM, Strausz OP. Ruthenium-ions-catalyzed oxidation of an immature asphaltene: structural features and biomarker distribution. Energy Fuels 1999;13(2):266–77.

[523] Peng P, Morales-Izquierdo A, Lown EM, Strausz OP. Chemical structure and biomarker content of Jinghan asphaltenes and kerogens. Energy Fuels 1999;13(2):248–65.
[524] Strausz OP, Mojelsky TW, Faraji F, Lown EM, Peng P. Additional structural details on Athabasca asphaltene and their ramifications. Energy Fuels 1999;13(2):207–27.
[525] Strausz OP, Mojelsky TW, Lown EM, Kowalewski I, Behar F. Structural features of Boscan and Duri asphaltenes. Energy Fuels 1999;13(2):228–47.
[526] Strausz OP, Peng P, Murgich J. Covalent structure and molecular architecture of Athabasca asphaltene. ACS Div Fuel Chem 1999;44(4):743–5. Preprints.
[527] Peksoz A, Akay SK, Kaya Y, Ovalioglu H, Kaynak G, Yalciner A. Analytical information on the asphaltenes from a few standard characterization techniques. Energy Sources Part A Recov Util Environ Eff 2011;33(15):1474–81.
[528] Leyva C, Ancheyta J, Berrueco C, Millán M. Chemical characterization of asphaltenes from various crude oils. Fuel Proc Technol 2013;106:734–8.
[529] Mckay JF, Latham DR, Haines WE. Composition of petroleum heavy ends 3. Comparison of the composition of high-boiling petroleum distillates and petroleum >675 °C residues. Fuel 1981;60(1):27–32.
[530] Jewell DM, Ruberto RG, Albaugh EW, Query RC. Distribution and structural aspects of sulfur compounds in petroleum residuals (Kuwait). Ind Eng Chem Fund 1976;15(3):206–11.
[531] Savel'ev VV, Golovko AK, Gorbunov LV, Kam'yanov VF, Galvalizi C. High-sulfur asphaltites of Argentina. Khim Tverd Topl (Moscow Russian Federation) 2007;(2):33–41.
[532] Liu P, Shi Q, Chung KH, Zhang Y, Pan N, Zhao S, et al. Molecular characterization of sulfur compounds in Venezuela crude oil and its SARA fractions by electrospray ionization Fourier transform ion cyclotron resonance mass spectrometry. Energy Fuels 2010;24(9):5089–96.
[533] Pomerantz AE, Seifert DJ, Bake KD, Craddock PR, Mullins OC, Kodalen BG, et al. Sulfur chemistry of asphaltenes from a highly compositionally graded oil column. Energy Fuels 2013;27(8):4604–8.
[534] Savel'ev V, Golovko A, Gorbunova L, Kamyanov V, Galvalizi C. High-sulfurous Argentinian asphaltites and their thermal liquefaction products. Oil Gas Sci Technol Rev IFP 2008;63(1):57–67.
[535] Mitra-Kirtley S, Mullins OC, Ralston CY, Pareis C. Sulfur characterization in asphaltene, resin, and oil fractions of two crude oils. In: Proc. ACS Fall Meeting, Resid Characterization Upgrading Symp. New Orleans, LO, USA; August 22, 1999. p. 763–7.
[536] George GN, Gorbaty ML. Sulfur K-edge x-ray absorption spectroscopy of petroleum asphaltenes and model compounds. JACS 1989;111(9):3182–6.
[537] Mitra-Kirtley S, Mullins OC, Van Elp J, George SJ, Chen J, Cramer SP. Determination of the nitrogen chemical structures in petroleum asphaltenes using XANES spectroscopy. JACS 1993;115(1):252–8.
[538] Chiaberge S, Guglielmetti G, Montanari L, Salvalaggio M, Santolini L, Spera S, et al. Investigation of asphaltene chemical structural modification induced by thermal treatments. Energy Fuels 2009;23(9):4486–95.
[539] Zhao Y, Wei F. Simultaneous removal of asphaltenes and water from water-in-bitumen emulsion. II. Application feasibility. Fuel Proc Technol 2008;89(10):941–8.
[540] Iwamatsu E, Hayashi E, Sanada Y, Ali MA, Ahmed S, Hamid H, et al. Proton magnetic resonance thermal analysis of petroleum derived asphaltene with Co smectite and CoMo alumina based catalysts. J Jpn Pet Inst 2002;45(3):135–43.
[541] Andersen SI. Separation of asphaltenes by polarity using liquid-liquid extraction. Pet Sci Technol 1997;15(1–2):185–98.
[542] Waldo GS, Mullins OC, Penner-Hahn JE, Cramer SP. Determination of the chemical environment of sulphur in petroleum asphaltenes by X-ray absorption spectroscopy. Fuel 1992;71(1):53–7.
[543] Sarret G, Mongenot T, Connan J, Derenne S, Kasrai M, Michael Bancroft G, et al. Sulfur speciation in kerogens of the orbagnoux deposit (upper kimmeridgian, jura) by XANES spectroscopy and pyrolysis. Org Geochem 2002;33(8):877–95.

[544] Wiltfong R, Mitra-Kirtley S, Mullins OC, Andrews B, Fujisawa G, Larsen JW. Sulfur speciation in different kerogens by XANES spectroscopy. Energy Fuels 2005;19(5):1971−6.
[545] Ignasiak T, Kemp-Jones AV, Strausz OP. Properties of asphaltenes from various Alberta crude oils. ACS Div Fuel Chem Prepr 1977;22(3):126−31.
[546] Boduszynski MM, Aickay JF, Latham DR. Asphaltenes, where are you?. In: Proc. Asphalt Paving Technol. Louisville, KN, USA, vol. 49; February 18−20, 1980. p. 123−43.
[547] Holmes SA. Nitrogen functional groups in Utah tar sand bitumen and produced oils. AOSTRA J Res 1986;2:167−75.
[548] Dutta PK, Holland RJ. Acid—base characteristics of petroleum asphaltenes as studied by non-aqueous potentiometric titrations. Fuel 1984;63(2):197−201.
[549] Wilhelms A, Patience RL, Larter SR, Jørgensen S. Nitrogen functionality distributions in asphaltenes isolated from several oils from different source rock types. Geochim Cosmochim Acta 1992;56(10):3745−50.
[550] Farhat Ali M, Saleem M. Characterization of asphaltenes from Saudi Arabian crude oils by nonaqueous potentiometric titration and elemental analysis. Fuel Sci Technol Intern 1994;12(5):675−91.
[551] Sharma BK, Sharma CD, Bhagat SD, Erhan SZ. Maltenes and asphaltenes of petroleum vacuum residues: physico-chemical characterization. Pet Sci Technol 2007;25(1 & 2):93−104.
[552] Sharma BK, Sharma CD, Tyagi OS, Bhagat SD, Erhan SZ. Structural characterization of asphaltenes and ethyl acetate insoluble fractions of petroleum vacuum residues. Pet Sci Technol 2007;25(1 & 2):121−39.
[553] Chen X, Shen B, Sun J, Wang C, Shan H, Yang C, et al. Characterization and comparison of nitrogen compounds in hydrotreated and untreated shale oil by electrospray ionization (ESI) Fourier transform ion cyclotron resonance mass spectrometry (FT-ICR MS). Energy Fuels 2012;26(3):1707−14.
[554] Liao Y, Shi Q, Hsu CS, Pan Y, Zhang Y. Distribution of acids and nitrogen-containing compounds in biodegraded oils of the Liaohe basin by negative ion ESI FT-ICR MS. Org Geochem 2012;47:51−65.
[555] Smith DF, Rodgers RP, Rahimi P, Teclemariam A, Marshall AG. Effect of thermal treatment on acidic organic species from Athabasca bitumen heavy vacuum gas oil, analyzed by negative-ion electrospray fourier transform ion cyclotron resonance (FT-ICR) mass spectrometry. Energy Fuels 2009;23(1):314−9.
[556] Tong J, Liu J, Han X, Shi Q, Jiang X. Influence of heating rate on basic nitrogen-containing species content in Dachengzi shale oil studied by positive-ion electrospray ionization FT-ICR mass spectrometry. Oil Shale 2013;30(1):76−89.
[557] Moschopedis SE, Speight JG. Oxygen functions in asphaltenes. Fuel 1976;55(4):334−6.
[558] Siddiqui MN. Infrared study of hydrogen bond types in asphaltenes. Pet Sci Technol 2003;21(9 & 10):1601−15.
[559] Chung KH, Xu C, Hu Y, Wang R. Supercritical fluid extraction reveals resid properties. Oil Gas J 1997;95(3):66−9.
[560] Sanchez-Berna AC, Camacho-Moran V, Romero-Guzman ET, Yacaman MJ. Asphaltene aggregation from vacuum residue and its content of inorganic particles. Pet Sci Technol 2006;24(9):1055−66.
[561] Mullins OC, Sheu EY, Hammami A, Marshall AG. Asphaltenes, heavy oils, and petroleomics. New York: Springer-Verlag Inc.; 2006. 800 pp.
[562] Mullins OC, Sheu EY. Structures and dynamics of asphaltenes. New York: Plenum Press; 1998. 438 pp.
[563] Mullins OC. The asphaltenes. Annu Rev Anal Chem 2011;4:393−418.
[564] Speight JG. The molecular nature of petroleum asphaltenes. Arab J Sci Eng 1994;19(2 B):335−53.
[565] Speight JG. The structure of Athabasca asphaltenes and the conversion of these materials to useful by-products. ACS Div Pet Chem Prepr 1971;13(2):76−82.
[566] Speight JG, Pancirov RJ. Structural types in petroleum asphaltenes as deduced from pyrolysis/gas chromatography/mass spectrometry. Liq Fuels Technol 1984;2(3):287−305.
[567] Mcintyre DD, Montgomery DS, Strausz OP. Carboxylic acids bound to the mineral matter in heavy cretaceous oils of Alberta. AOSTRA J Res 1986;2:251−65.

[568] Speight JG, Pancirov RJ. Some aspects of the structure of petroleum asphaltenes. In: Proc. ACS Div. Pet. Chem. Preprints. Washington, DC, USA, vol. 28. ACS; 1983. p. 1319–32.
[569] Yen TF, Wen CS, Kwan JT, Chow E. The nature of shale oil-derived asphaltenes. In: Strausz OP, Lown EM, editors. Oil sand and oil shale chemistry. New York: Verlag; 1978. p. 309–24.
[570] Danil'yan TD, Rogacheva OV, Yangurazoya AR, Mindiyarov KG, Gimaev RN. Structure of high-molecular-weight compounds in the thermal degradation of petroleum residues. Solid Fuel Chem 1987;21(4):83–6.
[571] Calemma V, Iwanski P, Nali M, Scotti R, Montanari L. Structural characterization of asphaltenes of different origins. Energy Fuels 1995;9(2):225–30.
[572] Bergmann U, Mullins O. Carbon X-ray raman spectroscopy of pahs and asphaltenes. In: Mullins O, Sheu E, Hammami A, Marshall A, editors. Asphaltenes, heavy oils, and petroleomics. New York: Springer; 2007. p. 139–55.
[573] Bergmann U, Mullins OC, Cramer SP. X-ray raman spectroscopy of carbon in asphaltene: light element characterization with bulk sensitivity. Anal Chem 2000;72(11):2609–12.
[574] Fergoug T, Bouhadda Y. Determination of hassi messaoud asphaltene aromatic structure from 1H & 13C NMR analysis. Fuel 2014;115:521–6.
[575] Abdallah WA, Yang Y. Raman spectrum of asphaltene. Energy Fuels 2012;26(11):6888–96.
[576] Bouhadda Y, Bormann D, Sheu E, Bendedouch D, Krallafa A, Daaou M. Characterization of algerian hassi-messaoud asphaltene structure using Raman spectrometry and X-ray diffraction. Fuel 2007;86(12–13):1855–64.
[577] Bouhadda Y, Fergoug T, Sheu EY, Bendedouch D, Krallafa A, Bormann D, et al. Second order Raman spectra of Algerian hassi-messaoud asphaltene. Fuel 2008;87(15–16):3481–2.
[578] Winans RE, Hunt JE. An overview of resid characterization by mass spectrometry and small angle scattering techniques. ACS Div Fuel Chem 1999;44(4):725–7. Preprints.
[579] Ruiz-Morales Y, Mullins OC. Measured and simulated electronic absorption and emission spectra of asphaltenes. Energy Fuels 2009;23(3):1169–77.
[580] Zhao S, Kotlyar LS, Woods JR, Sparks BD, Hardacre K, Chung KH. Molecular transformation of Athabasca bitumen end-cuts during coking and hydrocracking. Fuel 2001;80(8):1155–63.
[581] El-Mohamed S, Achard MF, Hardouin F, Gasparoux H. Correlations between diamagnetic properties and structural characters of asphaltenes and other heavy petroleum products. Fuel 1986;65(11): 1501–4.
[582] Rubinstein I, Spyckerelle C, Strausz OP. Pyrolysis of asphaltenes: a source of geochemical information. Geochim Cosmochim Acta 1979;43(1):1–6.
[583] Rubinstein I, Strausz OP. Thermal treatment of the Athabasca oil sand bitumen and its component parts. Geochim Cosmochim Acta 1979;43(12):1887–93.
[584] Savage PE, Klein MT, Kukes SG. Asphaltene reaction pathways. 1. Thermolysis. Ind Eng Chem Proc Des Dev 1985;24(4):1169–74.
[585] Artok L, Su Y, Hirose Y, Hosokawa M, Murata S, Nomura M. Structure and reactivity of petroleum-derived asphaltene. Energy Fuels 1999;13(2):287–96.
[586] Yen TF. Structural differences between asphaltenes isolated from petroleum and from coal liquid. In: Bunger JW, Li NC, editors. Adv. chem. series 195: chemistry of asphaltenes. ACS; 1982. p. 39–51.
[587] Xu Y, Koga Y, Strausz OP. Characterization of Athabasca asphaltenes by small-angle X-ray scattering. Fuel 1995;74(7):960–4.
[588] Sheremata JM, Gray MR, Dettman HD, Mccaffrey WC. Quantitative molecular representation and sequential optimization of Athabasca asphaltenes. Energy Fuels 2004;18(5):1377–84.
[589] Murgich J, Rodriguez MJ, Aray Y. Molecular recognition and molecular mechanics of micelles of some model asphaltenes and resins. Energy Fuels 1996;10(1):68–76.
[590] Parkash S, Moschopedis S, Speight J. Physical properties and surface characteristics of asphaltenes. Fuel 1979;58(12):877–82.

[591] Pacheco-Sanchez JH, Alvarez-Ramırez F, Martınez-Magadan JM. Morphology of aggregated asphaltene structural models. Energy Fuels 2004;18:1676–86.

[592] Michael G, Al-Siri M, Khan ZH, Ali FA. Differences in average chemical structures of asphaltene fractions separated from feed and product oils of a mild thermal processing reaction. Energy Fuels 2005;19:1598–605.

[593] Gonzalez G, Neves GBM, Saraiva SM, Lucas EF, De Sousa MDA. Electrokinetic characterization of asphaltenes and the asphaltenes-resins interaction. Energy Fuels 2003;17:879–86.

[594] Liao Z, Geng A, Graciaa A, Creux P, Chrostowska A, Zhang Y. Different adsorption/occlusion properties of asphaltenes associated with their secondary evolution processes in oil reservoirs. Energy Fuels 2006;20(3):1131–6.

[595] Siskin M, Kelemen SR, Eppig CP, Brown LD, Afeworki M. Asphaltene molecular structure and chemical influences on the morphology of coke produced in delayed coking. Energy Fuels 2006;20(3):1227–34.

[596] Sabbah H, Morrow AL, Pomerantz AE, Zare RN. Evidence for island structures as the dominant architecture of asphaltenes. Energy Fuels 2011;25(4):1597–604.

[597] Camacho-Bragado GA, Santiago P, Marin-Almazo M, Espinosa M, Romero ET, Murgich J, et al. Fullerenic structures derived from oil asphaltenes. Carbon 2002;40(15):2761–6.

[598] Jovanovic JA. Models of an asphaltene aggregate and a micelle of the petroleum colloid system. Hem Ind 2000;54(6):270–5.

[599] Mojelsky TW, Montgomery DS, Strausz OP. The side chains associated with the undistillable aromatic and resin components of Athabasca bitumen. AOSTRA J Res 1986;2:177–84.

[600] De Oliveira LP, Vazquez AT, Verstraete JJ, Kolb M. Molecular reconstruction of petroleum fractions: application to vacuum residues from different origins. Energy Fuels 2013;27(7):3622–41.

[601] Wiehe IA. The pendant-core building block model of petroleum residua. Energy Fuels 1994;8(3):536–44.

[602] Wiehe IA. Process chemistry of petroleum macromolecules. Boca Raton, FL: CRC Press; 2008. 456 pp.

[603] Zhang L, Hou Z, Horton SR, Klein MT, Shi Q, Zhao S, et al. Molecular representation of petroleum vacuum resid. Energy Fuels 2014;28(3):1736–49.

[604] Diallo MS, Cagin T, Faulon JL, Goddard Iii WA. Thermodynamic properties of asphaltenes: a predictive approach based on computer assisted structure elucidation and atomistic simulations. In: Yen TF, Chilingarian GV, editors. Asphaltenes asphalts. 2. Developments in pet. Science. Los Angeles, CA: Elsevier; 2000. p. 103–27.

[605] Speight JG, Wernick DL, Gould KA, Overfield RE, Rao BML, Savage DW. Molecular weight and association of asphaltenes: a critical review. Rev Inst Francais Pet 1985;40(1):51–61.

[606] Reerink H. Size and shape of asphaltene particles in relationship to high-temperature viscosity. Ind Eng Chem Prod Res Dev 1973;12(1):82–8.

[607] Moschopedis SE, Fryer JF, Speight JG. Investigation of asphaltene molecular weights. Fuel 1976;55(3):227–32.

[608] Koots JA, Speight JG. Relation of petroleum resins to asphaltenes. Fuel 1975;54(3):179–84.

[609] Griffin RL, Simpson WC, Miles TK. Influence of composition of paving asphalt on viscosity, viscosity-temperature susceptibility, and durability. J Chem Eng Data 1959;4(4):349–54.

[610] Katz M. Alberta bitumen: I. The composition of blown Alberta bitumen. Can J Res 1934;10:435–51.

[611] Labout JWA. Constitution of asphaltic bitumen. In: Pfeiffer JP, editor. The properties of asphaltic bitumen. Amsterdam: Elsevier; 1950. p. 13–48.

[612] Malhotra VM, Buckmaster HA. 34 GHz EPR study of vanadyl complexes in various asphaltenes. Statistical correlative model of the coordinating ligands. Fuel 1985;64(3):335–41.

[613] Ceballo CD, MartNez R, Escalona A. Characterization of petroleum asphaltenes by thermal field flow fractionation using modified mobile phases. Pet Sci Technol 1999;17(7):783–810.

[614] Mullins OC, Nez-Haya B, Marshall AG. Contrasting perspective on asphaltene molecular weight. This comment vs the overview of A. A. Herod, K. D. Bartle, and R. Kandiyoti. Energy Fuels 2008;22(3):1765–73.

[615] Behrouzi M, Luckham PF. Limitations of size-exclusion chromatography in analyzing petroleum asphaltenes: a proof by atomic force microscopy. Energy Fuels 2008;22(3):1792−8.
[616] Guzman A, Bueno A, Carbognani L. Molecular weight determination of asphaltenes from Colombian crudes by size exclusion chromatography (SEC) and vapor pressure osmometry (VPO). Pet Sci Technol 2009;27(8):801−16.
[617] Hansen BE, Malmros O, Turner NR, Stenby EH, Andersen SI. Size exclusion chromatography in the analysis of pitch. In: Proc. Light Metals. New Orleans, LA; 2001. p. 559−64.
[618] Mullins OC. Review of the molecular structure and aggregation of asphaltenes and petroleomics. SPE J 2008;13(1):48−57.
[619] Al-Jarrah MMF, Apikian RL. Investigation of asphaltenes association by a vapour pressure osmometric method. J Chem Technol Biotechnol 1987;39(4):231−6.
[620] Kok WT, TüDöS AJ, Grutters M, Shepherd AG. Characterization of asphaltenes by nonaqueous capillary electrophoresis. Energy Fuels 2011;25(1):208−14.
[621] Yarranton HW, Hussein H, Masliyah JH. Water-in-hydrocarbon emulsions stabilized by asphaltenes at low concentrations. J Colloid Interface Sci 2000;228(1):52−63.
[622] Larsen JW, Li S. Determination of bitumen molecular weight distributions using 252cf plasma desorption mass spectrometry. Energy Fuels 1995;9(5):760−4.
[623] Acevedo S, Gutierrez LB, Negrin G, Pereira JC, Mendez B, Delolme F, et al. Molecular weight of petroleum asphaltenes: a comparison between mass spectrometry and vapor pressure osmometry. Energy Fuels 2005;19(4):1548−60.
[624] Mathews JP, Eser S, Parviz R. Preliminary results on the molecular structures of the Athabasca and Cold Lake asphaltenes. ACS Div Fuel Chem 1999;44(4):783−5. Preprints.
[625] Speight JG. Asphaltene characterization and use in understanding processes. In: Proc. ACS Div. Pet. Chem. Preprints. San Diego, CA, USA, vol. 39. ACS; 1994. p. 200−3.
[626] Loegel TN, Danielson ND, Borton DJ, Hurt MR, Kenttämaa HI. Separation of asphaltenes by reversed-phase liquid chromatography with fraction characterization. Energy Fuels 2012;26(5):2850−7.
[627] Sheu EY. Petroleum asphaltene-properties, characterization, and issues. Energy Fuels 2002;16(1):74−82.
[628] Schneider MH, Andrews AB, Mitra-Kirtley S, Mullins OC. Asphaltene molecular size by fluorescence correlation spectroscopy. Energy Fuels 2007;21(5):2875−82.
[629] Mckenna AM, Rodgers RP, Hendrickson CL, Blakney GT, Savory JT, Kaiser NT, et al. The compositional continuum of petroleum: detailed molecular characterization of heavy crude oils and asphaltenes by ultrahigh resolution FT-ICR mass spectrometry. In: Proc. AIChE Annual Meeting. Salt Lake City, UT; 2010. 1 pp.
[630] Andrews AB, Guerra RE, Mullins OC, Sen PN. Diffusivity of asphaltene molecules by fluorescence correlation spectroscopy. J Phys Chem A 2006;110(26):8093−7.
[631] Acevedo S, Rodriguez P, Labrador H. An electron microscopy study of crude oils and maltenes. Energy Fuels 2004;18(6):1757−63.
[632] Chen Z, Zhao S, Xu Z, Gao J, Xu C. Molecular size and size distribution of petroleum residue. Energy Fuels 2011;25(5):2109−14.
[633] Storm DA, Sheu E, De Tar MM. Small angle x-ray scattering study of asphaltenes. In: Proc. ACS Nat. Meeting. San Francisco, CA, vol. 37; 1992. p. 833−43.
[634] Kim HG, Long RB. Characterization of heavy residuum by a small angle X-ray scattering technique. Ind Eng Chem Fund 1979;18(1):60−3.
[635] Andrews AB, Shih WC, Mullins OC, Norinaga K. Molecular size determination of coal-derived asphaltene by fluorescence correlation spectroscopy. Appl Spect 2011;65(12):1348−56.
[636] Wang Z, Guo A, Zhang L, Ren R, Que G. Molecular sizes and micellar sizes of asphaltenes in vacuum residues. ACS Div Pet Chem 2003;48:175−8. Preprints.
[637] Fujii M, Yoneda T, Sanada Y. Study on the micelle size of asphaltenes in vacuum residue by dielectric loss. ACS Div Fuel Chem 1999;44(4):738−41. Preprints.

[638] Acevedo S, Guzman K, Ocanto O. Determination of the number average molecular mass of asphaltenes (MN) using their soluble A2 fraction and the vapor pressure osmometry (VPO) technique. Energy Fuels 2010;24(3):1809–12.

[639] Lin MY, Gang H. A study of heavy oil asphaltene by neutron scattering. In: Proc. 7th UNITAR Intern. Conf. 'Heavy Crude Tar Sands'. Beijing, China, vol. 1; October 27–30, 1998. 6 pp.

[640] Dolomatov MY, Dezortsev SV, Bakhtizina RZ, Shutkova SA, Shulyakovskaya DO, Paymurzina NK. About possible formation of organic semiconductors from paramagnetic phase of asphaltenes. In: Proc. ElecMol'12. Grenoble, France; 2012. p. 160–4.

[641] Pollack SS, Yen TF. Structural studies of asphaltics by X-ray small angle scattering. Anal Chem 1970;42(6):623–9.

[642] Tanaka R, Sato E, Hunt JE, Winans RE, Sato S, Takanohashi T. Characterization of asphaltene aggregates using X-ray diffraction and small-angle X-ray scattering. Energy Fuels 2004;18(4):1118–25.

[643] Tanaka R, Hunt JE, Winans RE, Thiyagarajan P, Sato S, Takanohashi T. Aggregates structure analysis of petroleum asphaltenes with small-angle neutron scattering. Energy Fuels 2003;17(1):127–34.

[644] Storm DA, Sheu EY, Detar MM. Macrostructure of asphaltenes in vacuum residue by small-angle X-ray scattering. Fuel 1993;72(7):977–81.

[645] Thiyagarajan P, Hunt JE, Winans RE, Anderson KB, Miller JT. Temperature-dependent structural changes of asphaltenes in 1-methylnaphthalene. Energy Fuels 1995;9(5):829–33.

[646] Ravey JC, Ducouret G, Espinat D. Asphaltene macrostructure by small angle neutron scattering. Fuel 1988;67(11):1560–7.

[647] Gawrys KL, Kilpatrick PK. Asphaltenic aggregates are polydisperse oblate cylinders. J Colloid Interface Sci 2005;288:325–34.

[648] Storm DA, Barresi RJ, Sheu EY. Flocculation of asphaltenes in heavy oil at elevated temperatures. Fuel Sci Technol Intern 1996;14(1–2):243–60.

[649] Aske N, Orr R, Sjöblom J, Kallevik H, Øye G. Interfacial properties of water – crude oil systems using the oscillating pendant drop. Correlations to asphaltene solubility by near infrared spectroscopy. J Dispersion Sci Technol 2004;25(3):263–75.

[650] Ching MJTM, Pomerantz AE, Andrews AB, Dryden P, Schroeder R, Mullins OC, et al. On the nanofiltration of asphaltene solutions, crude oils, and emulsions. Energy Fuels 2010;24(9): 5028–37.

[651] Rajagopal K, Silva SMC. An experimental study of asphaltene particle sizes in n-heptane-toluene mixtures by light scattering. Braz J Chem Eng 2004;21(4):601–9.

[652] Czarnecki J, Moran K. On the stabilization mechanism of water-in-oil emulsions in petroleum systems. Energy Fuels 2005;19(5):2074–9.

[653] Layrisse I, Rivas H, Acevedo S, Medina R, Sanchez M, Utrera M. Composition and physico-chemical characteristics of extra heavy crudes [composicion y caracteristicas fisico-quimicas de crudos extrapesados]. Rev Tec Intevep 1984;4(1):3–18.

[654] Marquez HM, Rivas H, Layrisse I, Acevedo S, Escobar G, Gutierrez LB, et al. Composition and physicochemical characteristics of natural surfactants from extra heavy crude oils. Rev Tec Intevep 1987;7(2):117–22.

[655] Varadaraj R, Brons C. Molecular origins of heavy oil interfacial activity part 1: fundamental interfacial properties of asphaltenes derived from heavy crude oils and their correlation to chemical composition. Energy Fuels 2007;21(1):195–8.

[656] Varadaraj R, Brons C. Molecular origins of heavy crude oil interfacial activity part 2: fundamental interfacial properties of model naphthenic acids and naphthenic acids separated from heavy crude oils. Energy Fuels 2007;21(1):199–204.

[657] Varadaraj R, Brons C. Molecular origins of crude oil interfacial activity part 3: characterization of the complex fluid rag layer formed at crude oil–water interfaces. Energy Fuels 2007;21(3):1617–21.

[658] Kilpatrick PK, Spiecker PM. Asphaltene emulsions. In: Sjoblom J, editor. Encyclopedia of emulsion technology. New York: Dekker; 2001. p. 707–30.

[659] Mclean JD, Spiecker PM, Sullivan AP, Kilpatrick PK. The role of petroleum asphaltenes in the stabilization of water-in-oil emulsions. In: Sheu E, Mullins O, editors. Structure dynamics of asphaltenes. New York: Plenum; 1998. p. 377–422.

[660] Reisber J, Doscher TM. Interfacial phenomena in crude oil–water systems. ACS Div Pet Chem 1956;21:43–9. Preprints.

[661] Strassner JE. Effect of pH on interfacial films and stability of crude oil-water emulsions. J Pet Technol 1968;20:303–12.

[662] Yang F, Tchoukov P, Pensini E, Dabros T, Czarnecki J, Masliyah J, et al. Asphaltene subfractions responsible for stabilizing water-in-crude oil emulsions. Part 1: Interfacial behaviors. Energy Fuels 2014;28(11):6897–904.

[663] Yang F, Tchoukov P, Dettman HD, Teklebrhan RB, Liu L, Dabros T, et al. Asphaltene subfractions responsible for stabilizing water-in-crude oil emulsions: Part 2. Molecular representations and molecular dynamic simulations. Energy Fuels 2015;29(8):4783–94.

[664] Shengqun W, Jianjun L, Liyan Z, Zhenghe X, Masliyah J. Colloidal interactions between asphaltene surfaces in toluene. Energy Fuels 2009;23(2):862–9.

[665] Mclean JD, Kilpatrick PK. Effects of asphaltene aggregation in model heptane–toluene mixtures on stability of water-in-oil emulsions. J Colloid Interface Sci 1997;196:23–34.

[666] Nour AH, Suliman A, Hadow MM. Stabilization mechanisms of water-in-crude oil emulsions. J Appl Sci 2008;8:1571–5.

[667] Spiecker PM, Gawrys KL, Trail CB, Kilpatrick PK. Effects of petroleum resins on asphaltene aggregation and water-in-oil emulsion formation. Colloids Surf A Physicochem Eng Asp 2003;220:9–27.

[668] Elsharkawy AM, Al-Sahhaf TA, Fahim MA, Yarranton HW. Characterization of asphaltenes and resins separated from water-in-crude oil emulsions formed in Kuwaiti oil fields. In: Proc. AIChE Annual Meeting. Cincinnati, OH; 10/30–11/4, 2005. p. 697–718.

[669] Elsharkawy A, Al-Sahhaf T, Fahim M. Characterization of asphaltenes and resins separated from water-in-oil emulsions. Pet Sci Technol 2008;26(1–3):153–69.

[670] Syunyaev RZ, Balabin RM, Akhatov IS, Safieva JO. Adsorption of petroleum asphaltenes onto reservoir rock sands studied by near-infrared (NIR) spectroscopy. Energy Fuels 2009;23(3):1230–6.

[671] Betancourt SS, Ventura GT, Pomerantz AE, Viloria O, Dubost FX, Zuo J, et al. Nanoaggregates of asphaltenes in a reservoir crude oil and reservoir connectivity. Energy Fuels 2009;23(3):1178–88.

[672] Dalmazzone C, Noïk C, Glénat P, Dang F. Development of a methodology for the optimization of dehydration of extraheavy-oil emulsions. SPE J 2010;15(3):726–36.

[673] Hemmingsen PV, Silset A, Hannisdal A, Sjöblom J. Emulsions of heavy crude oils: influence of viscosity, temperature, and dilution. J Dispers Sci Technol 2005;26(5):615–27.

[674] Kralova I, Sjöblom J, Øye G, Simon S, Grimes BA, Paso K. Heavy crude oils/particle stabilized emulsions. Adv Colloid Interface Sci 2011;169(2):106–27.

[675] Poteau S, Argillier JF, Langevin D, Pincet F, Perez E. Influence of pH on stability and dynamic properties of asphaltenes and other amphiphilic molecules at the oil-water interface. Energy Fuels 2005;19(4):1337–41.

[676] Hung J, Ortega D, Castillo J, Acevedo S. Studies of kinetics of coalescence of water-in-furrial crude oil emulsions by confocal microscopy. J Dispers Sci Technol 2007;28(5):717–23.

[677] Rahmani S, Mccaffrey W, Elliott Ja W, Gray MR. Liquid-phase behavior during the cracking of asphaltenes. Ind Eng Chem Res 2003;42(17):4101–8.

[678] Acevedo S, Borges B, Quintero F, Piscitelly V, Gutierrez LB. Asphaltenes and other natural surfactants from Cerro Negro crude oil. Stepwise adsorption at the water/toluene interface: film formation and hydrophobic effects. Energy Fuels 2005;19:1948–53.

[679] Zhang LY, Lopetinsky R, Xu Z, Masliyah JH. Asphaltene monolayers at a toluene/water interface. Energy Fuels 2005;19:1330–6.

[680] Yang X, Verruto VJ, Kilpatrick PK. Dynamic asphaltene-resin exchange at the oil/water interface: time-dependent w/o emulsion stability for asphaltene/resin model oils. Energy Fuels 2007;21(3):1343–9.
[681] Freer EM, Radke CJ. Relaxation of asphaltenes at the toluene/water interface: diffusion exchange and surface rearrangement. J Adhes 2004;80(6):481–96.
[682] Ese MH, Sjöblom J, Djuve J, Pugh R. An atomic force microscopy study of asphaltenes on mica surfaces. Influence of added resins and demulsifiers. Colloid Polym Sci 2000;278(6):532–8.
[683] Ese MH, Yang X, Sjöblom J. Film forming properties of asphaltenes and resins. A comparative Langmuir-Blodgett study of crude oils from North Sea, European continent and Venezuela. Colloid Polym Sci 1998;276(9):800–9.
[684] Li M, Xu M, Ma Y, Wu Z, Christy AA. Interfacial film properties of asphaltenes and resins. Fuel 2002;81(14):1847–53.
[685] Ostlund JA, Nyden M, Auflem IH, Sjoblom J. Interactions between asphaltenes and naphthenic acids. Energy Fuels 2003;17(1):113–9.
[686] Mingyuan L, Christy AA, Sjoblom J. Water-in-crude oil emulsions from the Norwegian continental shelf part VI – diffuse reflectance fourier transform infrared characterization of interfacially active fractions from North Sea crude oil. In: Sjoblom J, editor. Emulsions – a fundamental practical approach. The Netherlands: Kluwer Academic Publishers; 1992. p. 157–72.
[687] Adams JJ. Asphaltene adsorption, a literature review. Energy Fuels 2014;28(5):2831–56.
[688] Liao Z, Zhou H, Graciaa A, Chrostowska A, Creux P, Geng A. Adsorption/occlusion characteristics of asphaltenes: some implication for asphaltene structural features. Energy Fuels 2005;19(1):180–6.
[689] Moustafa NE, Shara SI, Abdalah R. Asphaltenes as adsorbents in GC. II. Separation of some homologous series and isomeric compounds. Pet Sci Technol 2006;24(1):31–40.
[690] Pokonova YV. Carbon adsorbents with resins and asphaltenes. Chem Technol Fuels Oils (Translation of Khimiya i Tekhnologiya Topliv i Masel) 2006;42(4):313–6.
[691] Pokonova YV, Oleinik MS. Carbon adsorbents from petroleum resids. Chem Technol Fuels Oils (Translation of Khimiya i Tekhnologiya Topliv i Masel) 2001;37(1):44–7.
[692] Pokonova YV, Potashov VA. Asphaltene concentrates as bases for carbon adsorbents. Chem Technol Fuels Oils (Translation of Khimiya i Tekhnologiya Topliv i Masel) 2002;38(3):197–206.
[693] Marques LCC, Pereira JO, Bueno AD, Marques VS, Lucas EF, Mansur CRE, et al. A study of asphaltene-resin interactions. J Braz Chem Soc 2012;23(10):1880–8.
[694] Acevedo S, Castro A, Vásquez E, Marcano F, Ranaudo MA. Investigation of physical chemistry properties of asphaltenes using solubility parameters of asphaltenes and their fractions A1 and A2. Energy Fuels 2010;24(11):5921–33.
[695] Khvostichenko DS, Andersen SI. Interactions between asphaltenes and water in solutions in toluene. Energy Fuels 2008;22(5):3096–103.
[696] Sheu EY, Storm DA, Shields MB. Adsorption kinetics of asphaltenes at toluene/acid solution interface. Fuel 1995;74(10):1475–9.
[697] Auflen IH. Influence of asphaltene aggregation and pressure on crude oil emulsion stability. DOKTOR INGENIØR Thesis from Norwegian University of Science and Technology. Chem. Eng; June 2002. 58 pp.
[698] Lordo SA. Identifying and mitigating opportunity crude impacts. In: Proc. NPRA Annual Meeting. San Antonio, TX, vol. 1; March 22–24, 2009. p. 437–53. AM-09-43.
[699] Lordo S, Garcia JM, Garcia-Swofford S. Desalter acidification additives and their potential impacts on crude units. In: Proc. Corrosion. New Orleans, LO; 2008. p. 085561–0855612.
[700] Lordo S. Impact and management of opportunity crudes on the desalting and wastewater process. In: Proc. Opportunity Crudes Conf.: Challenges, Benefits. Houston, TX. Hydrocarbon Publishing Co.; April 28–May 2, 2008 [Unpublished presentation].
[701] Yan N, Gray MR, Masliyah JH. On water-in-oil emulsions stabilized by fine solids. Colloids Surf A Physicochem Eng Asp 2001;193(1–3):97–107.

[702] Yan N, Masliyah JH. Characterization and demulsification of solids-stabilized oil-in-water emulsions part 2. Demulsification by the addition of fresh oil. Colloids Surf A Physicochem Eng Asp 1995;96(3):243–52.

[703] Yan N, Masliyah JH. Characterization and demulsification of solids-stabilized oil-in-water emulsions part 1. Partitioning of clay particles and preparation of emulsions. Colloids Surf A Physicochem Eng Asp 1995;96(3):229–42.

[704] Zaki NN, Maysour NWS, Abdel-Azim A. Polyoxyalkylenated amines for breaking of water-in-oil emulsions stabilized by asphaltenes and clay. Pet Sci Technol 2000;18(9–10):1009–25.

[705] Tambe DE, Sharma MM. Factors controlling the stability of colloid-stabilized emulsions: II. A model for the rheological properties of colloid-laden interfaces. J Colloid Interface Sci 1994;162(1):1–10.

[706] Tambe DE, Sharma MM. Factors controlling the stability of colloid-stabilized emulsions: I. An experimental investigation. J Colloid Interface Sci 1993;157(1):244–53.

[707] Sztukowski DM. Asphaltene and solids-stabilized water-in-oil emulsions. Ph.D. thesis from University of Calgary. Department of Chem. Pet. Eng; February 2005. 333 pp.

[708] Kotlyar LS, Sparks BD, Woods JR, Chung KH. Solids associated with the asphaltene fraction of oil sands bitumen. Energy Fuels 1999;13(2):346–50.

[709] Sztukowski DM, Yarranton HW. Oilfield solids and water-in-oil emulsion stability. J Colloid Interface Sci 2005;285(2):821–33.

[710] Gafonova OV, Yarranton HW. The stabilization of water-in-hydrocarbon emulsions by asphaltenes and resins. J Colloid Interface Sci 2001;241(2):469–78.

[711] Erdman JG, Harju PH. Capacity of petroleum asphaltenes to complex heavy metals. J Chem Eng Data 1963;8(2):252–8.

[712] Grijalva-Monteverde H, Arellano-Tanori OV, Valdez MA. Zeta potential and Langmuir films of asphaltene polar fractions. Energy Fuels 2005;19(6):2416–22.

[713] Acevedo S, Ranaudo MA, Escobar G, Gutiérrez L, Ortega P. Adsorption of asphaltenes and resins on organic and inorganic substrates and their correlation with precipitation problems in production well tubing. Fuel 1995;74(4):595–8.

[714] Alboudwarej H, Pole D, Svrcek WY, Yarranton HW. Adsorption of asphaltenes on metals. Ind Eng Chem Res 2005;44:5585–92.

[715] Xie K, Karan K. Kinetics and thermodynamics of asphaltene adsorption on metal surfaces: a preliminary study. Energy Fuels 2005;19(4):1252–60.

[716] Rudrake A. Investigation of asphaltene-metal interactions. MSc thesis from Queen's University. Department of Chem. Eng; February 2008. 151 pp.

[717] Acevedo S, Castillo J, Fernández A, Goncalves S, Ranaudo MA. A study of multilayer adsorption of asphaltenes on glass surfaces by photothermal surface deformation. Relation of this adsorption to aggregate formation in solution. Energy Fuels 1998;12(2):386–90.

[718] Nassar NN, Husein MM, Al-Jabari ME. Investigation on asphaltene deposition onto different surfaces of ultradispersed catalysts. In: Proc. 8th World Congress of Chem. Eng. Montreal, QC; August 23–27, 2009. 5pp.

[719] Xing C, Hilts RW, Shaw JM. Sorption of Athabasca vacuum residue constituents on synthetic mineral and process equipment surfaces from mixtures with pentane. Energy Fuels 2010;24(4):2500–13.

[720] González G, Moreira MBC. The adsorption of asphaltenes and resins on various minerals. In: Yen TF, Chilingarian GV, editors. Asphaltenes asphalts. 1. Developments in pet. Science. Elsevier; 1994. p. 207–31.

[721] Acevedo S, Ranaudo MA, García C, Castillo J, Fernández A. Aggregation and adsorption of asphaltenes: a kinetic study. In: Proc. 224th ACS Nat. Meeting. Boston, MA, vol. 47; August 18–22, 2002. p. 529–30.

[722] Abu Tarboush BJ, Husein MM. Adsorption of asphaltenes from heavy oil onto in situ prepared NiO nanoparticles. J Colloid Interface Sci 2012;378(1):64–9.

[723] Acevedo S, Ranaudo MA, Garcia C, Castillo J, Fernandez A, Caetano M, et al. Importance of asphaltene aggregation in solution in determining the adsorption of this sample on mineral surfaces. Colloids Surf A Physicochem Eng Asp 2000;166(1−3):145−52.
[724] Castillo J, Ranaudo MA, Fernández A, Piscitelli V, Maza M, Navarro A. Study of the aggregation and adsorption of asphaltene sub-fractions A1 and A2 by white light interferometry: importance of A1 sub-fraction in the aggregation process. Colloids Surf A Physicochem Eng Asp 2013;427:41−6.
[725] Dubey ST, Waxman MH. Asphaltene adsorption and desorption from mineral surfaces. SPE Reserv Eng (Soc Pet Eng 1991;6(3):389−95.
[726] Dudášová D, Simon S, Hemmingsen PV, Sjöblom J. Study of asphaltenes adsorption onto different minerals and clays. Part 1. Experimental adsorption with UV depletion detection. Colloids Surf A Physicochem Eng Asp 2008;317(1−3):1−9.
[727] Gonzalez MF, Stull CS, Lopez-Linares F, Pereira-Almao P. Comparing asphaltene adsorption with model heavy molecules over macroporous solid surfaces. Energy Fuels 2007;21(1):234−41.
[728] Jada A, Debih H, Khodja M. Montmorillonite surface properties modifications by asphaltenes adsorption. J Pet Sci Eng 2006;52(1−4):305−16.
[729] López-Linares F, Carbognani L, Sosa-Stull C, Pereira-Almao P, Spencer RJ. Adsorption of virgin and visbroken residue asphaltenes over solid surfaces. 1. Kaolin, smectite clay minerals, and Athabasca siltstone. Energy Fuels 2009;23(4):1901−8.
[730] Marchal C, Abdessalem E, Tayakout-Fayolle M, Uzio D. Asphaltene diffusion and adsorption in modified NiMo alumina catalysts followed by ultraviolet (UV) spectroscopy. Energy Fuels 2010;24(8):4290−300.
[731] Marczewski AW, Szymula M. Adsorption of asphaltenes from toluene on mineral surface. Colloids Surf A Physicochem Eng Asp 2002;208(1−3):259−66.
[732] Nassar NN, Hassan A, Carbognani L, Lopez-Linares F, Pereira-Almao P. Iron oxide nanoparticles for rapid adsorption and enhanced catalytic oxidation of thermally cracked asphaltenes. Fuel 2012;95:257−62.
[733] Pernyeszi T, Dékány I. Sorption and elution of asphaltenes from porous silica surfaces. Colloids Surf A Physicochem Eng Asp 2001;194(1−3):25−39.
[734] Pernyeszi T, Patzkó A, Berkesi O, Dékány I. Asphaltene adsorption on clays and crude oil reservoir rocks. Colloids Surf A Physicochem Eng Asp 1998;137(1−3):373−84.
[735] Sakanishi K, Saito I, Watanabe I, Mochida I. Dissolution and demetallation treatment of asphaltene in resid using adsorbent and oil-soluble Mo complex. Fuel 2004;83(14−15):1889−93.
[736] Sakanishi K, Yamashita N, Whitehurst DD, Mochida I. Depolymerization and demetallation treatments of asphaltene in vacuum residue. Catal Today 1998;43(3−4):241−7.
[737] Murrell LL, Grenoble DC, Long RB. Separating basic asphaltenes using transition metal oxide acid catalysts Patent No. US4424114. Exxon Research and Engineering Co.; January 03, 1984.
[738] Carbognani L. Effects of iron compounds on the retention of oil polar hydrocarbons over solid sorbents. Pet Sci Technol 2000;18(3 & 4):335−60.
[739] Almarri MS, Ma X, Song C. Adsorption pretreatment of crude oil and its effect on subsequent hydrotreating process. In: Proc. 230th ACS Nat. Meeting. Washington, DC. USA. PETR-067; August 28−September 1, 2005. 1 pp.
[740] Bantignies J-L, Dit Moulin CC, Dexpert H. Asphaltene adsorption on kaolinite characterized by infrared and X-ray absorption spectroscopies. J Pet Sci Eng 1998;20(3−4):233−7.
[741] Dudášová D, Flåten GR, Sjöblom J, Øye G. Study of asphaltenes adsorption onto different minerals and clays. Part 2. Particle characterization and suspension stability. Colloids Surf A Physicochem Eng Asp 2009;335(1−3):62−72.
[742] Bauget F, Langevin D, Lenormand R. Effects of asphaltenes and resins on foamability of heavy oil. In: Proc. SPE Annual Technical Conf. Exhibition. New Orleans, LA; September 30−October 3, 2001. p. 1513−22.

[743] León O, Contreras E, Rogel E, Dambakli G, Acevedo S, Carbognani L, et al. Adsorption of native resins on asphaltene particles: a correlation between adsorption and activity. Langmuir 2002;18(13):5106–12.

[744] León O, Contreras E, Rogel E, Dambakli G, Espidel J, Acevedo S. The influence of the adsorption of amphiphiles and resins in controlling asphaltene flocculation. Energy Fuels 2001;15(5):1028–32.

[745] Pereira JC, Lopez I, Salas R, Silva F, Fernandez C, Urbina C, et al. Resins: the molecules responsible for the stability/instability phenomena of asphaltenes. Energy Fuels 2007;21(3):1317–21.

[746] Ye ZB, Cheng L, Xiang WT, Liu YG, Li JH. Characterization of resins and the adsorption of resins on asphaltene particles. Xinan Shiyou Daxue Xuebao/J Southwest Pet Univ 2010;32(6):147–54.

[747] Acevedo S, Ranaudo MA, Castillo J, Caetano M, Fernández A. Use of laser techniques for the study of asphaltene aggregation and adsorption. ACS Div Fuel Chem 1999;44(4):733–5. Preprints.

[748] Abraham T, Christendat D, Karan K, Xu Z, Masliyah J. Asphaltene-silica interactions in aqueous solutions: direct force measurements combined with electrokinetic studies. Ind Eng Chem Res 2002;41:2170–7.

[749] Acevedo S, Ranaudo MA, García C, Castillo J, Fernández A. Adsorption of asphaltenes at the toluene-silica interface: a kinetic study. Energy Fuels 2003;17(2):257–61.

[750] Nassar NN, Hassan A, Pereira-Almao P. Effect of surface acidity and basicity of aluminas on asphaltene adsorption and oxidation. J Colloid Interface Sci 2011;360(1):233–8.

[751] Nassar NN, Hassan A, Pereira-Almao P. Effect of the particle size on asphaltene adsorption and catalytic oxidation onto alumina particles. Energy Fuels 2011;25(9):3961–5.

[752] Daughney CJ. Sorption of crude oil from a non-aqueous phase onto silica: the influence of aqueous pH and wetting sequence. Org Geochem 2000;31(2–3):147–58.

[753] Franco CA, Nassar NN, Ruiz MA, Pereira-Almao P, Cortés FB. Nanoparticles for inhibition of asphaltenes damage: adsorption study and displacement test on porous media. Energy Fuels 2013;27(6):2899–907.

[754] Taylor SE. The electrodeposition of asphaltenes and implications for asphaltene structure and stability in crude and residual oils. Fuel 1998;77(8):821–8.

[755] Tukhvatullina AZ, Barskaya EE, Kouryakov VN, Ganeeva YM, Yusupova TN, Romanov GV. Supramolecular structures of oil systems as the key to regulation of oil behavior. J Pet Environ Biotechnol 2013;4(4):8. http://dx.doi.org/10.4172/2157-7463.1000152.

[756] Schabron JF, Speight JG. The solubility and three-dimensional structure of asphaltenes. Pet Sci Technol 1998;16(3–4):361–75.

[757] Pacheco-Sanchez JH, Mansoori GA. Tricritical phenomena in asphaltene/aromatic hydrocarbon systems. Rev Mex Física 2013;59(6):584–93.

[758] Rogel E. Molecular thermodynamic approach to the formation of mixed asphaltene-resin aggregates. Energy Fuels 2008;22(6):3922–9.

[759] Soorghali F, Zolghadr A, Ayatollahi S. Effect of resins on asphaltene deposition and the changes of surface properties at different pressures: a microstructure study. Energy Fuels 2014;28(4):2415–21.

[760] Alboudwarej H, Beck J, Svrcek WY, Yarranton HW, Akbarzadeh K. Sensitivity of asphaltene properties to separation techniques. Energy Fuels 2002;16(2):462–9.

[761] Leontaritis KJ, Mansoori GA. Asphaltene deposition during oil production and processing: a thermodynamic colloidal model. In: Proc. SPE Intern. Symp. Oilfield Chem. San Antonio, TX. SPE 16258; February 4–6, 1987. 10 pp.

[762] Carnahan N, Salager JL, Anton R. Effect of resins on stability of asphaltenes. In: Proc. Offshore Technol. Conf. Houston, TX, USA. OTC-19002; April 30–May 3, 2007. 9 pp.

[763] Daaou M, Modarressi A, Bendedouch D, Bouhadda Y, Krier G, Rogalski M. Characterization of the nonstable fraction of hassi-messaoud asphaltenes. Energy Fuels 2008;22(5):3134–42.

[764] Carbognani L, Orea M, Fonseca M. Complex nature of separated solid phases from crude oils. Energy Fuels 1999;13:351–8.

[765] Chang CL, Fogler HS. Stabilization of asphaltenes in aliphatic solvents using alkylbenzene-derived amphiphiles. 2. Study of the asphaltene-amphiphile interactions and structures using fourier transform infrared spectroscopy and small-angle X-ray scattering techniques. Langmuir 1994;10(6):1758–66.
[766] De Boer RB, Leerlooyer K, Eigner MRP, Van Bergen ARD. Screening of crude oils for asphalt precipitation: theory, practice, and the selection of inhibitors. In: Proc. SPE Production and Facilities. Richardson, TX, United States, vol. 10. Society of Petroleum Engineers (SPE); 1995. p. 55–61.
[767] Gonzalez DL, Vargas FM, Hirasaki GJ, Chapman WG. Modeling study of CO_2-induced asphaltene precipitation. Energy Fuels 2008;22(2):757–62.
[768] Verdier S, Carrier H, Andersen SI, Daridon JL. Study of pressure and temperature effects on asphaltene stability in presence of CO_2. Energy Fuels 2006;20(4):1584–90.
[769] Correra S, Donaggio F, Capuano F, Onorati N. Temperature effect on asphaltene instability. In: Proc. 4th Intern. Conf. on Pet. Phase Behaviour Fouling. Trondheim, Norway; June 23–26, 2003. 14 pp.
[770] Rogel E, Leon O, Contreras E, Carbognani L, Torres G, Espidel J, et al. Assessment of asphaltene stability in crude oils using conventional techniques. Energy Fuels 2003;17(6):1583–90.
[771] Hung J, Castillo J, Reyes A. Kinetics of asphaltene aggregation in toluene-heptane mixtures studied by confocal microscopy. Energy Fuels 2005;19(3):898–904.
[772] Castillo J, Caetano M, Hung J. Kinetics of asphaltene nanoaggregation in crude oil. In: Proc. International Conference on Nanoscience, ICON. Choroni, Venezuela, May 7–11, 2006. 2 pp.
[773] Gaestel CR, Smadja R, Lamminan KA. Contribution of consistency properties to bitumen function. Eur Roads Rev 1971;(466):85–94.
[774] Schabron JF, Pauli AT, Rovani JF. Petroleum residua solubility parameter/polarity map: stability studies of residua pyrolysis. Report No. WRI-99-R004. Laramie, WY, USA: Western Research Institute; April 1999. 30 pp.
[775] Zhang LL, Yang GH, Que GH. The conglomerating characteristics of asphaltenes from residue during thermal reaction. Fuel 2005;84(7–8):1023–6.
[776] Brons G, Yu JM. Solvent deasphalting effects on whole Cold Lake bitumen. Energy Fuels 1995;9(4):641–7.
[777] Watson KM, Nelson EF. Improved methods for approximating critical and thermal properties of petroleum fractions. Ind Eng Chem 1933;25(8):880–7.
[778] Shi TP, Xu ZM, Cheng M, Hu YX, Wang RA. Characterization index for vacuum residua and their subfractions. Energy Fuels 1999;13(4):871–6.
[779] Zhang L, Li S, Han L, Sun X, Xu Z, Shi Q, et al. Coking reactivity of laboratory-scale unit for two heavy petroleum and their supercritical fluid extraction subfractions. Ind Eng Chem Res 2013;52(16):5593–600.
[780] Zhao S, Xu Z, Xu C, Chung KH. Feedstock characteristic index and critical properties of heavy crudes and petroleum residua. J Pet Sci Eng 2004;41(1–3):233–42.
[781] Ovalles C, Rogel E, Lopez J, Pradhan A, Moir M. Predicting reactivity of feedstocks to resid hydroprocessing using asphaltene characteristics. Energy Fuels 2013;27(11):6552–9.
[782] Chang J, Fujimoto K, Tsubaki N. Effect of initiative additives on hydro-thermal cracking of heavy oils and model compound. Energy Fuels 2003;17(2):457–61.
[783] Guo A-J, Ren Z-H, Tian L-Y, Wang Z-X, Li K-Q. Characterization of molecular change of heavy oil under mild thermal processing using FT-IR spectroscopy. J Fuel Chem Technol 2007;35(2):168–75.
[784] Alhumaidan FS, Hauser A, Rana MS, Lababidi HMS. Impact of thermal treatment on asphaltene functional groups. Energy Fuels 2016:12 (Published online March 08), Article ASAP.
[785] Friesen WI, Michaelian KH, Long Y, Dabros T. Effect of solvent-to-bitumen ratio on the pyrolysis properties of precipitated Athabasca asphaltenes. Energy Fuels 2005;19:1109–15.
[786] Calemma V, Rausa R. Thermal decomposition behaviour and structural characteristics of asphaltenes. J Anal Appl Pyrolysis 1997;40-41:569–84.
[787] Douda J, Alvarez R, Bolaños JN. Characterization of Maya asphaltene and maltene by means of pyrolysis application. Energy Fuels 2008;22(4):2619–28.

[788] Hauser A, Bahzad D, Stanislaus A, Behbahani M. Thermogravimetric analysis studies on the thermal stability of asphaltenes: pyrolysis behavior of heavy oil asphaltenes. Energy Fuels 2008;22(1):449–54.
[789] Zhang C, Lee CW, Keogh RA, Demirel B, Davis BH. Thermal and catalytic conversion of asphaltenes. Fuel 2001;80(8):1131–46.
[790] Yasar M, Trauth DM, Klein MT. Asphaltene and resid pyrolysis. 2. The effect of reaction environment on pathways and selectivities. Energy Fuels 2001;15(3):504–9.
[791] Savage PE, Klein MT, Kukes SG. Asphaltene reaction pathways. 3. Effect of reaction environment. Energy Fuels 1988;2(5):619–28.
[792] Rahimi PM, Gentzis T. Thermal hydrocracking of Cold Lake vacuum bottoms asphaltenes and their subcomponents. Fuel Process Technol 2003;80(1):69–79.
[793] Storm DA, Barresi RJ, Sheu EY, Bhattacharya AK, Derosa TF. Microphase behavior of asphaltic micelles during catalytic and thermal upgrading. Energy Fuels 1998;12(1):120–8.
[794] Rahmani S, Gray MR. Dependence of molecular kinetics of asphaltene cracking on chemical composition. Pet Sci Technol 2007;25(1–2):141–52.
[795] Trejo F, Rana MS, Ancheyta J. Thermogravimetric determination of coke from asphaltenes, resins and sediments and coking kinetics of heavy crude asphaltenes. Catal Today 2010;150(3–4):272–8.
[796] Wang J, Anthony EJ. A study of thermal-cracking behavior of asphaltenes. Chem Eng Sci 2003;58(1):157–62.
[797] Rahmani S, Mccaffrey WC, Dettman HD, Gray MR. Coking kinetics of asphaltenes as a function of chemical structure. Energy Fuels 2003;17:1048–56.
[798] Rahimi PM, Teclemariam A, Taylor E, Debruijn T, Wiehe IA. Determination of coking onset of petroleum feedstocks using solubility parameters. In: Proc. SPE/EPA/DOE Exploration Production Environmental Conference. San Antonio, TX, vol. 48; 2003. p. 103–5.
[799] Murray D. Bitumen quality and processability. In: Proc. AI-EES Partial Upgrading Workshop; May 29, 2014. 40 pp. (Available from: http://www.ai-ees.ca/media/13006/4_ccqta__dave_murray_for_aiees_may29_2014.pdf.
[800] Gray MR. Consistency of asphaltene chemical structures with pyrolysis and coking behavior. Energy Fuels 2003;17:1516–9.
[801] Schabron JF, Speight JG. Correlation between carbon residue and molecular weight. ACS Div Fuel Chem 1997;42(2):386–9. Preprints.
[802] Butz T, Oelert HH. Application of petroleum asphaltenes in cracking under hydrogen. Fuel 1995;74(11):1671–6.
[803] Michels R, Langlois E, Ruau O, Mansuy L, Elie M, Landais P. Evolution of asphaltenes during artificial maturation: a record of the chemical processes. Energy Fuels 1996;10:39–48.
[804] Fan H, Liu Y, Zhang L, Zhao X. The study on composition changes of heavy oils during steam stimulation processes. Fuel 2002;81(13):1733–8.
[805] Ancheyta J, Centeno G, Trejo F, Marroquin G. Changes in asphaltene properties during hydrotreating of heavy crudes. Energy Fuels 2003;17(5):1233–8.
[806] Bartholdy J, Andersen SI. Changes in asphaltene stability during hydrotreating. Energy Fuels 2000;14(1):52–5.
[807] Buch L, Groenzin H, Buenrostro-Gonzalez E, Andersen SI, Lira-Galeana C, Mullins OC. Molecular size of asphaltene fractions obtained from residuum hydrotreatment. Fuel 2003;82(9):1075–84.
[808] Bartholdy J, Lauridsen R, Mejlholm M, Andersen SI. Effect of hydrotreatment on product sludge stability. Energy Fuels 2001;15(5):1059–62.
[809] Zou R, Liu L. Role of asphaltenes in petroleum cracking and refining. In: Yen TF, Chilingarian GV, editors. Asphaltenes asphalts. 1. Developments in pet. Science. Elsevier; 1994. p. 339–63.
[810] Solari RB. In: Teh Fu Y, George VC, editors. Asphaltene hydroconversion in Asphaltenes asphalts. 2. Developments in pet. Science. Elsevier; 2000. p. 149–71.
[811] Ancheyta J, Centeno G, Trejo F, Speight JG. Asphaltene characterization as function of time on-stream during hydroprocessing of Maya crude. Catal Today 2005;109(1–4):162–6.

[812] Trejo F, Ancheyta J, Morgan TJ, Herod AA, Kandiyoti R. Characterization of asphaltenes from hydrotreated products by SEC, LDMS, MALDI, NMR, and XRD. Energy Fuels 2007;21(4):2121−8.
[813] Sun YD, Yang CH, Zhao H, Shan HH, Shen BX. Influence of asphaltene on the residue hydrotreating reaction. Energy Fuels 2010;24(9):5008−11.
[814] Seki H, Kumata F. Structural analyses of petroleum asphaltenes and resins after HDM and subsequent HDS treatments. ACS Div Fuel Chem 1999;44(4):786−90. Preprints.
[815] Kim JW, Longstaff DC, Hanson FV. Catalytic and thermal effects during hydrotreating of bitumen-derived heavy oils. Fuel 1998;77(15):1815−23.
[816] Callejas MA, Martinez MT. Hydroprocessing of a Maya residue. Intrinsic kinetics of sulfur-, nitrogen-, nickel-, and vanadium-removal reactions. Energy Fuels 1999;13(3):629−36.
[817] Hunt JE, Winans RE, Miller JT. Characterization of asphaltenes from processed resids. In: Proc. 213th ACS Nat. Meeting. San Francisco, CA; April 13−17, 1997. 7 pp.
[818] Chung KH, Xu C. Narrow-cut characterization reveals resid process chemistry. Fuel 2001;80(8): 1165−77.
[819] Steele WV, Chirico RD. Hydrodenitrogenation: an increasingly important part of catalytic hydroprocessing interlocking of thermodynamics and kinetics. Report No. NIPER-570. Bartlesville, OK, USA: National Institute for Petroleum and Energy Research; April 1992. 37 pp.
[820] Weeks RW, Mcbride WL. Chemistry and processability of crude oil asphaltenes as studied by ultracentrifugation. In: Proc. ACS Nat. Meeting. Washington, DC; 1980. 34 pp.
[821] Wandas R. Structural characterization of asphaltenes from raw and desulfurized vacuum residue and correlation between asphaltene content and the tendency of sediment formation in h-oil heavy products. Pet Sci Technol 2007;25(1 & 2):153−68.
[822] Trejo F, Ancheyta J, Rana MS. Structural characterization of asphaltenes obtained from hydroprocessed crude oils by SEM and TEM. Energy Fuels 2009;23(1):429−39.
[823] Ancheyta J, Betancourt G, Marroquin G, Centeno G, Castaneda LC, Alonso F, et al. Hydroprocessing of Maya heavy crude oil in two reaction stages. Appl Catal A General 2002;233(1−2):159−70.
[824] Zhao S, Sparks BD, Kotlyar LS, Chung KH. Correlation of processability and reactivity data for residua from bitumen, heavy oils and conventional crudes: characterization of fractions from super-critical pentane separation as a guide to process selection. Catal Today 2007;125(3−4): 122−36.
[825] Xu H, Xu Z, Sun X, Xu C, Chung KH, Zhao S. Antisolvent separation of thermally cracked vacuum resid. Energy Fuels 2013;27(12):7885−95.
[826] Andersen SI, Duong D, Bartholdy J. Changes in asphaltene structure and stability during hydrotreating. In: Proc. AIChE Spring Nat. Meeting. Orlando, FL; 2006. 1 pp.
[827] Carbognani L, Buenrostro-Gonzalez E. Oxidation evidenced during hydrocarbon preparative fraction isolation. Intrinsic stability aspects affecting crude oils. Energy Fuels 2006;20(3):1137−44.
[828] Kök MV, Karacan Ö, Pamir R. Kinetic analysis of oxidation behavior of crude oil SARA constituents. Energy Fuels 1998;12(3):580−8.
[829] Herrington PR. Effect of concentration on the rate of reaction of asphaltenes with oxygen. Energy Fuels 2004;18:1573−7.
[830] Zhan GZ, Shaohui G, Suoqi Z, Guangxu Y, Lanqi S, Libo C. Alkyl side chains connected to aromatic units in dagang vacuum residue and its supercritical fluid extraction and fractions (SFEFS). Energy Fuels 2009;23(1):374−85.
[831] Pacheco LDC, Coelho RR. Caracterização das cadeias naftênicas dos asfaltenos com o uso de modelagem molecular e técnicas de IV. In: Proc. XIV Jornada de Iniciação Científica. Rio de Janeiro, Brazil; 2006. 7 pp.
[832] Zhou X, Zhang Y, Zhao S, Chung KH, Xu C, Shi Q. Characterization of saturated hydrocarbons in vacuum petroleum residua: redox derivatization followed by negative-ion electrospray ionization Fourier transform ion cyclotron resonance mass spectrometry. Energy Fuels 2014;28(1):417−22.

[833] Ali MF, Siddiqui MN, Al-Hajji AA. Structural studies on residual fuel oil asphaltenes by RICO method. Pet Sci Technol 2004;22(5–6):631–45.
[834] Sato T, Adschiri T, Arai K, Rempel GL, Ng FTT. Upgrading of asphalt with and without partial oxidation in supercritical water. Fuel 2003;82(10):1231–9.
[835] Sato T, Mori S, Watanabe M, Sasaki M, Itoh N. Upgrading of bitumen with formic acid in supercritical water. J Supercrit Fluids 2010;55(1):232–40.
[836] Siddiqui MN. Alkylation and oxidation reactions of Arabian asphaltenes. Fuel 2003;82(11):1323–9.
[837] Cosultchi A, Ascencio-Gutiérrez JA, Reguera E, Zeifert B, Yee-Madeira H. On a probable catalytic interaction between magnetite (Fe_3O_4) and petroleum. Energy Fuels 2006;20(3):1281–6.
[838] Nassar NN, Hassan A, Pereira-Almao P. Comparative oxidation of adsorbed asphaltenes onto transition metal oxide nanoparticles. Colloids Surf. A Physicochem. Eng Asp 2011;384(1–3):145–9.
[839] Boukir A, Guiliano M, Asia L, El Hallaoui A, Mille G. A fraction to fraction study of photo-oxidation of BAL 150 crude oil asphaltenes. Analusis 1998;26(9):358–64.
[840] Salmanova CK, Akhmedbekova SF, Mamedov AP, Kyazimov SM, Abdulova SN. Transformations of resins and asphaltenes in photoirradiation. Chem Technol Fuels Oils 2007;43(5):415–21.
[841] Platonov VV, Valdez JC, Klyavina OA, Vol-Epshtein AB, Shpil'berg MB, Chizhevskii AA, et al. Mechanical degradation of petroleum asphaltenes from the mantansas field, Cuba. Khimiya Tverd Topl (Solid Fuel Chem.) 1989;(4):102–8.
[842] Sato T, Tomita T, Trung PH, Itoh N, Sato S, Takanohashi T. Upgrading of bitumen in the presence of hydrogen and carbon dioxide in supercritical water. Energy Fuels 2013;27(2):646–53.
[843] Ignasiak T, Ruo TCS, Strausz OP. Investigation of Alberta [Canada] oil sand asphaltenes using thermal degradation reactions. ACS Div Fuel Chem 1979;24(2):178–83. Preprints.
[844] Ignasiak T, Bimer J, Samman N, Montgomery DS, Strausz OP. Lewis acids assisted degradation of Athabasca asphaltene. In: Bunger JW, Li NC, editors. Adv. chem. series 195: chemistry of asphaltenes. ACS; 1982. p. 183–201.
[845] Ramirez-Corredores MM, Borole AP. Biocatalysis in oils refining. Studies surf. sci. catal, vol. 164. Amsterdam Elsevier; 2007. 406 pp.
[846] Shong RG. Bioprocessing of crude oils. In: Proc. ACS Spring Meeting. Anaheim, CA, USA, vol. 44; 1999. p. 1–4.
[847] Ramirez M. Biorefining of crude oil. In: Schwartz S, Masciangioli T, Boonyaratanakornkit B, editors. Bioinspired chemistry for energy: a workshop. National Academies Press; 2008. 3 pp.
[848] Le Borgne S, Quintero R. Biotechnological processes for the refining of petroleum. Fuel Proc Technol 2003;81(2):155–69.
[849] Pineda-Flores G, Mesta-Howard AM. Petroleum asphaltenes: generated problematic and possible biodegradation mechanisms. Rev Latinoam Microbiol 2001;43(3):143–50.
[850] Kirkwood KM, Foght JM, Gray MR. Prospects for biological upgrading of heavy oils and asphaltenes. In: Vazquez-Duhalt R, Quintero-Ramirez R, editors. Studies surf. sci. catal. Amsterdam: Elsevier B.V; 2004. p. 113–43.
[851] Pineda-Flores G, Boll-Argüello G, Lira-Galeana C, Mesta-Howard AM. A microbial consortium isolated from a crude oil sample that uses asphaltenes as a carbon and energy source. Biodegradation 2004;15(3):145–51.
[852] Dunn K, Yen TF. A plausible reaction pathway of asphaltene under ultrasound. Fuel Proc Technol 2001;73(1):59–71.
[853] Poveda JC, Molina D, Martínez H, Florez O, Campillo B. Molecular changes in asphaltenes within h2 plasma. Energy Fuels 2014;28(2):735–44.
[854] Christy AA, Dahl B, Kvalheim OM. Structural features of resins, asphaltenes and kerogen studied by diffuse reflectance infrared spectroscopy. Fuel 1989;68(4):430–5.
[855] Lindeloff N, Heidemann RA, Andersen SI, Stenby EH. A thermodynamic mixed-solid asphaltene precipitation model. Pet Sci Technol 1998;16(3–4):307–21.

[856] Kohse BF, Nghiem LX, Maeda H, Ohno K. Modelling phase behaviour including the effect of pressure and temperature on asphaltene precipitation. In: Proc. SPE Asia Pacific Oil and Gas Conference and Exhibition, Winning Technology and Teamwork, APOGCE 2000. Brisbane, QLD; October 16–18, 2000. p. 883–95.
[857] Nghiem LX, Coombe DA. Modeling asphaltene precipitation during primary depletion. SPE J 1997;2(2):170–6.
[858] Nghiem LX, Coombe DA, Farouq Ali SM. Compositional simulation of asphaltene deposition and plugging. In: Proc. SPE Technical Conference and Exhibition. Part Omega. Richardson, TX, United States. New Orleans, LA, USA. Soc Pet Eng (SPE). Sigma; September 27–30, 1998. p. 129–40.
[859] Nghiem LX, Hassam MS, Nutakki R, George AED. Efficient modelling of asphaltene precipitation. In: Proc. SPE Annual Technical Conference and Exhibition. Part 3 (of 5). Richardson, TX, United States. Houston, TX, USA. Society of Petroleum Engineers (SPE). Sigma; October 3–6, 1993. p. 375–84.
[860] Nghiem LX, Kohse BF, Ali SMF, Doan Q. Asphaltene precipitation: Phase behaviour modelling and compositional simulation. In: Proc. SPE Asia Pacific Conference on Integrated Modelling for Asset Management "Rising to the Challenges of Enhancing our Assets". Yokohama, Japan; April 25–26, 2000. p. 283–96.
[861] Mansoori GA. A unified perspective on the phase behaviour of petroleum fluids. Intern J Oil, Gas Coal Technol 2009;2(2):141–67.
[862] Alvarez-Ramirez F, Ramirez-Jaramillo E, Ruiz-Morales Y. Calculation of the interaction potential curve between asphaltene-asphaltene, asphaltene-resin, and resin-resin systems using density functional theory. Energy Fuels 2006;20(1):195–204.
[863] Andrews AB, Mcclelland A, Korkeila O, Demidov A, Krummel A, Mullins OC, et al. Molecular orientation of asphaltenes and pah model compounds in langmuir-blodgett films using sum frequency generation spectroscopy. Langmuir 2011;27(10):6049–58.
[864] Klee T, Masterson T, Miller B, Barrasso E, Bell J, Lepkowicz R, et al. Triplet electronic spin states of crude oils and asphaltenes. Energy Fuels 2011;25(5):2065–75.
[865] Kowalewski I, Vandenbroucke M, Huc AY, Taylor MJ, Faulon JL. Preliminary results on molecular modeling of asphaltenes using structure elucidation programs in conjunction with molecular simulation programs. Energy Fuels 1996;10(1):97–107.
[866] Mujica V, Nieto P, Puerta L, Acevedo S. Caging of molecules by asphaltenes. A model for free radical preservation in crude oils. Energy Fuels 2000;14(3):632–9.
[867] Pan YQ, Gao JS, Wang DX. Accurate calculating method for molecular dimensions of heavy oil character molecules. Shiyou Xuebao, Shiyou Jiagong Acta Pet Sin Pet Proc Sect 2007;23(4):63–7.
[868] Rosales S, Machín I, Sánchez M, Rivas G, Ruette F. Theoretical modeling of molecular interactions of iron with asphaltenes from heavy crude oil. J Mol Catal A Chem 2006;246(1–2):146–53.
[869] Ruiz-Morales Y, Mullins OC. Polycyclic aromatic hydrocarbons of asphaltenes analyzed by molecular orbital calculations with optical spectroscopy. Energy Fuels 2007;21(1):256–65.
[870] Branco VAM, Mansoori GA, De Almeida Xavier LC, Park SJ, Manafi H. Asphaltene flocculation and collapse from petroleum fluids. J Pet Sci Eng 2001;32:217–30.
[871] Carauta ANM, Seidl PR, Chrisman ECAN, Correia JCG, Menechini PD, Silva DM, et al. Modeling solvent effects on asphaltene dimers. Energy Fuels 2005;19(4):1245–51.
[872] Murgich J, Rogel E, Leon O, Isea R. A molecular mechanics-density functional study of the adsorption of fragments of asphaltenes and resins on the (001) surface of Fe_2O_3. Pet Sci Technol 2001;19(3):437–55.
[873] Murgich J, Strausz OP. Molecular mechanics of aggregates of asphaltenes and resins of the Athabasca oil. Pet Sci Technol 2001;19(1):231–43.
[874] Ortega-Rodrıguez A, Cruz SA, Gil-Villegas A, Guevara-Rodrıguez F, Lira-Galeana C. Molecular view of the asphaltene aggregation behavior in asphaltene-resin mixtures. Energy Fuels 2003;17: 1100–8.

[875] Kuznicki T, Masliyah JH, Bhattacharjee S. Molecular dynamics study of model molecules resembling asphaltene-like structures in aqueous organic solvent systems. Energy Fuels 2008;22(4):2379–89.
[876] Correra S, Merino-Garcia D. Simplifying the thermodynamic modeling of asphaltenes in upstream operations. Energy Fuels 2007;21(3):1243–7.
[877] Merino-Garcia D, Correra S. A shortcut application of a flory-like model to asphaltene precipitation. J Dispers Sci Technol 2007;28(3):339–47.
[878] Mohammadi AH, Richon D. A monodisperse thermodynamic model for estimating asphaltene precipitation. AIChE J 2007;53(11):2940–7.
[879] Painter P, Veytsman B, Youtcheff J. Association model for the phase behavior of asphaltenes. Energy Fuels 2014;28(4):2472–80.
[880] Akbarzadeh K, Alboudwarej H, Svrcek WY, Yarranton HW. A generalized regular solution model for asphaltene precipitation from n-alkane diluted heavy oils and bitumens. Fluid Phase Equilibria 2005;232(1–2):159–70.
[881] Akbarzadeh K, Rahimi P, De Bruijn T, Irwin AW, Yarranton HW. Application of a generalized regular solution model in the oil compatibility determination. In: Proc. ACS Nat. Meeting. San Diego, CA, vol. 50; 2005. p. 251–4.
[882] Alboudwarej H, Akbarzadeh K, Beck J, Svrcek WY, Yarranton HW. Regular solution model for asphaltene precipitation from bitumens and solvents. AIChE J 2003;49(11):2948–56.
[883] Laux H, Rahimian I, Schorling P. Colloidal stability of crude oils and crude oil residues. Pet Sci Technol 1999;17(3):349–68.
[884] Tejeda P, Lira-Galeana C, Rodriguez F. Analysis of a thermodynamic micellization model for asphaltene precipitation. In: Proc. Latin American Caribbean Pet. Eng. Conf. Rio de Janeiro, Brazil; August 30–September 3, 1997. 10 pp.
[885] Tharanivasan AK. Asphaltene precipitation from crude oil blends, conventional oils, and oils with emulsified water. Ph.D. thesis from University of Calgary. Department of Chem. Pet. Eng; 2012. 211 pp.
[886] Tharanivasan AK, Svrcek WY, Yarranton HW, Taylor SD, Merino-Garcia D, Rahimi PM. Measurement and modeling of asphaltene precipitation from crude oil blends. Energy Fuels 2009;23(8):3971–80.
[887] Tharanivasan AK, Yarranton HW, Taylor SD. Application of a regular solution-based model to asphaltene precipitation from live oils. Energy Fuels 2011;25(2):528–38.
[888] Beck J, Svrcek WY, Yarranton HW. Hysteresis in asphaltene precipitation and redissolution. Energy Fuels 2005;19:944–7.
[889] Horvath-Szabo G, Masliyah JH, Elliott Ja W, Yarranton HW, Czarnecki J. Adsorption isotherms of associating asphaltenes at oil/water interfaces based on the dependence of interfacial tension on solvent activity. J Colloid Interface Sci 2005;283(1):5–17.
[890] Ortega-Rodriguez A, Duda Y, Guevara-Rodriguez F, Lira-Galeana C. Stability and aggregation of asphaltenes in asphaltene-resin-solvent mixtures. Energy Fuels 2004;18:674–81.
[891] Vargas FM, Gonzalez DL, Creek JL, Wang J, Buckley J, Hirasaki GJ, et al. Development of a general method for modeling asphaltene stability. Energy Fuels 2009;23(3):1147–54.
[892] Vargas FM, Gonzalez DL, Hirasaki GJ, Chapman WG. Modeling asphaltene phase behavior in crude oil systems using the perturbed chain form of the statistical associating fluid theory (PC-SAFT) equation of state. Energy Fuels 2009;23(3):1140–6.
[893] Gonzalez DL, Hirasaki GJ, Creek J, Chapman WG. Modeling of asphaltene precipitation due to changes in composition using the perturbed chain statistical associating fluid theory equation of state. Energy Fuels 2007;21(3):1231–42.
[894] Merino-Garcia D, Andersen SI. Thermodynamic characterization of asphaltene-resin interaction by microcalorimetry. Langmuir 2004;20(11):4559–65.
[895] Tavakkoli M, Panuganti SR, Taghikhani V, Pishvaie MR, Chapman WG. Precipitated asphaltene amount at high-pressure and high-temperature conditions. Energy Fuels 2013;28(3):1596–610.

[896] Ting PD, Hirasaki GJ, Chapman WG. Modeling of asphaltene phase behavior with the SAFT equation of state. Pet Sci Technol 2003;21(3–4):647–61.
[897] Wu J, Prausnitz JM, Firoozabadi A. A molecular-thermodynamic framework for asphaltene-oil equilibria. Report No. LBNL–40045; Other: ON: DE97004175; TRN: AHC29708%%17. Berkeley, CA: Lawrence Berkeley National Lab; February 1997. 39 pp.
[898] Wu J, Prausnitz JM, Firoozabadi A. Molecular-thermodynamic framework for asphaltene-oil equilibria. AIChE J 1998;44(5):1188–99.
[899] Agrawal P. Measurement and modeling of the phase behavior of solvent diluted bitumens. MSc thesis from University of Calgary. Department of Chem. Pet. Eng; March 2012. 133 pp.
[900] Akbarzadeh K, Ayatollahi S. Equations lead to asphaltene deposition predictions. Oil Gas J 2002;100(44):51–2. 54.
[901] Mullins OC, Andrews AB, Pomerantz AE, Dong C, Zuo JY, Pfeiffer T, et al. Impact of asphaltene nanoscience on understanding oilfield reservoirs. In: Proc. SPE Annual Technical Conf. Exhibition. Denver, CO, vol. 3; 2011. p. 2197–209.
[902] Mullins OC, Pomerantz AE, Zuo JY, Andrews AB, Hammond P, Dong C, et al. Asphaltene nanoscience and reservoir fluid gradients, tar mat formation, and the oil-water interface. In: Proc. SPE Annual Technical Conf. Exhibition. New Orleans, LA, vol. 3; 2013. p. 2387–400.
[903] Mullins OC, Sabbah H, Eyssautier J, Pomerantz AE, Barré L, Andrews AB, et al. Advances in asphaltene science and the Yen-Mullins model. Energy Fuels 2012;26(7):3986–4003.
[904] Nikookar M, Omidkhah MR, Pazuki GR. Prediction of density and solubility parameter of heavy oils and SARA fractions using cubic equations of state. Pet Sci Technol 2008;26(16):1904–12.
[905] Pan H, Firoozabadi A. Thermodynamic micellization model for asphaltene precipitation from reservoir crudes at high pressures and temperatures. SPE Prod. Facilities 2000;15(1):58–65.
[906] Salarieh M, Haqparast S, Kharrat R. Experimental and modeling study of asphaltene precipitation in petroleum fluids. In: Proc. 4th Intern. Conf. on Thermal Eng.: Theory Applications. Abu Dabi, UAE; January 12–14, 2009. 12 pp.
[907] Sabbagh O, Akbarzadeh K, Badamchi-Zadeh A, Svrcek WY, Yarranton HW. Applying the PR-EoS to asphaltene precipitation from n-alkane diluted heavy oils and bitumens. Energy Fuels 2006;20(2):625–34.
[908] Tian L, Wang J, Shen B, Liu J. The modeling of residue solvent deasphalting product molecular weight distribution with continuous thermodynamics. Pet Sci Technol 2010;28(14):1466–75.
[909] Victorov AI, Firoozabadi A. Thermodynamic micellization model of asphaltene precipitation from petroleum fluids. AIChE J 1996;42(6):1753–64.
[910] Manshad AK, Edalat M. Application of continuous polydisperse molecular thermodynamics for modeling asphaltene precipitation in crude oil systems. Energy Fuels 2008;22(4):2678–86.
[911] Correra S, Merino-Garcia D. Physical models of asphaltene solution behavior. In: Safieva RZ, Syunyaev RZ, editors. Physical and chemical properties of petroleum disperse systems and oil-and-gas technologies. Moscow-Izhevsk, Russia: ICS Publications; 2007. 15 pp.
[912] Subramanian S, Simon S, Sjöblom J. Asphaltene precipitation models: A review. J Dispersion Sci Technol 2016;37(7):1027–49.
[913] Coelho RR, Hovell I, De Mello Monte MB, Middea A, De Souza AL. Characterisation of aliphatic chains in vacuum residues (VRS) of asphaltenes and resins using molecular modelling and FTIR techniques. Fuel Proc Technol 2006;87(4):325–33.
[914] Dolomatov MY, Dezortsev SV, Bakhtizin RZ, Shulyakovskaya DO, Dolomatova MM, Kharisov BR, et al. Asphaltenes as objects of nanoelectronics. In: Proc. Trends in Nanotechnology Intern. Conf. Seville, Spain; September 9–13, 2013. 2 pp.
[915] Unger FG, Andrejeva LN. Grundlegende aspekte der chemie der erdöle. Natur der harze und asphaltene (fundamental aspect of oil chemistry. Nature of tars and asphaltenes). Erdol Erdgas Kohle 1994;47(1):18–23.

[916] Boek ES, Yakovlev DS, Headen TF. Quantitative molecular representation of asphaltenes and molecular dynamics simulation of their aggregation. Energy Fuels 2009;23(3):1209–19.
[917] Headen TF, Boek ES. Molecular dynamics simulations of asphaltene aggregation in supercritical carbon dioxide with and without limonene. Energy Fuels 2011;25(2):503–8.
[918] Boek ES, Wilson AD, Padding JT, Headen TF, Crawshaw JP. Multi-scale simulation and experimental studies of asphaltene aggregation and deposition in capillary flow. Energy Fuels 2010;24(4):2361–8.
[919] Boek ES, Headen TF, Padding JT. Multi-scale simulation of asphaltene aggregation and deposition in capillary flow. Faraday Discuss 2009;144:271–84.
[920] Headen TF, Boek ES. Potential of mean force calculation from molecular dynamics simulation of asphaltene molecules on a calcite surface. Energy Fuels 2011;25(2):499–502.
[921] Laux VH, Rahimian I, Neumann HJ. Particle distribution, colloid stability and rheological behaviour of crude oil residues and bitumen [partikelgroessenverteilung, kolloidstabilitaet und rheologie von erdoeldestillationsrueckstaenden und bitumen]. Erdol Erdgas Kohle 1993;109(9):368–72.
[922] Browarzik D, Laux H, Rahimian I. Asphaltene flocculation in crude oil systems. Fluid Phase Equilibria 1999;154(2):285–300.
[923] Laux H, Rahimian I, Butz T. Phase behaviour of colloidal crude oil system. Rev Inst Francais Pet 1997;52(2):226–7.
[924] Laux H, Rahimian I, Butz T. Theoretical and practical approach to the selection of asphaltene dispersing agents. Fuel Proc Technol 2000;67(1):79–89.
[925] Laux H, Kopsch H, Rahimian I. About formation of coke in the cracking of crude oil residues (part 1): Theoretical foundation and results of thermoanalysis [zur koksbildung bei der spaltung von erdoelrueckstaenden (1) theoretische grundlagen und ergebnisse der thermoanalyse]. Erdol Erdgas Kohle 1995;111(7–8):325–9.
[926] Laux H, Kopsch H, Rahimian I. About formation of coke in the cracking of crude oil residues (3) – modelling [zur koksbildung bei der spaltung von erdolruckstanden]. Erdol Erdgas Kohle 1996;112(7–8):319–21.
[927] Rahimian I, Laux H, Oschmann HJ. Cloud and pour point dependences of crude oils [trubungs- und stockpunktabhangigkeiten in erdolen]. Erdol Erdgas Kohle 1997;113(2):80–3.
[928] Barcenas M, Orea P, Buenrostro-González E, Zamudio-Rivera LS, Duda Y. Study of medium effect on asphaltene agglomeration inhibitor efficiency. Energy Fuels 2008;22(3):1917–22.
[929] Yu C, Zhang L, Guo X, Xu Z, Sun X, Xu C, et al. Association model for nickel and vanadium with asphaltene during solvent deasphalting. Energy Fuels 2015;29(3):1534–42.
[930] Fahim MA, Al-Sahhaf TA, Elkilani AS. Prediction of asphaltene precipitation for Kuwaiti crude using thermodynamic micellization model. Ind Eng Chem Res 2001;40:2748–56.
[931] Wang JX, Buckley JS, Burke NE, Creek JL. A practical method for anticipating asphaltene problems. SPE Prod Facil 2004;19(3):152–60.
[932] Pan H. A thermodynamic micellization model for asphaltene precipitation: Part I: Micellar size and growth. SPE Prod Facil 1998;13(2):118–27.
[933] Park SJ, Ali Mansoori G. Aggregation and deposition of heavy organics in petroleum crudes. Energy Sources 1988;10(2):109–25.
[934] Nabzar L, Aguilera ME. The colloidal approach. A promising route for asphaltene deposition modelling. Oil Gas Sci Technol Rev IFP 2008;63(1):21–35.
[935] Hu YF, Li S, Liu N, Chu YP, Park SJ, Mansoori GA, et al. Measurement and corresponding states modeling of asphaltene precipitation in Jilin reservoir oils. J Pet Sci Eng 2004;41(1–3):169–82.
[936] Kohse BF, Nghiem LX. Modeling asphaltene precipitation and deposition in a compositional reservoir simulator. In: Proc. SPE SPE/DOE 14th Symp. on Improved Oil Recovery Tulsa, Oklahoma; April 17–21, 2004. Paper SPE 89437, 11 pp.
[937] Kraiwattanawong K, Fogler HS, Gharfeh SG, Singh P, Thomason WH, Chavadej S. Thermodynamic solubility models to predict asphaltene instability in live crude oils. Energy Fuels 2007;21(3):1248–55.

[938] Vargas FM, Chapman WG. Application of the one-third rule in hydrocarbon and crude oil systems. Fluid Phase Equilibria 2010;290(1−2):103−8.
[939] Gonzalez DL, Vargas FM, Mahmoodaghdam E, Lim F, Joshi N. Asphaltene stability prediction based on dead oil properties: experimental evaluation. Energy Fuels 2012;26(10):6218−27.
[940] Panuganti SR, Tavakkoli M, Vargas FM, Gonzalez DL, Chapman WG. SAFT model for upstream asphaltene applications. Fluid Phase Equilibria 2013;359:2−16.
[941] Punnapala S, Vargas FM. Revisiting the PC-SAFT characterization procedure for an improved asphaltene precipitation prediction. Fuel 2013;108:417−29.
[942] Alhammadi AA, Vargas FM, Chapman WG. Comparison of cubic-plus-association and perturbed-chain statistical associating fluid theory methods for modeling asphaltene phase behavior and pressure-volume-temperature properties. Energy Fuels 2015;29(5):2864−75.
[943] Giraldo J, Nassar NN, Benjumea P, Pereira-Almao P, Cortés FB. Modeling and prediction of asphaltene adsorption isotherms using Polanyi's modified theory. Energy Fuels 2013;27(6):2908−14.
[944] Tavakkoli M, Panuganti SR, Vargas FM, Taghikhani V, Pishvaie MR, Chapman WG. Asphaltene deposition in different depositing environments: part 1. Model oil. Energy Fuels 2014;28(3): 1617−28.
[945] Tavakkoli M, Panuganti SR, Taghikhani V, Pishvaie MR, Chapman WG. Asphaltene deposition in different depositing environments: part 2. Real oil. Energy Fuels 2014;28(6):3594−603.
[946] Acevedo S, Caetano M, Ranaudo MA, Jaimes B. Simulation of asphaltene aggregation and related properties using an equilibrium-based mathematical model. Energy Fuels 2011;25(8): 3544−51.
[947] Sahimi M, Rassamdana H, Dabir B. Asphalt formation and precipitation: Experimental studies and theoretical modelling. SPE J 1997;2(2):157−69.
[948] Fahim MA. Prediction of asphaltene precipitation from empirical models. Pet Sci Technol 2007;25(12):1605−12.
[949] Redelius P. Solubility parameters and bitumen. Fuel 2000;79:27−35.
[950] Carauta ANM, Seidl PR, Correia JCG. Computational simulation of asphaltene aggregation. In: Proc. 2nd Mercosur Congress on Chem. Eng., 4th Mercosur Congress on Proc. Systems Eng. Rio de Janeiro, Brazil; August 14−18, 2006. 7 pp.
[951] Wiehe IA, Kennedy RJ. Application of the oil compatibility model to refinery streams. Energy Fuels 2000;14(1):60−3.
[952] Redelius P. Bitumen solubility model using Hansen solubility parameter. Energy Fuels 2004;18:1087−92.
[953] Wang J, Buckley JS. Improved modeling of the onset of asphaltene flocculation. In: Proc. 2nd Intern. Conf. on Pet. Phase Behavior Fouling. Copenhagen, Denmark; August 1, 2000. p. 1−6. Paper #18.
[954] Verdier S. Experimental study and modelling of asphaltene precipitation caused by gas injection. Ph D. thesis from Technical University Of Denmark. Department of Chem. Eng; April 2006. 281 pp.
[955] Rogel E. Studies on asphaltene aggregation via computational chemistry. Colloids Surf. A: Physicochem. Eng Asp 1995;104(1):85−93.
[956] Rahmani NHG, Masliyah JH, Dabros T. Characterization of asphaltenes aggregation and fragmentation in a shear field. AIChE J 2003;49(7):1645−55.
[957] Solaimany-Nazar AR, Rahimi H. Dynamic determination of asphaltene aggregate size distribution in shear induced organic solvents. Energy Fuels 2008;22(5):3435−42.
[958] Solaimany-Nazar AR, Rahimi H. Investigation on agglomeration-fragmentation processes in colloidal asphaltene suspensions. Energy Fuels 2009;23(2):967−74.
[959] Khoshandam A, Alamdari A. Kinetics of asphaltene precipitation in a heptane-toluene mixture. Energy Fuels 2010;24(3):1917−24.
[960] Laux H, Rahimian I, Browarzik D. Flocculation of asphaltenes at high pressure. I. Experimental determination of the onset of flocculation. Pet Sci Technol 2001;19(9−10):1155−66.

[961] Laux H, Butz T, Rahimian I. Thermogravimetric investigation of the influence of colloidal phase on the processing properties of crude oil residues. Oil Gas Sci Technol Rev IFP 2000;55(3):315–20.
[962] Angle CW, Hamza HA, Dabros T. Size distributions and stability of toluene diluted heavy oil emulsions. AIChE J 2006;52(3):1257–66.
[963] Pal R, Vargas F. On the interpretation of viscosity data of suspensions of asphaltene nano-aggregates. Can J Chem Eng 2014;92(3):573–7.
[964] Luo P, Gu Y. Effects of asphaltene content on the heavy oil viscosity at different temperatures. Fuel 2007;86(7–8):1069–78.
[965] Victorov AI, Smirnova NA. Thermodynamic model of petroleum fluids containing polydisperse asphaltene aggregates. Ind Eng Chem Res 1998;37(8):3242–51.
[966] Gabitto J, Barrufet M. Experimental and theoretical determination of heavy oil viscosity under reservoir conditions. Report No. OSTI ID: 816597 (Final Report). Tulsa, OK, USA: A&M University, National Petroleum Technology Office; May 2003.
[967] Sattarin M, Modarresi H, Bayat M, Teymori M. New viscosity correlations for dead crude oils. Pet Coal 2007;49(2):33–9. Available from: http://www.vurup.sk/sites/vurup.sk/archivedsite/www.vurup.sk/pc/vol49_2007/issue2/pdf/pc_2_2007_sattarin.pdf.
[968] Barrufet MA, Setiadarma A. Reliable heavy oil-solvent viscosity mixing rules for viscosities up to 450 K, oil-solvent viscosity ratios up to 4×10^5, and any solvent proportion. Fluid Phase Equilibria 2003;213(1–2):65–79.
[969] Rane JP, Zarkar S, Pauchard V, Mullins OC, Christie D, Andrews AB, et al. Applicability of the Langmuir equation of state for asphaltene adsorption at the oil–water interface: coal-derived, petroleum, and synthetic asphaltenes. Energy Fuels 2015;29(6):3584–90.
[970] Pina A, Mougin P, Béha E. Characterisation of asphaltenes and modelling of flocculation – state of the art. Oil Gas Sci Technol Rev IFP 2006;61(3):319–43.
[971] Carnahan NF, Salager J-L, Anton R, Davila A. Properties of resins extracted from Boscan crude oil and their effect on the stability of asphaltenes in Boscan and Hamaca crude oils. Energy Fuels 1999;13:309–14.
[972] Gray MR, Assenheimer G, Boddez L, Mccaffrey WC. Melting and fluid behavior of asphaltene films at 200–500°C. Energy Fuels 2004;18:1419–23.
[973] Ge Q, Yap YF, Vargas FM, Zhang M, Chai JC. Numerical modeling of asphaltene deposition. Comput Therm Sci 2013;5(2):153–63.
[974] Kurup AS, Vargas FM, Wang J, Buckley J, Creek JL, Subramani HJ, et al. Development and application of an asphaltene deposition tool (adept) for well bores. Energy Fuels 2011;25(10):4506–16.
[975] Marques LCC, Monteiro JB, González G. Asphaltenes flocculation in light crude oils: A chemical approach to the problem. J Dispers Sci Technol 2007;28(3):391–7.
[976] Marcano F, Antonieta Ranaudo M, Chirinos J, Castillo J, Daridon JL, Carrier H. Study of asphaltenes aggregation in toluene/n-heptane/CO_2 mixtures under high-pressure conditions. Energy Fuels 2013;27(8):4598–603.
[977] Jamaluddin AKM, Creek J, Kabir CS, Mcfadden JD, D'cruz D, Manakalathil J, et al. Laboratory techniques to measure thermodynamic asphaltene instability. J Can Pet Technol 2002;41(7):44–52.
[978] Akbarzadeh K, Hammami A, Kharrat A, Zhang D, Allenson S, Creek J, et al. Asphaltenes – problematic but rich in potential. Oilfield Rev 2007;19(2):22–43.
[979] Mozaffarian M, Dabir B, Sohrabi M, Rassamdana H, Sahimi M. Asphalt flocculation and deposition IV. Dynamic evolution of the heavy organic compounds. Fuel 1997;76(14–15):1479–90.
[980] Rassamdana H, Sahimi M. Asphalt flocculation and deposition: II. Formation and growth of fractal aggregates. AIChE J 1996;42(12):3318–32.
[981] Dabir B, Nematy M, Mehrabi AR, Rassamdana H, Sahimi M. Asphalt flocculation and deposition: III. The molecular weight distribution. Fuel 1996;75(14):1633–45.

[982] Rassamdana H, Mirzaee N, Mehrabi AR, Sahimi M. Field-scale asphalt precipitation during gas injection into a fractured carbonate reservoir. In: Proc. SPE Annual Western Regional Meeting. Long Beach, CA, USA. Society of Petroleum Engineers (SPE); June 25–27, 1997. p. 519–34.

[983] Sahimi M, Mehrabi AR, Mirzaee N, Rassamdana H. The effect of asphalt precipitation on flow behavior and production of a fractured carbonate oil reservoir during gas injection. Transp Porous Media 2000;41(3):325–47.

[984] Wu Q, Seifert DJ, Pomerantz AE, Mullins OC, Zare RN. Constant asphaltene molecular and nanoaggregate mass in a gravitationally segregated reservoir. Energy Fuels 2014;28(5):3010–5.

[985] Carbognani L. Dissolution of solid deposits and asphaltenes isolated from crude oil production facilities. Energy Fuels 2001;15(5):1013–20.

[986] Le Page JF, Chatila SG, Davidson M. Raffinage et conversion des produits lourds du petrole. Paris: Editions Technip; 1990. 190 pp.

[987] Wiehe I. Blending compatibility and incompatibility of opportunity crudes. In: Proc. Opportunity Crudes Conf.: Challenges, Benefits. Houston, TX; 30 April–2 May, 2008 [Unpublished presentation].

[988] Rogel E, León O, Torres G, Espidel J. Aggregation of asphaltenes in organic solvents using surface tension measurements. Fuel 2000;79(11):1389–94.

[989] Andersen SI. Flocculation onset titration of petroleum asphaltenes. Energy Fuels 1999;13:315–22.

[990] Mason TG, Lin MY. Time-resolved small angle neutron scattering measurements of asphaltene nanoparticle aggregation kinetics in incompatible crude oil mixtures. J Chem Phys 2003;119(1):565–71.

[991] Garcia MC, Carbognani L. Asphaltene-paraffin structural interactions. Effect on crude oil stability. Energy Fuels 2001;15(5):1021–7.

[992] Yanfei W, Jian C, Shengsheng J, Benxian S. Investigation on the compatibility and incompatibility of vacuum residua with catalytic cracking bottom oil. Energy Fuels 2003;17:344–7.

[993] Wiehe I. Crude oil compatibility. In: Proc. COQG Meeting. Houston, TX. USA; October 2, 2003. 24 pp.

[994] Wiehe I. Crude oil incompatibility and near incompatibility. In: Proc. MUG 36. HaverlyTechnical Conference. Vancouver, Canada; September 15–18, 2002. 25 pp.

[995] Rahimi P, Alem T. Crude oil compatibility and diluent evaluation for pipelining. In: Proc. Crude Oil Quality Association Meeting. New Orleans, LA. USA; February 10–11, 2010. 36 pp. (Available from: http://www.coqa-inc.org/docs/default-source/meeting-presentations/20100211_Rahimi_Compatibility.pdf.

[996] Wiehe IA, Kennedy RJ. Process for blending potentially incompatible petroleum oils. Patent No. US5871634 (Also published as WO9826026, EP0948580, CA2271957, TW460568B). Exxon Research and Engineering Co.; February 16, 1999.

[997] Rathore V, Brahma R, Thorat TS, Rao PVC, Choudary NV. Assessment of crude oil blends. PTQ 2011;16(4):6.

[998] Dion M. Challenges and solutions for processing opportunity crudes. In: Proc. AFPM Annual Meeting. Orlando, FL, USA; March 23–25, 2014. 13 pp.

[999] Pauli AT. Asphalt compatibility testing using the automated Heithaus titration test. ACS Div Fuel Chem 1996;41(4):1276–80. Preprints.

[1000] Pauli AT, Robertson RE, Branthaver J, Schabron JF. Automated flocculation titrimeter system, Patent No. WO2002063292. The University of Wyoming Research Corp.; August 15, 2002.

[1001] Dickakian G. Crude oil fouling control by a fouling analyzer. In: Proc. ASME Heat Transfer Division, HTD. Philadelphia, PA, USA, vol. 108. ASME; 1989. p. 331–6.

[1002] Aguiar JIS, Garreto MSE, González G, Lucas EF, Mansur CRE. Microcalorimetry as a new technique for experimental study of solubility parameters of crude oil and asphaltenes. Energy Fuels 2014;28(1):409–16.

[1003] Pauli AT, Branthaver JF. Relationships between asphaltenes, Heithaus compatibility parameters, and asphalt viscosity. Pet Sci Technol 1998;16(9–10):1125–47.

[1004] Schermer WEM, Melein PMJ, Van Den Berg FGA. Simple techniques for evaluation of crude oil compatibility. Pet Sci Technol 2004;22(7–8):1045–54.

[1005] Pauli AT, Schabron JF. Stability and compatibility testing of petroleum and asphalt. Am Lab 2003;35(19):8–10.
[1006] Alomair OA, Almusallam AS. Heavy crude oil viscosity reduction and the impact of asphaltene precipitation. Energy Fuels 2013;27(12):7267–76.
[1007] Mansoori GA, Vazquez D, Shariaty-Niassar M. Polydispersity of heavy organics in crude oils and their role in oil well fouling. J Pet Sci Eng 2007;58(3–4):375–90.
[1008] Wiehe IA, Kennedy RJ, Dickakian G. Fouling of nearly incompatible oils. Energy Fuels 2001;15(5):1057–8.
[1009] Wiehe I. Mitigation of asphaltene fouling with the oil compatibility model. In: Proc. Intern. Conf. on Refinery Proc. Houston, TX, USA, vol. 4; April 24, 2001. p. 300–6.
[1010] Wiehe IA. Petroleum fouling: causes, tools, and mitigation methods. In: Proc. AIChE Spring Nat. Meeting. Orlando, FL; 2006. 1 pp.
[1011] Wiehe IA. The chemistry of petroleum fouling. In: Proc. AIChE Spring Nat. Meeting. Houston, TX, USA; April 24, 2001. 7 pp.
[1012] Jacobs IC, Thorne MA. Asphaltene precipitation during acid stimulation treatments. In: Proc. 7th SPE Symp. on Formation Damage Control. Lafeyette, LA, USA. USA: Soc of Petroleum Engineers of AIME; February 26–27, 1986. p. 131–6. SPE 14823.
[1013] Leontaritis KJ, Ali Mansoori G. Asphaltene deposition: A survey of field experiences and research approaches. J Pet Sci Eng 1988;1(3):229–39.
[1014] Saleh ZS, Sheikholeslami R, Watkinson AP. Blending effects on fouling of four crude oils. In: Proc. 6th Intern. Conf. on Heat Exchanger Fouling Cleaning Challenges Opportunities. Kloster Irsee, Germany; June 5–10, 2006. p. 37–46.
[1015] Andersen SI, Speight JG. Thermodynamic models for asphaltene solubility and precipitation. J Pet Sci Eng 1999;22(1–3):53–66.
[1016] Gawel I, Bociarska D, Biskupski P. Effect of asphaltenes on hydroprocessing of heavy oils and residua. Appl Catal A Gen 2005;295(1):89–94.
[1017] Carbognani L, Espidel J, Izquierdo A. Characterization of asphaltenic deposits from oil production and transportation operations. In: Yen TF, Chilingarian GV, editors. Asphaltenes asphalts. 2. Developments in pet. Science. Amsterdam: Elsevier; 2000. p. 335–62.
[1018] Storm DA, Decanio SJ, Edwards JC, Sheu EY. Sediment formation during heavy oil upgrading. Pet Sci Technol 1997;15(1–2):77–102.
[1019] Gentzis T, Parker RJ, Mcfarlane RA. Microscopy of fouling deposits in bitumen furnaces. Fuel 2000;79(10):1173–84.
[1020] Gentzis T, Rahimi P. Processability and thermal behaviour of Athabasca bitumen with varying asphaltenes concentration. ACS Div Fuel Chem 1999;44(4):810–6. Preprints.
[1021] Schabron JF, Pauli AT, Rovani Jr JF. Non-pyrolytic heat induced deposition from heavy oils. Fuel 2001;80(7):919–28.
[1022] Stark JL, Asomaning S. Crude oil blending effects on asphaltene stability in refinery fouling. Pet Sci Technol 2003;21(3–4):569–79.
[1023] Stark JL, Falkler T, Dennie J. New technology for high temperature furnace fouling control. In: Proc. NPRA Annual Meeting. San Antonio, TX; 2007. p. 437–48. Paper #AM-07–61.
[1024] Falkler TJ, Stark JL. Index correlates asphaltene stability to coker furnace fouling potential. Oil Gas J 2004;102(34):72–4.
[1025] Sinnathambi CM, Nor NM. Relationship between SARA fractions and crude oil fouling. J Appl Sci 2012;12(23):2479–83. Available from: http://docsdrive.com/pdfs/ansinet/jas/2012/2479-2483.pdf.
[1026] Asomaning S, Watkinson AP. Deposit formation by asphaltene-rich heavy oil mixtures on heat transfer surfaces. Understanding heat exchanger fouling and its mitigation. 1999. p. 283–90.
[1027] Bennett CA, Appleyard S, Gough M, Hohmann RP, Joshi HM, King DC, et al. Industry-recommended procedures for experimental crude oil preheat fouling research. Heat Transf Eng 2006;27(9): 28–35.

[1028] Al-Atar E. Effect of oil compatibility and resins-asphaltene ratio on heat exchanger fouling. MSc thesis from University of British Columbia. Chem. Bioresource Eng; March 1, 1997. 124 pp.

[1029] Wang S, Chung K, Gray MR. Role of hydrotreating products in deposition of fine particles in reactors. Fuel 2001;80(8):1079−85.

[1030] Nelson MA. Field dispersant test for determining the fouling tendency of a crude oil process, Patent No. US5156975. Nalco Chemical Company; October 20, 1992.

[1031] Pugh SJ, Hewitt GF, Muller-Steinhagen H. Heat exchanger fouling in the pre-heat train of a crude oil distillation unit − the development of a user guide. In: Proc. Heat Exchanger Fouling Fundamental Approach & Technical Solutions. Davos, Switzerland; July 8−13, 2002. p. 201−12.

[1032] Lemke HK. Fouling in refinery equipment − an overview. In: Proc. 1st Int'l Conference on Petroleum Phase Behavior and Fouling, AIChE Spring Meeting. New York, NY. USA; March 1999. p. 375−82.

[1033] Gleim WKT. Role of asphaltene in refining. In: Chilingarian GV, Yen TF, editors. Asphaltenes asphalts. 1. Developments in pet. Science. Elsevier; 1978. p. 243−58.

[1034] Ancheyta J, Betancourt G, Centeno G, Marroquín G. Catalyst deactivation during hydroprocessing of Maya heavy crude oil. (II) Effect of temperature during time-on-stream. Energy Fuels 2003;17(2):462−7.

[1035] Ancheyta J, Betancourt G, Centeno G, Marroquín G, Alonso F, Garciafigueroa E. Catalyst deactivation during hydroprocessing of Maya heavy crude oil. 1. Evaluation at constant operating conditions. Energy Fuels 2002;16(6):1438−43.

[1036] Mochida I, Zhao ZZ, Sakanishi K. Catalyst deactivation during the hydrotreatment of heavy oils. ACS Div. Pet Chem 1988;33(2):360. Preprints (1 pp.).

[1037] Rana MS, Ancheyta J, Sahoo SK, Rayo P. Carbon and metal deposition during the hydroprocessing of Maya crude oil. Catal Today 2014;220−222:97−105.

[1038] Seki H, Yoshimoto M. Deactivation of HDS catalyst in two-stage RDS process II. Effect of crude oil and deactivation mechanism. Fuel Proc Technol 2001;69(3):229−38.

[1039] Schmidt M. Deactivation behaviour. Hydrocarbon Eng; March 2011. 3 pp.

[1040] Rastegari K, Svrcek WY, Yarranton HW. Kinetics of asphaltene flocculation. Ind Eng Chem Res 2004;43(21):6861−70.

[1041] Thakur DS, Thomas MG. Catalyst deactivation in heavy petroleum and synthetic crude processing: a review. Appl Catal 1985;15(2):197−225.

[1042] Furimsky E, Massoth FE. Deactivation of hydroprocessing catalysts. Catal Today 1999;52(4):381−495.

[1043] Dufresne P. Hydroprocessing catalysts regeneration and recycling. Appl Catal A General 2007;322(Suppl.):67−75.

[1044] Eijsbouts S, Battiston AA, Van Leerdam GC. Life cycle of hydroprocessing catalysts and total catalyst management. Catal Today 2008;130(2−4):361−73.

[1045] Oelderik JM, Sie ST, Bode D. Progress in the catalysis of the upgrading of petroleum residue: A review of 25 years of R&D on shell's residue hydroconversion technology. Appl Catal 1989;47(1):1−24.

[1046] Villasana Y, Luis MA, Méndez FJ, Labrador H, Brito JL. Upgrading and hydrotreating of heavy oils and residua. In: Govil JN, editor. Energy science and technology series. Studium Press LLC; 2015. p. 304−28.

[1047] Marugán J, Calles JA, Dufour J, Giménez-Aguirre R, Peña JL, Merino-García D. Characterization of the asphaltene onset region by focused-beam laser reflectance: A tool for additives screening. Energy Fuels 2009;23(3):1155−61.

[1048] Ostlund JA, Nyden M, Fogler HS, Holmberg K. Functional groups in fractionated asphaltenes and the adsorption of amphiphilic molecules. Colloids Surf A Physicochem Eng Asp 2004;234(1−3):95−102.

[1049] Lindemuth PM, Lessard RB, Lozynski M, Kremer M. Regarding asphaltene. Dispersants as desalting benefactors. In: Proc. 3rd International Conference on Refinery Processing − Recent Advances in Desalting. Atlanta, GA. USA; March 5−9, 2000. 11 pp.

[1050] Daniel-David D, Le Follotec A, Pezron I, Dalmazzone C, Noik C, Barre L, et al. Destabilisation of water-in-crude oil emulsions by silicone copolymer demulsifiers. Oil Gas Sci Technol Rev IFP 2008;63(1):165–73.

[1051] Xia L, Lu S, Cao G. Stability and demulsification of emulsions stabilized by asphaltenes or resins. J Colloid Interface Sci 2004;271(2):504–6.

[1052] Stark JL, Asomaning S. Synergies between asphaltene stabilizers and demulsifying agents giving improved demulsification of asphaltene-stabilized emulsions. Energy Fuels 2005;19(4):1342–5.

[1053] Liu G, Xu X, Gao J. Study on the compatibility of asphaltic crude oil with the electric desalting demulsifiers. Energy Fuels 2003;17(3):543–8.

[1054] Liu G, Xu X, Gao J. Study on the compatibility of high-paraffin crude oil with electric desalting demulsifiers. Energy Fuels 2003;17(3):625–30.

[1055] Correra S. Asphaltene problem: Will it be understood before the end of the oil? (living and dead theories for asphaltene stability). In: Proc. Adv. Flow Assurance Technol. San Donato Milanese, Italy; February 23, 2006. 7 pp.

CHAPTER 3

Metal Compounds

1. Introduction

Contaminating inorganic trace elements are present in crude oils and source-rock bitumen, as has been mentioned in Chapter 1. Certain elements can form heteroatomic organic compounds, while other inorganic elements are present in the form of both organic and inorganic compounds. The abundances of these trace elements vary widely among the different crude oils. Abundance depends on the source rock's (1) diagenesis, (2) depositional environment, (3) geological setting, (4) contained organic material, (5) maturity, and (6) chemical composition [1]. These trace elements, the compound types, and the chemical and physical species were described by Filby [1] and are summarized in Table 3.1.

Heavy as well as light metals are present in the crude oil, e.g., nickel, vanadium, iron, cadmium, mercury, aluminum, sodium, calcium, and magnesium [2], among others. Abundance of these metals ranges from less than a part per billion (ppb) to more than

Table 3.1: Trace Elements Found in Crude Oil and Bitumen

Chemical and Physical Species	Elements Found	Compound Types
Elemental species dissolved in oil or associated with polar constituents	S, Se, Hg	S°, Se°, Hg°
Discrete and extractable metal-organic complexes	Ni, V, Fe	Porphyrin complexes; metal chlorins; metal naphthenates; asphaltene moieties
Metal complexes or metal ions associated with or bonded in polar components of oils	As, Se	Possibly in molecular structure
Organometallic complexes with metal—carbon bonds	As, Hg	Methyl and phenyl arsenic acids; (methyl mercury compounds?)
Entrained formation waters	Na, K, Ca, Mg, I, Cl, Br, S, O	Na^+, K^+, Ca^{2+}, Mg^{2+}, I^-, Cl^-, Br^-, SO_4^{2-}
Entrained mineral matter	Al, K, Si, Mg, Ca, Na, Sc, rare earth	Clay minerals
Drilling/production contaminants	Ba, As, Hg	$BaSO_4$ in drilling muds; As in bactericides; Hg in oils flashed using liquid Hg

The Science and Technology of Unconventional Oils. http://dx.doi.org/10.1016/B978-0-12-801225-3.00003-6

several thousand parts per million (ppm). Nickel and vanadium are usually the most abundant metals within the organic environment and in concentration ranges of up to 200 ppm for Ni and 2800 ppm for V. Abundance of other trace elements is generally less than 100 ppm. The concentrations of Fe, Al, Na, Ca, and Mg have also been reported to vary in deposits as a function of well depth [3] and also among asphaltene subfractions [4,5]. In particular, the heteroatomic, sulfur, and nitrogen moieties are accumulated in the heaviest fraction of crude oil together with the molecules bearing heavy metals [4–6]. Examples of the V, Ni, and Fe contents in different types of oils are presented in Table 3.2 from data reported in Refs. [7–12].

From the three most common heavy metals in crude oils, regarding unconventional oils (UOs), Ni and V are the most abundant in bitumen and heavy and extra-heavy crudes, while Fe is in larger proportions in high-acid crudes (HACs) and shale oils (in which Ca is present in a major concentration as well). Heavy oils and bitumen are characterized by the presence of large heavy molecules, in which atoms of heavy metals are either associated or embedded. Most abundant heavy metals concentrate in the bottom of the barrel (BotB) fraction. Both the asphaltene as well as the resin (maltene) fractions contain part of the

Table 3.2: Most Abundant Metals in Crude Oil

		Metals (ppm)			
Oil Type	**Location**	**Ni**	**V**	**Fe**	**References**
Condensate	Kapuni, Taranaki Basin, NZ	<1	<1	6	[7]
Light oil	Tariki, Taranaki Basin, NZ	1.2	0.4	0	[7]
Crude oil	Gilby, Mannville Group, Alberta Basin	11	32	0.8	[8]
Heavy oil	Boscan, Venezuela	97	1200	16	[9]
	Zuata, Venezuela	155	820	2	[10]
	Tia Juana, Venezuela	100	765	2	[10]
Bitumen	New Albany shale, Indiana	2560	2700	353	[9]
Bitumen	Green River shale, Mahogany Zone, Colorado	59.1	41.2	79.0	[12]
	Asphalt Ridge, county asphalt pit, Vernal, Utah	88.1	5.9	N/R	[11]
	Sunny Side, Utah	61.6	5.9	N/R	[11]
	Big Clifty, Logan county, Kentucky	75.0	169.9	N/R	[11]
	Arroyo Grande, Pismo Beach, California	137.3	76.1	N/R	[11]
	McKittrick, Diatomite, California	62.4	39.4	N/R	[11]
	Commercial Kentucky	66.5	162.8	N/R	[11]
	Athabasca, Canada	81.4	216.8	N/R	[11]

N/R, not reported.

metal compounds; typically, metals in resins are in a lesser concentration [13–16] than those in the asphaltenes. Some examples are given in Table 3.3.

Other chemical elements contaminating crude oils and of interest in refining concern nonmetals such as arsenic and selenium. The presence of As (as well as halides) has been known for a long time and these compounds are mainly distributed in light fractions. The presence of significant concentrations of selenium in the refineries' wastewaters [17] has brought attention to this element. The Se compounds in the crude are mainly in the form of various selenides. Since Se chemistry is similar to that of S, Se profile will follow the pathways of sulfur. Likewise, the abundance of Se in crude is usually proportional to that of S, at an estimated ratio of 1:1000. As and Se represent environmental burdens that need to be taken care of during refining. This need for additional treatment will increase operating costs to meet compliance. Nevertheless, they do not affect operability and processability of the refining processing units. The same situation pertains to another heavy metal, mercury, which is also mainly present in the light cuts.

This chapter will review the knowledge developed around the heavy metal contaminants present in residues and heavy crude oils (as will become evident later, mostly Ni and V, and to some extent Fe). Light metals, such as Na, Ca, K, Si, etc., which are mostly present in the form of inorganic salts, will not be considered here for two reasons: these metals are not indigenous to the oil and are typically removed by the well-known desalting processes at the refining front end (see, for instance, Ref. [18–20]). The particular case of organic salts, typically of Ca, will be discussed in Chapter 4 since these compounds are mainly concerned with naphthenates.

Since the most abundant metal contaminants are concentrated in the asphaltic fraction, the reader is invited to visit the previous chapter, in which asphaltenes where considered and thoroughly discussed.

Table 3.3: Metal Content in the Bottoms Fractions of Various Crude Oils

	Resins		Asphaltenes	
Resid	Ni	V	Ni	V
Maya	44	242	286	1602
Arabian Heavy	30	81	170	594
Beta	124	147	429	749
Kern River	149	75	492	361
Boscan	160	1595	375	4410
Hondo	105	198	474	979
Heavy Maya	21	77	202	1141
Cerro Negro	26	110	360	1600
Wilmington	32	17	430	360
Prudhoe Bay	4	10	96	250
Furrial			164	845

2. Origin

Petroleum formation from source rocks is widely accepted to take place at temperatures in the range 50–200°C. This process is extremely slow and needs millions of years to complete. In the case of metals, the present compounds are formed by metal ions or groups bound to organic moieties. The incorporation of metals into the crude oil has been linked to the origin and aging processes associated with hydrocarbon formation [21] through different sources:

- Assimilation into the organic matrix during diagenesis of the biological material in the source rocks from surrounding clay minerals or interstitial aqueous solution;
- Metal complex diagenesis of the original biological material;
- During primary or secondary migration by uptake from an aqueous phase or mineral phases;
- From formation waters or from reservoir rock minerals.

Thus an understanding of the reactions resulting in the incorporation of trace metals has been the objective of many research projects. A characterization of the source rock in La Luna Formation (Maraca section), Maracaibo Basin, in Venezuela established correlations among bitumen, total organic carbon, total sulfur, major elements (Si, Al, Fe, Mg, Mn, Ca, Ti, Na, K, P), trace elements (V, Ni, Co, Cr, Cu, Zn, Mo, Ba, U, Th), and rare-earth elements (La, Ce, Nd, Sn, Eu, Th, Yb, Lu) of the carbonate-free fraction. From the observed correlations it was suggested that sedimentation occurred under dysoxic conditions for the chert layers and under euxinic conditions for argillaceous limestone. During the sequence corresponding to the argillaceous limestone, the sedimentation conditions of several intervals under euxinic conditions varied related to a high contribution and/or preservation of the organic matter at sedimentation intervals under relatively lower anoxic conditions and in the presence of H_2S [22].

The definition of indices based on metals concentrations has been proposed for distinguishing the geologic age of oils. Certain metal concentration ratios were examined and V, Ni, and Mg seem to provide better correlations, indicating that the mechanism of other metals' incorporation into the oil resembles that of V and Ni [23]. In other instances, S, V, and Ni concentrations and ratios showed good correlations as age indicators [24–26].

Metals were first identified in crude oils and bitumen in the form of porphyrins. Early last century petroporphyrins were probably the first compounds that were used by Treibs [27] for building evidence, which linked crude oil to a biological origin. During initial stages of diagenesis, metalloporphyrins are formed from surrounding rocks, thus Ni and V relative abundances would depend of the environment composition. Treibs also proposed the reactions sequence that may have given rise to porphyrins [particularly the most abundant types, i.e., deoxophylloerythroetioporphyrin (DPEP) and etioporphyrin (Etio)] from chlorophyll a. This sequence is initiated by demetallization of chlorophyll, followed by subsequent hydrolysis/saponification, reduction, aromatization, decarboxylation, and chelation of rock metals [27−29]. This reaction network is exemplified for DPEP in Fig. 3.1 [30].

The Treibs reaction sequence may vary depending on the depositional environment and the presence of other porphyrin precursors such as type of chlorophyll (b or c), heme, bacteriochlorophylls a−e, and other biological pigments. These latter precursors were categorized into five classes: (1) compounds with carbon skeletons related to selected types of precursors (e.g., DPEP), (2) compounds with carbon skeletons related to specific precursors (derivatives of chlorophyll b and c, and of bacteriochlorophyll d), (3) compounds exclusively related to nonspecific precursors (e.g., cytochrome derivatives), (4) compounds with carbon skeletons not obviously related to known precursors [e.g., rhodoporphyrins (Rhodo) of bacterial origin], and (5) compounds with a structure occurring naturally but that may have been chemically modified during isolation or purification procedures (this category addresses the implications to characterization work and the implications and/or interpretation of the genesis of the metal compounds) [31]. The presence of carboxylated DPEP and Etio porphyrins in the kerogen was used as a

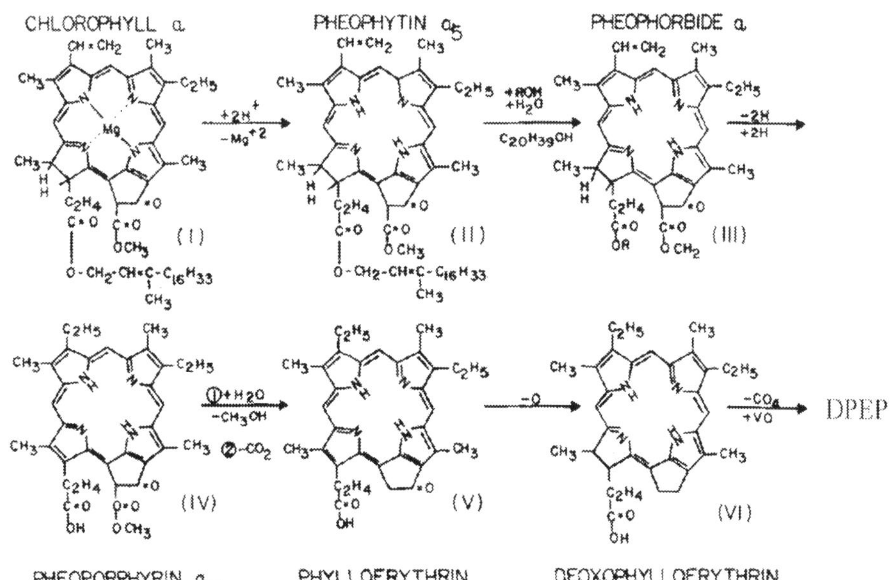

Figure 3.1

Treibs' reaction sequence for deoxophylloerythroetioporphyrin (DPEP) formation from chlorophyll a. *Reproduced from Yen TF, Chilingarian GV. Structural parameters from asphaltenes and their geochemical significance. In: Yen TF, Chilingarian GV, editors. Asphaltenes asphalts. 1. Developments in pet. science. Elsevier; 1994. p. 159−78, with permission from Elsevier.*

suggestion of the direct incorporation of these porphyrins into the oil upon simple decarboxylation. A higher proportion of Etio in the kerogen was also hypothesized as the reason for an increase in its proportion upon maturation [32].

A significant amount of nickel and vanadyl (VO) geoporphyrins exhibiting more condensed tetrapyrrolic cores than just chlorophyll-derived DPEP and Etio were found in an onshore Californian crude oil [33]. The simulation of kerogen catagenesis at bench scale using pyrolysis caused the effective liberation of Ni and V porphyrins in a proportional amount to that originally present on the kerogen as organically bound metals [34].

The origin of the less abundant porphyrins has been a subject of speculation [31]. Besides originating from different precursors, reactions of the most abundant porphyrins to convert into those less abundant have also been proposed. Reactions proposed include alkylation, transalkylation, decarboxylation, etc. [35,36]. The alkylation and transalkylation patterns of Boscan porphyrins were studied for the purpose of clarification of the Rhodo-porphyrin structure and origin. It was speculatively suggested that both reactions might occur during diagenesis and maturation [35]. An unexpected alkylation (methylation and propylation) of the ring E of chlorophyll occurred in the presence of CH_2X_2 ($X = Cl, Br$) [37]. The presence of significant concentrations of chlorin- and phylloerythrin-type compounds in immature fields, together with a high proportion of DPEP and Etio porphyrins [38] may be indicative of slower maturation processes from other precursors.

Another hypothesis is that petroleum VO porphyrins were released from kerogen in a two-step process. At low temperature, kerogen weaker bonds were broken and asphaltenes were produced. Upon temperature increases, asphaltenes decomposed releasing the porphyrin [39]. The finding of Etio porphyrins entrapped in Green River kerogen was taken as a substantiation of a later release of fossil molecules by thermal or thermocatalytic cracking of those that were originally trapped in the kerogen network or alternatively attached to the kerogen by chemical bonds [40]. A study on the immature La Luna and Serpiano source rocks concluded that such mechanism would only contribute to a minor extent [41]. This hypothesis does not have the acceptance of the other one. The reasoning for the rejection of that hypothesis is that if porphyrins were originated through degradation of kerogen, then it would be expected under geological conditions that porphyrin content would start to increase with the onset of petroleum generation (125–130°C) and would reach a maximum at higher maturation (145–160°C) [42]. Instead, maturation decreases porphyrin concentration.

A geochemical characterization of Saudi Arabian crude oils indicated that the content of VO porphyrins decreases with the depth of the well, coinciding with well temperature increases. This was explained to be caused by the instability of metalloporphyrins at higher temperatures (as suggested in Ref. [39]). As already mentioned, the abundance of VO porphyrins also decreases with increasing maturity of the crude oil. This was also explained to be caused by the length of time the oil has been trapped in deeper horizons and been exposed to higher thermal action. Hence the content of VO porphyrins was suggested to be governed by the thermal stress of depth to which the evolving sources were subjected [43]. Although temperature has a well-established role in chemical reactions, it has been suggested that chemical control should explain the presence of vanadium and nickel while excluding other metals. In addition to redox properties, sensitivity to acids and tendencies to exchange inorganic ligands for organic ligands must also be of importance in the formation of the metal-organic compounds [44].

Geochemical studies indicated that regardless of kerogen type, S and metal content decrease, while H/C ratio increases upon maturation. Meanwhile, maturation also brings a negative effect on stability and coking tendency of the residue fraction. The molecular structures change with aging, particularly aromaticity, i.e., asphaltenes becoming more aromatic, while aromaticity decreases in resins. This is a consequence of thermodynamic control. The rationale being that C-interaction with heteroatoms is either weaker than C–C bonds or complex organometallic compounds condense into even larger structures that consolidate within the rocks.

The degree of transformation and the kinetics of this transformation of the kerogen occurring during maturation depend on the type of kerogen. Although metal content typically decreases with maturation, it is less evident in kerogens of type I than in those of type II. Asphaltene content is also expected to increase with a decrease in maturation, following the same trend as metals. In Fig. 3.2, the data from Ref. [45] has been plotted for both types of kerogens and all degrees of maturation. The expected trend is more pronounced for crudes originating from type II kerogens.

Nevertheless, when a selected group of crudes is grouped according to a similar type of maturation, type I and type II kerogens show the expected trend (Fig. 3.3).

The well-accepted correlation of porphyrin content and maturation has enabled the establishment of methodologies for the determination of oil age. Porphyrins are then considered biomarkers for oil age and the ΣDPEP/ΣEtio ratio is one of the most accepted indicators for this purpose. The study of the origin of these biomarkers has been paramount in the understanding of the origin of the oil itself. One study considered asphaltenes from the Dead Sea Basin and concluded that kerogens with high porphyrin content should have had a marine origin [46]. A more detailed comparison of the metals' (V and Ni) profiles with other biomarkers susceptible to biodegradation provided deeper insight into the origin, types, and grouping of crude oil mixtures [47–49].

Electron probe microanalysis was employed for a study of Venezuelan source rocks for stylolites (Querecual Formation), framboids (La Luna Formation), and kerogen (La Luna and Querecual Formations). The element concentrations and distributions obtained in the

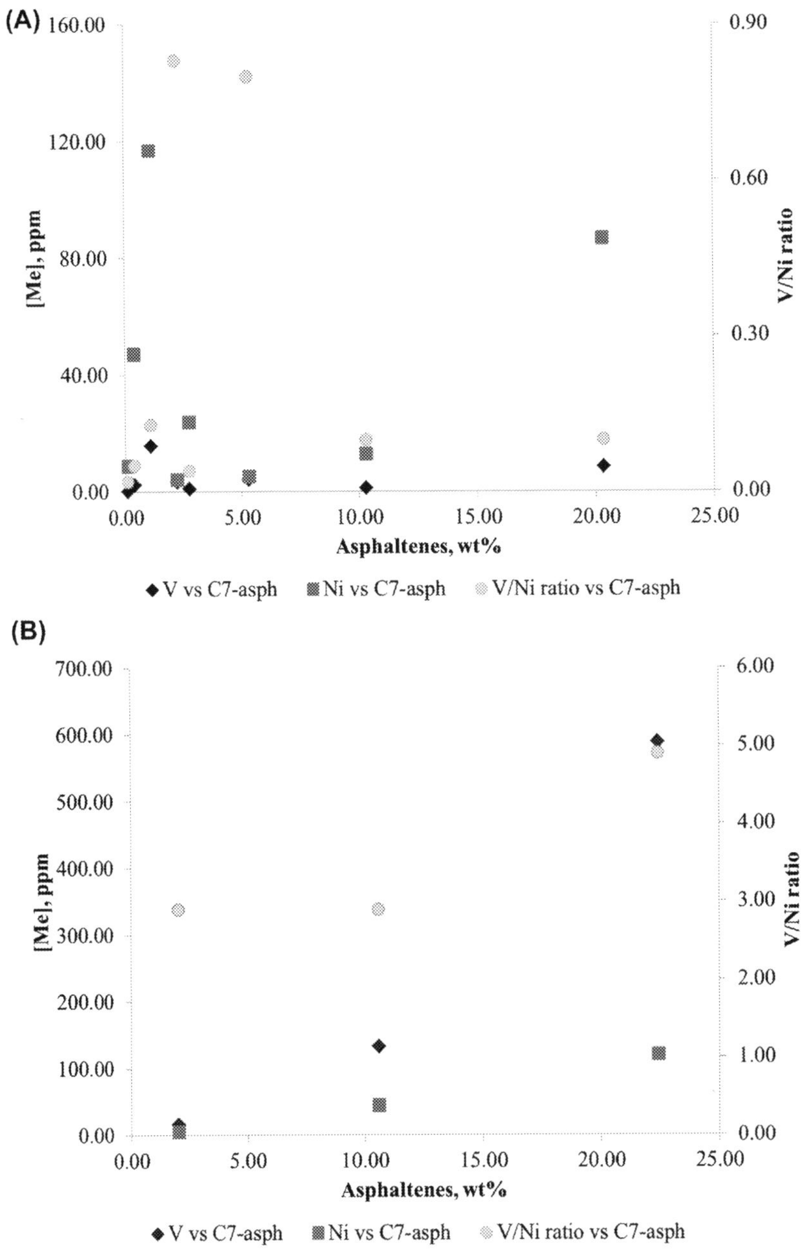

Figure 3.2

Metals and asphaltenes. (A) Kerogen type I, (B) kerogen type II.

stylolite and in the rock showed that V and to a lesser extent Ni were associated with the organic matter [50]. Chemically speaking, this might be taken as a more rapid incorporation of vanadium into the organic substance and/or a higher reactivity of the

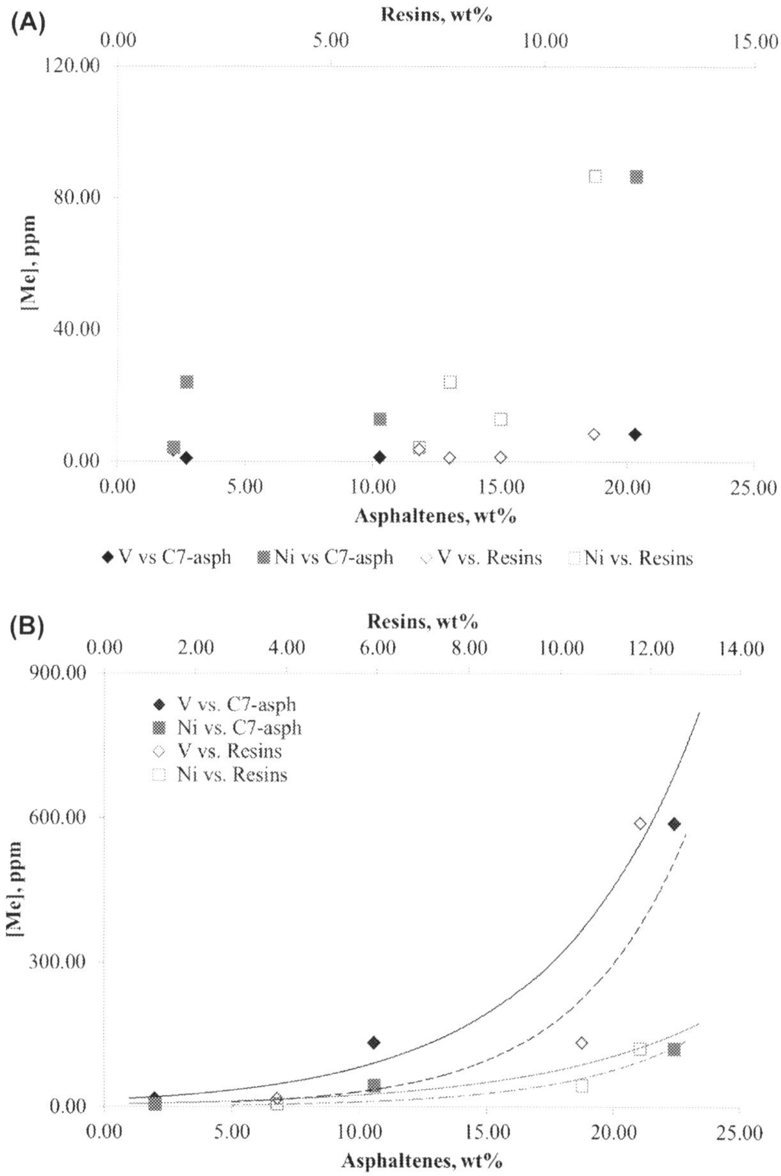

Figure 3.3
Metals and asphaltenes. (A) Kerogen type I—mature, (B) kerogen type II—immature.

vanadium source mineral matter toward the organic components of the kerogen. Other metals were also investigated in these crudes, namely, Fe, Mo, Ca, and Na, and a relationship with maturity was targeted but was not found [45]. The reason is believed to be that at least part of these metals originates from other sources (seawater, corrosion products, etc.) and those are not indigenous to the crude.

Apart from metalloporphyrins, any other Ni and V species occurring in oils and source-rock bitumen is categorized simply as nonporphyrin, since both nature and structure remain unknown. The nonporphyrin complexes of any trace elements in crude oils originate either through primary or secondary processes. Catagenesis of the kerogen is the primary process, involving leaching of metal and formation of metal-organic complexes by mineral–kerogen reactions. Secondary processes occur during maturation, biodegradation, or migration by interaction of mineral matter or formation waters with organic compounds in the oil [1]. The isothermal heating at laboratory scale of three different kerogen samples at 150 and 250°C for 1–20 days in the presence of air O_2 converted the kerogen VO porphyrins into VO nonporphyrin [51]. This route of nonporphyrin formation from porphyrins indicates that nonporphyrins may be regarded as evolutionary products of porphyrin maturation.

A study on vanadyl $\left(VO_2{}^+\right)$ nonporphyrins in Western Kentucky No. 9 coal concluded that the vanadylation occurred during the diagenetic (peat-forming) stage in the Pennsylvanian swamp. An unusual V enrichment of the top 15 cm, relative to the lower parts of the coalbed, was interpreted to be caused by a high V-content in the past swamp water, predominantly from the weathered/leached V of a volcanic ash source. This volcanic ash was probably derived from basalts because of the observed association of V/Cr (together with their enrichment in the top of the seam) and Ni. Thus the high V-content was thought to be caused by the high abundance of organic (humic) materials and low rate of deposition. The study predicted that the oxidation potential and pH of the ancient peat interstitial water were approximately −0.2 to −0.3 V and 5–6, respectively, based on the chemistry of $VO_2{}^+$, FeS_2, and $CrOH_2{}^+$ [52].

3. Isolation

Metal compounds concentrate in the BotB fraction. The initial attempts for characterizing these compounds by direct examination of the BotB fraction showed the complexity of the problem. The need for isolation, recovery, separation, and/or purification became evident.

It is expected that the metal compounds would be soluble in the most polar solvents though still some partitioning between the solvent and the oil fraction would limit the extraction effectiveness. In fact, metal compounds naturally present in the crude oil are very difficult to extract from the asphaltene fraction, regardless of the polarity of used solvent [53,54]. Furthermore, the low solubility of the heavy asphaltenes and resins would limit the accessibility of the solvent to the metal compounds.

More stringent conditions have to be applied. Thus strong polar solvents at boiling conditions have been required for extraction and sometimes results are not very effective. Extraction proceeds according to solubility differences between resins or asphaltenes and metal compounds. Asphaltenes and resins are a solubility class of compounds. Asphaltenes are a crude oil fraction that is insoluble in light paraffins (usually heptane, but also any $n-C_6^-$ alkane) and is soluble in toluene, while resins are insoluble in lower molecular weight (MW) *n*-alkanes and soluble in higher MW *n*-alkanes. More specifically, resins are a polar fraction of petroleum that is soluble in *n*-alkanes and aromatic solvents and insoluble in ethyl acetate. These solubility differences between asphaltenes and resins, because of the higher and broader solubility of resins, make the selection of an extracting solvent for the metal compounds a difficult challenge. Considering the antisolvent/solvent (heptane/toluene) pair used for the operational definition of asphaltenes, metalloporphyrins will meet the same criterion, i.e., they will exhibit the same solubility/insolubility behavior. Therefore if the asphaltenes or the resins are first dissolved in a given solvent, then a more polar solvent can be used for the separation of the metal compounds, such as dimethylformamide [55], methanol/acetone mixtures [56,57], methanol and tetrahydrofuran [58], methylene chloride, or chloroform [59]. A combination of pyridine or aniline would solubilize metal compounds from the BotB fraction. From this extracted mixture of compounds certain nonporphyrinic types of compounds can be extracted using hydrocarbon solvents [60]. These methods can be combined with the saturate, aromatic, resin, and asphaltene (SARA) fractionation (see Chapter 2 for a description of this method) and then Soxhlet extraction [59] can be applied to the SARA fractions with combinations of the convenient solvents, according to porphyrin or nonporphyrin compounds.

A preparative liquid chromatography (LC) or high-performance liquid chromatography (HPLC) system has been used for purification purposes of the isolated porphyrin mixture [55]. Preparative columns of alumina and silica are used in sequence and mixtures of solvents are used to elute the different compounds. The sequence hexane, hexane–toluene, toluene, toluene–methylene chloride, and methylene chloride was used by Doukkali et al. for the separation of V and Ni porphyrins from a shale oil vacuum resid (VR) [59]. Irreversible poisoning of the packing materials causes lack of reproducibility and poor effectiveness in the chromatographic separation of the nonporphyrinic compounds [58]. Other columns have been suggested, e.g., aminopropyl column [61].

A separation method was specifically applied for obtaining VO porphyrins and characterizing these using an oxidation method [35]. The porphyrins could be separated from the resins fraction using silica gel functionalized with alkylsulfonic acid and eluting with toluene, followed by 2% ethyl acetate in toluene. Further separation of Rhodo from nonrhodo-porphyrins was carried out by thin layer chromatography using precoated silica plates [35]. The details of the separation method developed by Barwise and Whitehead were given in a British patent application [62], which was subsequently withdrawn, breaking away from any possible access to the information.

Metal removal with strong acid solutions was thought as a way of limiting the contamination of the metal compounds with the organic polar compounds in the oil matrix. The porphyrin mixture could then be processed with different solvents and/or eluted from an adsorbent column. A method using a silica gel column and mixtures of benzene, chloroform, and isopropanol was reported as capable of separating Etio-, DPEP-, and Rhodo-type porphyrins in an effective way [63]. Separation of the VR into acid, base, and neutral (ABN) fractions allows the recovery of metal compounds according to chemical nature. The ABN separation is carried out by sequential extraction or LC using either solid adsorbents or ion-exchange resins such as macroreticular styrenedivinylbenzene polymer bonded with strong base or acid groups. A nonaqueous ion-exchange liquid chromatographic procedure used BioRad MP1 and MP50 resins and reported some drawbacks in the method: contamination of the acid fraction by the resins' material, precipitation/fouling of organics on the column, solvents selection, and poor quality of the fractions [64]. Nonetheless, the ABN fractions were further subfractionated by preparative HPLC. The neutral and weak acid fractions were eluted from a 25×2.5 (i.d.) cm column packed with (4-pyridyl)ethyl silica, with three mobile phases: *n*-heptane, benzene, and 70:30 benzene:ethanol. The strong acid and strong base fractions were enriched with the nonporphyrinic metal compounds whereas the weak base or neutral fractions concentrated the bulk of the porphyrinic complexes. A brief description of the whole separation LC/HPLC scheme is presented in Fig. 3.4 [65,66]. Repetition of the fractionation basis on the subfractions was used in an attempt to concentrate the metal compounds in a given subfraction. This attempt was more successful for V than it was for Ni [65]. This separation scheme, though containing some pitfalls, set the basis for many improved versions of the principle, which later served for the isolation of simplified chemical families and facilitated the titanic efforts needed for not only the identification of the metal compounds, but also the definition of compound classes and types, or pseudotypes, and the understanding of properties and chemical behavior.

The porphyrin mixture was separated from three different feeds (Boscan 9K3 oil, La Luna 36E shale, and Carneros 31042 oil) by preparative LC through first alumina and then silica columns eluting with a hexane—methylene chloride mixture. Subsequently, the porphyrin-containing fractions were subfractionated by silica thin layer chromatography (TLC) using toluene—hexane as the mobile phase. The Ni—porphyrin subfractions were further subfractionated by another silica TLC using toluene—methylene chloride as the mobile phase. Etio-isomers were separated using an octadecyl-silica (ODS) (NARP) column, using a mixture of chloroform, acetonitrile, and water (5:90:5, v/v/v) as solvent under isocratic conditions [67—69].

The integration of online element detection with LC gave the opportunity to distinguish V from Ni compounds. Thus for this purpose, element-selective graphite furnace atomic absorption detection has been used in combination high-performance liquid

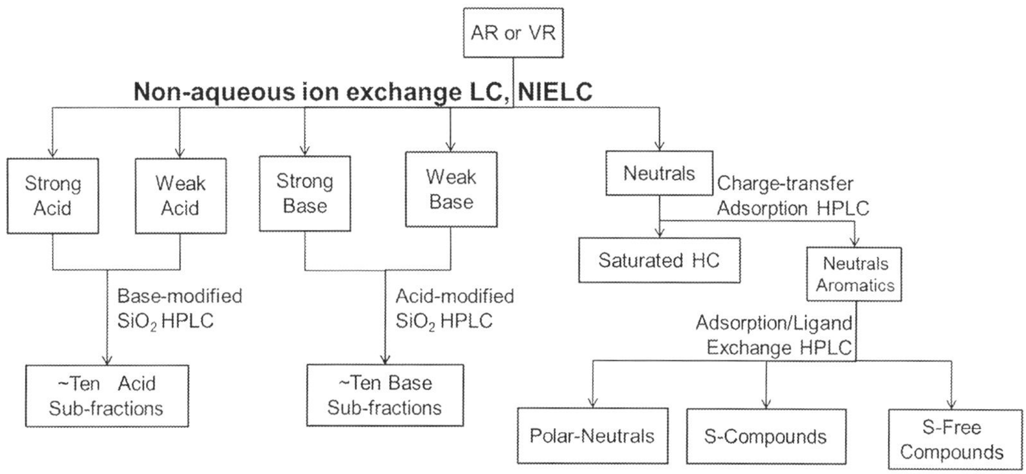

Figure 3.4
Liquid chromatography/high-performance liquid chromatography (LC/HPLC) separation scheme of metal compounds.

chromatography systems (HPLC/GFAA) [70]. Based on these techniques, reversed-phase HPLC/GFAA has been proven useful for the separation and identification of V and Ni nonporphyrinic compounds. The VO and nickel compounds were extracted from the crude oil using pyridine as solvent (pyridine:water 4:1 solution). The reversed phase HPLC was run using an ODS column with a solvent gradient starting with 100% methanol:water (3:1 v/v) to 30% tetrahydrofuran (THF) from 1 to 3 min, followed with 30–70% THF from 22 to 24 min, and finally 100% THF from 25 to 27 min. The column was restored after each run by ramping to 100% methanol:water (3:1 v/v) over 3 min and holding until over 10 column volumes of solvent were eluted. Whereas fractionation using silica or alumina columns would retain the nonporphyrins, the reversed phase ODS column allowed typical metal recoveries in excess of 97% [15].

A vacuum sublimation method was proposed for the isolation of metal compounds from asphaltene fractions. At a pressure of 10^{-2} mm Hg and between 80 and 240°C, four fractions could be collected: (1) a light yellow oil fraction (at 90–120°C), (2) an orange oil fraction (130–160°C), (3) a dark red gluey semisolid fraction (170–200°C), and (4) a red brittle semisolid fraction (210–240°C). Roughly 50% of the total metals were left in the black residue remaining after sublimation. The V-content increased with increasing sublimation temperature, while the Ni-content decreased and (V and Ni) porphyrins were detected in all these sublimed fractions [71].

Short-path distillation has been reported as a method for separating an atmospheric residue (AR) into a porphyrin-containing distillate fraction that boils between 454 and 677°C (98% porphyrinic metals) from the nonporphyrin-containing residue that boils above 677°C (2% porphyrinic metals) [72].

4. Description and Properties

Organometallic complexes and salts are present in UOs. Porphyrin and nonporphyrin complexes of transition metals and more particularly of V and Ni are the more representative compounds. Organic and inorganic salts are also present; the latter are not considered here for reasons given earlier. Organic salts are more abundant in HACs than in heavy oils and bitumen.

4.1 Molecular Structure

Molecular characterization of metal compounds has not been an easy task. Although isolation and purification of these compounds have received much attention, there still exists the possibility and uncertainty of structural modifications during the course of these processes.

The metal-bearing molecules usually contain sulfur, oxygen, and more particularly nitrogen. While V and Ni are typically present in the form of organometallic compounds (minor amounts of iron as well), the other light metals (Al, Na, Ca, and Mg, including Fe) consist of inorganic and/or organic salts.

Early work showed the complexity of the characterization problem, for instance, in the case of Boscan crude oil, VO−Etio was not found in a 1967 work [73], and two molecularly different vanadium porphyrin compounds differing in polarity were proposed in 1989 [74]. Asphaltene fractionated in two fractions differing in polarity, concentrates the metal (Fe, Ni, and V) compounds in the more polar fraction, with a partition of about 2:1 [75]. Early electron paramagnetic resonance characterization of asphaltenes found out that vanadium was present in the oxidation state of +4 [76]. The elucidation of the molecular structures of the metal compounds has required titanic isolation and characterization efforts. Nevertheless, this task is yet to be completed.

Porphyrins are highly aromatic large molecules with discontinued ring conjugation. The structure of porphyrin is based on porphin, which is an aromatic compound formed by four pyrrole rings linked by unsaturated CH bridges. The porphin molecule (Fig. 3.5A) appears as a chelating agent in the metal complexes. In the porphyrinic compounds, the metals are embedded in the core of the porphyrin compound (see examples in Fig. 3.5B and C showing the case of tetraethyltetramethylporphin (TETMP) complexes of Ni and VO, respectively) by the elimination of the two hydrogen atoms from opposite amide groups.

Some of the common porphyrin structures have already been mentioned. The core structures of synthetically obtained porphyrin molecules are shown in Fig. 3.6. Examples shown in this figure include 2,7,12,18-tetraethyl-3,8,13,17-tetramethyl-21H,23H-porphine (Etio), diethylpyrrole-bridged bisporphyrin, tetraphenylporphyrin (TPP), and 10-[2-(4-carboxyphenyl)ethenyl]-5,15-bis(2,4,6-trimethylphenyl)porphyrin. Among these, only Etio

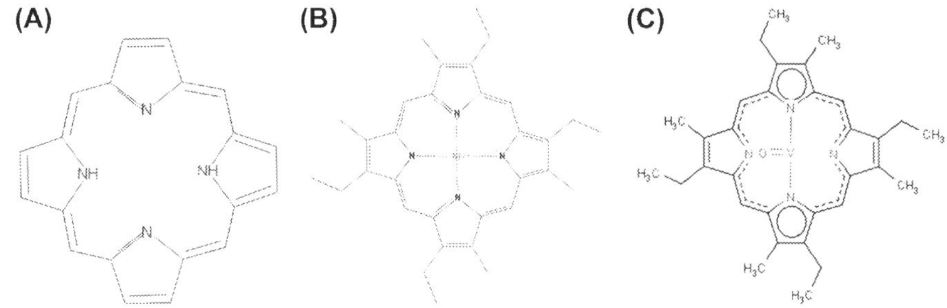

Figure 3.5
(A) Porphyrin molecule, (B) Ni-tetraethyltetramethylprophin (TETMP) complex, and (C) vanadyl-TETMP complex.

has been found in crude oils. The identification of the others in crude samples has been difficult.

Instead, petroporphyrins are a complex mixture of five types of porphyrins, each type consisting of a homologous series. These complex mixtures differ in the relative population of each series, MW and MW distribution (MWD), and number of isomers present. The variations observed also depend on the source (crude oil). As an example, the characteristics of the MWD of petroporphyrins for various crude oils is presented in Table 3.4 [77]. For this particular example, the MW of the most abundant porphyrin in

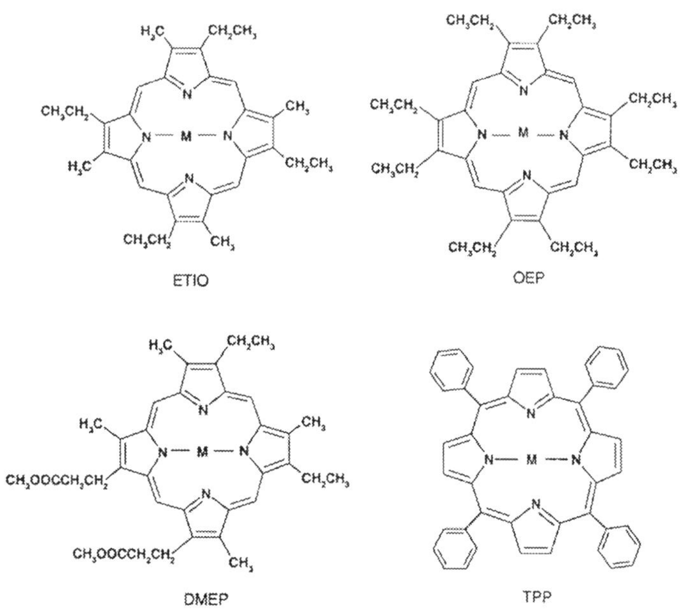

Figure 3.6
Core structures of synthesized porphyrins. DMEP, *10-[2-(4-carboxyphenyl)ethenyl]-5,15-bis(2,4,6-trimethylphenyl)porphyrin;* Etio, *etioporphyrin;* OEP, *octoethylprophyrin;* TPP, *tetraphenylporphyrin.*

Table 3.4: Characteristics of the Molecular Weight Distribution of Petroporphyrins

	Molecular Weight/z	Width (amu)	CH_2—Units (#)
Boscan asphaltene	476	70	5
Boscan asphaltene (DPEP)	462	42	3
Boscan asphaltene (Etio)	450	70	5
Mara asphaltene	450	70	5
Mara gas oil	436	56	4
Wilmington asphaltene	436	42	3
Gilsonite	476	No envelope	

DPEP, deoxophylloerythroetioporphyrin; *Etio*, etioporphyrin.

Boscan asphaltene is heavier than that in a Mara asphaltene and this one is heavier than that in a Mara gas oil; meanwhile the broadness of the MWD is equal for Mara and Boscan asphaltenes, but narrower for the Mara gas oil, which is also the case comparing the Etio series with the DPEP series of Boscan asphaltene.

Vanadium, as the most abundant metal contaminant, has received the most attention. Studies of V porphyrins indicated that although the rearrangements of the porphin substituents to generate an infinite number of porphyrinic compounds seem probable, the variation found in oils have been proven to be very limited. For instance, the cyclicity of the substituents at the porphin nucleus of porphyrins is not very common. In fact, most porphyrins (80–95%) are alkyl-porphyrins (Fig. 3.7A, etio-) and monocycloalkyl-porphyrins (Fig. 3.7B, DPEP), which are present in homologous series differing in the length of the C-chain of the substituents, i.e., $MW_{series\ member} = MW_{core} + 14n$ [78]. Thus the MW (in amu) of each member of the (metal-free) Etio series is $310 + 14n$ and it is $308 + 14n$ for the DPEP series [36].

Consistently with their molecular structures, the results from probe mass spectrometry (MS) and nuclear magnetic resonance experiments indicated that DPEP porphyrins are more polar than Etio, and for a homologous series, the larger the C_n, the lower the polarity [69]. These DPEP and Etio porphyrins of Ni^{2+} or VO^{2+} are most predominantly found in crude oils, although smaller amounts of rhodo, benzo, and other complex porphyrins are also found [1,59]. This small proportion of porphyrins (5–10%) contains more condensed substituents; examples include benzoporphyrins (Fig. 3.7C, Rhodo–etio with $MW = 458 + 14n$ [36]), cycloalkyl-monobenzoporphyrins (Fig. 3.7D, Rhodo–DPEP with $MW = 456 + 14n$ [36]), and bicycloalkyl-porphyrins (Fig. 3.7E, Di-DPEP with $MW = 306 + 14n$). These latter types of porphyrins seem to be exclusively found in crude oils with no structural equivalents in living beings and/or of no biological origin.

In summary, the five homologous series of porphyrin types are: (1) alkyl-porphyrins, (2) monocycloalkyl-porphyrins, (3) bicycloalkyl-porphyrins, (4) cycloalkyl-monobenzoporphyrins, and (5) benzoporphyrins. Each type is formed by a continuous

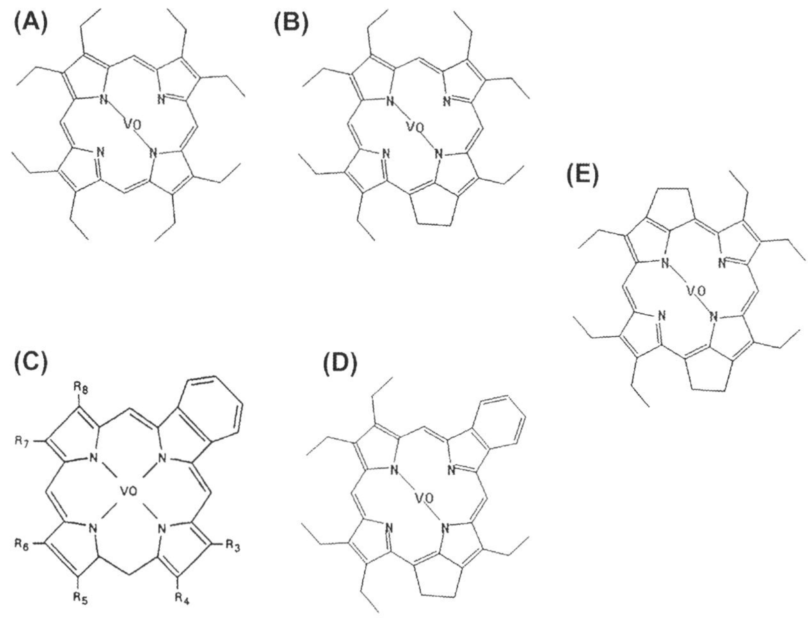

Figure 3.7
Most common V porphyrins found in crude oil. (A) Etio, (B) deoxophylloerythroetioporphyrin (DPEP), (C) Benzo-etio, (D) Rhodo-DPEP, (E) Di-DPEP.

isobaric series in oil. Each series starts with members containing C_5—C_7 alkyl side substituents and ending with C_{25}—C_{30}-type branches, the most common being C_{10}—C_{13}. A minor proportion of VO porphyrins contains a greater number of condensed rings in the structure, such as naphthene-aromatic moiety, bicycloalkyl-benzoporphyrins, and dibenzoporphyrins with adjacent and/or opposite benzene rings [78].

Confirmation of homologous series can be achieved by obtaining a linear correlation in the so-called Kovats' plots [79], in which the C-number is plotted as a function of the logarithm of the retention time [67,68,80]. This kind of plot has been used to characterize series and identify series members. Construction of a Kovats' plot of a mixture of petroporphyrins requires separation, isolation, purification, and demetallization of the porphyrinic fraction. A combination of chromatographic techniques and methodologies (preparative and analytical) needs to be employed. The demetallization step is typically done by acid extraction using alkyl-sulfonic acid [35,36,77], though hydrotreatment (HDT) has also been considered [65]. However, an apparently homologous series according to Kovats' plots might be a pseudohomologous series or not even homologous at all. The use of MS has revealed that a group of related compounds were characterized by formulae in which the preceding and succeeding members differ by one methylene group, whereas the methylene group was not confined to a common alkyl chain, contrary to what is required to be a homologous series. For example, as a representative of the alkyl-porphyrin series, the etio-homologous series should have an MW of 310 amu + 14n, where n is an integer. However, for a given species to be part of a homologous series, the incremental $-CH_2-$ has to belong to the same alkyl substituting chain in the series [36,77]. The credibility of the structural implications derived from the characterization of treated samples trembles on the possible modifications that such treatments may have caused [77].

Derivatization by incorporation of silicon into the free-base porphyrin has enabled the application of gas chromatography–mass spectrometry (GC–MS) and elucidation of isomeric structures. A comparison of different derivatization routes and reagents indicated that the bis(trialkylsiloxy)silicon (IV) derivatives can be analyzed by GC–MS on nonpolar stationary phases. Both identification of individual components (series and structural type) and classification of the porphyrins into the homologous or pseudohomologous series are possible with careful calibration of the GC and MS spectral data, which are combined into Kovats'-type plots [67–69]. A large number of individual porphyrins were found on the application of this methodology on Boscan oil and its source rock, La Luna shale. The Kovats' plots indicated a high degree of similarity between the homologous and pseudohomologous series found in the crude and its source rock [67].

High-resolution MS has been used for the detection and identification of porphyrins. For these purposes, atmospheric pressure photoionization (APPI) and Fourier transform ion cyclotron resonance mass spectrometry (FTICR-MS) were used on VO porphyrins and sulfur-containing vanadyl (VOS) porphyrins. A series with a wide carbon number range ($C_{26}-C_{52}$) and Z-number range (−28 to −54) was considered. Under APPI conditions, extremely low fragmentation occurs and primarily molecular ions (M+.) are detected. Thus this methodology allows the molecular characterization of the whole series, with no need for predemetallization or chromatographic separation. DPEP was the most prevalent structure found, followed by Etio- and then Rhodo-DPEP. Although the employed technique indicated the presence of sulfur in the porphyrinic sample, whether it is part of the molecular structure and how it is bound to the molecule remain unknown and require further characterization work. The observed Z-distribution of VO porphyrins suggested that the molecular growth of the ring structures was caused by incremental benzene and naphthene units. The recorded carbon number distributions of VO porphyrins was bimodal [81]. This bimodal distribution in terms of C-number is consistent with the bimodal molecular size observed by Reynolds [82]. The bimodal size distribution was not observed from Fe compounds; instead it was unimodal with a maximum at significant large molecular size.

The most abundant Ni porphyrins family is of monocycloalkyl type (presumed to be DPEP) followed by Etio- (of bicycloalkyl type), and then Rhodo-DPEP (monocycloalkyl type). A similar study using APPI–FTICR-MS on a series of Z-numbers ranging from

−28 to −44 and a carbon number ranging from 26 to 41 was carried out. In the Ni mixture, the Etio/DPEP ratio and Rhodo-etio/Rhodo-DPEP ratio were determined to be higher compared to those typically observed on VO−porphyrin mixtures. As already mentioned, in studied nickel and VO porphyrins, the higher degree of condensation of tetrapyrrolic cores, compared to chlorophyll-derived DPEP and etioporphyrins, indicates a change in origin or evolutive transformation [33]. This technique also served to confirm the coordination sphere of the VO porphyrins, the MW distribution of the series and the double bond equivalent (DBE) of the molecular structures [83]. DBE together with carbon number are specific parameters for each member and type of porphyrin family. In the case of Gudao oil, a crude containing a higher proportion of Ni over V, the Etio homologs spread through the C_{25}−C_{34} and maximized at C_{30} while the DPEP homologs comprised a C_{28}−C_{37} range and maximized at C_{31}. A slightly higher population of DPEP resulted in a ΣDPEP/ΣEtio ratio of 1.04 [84]. This ratio has been associated with the geological age of the crude oil or bitumen and two more Chinese crude oils were characterized with that purpose [85]. This ratio is evaluated spectroscopically, using either MS or UV−Vis spectra.

Actually, a primary distinguishing characteristics of these molecular structures is their intense absorption UV−Vis spectra. The most intense absorption, named the Soret band, occurs around 400 nm, and two additional visible bands, the α and β bands, appear in metalloporphyrin spectra in the region of 500−580 nm (Table 3.5) [36,60,70,74,84,86,87]. Peripheral substitution does not affect significantly the Soret band but rather changes the visible range peaks [87,88]. The spectra are also modified by solvent effects [87,89] and by coordination and association [90−93]. The positions of the UV−Vis bands can distinguish between Ni and VO porphyrins, but the spectra of closely related positional isomers of the same metal are identical. For these reasons, the Soret band is typically used for porphyrin quantification, according to the method proposed by Sugihara and Bean [94].

Table 3.5: UV−Vis Spectral Bands of Petroporphyrins (nm)

	Soret Band	β Band	α Band
Vanadyl Porphyrin			
Etio	400−406	532.8	570.7
DPEP	410.5	533.3	573.0
Rhodo-Etio	414.0	545	579−587
Rhodo-DPEP	590	620	632
Nickel Porphyrins			
Etio	390	511	542
DPEP	410	515	550
Rhodo-Etio	394	513	553
Di-DPEP	398	526	568

DPEP, deoxophylloerythroetioporphyrin; *Etio*, etioporphyrin; *Rhodo*, rhodoporphyrin.

Since the Soret band can be hidden or masked by electronic interactions of the porphyrins with other components in the oil, nowadays researchers refer to the metal compounds as Soret active and inactive, or Soret and non-Soret compounds. The optical spectra of porphyrins and metalloporphyrins have been reviewed in great detail and explained in terms of the molecular electronic structure, including theoretical calculations [95].

Several structural motifs have been identified for petroporphyrins, e.g., the relative position of the ethyl substituents in Etio give rise to four isomers distinguished with roman numbers from I to IV (see this and other examples in Fig. 1 of Ref. [96]). Differences in the resonance Raman scattering spectra of isomeric Etio-I and Etio-III porphyrins of both VO and nickel have been established. In the case of the υ_6 vibration mode in Etio-I and Etio-III spectra, an intense single peak appears at 803 and 821 cm^{-1}, respectively, for VO porphyrins. Most of the vibrational modes gave rise to peaks that can differentiate the Etio-I isomer from the Etio-III isomer (see Table 1 in Ref. [96]). Additionally, the high-frequency (>1400 cm^{-1}) vibrations caused by pyrrole $C_\beta C_\beta$ and methine bridge $C_\alpha C_m$ stretching modes for (VO)Etio differed to those of NiEtio, which were 14–28 cm^{-1} lower in frequency [96]. Similar studies were carried out to distinguish among Ni porphyrins: Etio, DPEP, and Rhodo families and family's members [97]; regioisomers of tetrahydrobenzoetioporphyrin [98]; DPEP isomers [88,99,100]; and Rhodo–Etio [101].

The average MW of the DPEP and Etio series from a number of different crude oils and bitumen is collected in Table 3.6, taken from data reported in Ref. [36]. Although a higher MW of series Etio than that of the DPEP series is expected, the fact that the ratio is always closed to 1 when the average MW of the total mixtures increased in 30 amu (two methylene groups) indicated that the degree of alkyl substitution is kept through both series equally. A different explanation was given by the authors in Ref. [36], the spectral analysis indicated variations of the DPEP and Etio relative population. The DPEP/Etio ratio varied from 0.27 (Baxterville) to 5.8 (Rozel Point) within the examined samples. Thus based on the average MWs, Baxterville should resemble the properties of Etio series better (MW Etio = 444 amu and MW mixture = 447 amu) and the Rozel Point would rather resemble that of the DPEP series (MW DPEP = MW mixture = 475 amu).

Ultrahigh-resolution atmospheric pressure photoionization Fourier transform ion cyclotron resonance mass spectrometry (APPI FT-ICR-MS) allowed the first direct identification of VO porphyrins at the molecular level from asphaltenes of a South American heavy crude oil and of Athabasca bitumen, without prior isolation or pretreatment. Relative abundance of the porphyrins determined by UV–Vis coincided with that obtained by MS, confirming the validity of the identity assignment (Fig. 3.8). The porphyrin distribution in the asphaltene fraction of Athabasca bitumen showed the prevalence of the DPEP family over the other families (Fig. 3.8A), while on the whole, South American crude oil DPEP and Etio populations were very similar (Fig. 3.8B) [102].

Table 3.6: Average Molecular Weight (MW) (amu) of Deoxophylloerythroetioporphyrin (DPEP) and Etioporphyrin (Etio) Homologous Series in Various Oils, Source Rocks, and Bitumen

Oil/Bitumen/Rock	DPEP Series	Etio Series	DPEP/Etio Ratio	Total Mixture
Baxterville	461	444	1.04	447
Aghajari	455	442	1.03	448
Wilmington	454	443	1.02	448
Bel-Ridge	458	448	1.02	454
Santiago	458	446	1.03	455
Melones	468	457	1.02	462
Athabasca tar sands	467	457	1.02	463
Colorado oil shale	466	457	1.02	464
Mara	472	456	1.04	464
Burgan	473	466	1.02	469
Rozel Point	475	477	1.00	475
Boscan	482	469	1.03	476
Gilsonite	479	469	1.02	476

This methodology has permitted the simultaneous identification of Ni and V porphyrins directly in the BotB fraction. The prevalence of DPEP was also observed in the case of Ni for Californian Carpinteria crude oil asphaltenes. There, additional environments were found for a sample of Il Duomo asphalt volcano corresponding to $N_4O_2V_1$, $N_4O_3V_1$, and $N_4O_4V_1$ [103]. These environments have also been observed in Venezuelan Orinoco heavy oil and the full series have been characterized as responding to the general formulas $C_nH_mN_4VO_2$, $C_nH_mN_4VO_3$, and $C_nH_mN_4VO_4$ [104]. This unusual higher coordination of vanadium and the additional ligands other than N should be taken as a reason to categorize the metal compounds as nonporphyrins; however, the fractions where these compounds were present exhibited UV–Vis porphyrin characteristic bands.

Molecular structural and other characteristics of the Rhodo-type porphyrin were summarized as:

1. Minor proportion of the other petroporphyrins;
2. Two (pseudo) homologous series appear to exist: (1) $456 + 14n$ and (2) $458 + 14n$;
3. Only C, H, and N present an unsaturated side chain with six-H in series (1) and eight-H in series (2);
4. Solubility in aromatic hydrocarbons is greater than it is for Etio;
5. The visible spectral shift to 632 μm versus 620 μm for Etio (indicating presence of e-withdrawing groups);
6. Absence of carbonyl group; and
7. Resistance to bromination and to catalytic hydrogenation [36].

Although initially metals present in the heavy oils were associated with the porphyrin structures, it is now accepted that porphyrins are only a fraction of the metal-containing

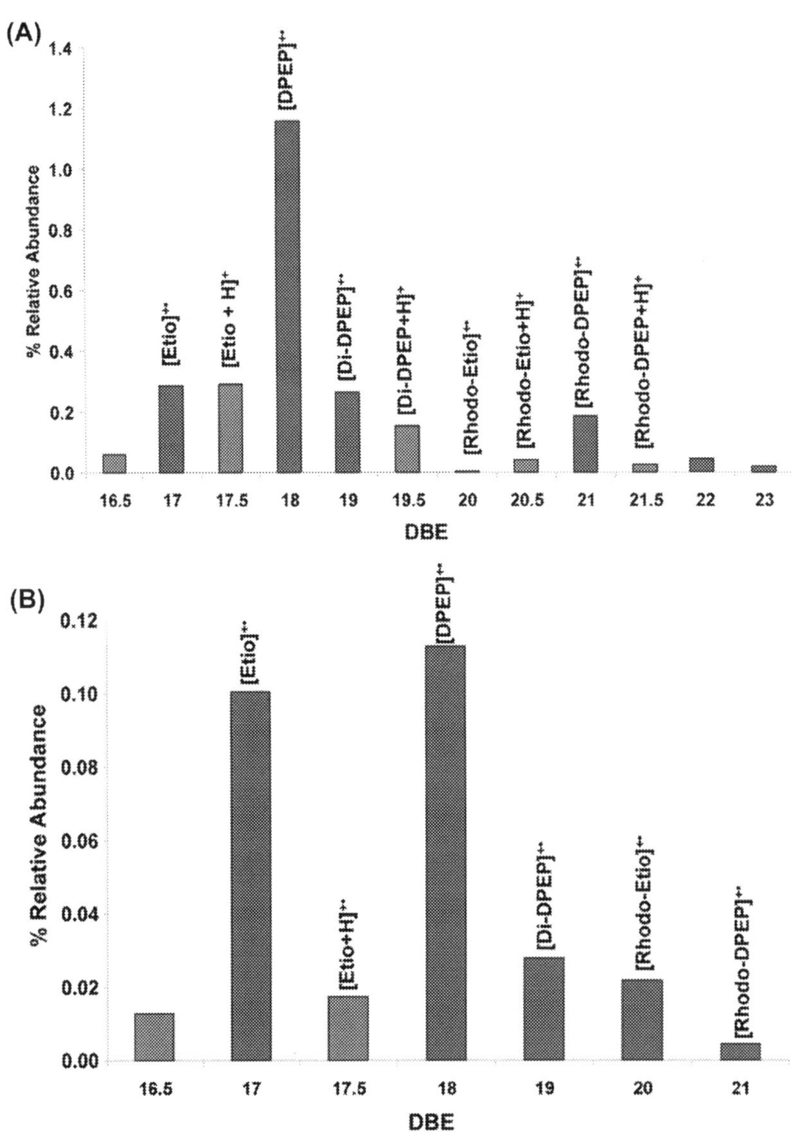

Figure 3.8

Distribution for vanadyl porphyrins in (A) Athabasca asphaltenes, (B) South American heavy crude oil. *DBE*, Double bond equivalent; *DPEP*, deoxophylloerythroetioporphyrin; *Etio*, etioporphyrin; *Rhodo*, rhodoporphyrin. *Reproduced from Mckenna AM, Purcell JM, Rodgers RP, Marshall AG. Identification of vanadyl porphyrins in a heavy crude oil and raw asphaltene by atmospheric pressure photoionization Fourier transform ion cyclotron resonance (FT-ICR) mass spectrometry. Energy Fuels 2009;23(4):2122–28, with permission from ACS Publications.*

molecules. Apparently, V is equally distributed in porphyrinic and some other nonporphyrinic cores, whereas only about 30% of the total Ni seems to be porphyrinic [74]. Although there are controversial discussions on the real existence of any nonporphyrinic structures [72,105,106], separation, extraction, and characterization studies have identified molecular features that could not be found in porphyrins. Therefore, another molecular subclass became known as nonporphyrins. They are supposed to be heavier and larger than porphyrins [15,70,107]; similarly they exhibit interrupted conjugated and aromatic cycles [54]. The SEC-calibrated MW of V and Ni nonporphyrins, separated by reversed phase ODS HPLC, ranged from above 2000 Da to less than 400 Da. Although in certain asphaltenes the concentration of nonporphyrins could be larger than that of porphyrins, the opposite was found and made the separation and identification of the nonporphyrins in this fraction more difficult. HPLC/GFAA allowed the separation and determination of the metal concentration in nonporphyrins, but the model compounds considered for retention time calibration did not provide insights into the identification of these compounds [15].

Nonporphyrins are generally more polar than porphyrins [15] and have elements in their first coordination spheres other than the four nitrogen atoms existing in those of porphyrins [108,109]. Thus the most prominent differentiating feature between porphyrinic and nonporphyrinic compounds is that in the former the first coordination sphere of the metal is formed by N atoms, while in the latter S and O might be present [110]. The existence of coordination spheres other than porphyrin N_4 was proven by electron paramagnetic resonance (EPR). NOS_2, N_3S, S_2O_2, and N_3O were found in the vanadium coordination spheres of a separated nonporphyrin fraction. Linear correlations were found on the EPR parameters that define the arrangements of the metal coordination sphere of nonporphyrin compounds [108,111,112].

A summary of the distinguishing structural features of nonporphyrins includes: (1) the coordination sphere contains heteroatoms other than nitrogen, (2) this coordination sphere is crude dependent, and (3) part of the metal nonporphyrin compounds are small molecules [109]. The characterization of the isolated nonporphyrin fraction of several crude oils and bitumen (Boscan, Cerro Negro, Wilmington, and Prudhoe Bay) identified part of the Ni species as naphthenates [113].

The paramagnetic nature of the VO group makes EPR (also known as electron spin resonance) a technique of choice for the characterization of its local environment. Typically, the spectra of asphaltene fractions as well as subfractionated samples are compared with model compounds. The values of the spin-Hamiltonian parameters g_0 and A_0 are determined by the coordination sphere surrounding the paramagnetic center. The g_0 value has been postulated as a better indicator of the coordinating ligands than the A_0 value [112]. Based on a linear correlation found between the difference in g_0 and the electronegativity of the surrounding elements in the coordination sphere, it was postulated that possible coordination of the VO group in a nonporphyrin (UV–Vis non-Soret) fraction from West Mara crude might be N_2OS, S_2O_2, or N_3O [112]. EPR studies indicated that the spin density distribution of the VO ion is affected by the ligands in the second coordination sphere, namely, substitutions in the methine bridges of porphyrins with respect to the substitutions in the pyrrole rings, giving rise to excited electronic configurations [78]. High-precision EPR spectra from various crude oils and bitumen were characterized, namely, Athabasca, Peace River, Wabasca, Boscan, Gato Ridge, and Circle Cliff. The number of model compounds was broadened and statistically tested to determine significance. Although g_0 was found to be a better indicator for the coordinating ligands, it could only distinguish with reasonable certainty between N_4, S_4, and S_2O_2 groups if high precision or high frequency was used. In this latter case, a careful separation first between porphyrins and nonporphyrins and then among the porphyrin series was exercised prior to spectroscopic characterization [114].

EPR has provided a geometrical view for porphyrins and nonporphyrins. Porphyrins were found to exhibit a tetragonal (C_{4V}) distorted average coordination geometry. The distortion increases with the shortening of the V=O bond and with the increase of the ligand distance in the basal plane. Parametrization of the EPR spectra can be used to estimate the covalent or ionic character of the V–N ligand bonds, the V=O bond strength, and the delocalization of the unpaired electrons V^{4+} onto the coordinative nitrogen bond [115].

Nonporphyrins of vanadium appear to be square planar complexes with idealized C_{4v} to distorted C_{2v} symmetry, with possible coordination in the axial position, such as that shown by oxovanadium complexes [116,117]. The EPR spectra of VO^{2+} nonporphyrins from Western Kentucky No. 9 coal indicated that the VO_2^+ ion was surrounded by an environment with approximately axial symmetry and chelated possibly by carboxylic/phenolic oxygen ligand donor atoms [52].

In the nonporphyrinic compounds, nickel is ordinarily present as Ni^{2+} as it is in the porphyrins. Meanwhile, vanadium typically present as a VO^{2+} cation in porphyrins, in nonporphyrins, it would probably be in the forms of $VO(N_2OS)$, $VO(S_2O_2)$, or $VO(N_3O)$ [112]. However, the existence of S atoms close to the VO cation has been ruled out by some authors [118]. These different local arrangements of the metals leads to fairly planar structures in the case of Ni, but a square pyramid in the case of V, with the VO cation at the top and the oxygen atom facing out.

The evidence from X-ray absorption spectroscopy is quite different. Extended X-ray absorption fine structure (EXAFS) and X-ray absorption near edge spectroscopy (XANES) spectra of a non-Soret fraction of Boscan, Cabimas, and Aramco asphaltenes showed that the vanadium within this fraction was of the porphyrin macrocycle type, i.e., oxovanadyl type coordinated to four nitrogen atoms [105,106]. Similar results were found with the non-Soret asphaltene fraction of Maya oil. The lack of Soret band might be caused by a masking effect of the asphaltene and/or the presence of other interacting ligands. In this case, octahedral coordination was also observed [119]. X-ray photoelectron spectroscopy of asphaltene fractions from the same crude oils also indicated that the binding energy of V was closer to that of compounds with N_4 coordination than to any other S-containing arrangement [120]. Examination of unconverted asphaltenes from a hydroprocessed crude oil by EXAFS showed changes in the coordination sphere of the vanadium center. While the untreated asphaltenes did not show any differentiating feature from porphyrinic coordination, the treated unconverted asphaltenes changed to a six-ligand distorted octahedron coordination of N_4O_2 type [119]. X-ray absorption results consistently indicate that V is present as a VO group, coordinated to four N atoms, regardless of porphyrin or nonporphyrins compounds. In summary, a lack of agreement around the chemical nature of the V-coordination sphere in nonporphyrins still prevails among different research groups.

A higher proportion of nonporphyrins over porphyrins were found to be present in bitumen from various tar sand fields. The metal (V and Ni) concentration in the bitumen of these tar sands was already reported in Table 3.2, and the percentage of the total metal present as porphyrin was always below 25% (Table 3.7) [11].

The fact that some of the metal moieties are associated or incorporated into the asphaltene fraction and cannot be separated chromatographically has been interpreted as these compounds being molecularly embedded into the asphaltene molecule. Whether this metal moiety is of porphyrin or nonporphyrin type remains inconclusive. A thermal treatment of a nonporphyrin fraction showed reduction of molecular size, but no appearance of spectral characteristic of porphyrins. These pioneering results indicated molecular differences between porphyrins and nonporphyrins and confirmed that nonporphyrins were not asphaltene-encapsulated porphyrins [72]. However, this argument is based on the assumption that the thermal demetallization treatment only removes the metal leaving intact the organic core.

Table 3.7: Metal Percentage Present as Petroporphyrin in Tar Sand Bitumen [11]

Bitumen	Nickel	Vanadium
Asphalt Ridge	24	0
Sunnyside	1	0
Big Clifty	3	13
Arroyo Grande	21	16
McKittrick	13	21
Commercial Kentucky	1	10
Athabasca	2	12

Thus metals can be occluded or trapped in larger agglomerative networks, giving apparent large size and weight [15,16,70]. The trapping or occlusion capacity of asphaltenes is so large that the quantity of materials (both metals and resins) may represent 80−90% of their total amount present in the oil [16]. This can be caused by the ligating effect of asphaltenic sheets [54] or to agglomeration caused by donor−acceptor interactions [108]. Consequently, a separated nonporphyrinic fraction may contain not only asphaltenes, but also porphyrinic molecules. Complete characterization of porphyrin and nonporphyrin metals associated with asphaltenes has not been easy and their molecular structure identification remains unknown. The intermolecular association of metal compounds with the asphaltene molecules is one of the complicating features that introduce many obstacles in the study of each of these moieties.

The real molecular structure of asphaltenes also remains undiscovered; some hypothetical models have been proposed, which seem to explain some of the experimental observations. Four molecular models have been proposed as hypothetical structures [121] for modeling and theoretical calculation purposes. In these models, metal atoms are not currently included, probably because the theoretical methods used for modeling (particularly molecular mechanics methods) limited the type of atoms and the number of atoms (molecular size) that may be calculated. Characterization results and molecular modeling studies led to other structural models [122,123]. It is important to point out that in this case, proposed structures are derived from structural parameters calculated from the characterization results, so they represent average models rather than an identified molecule. The correlation between metal content and asphaltene content, and the similarities on agglomerating capabilities, has led to assumptions on the chemical structural location of the metals, such as the hypothetical model of Strausz et al. [124] (Fig. 3.9). It is known that the porphyrins concentrate in the BotB fraction, but whether they are an integral part of the asphaltene structure may still need confirmation [125].

Sixteen new series of vanadium compounds have been identified in a Venezuelan heavy crude oil. Some have the basic core of porphyrins, some have a higher degree of condensed ring substitution, and others will fall into the nonporphyrins category, i.e., other than N_4 coordination. These series are represented by the general formulas [126]:

$C_nH_{2n-40}N_4V_1O_1$ $(36 \leq n \leq 58)$	$C_nH_{2n-42}N_4V_1O_1$ $(37 \leq n \leq 57)$	$C_nH2_{n-44}N_4V_1O_1$ $(38 \leq n \leq 59)$	$C_nH_{2n-46}N_4V_1O_1$ $(43 \leq n \leq 54)$
$C_nH_{2n-48}N_4V_1O_1$ $(45 \leq n \leq 55)$	$C_nH_{2n-38}N_4V_1S_1O_1$ $(36 \leq n \leq 41)$	$C_nH_{2n-40}N_4V_1S_1O_1$ $(35\ n \leq 51)$	$C_nH_{2n-42}N_4V_1S_1O_1$ $(36 \leq n \leq 54)$
$C_nH_{2n-44}N_4V_1S_1O_1$ $(41 \leq n \leq 55)$	$C_nH_{2n-46}N_4V_1S_1O_1$ $(39 \leq n \leq 55)$	$C_nH_{2n-27}N_5V_1O_1$ $(29 \leq n \leq 40)$	$C_nH_{2n-29}N_5V_1O_1$ $(34 \leq n \leq 42)$
$C_nH_{2n-33}N_5V_1O_1$ $(31 \leq n \leq 38)$	$C_nH_{2n-35}N_5V_1O_1$ $(32 \leq n \leq 41)$	$C_nH_{2n-27}N_5V_1O_2$ $(32 \leq n \leq 41)$	$C_nH_{2n-29}N_5V_1O_2$ $(33 \leq n \leq 42)$

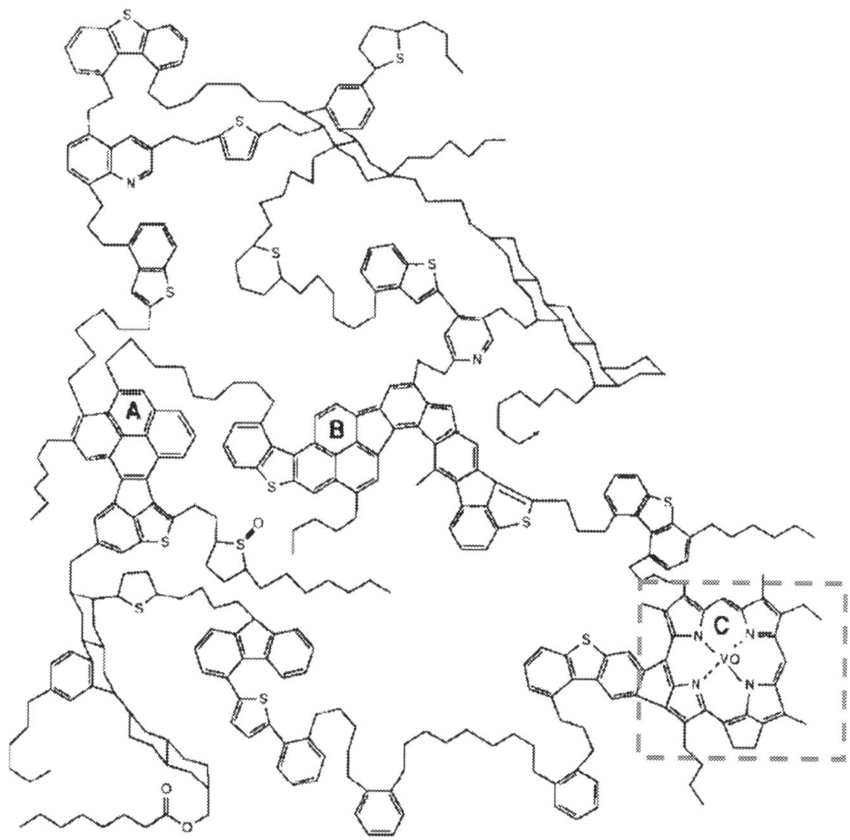

Figure 3.9

Hypothetical model structure of an asphaltene molecule. *Reproduced from Strausz OP, Mojelsky TW, Lown EM. The molecular structure of asphaltene: an unfolding story. Fuel 1992;71:1355–63, with permission from ACS Publications.*

4.2 Characteristics

Significant advances in analytical instrumentation have taken place during the last two decades, which have improved both sensitivity and selectivity of the determinations, enabling direct examination and identification of the metal compounds. Notwithstanding, the pioneering efforts in isolation not only serve as a simplification of a complex problem but also facilitate the comparison with synthetic standards and model compounds. Any characterization methodology starts with the determination of elements and the understanding of the effects that the organic matrix may have. Review articles on metals determination have been published [127–133], and sample preparation, diluents, effect of dilution, techniques, equipment, and methodologies can be found in those references.

The combination of several quantitative methods is required to characterize metal components in heavy oils. The presence of microheterogeneities within asphaltene solutions creates disparities that affect the quantitative determination of metal elements. This microheterogeneity has been proven by EXAFS and UV–Vis spectroscopy, with a limited number of samples. Therefore further examination of a larger range of samples and in more detail should be undertaken [134].

Opportunity crudes are immature oils, which are characterized with a V/(Ni or Fe) ratio that typically is greater than one. Nevertheless, some opportunity crudes exhibit unusual metallic composition, with excess content of Ni or Fe relative to V. Such is the case exhibited by Moslavina basin crude oil, which shows Fe-content higher than that of Ni or V. For this crude oil, Fe concentration is more than 10-fold that of V, and the V-content of its resins and asphaltene fractions is equal (28 ppm) [135].

The main characteristic to bear in mind when considering heteroatoms present in the bottoms fraction of the heavy oil is the interdependence among them, as clearly shown in Fig. 3.10 [136], and within the limitations set by the origin of the crude oil and of the metal compounds, as pointed out in the previous section. This association with the heavy ends was originally suspected based on elemental quantitative analysis. Metal concentration typically correlates with sulfur content, though not all the crudes contain both compounds. It has been observed that crudes with high vanadium content are high in sulfur while those with high nickel content are high in nitrogen [44].

Quantitatively, sulfur and nitrogen are present in the percentage range, while metals could reach up to 2000 ppm. Asphaltene fraction can be as high as 20–25% in extraheavy oils. The profile of heteroatoms as a function of the atmospheric equivalent boiling point of the crude oil fraction was examined by Boduszynski, whose *continuous model* of the crude oil has already been described in Chapter 1. Four ARs were fractionated and the metal content was determined for each fraction. The results are shown in Fig. 3.11.

In both cases, the metal profile showed a bimodal distribution pattern, though Ni pattern is less pronounced than that observed for vanadium. The distribution of the other heteroatoms (S, N, and O) showed similar patterns, in some instances with a lower definition. This similarity was interpreted as being caused by all heteroatoms being constituents of the same molecular structures [137]. Size exclusion chromatography (SEC) [coupled with induced coupled plasma (ICP) detection] showed similar results when the metal profile was examined as a function of molecular size (Fig. 3.12) [138].

The distribution of other metals in the SARA fractions of a Nigerian crude oil (with low metal content) showed a different profile, having the higher metal contents in the aromatic fraction [139]. Aromaticity of porphyrins increases with aging/maturation, leading to more stable π-systems. The presence of the π-system induces intermolecular

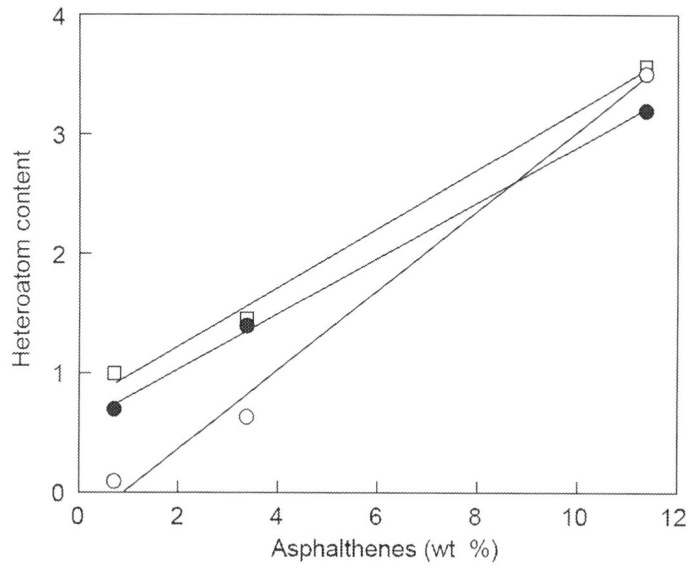

Figure 3.10
Heteroatom and asphaltenes content. *Open circles*: metals (Ni + V) in ppm × 10^2; *black circles*: nitrogen in wt% × 10^{-1}; and *open squares*: sulfur in wt%. *Reproduced from Pina A, Mougin P, Béha E. Characterisation of asphaltenes and modelling of flocculation — state of the art. Oil Gas Sci Technol Rev IFP 2006; 61(3):319—43, with permission from IFP.*

interactions that may result in dimerization and aggregation [140—144]. ESIFT-ICR showed the presence of dimers with a greater relative abundance of monomers in porphyrins isolated from Cerro Negro bottoms fraction [145]. Self-association of VO porphyrins, both model compounds as well as petroleum-extracted, was evidenced by EXAFS and XANES [140]. This study also considered the adsorption of the porphyrins on a commercial NiMo/Al_2O_3 catalyst and on the Al_2O_3 support, observing that the interactions between the porphyrin with the bare support were much weaker than with the catalyst. Upon adsorption the macrocyclic cavity was found to become slightly larger. This structural expansion is compensated by a shift of the metal group closer to the plane of the porphyrin [140].

The polar nature of the metal compounds is clearly evidenced by their adsorption behavior. The oxidation states for Ni (+2) and V (+4) are forced to different coordination environments for each of these metals. In the case of Ni, all the bonds can be satisfied within a plane, while for V, the oxygen atom of the VO group is projected out of the plane [108]. Consequently, the average V-containing molecule is more polar than the Ni-containing molecules. Furthermore, Ni accessibility is more restricted than that of V [74,107]. The percolation through a bauxite matrix illustrates the surface properties difference between both compounds. It showed that Ni eluted more quickly than V.

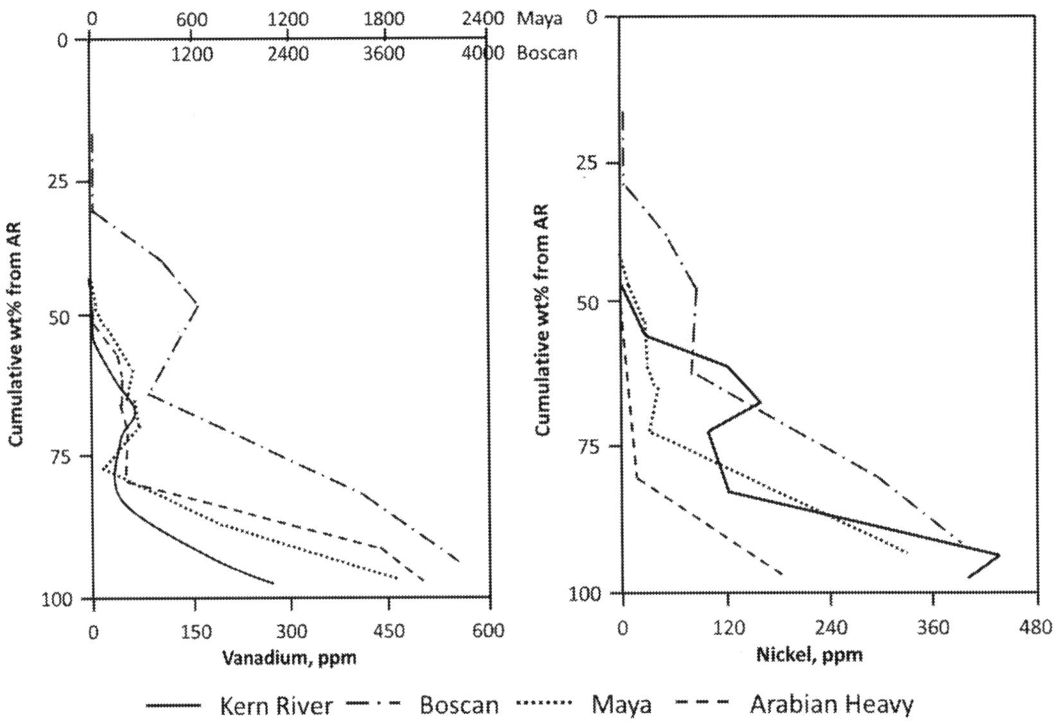

Figure 3.11
Metal distribution in atmospheric residue (AR) as a function of atmospheric equivalent boiling point (AEBP). *Reproduced from Boduszynski MM. Composition of heavy petroleums. 1. Molecular weight, hydrogen deficiency, and heteroatom concentration as a function of atmospheric equivalent boiling point up to 1400°F (760°C). Energy Fuels 1987;1(1):2–11, with permission from ACS Publications.*

Furthermore, the Ni-adsorbed species could be desorbed with a less polar solvent than that used for desorbing the V-adsorbed compounds [146].

The adsorption of VO porphyrins on Mo/alumina catalyst at room temperature showed two modes of interaction between the metal compound and the solid surface: (1) coordination of the V=O group on basic surface sites, the latter acting as electron donor centers, and (2) coordination of the V=O group on acidic surface sites, the latter being the electron acceptor site. Additional interaction of the π-electron system of the porphyrin aromatic core and the surface indicated the porphyrin molecule laid flat on the surface [147]. This stereochemical configuration and the multiple interaction modes imply that saturation or full surface coverage could be reached easily and poisoning effects could be strong. A donor–acceptor interaction was found to be dominant for the binding of V and Ni porphyrins to γ-Al_2O_3, SiO_2–Al_2O_3, MoO_3/Al_2O_3, NiO/Al_2O_3, and to Co– and Ni–Mo/Al_2O_3 catalysts in their oxide and sulfide forms. The electron donor was considered to be the delocalized π-system of the porphyrin ring, while the Bronsted and Lewis sites on the catalyst

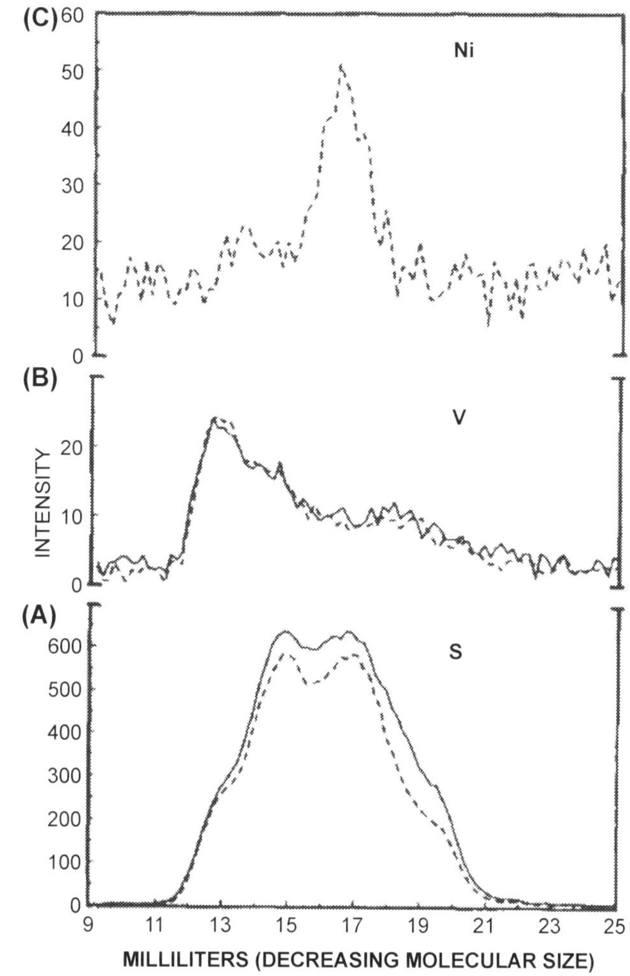

Figure 3.12

(A) Sulfur, (B) vanadium, and (C) nickel molecular size distribution in Arabian heavy crude (——) and Arabian heavy vacuum resid (VR) (——). *Reproduced from Hausler DW. Molecular size distribution of specific elements in petroleum crudes and 650 F + residua by size exclusion chromatography with inductively coupled plasma spectrometry detection. Spectrochim Acta Part B At Spect 1985; 40(1–2):389–96, with permission from Pergamon Press Ltd.*

surface constituted the acceptor. Either oxidation or hydrogenation of the porphyrin ring at the mesocarbons could be the initial stage in the decomposition of the bound porphyrins, depending on the reaction conditions [148].

Additionally, this π-system will contribute to the development of strong interactions with π-systems in the organic matrix. This might be the reason why these metal compounds concentrate in the fraction of the crude oil with the more dense aromatic population.

Their solubility in hydrocarbon media is determined by the combined action and effects of aromaticity, MW, and polarity [149].

Solubility of metal compounds is generally low, probably better for the porphyrins than for the nonporphyrins. Metalloporphyrins are soluble in polar solvents such as γ-butyrolactone, acetonitrile, phenol, furfural, 2-pyrrolidone, dimethylformamide, pyridine−water mixtures, ethylene carbonate, propylene carbonate, ethylene trithiocarbonate, and dimethylsulfoxide. VO porphyrins have very low solubility in common organic solvents. The best solvents were chlorinated compounds (e.g., chloroform, dichloromethane, etc.); solubility was low in any other solvents [87]. Solvent extraction and preparative chromatography in the range of 25−100°C using solvent gradients or taking advantage of the difference in solubility for the various species were described in Section 3. Solubility in water is extremely low; however, the presence of surfactants changed this behavior. The partition from the oil into an aqueous phase was 1:1 in the presence of Brij-35 surfactant (polyoxyethylene 23 lauryl ether). This may raise concerns of the toxicity effects during remediation methods that use surfactants and also on aquifers in situations where natural humid surfactants are present in the oil wells [150].

SEC methodology integrated with ICP detection was used for the determination of the elemental profiles in petroporphyrins. Results indicated that a greater percentage of the V in the crudes is present in the porphyrin size range than the percentage of Ni species [151], in agreement with results previously mentioned [74]. Consequently, distinguishing properties between porphyrins and nonporphyrin compounds will markedly affect the behavior of V versus that of Ni species.

A differentiating characteristic among porphyrins and nonporphyrins is their volatility. This significantly higher volatility of porphyrins compared to nonporphyrins has advantageously served as a separation tool prior to characterization [152−154]. Only thermally stable species were found to precipitate with the asphaltenes, while in the resins and polar fractions both stable and volatile species could be found [155]. The quantification of the thermally stable portion is questioned because of coke formation that might trap metal species during pyrolysis. This effect is larger the higher the asphaltene content of the crude [154].

The interaction of the metal compounds with asphaltenes has caught the attention of many researchers and it is not well understood yet. Polarity in asphaltene molecules is caused by the presence of heteroatomic moieties (among them metals) [156]. The most polar fraction from an asphaltene sample was isolated and the solubility in 10% dodecyl benzene sulfonic acid in heptane solution was tested. The solubility of such fraction was greatly enhanced in the presence of a strong chelating agent (ethylenediamine tetraacetic acid, EDTA), indicating that the elimination or mitigation of the metal interactions increased the rate of dissolution. In this regard, metal content has a direct effect on asphaltene dissolution in amphiphile/alkane solvents [156]. It is possible that metallocompounds present in the asphaltenic fraction coordinate with the heteroatoms of the asphaltene molecules. However, because of steric hindrance of the porphyrin core to the metal center in this additional association, only those heteroatoms present at a pendant position will be accessible for a coordinative interaction [157]. The presence or absence of carboxylic acid side chains of porphyrins has been considered to be the cause for the most distinct differences in their polarity [158–160]. The presence of the carboxylic group has been associated with the very strong interaction between porphyrins and asphaltenes that obstructs or prevents their separation and isolation from the residue fraction [161–163].

As pointed out earlier, in the Strausz model of the asphaltene molecule, a porphyrinic moiety is part of the molecular structure (Fig. 3.9). Other authors have suggested (1) an active role of the metal groups in asphaltene agglomeration, (2) occlusion or trapping during asphaltene precipitation, and (3) adsorption on the asphaltene solid surface. Both self-association and association with and into the asphaltene agglomerates were evidenced by EPR studies [164].

Dispersion of asphaltene agglomerates under the action of ultrasound released maltenes and metal compounds, which was interpreted as asphaltenes conversion. The ultrasonic treatment can be enhanced by the presence of an adsorbent that would retain the metal compounds [165]. Whether the observed maltenes are the result of asphaltenes cracking or these were occluded during asphaltenes precipitation is a subject of discussion. In Chapter 6, ultrasonic upgrading will be discussed and as mentioned there, the action of ultrasound was observed to be temporary, indicating that no real conversion occurred. The disaggregation of the asphaltenes agglomerates in the presence of a metal adsorbent benefits the treatment results by precluding the interactions that metal compounds exerted on asphaltenes interconnection. A similar effect to that caused by ultrasound was observed by dissolving the asphaltenes in the presence of either activated carbon adsorbent and/or an oil-soluble Mo dithiocarbamate complex. In both of these cases, metal compounds and asphaltenes will interact with the additives, causing a greater upgrading effect of the VR [166].

Metal porphyrins adsorption on asphaltenes has been studied considering Ni [167] and V [168] Etio model compounds. In both cases, a very rapid initial adsorption is followed by a slowing down effect toward reaching the equilibrium. An initial monolayer adsorbate distribution is suspected to be followed by the formation of a multilayer adsorption process.

Theoretical calculations predicted a caging effect that would organize asphaltenes surrounding free radicals and creating a protective shield that stabilized these species. This caging effect would explain the aggregation behavior of asphaltenes as well as the interaction with VO porphyrins. The paramagnetic nature of the VO group and the presence of shielded free radicals within the asphaltene structure were well predicted by the theoretical calculations, which also incorporated the V—asphaltene interaction as an explanation of asphaltene aggregation [169]. Molecular interactions were also evidenced by the changes in the paramagnetic properties as shown in the EPR spectrum. The dynamic effect of temperature on association behavior was studied of asphaltenes of different origin. Formation of aggregates was considered to be caused by a considerable energy of the exchange interaction between the paramagnetic centers of the dispersed phase. The energy of resonance interaction of the free radicals of the dispersed phase with the diamagnetic molecules of the dispersion medium leads to the formation of solvate shells. The destruction of the asphaltene aggregates could be evidenced by the appearance of isotropy of the hyperfine structure of the EPR spectra of the VO complexes [170]. A study of the asphaltene aggregation kinetics in toluene/*n*-heptane mixtures indicated that the colloidal properties of asphaltenes depend on vanadium content. A universal model of asphaltene aggregation was proposed, including the V—asphaltene interaction [171]. These findings confirmed what was speculatively proposed in the early years: the implication of vanadium in asphaltene aggregation. In general, it could be stated that the interaction of the metal compounds with the asphaltenes might explain the formation of asphaltene micelles. According to the Yen model (see Chapter 2), metal-bearing molecules played an important role in the construction of molecular micellar arrangements (and consequently in asphaltene agglomeration), by interconnecting asphaltene molecules. Part of the micelle model of asphaltenes according to Yen [172—174] is depicted in Fig. 3.13.

However, when model porphyrins and aromatic ring compounds were allowed to interact, neither UV—Vis nor fluorescence spectroscopy could detect any interactions in either the ground or excited states. The lack of detectable binding may be caused by the considered model compounds themselves or may suggest that the porphyrins—asphaltenes association could involve other functionality rather than the $\pi-\pi$ stacking interactions of aromatic rings [175].

These micelles themselves have surfactant properties that are known to stabilize (water and oil) emulsions [176]. Furthermore, porphyrins were thought to have surfactant capabilities and stabilize oil—water emulsions [177]. The mixing with seawater and the addition of nickel porphyrins to nonemulsion-forming crude oils resulted in the formation of a stable reddish brown water-in-oil emulsion [178]. Instead, the lack of surfactant capabilities of V porphyrins was confirmed by separating V and Ni porphyrins from an emulsion-forming crude oil (Monterey) and testing them on emulsion formation with a crude oil that does not form emulsions (South Louisiana). Similarly, when synthetic pure Ni porphyrins were tested no emulsion was formed, indicating that other oil component(s) are needed. This suggested a concerted action of Ni porphyrins and asphaltenes, most likely via micelle formation, as explained earlier [178].

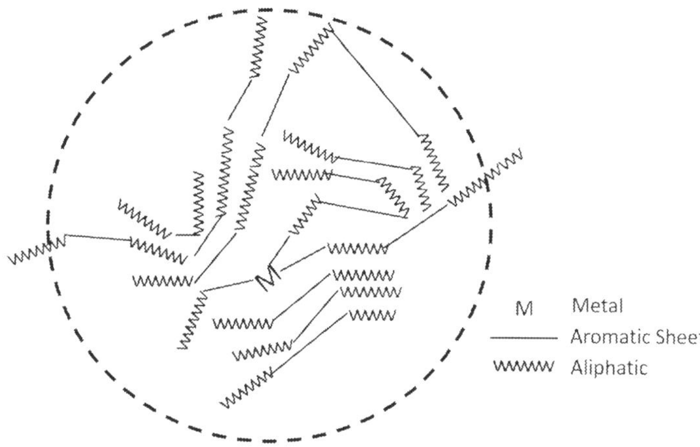

Figure 3.13
Interconnecting role of metal-bearing molecules in asphaltene micelles formation, according to the Yen model. *Reproduced from Dickie JP, Yen TF. Macrostructures of the asphaltic fractions by various instrumental methods. Anal Chem 1967;39(14):1847–52, with permission from ACS Publications.*

The vanadium porphyrins are more hydrophilic than nickel porphyrins, because of the higher polarity of the oxovanadium cation and VO moiety present in vanadium porphyrins compared to the nickel ion [179].

Asphaltene fractionation into acid and basic fractions showed very similar metal partitioning in each fraction for both Ni and V compounds. The characterization of these fractions by UV–Vis spectroscopy and EPR showed that the porphyrins concentrated in the basic fraction, while the acid fraction did not show the Soret band [180].

4.3 Reactivity/Stability

Generally, Ni and V are the most abundant of the trace metals and have negative effects on refining processes [181–183], as will be discussed in the next section. Furthermore, transportation fuels cannot contain these or any other heavy metals. For these and other reasons their removal becomes mandatory. Consequently, the most studied reactivity is that concerning routes to demetallization.

The reactivity and processability of vacuum residua have been associated to crude maturity, size distribution of HC and of metal compounds, sulfur content, and asphaltene characteristics (aromaticity, H/C ratio, etc.) [184].

Geochemical efforts have been carried out to understand and classify crudes in terms of their maturity and origin: kerogen I and II types [45]. As already mentioned, the effect of maturation on chemical composition is an increase in the H/C ratio, and a decrease in the concentration of heteroatoms (S, N, O, and metals). The V/Ni ratio is an indication of kerogen type (type I < 1; type II > 1). The apparent reactivity of metal compounds is larger in kerogen II (containing more V) than in kerogen I (containing more Ni), probably because of the intrinsic reactivity of Ni and V compounds themselves (V > Ni). Nevertheless, metal reactivity may be also affected by accessibility caused by steric hindrance of the organic complex core. These properties may suggest that processability is facilitated in more mature crude oils, which may appear to be a contradiction of what was stated in Ref. [184]: "A very mature crude can be considered as if it has been already processed. Therefore, its residue will not, relative speaking, yield much distillates after further treatment."

The negative impact of the metal compounds on refining processes had imposed high pressure on the study of demetallization reaction. Anticipation of processing characteristics and potential problems is desirable. Since metals are concentrated in the asphaltene fraction, deasphalting produced a low metal-containing product (the deasphalted oil, DAO) and a metal-concentrated pitch (mainly asphaltenes and resins). The metal compounds in the pitch are known to be either highly agglomerated, trapped, or otherwise strongly bound to the pitch constituents, which lower their reactivity and hamper further processing [14].

4.3.1 Hydroprocessing

Conventional hydroprocessing (HDP) technologies remove the metals via hydrogenation and hydrogenolysis reactions. These reactions are catalyzed by transition metal sulfides. In commercially available catalysts, molybdenum or tungsten is promoted by nickel or cobalt and these metals are supported on alumina. As in the particular case of (hydro) demetallization (HDM), unpromoted molybdenum sulfide has been proven to be very active, indicating that the metal compounds' reactivity toward hydrogenation is high. The reactivity toward HDP seems to be dictated by the relative molecular size of the metal compounds' with respect to the catalysts' pore diameter, but also for the reactivity of the sulfur compounds and asphaltenes, which will affect the catalyst activity [10]. This has been widely confirmed; an example is given by the catalytic HDM of AR and DAO, which indicated that the conversion of porphyrin and nonporphyrin molecules was limited by size accessibility of those molecules and the pore size of the catalyst. A study observed that metal compounds in the resins fraction looked more reactive than those in the asphaltene fraction, which may have to be related to size accessibility/catalyst pore size [185]. SEC was used to quantify the refractory fractions for demetallization (DM of V compounds, i.e., hydrodevanadation) under desulfurization (DS) conditions. It was found that the major proportion of reactive V compounds were of intermediate size. Roughly, the same proportion of refractory V compounds was found for large and small molecular size fractions [186]. The use of SEC-ICP/MS for monitoring the HDM of residues confirmed a lower conversion of the larger molecules, i.e., metals present in high MW compounds were the most refractory to conversion. However, since only light metal complexes were easily converted during HDT, it was interpreted as the impact of molecular size on reactivity [187].

The same mechanism seems to take place for all kinds of metal compounds, which involves a first hydrogenation step, leading to the formation of chlorins. The complete mechanism for the reduction of the VO porphyrins to VO chlorins is depicted in Fig. 3.14. This mechanism has been proposed for the thermal demetallization of the VO porphyrins, based on the spectroscopic determination of reaction intermediates, using the data for reactions run at different levels of severity [188].

These reactions and the catalytic chemistry involved have been the subject of books, reviews, and reports [189–208]. Hydrogenation occurs easily on the isolated metal compounds, indicating that the rate of demetallization might be a function of diffusion through or denaturing of the polar medium around the metal centers and not of the coordination sphere around the metal ion [72]. In cases where porphyrins are monitored through the Soret band, hydrogenation makes porphyrins appear more reactive than nonporphyrins. Differences in reactivity should exist and there is evidence [185]; however, a more systematic study is still needed.

However, it is worth mentioning that the degree of hydrogenation required for demetallization to occur depends on the porphyrin or complexing core, the reaction

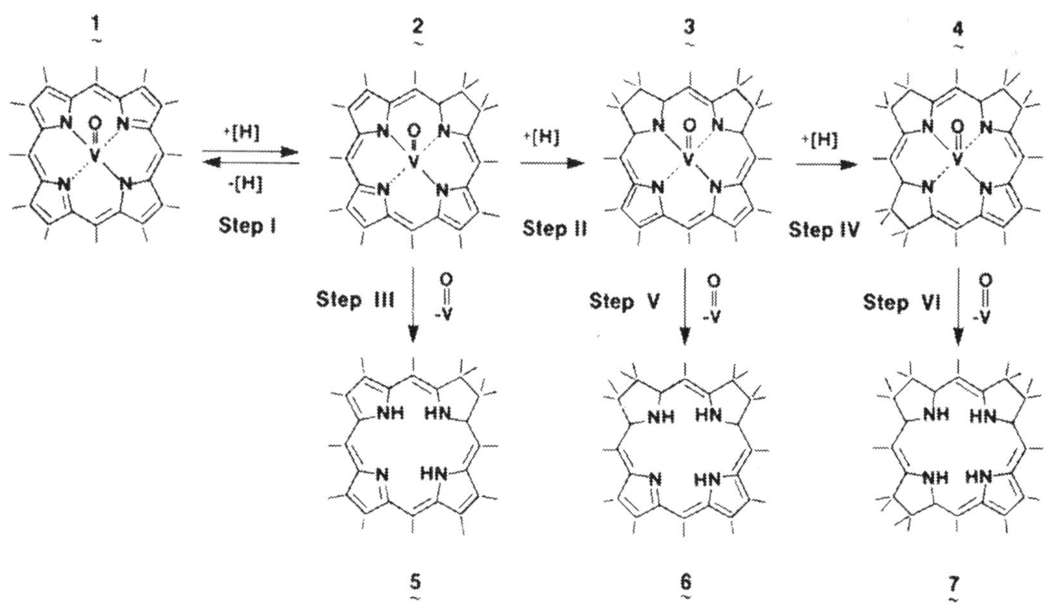

Figure 3.14
Demetallization mechanism of V porphyrins. *Reproduced from Reynolds JG, Biggs WR, Bezman SA. Reaction sequence of metallopetroporphyrins during heavy residuum upgrading. In: Metal complexes in fossil fuels, ACS symposium series. Washington (DC): ACS; 1987, p. 205–19 , with permission from ACS Publications.*

conditions, and presence of a catalyst and/or of H_2S. These effects were investigated using model compounds and the role of S was postulated as weakening the metal–nitrogen bonds by coordinating to the metals [209].

The ABN fractions showed differences in the reactivity toward hydrodemetallization of their contained metal compounds. Wilmington and Mayan VRs were compared. In these VRs, the nonporphyrinic form represents the major proportion of the nickel and vanadium compounds. These compounds covered a broad range of MW and chemical behavior. Reactivity depended not only on process conditions, but also on the chemical type of the compound itself. The basic fractions were more reactive than the acid fractions. The MW distribution of the Maya basic fraction shifted to heavier MW compared to that of the Wilmington VR. Thus regardless of its increased proportion in porphyrinic metal, the strong base fraction of Maya was more refractory than the Wilmington counterpart. The strong acid fractions were the most refractory while for the weak acid fractions Maya was more refractory than Wilmington. Although the Mayan weak acid was converted, it was not demetallized at all, while the Wilmington was almost completely demetallized. The detection of metal porphyrins in the acid and basic fractions of hydrotreated samples was attributed to a masking effect of other molecules that were originally present in the feed, but reacted upon HDT. The porphyrins present in the neutral fraction were converted during HDT [210,211].

The way other heteroatomic compounds affect the catalytic reaction of the metal compounds varies depending on whether the heteroatomic compound interacts with the active sites of the catalyst or with the metal center. An example was given for HDT in the presence of S, O, and N compounds. It was found that quinoline, benzofuran, dibenzothiophene, and anthracene inhibited the conversion of the Me compounds by competitive adsorption on the active sites of the catalyst. Meanwhile, water and ammonia decrease the rate of metal removal by coordinating to the metal group in the porphyrins [212].

Although HDT will remove metals, the processing of crude oils with high metal concentrations has negative effects in most of the refining processes (this topic will be considered in the next section). Thus it is still desirable to have a cost-effective metal removing treatment, and accordingly the understanding of the state and identification of the metal compounds become relevant.

The characterization of the asphaltenic fraction present in a hydroprocessed Maya crude oil showed changes in composition, and at 440°C the structural properties of asphaltenes changed the most drastically [213]. As the reaction temperature is increased, asphaltene aromaticity, nitrogen, and metal content increased, and sulfur decreased. This difference in behavior was attributed to the structural localization of the heteroatoms in the molecule. Once the asphaltene aromatics are hydrogenated to some degree, they ceased to be asphaltene and passed into the maltene fraction, but metal and sulfur removal follows a different chemical path [214]. However, a similarity exists on the planar and rigid nature of the fused-rings structure, which loosens with its reaction with oxygen or hydrogen.

Chiyoda researchers also associated the reactivity of asphaltenes to the reactivity of the metal and sulfur compounds [215] by characterizing the involved molecular groups prior and after conversion through the Asphaltenic Bottoms Cracking (ABC) process [216]. In the ABC process, a VR is hydroconverted to supposedly total asphaltene extinction. During conversion, the asphaltene micelles are believed to decompose (Fig. 3.15) as vanadium is removed. The V remaining after the reaction does not form micelles but is combined with smaller molecules, facilitating its removal by further HDT. Their characterization studies indicated a 4-N first coordination sphere prior to processing (porphyrinic) and 4-S upon depositing on the catalyst surface (vanadium sulfide) [217].

In summary and regarding reactivity, the most significant feature affecting reactivity is the π-system and polarity of the metal compounds. Both features define the degree of association, the components of such interactions, and as a whole the accessibility of the metal centers. The most important issue is the interactions between metal compounds and asphaltenes and second the association. Metal compounds are thought to be occluded, adsorbed, interconnected, and/or even being integral part of the asphaltene molecule. For these reasons, the solution chemistry of asphaltene and of the metal compounds is intrinsically connected. Any treatment process should start by disrupting the association of asphaltenes with the metal compounds. A solvent system for this purpose has been suggested to be multifunctional to counteract the interactions by providing polar, hydrogen bonding, and aromatic functionalities [86]. Oppositely, improved versions of conventional deasphalting technologies try to incorporate solvent systems that favor the precipitation of

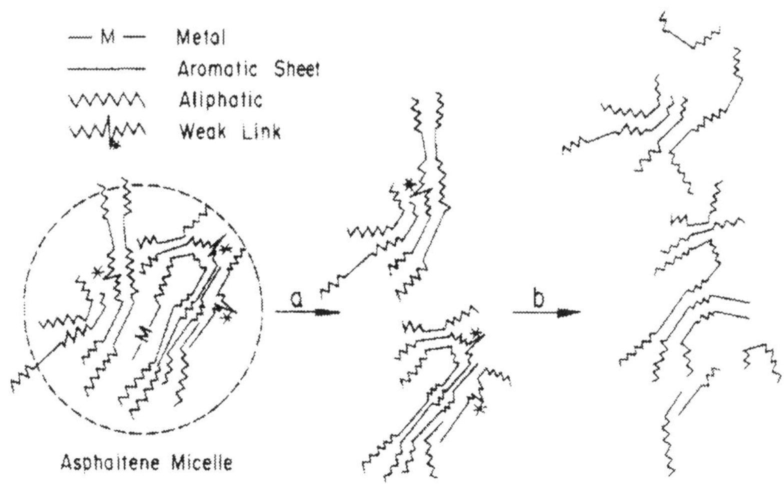

Figure 3.15
Chiyoda mechanism of asphaltene cracking. (A) Micelle destruction, (B) depolymerization caused by heteroatom removal. *Reproduced from Takeuchi C, Fukui Y, Nakamura M, Shiroto Y. Asphaltene cracking in catalytic hydrotreating of heavy oils. 1. Processing of heavy oils by catalytic hydroprocessing and solvent deasphalting. Ind Eng Chem Proc Des Dev 1983;22(2):236–42, with permission from ACS Publications.*

metal compounds together with asphaltenes. An example can be seen in Ref. [218] and the reader is invited to visit Chapter 5 for more on this topic.

4.3.2 Other Reactions

Metalloporphyrins are stable compounds, which suggests a low reactivity of these compounds. Most porphyrins are thermally stable up to 400°C; above this temperature thermolytic decomposition and demetallization have been observed [41]. In terms of chemical stability, the cycloalkane ring on the porphin nucleus of DPEP imposes a strain on the molecule that makes DPEP less stable than the etio molecule [30]. Consequently, DPEP reactivity could be expected to be greater than that of etio molecules. Changes in the coordination spheres of the metal center, both the first and the second spheres, alter its electronic configuration leading to differences in reactivity of the various metal compounds present in the oil.

The reaction between petroporphyrins and a variety of chemical reagents was screened for demetallization capabilities. These reagents included EDTA, disodium salt of EDTA (Na_2EDTA), nitrilotriacetic acid (NTA), disodium NTA (Na_2NTA), diethylenetriamine pentaacetic acid, *N*-(2-hydroxyethyl)ethylene diamine triaacetic acid (*N*-HEEDTA), trisodium *N*-HEEDATA (Na_3*N*-HEEDTA), benzoic acid, malic acid, malonic acid, maleic acid, maleic anhydride, fumeric acid, oxalic acid, triethylcitrate, hexamethylenetetraamine, phthalocyanine, tetraphenyl porphyrin, catechol, di-*t*-butyl, coumarin, formic acid, diethyldithiocarbamate sodium salt (Et_2dtcNa)(H_2O)$_3$, Et_2dtc diethylammonium salt (Et_2dtcEt_2NH_2), *t*-butyl hydroperoxide, cumene hydroperoxide, peroxyacetic acid, sodium sulfide nonahydrate (Na_2S*$9H_2O$), Fuller's earth, bentonite, montmorillonite, pyridine-*N*-oxide, 3-pyridylcarbinol-*N*-oxide, acetone oxime, titanium tetrakis-isopropoxide ($Ti(OCH(CH_3)_2)_4$), titanium oxide acetyl acetonate (TiO(acac)$_2$), fluorosulfonic acid (FSO_3H), trifluoro sulfonic acid (CF_3SO_3H), titanium dioxide, and tartaric acid and examined solvents included dimethyl furan (DMF), 1-methylnaphthalene (1-MN), CH_2Cl_2, toluene, and methanol. Some of the reacting systems were observed to work well in DMF: NTA, carboxylic acid compounds (malic, benzoic, maleic, and malonic), coumarin, hydroperoxides (*t*-butyl and cumene), and in 1-MN; montmorillonite and superacids (fluoro- and trifluoromethane sulfonic); and peroxyacetic acid in CH_2Cl_2. Meanwhile, there was no reaction with polyamino compounds, dithiocarbamates, catechol compounds, amino-*N*-oxides, S^{2-}, and oxotitanium compounds [219].

The reactivity of the VO group under redox conditions has been examined [220–223]. The reaction with thionyl (SO–), acetyl (C_2O_2–), and/or acyl (RCO–) halides converts the VO group into the corresponding dihalide:

$$[VO]^{2-} + 2RCOCl \rightarrow [VCl_2]^{2-} + (RCO)_2O$$
$$[VO]^{2-} + SOCl_2 \rightarrow [VCl_2]^{2-} + SO_2$$

These reactions were demonstrated by converting VO complexes [$VO(S_2CNR_2)_2$ [220] and other S-free complexes [221] with acetyl halide in methylene chloride; diketonates, ketiiminates, 8-quinolinolate, dialkyldithiocarbamates, and porphyrin complexes with carboxylic acid halides [222]; and porphyrins with SOX_2 and COX_2 [223]] to the corresponding dihalide complexes. A primary amine can be used to form stable imidovanadium complexes from the dihalide complexes [224]. When the primary amine is either followed by a long chain alkylamine or perfluoroalkylaniline, the produced derivatized imidovanadium (IV) complexes can be selectively adsorbed on C_{18}-silica gel or perfluorous ($-C_8F_{17}$) silica gel column. This reaction/adsorption sequence has been examined in an asphaltene fraction from Athabasca and Cold Lake bitumens and from heavy oil. Although there is room for optimization, preliminary results indicate selective demetallization as high as about 75% [225].

Partial reduction of the VO groups transforms it into an activated V^{3+} species, which undergoes aggregation to give di- or higher nuclearity products:

$$2[VO]^{2-} + 2e^- + 2H^+ \rightarrow [V_2O]^{4-} + H_2O$$

Porphyrins are known to be photosensitive. Photochemical degradation of Colombian and Arabian crude oils was followed by EPR and fluorescence spectroscopy. Photodegradation under solar irradiation was found to be a slow process that could only be detected after 100 h of irradiation [226]. However, the detection methods are not exclusively sensitive to porphyrins and degradation of other free radicals and/or fluorescent (e.g., polycyclic aromatic hydrocarbons) species could be responsible for the observed changes. Irradiation by the immersion of a high-pressure mercury UV lamp (100 W) in a porphyrin-containing solution may induce porphyrin degradation. The presence of protonic solvent, such as alcohols, accelerates the decomposition and facilitates demetallization. The porphyrin solutions are prepared using aromatic solvents. While model compounds react fast, residues or their bottoms fraction do not. The association of the metal compound, i.e., its trapping or interaction with asphaltene or maltene molecules, has been held responsible for slowing down or stopping the reaction from occurring. The use of 2-propanol favored the release of the porphyrin from binding asphaltic molecules and its reaction [227]. The quantification method employed in this work for evaluating demetallization was based on extracting the supposedly removed metals by acid and basic extraction. The possibility of the acid/base treatment assisting demetallization was not mentioned or discussed by the authors, nor were the blank experiments run.

Metalloformylbiliverdin formation resulted from the photooxygenation of metalloporphyrins and metallochlorins in a single step. Metallobilatrienes are fully conjugated compounds analogous to that of the phlorins, presumably of square planar coordination. The reaction starts by protonation of one of the inner N atoms. Acid demetallization of metallobiliverdins generally includes the irreversible addition to the ligand [228].

In fact, metals can be removed from the porphyrin by strong acids, according to the reaction:

$$\text{Me-Porph} + \text{HX} \rightarrow \text{MeX} + \text{Proph-H}$$

Examples of acid and basic treatments of metalloporphyrins and/or of residues have been given in the literature:

1. Hydrofluoric, hydrochloric, sulfuric, sulfonic, polyphosphoric, hydrobromic [229,230];
2. Hydrochloric or nitric [231];
3. H_3PO_3 and/or H_3PO_4 [232,233];
4. Nitric supported on activated carbon [234–236];
5. H_2SO_3 [237];
6. Ferric nitrate in nitric acid [238];
7. Magnesium sulfate in sulfuric acid [239];
8. Urea and sulfuric acid [240,241];
9. Hydroxides and carbonates [242,243];
10. Superacids, fluorosulfonic, and trifluoromethane sulfonic [219,244];
11. Trifluoroacetic acid and isoperfluorooctanoic acid [245].

Among all these treating acids and bases, hydrofluoric acid was the most effective. However, an oil-soluble acid would decrease the mass transfer limitations of biphasic systems. In the case of isoperfluorooctanoic acid, TPP-porphyrin demetallization was observed while with trifluoroacetic acid, vanadium oxidation followed the demetallization reaction [245].

The relative solubility of the metal compounds in acid (6M HCl) and basic (0.4M ammonium citrate) solutions was evaluated by first concentrating them onto an ion-exchange resin [246]. While Ni (acid) and Fe (basic) could be extracted, V could not be extracted by using any of the employed solutions.

Catalytic and electrochemical conversion of Boscan asphaltenes in lithium ethylenediamine, with Raney nickel, and electrolysis in lithium chloride–ethylene diamine effected reduction and resulted in vanadium reduction and decrease in S-content of the bottoms fraction [247]. Electrolysis of the metal compounds has also been reported to occur at a cathodic voltage from 0 to −3.0 V in aqueous solutions at basic pH [248]. The electrochemical conversion and more particularly the demetallization of crude oil is not very high. While model compounds can be easily demetallized, demetallization drops below 70% for crude oil extracts. The performance for the whole crude was reported in terms of charge efficiency, which was below 10% for V and Ni and only reached interesting levels for Fe (~80%) [249].

Metal compounds were found to react with cumene hydroperoxide at temperatures between 80 and 250°C. Besides metal removal, asphaltenes underwent polymerization, polycondensation, and oxidation reactions [250]. The reactivity toward hydrogen peroxide can be enhanced by using chloroperoxidase as biocatalyst. Enzymes are water soluble and any oil treatment has to be carried out at conditions that overcome the mass transfer limitations. The advantage of any biotreatment is the mild conditions employed (room temperature, atmospheric pressure, etc.). A ternary solvent system composed of toluene, isopropanol, and aqueous buffer solution was found to facilitate completeness of the reaction. The buffer was made up of 3 mM KH_2PO_4 pH 3.0, with 20 mM KCl. Although the treatment was very effective for the demetallization of porphyrins [251], the potential incorporation of Cl into the oil represented a drawback of the enzymatic treatment with that particular biocatalyst [252]. The interaction between metal ions and asphaltene molecules seems to occur by H-bonding between donors and acceptors embedded in the molecular cores. This kind of interaction plays a stabilizing role in water-in-oil emulsions by facilitating the interfacial activity of the asphaltene micelles [253]. Probably this interfacial activity may be used advantageously for enabling the development of a biocatalytic process.

Conversion into oil-insoluble compounds occurs when BotB fractions are subjected to reaction with $ZnCl_2$ and $TiCl_4$ in hydrogen at 285–485°C; coke formation was found to be very low (<3 wt%) [254]. Strong chlorinating compounds and their aqueous solutions also react with the metal compounds in the oil at temperatures from 40 to 300°C, leading to demetallization. Chloride-containing gases as well as inorganic salts are active reagents, e.g., Cl_2, $SOCl_2$, $FeCl_2$, $SnCl_2$, $ZnCl_2$, $TiCl_4$, $RuCl_3$, $CrCl_3$, and $CoCl_2$ [238,254,255].

Vanadium compounds were found to react with phosphorous [256] and P compounds [257,258], producing oil-insoluble vanadium compounds. In the case of organophosphorus compounds, steric limitations control the reactivity. Thus the demetallization effectiveness decreases in the order of $(RO)_3P > Ar_2POH > Ar_3P > Ar_3PO > (NH_4)_2HPO_4$ [259,260]. Reactions with organophosphate esters and phosphorous sulfides [261] also lead to demetallization.

Some crude oils have shown a prevalent content of nonporphyrinic compounds over the porphyrins, while others do not seem to have them at all. As already mentioned, the reported molecular differences may imply different reactivity of each of these subclasses of metal compounds. Certain evidence seems to indicate that porphyrins are more labile to both reducing [262] and oxidative demetallization [263] than the nonporphyrinic compounds. The thermal conversion of the VR from a Chinese crude oil, Daqing, showed a higher conversion of Ni porphyrins, including demetallization, in comparison with the nonporphyrinic Ni [264].

The effect of the surrounding medium and particularly of asphaltenes has also been mentioned in previous sections. The reactivity of the metal was thought to be determined by the macrostructure of the residuum. Whatever the chemistry involved in metal removal, it typically exceeds that of asphaltene conversion. However, the strong interactions between metal compounds and asphaltenes create a dependency between their observable reactivity and the presence of asphaltenes. The degree of agglomeration of asphaltenes is crude dependent and disaggregation is temperature dependent. Asphaltene aggregates from three different sources were characterized by X-ray diffraction and small-angle X-ray scattering and disaggregation was thermally induced. The starting temperature for disaggregation was different for each crude source and the lowest was about 180°C, while the other two were around 240°C [265]. Since only three crude oils were considered, a correlation cannot be derived; nevertheless, it is worth noticing that the lowest disaggregation temperature was found for the crude containing the lowest concentration of metals.

Chemical reactivity of original asphaltenes is low, and furthermore the reactivity is even lower after the molecules have been partially converted, regardless of whether the conversion has been carried out thermally or catalytically. Studies of asphaltene reactivity showed that the final structure of the product depends on the operating conditions rather than on the presence or not of a catalyst. The effect of thermal processing on a vacuum residue was to crack selectively the aliphatic or naphthenic side chains of the molecule, but leaving the highly condensed aromatic core structure almost intact [266]. These aromatic moieties are responsible for the $\pi-\pi$ interactions with metal compounds. Additionally, thermal reactions, HDP of residues [110], and other results indicated that:

1. Demetallization is a function of temperature [110] and its rate is different for the VR than it is for the porphyrins and for the nonporphyrin compounds [262];
2. Product distribution is a function of temperature, but not of reaction time;
3. Porphyrins react preferentially over nonporphyrins [263,267];
4. The porphyrin demetallization mechanism goes through hydrogenated intermediates;
5. Downstream processing may be more amenable to catalytic upgrading because of the removal and size reduction of the metal-containing molecules.

In Chapter 5 more details will be provided on the commercially available technologies for the removal of metal compounds, while in Chapter 6 the emerging concepts, ideas, and technologies will be considered.

5. Impact in the Oil Industry

The presence of metal moieties together with asphaltenes in the refinery streams is mainly responsible for the processing limitations faced by heavy crude oils and bitumen. Metals are present at levels well below 1%, but this amount is more than enough for creating problems in petroleum refining and upgrading [110,125].

As already mentioned, metals are present in the form of organic and inorganic compounds. They deposit in some process equipment blocking and dropping the pressure of the system, and they have also been attributed to be responsible for corrosion and destruction of refractory bricks in industrial furnaces [268]. Catalyst poisoning, polymerization, coke formation, precipitation, emulsification, foaming, etc. are some of these problems.

Besides Ni and V originally present in the crude oil, other inorganic metal compounds can also be present in the feedstock, because of their use in upstream processes or their deficient performance in those processes, e.g., Ca, Pb, Si, Fe, As, Hg, Na, etc. Sodium is present in the form of chloride or caustic carried over from previous (upstream) refining processes. Sodium is particularly dangerous to the fluidized catalytic cracking (FCC) catalysts, since it literally destroys the zeolite component of the catalyst during the regeneration stage. It is also implicated in the formation of sodium vanadate, which is a well-known corrosive compound. Silicon is added in coking and visbreaking as part of antifoaming additives. Arsenic is another contaminating element, which might be present in the feed. All metal contaminants will deposit on the top part of the catalytic bed and may contribute to the buildup of a significant pressure drop in fixed bed processes.

As mentioned in the Introduction section, light metals present in the form of inorganic salt (suspended brine) are removed in the front end of the refinery by a desalting process. In desalting, the crude oil is deliberately emulsified with fresh water for improving contact between the brine droplets and the fresh water and consequently washing off the salts. This contact induces coalescences to dehydrate the mixed oil/water system. The emulsion is subsequently broken using electrical precipitators/dehydrators. Electrostatic dehydrators apply modulation of the electrostatic field to maximize mixing and coalescence to remove brine from the feed oil and recover the added chemicals. The presence of Fe, Ni, and V has been shown to interfere with the desalting efficiency; Fe particularly affected the removal of Ca. Advantageously, specific emulsifiers have been developed for the partial removal of the iron compounds [269]. Additionally, another factor limiting desalting efficiency is the presence of solid particulates in the desalter, which also plays an important role in emulsion stabilization [270,271].

In the case of metal-containing heavy oils, the role of metals in asphaltene association could provoke and stabilize the formation of asphaltene particles that may concentrate in the oil/water interface. Solutions include the use of additives, dispersants, emulsifiers [272], and/or the design of improved desalters with more than one stage [273]. Preventing desalter upsets requires modeling and simulations to be run aprioristically on crude slate blending, compatibility, and desalting characteristics [272].

In dealing with vanadium containing by-products (fuels, contaminated catalysts, etc.), one should keep in mind the issues created by combustion of these products. Upon burning, vanadium will be converted into V_2O_5 that will cause severe corrosion of turbine blades and enhanced deposition of vanadium-containing salts in exhaust systems. Vanadium-induced corrosion is more important at temperatures above 680°C, at which point vanadium pentoxide melts. Sodium vanadate is easily formed at combustion conditions and it is an aggressive low melting point compound that accelerates deposit formation and high-temperature corrosion of engine components [274].

The foremost consequence of the most abundant heavy metals naturally occurring in petroleum (V, Ni, and Fe) is the deactivation of conversion catalysts [both hydrocracking (HCK) and FCC catalysts], and to a lesser extent HDT catalysts. These metal compounds deposit on the catalysts, deactivating it by fouling and pore plugging. They also catalyze undesirable reactions, such as dehydrogenation, coking, etc. In recent times, HCK and FCC catalysts have been developed to tolerate higher amounts of metals. Therefore the level of demetallization is imposed by the feedstock specification of the downstream unit, which would be fed. Consequently, the removal of metals from crude oil or its fractions is envisioned as a need for improving the processability of the heavy stocks. Therefore highly asphaltenic resources must be rigorously treated to convert them into upgraded crude or into clean fuels. A comparison of the visbreakability of an entire crude versus DAO showed that under the same conditions the former was 70% demetallized, while the latter was 96.5% demetallized [146]. A more refractory nature of Ni relative to V was also shown.

The impact of metals in refining has pushed the definition of new methodologies and the development of techniques for coping with the high metals-containing crude oils. New scheduling methods, new linear programming simulations, optimization of operating conditions, etc. have been implemented. The optimization of the vacuum distillation unit to produce the required metal profile in the heavy fractions led to minimizing the impact of the metals in the downstream unit and the metal content in the most valuable products [275].

5.1 Effects on FCC

Since HDT is a conventional process that will remove metal contaminants, through HDM reactions the pretreatment of streams prior to being fed to a process sensitive to these contaminants is recommended. However, complete removal cannot be achieved. Refineries processing heavy oils have to deal with the remaining presence of trace metals, although HDT of the FCC feed is broadly employed. Metals cause adverse effects in both conversion and yield.

Metals deposit on the FCC catalysts and accumulate quantitatively through time on stream, up to levels of about 10,000 ppm (~2000–3000 ppm Ni and ~6000–8000 ppm V), at current make-up rates. The irreversible deactivation caused by metals dictates the rate at which fresh catalyst has to be added to the reactor and so it is directly related to increases in operating costs.

The 1980s and 1990s marked titanic efforts for dealing with the increasing proportion of heavy oils in the refinery diet, which led to increasing the amount of heavy residues fed to the FCC unit (FCCU). At that time the increasing operating cost caused by continuous catalyst additions obliged refiners to try to increase conversion, but unfortunately technical means were scarce. A considered option was to remove the deposited metals from the deactivated catalyst, ex situ the unit, and to recycle the demetalated catalyst [276–279]. Overall, the technical benefits were marginal and the process was burdensome.

Currently, environmental regulations not only impose rough measures for the reduction of contaminants in the transportation fuels, but also have limited their presence in fuel oil as well. Thus nowadays FCC operation mode has moved to maximize bottoms conversion [280]. Catalyst manufacturers, catalysis researchers, and refiners were challenged with the task of mitigating the problems that metals were causing to the performance and operating costs by passivating the deleterious effects on the catalyst [281].

Years of research were devoted to finding ways to artificially deposit the metals on the FCC catalysts, not only to understand the mechanisms of interaction, but also to simulate in an expeditious manner the effects of metals accumulation on performance. Impregnation of inorganic as well as organic salts was unsuccessfully attempted. Some examples can be seen in Refs. [282,283]. Soon it was realized that these deactivation methodologies were not mimicking accurately the actual changes occurring in the catalyst in the FCCU. New deactivation methods have been defined for a better representation of real conditions, such as using a recirculating unit and running as many reaction–regeneration cycles as needed to accumulate similar metal levels as in equilibrium catalyst (E-cat) from commercial units [284]. The way of minimizing the number of cycles is by increasing the metal content of the vacuum gas oil (VGO), if possible with metal compounds extracted from real feeds.

Metals are thought to affect the FCC catalyst in two different ways. Physically, metal deposits may obstruct the access of reacting molecules to the zeolite pore structure. Chemically, the superficially exposed metals exhibit intrinsic catalytic activity for dehydrogenation reactions, which will increase hydrogen, olefins, and coke yields during operations. Ni and V act neither synergistically nor proportionally; instead Ni produces three to four times as much hydrogen as an equivalent amount of V [285]. This increased dehydrogenation exhibited by Ni is also reflected in an increase in coke yield.

Conversion and throughput in commercial FCCUs are limited by either coke yield or gas yield. The former is determined by regenerator metallurgy and air rate, while the latter is determined by the gas plant and compressor capacities. This compressor capacity is also highly impacted by the difference in compressibility between hydrogen and hydrocarbons. Hence for processing high metal-containing feeds, the FCCUs and the gas plant have to be designed to account not only for the total metal content, but also for an Ni/V activity ratio of 4. In terms of operations and for a more efficient processing, knowledge of catalyst metal resistance and feedstock properties should be considered [285].

In the case of vanadium, these deactivating actions are additionally aggravated by the formation of vanadic acid in the regenerator where V is first oxidized and then the oxide reacts with the water formed during coke combustion:

$$2[-VO] + 3/2O_2 \rightarrow V_2O_5$$
$$V_2O_5 + 3H_2O \rightarrow 2H_3VO_4$$

Vanadic acid has been found to destroy the zeolite crystal structure by hydrolysis of its $SiO_2-Al_2O_3$ framework.

One of the passivating alternatives involved the development of metal traps, which were materials that chemically interacted with the metal rendering catalytically inactive solid compounds. Besides a high storage capacity, the metal—trap interaction should be strong and should occur at a faster rate than that taking place between the metal and the zeolite. The strength of the binding of the metal to the trap will determine the effectiveness of the trap. Initially, in vanadium mobility, both intraparticle and interparticle mobility was observed and then nickel mobility was proven. Interparticle mobility is thought to be enabled by the action of steam. Obviously, the desired mobility is from the zeolite to the trap, though this is not always the case [286]. In other words, an irreversible trapping effect is desired.

First to appear in the FCC market were the vanadium traps [287]. Two classes of metal traps are available. Those that are incorporated as an integral part of the catalyst particles and those that come in separate particles; the "dual-particle" system has to be fed to the riser.

Regarding trap materials, Ashland Oil considered oil and aqueous soluble compounds of aluminum or silicon, such as silicic acid, aluminum isopropoxide, aluminum acetate, aluminum alcoholates, silanes, and silicates comprising an alkyl or aryl group [288]. This company also considered the coating of catalyst particles with nonionic lanthanum or lanthanum-rich rare-earth metals [287].

Chevron formulated traps [287] based on sepiolite [289], dolomite/sepiolite [290], aluminas [291], magnesium oxide [292], and titania and zirconia [293]. Chevron conducted a commercial demonstration trial with one of this formulated traps by feeding the trap at 4% of make-up rate. The trapping capacity was defined as the V/trap weight ratio (determined from the scanning electron microscopy examination of E-cat samples) and was determined to be 17/1 [294]. An investigated option was to coat the Y-zeolite particles with an AlO_x layer either by depositing an $[Al_{13}O_4(OH)_{24}(H_2O)_{12}]^{7+}$($[Al_{13}]$) complex or by anchoring alumoxane through in situ hydrolysis of triisobutylaluminum. The first method was more effective than the second in capturing V from the feed [295].

Other traps were designed for trapping the vanadic acid by an acid–base reaction to produce the corresponding basic metal vanadate. Basic materials, such as barium titanate, calcium titanate, calcium carbonate, and strontium titanate, have been proven at laboratory scale [286]. The most common employed materials as V-traps and commercially available through catalyst vendors are magnesia and titania.

While some of the V-traps would pick nickel up, their effectiveness for Ni is not adequate. BASF developed a specialty alumina with Ni-trapping capabilities. This trap could be incorporated into the catalyst particle. Consequently, the distribution of the trap relative to the active zeolite is crucial. Two specific manufacturing technologies were developed and are employed for meeting the careful design of this family of residue FCC catalysts. Details are given in Refs. [296–298].

Most of the time, traps have been evaluated for their storage capacity and stability during reaction conditions. However, refiner objectives vary, and although related to the storage and stability of the trap, they are more closely associated with the economic performance parameters. The evaluation at pilot plant level needs to consider such objectives, examples are:

1. Incorporation of a higher metal feed or process more residue;
2. Keep feed and catalyst constant and allow conversion (and yields) to change;
3. Reduce fresh catalyst addition.

The ultimate test is the commercial test, under which catalyst performance should not be affected significantly. In commercial FCC operation, when traps are run at constant conversion, higher metal levels are let to accumulate in the E-cat, either by running higher-metals contaminated feeds or by lowering catalyst additions. If this mode of operation is applied in the absence of a trap, yields will change, but the octane in the gasoline has been observed to increase because of the dehydrogenation activity of the metals [299]. In the presence of the trap, at a similar metal level on the E-cat as per the case of no trap, no effect on octanes would be observed since a major part of the metals would be inactivated by the trapping.

In some instances, where bottoms-cracking additives (e.g., Refs. [300,301]) are used, the metal trap might be incorporated with such additive. A bottoms-cracking additive might be formulated similarly to an FCC catalyst, in which a high activity (high acidity) matrix is combined with the other components [302]. A high surface area matrix was found to be capable of trapping vanadium effectively [303]. Another modification of the matrix includes the creation of reactive defects, such as those generated on kaolin by acid or caustic modifications [304].

An E-cat pretreatment with hydrogen or methane was a proposed option to cope with deactivation caused by metal poisoning. These pretreatments decreased hydrogen and coke yields while gasoline yield increased. The higher the metal accumulated on the catalyst, the higher the benefits found with the pretreatment. These benefits were explained to be caused by an increase in the hydrogen transfer reactions by the reduction of the loaded metals [305].

Whereas some of the additives employed in FCC can be envisioned as cocatalysts that promote desired reactions (octane enhancers, olefins yield, aromatic yields, etc.), the metal passivating traps are supposed to avoid or minimize the deactivation effect of the metals. In this way, traps would diminish the undesired reactions promoted by metals, prolonging performance indicators (conversion and yields).

A trap is not supposed to possess any catalytic activity and would represent a diluent of the catalyst within the reactor [301]. In a way, the trap would affect activity and performance, since in its absence, the catalyst will deactivate faster. Thus the decrease in activity caused by dilution would be much less than that caused by metal deactivating effect.

In some instances, a sort of complementary effect has been found with the metalated catalyst and the action of certain additives. This is the case, for instance, of ZSM-5 additives, particularly when used for boosting the olefins yield [306]. As already mentioned, olefins yield is also increased by the dehydrogenating activity of both Ni and V, much more with Ni than with V.

As mentioned previously, a high metal content is typically accompanied by high S-content. Hence when processing high metal-containing feeds, the already stretched targets of the FCC process are more stressed by including S-reduction of the produced naphtha. On the other hand, the presence of S in the feed has been found to be detrimental to the performance of the metal traps. Different additives are commercially available for this purpose [286]. It was observed that when V and Ni were impregnated on the FCC catalyst the sulfur of the naphtha could be reduced [307].

5.2 Effects on HDP

As already anticipated, VGO, AR, and VR may contain traces (ppm levels) of organometallic compounds (typically, nickel and vanadium). Additionally, iron may be present as inorganic compounds (corrosion products) or as organometallic compounds.

Although HDP catalysts are active for demetallization reactions of the organometallic contaminants present in the crude oil, they are actually poisoned by the metals [308,309]. Under HDT conditions, the organometallics decompose, transform into their sulfided form, and deposit on the HDT catalyst, causing its deactivation, particularly of the hydrodesulfurization (HDS) and hydrodenitrogenation (HDN) activity [310]. Typically, the inorganic form of iron fouls and accumulates on the top of the bed causing an even more dangerous plugging that leads to high pressure drop [208,311]. This irreversible deactivation caused by metals deposition degrades performance by occupying active sites and blocking pores. However, this fouling may also cause a second cause of irreversible deactivation, and that is a chemical reaction with the active phase yielding mixed sulfides [312], and thermally or chemically induced sintering (decrease of the surface area) of the metal phase or of the support material [313,314]. In summary, the two mechanisms believed to be the cause of irreversible deactivation are [313,315,316]:

- Pore blocking: metals accumulate as agglomerated structures inside the pores or at the pore mouth, restricting accessibility to reacting molecules, increasing coke formation, and building up pressure drops.
- Fouling on the active phase and poisoning of the active sites: physical deposition and chemical solid-state interactions lead to the formation of mixed sulfides and sintering.

The metal–catalyst interactions are so strong that their deposition starts at the external surface of the particles, thus, typically, pore blocking is mostly mouth plugging. This pore-mouth obstruction limits the access of reacting molecules to the active sites, restricting the overall performance of the catalyst. Similarly, the metals start depositing from the top layers of the catalytic bed, and this loading moves downward continuously, unless the plugs at reactor inlet cause a pressure drop that requires stopping the run [208,311].

Research has focused on the understanding of metal issues by addressing all possible fronts: the relative reactivity of the different metal compounds, the interaction mechanisms, the deactivation phenomenon, the modeling of kinetic effect, activity/deactivation, reactor performance, etc.

5.2.1 Interaction Mechanisms

The decreasing deposition profile of vanadium and nickel from top to bottom of the bed matched well with the observed carbon increases. Both hard and soft coke amounts increased from the top toward the bed outlet. Furthermore, a tendency of metals to deposit in pores of midsize and coke to fill big pores was also observed. The observed changes in activity and overall performance were associated with the metal deposition [317]. While vanadium tends to deposit in the external surface, Ni is found to distribute more evenly throughout the internal surface [318].

Catalyst fluorination was found to inhibit the deactivation effects of metal deposition. An optimum of fluorine surface concentration was determined to be F/Al = 0.1. The action of fluorine was considered to be the prevention of metals to interact with the internal surface area of the catalyst and keep the metals in the exterior of the particles in a highly dispersed state [319].

Occasionally, a low metal content in VGO can be handled by using either protective guard beds, sacrificed catalyst, or an appropriate (pressure-drop control) catalyst at the top of the active catalyst bed. The condition for the guard bed is to exhibit a very strong interaction with the metal compounds, especially stronger than those existing with the other compounds present in the crude oil feed.

5.2.2 Relative Reactivity of Metal Compounds

As already mentioned, Ni removal is always more difficult than that of V. The rate constant for Ni removal was found to be about 60% of that attained for V on a wide range of crude oils, catalysts, and processing conditions. This was explained in terms of a stronger interaction of V compounds with the catalysts that compete with the Ni compounds. However, the lower concentration of Ni in the feed drives unfavorably competition in terms of the interactions with the active sites of the catalyst [146]. The higher reactivity of the vanadium has also been explained as being caused by its coordination to an oxygen ion in the VO group [318].

5.2.3 Deactivation Phenomenon

Along the reaction run, changes in pore size distribution are caused by the deposition of both metals and coke. Coke seems to deposit as a whole at the start of the run, at a level of 18–20% (based on catalyst mass) [317]. Instead, metals gradually build up from the external surface of the particle through the inner surface, and from the top of the bed downward to the outlet, up to an average level of about 25% [320]. An increase in mesopores in the range of 100–250 Å has been observed with time on stream and it has been explained as being caused by partial blockage of larger pores (>250 Å) [318,321,322], which are believed to be the most involved in the metals accumulation process [323]. The gradual deposition of the metals along the length of the catalyst bed within the reactor and preferentially at the top was thought to be a consequence of the big molecular volume rather than of a higher reactivity of the metal-containing molecules [324].

Catalysts with broader and higher tolerance to metals have been optimized following two approaches: pore engineering and/or modifications of the metal–catalyst interaction. Pore engineering involves increasing pore volume, optimizing the proportions of larger pores (meso- and macropores), and optimizing surface area/pore size distribution. Meanwhile, modification of the metal–catalyst interaction could be pursued either at catalyst level or on the catalytic bed. The molecular design of the catalyst requires allocating the HDS/HDN active sites away from the demetallization sites. The catalytic bed approach works similarly by loading catalyst with a gradual increase in activity, i.e., there would be an activity profile (gradient) throughout the catalyst bed.

5.2.4 Kinetic Effects, Activity/Deactivation, Reactor Performance, and Modeling

Modeling efforts have been pursued with several objectives in mind. As mentioned previously for the FCC case, artificial deposition has not been a success for modeling deactivation [325,326]. In the case of Ni and V, artificial deposition has indicated that hydrogenation is not significantly affected by nickel sulfide deposits, but its rate constant decreases with vanadium sulfide deposits. A similar situation was observed for hydrogenolysis. Nickel sulfide deposits were almost inactive for cracking but vanadium sulfides showed a relatively high activity, suggesting an acid nature of these surface sites. Additionally, the vanadium sites were poisoned by quinoline [326]. Although the activity and catalytic behavior of Ni and V sulfides has been reported in different studies (e.g., Refs. [327–329]), the intrinsic activity of the metal compounds present in crude oil as such has not received much attention. Nonetheless, the hydrogenation capabilities of V and Ni porphyrins have been reported [330].

The kinetics of the different reactions occurring on HDP is affected by the metal deposits. Each reaction is affected differently, therefore product distribution changes with time on stream. The incorporation of the kinetics of HDM and of coke formation in an HDP model for the mathematical evaluation of metal and coke deposition was considered. In this case, HDM was assumed to occur through a network of parallel and consecutive reactions. Deposition will restrict the diffusion of larger molecules throughout the pore system of the catalyst. Deactivation was then modeled as an increasingly diffusion-limited effect [331]. The HDM reaction network predicted a lower metal deposition at the top of the reactor, indicating the invalidity of such network.

Modeling has been also justified as a way to save resources, by predicting performance at different deactivation levels [309]. The difficulties encountered by the simultaneous deposition of coke and metals made it impossible to uncouple their individual effects on deactivation under real conditions. Moreover, the main cause of coke formation is asphaltenes and their associative behavior. Metals are intimately associated with the asphaltenes. For these reasons, coke and metal deposition together are widely accepted as the causes for deactivation of HDP catalysts when processing heavy feeds [332]. However, the comparison of HDT performance for low metal- and high metal- containing feeds has provided certain insights. It has been found that the life cycle was drastically shortened by the high metal-containing feed, for which an accelerated coke deposition was observed on increasing metals accumulation [333].

The modeling of reactor performance has required an increasing incorporation of deactivation details. The simplest model that considered the changes in activity as exclusively caused by a purely pore-plugging model has been evolving through the years, to incorporate the poisoning effect of metals, their profile through the catalytic bed, the intrinsic reactivity of the crude oil molecules, the inherent activity of the deposited metals, the changes on catalytic functionality caused by the presence of contaminants, etc. The reader interested in further details on the modeling of activity, deactivation, and reactor performance is directed to the work of Ancheyta's group and the references cited therein, representative examples are Refs. [310,317,318,320,322,334–352].

6. General Remarks

The increase in the proportion of heavy oils into the refinery diet, together with the increasingly stringent environmental requirements, as a whole makes the fuel manufacturing process one of enormous complexity. The presence of metal compounds represents one of these complications.

More than 60 years of research devoted to the trace metals present in crude oils have contributed to a continuous growth of the associated knowledge. The task has not been easy and is still lacking completion. Two major difficulties have been faced by researchers dealing with the characterization of the metal compounds. The most abundant metals, Ni and V, occur in two different forms: porphyrins and the rest, grouped as nonporphyrins. Porphyrins are well-known compounds since they are naturally occurring complexes and chelating organic molecules, chlorophyll being the best example. However, very little is known about the nonporphyrin compounds and controversial results cloud the scenario of their existence.

The organic mixture where metal compounds coexist is highly complex, requiring separation, isolation, and purification prior to any attempt for their examination under high-resolution or highly sensitive techniques. Separation alone has been proven not to be effective enough in simplifying the system. Nevertheless, the studies made on the fractions provide some information, such as total metal content, porphyrins content, and types, but hardly allow definitive identification. Further (sub)fractionation and isolation methods have to be and have been developed. Such approaches lead to partitioning of the different types of metal compounds into several subfractions, each of them requiring independent characterization. However, in the best case scenario of isolating and purifying the fractions into simpler mixtures, a second major concern arises as to whether any unwanted chemistry took place during the process. These concerns are based on the fact that separation and purification of these compounds for subsequent studies require stringent extraction conditions that might change or modify the molecular structure. Thus the validity of such results has been questioned and consequently attention should be paid to the presence of any experimental artifacts or any molecular structural modifications. Fortunately, analytical techniques have been advancing rapidly and at least demetallization is no longer required in advance of the examination of the metal compounds.

Multiple reactions have been examined for the demetallization of model compounds and of isolated metal compounds from crude oils. Examples are selective oxidation [240,267], biological demetallization [251,353,354], electrolysis [181,248], adsorption [147,167,168], photochemical reaction [227], and (reactive-)extraction [180,222,230,232,241,254,260]. However, the economic benefits of these reactions do not seem to break the commercialization barrier and neither the technical effectiveness is high enough.

The lack of knowledge of the chemical nature of all the metal compounds, specifically the nonporphyrins, and of the understanding of the reactivity has precluded new ideas from emerging. Furthermore, the complicating effect of metals—asphaltenes interactions, the possibility of metal compounds being trapped, shielded, or occluded by asphaltenes, or even being a structural part of the asphaltene molecules make any approach to study these compounds a very difficult one.

Further research is needed. The future focus may be on the elucidation of the molecular structure and the physical form of the metal compounds. Clarification of the interactions, aggregation, adsorption, and chemically bound molecular entities in the crude oil matrix is desirable. Once these are achieved, then the understanding of metal compounds reactivity would become easier and ideas for a selective metal removal process would flourish.

Probably the inclusion of new units dedicated to change the properties of the new crude blend diet into the specifications of the currently operating units might have to be considered. Processes specifically for demetallization are not commercially available; instead, metals are removed by either deasphalting, conversion (thermal or catalytic), or HDP. These processes either currently commercially practiced or ready for commercial licensing will be described in Chapter 5. A selective demetallization process may contribute to get more value from the crude oil and increase refining margins. Few new ideas are emerging and will be discussed in Chapter 6.

References

[1] Filby RH. Origin and nature of trace element species in crude oils, bitumens and kerogens: implications for correlation and other geochemical studies. Geol Soc Lond Spec Pub 1994;78:203–19.

[2] Olajire AA, Oderinde RA. Trace metals in Nigerian crude oils and their heavy-end distillates. Bull Chem Soc Jpn 1993;66(2):630–2.

[3] Chouparova E, Lanzirotti A, Feng H, Jones KW, Marinkovic N, Whitson C, et al. Characterization of petroleum deposits by synchrotron radiation-based microanalyses. Energy Fuels 2004;18:1199–212.

[4] Spiecker PM, Gawrys KL, Kilpatrick PK. Aggregation and solubility behavior of asphaltenes and their subfractions. J Colloid Interface Sci 2003;267(1):178–93.

[5] Yang X, Hamza H, Czarnecki J. Investigation of subfractions of Athabasca asphaltenes and their role in emulsion stability. Energy Fuels 2004;18(3):770–7.

[6] Mclean JD, Kilpatrick PK. Comparison of precipitation and extrography in the fractionation of crude oil residua. Energy Fuels 1997;11:570–85.

[7] Frankenberger A, Brooks RR, Varela Alvarez H, Colleen JD, Filby RH, Fitzgerald SL. Classification of some New Zealand crude oils and condensates by means of their trace element contents. Appl Geochem 1993;9:65–72.

[8] Hitchon B, Filby RH. Trace elements in Alberta crude oils. Report No. 02. Edmonton (Canada): Alberta Research Council; 1983.

[9] Mercer GE, Regner AJ, Filby RH. Trace element distributions in kerogen, bitumen, and pyrolysates isolated from the New Albany shale. Proc ACS Div Fuel Chem Prepr 1992;37:1761−8.

[10] Carbognani L, Garcia C, Izquierdo A, Di Marco MP, Perez C, et al. Correlation between physical chemical properties and reactivity in hydroprocessing of Venezuelan heavy ends. ACS Div Pet Chem Prepr 1987;32(2):406−12.

[11] Reynolds JG, Jones EL, Bennett JA, Briggs WR. Characterization of nickel and vanadium compounds in tar sand bitumen by petroporphyrin quantitation and size exclusion chromatography coupled with element specific detection. In: Proc. 195th ACS Nat. Meeting, Toronto, Canada; June 5−11, 1988. 17 pp.

[12] Van Berkel GJ. The role of kerogen in the evolution of the nickel and vanadyl porphyrins [Ph.D. thesis] from Washington State University. December 1986.

[13] Roussel J-C, Boulet R. Composition of crude oils and petroleum products. Pet Refin 1995;1 (Crude Oil, Petroleum Products, Process Flowsheets), 1−15, 453−460.

[14] Reynolds JG. Effects of asphaltene precipitation on the size of vanadium-, nickel-, and sulfur-containing compounds in heavy crude oils and residua. In: Yen TF, Chilingarian GV, editors. Asphaltenes asphalts. 1. Developments in pet. science. Amsterdam: Elsevier; 1994. p. 233−48.

[15] Fish RH, Komlenic JJ, Wines BK. Characterization and comparison of vanadyl and nickel compounds in heavy crude petroleums and asphaltenes by reverse-phase and size-exclusion liquid chromatography/graphite furnace atomic absorption spectrometry. Anal Chem 1984;56(13):2452−60.

[16] Acevedo S, Guzmán K, Labrador H, Carrier H, Bouyssiere B, Lobinski R. Trapping of metallic porphyrins by asphaltene aggregates: a size exclusion microchromatography with high-resolution inductively coupled plasma mass spectrometric detection study. Energy Fuels 2012;26(8):4968−77.

[17] Laura Reyes H, García-Ruiz S, Gisele Tonietto B, Marcus Godoy J, Ignacio García Alonso J, Sanz-Medel A. Quantification of selenium species in petroleum refinery wastewaters using ion chromatography coupled to post-column isotope dilution analysis ICP-MS. J Braz Chem Soc 2009;20(10):1878−86.

[18] Sams GW. Challenges of desalting Canadian crudes. In: Proc. Crude Oil Quality Association Meeting, New Orleans, LA, USA; February 10−11, 2010. p. 23. Available from: http://www.coqa-inc.org/docs/default-source/meeting-presentations/20100211_Sams_Desalting.pdf.

[19] Sellman E, Sams G, Pavan Kumar Mandewalkar S. Improved dehydration and desalting of mature crude oil fields. In: Proc. 18th Middle East Oil Gas Show Conf.: Transforming the Energy Future, Manama, Bahrain, vol. 2; March 10−13, 2013. p. 1133−45.

[20] Sams G. New electrostatic technology for desalting crude oil. In: Proc. NPRA Annual Meeting, Salt Lake City, UT; March 19−21, 2006. p. 33. AM-06-30.

[21] Fausnaugh JM. Nickel and vanadium: using trace metals for anomaly identification and characterization. U. S Geochem 2002:3. Available from: http://www.geotech.org/files/69677365.pdf.

[22] Lo Mónaco S, López L, Rojas H, Garcia D, Premovic P, Briceño H. Distribution of major and trace elements in La Luna Formation, Southwestern Venezuelan basin. Org Geochem 2002;33(12):1593−608.

[23] Abu-Elgheit M, Khalil SO, Barakat AO. Characterization of crude oils by new trace-metal indexes. ACS Div Pet Chem Prepr 1979;24(3):793−7.

[24] López L, Lo Mónaco S. Geochemical implications of trace elements and sulfur in the saturate, aromatic and resin fractions of crude oil from the mara and mara oeste fields, Venezuela. Fuel 2004;83(3):365−74.

[25] López L, Lo Mónaco S, Galarraga F, Lira A, Cruz C. V/Ni ratio in maltene and asphaltene fractions of crude oils from the west Venezuelan basin: correlation studies. Chem Geol 1995;119(1−4):255−62.

[26] Ali MF, Bukhari A, Saleem M. Trace metals in crude oils from Saudi Arabia. Ind Eng Chem Prod Res Dev 1983;22(4):691−4.

[27] Treibs A. Uber das vorkommen von chlorophyllderivaten en eines olschiefer aus der oberen triss (occurrence of chlorophyll derivatives in an oil shale from the upper triassic). Ann Chem Liebigs 1934;509:103−14.

[28] Treibs A. Chlorophyll and hemin derivatives in bituminous rock, petroleums, coals, and phosphate rock. Ann Chem Liebigs 1935;517:172–96.
[29] Treibs A. Chlorophyll and hemin derivatives in organic mineral substaces. Angew Chem 1936;49:682–6.
[30] Yen TF, Chilingarian GV. Structural parameters from asphaltenes and their geochemical significance. In: Yen TF, Chilingarian GV, editors. Asphaltenes asphalts. 1. Developments in pet. science. Elsevier; 1994. p. 159–78.
[31] Chicarelli MI, Kaur S, Maxwell JR. Sedimentary porphyrins: unexpected structures, occurrence and possible origins. In: Filby RH, Branthaver JF, editors. Metal complexes in fossil fuels. Washington (DC): ACS; 1987. p. 40–67.
[32] Huseby B, Ocampo R, Bauder C, Callot HJ, Rist K, Barth T. Study of the porphyrins released from the Messel oil shale kerogen by hydrous pyrolysis experiments. Org Geochem 1996;24(6–7):691–703.
[33] Qian K, Edwards KE, Mennito AS, Walters CC, Kushnerick JD. Enrichment, resolution, and identification of nickel porphyrins in petroleum asphaltene by cyclograph separation and atmospheric pressure photoionization fourier transform ion cyclotron resonance mass spectrometry. Anal Chem 2010;82(1):413–9.
[34] Van Berkel GJ, Filby RH. Generation of nickel and vanadyl porphyrins from kerogen during simulated catagenesis. In: Proc. 191st ACS Nat. Meeting, Metal Complexes in Fossil Fuels: Geochem., Characterization Symp., New York, NY, USA; April 13–18, 1987. p. 110–34.
[35] Barwise AJG, Whitehead EV. Separation and structure of petroporphyrins. Phys Chem Earth 1980;12(C):181–92.
[36] Baker EW, Yen TF, Dickie JP, Rhodes RE, Clark LF. Mass spectrometry of porphyrins. II. Characterization of petroporphyrins. JACS 1967;89(14):3631–9.
[37] Jeandon C, Ocampo R, Callot HJ. Unexpected methylation and propylation of porphyrin E-ring during the hemisynthesis of deoxophylloerythroetioporphyrin (DPEP). Tetrahedron 1997;53(47):16107–14.
[38] Ekstrom A, Dale L, Loeh H. An examination of the petroporphyrins found in oil shale from the julia creek deposit of the toolebuc formation. In: Proc. ACS Nat. Meeting, Div. Fuel Chem. Symp. Geochem. Chem. of Oil Shale, Seattle, WA, USA; March 25, 1983. 11 pp.
[39] Baker EW, Palmer SE. Petroporphyrins. In: Dolphin D, editor. The porphyrins. New York: Academic Press; 1978. p. 485–551.
[40] Barakat AO, Yen TF. Nature of porphyrins in kerogen. Evidence of entrapped etioporphyrin species. Energy Fuels 1989;3(5):613–6.
[41] Premović PI, Jovanović LS, Nikolić GS. Thermal stability of the asphaltene/kerogen vanadyl porphyrins. Org Geochem 1996;24(8–9):801–14.
[42] Premović PI, Jovanović LS. Are vanadyl porphyrins products of asphaltene/kerogen thermal breakdown? Fuel 1997;76(3):267–72.
[43] El-Sabagh SM. Occurrence and distribution of vanadyl porphyrins in Saudi Arabian crude oils. Fuel Proc Technol 1998;57(1):65–78.
[44] Sugihara JM, Branthaver JF, Wu GY, Weatherbee C. Research on metal compounds in petroleum. In: Proc. Symp. on Heterocompounds in Pet., Houston, TX, USA. ACS div petrol chem, vol. 15 (2C); February 22–27, 1970. p. 5–11.
[45] Van Den Berg FGA, Heijnis RMA, Stamps PA, Kramer PA. A geochemical framework for understanding petroleum residue properties. Pet Sci Technol 2003;21(3 & 4):449–60.
[46] Premović PI, Tonsa IR, Pavlović MS, López L, Lo Monaco S. Classification of the asphalts and their source rock from the dead Sea basin (Israel): the asphaltene/kerogen vanadyl porphyrins. Fuel 1998;77(15):1769–76.
[47] López L, Lo Mónaco S, Volkman JK. Evidence for mixed and biodegraded crude oils in the socororo field, eastern Venezuela basin. Org Geochem 2015;82:12–21.
[48] Lópe L, Lo Mónaco S. Geochemistry of crude oils from the orinoco oil belt, Venezuela. Rev la Fac Ing 2010;25(2):41–50.

[49] López L. Study of the biodegradation levels of oils from the Orinoco Oil Belt (Junin area) using different biodegradation scales. Org Geochem 2014;66:60–9.

[50] Mónaco SL, López L, Rojas H, Lugo P, García D, Gastiel J. Applications of electron microprobe analysis (EPMA) in the study of Venezuelan source rocks: La Luna and Querecual Formations. Fuel 2007;86(5–6):641–8.

[51] Premović PI, Tonsa IR, Pajović MT, Lopez L, Lo Monaco S, Dordević DM, et al. Electron spin resonance study of the kerogen/asphaltene vanadyl porphyrins: air oxidation. Fuel 2001;80(5):635–9.

[52] Premović PI, Nikolić ND, Pavlović MS, Jovanović LS, Premović MP. Origin of vanadium in coals: parts of the Western Kentucky (USA) No. 9 coal rich in vanadium. Geol Soc Lond Special Pub 1997;125:273–86.

[53] Baker EW. Porphyrins. In: Eglinton G, Murphy MTJ, editors. Org. Geochem. London (UK): Longmans Ltd; 1969. p. 464–97.

[54] Yen TF. The role of trace metals in petroleum. Ann Arbor: Ann Arbor Science; 1975.

[55] Yang Z, Cheng Z. Isolation, identification and correlation of porphyrins from some continental crude oils, source rocks, oil shales and coals in China. In: Yen TF, Moldowan JM, editors. Geochemical biomarkers. Chur (Switzerland): Harwood Academic Publishers; 1988. p. 115–32.

[56] Tooulakou D, Filby RH. Separation of nickel and vanadium complexes from the Athabasca oil sand asphaltenes. In: Yen TF, Moldowan JM, editors. Geochemical biomarkers. Chur (Switzerland): Harwood Academic Publishers; 1988. p. 221–40.

[57] Galimov RA, Romanov GV. Extraction of primary concentrated petroporphyrin product. In: Proc. 7th UNITAR Intern. Conf. 'Heavy crude tar sands.' Beijing, China; October 27–30, 1998. 5 pp.

[58] Spencer WA, Galobardes JF, Curtis MA, Rogers LB. Chromatographic studies of vanadium compounds from Boscan crude oil. Sep Sci Technol 1982;17(6):797–819.

[59] Doukkali A, Saoiabi A, Zrineh A, Hamad M, Ferhat M, Barbe JM, et al. Separation and identification of petroporphyrins extracted from the oil shales of Tarfaya: geochemical study. Fuel 2002;81(4):467–72.

[60] Millson MF, Montgomery DS, Brown SR. An investigation of the vanadyl porphyrin complexes of the athabasca oil sands. Geochim Cosmochim Acta 1966;30(2):207–21.

[61] Xu H, Lesage S. Separation of vanadyl and nickel petroporphyrins on an aminopropyl column by high-performance liquid chromatography. J Chromatogr A 1992;607(1):139–44.

[62] Barwise AJG, Whitehead EV. Separation process. Patent No. British Application No. 35626/78. 1979. Withdrawn.

[63] Howe WW. Improved chromatographic analysis of petroleum porphyrin aggregates and quantitative measurement by integral absorption. Anal Chem 1961;33(2):255–60.

[64] Green JB, Hoff RJ, Woodward PW, Stevens LL. Separation of liquid fossil fuels into acid, base and neutral concentrates: 1. An improved nonaqueous ion exchange method. Fuel 1984;63(9):1290–301.

[65] Pearson CD, Green JB. Vanadium and nickel complexes in petroleum resid acid, base, and neutral fractions. Energy Fuels 1993;7(3):338–46.

[66] Pearson CD, Green JB. Vanadium and nickel complexes in petroleum resid acid, base, and neutral fractions. Report No. NIPER-536. Bartlesville, OK, USA: National Institute for Petroleum and Energy Research; January 1993. 32 pp.

[67] Gill JP, Evershed RP, Eglinton G. Comparative computerised gas chromatographic-mass spectrometric analysis of petroporphyrins. J Chromatogr A 1986;369(C):281–312.

[68] Gill JP, Evershed RP, Chicarelli MI, Wolff GA, Maxwell JR, Eglinton G. Computerised capillary gas chromatographic-mass spectrometric studies of the petroporphyrins of the gilsonite bitumen (Eocene, U.S.A.). J Chromatogr A 1985;350(C):37–62.

[69] Marriott PJ, Gill JP, Evershed RP, Hein CS, Eglinton G. Computerised gas chromatographic-mass spectrometric analysis of complex mixtures of alkyl porphyrins. J Chromatogr A 1984;301(C):107–28.

[70] Fish RH, Komlenic JJ. Molecular characterization and profile identifications of vanadyl compounds in heavy crude petroleums by liquid chromatography/graphite furnace atomic absorption spectrometry. Anal Chem 1984;56(3):510–7.

[71] Marquez N, Ysambertt F, De La Cruz C. Three analytical methods to isolate and characterize vanadium and nickel porphyrins from heavy crude oil. Anal Chim Acta 1999;395(3):343–9.
[72] Reynolds JG, Biggs WR. Analysis of residuum demetalation by size exclusion chromatography with element specific detection. Fuel Sci Technol Intern 1986;4(5):593–619.
[73] Rosscup RJ, Pohlmann HP. Fractionation and characterization of petroleum residues using gel permeation chromatography. I. Molecular size distribution of hydrocarbons and metalloporphyrins in petroleum. ACS Div Pet Chem 1967;12(2):A103–9.
[74] Barbooti MM, Said EZ, Hassan EB, Abdul-Ridha SM. Separation and spectrophotometric investigations of the distribution of nickel and vanadium in heavy crude oils. Fuel 1989;68(1):84–7.
[75] Nalwaya V, Tangtayakom V, Piumsomboon P, Fogler S. Studies on asphaltenes through analysis of polar fractions. Ind Eng Chem Res 1999;38(3):964–72.
[76] Saraceno AJ, Fanale DT, Coggeshall ND. An electron paramagnetic resonance investigation of vanadium in petroleum oils. Anal Chem 1961;33(4):500–5.
[77] Baker EW. Mass spectrometric characterization of petroporphyrins. JACS 1966;88(10):2311–5.
[78] Gilinskaya LG. EPR spectra of V(IV) complexes and the structure of oil porphyrins. J Struct Chem 2008;49(2):245–54.
[79] Kovats E. Gas-chromatographische charakterisierung organischer verbindungen. Teil 1: Retentionsindices aliphatischer halogenide, alkohole, aldehyde und ketone. Helv Chim Acta 1958;41(7):1915–32.
[80] Hajibrahim SK, Quirke JME, Eglinton G. Petroporphyrins. V. Structurally-related porphyrin series in bitumens, shales and petroleums – evidence from HPLC and mass spectrometry. Chem Geol 1981;32(1–4):173–88.
[81] Qian K, Mennito AS, Edwards KE, Ferrughelli DT. Observation of vanadyl porphyrins and sulfur-containing vanadyl porphyrins in a petroleum asphaltene by atmospheric pressure photonionization fourier transform ion cyclotron resonance mass spectrometry. Rapid Commun Mass Spectrom 2008;22(14):2153–60.
[82] Reynolds JG. Trace metals in heavy crude oils and tar sand bitumens. In: Proc. Eastern Oil Shale Symp. Lexington, Kentucky; November 6–8, 1990. 15 pp.
[83] Kekäläinen T, Pakarinen JMH, Wickström K, Lobodin VV, Mckenna AM, Jänis J. Compositional analysis of oil residues by ultrahigh-resolution fourier transform ion cyclotron resonance mass spectrometry. Energy Fuels 2013;27(4):2002–9.
[84] Xu H, Yu D, Que G. Characterization of petroporphyrins in Gudao residue by ultraviolet-visible spectrophotometry and laser desorption ionization-time of flight mass spectrometry. Fuel 2005;84(6):647–52.
[85] Xu H, Que G, Yu D, Lu JR. Characterization of petroporphyrins using ultraviolet-visible spectroscopy and laser desorption ionization time-of-flight mass spectrometry. Energy Fuels 2005;19(2):517–24.
[86] Dechaine GP. Solubility and diffusion of vanadium compounds and asphaltene aggregates [Ph.D. thesis] from University of Alberta, Department of Chem. Mat. Eng. January 26, 2010. 246 pp.
[87] Freeman DH, Saint Martin DC, Boreham CJ. Identification of metalloporphyrins by third-derivative UV/vis diode array spectroscopy. Energy Fuels 1993;7(2):194–9.
[88] Rankin JG, Cantu R, Czernuszewicz RS, Lash TD. Fingerprinting petroporphyrin structures with vibrational spectroscopy. Part 5. Structural influences of the porphyrin 13-alkyl substituent on Resonance raman scattering from nickel(II) cycloalkanoporphyrins. Org Geochem 1999;30(4):201–28.
[89] Ferrer EG, Baran EJ. Electronic and photoelectron spectra of vanadyl(IV) tetraphenylporphyrin. J Electron Spectrosc 1991;57(2):189–97.
[90] White WI. Aggregation of porphyrins and metalloporphyrins. In: Dolphin D, editor. The porphyrins. NewYork (NY, USA): Academic Press; 1978. p. 303–39.
[91] Walker FA, Hui E, Walker JM. Electronic effects in transition metal porphyrins. I. Reaction of piperidine with a series of para- and meta-substituted nickel(II) and vanadium(IV) tetraphenylporphyrins. JACS 1975;97(9):2390–7.

[92] Bencosme CS, Romero C, Simoni S. Axial interaction of vanadyltetraphenylporhyrin with lewis bases. Inorg Chem 1985;24(10):1603–4.
[93] Bonnett R, Brewer P, Noro K, Noro T. Chemistry of vanadyl porphyrins. Tetrahedron 1978;34(3):379–85.
[94] Sugihara JM, Bean RM. Direct determination of metalloporphyrins in Boscan crude oil. J Chem Eng Data 1962;7(2):269–71.
[95] Gouterman M. Optical spectra and electronic structure of porphyrins and related rings. In: Dolphin D, editor. The porphyrins. Academic Press; 1978. p. 1–165.
[96] Rankin JG, Czernuszewicz RS. Fingerprinting petroporphyrin structures with vibrational spectroscopy. 1. Resonance raman spectra of nickel and vanadyl etioporphyrins I and III. Org Geochem 1993;20(5):521–38.
[97] Rankin JG, Czernuszewicz RS, Lash TD. Fingerprinting petroporphyrin structures with vibrational spectroscopy. 2. Resonance raman marker bands for the exocyclic rings of nickel tetrahydrobenzoporphyrins and nickel cycloalkanoporphyrins. Org Geochem 1995;23(5):419–27.
[98] Rankin JG, Czernuszewicz RS, Lash TD. Fingerprinting petroporphyrin structures with vibrational spectroscopy. 3. Resonance raman characterization of regioisomers of nickel(II) tetrahydrobenzoporphyrin. Inorg Chem 1995;34(11):3025–37.
[99] Czernuszewicz RS, Rankin JG, Lash TD. Fingerprinting petroporphyrin structures with vibrational spectroscopy. 4. Resonance raman spectra of nickel(II) cycloalkanoporphyrins: structural effects due to exocyclic ring size. Inorg Chem 1996;35(1):199–209.
[100] Czader A, Czernuszewicz RS. Fingerprinting petroporphyrin structures with vibrational spectroscopy. Part 7. Calculations using density functional theory of the molecular structures and structure-sensitive vibrational modes of type II nickel(II) cycloalkanoporphyrins. Org Geochem 2007;38(2):250–66.
[101] Boggess JM, Czernuszewicz RS, Lash TD. Fingerprinting petroporphyrin structures with vibrational spectroscopy. Part 6: resonance raman characterization of regioisomers of nickel(II) benzoetioporphyrin. Org Geochem 2002;33(9):1111–26.
[102] Mckenna AM, Purcell JM, Rodgers RP, Marshall AG. Identification of vanadyl porphyrins in a heavy crude oil and raw asphaltene by atmospheric pressure photoionization Fourier transform ion cyclotron resonance (FT-ICR) mass spectrometry. Energy Fuels 2009;23(4):2122–8.
[103] Mckenna AM, Williams JT, Putman JC, Aeppli C, Reddy CM, Valentine DL, et al. Unprecedented ultrahigh resolution FT-ICR mass spectrometry and parts-per-billion mass accuracy enable direct characterization of nickel and vanadyl porphyrins in petroleum from natural seeps. Energy Fuels 2014;28(4):2454–64.
[104] Zhao X, Liu Y, Xu C, Yan Y, Zhang Y, Zhang Q, et al. Separation and characterization of vanadyl porphyrins in Venezuela Orinoco heavy crude oil. Energy Fuels 2013;27(6):2874–82.
[105] Goulon J, Esselin C, Friant P, Berthe C, Muller JF, Poncet J-L, et al. Structural characterization by X-ray absorption spectroscopy (EXAFS/XANES) of the vanadium chemical environment in various asphaltenes. In: Proc. Intern. Symp. on Characterization of Heavy Crudes Pet. Residues, Lyon, Francevol. 40. Technip; 1984. p. 158–63.
[106] Goulon J, Retournard A, Friant P, Goulon-Ginet C, Berthe C, Muller JF, et al. Structural characterization by X-ray absorption spectroscopy (EXAFS/XANES) of the vanadium chemical environment in Boscan asphaltenes. J Chem Soc Dalton Trans 1984:1095–103.
[107] Biggs WR, Brown RJ, Fetzer JC, Reynolds JG. Characterization of vanadium compounds in selected crudes. I. Porphyrin and non-porphyrin separation. Liq Fuels Technol 1985;3:397–422.
[108] Reynolds JG, Biggs WR, Fetzer JC. Characterization of vanadium compounds in selected crudes. II. Electron paramagnetic resonance studies of the first coordination spheres in porphyrin and non-porphyrin fractions. Liq Fuels Technol 1985;3(4):423–48.
[109] Reynolds JG, Gallegos EJ, Fish RH, Komlenic JJ. Characterization of the binding sites of vanadium compounds in heavy crude petroleum extracts by electron paramagnetic resonance spectroscopy. Energy Fuels 1987;1(1):36–44.

[110] Reynolds JG, Biggs WR. Application of size exclusion chromatography coupled with element-specific detection to the study of heavy crude oil and residua processing. Acc Chem Res 1988;21:319–26.
[111] Dickson FE. Use of electron spin resonance to characterize the vanadium(IV)-sulfur species in petroleum. Anal Chem 1972;44(6):978–81.
[112] Dickson FE, Petrakis L. Application of electron spin resonance and electronic spectroscopy to the characterization of vanadium species in petroleum fractions. Anal Chem 1974;46(8):1129–30.
[113] Fish RH, Reynolds JG, Gallegos EJ. Molecular characterization of nickel and vanadium nonporphyrin compounds found in heavy crude petroleums and bitumens. In: Proc. 191st ACS Nat. Meeting, Metal Complexes in Fossil Fuels: Geochem., Characterization Symp. New York, NY, USA; April 13–18, 1987. p. 332–49.
[114] Malhotra VM, Buckmaster HA. 34 GHz EPR study of vanadyl complexes in various asphaltenes. Statistical correlative model of the coordinating ligands. Fuel 1985;64(3):335–41.
[115] Espinosa M, Campero A, Salcedo R. Electron spin resonance and electronic structure of vanadyl-porphyrin in heavy crude oils. Inorg Chem 2001;40(18):4543–9.
[116] Stoklosa HJ, Wasson JR, Mccormick J. Electronic structure of bis(2-methyl-8-quinolinolato) oxovanadium(IV). Inorg Chem 1974;13(3):592–7.
[117] Vigee G, Selbin J. Oxovanadium(IV) complexes of dithiocarbamates. J Inorg Nucl Chem 1969;31(10):3187–94.
[118] Zhang G, Boduszynski MM. The binding structure of vanadium in Arabian heavy petroleum: an X-ray absorption investigation. In: Proc. Intern. Conf. Pet. Refin. Petrochem. Proc., Beijing, China, vol. 2. International Academic Publishers; September 11–15, 1991. p. 821–6.
[119] Miller JT, Fisher RB, Van Der Eerden AMJ, Koningsberger DC. Structural determination by XAFS spectroscopy of non-porphyrin nickel and vanadium in Maya residuum, hydrocracked residuum, and toluene-insoluble solid. Energy Fuels 1999;13(3):719–27.
[120] Berthe C, Muller JF, Cagniant D, Grimblot J, Bonnelle JP. Characterisation des fractions lourdes du pétrole par spectroscopie IR, UV-visible, spectroscopie photoélectronique esca et spectrométrie lamma en relation avec l'environnement du vanadium. In: Proc. Characterization of Heavy Crude Oils Pet. Residues: Intern. Symp. on Characterization of Heavy Crude Oils Pet. Residues. Lyon, France; June 25–27, 1984. p. 164–8.
[121] Andreatta G, Goncalves CC, Buffin G, Bostrom N, Quintella CM, Arteaga-Larios F, et al. Nanoaggregates and structure-function relations in asphaltenes. Energy Fuels 2005;19:1282–9.
[122] Leon O, Rogel E, Espidel J, Torres G. Asphaltenes: structural characterization, self-association, and stability behavior. Energy Fuels 2000;14(1):6–10.
[123] Siskin M, Kelemen SR, Eppig CP, Brown LD, Afeworki M. Asphaltene molecular structure and chemical influences on the morphology of coke produced in delayed coking. Energy Fuels 2006;20(3):1227–34.
[124] Strausz OP, Mojelsky TW, Lown EM. The molecular structure of asphaltene: an unfolding story. Fuel 1992;71:1355–63.
[125] Branthaver JF. Influence of metal complexes in fossil fuels on industrial operations. In: Filby RH, Branthaver JF, editors. Metal complexes in fossil fuels. Washington (DC): ACS; 1987. p. 188–204.
[126] Zhao X, Shi Q, Gray MR, Xu C. New vanadium compounds in Venezuela heavy crude oil detected by positive-ion electrospray ionization fourier transform ion cyclotron resonance mass spectrometry. Sci Rep 2014;4:6.
[127] Podobaeva TP, Gonsales MG, Polyakova AA, Kogan LO, Vasilenko GV. Determination of metal porphyrins in asphaltenes. Chem Technol Fuels Oils Transl Khim Tekhnol Topl Masel 1980;16(5):351–3.
[128] Zaki NS, Barbooti MM, Baha-Uddin SS, Hassan EB. Determination of trace metals and their distribution in heavy crude oil distillates (350°C+) by atomic absorption spectrophotometry. Appl Spect 1989;43(7):1257–9.

[129] Lord III CJ. Determination of trace metals in crude oil by inductively coupled plasma mass spectrometry with microemulsion sample introduction. Anal Chem 1991;63(15):1594–9.

[130] Amorim Fa C, Welz B, Costa ACS, Lepri FG, Vale MGR, Ferreira SLC. Determination of vanadium in petroleum and petroleum products using atomic spectrometric techniques. Talanta 2007;72(2):349–59.

[131] Duyck C, Miekeley N, Porto Da Silveira CL, Aucélio RQ, Campos RC, Grinberg P, et al. The determination of trace elements in crude oil and its heavy fractions by atomic spectrometry. Spectrochim Acta Part B At Spect 2007;62(9):939–51.

[132] Caumette G, Lienemann C-P, Merdrignac I, Bouyssiere B, Lobinski R. Element speciation analysis of petroleum and related materials. J Anal At Spectrom 2009;24(3):263–76.

[133] Khuhawar MY, Mirza MA, Jahangir TM. Determination of metal ions in crude oils. In: Abdel-Raouf ME-S, editor. Crude oil emulsions – composition stability characterization; 2012. p. 121–44. Chap. 7: Rijeka, Croatia. Available from: http://www.intechopen.com/books/crude-oil-emulsions-composition-stability-and-characterization/determination-of-metal-ions-in-crude-oils.

[134] Senglet N, Williams C, Faure D, Des Courières T, Guilard R. Micro-heterogeneity study of heavy crude petroleum by U.V.-visible spectroscopy and small angle X-ray scattering. Fuel 1990;69(1):72–7.

[135] Ugarkovic D, Premerl D. Trace-metal distribution in Moslavina basin crude oil and oil products. Fuel 1987;66(10):1431–5.

[136] Pina A, Mougin P, Béha E. Characterisation of asphaltenes and modelling of flocculation – state of the art. Oil Gas Sci Technol Rev IFP 2006;61(3):319–43.

[137] Boduszynski MM. Composition of heavy petroleums. 1. Molecular weight, hydrogen deficiency, and heteroatom concentration as a function of atmospheric equivalent boiling point up to 1400°F (760°C). Energy Fuels 1987;1(1):2–11.

[138] Hausler DW. Molecular size distribution of specific elements in petroleum crudes and 650 F + residua by size exclusion chromatography with inductively coupled plasma spectrometry detection. Spectrochim Acta Part B At Spect 1985;40(1–2):389–96.

[139] Muhammad I, Tijjani N, Dioha IJ, Musa A, Sale H, Lawal AM. SARA separation and determination of concentration levels of some heavy metals in organic fractions of nigerian crude oil. Chem Mat Res 2013;3(4):7–14.

[140] Loos M, Ascone I, Friant P, Ruiz-Lopez MF, Goulon J, Barbe JM, et al. Vanadyl porphyrins: evidence for self-association and for specific interactions with hydroprocessing catalysts shown from XAFS and ESR studies. Catal Today 1990;7(4):497–513.

[141] Lemtur A, Chakravorty BK, Dhar TK, Subramanian J. Electron spin resonance of dimeric species from the radical cations of vanadyl porphyrins. J Phys Chem 1984;88(23):5603–8.

[142] Banci L. Homo- and heterodimer formation in metalloporphyrins. Inorg Chem 1985;24(5):782–6.

[143] Thanabal V, Krishnan V. Cation-induced crown porphyrin dimers of oxovanadium(IV). Inorg Chem 1982;21(10):3606–13.

[144] Maiya GB, Krishnan V. Electrochemical redox properties and spectral features of supermolecular porphyrins. Inorg Chem 1985;24(20):3253–7.

[145] Rodgers RP, Hendrickson CL, Emmett MR, Marshall AG, Greaney M, Qian K. Molecular characterization of petroporphyrins in crude oil by electrospray ionization fourier transform ion cyclotron resonance mass spectrometry. Can J Chem 2001;79(5–6):546–51.

[146] Larson OA, Beuther H. Processing aspects of vanadium and nickel in crude oils. ACS Div Pet Chem 1966;11(2):B95–104.

[147] Knozinger H, Cordischi D, Vielhaber B. Adsorption of oxovanadium(IV) porphyrins on oxides and molybdenum-containing catalysts. Catal Today 1990;7(4):447–66.

[148] Mitchell PCH, Scott CE. Interaction of vanadium and nickel porphyrins with catalysts, relevance to catalytic demetallization. Catal Today 1990;7(4):467–77.

[149] Acevedo S, Ranaudo MA, Escobar G, Gutierrez LB, Gutierrez X. A unified view of the colloidal nature of asphaltenes. In: Sheu EY, Mullins OC, editors. Asphaltene constituents: fundamentals applications. New York: Plenum Press; 1995. p. 131–54.

[150] Cantu R, Stencel JR, Czernuszewicz RS, Jaffe PR, Lash TD. Surfactant-enhanced partitioning of nickel and vanadyl deoxophylloerythroetioporphyrins from crude oil into water and their analysis using surface-enhanced resonance raman spectroscopy. Environ Sci Technol 2000;34(1):192–8.
[151] Biggs WR, Brown RJ, Fetzer JC. Elemental profiles of hydrocarbon materials by size-exclusion chromatography/inductively coupled plasma atomic emission spectrometry. Energy Fuels 1987;1(3):257–62.
[152] Vale MGR, Damin ICF, Klassen A, Silva MM, Welz B, Silva AF, et al. Method development for the determination of nickel in petroleum using line-source and high-resolution continuum-source graphite furnace atomic absorption spectrometry. Microchem J 2004;77(2):131–40.
[153] Damin ICF, Vale MGR, Silva MM, Welz B, Lepri FG, Dos Santos WNL, et al. Palladium as chemical modifier for the stabilization of volatile nickel and vanadium compounds in crude oil using graphite furnace atomic absorption spectrometry. J Anal At Spectrom 2005;20(12):1332–6.
[154] Lepri FG, Welz B, Borges DLG, Silva AF, Vale MGR, Heitmann U. Speciation analysis of volatile and non-volatile vanadium compounds in Brazilian crude oils using high-resolution continuum source graphite furnace atomic absorption spectrometry. Anal Chim Acta 2006;558(1–2):195–200.
[155] Vale MGR, Silva MM, Damin ICF, Sanches Filho PJ, Welz B. Determination of volatile and non-volatile nickel and vanadium compounds in crude oil using electrothermal atomic absorption spectrometry after oil fractionation into saturates, aromatics, resins and asphaltenes. Talanta 2008;74(5):1385–91.
[156] Kaminski TJ, Fogler HS, Wolf N, Wattana P, Mairal A. Classification of asphaltenes via fractionation and the effect of heteroatom content on dissolution kinetics. Energy Fuels 2000;14(1):25–30.
[157] Dechaine GP, Gray MR. Chemistry and association of vanadium compounds in heavy oil and bitumen, and implications for their selective removal. Energy Fuels 2010;24(5):2795–808.
[158] Moore J, Dunning H. Interfacial activities and porphyrin contents of oil-shale extracts. Ind Eng Chem 1955;47:1440–4.
[159] Dunning HN, Moore JW, Myers AT. Properties of porphyrins in petroleum. Ind Eng Chem 1954;46(9):2000–7.
[160] Dunning HN, Moore JW, Denekas MO. Interfacial activities and porphyrin contents of petroleum extracts. Ind Eng Chem 1953;45(8):1759–65.
[161] Dunning HN, Moore JW. Decomposition of metal-porphyrin complexes by gamma irradiation. Ind Eng Chem Res 1959;51:161–4.
[162] Dunning H, Moore J. Porphyrin research and origin of petroleum. Bull Amer Assoc Pet Geol 1957;41:2403–12.
[163] Dunning HN, Raban NA. Porphyrin-metal complexes in petroleum stocks. Ind Eng Chem 1956;48(5):951–5.
[164] Tynan EC, Yen TF. ESR study of vanadium occurring in petroleum asphaltenes in solution. ACS Div Pet Chem Prepr 1967;12(2):A89–101.
[165] Sakanishi K, Yamashita N, Whitehurst DD, Mochida I. Depolymerization and demetallation treatments of asphaltene in vacuum residue. Catal Today 1998;43(3–4):241–7.
[166] Sakanishi K, Saito I, Watanabe I, Mochida I. Dissolution and demetallation treatment of asphaltene in resid using adsorbent and oil-soluble Mo complex. Fuel 2004;83(14–15):1889–93.
[167] Chen FF, Xu ZM, Sun XW, Zhao SQ. Adsorption kinetics of nickel etioporphyrin on VTB-asphaltene in N-pentane solution. Adv Mat Res 2013;781–784:24–9.
[168] Chen F, Liu Q, Xu Z, Sun X, Shi Q, Zhao S. Adsorption kinetics and thermodynamics of vanadyl etioporphyrin on asphaltene in pentane. Energy Fuels 2013;27(11):6408–18.
[169] Mujica V, Nieto P, Puerta L, Acevedo S. Caging of molecules by asphaltenes. A model for free radical preservation in crude oils. Energy Fuels 2000;14(3):632–9.
[170] Danil'yan TD, Rogacheva OV, Unger FG, Gimaev RN, Akhmetov SA. Influence of the paramagnetism of asphaltenes on their phase state in dispersed petroleum systems. Colloid J USSR 1987;49(4):666–70.
[171] Yudin IK, Nikolaenko GL, Gorodetskii EE, Markhashov EL, Frot D, Briolant Y, et al. Universal behavior of asphaltene aggregation in hydrocarbon solutions. Pet Sci Technol 1998;16(3–4):395–414.

[172] Yen TF, Erdman JG, Pollack SS. Investigation of the structure of petroleum asphaltenes by X-Ray diffraction. Anal Chem 1961;33(11):1587–94.
[173] Dickie JP, Yen TF. Macrostructures of the asphaltic fractions by various instrumental methods. Anal Chem 1967;39(14):1847–52.
[174] Dickie JP, Haller MN, Yen TF. Electron microscopic investigations on the nature of petroleum asphaltics. J Colloid Interface Sci 1969;29(3):475–84.
[175] Yin C-X, Tan X, Mullen K, Stryker JM, Gray MR. Associative $\pi - \pi$ interactions of condensed aromatic compounds with vanadyl or nickel porphyrin complexes are not observed in the organic phase. Energy Fuels 2008;22(4):2465–9.
[176] Lee RF. Agents which promote and stabilize water-in-oil emulsions. Spill Sci Technol Bull 1999;5(2):117–26.
[177] Canevari GP. The formulation of an effective emulsifier for oil spill emulsions. Mar Pollut Bull 1982;13:49–54.
[178] Lee RF. Isolation and identification of compounds and mixture which promote and stabilize water-in-oil emulsions. Report No. MSRC 95-002. Washington (DC, USA): Marine Spill Response Corporation; 1995. 62 pp.
[179] Lou SS, Zuo Y, Zhao JM. Removal of metallic impurities from residue oil. Petrochem Technol 2002;31(5):357–60.
[180] Gould KA, Long RB. A new technique for the acid/base separation of petroleum and coal-derived fractions. Fuel 1986;65(4):572–6.
[181] Ovalles C, Rojas I, Acevedo S, Escobar G, Jorge G, Gutierrez LB, et al. Upgrading of Orinoco belt crude oil and its fractions by an electrochemical system in the presence of protonating agents. Fuel Proc Technol 1996;48(2):159–72.
[182] Pena ME, Manjarrez A, Campero A. Distribution of vanadyl porphyrins in heavy crude oil a Mexican offshore. Fuel Proc Technol 1996;46:171–82.
[183] Ali MF, Perzanowski H, Bukhari A, Al-Haji AA. Nickel and vanadyl porphyrins in Saudi Arabian crude oils. Energy Fuels 1993;7(2):179–84.
[184] Izquierdo A, Carbognani L, Leon V, Parisi A. Characterization of Venezuelan heavy oil vacuum residua. Fuel Sci Technol Intern 1989;7(5–6):561–70.
[185] Galiasso R, Garcia J, Caprioli L, Pazos JM, Soto A. Reactions of porphyrinic and non porphyrinic molecules during hydrodemetalation of heavy crude oils. ACS Div Pet Chem Prepr 1985;30(1):50–61.
[186] Barbier J, Lienemann C-P, Le Masle A, Chatron-Michaud P, Guichard B, Digne M. New insights into resid desulfurization processes: molecular size dependence of catalytic performances quantified by size exclusion chromatography-ICP/MS. Energy Fuels 2013;27(11):6567–74.
[187] Barbier J, Marques J, Caumette G, Merdrignac I, Bouyssiere B, Lobinski R, et al. Monitoring the behaviour and fate of nickel and vanadium complexes during vacuum residue hydrotreatment and fraction separation. Fuel Proc Technol 2014;119:185–9.
[188] Reynolds JG, Biggs WR, Bezman SA. Reaction sequence of metallopetroporphyrins during heavy residuum upgrading. In: Metal complexes in fossil fuels. ACS symposium series. Washington (DC): ACS; 1987. p. 205–19.
[189] Harnett CG. Heavy oil processing. Proc Refin Dep Am Pet Inst 1972;52:703–18.
[190] Johnson CA, Maruhnic P, Nongbri G, Wolk RH. Catalyst development for demetallization of petroleum residua. In: Proc. AIChE Nat. Meeting, Houston, TX, USA; March 16–20, 1975. A, 347 (1 p.).
[191] Weekman VW. Hydroprocessing reaction engineering. In: Proc. 4th Intern. Symp. Chem. React. Eng., Heidelberg, Germany; April 6–9, 1976. p. 615–46.
[192] Kobayashi S, Kushiyama S, Aizawa R, Koinuma Y, Inoue K, Shimizu Y, et al. Kinetic study on the hydrotreating of heavy oil. 1. Effect of catalyst pellet size in relation to pore size. Ind Eng Chem Res 1987;26(11):2241–5.
[193] Kobayashi S, Kushiyama S, Aizawa R, Koinuma Y, Inoue K, Shimizu Y, et al. Kinetic study on the hydrotreating of heavy oil. 2. Effect of catalyst pore size. Ind Eng Chem Res 1987;26(11):2245–50.

[194] Heck RH, Degnan TF. Fundamentals of resid upgrading. In: AIChE Symp. Series, vol. 85 No. 273. New York: American Institute of Chemical Engineers; 1989. 77 pp.
[195] Le Page JF, Chatila SG, Davidson M. Raffinage et conversion des produits lourds du petrole. Paris: Editions Technip; 1990. 190 pp.
[196] Adarme R, Sughrue EL, Johnson MM, Kidd DR, Phillips MD, Shaw JE. Demetallization of asphaltenes: thermal and catalytic effects with small-pore catalysts. In: Proc. ACS Nat. Meeting, Div. Pet. Chem. Symp. on Resid Upgrading, Washington, DC, USA, vol. 35; August 26–31, 1990. p. 614–8.
[197] Gray RM. Upgrading petroleum residues and heavy oils. Chem. industries series, vol. 56., New York: Marcel Dekker/CRC Press; 1994. 368 pp.
[198] Ingram C. Heavy oil upgrading. Report No. DOE/MT/95014-2. Atlanta (GA, USA): Clark Atlanta University; September 30, 1996. 17 pp.
[199] Speight JG. The desulfurization of heavy oils and residua. New York (NY, USA): Marcel Dekker, CRC Press; 1999. 458 pp.
[200] Rahimi PM, Gentzis T. The chemistry of bitumen and heavy oil processing. In: Hsu CS, Robinson PR, editors. Practical Adv. Pet. Proc. New York: Springer; 2006. p. 597–634.
[201] Ali MF, Abbas S. A review of methods for the demetallization of residual fuel oils. Fuel Proc Technol 2006;87(7):573–84.
[202] Delbianco A, Montanari R. Emerging technologies for the conversion of residues. In: Treccani P, editor. Encyclopaedia of hydrocarbons. Roma: Istituto Della Enciclopedia Italiana, Fondata Da Giovanni Treccani S.P.A., Marchesi Grafiche Editoriali S.P.A.; 2007. p. 137–70.
[203] Treccani P. Emerging technologies for the conversion of residues. In: Poli R, editor. Encyclopaedia of hydrocarbons. Roma (Italy): Istituto Della Enciclopedia Italiana; 2008. p. 137–60.
[204] Speight JG. The refinery of the future. Amsterdam: Elsevier; 2010. 416 pp.
[205] Fahim MA, Alsahhaf TA, Elkilani A. Fundamentals of petroleum refining. Oxford (Great Britain): Elsevier; 2010. 496 pp.
[206] Abbas S, Maqsood ZT, Ali MF. The demetallization of residual fuel oil and petroleum residue. Pet Sci Technol 2010;28(17):1770–7.
[207] Magomedov RN, Popova AZ, Maryutina TA, Kadiev KM, Khadzhiev SN. Current status and prospects of demetallization of heavy petroleum feedstock (review). Pet Chem 2015;55(6):423–43.
[208] Fujita K, Abe S, Inoue Y. New developments in resid hydroprocessing. PTQ 2001;6(4):51–8.
[209] Bonne RLC, Van Steenderen P, Moulijn JA. Hydrogenation of nickel and vanadyl tetraphenylporphyrin in absence of a catalyst: a kinetic study. Appl Catal A General 2001;206(2):171–81.
[210] Pearson CD, Green JB. Comparison of processing characteristics of mayan and wilmington heavy residues: 1. Acid-base-neutral fractionation and characterization. Fuel 1989;68(4):456–64.
[211] Pearson CD, Green JB. Comparison of processing characteristics of Mayan and Wilmington heavy residues: 2. Characterization of vanadium and nickel complexes in acid-base-neutral fractions. Fuel 1989;68(4):465–74.
[212] Janssens JP, Xu Y, Witte G, Van Langeveld AD, Sie ST, et al. Competitive effects of hetero-atom containing compounds in the hydrodemetallisation of vanadyl-tetraphenyl-porphyrin. Fuel 1998;77(12):1367–75.
[213] Ancheyta J, Centeno G, Trejo F, Marroquin G. Changes in asphaltene properties during hydrotreating of heavy crudes. Energy Fuels 2002;17:1223–8.
[214] Ware JC, Schramm CM. NMR analysis of heavy crude oil resids. ACS Div Fuel Chem 1985;30(4):104–12.
[215] Asaoka S, Nakata S, Shiroto Y, Takeuchi C. Asphaltene cracking in catalytic hydrotreating of heavy oils. 2. Study of changes in asphaltene structure during catalytic hydroprocessing. Ind Eng Chem Proc Des Dev 1983;22(2):242–8.
[216] Takeuchi C, Fukui Y, Nakamura M, Shiroto Y. Asphaltene cracking in catalytic hydrotreating of heavy oils. 1. Processing of heavy oils by catalytic hydroprocessing and solvent deasphalting. Ind Eng Chem Proc Des Dev 1983;22(2):236–42.

[217] Asaoka S, Nakata S, Shiroto Y, Takeuchi C. The characteristics of metal complexes before and after hydrotreating. ACS Div Pet Chem 1986;31(2):597−605.
[218] Onukwuli OD, Onyia IM, Ekumankama EO, Okeke SI. Solvent demetallization of atmospheric and vacuum residues. Pet Sci Technol 1999;17(1 & 2):37−49.
[219] Reynolds JG. Removal of nickel and vanadium from heavy crude oils by exchange reactions. In: Proc. 227th ACS Nat. Meeting, Anaheim, CA, USA, vol. 49; March 28−April 1, 2004. p. 79−80.
[220] Jones EL, Reynolds JG, Huffman JC, Christou G. Deoxygenation of oxovanadium(IV) complexes under mild conditions: synthesis and structural characterization of cis- dihalobis(diakyldithiocarbamato) vanadium(IV). Polyhedron 1991;10(15):1817−25.
[221] Reynolds JG, Christou G, Huffman JC. Removal of vanadium in heavy crude oil refining. In: Proc. 205th ACS Nat. Meeting, Denver, Co, USA; March 31−April 2, 1993. 1951 (1 pp.).
[222] Reynolds JG, Jones EL, Huffman JC, Christou G. Deoxygenation of oxovanadium(IV) complexes under mild conditions : conversion of vanadyl species to the corresponding dihalides with carboxylic acid halides. Polyhedron 1993;12(4):407−14.
[223] Richard P, Poncet JL, Barbe JM, Guilard R, Goulon J, Rinaldi D, et al. Vanadium(IV) porphyrins: synthesis and spectrochemical characterization of dihalogenovanadium(IV) porphyrins. Extended X-ray absorption fine structural study of dibromo(2,3,7,8,12,13,17,18-octaethylporphyrinato) vanadium(IV). J Chem Soc Dalton Trans 1982;(8):1451−6.
[224] Schumann H. Spectroscopy, molecular structure, electrochemistry, and reactivity of vanadium(IV,V) imido complexes of 5,7,12,14-tetramethyldibenzo[b.i][1,4,8,11]tetraazacyclotetradecine. Inorg Chem 1996;35(7):1808−13.
[225] Yin CX, Stryker JM, Gray MR. Separation of petroporphyrins from asphaltenes by chemical modification and selective affinity chromatography. Energy Fuels 2009;23(5):2600−5.
[226] Guedes CLB, Di Mauro E, De Campos A, Mazzochin LF, Bragagnolo GM, De Melo FA, et al. EPR and fluorescence spectroscopy in the photodegradation study of Arabian and Colombian crude oils. Intern J Photoenergy 2006;2006:6. Article #48462.
[227] Shiraishi Y, Hirai T, Komasawa I. A novel demetalation process for vanadyl- and nickelporphyrins from petroleum residue by photochemical reaction and liquid-liquid extraction. Ind Eng Chem Res 2000;39(5):1345−55.
[228] Wasser PKW, Fuhrhop JH. Photooxygenation of metalloporphyrins and metallochlorins. Ann N Y Acad Sci 1973;206:533−48.
[229] Maxwell JR, Pillinger CT, Eglinton G. Organic geochemistry. Q Rev Chem Soc 1971;25(4):571−628.
[230] Tilley GL, Young DC. Demetallization of residual oils with polyphosphoric acids. Patent No. US3622505. Union Oil Co.; November 23, 1971.
[231] Xu Z, Tan L, Yu L. Method for removing metal from hydrocarbon oil by using inorganic acid. Patent No. CN1431276. China Petrochemical Corp., Peop. Rep. China; Research Institute of Petroleum Processing; July 23, 2003.
[232] Kukes SG, Battiste DR. Demetalization of heavy oils with phosphorous acid. Patent No. US4522702. Phillips Petroleum Co.; June 11, 1985.
[233] Eidem PK. Reducing the metals content of petroleum feedstocks. Patent No. US4752382. Chevron Research Co.; June 21, 1988.
[234] Blytas GC. Demetallizing of heavy hydrocarbons. Patent No. CA1117062. Shell Canada Ltd; January 26, 1982.
[235] Blytas GC. Process for demetallizing heavy hydrocarbons. Patent No. US4116820. Shell Oil Co.; September 26, 1978.
[236] Blytas GC. Removal of metal compounds from hydrocarbons. Patent No. US4048061. Shell Oil Co.; September 13, 1977.
[237] Bearden R, Stuntz GF. Reducing metal content of oil feeds. Patent No. US4537676. Exxon Research and Engineering Co.; August 27, 1985.

[238] Michlmayr MJ. Upgrading metal-contaminated petroleum oils containing vanadium and/or nickel. Patent No. US4039432. Chevron Research Co.; August 2, 1977.
[239] Garwood EP. On-site purification of problem petrolic liquid fuels. Patent No. US3664802. Chevron Res Co.; May 23, 1972.
[240] Young DC. Acid catalyzed oxidative reactions. Patent No. US5105040. Union Oil Co.; April 14, 1992.
[241] Young DC. Demetallizing organometallic compounds using sulphuric acid-urea adducts. Patent No. US5099014. Union Oil Co.; March 24, 1992.
[242] Greaney MA, Polini PJ. Method for demetallating petroleum streams (LAW772). Patent No. US6007705 (Also published as WO0037588, JP2002533527 ,EP1149141, DE69903195, CA2353951, AU2156200). Exxon Research and Engineering Co.; December 28, 1999.
[243] Greaney MA, Bearden R, Kerby MC. Method for decreasing the metals content of petroleum streams. Patent No. US6013176 (Also published as WO0037589, JP2002533528, EP1151059, DE69911654, CN1330698, CA2350707, AU1934200). Exxon Research & Engineering Co.; January 11, 2000.
[244] Kotova AV, Emelyanova SV. Separation of vanadium from crude oils and petroleum products by means of aqueous sulfonic acid solutions. Report No. AD0633830. Washington (DC): Department of the Navy; 1966. 6 pp.
[245] Lerda D, Bolzacchini E, Montanari L. EPR, UV-vis and HPLC studies of the behaviour of vanadyl-mesotetraphenyl-porphyrin in the presence of perfluorinated acids. Fuel 1992;71(10):1153–9.
[246] Gonzalez MC, Rodriguez AR. Preconcentration and recovery of metal ions from petroleum and its by-products using a macroporous ion-exchange resin. Analusis 1990;18(7):433–9.
[247] Sugihara JM, Okada T, Branthaver JF. Reductive desulfuration on vanadium and metalloporphyrin contents of fractions from Boscan asphaltenes. J Chem Eng Data 1965;10:190–4.
[248] Greaney MA, Kerby Jr MC, Olmstead WN, Wiehe IA. Method for demetallating refinery feedstreams. Patent No. US5529684 (Also published as WO9620133, EP0800567). Exxon Research and Engineering Co.; June 25, 1996.
[249] Welter K, Salazar E, Balladores Y, Márquez OP, Márquez J, Martínez Y. Electrochemical removal of metals from crude oil samples. Fuel Proc Technol 2009;90(2):212–21.
[250] Bukowski A, Gurdzinska E, Gardzinski M, Wawszczak K. Removal of'heavy metals from petroleum fractions. Patent No. PL100790. Politechnika Warszawska; November 30, 1978.
[251] Fedorak PM, Semple KM, Vazquez-Duhalt R, Westlake DWS. Chloroperoxidase-mediated modifications of petroporphyrins and asphaltenes. Enzyme Microb Technol 1993;15:429–37.
[252] Vázquez-Duhalt R, Ayala M, Márquez-Rocha FJ. Biocatalytic chlorination of aromatic hydrocarbons by chloroperoxidase of caldariomyces fumago. Phytochemistry 2001;58(6):929–33.
[253] Singh S, Mclean JD, Kilpatrick PK. Fused ring aromatic solvency in destabilizing water-in-asphaltene-heptane-toluene emulsions. J Dispers Sci Technol 1999;20(1 & 2):279–93.
[254] Young DA. Demetallization of petroleum feedstocks with zinc chloride and titanium tetrachloride catalysts. Patent No. US4148717. Union Oil Company of California; April 10, 1979.
[255] Gleim WKT. Hydrorefining metal-containing sour crudes treated with a lewis acid and a molybdenum chloride. Patent No. US3483117. Universal Oil Products Co.; December 9, 1969.
[256] Wolff WF, Rosscup RJ, Ridgway JA. Retardation of coke formation. Patent No. US4419225. Standard Oil Co.; March 7, 1972.
[257] Kukes SG. Demetallization of hydrocarbon containing feed streams. Patent No. US4457835. Phillips Petroleum Co.; July 3, 1984.
[258] Kukes SG, Hawley GR. Demetallization of hydrocarbon containing feed streams. Patent No. US4557824. Phillips Petroleum Co.; December 10, 1985.
[259] Mann DP, Kukes SG, Coombs DM. Metals removal from oils with a light hydrocarbon and an organophosphorous compound. Patent No. US4518484. Phillips Petroleum Co.; May 21, 1985.
[260] Kukes SG. Demetallization of hydrocarbon-containing feed streams with phosphorus compounds. Patent No. US4529503. Phillips Petroleum Co.; July 7, 1985.

[261] Kukes SG, Johnson MM, Nowack GP. Demetallization of heavy oils. Patent No. US4421638. Phillips Petroleum Co; December 20, 1983.

[262] Rankel LA, Rollmann LD. Catalytic activity of metals in petroleum and their removal. Fuel 1983;62(1):44–6.

[263] Gould KA. Oxidative demetallization of petroleum asphaltenes and residue. Fuel 1980;59(10):733–6.

[264] Wen P, Huang Q, Shi B, Yu D, Que G. Research on porphyrin and non-porphyrin nickel distribution in daqing vacuum residue and its cracked residue. In: Proc. ACS Div. Pet. Chem. Preprints, Boston, MA, vol. 47; 2002. p. 295–7.

[265] Tanaka R, Sato E, Hunt JE, Winans RE, Sato S, Takanohashi T. Characterization of asphaltene aggregates using X-ray diffraction and small-angle X-ray scattering. Energy Fuels 2004;18(4):1118–25.

[266] Michael G, Al-Siri M, Khan ZH, Ali FA. Differences in average chemical structures of asphaltene fractions separated from feed and product oils of a mild thermal processing reaction. Energy Fuels 2005;19:1598–605.

[267] Sugihara JM, Branthaver JF, Willcox KW. Oxidative demetallation of oxovanadium(IV) porphyrins. In: Yen TF, editor. The role of trace metals in pet.; 1975. p. 183–93.

[268] Abu-Elgheit MA, Elkholy IS, Barakat AO. Distribution of trace metals in petroleum and its heavy refinery products. J Chem Technol Biotechnol 1981;31(12):760–4 (1979–1982).

[269] Xinru X, Jingyi Y, Ying J, Jinsheng G. Effects of process conditions on desalting and demetalization of crude oil. Pet Sci Technol 2006;24(11):1307–21.

[270] Gafonova OV, Yarranton HW. The stabilization of water-in-hydrocarbon emulsions by asphaltenes and resins. J Colloid Interface Sci 2001;241(2):469–78.

[271] Sztukowski DM, Yarranton HW. Oilfield solids and water-in-oil emulsion stability. J Colloid Interface Sci 2005;285(2):821–33.

[272] Stark JL, Nguyen J, Kremer LN. New method prevents desalter upsets from blending incompatible crudes. Oil Gas J 2002;100(11):89–91.

[273] Lordo SA. New desalting chemistry for heavy/high solids crudes (Unpublished Presentation at Joint Canadian Crude Quality Technical Association/Crude Oil Quality Association Meeting, New Orleans, LA, Februrary 10–11). 2010. p. 32. Available from:, http://www.coqa-inc.org/20100211_Lordo_Solids_in_Crude.pdf.

[274] Eliaz N, Shemesh G, Latanision RM. Hot corrosion in gas turbine components. Eng Fail Anal 2002;9(1):31–43.

[275] Golden SW. Managing vanadium from high metals crude oils. PTQ 2003:26–30.

[276] Pavel SK, Elvin FJ. New FCC Demet® process to produce demetallized spent catalyst with high activity and low hydrothermal deactivation rates. In: Proc. NPRA Annual meeting. San Antonio, TX, USA; March 17–19, 1996. Paper AM-96-45, 18 pp.

[277] Pavel SK. New FCC demetallization procedures. Chem Ind (Dekker) 1998;74:291–336 (Fluid Cracking Catalysts).

[278] Henry E. Cracking catalyst demetallization. Patent No. US3147228. Sinclair Research Inc.; September 1, 1964.

[279] Elvin FJ, Pavel SK. Metals removal of FCC catalyst operating in refinery. Oil Gas J 1991;89(29):95–8.

[280] Nicolin A, Evans M. Trial run of FCCU additive reduces fuel-oil production in heavy feed. Oil Gas J 2000;98(35):44–8.

[281] Krishna AS, Koch RW, Bott DJ. Economics of resid upgrading via delayed coking versus fluid catalytic cracking. In: Proc. AIChE Spring Nat. Meeting. Anaheim, CA, USA; 1984. 1 pp.

[282] Thomas Habib Jr E, Owen H, Synder PW, Streed CW, Venuto PB. Artificially metals-poisoned fluid catalysts. Performance in pilot plant cracking of hydrotreated resid. Ind Eng Chem Prod Res Dev 1977;16(4):291–6.

[283] Mitchell BR. Metal contamination of cracking catalysts. 1. Synthetic metals deposition on fresh catalysts. Ind Eng Chem Prod Res Dev 1980;19(2):209–13.

[284] Ravichander N, Chiranjeevi T, Gokak DT, Voolapalli RK, Choudary NV. FCC catalyst and additive evaluation-a case study. Catal Today 2009;141(1–2):115–9.
[285] Chester AW. Studies on the metal poisoning and metal resistance of zeolitic cracking catalysts. Ind Eng Chem Res 1987;26(5):863–9.
[286] Dougan TJ, Alkemade U, Lakhanpal B, Boock LT. New vanadium trap proven in commercial trails. Oil Gas J 1994:10.
[287] Beck HW, Lochow J, Charles F, Nibert CW. Passivation of vanadium accumulated on catalytic solid fluidizable particles. Patent No. US4515683 (Also published as EP0140007, NO843643, NO167586, JPS6096687, JPS6335197, CA1212367). Ashland Oil Inc.; May 7, 1985.
[288] Hettinger WP. Passivating heavy metals in carbo-metallic oil conversion. Patent No. US4454025. Ashland Oil Inc.; June 12, 1984.
[289] Van Nordstrand RA. Catalytic cracking with sepiolite. Patent No. US4266672. Chevron Research Company; May 12, 1981.
[290] Kennedy JV, Jossens LW. Catalytic cracking process with vanadium passivation and improved. Patent No. US5002653. Chevron Research Company; March 26, 1991.
[291] Mitchell BR, Swift HE. Process for the cracking of high metals content feedstocks. Patent No. US3944482. Gulf Research & Development Co.; March 16, 1976.
[292] Occelli ML, Kennedy JV. Process for cracking high metals content feedstocks. Patent No. US4944865. Chevron Research Co.; July 31, 1990.
[293] Mitchell BR, Vogel RF. Vanadium passivation in a hydrocarbon catalytic cracking process. Patent No. US4451355. Gulf Research & Development Co; May 29, 1984.
[294] Kennedy JV, Kuehelr CW, English AR. Commercialization of the Chevron FCC vanadium trap. In: Proc. INTERCATS' Symp. on additives used in the FCC Proc. Jakarta, Singpore; 1994. p. 19–28.
[295] Catana G, Grünert W, Van Der Voort P, Vansant EF, Schoonheydt RA, Weckhuysen BM. Alox coating of ultrastable zeolite y: a possible method for vanadium passivation of FCC catalysts. J Phys Chem B 2000;104(39):9195–202.
[296] Komvokis V, Keeley C, Challis S. Increase refinery profitability using innovative FCC catalysts to process residue feedstock – part 1: use catalyst fundamentals to select the best catalysts. PTQ Digit Refin 2014:6 (Q3).
[297] Jollien YA, Mayol J, Komvokis V, Keeley C. Multistage reaction catalyst with advanced metals tolerance: the metal trapping capabilities of a catalyst gave a refiner greater flexibility to upgrade heavy residue feed with increased metals content in its RFCC unit. PTQ 2013;18(5):131–7.
[298] Mclean J, Hoffer B, Smith G, Stockwell D, Shackleford A. Multi-stage reaction catalysts. PTQ 2011;16(3):109–17.
[299] Myrstad T. Effect of vanadium on octane numbers in FCC-naphtha. Appl Catal A General 1997;155(1):87–98.
[300] W ET, Demmel EJ, Steves CA, Johnson CR. Enhanced bottoms cracking using a solid particle additive in a commercial FCC unit. In: Proc. NPRA Annual Meeting. San Antonio, TX, USA; March 21–23, 1993. 17 pp.
[301] Mitchell MM, Moore HF, Goolsby TL. Improvement of FCC catalyst performance with bottom- cracking additives. In: Proc. AIChE Spring Nat. Meeting. Orlando, FL, USA; March 18–22, 1990. 9 pp.
[302] Evans M, Brown M. Commercial experience with BCA-105 bottoms cracking additive. In: Proc. AIChE Spring Nat. Meeting, New Orleans, Louisiana, USA; February 26–29, 1996. 9 pp.
[303] Engelhard Corporation. Improve catalyst resistance to feedstock poisons by increasing matrix surface area. 2008. 4 pp. Available from: http://www.refiningonline.com/engelhardkb/crep/TCR3_15.htm.
[304] Liu C, Deng Y, Pan Y, Zheng S, Gao X. Interactions between heavy metals and clay matrix in fluid catalytic cracking catalysts. Appl Catal A General 2004;257(2):145–50.
[305] Bayraktar O, Kugler EL. Effect of pretreatment on the performance of metal-contaminated fluid catalytic cracking (FCC) catalysts. Appl Catal A General 2004;260(1):119–24.

[306] Wallenstein D, Kanz B, Haas A. Influence of coke deactivation and vanadium and nickel contamination on the performance of low ZSM-5 levels in FCC catalysts. Appl Catal A General 2000;192(1):105–23.
[307] Myrstad T, Seljestokken B, Engan H, Rytter E. Effect of nickel and vanadium on sulphur reduction of FCC naphtha. Appl Catal A General 2000;192(2):299–305.
[308] Ledoux MJ, Hantzer S. Hydrotreatment catalyst poisoning by vanadium and nickel porphyrin: ESR and NMR. Catal Today 1990;7(4):479–96.
[309] Reyes L, Zerpa C, Krasuk JH. Catalyst deactivation in HDM of heavy deasphalted oils. Stud Surf Sci Catal 1994;88:85–95 (Catalyst Deactivation 1994).
[310] Maity SK, Pérez VH, Ancheyta J, Rana MS. Catalyst deactivation during hydrotreating of Maya crude in a batch reactor. Energy Fuels 2007;21(2):636–9.
[311] Abe S, Fujita K, Takada M, Inoue Y, Plantenga FL, Leliveld RG. New developments in resid hydroprocessing. Stud Surf Sci Catal 2002;145:491–4.
[312] Wei J, Wei RG. Metal deposition profiles in hydrodemetalation. Chem Eng Commun 1982;13(4–6):251–60.
[313] Matos EM, Guirardello R. Modeling and simulation of hydrodemetallation and hydrodesulfurization processes with transient catalytic efficiency. Braz J Chem Eng 2000;17(2):171–9.
[314] Ying Z-S, Gevert B, Otterstedt J-E, Sterte J. Hydrodemetalation of residual oil with catalysts using fibrillar alumina as carrier material. Appl Catal A General 1997;153(1–2):69–82.
[315] Minaev VZ, Zaidman NM, Spirina GA, Samakhov AA, Lur'e MA, Lipovich VG, et al. Effect of pore structure of alumina-cobalt-molybdenum catalyst on activity and stability in hydrodesulfurization of heavy feedstocks. Chem Technol Fuels Oils Transl Khim Tekhnol Topl Masel 1975;11(6):436–9.
[316] Tamm PW, Harnsberger HF, Bridge AG. Effects of feed metals on catalyst aging in hydroprocessing residuum. Ind Eng Chem Proc Des Dev 1981;20(2):262–73.
[317] Torres-Mancera P, Rayo P, Ancheyta J, Marroquín G, Centeno G, Alonso F. Catalyst deactivation pattern along a residue hydrotreating bench-scale reactor. Catal Today 2014;220-222:153–8.
[318] Rana MS, Ancheyta J, Sahoo SK, Rayo P. Carbon and metal deposition during the hydroprocessing of Maya crude oil. Catal Today 2014;220–222:97–105.
[319] Sarbak Z, Andersson SLT. Effect of metal-organic compounds on thiophene hydrodesulphurization over sulphided forms of fluoride-containing $CoMo/Al_2O_3$ catalysts. Appl Catal 1991;69(1):235–51.
[320] Ancheyta J, Betancourt G, Centeno G, Marroquín G, Alonso F, Garciafigueroa E. Catalyst deactivation during hydroprocessing of Maya heavy crude oil. 1. Evaluation at constant operating conditions. Energy Fuels 2002;16(6):1438–43.
[321] Khang SJ, Mosby JF. Catalyst deactivation due to deposition of reaction products in macropores during hydroprocessing of petroleum residuals. Ind Eng Chem Proc Des Dev 1986;25(2):437–42.
[322] Ancheyta J, Betancourt G, Centeno G, Marroquín G. Catalyst deactivation during hydroprocessing of Maya heavy crude oil. (II) Effect of temperature during time-on-stream. Energy Fuels 2003;17(2):462–7.
[323] Newson E. Catalyst deactivation due to pore-plugging by reaction products. Ind Eng Chem Proc Des Dev 1975;14(1):27–33.
[324] Bridge AG, Reed EM, Tamm PW, Cash DR. Chevron hydrotreating processes desulfurize Arabian heavy residua. AIChE Symp Ser 1975;71(148):225–33.
[325] Yumoto M, Kukes SG, Klein MT, Gates BC. Catalytic hydroprocessing of aromatic compounds: effects of metal sulfide deposits formed in commercial residuum hydroprocessing. Ind Eng Chem Res 2001;40(1):131–5.
[326] Yumoto M, Kukes SG, Klein MT, Gates BC. Catalytic hydroprocessing of aromatic compounds: effects of nickel and vanadium sulfide deposits on reactivities and reaction networks. Ind Eng Chem Res 1996;35(9):3203–9.
[327] Ramirez de Agudelo MM, Gonzalez Reyes L, Gajardo P. The role of transition metal sulfides in hydrotreatment – first row transition metal sulfides. Rev Tec INTEVEP 1984;4(1):19–26.

[328] Ramirez de Agudelo MM, Reyes LG, Gajardo P. The role of transition metal sulfides in hydrotreatment. Part II: first row transition metal sulfides. In: Proc. AIChE Nat. Meeting. Houston, TX, USA; 1983. 8 pp.
[329] Sawyer WH, Bearden R, Chianelli RR, Winter WE. Slurry hydroprocessing process. Patent No. US4952306 (Also published as JPH03139595, EP0419266, CA2025045). Exxon Research and Engineering Co.; August 28, 1990.
[330] Carnell ID. The effects of various porphyrin species on the decomposition and hyerogenation of hydrocarbons [Ph.D. thesis] from University of Birmingham Chem Eng July 1989. 505 pp.
[331] Pazos JM, Gonzalez JC, Salazar-Guillen AJ. Effect of catalyst properties and operating conditions on hydroprocessing of high-metals feeds. Ind Eng Chem Proc Des Dev 1983;22(4):653–9.
[332] Maity SK, Pérez VH, Ancheyta J, Rana MS, Centeno G. Effect of asphaltene contained in feed on deactivation of Maya crude hydrotreating catalyst. Pet Sci Technol 2007;25(1–2):241–9.
[333] Hauser A, Marafi A, Almutairi A, Stanislaus A. Comparative study of hydrodemetallization (HDM) catalyst aging by Boscan feed and Kuwait atmospheric residue. Energy Fuels 2008;22(5):2925–32.
[334] Ancheyta J, Sánchez S, Rodríguez MA. Kinetic modeling of hydrocracking of heavy oil fractions: a review. Catal Today 2005;109(1–4):76–92.
[335] García-López AJ, Cuevas R, Ramírez J, Ancheyta J, Vargas-Tah AA, Nares R, et al. Hydrodemetallation (HDM) kinetics of Ni-TPP over Mo/Al_2O_3-TiO_2 catalyst. Catal Today 2005;107–108:545–50.
[336] Trejo F, Ancheyta J. Kinetics of asphaltenes conversion during hydrotreating of Maya crude. Catal Today 2005;109(1–4):99–103.
[337] Trejo F, Ancheyta J, Centeno G, Marroquín G. Effect of hydrotreating conditions on Maya asphaltenes composition and structural parameters. Catal Today 2005;109(1–4):178–84.
[338] Rana MS, Ancheyta J, Maity SK, Rayo P. Hydrotreating of maya crude oil: I. Effect of support composition and its pore-diameter on asphaltene conversion. Pet Sci Technol 2007;25(1–2):187–99.
[339] Rana MS, Ancheyta J, Maity SK, Rayo P. Hydrotreating of maya crude oil: II. Generalized relationship between hydrogenolysis and hdas. Pet Sci Technol 2007;25(1–2):201–13.
[340] Sánchez S, Ancheyta J. Effect of pressure on the kinetics of moderate hydrocracking of Maya crude oil. Energy Fuels 2007;21(2):653–61.
[341] Trejo F, Ancheyta J, Sánchez S, Rodríguez MA. Comparison of different power-law kinetic models for hydrocracking of asphaltenes. Pet Sci Technol 2007;25(1–2):263–75.
[342] Maity SK, Ancheyta J, Alonso F, Vazquez JA. Study of accelerated deactivation of hydrotreating catalysts by vanadium impregnation method. Catal Today 2008;130(2–4):405–10.
[343] Centeno G, Ancheyta J, Alvarez A, Marroquín G, Alonso F, Castillo A. Effect of different heavy feedstocks on the deactivation of a commercial hydrotreating catalyst. Fuel 2012;100:73–9.
[344] Maity SK, Blanco E, Ancheyta J, Alonso F, Fukuyama H. Early stage deactivation of heavy crude oil hydroprocessing catalysts. Fuel 2012;100:17–23.
[345] Martínez J, Ancheyta J. Kinetic model for hydrocracking of heavy oil in a CSTR involving short term catalyst deactivation. Fuel 2012;100:193–9.
[346] Mederos FS, Ancheyta J, Elizalde I. Dynamic modeling and simulation of hydrotreating of gas oil obtained from heavy crude oil. Appl Catal A General 2012;425–426:13–27.
[347] Rodríguez MA, Elizalde I, Ancheyta J. Modeling the performance of a bench-scale reactor sustaining HDS and HDM of heavy crude oil at moderate conditions. Fuel 2012;100:152–62.
[348] Ramírez S, Martínez J, Ancheyta J. Kinetics of thermal hydrocracking of heavy oils under moderate hydroprocessing reaction conditions. Fuel 2013;110:83–8.
[349] Alvarez A, Castañeda LC, Ancheyta J. On the application of petroleum feedstock modeling techniques for developing molecule-based models of hydrocarbon conversion processes. Catal Today 2014;220–222:198–207.
[350] Elizalde I, Ancheyta J. Application of a three-stage approach for modeling the complete period of catalyst deactivation during hydrotreating of heavy oil. Fuel 2014;138:45–51.

[351] Elizalde I, Ancheyta J. Modeling the deactivation by metal deposition of heavy oil hydrotreating catalyst. Catal Today 2014;220–222:221–7.
[352] Trejo F, Rana MS, Ancheyta J, Chávez S. Influence of support and supported phases on catalytic functionalities of hydrotreating catalysts. Fuel 2014;138:104–10.
[353] García-Arellano H, Buenrostro-Gonzalez E, Vazquez-Duhalt R. Biocatalytic transformation of petroporphyrins by chemical modified cytochrome C. Biotechnol Bioeng 2004;85(7):790–8.
[354] Xu G-W, Mitchell KW, Monticello DJ. Process for demetalizing a fossil fuel. Patent No. US5624844 (Also published as EP828805, WO9638517). Energy Biosystems Corp.; April 29, 1997.

CHAPTER 4

Acidity in Crude Oils: Naphthenic Acids and Naphthenates

1. Introduction

Among the enormous amount of compounds present in crude oils, carboxylic acids are a part of them. A diversity of organic acids is present in crude oil, including fatty acids, but sulfur compounds and other heteroatomic compounds are also acidic in nature. Naphthenic (cycloaliphatic)-type organic acids, i.e., naphthenic acids (NAs), represent the major proportion of acid compounds in crude oil.

Acidity in crude oils is measured as the *total acid number* (TAN) obtained by titrating with potassium hydroxide, and thus it is expressed in mg KOH/g units (a standard method is the ASTM D664 [1]). In other words, TAN is the amount of KOH in milligrams that neutralizes the acids contained in 1 g of oil sample. The potentiometric method ASTM D664 (IP177/96) is the most commonly used titration method. Another method, ASTM D974 (IP 139/86), is a colorimetric titration and is mainly used on petroleum products [2]. In a typical procedure, a known amount of crude oil sample (in grams, W_s) is dissolved in an organic solvent and is titrated with a KOH solution of known normal concentration $[N]$. The titrant volume required for neutralization (V_{eq}) is used to evaluate TAN as:

$$\text{TAN} = 56.1\frac{V_{eq} * [N]}{W_s}$$

The *naphthenic acid number* (NAN) is measured by extraction of the acids by liquid chromatography followed by Fourier transform infrared spectroscopy (FTIR) analysis (see Refs. [3–6]). The FTIR instrument is calibrated with a standard NA, for evaluating the wt %, which is then converted to an acid number assuming a molecular weight (MW) of 250 for whole crude and 300 for cuts (see, for instance, Ref. [7]). A third method is *naphthenic acid titration* (NAT), which consists of extraction of the acids by chromatography and then titration (titration is carried out as per ASTM D664). A comparison of TAN and NAT values for a high-acid crude (HAC) (Grana) and its boiling point fractions is presented in Fig. 4.1. As can be seen, the two values are quite similar for the light distillates but differ more toward the bottom of the barrel (BotB) fractions. In fact, the differences are more pronounced as the boiling point (MW) of the fraction increased [8].

The Science and Technology of Unconventional Oils. http://dx.doi.org/10.1016/B978-0-12-801225-3.00004-8

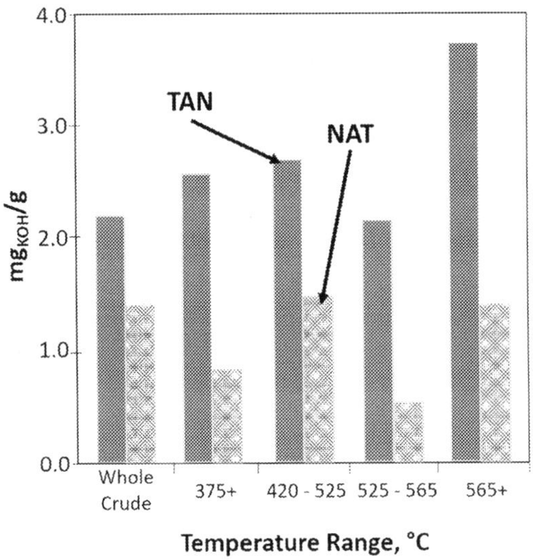

Figure 4.1
TAN and NAT values of Grana crude oil and its boiling point fractions (data taken from Ref. [8] emphasizing fractions with major differences).

A comparison between these methods and the effect of the presence of additives in the TAN values has been presented and discussed by Tebbal [9]. The discrepancies could be because ASTM methods would titrate NAs, phenols, carbon dioxide, hydrogen sulfide, mercaptans, and other acidic compounds present in the oil. However, since NAT is based on NA extraction, it could also be that the extraction method is not 100% effective.

Although TAN continues to be the only standard method, its failure in anticipating the potential impact of HAC processing and the lack of proportionality with the actual NA concentration has prompted the oil industry and related service companies to develop new measuring methods. None of these are publically available; only their merits have been emphasized [10]. The UOP procedures (UOP 565 and UOP 587) require the removal of the S-compounds prior to analyzing the acid number. UOP 565 is a potentiometric method recommended for petroleum products and petroleum distillates [11] and UOP 587 is a colorimetric method, limited to light-colored distillates [12]. Baker Petrolite developed a proprietary method (SCAN) that claimed to determine a "Specific Carboxylic Acid Number" [13]; no further report describes it. Near- and mid-infrared methodologies have been developed for online monitoring of acidity [14,15].

These test methodologies can have many drawbacks, especially when applied to crude oil. It was realized early on that the general use of the acid number as a criterion for the purchase or evaluation of NAs does not give sufficient detailed information to the consumer [16]. Inorganic acids, esters, phenolic compounds, S-compounds, lactones, resins, salts (some of the metals present in HACs may associate with acids and form naphthenate salts), and additives such as inhibitors and detergents interfere with any of aforementioned acidity methods.

A long-standing belief is the association between TAN and corrosion. This belief has caused more misunderstanding and controversy than providing a pathway for finding solutions to any problems [10]. The Canadian Crude Quality Technical Association (CCQTA) examined the correlation between the TAN of Athabasca bitumen and its corrosivity and encountered problems in applying the standard method [17]. The suspected problems included the following:

- Some acidic compounds, which were soluble in toluene/isopropanol mixture (the titration solvent), induced time-dependent asphaltene precipitation;
- Effects that depended on synergies between sample size and KOH concentration;
- Bitumen coating the electrodes; and
- The high viscosity of bitumen was inhibiting its effective dissolution.

The high viscosity of bitumen and heavy oils, as well as their tendency to precipitate asphaltenes, requires special consideration when determining TAN by ASTM D664. CCQTA then recommended special handling of the samples when determining TAN of heavy oils with ASTM 664. Sample handling may be modified to account for these issues. Some precautionary steps include predilution with toluene (and sonication) to ensure complete dissolution of the bitumen and carrying out titration without delay to avoid asphaltene precipitation. This additional step chooses toluene as diluent to ensure complete asphaltene solubility and for preventing its precipitation. They also recommended immediate titration following the dilution step [17].

The presence of acidity in crude oils is regarded as an impoverishment of the quality. This quality assessment is a consequence of the difficulties faced by their processing. Thus crude oils with a TAN value greater than around 0.5 mg KOH/g are collectively named *high-acid* (or *TAN*) *crudes* (HACs). At this point, this value of 0.5 mg KOH/g would appear somehow arbitrary; the following paragraphs will bring some clarity to it. Since TAN is a measurement of all acidic species present in the crude, there is no direct correlation between the NA concentration and the TAN number. The TAN test measures all "mobile protons" including esters, phenols, lactones, resins, and some additives (inhibitors and surfactants) that might be present. Many crudes with high TAN numbers can have a low NA content and vice versa. There is no NA standard for the three methods and the NAN changes depending on the acid standard used. In reality, NAs constitute at least 50% by weight of the total acidic compounds in crude oil; nonetheless TAN would overestimate their concentration.

Since most refineries were not prepared for processing this type of crude, refiners demand discounted prices for purchasing. In fact, there is a lack of refining capacity for processing crude oils considered acidic [18]. The price differential is then proportional to the acidity of the oil above a certain threshold level [19–26]. This threshold value has been set at 0.5 mg KOH/g and though it may sound arbitrary, it is the maximum value of TAN that the vast majority of refineries are suited for processing. The impact of TAN in the price differential was discussed in Chapter 1. The existence of this price differential makes HACs, another type of opportunity crudes, however in this case their processing options are limited by the existing refining capabilities for feeding crudes with TAN > 0.5 mg KOH/g.

In terms of API gravity, most HACs are light to median crude oils and contain low levels of sulfur. Besides API gravity, S-content is also regarded as a quality indicator and associated with the *sourness* of the oil. Thus high S-containing crude oils are categorized as *high sour* oils. Nevertheless, there are also examples of heavy and acid, and, worse than these, heavy-sour-acid, crude oils. Since the negative effects on refining of TAN synergize with the negative effects of S, the latter crudes have the greatest price differentials. Chapter 1 gave a description of HACs by country of origin, current size of market, and expected growth.

More details of the isolation, characterization, and properties of NAs follow in the next sections. The impact on refining processes and the current practices for the mitigation of the problems caused by NAs will be discussed at the end of this chapter. However, the concerns with NAs go beyond refining, up to the production well. For instance, one of the oil production methods from tar sands involves the separation of the bitumen from the sands using a caustic hot water flotation process, which produces large volumes of fluid wastes including process-affected water and a relatively stable suspension of solids and unrecovered bitumen called fine tailings. These tailings are highly contaminated with NAs and represent an ecological hazard. NAs are toxic components not only in oil sands extraction waters, but also in refinery wastewaters. A great deal of research effort has been focused on the environmental fate, transport, degradation, isolation of specific toxic NAs, and epidemiology. The interested reader is directed to a review that provides a comprehensive look at the microbiological degradation and adsorptive properties of NAs in aquatic environments, as well as detailed information regarding the origin of NAs in tailings ponds, chemistry and toxicological considerations, current analytical methods for aquatic sampling, and areas of future remediation research [27].

2. Origin and Nature

An enormous variety of acid compounds are present in crude oil, including NAs, fatty acids, saturated acids, aromatic acids, phenols, mercaptans, etc. Carboxylic acids as low in MW as acetic acid, saturated, and unsaturated acids based on single and multiples of five- and six-membered rings can be found in crude oils. NAs occur naturally in crude oils and in oil sands bitumen.

Acidity in crude oils originates from different sources; one of them is in-reservoir biodegradation of hydrocarbons in fossil deposits. The origin of carboxylic acids in petroleum is not completely understood. It is believed that either the deposit has not undergone sufficient catagenesis or it has been biodegraded by bacteria [28]. Carboxylic acids have been found in deposits of naturally biodegraded oil [29,30] and in crude oil that was biodegraded in laboratory experiments [31,32]. The analysis of biodegraded oils showed higher total acid and total base contents. The results indicate that the acidic constituents in biodegraded oils are a product of the biodegradation, as the composition is very different from the nonbiodegraded oils [33]. NAs in the Athabasca oil sands in Canada seem to have been produced by biodegradation of mature petroleum [28].

Two alternative routes were proposed to explain the origin of acid compounds in San Joaquin Californian crude oil. According to the first hypothesis, various degrees of successive oxidation of the parent oil species would explain the presence of a series of oxygen-only types such as two, four, and six oxygen-containing species. As oxidation proceeds, the polarity of species increases leading to higher degrees of saturation on those species containing higher oxygen (lower degrees of aromatization). Further migration is facilitated by increasing addition of acid groups via oxidation. Components with many or bulky substituents and high aromaticity would be hindered for migration because of a chromatographic effect of the rock minerals. The second hypothesis is based on the degradation pathway; however, there is conflicting information as to whether aerobic organisms preferentially attack unsubstituted alkanes or aromatics. It is possible that both kinds of organisms can exist at the same time. It seems that aromatic functionalities are the most degraded portions and that the organisms cleaved off these unsubstituted fractions, leaving behind compounds with higher heteroatom-to-carbon ratios. These fragments could reconfigure, creating highly substituted compounds, with low aromaticity [34].

The oil TAN values were found to increase with decreasing reservoir depth. This variation was attributed to biodegradation at or near the oil—water contact and diffusive mixing of biodegraded oil with nonbiodegraded oil [35] supplied through either single/episodic recharge or continuous charging of the shallow reservoir [36]. Consequently, there is the possibility of distinguishing acidic compounds newly formed by way of in-reservoir biodegradation and those contributed directly from oil-degrading bacteria [35]. Additionally, the extensive variation in the oil acidity at different sites in shallow reservoirs has been proposed to be likely controlled by local geological factors, such as the presence and thickness of a water leg, reflecting differences in accessibility, and availability of nutrients to the reaction site [37—39]. Another explanation given for large horizontal and vertical variations in TAN was proposed for the case of the Yabus and Samma formations of the Great Palogue Field of the Melut Basin in Sudan. A multiple-phase oil-charging model based on the molecular-level source and maturity constraints of the field would account for the two types of HACs that could be recognized in the Melut Basin [40].

As mentioned in the previous chapter, the understanding of the origin of the heteroatomic species in crude oil connects both ends: the specific compound and the oil itself. If that is the case, then such species becomes a biomarker. We have described the paramount importance of porphyrins as biomarkers, enabling geochemists to relate crude oils to their parent kerogen and thus draw a genetic map for the origin of a crude oil (see Chapter 3 and references therein). NAs as biomarkers have also been used as indicators of oil maturity [30,41,42], biodegradation [29,30,43], and geographic location [44–46].

Traditional hydrocarbon biomarker analyses have been used to determine the degree of biodegradation in two reservoir and two surface oils. These data were then correlated to the distribution and type of acidic NSO species (rings plus double bonds) selectively ionized and mass resolved by negative-ion electrospray Fourier transform ion cyclotron resonance mass spectrometry (ESI FT-ICR-MS). The biodegraded reservoir crude oil and surface oil samples exhibited an increase in relative abundance of O_2 species, a decrease in acyclic fatty acids, an increase in multiring NAs, and a decrease in C_{32} hopanoic acids compared to the nonbiodegraded reservoir crude oil. However, one surface sample also exhibited biomarker signatures indicative of a nondegraded oil [43]. Clear differences in the distribution of NSO compound classes, types (number of rings plus double bonds within a class), and number of alkyl carbons were observed when comparing Smackover oils of different levels of thermal maturity. With increasing thermal stress, the relative amount of sulfur- and oxygen-containing compounds decreases, condensation and aromatization of the polar cores increase, and the number of alkyl carbons decreases, reflecting the distribution of saturated hydrocarbons [42].

The study of the degradation mechanism of petroleum hydrocarbon by *Brevibacillus brevis* and *Bacillus cereus* indicated that some fatty acids could have been generated by biodegradation. The alkyl acids, especially those with linear and saturated alkyl acids, are dominant in the newly generated acids. A certain amount of naphthenic, alkenyl, and aromatic acids are also generated in the degraded samples. Biooxidation is the main degradation pathway of crude oil by *B. brevis* and *B. cereus*. The unconventional subterminal oxidation also existed, and *B. brevis* and *B. cereus* converted single long-chain hydrocarbons into short-chain fatty monoacids or alcohols [47]. They also found that neither nC_{17} nor nC_{18} in oil was degraded or generated when the heavy hydrocarbon was degraded.

3. Isolation and Identification

NAs can be defined as a complex mixture of carboxylic acids with the general formula $C_nH_{2n+Z}O_2$, where n indicates the carbon number and Z specifies the hydrogen deficiency resulting from ring formation. More details will be given in Section 3.3.

The methods for acidity evaluation (TAN, NAN, and NAT) mentioned earlier are quantification methods. Early attempts to separate and identify NAs were reported [48,49].

A routine, rapid, and quantitative method was developed for the analysis of aliphatic and NAs in crude oils, based on their isolation using nonaqueous ion-exchange solid-phase extraction (SPE) cartridges. The isolated acid fractions were derivatized by methylation and analyzed by gas chromatography—flame ionization detector and gas chromatography—mass spectrometry (GC—MS) [3]. This isolation method is an example where selectivity and purity of the resultant product were important, which is the case for analytical quantitative purposes.

HACs are the source for the commercial production of NAs and extraction methodologies have been developed. These methods try to maximize yields and not purity. Methodologies for the whole crude oil differ from those applied to oil products and will be described in the next sections.

3.1 Crude Oils

Although the separation of the NA mixture from the hydrocarbon matrix is a relatively simple process, their direct individual extraction from crude oil is not feasible. The commercially used method is based on neutralization with sodium hydroxide to form the naphthenate soaps, from which the acids are reconstituted by acid wash. An example has been reported for a Gulf Coast crude. Thus a mixture of NAs was extracted from the crude oil by neutralizing with soda, distilling out the oil, liberating the acids, and redistilling. The fractions from the front and tail ends were discarded. Further purification was effected by neutralizing this material with strong alkali and subjecting the soaps to a pressure extraction with a liquid propane—butane mixture until no more extract was obtained. The extracted soaps were then acidified with sulfuric acid, dissolved in ether, and washed until the aqueous layer was free of mineral acid. The acids were then dried and freed of ether by evaporation under diminished pressure. This method rendered a mixture with an acid number of 174 mg KOH/g [50]. To obtain more homogeneous fractions and to purify further, the acid mixture can be redistilled at a pressure of less than 0.005 bar in a molecular distiller. This scheme is presented in Fig. 4.2.

More complex separation schemes have been defined, particularly for analytical characterization purposes. Both liquid/liquid and liquid/solid extractions have been applied. SPE using ion-exchange resins has been proven to be an effective technique for separating NAs from simulated groundwater and river waters. The use of cartridges loaded with ENV+ (a cross-linked polystyrene-based polymer, from Biotage), C18, and Oasis the separation and quantification of NAs was achieved. ENV+ was found to be more efficient than Oasis and C18 [51].

The extraction scheme shown in Fig. 4.3 is based on high-performance liquid chromatography (HPLC) and involves separation on a dual basic alumina/acidic alumina column (HPLC-BA/AA) [52]. Conceptually, this separation step first separates the sample

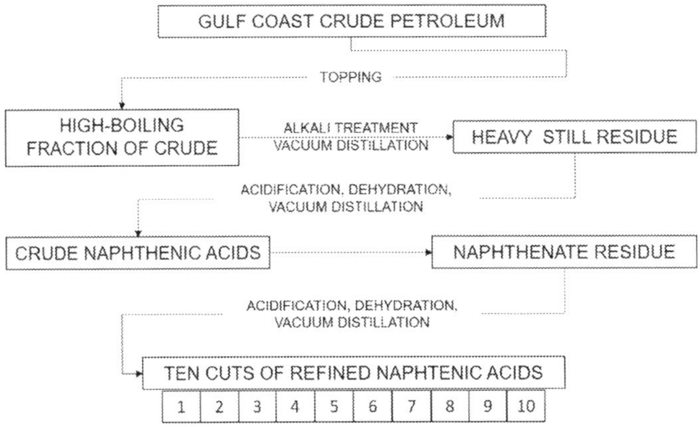

Figure 4.2

Separation of naphthenic acids (NAs) from crude oil. *Reproduced from Goheen GE, Conversion of naphthenic acids to naphthene hydrocarbons. Chemical constitution. Ind Eng Chem 1940;32(4):503–08, with permission from ACS Publications.*

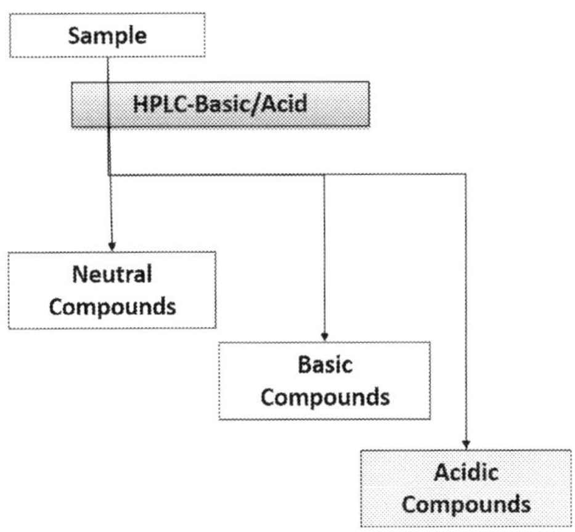

Figure 4.3

Separation scheme for naphthenic acid (NA) analysis. *HPLC*, High-performance liquid chromatography. *Reproduced from Boduszynski MM. Composition of heavy petroleums. 2. Molecular characterization. Energy Fuels 1988;2(5):597–613, with permission from ACS Publications.*

on the basis of acidity and basicity, and uses a basic/acidic alumina instead of anion- and cation-exchange resins, as other methods do [53–57]. Although the latter resin-based methods were believed to be more selective than alumina (because of simpler in-nature interactions with solute molecules), they required tedious preparations, were difficult to reproduce from batch to batch, and were reported to introduce artifacts caused by the deterioration of resins. The HPLC-BA/AA method uses fresh, "as received," basic and acidic alumina for each separation. The separation requires less than 2 h to complete. The fractions are operationally very well defined because of the use of automated HPLC equipment. The method has been developed primarily for separations of distillable cuts covering a boiling point range from about 345 to 700°C atmospheric equivalent boiling point (AEBP). However, it has also been used for sequential elution fractionation (SEF) fractions, which were derived from "no distillable" residues. The solubility SEF-1 fraction had a 50% AEBP of approximately 750°C as determined by Simdis methods. The results obtained for fractions boiling up to about 700°C AEBP showed material recoveries of 98.5 wt% or better. Losses for the SEF fraction did not exceed about 8 wt%. The separation of nitrogen-containing species into "acidic," "basic," and "pyrrolic" compounds using the HPLC-BA/AA method was excellent, with nitrogen balance of 95% or better [52].

Aminopropyl silica (APS) supplied by Baker was used as the solid phase for extracting the acid components from a diluted sample of the total crude oil (South American heavy oil). The solvent employed was a 70:30 toluene/methanol solution. The acid-loaded APS was then Soxhlet extracted with 30% acetic acid in toluene. The extract was water washed to remove residual acetic acid and rotovapped to remove solvent. The residue was reextracted with hexane. The "acid fraction" was obtained as the hexane-soluble fraction [58]. A modified version of this type of aminopropyl silica was used to isolate and separate NAs into discrete MW ranges [59].

Solid-phase extrography on $KOH-SiO_2$ was used to subfractionate an atmospheric residue of Liaohe crude oil from an oil field in Bohai Basin, China. A first work of the Chinese group identified two monooxygenated compounds ($C_{27}H_{48}O$ and $C_{28}H_{50}O$) as the major acid compounds present [60]. The extrographic subfractions allowed the identification of these $C_{27}H_{48}O$ and $C_{28}H_{50}O$ compounds as isoprenoidyl phenols. An additional mass peak with a molecular formula of $C_{27}H_{46}O_2$ was identified as δ-tocopherol, a phenolic compound with vitamin E activity [61].

The effectiveness of a Sudanese muscovite clay activated with NaOH was very poor for the separation of NAs from Nile blend crude oil and only slightly better for Fula crude oil. Effectiveness was measured as TAN reduction, which was 27% for Fula and just about 1% for Nile blend, though this last result could be considered statistically irrelevant [62].

A liquid/liquid extraction procedure was applied to Maya crude oil for the recovery of the acid components. A sample of crude was dissolved in (50:50) acetonitrile:methanol solution, from which the produced black residue was discarded. A golden supernatant was then filtered and blown down to dryness. The dried solid was reconstituted in methanol containing 0.5% ammonia [63].

Ethanolamine has been used for extracting NAs from the Penglai HAC and their composition and structures were analyzed by elemental analysis and characterized by infrared, MS, and nuclear magnetic resonance spectroscopic methods [64]. The NAs in Penglai crude oil were mainly monobasic acids containing 1 or 2 rings. The carbon number ranges from 9 to 28 and the average MW was 278 amu, responding to an average empirical formula of $C_{18}H_{30}O_2$.

Tetraalkylammonium and tetraalkylphosphonium hydroxide ionic liquids (ILs) were employed for the isolation and recovery of NAs from an HAC model oil. This model oil consisted of a commercial mixture of NAs dissolved in dodecane. Complete NA extraction was achieved at a very low IL/oil ratio. The NAs were recovered from the IL by extraction with an aqueous solution of an inorganic acid, leaving the IL ready for recycling and/or reuse [65]. A similar study was carried out using 1-*n*-butyl-3-methyl imidazolium IL with three types of anions, namely, octylsulfate, trifluoromethanesulfonate, and thiocyanate, which showed removal capability of up to 99%. Theoretical calculations indicated that the polarization charge density was responsible for the interaction between the anion and the carboxylic acids [66]. Nevertheless, effectiveness of the methodology was not proven to work with real HACs in both cases.

The difficulties encountered in isolating individual NAs present in HACs make their identification a daunting task, as can be deduced from Fig. 4.4 in which an MW

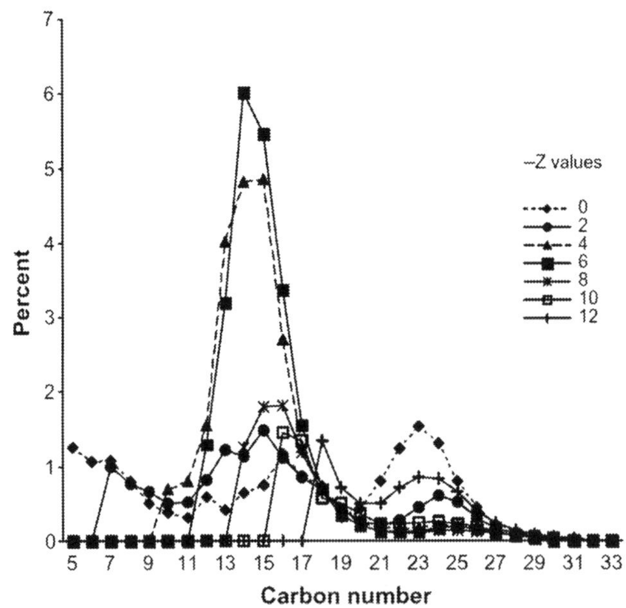

Figure 4.4

Carbon number distribution of ring types (based on Z number). *Reproduced from Rogers VV, Liber K, Mackinnon MD. Isolation and characterization of naphthenic acids from Athabasca oil sands tailings pond water. Chemosphere 2002;48:519–27, with permission from Elsevier.*

distribution obtained by MS is shown (published in Ref. [27] from data of Ref. [67]). However, efforts continued in this area targeting a better understanding of the differences in composition of NA preparations and effects on observed MW distributions. There is ample evidence that NAs from different sources have different compositions based on carbon and Z numbers, as will become evident later (Section 3.3).

The detection, identification, and characterization of NAs are of considerable importance. Fedorak [68] summarized the steps required for analytical purposes. This scheme for the separation and characterization of NAs was practiced in Alberta University and can be seen in Fig. 4.5.

Fedorak's scheme (Fig. 4.5) has been applied for the analysis of bitumen's NAs. Eight NA preparations (four from commercial sources and four from the oil sands operations) were derivatized and analyzed by GC–MS. The composition of each mixture was summarized as a three-dimensional plot of the abundance of specific ions (corresponding to NAs) versus carbon number (ranging from 5 to 33) and Z family (ranging from 0 to −12). The data in these plots were divided into three groups according to carbon number (group 1

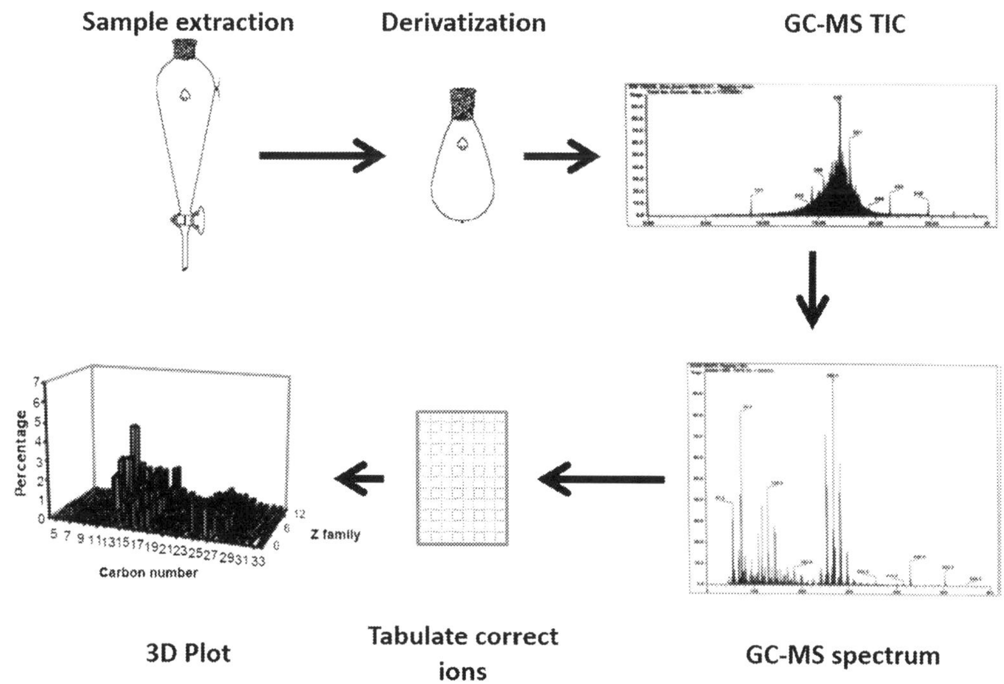

Figure 4.5

Method for naphthenic acid (NA) separation and analysis. *GC-MS TIC*, Gas chromatography–mass spectrometry total ion chromatogram. *Reproduced from Fedorak PM. Overview of naphthenic acids analyses at the University of Alberta. In: Proc. CONRAD/OSERN Symp. Coast Terrace Inn, Edmonton, Alberta; May 12–13, 2003. 50 pp., with permission from CONRAD/OSERN.*

contained carbon numbers 5–14, group 2 contained carbon numbers 15–21, and group 3 contained carbon numbers 22–33). A statistical *t*-test, using arcsine-transformed data, was applied to compare corresponding groups in samples from various sources. Results of the statistical analyses showed differences between various commercial NA preparations and between NAs from different oil sands ores and tailings ponds. This statistical approach can be applied to data collected by other MS methods [69].

The extraction scheme shown in Fig. 4.6 was applied to fractionate the acidic compounds from a North Sea HAC into differing acidity fractions using (KOH) pH-adjusted solutions of ethanol in distilled water. These fractionation conditions extracted 88% of the total acids in the crude oil, indicating that complete extraction would require a very strong and basic solution (pH > 14). Around 90% of these acidic compounds consisted of carboxylic acids, with MWs in the range 300–800 amu. The fraction extracted at pH 7 was the largest fraction. The acidic compounds remaining in the oil had MW > 600 amu and very low solubility in the 70% ethanolic aqueous phase [70].

The effectiveness of the separation could be visualized by changes in oxygen concentration with decreasing volatility (AEBP) and solubility (SEF) of oil components. In two California crude oils, Offshore California and Kern River, the oxygen content increased with increasing

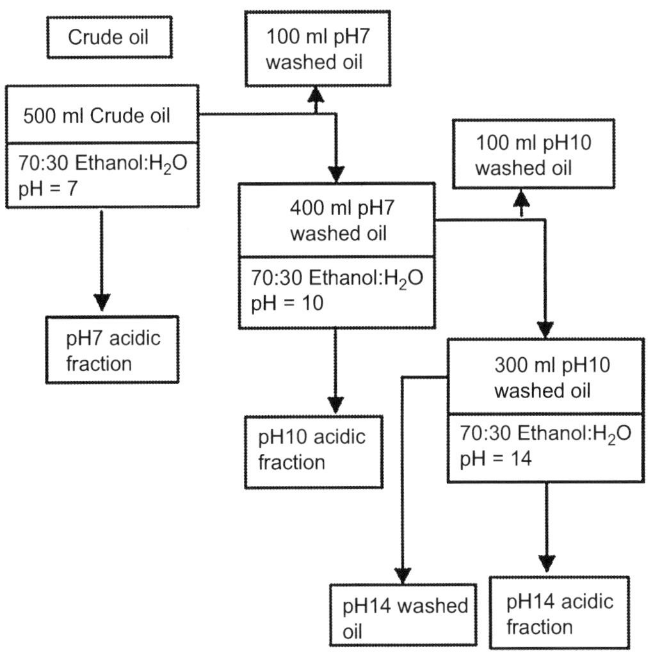

Figure 4.6

Fractionation of acidic compounds. *Reproduced from Hemmingsen PV, Kim S, Pettersen HE, Rodgers RP, Sjöblom J, Marshall AG. Structural characterization and interfacial behavior of acidic compounds extracted from a North Sea oil. Energy Fuels 2006;20(5):1980–7, with permission from ACS Publications.*

AEBP in a fairly similar fashion for both residues, from approximately 0.3–0.4 wt% for distilled fractions to 1.2–1.35 wt% for SEF-3 fraction [71].

3.2 Oil Products and Other Streams

The NAs of commercial importance are concentrated in the refinery process streams boiling between 204 and 371°C. These lighter NAs can be extracted from light gas oil and kerosene fractions by use of a caustic solution. The use of dilute caustic solution of 7–10% was preferred for the separation of NAs based on the emulsifying characteristics of NAs. When strong caustic solutions are used, the solubility of hydrocarbon oils in the aqueous phase increases with the assistance of the salted-out sodium naphthenates. The aqueous layer containing the naphthenates is separated from the hydrocarbon layer and treated with dilute mineral acids to release the free acids, which are separated from the aqueous layer, dried, and distilled [72]. Irradiation with microwaves have been found to enhance the neutralization reaction rate. Optimum conditions for this process to achieve about 92% acidity reduction were: ratio of caustic solution to oil = 0.23:1; pressure of 0.11 MPa; irradiation time of 5 min; microwave power of 375 W; and postreaction resting time of 25 min [73].

The characterization of petroleum acids in three diesel fractions of Chinese crude oils with high acid number (Penglai, Doba, and Liaohe) was carried out by first obtaining the NAs using caustic extraction and then derivatizing by methyl esterification. The content of 1–3-ring NAs represented more than 60% of the total organic carboxylic acids in the diesel fractions [74]. The light (C_8^-) carboxylic acids present in crude oil are believed to be monocyclic aliphatic with a C5 or C6 ring and predominate in a hydrocarbon fraction in the range of C_{14}^+, indicating that the more complex structures might correspond to those present in heavier fractions [74–77].

In the vacuum resid (VR), the isolated NAs were found to be part of the resin fraction [78]. Another attempt to identify the acid–base compounds present in the asphaltene fraction of a VR was based on a separation scheme developed by Gould and Long [79] and shown in Fig. 4.7.

An (3–5%) ammonia solution in ethylene glycol can be used to extract NAs from HAC vacuum fractions at 50–60°C. After mixing, two phases will spontaneously separate, with the NAs being extracted into the ethylene glycol solution. NAs can be recovered decomposing the NA–ammonia salt by heating. Petroleum ether could be used for purification. The optimal extraction conditions were a contact time of more than 10 min, with a reagent/oil ratio of more than 0.3 (wt/wt), rendering an acid removal greater than 85% [80]. More sophisticated extractants, such as 2-methylimidazole IL in ethanol (20% wt/wt), may improve extraction; in fact, this IL only extracted 67% of NAs [81]. Meanwhile, the neutral and acid compounds of (Cold Lake) asphaltenes could be

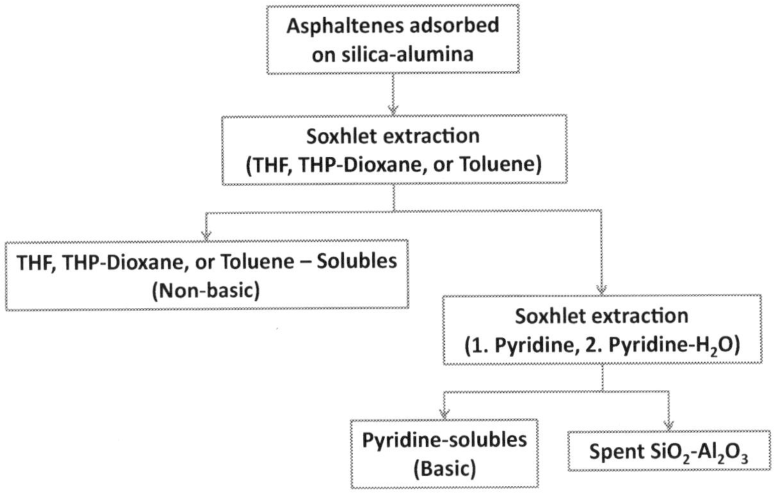

Figure 4.7

Acid–base separation scheme. *THF*, tetrahydrofuran; *THP*, tetrahydropyran. *Reproduced from Gould KA, Long RB. A new technique for the acid/base separation of petroleum and coal-derived fractions. Fuel 1986;65(4):572–6, with permission from Elsevier.*

separated using a KOH–SiO_2 column and then eluting with dichloromethane and a mixture of dichloromethane–formic acid mixture, respectively [82].

Measuring the concentrations of NAs in environmental samples and determining the chemical composition of such mixtures are huge analytical challenges. New analytical methods are being applied to these problems. Progress is being achieved in a better understanding of the mixture, emphasizing the challenges of identifying compounds, which are chemically similar. Clemente et al. have reviewed a variety of analytical methods and their application in assessing biodegradation of NAs in environmental samples [27].

The fundamental principles and characteristic features of NA analysis have been reviewed by Conrad Environmental Aquatic Technical Advisory Group (CEATAG) [72] as well. The analytical techniques for the quantitative analyses of NAs in aqueous solutions include FTIR spectroscopy, GC, ESI-MS, and HPLC. While MS is the preferred method used to determine the molecular composition of NAs, the combination GC–MS is also widely applied for semivolatile compounds.

One of the current quantitative analyses of NAs [68] is based on FTIR and consists of:

- The aqueous sample is filtered, acidified, and extracted with CH_2Cl_2;
- The extract is concentrated and analyzed by FTIR spectroscopy;
- The absorbance of the carboxyl group is calibrated to assess NA concentration.

3.3 Molecular Analysis

The physicochemical properties of NAs depend on their structural configuration, highlighting the importance of the molecular characterization of these compounds. The focus of the efforts differs between upstream (E&P) and downstream (refining), which in turn is conditioned by the impact that NAs have on each sector's operations. In refining, huge dedication has been placed on the understanding of the effect of NAs' molecular structure on corrosion. The dependence of corrosion capabilities of NAs on the molecular sizes, structures of the ring type, and carbon number distributions has been generally observed, making desirable a better understanding of their molecular structure. Early characterization (1940) of the NAs present in the lubricating oil portion of a Gulf Coast petroleum showed an MW range of about 220–440, corresponding to 14–29 carbon atoms per molecule. Hydrogen deficiency below the fatty acid series was 4–10 atoms per molecule, corresponding to average type formulas $C_nH_{2n-4}O_2$ through $C_nH_{2n-10}O_2$ [16,50,83].

The potentiality of using MS in combination with analytical separation techniques (e.g., chromatography) was realized early on. Two methods of ionization were examined for the MS analysis, namely, fluoride ion chemical ionization (CI) and electron impact (EI). Still, ion fragmentation was significant and complicated the spectra beyond possible interpretation. NA derivatization to its corresponding tertbutyldimethylsilyl analogs decreased the extent of molecular fragmentation, enabling identification [84]. A method based on positive-ion CI-MS using isobutane reagent gas to produce $(M+15)^+$ ions was applied to the analysis of NA esters. Since the complex mixture of NAs cannot be separated into individual components, only the determination of relative distribution of acids classified in terms of hydrogen deficiency was possible. The identities and relative distribution of fatty and mono-, di-, tri-, and higher polycyclic acids were obtained from the intensities of the $(M+15)^+$ ions according to Z-series formula $C_nH_{2n+z}O_2$ of NAs. The components are characterized on the basis of group-type and carbon number distributions. A comparison of the fast atom bombardment (FAB) and CI results showed that the group-type distributions obtained by both methods agree surprisingly well [85]. To simplify the complexity of the mixture, NAs were separated by extraction with column chromatography using an anion-exchange resin [54]. CI-MS of chromatographically extracted NAs from a Chinese VR showed that the type of NAs can be classified into fatty, mono-, bi-, tri-, up to hexacyclic. The application of these methodologies of extraction and CI-MS characterization indicated that the MW distribution of NAs extended between 198 and 540 amu, which corresponded to a carbon number distribution of about C_{12}–C_{37}. NA distribution was suggested as a tool for fingerprinting oil–oil and oil–source correlations [86].

A complex set of techniques comprising CI, liquid secondary ion MS (fast ion bombardment), atmospheric pressure chemical ionization (APCI), and ESI in both positive- and negative-ion modes was used for the determination of MW distribution of acids without derivatization [87]. Negative-ion APCI using acetonitrile as a mobile phase yields the cleanest spectra with good sensitivity among the ionization techniques evaluated. The selectivity of negative-ion APCI for NAs has also been demonstrated by comparing results for a whole crude oil with those for the isolated acid fraction. APCI also holds a great potential for online LC/MS for separating acid mixtures by HPLC and detecting with MS characterization.

The use of MS to investigate the NA mixture present in crude oils and to ascertain the nature of these species requires employing an ionization technique that does not result in fragmentation. Ensuring the detection only of molecular species provides useful information about the sample constitution. ESI-MS was proven to be a convenient way for NA analysis, in comparison with the extensive fragmentation caused by electron ionization MS. Model compounds, mixtures, and NAs extracted from Athabasca bitumen were analyzed and pseudoquantitatively determined. The calibration obtained with model compounds could be used under certain limitations for the native NAs in the bitumen [88]. However, the low resolution of ESI-MS provided little compositional information or accurate measurement of the MW distribution of a heavy vacuum gas oil (HVGO) acidic fraction of Athabasca bitumen because of the presence of multimers [89].

A chip-based nanoelectrospray system enabled microscale (<200 mg) and high-throughput (20 samples/h) measurement. A mass-dependent collision-induced dissociation technique was developed to eliminate dimer formation while minimizing the fragmentation of low MW acids in the ESI process. Stearic acid is used as an internal standard to calibrate ESI-MS response factors for quantification purposes. With the use of structure–property correlations, boiling point distributions of TAN values could be calculated from the composition [90]. A comparison with field desorption (FD) MS showed that ESI is selective toward the ionization of polar molecules (mostly basic and acidic molecules) while FD is a more universal ionization method. Although general agreement was found between the two techniques, small variation in the MW distribution of polar and nonpolar molecules was observed. ESI yields lower average MWs and the difference increases with the boiling point [91].

Among the new, powerful MS methods that have been applied to the analysis of NAs, ESI high-field asymmetric waveform ion mobility spectrometry (FAIMS) shows strong potential to provide new insights into the compositions of NAs. The ability of FAIMS to separate ions in the gas phase provides an opportunity to select certain ions and study them in detail with various MS methods [92]. This was demonstrated by the identification of specific compounds in an NA preparation. In addition to facilitating the identification of compounds in complex mixtures, ESI-FAIMS-MS can provide quantitative results [27]. The ESI-FAIMS-MS method was combined with quadrupole, time-of-flight (TOF), and tandem MS to characterize commercial and naturally occurring NA mixtures [92]. This new method provides quantitatively reliable mass and isomer distributions of NA components in ~3 min without extensive sample preparation. ESI-FAIMS-MS seems to be especially useful for characterization of fragile ions that cannot be detected by other methods. A unique part of this technique is separation of structural isomers that proved to be critical in determination of elemental composition and in structure elucidation. Although tandem MS of NA ions separated by FAIMS might provide more information about the structure of NAs than any other method, no data on specific NA molecules present in the studied Athabasca sample were provided.

During the last few years, great effort has been concentrated in the molecular characterization of the most complex compounds of heavy oils and bitumen. Better resolution of the carbon profile has been achieved by petroleomics-related methodologies [93,94]. FT-ion cyclotron resonance (ICR)-MS is a technique with inherently ultrahigh mass accuracy and resolution, affording unequivocal mass assignments. The suitability of FT-ICR-MS has been demonstrated for both environmental and oil sands tailings pond samples [95]. The combination of ion mobility TOF-MS and ultrahigh-resolution FT-ICR-MS has allowed the identification of isomeric homolog series of NAs in petroleum samples. The latter technique (FT-ICR-MS) served to validate molecular formula assignments by the former (TOF-MS) and to determine whether or not isobaric ions of different mobility were isomers [96]. NAs comprise a complex mixture of alkyl-substituted acyclic and cycloaliphatic carboxylic acids, with the general chemical formula $C_nH_{2n+Z}O_2$, where n indicates the carbon number and Z is zero or a negative, even integer that specifies the hydrogen deficiency resulting from ring formation. The absolute value of Z divided by 2 gives the number of rings in the compounds. The rings may be fused or bridged. The acyclic components are highly branched. Fig. 4.8 shows molecular representations of NA structures [4]. Besides the carboxylic acid group, cyclic NAs are believed to be substituted with alkyl groups (R in Fig. 4.8).

Open chain acids with multiple bonds (such as olefinic compounds) and unsaturated rings (such as aromatic compounds) are not included in this structural definition of the NA group of compounds that might also be present in the fractions. The relationship between the number of C atoms in the molecule and the Z value defines a way of classifying the possible family of compounds [97]. In Table 4.1, these families are described in terms of certain molecular features.

Negative-ion mode nanospray FT-ICR-MS has been applied to the analysis of crude oil samples, providing insight into the different acidic species that were present. The use of the negative-ion mode allows the selective observation of NAs while the inherent high mass accuracy and ultrahigh resolution of FT-ICR-MS ensures potentialities for the characterization of NAs within a crude oil sample. Determination of the nature of the NAs present provides vital information, such as the molecular sizes and composition of acids,

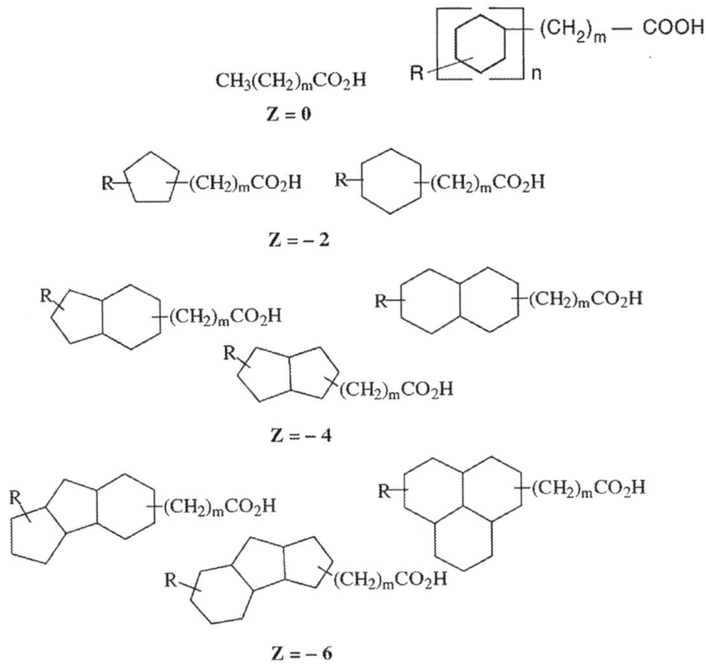

Figure 4.8

Molecular representation of naphthenic acid (NA) structures. R: alkyl chain, Z: hydrogen deficiency, and m: number of CH_2 units. *Reproduced from Clemente JS, Fedorak PM. A review of the occurrence, analyses, toxicity, and biodegradation of naphthenic acids. Chemosphere 2005;60(5):585–600, with permission from Elsevier.*

which may be used in the battle against corrosion and also used to fingerprint samples from different oil fields [98].

The ESI-FT-ICR MS of acidic NSO compounds in crude oils collectively showed 14,000 masses, spanning 18 different heteroatomic classes, from which identification could be derived as well [46]. The mass spectra of NAs obtained using a variant of ESI coupled with an FT-ICR-MS was greatly affected by solvent effects, reflecting a dependence of the spectra on solubility. For example, 1-octanol (similar solvent to fatty tissue) was compared to polar solvents such as methanol or acetonitrile, which were used as a surrogate to indicate the more bioavailable or toxic components of NAs in natural waters. The spectral changes observed with different solvents (1:1 water/acetonitrile, 1:1 dichloromethane/acetonitrile, and 1:1 1-octanol/acetonitrile) were reported. Monocarboxylic compounds ($C_nH_{2n+z}O_2$) in the $z = -4$, -6, and -12 (2-, 3-, and 6-ring NAs, respectively) families in the carbon number range of 13–19 were prevalent, regardless of the solvent systems [99].

The advances in FT-ICR-MS have enabled direct characterization of NAs in bitumen, eliminating the need for prefractionation or isolation. However, low abundant species

Table 4.1: Molecular Features of Naphthenic Acid (NA) Families [97]

	Z	0	−2	−4	−6	−8	−10	−12
C-number	Number of Rings	Open Chain	1	2	3	4	5	6
10		172	170	168	166	164	162	160
11		186	184	182	180	178	176	174
12		200	198	196	194	192	190	188
13		214	212	210	208	206	204	202
14		228	226	224	222	220	218	216
15		242	240	238	236	234	232	230
16		256	254	252	250	248	246	244
17		270	268	266	264	262	260	258
18		284	282	280	278	276	274	272
19		298	296	294	292	290	288	286
20		312	310	308	306	304	302	300
21		326	324	322	320	318	316	314
22		340	338	336	334	332	330	328
23		354	352	350	348	346	344	342
24		368	366	364	362	360	358	356
25		382	380	378	376	374	372	370

could not be detected; for this reason a comprehensive analysis can be obtained only if prefractionation or NA isolation is performed [100].

The characterization of a South American heavy crude oil sample and its extracted acid fraction by negative-ion ESI high-field FT-ICR-MS showed similar spectral profile for both samples, indicating that the identified carboxylic acids were not modified by the extraction protocol. NAs extracted from Athabasca River Basin, Alberta, Canada, were characterized by FT-ICR-MS and results were processed with data visualization techniques. The acids present responded to the empirical formula $C_nH_{2n+z}O_x$, where x varied from 2 to 5, indicating that not all of them were monoprotic or the presence of other oxygenated functional groups. Furthermore, the presence of such species was postulated as a partial explanation of the discrepancies between TAN and acid content [101]. The characterization (negative-ion ESI FT-ICR-MS) of the maltenes and asphaltenes fractions of five biodegraded tar sand bitumens obtained from a tar sand column of a tar sand mining plant in Athabasca, Canada, showed that NAs were concentrated in the maltenes fraction [102].

ESI high-field FT-ICR-MS could selectively ionize and identify NAs without interference from the hydrocarbon background. When combined with a prechromatographic separation, ESI FT-ICR-MS reveals an even more detailed acid composition, otherwise unrevealed. Furthermore, the minor compounds undetected in the whole crude became evident and well resolved in the extracted fraction. The powerful mass resolving power distinguishes as many as 15 distinct chemical formulas within a 0.26 amu mass window. The ultrahigh resolution and mass accuracy (typically subppm) associated with the FT-ICR-MS affords a high degree of confidence in the mass assignments, and therefore the instrument is well suited to the analysis of complex mixtures such as petroleum-related samples. This analysis and the whole crude–acid fraction comparison inferred the identification of S-containing NAs. The proposed identified molecular structures are presented in Fig. 4.9. More than 100 acid homologs were identified within about 3000 chemical formulas containing O_2, O_3, O_4, O_2S, O_3S, and O_4S groups with carbon numbers ranging from 15 to 55, with cyclic (1–6-ring) and aromatic (1–3-ring) structures. Additionally, aromatic NAs were found to be concentrated in the asphaltenes fraction [58]. To avoid thermal decomposition of NAs of high boiling cuts, a new technology (electrospray TAN or E-TAN) was developed, based on ESI-MS and it did not involve physical distilling of the sample [103].

The possible presence of aromatic rings as part of the molecular structure was suggested. If aromatic rings were absent, at least five naphthenic ring closures should be possible in the acids with largest hydrogen deficiency. The acids were at least substantially monobasic [83]. A complex array of techniques have demonstrated that aromatics make up 3.1% of the considered California crude oil acids [87].

At the University of Warwick, a Bruker BioAPEX II 9.4 T FT-ICR-MS was used for the analysis of NAs, crude oil extracts, and fuel-related samples. A nanospray ion source was used for sample introduction and ionization, which resulted in high sensitivity and minimized fragmentation. A portion of a well-resolved spectrum, indicating the achieved sensitivity and resolution, is shown in Fig. 4.10 [104].

The same methodology was applied to commercial mixtures of NAs and NAs extracted from a Syncrude. The carbon profile obtained for the latter sample is shown in Fig. 4.11A [104]. The different profiles exhibited by HACs of various locations and so various origins can be concluded for a collective comparison of profiles shown in Fig. 4.11A and B [98,104]. Thus the NA content and profile of an oil would identify a particular oil sample and link it to a particular oil field. This work seems to indicate that NA profiles can act as fingerprints for different oil fields [98,104].

The negative ESI FT-ICR-MS-detailed characterization of acids and neutral nitrogen compounds in a Chinese HAC indicated that the most abundant O1 and O2 class species correspond to biologic skeleton structures, such as hopanoic acids, secohopanoic acids, and sterols. The distributions of acidic compounds within 39 narrow distillate fractions showed a preferential presence of acidic compounds in the vacuum distillate fractions. While hopanoic acids and secohopanoic acids were more abundant in the 500°C$^+$ fractions, the sterol-like compounds were enriched in the 460–500°C fractions. The N1 class species were grouped together with these O1 and O2 species as being responsible for the high TAN of the crude oil. These N1 species were centered at DBE values of 9, 12,

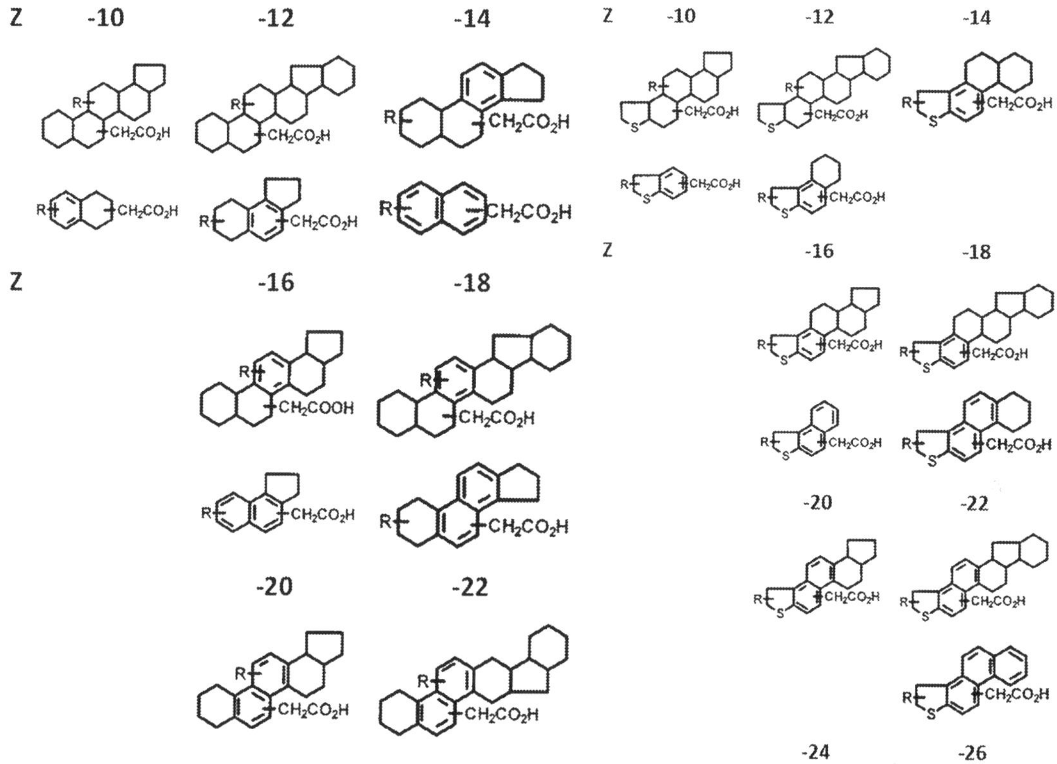

Figure 4.9

Identified naphthenic acids (NAs) and S-containing NAs. *Reproduced from Qian K, Robbins WK, Hughey CA, Cooper HJ, Rodgers, RP, Marshall AG. Resolution and identification of elemental compositions for more than 3000 crude acids in heavy petroleum by negative-ion microelectrospray high-field Fourier transform ion cyclotron Resonance mass spectrometry. Energy Fuels 2001;15(6):1505–11, with permission from ACS Publications.*

15, and 18 and were considered likely to be pyrrolic compounds with various numbers of aromatic rings [60].

Acidic and basic characterization of a light distillate fraction exhibited monoaromatic and low DBE polycyclic species. Low MW polycyclic and monoaromatic carboxylic acids and oxygen–sulfur (S_xO_y) species solely populate the negative-ion mass spectrum of the light distillate. In contrast, the middle and heavy distillates are composed of high MW and high DBE polyaromatic species such as pyrrolic and phenolic species, in addition to polyaromatic carboxylic acids. Light and middle distillate carboxylic acids and O_xS species fractionate by alkyl chain length and not by DBE value [105].

The acid fraction extracted from the liquid/liquid (acetonitrile–methanol) extraction of a Maya crude was characterized by ESI(−) MS and ESI(−) MS/MS. The fragmentation patterns were determined by ESI(−) MS/MS and a typical fragment from the loss of CO_2

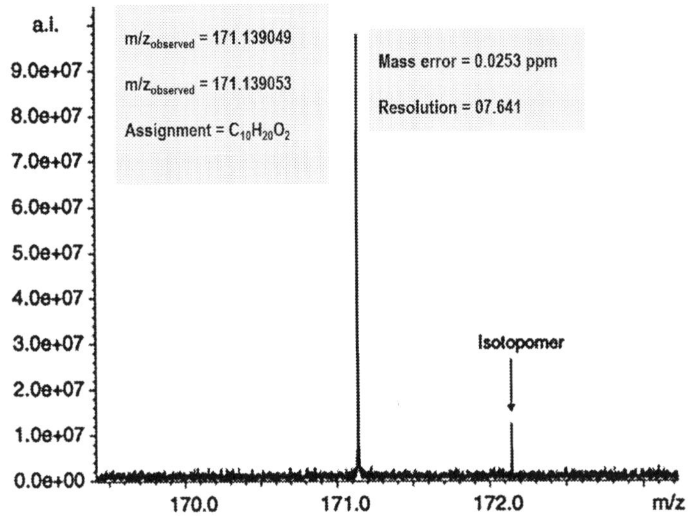

Figure 4.10

Portion of a resolved mass spectrum of naphthenic acid (NA) extract. *Reproduced from Barrow MP, Headley JV, Peru KM, Derrick PJ. Fourier transform ion cyclotron resonance mass spectrometry of principal components in oilsands naphthenic acids. J Chromatogr A 2004;1058(1–2):51–9, with permission from Elsevier.*

was observed. Additionally, a loss of 64 amu as well as an intense peak at *m/z* 80 were also observed. The *m/z* 80 fragment was observed in all spectra in which the peak was above the low mass cutoff of the ion trap. However, no peaks of *m/z* 64 were detected in the lower *m/z* MS/MS scans, indicating that it was a neutral fragment. These results were interpreted as a neutral loss of SO_2 (64 amu) and the $(SO_3)^-$ ion (80 amu), indicating the presence of a homologous series of alkyl-sulfonic acids [63]. Although these SO_2 and SO_3 moieties have been observed in a South American heavy crude [58] and in a California crude [34], they were not previously [34] associated with the presence of sulfonic acids.

In summary, high mass accuracy and high resolution are a prerequisite for full characterization of different ions of very similar mass, but same class in complex mixtures. Although magnetic sector and FT-ICR instruments have been most commonly applied to such investigations, TOF-MS has also been used to analyze crude oil samples. Techniques such as APCI in negative-ion mode can produce very clean spectra with good sensitivity compared to other techniques. ESI, however, is becoming the ionization technique of choice for the MS of NAs in crude oil samples. ESI has the advantage of being an ionization technique that imparts little energy to the nascent ions and results in the vaporization and ionization of a sample while minimizing fragmentation. The combination of the ultrahigh mass accuracy, ultrahigh resolution, and selective observation of the deprotonated NAs make negative-ion mode FT-ICR-MS an attractive technique for the characterization of the NAs within a crude oil. The ability to determine the empirical

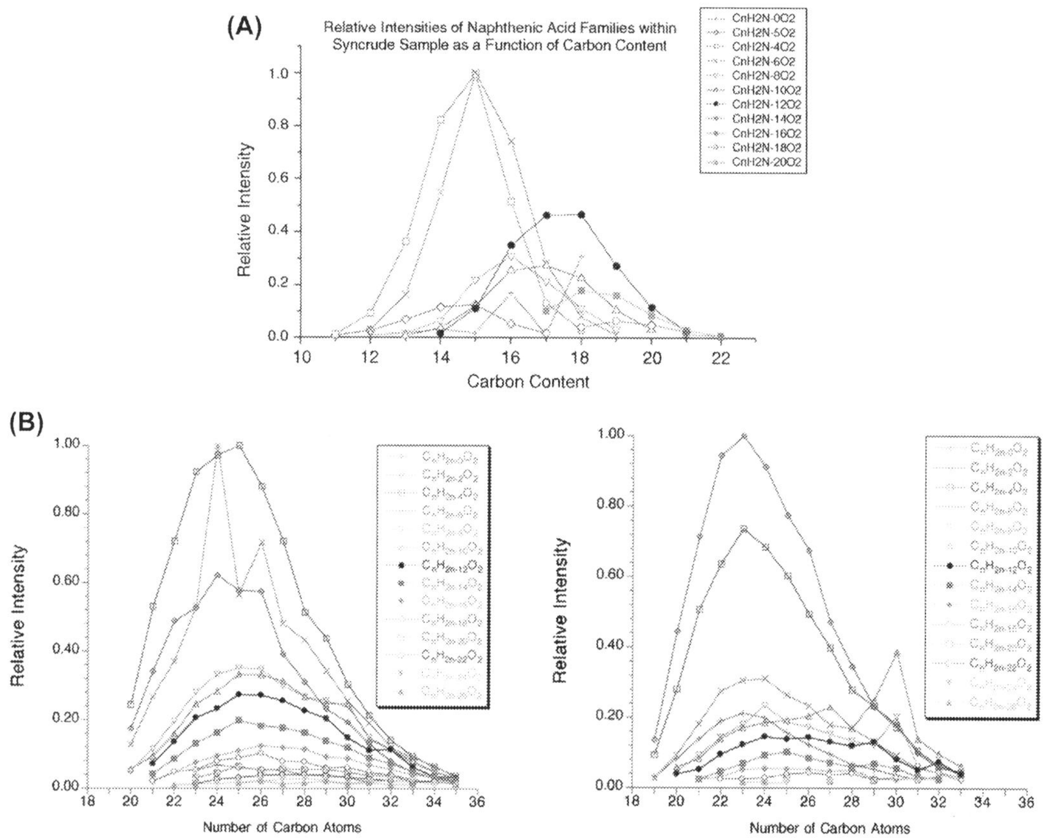

Figure 4.11

Carbon profile of naphthenic acids (NAs) extracted from: (A) Syncrude and (B) crudes from two different fields. *Reproduced from (A) Barrow MP. Petroleomics; 2008. Available from: http://www.warwick.ac.uk/staff/M.P.Barrow/petroleum.html, with permission from Elsevier; (B) Barrow MP, Mcdonnell LA, Feng X, Walker J, Derrick PJ. Determination of the nature of naphthenic acids present in crude oils using nanospray Fourier transform ion cyclotron resonance mass spectrometry: the continued battle against corrosion. Anal Chem 2003;75(4):860–6, with permission from ACS Publications.*

formulas of the acidic species contained within a given crude oil is of relevance to the continued fight against corrosion within the petroleum industry and environmental concerns [99]. Molecular structures differ between biodegraded oils and nondegraded oils. The former showed MW ranging between 300 and 500 amu and were more carboxylic and aliphatic, meanwhile the latter were more phenolic [33]. Furthermore, NAs in the considered Californian crude oil exhibit a wide range of MWs and MW distributions. The molecular weight distribution by family can be seen in Fig. 4.12 [87].

The environmental importance of the identification and characterization of water-soluble acid compounds has been mentioned already. A comprehensive two-dimensional GC × GC/TOF-MS system was applied for the characterization of two commercial

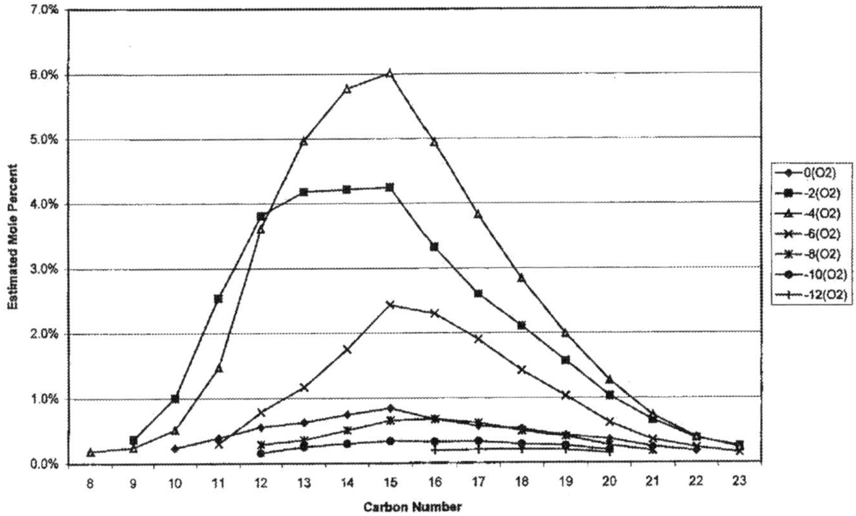

Figure 4.12

Carbon number distribution of naphthenic acids (NAs) from a Californian crude oil. *Reproduced from Hsu CS, Dechert GJ, Robbins WK, Fukuda EK. Naphthenic acids in crude oils characterized by mass spectrometry. Energy Fuels 2000;14(1):217–23, with permission from ACS Publications.*

mixtures of NAs (Fluka and Acros) and an NA sample extracted from a Syncrude tailings. Contour plots of chromatographic distributions of different *Z* homologous series of the Fluka, Acros, and Syncrude NAs were constructed using fragment ions that were characteristic of the NAs' molecular structures. Well-ordered patterns were observed for NAs of $Z = 0$ and -2, which corresponded to acyclic acids and monocyclic acids, respectively. For NAs of $Z = -4$, -6, and -8, specific zones were observed that would allow the pattern recognition of these NAs obtained from different origins. As expected, GC retention times increase with the number of carbon atoms and the number of rings in the molecules. Little signal was obtained for NAs with *Z* numbers of -10 or lower. Deconvoluted mass spectra of various NA isomers were derived from the reconstructed GC × GC chromatogram, permitting detailed structural elucidations for NAs in the future [106].

A laboratory bench procedure was developed to efficiently extract NAs from bulk volumes of tailings pond water from Athabasca oil sands. ESI-MS suggested a highly heterogeneous mixture of NAs. Monocyclic, polycyclic, and acyclic acids with MWs primarily between 220 and 360 were present at nearly identical proportions. Biphenyls, naphthalenes, and phenanthrene/anthracene were the most prominent impurities detected, but their levels were low (13 μg/L) even in a concentrated solution of the NAs (8549 mg/L). NAs stored at 4°C at this concentration were stable, exhibiting no significant change in concentration over a 10-month period [67].

Water-soluble (23°C) crude oil NSO nonvolatile acidic, basic, and neutral crude oil hydrocarbons have been identified by negative-ion ESI and continuous flow FD FT-ICR-MS at an average mass resolving power of $m/m_{50\%} = 550{,}000$. More than 7000 singly charged acidic species were identified in South American crude oil. Although water solubility is not restricted to the acid compounds, surprisingly, many were water soluble and much more so in pure water than in seawater. The truncated *m/z* distributions for water-soluble components exhibit preferential influences of MW, size, and heteroatom class on hydrocarbon solubility. Acidic water-soluble heteroatomic classes detected at >1% relative abundance include O, O_2, O_3, O_4, OS, O_2S, O_3S, O_4S, NO_2, NO_3, and NO_4. Parent oil class abundance does not directly relate to abundance in the water-soluble fraction. Acidic oxygen-containing classes were most prevalent in the water-soluble fractions, whereas acidic nitrogen-containing species were least soluble. In contrast to acidic nitrogen-containing heteroatomic classes, basic nitrogen classes were water soluble. Water-soluble heteroatomic basic classes detected at >1% relative abundance included N, NO, NO_2, NS, NS_2, NOS, NO_2S, N_2, N_2O, N_2O_2, OS, O_2S, and O_2S_2 [107]. Thus MS identification of molecular structures has confirmed that contaminants present in tailings go beyond NAs and include also neutrals and basic compounds.

4. Physicochemical Properties

A selected group of physical and chemical characteristics of HACs has been collected in Table 4.2 [108]. Some examples of Kuwaiti sour crudes [109] have been included, which are known to exhibit some features affecting the behavior of HACs that will be considered in the following sections.

Additionally, a review of literature, up to 1996, on NAs was reported by CEATAG to allow a description of their production and uses; physical and chemical properties; sources and environmental concentrations; environmental fate and persistence; and toxicity to biota. A summary of the reported physical and chemical properties of NAs is shown in Table 4.3 [72].

4.1 Characterization Results

In general, the physicochemical characterization of crude oils has shown that their indigenous components, like asphaltenes, resins, and NAs, are surface active. Thus besides the problems caused by their reactivity, their surface activity together with their solubility in water bring forth other sources for trouble. Solubility, as any other physicochemical property, depends on molecular structure. NAs variety goes from water soluble all the way through to oil soluble, which introduces a complicating factor in designing methodologies for separation, fractionation, isolation, and purification. Since NA pK_a values fall in the range of approximately 5–6, at pH above pK_a the acids will be ionized and remain in the solution, but at pH below the pK_a (acid solutions) solubility is hindered.

Table 4.2: Physical and Chemical Properties of High-Acid Crudes (HACs)

Crude Oil	Doba	Duri	Eocene	Hange	KEC	Kuito	Liaohe	Liuhua	Marlin	Penglai	Ratawi- Burgan
Density, g/cm^3 (at 20°C)	0.9309	0.921	0.92–0.98	0.8829	082–0.91	0.9283	0.9338	0.9208	0.9326	0.9276	0.91–0.95
Viscosity, mm^2/s (at 50°C)	141.9	179.5	–	14.64	–	54.83	509.4	67.96	78.62	97.91	–
CCR, %	5.47	7.61	9–13	5.28	4–6	6.54	8.98	6.7	6.78	6.11	9–12
Sulfur, %	0.16	0.221	4–4.5	0.661	2.2–3.0	0.691	0.2	0.243	0.68	0.32	3.8–4.4
N, ppm	1874–2300	2235	1500–2500	2538	1200–1800	4439	4700	2569	3700	3000	1600–2500
TAN, mg KOH/g	3.55–5.05	1.4	–	0.51	–	1.9	2.12	1.12	1.27	3.57	–
Asphaltenes, %	1.19	5.42	3–7	1.63	1.5–3.5	1.55	–	4.12	–	–	3–6
Resins, %	19.87	19.14	–	12.19	–	28.34	18.54	16.56	16.98	17.7	–
Wax, %	10.36	13.55	–	1.51	–	1.52	7.5	6.37	2.92	6	–
				Metals, ppm							
Fe	11.02	12.51	–	7.89	–	14.16	–	2.45	–	–	–
Ni	7.13–10.53	36.61	15–30	8.41	5–18	41.96	46.6	5.68	16.6	31.34	20–30
Cu	0.03	0.34	–	<0.03	–	<0.03	–	0.04	0.16	–	–
V	0.13–0.62	1.43	40–60	14.46	20–50	39.12	0.17	0.41	24.7	1.17	50–70
Ca	199.33	12.22	–	0.33	–	1.18	–	11.5	–	–	–
Mg	2.42	1.78	–	<0.03	–	0.47	–	1.28	–	–	–
Pb	0.18	1.55	–	<0.10	–	<0.10	–	0.12	–	–	–
Na	2.92	8.34	–	1.02	–	13.63	–	8.24	–	–	–
As	<0.10	–	–	<0.10	–	0.81	–	–	–	–	–
Light yield, %	20.33	21.83	–	44.75	–	34.72	–	26.77	–	–	–
Total yield, %	54.18	51.19	–	72.07	–	67.1	–	63.92	–	–	–
References	[108]	[108]	[109]	[108]	[109]	[108]	[108]	[108]	[108]	[108]	[109]

CCR, Conradson Carbon Residue.

Table 4.3: Physical and Chemical Properties of Naphthenic Acids (NAs)

Property	
CAS#	677090-13-4
Appearance	Brown or black liquid
Odor	Musty, hydrocarbon
pH	>10
Boiling point	~104°C
Freezing point	−17 to 10°C
RVP	0.36 at 221°C
	2.2 at 54.5°C
Specific gravity	1.00−1.10 at 15.6°C
pK_a	5.2 at 25°C
Solubility in water	Soluble

RVP, Reid vapor pressure.

NAs are generally quite stable compounds. As carboxylic acids, NAs are weak acids, whose pH depends on their concentration and on the proton−oxygen bond strength. As an example, aqueous solutions of acetic acid exhibit pH values in the range of 2−3. NAs have dissociation constants that range between 10^{-5} and 10^{-6} [110], which is typical of most carboxylic acids (e.g., acetic acid = $10^{-4.7}$, propionic acid = $10^{-4.9}$, palmitic acid $10^{-8.7}$). Their polarity and nonvolatility increases with increasing MW. Individual compounds within the group are expected to have varying physical, chemical, and toxicological propensity [111].

Infrared cannot provide discriminated information on the different families or individual compounds. The FTIR spectra of NAs show an absorbance peak in the region of 1740−1750 cm^{-1} for the monomeric form of the acid, and at 1700−1715 cm^{-1} for the dimer. In commercially available NAs (Kodak Standard P2388; Acid # 225−260), these two peaks were observed at 1743 cm^{-1} (monomer) and 1706 cm^{-1} (dimer). On the other hand, FTIR can be used for quantifying NA concentration. Meanwhile, fast atom bombardment mass spectrometry (FABMS) has also shown to be successfully applied in the analysis of NAs. Unfortunately, FABMS only details the various NA groups within a mixture. It does not provide quantification of the compounds [72].

As shown earlier, NA concentration and profile vary depending on the source of oil. Although there is no correlation with API gravity, the most acidic crude oils are also heavy (API <20). Fig. 4.13 shows the TAN values of a large number of crude oils as a function of their API gravity.

The concentration of heteroatomic compounds in heavy crude oils typically increases toward the BotB. While in light acid crude oils the acid profile varies with the boiling

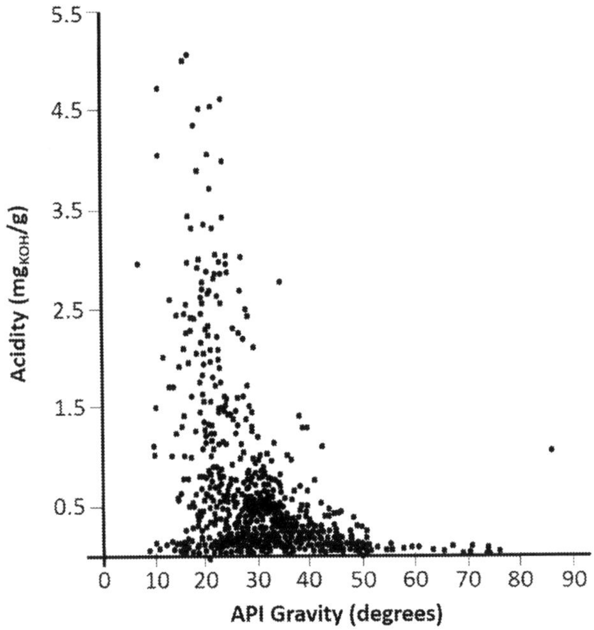

Figure 4.13
Acidity of crude oils according to their API gravity. *TAN*, Total acid number.

point fraction [112], in heavier HACs, the acidity profiles look similar and maximize toward the BotB boiling point fractions [8].

This is reflected in the acidity profile of Alba crude oil shown in Fig. 4.14. Clearly, an increasing NA concentration toward heavier fractions is demonstrated [113].

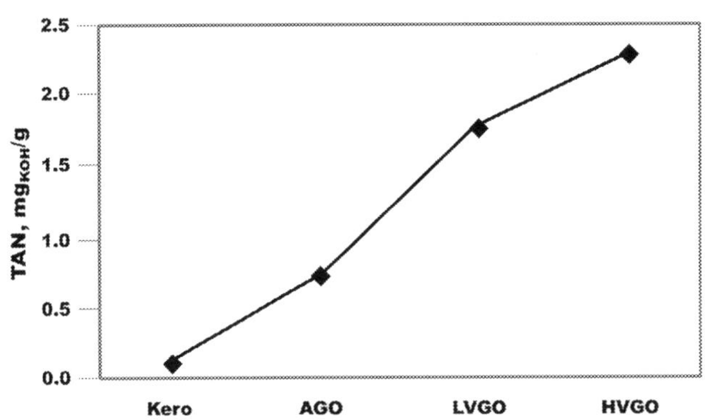

Figure 4.14
Acid profile of Alba crude oil. *AGO*, Atmospheric gas oil; *HVGO*, heavy vacuum gas oil; *LVGO*, light vacuum gas oil; *TAN*, total acid number. *Reproduced from Skippins J, Johnson D, Davies R. Corrosion-mitigation program improves economics for processing naphthenic crudes. Oil Gas J 2000;98(37):64–8, with permission from Pennwell.*

The results of the physicochemical characterization of the separated and fractionated NAs from a Gulf Coast HAC, according to the scheme shown in Fig. 4.2, were summarized as:

- Specific gravity of the NAs from the Gulf Coast decreased with increasing MW, similarly to NAs from a Russian crude oil but opposite to what was observed for NAs of a Rumanian crude.
- Higher boiling point NAs were monobasic.
- The average empirical formulas indicated that the composition ranges from $C_nH_{2n-4}O_2$ up to $C_nH_{2n-10}O_2$, and the average number of carbon atoms in the molecule varies from 14 to 29.
- Negligible bromine numbers indicated the high C/H ratios are not caused by the presence of unsaturated fatty acids. Whether the molecular structure includes 2 and 5 carbon rings or aromatic rings remained inconclusive [83].

The most prolific hydrocarbon resource in central Sudan, M Basin is one of the most productive zones in the whole Africa shear zone. Most of these crude oils are within the 20−34 degree API range, though a few are heavy oil with <20 degrees API. The crude oils are characterized by a low sulfur content, high asphaltene, high wax, high pour point, and high TAN especially in Bentiu reservoirs. The primary geochemical analyses indicate that biodegradation is the key genetic factor for high TAN crudes. It was found that the organic acids from high TAN crudes are mainly distributed in the heavy fractions distilling above 300°C. The molecules are composed with mono-, bi-, and tri-cycles. As fraction becomes heavier and acid value increases, the average molecular weight of NAs increases, its distribution gets wider, and the carbon number range increases [114].

4.2 Surface Activity and Behavior

NA surface activity deserves particular attention, since it is responsible for many of the problems caused during refining (and production) of HACs. One interesting feature of the acidic fractions of HACs is their ability to act as antiagglomerants/dispersants, surfactants, emulsifiers, etc.

4.2.1 Self-Aggregation

The surface activity of the acid compounds present in crude oils provokes their accumulation at interfaces between oil and water [115]. The solution-phase self-aggregation of NAs was demonstrated by negative-ion ESI high (FT-ICR) and low [linear trap quadrupole, (LTQ)] resolution MS for a whole crude oil, whole bitumen, and bitumen distillate fractions. Ordinarily monoacids are micelle-forming compounds in solutions above their pK_a values [116]. Aggregation was found to be concentration dependent. At high concentrations (larger than 1 mg/mL), the disruption of noncovalent interactions between heteromultimers by low-energy collision-activated dissociation yields LTQ dissociation mass spectra with MW distributions identical to those observed by FT-ICR-MS analysis at lower concentrations for purely monomeric species. These materials can exist as aggregates in solution even at high dilution (<0.1 mg/mL). Multimerization of polar acidic species in the Athabasca bitumen and bitumen distillates was found to be concentration and boiling point dependent. Interestingly, the lowest boiling distillation cut (375–400°C) displays the highest aggregation tendency, with dimers at concentrations as low as 0.05 mg/mL. The distillation cuts with higher boiling points display a decreased aggregation tendency with an increasing cutpoint. Acidic heteroatomic classes for the distillation cut multimers include O_4, S_1O_4, O_3, S_1O_3, N_1O_2, and $N_1S_1O_2$. The most abundant multimers for the 375–400°C distillation cut are O_4 species, whereas the 450–475°C cut contains N_1O_2 multimers in the highest relative abundance. Changes in multimer heteroatom content as a function of the monomer composition and distillation cut suggest that aggregation depends upon the chemical functionalities of the monomer species [117]. MS detection and characterization of the NAs multimers have been troublesome. FT-ICR-MS was used to study the tendency of NAs to form dimers in solution. It was found that increasing the residence time in the hexapole ion trap resulted in a higher number of collisions, which increased the chances for dissociation [118].

The interfacial tension (IFT) not only depends on the chemistry of the hydrophilic group, and size of the hydrophobic group, but also on their molecular structure and their ability to self-assemble. This associative trend represents an additional challenge for the determination of molecular structure. The correlation between IFT and pH was studied for several NA systems. A decrease in IFT was seen at high pH; this was explained by an increase in the total concentration of ionized NAs at the interface. The fraction of ionized NAs at the interface was lower than the fraction in the bulk phases. This effect is explained by the fact that interfacial pH is lower than the bulk pH because of electrical double layer effects. Partitioning and dissociation constants for NAs from crude oils and several NA model compounds were reported as well. The pK_a for NAs from crude oil was found to be 4.9. The logarithm of the partition coefficients varies linearly with the number of carbon atoms in the NA molecules. The NAs with 3 rings were found to be more hydrophilic than acids with 1 or 2 rings. The data could be used to predict the content of different NAs in produced water at a particular pH, given the acid content and mole weight distribution in the crude oil. A model for the total acid concentration in water was derived to account for the dissociation in water and partitioning of nondissociated acid. Experiments with synthetic NAs were performed and the model proved to describe the experimental points at $pH < 8$. At higher pH there was a deviation between the model and the experimentally determined concentrations. This was suggested to be caused by formation of ordinary and reversed micelles [116].

Both composition and chemical structure of NAs determine the nature of molecular aggregation at the hydrocarbon—water interface and impact interfacial properties and emulsion stabilization performance. Crude oil NAs containing about 80% primary acids and a higher proportion of 250—425 MW species were the most effective in reducing the IFT and stabilizing water-in-oil emulsions. Medium effects on interfacial properties were investigated using a model hydrocarbon, 5:1 hexadecane:toluene, and NA-free crude oil as the hydrocarbon phase. Crude oil NAs reduce the IFT of the NA-free crude oil to 18.5 dyn/cm, whereas they reduce the IFT of the model oil to 28 dyn/cm [119].

4.2.2 Effects on Asphaltenes

The enhanced surface activity of the NAs in the crude oil medium is attributed to the interaction between crude oil NAs and crude oil asphaltenes [119]. The presence of NAs increases the solubility of asphaltenes and this was considered to be caused by a solvent effect rather than by specific interactions [120]. A study of the dynamic IFT of interfaces between water and a model oil (toluene) in which variable amounts of asphaltenes are solubilized has shown that pH has a strong influence on interfacial properties of asphaltenes at the oil—water interface. At high or low pH, asphaltene functional groups become charged, enhancing their surface activity. An interaction between asphaltenes and maltenes that facilitates molecular arrangement at the interface was detected when natural NAs are contained in maltenes. Micrometric drop additions indicate that very little coalescence of water droplets is observed at high or low pH. The microscopic properties of the interface and the macroscopic behavior of the emulsion were determined to be correlated [121].

The dispersive effects of amphiphiles and NAs on asphaltene aggregation in model heptane—toluene mixtures were studied by following the disintegration of asphaltene aggregates at 1600 nm, using near-infrared spectroscopy. The disaggregating capabilities of NAs, as well as of the various chemical additives studied, were confirmed [122]. The dispersing capability of NAs is not observed with waxes, therefore no decrease of pour point would be observed for waxy fractions of HACs. While asphaltenes were capable of a large decrease in the pour point of a wax solution in a solvent mixture of *n*-decane, toluene, and cyclohexane, NAs did not show the same effect [123].

The NA content of deasphalted oils was observed to influence the interfacial properties of the asphaltenes at the crude oil—water interface. As a first hypothesis the basic nitrogen-containing asphaltenes and crude oil NAs were supposed to interact with each other synergistically, leading to NA—asphaltene acid—base complexes. The high propensity for aggregation of these complexes at the hydrocarbon—water interface leads to the observed high interfacial activity of asphaltenes in crude oils [124].

The fundamental interfacial properties at the hydrocarbon−water interface were determined for five different samples of acid compounds: cholanic acid, abietic acid, and three crude oil-extracted NAs. The percentage of primary NAs and proportion of low MW NAs influence interfacial activity and emulsion stability of water-in-oil emulsions. The interactions between two kinds of NAs (5-β-(H)-cholanoic acid and decahydro-1-naphthalenepentanoic acid) and asphaltenes separated from two different North Sea oils were found to take place upon mixing. It also appeared as if the tendency of the acids to interact with asphaltenes depended on the asphaltene type (in this case, the crude oil from which they were separated). Furthermore, an increase in asphaltenes concentration caused a dramatic decrease in diffusion coefficients, suggesting that the investigated asphaltenes were prone to self-associate even in toluene [125]. However, these results are somehow conflicting with those found when the same group studied the adsorption of amphiphilic molecules on asphaltenes. It was observed that amphiphilic molecules with acidic functional groups adsorbed onto the asphaltenes from the unstable subfraction to a larger extent than amphiphilic molecules with basic functional groups did. The more acidic amphiphilic molecules were also observed to be the most effective ones in stabilizing asphaltenes [126]. A theoretical study based on molecular mechanics calculations supported those experimental results. The derived structure−property relationships for the relative contributions and importance of various functional groups, such as C=C, C=O, COOR, and COOH, indicated that as polar functionality was incorporated into the acid structure (e.g.,, C−C to C=C in the case of hydrogenated methyl abietate to methyl abietate) an increase was predicted for the interaction energy and a delay in the precipitation onset was observed [127].

The nonbasic fraction obtained by the fractionation scheme shown in Fig. 4.7 [79] contains both the acid and neutral compounds. Their elemental analysis is shown in Table 4.4.

Table 4.4: Elemental Composition of Basic and Nonbasic Fractions of Asphaltenes From Canadian Samples

	Cold Lake		Wyodak	
	Nonbasic	Basic	Nonbasic	Basic
%C	80.7	80.6	87.7	84.1
%H	7.8	7.6	6.4	6.3
%N	1.1	1.4	0.6	3.2
%S	7.8	7.6	0.6	0.6
%O	nd	nd	3.9	5.5
V (ppm)	813	770	nd	nd
Ni (ppm)	322	350	nd	nd

nd, Not detected.

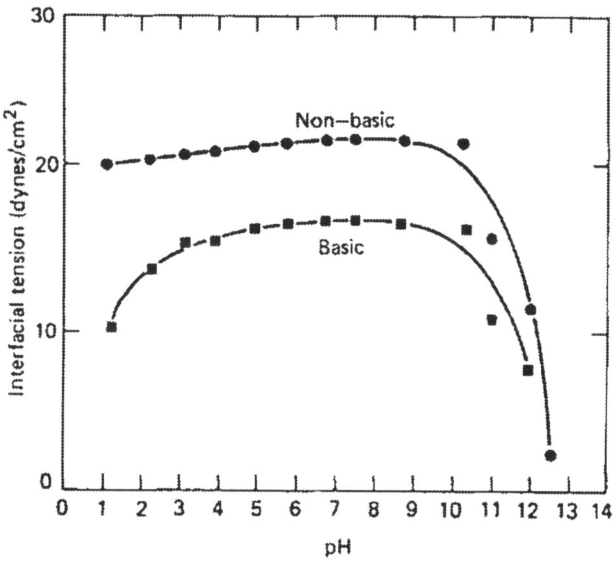

Figure 4.15
Interfacial tension of (coal-derived) asphaltene fractions. *Reproduced from Gould KA, Long RB. A new technique for the acid/base separation of petroleum and coal-derived fractions. Fuel 1986;65(4):572–76, with permission from Elsevier.*

The presence of acid compounds in the nonbasic fraction of coal-derived asphaltenes was reflected from the changes in interfacial surface tension at higher pH, as shown in Fig. 4.15. At high pH the acid compounds form a corresponding surface-active salt. The Soret band of each fraction was also measured for porphyrins and found that porphyrins are concentrated in the nonbasic fraction. It means that the excess N in this basic fraction might be associated with the presence of the porphyrinic compounds [79].

4.2.3 Emulsion Stability

The emulsion stability of water-in-oil emulsions of a North Sea crude oil and the alkaline-(at pH 7, 10, and 14 according to scheme presented in Fig. 4.6) washed crude oils was determined by the critical electric field cell method. The removal of acidic compounds from crude oils increases both the IFT and the water-in-oil emulsion stability, indicating that such indigenous acidic compounds destabilize water-in-oil emulsions [70]. The interfacial properties of the acidic fractions and the alkaline-washed oils were compared. The original crude oil, the acidic fractions, and the pH-washed oils have been characterized by ESI FT-ICR-MS and by FTIR spectroscopy. The smallest/lightest compounds were extracted at pH 7, which were found to be water soluble to a degree that the increase of IFT with time was caused by the dissolution of some of these acidic compounds into the aqueous phase [70].

Nonvolatile polar acidic and basic emulsion stabilizers were identified by ESI FT-ICR-MS in nine geographically distinct light, medium, and heavy oils. Although oil class distributions were unique, oils of similar API gravity exhibited similar relative abundances for the O_2 and O_4S classes. While heavy oils were high in O_2 and low in O_4S, light oils followed the opposite trend. However, regardless of the O_2 and O_4S class abundance in the parent oil, these O_2 and O_4S species were the two most abundant classes in the emulsion interfacial material, preferentially adsorbed at the interface. Negative-ion nitrogen-containing classes did not have a high affinity for emulsion interface adsorption. However, all positive-ion nitrogen-containing species adsorbed to the oil—water interface [128].

The role of NAs in emulsion stabilization of low-TAN/high-asphaltene oils was studied by focusing on the interfacial material. The characterization of this material by negative-ion ESI FT-ICR-MS indicated the presence of two groups of acidic compounds bound with different strengths: strongly and loosely bound. Among the identified indigenous acid compounds besides NAs, saturated (fatty) monoprotic acids and aromatic diprotic acids were also present. These acids exhibited MW in the range from 200 to 700 amu [129,130]. The interfacial rheology indicated the formation of a two-dimensional gel from the coprecipitation of the asphaltenes and organic acids into a very cohesive interface. The disruption of this interface required not only the drainage of individual molecules but also a collective yield of the gel. Different roles were proposed for the various types of organic acids found in the interfacial material:

1. Unsaturated monoprotic acids (naphthenic and aromatic) having higher affinity with bulk oil did not contribute significantly to the interface though they may coprecipitate with asphaltenes and contribute to the growth of the two-dimensional gel and favor stable emulsions.
2. On the contrary, fatty acids specifically adsorbed at the interface with paraffins led to the formation of a strong interface and to soap emulsions at near neutral pH values. Thus for high-asphaltene oils, asphaltenes would stabilize the interface at acidic pH and fatty acids would remove asphaltenes from the interface at neutral pH in a sort of competition among fatty acids and asphaltenes. Consequently, at near neutral pH, fatty acids would destabilize emulsions.
3. Diprotic acids are adsorbed at the interface regardless of their saturation degree and would favor asphaltenes adsorption forming very tight emulsions at near neutral pH values [130].

Surface-active derivatives can be separated or chemically derived from Athabasca (Canada) bitumen. These compounds have the ability to lower the surface tension of aqueous solutions as well as substantially reduce the IFTs of aqueous—organic systems. For this reason, it is suggested to use this beneficial effect on bitumen recovery processes [131]. In fact, acids in some cases are able to convert hydrate-forming systems from aggressively plugging materials into dispersed systems. Acids SPE from three low-plugging tendency oils were able to change the wettability of hydrates toward a more oil-wet state. A given subfraction from each of the considered oils exhibited particularly promising capabilities, well in excess of the whole acid extract. This subfraction was not characterized and no additional information on their composition was reported. However, it was suggested that the observed strong hydrate surface activity effects may be a result of a change in the acid monomer—multimer equilibrium state after resolving the acid matrix into subfractions. If acid monomers were the predominant form of species present in the subfractions, a higher hydrate surface activity would be expected [132]. Nevertheless, in this regard naturally occurring surfactants, specifically short-chain carboxylic acids and phenols of $MW < 400$, are responsible for stabilizing foams, as pointed out by Callaghan and coworkers [133]. This work represents an attempt for understanding the factors controlling crude oil foam stability and the identification of components responsible for foam stabilization. Their work also tried to demonstrate that a similar suite of compounds was responsible for the stabilization of a wider range of crude oil foams.

4.2.4 Naphthenates and Their Surface Activity

The emulsion-stabilizing mechanisms depend on which form of the acid/salt is present and/or the dominant species (NA/naphthenate ratio) [134—136]. Electrostatically stabilized water continuous emulsions are formed when deprotonated (charged) naphthenates dominate the system. Intermediate to high-acid/salt ratios will give rise to the formation of a lamellar liquid crystalline phase, which adsorbs to dispersed water droplets in a water-in-oil emulsion [137].

Most of the metal salts of NAs are soluble in water. The study of the interactions between NAs and divalent metal cations (Mg, Ca, Sr, and Ba) showed changes in dynamic IFT. These cations are all common in coproduced formation water. Dynamic IFT strongly depends on NA structure, type of divalent cation, and the concentration of the compounds as well as the pH of the aqueous phase. A permanent decline of the IFT was attributed to electrostatic attraction forces across the interface between the cations in the aqueous phase and the carboxylic groups at the oil—water interface, which cause a higher interfacial density of NA monomers and the formation of positively charged monoacid complexes that possess high interfacial activity [138].

A C_{80} NA-ARN and its calcium naphthenate have been characterized with respect to their interfacial properties at both the oil—water and the air—water interfaces as well as their thermal properties. The interfacial activity and film properties of the compounds were examined by the oscillating drop method. As per monoprotic NAs, ARN also exhibit a dynamic self-association phenomenon. An elastic, solid-like film can be formed that leads to the formation of crystalline structures in the interfacial film, upon compression. The effect can be more pronounced for calcium ARN naphthenates, this film is also elastic and solid-like but on the contrary it forms a continuous amorphous film on the substrate [139]. While the ARN films are unstable, the naphthenates' films are more robust [140]. Langmuir–Blodgett films of the acid and its naphthenate were deposited on conducting substrates and subsequently analyzed by scanning electron microscopy. It was found that the pure acid formed crystalline structures in the interfacial film on compression. The calcium naphthenate, on the other hand, forms a continuous amorphous film on the substrate [139].

The degree of hydration of divalent cations affects their reactivity with NAs. Strongly hydrated cations, like Mg^{2+}, will be less preferable toward the oil–water interface than less hydrated cations, like Ba^{2+}. The accumulation of the metal naphthenate at the oil–water interface leads to the formation of films. Film stability against compression was found to depend on NA structure, type of divalent cation, and the pH of the aqueous phase. The NAs with more branched structures formed less stable films [140].

The surface activity of monoacids will lead to the formation of particles at the water–oil interface, which will disperse either in the oil or the water phase depending on the solubility of the product. Meanwhile C_{80} NAs form a continuous sticky network at the interface, insoluble both in water and oil. This network will function as glue and will under production conditions bind up sand, scale, and other crude oil components [134]. The aggregation of fully ionized C_{80} ARN leads to the formation of very large structures in pure D_2O and NaCl 20 mM, particles are so large that their exact dimensions could not be determined by SANS [141].

Calcium naphthenate deposits were first recovered on a North Sea field. The sample contained massive proportions of a high MW tetraprotic acid compound, at that time coined ARN. The ARN acid is a family of 4-protic carboxylic acids with MWs in the range 1227–1235 amu. The MW of the homologous ARN acids series are 1227, 1229, 1231, 1233, and 1235 (basic structures) + $n \times 14$ ($n =$ number of additional CH_2 groups in hydrocarbon skeleton). The most populated ARN acid with an MW of 1232 has $C_{80}H_{142}O_8$ as its empirical formula [142]. The molecule consists of four-branched, aliphatic molecules with one terminal carboxylic group attached to each branch, with four to eight sites of unsaturation (or 4–8 cyclopentyl rings). Four members have been identified in the series that differ by the number of cyclopentyl rings: 4 ($C_{80}H_{146}O_8$), 5 ($C_{80}H_{144}O_8$), 7 ($C_{80}H_{140}O_8$), and 8 ($C_{80}H_{138}O_8$).

ARNs are also characterized by the absence of quaternary carbons, double C-bonds, and aromatic rings. This structure led to the suggestion that ARN acid was a metabolic product of Archaea microorganisms [143].

Examination by high-temperature GC of the methyl esters of the so-called ARN NAs from crude oils of North Sea UK, Norwegian Sea, and West African oil fields revealed the distributions of resolved 4–8 ring C_{80} tetra acids and trace amounts of other acids. While all three oils contained apparently the same major acids, the proportions of each compound differed from crude to crude, possibly reflecting the growth temperatures of the archaebacteria from which the acids are assumed to have originated. An HPLC method for the isolation of the individual acids was described. The results revealed structures of the 4-, 5-, 7-, and 8-ring acids, which were tentatively assigned by comparison with the known 6-ring acid and related natural products. ESI-MS indicated the presence of previously unreported acids, tentatively identified as C_{81} and C_{82} formed by 7- and 8-ring analogs [144].

The characterization of a Statoil ARN NA deposit (containing >40 wt% NAs [145]) showed a broad MW distribution, with an average MW of 330. A first molecular structure was proposed (Fig. 4.16A). This structure was the result of ConocoPhillips and StatoilHydro research, which associated such molecular structure as responsible for the deposition of calcium naphthenate in oil production facilities [146]. A refined structure from Lutnaes et al. of the Norwegian University of Science and Technology (NTNU) is also included in that figure (Fig. 4.16B [134,147]).

The quantitative determination of ARN in crude oil is of great importance in the assessment of the potential risk of deposits formation. The very low concentration of ARN in crude oil is the main challenge to overcome. Statoil [148] and Total [149] as well as the NTNU [150] have devoted great efforts to this endeavor. A synthetic model acid compound (BP10) similar to C_{80} ARN was employed for studying the possible factors controlling naphthenate formation. The structure of BP10 has been included as Fig. 4.16C; this molecule with a benzophenone core has an MW of 983 amu. The affinity of this molecule with different cations for forming the corresponding salt follows the order $Ca^{2+} > Ba^{2+} \approx Sr^{2+} > Mg^{2+} >> Na^{+}$. This order seems to correspond with the number of hydration water molecules complexed to the cation in solution, in combination with the cation size, which also reflects the shielding density from water molecules [151]. The comparison of the interactions of the same cations with ARN and BP10 indicated that film formation was dependent on the ratio M^{2+}/acid. The main deviations were in the solid state, where BP10 salts showed crystalline transitions, while the ARN naphthenates did not [152]. The reaction of C_{80} ARN and BP10 with Ca^{2+} showed that both compounds precipitated within similar pH ranges; however, the dependence of acid concentration with pH differed [153]. The capacity of BP10 to model the behavior of ARNs has been the subject of several studies. A comparison of the key physicochemical properties has been collected by Sjöblom et al. and is summarized in Table 4.5 ([154] and references therein).

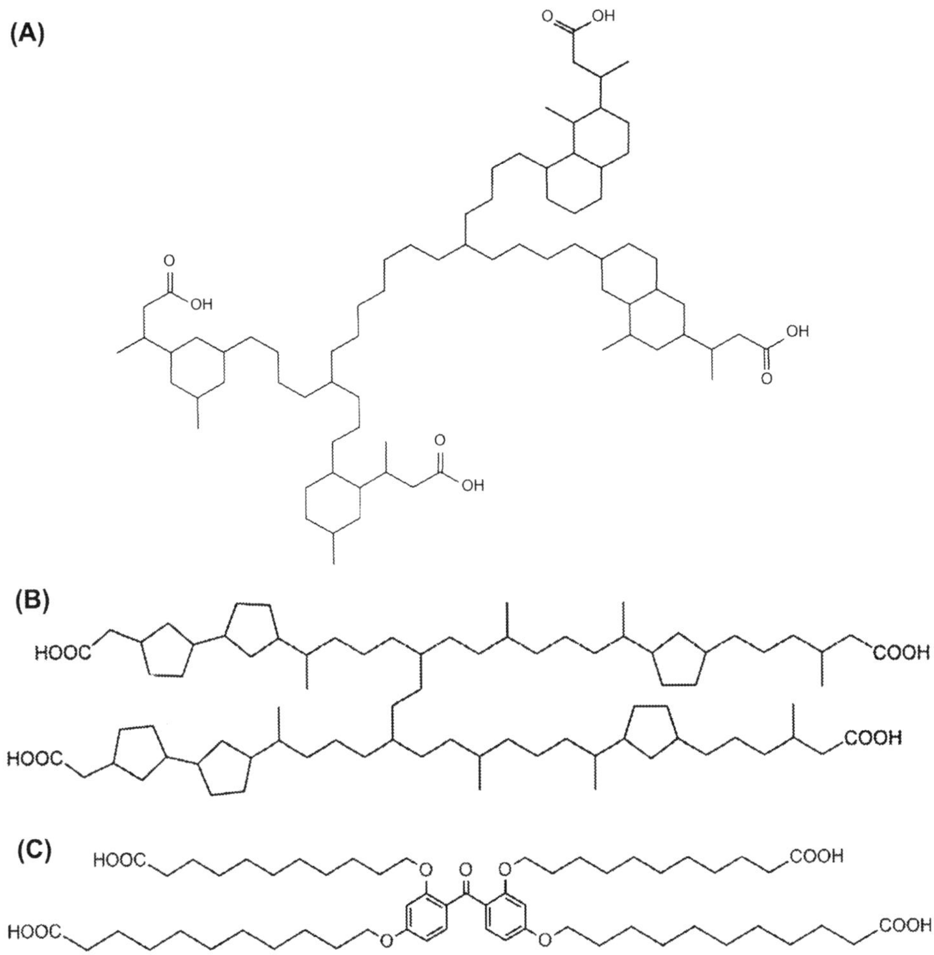

Figure 4.16

ARN molecular structure: (A) original [145], (B) refined, and (C) BP10 model compound. *Reproduced from Sundman O, Simon SB, NordgåRd EL, SjöBlom J. Study of the aqueous chemical interactions between a synthetic tetra-acid and divalent cations as a model for the formation of metal naphthenate deposits. Energy Fuels 2010;24(11):6054–60, with permission from ACS Publications.*

A review on the chemistry of these tetrameric acid species has been published, in which more than 90 references have been cited, indicating the great interest these species have attracted since their discovery [154]. The detection of ARN acids in crudes is of paramount importance, especially for designing future installations and allowing for mitigation means. The proposed ARN detection strategy involves a concentration step and a high-pressure LC-MS analysis [155]. Using this methodology, ARN has been detected in a large number of crude oils, most of which do not show any operational

Table 4.5: Comparison of Physicochemical Properties of ARN and BP10 Model Compound [154]

Property	Similarity	Comments
Critical micellar concentration	Yes	Same order of magnitude: BP10 < C_{80}-ARN
Solubility in oil	Partially	BP10: low solubility in toluene, more soluble in $CHCl_3$ than C_{80}-ARN
Solubility in water	Different	BP10 more soluble than C_{80}-ARN
Langmuir films	Some	BP10 forms the Ca film at higher pH than C_{80}-ARN
Emulsions	Yes	Generally oil-in-water
Interfacial activity	Yes	Similar interfacial tension
Glass transition temperature	No	No in BP10, Yes for C_{80}-ARN

difficulties. A physicochemical model is then proposed for interpreting these results. It incorporates the formation, in specific conditions, of a polymer-like structure through ARN association [156]. Although it was found that pH was the most important variable in naphthenate formation, it failed to explain the lower average MW of acids in the deposit than in the crude (330 vs. 430 amu), the low dosage of naphthenate inhibitor sufficient to inhibit deposition at the field, and the undetectable change of Ca^{2+} and acids concentrations in the uninhibited system [17].

While great attention has been given to the characterization of ARNs, sodium naphthenates have been studied in a lesser extent. The high-resolution FT-ICR-MS characterization of a deposit containing mixed naphthenates indicated significant differences in the chemical composition of Na- and Ca-naphthenates. Na-naphthenate originates from lower MW (C_{15}—C_{35}) monoprotic NAs. Additionally, low MW tetraprotic ARN acids with C_{60}—C_{77} hydrocarbon skeletons were identified in the Ca-naphthenate deposit [157].

A summary of the contribution of the amphiphiles present in HACs on the formation and stabilization of emulsions depending upon the water content and initial pH of the aqueous phase has been identified by Arla et al. [158] as:

1. The light NAs and naphthenates are able to form and stabilize emulsions;
2. The acids contained in the intermediate fraction seem to dictate the type of emulsions formed with the crude oil;
3. The naphthenates generated from the intermediate fraction are responsible for the formation and stabilization of oil-in-water emulsions formed at high pH and water content; and

4. The heaviest amphiphiles contained in the crude oil, i.e., resins and asphaltenes, played a major role in the long-term stability of the oil-continuous emulsions.

4.2.5 Adsorptivity

The affinity of NAs with mineral and other materials has been evidenced by the study of the adsorption on magnetite [159], calcite, alumina, and silica [160], acid montmorillonite [161], activated alumina and bentonite [162], and on β-cyclodextrin (β-CD) copolymers [163,164]. The first two studies demonstrated an increased affinity for Fe and Ca, in comparison with that for Si or Al [159,160]. The comparison between clays (having a negative surface charge) with activated alumina (having amphoteric character and a positive surface charge) indicated an NA preference either for basic sites or for positive surface charge [161,162].

Adsorption on clays has been proposed to occur in two steps. The first step involves either hydrogen bonding between the carboxyl group of the NAs and surface hydroxyl groups or an electrostatic charge–dipole interaction between the clay and the NAs' strong polar group. The second step was caused by van der Waals forces between the large hydrophobic group of the adsorbed NA molecule and another one in solution [165].

In general, adsorption capacity and selectivity of any adsorbent are higher when tested with synthetic mixtures or model NA compounds than what is observed with real feeds.

5. Reactivity/Stability

Most of the reactivity of NAs is of acid–base type, though redox behavior has also been observed. Acidity (pK_a) of NAs is typically lower than their carboxylic counterpart. Thus neutralization reactions (producing the corresponding naphthenate and water) are fast with strong bases, but slow down with the pK_b of the neutralizing base. Lewis bases, basic oxides, carbonates, and bicarbonates will react in a similar way, while the latter two additionally will release CO_2.

Neutralization with ammonia and organic bases is slow at room temperature, but could proceed significantly faster at higher temperatures (50–350°C). Some examples have been reported such as the reactions with tetraalkylammonium hydroxide [166], polypropylene amine and anionic exchange resins having amino groups [167], and with aqueous ammonia, diethylamine, and diethylenetriamine [168,169]. Increasing the alkalinity of the media and decreasing steric hindrance of the base increase reactivity. In the case of aqueous solutions, a dosage of demulsifiers is required.

There might be more than 1000 NA species; some are very reactive, others are relatively inert, and different species distil at different temperatures and can concentrate in specific areas in the refinery.

5.1 Reaction With Metals: Corrosion

In principle, corrosion is the oxidation reaction of metals (more particularly of iron) and it is typically followed by the hydration of the oxide. The formed rust could on occasions be easily detached from the bulk, exposing fresh metals to the oxidizing atmosphere. Instead, acid corrosion follows a different chemistry induced by the presence of free protons and/or highly oxidizing anions.

The NAs' main featuring reactivity concerns corrosion, which is their most studied reaction. As carboxylic acids, NAs react with reactive metals to produce the corresponding naphthenate salt and hydrogen (Rxn. 4.1, showing the Fe case). Rxn. 4.1 is a typical metal–acid reaction (except noble metals), which could occur rapidly with strong inorganic acids but much slower with NAs.

$$2R[CH_2]_nCOOH + Fe \rightarrow (R[CH_2]_n\ COO)_2Fe + H_2 \tag{4.1}$$

Recent research has shown that the corrosivity of NAs is related to their molecular mass and that the TAN, traditionally used as an indicator of the NA content of oil, is not a reliable measurement of corrosivity of the crude, as it was first believed. In this regard, two types of NAs have been distinguished: α and β, the former corresponding to low MW NAs and exhibiting very high corrosivity and the latter having high MW and low corrosivity [170,171] (explained as caused by steric hindrance [172]). A summary of the distinguishing features between these two acid classes has been given by Kane and Chambers [173] and is reproduced here:

α NAs	β NAs
• low MW, ~125–425 amu	• High MW, ~325–900 amu
• Moderate to high solubility in water, moderate to low solubility in oil	• Low solubility in water, high solubility in oil
• Carboxyl group readily ionizes in aqueous solutions	• Carboxyl group poorly ionizes in aqueous solutions
• Neutralizes to form salts	• Difficult to neutralize
• Iron naphthenates—highly soluble in oil	• Iron naphthenates—difficult to form
• True boiling point up to 385°C	• True boiling point ~357–816°C typically above average crude true boiling point
• No protective film formation	• Formation of protective/inhibitive film
• NA corrosion follows classical model	• No classical NA corrosion model

According to this α/β classification, the former acids will form an oil-soluble product that will erode the metal surface. Meanwhile, the latter acids will render a film on the surface of the metal. An oxide film has been observed to form upon corrosion with model acid compounds as well as extracted NAs. Magnetite was identified as the more prominent iron oxide in the film. However, whether this oxide film has protective or inhibiting action was not proven [174,175].

The difference in corrosivity between Isthmus and Maya crude oil was explained as being caused by a higher proportion of low MW acids in the former, a more mature crude oil (lighter). Consequently, the corrosivity observed in the crude distillation units (CDUs) of the Salina Cruz Refinery, when distilling blends with higher relations of the Isthmus crude oil, may occur at boiling point temperatures below 200°C when the distilled fractions are rich in low MW NAs and S-content is nearly zero, preventing the formation of the S-protective film [176].

This difference in reactivity has been suggested as a way of separating reactive NAs from nonreactive NAs. Reaction with Fe powder would form the naphthenates from the reactive NAs, leaving the nonreactive NAs intact. However, it has been noticed that water and naphthenate decomposition are complicating factors for this reaction. The recommended conditions for the reaction were a fixed bed reactor loaded with the (15 g or 5 cc) iron powder, 250–350°C, flow rate of 0.17–0.5 cc/min, liquid hourly space velocity of 2–6 h^{-1}, and residence time of 10–30 min. After reaction a two-stage ion-exchange resin extraction will separate the formed Fe-naphthenate from the nonreactive NAs. The complete process scheme is shown in Fig. 4.17 [170].

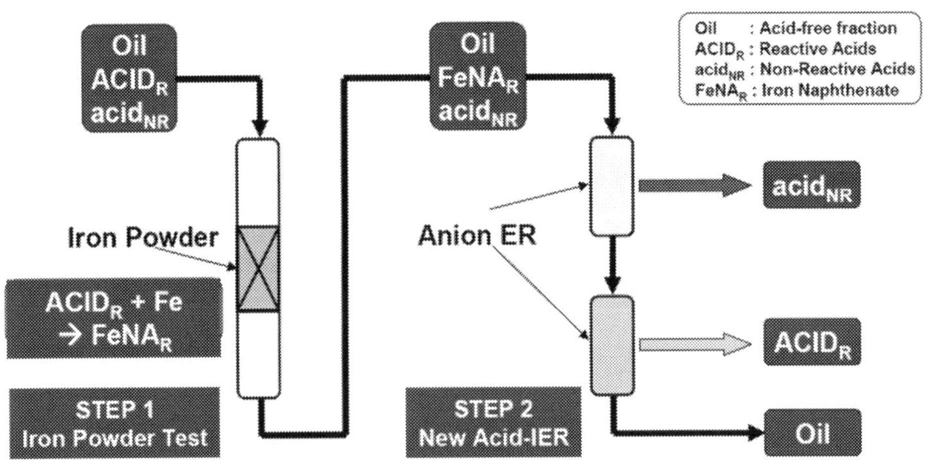

Figure 4.17

Separation of reactive and nonreactive naphthenic acids (NAs). *ER*, Exchange resin; *IER*, ion-exchange resin. *Reproduced from Rahimi P, Kayukawa T, Ryan R, Alem T. Comparison of the reactivity of naphthenic acids in athabasca bitumen and san Joaquin valley. In: Proc. Crude Oil Quality Association Meeting. New Orleans, LA. USA, February 10–11; 2010. 48 pp. Available from: http://www.coqa-inc.org/docs/default-source/meeting-presentations/20100211_rahimi_athabasca_-napacids.pdf?sfvrsn=2, with permission from COQA.*

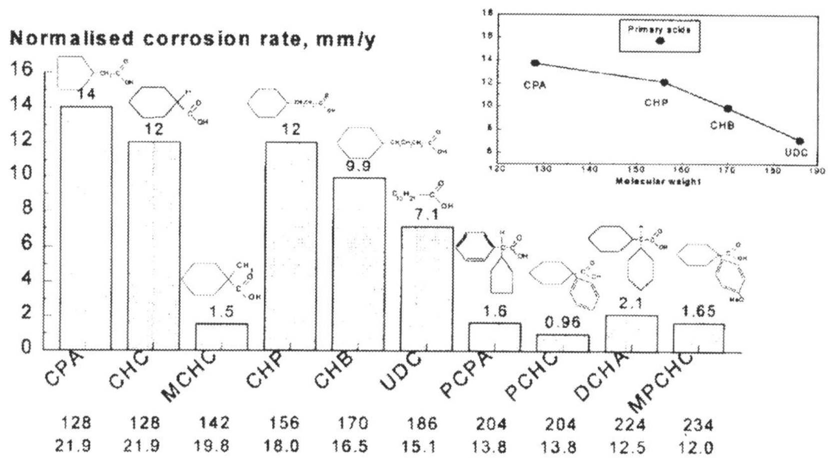

Figure 4.18

Effect of acid structure and molecular weight on corrosion of carbon steel in mineral oil. *Reproduced from Lewis KR, Daane ML, Schelling R. Processing corrosive crude oils. In: Proc. Corrosion/99. San Antonio, TX, USA; April 2–30, 1999. Paper #377. 10 pp., with permission from NACE International.*

Turnbull et al. reported that increasing alkyl chain lengths increases corrosivity when the chain has less than three methylene groups, after which corrosivity decreases [177]. Lewis et al. studied the impact of NA MW and structure on corrosion. They found that at constant temperature the lighter MW acids have a relatively higher corrosivity. Fig. 4.18 shows the decreases in corrosion rate as the acid structures shift from primary to tertiary carbon atom [178]. From a practical standpoint the applicability of this type of result is limited because of the extensive and detailed amount of analysis required to determine the types of acids present in any given crude oil or fraction. Nevertheless, it is indicative that molecular structure counts and not all the NAs are corrosive.

NAs with lower MW and fewer ring structures were found to be more corrosive in five studied NA samples for which the acidic species were saturated aliphatic and NAs, and the amount of aromatic and unsaturated acids was negligible [179]. A set of 10 crude oils with different TAN, corrosivity, and from various source-rock materials were analyzed by ultrahigh-resolution FT-ICR-MS. A variety of graphical methods were used to visually display each crude oil and revealed the lack of correlation between basic and acidic species to corrosive character; neither was there any correlation between a given specific class or type of acidic species and corrosive character. Furthermore, a lack of correlation was found when examining the relative abundance of the O_2-containing acids relative to all species versus TAN. Similarly, the abundance of the NAs as the acid percent relative abundance to the percent relative abundance of all other carboxylic acids versus TAN did not show any correlation. It was also affirmed that the MW distribution played no role in the corrosive nature of a particular crude [45]. In the particular case of the distillates of a

Mexican crude, a relationship between TAN and corrosion was found, while no association was observed between the concentrations of NAs and TAN in the HVGO [180].

NAs are responsible for a specific type of corrosion in specific areas of the refining processes and NA concentration is not necessarily a function of TAN magnitude [181]. The belief that crude corrosivity increases with increasing crude oil TAN has been proven not to be quite accurate, as mentioned before [173,182–185]. Clearly then, NAs are not the only acid compounds present in crude oils and also they are not the only species responsible for corrosion. For this reason, TAN only correlates with corrosivity when different concentrations of the same acids are tested. Nevertheless, this is just one of the reasons. The intrinsic extent of corrosion by NAs depends not only on TAN, but also on the availability of the carboxylic acid group to form metal complexes and the type of compounds present. Carboxylic corrosion starts at temperatures above 115°C, though the rate of reaction is very slow at this temperature. Corrosion might stop at temperatures above their decomposition temperature, which is not exactly known and depends on their molecular structure [186].

Thermal conversion of NAs in an Athabasca HVGO was followed by ESI FT-ICR-MS between 300 and 400°C [187]. However, the decomposition temperature range extends broadly. It has been reported to start at 370°C [188], becomes appreciable only above 400°C, and shifts to between 450 and 550°C for heavier compounds [189,190]. Consequently, no corrosion damage has been observed at temperatures above 400°C, which was explained as being caused by the decomposition of NAs accompanied by the formation of coke at the metal surface [182]. Additionally, the decrease in TAN (and in corrosivity [190]) observed on HACs in the 300 and 400°C temperature range [187] has been explained by the evaporation of lighter NAs [191].

The corrosiveness of crude oil fractions containing various amounts of NAs was tested for different distillate fractions. They tested the inhibiting behavior of nonacidic kerosene compared to commercial inhibitors. Nonacidic kerosene became especially effective at a volume ratio of 1:3. Heavier acid fractions showed an opposite corrosion behavior compared to the lighter ones, particularly to the three acidic kerosene fractions. In this case, the heavier fractions were less corrosive than the lighter ones. It seems that MW might be as important as molecular structure in determining the reactivity/aggressiveness of NAs [192]. The vast difference in relative abundance of the carboxylic acids suggests that although these compounds contribute to corrosion, other species, such as the nonpolar sulfur compounds, also play a key role in the corrosive nature of a crude oil.

Thus the complexity introduced by diversity in the molecular structures of NAs is derived from the following effects:

1. The boiling point difference creates a concentration profile throughout the distillation cuts;

2. The inherent reactivity of different molecules exhibits steric hindrance in some and a superior inhibition behavior in others;
3. The differences in decomposition temperatures and reactions leads either to the formation of more acidic light carboxylic acids or to CO_2 formation rendering a nonacid hydrocarbon; and
4. High boiling point NAs are exposed to conditions for high-temperature corrosion.

The reaction between iron and NAs has been suggested as one of the possible corrosion mechanisms and as a way for a quick test for NA corrosivity [193,194]. This reaction (shown in Rxn. 4.1) produces H_2, along with iron naphthenate [195]. The CO_2 detected during the experiment was supposedly the product of the acid thermal decomposition [196]. However, reaction was carried out at temperatures below 340°C, at which NA thermal decomposition is low.

The hydrogen produced in Rxn. 4.1 may induce the decomposition of sulfoxide compounds through a water-producing reaction as shown in Rxn. 4.2. Similarly, sulfoxides might react with mercaptans through Rxn. 4.3 and contribute to additional water formation [197]. This formed water will contribute to keeping the NAs in a dissociative state and becoming more reactive [195], particularly for reaction with metals as per Rxn. 4.1. The determination of sulfoxide presence and content in a crude oil and/or its distillates is necessary for forecasting possible contribution of Rxn. 4.2 in NA corrosion problems. These sulfoxides might be produced by a mild oxidation with air during storing of certain crudes.

$$R_2SO + H_2 \rightarrow R_2S + H_2O \tag{4.2}$$

$$R_2SO + RSH \rightarrow R_2S + RSSR + H_2O \tag{4.3}$$

NAs are just one source of the corrosive properties of crude or crude fractions. Other naturally found crude components that can contribute to corrosion in the refining process are hydrogen sulfide, mercaptans, organic sulfides, mineral acids, phenols, and carbon dioxide. Thus three types of corrosion in the presence of NAs have been distinguished [198]:

- Type I (NA dominated) is a pure NA corrosion scenario in the absence of sulfur compounds or with very little or no effect if they are present;
- Type II (S dominated) in which sulfidic corrosion is accelerated by the presence of naphthenic acid (see Ref. [109]);
- Type III in which NA corrosion is inhibited to some extent by the presence of H_2S.

From these corrosion types, the two dominant corrosion mechanisms can be easily deduced:

- NA corrosion, in which soluble corrosion products are formed; and
- Sulfidic corrosion, in which corrosion products would form a film under most cases [173,183].

These two mechanisms were suggested either to act independently with one dominating the corrosion behavior or to interact synergistically adding to the complexity. In both cases velocity as flow/turbulence highly affects the predominance of one mechanism over the other by inducing wall shear stress. Furthermore, the resulting chemical, structural, and morphological changes vary among the undergoing dominant corrosion type [199,200]. A micrography study showed the dominance of Type I corrosion under the effects of an oil with 5.0 TAN, 0.38% total sulfur, a temperature of 343°C, and 24 h exposure; and of Type II under the effects of an oil with 0.1 TAN, 0.38% total sulfur, a temperature of 343°C, and 96 h exposure [199].

Corrosion based on those first two mechanisms is relatively easy to predict. However, the complex interrelations between S, NAs, and operating conditions (e.g., shear stress and metallurgy) make corrosion prediction very difficult [201].

High corrosion rates with reactive sulfur species have been used to explain "NA corrosion" of some systems when processing low-TAN crudes [109] since there is evidence from laboratory tests that reactive sulfur species and NAs synergize to increase corrosion in the liquid phase [202–205]. These possible synergies among the different corrosive compounds require a more systematic study and comparison with real field or operational situations [206]. Rxn. 4.2 explains one of the possible synergy mechanisms with S-compounds, since, for instance, $CH_2{-}S$ bonds (sulfides) start cracking to form H_2S at temperatures as low as 200°C.

Similar to the situation with NA corrosion, sulfidic corrosion varies with molecular structure. Thus predictions with model compounds or synthetic mixtures typically underestimate corrosion rates, even on the basis of breakdown into type of reactive sulfur species closer to what is present in the crude oil [178,207]. Exxon has recognized that sulfidic corrosivity of a crude oil or fraction depends not only on total S concentration, but more specifically on that of reactive sulfur compounds, and also on the physical nature of the refinery stream (stream velocity and physical state) and metal temperatures [208,209]. H_2S reacts with iron to form iron sulfide scale at temperatures above 260°C. The reaction rate depends on H_2S concentration, temperature, stream flow rate, and steel composition. The product of sulfidic corrosion is iron sulfide, which is believed to form a protective film on the surface of steel that prevents further corrosion [210]. Other metal sulfides have been observed upon high-temperature sulfidic corrosion of alloys [200]. Another synergy between NAs and sulfur compounds is shown in Rxn. 4.4 and Rxn. 4.5. The reaction shown in Rxn. 4.4 is a reversible reaction and its reversed reaction is shown as reaction Rxn. 4.5. According to reaction Rxn. 4.4, the presence of NAs will lead to the formation of the oil-soluble naphthenate, from the iron sulfide protective film, formed according to Rxn. 4.6.

$$2R[CH_2]_nCOOH + FeS \rightarrow (R[CH_2]_nCOO)_2Fe + H_2S \tag{4.4}$$

$$(R[CH_2]_nCOO)_2Fe + H_2S \rightarrow 2R[CH_2]_nCOOH + FeS \tag{4.5}$$

$$Fe + H_2S \rightarrow FeS + H_2 \tag{4.6}$$

In fact, metallographic measurements of sulfide film thickness before and after exposure to a high-TAN environment revealed that NA has the ability to chemically dissolve sulfide films on the metal surface for all alloys examined, thus leading to increasing corrosive attack [211]. This reaction (Rxn. 4.4) not only contributes to the erosion of the steel wall, but also would carry off heavy metals into the oil stream. In the presence of H_2S or reactive-S species that produce it upon decomposition, Me-naphthenates will not form a protective layer on the metal walls (Rxn. 4.5) and because solubility in the oil phase. The understanding of the nature of films formed in alloyed materials is vital for an explanation of the effects of the different compounds and components in the oil and of the operations and service conditions. For instance, the addition of 0.1% H_2S to the HAC or fraction containing NAs tended to decrease the corrosion rate, most likely because of formation of stable iron sulfides in competition with the soluble iron naphthenates. The degree of corrosion inhibition of HVGO compared to that of mineral oil was thought to be caused by the presence of sulfur compounds generating H_2S and/or to the presence of naphthenoaromatic acids or large S-containing molecules with good adsorbability on iron and steel [177].

Whether the oxide film possibly formed through NAs corrosion has a protective action against sulfidic corrosion, whether it is formed in the copresence of S-compounds and NAs, or what kind of interaction may exist between the two probable films at all remains unknown [175].

In conclusion, active sulfur could inhibit or could also promote corrosion. The effect of sulfur compounds in NA corrosion involves reduction by the protons provided by the corrosion process. When the reduction product is H_2S, the production of the protective sulfide layer could prevent NA attack. However, when the reduction product is H_2O an increased deprotonation of the NAs would contribute to the acceleration of the corrosion rate [197].

A combination of infrared and LC/ESI-MS methodologies for analyzing NAs from comparative crude oils and gas oil was applied before and after corrosivity tests. The changes observed on the acidity profile and on the carbon distribution indicated that NAs with 1–3 naphthenic rings and C_{14}–C_{17} C-number preferentially undergo dissolution as iron naphthenates (Rxn. 4.4), ring decomposition into more aliphatic and eventually NAs are released into the fluid (Rxn. 4.5) [212].

Passivation chemistry is based on creating a more resilient layer to NA action than the FeS counterpart. One of these chemicals is based on phosphate ester compounds, which decompose forming phosphoric acid. The acid then complexes with the FeS layer present, creating a new layer that serves for improved protection in an NA environment. This

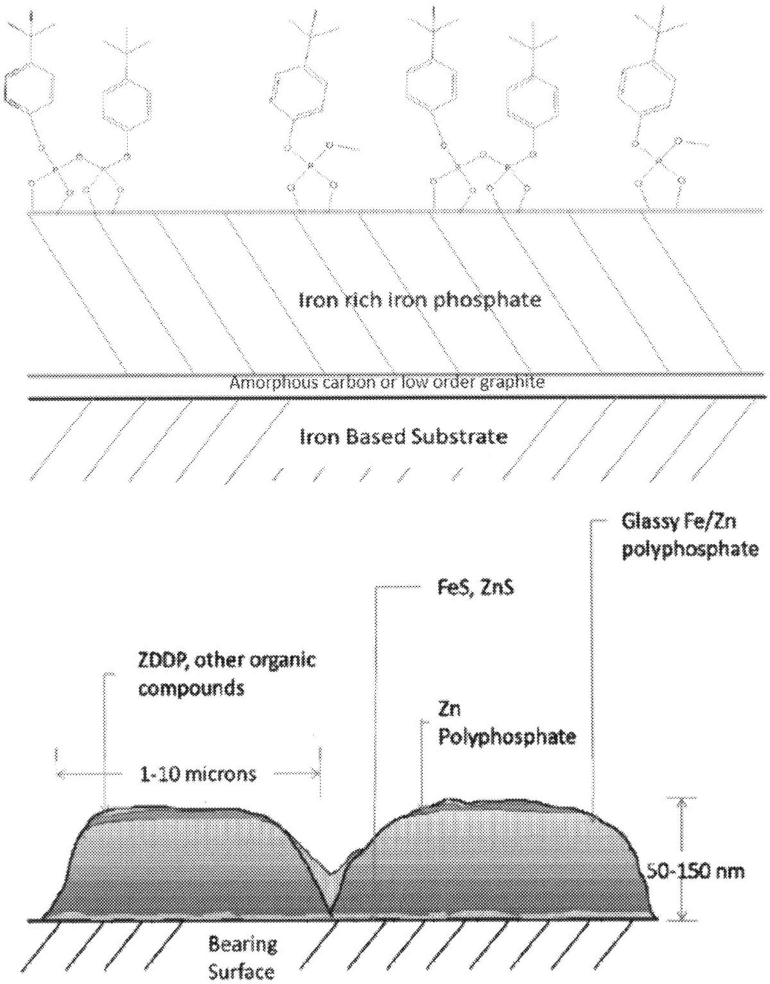

Figure 4.19

Passivating action of P-based additives. *ZDDP*, Zinc dialkyldithiophosphate. *Reproduced from Johnson DW, Hils J. Phosphate esters, thiophosphate esters and metal thiophosphates as lubricant additives. Lubricants 2013;1:132–48, with permission from MDPI AG.*

action mechanism is presented in Fig. 4.19; drawing (A) shows the complexing action and (B) the creation of the protective layer [213].

5.2 Decarboxylation and Decomposition

The decomposition of aliphatic carboxylic acids and fatty acids may involve both decarboxylation and cracking, depending on the C-chain. The minimum temperature at which decarboxylation of stearic acid was observed was 225°C [214]. Thermal cracking of alkyl chains in NAs may start at temperatures as low as 300°C, forming lower MW NAs and probably releasing light carboxylic acids as well. A particular case of decomposition is decarboxylation that eliminates a CO_2 molecule from NAs rendering a hydrocarbon molecule (Rxn. 4.7). This reaction occurs at temperatures above 400°C. The formed CO_2 in the presence of water would produce carbonic acid that may contribute to corrosion [199].

$$RCOOH \rightarrow RH + CO_2 \tag{4.7}$$

Catalytic decarboxylation has been studied on various catalytic compositions. In organic and biochemical processes, this chemical reaction has been applied for organic synthesis employing homogeneous Cu-based catalysts [215] and tungsten complexes [216].

Basic as well as acid heterogeneous systems have also been examined. Oxides (MgO, CaO, Ag_2O, Cu_2O, and MnO_2), metal supported (Pt/SiO_2, Ni/Al_2O_3, and Cu/SiO_2), and ZSM-5 catalysts were theoretically as well as experimentally studied. Model compounds and their mixtures were tested. Although considered catalysts acted through different reaction mechanism, all catalysts showed high catalytic decarboxylation activities at relatively low temperatures of 200–300°C. Alkaline earth metal oxides catalyzed the decarboxylation reaction via a ketonization mechanism through the coupling of 2 mol of carboxylic acid and producing only 1 mol of CO_2. A free radical mechanism was proposed as the most probable route for Ag_2O and Cu_2O. Meanwhile, on the zeolite, decarboxylation would occur by C–C bond cracking. Regardless of the good results with model compounds, passing to the crude oil brought catalyst poisoning problems. Poisoning was explained as being caused by S-compounds and coking [217]. Zirconia promoted catalytic decarboxylation of acetic acid in supercritical water at 400°C and 25–40 MPa [218]. A performance comparison among catalysts of oxides of cerium, zinc, and tin supported on alumina placed cerium oxide on the top position when evaluated for the processing of a Korean HAC of TAN above 8 mg KOH/g. However, ceria was not capable of bringing TAN below 1. An improvement attempt was considered by promotion with Mn, Co, Ni, and Cu. This latter catalyst enhanced performance to the desired levels and it was assigned to the increase in both basicity and surface area [219].

Catalytic decarboxylation of benzoic acid also occurs at about 400°C on REY-zeolites [220]. The reaction mechanisms occurring on Bronsted and Lewis acid sites have been proposed, based on theoretical molecular simulations. In Fig. 4.20, these two mechanisms are shown. Based on the theoretical results a catalyst with a relative population of about 60% Lewis acid sites and about 40% Bronsted was prepared and tested with an HAC with TAN of 12.52 mg KOH/g at a volume space velocity of 8 h^{-1}, ratio of catalyst to oil of 7.5, and temperature of 460°C, obtaining 97% decarboxylation [221].

An increase in the density of basic sites on the surface of MgO increased the decarboxylation reaction of NAs, indicating that deprotonation of the acid might be a

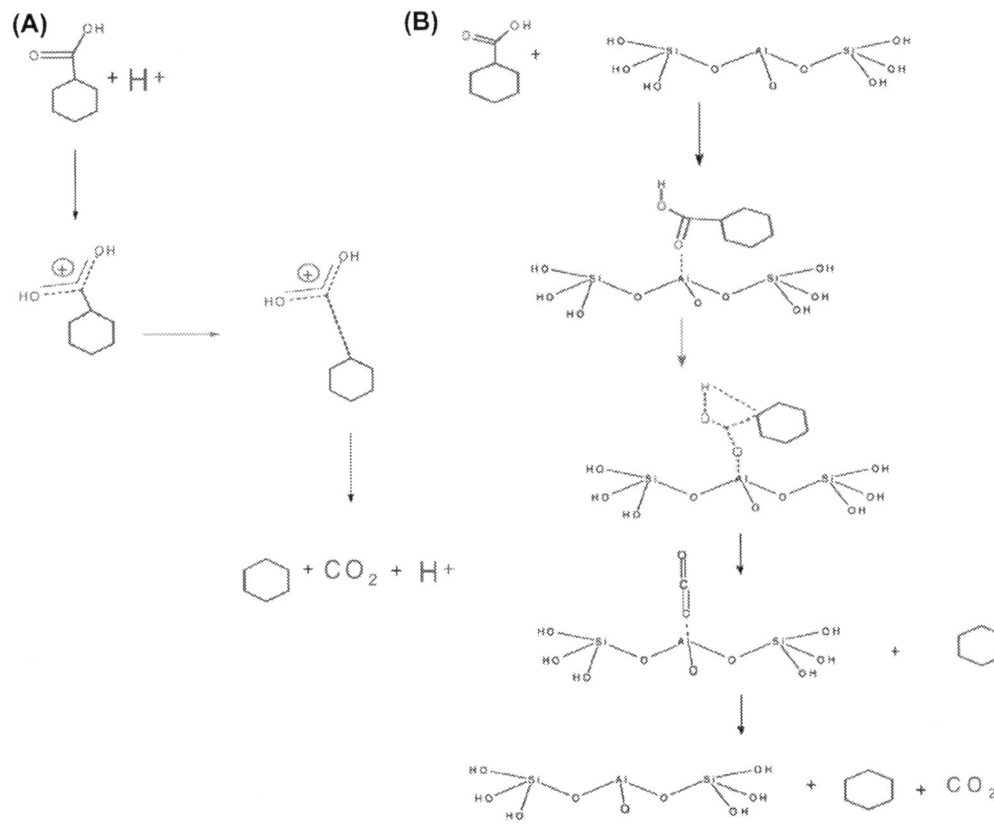

Figure 4.20

Catalytic decarboxylation mechanisms on: (A) Bronsted acid and (B) Lewis acid. *Reproduced from Fu X, Dai Z, Tian S, Long J, Hou S, Wang X. Catalytic decarboxylation of petroleum acids from high acid crude oils over solid acid catalysts. Energy Fuels 2008;22(3):1923–29, with permission from ACS Publications.*

controlling step of the reaction mechanism [222]. A thermal treatment on MgO led to similar results. In this case, the surface properties of MgO were modified by calcination and/or hydration followed by dehydration under vacuum. Although calcination at high temperatures reduced the Brunauer–Emmett–Teller surface area and number of basic sites, it gave the highest decarboxylation activity. However, these treatments increased the direct reaction of MgO with NAs, leading to the bulk formation of Mg-naphthenate that dissolved in the treated oil [223]. The role of MgO in the system is considered to be multiple: (1) adsorption of acidic compounds via acid/base neutralization and (2) promotion of decarboxylation and cracking reactions at increased temperatures [224]. Catalyst structure and basic strength have been found to play an important role in the catalytic decarboxylation of alkaline-earth metal oxide catalysts [225].

Powders of Li_2O, CaO, $CaCO_3$, BaO, and MgO were tested at 385°C in a batch reactor, and their effectiveness in the liquid phase decarboxylation reaction was determined based on acid (TAN) reduction. CO_2 production rate was independent of the NA concentration and the corresponding naphthenates of the catalyst metal were detected on the spent catalyst [226,227]. These two observations (in agreement with the work of Ref. [222]) are indicative of a zero order of reaction, which is directly evidenced by the no-dependence on reactant concentration or a catalyst surface saturation on the reactive species.

Finally, depending on the conditions and reactivity/acidity of NAs, on basic catalytic materials such as alkaline and alkaline-earth oxides, catalytic decarboxylation, neutralization, and cracking may coexist leading to catalyst deactivation by irreversible changes in the crystalline structure and/or active phase composition. ZnO and in some instances MgO do not undergo changes in their crystalline phases during catalytic decarboxylation of NAs [228].

Symmetrical ketones can be formed upon catalytic (e.g., MnO_2) thermal decomposition of aliphatic carboxylic and fatty acids [229–231], though there are no reports of this reaction being observed on NAs. Instead, thermal dehydration of NAs has been reported to be a way of reducing acidity by producing an anhydrite (Rxn. 4.8) [170].

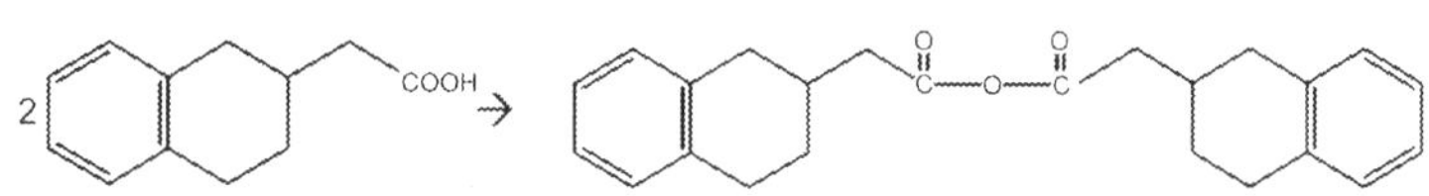

(4.8)

Reactivity toward degradation is also a function of NA structure. While a single-ring NA model compound (*trans*-4-methyl-1-cyclohexane carboxylic acid) could be appreciably degraded by *Variovorax paradoxus* and *Pseudomonas putida* [232,233], only the light acid compounds were metabolized from a Merichem commercial preparation [234]. Biodegradation of commercial NAs revealed that the mixture contained a substantial labile fraction, which was rapidly biodegraded, and a recalcitrant fraction composed of highly branched compounds. Conversely, NAs in oil sands process-affected water (OSPW) were predominantly recalcitrant and degraded slowly by first-order kinetics. Carbon number (n) had little effect on the rate of biodegradation, whereas a general structure persistence relationship was observed indicating that increased cyclization (Z) decreased the biodegradation rate for NAs in both mixtures. Time to 50% biodegradation ranged from 1 to 8 days among all NAs in the commercial mixture, whereas half-lives for OSPW NAs ranged from 44 to 240 days, likely a result of a relatively high alkyl branching among OSPW NAs [235].

Ozonation has been reported as a more effective treatment than biodegradation for the decomposition of recalcitrant NAs present in OSPW, minimizing the toxicity of the water phase [236,237]. However, any potential implementation of this type of treatment would also require an economical assessment in terms of the magnitude of future tailings and reclamation requirements of the industry.

5.3 Esterification

NAs will react with alcohols to form esters in a slow and reversible reaction (Rxn. 4.9). Reactivity toward esterification has been reported to depend not only on molecular structure but also on other physicochemical parameters of the NAs. Lighter acids (lower MW and boiling point) were found more reactive. Produced esters were less corrosive than parent NAs, decreasing corrosion rate of the crude oil in more than 90% after an esterification treatment [190].

$$\mathrm{RCOOH} + \mathrm{R'OH} \rightarrow \mathrm{RCOOR'} + \mathrm{H_2O} \tag{4.9}$$

The kinetics of esterification of NAs present in a VR from heavy Colombian oil was studied in the temperature range of 150–250°C. At the lower temperatures esterification proceeded very slowly but became appreciable at 250°C [190]. Meanwhile, on a SnO catalyst it was already measurable at 180°C [238]. In both cases, the kinetics of the noncatalyzed and of the catalytic reactions followed a second-order reaction rate equation [190,238]. An SnO/Al_2O_3 catalyst was employed for esterification with methanol of NAs in a diesel cut [239], an HVGO [240], and in a whole oil [241]. Esterification rate was found to be affected by temperature, methanol/oil ratio, and space velocity. A high reaction temperature and methanol/oil ratio or low space velocity were favorable for the esterification reaction. The optimum reaction temperature was 280°C for the diesel feed [239], while it was 300°C for the HVGO [240]. The crude oil TAN was reduced from 2.8 mg KOH/g to a value below 0.5 mg KOH/g. In this case, dewatering was found to favor esterification [241].

5.4 Other Chemical Properties

The characterization of NAs is also of interest to geochemical studies, particularly migration and biodegradation, and to refinery wastewater treatment for environmental compliance. The differences in relative abundance of the carboxylic acids do support the use of these carboxylic acids as biomarkers for specific geographical location as well as locaters of specific regions in a reservoir [242].

The characterization of NAs in water tailings provides additional information on biodegradation, along with their environmental/toxicity impact. GC-EI-MS was used to characterize NAs in nine water samples derived from oil sands extraction processes. For each sample, the relative abundances for up to 156 base peaks, with each representing at least one NA structure, were determined. Plotting the relative abundances of NAs as three-dimensional bar graphs showed differences among samples (e.g., Fig. 4.21). The relative

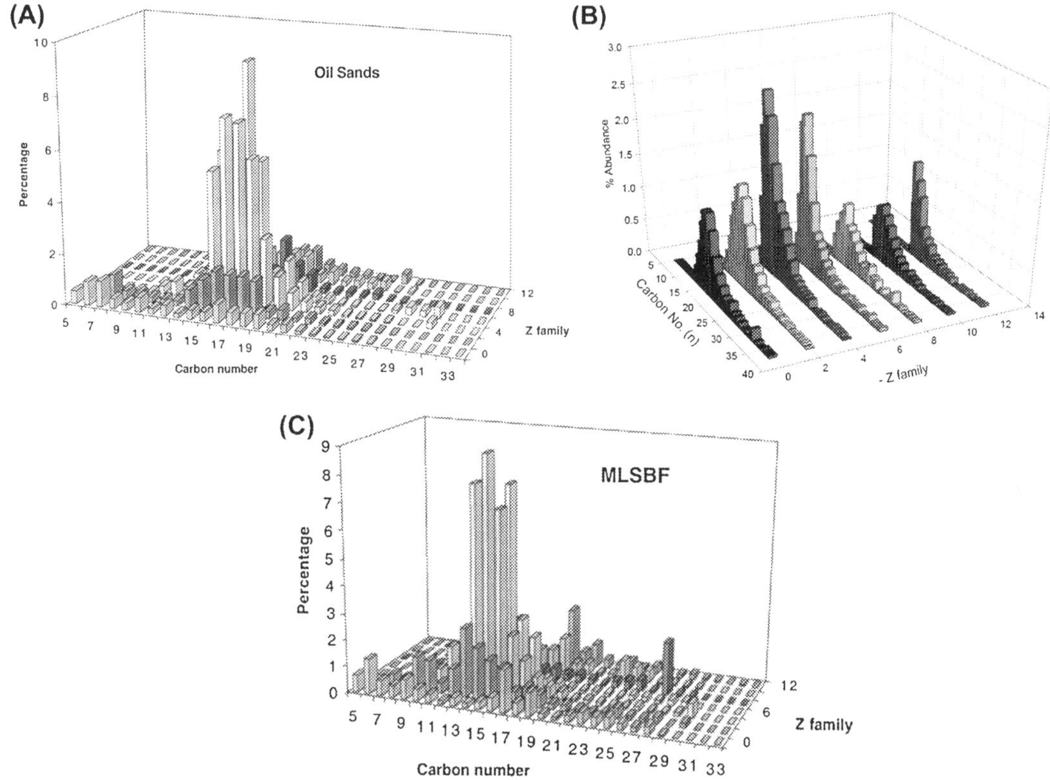

Figure 4.21

Carbon numbers and *Z* families distribution of naphthenic acids (NAs): (A) sand and (B) contaminated water. *MLSBF*, Mildred lake settling basin. *Reproduced from Green JB, Hoff RJ, Woodward PW, Stevens LL. Separation of liquid fossil fuels into acid, base and neutral concentrates: 1. An improved nonaqueous ion exchange method. Fuel 1984;63(9):1290–301, with permission from Elsevier.*

abundance of NAs with carbon numbers $\geq$21 to those in the "C22$^+$ cluster" (sum of all NAs with carbon numbers $\leq$22 in *Z* families 0 to -12) proved useful for comparing the water samples that had a range of toxicities. A decrease in toxicity of OSPW accompanied an increase in the proportion of NAs in the "C22$^+$ cluster," likely caused by biodegradation of NAs with carbon numbers <21. In addition, an increase in the proportion of NAs in the "C22$^+$ cluster" was accompanied by a decrease in the total NAs in the OSPW, again suggesting the selective removal of NAs with carbon numbers of $\geq$21. This work was the first investigation in which changes in the fingerprint of the NA fraction of OSPW from the operations have corresponded with the measured toxicity in these waters [243].

Biodegradation and toxicity of NAs are two correlated topics that have caught the attention of many researchers. A common agreement is the lack of a detailed and comprehensive characterization of these compounds, which obstructs progress in the area and achievement of solutions to the problems [236,244–251].

6. Impact in the Oil Industry

Refineries investigate the possibility of running difficult "opportunity crudes" for optimizing profitability. Crude oils from Venezuela, California, Russia, Canada, and the Gulf Coast have been identified as NA-bearing crudes. New crudes from the Far East and Africa are also indicating high TAN in crude assays. Purchased cat-cracker feeds may also have significant NA content. Over the past few years, it has become evident that NA corrosion can occur at much lower TAN levels than the previously reported crude TAN level of 0.5.

6.1 Impact in Oil Production

The challenges posed by acid compounds go beyond the petroleum industry, since these are a growing concern for the aquatic environment as well. An assessment of the impact of unconventional crude oils on operational and product quality has been reported. Explanations for the way in which the crude oil supply chain and the sources of many crude constituents affect production were included [252].

As mentioned earlier, NA molecular structure dictates its interfacial activity. The more interfacially active NAs may form salts in water above their pK_a (slightly acidic to basic pH). In turn, naphthenates promote the formation of tight emulsions or rag layers. Directly, naphthenate salts may cause emulsions formation that indirectly leads to oil carryover into water treatment systems and water carryover into oil systems. Additionally, some NAs may coordinate with divalent metal ions to form polymeric salts such as ARN-naphthenates, which lead to organic deposition [253].

In oil sands processing, hot caustic solutions are used to separate the bitumen from the sand. Because of the high pH, NAs preferentially partition to the aqueous phase, and are discharged with the water into the tailings ponds. NAs are believed to account for the high acute toxicity of the tailings waters [254]. These upstream situations lead to large amounts of NAs being released from the oil sands extraction process, and so continue to stimulate any research to understand the composition, toxicity, and biodegradation of the complex acid mixtures. A comprehensive review on the environmental fate and persistence of NAs, particularly in the Canadian scenario, has been published by CEATAG [72].

The reverse emulsion produced under steam-assisted gravity drainage conditions has to be treated for separation of the water and oil. The various heat exchangers used to heat the reverse emulsion and cool the produced water all experience severe fouling. All deposits contained both inorganic and organic components, with high Mg^{+2} and Si^{+4} content. Additional deposition of other sand/silicates was also found. Significant concentrations of organic acids and organic acid salts were observed in the organic portion of these deposits [255]. A detailed compositional analysis of the organic acids was performed by ultrahigh-resolution ESI(−) FT-ICR-MS and revealed the MW distributions, heteroatom content, aromaticity, and carbon number for several thousand organic acids in each sample [256]. High mass accuracy combined with Kendrick mass defect analysis allowed for the assignment of a unique elemental composition, $C_cH_hN_nO_oS_s$, to each mass spectral peak. Three-dimensional mass spectral images generated from the molecular formulas revealed compositional differences that may be rationalized on the basis of sample processing. Early-stage deposits showed lower overall heteroatom content, higher average carbon number, and lower aromaticity for the produced bitumen compared to later-stage samples that show increased water solubility. MS/MS by infrared multiphoton dissociation of trapped ions indicated dicarboxylic acid structures for predominant $C_cH_hO_4$ and dicarboxylic acid structures with thiophenic sulfur predominating in $C_cH_hSO_4$ species [256].

During production of an HAC from a West Africa field, high pH connate waters cause severe problems during the separation of oil and water. The neutralization products of the acids in the crude oil, with basic components in the water (metal naphthenates), tend to form very stable water-in-oil emulsions and insoluble sticky naphthenate sludge (Fig. 4.22A), which may contain water, oil, sand, scale, and other solids. These phenomena considerably hamper oil production and result in high treatment costs, for instance, treating with acetic acid [257]. Nevertheless, acetic acid brings serious disadvantages, such as increasing water production, corrosion, and water-in-oil emulsion formation, representing an expensive and logistically difficult mitigating solution [145]. The incorporation of this chemical or any other at production level is rarely reported to the refinery purchasing the crude oil. For this reason not only a comprehensive crude assay is necessary, but also a detail speciation of inorganics present in it.

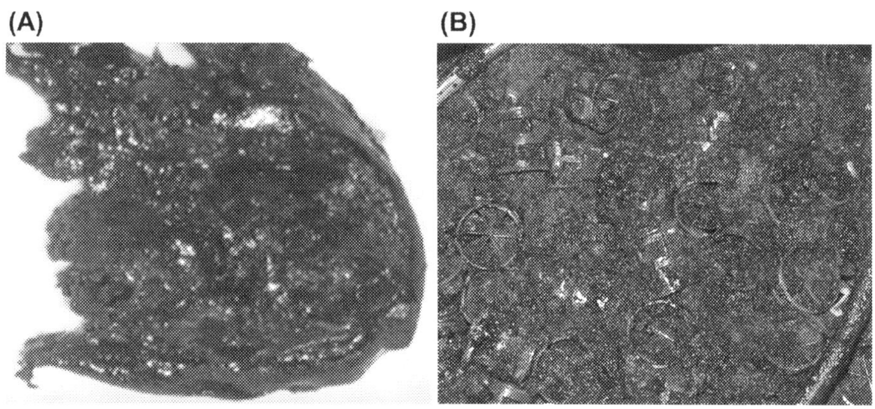

Figure 4.22
(A) Calcium naphthenate sludge and (B) clogged flash drum. *Reproduced from Mediaas H, Grande KV, Vindstad JE. Efficient management of calcium naphthenate deposition at oil fields. In: Proc. Tekna – separation Technol. Stavanger, Norway; September 26–27, 2007. 1 pp., with permission from Statoil.*

During production, the high concentration of NAs in HACs frequently has caused the formation and precipitation of naphthenate soaps by their interaction with the inorganic cations present in the reservoir brine. Depending on the level and quantity of the precipitate, formation damage and well productivity loss in the petroleum reservoirs might be caused. A "pore-throat plugging" mechanism that would restrict the flow of fluids through the formation and eliminate the hydraulic flow tubes was proposed as responsible for formation damage. A lab study had proven that formation damage can be caused by naphthenate soap precipitation [258].

The problems faced upstream during production and transportation of HACs can be summarized as:

- Stabilization of water-in-oil emulsions by accumulation at interfaces of amphiphilic NAs causing enhanced separation problems;
- Release of carbon dioxide and pressure drop during fluid transportation from the reservoir;
- Higher concentration of dissociated NAs at the oil—water interface, caused by decreased pH of the coproduced water;
- Increased formation of metal naphthenates by reaction of dissociated NAs with metal cations in the water;
- Precipitation and accumulation of naphthenates at the oil—water interface, caused by low interfacial affinity and low solubility in water;
- Adsorption and adhesion of metal naphthenates on pipelines and/or on the walls of process unit surfaces (Fig. 4.22B).

6.2 Technical Issues in Refining

The problems caused by NAs in refining are reflected in both operations and quality of the final products [259]. Availability, reliability, and operations in refining can be undesirably impacted by HACs with problems in the tankage, desalter upsets, corrosion in unexpected areas, fouling, catalyst poisoning, off-specs and degradation of finished products, and environmentally harmful discharges. The potential impact from upstream (oil production) additives and from crude properties on downstream operations are the continuous subject of review and discussion (e.g., [260]).

HACs, which are often available at discounted prices, may offer relative value compared to other grades. Because of their economic attractiveness, there is rising interest in the processing opportunity crudes, particularly those with high TAN. However, the understanding of the operability issues of opportunity crude oils is paramount to capitalizing on the crude value without impacting unit reliability and operation. Since acidity in crude oils has long been recognized as a problem for refining, the knowledge of the detailed chemical composition of the responsible acids can facilitate identification of problematic crude oils and potentially lead to improved processing options for HACs. Cost-effective solutions are needed to cut operating costs and boost margins.

While light organic acids do cause overheads corrosion and other acids can cause other problems, the group of acids that cause the most corrosion in refineries are NAs. As well as corrosion (especially at high temperatures), NAs are directly involved in other operational problems such as foaming in the desalter, off-specification products, e.g., JP-4 and kerosene, odor problems in products, scaling, and deactivation of downstream catalysts. Additionally, NAs are also indirectly involved in other operability issues including fouling and carryovers. Details and examples will be discussed in the following paragraphs.

6.2.1 Impact on Tank Farm

During crude storage in the refinery, the previously described physicochemical properties of NAs act together to create problems. The presence of NAs makes the removal of water more difficult in HACs. Additionally, emulsions and sediments are prone to form.

Foaming is another NA-associated issue. Crude oil foams, a major problem for operators of gas/oil separation plants in the upstream sector, cause loss of crude in the separated gas stream and consequent loss of revenue. Downstream it can pose possible damage to compressors [133].

6.2.2 Impact on Desalter

A desalter in the refinery is part of the crude unit and prepares the feed for the atmospheric distillation tower. The purposes of the desalter are:

1. Remove salts, particularly chlorides of calcium, magnesium, and sodium that would corrode the tower overhead. The hydrolysis of these salts will produce hydrochloric acid during the distillation process;
2. Remove solids and sediment that cause erosion or abrasion of equipment and may deposit and accumulate in the preheat exchanger train, plugging or fouling the tubes and leading to higher energy consumption; and
3. Prevent water slugs from tankage to be charged directly to the distillation tower.

Although the desalting process is based on the solubility of the salts in water, actual oil—water separation is based on density differences. The oil is washed with water and let to settle in a tank or decanter. Ammonia may be used to reduce corrosion and caustic or acid may be added to adjust the pH of the water wash. The desalted crude is continuously drawn from the top of the settling tanks and sent to the atmospheric distillation unit. The heavier the oil, the more difficult desalting (water—oil separation) is. Thus in the desalter, HACs are more difficult to desalt but even more so are the heavier HACs. Heavier crudes require longer residence time because the gravity difference between the oil and water is reduced. Furthermore, chemical additives for emulsion breaking have to be added to prevent the effect on oil undercarry in the water, which is caused by inadequate residence time. Surface activity of NAs leads to a greater tendency for emulsion formation that turns into oil undercarry (wastewater treatment plant issue) and water carryover. Thus the dosage of primary demulsifiers has to be increased to maintain any level of performance. Similarly, downstream of the desalter the usage of caustic soda also has to be increased [261]. The presence of NAs in crude oils induce the formation of organic metal salts with metals present in the formation, brought in by the production water, from additives and/or any other contamination. Organic metal salts promote the formation of stable emulsions in the desalter and are carried over into the desalted crude. Ultimately, these salts are later found in the heavy products, which are further upgraded. Consequently, a desalted HAC contains higher salt and water than a conventional crude oil. Therefore imported HACs might present larger concentrations of salts and water, as well as production additives. The presence of numerous additives in the raw oil makes the selection of refining additives more difficult [262]. These high levels of organic salts in HACs consist most commonly of calcium and sodium naphthenates. Several operating challenges derive from the presence of these salts. These compounds can increase conductivity of crude blends and promote a tendency to form stable emulsions. Both of these outcomes penalize (electric) desalter operation. The naphthenate salts are much less water soluble and hence difficult to remove in a desalter. This implies a larger use of water, a more careful control of the pH and conductivity, a larger use of chemical additives, and an increased amount of caustic soda [263].

If the naphthenates are not adequately removed, they can cause quality issues with residual fuel (off-spec. for metals), increased metal loading to coke (resulting in lower grade), catalyst poisoning on fluidized catalytic cracking (FCC) feeds, crude preheat and tower fouling, and activation of coker furnace fouling. Moreover, all these inefficiencies in the desalter cause increases of strong acid levels in the crude unit overheads (primarily HCl).

The stabilization of water-in-oil emulsions represents a serious challenge in petroleum production and refining since they can cause problems in the topside and subside of separators. A variety of acids and their corresponding soaps/salts have been thoroughly studied to investigate how these compounds interact in aqueous solutions at different pH, and how these association structures relate to emulsion formation and stability. As already mentioned, the formation of aggregates and hence the stabilizing properties of this class of material are strongly sensitive to the relative proportion of uncharged acid and charged soap anion. A review of experimental results on NA/naphthenate stabilized emulsions utilizing several model compounds, including heptylbenzoic acid, *trans*-4-

pentylcyclohexane-carboxylic acid, and 5-β-cholanic acid, confirmed the high stability of formed emulsions [137].

Stable emulsions not only have adverse impact in desalting but also cause problems in wastewater treatment plants. Besides the water-in-oil emulsions, solids-in-oil dispersions, oil-in-oil dispersions, and oil-in-water-in-oil multiple emulsions may coexist in the continuous oil phase of the complex fluid rag layer. Strong interactions between these phases impart viscoelastic properties to this rag layer. In heavy-sour HACs with high content of S, N, metals, and Conradson Carbon Residue, these heteroatomic and heavy molecules will concentrate in the complex fluid rag layer compared to the bulk crude oil, most likely induced by the presence of highly surface-active asphaltenes and NAs. Both of these types of compounds also concentrate into the rag layer. In summary, asphaltenes and NAs not only stabilize the water and solids in the crude oil-continuous phase of the rag layer resulting in a complex fluid rag layer, but also concentrate in this layer [119,264].

6.2.3 Impact on Fouling

While poor desalting can increase salt contents in atmospheric and vacuum residues, their subsequent processing in the downstream units can lead to increased fouling of heaters and exchangers in those units. The formation of sediments has been already mentioned in relation to storage of crude oil in the tank farm of the refinery. In fact, the indigenous acid compounds play an important role in the destabilization of water droplets. Surface activity of NAs may hinder asphaltenes and other resins from forming stable films around the droplets. NAs are also known to have a solubilizing effect on asphaltenes, and removing them from a crude oil may cause the asphaltene to aggregate more easily at the water–oil interface [70].

Generally, the presence of surface-active/dispersant NAs in HACs would be beneficial to prevent fouling of the CDU and vacuum distillation unit (VDU) preheaters since asphaltenes tend to be solubilized. In such cases, recovery or stabilization at furnace inlet temperature might be observed when processing. However, the presence of other inorganic contaminants (sediments/corrosion products/filterable solids) could increase fouling rates. Furthermore, the associative and interaction behavior of NAs leads to polymeric fouling. HACs could often be destabilized by mixing with paraffinic crudes [261].

Another type of fouling on fixed bed reactors is the deposition of corrosion products, i.e., metal ores formed upstream of the unit in question. This type of fouling would increase in the absence of corrosion mitigating actions, and in general of any other corrosion control aid.

Naphthenate formation and more particularly calcium naphthenate (shown already in Fig. 4.22A) is a common problem in production and refining. Formed naphthenates foul, deposit, and accumulate in the units, plugging the lines or catalytic beds, creating pressure drops or blocking the ordinary flow of fluids (Fig. 4.22B). The unit facing the largest problem is the crude preheater [145].

6.2.4 Impact on Crude Unit

Besides the fouling of the preheat train of the crude unit, NA corrosion highly affects the other elements of the crude systems. During processing of HAC, formation of lower MW organic acids, which are more volatile, will go overhead in the CDU. The corrosion of the overhead systems of both the CDU and VDU has been observed at temperatures below 220°C. Light organic acids can be formed by the decomposition of NAs in the vacuum feed furnace. Although corrosion from NAs has been observed at temperatures as low as 180°C [192,265,266], the most important corrosion is that occurring at higher temperatures, which would have fastest consequences [267]. At temperatures in excess of 180°C and particularly in the distillation units, the more predominant corrosion was found to occur [268]. High-temperature corrosion of NAs affects both crude units CDU and VDU, and also the coker and visbreaker if installed in the refinery. The affected areas of NA corrosion in the atmospheric and vacuum distillation units are exemplified in Fig. 4.23 from Ref. [112].

Fig. 4.23 shows one of the main characteristics of NA corrosion, which is having a more localized attack and particularly at areas of high velocity and, in some cases, where condensation of concentrated acid vapors can occur in CDUs. Thus feedstock heaters, furnaces, transfer lines, feed and reflux sections of columns, columns, heat exchangers, and condensers are all susceptible to NA corrosion. Similarly, while in the drums of the coker unit, coke forms a protective layer; other elements such as furnace, soaking drums, lower part of the tower, high-temperature exchangers, and fractionation tower and its side pipes are subjected to high temperature corrosion (above 475°C). The attack is oftentimes

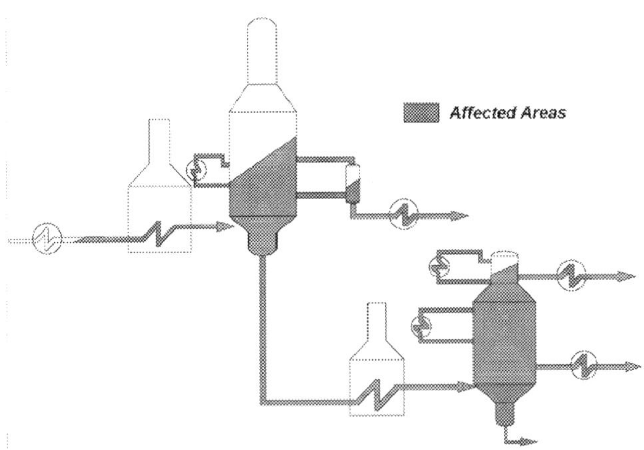

Figure 4.23
Refinery units more susceptible to naphthenic acid (NA) corrosion. *Reproduced from Haynes, D. Naphthenic acid bearing refinery feedstocks and corrosion abatement. In: Proc. AIChE — Symp. Chicago, IL, October 2006. Nalco Co. Proprietary Knowledge. 30 pp., with permission from COQA.*

described as lacking surface corrosion products thereby exposing fresh metal. In fact, NA corrosion becomes apparent in the form of pitting and impingement.

NA corrosion is not well understood, probably because of the many variables and parameters involved, such as the NAs' chemical nature (Figs. 4.8 and 4.9), molecular structure (see NA structures in Figs. 4.11 and 4.12; see corrosivity of different structures in Fig. 4.18), concentration (e.g., for TAN in crude oils see Fig. 4.13), boiling point distribution (Fig. 4.14), decomposition temperature (see Section 5.2), and the possible synergy with other compounds present in the oil, the effect of temperature on corrosion, and process conditions such as stream flow velocity (in reality, wall shear stress), degree of vaporization, and metallurgy (alloy composition) of the unit. For instance, conditions of low vaporization and medium fluid velocity exist in the side-cut piping. Under these conditions, an increase in velocity increases corrosion rates up to the point where impingement starts and corrosion is accelerated dramatically. Petkova et al. considered the critical factors causing HAC corrosion aggressiveness as density, content of water-soluble salts, total sulfur content, organically bonded chlorine, TAN/NAN, and metals contents (Ni, V, Fe, Hg, Se, etc.) [269].

Slavcheva et al. have reviewed NA corrosion in oil refining, indicating that it may involve the chelation of the metal ion by the carboxylate with the formation of hydrogen gas [206], as shown in Rxn. 4.1. NA concentration may be a serious source of corrosion for one process and have relatively benign effects for another. The density and viscosity of the liquid and the vapor in a pipe, the degree of vaporization in a pipe, and the pipe diameter are all factors affecting the corrosive activity of NAs. High-throughput rates, together with operating temperatures between 220 and 400°C, favor corrosion. Above NA decomposition temperatures (>400°C), a film is formed on the metal surface that protects the alloy [177]. The NA corrosion behaviors of 5Cr-steel and that of 9Cr-steel were similar in nature with respect to TAN and flow velocity limits. Both exhibited the onset of impingement attack at about a TAN value of 1.5 or higher, depending on the flow velocity [211]. The comparison of the corrosivity of an HAC, the corresponding desalted HAC, fractions of the HAC, AR, nonacid crude oil, viscosity-decreased VR, and fractions of VR showed increased corrosivity with increasing acidity, increasing cutting temperature of the fraction, and increasing temperature of the corrosion test [270].

The damage caused by NAs is in the form of unexpectedly high corrosion rates and localized attack on alloys that would normally be expected to resist sulfidic corrosion (particularly steels with more than 9% Cr). In some cases, even very highly alloyed materials (i.e., 12Cr, 316SS, and 317SS, and in some rare cases even 6Mo stainless alloys) have been reported to exhibit sensitivity to corrosion [211].

The presence of low MW carboxylic acids in the tower overheads is partially responsible for corrosion in this area. There are two possible sources of acetic acid, one is NA decomposition and the other is upstream additives. Decomposition during visbreaking may explain the corrosion observed at the top trays of the fractionation column of the visbreaker unit. That was the case when treating a blend of heavy oils with 0.24 TAN, which was unexpectedly more corrosive at such low temperature (~200°C) [271]. The precipitation, accumulation, deposition, and fouling of metal naphthenates upstream of the crude unit (including pipelines and desalter) could have been combated using acetic acid, of which the refiner may be unaware.

Desalter upsets have already been mentioned as responsible for the increase in chloride concentration and for the presence of a weak acids level. In addition to NAs themselves, overhead corrosion is caused also by the presence of mineral salts, such as magnesium, calcium, and sodium. More important among the mineral salts, chlorides are the most dangerous since their hydrolysis produces volatile hydrochloric acid. A highly corrosive environment is created in the overhead exchangers by the presence of chlorides, which can override the desalter when processing HACs. Additionally, NAs also accelerate the hydrolysis reactions of the inorganic salts.

Experience with NA corrosion at low TAN levels is most often associated with sweet crudes (less than 0.5% sulfur). Details have been given on the corrosion damage caused by sweet, low TAN crudes in a crude/vacuum unit designed for sour crude service. The crudes were primarily from West Africa. Several metallurgical upgrades were needed to improve plant reliability including furnace repairs, clad vacuum furnace headers and transfer lines, and vacuum tower cladding and internals work [272].

6.2.5 Impact on Finished Products

The presence of NAs in HAC together with their effects on the front end units of the refinery shifts products off-specifications. NA boiling ranges start near to the heavy naphtha end of the boiling range (above 200°C). Consequently, during fractionation higher MW NAs may be taken off with the side cuts, causing some stability problems. NAs appear more in the kerosene and upper streams (straight-run gas oil through bottoms). They start decomposing appreciably at 400°C, from which the problem changes, as mentioned earlier [273].

In terms of limitations, most refineries set limits around the 0.3−0.5 mg KOH/g range for crudes to be processed; only a few have been upgraded to tolerate much higher levels up to about a TAN of 2.5−3 mg KOH/g. No TAN limit is in place for overheads; the corrosion rate is controlled mostly by pH, and other limits effecting high TAN crude processing are: P, Ca, metals, and overhead fouling. Some refineries have side draw limits in addition to TAN in crude oil, an example is the Kero TAN limit, which is determined based on either from Jet specifications or from the feed limits or specifications of the downstream unit. This particular example might be based on the fact that the refinery production of jet fuel, in some refineries is done by simply withdrawing a side stream product from the crude oil fractionator, which requires no additional treating or cleanup. In other cases (e.g., higher TAN), caustic treating followed by water washing, salt drying, and clay filtration are required. Others pursue more complicated processing to fully meet jet specifications when feeding more contaminated or lower-quality streams.

Typically, HACs produce a low cetane number diesel. Diesel fractions derived from Athabasca bitumen, syncrude oil, and in general heavy-sour HACs have cetane numbers below 30. The presence of high sulfur, nitrogen, and heavy metals contents contributes to the difficulties in meeting ultralow-sulfur diesel on road specifications on bitumen-derived diesel [8,274]. In the case of Sudanese Fula crude oil, the presence of iron (both indigenous and corrosion product) has been indirectly linked to causing the diesel off-specs [275]. The Brazilian Marlim is a heavy-acid-sour naphthenic crude oil, rich in heavy fractions with low sulfur content and high levels of bottom sedimentation water, nitrogenated compounds, and TAN. Its derived fuel oil and coke are characterized by low metal and sulfur content levels. Meanwhile, Marlim's gasoline is highly naphthenic, with competitive octane numbers. Instead, diesel and lube oil have poor quality. Nevertheless, this oil can be easily handled and stored, even at low temperatures, without leaving undesirable paraffin deposits in pipelines and tanks [276].

The ESI FT-ICR-MS characterization of acidic compounds in VGOs and their dewaxed oils indicated that the dominant polycyclic NAs present in the VGO were easily eliminated by the dewaxing process. However, the single-ring NAs survived the process and contaminated the potential lube base oil [277].

The presence and/or formation of Ca-naphthenate during processing will affect the quality of fuel oil and coke products. Ca contamination of these products precludes their use in burners for the former cut and as a raw material for anode production for the latter product [278]. Indeed, calcium is detrimental to coke quality. Aluminum manufacturers require premium-grade (needle) coke to have less than 300 ppm iron or calcium. This can equate to as little as 20–25 ppm oil-soluble metal in the crude. Often, opportunity crudes contain over 300 ppm of these metals, which must be removed.

Ca-naphthenate also has a deactivating effect of cracking (FCC) and hydrocracking (HCK) catalysts [278]. Even though the quality of the finished products will not be affected in the case of the FCC catalysts, the yields will be negatively affected. In the case of HCK catalysts, both yields and quality will be equally affected by the excess Ca in the feed.

6.3 Economical Aspects of Acidity in Refining

The economic view of processing HACs differs from one refinery to another. Thus some predict the economics of opening the TAN limits based on purchases of corrosion-mitigating chemicals. Others predict the economics of implementing projects for corrosion monitoring and control, passivation, or upgrading refining units. Crude availability seems to restrict the options in certain regions and case-by-case decisions have to be taken.

Analysis of the economics of running HACs is the first step prior to any attempt of their purchase. The case of Liaohe Petrochemical Co., which has been engaging in refining of various kinds of heavy-sour crude oils, with no consideration of the problem of equipment corrosion, is an illustration of a not-to-follow model. The severe corrosion of equipment affected normal operation and resulted in great economic loss that greatly overcame the discounts of the opportunity crudes processed [279].

As already mentioned, NA corrosion is characterized by deep pitting and metal impingement. These effects can be seen even in highly alloyed metallurgies, which are normally resistant to corrosive attack from sulfidic species. The corrosive potential from NAs is a function not only of concentration, but also of temperature, wall shear stress (along with Reynolds number), and the presence of other synergizing compounds. When two refineries process the same blend of crude oil, corrosion may be observed in one refinery and not in the other. Thus corrosion relates not only to the nature of the crude oils, but also to the operational conditions in a refinery.

The cost of corrosion can be defined in different ways depending on what is included and which units are affected. The cost of corrosion has been defined as the corrosion fraction of design, manufacturing, operation and maintenance, technology development, and asset value loss [280]. Changes in the cost of corrosion could be addressed by examining changes in corrosion control practices over the last few decades. This allows placing current practices into perspective within the sector's history and demonstrating achievements to date. From economic analysis, it has been stated that the current cost of corrosion may be lowered by implementing optimal corrosion management practices.

Sources of direct costs could be divided into two categories: (1) costs of design, manufacturing, and construction and (2) costs of management. The first one includes material selection, such as stainless steel to replace carbon steel; additional material, such as increased wall thickness for corrosion allowance; material used to mitigate or prevent corrosion, such as coatings, sealants, corrosion inhibitors, and cathodic protection; and their application, including the cost of labor and equipment. The latter involves corrosion-related inspection, corrosion-related maintenance, repairs due to corrosion, replacement of corroded parts, inventory of backup components, rehabilitation, and loss of productive time [280].

The costs associated with conventional mitigation strategies in refining have been summarized by Skipping et al. [113,281] as:

- The cost of corrosion monitoring hardware, inspection tests, chemical treatment, and mechanical upgrades as preparation means for processing of higher-TAN crude slates varies from one refinery to another.

- For a 120,000-b/d unit, the cost of purchasing coupons for an atmospheric distillation column and vacuum unit is 005–0.03¢/bbl for 6–10 coupons. Probes cost 0.04–0.28¢/bbl.
- Costs associated with nondestructive testing and metal thickness measurements vary from one refinery to another. Processing HACs may require considerable measurements initially. The number of inspections decreases when the refinery gains confidence in the reliability of inspection results.
- The typical initial frequency of ultrasonic measurements on areas susceptible to NA corrosion is monthly. It later declines to 6-month intervals. These measurements may require additional scaffolding to reach the more remote locations. One-time initial costs associated with building scaffolding may also be high.
- The approximate incremental annual cost of using a certified technician to take ultrasonic measurements is $20,000, or 0.11–0.45¢/bbl of opportunity crude.
- The costs of analytical tests for measuring metals content, TAN, NAN, and sulfur are insignificant when compared to the cost per barrel of HAC.
- Early detection of corrosion will allow corrosion mitigation using chemical treatment. (Several refiners has been practicing this method of mitigation for more than 25 years.) The cost associated with a chemical treatment is a function of existing metallurgy and operating conditions for the equipment in question.
- The cost of corrosion inhibitors for mitigating NA corrosion ranges from 1 to 10¢/bbl of HAC, depending upon the extent of areas requiring protection and the percentage of naphthenic crude in the crude slate.
- If the refinery has a long-term commitment to process HAC oil, upgrading metallurgy in susceptible areas of corrosion may be economically feasible. For materials selection, the life-cycle cost must be considered in addition to purchase price. The typical cost of a refinery metallurgy upgrade is $12–20 million, which is a payback period of 2–3 years based on a run rate of 40,000 b/d of the HAC in a 120,000 b/d refinery.

Another review on the corrosion costs in refining was published later [282]. By the time that study was published (2006) the total cost of corrosion control in US refineries (163 refineries) was estimated at $3.692 billion. Of this total, maintenance-related expenses were estimated at $1.767 billion annually, vessel turnaround expenses accounted for $1.425 billion annually, and fouling costs were approximately $500 million annually. This reported analysis concerns mostly the processing of conventional crude oils.

Nalco/Exxon Energy Chemicals LP, Sugar Land, Texas, estimated the refinery costs for corrosion mitigation to process HACs using Alba crude (TAN = 1.4 mg KOH/g, produced by Chevron in the North Sea) [192]. The estimated associated costs for running this HAC was 1.15−10.73¢/bbl, giving an economical advantage of 54.27−63.85¢/bbl compared to processing Brent.

A review of the economy of processing HACs, with respect to anticipated impacts and mitigation strategies, was recommended as a reasonable practice to ensure that the HAC being purchased will yield the return on investment. The design of the appropriate mitigation strategy and whether to include crude blending, metallurgy upgrades, and/or chemical inhibition for those areas affected by NA attack should be properly addressed and assessed. A first step may be to try to define a reliable prediction of possible corrosion rates in the refining circuit, based on general rules for TAN and total sulfur observed corrosion trends. A brief review of three available methodologies, API 581, Modified McConomy plots, and Predict-Crude, has been provided in Ref. [183].

Any given strategy for safely processing HACs has to be based on a risk assessment of the susceptibility of the plant and its equipment to undergo naphthenic corrosion. Since corrosion depends on many variables and not only feed composition, a holistic evaluation is required. These variables include: stream composition (TAN, NAN, S/reactive S, etc.), flow rate, two-phase flow, turbulence, vaporization or condensation, metallurgy, other "process unit" conditions (temperature, residence time, environment, etc.), and side-cut/stream stability. A dynamic loop as that shown in Fig. 4.7 of Ref. [283] has been proposed and tested for exposing coupons of different metallurgy. Experiments can be designed for defining both the most critical parameters and the most susceptible areas of the refinery unit where monitoring should be emphasized. Even additives effectiveness and injection points could also be evaluated.

Devices for monitoring corrosion online, by measuring wall thickness, have been known and used upstream for a long time. Portable nonintrusive versions were needed for refining units. Devices are based either on ultrasonics or radiography [263]. The selection of corrosion-monitoring methods, techniques, and technology, such as visual inspection, electrical resistance probes and weight loss coupons, field signature method (FSM), ultrasonic thickness measurements, digital radiography, hydrogen permeation and pulsed eddy current, and the level of improvement achieved in the ability to manage corrosion and processing the HACs have to be assessed when planning on purchasing HACs [13,284]. Details of an example of FSM application can be seen in [285,286]. Another example is the Radioactive Tracer Technology developed by the South West Research Institute for real-time assessment of corrosion. A test was validated with coupons that were preactivated (thermal nuclear activation) and placed in a flow loop where corrosion was evaluated at refining units conditions [287,288]. Once the monitoring protocol has been established, continuous improvement should be practiced, with clear optimization objectives.

ChevronTexaco proposed an economic valuation for selecting and running HACs [289]. A first step would consider the basis of crude selection (see, e.g., Table 4.6). The second step

Table 4.6: Example of Crude Valuation for Selection Purposes [289]

	Alba	H	Kuito	S	U	G
Gravity API	19.4	27.3	19.0	23.8	31.3	22.2
Sulfur (wt%)	1.3	2.9	0.7	4.0	1.4	3.0
Acid number	**1.4**	0.3	**2.1**	0.3	0.1	0.3
TBP yields (vol%)						
Butanes and lighter	0.1	2.9	0.2	1.7	1.8	0.6
Light gasoline	0.2	5.7	0.6	7.0	5.5	3.0
Light naphtha	1.0	8.4	5.2	9.6	10.2	8.1
Heavy naphtha	2.8	8.2	5.0	7.2	8.8	6.8
Kerosene	6.6	8.1	8.9	7.4	9.6	6.8
LGO	17.5	13.9	16.1	12.8	15.6	12.9
LVGO	15.4	11.0	16.3	11.9	14.2	12.9
HVGO	26.9	14.8	23.0	17.7	17.0	20.2
VR	29.6	27.0	24.6	24.7	17.3	28.8
Run it?	**No!**	Yes	**No!**	Yes	Yes	Yes

HGVO, Heavy vacuum gas oil; *LGO*, light gas oil; *LVGO*, light vacuum gas oil; *TBP*, true boiling point; *VR*, vacuum resid.

would be to challenge a decision based on acid number exclusively, as shown in the given example. In this example, the crude oils were not identified. However, HAC properties and valuation data were reported and Alba and Kuito could be identified.

Crude valuation of HACs needs to include the balance between crude discounts and mitigation costs. In this regard, it is important to count on a grade with a reasonable price history and a consistent valuation methodology. Discount ranges recorded for HACs in the period 2000–2003 are shown in Table 4.7 [262].

It seems that compared to mitigation costs, discounts make HAC a good option for refining. Nevertheless, corrosion contributing the most to refining costs is not the only

Table 4.7: Discount Ranges for Selected High-Acid Crudes (HACs) (2000–03) [262]

HACs	Discount
Alba	−5.00 to −2.00
Captain	−5.00 to −1.50
Gryphon	−2.50 to −1.50
Troll	−1.50 to +0.75
Heidrun	−2.50 to −0.50
Kuito	−5.50 to −2.00
Ceiba	−5.00 to −2.50
Lokele	−4.40 to −3.20
Marlim	−5.50 to −2.50

problem brought by NAs and causing increases in refining expenses. An example is the need to bring the diesel to meet on-road specifications. Allowance has to be made either for the use of cetane improver or amount of higher cetane gas oil available in the refinery pool for blending [261]. Other sources of unexpected costs are nonprogrammed stops caused by catalyst deactivation and unplugging or cleaning fouled deposits and scales.

The industry faces a trend toward refining more HACs (which are thought to be refined at a higher margin because of their discounted price) since the early 1990s. Obviously, there would be increases of potential corrosion problems and therefore refining costs. Thus, to strength the opportunity for good margins, new cost-effective technologies have to be developed and implemented on time to take full advantage of the market trend.

7. Current Practices

Although efforts on R&D activities concerning NAs and HACs have been substantial, technologies for their abatement has not become commercially available. As seen earlier, the problems caused by HAC involve not only refining operation, but also upstream operations that raise environmental concerns and result in environmental and social problems. It is clear that there exist plenty of opportunities for technology development and adaptation to the different segments scenarios within the value chain of the oil industry (production, transportation, and refining). So far, each sector deals with the mitigation of the derived issues, but do not tackle directly the source of such problems. As will become evident from the following paragraphs, there is no unique and universal solution for all the problems. Each sector, including refining, needs to face its own situation and find a suitable mitigation strategy, which may at best extend the operability period of time from months probably to years, providing part of the expected benefits for the opportunity crudes' discounts [290].

The properties and composition of HACs diverge from one crude to another and refinery configurations and capabilities also vary from one site to another. In consequence, there is no single solution to best suit all sites and an answer has to be designed on a case-by-case basis. The many and great variety of problems caused by HACs during refining impose complicated and interrelated challenges.

For these reasons, the purchase and processing of HACs are limited to:

- Refineries with specialized metallurgy;
- Large refineries that can dilute acidic crude and use it as trim;
- Those refiners with sufficient specialized facilities to handle HACs when discounted;
- Specialized, nonprocessing utilizations;
- Risk-adverse refiners that will only occasionally experiment with high-TAN grades;
- Large and sophisticated refineries buying discounted acidic crude in the hope of a double-dip in profits, using residual from high-TAN grades as cat-cracker feedstock [18].

In the case of Petrobras' heavy-acid-sour crude oils, prior to 2000, the destiny of these crudes was mainly domestic. Their internationalization strategy after 2000 was aiming to place a larger slice into the oil market, by building in their technical support to customer refiners that were processing these oils for the first time [276].

However, as discussed earlier, processing HACs does not have to reduce refinery reliability; a systematic approach for assessing corrosion risk and opportunities could and has to be used. Currently, refiners purchasing HACs are using mitigating methods that allow safe processing, namely, blending (dilution with a nonacid crude), chemical additives for compensating NAs' negative effects, specialized materials in the process units (alloy steels), and extractive neutralization [182,291–294].

7.1 Corrosion

The most detrimental effect of NAs is corrosion, and so refiners address it through mitigation approaches when processing acid/sour crude oils. Coping with the corrosive actions of NAs includes: blending (dilution with nonacid crude), anticorrosive chemical additives, and upgrading the units with specialized materials. Selecting one of these options or an adequate combination of them depends on the refinery situation and on the economical assessment. Three different strategies have been proposed for the comparative evaluation of options: (1) fixed corrosion rate, (2) limits imposed by the conditions of existing equipment, and (3) balancing the limitations based on existing equipment with replacement costs [292].

For the blending option, the refiner would set the tolerance in terms of NAs and S and will need to pay the full cost of high-quality crudes to dilute the HACs or blend sour crudes with a low-S HAC. It only requires adequate planning and scheduling, but counts on crude availability and does not take full advantage of the discount.

The use of anticorrosive additives requires more dedication from the refiner and advice from the experts. These chemicals are known to be phosphorous-only based, sulfur-only based, or sulfur and phosphorous based. Some fouling issues and catalyst deactivation can be observed from use of phosphorous-based additives; however, sulfur-based additives do not achieve the desired passivation in many cases. Furthermore, installation of proper metering devices is required for avoiding overdosing. Careful selection of the additive on the basis of the crude composition and properties, and more particularly the presence of or need for other additives, is typically desired. In this regard, the selection of the diluent stream, the injection conditions, and devices have to be suitable for the operation and operating conditions of the corresponding unit. Additive vendors usually offer an integral program for monitoring and control of corrosion, sometimes including laboratory services (see examples in Ref. [295–297]). The monitoring part is typically twofold: feed quality monitoring and corrosion monitoring [298]. In this regard, the refiner is better suited for the requirements on feed quality sampling and analysis, and also for the identification of the most vulnerable pieces of the units. On the other hand, collaboration with a technical expert from the vendors is needed for assessing the best location for additives injection points.

BP briefly described the different approaches followed in its refineries: (1) a controlled stepwise increase of HAC in the diet fed to Grangemouth refinery (France) improved margins significantly; (2) Nerefco site (Netherlands) used FSM corrosion monitoring devices and a passivating chemical, supplied by Ondeo Nalco; and (3) Texas City refinery (USA) was considering a new corrosion monitoring and control system. It was indicated that operating costs for the corrosion monitoring and chemical injection systems were relatively low compared with the high payback derived from the opportunity given by discounted acid crudes. It was clear for this company that not all its refineries could profit from using HACs [299].

A computational fluid-dynamic (CFD) methodology has been proposed for anticipating the most susceptible elements for NA attack. CFD will provide the refiner with a map of specific locations and prioritized details of corrosion locations within the refinery. An appropriate and preventing plan for monitoring and controlling the effects of NAs might be designed focused on those locations [300]. Although CFD could allow the incorporation of other dynamic factors into the corrosion modeling, the problem with most modeling approaches is the use of data obtained with model compounds, extracted NA compounds or real feeds far from operating conditions, limited field data, etc.

Finally, the use of upgraded metallurgy as internal lining of the affected unit is a more permanent solution and might take better advantage of the HAC discount. Of course, this would require higher investment and longer downtime of the unit. Nevertheless, it might be the only long-term solution for the processing of heavy-sour HACs, such as some of the Venezuelan crude oils [301].

7.2 Other Acidity-Related Issues

Opportunity HAC oils can have associated processing problems in addition to naphthenic corrosion, as has been discussed in the previous sections. Desalter upsets, preheat and reactor bed fouling, side-cut stability, and cetane issues can all be present. Mitigation strategies are commercialized for all ancillary problems and the refiner is typically aware of these and consults the vendors for recommendations and advice in a timely manner.

7.2.1 Desalter

In the first place, dealing with the processing difficulties associated with the high density and viscosity of (heavy) HACs requires equipment considerations for electric desalters. The desalting tanks have to be designed at three levels: level 1 is the high-velocity level, and levels 2 and 3 are low-velocity tanks specially designed with back flush facilities inside the electric desalting tube. Since in these cases oil—water separation will be more difficult and emulsification increases, it is recommended to test dewatering and demulsifying agents prior to feeding and during operations to consider the supplemental addition of reverse demulsifiers on the basis of the previously added amount of demulsifiers [302].

Abatement of the formation and presence of naphthenates is another target for the desalter. Calcium naphthenate is an organic metal salt, which is not effectively removed by the desalter. Specialized additives have been developed that can significantly reduce the concentration of these undesirable species in the desalting process [303]. Typically, oil-soluble metal compounds from crude are removed by acetic acid treatment, but its high solubility in the hydrocarbon phase causes accelerated crude unit overhead corrosion activity. A new chemistry is claimed to reduce oil-soluble Ca compounds from 100 ppm to less than 5 ppm, as well as capabilities for removing other metals [304]. The key components of this Ca-removal technology include the use of a proprietary complexing agent that can increase the solubility of Ca-naphthenate salts, and the implementation of appropriate demulsifier chemistries and desalter optimization strategies to minimize the emulsion-stabilizing effects of Ca-naphthenates. Ca-removal efficiencies averaged over 85%. Doba crude blends, for instance, could be processed without experiencing fouling in the exchangers of desalter effluent water and without significant impact on wastewater treatment plant operations [305]. Many chelating additives can react with Ca [305] (or any other metal contaminant prone to form oil-soluble/water-insoluble naphthenates) though only a few would do it without having side effects throughout refining. A field test comparison in various US and European refineries processing West African crudes such as Doba [306] and North Sea crudes such as Harding and Heidrun [307] indicated a 95% effectiveness on Ca and more than 70% on Fe for the Baker Petrolite technology [308]. Additionally, Dorf Ketal has also developed proprietary Ca-removing agents, namely, two acid-based formulations and one that is acid free. The acid-based products are water-soluble, organic acid-based additives that hydrolyze Ca-naphthenate to form water-soluble calcium salts. The latter acid-free additive forms a water-soluble Ca adduct that is biodegradable and noncorrosive. It could be used as a stand-alone additive that would not require corrosion or scale inhibitors [108].

The role of Ca-naphthenate and NAs in the stabilization of water—oil emulsion [135,136] and the effect of different pH levels and water contents [309] have been reported. Near infrared spectroscopy was utilized to study the ability of NAs and other amphiphiles to disintegrate asphaltene particles. It was shown that the structure of the NAs is important and that a commercial mixture of different surfactants had the best disintegration effect. The technique is applicable for screening the efficiency of various additives as disaggregating chemicals. Emulsion stability was studied by means of Langmuir technique, critical electric field, and bottle tests. The Langmuir study showed that the presence of Ca^{2+} ions at elevated pH gave rise to stable naphthenate monolayers. It was demonstrated that water–oil emulsions could be stabilized by a combination of multilayer (D-phase) and asphaltene particles. Critical electric field was used to determine the emulsion stability of these systems and it was shown that a combination of 60% asphaltene particles and 40% D-phase gave the most stable of the water–oil emulsions [309].

A complex, rigid layer formed at the water droplet surface is partly responsible for the stability of water-in-crude oil emulsions. This layer is likely made of large molecules such as asphaltenes, exhibiting rigid interfacial skins and a complete lack of cohesiveness. The surface of this layer is quite rigid in highly diluted bitumen, where water droplet surfaces are relatively free of low MW surfactant species. Upon addition of low MW sodium naphthenates, the properties of this layer were dramatically changed. At sufficient concentrations of added Na-naphthenates, larger macromolecules residing in heptol-diluted bitumen were either hindered from adsorbing or competitively displaced from the water droplet surface. These macromolecules may act as mechanical barriers to flocculation and potentially confer interfacial rigidity to droplets emulsified in diluted bitumen. The dynamic IFT data suggested long-term irreversible adsorption of the adsorbed macromolecules or their slow reorganization into lower energy configurations [310].

While Ca-naphthenate deposition can be prevented by the use of acetic-based demulsifiers as well as nonacid demulsifiers, acidification of Na-naphthenate emulsions breaks the emulsion and converts the fatty acid-like NAs into their original acid form, therefore mitigating emulsion formation [157]. The presence of Na-naphthenate is also common when caustic solution is injected upstream of the desalters to neutralize NAs, and/or is injected downstream of the desalters for corrosion control. A refinery using these practices experienced serious fouling and corrosion problems in the second preheat train in the crude unit. Besides a better control of the amount and frequency of the injection, the second injection point was relocated for exercising a better neutralization of the overhead corrosive moieties [311].

In normal operations, overheads water is recycled to optimize use of the wastewater treatment facilities. However, with HACs, if overheads water is recycled, organic acids will accumulate in the wash water and potentially repartition into the crude oil. Although lower pH generally destabilizes emulsions and higher pH results in saponification of acids, as mentioned before, the low pHs may drive organic acids into the oil phase [289].

In summary, management of naphthenates in the desalter requires treatment with at least three specialized additives: emulsion breakers, complexing agent, and scale inhibitor (see example for Ca in Ref. [278]). Furthermore, the recalcitrant presence of solids would call for the supplemental installation of deoilers behind electric desalting tanks, the addition of antisludging agents, dispersants or washing, and changes in operation modes [302].

Overall, the variables and parameters to pay attention to when processing opportunity crudes in general have been summarized by Collins and Barletta [312] as:

- Operating temperature
- Amount and quality of water
- Water and oil mixing
- Mud washing to remove solids
- Brine cooling heat exchanger design
- Chemical treatment
- Desalter design
- Transformer size

7.2.2 Product Specifications

In the case of petroleum products, there are current finishing practices to meet acidity specifications, mostly based on extractive neutralization. The Merichem Napfining System is one of the commercially available technologies for the treatment of finished products and even light crudes [313]. An example of a process scheme has been shown in Fig. 4.1 of Ref. [314] for the treatment of jet fuel from HAC. This chemical treatment for kerosene fractions consisted of the removal of the naturally occurring contaminants with sodium hydroxide (caustic) solutions.

However, formation of naphthenates has been reported to take place during alkaline washing of the diesel blendstock. These naphthenates induce emulsion formation and under this circumstance, a microwave treatment was recommended for breaking the emulsion and recovering the washed stock [315].

NAs are the main contributors to acidity of jet fuels from HACs; however, in some instances mercaptans are the main acidic compounds. It is interesting to note that jet fuel fractions that are derived from crude oils containing large amounts of NAs seldom contain significant quantities of mercaptans. Likewise, the opposite is also true. This wet treatment of jet fuel may consist of two steps: total acidity reduction and mercaptan oxidation. A weak caustic prewash, which is designed specifically to extract strongly acidic compounds such as H_2S, but in particular NAs from the jet fuel, is used when total acidity must be reduced. The typical neutralization number specifications vary from 0.005 mg NaOH/g of hydrocarbon to as high as 0.10. Some of the characteristics of this process that were designed for the National Petroleum Refiners of South Africa (Pty) Ltd. are collected in Table 4.8 [314]. Besides this technology (NapfiningSM) to remove NAs from jet fuels and diesel streams by caustic washing, Merichem has announced their proprietary technology development to remove NAs from the whole crude oil. In this way, its current position as the world's largest producer of refined NAs (plant is located in Tuscaloosa, Alabama) might become consolidated [316].

Table 4.8: Characteristics of the Merichem Napfining Process

Flow Rate, m^3/h	
Design	80
Normal	80
Minimum	30
Inlet Impurities	
Acid number, mg KOH/gm	0.06 (max.)
H_2S as S, ppb (wt)	1000 (max.)
Elemental sulfur, ppb (wt)	1288
Water content, ppm (wt)	100–150
RSH-S, ppm (wt)	130
Total sulfur, ppm (wt)	1000
Treating Reagents	
Sodium hydroxide, °Be	7.0
Caustic spending, %	50–60
Guaranteed Product Specifications	
Acid number, mg KOH/gm	0.004 (max.)
Sodium, ppm (wt)	5 (max.)
H_2S as S, ppb (wt)	500 (max.)
Elemental sulfur, ppb (wt)	197
Battery Limit Conditions	
Temperature, °C	38
Pressure, kg/cm^2g	6.2
Design temperature, °C	66
Design pressure, kg/cm^2g	8.0

7.2.3 Oil Production

Calcium naphthenate deposition may have serious economic consequences in crude oil production, besides environmental and safety issues. On one side, it would reduce operational regularity, and on the other, it would increase maintenance costs. The main component of the deposits is typically the tetra acid (ARN) naphthenate compound. Calcium naphthenate may be managed if addressed in the design process. Two fundamentally different ways are used for managing calcium naphthenate deposition: pH regulation and chemical inhibition [142].

Indeed, the NA responsible for solids deposition is the archaeal C_{80} isoprenoid ARN described previously (Figs. 4.16 and 4.21A). Mitigation techniques are based on the fact that NAs are interfacial active. The structure of the acid strongly affects the IFT and the rate at which the IFT changes with a change in pH [253].

A model study at laboratory scale indicated that the formed calcium naphthenate may be strongly retained inside the porous medium in a well of low permeability. According to these findings, propagation of calcium naphthenate agglomerates through the reservoir and its accumulation near the well bore should only be of concern in field cases with rock permeability above ~1 darcy [317].

7.2.4 Contaminated Waters

Technologies for the treatment of NA-contaminated waters are practiced more in the upstream segment than they are in the downstream segment. SPE has been proven to be a good option for treating production waters [51].

8. General Remarks

HACs are a subclass of opportunity crudes characterized by containing high concentrations of acid species. Besides NAs, other acidity-bearing moieties are also present. Acidity standardized methods (e.g., TAN) account for all acid species present and not exclusively NAs. Therefore TAN is not a direct measurement of NA concentration. For this reason, HACs having the same TAN exhibit different properties.

HACs containing high levels of NAs have adverse consequences in the refining processes and on the refined products, particularly on yields and quality. Furthermore, NAs cause corrosion and corrosion increases are triggered by S-compounds in a synergistic way. However, high corrosivity has also been found with crudes with low sulfur and higher TAN values. The unit most affected by corrosion is the CDU, particularly in areas of high temperature. Additionally, these compounds negatively affect other refining units, causing poor desalter performance, fouling, catalyst poisoning, product degradation, and environmentally troubled discharges.

Other problems found with HACs concern naphthenates formation. Besides their precipitation and accumulation, the surface activity of these compounds and their role in water–oil emulsion stabilization are aggravated by the presence of asphaltenes. Therefore the most troublesome crudes are those categorized as heavy-sour-acid crude oils. In this regard, a high TAN value would serve as an indication of corrosivity of the crude, but it will not provide information on its extent or anticipation for any other problems that may be found. These problems make the processing of HACs technically and economically challenging and probably not incentivized enough by the price discount. Nevertheless, refiners have found ways for the safe processing of HACs. Four key management elements have been defined:

1. Preestimation or assessment of risks;
2. Constant monitoring of feed properties and vulnerabilities;
3. Careful design of the mitigation strategy (blending, passivation, or metallurgy upgrade); and
4. Periodic revision of technical and economic value.

Oil refining and upstream operators are conscious of all the problems caused by HAC, which give rise to concerns and result in operative-technical, environmental, and social problems. All sectors concurred in sponsoring and promoting long-lasting efforts on R&D. So far, the industry has been dealing with the mitigation of the derived issues on each affected sector, rather than directly tackling the source of such problems. Consequently, technologies for NA abatement have not become commercially available. Since the removal or conversion of these compounds has no direct incidence on the value of the refining products, producers cannot economically justify the additional abatement costs. Similarly, refiners sometimes overlook the mitigation costs by justifying operability, reliability, and availability because the value of the refined products is market controlled.

The number of operating refineries in the United States has dropped from 324 in 1981 to about 160. In this period a trend toward refining more HACs has been faced by the industry. The potential increase in margin is accompanied by corrosion problems but it is also a possible extension of the economic life of some existing refineries. In consequence, refining margins might benefit from a cost-effective more direct solution of NA-derived issues. The recent fall of oil prices has benefited to a larger extent refining operations and margins. Unfortunately, this has not been translated into larger investments and, more particularly, refining R&D activities are still in a declining mode though technically and economically NA abatement remains unsolved.

References

[1] ASTM, Standard test method for acid number of petroleum products by potentiometric titration. West Conshohocken, PA. USA: ASTM International; 2011. Available from: www.astm.org.

[2] ASTM, Standard test method for acid and base number by color-indicator titration. West Conshohocken, PA, USA: ASTM International; 2002. Available from: www.astm.org.

[3] Jones DM, Watson JS, Meredith W, Chen M, Bennett B. Determination of naphthenic acids in crude oils using nonaqueous ion exchange solid-phase. Anal Chem 2001;73(3):703−7.

[4] Green JB, Reynolds JW, Yu SKT. Liquid chromatographic separations as a basis for improving asphalt composition-physical property correlations. Fuel Sci Technol Intern 1989;7(9):1327−63.

[5] Green JB, Stierwalt BK, Thomson JS, Treese CA. Rapid isolation of carboxylic acids from petroleum using high-performance liquid chromatography. Anal Chem 1985;57(12):2207−11.

[6] Ramijak Z, Solc A, Arpino P, Schmitter J-M, Guiochon G. Separation of acids from asphalts. Anal Chem 1977;49(8):1222−5.

[7] Matsushita K, Yokota H, Fujita K, Tsukamoto A. Method of acid value determination by infrared absorption. Patent No. US5420041. Kurashiki Boseki Kabushiki Kaisha; May 30, 1995.

[8] Lu T. Challenges faced in high TAN crude processing. In: Proc. Intern. Conf. Petrofed. New Delhi, India, April 16, 2012. 36 pp.

[9] Tebbal S. Naphthenic acid analysis TAN − nan − nat- MS. In: Proc. COQG Meeting. Edmonton, Canada, September 29, 2005, COQG Meeting. 14 pp.

[10] Coe C. A case study in TAN determination: devils tower (gom) crude. 2007 [Unpublished] Available from: http://www.coqa-inc.org/Devils_Tower_TAN_SAT_Feb_2007.pdf.

[11] UOP, acid number and naphthenic acids by titration. West Conshohocken, PA, USA: ASTM; 2002. Available from: http://www.astm.org/Standards/UOP565.htm.
[12] UOP, acid number and naphthenic acids by colorimetric titration. 1992. West Conshohocken, PA, USA. Available from: http://www.astm.org/DATABASE.CART/HISTORICAL/UOP587-92.htm.
[13] Rechten R. Naphthenic acid corrosion control strategies. In: Proc. AIChE Nat. Meeting. Chicago, IL, July, 2006. Unpublished presentation. 24 pp.
[14] Ramamoorthy P. Controlling acid concentration in a hydrocarbon process. Patent No. US5681749. Chevron U.S.A. Inc.; October 28, 1997.
[15] Da Silva MT, De Andrade GH, Lopes MIC, Braga CMS. Method for determination of the total acid number and naphthenic acid number of petroleum, petroleum cuts and petroleum emulsions of water-in-oil type by mid-infrared spectroscopy. Patent No. US8222605 (Also published as EP2131181, US20090294672). Petroleo Brasileiro S.A.—Petrobras; July 17, 2012.
[16] Klotz JRM, Littmann ER. Naphthenic acids Aruba acid. Ind Eng Chem 1940;32(4):590−1.
[17] Fuhr B, Banjac B, Blackmore T, Rahimi P. Applicability of total acid number analysis to heavy oils and bitumens. Energy Fuels 2007;21(3):1322−4.
[18] Processing news refiner demand for discounted feeds will increase trade of high-acid crudes. Oil Gas J 2008;106(11):52−5.
[19] Bacon R, Tordo S. Crude oil price differentials and differences in oil qualities: a statistical analysis. Report No. WB.34906. Technical paper No. 81. Washington, DC, USA: ESMAP; October 01, 2005. 40 pp.
[20] Fattouh B. The dynamics of crude oil price differentials. Report No. Oxford, UK: Centre for Financial and Management Studies, SOAS and Oxford Institute for Energy Studies; January 2008, ISBN 978-1-901795-70-7. 42 pp.
[21] Tahmassebi H. Crude oil and product value differentials: historical perspective and outlook. Energy 1986;11(4−5):343−59.
[22] Bendiner J, Burleton D, Preston L. Drilling down on crude oil price differentials. TD Economics; March 2013. 6 pp. Available from: http://www.td.com/document/PDF/economics/special/DrillingDownOnCrudeOilPriceDifferentials.pdf.
[23] Fattouh B, Sen A. New swings for West African crudes. The Oxford Institute for Energy Studies: Oxford Energy Comment; August 2014. 7 pp.
[24] Bacon R, Tordo S. Crude oil prices: predicting price differentials based on quality. Viewpoint, The World Bank; 2004 (Note #275), 4 pp.
[25] Switzer M. Explaining oil price differentials. Calgary, Canada: Best of BNN; November 21, 2013. 1 pp.
[26] Katusa M. The tricky calculus of oil price differentials. Calgary, Canada: The World of Energy; January 29, 2014. 1 pp.
[27] Clemente JS, Fedorak PM. A review of the occurrence, analyses, toxicity, and biodegradation of naphthenic acids. Chemosphere 2005;60(5):585−600.
[28] Tissot BP, Welte DH. Petroleum formation and occurrence. New York: Springer-Verlag; 1978.
[29] Nascimento LR, Reboucas LMC, Koike L, Reis FDM, Soldan L, Cerqueira JR, et al. Acidic biomarkers from Albacora oils, Campos basin, Brazil. Org Geochem 1999;30:1175−91.
[30] Meredith W, Kelland SJ, Jones DM. Influence of biodegradation on crude oil acidity and carboxylic acid composition. Org Geochem 2000;31(11):1059−73.
[31] Roques DE, Overton EB, Henry CB. Using gas chromatography/mass spectrometry fingerprint analyses to document process and progress of oil degradation. J Environ Qual 1994;23:851−5.
[32] Watson JS, Jones DM, Swannell RPJ, Van Duin ACT. Formation of carboxylic acids during aerobic biodegradation of crude oil and evidence of microbial oxidation of hopanes. Org Geochem 2002;33(10):1153−69.
[33] Barth T, Høiland S, Fotland P, Askvik KM, Pedersen BS, Borgund AE. Acidic compounds in biodegraded petroleum. Org Geochem 2004;35(11−12 Spec. Iss.):1513−25.

[34] Tomczyk NA, Winans RE, Shinn JH, Robinson RC. On the nature and origin of acidic species in petroleum. 1. Detailed acid type distribution in a California crude oil. Energy Fuels 2001;15(6):1498–504.

[35] Li M, Cheng D, Pan X, Dou L, Hou D, Shi Q, et al. Characterization of petroleum acids using combined FT-IR, FT-ICR–MS and GC–MS: implications for the origin of high acidity oils in the muglad basin, Sudan. Org Geochem 2010;41(9):959–65.

[36] Huang H, Larter SR, Bowler BFJ, Oldenburg TBP. A dynamic biodegradation model suggested by petroleum compositional gradients within reservoir columns from the Liaohe basin, ne China. Org Geochem 2004;35(3):299–316.

[37] Larter S, Wilhelms A, Head I, Koopmans M, Aplin A, Di Primio R, et al. The controls on the composition of biodegraded oils in the deep subsurface – Part 1: biodegradation rates in petroleum reservoirs. Org Geochem 2003;34(4):601–13.

[38] Larter S, Huang H, Adams J, Bennett B, Jokanola O, Oldenburg T, et al. The controls on the composition of biodegraded oils in the deep subsurface: Part II – geological controls on subsurface biodegradation fluxes and constraints on reservoir-fluid property prediction. AAPG Bull 2006;90(6):921–38.

[39] Bennett B, Adams JJ, Gray ND, Sherry A, Oldenburg TBP, Huang H, et al. The controls on the composition of biodegraded oils in the deep subsurface – Part 3. The impact of microorganism distribution on petroleum geochemical gradients in biodegraded petroleum reservoirs. Org Geochem 2013;56:94–105.

[40] Dou L, Cheng D, Li M, Xiao K, Shi B, Li Z. Unusual high acidity oils from the great palogue field, melut basin, Sudan. Org Geochem 2008;39(2):210–31.

[41] Jaffe R, Gardinali PR. Generation and maturation of carboxylic acids in ancient sediments from the maracaibo basin, Venezuela. Org Geochem 1990;16(1–3):211–8.

[42] Hughey CA, Rodgers RP, Marshall AG, Walters CC, Qian K, Mankiewicz P. Acidic and neutral polar NSO compounds in smackover oils of different thermal maturity revealed by electrospray high field Fourier transform ion cyclotron resonance mass spectrometry. Org Geochem 2004;35(7):863–80.

[43] Hughey CA, Galasso SA, Zumberge JE. Detailed compositional comparison of acidic NSO compounds in biodegraded reservoir and surface crude oils by negative ion electrospray Fourier transform ion cyclotron resonance mass spectrometry. Fuel 2007;86(5–6):758–68.

[44] Jaffé R, Gallardo MT. Application of carboxylic acid biomarkers as indicators of biodegradation and migration of crude oils from the maracaibo basin, western Venezuela. Org Geochem 1993;20(7):973–84.

[45] Klein GC. Petroleomics: applications in the fingerprinting of the acidic and basic crude oil components detected by electrospray ionization Fourier transform ion cyclotron Resonance mass spectrometry [Doctor of Philosophy thesis from Florida State University, Department of Chemistry and Biochemistry]. October 12, 2005. 145 pp.

[46] Hughey CA, Rodgers RP, Marshall AG, Qian K, Robbins WK. Identification of acidic NSO compounds in crude oils of different geochemical origins by negative ion electrospray Fourier transform ion cyclotron resonance mass spectrometry. Org Geochem 2002;33(7):743–59.

[47] Huang X, Wu X, Hou Z. Mechanism of degradation for petroleum hydrocarbon by Brevibacillus brevis and Bacillus cereus. Shiyou Xuebao Acta Pet Sin 2006;27(5):92–5.

[48] Seifert WK, Teeter RM. Preparative thin-layer chromatography and high resolution mass spectrometry of crude oil carboxylic acids. Anal Chem 1969;41:786–95.

[49] Seifert WK, Teeter RM, Howells WG, Cantow MJR. Analysis of crude oil carboxylic acids after conversion to their corresponding hydrocarbons. Anal Chem 1969;41:1638–47.

[50] Goheen GE. Conversion of naphthenic acids to naphthene hydrocarbons. Chemical constitution. Ind Eng Chem 1940;32(4):503–8.

[51] Headley J, Du J-L, Peru K, Janfada A. Solid phase extraction (SPE) and quantification of naphthenic acid in milli-q, simulated groundwater and river waters [Unpublished Presentation]. 2007. Available from: http://www.traceorganic.com/2007/presentations/Jing-LongDu_Solid%20Phase%20Extraction%20(SPE)%20and%20Quantification%20of%20%20Naphthenic%20Acid.pdf. 30 pp.
[52] Boduszynski MM. Composition of heavy petroleums. 2. Molecular characterization. Energy Fuels 1988;2(5):597–613.
[53] Gaikar VG, Maiti D. Adsorptive recovery of naphthenic acids using ion-exchange resins. React Funct Polym 1996;31(2):155–64.
[54] Lu Z, Zhai Y, Zhao S, Zhuang L. A new method for determination of naphthenic acids in crude oil. China Pet Proc Petrochem Technol 2004;3:39–43.
[55] Saab J, Mokbel I, Razzouk AC, Ainous N, Zydowicz N, Jose J. Quantitative extraction procedure of naphthenic acids contained in crude oils. Characterization with different spectroscopic methods. Energy Fuels 2005;19(2):525–31.
[56] Zhuang L, Lu Z, Tian Y, Zhai Y. Process for separation of petroleum acids from crude oil. China Pet Proc Petrochem Technol 2005;(2):37–9.
[57] Green JB, Hoff RJ, Woodward PW, Stevens LL. Separation of liquid fossil fuels into acid, base and neutral concentrates: 1. An improved nonaqueous ion exchange method. Fuel 1984;63(9):1290–301.
[58] Qian K, Robbins WK, Hughey CA, Cooper HJ, Rodgers RP, Marshall AG. Resolution and identification of elemental compositions for more than 3000 crude acids in heavy petroleum by negative-ion microelectrospray high-field Fourier transform ion cyclotron Resonance mass spectrometry. Energy Fuels 2001;15(6):1505–11.
[59] Rowland SM, Robbins WK, Corilo YE, Marshall AG, Rodgers RP. Solid-phase extraction fractionation to extend the characterization of naphthenic acids in crude oil by electrospray ionization Fourier transform ion cyclotron resonance mass spectrometry. Energy Fuels 2014;28(8):5043–8.
[60] Shi Q, Zhao S, Xu Z, Chung KH, Zhang Y, Xu C. Distribution of acids and neutral nitrogen compounds in a chinese crude oil and its fractions: characterized by negative-ion electrospray ionization Fourier transform ion cyclotron resonance mass spectrometry. Energy Fuels 2010;24(7):4005–11.
[61] Zhang Y, Shi Q, Li A, Chung KH, Zhao S, Xu C. Partitioning of crude oil acidic compounds into subfractions by extrography and identification of isoprenoidyl phenols and tocopherols. Energy Fuels 2011;25(11):5083–9.
[62] Saad OM, Gasmelseed GA, Hamid AHM. Separation of naphthenic acid from sudanese crude oil using local activated clays. J Appl Ind Sci 2014;2(1):14–8.
[63] Rudzinski WE, Oehlers L, Zhang Y, Najera B. Tandem mass spectrometric characterization of commercial naphthenic acids and a Maya crude oil. Energy Fuels 2002;16(5):1178–85.
[64] Shi LJ, Shen BX. Composition and distribution of naphthenic acids in penglai crude oil with high acid number. Hua Dong Li Gong Da Xue J East China Univ Sci Technol 2007;33(3):314–7.
[65] Nasir Shah S, Mutalib MIA, Pilus RBM, Lethesh KC. Extraction of naphthenic acid from highly acidic oil using hydroxide-based ionic liquids. Energy Fuels 2015;29(1):106–11.
[66] Kamarudin H, Mutalib MIA, Man Z, Azmi M, Extraction of naphthenic acids from liquid hydrocarbon using imidazolium ionic liquids. In: Proc. Intern. Conf. on Environment Sci. Eng. Rome, Italy, September 26–27. IPCBEE, 2012. vol. 32, 17–23.
[67] Rogers VV, Liber K, Mackinnon MD. Isolation and characterization of naphthenic acids from Athabasca oil sands tailings pond water. Chemosphere 2002;48:519–27.
[68] Fedorak PM, Overview of naphthenic acids analyses at the University of Alberta. In: Proc. CONRAD/OSERN Symp. Coast Terrace Inn, Edmonton, Alberta, May 12–13, 2003. 50 pp.
[69] Clemente JS, Prasad NGN, Mackinnon MD, Fedorak PM. A statistical comparison of naphthenic acids characterized by gas chromatography-mass spectrometry. Chemosphere 2003;50(10):1265–74.

[70] Hemmingsen PV, Kim S, Pettersen HE, Rodgers RP, Sjöblom J, Marshall AG. Structural characterization and interfacial behavior of acidic compounds extracted from a North Sea oil. Energy Fuels 2006;20(5):1980−7.

[71] Boduszynski MM. Composition of heavy petroleums. 1. Molecular weight, hydrogen deficiency, and heteroatom concentration as a function of atmospheric equivalent boiling point up to 1400°F (760°C). Energy Fuels 1987;1(1):2−11.

[72] Conrad Environmental Aquatic Technical Advisory Group (Ceatag). Naphthenic acid: background information. Report No. Discussion Report, CONRAD. June 1998. 89 pp.

[73] Huang M, Zhao S, Li P, Huisingh D. Removal of naphthenic acid by microwave. J Clean Prod 2006;14(8):736−9.

[74] Huang S, Tian S, Liu Z, Zhu C. Determination of the structure and composition of petroleum acids in the diesel fractions of crude oil with high acid number. Pet Proc Petrochem 2007;38(4):51−5.

[75] Rowland SJ, West CE, Scarlett AG, Jones D, Boberek M, Pan L, et al. Monocyclic and monoaromatic naphthenic acids: synthesis and characterisation. Environ Chem Lett 2011;9(4):525−33.

[76] Wang Z, Tu Y. Comparison of carboxylic acids in some crude oils and their diesel distillates and vgos: characterized by negative-ion electrospray ionization Fourier transform ion cyclotron resonance mass spectrometry (negative-ion esift-ICR MS). China Pet Proc Petrochem Technol 2011;(3):8−15.

[77] Damasceno FC, Gruber LDA, Geller AM, Vaz De Campos MC, Gomes AO, Guimaraes RCL, et al. Characterization of naphthenic acids using mass spectroscopy and chromatographic techniques: study of technical mixtures. Anal Methods 2014;6(3):807−16.

[78] Aske N. Characterisation of crude oil components, asphaltene aggregation and emulsion stability by means of near infrared spectroscopy and multivariate analysis [DOKTOR INGENIØR thesis from Norwegian University of Science and Technology, Department of Chem. Eng.]. June 2002. 51 pp.

[79] Gould KA, Long RB. A new technique for the acid/base separation of petroleum and coal-derived fractions. Fuel 1986;65(4):572−6.

[80] Wang Y, Chu Z, Qiu B, Liu C, Zhang Y. Removal of naphthenic acids from a vacuum fraction oil with an ammonia solution of ethylene glycol. Fuel 2006;85(17−18):2489−93.

[81] Shi LJ, Shen BX, Wang GQ. Removal of naphthenic acids from Beijiang crude oil by forming ionic liquids. Energy Fuels 2008;22(6):4177−81.

[82] Khulbe KC, Mann RS, Macphee JA. Separation of acidic fraction from the Cold Lake bitumen asphaltenes and its relationship to enhanced oil recovery. Fuel Proc Technol 1996;46(1):63−9.

[83] Harkness RW, Bruun JH. Naphthenic acids from gulf coast petroleum composition of higher boiling acids. Ind Eng Chem 1940;32(4):499−502.

[84] St John WP, Rughani J, Green SA, Mcginnis GD. Analysis and characterization of naphthenic acids by gas chromatography−electron impact mass spectrometry of tert.-butyldimethylsilyl derivatives. J Chromatogr A 1998;807(2):241−51.

[85] Lu Z, Tian S, Zhai Y, Ding Y, Zhuang L. Determination of naphthenic acids in crude oil by chemical ionization mass spectrometry. Chin J Geochem 2005;24(1):68−72.

[86] Lu ZB, Chen QL, Li Q, Wang XQ. Determination of petroleum acids in vacuum residue by ci-MS. Shiyou Huagong Gaodeng Xuexiao Xuebao J Petrochem Univ 2007;20(4):13−5.

[87] Hsu CS, Dechert GJ, Robbins WK, Fukuda EK. Naphthenic acids in crude oils characterized by mass spectrometry. Energy Fuels 2000;14(1):217−23.

[88] Lo CC, Brownlee BG, Bunce NJ. Electrospray-mass spectrometric analysis of reference carboxylic acids and Athabasca oil sands naphthenic acids. Anal Chem 2003;75(23):6394−400.

[89] Rahimi P, Rodgers R, Marshall AG, Teclemariam A, Akbarzadeh K, De Bruijn T. Detailed molecular characterization of naphthenic acids in Athabasca bitumen. ACS Div Pet Chem Prepr 2005;50(2):266−8.

[90] Qian K, Edwards KE, Dechert GJ, Jaffe SB, Green LA, Olmstead WN. Measurement of total acid number (TAN) and TAN boiling point distribution in petroleum products by electrospray ionization mass spectrometry. Anal Chem 2008;80(3):849−55.

[91] Qian K, Edwards KE, Siskin M, Olmstead WN, Mennito AS, Dechert GJ, et al. Desorption and ionization of heavy petroleum molecules and measurement of molecular weight distributions. Energy Fuels 2007;21(2):1042–7.
[92] Gabryelski W, Froese KL. Characterization of naphthenic acids by electrospray ionization high-field asymmetric waveform ion mobility spectrometry mass spectrometry. Anal Chem 2003;75(17):4612–23.
[93] Rodgers RP, Schaub TM, Marshall AG. Petroleomics: MS returns to its roots. Anal Chem 2005;77(1):20–7.
[94] Senko MW, Hendrickson CL, Pasa-Tolic L, Marto JA, White FM, Guant S, et al. Electrospray ionization Fourier transform ion cyclotron resonance at 9.4 t. Rapid Commun Mass Spectrom 1996;10(14):1824–8.
[95] Barrow MP, Headley JV, Peru KM, Derrick PJ. Fourier transform ion cyclotron resonance mass spectrometry of principal components in oilsands naphthenic acids. J Chromatogr A 2004;1058(1–2):51–9.
[96] Lalli PM, Corilo YE, Rowland SM, Marshall AG, Rodgers RP. Isomeric separation and structural characterization of acids in petroleum by ion mobility mass spectrometry. Energy Fuels 2015;29(6):3626–33.
[97] Headley JV, Mcmartin DW. A review of the occurrence and fate of naphthenic acids in aquatic environments. J Environ Sci Health Part A Toxic Hazard Subst Environ Eng 2004;39(8):1989–2010.
[98] Barrow MP, Mcdonnell LA, Feng X, Walker J, Derrick PJ. Determination of the nature of naphthenic acids present in crude oils using nanospray Fourier transform ion cyclotron resonance mass spectrometry: the continued battle against corrosion. Anal Chem 2003;75(4):860–6.
[99] Headley JV, Peru KM, Barrow MP, Derrick PJ. Characterization of naphthenic acids from Athabasca oil sands using electrospray ionization: the significant influence of solvents. Anal Chem 2007;79(16):6222–9.
[100] Smith DF, Rahimi P, Teclemariam A, Rodgers RP, Marshall AG. Characterization of Athabasca bitumen heavy vacuum gas oil distillation cuts by negative/positive electrospray ionization and automated liquid injection field desorption ionization Fourier transform ion cyclotron resonance mass spectrometry. Energy Fuels 2008;22(5):3118–25.
[101] Barrow MP, Headley JV, Peru KM, Derrick PJ. Data visualization for the characterization of naphthenic acids within petroleum samples. Energy Fuels 2009;23(5):2592–9.
[102] Pan Y, Liao Y, Shi Q, Hsu CS. Acidic and neutral polar NSO compounds in heavily biodegraded oils characterized by negative-ion ESI FT-ICR MS. Energy Fuels 2013;27(6):2960–73.
[103] Qian K, Edwards KE, Dechert GJ, Jaffe SB, Green LA, Olmstead W. Measurement of distributed total acid numbers by electrospray ionization mass spectrometry. In: Proc. ACS Nat. Meeting. Boston, MA, 2007. 1 pp.
[104] Barrow MP. Petroleomics. 2008. Available from: http://www.warwick.ac.uk/staff/M.P.Barrow/petroleum.html.
[105] Stanford LA, Kim S, Rodgers RP, Marshall AG. Characterization of compositional changes in vacuum gas oil distillation cuts by electrospray ionization Fourier transform-ion cyclotron resonance (FT-ICR) mass spectrometry. Energy Fuels 2006;20(4):1664–73.
[106] Hao C, Headley JV, Peru KM, Frank R, Yang P, Solomon KR. Characterization and pattern recognition of oil-sand naphthenic acids using comprehensive two-dimensional gas chromatography/time-of-flight mass spectrometry. J Chromatogr A 2005;1067(1–2):277–84.
[107] Stanford LA, Kim S, Klein GC, Smith DF, Rodgers RP, Marshall AG. Identification of water-soluble heavy crude oil organic-acids, bases, and neutrals by electrospray ionization and field desorption ionization Fourier transform ion cyclotron resonance mass spectrometry. Environ Sci Technol 2007;41(8):2696–702.
[108] Qing W. Processing high TAN crude: Part I. PTQ 2011;16(Q1):35–43.
[109] Shalaby HM. Refining of Kuwait's heavy crude oil: materials and challenges. In: Proc. Workshop on Corrosion Protection of Metals, Arab School for Sci. Technol. Kuwait, December 3–7, 2005. 15 pp.

[110] Brient JA. Commercial utility of naphthenic acids recovered from petroleum distillates. In: Proc. 215th ACS Nat. Meeting. Dallas, TX, April 2, 1998. Book of Abstract, PETR Part 2. 1 pp.

[111] Lo CC, Brownlee BG, Bunce NJ. Mass spectrometric and toxicological assays of Athabasca oil sands naphthenic acids. Water Res 2006;40(4):655−64.

[112] Haynes D. Naphthenic acid bearing refinery feedstocks and corrosion abatement. In: Proc. AIChE − Symp. Chicago, IL, October 2006. Nalco Co. Proprietary Knowledge. 30 pp.

[113] Skippins J, Johnson D, Davies R. Corrosion-mitigation program improves economics for processing naphthenic crudes. Oil Gas J 2000;98(37):64−8.

[114] Cheng DS, Dou LR, Li YY, Li Z. Component and distribution of organic acid oil with high TAN, m basin, Sudan. Shiyou Kantan Yu Kaifa Pet Explor Dev 2006;33(6):762−5.

[115] Speight JG. Chapter 1-naphthenic acids in petroleum. In: Speight JG, editor. High acid crudes. Boston: Gulf Professional Publishing; 2014. p. 1−29.

[116] Havre TE, Sjöblom J, Vindstad JE. Oil/water-partitioning and interfacial behavior of naphthenic acids. J Dispers Sci Technol 2003;24(6):789−801.

[117] Smith DF, Schaub TM, Rahimi P, Teclemariam A, Rodgers RP, Marshall AG. Self-association of organic acids in petroleum and Canadian bitumen characterized by low- and high-resolution mass spectrometry. Energy Fuels 2007;21(3):1309−16.

[118] Da Campo R, Barrow MP, Shepherd AG, Salisbury M, Derrick PJ. Characterization of naphthenic acid singly charged noncovalent dimers and their dependence on the accumulation time within a hexapole in Fourier transform ion cyclotron Resonance mass spectrometry. Energy Fuels 2009;23(11):5544−9.

[119] Varadaraj R, Brons C. Molecular origins of heavy crude oil interfacial activity Part 2: fundamental interfacial properties of model naphthenic acids and naphthenic acids separated from heavy crude oils. Energy Fuels 2007;21(1):199−204.

[120] Carbonezi CA, Almeida LCD, Araujo BC, Lucas EF, González G. Solution behavior of naphthenic acids and its effect on the asphaltenes precipitation onset. Energy Fuels 2009;23(3):1249−52.

[121] Poteau S, Argillier JF, Langevin D, Pincet F, Perez E. Influence of pH on stability and dynamic properties of asphaltenes and other amphiphilic molecules at the oil-water interface. Energy Fuels 2005;19(4):1337−41.

[122] Auflem I, Havre T, Sjöblom J. Near-IR study on the dispersive effects of amphiphiles and naphthenic acids on asphaltenes in model heptane-toluene mixtures. Colloid Polym Sci 2002;280(8):695−700.

[123] Oliveira GE, Mansur CRE, Lucas EF, González G, De Souza WF. The effect of asphaltenes, naphthenic acids, and polymeric inhibitors on the pour point of paraffins solutions. J Dispers Sci Technol 2007;28(3):349−56.

[124] Varadaraj R, Brons C. Molecular origins of heavy oil interfacial activity Part 1: fundamental interfacial properties of asphaltenes derived from heavy crude oils and their correlation to chemical composition. Energy Fuels 2007;21(1):195−8.

[125] Ostlund JA, Nyden M, Auflem IH, Sjöblom J. Interactions between asphaltenes and naphthenic acids. Energy Fuels 2003;17(1):113−9.

[126] Ostlund JA, Nyden M, Fogler HS, Holmberg K. Functional groups in fractionated asphaltenes and the adsorption of amphiphilic molecules. Colloids Surf A Physicochem Eng Asp 2004;234(1−3):95−102.

[127] Heaps DT, Madasu PK, Magers DH, Buchanan JP. Investigation of the precipitation behavior of asphaltenes in the presence of naphthenic acids using light scattering and molecular modeling techniques. Energy Fuels 2012;26(3):1862−9.

[128] Stanford LA, Rodgers RP, Marshall AG, Czarnecki J, Wu XA, Taylor S. Detailed elemental compositions of emulsion interfacial material versus parent oil for nine geographically distinct light, medium, and heavy crude oils, detected by negative- and positive-ion electrospray ionization Fourier transform ion cyclotron resonance mass spectrometry. Energy Fuels 2007;21(2):973−81.

[129] Muller H, Pauchard VO, Hajji AA. Role of naphthenic acids in emulsion tightness for a low total acid number (TAN)/high asphaltenes oil: characterization of the interfacial chemistry. Energy Fuels 2009;23(3):1280−8.

[130] Pauchard V, Sjöblom J, Kokal S, Bouriat P, Dicharry C, Müller H, et al. Role of naphthenic acids in emulsion tightness for a low-total-acid-number (TAN)/high-asphaltenes oil. Energy Fuels 2009;23(3):1269–79.

[131] Moschopedis SE, Schulz KF, Speight JG, Morrison DN. Surface-active materials from athabasca oil sands. Fuel Proc Technol 1980;3(1):55–61.

[132] Erstad K, Høiland S, Fotland P, Barth T. Influence of petroleum acids on gas hydrate wettability. Energy Fuels 2009;23(4):2213–9.

[133] Callaghan IC, Mckechnie AL, Ray JE, Wainwright JC. Investigation of the foaming properties of crude oil. Part 1. Identification of the crude oil components responsible for foaming. Soc. Pet. Eng. AIME, 1983, SPE-12342. 5 pp.

[134] Hanneseth A-MD. An experimental study of tetrameric naphthenic acids at w/o interfaces: reactivity, inhibition and emulsion formation [Ph.D. thesis from Norwegian University of Science and Technology, Chem. Eng.]. January 2009. 70 pp.

[135] Hanneseth AMD, Fossen M, Selsbak C, Sjöblom, J. Naphthenic acid/naphthenate stabilized emulsions and the influence of crude oil components. In: Proc. 8th International Conference on Petroleum Phase Behavior and Fouling. Pau, France, June 10–14, 2007. 1 pp.

[136] Hanneseth AMD, Selsbak C, Sjöblom J. Behavior and stability of naphthenic acid/naphthenate stabilized emulsions. Mixed c80-tetraacid and stearic acid stabilization. J Dispers Sci Technol 2010;31(6):770–9.

[137] Ese M-H, Kilpatrick PK. Stabilization of water-in-oil emulsions by naphthenic acids and their salts: model compounds, role of pH, and soap to acid ratio. J Dispers Sci Technol 2004;25(3):253–61.

[138] Brandal Ø, Hanneseth AMD, Sjöblom J. Interactions between synthetic and indigenous naphthenic acids and divalent cations across oil-water interfaces: effects of addition of oil-soluble non-ionic surfactants. Colloid Polym Sci 2005;284(2):124–33.

[139] Magnusson H, Hanneseth AMD, Sjöblom J. Characterization of c80 naphthenic acid and its calcium naphthenate. J Dispers Sci Technol 2008;29(3):464–73.

[140] Brandal Ø. Interfacial (o/w) properties of naphthenic acids and metal naphthenates, naphthenic acid characterization and metal naphthenate [Doctor Ingenier thesis from Norwegian University of Science and Technology, Department of Chem. Eng.]. February 2005. 151 pp.

[141] Simon S, Knudsen KD, Nordgård E, Reisen C, Sjöblom J. Aggregation of tetrameric acids in aqueous media studied by small-angle neutron scattering. J Colloid Interface Sci 2013;394:277–83.

[142] Baugh TD, Grande KV, Mediaas H, Vindstad JE, Wolf NO. The discovery of high-molecular-weight naphthenic acids (ARN acid) responsible for calcium naphthenate deposits. In: Proc. SPE 7th Intern. Symp. Oilfield scale: Pushing the Boundaries of Scale Control. Aberdeen, UK; 2005. pp. 9–15.

[143] Lutnaes BF, Brandal Ø, Sjöblom J, Krane J. Archaeal c80 isoprenoid tetraacids responsible for naphthenate deposition in crude oil processing. Org Biomol Chem 2006;4(4):616–20.

[144] Smith BE, Sutton PA, Lewis CA, Dunsmore B, Fowler G, Krane J, et al. Analysis of 'RN' naphthenic acids by high temperature gas chromatography and high performance liquid chromatography. J Sep Sci 2007;30(3):375–80.

[145] Mediaas H, Grande KV, Vindstad JE. Efficient management of calcium naphthenate deposition at oil fields. In: Proc. TEKNA – Separation Technol. Stavanger, Norway, September 26–27, 2007. 1 pp.

[146] Naphthenate chemistry breakthrough. Statoil-Hydro; 2007. Available from: http://www.statoilhydro.com/en/technologyinnovation/RefiningAndProcessing/OilRefining/NaphthenateChemistryBreakthrough/Pages/default.aspx.

[147] Lutnaes BF, Krane J, Smith BE, Rowland SJ. Structure elucidation of c80, c81 and c82 isoprenoid tetraacids responsible for naphthenate deposition in crude oil production. Org Biomol Chem 2007;5(12):1873–7.

[148] Mediaas H, Grande K, Kummernes H, Vindstad JE. Method for isolation and quantification of naphthenate forming acids ("ARN-acids"). Statoil Petroleum As; March 18, 2014. Patent No. US8674161 (Also published as CA2766384, CN102597176, EP2445994, US20120190907, WO2010151139).

[149] Passade-Boupat N, Gonzalez MR, Brocart B, Hurtevent C, Palermo T. Risk assessment of calcium naphtenates and separation mechanisms of acidic crude oil. In: Proc. SPE 14th Intern. Oilfield Scale. Aberdeen, UK, May 30–31, 2012. pp. 443–454.
[150] Nenningsland AL. Extraction, quantification and study of interfacially active petroleum compounds [Ph.D. thesis from Norwegian University of Science and Technology (NTNU), Chemical Engineering]. 2012.
[151] Malhotra VM, Buckmaster HA. 34 GHz EPR study of vanadyl complexes in various asphaltenes. Statistical correlative model of the coordinating ligands. Fuel 1985;64(3):335–41.
[152] Nordgård EL, Magnusson H, Hanneseth AMD, Sjöblom J. Model compounds for c80 isoprenoid tetraacids. Part II. Interfacial reactions, physicochemical properties and comparison with indigenous tetraacids. Colloids Surf A Physicochem Eng Asp 2009;340(1–3):99–108.
[153] Simon S, Reisen C, Bersås A, Sjöblom J. Reaction between tetrameric acids and ca^{2+} in oil/water system. Ind Eng Chem Res 2012;51(16):5669–76.
[154] Sjöblom J, Simon S, Xu Z. The chemistry of tetrameric acids in petroleum. Adv Colloid Interface Sci 2014;205:319–38.
[155] Simon S, Nordgård E, Bruheim P, Sjöblom J. Determination of c80 tetra-acid content in calcium naphthenate deposits. J Chromat A 2008;1200(2):136–43.
[156] Brocart B, Bourrel M, Hurtevent C, Volle JL, Escoffier B. ARN-type naphthenic acids in crudes: analytical detection and physical properties. J Dispers Sci Technol 2007;28(3):331–7.
[157] Mapolelo MM, Stanford LA, Rodgers RP, Yen AT, Debord JD, Asomaning S, et al. Chemical speciation of calcium and sodium naphthenate deposits by electrospray ionization FT-ICR mass spectrometry. Energy Fuels 2009;23(1):349–55.
[158] Arla D, Sinquin A, Palermo T, Hurtevent C, Graciaa A, Dicharry C. Influence of pH and water content on the type and stability of acidic crude oil emulsions. Energy Fuels 2007;21(3):1337–42.
[159] Balmasova OV, Ramazanova AG, Korolev VV. Adsorption of naphthenic acid on magnetite at different temperatures. Russ J Phys Chem A 2015;89(3):487–90.
[160] Keleşoğlu S, Volden S, Kes M, Sjöblom J. Adsorption of naphthenic acids onto mineral surfaces studied by quartz crystal microbalance with dissipation monitoring (qcm-d). Energy Fuels 2012;26(8):5060–8.
[161] Silva JP, Costa ALH, Chiaro SSX, Delgado BEPC, De Figueiredo MAG, Senna LF. Carboxylic acid removal from model petroleum fractions by a commercial clay adsorbent. Fuel Proc Technol 2013;112:57–63.
[162] Silva JP, De Senna LF, Do Lago DCB, Da Silva Jr PF, Dias EG, De Figueiredo MAG, et al. Characterization of commercial ceramic adsorbents and its application on naphthenic acids removal of petroleum distillates. Mat Res 2007;10(2):219–25.
[163] Headley JV, Peru KM, Mohamed MH, Wilson L, Mcmartin DW, Mapolelo MM, et al. Electrospray ionization Fourier transform ion cyclotron resonance mass spectrometry characterization of tunable carbohydrate-based materials for sorption of oil sands naphthenic acids. Energy Fuels 2013;27(4):1772–8.
[164] Headley JV, Peru KM, Mohamed MH, Wilson L. Electrospray ionization mass spectrometry characterization of sorption of oilsands naphthenic acids to engineered polymers. In: Proc. Asms Conference nn Mass Spectrometry and Allied Topics. Denver, Co, USA, June 2008. vol. 56, 1495. 1 pp.
[165] Zou L, Han B, Yan H, Kasperski KL, Xu Y, Hepler LG. Enthalpy of adsorption and isotherms for adsorption of naphthenic acid onto clays. J Colloid Interface Sci 1997;190(2):472–5.
[166] Sartori G, Savage DW, Ballinger BH. Process for neutralization of petroleum acids. Patent No. GB2304729 (Also published as US5683626). Exxon Research and Engineering Co.; March 26, 1997.
[167] Sartori G, Savage DW, Ballinger BH, Dalrymple DC. Method of decreasing acidity of crude oils and fractions. Patent No. EP0924286 (Also published as US6121411, NO985879, NO318135, ES2267161, DK0924286, DE69834896, CA2252040, AU9714198, AU743069). Exxon Research and Engineering Co.; June 23, 1999.
[168] Lu R, Xu X, Yang J, Gao J. Reduction of total acid number of crude oil and distillate. Energy Sources Part A Recovery, Util Environ Eff 2007;29(1):47–57.

[169] Lu RJ, Xu XR, Yang JY, Gao JS. Reduction of the total acid number of crude oil with basic solution. Pet Sci Technol 2007;25(11):1415–27.

[170] Rahimi P, Kayukawa T, Ryan R, Alem T. Comparison of the reactivity of naphthenic acids in athabasca bitumen and san Joaquin valley. In: Proc. Crude Oil Quality Association Meeting. New Orleans, LA, USA, February 10–11, 2010. 48 pp. Available from: http://www.coqa-inc.org/docs/default-source/meeting-presentations/20100211_rahimi_athabasca_-napacids.pdf?sfvrsn=2.

[171] Messer B, Tarleton B, Beaton M. Compositions, configurations, and methods of reducing naphthenic acid corrosivity. Patent No. US8118994 (Also published as WO2005040313, US2008164137, JP2007508437, EP1678278, CN102382681, CN1894389, CA2541997). Fluor Corp.; February 21, 2012.

[172] Zhang Q, Tian S, Huang S, Wang J. Relationship between naphthenic acid structure and its corrosion performance. Shiyou Xuebao Shiyou Jiagong Acta Pet Sin Pet Proc Sect 2012;28(4):652–6.

[173] Kane RD, Chambers B. High temperature crude oil corrosivity: where sulfur & naphthenic acid chemistry & metallurgy meet. In: Proc. Corrosion Solutions 8th International Conference. Lake Louise, Alberta, Canada, September 25–30, 2012. iCorrosion LLC & Honeywell Corrosion Solutions. Available from: https://www.atimetals.com/news/corrosion-conference/Documents/CSC11-pdfs/presentation_4a_kane.pdf. 20 pp. presentation.

[174] Jin P, Nesic S, Wolf HA. Analysis of corrosion scales formed on steel at high temperatures in hydrocarbons containing model naphthenic acids and sulfur compounds. Surf Interface Anal 2015;47(4):454–65.

[175] Jin P. Mechanism of corrosion by naphthenic acids and organosulfur compounds at high temperatures [Ph.D. thesis from Ohio University, Chem. Eng.]. 2013. 233 pp.

[176] Laredo GC, Lopez CR, Alvarez RE, Cano JL. Naphthenic acids, total acid number and sulfur content profile characterization in isthmus and Maya crude oils. Fuel 2004;83(11–12):1689–95.

[177] Turnbull A, Slavcheva E, Shone B. Factors controlling naphthenic acid corrosion. Corrosion 1998;54:922–30.

[178] Lewis KR, Daane ML, Schelling R. Processing Corrosive Crude Oils. In: Proc. Corrosion/99. San Antonio, TX, USA, April 25–30, 1999. Paper #377. 10 pp.

[179] Qu D-R, Zheng Y-G, Jiang X, Ke W. Correlation between the corrosivity of naphthenic acids and their chemical structure. Anticorros Methods Mat 2007;54(4):211–8.

[180] Laredo GC, Lopez CR, Alvarez RE, Castillo JJ, Cano JL. Identification of naphthenic acids and other corrosivity-related characteristics in crude oil and vacuum gas oils from a Mexican refinery. Energy Fuels 2004;18(6):1687–94.

[181] COQG. Naphthenic acids: an overview of their contributions to corrosion effects. In: Proc. Crude Oil Quality Association Meeting. January 27, 2005. 19 pp.

[182] Tebbal S, Kane RD, Hau JL, Mirabal EJ. Critical factors affecting crude corrosivity. PTQ 1997:85–9. Spring.

[183] Kane RD, Chambers B. High temperature crude oil corrosivity: where sulfur & naphthenic acid chemistry & metallurgy meet. In: Proc. Corrosion Solutions Conference. iCorrosion LLC & Honeywell Corrosion Solutions; February 2011. p. 138–43. Available from: http://www.atimetals.com/news/corrosion-conference/Documents/CSC11-pdfs/paper_4a_kane.pdf.

[184] Dettman HD, Li N, Wickramasinghe D, Luo J. The influence of naphthenic acid and sulfur compound structure on global crude corrosivity under vacuum distillation conditions. In: Proc. Crude Oil Quality Association Meeting. New Orleans, LA, USA, February 10–11, 2010. 29 pp. Available from: http://www.coqa-inc.org/docs/default-source/meeting-presentations/20100211_Dettman_Napacid_Corrosivity.pdf.

[185] Dettman HD, Li N, Wickramasinghe D, Xu Z, Chen XN, Elliott GRD, et al. The influence of naphthenic acid and sulfur compound structure on global crude corrosivity under vacuum distillation conditions. In: Proc. Corrosion Salt Lake City, UT, March 11–15, 2012. vol. 4. pp. 2594–2608.

[186] Randolph B, Vetters E. Corrosion issues with opportunity crudes – more than TAN. In: Proc. opportunity crudes 2008: challenges, benefits. Houston, TX, April 28–May 2, 2008. Hydrocarbon Publishing Co. Presentation. 22 pp.

[187] Rahimi P, Rodgers RP, Marshall AG, Schaub TM, Smith DF, Alem T. Thermal degradation of naphthenic acid in Athabasca bitumen. In: Proc. 231th ACS Nat. Meeting. Atlanta, GA. USA, May 26−30, 2006. vol. 231. 1 pp.

[188] Blum SC, Olmstead WN, Bearden R. Thermal decomposition of naphthenic acids. Patent No. US5820750 (Also published as WO9625471, SG68588, MX9603335, KR100451614, EP0809683, CN1175972, CA2212775, AU708575). Exxon Research and Engineering Co.; October 13, 1998.

[189] Ritchie RGS, Roche RS, Steedman W. Non-isothermal programmed pyrolysis studies of oil sand bitumens and bitumen fractions. 1. Athabasca asphaltene. Fuel 1985;64(3):391−9.

[190] Quiroga-Becerra H, Mejía-Miranda C, Laverde-Cataño D, Hernández-López M, Gómez-Sánchez M. A kinetic study of esterification of naphthenic acids from a colombian heavy crude oil. CTyF Cienc Tecnol Futuro 2012;4(5):21−31.

[191] Smith DF, Rodgers RP, Rahimi P, Teclemariam A, Marshall AG. Effect of thermal treatment on acidic organic species from Athabasca bitumen heavy vacuum gas oil, analyzed by negative-ion electrospray Fourier transform ion cyclotron resonance (FT-ICR) mass spectrometry. Energy Fuels 2009;23(1):314−9.

[192] Groysman A, Brodsky N, Penner J, Goldis A, Savchenko N. Corrosiveness of acidic crude oil and its fractions. Mater. Perform April 2005:34−9.

[193] Yepez O, Vera J. Method of determining the corrosiveness of naphthenic acid in crude oil refinery streams. Patent No. US6294387 (Also published as EP1039290, CA2300608). Intevep S.A; 2001.

[194] Hau JL, Yepez O, Specht MI, Lorenzo R. The iron powder test for naphthenic acid corrosion studies. In: Proc. Corrosion 99. San Antonio, TX, USA, April 25−30, 1999. NACE International Paper #379. 16 pp.

[195] Alvisi PP, Lins VFC. An overview of naphthenic acid corrosion in a vacuum distillation plant. Eng Fail Anal 2011;18(5):1403−6.

[196] Yepez O. On the chemical reaction between carboxylic acids and iron, including the special case of naphthenic acid. Fuel 2007;86(7−8):1162−8.

[197] Yepez O. Influence of different sulfur compounds on corrosion due to naphthenic acid. Fuel 2005;84(1):97−104.

[198] Craig Jr. HL. Naphthenic acid corrosion in the refinery. In: Proc. CORROSION/95. Houston, Texas, USA 1995. Paper No. 333. pp. 120−141.

[199] Kanukuntla V, Qu D, Nesic S, Wolf A. Experimental study of concurrent naphthenic acid and sulfidation corrosion. In: Proc. 17th international corrosion Congress: Corrosion Control in the Service of Society. Las Vegas, NV 2008. vol. 2. pp. 876−897.

[200] Qu DR, Zheng YG, Jing HM, Yao ZM, Ke W. High temperature naphthenic acid corrosion and sulphidic corrosion of q235 and 5cr1/2mo steels in synthetic refining media. Corros Sci 2006;48(8):1960−85.

[201] Kane RD, Chambers B. Refining high acid crudes: when is an opportunity not an opportunity? In: Proc. European Corrosion Congress: Managing Corrosion for Sustainability, EUROCORR. Edinburgh, UK, September 7−11, 2008. vol. 283. 1 pp.

[202] Huang BS, Yin WF, Sang DH, Jiang ZY. Synergy effect of naphthenic acid corrosion and sulfur corrosion in crude oil distillation unit. Appl Surf Sci 2012;259:664−70.

[203] Bota G, Nesic S. Naphthenic acid challenges to iron sulfide scales generated in-situ from model oils on mild steel at high temperature. In: Proc. Corrosion Orlando, FL, March 17−21, 2013. Paper #2512. 13 pp.

[204] Huang BS, Sang DH, Jiang ZY, Liu QY. Coupled erosive behavior of naphthenic acid and sulfur at high temperatures. Cailiao Gongcheng J Mat Eng 2011;(9):39−44. +49.

[205] Dong Z, He J, Guo X, Zhang Y, Han J. High temperature interactive corrosion of naphthenic acid and organic sulphide on cr5mo steel in synthetic refining mixture. J Chin Soc Corros Prot 2011;31(3):219−24.

[206] Slavcheva E, Shone R, Turnbull A. Review of naphthenic acid corrosion in oil refining. Br Corros J 1999;34:125−31.

[207] Chambers BD, Srinivasan S, Kane R, Blades MA. An experimental method for evaluation of crude corrosivity − naphthenic acid and sulfidic corrosion of oil fractions. In: Proc. Corrosion. Salt Lake City, UT, March 11−15, 2012. vol. 6. pp. 4553−4566.

[208] Cui X, Ning Z. Sulfur corrosion and prevention in petroleum processing. Pet Refin Eng 1999;29(8):61–7.
[209] Li JH, Zhao SL, Li P, Zhao ZH, Wang C. Corrosion and protection of sulfur containing oil tanks. Corros Prot 2005;26(8):344–6.
[210] Bagdasarian A, Feather J, Hull B, Stephenson R, Strong R. Crude unit corrosion and corrosion control. In: Proc. Corrosion. Houston, TX 1996. NACE International Paper 615. 15 pp.
[211] Kane RD, Cayard MS. A comprehensive study on naphthenic acid corrosion. In: Proc. Corrosion. Denver, CO, April 7–11, 2002. 16 pp.
[212] Flego C, Galasso L, Montanari L, Gennaro ME. Evolution of naphthenic acids during the corrosion process. Energy Fuels 2014;28(3):1701–8.
[213] Johnson DW, Hils J. Phosphate esters, thiophosphate esters and metal thiophosphates as lubricant additives. Lubricants 2013;1:132–48.
[214] Maher KD, Kirkwood KM, Gray MR, Bressler DC. Pyrolytic decarboxylation and cracking of stearic acid. Ind Eng Chem Res 2008;47(15):5328–36.
[215] Darensbourg DJ, Holtcamp MW, Longridge EM, Khandelwal B, Klausmeyer KK, Reibenspies JH. Role of the metal center in the homogeneous catalytic decarboxylation of select carboxylic acids. Copper(I) and zinc(II) derivatives of cyanoacetate. JACS 1995;117(1):318–28.
[216] Darensbourg DJ, Chojnacki JA, Atnip EV. The catalytic decarboxylation of cyanoacetic acid: anionic tungsten carboxylates as homogeneous catalysts. JACS 1993;115(11):4675–82.
[217] Zhang A, Ma Q, Wang K, Tang Y, Goddard WA. Improved processes to remove naphthenic acids. Report No. Final Technical Report Grant #DE-FC26-02NT15383. Pasadena, CA, USA: California Institute of Technology; December 09, 2005. 96 pp.
[218] Watanabe M, Inomata H, Smith Jr RL, Arai K. Catalytic decarboxylation of acetic acid with zirconia catalyst in supercritical water. Appl Catal A General 2001;219(1–2):149–56.
[219] Shukri NM, Bakar WAWA, Jaafar J, Majid ZA. Removal of naphthenic acids from high acidity Korean crude oil utilizing catalytic deacidification method. J Ind Eng Chem 2015;28:110–6.
[220] Takemura Y, Nakamura A, Taguchi H, Ouchi K. Catalytic decarboxylation of benzoic acid. Ind Eng Chem Prod Res Dev 1985;24(2):213–5.
[221] Fu X, Dai Z, Tian S, Long J, Hou S, Wang X. Catalytic decarboxylation of petroleum acids from high acid crude oils over solid acid catalysts. Energy Fuels 2008;22(3):1923–9.
[222] Mejía C, Quiroga H, Vásquez C, Peña D. Influence of basic catalysts on the decarboxylation kinetics of naphthenic acids. Rev Tec Ing Univ Zulia 2010;33(3):245–53.
[223] Oh HY, Park JH, Rhee YW, Kim JN. Decarboxylation of naphthenic acid using alkaline earth metal oxide. J Ind Eng Chem 2011;17(4):788–93.
[224] Zhang A, Ma Q, Wang K, Liu X, Shuler P, Tang Y. Naphthenic acid removal from crude oil through catalytic decarboxylation on magnesium oxide. Appl Catal A General 2006;303(1):103–9.
[225] Ding L, Rahimi P, Fan Z. Naphthenic acid removal from HVGO by alkaline earth metal catalysts. In: Proc. 238th ACS Nat. Meeting. Washington, DC, August 16–10, 2009. 1 pp.
[226] Yang L, Nelson AR, Gray MR. Liquid-phase catalytic decarboxylation of naphthenic acids. In: Proc. AIChE Spring Nat. Meeting. New Orleans, LA 2008. 1 pp.
[227] Yang L, Gray MR. Liquid-phase decarboxylation of naphthenate acids in oil. In: Proc. 235th ACS Nat. Meeting. New Orleans, LA, April 6–10, 2008. 1 pp.
[228] Ding L, Rahimi P, Hawkins R, Bhatt S, Shi Y. Naphthenic acid removal from heavy oils on alkaline earth-metal oxides and ZnO catalysts. Appl Catal A Gen 2009;371(1–2):121–30.
[229] Müller-Erlwein E. Heterogen katalysierte ketonisierung von laurin- und stearinsäure in flüssigphase. Chem Ing Tech 1990;62(5):416–7.
[230] Leung A, Boocock DGB, Konar SK. Pathway for the catalytic conversion of carboxylic acids to hydrocarbons over activated alumina. Energy Fuels 1995;9(5):913–20.
[231] Vonghia E, Boocock DGB, Konar SK, Leung A. Pathways for the deoxygenation of triglycerides to aliphatic hydrocarbons over activated alumina. Energy Fuels 1995;9(6):1090–6.

[232] Paslawski JC, Headley JV, Hill GA, Nemati M. Biodegradation kinetics of trans-4-methyl-1-cyclohexane carboxylic acid. Biodegradation 2009;20(1):125–33.

[233] Paslawski J, Nemati M, Hill G, Headley J. Biodegradation kinetics of trans-4-methyl-1-cyclohexane carboxylic acid in continuously stirred tank and immobilized cell bioreactors. J Chem Technol Biotechnol 2009;84(7):992–1000.

[234] Biryukova OV, Fedorak PM, Quideau SA. Biodegradation of naphthenic acids by rhizosphere microorganisms. Chemosphere 2007;67(10):2058–64.

[235] Han X, Scott AC, Fedorak PM, Bataineh M, Martin JW. Influence of molecular structure on the biodegradability of naphthenic acids. Environ Sci Technol 2008;42(4):1290–5.

[236] Pereira AS, Islam MS, Gamal El-Din M, Martin JW. Ozonation degrades all detectable organic compound classes in oil sands process-affected water; an application of high-performance liquid chromatography/obitrap mass spectrometry. Rapid Commun Mass Spectrom 2013;27(21):2317–26.

[237] Scott AC, Zubot W, Mackinnon MD, Smith DW, Fedorak PM. Ozonation of oil sands process water removes naphthenic acids and toxicity. Chemosphere 2008;71(1):156–60.

[238] Wang Y-Z, Liu Y-P, Liu C-G. Kinetics of the esterification of low-concentration naphthenic acids and methanol in oils with or without SnO as a catalyst. Energy Fuels 2008;22(4):2203–6.

[239] Wang Y, Sun X, Liu Y, Liu C. Removal of naphthenic acids from a diesel fuel by esterification. Energy Fuels 2007;21(2):941–3.

[240] Wang YZ, Liu YP, Liu CG. Removal of naphthenic acids of a second vacuum fraction by catalytic esterification. Pet Sci Technol 2008;26(12):1424–32.

[241] Wang YZ, Li JY, Sun XY, Duan HL, Song CM, Zhang MM, et al. Removal of naphthenic acids from crude oils by fixed-bed catalytic esterification. Fuel 2014;116:723–8.

[242] Klein GC, Rodgers RP, Teixeira M.A.G, Teixeira AMRF, Marshall AG. Petroleomics: electrospray ionization FT-ICR mass analysis of NSO compounds for correlation between total acid number, corrosivity, and elemental composition. In: Proc. ACS Div. Fuel Chem. Preprints. San Antonio, TX 2003. vol. 48. pp. 14–15.

[243] Holowenko FM, Mackinnon MD, Fedorak PM. Characterization of naphthenic acids in oil sands wastewaters by gas chromatography-mass spectrometry. Water Res 2002;36(11):2843–55.

[244] Mahdavi H, Prasad V, Liu Y, Ulrich AC. In situ biodegradation of naphthenic acids in oil sands tailings pond water using indigenous algae-bacteria consortium. Bioresour Technol 2015;187:97–105.

[245] Brown LD, Ulrich AC. Oil sands naphthenic acids: a review of properties, measurement, and treatment. Chemosphere 2015;127:276–90.

[246] Mckenzie N, Yue S, Liu X, Ramsay BA, Ramsay JA. Biodegradation of naphthenic acids in oils sands process waters in an immobilized soil/sediment bioreactor. Chemosphere 2014;109:164–72.

[247] West CE, Scarlett AG, Tonkin A, O'carroll-Fitzpatrick D, Pureveen J, Tegelaar E, et al. Diaromatic sulphur-containing 'naphthenic' acids in process waters. Water Res 2013;51:206–15.

[248] Wang B, Wan Y, Gao Y, Yang M, Hu J. Determination and characterization of oxy-naphthenic acids in oilfield wastewater. Environ Sci Technol 2013;47(16):9545–54.

[249] Toor NS, Han X, Franz E, Mackinnon MD, Martin JW, Liber K. Selective biodegradation of naphthenic acids and a probable link between mixture profiles and aquatic toxicity. Environ Toxicol Chem 2013;32(10):2207–16.

[250] Sohrabi V, Ross MS, Martin JW, Barker JF. Potential for in situ chemical oxidation of acid extractable organics in oil sands process affected groundwater. Chemosphere 2013;93(11):2698–703.

[251] Collier TK, Anulacion BF, Arkoosh MR, Dietrich JP, Incardona JP, Johnson LL, et al. Effects on fish of polycyclic aromatic hydrocarbons (pahs) and naphthenic acid exposures. In: Tierney KB, editor. Org. chem. toxicology of fishes. London: Elsevier; 2013. p. 195–255.

[252] Kremer LN. Crude oil and quality variations. PTQ 2004 Autumm:87–93.

[253] Rodriguez RA, Ubbels SJ. Understanding naphthenate salt issues in oil production. World Oil 2007;228(8):143–6.

[254] Mackinnon MD, Boerger H. Description of two treatment methods for detoxifying oil sands tailings pond water. Water Pollut Res J Can 1986;21(4):496–512.
[255] Jennings DW, Shaikh A. Heat-exchanger deposition in an inverted steam-assisted gravity drainage operation. Part 1. Inorganic and organic analyses of deposit samples. Energy Fuels 2007;21(1):176–84.
[256] Schaub TM, Jennings DW, Kim S, Rodgers RP, Marshall AG. Heat-exchanger deposits in an inverted steam-assisted gravity drainage operation. Part 2. Organic acid analysis by electrospray ionization Fourier transform ion cyclotron resonance mass spectrometry. Energy Fuels 2007;21(1):185–94.
[257] Hurtevent C, Ubbels S. Preventing naphthenate stabilised emulsions and naphthenate deposits on fields producing acidic crude oils. In: Proc. SPE 8th intern. symp. oilfield scale. Aberdeen, UK, May 31–June 1, 2006. pp. 69–75.
[258] Sarac S, Civan F. Experimental investigation and modeling of naphthenate soap precipitation kinetics in petroleum reservoirs. In: Proc. SPE intern. symp. oilfield chem. Houston, TX, February 28–March 2, 2007. pp. 280–288.
[259] Yeung TW. Evaluating opportunity crude processing. PTQ 2006;11(Q4):93–6.
[260] Lordo S. Impact and management of opportunity crudes on the desalting and wastewater process. Proc. Opportunity crudes Conf.: challenges, benefits. Houston, TX, April 28–May 2, 2008. Hydrocarbon Publishing Co. [Unpublished presentation].
[261] [Unpublished Presentation at Refining High Acid Crudes II] Nalco and exxon energy chemicals inhibition of napthenic acid corrosion and other related issues. 2002. 72 pp.
[262] Skippins J, Bell K, Kronk J, Bagdasarian A, Johnson D. High acid crude. In: Proc. Crude Oil Quality Association Meeting. New Orleans, LA, USA, May 2002. ChevronTexaco 180 pp.
[263] Ayello F, Robbins W, Richter S, Nešić S. Crude oil chemistry effects on inhibition of corrosion and phase wetting. In: Proc. Corrosion, NACE Intern. Houston, TX, USA, March 13–17, 2011. 19 pp.
[264] Varadaraj R, Brons C. Molecular origins of crude oil interfacial activity part 3: characterization of the complex fluid rag layer formed at crude oil – water interfaces. Energy Fuels 2007;21(3):1617–21.
[265] Kapusta SD, Ooms A, Smith A, Van Den Berg F, Fort W. Safe processing of acid crudes. In: Proc. Corrosion. New Orlean, LA 2004. Paper No. 04637. 19 pp.
[266] Groysman A, Brodsky N, Pener J, Shmulevich D. Low temperature naphthenic acid corrosion study. In: Proc. Corrosion. Nashville, TN, USA, March 11–15, 2007. NACE International Paper #NACE-07569. 20 pp.
[267] Jiang K, Chen X, Yang T, Qin Z. Experiment study of high-temperature and high-flow rate naphthenic acid corrosion. In: Proc. ASME Pressure Vessels and Piping Conf. Baltimore, MD, 2011, vol. 6. pp. 1403–1409.
[268] Ghoshal S, Sainik V. Monitor and minimize corrosion in high-TAN crude processing. Hydrocarb Process 2013;92(3):35–8.
[269] Petkova N, Angelova M, Petkov P. Establishing the reasons and type of the enhanced corrosion in the crude oil atmospheric distillation unit. Pet Coal 2009;51(4):286–92.
[270] Lou S, Zhao D, Wang H, Du J. Decreasing the acid content in high-acid crude oil. Pet Sci Technol 2009;27(1):111–21.
[271] O'Kane JM, Rudd TF, Cooke D, Dean FWH, Powell SW. Detection and monitoring of naphthenic acid corrosion in a visbreaker unit using hydrogen flux measurements. In: Proc. Corrosion. San Antonio, TX, USA, March 14–18, 2010. NACE International Paper No. 10351. 15 pp.
[272] Nugent MJ, Dobis JD. Experience with naphthenic acid corrosion in low TAN crudes. In: Proc. Corrossion/98. Houston, TX 1998. NACE. Paper No. 577. 8 pp.
[273] Kumar R, Thorat TS, Chithra V, Rathore V, Rao PVC, Choudary NV. Processing opportunity crude oils – catalytic process for high-acid crudes. Hydrocarb World 2009;4(2):64–8.
[274] Dion M. Challenges and solutions for processing opportunity crudes. In: Proc. AFPM Annual Meeting. Orlando, FL, USA, March 23–25, 2014. 13 pp.
[275] Afaf GA, Badia HA, Hassan EE. Corrosion treatment of high TAN crude: case of fula crude, Sudan. Annual Conf. For basic Eng. Sci. 2012. pp. 1–10. Available from: http://research.uofk.edu/multisites/UofK_research/images/stories/research/PDF/BESBC/corrosion%20treatment%20of%20high%20tan%20crude%20%20case%20of%20fula%20crude%20sudan.pdf.

[276] Szklo AS, Machado G, Schaeffer R, Felipe Simoes A, Barboza Mariano J. Placing Brazil's heavy acid oils on international markets. Energy Policy 2006;34(6):692–705.
[277] Li X, Zhu J, Wu B, Mao X. Characterization of acidic compounds in vacuum gas oils and their dewaxed oils by Fourier transform-ion cyclotron resonance mass spectrometry. Energy Fuels 2012;26(9):5646–54.
[278] Bieber S. A new process for removing calcium from crude oils containing calcium naphthenates. In: Proc. Crude Oil Quality Association Meeting. Chicago, IL, May 26, 2005. 14 pp. Available from: http://www.coqa-inc.org/docs/default-source/meeting-presentations/20050526calcium-removal.pdf?sfvrsn=2.
[279] Zhang S. Processing of Liaohe highly sour crude. China Pet Proc Petrochem Technol 2006;(4):7–13.
[280] Corrosion Cost Technologies. What are corrosion costs? 2006. Available from: http://www.corrosioncost.com/downloads/pdf/index.htm.
[281] Skippins J, Johnson D, Davies R, Evaluation of the economics for the processing of naphthenic crudes. In: Proc. Intern. Conf. Corrosion in Refinery Petrochemical Power Generation Plants. Venice, Italy, May 18–19, 2000. pp. 275–286.
[282] Corrosion Cost Technologies. Petroleum refining cost of corrosion. 2006. Available from: http://www.corrosioncost.com/downloads/pdf/index.htm.
[283] Johnson D, Mcateer GR, Zuk H. Mitigating corrosion from naphthenic acid streams. PTQ 2002 Winter:79–85.
[284] Mathers R. Mitigating circumstances. Hydrocarb Eng 2006;11(8):27–30.
[285] Mathers RE. Corrosion monitoring improvements. PTQ Revamps & Operations; 2005.
[286] Roxar. Corrosion monitoring SenCorr sensors. 2008. Available from: http://www.roxar.com/category.php?categoryID=1150.
[287] Eberle DC. Real time oil corrosivity measurement using radioactive tracer technology [Unpublished Presentation]. 2003. 25 pp. Available from: http://www.swri.org/3pubs/divhome.htm.
[288] Eberle DC. Update on crude oil corrosivity measurement techniques using radioactive tracer technology [Unpublished Presentation]. 2004. 25 pp. Available from: http://www.swri.org/3pubs/divhome.htm.
[289] Shafizadeh A, Mcateer G, Sigmon J. High acid crudes. In: Proc. Crude Oil Quality Association Meeting. New Orleans, January 30, 2003. ChevronTexaco. Available from: http://www.coqg.org/20030130High%20Acid%20Crudes.pdf. 50 pp.
[290] Process notes Opportunity knocks. PTQ; 2011. 1 pp.
[291] Hodges M, Gould C. Smart, integrated smart, integrated approach to capturing acid crude value. In: Proc. AIChE Spring Nat. Meeting, 9th Topical Conf. On Refinery Proc. Orlando, FL, April 25, 2006. 26 pp.
[292] Vetters E, Clarida D. Maintaining reliability when processing opportunity crudes. PTQ 2013;18(5):59–67.
[293] Qing W. High TAN crude and its processing [Unpublished Presentation]. 2010. 61 pp. Available from: http://www.pecj.or.jp/japanese/overseas/conference/pdf/conference04-07.pdf.
[294] Afaf GA, Badiab HA, Hassan EE. Corrosion management methods of high TAN crude case study: Fula crude oil, Sudan. Am Sci Res J Eng Technol Sci 2015;11(1). Available from: http://asrjetsjournal.org/index.php/American_Scientific_Journal/article/view/530/481. 7 pp.
[295] Basconi J. In: Hughes B, editor. Baker petrolite develops technologies for challenging feedstocks. Sugar Land, Texas: Baker Petrolite; 2007.
[296] Rechtien R. Crude unit corrosion control programme. PTQ 2007;Q4:55–60.
[297] Poole K. Laboratory evaluation of high temperature corrosion inhibitors [Unpublished Data report]. 2007. p. 8 [Available from: BP intranet proprietary information].
[298] Petrolite B. In: Hughes B, editor. Smartguard naphthenic acid corrosion control program; 2010 [Sugar Land, TX, USA].
[299] Knot T. Passing the acid test. Fronteirs; December 26, 2001. 1 pp.
[300] Cross C. High-acid crude processing enabled by unique use of computational fluid dynamics: a methodology for applying CFD enables faster identification of pipe elements that are the most vulnerable to attack by high-acid crudes. PTQ 2013;18(5):39–49.

[301] Hopkinson BE, Penuela LE. Naphthenic acid corrosion by venezuelan crudes. In: Proc. Corrosion. New Orleasn, LA, USA 1997. NACE International Paper #NACE-502. 15 pp.
[302] Qing W. Processing high TAN crude: Part II. PTQ 2011;16(1):7.
[303] Kremer L, Nguyen J, Weers J. Removal of calcium and other metal species from crude oil in the desalting process. In: Proc. AIChE Spring Nat. Meeting. New Orleans, LA, April 25–29, 2004. pp. 1273–1279.
[304] Green K. Improve processing of 'opportunity crudes'. Hydrocarb Proc 2004;83(6 Section 1):53–5.
[305] Weers JJ, Bieber S. Calcium removal from high TAN crudes. PTQ 2005;Q3:5.
[306] Weers J, Nguyen J. A new metals removal process for Doba crude oil. In: Proc. 9th ERTC Annual Meeting. Prague, Czech Republic, February 15, 2004. 10 pp.
[307] Bieber S, Fahey B, Renbin J, Hongbin T, Tonghua L. Successful strategies for processing high calcium, high TAN crude oils. In: Proc. 9th annual Asian Refining Technol. Conf. Kuala Lumpur, Malaysia, March 7–9, 2006. 22 pp.
[308] Weers J. New chemical process removes crude oil contaminants. In: Proc. NPRA Annual Meeting. Salt Lake City, UT, March 19–21, 2006 AM-06-31. 10 pp.
[309] Havre TE. Formation of calcium naphthenate in water/oil systems, naphthenic acid chemistry and emulsion stability [DOKTOR INGENIØR thesis from Norwegian University of Science and Technology, Department of Chem. Eng.]. October 2002. 168 pp.
[310] Moran K, Czarnecki J. Competitive adsorption of sodium naphthenates and naturally occurring species at water-in-crude oil emulsion droplet surfaces. Colloids Surf A Physicochem Eng Asp 2007;292(2–3):87–98.
[311] Eid MS. Moving caustic injection point improves crude unit operations. Oil Gas J 2008;106(15):60–5.
[312] Collins T, Barletta T. Desalting heavy Canadian crudes. PTQ 2012;3:23–6.
[313] Suarez F, Wizig H, Youchao L, Zhang J. Light crude oil treatment. PTQ 2004 Autumn:21–5.
[314] Forero P, Suarez FJ. Caustic treatment of jet fuel streams. PTQ 1996 Winter:43–6.
[315] Zhao S, Li P, Kong L. Microwave demulsification in removing naphthenic acids from diesel oil. Pet Sci Technol 2007;25(12):1583–92.
[316] Kittrell N. Removing acid from crude oil. In: Proc. Crude Oil Quality Association Meeting. New Orleans, LO, February 2006. 19 pp.
[317] Holt T, Olav Jøsang L, Sandengen K. Calcium naphthenate propagation during flow in a porous medium. Energy Fuels 2013;27(11):6440–6.
[318] Sundman O, Simon SB, NordgåRd EL, SjöBlom J. Study of the aqueous chemical interactions between a synthetic tetra-acid and divalent cations as a model for the formation of metal naphthenate deposits. Energy Fuels 2010;24(11):6054–60.

CHAPTER 5

Bottom of the Barrel Upgrading Technologies

1. Introduction

The high proportion of troublesome species in unconventional oils not only represents operability problems for the refiner, but is also a source of margin losses. One of these margin losses and probably the most difficult to quantify is the disposition of stranded streams in the refinery. These stranded streams are categorized as low-value products either because they do not meet market specifications or because they fall into that category by virtue of their composition. These streams sometimes oblige the refiner to sacrifice high-value products for blending them off and making more saleable products. Vacuum resid (VR) is probably the single largest stranded stream for most refineries [1]. The VR proportion is higher for heavy and extraheavy oils and bitumen than it is in lighter crudes, hence the quality of these crudes needs to be upgraded. Thus upgrading objectives are (1) the conversion of high molecular weight (MW) compounds into lower MW compounds, which are less viscous and dense, and/or (2) the removal or diminishing of the concentration of contaminants. Upgraded crude oil must be a synthetic crude oil (SCO, Syncrude), comparable to conventional crudes, pipeline transportable, and suitable for processing in conventional refineries.

Simple refineries cannot upgrade heavy crude oils, nor bitumen; in fact, the complexity of a refinery is adjusted to cope with the API gravity and the level of crude oil contamination. A large volume of stranded VR and/or low-quality fuel oil (FO) could typically accumulate in those simple refineries that are not equipped for the processing of VRs from opportunity crudes. Upgrading the whole crude oil or the VRs may signify a comprehensive solution for the abatement of the problematic species.

The presence of heteroatomic functional groups that replace hydrogen atoms in the hydrocarbon and of highly conjugated and aromatic systems results in an H/C ratio as low as 0.8. The H/C ratio is directly associated with the fuel value of the hydrocarbon mixture. To improve the quality of the bottom of the barrel (BotB) fraction requires not only removing the heteroatomic moieties, but also increasing the H/C ratio to values in the order of 1.6. The H/C ratio of examples of different types of crude oils was given in Chapter 1; here the H/C ratio of other hydrocarbonaceous materials and compounds is

The Science and Technology of Unconventional Oils. http://dx.doi.org/10.1016/B978-0-12-801225-3.00005-X

presented in Table 5.1. This change in the H/C ratio could be achieved either by increasing the hydrogen content or by decreasing the carbon content.

Oil residues and specific BotB compounds can be converted to lighter or better quality products by thermal and/or catalytic processing in the absence or presence of hydrogen pressure using conventional technologies. A variety of technologies is available commercially. These were developed for conventional crude oils, containing the contaminant compounds in a much lesser concentration. Improvements, modifications, and adaptations have been added to try to cope with the issues brought by these compounds. According to the followed approach, these processes fall into two broad categories: the *C-rejection processes* (those that eliminate part of the C-atoms from the molecules) and the *H-addition processes* (those that increment their relative H-content). Examples of such processes include visbreaking, hydrovisbreaking, coking, fluid cracking, hydrocracking (HCK), and hydroprocessing (HDP) in general. The presence of asphaltenes, metal compounds, and naphthenic acids (NAs), whose concentration is highest in opportunity crudes, negatively affects the performance and operability of these processes. Specific technologies were developed and are commercially operative or available for licensing for the removal of asphaltenes. Since the metal compounds concentrate in the asphaltic fraction, some of these technologies have been either adapted or intensified for the simultaneous removal of asphaltenes and metal compounds. However, these technologies are not effective enough and the treated product still needs quality improvements. Apart from asphaltenes, there are no specific technologies for the removal of metals or NAs, commercially available. Instead, these compounds are dealt differently throughout refineries, depending on their configurations. Refinery streams containing metals and/or NAs are treated via HDT prior to reaching a process where their effect would be more severely negative.

Table 5.1: Examples of H/C Ratio of Different Materials and Compounds

Material or Compound	H/C Ratio
Bitumen	0.8
Benzene	1.0
Asphaltenes	1.0
Heavy crude	1.5
Shale oil	1.5
Light crude	1.8
Transportation fuels	1.9
Liquefied petroleum gas	2.0
Hexane	2.3
Butane	2.5
Methane	4.0

The commercially available technologies are mature process technologies that have been in refining operations for more than 50 years. Most of the original patents have expired and leave open new developmental opportunities and freedom to design and operate similar processes. Some of the commercially available processes have been mentioned in previous chapters when used in an integrated manner for abating specific species (asphaltenes, metals, or acids).

In this chapter, a brief description of these process technologies, the negative impact caused by BotB processing, and the possible amendments to these technologies will be given. These conventional technologies are not considered in detail since abundant literature—books (e.g., Refs. [2–13]), reviews and reports (e.g., Refs. [14–32]), articles, patents, and brochures—has been published throughout the long history of oil refining.

Selection of the more convenient technology option is not an easy task and depends on refinery configuration and diet changes. Changes in the diet affect VR quality for two reasons: first, oil quality itself and second, the refiner would fractionate the crude to comply with the feed specifications required by the downstream units, depending on refinery configuration, and installed units [fluidized catalytic cracking (FCC), isomerization, reforming, etc.]. For instance, the final boiling point of a given stream would be the lowest boiling point of a contaminant/impurity molecule that need to be kept away from the next downstream unit, in which such stream would be fed. A list of considerations regarding technology selection has been compiled in Ref. [1]:

1. Demonstration at large scale
2. Demonstration of feed type flexibility
3. Feed quality flexibility
4. Conversion level on a once-through basis
5. Products quality (proximity to finished products)
6. Technology reliability
7. Economics

Opportunity crudes are marketed to the refiners with discounted prices, trying to create incentives and processing advantages. Conventional refining soon realized these crudes brought more problems than opportunities. The real opportunities were open to R&D, and the search for new catalysts and processes started long ago and continues today. The improved and modified technologies for the processing of residues from heavy oil, extraheavy oil, and bitumen are commercially available from several licensors and vendors.

2. Carbon Rejection

As already mentioned, these processes consist in the thermal elimination of part of the C-atoms from the molecules, mostly in the absence of a catalyst. Unfortunately, thermal processes lack selectivity but take advantage of the low reactivity of the BotB compounds. Thus the content of metal compounds and/or coke-forming species does not constrain limitations for the processes, as they do in the case of catalytic upgrading processes. Typically, a correlation between Conradson Carbon Residue (CCR) and coke yield is observed.

2.1 Deasphalting

In refining, the existing deasphalting (DA) processes, under operation, are all based on the asphaltene solubility behavior. Precipitation is the most recalled behavior of asphaltenes in crude oils. An excess volume of paraffin precipitates asphaltenes while the maltene fraction may remain soluble.[1] Consequently, this type of solvent deasphalting process is known under the acronym SDA. Thus for DA, flocculation and precipitation are carried out by contacting the asphaltene-containing residue with a paraffinic solvent (in reality, an antisolvent) under strong mixing, at set time, temperature, and pressure conditions, which depend on the solvent used. Clearly from these statements, process variables for a given antisolvent include the ratio of solvent to oil, the contact time, the temperature, the pressure, and the number of stages. In general, the mixture of DA feed and antisolvent is pumped to an extraction tower at temperature ranges of 60–120°C and pressures from 25 to 43 bar. Separation occurs in this extraction tower containing contactor devices. The liquid product is separated from the solids and stripped from solvent.

Currently, SDA units have simplified the solvent recovery train by operating the removal unit at the solvent supercritical conditions. In this way, conventional fractionation towers are not necessary. Additionally, the application of extraction under solvent critical conditions advantageously enhances asphalt phase separation because of solvent vapor density and diffusivity and facilitates turbulent mixing with the feed that contributes to improve mass transfer. The industrial supercritical units utilize light hydrocarbons, such as propane, *n*-butane, *i*-butane, *n*-pentane, *i*-pentane, or a mixture of these, as the antisolvent [33].

VRs are the typical feed, particularly those from heavy oils, bitumen, and tar sands. In some instances, DA of atmospheric residues (ARs) is applied instead of the VR cut, depending on the deasphalted oil (DAO) required quality. Regardless of the feed, the basis of the extraction process is an intimate contact between the solvent and the feed. To accomplish the required degree of contact an efficient mixing of the phases has to take place in a manner that exposes the maximum surface area of each phase to the other.

[1] At bench scale, and for research purposes, maltenes are further separated into oil and resin, for instance, by selective adsorption on supports such as clay or silica. The oil fraction is the portion eluted by nonpolar solvents, while the resin is eluted with strongly polar solvents. [Refer to Chapter 2 for detailed discussion].

Enough equilibrium timing allows an efficient separation by letting the heavy phase settling from the lighter continuous phase. Factors affecting separation include particle and/or droplet size, the relative density of the two phases, and the space available for settling.

The first approach for the liquid–liquid extraction device was a mixer settler vessel, where the feed was mixed with propane and continuously decanted at a temperature approximately 38°C below the critical temperature of the solvent. Solvents ranged from propane to pentane and their mixtures. From this first approach, operation in a multiple stage countercurrent device was developed to take advantage of using a temperature gradient between the final extraction stage and the feed stage. Later, the original multiple stages were replaced by a countercurrent extraction column, using a variety of internal designs. Efficiency of the separation was approached by designing better mixer–extractors, for instance, Universal Oil Products (UOP) uses tilted parallel plates and KBR packed columns (Rosemax). UOP have designed a new set of internals for the extractors, for which information on commercial experience is lacking.

A top raffinate or DAO and a bottom pitch (removed asphaltenes) are the two produced cuts from the unit. Besides the DAO and pitch products, some of the commercial units are also set up to produce an intermediate fraction composed of resins. The obtained DAO is lighter than the original residue, contains fewer metals, and has a lower CCR. The choice of solvent is a compromise between DAO quality and yield. For a given solvent, improving DAO quality requires increasing the number of stages, using a heavier solvent, having a larger solvent/feed ratio, increasing the temperature, etc.

The DAO could be blended into the VGO and fed as part of the conversion feed [for fluidized catalytic cracking (FCC) or hydrocracking (HCK)] [34] or as a blending stock for lube oil production. The quality of the DAO determines its downstream use and may preclude certain applications, such as lube oil production. Originally, DA was designed for lube oils manufacturing. However, the DAO used for lube stocks production requires limited and specific residues and SDA has to be operated at particular conditions. The more recent lube specifications require different manufacturing methods and leave the DAO supply open for uses and applications.

The pitch fraction was thought to be used for asphalt production or as a refinery fuel. Depending on the employed solvent, the pitch contains all the asphaltenes and most or just part of the high MW resins and other molecules. The pitch may vary in its hardness, but it is a brittle solid of zero penetration at room temperature and a liquid of very high viscosity at temperatures above its softening point. The contaminants, metals, sulfur, nitrogen, etc. are concentrated in the pitch. Sulfur content may be as much as 20–100% higher than that of the VR from which it was extracted. As explained earlier, it is not that difficult to feed the DAO product of the SDA units into other refining conversion processes, while the pitch is generally considered a disposable material. It could be fed into a coker unit or further processed through additional extraction either to increase DAO yield or to separate the resins from the asphaltenes. Pitch disposition may be regarded as the main drawback of SDA, since pitch has to be kept hot for maintaining its fluid properties. Feeding these bottoms to gasification improves both processes' economics, but introduces interdependencies between the units. Technologies for pitch solidification are also available [26] and facilitate pitch transportation and distribution, but probably it would impact economics and product marketing. Solidifying processes include flaking and pelletizing. A pelletizing process is offered by KBR, in which the hot pitch is fed through a specially designed distributor into a water-containing vessel (Aquaform [35]). Sandvik offers different solidification alternatives, which have been proved to work in a large number of refineries. Solidification into flakes is carried out in a steel belt cooling system, while pastilles are made in the Rotoform system. For asphaltenes, pastilles give a lower dusting option; pastilles are also safer and environmentally friendlier.

Several uses of the pitch have been suggested, e.g., in the cement industry (in combination with coal and petcoke, more typically employed), further conversion (via visbreaking, HCK, or coking), and/or quality upgrade (via feed pretreatment, e.g., HDT). Some of these applications are available commercially, but others need development. The pitch used in heat or power generation either in the refinery itself or nearby installations has been suggested. In this case, a liquid fuel blend is preferred and prepared in the form of: heavy fuel oil (HFO), hot undiluted liquid, emulsions, or slurries/suspensions. Nevertheless, the current market for high-sulfur FO impedes the inclusion of heavy and/or extra-heavy oil-derived pitch in the FO pool. For this application, the inclusion of SDA bottoms to any other conventional fuel will increase the sulfur emissions from the power station. For this reason, application is limited to fluidized-bed boilers equipped with SO_x capture (e.g., limestone) from the combustion gases or boilers with stack-gas clean-up.

Each of the probable uses of the produced bottom pitch (asphalt cements, asphalt emulsions, industrial asphalts, binders, energy, etc.) requires a proper selection of the VR quality, but still the demand for the product is typically lower than what the SDA units might offer. Therefore when a potential SDA application is identified, special care has to be taken in considering the use/disposal/marketing alternatives for the bottoms. Furthermore, in the future, two more adverse factors act together to impose tighter constraints: (1) increased proportion of heavy oils in the refinery diet that increase contaminants content in the SDA feed and (2) more stringent environmental regulations.

SDA is offered commercially by four licensors [26]. The best known are the Residuum Oil Supercritical Extraction (ROSE) process licensed by M.W. Kellogg (formerly by Kerr-McGee) and the Demex process licensed by UOP. A third technology is offered by Axens/Institut Francais Du Petrol (IFP), their Solvahl process. In addition, most of Petrobras's old units have been redesigned into its Advanced Supercritical Solvent Deasphalting Process PASD [36]. These SDA processes, commercially available through the various licensors, differ in the used solvent, the mixing conditions, and the method for recovery of the separated fractions. Typical temperatures range from 50 to 235°C with pressures of 20–40 bar.

The SDA-installed capacity of ROSE and Demex units, by March 2003 [26], was nearly 500 kbpd corresponding to 24 trains (367,550 bpd in the United States by 2013 [37]), in a proportion of 75:25 (ROSE:Demex) through the years. Therefore commercial experience favors the ROSE technology. During the 1980s seven units were shut down, five of them ROSE units and the other two Demex units. These latter two units were later converted to ROSE units. Operationally, the main difference between these two technologies lies in the solvent-to-oil ratio employed. For ROSE (8:1) it is supposed to be an economical breakpoint, while for Demex (5:1) it is a capital costs optimization (based on size of the unit). Both technologies show limitations on solvent cooling. ROSE recommends siting the recirculation pump in the cleanest and coldest place of the refinery and operating it at the minimum flow. Furthermore, they include reboilers on each stripper and a pump-around. In Demex, the solvent recirculation pump is placed downstream of the extractor and has to handle a high DAO-containing stream (in hot duty). There are no reboilers or pump-around, but instead a DAO wash zone.

In terms of reliability, safety, and environmental concerns, the SDA processes are considered very reliable, safe, and environmentally handy. The stream factors are above 95% and the turnaround timing can be arranged to coincide with that of the conversion units. A very limited number of incidents and accidents have been reported with no fatalities involved. If one considers all the products marketable, then the only effluent/waste stream is sour water (with entrained asphalt), which can be easily handled. However, as previously mentioned the quality of the pitch depends on the quality of the feed and it could become a disposal problem for the refiner. The market growth for this bottom fraction would benefit from its use as cofeed material for power stations. This market looks promising if "zero emissions"-integrated gasification combined cycle and other cleaner gasification techniques succeed in their commercial implementation.

The heat demand on the solvent recovery units of the early processes heavily affected the economics of the technologies. In the 1970s the use of supercritical conditions in these units was introduced as a means to improve the efficiency of the process and therefore the overall economics. In recent times, critical solvent DA became the preferred technology when choosing new units. In 1985, the idea of using the solvent as a gaseous phase was patented, but it seems it never reached commercial scale [38]. In an attempt to improve the capabilities of this type of process, the oil was first diluted in a totally miscible solvent and then treated with the gaseous antisolvent (e.g., CO_2) [39].

The need for cost-effective solutions for marketing the bottoms fraction remains. However, SDA is nowadays seen as a way of minimizing low-quality FO production, while increasing higher-quality distillates and in consequence, increasing the average refinery basket price also increases.

2.1.1 Residual Oil Supercritical Extraction

The ROSE technology was originally developed by Kerr-McGee Refining Co. [40–43]. In this process, solvent and feed are premixed and sent to the precipitating vessel. Most of the earlier developmental stages were based on pentane as solvent and base lubes as the end product [44,45]. Solvents and conditions were broadened to diversify the downstream use of the products, particularly underpinning the energy efficiency of the supercritical conditions [46–48]. Nowadays, Halliburton KBR is the process licensor. By July 1994, the estimated total capacity of the ROSE units was just over 200 kbbd [33], and by 2004, 33 units of about 600 kbbd were licensed or designed [49], from which by 2006, 29 units were installed [50]. In 2008, it was reported that 48 units were licensed or designed for a total capacity of 900 kbbd, and 25 units were operating [51].

Currently, the bottoms (asphalt stream, pitch, or asphaltenic residue) are stripped with steam or heated for solvent recovery. Meanwhile, the top is heated to reach solvent supercritical temperatures to form two phases: the DAO and the solvent. Most of the solvent is recycled directly, but part of it is stripped with steam prior to recycling. The heat of the supercritically heated solvent is exchanged to preheat the incoming top of the asphalt separator (see block diagram in Fig. 5.1).

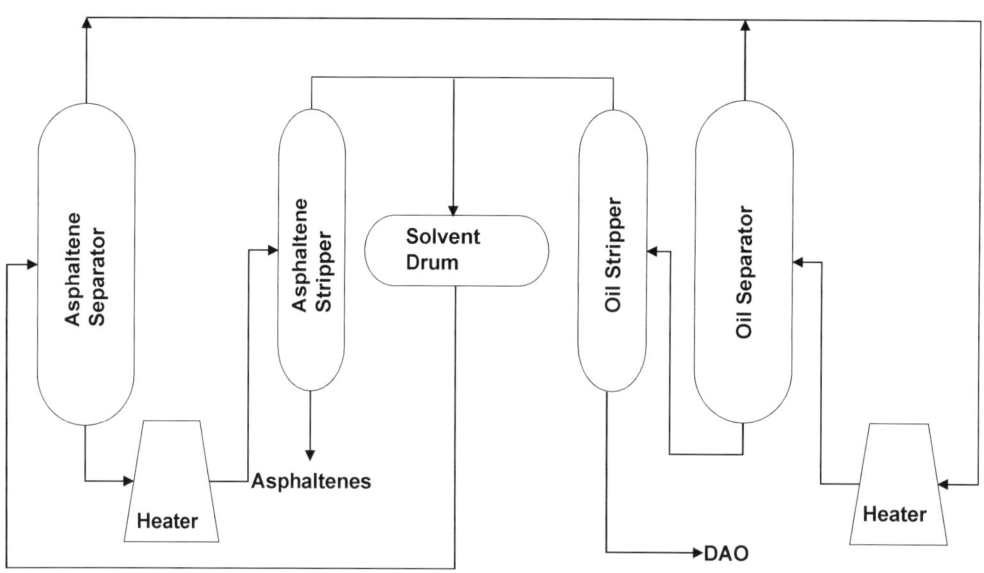

Figure 5.1
Residuum Oil Supercritical Extraction (ROSE) schematics. *DAO*, deasphalted oil.

Commonly used C_3-C_6 solvents and blends are mainly recovered by vaporizing the supercritical extractant, reducing utility requirements and costs. A two-stage unit is employed for separation into two products (DAO and asphaltenes); a third stage is added if resins are desired as a third product [49].

The trends between yield and the DAO quality obtained with the ROSE process of a typical AR are shown in Fig. 5.2 [52]. As can be expected from the discussion in previous sections and Chapter 2, quality and yield are intrinsically associated. The higher the DAO yield (obtained with the heavier paraffin solvent), the lower its quality, i.e., higher metals, sulfur, nitrogen, and CCR. Hydrotreating capacity and downstream integration of the SDA unit will determine the quality requirements for the DAO and/or the end use of the pitch [51]. If there is an HDT FCC-feed pretreatment unit in the refinery, a typical solvent:oil ratio of 8:1 would optimize the DAO yield for FCC, using ROSE technology [44,45].

2.1.2 UOP—Foster Wheeler Demex

The UOP Demex technology was oriented toward the production of (low metals) FCC feed, and it is still recommended for such purpose [53]. Consequently, UOP focused its development in maximizing DAO yield by using a C_4^+ type of solvent, for which supercritical solvent—DAO separation technology would advantageously recover the solvent. UOP also devoted efforts in trying to minimize solvent:oil ratio, affecting DAO quality as little as possible. Meanwhile, Foster Wheeler (FW) SDA technology [Low Energy Deasphalting (LEDA)] emphasized its development on yielding a very high-quality DAO for production of lube oil feedstocks, using propane as solvent and focusing on the extraction techniques. Additionally, FW made available its detailed design and

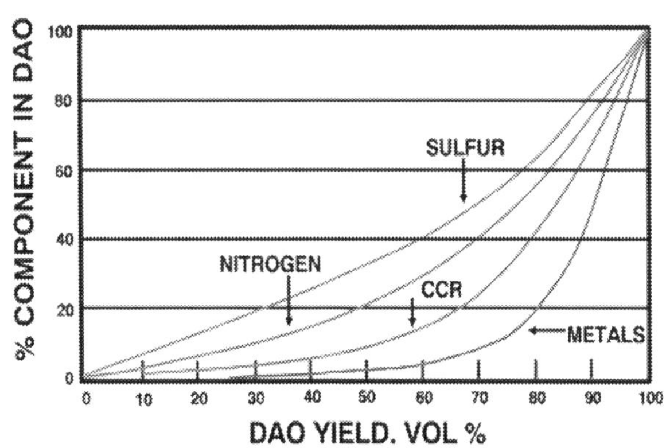

Figure 5.2
Yield versus quality of Residuum Oil Supercritical Extraction (ROSE) deasphalted oil (DAO). *CCR*, conradson carbon residue. *Reproduced from Patel V, Iqbal R, Eng O. To vacuum or not to vacuum. Hydrocarb Eng. June 2007;83—87, with permission from Palladian Publications.*

construction experiences from a multitude of SDA projects [54]. Thus back in 1996, FW LEDA and UOP (Demex) joined their SDA technologies into the Demex process (Fig. 5.3).

For this reason, the resins are separated from the DAO at the top of the asphalt extractor. Thus the resins-containing DAO—solvent mixture is fed to a small heated extra vessel that separates them from the DAO—solvent mixture [55,56]. The resins are drained from the bottom and recycled with the feed. The top of this vessel is the DAO—solvent mixture, which is heated to above the solvent critical temperature to recover the solvent. In this process, the performance of the extraction section depends on the used internals [57], for which four options are offered:

- Rotating disc contactors (RDC): A vertical tower comprising individual compartments divided by baffles attached to the walls of the vessel. The discs are centered in each compartment and rotate at a given speed, which depends on the feed rate. The dispersed feed droplets flow downward. The oil from the feed dissolves in the upward-flowing solvent, while the heavier fraction remains in the droplets, settling at the bottom.

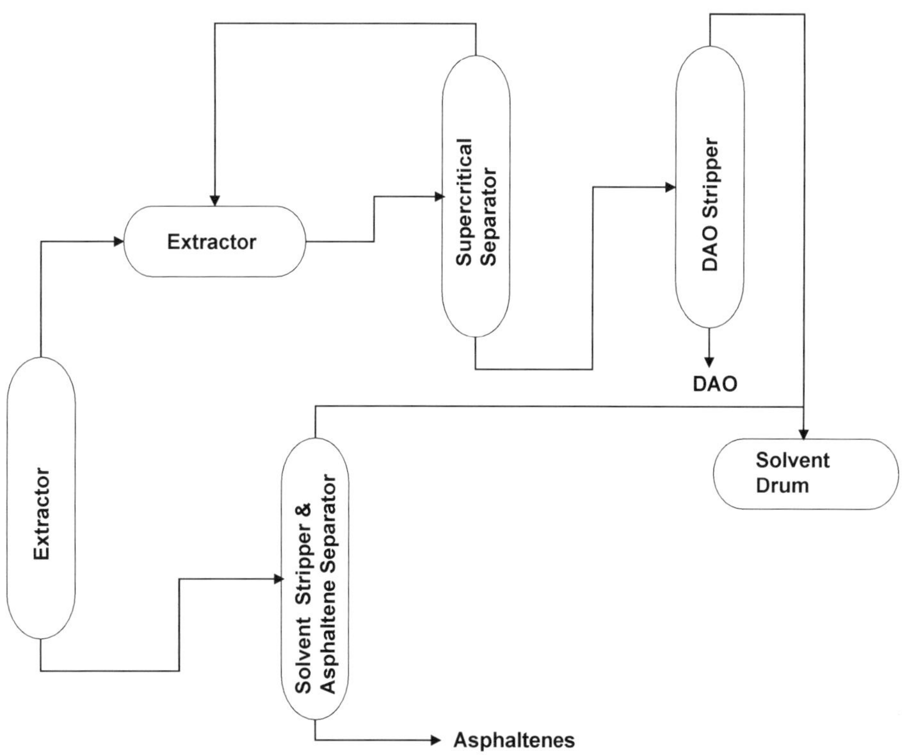

Figure 5.3
Demex schematics. *DAO*, deasphalted oil.

- Parallel Interceptor Plates (PIP) Assemblies: A series of compactly arranged inclined plates are provided as surfaces for the oil particles to coalesce against, become buoyant, and flow accordingly to density differences from the surrounding liquid [58,59].
- Proprietary trays [60,61].
- Proprietary packings.

More details on the PIP Assemblies and RDC systems can be found in Ref. [62]. The selection of internals is a function of the size of the unit and the desired DAO quality. There are also several options available for solvent–DAO separation (solvent recovery), such as single-stage recuperative evaporation, multiple effect evaporation, single-stage supercritical solvent recovery, and two-stage supercritical solvent recovery (in this latter case, the resin product can be separated). Units from 1983 use supercritical solvent recovery techniques (earlier units used conventional subcritical recovery methods). A typical 5:1 solvent:oil ratio is used with a resin recycle. To recover the resins an extra vessel, heat exchanger, and set of pumps are required [57].

As before, maximizing DAO yield can be achieved by using heavier solvents [63], in which case its quality will decrease. Table 5.2 collects data for the Demex treatment of VRs derived from different crudes. As explained earlier, yield and quality compromise depends on the HDT capacity and FCC or HCK feed specification in the refinery.

By 2003, UOP had designed 13 units, while FW did 40 [57]. In the ASCOT (Asphalt Coking Technology), Foster Wheeler licensed a DA–coking configuration (increasing liquid yield and energy utilization), which uses their LEDA–Selective Yield Delayed Coking (SYDEC) technologies. In the LEDA process, C_2–C_7 blends or light naphtha are used as solvents and the extraction takes place in a rotating disk contactor [64].

2.1.3 Axens/IFP Solvahl

The feeds for the Solvahl process (Fig. 5.4) can be either atmospheric or vacuum residue. The feed and solvent are mixed at low temperature, while the extractor is operated at high temperature and low vertical velocity. The extractor configuration and the hot DAO

Table 5.2: Effect of Solvent Type on Deasphalted Oil (DAO) Yield and Quality for the Demex Process [63]

Solvent	C_3	C_4	C_5	C_4			
Crude	Arabian Heavy			Maya	Arabian Light	North Slope	Lagomedio/Iranian Light
Yield (%)	10	35	60	50	60	60	70
Sulfur (%)	2.5	3.6	4.0	Δ^a 1.2	Δ^a 0.8	Δ^a 1.0	Δ^a 0.2
API	21	15	12	–	–	–	–

[a]Sulfur content improvement (in percent points) from the feed to the DAO.

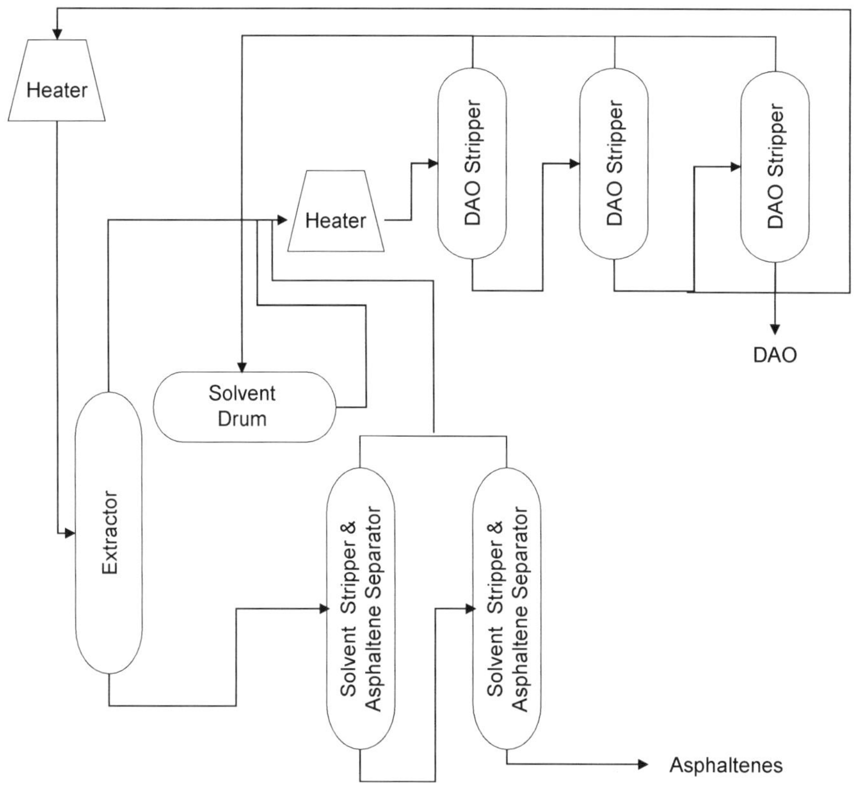

Figure 5.4
Solvahl schematics. *DAO*, deasphalted oil.

pump-around are the distinguishing features of this technology [65]. The solvent is recovered from the mixture of DAO and asphalt in different stages and is accumulated in a drum.

Yield optimization is defined in terms of the acceptable metal content of the downstream application, whether it is a HCK unit or an FCC unit. The solvent is a heavy hydrocarbon used at low solvent:oil ratio (4:1–6:1) to maximize DAO yield (Table 5.3). This table includes details of the quality of the Solvahl DAO.

In terms of commercial activities, Axens is reported to have licensed 13 units by 2015 [66].

2.1.4 Advanced Supercritical Solvent Deasphalting Process

In 1983 Petrobras–Cenpes completed a basic design package for a grassroots unit of their developed SDA technology (PASD), which was started up in 1989 in the São José dos Campos Refinery (REVAP). Process improvements include an advanced scheme with heat exchange matches for energy optimization. This feature enabled these units to improve the

Table 5.3: Effect of Solvent Molecular Weight (MW) on Deasphalted Oil (DAO) Properties

		DAO	
	Feed	C_4 Solvent	C_5 Solvent
Yield, wt%	—	**70.1**	**85.5**
Specific gravity	1.003	0.959	0.974
Viscosity @ 100°C, cSt	345	63	105
Conradson carbon, wt%	16.4	5.3	7.9
Asphaltenes, wt%	4.2	0.05	0.05
Nickel, ppm	19	2	7
Vanadium, ppm	61	2.6	15.5
Sulfur, wt%	4.05	3.3	3.65
Nitrogen, ppm	2875	1950	2170

benefit/cost ratio of energy recovery in comparison to conventional solvent recovery systems. Additionally, the unit throughput was increased by incorporating an advanced control system. The REVAP refinery unit was designed to process 38 kbpd of residue, for producing FCC feed at a solvent:oil ratio of 5:1. The introduction of the advanced control system in REVAP resulted in an improved throughput of the unit to 42 kbpd [36]. A simplified process schematic is presented in Fig. 5.5.

That unit incorporated and commercially demonstrated the suggested improvements. Optimization studies included the effect of solvent:oil ratios, which is reported to be in the 5:1−12:1 range [67], depending on yield and quality requirements. Later in 1999, technology was rolled out into Petrobras's RLAM refinery, which was revamped for bright stock production, having to increase the solvent:oil ratio from 9:1 to 12:1 simultaneously, improving DAO quality and yield (from 54% to 58%) [67].

2.1.5 MDS Process

The MDS process, developed by Cosmo Oil, is a basic solvent DA process [68,69], in which feed and solvent are mixed and fed into a DA tower. The feed includes AR, VR, oil sands, bitumen, visbroken tar, etc. Most of the asphaltenes extraction takes place in the upper half of the tower.

The mixed stream of DAO and solvent flows upstream and exits from the top of the tower. Flocculated asphaltenes flow downward by coming in contact with the countercurrent solvent that simultaneously stripped the DAO. The solvent and DAO mixture is heated through a heating furnace, and solvent is recovered in a flash tower under pressure. Further heating and stripping recover a lesser amount of solvent still present in the separated DAO. Asphalt is withdrawn from the bottom of the extractor and is stripped for recovering any remaining solvent. The recovered solvent needs to be cooled and condensed into liquid, prior to recycling it to the extraction tower.

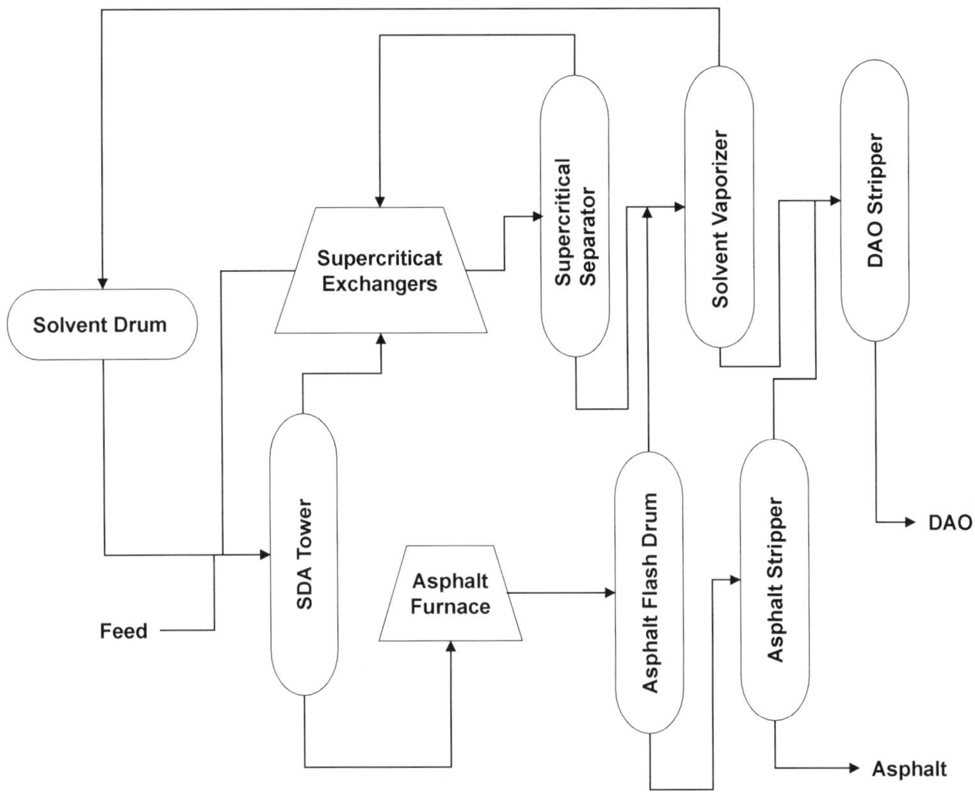

Figure 5.5

PASD simplified scheme. *DAO*, deasphalted oil; *SDA*, solvent deasphalting.

2.1.6 Other Processes

Other companies have developed and practiced processes commercially in their own refineries, for which very limited information can be accessed. The Research Institute of Petroleum Processing (RIPP) developed a process, which is known to be in operation in China, but no information on its technical details was found within the publically accessible literature. A patent for an integrated SDA-coking process [70] was granted to RIPP and China Petroleum & Chemical Company (now both integrated into Sinopec). Although there is no direct connection between this patent and the RIPP DA process, it was the only SDA-regarding reference found. RIPP-Sinopec has devoted more recent efforts to the depth of the vacuum distillation cuts. The SDA process can be fed either with AR or VR, and thus the depths of the cuts from the crude distillation unit (CDU) and the vacuum distillation unit (VDU) depend on the product objectives. For instance, the CDU and VDU can be operated to maximize lube oil production at the required lube viscosity, if the cuts from the distillation units (DUs) and the propane SDA are configured accordingly. A different case arises when paving asphalt is the desired product. Then, the

CDU and VDU are operated and cuts are tuned to satisfy asphalt performance requirements. A more complex case is the processing of high-sulfur and high-metal crude, in which case the VDU cut has a more impacting effect on downstream process units. VR with high sulfur and metal contents is fed either to the SDA unit or delayed coking process. The HDT requirement of the liquid product would depend on the HDT upstream unit (the VR treating unit) and the feed specifications for the HDT downstream processing unit. In summary, optimization of the operating conditions of the SDA unit takes into consideration the upstream SDA unit (DUs) and the downstream SDA unit (e.g., HDT), and the boundary conditions are the specifications of the feed for the downstream SDA unit, ultimately defined by the specification of the finished product. These optimization efforts maximize yields and quality via operating conditions of all units in an integrated manner. This principle is basic, for any integrating purpose and somehow links the previously cited patent and the RIPP installed/commercialized process.

More recently and also from China, the State Key Laboratory of Heavy Oil Processing at the China University of Petroleum, Beijing, has announced having demonstrated its patented Selective Asphaltene Extraction (SELEX-Asp) process that removes only the solid asphaltenes from heavy crude or residuum (Fig. 5.6) and recovers all desired components from residuum [71]. The asphaltene-free heavy crude or residuum produced by this process was claimed to be suitable for processing under conventional refining operations.

Additional studies were carried out for using the supercritical extraction system to further subfractionate the BotB in (10) narrow cuts [72]. The physicochemical properties of these fractions and their impact on processability were already discussed in Chapter 2 [73,74]. However, from a technology viewpoint, these studies indicated that integration and operational optimization of the CDU, VDU, and SDA are required and clearly depend on the desired targets for quality, yields, and refinery configuration downstream these units. According to the US Application Serial Number included in the process scheme depicted in Ref. [71], a US patent has been granted to the China University of Petroleum that discloses the SELEX-Asp process [75]. This process (shown in Fig. 5.7) was described as a separation method and system, in which granulation of the DA residue is coupled to the extraction system. This postextraction granulation is used to achieve deeper separation from the heavy oil. A dispersing solvent is introduced into the asphalt phase after SDA extraction and the asphalt phase undergoes rapid phase change in a gas—solid separator. Thus the solid particles disperse while the solvent vaporizes, resulting in low-temperature separation of asphalt and solvent with adjustable size of the asphalt particles. This method also includes a three-stage separation of the heavy oil feedstock, in which the DAO phase separated from heavy oil is further treated with supercritical solvent and results in the recovery of the resins from the DAO, maximizing its yield and quality (as originally conceived and depicted in Fig. 5.6). The differentiating features of this process are the use of atmospheric pressure and a low-temperature gas—solid separator instead of

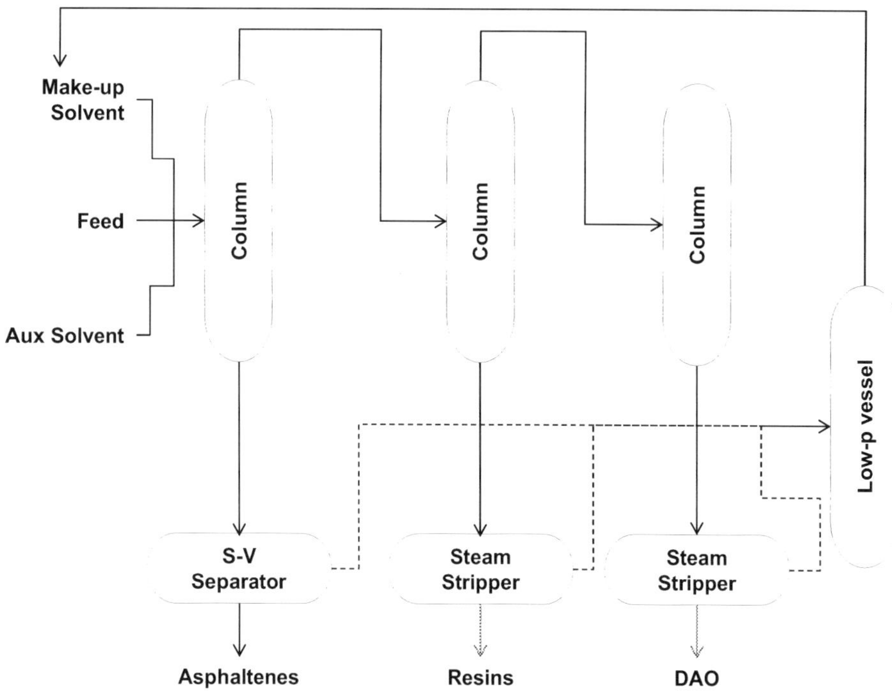

Figure 5.6
Selective Asphaltene Extraction (SELEX-Asp) process simplified scheme. *DAO*, deasphalted oil; *S-V*, solid-vapor.

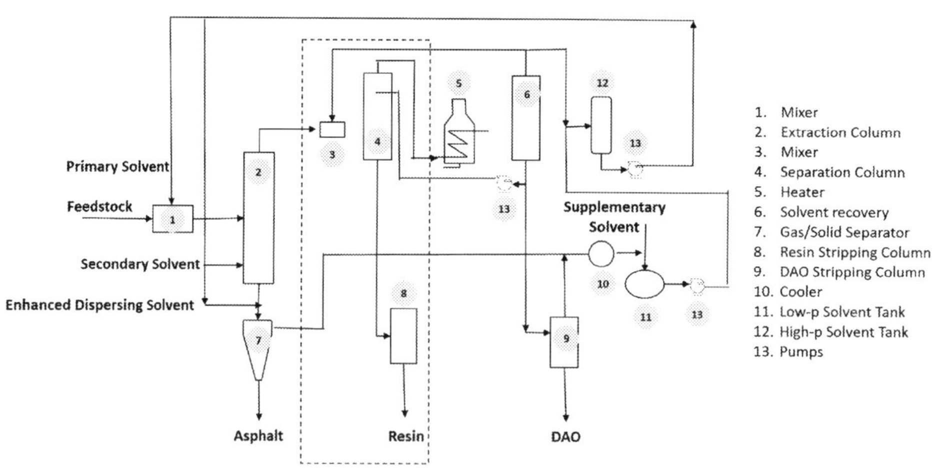

Figure 5.7
Deasphalting process disclosed by China University of Petroleum [75].

high-temperature and high-pressure furnace for feed preheating and/or heat exchange equipment at the inlet of the resin separator column. The result is a simplified process flow with reduced investment. PetroChina built a 500-bpd field demonstration unit at the Panjin refinery that performed to fulfill the expectations, by producing predicted yields and quality of both DAO and (free-flowing) asphaltenes [76].

Froth treatment technology introduced in the oil sand industry by Suncor uses naphtha as diluent. The froth treatment process developed by Syncrude Canada Ltd.[2] uses hydrotreated naphtha and is available to members of a Froth Treatment Consortium to be applied in other oil sands plants in Canada. It has been considered an integrated step in the Athabasca oil sands bitumen recovery process, with the main objectives of separating the organic compounds from the inorganic solids and water [77]. As has been discussed in Chapter 2, a great deal of research efforts have been targeting the understanding of the formation, stabilization, and properties of the water-in-oil emulsions in this kind of organic bitumen/water/inorganic solids system, for which the asphaltenes are known to play a stabilizing effect. Hence removal of the asphaltenes would facilitate breaking the water-in-oil emulsion and the removal of inorganics. This process can use either a paraffinic (PFT) or a naphthenic (NFT) solvent to produce a bitumen product essentially free of solids and water [78]. In general, the solvent-to-bitumen mass ratio is about 0.6 to 0.75. While PFT would precipitate asphaltenes, NFT will not, thus NFT is preferred by the producing unit and PFT by the downstream processor. In practice PFT application has been limited to Albian (Shell) and Kearl Lake (Imperial Oil), while NFT is widely applied in most of the Athabasca projects. PFT applied by Albian Sands Energy (155,000 bbd) yields a clean bitumen product (<0.1 wt% water and <0.1 wt% solids) at a solvent-to-bitumen ratio above a critical value, which corresponds to the onset of asphaltene precipitation. The most critical feature of the technology is the solvent recovery unit and the disposal of by-products (tailings). Once the bitumen froth is separated from the water and mineral solid fractions, these become the froth treatment tailings, which typically contain 2 wt% hydrocarbon solvent, 4.5 wt% bitumen, 17 wt% mineral solid particles, in 76.5 wt% water. The composition of these tailings indicates a significantly negative potential impact on the environment. Testing the incorporation of the PFT-treated bitumen with the FCC typical feed was proven without any detrimental effects on yields or product quality [79]. Compact online analysis has been developed for monitoring operations and/or product quality in the field unit [80]. This process enables businesses to remotely locate the upgrading conversion plant [81]. Further development of the froth treatment technology addresses the needs for oil sands exploitation such as maximizing bitumen recovery,

[2] Technology development team: Dr. Robert Tipman, Senior Process Consultant, Shell Canada Limited, Dr. William Power, Head Process Chemist—Oil Sands, Shell Canada Limited, Dr. Yicheng Long, Team Leader—Froth Treatment Pilot Plant, CANMET Energy Technology Centre, Dr. Tadeusz Dabros, Team Leader—Multiphase Systems Group, CANMET Energy Technology Centre.

minimizing the water and solids content of the produced bitumen, and minimizing overall energy consumption. Contrary to the first need (maximizing bitumen recovery) that favors NFT over PFT, the latter introduces quality benefits. Hence published work on NFT showed the focus on solids removal (e.g., Refs. [82–86]). On the other hand, some PFT additional work addresses further hydrocarbon recovery and inorganic contaminant separation from the asphaltene-containing fraction and asphaltene precipitation from the bitumen froth (e.g., Refs. [82,86–93]).

BP also developed a process, Selective Solvent Deasphalting, that was conceived to maximize the DAO yield from thermally cracked feeds prior to its posttreatment in a hydrogen adding process. It uses a C_7 solvent at low solvent-to-oil ratio. The technological advantages may be seen as the minimization of the hydrogen consumption in the HDP units and the solid state of the rejected asphalt [94,95].

Intevep (Venezuela) developed a similar process using C_6 solvent and a different approach for separating the asphalt [96]. Other developmental work seems to have changed gear toward froth-type treatments [97].

2.1.7 Improvements of Available Technologies

A worldwide search retrieved more than 800 patents beginning in 1951, though commercial applications of SDA started in the 1930s, indicating that this search is far from complete, particularly lacking of patents of early processes. The work prior to 1980 concerned improvements to the basics of a conventional liquid–liquid extraction, such as mixing, contacting, solvent composition, solvent recovery systems, temperature and pressure conditions, etc. Most of the oil companies and engineering firms filed patents that differed from each other on those features, e.g., Standard Oil [98,99], Foster Wheeler [100–103], Exxon [104–110], Kerr-McGee [40,41], Mobil [111], Texaco [112–115], Chevron [116], Phillips Petroleum [117–119], UOP [120–125], Gulf [126], Shell [127], and others [128–130].

By 1980, supercritical extraction emerged and rapidly became commercial. It did not take long to be adopted by SDA and Kerr-McGee was awarded a patent for this application [131]. The already discussed available technologies are based on subcritical/supercritical extraction and supercritical solvent recovery. During the 1980s, most of the actors with original liquid–liquid extraction process technologies also filed for patents that adapted or incorporated the sub- or supercritical conditions to their current know-how [39,94–96,132,133]. However, sub- and supercritical conditions were not enough and further improvements were needed.

Although since 1990 SDA-awarded patents have been focused on increasing efficiency and solvent-related costs, there is no reported evidence that these have been commercially implemented.

2.1.7.1 Solvents

As mentioned, paraffinic solvents are preferred for SDA, and an increase in the number of C-atoms in the alkane chain increases DAO yield and decreases its quality. Mixtures of paraffins and paraffin-enriched naphtha have been used for particular applications, in addition to the froth treatment discussed in the previous section. For instance, a mixture of naphtha and light aliphatic hydrocarbon, such as butane, was used to destabilize bitumen—water emulsions, resulting in the removal of most of the water and minerals, and in the precipitation of some of the asphaltenes [134].

Several polar solvents have been suggested in the patent literature as improvements to the yield, quality, selectivity, process, and/or a way to intensify the DA process by promoting other capabilities (e.g., demetallization, DM) or facilitating other operations; some examples follow.

A DA method for using a water-soluble (polar) solvent has been proposed. In this process, a solvent—crude oil solution is prepared and then water is gradually added at an elevated temperature. This temperature has to be lower than the boiling point of the polar solvent. The DAO fraction is separated by a stepwise prefractionation prior to adding the water. Temperature is lowered at 15—35°C steps and fractions released in each step are simultaneously withdrawn up to a solution temperature of 20—25°C [135].

Superheated water (300—425°C) [136] and also superheated water containing a soluble carbonate salt [137] have been claimed as extracting solvent for the SDA process.

An unusual SDA solvent, *N*-methyl-2-pyrrolidone (NMP) was found to improve the quality of the base lube final product. The actual solvent comprises NMP, 0—5 v% water, and a light hydrocarbon. An asphalt-containing mineral oil is simultaneously deasphalted and extracted in a combination zone and produces a raffinate and extract phase. The raffinate phase contains the desired oil, most of the hydrocarbon solvent, and some NMP. The light hydrocarbon is removed from this raffinate phase by reducing pressure and the remaining NMP by chilling, since lowering the temperature makes the oil and NMP immiscible [138].

As per the Demex process, SDA conditions could be optimized to maximize DM, but only to a certain extent. Incorporation of polar solvents can be seen as a particular case of process intensification for carrying out two extraction processes in a single unit. Additional details of these intensified approaches will be given in Section 2.1.7.5, and some new ideas in Section 3 of Chapter 3. Two other unusual solvent systems have been claimed for promoting DM under SDA processing. One is dimethylcarbonate in the presence of an overpressure of CO_2 [139] and the other is sulfuric acid [140]. For the same purposes, some additives have been recommended, e.g., an additive containing a polyoxyalkylene moiety that is insoluble in the DA solvent [141]. An increase in DM efficiency was claimed when organophosphate and phosphite compounds were used as DA additives [142]. Organic phosphate esters and light solvents (*n*-pentane) removed metals from heavy residues by hot extraction (about 200°C, 62 bar). Similarly, dimethylphosphite (0.057:1 agent:oil ratio) could also be used.

2.1.7.2 Solvent Recovery Systems

SDA is currently carried out at sub- or supercritical conditions. Thus solvent recovery is typically based on supercritical methodologies. In addition to those supercritical methods, other modifications have been suggested either to revamp old units or just as incremental improvements.

An apparatus for solvent recovery from the DAO/solvent mixture includes a pipeline arranged to produce a stream of fluid, in which the solvent would reach a supercritical condition. A horizontally disposed inlet line receives the stream of fluid and is designed to effect two-phase laminar flow at the outlet of the inlet line. The two-phase laminar flow has a horizontally disposed interface that divides the stream into an upper part containing the lighter supercritical solvent that floats on a lower part containing the heavier DAO [143].

A heat-transfer agent was recommended to be added to the coil located at the top of the extraction column of the solvent recovery unit for reducing power requirements and equipment corrosion. Additionally, the recovered solvent is stripped, compressed (50−60 bar), and preheated prior to be recycled to the SDA unit [144].

A two-stage solvent recovery method was described as consisting of heating the solution at different combinations of temperature and pressure. In the first stage, temperature is in the range of 80−90°C, at a pressure of 20−30 bar; then, it goes up to of 210°C under a pressure of not less than 17.5 bar. Part of the separated solvent from the first stage is used for refluxing in the second stage [145].

A liquid−gas jet apparatus was designed for solvent recovery. In this case, the mixture to be separated is forced into the liquid−gas jet apparatus and then fed into a separator and back to the extraction column [146,147].

Solvent recovery improvements have also been concentrated in the use of supercritical conditions. In this regard, the spent solvent was recovered from the DAO solution in a separator operated at solvent supercritical conditions. Then, the vaporized solvent is liquefied with a jet compressor [148]. A slight variation of this system uses the solvent flow obtained from the vaporizers as actuating fluid for the jet compressor. In this case also, solvent is recovered from the DAO under supercritical methodology, while it is stripped and flash vaporized from the pitch [149].

A recovery method in which the solvent is heated with no change in pressure prior to introduction in the separator has been described for accomplishing the vaporization of the solvent. Operating conditions for propane (subcritical conditions) were temperatures of 80–100°C and pressures of 37–40 bar [150].

2.1.7.3 Process Units and Conditions

A short vapor residence time process in a horizontal moving bed of fluidized and/or stirred hot particles was defined for improving DAO quality. The vapor-phase product stream ($510°C^-$) from the process unit is passed to a soaker drum where it undergoes oligomerization and/or polymerization reactions forming higher MW compounds. The heavier high-boiling point fraction is separated and recycled to the thermal conversion unit [151]. A mechanical means to fluidize a bed of hot particles was preferred [152]. An incremental improvement to enhance process heat efficiency was suggested, in which a portion of the DAO stream is heated, and this portion indirectly heats the two primary liquid streams (residue and solvent) [153].

An improvement in the hydrodynamics of the SDA unit is achieved by bubbling oil field gas when the solvent is either a C_5–C_{16} alkane or gasoline. Gas bubbling has a simultaneous dehydration effect on the crude [154].

The use of specific packing in the extraction tower and in the solvent recovery units has been recommended as a way of improving efficiency and minimizing energetic demands. Particular combinations of extraction-coalescing boards were recommended for a C_4 SDA extraction tower, namely, ZUPAC2 in the extracting section and NPPAC2 coalescing board in the deposition section, and in another configuration ZUPAC1 and ZUPAC2 were suggested from bottom to top [155]. In regard to solvent recovery in a supercritical C_4 SDA, packing the tower with NPPAC coalescing boards increased the recovering rate. In this process, solvent is introduced to the tower and heated up to a temperature of about 170–180°C at a pressure of 40–44 bar [156].

A *sui generis* extraction tower has been claimed in a Chinese patent, for which only an abstract could be accessed [157]. According to this abstract, the preheated feed is countercurrently introduced from the top of the SDA tower, while solvent comes from the lower portion of the tower. The excess of recovered solvent is recycled. Surprisingly, DAO is discharged through the bottom of the tower fed into a heating oven and fractionated. Neither the asphaltenes collection method nor their collection location was described in the abstract.

Another improvement aiming to facilitate solvent recovery and further enhance heat efficiency was claimed in Refs. [158,159]. In this process the hydrocarbon stream is fractionated at a cut-off point that is about 25°C below the T_c of the solvent to be used in the SDA unit. The bottom fraction (boiling at temperatures greater than T_c of −25°C) is fed to the SDA unit. The precipitated asphaltenes are considered to be solvent free. The DAO produced (also considered to be solvent free) is vacuum fractionated to yield DAO distillates and DAO residue. The DAO residue is then thermally cracked and the product is fractionated as well. A thermally cracked residue is separated from the thermally cracked distillates and recycled to the SDA unit. In these patents the solvent recovery system seemed to be an integral part of the SDA unit but there was no description of it included.

The recovery of the resin phase from the DAO—solvent mixture as a fraction for gasification feeds requires modifications of the SDA process conditions and method. Besides, more efficient heat integration can be accomplished by heat integrating the SDA and the gasification units. For this purpose, the resin is first precipitated by heating a primary DAO—solvent mixture. The remaining solvent is separated from the DAO solution by vaporization using additional heating. A higher quality resin-free DAO is claimed to be produced. The recovered resin or the total asphaltenic fraction can then be used as feedstock in the gasification process [160,161].

Decanted oil (DCO) from the FCC units as a refinery intermediate stream represents a feed where two problems coexist: (1) it consists of the most refractory organic compounds present in the VGO that resisted conversion throughout the FCC process, and (2) it is contaminated with inorganics, mainly comprising catalyst fines. The heavier organic compounds might be unconverted asphaltenes and resins that could be removed in an SDA unit, in which the inorganic solids would be retained also in the precipitated phase. A multistage solvent extraction process addresses these problems [162]. In this type of process, hydrotreated residues can also be processed simultaneously with the DCO. A nonaromatic C_3—C_7 hydrocarbon solvent is run for extraction at or below the critical condition. In the first stage asphaltenes/catalyst fines are dropped and the DAO—solvent mixture is recovered. Resins and/or solvent are separated from the DAO solution in a second (or additional) stage. A portion of the DAO from the second stage is recycled to the first stage.

Finally, a different application for regenerating spent motor oil is described in a Russian patent. A two-stage propane SDA process is recommended, in which a conventional VR is fed in the first stage, while the motor oil is incorporated into the second [163].

2.1.7.4 Deasphalter Design

A German patent claimed to have designed a new apparatus for achieving complete DA of crude oils, which is described as encompassing a continuous circulation of the solvent through the precipitating vessel [164]. However, the patent drawings show an extractor that closely resembles a Soxhlet system.

An apparatus comprising an array of collectors—coalescers and tray system to improve liquid—liquid contact and effectiveness of the DA process was described in Refs. [165,166]. The extraction tower disposes sealed sieve trays with pipes in either the downcomer or upcomer zones. The array suits two purposes: it minimizes the potential of dispersing the continuous phases and avoids its breaking into small droplets. The result is a homogeneous delivery of the continuous phase [165]. The sieve trays are vertically assembled spaced apart and include vertical weirs between groups of sieve holes (preferably rows). A sloped collector—coalescer plate (per sieve tray) provides a coalescence surface for dispersed phase droplets and directs the dispersed phase to the rows of sieve holes on the sieve trays. Additionally, the collector—coalescer plate prevents the dispersed phase droplets from impacting directly on the sieve holes where new drops of dispersed phase are being formed. The whole arrangement of sieve trays, pipes, sieve holes, weirs, and collector—coalescer plate provides for steady controlled flow of both continuous and dispersed phases throughout the apparatus [166].

2.1.7.5 Intensification

Originally, SDA was meant particularly for the manufacturing of base lubes, which greatly depended on the quality of the starting feed or source of crude oil. The incorporation of heavier stocks in the refinery diet gave new purposes to the SDA unit, the mitigation of downstream problems. First, the reduction of CCR in the DAO allowed its addition to the typical FCC feed. However, CCR reduction was not enough for this application and soon the metal contamination led to increases in operating costs. The need for intensifying the DA process by promoting a better DM was identified easily. Finding new solvents for this purpose was an approach already discussed in the Solvent Improvements section. However, in some instances the use of these new solvents required modifications to the existing processes. Additional to DM and CCR reduction, the intensification for including the removal of any other heteroatomic compounds during DA has been approached as well.

Three inventions have been developed for the use of organic carbonates as additives for increasing DM under DA conditions. The last two have already been mentioned [139,140]. In the first patent dimethyl carbonate is used in a continuous process, fed with AR. This process comprises the same stages that will be described for the second patent [167]. The process of the second patent uses dimethyl carbonate as an additive in the feed. This mixture is contacted with a homogeneous liquid phase (the solvent, which is supercompressed carbon dioxide) to precipitate a solid residue, enriched in asphaltenes and asphaltenic metals. A first separation stage removes the solids from the liquid phase, which is cooled to separate an oil-rich refined liquid phase from the extracted liquid phase containing the organic carbonate. Another option for the separation of the extracted and refined liquid phases is to add a liquid solvent, which is more polar than the carbonate. This second option may not require any cooling. Further cooling of the entire system to a temperature in which three phases are formed is employed to recover the gas phase CO_2 by venting. The partly demetallized DAO is separated from the intermediate phase and the spent dimethyl carbonate is recycled [139]. A completely different process and equipment are needed when aqueous sulfuric acid is used as solvent. In this case, only a batch operation has been described in which the aqueous sulfuric acid at 45–98 wt% is contacted with the feed; the acid is added in an amount of 5–175 wt% relative to the crude. Temperature is kept at 20°C for more than 10 min. The refined oily phase practically devoid of metals and asphaltenes spontaneously separates from the acid sludge [140]. The handling of sulfuric acid represents potential operational hazards and together with the disposal to such acid sludge creates strong drawbacks for this process technology.

Another additive, mentioned earlier, is based on a polyoxyalkylene moiety. The DA process including this additive for enhancing DM requires either the use of a solvent in which the additive is insoluble or applying process conditions that preclude the additive to solubilize into the solvent [141].

Another intensification approach is the global optimization of energy efficiency attained by integration of SDA to gasification, which also facilitates and adds value to the disposal of asphaltenes by their incorporation into the gasification unit. The asphaltenes produced in an SDA unit are fed into a gasification unit. The heavy crude feed is heated to 300–600°C to flash off any residual/dissolved gas. Steam is admixed with the heated crude and alkaline earth metal salts (e.g., magnesium chloride or calcium chloride) are also added. These salts react with water and form metal hydroxides and hydrochloric acid. Asphaltenes and other solids are subsequently separated from the DAO; and the hydrochloric acid from the crude in a flash drum and is neutralized in the vapor phase [160].

Incorporation of surface active additives into the SDA process could facilitate operations and increase efficiency. For instance, phase separation accelerators (alkyl and alkylaryl alkoxylates, alkylaryl sulfonates, and heteroatom punctuated fatty polymers) facilitate separation of resins and asphaltenes [168]. Another example is a nonionic surfactant that enhances asphaltene yield at lower solvent consumption. This process is operated with the addition of 50–5000 wppm (based on solvent) of a cationic surfactant (e.g., SNPKh-1004) to the hydrocarbon solvent at a solvent-to-feed volume ratio between 3:1 and 20:1 [169].

A DA process that eliminates the need for solvent is characterized by the use of a (not identified) inorganic material. The additive is incorporated at a certain proportion into a depressurized vessel containing the residue feed under stirring and heating. The stirring operation is stopped while the mixture cools down. The depressurized mix is further heated to a temperature, and held constant for a period of time until all the produced asphalt separates and can be discharged [170].

2.2 Moderate Conversion

2.2.1 Visbreaking

Initially (by 1930), visbreaking was conceived for the reduction of viscosity by thermally cracking large oil molecules into smaller ones at moderate conditions, in the absence of a catalyst. Viscosity is very sensitive to molecular size, thus an acceptable result can be attained by breaking a few bonds in large molecules and making them smaller. Meanwhile, in visbreaking, density does not change much, since neither carbon is removed nor hydrogen is added significantly. However, there is a slight decrease in the H/C ratio caused by the formation of light H-richer gases. Consequently, for heavy oils and bitumen, for which upgrading objectives include density and H/C ratio increases, visbreaking does not appear to be a good option.

Historically, residues were visbroken to meet specifications for heavy fuel oil (HFO) applications. A residue exceeding the minimum HFO specifications was used as feed at conversion levels just enough to obtain 15–30% transportation fuels and still have the heavy product meeting HFO specifications.

For transportation of heavy oils, viscosity reduction is needed. WRI offers this type of technology for pipeline readiness of heavy oils [171,172]. It is operated at temperatures between 430 and 490°C and nowadays remains the most inexpensive upgrading process, with a typical investment cost per barrel in the range of USD 900–1000. Although the visbroken product is relatively unstable, the reduction in viscosity and slight decrease in specific gravity, with a minimal investment, are the main advantages of such a simple process. Product stability is controlled by diminishing the severity of the process. To define operating conditions both stability and compatibility between the residue and its visbroken/cracked light products have to be evaluated.

The naphtha produced is highly olefinic and S-content is high. The light naphtha is typically treated in a Merox unit, while the heavy naphtha has to be hydrotreated and reformed prior to blending them into the gasoline pool. Gas oil is also high in S and the color unstable, so it has to be hydrotreated.

Additives have been suggested for improving performance; however, the commercial availability of these technologies is unknown. Examples include sorbents for retaining the removed metals [173–176], coke sorbent materials [177], alkali metal oxides for promoting DM [178,179], rare-earth-containing Y-zeolite [180], native solids in tar sands [181], and coal [182].

In any conversion unit, economy improves with conversion. However, the severity of the visbreaker is restricted by the stability of the final product [183]. In the case of high asphaltenic oils, the higher the conversion, the larger the proportion of light saturates in the cracked product. Furthermore, new asphaltenes are formed from maltenes, and the maltene-phase composition and behavior change. Eventually the equilibrium between asphaltenes and maltenes is disturbed to such an extent that part of the asphaltenes flocculates. The increasing proportion of saturates would make the asphaltenes destabilize, precipitate, and/or react [184,185].

There are two types of processes: coil (only heating) and soaker (heating followed by soaking); the stability is slightly improved in the latter. Vaporization and a mild cracking are induced in the feed within the heater of a coil visbreaker. The product stream formed in the coil heater is quenched prior to reaching the fractionator. The gas oil or the bottoms from the fractionator could be used for quenching. Gas and naphtha exit from the top of the fractionator. The other fractionator outlet streams are the gas oil and the tar (visbroken residue). The tar may be further fractionated to increase the yield of visbroken VGO. Incorporation of the Wood's technology improves gas oil recovery from the unit and reduces residue FO yield. In this improvement, a combination of high coil temperature and low fractionator pressure is used to delay the quench of the heater effluent after the vapor and liquid effluent have disengaged. This technology also reduces the capital costs since the vacuum flasher is no longer required.

For the soaker type, the outlet of the heater stream is fed into a vessel (the soaker) where further conversion takes place. Pressure and temperature are adjusted to control conversion level and/or unit throughput. The product is separated in an atmospheric fractionator designed as per customer basis to render as many cuts as desired (e.g., gas, liquefied petroleum gas, naphtha, kerosene, gas oil, waxes, residue, etc.). As mentioned before for the coil process system, the residue could also be further fractionated in a vacuum tower.

The total installed visbreaking capacity worldwide is in the order of 3 Mbpd. An example of commercially available technology is the Tervahl T process offered by IFP/Axens/ Technip [186].

The Tervahl process incorporates a stabilizing section in the process, in which purge gas is admixed for recovering the SCO.

Improved design versions of higher conversion visbreakers are available from Shell Global Solutions (soaker type, Shell Deep Thermal Conversion) and Foster Wheeler/ UOP (coil type + Wood's technology). These are claimed to render stable FO products.

The High Conversion Soaker Cracking Process, a high-severity soaker visbreaker, was jointly developed by Toyo Engineering Corporation and Mitsui Kozan Chemicals, Ltd. and is offered commercially [187−190]. The process operates at higher conversion than a typical visbreaker does, but lower than any coker. Still, occasionally, upsets cause coke formation and deposition that accumulate on visbreaker walls. Periodically (3−6 months), stops are required for decoking and cleaning.

In the Comprehensive Heavy Ends Reforming Refinery Process developed by Osaka Gas Co. [191−193], the feedstock is mixed with coal powder in a slurry-mixing vessel, heated and fed to the reactor. Products are subjected to fractionation, flashing, and vacuum flashing. No coke is reported to be produced.

2.2.2 Hydrovisbreaking and Aquaconversion

Visbreaking has also been improved by the incorporation of slight levels of hydrogenation. The inclusion of H_2 pressure has been tested in the absence of catalysts, e.g., the Tervahl H process offered by IFP (see, for instance, Refs. [194–200]). Catalysts, both supported [201,202] and dispersed [186], are offered. Tervahl HC soluble catalyst is the offered option for the Tervahl H process. Additional examples can be found in Refs. [203,204]. Hydrovisbreaking unconverted heavy feedstock and/or feeding hydrovisbroken heavy product to the coker maximizes the overall yield of high-value products and minimizes that of the low-value coke. Whether these technologies have been or are available commercially is unknown to the author, who believes that further development is still waiting. Most of the awarded patents in this area were filed in the late 1980s, early 1990s and must have expired by now, allowing freedom to operate of the newly developmental work.

Hydrogenation in the presence of steam under visbreaking conditions is the basis of Aquaconversion (AQC). The process was developed by Petroleos de Venezuela (PDVSA Intevep) [205–208] and is commercially licensed jointly by UOP, Intevep, and Foster Wheeler. The feed is mixed with a proprietary catalyst and water prior to entering the heater. The fire heater is then taken to the reaction temperature and the effluent is sent to the soaker. In AQC, mild hydrogenation of the free radicals formed thermally overturns condensation reactions from occurring. The proprietary catalyst is supposed to activate the water molecules dissociatively and transfer hydrogen to the feed oil. Asphaltene precipitation is repressed by that mild hydrogenation, allowing operation at higher conversion conditions. The increase in severity does not represent a drawback in product stability and no increase in asphaltenes has been observed. It is believed that the hydrogenation reactions are responsible for inhibiting the formation of thermal asphaltenes from maltenes [57]. The process also results in additional desulfurization (DS). PDVSA modified an existing 38,000-bpd soaker visbreaker at the ISLA refinery in Curacao for the commercial demonstration of the AQC process. Production of pipeline-ready SCO was targeted [30,209]. Commercial results showed higher improvement in gravity than that achievable by visbreaking with a total liquid yield of nearly 98%, indicating the possibility of easily revamping existing visbreakers to the new technology. These results also make this technology attractive for a surface field application since pipeline specifications could be met for a large variety of heavy crudes.

2.3 Coking

Coking processes are more thermally severe than visbreaking, using temperatures in the range of 450–500°C. Thus the light boiling products would exit overhead and leave the solid coke behind. Coking in the absence of H_2 moves a high amount of heteroatoms (and CCR) into the formed coke, making it a low-value by-product. Although heteroatom removal from the liquid product contributes to an increased H/C ratio of the liquid product, these products still have to be catalytically upgraded via hydrogen addition. Advantageously, coking removes a great deal of the poisons and fouling species that otherwise would deactivate downstream catalysts.

Favorable economics has made coking the choice for primary upgrading at field sites. Foster Wheeler, which offers SDA and delayed coking technologies, in addition to visbreaking has reported an economic comparison for these three processes, under similar terms. The evaluated net present value was very similar for visbreaking and SDA (USD 427 and 391 MM, respectively) while coking almost duplicated this value (USD 756 MM) [210]. The economics of delayed coking is driven by the differential between transportation fuels and high-sulfur residual FO. Regardless of the economic advantages, the trends of increasing supply of sour heavy crudes, of the reduction in the market for high-sulfur residual FO, and of decreasing high-sulfur coke prices have caused an oversupply of coke that refiners struggle to place in the market.

Originally, cokers were designed for complete conversion, but nowadays minimizing coke and producing heavy VGO (for further upgrade into higher value products) represent a better alternative. Additionally and to improve economy, refiners might consider incorporating gasification of coke to syngas to be used within the refinery, as it is the case of Reliance, the world's largest refinery [211]. Coking process technologies available commercially differ in the type of reactor used and fall into two main categories: semibatch reactors and fluid reactors.

2.3.1 Delayed Coking

Delayed coking was the first type of coker introduced in the refining market place. The preheated feed is flown into the bottom of the reactor, an isolated cylindrical drum. The volatile products exit through the overhead into a fractionator, while the solid coke product accumulates in the drum. The fractionator bottom cut is recycled to the drum. Once the drum is filled with coke (about 16 h), the feed is switched to a second drum. The formed coke is extracted from the filled drum by drilling a hole through the center and lateral cuts as well, with high-pressure water.

The product slate is controlled by the operating parameters. A reduction in recycle and operating pressure will decrease coke yield and the heavy coker gas oil conversion. Temperature increases do not have a strong influence on yields.

The price of the produced coke depends on the composition quality. Needle coke could be produced for making electrodes, but demands low levels of sulfur and metals. The anodes for the aluminum industry could be made from high-quality residues (higher costs), but the electrodes for the steel industry are even more demanding and only a more severely pretreated feed can be employed, such as FCC bottoms.

The Eureka process developed by Kureha Chemical Industry Company and Eureka Industry Co. [212,213] was originally operated at higher severity than a visbreaker, but lower than a coker. It was improved into the Advanced Eureka process by Fuji Oil and Chiyoda Corporation who currently license it jointly [214,215]. In this process, the thermal reaction is carried out in the presence of high-temperature steam, which is used for stripping the distillates. The fluid pitch residue can flow at high temperature and is recovered from the bottom of the drum. This pitch is then cooled and pelletized, to be sold as a finished solid fuel. The Eureka process is reported to be the basis for the ET-II (see p. 165 in Ref. [10]) and/or KKI process [216].

Numerous licensors offer delayed coking process technology, for instance, SYDEC by Foster Wheeler [217,218]. Foster Wheeler also licenses ASCOT, which integrates SDA with coking, increasing liquid yield and energy optimization [64]. The preextraction of the DAO, by SDA renders a greater liquid yield that otherwise would crack up to gas product and/or convert to coke, in the coker. Additionally, by-products from both processes, SDA and delayed coking, are minimized. The high- and low-level waste heat of the coker can be employed by heat integration with the SDA unit. A mixture of straight run naphtha and coker naphtha is used as solvent in the SDA unit. The low-boiling hydrocarbons are stripped off from a combined stream of DAO and heavy coker gas oil and are returned to the fractionator. The bottom stream of the fractionator is a mixture of the stripped DAO and heavy coker gas oil and could be sent to the cracking stock pool. The asphaltenic fraction is sent to the coker drum after recovery of any remaining solvent [64].

2.3.2 Fluid Coking

Two process technologies are licensed by ExxonMobil in this category, namely, Fluid Coking and Flexicoking. In a reactor vessel containing a hot fluidized bed of coke particles, the residue feed is sprayed in the case of Fluid Coking. The formed coke is removed from the bottom and transferred to the burner. Part of the coke is burnt in this other vessel to provide the process heat and the hot coke particles that are fluidized in the reactor. Again, the volatile products leave the vessel through the overhead to a fractionator and the heavy liquid product from the fractionator is recycled back to the reactor. The coke produced is typically a high-sulfur fuel grade coke of low value. Product yields depend on the feed and operational mode; when feeding Arab heavy AR, the recycle mode produces 12.9% gas, 14.4% naphtha, 10.2% light gas oil, 27.1% heavy gas oil, zero bottoms, and 35.4% coke [30].

Flexicoking incorporates a gasifier by adding a third vessel to a fluid coker. A mixture of steam and air is used to gasify part of the coke and to produce a low British thermal unit (BTU) syngas containing nitrogen and hydrogen sulfide. The hydrogen sulfide is removed and the low BTU gas can be burned as a clean fuel for power generation. The accumulation of nickel and vanadium in the gasifier is minimized by withdrawing around 3% of coke from the process.

2.4 Heavy-to-Light Crude Oil Upgrading Process

Heavy-to-light (HTL) is a thermal cracking upgrading process of the carbon rejection type, developed by Ivanhoe Energy [219]. Ivanhoe Energy's proprietary upgrading process is based on its RTP (commercial) technology [220–225], originally designed for the thermochemical conversion of biomass back in the early 1980s. The Ensyn Group developed the RTP biomass technology and has operated six commercial units in the United States and Canada for more than 25 years [226]. A pilot plant for crude oil processing was built in 1998 and in 2004 a 1000-bpd demonstration plant started up [227]. Ivanhoe Energy and Ensyn merged in 2005 making Ivanhoe Energy the sole owner of crude oil application technology. Prior to the transaction, Ensyn Group spun off the biomass business to its shareholders [228].

As per the RTP process, HTL employs conditions of fast pyrolysis (ultrashort contact time in a fluidized bed). More recent results include mechanical and physical modifications to the feeder system, separation stages, reactor design, process conditions, etc. [229–232]. The produced coke is converted to energy "on site." Several types of crude oils have been tested in the demonstration unit, confirming that the HTL process was capable of producing pipeline specification Syncrude by considerably reducing contaminants [233].

It is considered as a cost-effective process for rendering a stable synthetic oil product (Syncrude) along with by-product energy and/or steam. The production cost of this Syncrude is practically independent of natural gas and diluent prices. Consequently, a reduction of the negative environmental impact has been claimed [234]. HTL could capture an HTL price differential at a smaller scale than other technologies (as small as 10–20 kbpd), hence reducing capital costs [235]. For this reason, HTL can be easily modularized and deployed to remote locations, i.e., it can be located in the heavy oil field [235], to facilitate its integration with upstream operations [236,237]. Since the produced Syncrude meets pipeline specifications, HTL would enable its transportation from remote locations to the market place [238].

In 2008, Ivanhoe announced that it had acquired oil sands assets for the exploitation and processing of tar sands with HTL [239]. The reported historical advances of the technological development were reported in Ref. [238]. The realized scaling-up achievements are indicative of the commercialization pathway undertaken by this company and represent an example of a road to follow from R&D up to commercial scale.

The engineering and design of a 10-kbpd HTL facility for the Tamarack Project in Canada was announced in 2010 as a work in progress with AMEC, Ivanhoe's London-based contractor. The suspension of the project can be found on Ivanhoe's web page [228].

3. Hydrogen Addition

Hydroprocessing (HDP) has a long history in refining. Since the 1970s with the implementation of the Clean Air Act in the United States, a continuous decrease in S-specification in fuels has been implemented worldwide. Furthermore, the restrictions of use of lead octane booster additive (and any other metal additives) in gasoline was also implemented worldwide within a very short period of time. HDP processes and catalysts development has been a dynamically evolving practice. The stress imposed by fuel specifications has been aggravated by the changes in the quality of crude supply.

In general the HDP reactor can handle the presence of three evolving phases of reacting feed (liquid and H_2), catalyst, and products. The heavy liquid feed is converted to gases, vapors, liquid, and carbonaceous solids. In the case of residues of heavy/extraheavy oils and bitumen, the processing capability of HDP is limited by the presence of metals and asphaltenes. Metal deposition and accumulation in the catalyst pores are the main causes of catalyst deactivation during BotB HDP. Furthermore, metals removal is one of the fastest reactions under HDP conditions. Besides, other inorganic elements might be present in the feed (Si, Fe, Hg, Cd, etc.) and might be filtered off by the catalyst bed.

Diffusion of asphaltenes, the largest macromolecules of the residue, within the pore structure of the catalyst needs pore diameters in the range of 50−100 μm. Nevertheless, conversion of asphaltenes is by far the most difficult reaction in HDP. For residue conversion to volatile liquids of 50% or more, asphaltene agglomeration and deposition may occur. As discussed in Chapter 2, asphaltene agglomeration is promoted by insolubility in saturated hydrocarbons. Under HDP conditions conversion of asphaltenes starts by the trimming of alkyl substituting branches that upon hydrogenation will render light alkanes [240]. At a given point, insolubility will trigger agglomeration. Asphaltenes are soluble in aromatics, and finally colloidal stability is facilitated by the presence of the resins. Some aromatics might be hydrogenated. Resins are more reactive than asphaltenes and would be converted. The solubility and colloidal stability would be perturbed. In summary, the saturates, aromatics, resins, and asphaltenes ratios have significant impact on the formation of organic sediments.

Additionally, hydrogen addition processes need a hydrogen plant to generate the required high pressures. The process gas is recovered through a gas scrubbing system that removes the H_2S and NH_3 and recycles the H_2. The formed H_2S is fed to a Claus plant to be converted into sulfur. Catalyst management includes addition of fresh (or regenerated) catalyst and withdrawal (or separation) of spent catalyst. All these ancillary units and systems impose cost-impacting economics. Comparatively, C-rejection without all these units is less onerous.

3.1 Hydrotreatment

During hydrotreatment (HDT), two types of reactions take place, namely, hydrogenation and hydrogenolysis. In hydrogenolysis, C–C bonds are broken and under HDT conditions may be either hydrogenated or evolve toward coke formation. Similar reactions take place on C–S and C–N bonds (and in general C–X bonds, X = heteroatom) as well. Thus reaction products from heavy and sour crudes or fractions include H_2S and NH_3 that would irreversibly poison and deactivate noble metals, which are highly active for hydrogenation and hydrogenolysis reactions. Instead, sulfide catalysts of nonnoble transition metals are active enough and more stable to be preferred for the HDT of those crudes and their fractions. The conditions for breaking C–C bonds are more severe than those required for C–X bonds.

Hydrogenation and hydrogenolysis reactions can be catalyzed by the same active phases, i.e., transition metals and transition metal sulfides (nitrides and carbides also). While transition metals, particularly noble metals, are preferred for hydrogenation, sulfides are better in hydrogenolysis.

Commercially, HDT is applied in refining as a pretreatment of feedstocks for meeting specifications of the downstream process or as a finishing step for intermediate streams to produce blendstocks. Thus VGO pretreatment for FCC and aromatics saturation in diesel are respective examples of each of these applications. In the case of residues, HDT has been applied either way, as feedstock pretreatment for residue catalytic cracking (RCC) or HCK processes and as sulfur reduction for lower-sulfur FOs.

Sulfur reduction of residues was originally developed for the treatment of ARs for production of low-S FO. However, changes in crude oil quality, in product specifications, and in product demand have shifted the relative commercial importance of this application. All these impose additional processing than just HDT, thus nowadays residue HDT needs to be more directed into the pretreating of feedstocks for the conversion units. In this way, HDT serves as protection for the catalysts of other processing units and contributes to minimize catalyst replacement in those downstream units and in the case of coker to reduce coke yield and increase the quality of the fractions and by-products.

As mentioned, typically, the organometallic and asphaltene components in the feedstock determine the processing capabilities. Metals will deposit and deactivate the catalysts irreversibly. The higher the metal content of the residue, the shorter the catalyst life cycle. Process economics is deteriorated by the forced frequent stops necessary for unloading and reloading catalyst. Generally, the HDT applications are limited economically to residue feeds containing less than about 250 wppm Ni + V.

Technology licensors customize their applications to cope with the properties of the feedstock. The residue HDT process and catalyst technologies are licensed by several engineering companies. Axens-IFP licenses the T-Star Residue Desulfurization process, a fixed bed system based on a cost-effective permutable reactor system. One reactor would be in use, while the other is safely reloaded, increasing the on-stream factor [241]. An advanced process control system switches the valves that reroute the feed to alternate reactors, from deactivated to active bed.

Chevron Lummus licenses RDS and VRDS for AR and VR treatments, respectively [242–244]. The performance of some of the available catalysts was reported [245]. This company developed a system to cope with catalyst deactivation, the On-stream Catalyst Replacement (OCR), which facilitates catalyst unloading and reloading without having to stop the unit run [246]. In an OCR reactor, catalyst gradually moves downflow (plug flow) while feed goes upflow. Part of the spent catalyst is withdrawn from the bottom while a similar quantity is added at the top [247]. To avoid fouling or blockage during replacement operations, the main requirement for the particles is neither to break nor to agglomerate. In 1996, it was reported that OCR was commercially applied in three refineries in Japan, two for RCC feed pretreat and the other for production of low-S FO from AR, totaling 130 kbpd installed capacity [248].

ExxonMobil licenses RESIDfining [249], with four commercially available proprietary catalysts, RT2, RT228, RT621, and RT641, and an automatic catalyst bed technology to facilitate bypassing feed over deactivated bed and catalyst replacement.

Shell Global Solutions has evolved its initial residue hydrodesulfurization (LR-HDS) process to an integrated feed pretreatment for hydroconversion (HDC or hydrocracking, HCK) [250]. In the case of high metal containing feeds, it incorporates the Shell Bunker HDM Reactor in the front end [251–253], which facilitates spent catalyst replacement. In the reactor, screens separate the catalyst from the process fluids allowing continuous withdrawing and addition of catalyst particles. Opposite to Chevron's OCR, in Shell's bunker reactors, fresh or regenerated catalyst is added cocurrently with the feed through the top of the reactor. Since catalysts exhibit high metal removal activity, catalyst development was oriented to achieve high storing capacity [252]. In this way, the size of the bunker reactors is kept as small as possible and the run length can be as long as possible [250]. Catalysts are available through Criterion. Shell also developed a regeneration process for the DM catalyst (DCR), based on acid leaching of the deposited metal sulfides and coke burning of the spent catalysts [254]. A demonstration unit was built in Venezuela at PDVSA's Cardon refinery consisting of a series of four bunker-flow HDM reactors integrated with the DCR unit. Demonstration concluded in 1985 [251].

The UOP RCD Unionfining process provides DS, denitrogenization (DN), and DM for heavy feedstocks. A guard reactor is loaded with a hydrodemetallization (HDM) catalyst upstream from the rest of the HDT reactors [255]. Additionally to the expected contaminants removal, operating conditions allow a partial conversion to distillates as well [244]. The original development was based on Unocal catalysts [256] but currently Albemarle's residue hydrotreating catalysts are used [257].

Haldor Topsoe commercializes both equipment and catalysts for HDM, HDS, hydrodenitrogenization (HDN), and pressure drop control of residue HDP processes [258].

3.2 Hydroconversion and Hydrocracking

In HDC, bond breaking takes place by thermal cracking and hydrogenolysis and is accompanied by hydrogenation reactions as well. Coke formation, via condensation reactions, is suppressed by the hydrogenation of dangling bonds, free radicals, and coke precursors (polycyclic aromatics; dienes, olefins, etc.). Product slate and yields change drastically in comparison with thermal C-rejection processes and the heavier the residue, the larger the difference. Aromaticity of the product increases with severity [259]. For instance, a heavy crude with 40% AR, which might produce 50% liquid products by coking would render roughly, in terms of the whole crude, 65% of valuable products (gasoline through lubes), while HDC might render more than 85%; on the other side, an extraheavy crude with 60% AR would produce 50% valuable products by coking, while by HDC it would render about 85%. However, operating issues, such as hydrogen and catalyst consumption, formation of coke and sediments, poisoning of catalysts, energy consumption, and difficulty of asphaltenes to diffuse through small pores and capital factors, such as high-pressure vessels, presence of corrosive chemicals, ancillary recovery units, etc., make the HDC process less economical than coking.

Numerous hydrogen-addition technologies are currently marketed, trade names include Shell-HDM/HCON, IFP's HYVAHL and H-Oil, Chevron's RDS and LC-Fining, Veba-Combi-Cracking, UOP's Uniflex [formerly CANMET (Canada Centre for Minerals and Energy Technology)], PDVSA's HDH, and ExxonMobil's M-Coke.

As per any HDP process, catalyst life cycle is affected by formation of deposits, sediments, and precipitates. While inorganic deposits (mainly metals) accumulate irreversibly on the catalyst bed, organics (coke, measured indirectly as feed CCR and sediments) are considered to be either fouling or reversible poisons that can be eliminated by washing/flashing or burning. These characteristics led process technologists to examine different types of reactors and catalyst management systems. Selection of the reactor type depends on the level of contaminants present in the feedstock, roughly expressed in Fig. 5.8.

3.2.1 Fixed Bed

In fixed bed the HDC of residues and formation of deposits and sediments are the main problems. It is common practice in fixed bed HDC (and in general in HDP as mentioned in the previous section) to include a guard bed in front of the catalyst bed for protecting the catalyst from metal (and inorganics) deposition. Active guard beds are also commercially offered and depending on the residue composition, an HDT stage might be integrated upstream the HDC unit. When the guard bed does not offer sufficient DM capacity, the feed is switched to a second guard bed loaded with fresh catalyst/material and the spent guard bed is replaced in the first reactor. Run lengths longer than 1 year on fixed bed HDC units with VR feeds are very rare. The most relevant improvement of this approach is the on-stream replacement systems for unloading the deactivated

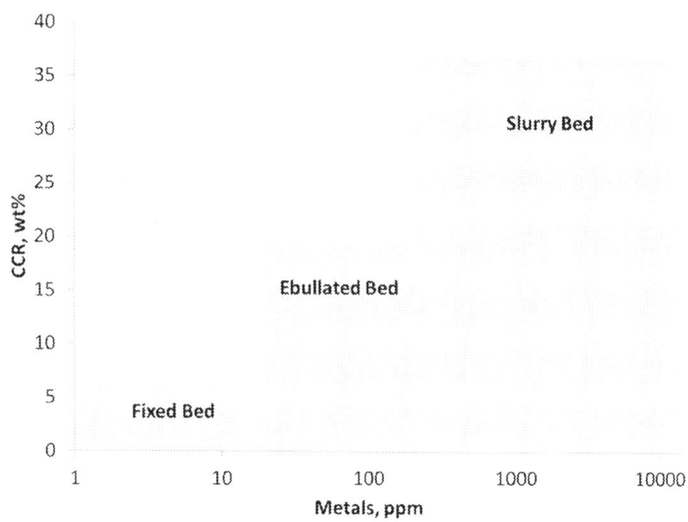

Figure 5.8
Selection of reactor type for hydroprocessing (HDP) of highly contaminated oils. *CCR*, conradson carbon residue.

(demetallizing) guard bed and reloading fresh material, without having to stop the operating run. The prominent fixed bed technologies are licensed by IFP (HYVAHL), Chevron (OCR) and Shell (HYCON).

The HYVAHL process licensed by IFP uses a "swing reactors" HDT stage prior to the HDC stage. Additionally, the IFP recommendation for heavier feeds is its integration of a upstream SDA unit. Thus the whole plant would consist of an SDA unit, followed by two-swinging HDT reactors, and then the HDC reactor. A detailed description of the process and catalysts has been reported in Ref. [260]. Low-S FO and about 30% distillates can be produced from ARs or VRs. Two units of about 25-kbpd capacity were reported to have been built in the United States and Korea.

Chevron's OCR system was discussed in the previous section as the most relevant improvement incorporated in residue HDT units. Nowadays it forms an integral part of Chevron's HDP technology packages for high metal-containing feeds, in multiple processing configurations, for new units, and/or revamping of existing ones [261].

Similarly, Shell's commercial experience with its HDM bunker technology paved the way for the development of the fixed bed HDC process, commercialized under the trade name HYCON. Thus HYCON integrates the HDM bunker technology with a multiple fixed bed VR hydrocracker [250]. A large suite of catalysts was also developed [251] and is commercially available through Criterion, allowing Shell to customize the application according to feed requirements. A 25-kbpd unit demonstrated the technology commercially at Shell's Pernis refinery in the Netherlands. Initial operations in the early 1990s presented significant problems that were corrected allowing a stable extended run that started in 1995 [248,250]. More information on HYCON development and experience can be found in Refs. 11–20 cited in Ref. [251].

Relative advantages of commercially available mobile bed reactors over conventional fixed bed are scarce in terms of process reliability, on-stream factors, and mechanical performance. For high metal-containing feedstocks, the increase in conversion is limited, a long-term run has not been commercially demonstrated, and the economics remains to be proved [25]. Additionally, some R&D challenges have to be addressed, such as the decrease in reactor efficiency caused by back mixing and high costs (capital and operating).

Typically, HDP fixed bed reactors are offered in highly integrated configurations, with other HDP and/or C-rejection processes (HYVALT, for instance). Open literature shows some examples; one of these has been described in Ref. [204], in which an integrated two-stage hydrotreating process comprised (1) hydrogen and a solid inorganic refractory material (preferably alumina) for at least partial removal of Ni and V, and (2) hydrovisbreaking of the intermediate product stream from step (1) in the presence of a decomposable compound (molybdenum dithiophosphate).

3.2.2 Ebullated Bed

In the ebullated bed reactor the catalyst is expanded or suspended by the pressure of the feed fluids (gas, vapors, and liquid), which are fed upflow allowing back mixing and minimizing Δp build-up. The main requirement for the catalyst is particle sizes of less than 1 mm, regardless of shape (pellets, spheres, or grains). The mechanically ebullated catalyst keeps a homogeneously mixed feed and allows on-stream replacement. The target is to operate efficiently under isothermal conditions. These features make operating cycles of ebullated bed generally longer than those of fixed bed.

Commercially, catalysts are offered by Albemarle, Criterion, and Axens (see corresponding websites). Albemarle's catalysts were reported in Ref. [262]. Catalyst replacement represents a large contributor to operating costs. Optimizing catalyst addition rate (CAR) has been one of the challenges faced by these technologies. CAR depends not only on catalytic properties, but also on feed properties/composition and products targeted. The impact of CAR on cost is counterbalanced by improved yields, product quality, and operability. An example of a possible optimization methodology has been proposed by Sayles and Bailor [263], who introduced theoretical means for calculating catalyst properties as a function of CAR, feed rate, and feed qualities. The approach was based on the assessment of the effect of catalyst metals content and catalyst physical properties on catalyst activity at different CARs. In practice CAR has to be equal to catalyst withdrawal rate to keep catalyst inventory constant. Furthermore, catalyst withdrawal should be based on catalyst deactivation rate. However, deactivation under HDC conditions of heavy oil residues is poorly understood. Metals are considered to deposit and to accumulate in the catalyst pore system, blocking the access of reacting molecules to the active sites. The catalytic changes are taken simply as physical and any chemical change is ignored. However, Ni and V sulfides are known to possess different catalytic functionality than the active HDP sites. Most of the proposed models are mainly phenomenological and based on pilot plant data, typically collected in the absence of catalyst addition (e.g., Ref. [264]).

Although conversion depends on feed composition, levels of 70% might be achievable for VRs [265]. Once again, formation and deposition of coke and of sediments limit the operability of hot and cold separators downstream the reactor [266]. In fact, the amount and level of deposits increases with the severity of the process. At moderate conversion (below 60%) problems are minimum [267]; however, at higher conversion levels, sedimentation and fouling start to build up. This might be one of the reasons why refiners have been cautious in operating at conversion levels that licensors claim are within the technology's capabilities [25]. Nevertheless, ebullated bed processes have charted remarkable experience in reducing operating costs at increased capacity by improving on-stream factors [248]. Actually, it is considered a mature technology with 17 operating units with capacity over 650 kbpd [268].

Chevron and ABB Lummus license the LC-Finning technology and by 2011 they had installed eight units worldwide, for a total capacity of nearly 400 kbpd [269]. Conversion and heteroatoms removal take place simultaneously, to render a full range of products. On the heavy side, it can produce FO and Syncrude, as well as feedstocks for residue fluidized catalytic cracking (RFCC), coker, visbreaking, or SDA. Additionally, gasoline, kero, jet, diesel, and lube base are also produced on the distillates side. Operation mode is adjusted to define product slate [270]; examples to maximize either middle distillates [271] or conversion [272] have been reported. CAR controls and maintains yields and/or product quality.

The H-Oil process is one of the two most marketed HDC process technologies. It was developed in the mid-1950s by Hydrocarbon Research Inc. [273]; since then, it has accumulated a large commercial experience with 12 units [274] that might exceed 360 kbpd of installed capacity. Currently, IFP-Axens licenses the process and commercializes the catalysts. In 2000, IFP reported the introduction of two improvements in the technology [264]. In the first one, determination of critical parameters (reactor hydrodynamics and liquid holdup) from the modeling of benchtop data would ensure scalability of the process, as confirmed by pilot plant results on residence–time distribution. The second was a new design for the hot high-pressure separator, which avoided excessive liquid carryover and gas carryunder. Process units could be designed for deep conversion of feeds with high metal content, and still achieve long operating cycle times at maximum liquid yields [275]. The limited vaporization of the feed (30–60 wt%) at severities required for high conversion affects the fraction of hydrocarbons entering the second reactor. Thus an interstage separator can be employed to eliminate gas streams from the first reaction zone and to enable increased conversion with the same reactor volume [275,276].

Ebullated beds have clear advantages over fixed bed, such as isothermal, good mixing (uniform distribution of liquid, gas, and catalyst throughout the reactor), heat transfer, reduced plugging and pressure drops, better catalyst performance, and lower catalyst deactivation. Fixed bed catalysts are millimetric sized, while ebullated bed catalysts are micrometric. Smaller catalysts will exhibit a larger proportion of pores on the external surface than an extruded catalyst would, resulting in less susceptibility to plugging by metal contaminants. Catalysts are designed with pores sizes to diffuse and accumulate metal compounds as much as possible (mean pore diameter of macropores in the range of 10–15 nm is typical). Coke and metals would deposit on a finely divided catalyst with a flatter radial profile than that on a pellet.

Despite all the advantages of ebullated bed systems, drawbacks are also present. For instance, back mixing promotes recombination reactions and favors more contact between deactivating/inhibiting compounds and catalyst. This is particularly the case for HDS, which has led to the need for increasing the number of these units for a more effective DS. Other weaknesses are complexity of the ebullating system, control systems, catalyst-associated costs (price, consumption/replacement rate, disposal, etc.), feed customization and changes in feedstocks, and unconverted bottoms/unstable FO (production, recycling, and disposition).

3.2.3 Slurry Bed

In this system for liquid phase hydrogenation processing, the slurry is formed by a mixture of residue and fine particulate either of a solid additive or of a solid catalyst, which is kept fluidized (fully expanded) by the conjoint motion of the gas and liquid fluids, in an upflow reactor within a hydrogen-rich environment. The slurry mixture is formed before entering the reactor. Generally, total holding time at conversion conditions is adjusted from about 10 min to about 200 min. A high-pressure, high-temperature separator is used to separate product and catalyst at the top of the reactor. The metal contaminants and formed coke deposit on the catalyst surface and are discharged from the reactor, avoiding any bed-plugging problem. Typically, vapor is disengaged from the slurry in the upper section of the reactor, and hydrogen and other permanent gases are removed in a high-pressure separator. The liquid product is condensed from the overhead vapor and fed to a distillation column. The produced fractions are blended with corresponding refinery intermediate streams for processing in downstream treatment facilities. The solids are recovered with the unconverted feed fraction at the bottom of the distillation tower or by SDA.

Slurry phase reactors were thought to overcome the issues and drawbacks of fixed and ebullated bed processes under conditions of high conversion of highly contaminated residues. Indeed, the dispersed catalyst particles can serve as nuclei for contaminants deposition. The higher activity or surface area of the dispersed catalyst constitutes the main advantage of the process. However, the better the dispersion, the more difficult product separation becomes. Hence catalyst dispersion and activity set the limits and compromises of the technology.

Two types of catalysts have been considered, namely, solid powders and dispersed systems, and within the latter oil-soluble and water-soluble compounds have been tested. Although oil-soluble catalysts have better dispersion, higher activity, and facile handling, the cost of water soluble is lower. Pretreatment is required for the water-soluble case; dispersing agents, emulsifiers, or dehydration are needed prior to processing. The preferred active metal in dispersed system is molybdenum. The use of dispersed catalysts in slurry-phase reactors was based on the idea of enabling the contact between large asphaltene molecules and a molecularly distributed active catalyst. Continuous addition of a diluted solution of the catalyst precursor compound into the mixture of feed and hydrogen facilitates the formation of a homogeneously dispersed catalyst. Consequently, a low amount of active phase, as low as 100 ppm, is needed. After reaction the catalyst remains with the solid product (if coke is produced) and the unconverted feed bottoms. Hence either for environmental or economic reasons any sort of recycle becomes mandatory.

Active metals of the solid powder catalysts are mainly Fe, Ni, and V, in the form of fine particulate powder that can be slurried/dispersed in the heavy feed before the reactor. Earlier developments used these fine powder catalysts in a once-through option and employed cheap naturally occurring materials, mineral ores, or even the contaminating metals of the heavy feeds. However, regardless of the catalyst type and depending on cost and catalyst concentration, regeneration and/or recycling is necessary. In anticipation of the technologies that will be briefly described in the next paragraphs, a summary of the employed and/or tested catalyst is collected in Table 5.4.

The dispersed catalysts, even from soluble precursors, tend to form large crystallites or agglomerated particles [307,308,337,338]. Although this might be seen as an enabler for separation and recovery, it also jeopardizes the effectiveness for reuse or recycling and threatens the required dispersion in the reactor. The agglomeration tendency is lower when fine powders are used; instead powder issues are associated with high attrition. In this regard, fine iron particles synthesized by laser or flame pyrolysis [339] have shown suitability for slurry operations.

To achieve single pass conversion in the range of 70–85%, typical operating reactor conditions are temperatures of 400–460°C, pressures of 100–200 bar, and liquid hourly space velocity (LHSV) of 0.5–2.0 h^{-1}. The lower severity ranges correspond to catalytic processes, while the higher side corresponds to the noncatalytic ones. Furthermore, the correlation observed between CCR and coke yield or refractory compounds seems to vanish for well-mixed slurry processes.

Research on slurry systems started in the first half of the 1900s by Veba Oel [340,341] and CANMET [342]. Research efforts extended through the rest of the last century into this century; recent trends in R&D include the assessment of new nanocatalysts, optimization of process conditions for more difficult and more contaminated feeds, product and catalyst separation and recovery, and in general any other option for the improvement of process economics.

Slurry reactor processes represent the ultimate technology: environmentally clean and friendly for asphaltene conversion and metals removal, and minimizing the drawbacks of

Table 5.4: Summary of Slurry-Phase Catalysts

Organization and Technology	Catalyst Precursor	References
	Fine Powders	
CanMet	Iron sulfate	[277]
UOP	Nonstochiometric vanadium sulfide	[278]
BP-KBR VCC	C-additive	[279,280]
Exxon Research and Engineering Co.	Supported and unsupported catalysts	[281]
	Alkali metal compounds or alkaline earth metal compounds	[282]
	Potassium sulfide	[283]
	Coking ashes	[284]
PDVSA HDH	Iron ores, iron minerals	[285]
PDVSA HDH-Plus	C-additive and Mo oil-soluble compounds	[286–291]
Chevron	Red mud, imogolite, or alumina	[292–296]
RIPI HRH	Me-oxide ashes and Mo and Ni aqueous emulsion	[297]
Idemitsu MRH	Spent FCC	[298]
Asahi Chemical Industries SOC	C-powder additive and Me-soluble compound of V, Cr, Fe, Co, Ni, Cu, Mo, or W	[299,300]
	Oil Soluble	
EST	Q2[MoxSy] (see text)	[301,302]
	Molybdenum naphthenate and recycle bottoms	[302–308]
Exxon Research and Engineering Co.	Molybdenum alicyclic or naphthenate	[309,310]
	Fe_2O_3 and Mo naphthenate, CrO_3 tert-butyl alcohol	[311–315] [313–315]
Alberta Oil Sands Technology and Research Authority	Iron pentacarbonyl or molybdenum 2-ethyl hexanoate	[316–321]
HTI HCAT	Mo, Ni acetylacetonates or 2-ethyl hexanoate	[322–324]
Chevron Inc.	Mo or W salts of fatty acids (C_7–C_{12})	[325]
	Water Soluble	
Chevron Inc.	Mo, Ni oxide with aqueous ammonia	[326–330]
Exxon Research and Engineering Co.	Phosphomolybdic acid ammonium, nickel carbonate ammonium, dimolybdate ammonium metatungstate,	[331]
	heptamolybdate, molybdenum oxalate	[332]
UOP	Mo, V, and Fe metal oxide or salt and heteropoly acid	[333–335]
PetroChina	Ni, Fe, Mo, and Fe–Co liquid catalyst	[336]

ENI EST, ENI Slurry Technology; *FCC*, fluidized catalytic cracking; *HTI*, Hydrocarbon Technologies Inc.; *MRH*, mild resid hydrocracking; *RIPI HRH*, Research Institute of Petroleum Industries Heavy Residue Hydroconversion; *SOC*, Super Oil Cracking; *UOP*, Universal Oil Products; *VCC*, VEBA Combi Cracking.

downstream processing of BotB. The main technical advantages of this type of process are that the unconverted bottom residues are virtually eliminated at the exceptionally high conversion levels achieved. Regardless of environmental and technical benefits, the economics of the slurry technology depends intensely on crude prices and price differentials. A threshold for economic feasibility has been established on price differentials of USD 16/bb. During 2014, an accelerated decline of crude oil prices helped maintain a sequel of low price differentials. The last few decades have been characterized by a volatile market that has precluded any possibility of installing a first commercial slurry reactor.

The reliability of slurry reactors at a commercial level still remains unproven. However, once the first commercial unit is on stream, most likely this technology would become the preferred option for high conversion, particularly in high oil prices and environmentally concerned scenarios. However, capital investment and operational cost reductions have to be addressed in future developments since this type of technology sorts out the transformation of low-value feeds into clean fuels, with the least environmental impact of all the other BotB processing options.

As already mentioned, commercial experience on slurry bed HDC processes is very small though the technologies, discussed later, are ready for licensing. Because of the high capital investment required, a large commercial application has not yet materialized though three applications are currently operating at a small commercial scale (CANMET, VCC, and EST). In fact, most of the technology licensors have accumulated in commercial demonstration tests that showed the potential of their processes, particularly for high conversion of residues with the poorest quality. Discussion of these technologies follows.

3.2.3.1 Uniflex (Formerly CANMET)

The CANMET Hydrocracking Process, developed at the Energy Research Laboratories of CANMET, was conceived in the 1970s for the high conversion (~90%) of residues under high DM conditions [343–346]. Besides, DS is in the order of 60% and DN is about 40%, under high severity conditions. Although product quality is not that good, a stable liquid is obtained and further HDP would be required. The liquid product is stripped and fractionated prior to HDP. The low DN and DS achieved in CANMET impose great challenges to the downstream units. First, HDP has to be used for the pretreatment of these intermediate products to comply with the specification of the other downstream processing/finishing units.

An additive was used for the inhibition of coke formation and for the conversion of high boiling point hydrocarbons into lighter products with low hydrogen consumption. This additive was preferably $FeSO_4$ and it is used at a relatively high dosage (1–5 wt%) [277].

The separation of catalyst and nonconverted bottom oil is somehow troublesome. Although the use of the CANMET technology for the coprocessing of coal was seen as a possibility for increasing coal utilization in conventional refineries, the low economic incentives and the lack of existence of commercial CANMET units preclude the realization of such an idea [347].

In the early 1980s, Petro-Canada built and operated a 5000-bpd demonstration plant in its Montreal East refinery [348]. This unit would be operated on an economical basis and would be shut down when there would be a better opportunity for the visbroken and/or the VRs employed as feed. A 100 vol% of distillate yield compared favorably to the 83 vol% typical of any coking process [349,350]. Typically, the remaining unconverted heaviest residue from the unit was sold to a local cement manufacturer that uses it as an energy source [248]. During the 1990s, incentivized by the high price differentials, the unit was operated with an average on-stream efficiency of about 97%.

In 2006 UOP joined Natural Resources Canada's efforts for the improvement of the CANMET Hydrocracking Process and in 2007 acquired the exclusive worldwide rights to license it. This work resulted in the Uniflex Process, which incorporated elements from UOP's Unionfining Process and Unicracking technologies. Although UOP had developed the Aurabon Process by 1980, there is no reported connection between CANMET and this process.

The Aurabon process, developed by Gleim and Gatsis [351,352], was described by Gleim [353] as the following eight steps:

1. Catalyst formation: H_2S converts the metals present in the feed to sulfides (V_xS_y);
2. Catalyst carrier: asphaltenes act both as a catalyst carrier and substrate;
3. Reaction: slurry operation where the catalyst sludge is adsorbed on the asphaltenes moves upward with the charge stock through the reactor at temperatures of 400–425°C (H_2 + 15% H_2S), pressure of 2000–3000 psi, and LHSV of 0.5–2 h^{-1};
4. Product separation: stripping and DA;
5. Catalyst recycle: deasphalted bottoms are recycled and mixed with the fresh feedstock;
6. Startup: process is initiated by adding about 10% vanadium sulfide into the feed.
7. Catalyst concentration: catalyst concentration is kept constant by a dragstream installed after the SDA unit and by controlling the conversion per pass. Asphaltene conversion is kept at the necessary level to keep the source of catalyst carriers needed in the reactor;
8. Product DS: an HDS unit is integrated downstream for liquid product treatment.

Although very little was published about the Aurabon process, it is worth mentioning the evaluated potentiality of its performance [62,354] and the favorable economics [354,355].

It is clear that long before the acquisition of CANMET rights, V-based catalysts were developed by UOP. A V_2O_5 catalyst based on an ammonium salt [278] and an oxovanadate [356] was reported to be tested by UOP. The active phase would interact with asphaltenes or with the organometallic compounds present in the feeds. UOP owned numerous patents for the use of this catalyst and other formulations [333–335] in slurry-phase processes [357–362].

In Uniflex, a proprietary nano-sized solid catalyst is used instead of the original CANMET additive [363]. A nanocatalyst advantageously offers a large surface area, which would inhibit the coalescence of coke precursors favoring their conversion. More recently, pursuing the development of a second-generation catalyst was announced.

As an H-addition process, Uniflex produces higher-quality distillate products at higher yields than a coking process. Additionally, the process enables replacement of more than 30 wt% coke yield that would be produced by a coker, with less than 10% FO [54,364,365]. A recycle of the heavy vacuum gas oil was recommended for maximizing the distillates yield and improving the quality of the bottoms [366,367]. Nevertheless, the quality of that FO would hinder its marketing in modern times. Other configurations are needed for improving bottoms quality and producing bunker or marine fuels that meet specifications [368]. A flow diagram for the Uniflex process is presented in Fig. 5.9.

3.2.3.2 Veba Combi Cracking

The Veba Combi Cracking (VCC) process was developed in Germany between 1950 and 1964 for the conversion of residual oils and is currently licensed by KBR [341,369].

The process (Fig. 5.10) consists of a liquid-phase hydrocracker followed by a gas-phase hydrofinishing step for rendering an SCO. Besides conversion of VR, it can also be used to process heavy crude oils for the production of SCOs or for the liquefaction of coal [279,280]. It was originally conceived employing an additive based on red mud (an iron-containing material disposed by the aluminum industry from the processing of bauxite) or a fine particulate coke obtained from Bovey coal. To achieve 95%+ conversion, highly severe conditions are utilized: 15–27 MPa reaction pressure, 5 wt% of additive, and temperatures in the range of 440–485°C. The liquid-phase reactor is operated upflow and does not include internals. A vacuum-flash distillation separator follows the reactor to remove the additive and unconverted materials from the lighter products, which are fed to a fixed bed HDT. The total product from the HDT unit is cooled, condensed, and separated from the permanent gases to recover the SCO. The whole gas stream is washed with a lean oil stream prior to be recycled to the conversion unit. The bottoms can be further

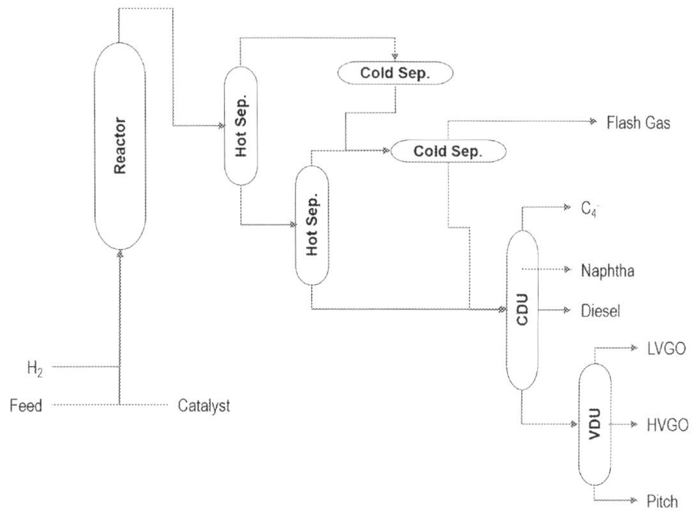

Figure 5.9
Simplified flow diagram of the Uniflex™ process, based on description from Refs. [363-368].

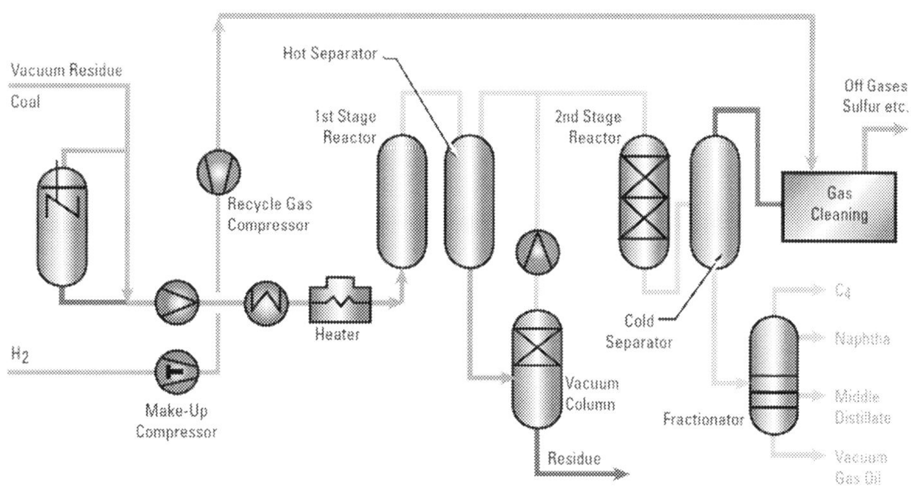

Figure 5.10
VCC process flow diagram. *Reproduced from KBR Technology, Veba combi-cracking technology. 2013;3 pp. Available from: http://www.kbr.com/Newsroom/Publications/Brochures/Veba-Combi-Cracking-Technology.pdf, with permission from KBR.*

processed through gasification, carbonization, coking, or applied as a binder [370–373]. VCC does not recycle the bottom product to withstand the previously mentioned overall conversion; neither does it require solvent addition for fouling control [341,369]. A joint development between Veba Oel and PDVSA Intevep optimized the particle size distribution of the additive. A binodal distribution was defined, including a fine grain portion with a grain size of less than 90 μm and a coarse grain portion with size in the range 100–1000 μm [374].

A 3500-bpd demonstration plant was built and tested in Bottrop, Germany, proving technical feasibility, expected yields, product quality, and reliability [371]. A variety of feeds were tested showing the flexibility of the technology for handling feeds with 2–80% asphaltenes, Ni, or V levels of about 2000 ppm and sulfur in the range 0.7–7% [30]. The produced SCO (~4000 bpd) was routed to the Scholven refinery as part of the daily diet.

In 1996, it was reported that Veba Oel adapted its 4000-bpd VCC plant to process plastic wastes from recycled domestic garbage. The main product served as petrochemical and refinery feedstocks, while the bottoms were disposed to feed cokers in nearby steel facilities [248]. Besides plastics, a variety of other waste materials were tested within that program: residues from degreasing plants, chlorinated used solvents, painting sludge, transformer oils, spent lube oils, cutting oils, spent catalysts, and carbon adsorbents.

After oil prices declined and a period of low prices ensued, Bottrop's unit was decommissioned and shut down in 2001. Then, the acquisition of Veba Oel by BP coincided with increasing market demand and better oil prices, moving BP to include the VCC process in its Vision 2030 portfolio and the Advanced Refining Program. A new 1-bpd pilot unit was built in BP's Naperville facility. In 2010, a BP and KBR alliance was created for VCC commercialization, licensing, and engineering of the technology. The current process, subject to commercialization, uses a noncatalytic, nonmetallic additive, which meets the marketed reduction on fouling tendencies and high asphaltene conversion [375]. In 2012 TAIF-NK announced the starting of a VCC unit construction in its Tatarstan refinery and in 2014 Shaanxi Yanchang Petroleum (Group) Co. and Shaanxi Yanchang Petroleum Anyuan Chemical Co., Ltd., both subsidiaries of Yanchang Petro, announced construction of one unit each.

3.2.3.3 ENI Slurry Technology

During the early 1990s, three affiliated companies, Enitecnologie S.p.A., Snamprogetti S.p.A., and ENI S.p.A., started the development of a process for the conversion of feeds with a high content of contaminants, i.e., heavy oils, VRs, and bitumen. This process evolved into the currently known ENI Slurry Technology (EST) process [376–380].

The EST process is operated at relatively low temperature (400–425°C) in the presence of a hydrogen/H_2S atmosphere (~160 bar) and a dispersed catalyst, which is recycled to the slurry reactor via SDA, together with an asphaltene recycle. In the EST process, the HCK reaction is terminated before the mixture becomes unstable and the asphaltenes precipitate. Thus the partially converted asphaltenes can remain dissolved in a more aromatic stream, allowing further processing. The hydroconverted product is fractionated and the bottoms are fed to the SDA unit. Unconverted asphaltenes and metals originally present in the feeds are less soluble in product oil, which can be easily separated in this SDA unit [381–386]. Since the liquid product is fractionated, the light and medium distillates are incorporated into the intermedia refinery streams for further processing. Meanwhile, the DAO is thought to be a suitable feedstock for FCC or HCK. EST has demonstrated residue conversion of 98–99%, HDS >80%, HDM >99%, and CCR removal >96% on a small pilot plant scale. The EST scheme [387] is presented in Fig. 5.11.

Initially, the precursor of the dispersed catalyst was an oil-soluble molybdenum compound, represented by the formula $Q_2[Mo_xS_y]$, where $x \geq 1$, $y \geq 4$, and Q was a phosphorus-containing salt of the formula $PPh_3{-}N{-}PPh_3$ (where Ph is the phenyl group) or PH_zR_{4-z}, in which R was an alkyl, cycloalkyl, aryl, aryl-alkyl group containing 1–20 C-atoms, and $z = 0{-}4$ [301,302]. From 1988 through 1999, a more systematic evaluation of several transition metal compounds carried out at lab and bench scale led to the selection of molybdenum as the best performer. Molybdenum naphthenate was selected based on its high conversion activity, high DS, and the highest DM, even at moderate conditions. The catalyst precursor is dissolved in the feed at a level of thousands of ppm molybdenum, prior to being fed to the reactor. On the catalyst, the layered structure of molybdenite was identified as a microcrystalline active phase of 20–40 Å average size, though agglomeration of crystallites into clusters of about 0.5–2 μm was observed during reaction. Typical S-content of the relevant feedstocks (3–5%) was enough for sulfurization of the dispersed catalyst; no additional sulfur was needed. Metals from the feed (V, Ni, Fe, etc.) were also converted to sulfides and were kept homogeneously dispersed in the slurry.

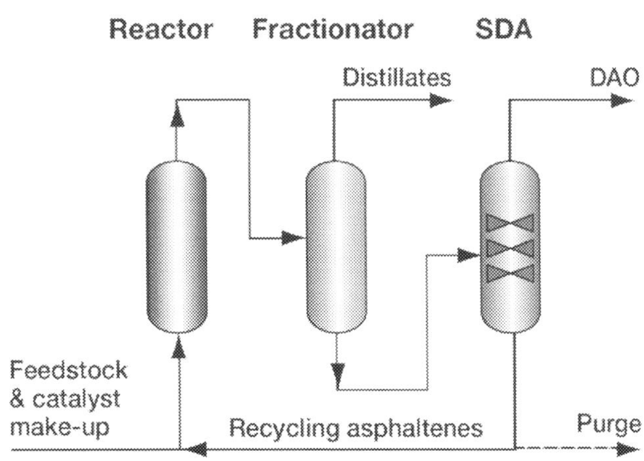

Figure 5.11

ENI Slurry Technology (EST) process scheme. *DAO*, deasphalted oil; *SDA*, solvent deasphalting. *Reproduced from Delbianco A, Panariti N, Correra S, Montanari R, Marchionna M, Rosi S. Petroleum residues upgrading: towards 100 percent conversion. In: Proc. AIChE spring meeting, 3rd intern. pet. phase behavior fouling. New Orleans, LA, USA; 2002. p. 520–525, with permission from PTQ.*

These sulfides act as catalysts but with significantly lower activity than the molybdenum main catalyst [302–308].

From 1999 to 2005 pilot plant testing went from 0.3 bpd up to 1200 bpd and Athabasca bitumen, Ural, Maya, Arabian heavy, Venezuelan Zuata, and other heavy crude oils were tested. The observed advantages of the EST process, such as feedstock flexibility, high conversion to distillates, negligible residue production, minimum hydrogen utilization/consumption, low catalyst consumption, and product slate flexibility, were reported [388]. In 2011, ENI announced the launching of a project for the construction of the first commercial unit of the EST process [389]. A 23-kbpd unit was built in ENI's Sannazzaro de' Burgondi refinery, which started up in October 2013 [390].

3.2.3.4 ExxonMobil M-Coke Process

Originally, Exxon Research and Engineering Co. (now ExxonMobil group) developed the Microcat Process. This process was taken as the basis of the M-Coke Technology [391,392]. The process is operated at comparatively moderate temperatures ($\sim$440°C) and pressures ($\sim$170 bar). A flash separation is applied to the fluid product to recover hydrogen and gases from the liquid product and separate the bottoms containing the spent catalyst. These latter bottoms are vacuum distillated to render a $565°C^-$ stream and a $565°C^+$ residue. Unconverted feed, microcatalyst, and most of the contaminating metals are left in that bottom residue. Conversion levels of about 90% can be achieved [314,315,393–397].

The catalyst is used in micrometric particle size to obtain a uniform dispersion throughout the feed within the reactor. One of the earliest awarded patents used coal to form the slurry with the feed [393]. Although the process is considered catalytic, one might think of coal acting only as an additive. A broad variety of catalyst formulations of HDC slurry catalysts was proposed and/or tested by ExxonMobil: copper chloride [398], transition metal (e.g., vanadyl phthalocyanine) and aluminum alkyl compounds [399], alkali metal compounds or alkaline earth metal compounds [282], potassium sulfide [283], alkali metals in the molten state [400], chromium-containing catalyst from a heated mixture of an aliphatic polyhydroxy compound (e.g., polyhydroxy alcohol, carbohydrate, chromic acid, a hydrocarbon material, and hydrogen sulfide) [311–313], coking ashes [284], nickel phthalocyanine or iron oxide [314,315], soluble metal compounds [331] (e.g., as colloids [401], phosphomolybdic acid, [332] and molybdenum naphthenate [309,310]), and a combination of supported and unsupported catalysts [281].

Experimentally at pilot plant level (1 bpd), the tested catalysts proved to exhibit certain activity and poor performance. As an example, the treatment of a metal-contaminated heavy oil (Safaniya AR with Ni 23 ppm, V 77 ppm, and Fe 5 ppm) in the presence of hydrogen or a hydrogen donor compound catalyzed by a mixture of a transition metal compound (either those present in the feed or a supplementary compound) and an aluminum alkyl compound (e.g., vanadyl phthalocyanine and 1.9 wt% triethyl aluminum at an Al alkyl/transition metal mole ratio of 5–20:1), at 435°C and 175 bar, resulted in DS of 30%, DM of 82%, coke yield <6%, and product API gravity of 23, in 60 min [399]. These results were only slightly better than those obtained without any catalyst.

Technically, there were limitations on catalyst recovery and recycling, which together with the high cost of the considered formulations created economic constraints. Regardless of the efforts devoted to recover the catalyst particles by microfiltration [402], a commercial demonstration never moved forward [403,404].

3.2.3.5 Intevep HDH Process

The R&D and Technology Center of PDVSA Intevep developed process technologies for the upgrading of heavy Venezuelan crudes, by recognizing that the extensive and complex operations required can only be favored by high-throughput "scalable" processes. To move in this direction, the HDH process integrates a slurry-phase *HDC* reactor, a vacuum *distillation* tower, and a *hydrotreating* train. The HDC slurry reactor is characterized by moderate reaction conditions and high liquid space velocities (420–480°C, 130–140 bar pressure, and weight hourly space velocity of 0.4–0.8 h^{-1}) [285]. The process can achieve conversion of about 92%, which can be increased to 97% with a recycle of the recovered solid-free residue. The finished products after the HDT stage possess enough quality to be blended with the intermediate streams and fed downstream to the other processing units [405]. Between 1984 and 1987, HDH was tested in Germany as part of collaborative work with Veba Oel, using two units of different capacity in a 3 bpd and 150 bpd pilot plant. After that, each company pursued separate developments [378].

The slurry process technology was originally based in an inexpensive (~USD 0.10/kg) natural ore catalyst. The low activity of the (iron) mineral-based catalyst compelled its use at very large amounts [96,406–409]. The idea of an inexpensive material was conceived as a cheap use-and-dispose strategy. However, not only a relatively complex system was required for the separation of the solids and unconverted bottoms, but also large amounts of a material contaminated with heavy metals (the spent catalyst) had to be disposed of Ref. [403]. Some catalyst management options were developed to overcome these issues, such as regeneration and recycling [410], application in the metallurgical [411], and extractive mining industries [412]. However, the fact that the natural ore was domestically mined also imposed geographical/logistic limitations for location of the HDH plants consuming those large quantities of minerals [404]. Alternatively, synthetic versions were also conceived [286–291], leading to the development of the improved process: HDH-Plus, currently licensed by IFP/Axens (Fig. 5.12 [413]). This new process employs molybdenum compounds dispersed in the feed by means of a catalytic emulsion. This most recent version of an Mo-containing catalyst might incorporate coke as support to improve the degree of dispersion and to make it easier to separate it from the bottom

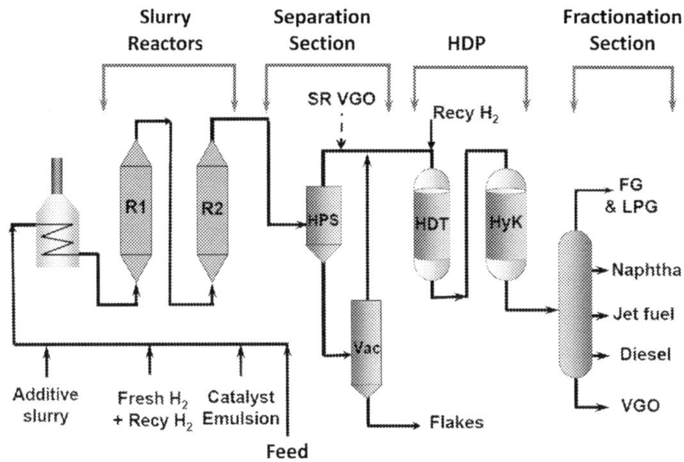

Figure 5.12

HDH-Plus integrated scheme. *FG*, fuel gas; *HDT*, hydrotreatment; *HPS*, high pressure separator; *HyK*, hydrocracking; *LPG*, liquefied petroleum gas; *SR VGO*, straight-run vacuum gasoil; *VGO*, vacuum gasoil. *Reproduced from Castillo E, Morel F. HDHPLUS®: integrated scheme with sequential hydroprocessing. In: Proc. Ertc 11th annual meeting. Paris, France; November 2006. 18 pp, with permission from ERTC.*

residue product. The particle size of this solid support material is adjusted to meet reactor fluid dynamics requirements [414] and to prevent foam formation. In this case, the support or solid-aid material is mixed with the feed in the feeding system and is fed to the heating system. The alliance with IFP/Axens introduced the sequential HDP units for the final stage of the technology [413,415–417]. The unconverted feed enters the vacuum tower where any remaining light material is recovered. The bottom residue from the vacuum flash tower is processed for its final application (e.g., solidification in a flaker unit [404]).

In 2005, PDVSA announced an expansion project for the Puerto La Cruz refinery (Venezuela) to 280 kbpd, which included the construction of the first commercial application of the HDH-Plus technology, supposedly starting up in 2010 [418]. It has been postponed through the years, but it seems finally a plant is being built. This HDH-Plus unit was designed to comprise two trains of 25 kbpd each. Several engineering firms have been contracted and roles assigned. Thus Toyo Engineering/Vepica (PMC) and JGC Corporation would be in charge of detail engineering, procurement support, and construction (EPC) management services, Hyundai-Wison would do EPC and start-up assistance, and Wison would also carry out the PC for site preparation [419].

3.2.3.6 Chevron CASH/VRSH Technology

Chevron has an extensive R&D history in the conversion of residues and heavy oils, dating back to the 1960s (e.g., Refs. [420–422]). Regarding HDC, early process research at Chevron seemed to have been focused on the once-through option, using cheap disposable materials, such as red mud, imogolite, or alumina [292–296].

In the late 1980s ChevronTexaco changed gear into dispersed catalyst systems. Chevron's Activated Slurry Hydrocracking (CASH) process was the result of a prolonged R&D program targeting high conversion and contaminant removal. Initially, water-soluble compounds were considered, and more particularly molybdenum and nickel [326–330]. Process development in a slurry-phase reactor at laboratory scale (0.1 bpd) continued its path into the 1990s, using catalysts based on molybdenum in the presence of nickel as a promoter [423–426]. The catalyst might be prepared in aqueous ammonia in which an oxide of a Group VIB metal (preferably molybdenum), in combination with a Group VIII metal (Co or Ni) compound, is mixed. A gas mixture of hydrogen sulfide and hydrogen might be used for sulfiding and to form the slurry. This particular combination of active phases performed advantageously on nitrogen removal. Catalyst concentration ranged from 500 ppm up to 1%. The incorporation of catalyst recycling from the bottom of the distillation column was employed for the prevention of catalyst deactivation [426]. CASH process conditions fall within typical ranges of slurry phase HDC. Further development reached the 30-bpd pilot plant level.

The basis of using a highly active catalyst was kept in the next generation of Chevron's heavy oil conversion process, Vacuum Residue Slurry Hydrocracking (VRSH). A wide variety of feedstocks has been tested at different pilot plants in Richmond, California, facilities. In this case, the key was the preparation of the catalyst (ISOSLURRY catalyst) into VGO or a light intermediate fraction like naphtha, i.e., oil-soluble compounds (e.g., a fatty acid salt [325]). The slurry is passed to a mixing zone into which the hydrocarbon (VGO or naphtha) solvent is added [427–452]. A sequential multistage reactor system is used with intermediate separation. The converted light fractions are removed and the unconverted heavier fraction containing the catalyst is moved to the next reactor. VRSH potential was to increase gasoline, diesel, and jet fuel yields from heavy and ultraheavy crude oils. Plans were announced for building a precommercial plant (3500 bpd) at Chevron's refinery in Pascagoula, Mississippi [453–455]. The target was to test the technical and economic viability of VRSH [456,457]. This demonstration plant appeared to have never been built [458].

More recently, the experience with LC-Fining has been incorporated in a hybrid concept with the VRSH process. A key feature of the hybrid system is capturing coke while suppressing sediments. While the ebullated bed catalyst provides high catalytic density with a slow turnover time, the slurry cocatalyst suppresses sediment formation in the thermal zones by continuously removing coke and coke precursors with a fast turnover time. Incorporation of the ISOSLURRY catalyst in existing LC-Fining units would facilitate the commercial demonstration of the catalytic system, while giving the refiner the opportunity to process more contaminated feeds at higher conversion levels [459].

3.2.3.7 Headwaters Technology Innovations, HTI HCAT/HC3 Process

In 1988 Alberta Research Council and Alberta Oil Sands Technology and Research Authority presented their initial activities toward the development of a slurry HDC process, the so-called (HC)3 process for the upgrading of highly contaminated feedstocks. A dispersed, cost-effective, oil-soluble catalyst precursor at ppm level forms an active colloidal (particle size ~0.1 nm) catalyst in situ, in the HDC reactor [316–321]. Iron and molybdenum compounds (e.g., iron pentacarbonyl or molybdenum 2-ethylhexanoate) were tested and Mo compounds were found to perform better. This molecular-size catalyst was chemically generated within the reactor system through a proprietary conditioning process [460]. The potential process scheme was reported in 1998 and is reproduced here in Fig. 5.13.

Later on, in 2003 Headwaters Heavy Oil, LCC, a subsidiary of Headwaters Inc. (currently devoted to light building products, heavy construction materials, and energy technologies), obtained the exclusive worldwide license to develop, market, and sublicense the (HC)3 technology [461]. According to Headwaters, in the (HC)3 approach, a large number of individual catalyst molecules diffuse onto the larger feedstock molecules and promote the conversion reactions by avoiding any geometrical, steric, and diffusional limitations. The advantages of the HC3 process have been reported to be: simple reactor system (two-phase gas–liquid), negligible catalyst deactivation, feedstock flexibility and high throughput, uniform product quality, and high conversion (~95%) [403]. In 2006 Headwaters announced the successful commercial demonstration of the (HC)3 process in cooperation with a host refinery and

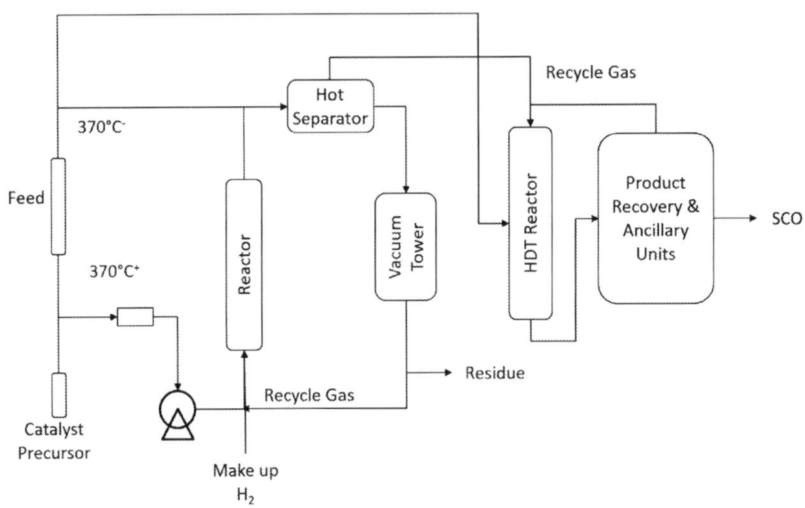

Figure 5.13
(HC)3 Proposed process scheme, based on description from Refs. [460–479].

with Advanced Refining Technologies LLC (a joint venture between W.R. Grace & Co. and Chevron Products USA) [462]. For this demonstration, only the performance of the dispersed catalyst was proven in an existing ebullated bed hydrocracker, at Neste Oil Porvoo refinery [463,464].

Headwaters Heavy Oil, LCC seems to have given rise to Hydrocarbon Technologies Inc. (HTI) [465] and the newly developed catalytic systems became HCAT [323,324,466–470]. Process development led to the application of HCAT in different reactor systems [322,471–475]. In 2011 and 2013 Criterion and Axens, respectively, joined efforts with HTI for testing HCAT in their ebullated bed processes and found additional advantages of this application mode. These advantages were: increasing residue conversion and throughput; enabling the processing of higher-fouling feedstocks; allowing HCK catalyst to perform under its capabilities; improving unit reliability; and reducing on-stream time sediment formation [274,476].

Further optimization studies included the effect of catalyst concentration on performance. An increase in catalyst dosage resulted in sediment formation reduction, increase in asphaltene conversion, improved HDS, and improvements in API gravity and viscosity of the product [477].

Three commercial tests were run from 2006 to 2013, two at Neste Oil Porvoo refinery and the other at Motiva Enterprises in Convent, Los Angeles, California, USA, confirming the economical improvements of the technology when used in ebullated bed reactor systems [478,479].

3.2.3.8 Heavy Residue Hydroconversion

The Research Institute of Petroleum Industries (RIPI) of Iran developed a processing concept in which a nanoparticulate catalyst diffuses onto the surface of the asphaltenes to promote their conversion. This concept is the basis of the Heavy Residue Hydroconversion (HRH) process [480,481]. The catalyst was described as a pseudohomogeneous catalyst [482], introduced to the reactor in a colloidal form of ammonium heptamolybdate, indicating it is a water-soluble dispersed catalyst [483]. The catalyst is prepared from an aqueous emulsion of a precursor compound of the active phase (Mo or Ni) in the heavy hydrocarbon. The stepwise addition of the catalyst emulsion to the reaction zone is supposed to lead to a higher catalyst activation that minimizes catalyst amount needed for a given conversion. This method is the result of a collaboration between RIPI and Russian researchers from the Academy of Science [297], which apparently continued during process development. Conversion was within the range of 90–95% at moderate conditions (100 bar of pressure and temperatures below 430–490°C). At this conversion range, 60–80% of sulfur is converted to H_2S, and 100% DM was observed. The product is recovered by using a high-/low-pressure separation train where sour gases, SCO, and residue are obtained. A system for catalyst recovery and regeneration was incorporated into the process [484]. The spent catalyst is separated via residue filtration in the recovery unit and then is sent to the regeneration section. A solvent is added to the resulting ash and this mixture is recycled to the reaction section. The metal oxide ashes from the feed are dried and sent outside the battery limit for their commercialization. Sulfur is recovered as ammonium sulfite and via air oxidation converted to crystallized ammonium sulfate. Catalyst recovery was established to be better than 95% and a regeneration unit allows for the recycling of the regenerated catalyst to the slurry reactor. The awarded patents are shared between RIPI and a Russian company (NTI Co.) [297,485].

The process was tested up to a pilot plant scale of 200 bpd. Three configurations were visualized: (1) wellhead, (2) new refinery scheme, and (3) retrofitted refinery [484]. The HRH process has been evaluated for wellhead applications. The simplicity of the HRH reacting system would facilitate field operation for producing a more marketable SCO. For this application, HRH would separate the SCO from the residue and solids using simple flash since distillation columns would either be unavailable, uneconomical, or impractical. Data obtained in pilot plants showed favorable economics, with an internal rate of return of over 30%. The refinery application has shown suitable incorporation of HRH products in downstream units, such as RFCC [485,486]. The construction of a 10-kbpd demonstration plant [484] and an unit for a 180-kbpd refinery [484] were announced in 2008; there is no evidence that these units were ever built.

In 2006 RIPI announced that Mobis Energy Inc. was the licensor for the HRH process in North America [481] and in 2009 Mobis Energy rebranded it as the Mobis HRH process [483]. Additionally, Zoominfo[3] described Mobis Energy as "an independent business development firm, building markets for transformational technologies in the oil & gas industry. In cooperation with the Topchiev Institute of Petrochemical Synthesis (TIPS), a division of the Russian Academy of Sciences, Mobis is offering the TIPS Residue Hydroconversion (TIPS) process to the North American market. Through an optimization of reactions on a molecular level, the TIPS slurry process achieves deep feedstock conversion, increasing yields of upgraded crude, while nearly eliminating unconverted by-products. The catalyst system, originally developed for direct coal liquefaction, is ideally suited to the challenging chemistry of heavy feeds, such as Alberta bitumen. Unique to TIPS is a recyclable, ultra-reactive catalyst providing up to 95% feed conversion at pressures between 900 and 1100 psig." Currently, Mobis Energy Inc. has no official website and no

[3] http://www.zoominfo.com/s/#!search/profile/company?companyId=356185137&targetid=profile.

additional information could be found about the TIPS process, other than a description, which resembles that of the HRH process.[4]

3.2.3.9 Mild Resid Hydrocracking

Idemitsu Kosan designed the mild resid hydrocracking (MRH) process for upgrading heavy feedstocks containing large amounts of metals and asphaltene in a mixed three-phase slurry reactor [487,488]. The reactor is fed downflow with slurry of feedstock and fine powder catalyst. Upflow hydrogen is fed from the bottom of the reactor for keeping the catalyst suspension in the reacting mixture. A slurry oil containing the catalyst, unconverted residue, and part of the VGO fraction are withdrawn at the reactor bottom section. The spent catalyst and coke are recovered by flashing off any remaining liquid product and fed to the regenerator.

Although spent FCC catalyst might be used, a proprietary catalyst R-HYC4 having high DS, DM, DA, and HCK activities is also offered [298]. Catalyst is added at a quite large proportion of the feed (20%). Mild operating conditions (60−80 bar and 380−420°C) result in conversions lower than 80% [489].

3.2.3.10 Super Oil Cracking Technology

A Japanese alliance formed by Asahi Chemical Industries [299,300], Nippon Mining Company [490,491], and Chiyoda Co [492,493] developed a slurry-phase technology, the Super Oil Cracking (SOC) process for the conversion of heavy residues. The dispersed catalyst consists of two components: a transition metal compound with hydrogenating activity (V, Cr, Fe, Co, Ni, Cu, Mo, or W) and an ultrafine particle powder. This catalytic system seems to reduce coke formation during conversion to levels of about 1%. The process using the dispersed system of Asahi Chemical Industries [299,300] has been described in Ref. [494] that corresponds to a patent assigned to Stone & Webster Engineering Corp., in which SOC was proposed to be integrated with ebullated bed processes (LC-Fining or H-Oil). Both, oil-soluble (π-compounds, alkoxy compounds, diketone, carbonyl, sulfonic or sulfinic acid, xanthinic acid, but preferably salts of aliphatic carboxylic acids) or water-soluble (carbonates, carboxylates, sulfates, nitrates, hydroxides, halide, and ammonium or alkali metal salts of transition metal acids) transition metal

[4] **TIPS RAS developments prepared or assigned for implementation: novel petroleum feed hydroconversion technology**. A novel world-unique petroleum feed hydroconversion technology was developed that increases refining depth to 90−92% compared to the traditional 70% of petroleum refining in Russia. The process is carried out in a flow unit at T = 420−450°C, P = 6.5−7.5 MPa in the presence of the new-type nanostructured catalyst forming in situ from the precursor. The main process products are C_5−C_{10} hydrocarbons—high-quality motor fuel components−and C_1−C_4 fraction and vacuum gasoil fractions—a valuable feed for the petroleum chemistry. The technology was tested at the pilot units; research data were developed and passed for the technical and economical industrial unit calculation for Taneco plc (Nizhnekamsk town). http://www.ips.ac.ru/index.php?option=com_content&task=view&lang=en&id=304

(preferably Mo) compounds can be used. Compounds of smaller MW are preferred because of the relative proportion of the transition metal. The ultrafine powder for the second component has an average particle size within the range of 5–1000 μm and is employed as a coke transport vehicle for preventing excessive coke deposition in the reactor. Both organic (carbon black) and inorganic (silicic acid, silicates, alumina, titania, etc.) powders have been recommended. A carbonaceous powder is preferred for being more lipophilic and wettable by hydrocarbon oils than the inorganic substances. Typically, the Mo compound (10 wt% or more as Mo) is dissolved in a convenient solvent before suspending it in the hydrocarbon feed together with the ultrafine powder of carbon black. With respect to the feed, total concentration of the catalytic system should be lower than 20 wt%, including 1–1000 ppm of metal and 0.005–10% fine powder. Any reactor (tubular, tower, or soaker) suitable for the processing of slurry phase could be used. SOC is operated at high severity with temperatures of 480°C and pressures as high as 220 bar for achieving conversion in the range of 90%. A 3500-bpd demonstration unit has been used for process proof-of-concept [495–497]. A high capital cost has been associated with this technology [404].

3.2.3.11 Other Slurry-Phase Processes

The Japanese company Kobe Steel Ltd. disclosed a slurry-phase process [498] that uses limonite ore as catalyst in the form of a fine particulate of less than 2 μm in size. As already mentioned, a mineral ore has been considered an economical solution for processes in a once-through mode; however, in this case a portion of the catalyst is recycled to the reactor, together with a portion of the unconverted bottoms [499–502].

In collaboration with the University of Petroleum China, PetroChina Company Limited has developed a slurry-phase process using a multimetallic liquid catalyst [336]. The preferred metals are nickel, iron, and molybdenum [503]. Both water-soluble and oil-soluble compounds have been tested [338,504–509]. An online mixer at the entrance of the reactor is used for the low-temperature sulfurization of the catalyst [510–512]. The process is carried out at moderate temperatures (420–460°C) and low pressures (80–120 bar). Pressure and gas holdup effects have been experimentally tested and simulated [513–516]. Product is recovered through a series of high-/low-pressure separators and the light fraction is hydrotreated in a fixed bed reactor. Recycle of the unconverted bottom is also considered [517]. Conversion ranges between 80% and 95% [404].

Although catalyst deactivation is comparatively low, the main causes of deactivation are coke and metals deposition, thermal aging, and attrition.

4. Integrated Processes and Combination of Available Technologies

There are numerous ways of combining the existing technologies for processing the most diversified types of crude oils and/or rendering the most valuable products. The combination of C-rejection with H-addition, and that of different categories of HDP (HDT, HCK, and HDC) processes, has already been mentioned and considered in previous sections. In this section, different ways of integrating SDA with the other refining processes will be reviewed.

SDA developmental activities have been paving a way to increase benefit in refinery operations, as is reflected from the numerous patents that keep being awarded. However, another way of interpreting this large number of patents is as a strategic measure for keeping alive ownership of old technologies in a new scenario with increasing share of heavy and extraheavy oils, bitumen, and tar sands. This will become more apparent throughout the following discussion. In these recent patents, process improvements are not significantly represented. Instead, these patents are more concerned with integration and/or incorporation of the old SDA processes with existing units.

Exergy and exergoeconomic analysis has been used to set the basis for optimization of a refinery configuration and detect opportunities for maximizing refining benefits [518]. In this referenced work, the analysis was only applied to the CDU and VDU; the application of these methodologies on defining the integration of other units would provide means for optimizing configuration alternatives. Additionally, models and simulations have been designed and run to predict operational conditions for the units of the integrated configurations (see, for instance, Refs. [519,520]).

In the case of SDA, there are also multiple ways of integrating this type of unit with other refining processes. In Fig. 5.14 some examples are shown. In this drawing and for clarity purposes the SDA unit was placed downstream the DUs; however, and as it will be discussed later, there are also possibilities of upstream integration.

As can be seen, an SDA unit could be fed either by the AR or the VR and the asphaltene fraction can be fed to a thermal conversion unit (TCU). The total liquid product or the bottoms from the TCU could also be fed to another SDA unit (or recycled to the first). The DAO products might be hydrotreated or not (depending on quality) and VGO fractions (combined or not with DAOs) fed into an FCC unit.

4.1 Separation/Fractionation Units

The efforts of Chinese organizations, namely, Sinopec, PetroChina, and China University of Petroleum, described earlier are clear examples of SDA integration with the fractionation units (see Section 2.1.6 and Refs. [70,75,157,521]). Nevertheless, there was prior work reported in the patent literature concerning integration and operation of the DUs for maximizing the DAO yield. In a patent [132], Exxon characterized the VDU as a short residence time unit that could be of a wiped-film type.

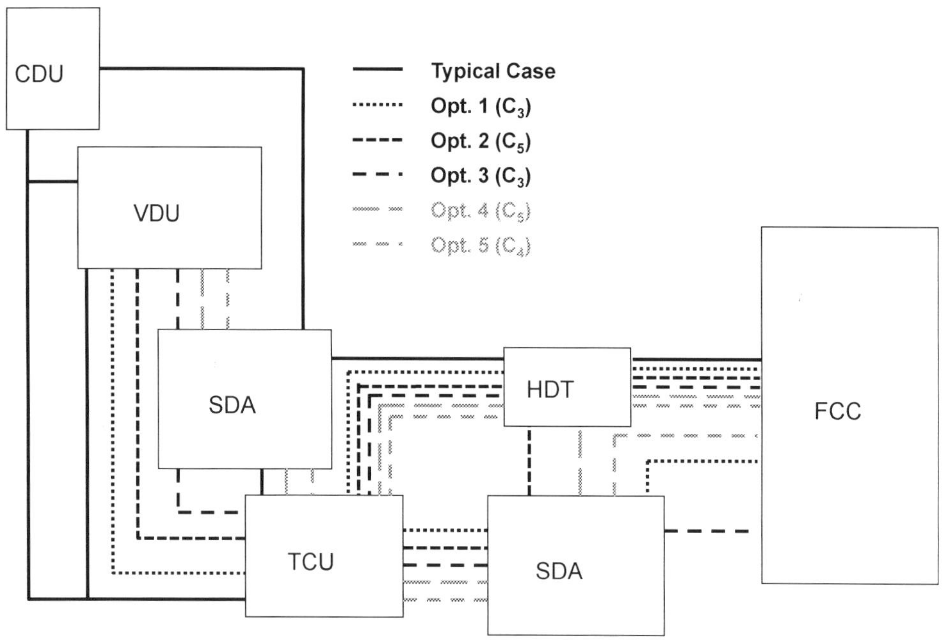

Figure 5.14
Examples of solvent deasphalting (SDA) integration with conventional refinery units. *CDU*, crude distillation unit; *FCC*, fluidized catalytic cracking; *HDT*, hydrotreatment; *SDA*, solvent deasphalting; *TCU*, thermal conversion unit; *VDU*, vacuum distillation unit.

ROSE has been recommended as a substitution of the VDU, as well as its integration with downstream units. The operating conditions of the SDA unit depends on the conditions of the CDU, and/or vice versa [50,52,522–524].

ORMAT, an Israeli company devoted to heat integration and power generation, defined a complex refining scheme that combined its newly developed SDA [143,153,525] with other refining units by the end of the 1990s [158,526,527]. In this process, a CDU produces two fractions: the distillate (considered to be asphaltene-free and metal-free fraction) and the residue fraction (containing sulfur, asphaltenes, and metals). An HDT unit produces a hydrogen donor component from the distillate and the residue fraction is treated in an SDA unit. A blend of the DAO (produced from the SDA-treated residue) and the hydrogen donor component is subjected to thermal cracking (TCU). The HDC is intended for suppressing the formation of thermal asphaltenes [528]. The thermally cracked product is recycled to the CDU and combined with the incoming crude oil [159,529]. In a later patent, this company substituted the HDT unit with a VDU for producing VR feed for the SDA unit, in the simplest configuration that keeps a TCU for cracking a blend made from the distillates of all DUs. More complicated configurations including one or more TCUs, HDTs, and distillation columns were claimed as well

[530,531]. For instance, the DAO can be fractionated and the top and bottoms sent to two separated TCUs; even conventional tanks were suggested as TCUs [532]. The asphaltic fraction, named A-fuel, is fed to a gasifier [533–535], which fulfills various objectives: power, hydrogen, and pitch use, etc. Both equipment and process schemes for the integration of SDA with fractionation via thermal cracking were disclosed by Ormat as well [529,536]. This company created a Canadian subsidiary, under Ormat Process Technologies, Inc. (OPTI), with the purpose of commercially demonstrating the technology with Canadian heavy oils, bitumen, and tar sands. The process technology became known as OrCrude. In 1999, an agreement was reached with Suncor and a 500-bpd unit was built and started up in 2001 for processing crude produced by this latter company using steam-assisted gravity drainage (SAGD), to which the steam produced in the gasifier can be incorporated [537]. The complete integrated scheme is shown in Fig. 5.15 [538].

Suncor's alliance did not last long and the same year Nexen took over its interests in OPTI. By 2006–7, phase I of Long Lake Project, a 72-kbpd OrCrude-based facility for producing 58.5 kbpd 39 degrees API SCO, was expected to start producing revenues. The stake jointly owned by OPTI and Nexen was assessed to contain 14 billion bbl, the production of which would not decline for over 40 years once started. In 2008, OPTI sold an additional 15% of its stake to Nexen to make an OPTI–Nexen 35–65% partnership [539], and in 2009 the launching of operations with the initiation of the SAGD production in Long Lake was announced [540]. However, at the same time, a slowing down of the plans for the project's second phase was also announced stating the global credit crisis as the main cause. Later that year the disappearance of OPTI was forecast by financial analysts [541]. Then, in 2011, CNOOC Luxembourg S.à r.l., an indirect wholly owned subsidiary of CNOOC Limited. acquired the whole company. In January 2016 the

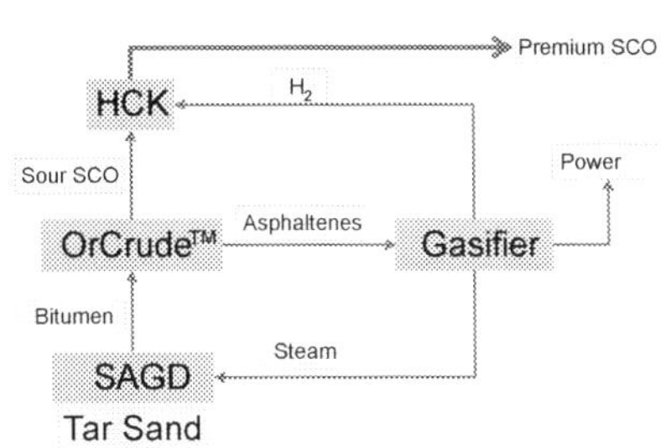

Figure 5.15
Nexen–OPTI upstream–downstream integrated scheme. *HCK*, hydrocracking; *SAGD*, steam-assisted gravity drainage; *SCO*, synthetic crude oil. *Reproduced from Birdgeneau J. The Canadian Long Lake project integrating SAGD, cogeneration and upgrading technologies to develop Athabasca bitumen. In: Proc. 19th world pet. congress, vol. 1. Madrid, Spain; June 29–July 3, 2008.p. 436–443, with permission from World Petroleum Council.*

shutdown of Long Lake Nexen–CNOOC facilities was announced,[5] because of an explosion in the HDC unit and a previous leak in the pipelines; resumption of operations has not been reported yet.

4.2 Conversion Units

The combination of SDA to conversion processes also includes the supercritical treatment of cracked feeds from thermal and catalytic units, i.e., upstream and downstream SDA integration. The integration of ROSE (e.g., Ref. [542]) technology with conversion units has been tested at both pilot [189,543,544] and commercial scales [545–547]. Integration of ROSE with a visbreaker was considered by the original developers of the technology [548]. Under these conditions, the stability of the products is important and in some cases a hydrogen addition process is required downstream. Although the need for hydrotreating the DAO prior to the conversion unit may be seen as an economical drawback, the imposed regulations will demand deep HDT for the fuels derived from heavy feedstocks prior to reaching the fastest-growing markets.

Shell disclosed two patents that combine SDA with thermal cracking, inverting the sequence in which the units are integrated. In the first patent, SDA is carried out in a first step (at conditions to yield up to 90% DAO), and the obtained DAO is thermally cracked [549]. Meanwhile, in the second patent, cracking is carried out first, with a 35 wt% conversion of the 520°C$^+$ fraction [550] and then the residual fraction is deasphalted. In a very old patent, Exxon also proposed a similar process, in which a visbroken residue is deasphalted [106]. In another process, the 510°C$^+$ residue fraction from a short vapor contact time thermal process unit is treated in an SDA unit and the pitch is recycled to the thermal unit by mixing it with the residue feed [551]. A variation of this process (already mentioned) includes a horizontal moving bed of fluidized and/or stirred hot particles in the short vapor residence time process unit [151].

Mobil disclosed two processes, which integrate SDA with catalytic cracking. In the first patent, relatively low efficiency DA is used to leave at least 10% of the asphaltenes and 10% of the solvent. The demetallized DAO is catalytically cracked and the solvent used in the SDA unit is derived and recovered in/from the FCC unit [552]. In the second patent, the whole crude is deasphalted by extracting with an aromatic solvent and then precipitated with a paraffinic solvent to obtain three fractions: deasphalted crude, resins, and asphaltenes. The deasphalted crude is distilled and the VGO fraction is fed into an FCC unit [553].

[5] See Bloomberg announcement at http://www.bloomberg.com/news/articles/2016-01-16/cnooc-s-nexen-stops-long-lake-oil-sands-work-after-fatal-blast, and Nexens' website at http://www.nexencnoocltd.com/en/Operations/OilSands/LongLakeExplosion.aspx

A combination process integrates a ROSE SDA with catalytic cracking. The SDA-produced DAO is either fed to a residue cracking unit (RCC) or to an FCC depending on the feed specifications of those units [554]. The asphaltenes fraction is blended with the extract from a lube plant and with the asphaltic fraction from the VDU, then the mixture is further processed through the asphalt units to meet road pavement specifications (Strategic Highway Research Program in the United States) [555].

A DM unit has been integrated with a residue conversion process (RCC/FCC) for reducing the relatively large amount of metal contaminants originally present in the VR. Two cracking units are used: the first catalytic cracking zone is operated at temperatures of 465–500°C with a metal-contaminated regenerated catalyst, and controlling temperature and performance by adjusting the catalyst-to-oil ratio. The distillate oil fractions containing significantly smaller amounts of metal contaminants are processed in a second catalytic cracking zone using regenerated, clean catalyst at a catalyst-to-oil ratio to keep the reaction temperature in the range of 475–525°C. The produced cycle oils are solvent extracted, and the raffinate from the solvent extraction step is recycled to the second catalytic cracking zone while the extract from the solvent extraction step is fed to the first catalytic cracking zone and mixed with hot, regenerated catalyst in a separate zone admixed at a catalyst-to-oil ratio such that temperature of the extract feed reaches 510–565°C [556].

Crude oil DM and liquefaction of coal is achieved by coprocessing through a two-stage hydrovisbreaking system (H_2 pressure of 50–1500 psi), integrated with an SDA unit. The oil and coal are visbroken separately at an optimum severity to minimize coke yield. The oil is heated in a visbreaking zone at c. 425–535°C for a residence time of 0.1–2 h. The effluent is mixed with particulate coal (1.5–10:1 wt ratio) and the mixture is heated in a second visbreaking zone at 370–455°C for a residence time of 0.1–2 h. Light ends are removed and the second-stage visbreaking effluent is subjected to SDA. The DAO is distilled to remove the DA solvent and yield a demetallized liquid hydrocarbon product. The asphaltic solids are subjected to flotation in an aqueous medium to separate organic solids from inorganic ash. DM was exemplified as being 76% [182].

4.3 Hydroprocessing

In general, integration with HDP is probably the most prolific application, considering that HDP includes the broad range of hydrogen-treatment processes, from a mild hydrogenation to HDT, mild HCK, HCK up to deep conversion (HDC). As mentioned earlier, depending on the SDA solvent and conditions, most of the time the produced DAO needs to be hydrotreated prior to being fed to an FCC unit, but in many cases refineries are already configured to hydrotreat the FCC feed. Some of the patents that explicitly include the HDT step have been discussed in the previous sections.

Two early and separate discoveries (pioneered by UOP and Exxon, prior to the ExxonMobil merge) led the way toward additional concepts for processes and integrated schemes. On one side, oil-soluble compounds of hydrogenating metals (see Section 3.2.3) were found to decompose and formed sulfides (with the S-compounds of the crude oil) under HDC conditions that could exhibit remarkable catalytic properties [310,313–315,358,360,394,395,557,558]. On the other side, it was found that the presence of coal particles synergized with the conversion of heavy oil [393,396]. Then, in an additional concept, the asphaltenes from the SDA unit were calcined into a sort of microcoke that in combination with a hydrogenation active phase was used as HDC catalyst in a slurry bed (slurry hydroconversion, SHC) [559]. This type of process (Fig. 5.16) could be used to upgrade the asphaltenic bottoms of the SDA unit and also the unconverted SHC fraction can be recycled to the SDA unit [560]. The additional incorporation of a coking unit (CKU) provides means for thermally converting the pitch and/or providing a metal-contaminated coke material suitable for the SHC or HDP catalytic system [561,562]. Even if the VR is first hydrotreated, the heavy bottoms can be converted in the CKU to provide a metal-enriched coke, useful as a catalyst. This type of SHC that incorporates an oil-soluble compound of hydrogenating metals and a coke or coke precursor was branded by Exxon as MicroCat-RC (see Section 3.2.3.4) technology and technical details were given in conference presentations [249,392].

The IFP disclosed a series of patents for integrating the SDA units into the hydrotreating/HCK operations in the refinery. The first two patents only differ in using HDT or HCK in the last stage. One example scheme is presented in Fig. 5.17A, corresponding to Ref. [563], for which the last stage is an HDT process unit. In another example, the last stage is an HCK unit [564]. A variation of this last patent includes an HDM stage in the multireactor–HCK first unit [565]. In another multireactor–HCK system [566], a portion of the heavier liquid

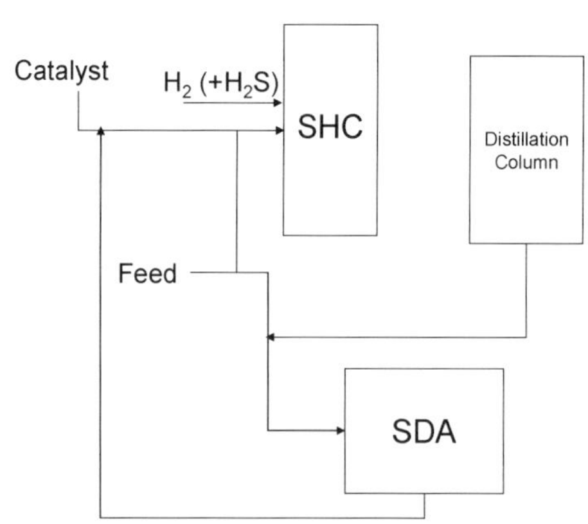

Figure 5.16

Solvent deasphalting (SDA) as a source of catalyst precursor and its integration with slurry hydroconversion. *SDA*, solvent deasphalting; *SHC*, slurry hydroconversion.

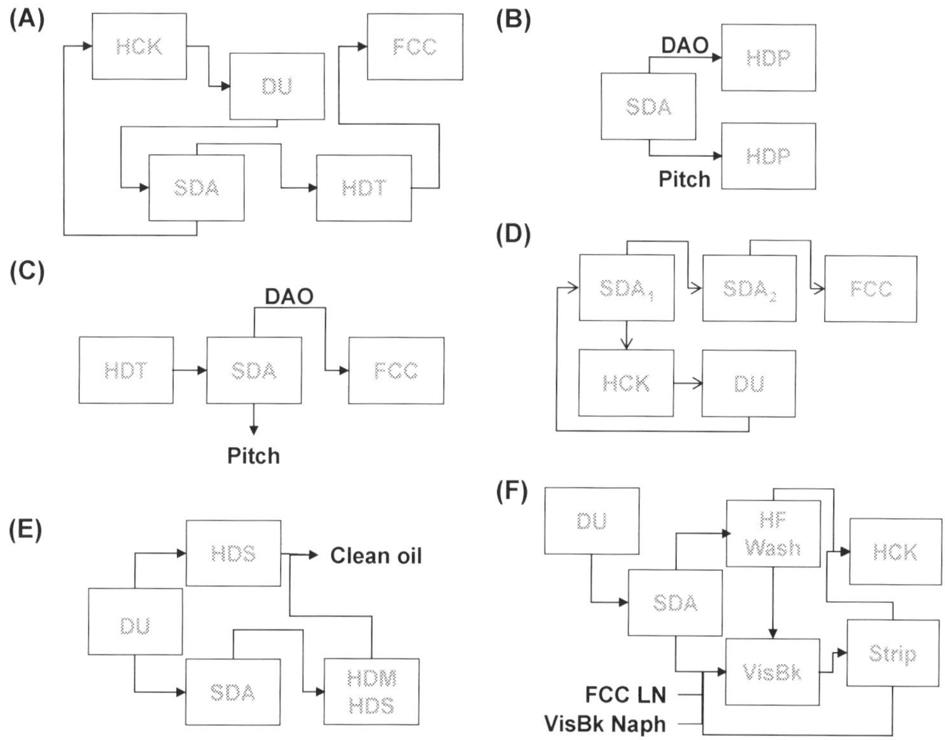

Figure 5.17

Summary of examples of hydroprocessing–solvent deasphalting (HDP–SDA) integrated schemes: (A) IFP fixed bed systems; (B) IFP ebullated bed systems; (C) Texaco; (D) Phillips Petroleum; (E) JGC Corporation; and (F) Chevron. *DAO*, deasphalted oil; *DU*, distillation unit; *FCC*, fluidized catalytic cracking; *HCK*, hydrocracking; *HDM*, hydrodemetallization; *HDS*, hydrodesulfurization; *HDT*, hydrotreatment; *HF*, hydrofluoric acid; *IFP*, Institut Francais Du Petrol; *LN*, light naphtha; *SDA*, solvent deasphalting; *VisBk*, visbreaking.

fraction recovered from the hydrotreater is sent to a catalytic cracking section. In another process that includes sequential HDM–HDS, initial hydrovisbreaking and/or final DA may act in a complementary manner [567]. In an integrated (SDA–HCK–RCC/FCC) scheme, the SDA unit is set to produce a DAO with an asphaltenes content below 3000 ppm. This DAO is hydrocracked in a bubbling bed reactor. The HCK product is fractionated and the residue from fractionation is then cracked [568]. Finally, two other IFP-awarded patents incorporate pitch HDP in an ebullated bed reactor downstream VR SDA unit [569,570]. First, the VR is sent to an SDA unit operated with C_4/C_5 solvent and then the produced DAO and asphaltene fractions are treated in separate (ebullated or fixed beds) HDP processes (Fig. 5.17B) under corresponding adequate conditions [571].

Phillips Petroleum disclosed an integrated scheme based on HDT–SDA–FCC (Fig. 5.17C). The heavy hydrocarbon oil or fraction is hydrotreated to remove part of the sulfur and metal contaminants. Additional removal of metal and sulfur contaminants is then accomplished in an SDA unit, which produces a DAO that is fed into an FCC [572]. In another series of patents for process configurations, two to four SDA steps are integrated with HDP and FCC in different ways [573–581]. The HDP unit might be an ebullated bed [574]. The product from the ebullated bed reactor is fractionated into three fractions: high, medium, and low boiling point. The high-temperature component is treated through an SDA unit to provide oil, resin, and asphaltene fractions [580]. An example is shown in Fig. 5.17D for a two-SDA-step process. In this example, the DAO from the first SDA step is fed to the second SDA step via a heat exchanger. The pitch from the first SDA step is fed to an HDP (HDT or HCK) unit and then to a fractionator (DU). The fractionator bottoms are recycled to the first SDA step. The bottoms from the second SDA step might be fed into a catalytic cracker or processed elsewhere [578–580]. The DCO from the FCC unit [573] might be fed to one of the SDA units [575–577].

A refining scheme from JGC Corporation comprises integration of DU, HDT, and SDA (Fig. 5.17E). The crude oil or heavy fraction is fractionated in two cuts: a distillate and a residue. The distillate is subjected to HDT to yield hydrotreated distillates. The residue fraction is fed to the SDA unit; the obtained DAO is the feed for an HDM–HDS unit, producing a refined oil that may be blended with the hydrotreated distillate or further processed downstream [582]. This process concept has been branded as GEfinery and details of its application for the manufacture of turbine fuels were reported [583]. Specifications of such a fuel require a V-content of less than 0.5 ppm, Na + K below 1 ppm, and viscosity lower than 20 cSt (@135°C).

Chevron disclosed an integrated pretreatment scheme for the production of hydrocracker feed from the higher boiling point fraction [421]. The VR is fed to an SDA unit, from which the DAO is washed with hydrofluoric acid and fractionated into two fractions, a light and a heavy. The heavy HF-treated DAO is combined with light FCC naphtha and visbroken naphtha and fed into a visbreaker. The visbroken product passes to a pitch stripper. The liquid visbroken product is combined with the light HF-treated DAO and fed to an HCK (Fig. 5.17F).

Texaco integrated an (C_5–C_7) SDA unit downstream a VR HCK for deasphalting the unconverted bottoms of the hydrocracked VR. The unconverted HCK bottoms are deasphalted at ambient or elevated temperature and a pressure sufficient to maintain the solvent in the liquid state. A performance improvement of the SDA–HCK system was observed by recycling the DAO to the HCK [584].

In Shell's awarded patent, an SDA–HDM–HCK process sequence produces a demetallized DAO in the first two stages that can be further processed through HCK under the desired level of conversion. Once the DAO is demetallized on an HDT bed of a HDM catalyst, it is blended with the distillate fractions feeding the hydrocracking unit [585].

ENI has branded the EST upgrading process described earlier (Section 3.2.3.3) [376,380,586], and later integrated it with an SDA unit in a similar way to that of the MicroCat-RC (Section 3.2.3.4), also discussed earlier (Fig. 5.16). An example of this process combines the use of catalytic SHC, distillation or flash topping (DU), and SDA [383,385]. One fraction of the feedstock is sent to SDA and the obtained asphalt is mixed with a hydrogenation catalyst and with the remaining fraction of the feed. This mixture is fed into the SHC reactor together with hydrogen or a mixture of hydrogen and H_2S. The hydroconverted product is sent to distillation or flash steps (in a DU). About 95% of the distillation residue containing the catalyst and possibly coke is recycled to the SDA unit. ENI differentiate from this patent since the SDA unit is used for catalyst recovery (and recycling) from the bottoms of the SHC effluent. The bottoms containing unconverted feed and spent catalyst is sent to a solvent treatment section for the separation of the product into a solid fraction and a liquid fraction. The liquid fraction is combined with the SHC liquid product and the solid fraction is treated and recycled [384]. A published patent application concerns the processing of coal-derived liquids with this type of technology and incorporates the use of the seven process units. It has been described as consisting of coal liquefaction (CL), flash, or distillation of the product obtained from the liquefaction (F), extraction with a solvent to remove the ashes (SDAsh), distillation to separate the solvent (RS), hydroconversion with catalysts in SHC, distillation or flash of the product obtained from the SHC reactor, and SDA [587]. This process sequence can be summarized as: CL− F−SDA−DU−SHC−DU−HDC−SDA.

A combined process includes a preconditioning stage in which the crude oil is heat soaked prior to (two-stage) HDT with a dispersed metal catalyst to reduce coking. The effluent of the hydrotreating process is filtered to recover catalytically active coke, which is recycled and mixed with the feed stream. A mild SDA step using heptane recovers metals in a reduced volume asphaltene fraction from the treated oil [588].

4.4 Coking and Gasification

As can be anticipated from the previous sections, the combination of a CKU with SDA can be practiced in either sequential case, i.e., SDA−CKU or CKU−SDA. Some examples are discussed next.

A low-conversion SDA is used to remove the asphaltenes from delayed coking feed, lowers the impurities content of the produced coke (needle grade), and increases the liquid yields [70]. Similarly, a ROSE SDA unit is employed to recover asphaltenes from a low-severity CKU that operates at a higher severity than a visbreaker, but lower than delayed coking [188].

The DAO yield was increased by deasphalting the fractionated residue from a CKU [133] or a fluid coking unit (FCKU) [589]. In this kind of process, the operability of the deasphalter was improved by mixing the residue with an aromatics solubilizing aid to prevent the formation of an immiscible third phase during DA [133]. The residue from the FCKU is separated in a low CCR stream and a high CCR stream. The first stream can be fed into an FCC unit, while the second is recycled into the FCKU [589].

An integrated oxidation/mild coking SDA process scheme performs with high levels of DM and low rejection of solids to provide an upgraded crude [590]. The oxidation stage (thermal or catalytic) adds 0.5–3.0 wt% oxygen to the crude. Then, the oxidized oil is fed to a coking stage at temperatures of 370–480°C, pressures of 15–35 bar, and LHSV of 3–5 h^{-1}. The formed paraffins in the coker induced deasphalting of the converted oil that is let to settle in a third stage.

A heavy crude oil processing scheme includes a distillation zone(s), an SDA unit, a high-pressure air partial oxidation gasifier, a shift reactor, and a purification step to produce 99.9% pure hydrogen. The integration consists of using the produced H_2 for the treatment of the hydrocarbons [591].

In most of the forthcoming patents, the SDA unit is employed as an asphaltenes purifying step for preparing gasification feeds, which requires the removal of inorganic solids, particularly catalyst fines.

Texaco disclosed a process involving heat integration of an SDA unit with a gasification process, mentioned earlier [160,161]. The solvent is recovered by vaporization from the DAO/solvent mixture using heat from the gasification unit. The vaporized solvent is used for heating the aforementioned solvent solution and pre-heating the resin-free DAO/solvent mixture. The heating step of solvent solution and the pre-heating step of the DAO/solvent mixture are usually done in series; i.e., the DAO/solvent mixture is first preheated. The vaporized solvent fraction is then used to provide the heat to the solvent solution prior to the resin separation step. A combined process comprises contacting an asphaltene-containing hydrocarbon liquid with an alkane solvent and removing the solids. Additional solvent is added, and the mixture is heated until asphaltenes precipitate into a separate phase. The asphaltenes are removed by heating with the steam produced in the gasification reaction [161] and the separated asphaltenes are then gasified [592]. In another patent, an asphaltene partial oxidation unit is integrated with an SDA unit by using part of the gasification heat for solvent recovery. The precipitated asphaltenes are gasified to synthesis gas [593].

4.5 Integration With Other Segments

SDA technology competitiveness could be enhanced with a more effective process. Probably that is the reason why integration with power generation or with the oil production segments has become increasingly considered in recent times, though the number of patents does not reflect the magnitude of the effort this area is actually receiving.

4.5.1 Upstream Integration

As mentioned earlier, the whole value chain of the industry is affected by the presence of asphaltenes in (heavy and extraheavy) crude oils, bitumen, and tar sands. Asphaltene deposition in production facilities, i.e., downhole locations, subsea systems, pipes, valves, and/or surface units, creates plugs, causing pressure build-up that disrupts operation and typically requires shutdown. In the crude oil at reservoir pressure and temperature, asphaltenes are undisturbedly solubilized or dispersed. During production, pressure drops and asphaltene solubility decreases, causing instability to increase, flocculation to start, and eventually asphaltenes to precipitate. All these problems have prompted R&D initiatives for removing and/or converting the asphaltenes as close as possible to upstream facilities.

Facing the problems from the upstream sector might facilitate (and probably resolve) the problems created for transportation and even for refining. However, oil fields are not equipped to perform demanding refining tasks and simple technologies for quick treatments are needed. Emerging concepts and new ideas will be discussed in the next chapter; in this section only those considering conventional existing technologies and involving at least one SDA unit will be considered.

Arco integrated the production from a subterranean formation through a wellbore with an SDA unit. The heavy oil is separated by distillation and the AR is fed to an SDA unit. The asphaltic residual portion is combusted in a fluidized bed to produce heat, steam, electricity, and/or synthesis gas [594]. Distillates can be used for dilution of the unconventional oils to facilitate transportation. This latter mode of SDA integration with the value chain of Canadian crudes exploitation, via synthetic/diluted bitumen (synbit or dilbit), was suggested as a more economical alternative compared to conventional coking upgrading [595]. The convenience of this type of upstream processing versus performing it within the refinery battery limits has been addressed [596].

A DA process was integrated with production facilities for the simultaneous DA and emulsion breaking of the wellhead emulsions (heavy oil—water). The separated asphaltene fraction serves as a vehicle for removing the heavy metals; meanwhile, the light metals (soluble salts) would be retained in the aqueous phase from the broken emulsion. The process comprises: (1) breaking the emulsion with a C_4—C_7 paraffinic solvent (solvent:oil ratio = 2:5) and coalescence of emulsion water and asphaltic as fast-settling particles, (2) separating the water—asphaltic particles from the bulk oil—solvent phase; (3) feeding the separated particles and residual solvent into a hot water bath to form stable asphaltic agglomerates with exclusion of emulsion water; (4) flashing off and recovering the residual solvent; (5) removing the agglomerates from the hot bath; and (6) recovering the DAO from the oil–solvent phase [597].

4.5.2 Integration With the Energy Grid

Since most of the gasification units are used for power generation purposes, some of the integrated processes discussed earlier might be considered as part of this section as well.

Heavy oils, bitumen, and pitch can be emulsified to facilitate transportation (see, for instance, the case of Orimulsion production) [598]. Energy generation is the main end use of such emulsions, though burners and boilers have to be modified and environmentally adapted [599].

The solvent vapor produced by a flash drum in an SDA unit can be fed to an organic Rankine cycle turbine for power generation [600]. The expanded solvent vapor is combined with a stream of solvent vapor and steam from the stripper of the SDA unit. Heat is extracted from the combined stream by an organic fluid that is vaporized and supplied to a Rankine cycle organic vapor turbine. The solvent is recovered from the cooled, combined stream and reused in the SDA unit [600].

An SDA unit is integrated with a generating power system. Residual FO is deasphalted with process steam, and three streams are produced: DAO, pitch, and DA condensate. DAO is burned in a pressurized oxygen-bearing gas to produce a pressurized hot gas stream. This pressurized hot gas stream is expanded in a turbine that produces shaft power and an expanded gas stream. The expanded gas stream is cooled by transferring heat to water and forming steam. A portion of the steam is used in the SDA unit [601].

In an SDA apparatus, the energetic content of solvent streams is employed in power generation. This power generator generates power in response to the flow of solvent, and includes a vaporizer for vaporizing solvent in the bypass line, an organic vapor turbine responsive to vaporized solvent for expanding the vaporized solvent, producing expanded vaporized solvent, and generating power, a condenser for condensing the expanded vaporized solvent and producing condensed solvent, and a pump for returning the condensed solvent to the solvent drum [525].

5. General Remarks on Upgrading Technologies

The molecular complexity of residues or heavy oils and the issues created upon refining processes have been discussed in previous chapters. The upgrading of these feeds represents enormous technical challenges to the refiner that ends up negatively impacting the margins. For conventional technologies, either economy or environment is severely jeopardized. Current existing technologies, operating commercially, render an unconverted or not-convertible heavy fraction product (oil residues). Initially, the bottom liquid fraction was marketed as FO, but this is no longer an option. The total production volume of this low-quality FO amounts around 9 million bpd, a number comparable to Saudi Arabia total oil production. Furthermore, more than 60 million tons/year of pet coke (coking by-product) are accumulated all over the world. Both of these products, coke and low-quality FO, negatively impact the environment, and for this reason they have been increasingly limited for use as fuel in power stations. In some instances, pet coke is disposed in landfills, but again this is becoming an inconvenience.

C-rejection and H-addition approach upgrading from different perspectives, with different processes, resulting in very different product slate and economics. In Table 5.5 a comparative summary is presented [2,602,603].

A special case of C-rejection is SDA, which can be regarded as a physical separation process rather than a chemical conversion process, such as those compared in Table 5.5. SDA is a mature technology that has been improved through time, particularly in terms of improving efficiency, not only for efficient removal at the DA step, but also for the solvent recovery stage. New solvents and/or substitute solvents and additives have been proposed, but none has been in use. The current practice has two modes of operation; the preferred mode would depend on refinery capabilities for FCC feed pretreating and/or HDP facilities (HDT and HCK). One mode is maximizing asphaltene yield to optimize DAO quality for cofeeding the FCC unit, in which case the preferred solvent is propane, but could be *n*-butane. The second mode is maximizing DAO yield, in which case high-quality asphaltenes are produced; the choice of precipitant should be *n*-heptane, but availability might lead to *n*-pentane. Nevertheless, even the *n*-heptane precipitate always contains

Table 5.5: Comparative Summary for C-Rejection and H-Addition Technologies [2,602]

	C-Rejection			H-Addition			
	Visbreaking	Delayed Coking	Fluid Coking	Fixed Bed	Moving Bed	Ebullated Bed	Slurry Bed
Objective	FO viscosity reduction	Conversion, coke make	Conversion	Conversion and heteroatoms removal			
Operability	Good	Good	Moderate	Good	Difficult	Difficult	Difficult
Maximum conversion $550^\circ C^+$, %	10	>90	>90	60–70	60–70	80	90
Temperature range, °C	470–485	450–520	510–570	400–450	400–450	420–450	450–480
Pressure range, psig	50–200	80–100	10	1800–2000	2500–3000	1500–3000	2000–3000
Tolerance for impurities	High	High	High	Low	Low	Average	High
Maximum (Ni + V) in feed, ppm	>700	>700	>700	120–400	500–700	>700	>700

chemisorbed and *n*-heptane-insoluble resinous and other maltene materials (carboxylic acids, fluorenones, fluorenols, polycyclic terpenoids, thiolane- and thiane-derived and acyclic sulfoxides, carbazoles, quinolines, vanadyl porphyrins, etc.) as well as low-MW asphaltene fragments that will impact downstream operations.

The cheapest of the C-rejection processes, visbreaking and hydrovisbreaking, result only in viscosity improvement. Feed quality is unchanged, since all contaminants remain in the product. For crude oil transportation purposes, a reduction in diluent requirements is beneficial, but they will not eliminate the need for dilution. A slight improvement in quality is achieved by hydrovisbreaking processes, and more specifically with the AQC process. The coking processes produce high levels of olefins and these streams have to be routed to HDT units in which catalysts with improved hydrogenation are needed and H-consumption would consequently increase.

Inclusion of technical considerations into the economical comparison between C-rejection and H-addition requires considering: (1) conversion in HDP, (2) availability (and consumption) of hydrogen, (3) coke disposal—market value, and (4) product margins (diesel vs gasoline market). The main difference in product yield is the production of (better-quality) FO in HDP versus low-grade coke in C-rejection as shown in the example of Fig. 5.18 (data were taken from Ref. [604] and refer to delayed coking and ebullated bed HDP).

Furthermore, the quality of all products from HDP is always superior to that of any C-rejection process [604,605]. Coking or C-rejection in general would become the

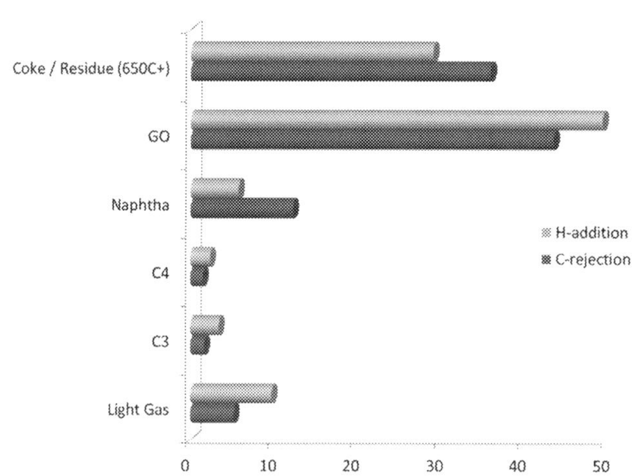

Figure 5.18

Yields comparison for C-rejection and H-addition technologies. *Reproduced from Sayles S, Romero S. Comparison of thermal cracking and hydrocracking yield distributions. PTQ 2011;(3):9 pp. Paper #1000070, with permission from PTQ.*

technology of choice within a low crude oil prices and high natural gas prices scenario. Low crude prices favor technologies with low investments and low operating cost. In Fig. 5.19, the operating cost for representative C-rejection and H-addition processes (as plotted in Ref. [403]) is presented. The current refiners' preference for C-rejection seems to be justified. However, under a low-grade petroleum coke surplus (or a decreasing demand), high crude oil prices, and low natural gas prices, a need for hydrogen addition would develop. In this regards, both CCR and metal content would determine the technology choice for HDP.

Currently, fixed bed, moving bed, and ebullated bed reactors are commercially used for heavy oils and VR upgrading [28,251,260,403,404,606–609]. The deactivation problems found with fixed bed systems were thought to be resolved by moving beds. Certainly, moving beds represented an improvement over fixed bed systems by offering an increase in conversion and liquid yields; however, back mixing, high operating/investment costs, complex operation, and sediments formation limited their application. These problems seem to have been solved by slurry-phase reactors, but before massive commercialization is seen, other issues need to be sorted out. An ebullated bed HDC technology is always more complicated than any slurry-phase reactor, while the latter is comparatively less prone to fouling. In general, slurry reactors offer other numerous advantages for the treatment of the more contaminated feeds, such as higher flexibility to product selectivity and yield, more flexible operation and higher reliability, high space velocity and

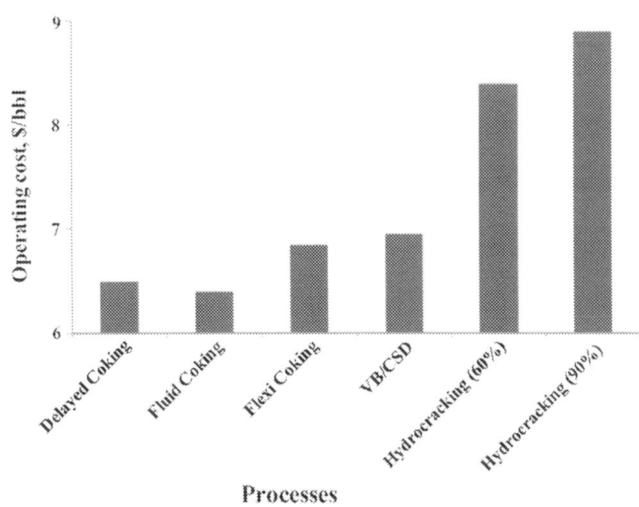

Figure 5.19

Operating cost for representative processes of C-rejection and H-addition. *VB/CSD*, visbreaking-combined with SDA. *Reproduced from Sahu R, Song BJ, Im JS, Jeon Y-P, Lee CW. A review of recent advances in catalytic hydrocracking of heavy residues. J Ind Eng Chem 2016:13 pp. http://dx.doi.org/10.1016/j.jiec.2015.01.011 [in press], with permission from Elsevier.*

conversion rates (higher throughput), no bed-plugging problems, minimum catalyst deactivation, and a broader adaptability to different feedstocks. Its integration with other conventional technologies results in an attractive product price structure. In fact, the flexibility of slurry-phase HDC allows the processing not only of heavy oils and residues, but also of streams of difficult disposal, limited application or low value, such as SDA pitch and FCC DCO. Nevertheless, the development of slurry-phase HDC processes has made it through small scale, but commercial operations have moved forward very slowly. Besides the high investment costs, other disadvantages of these processes are responsible for obstructing the path to commercialization, particularly the complex operability and the difficulties regarding catalyst recovery and recycling.

There is no doubt about the effectiveness of the conventional existing technologies. However, the negative technical/economical/environmental impact calls for more creative and innovative solutions. Licensors are stuck with offering stand-alone process units, combinations of these, and integrated versions of such combinations. Under the current scenario, it seems refiners are reluctant to invest in these and commercialization has moved enormously slowly. Nevertheless, it could be expected that sometime in the future more and heavier crude oil will be marketed, and refiners would be forced to incorporate the required conversion technology.

The increasing difficulties and the tougher environmental regulations limiting the specifications of transportation fuels and restricting the allocation of high-sulfur residues in the fuel market would drive R&D future trends to improving process economics. The evolution of existing technologies has been biased by the volatile behavior of oil prices. R&D interests peak and decline as refining investments come and go, regardless of the conscious perception that without basic and fundamental knowledge, technological advances would be just dreams. The refiner is seeking for cost-effective improvements or cost-reducing modifications. However, these achievements can only be gained by a better understanding of the process chemistry and its impact on the properties of the reacting media and of the products. Fundamental understanding is specifically needed in many more aspects, such as the molecular characterization of the reacting species; pathways, mechanisms, and kinetics of the chemical reactions; factors affecting product and reaction selectivity (coke and coke precursors formation, fouling species, poisons, etc.); process thermodynamics; fluid-phase behavior at process conditions; intermediate and final products stability and reactivity; property—structure relationships, etc.

It is clear that further research is needed toward the development of new or improved process technology. Although significant advances have been observed in areas such as molecular characterization, incorporation of this knowledge into existing technologies does not seem to have taken place. A large number of publications populate the literature reporting properties of feeds, catalytic properties (activity, selectivity, performance, deactivation, regeneration, etc.), kinetics, modeling, reaction mechanisms, operating conditions, yields, etc. Translating all these findings into capital savings, reduction of operating costs, and/or a new cost-effective technology is required sooner rather than later. The next chapter will go deeper into the revision of those findings and propose ideas that have not reached scaling up to the commercial stage.

References

[1] Montaghi M, Ulrich B, Subramanian A. Slurry-phase hydocracking possible solution to refining margins. Hydrocarb Proc 2011;2:37−43.

[2] Fahim MA, Alsahhaf TA, Elkilani A. Fundamentals of petroleum refining. Oxford, Great Britain: Elsevier; 2010. 496 pp.

[3] Gary JH, Handwerk GE. Petroleum refining: technology and economics. New York: CRC Press, Marcel Dekker; 2001. 441 pp.

[4] Gray RM. Upgrading petroleum residues and heavy oils. Chem. industries series, vol. 56. New York: Marcel Dekker/CRC Press; 1994. 368 pp.

[5] Heck RH, Degnan TF. Fundamentals of resid upgrading. In: American Institute of Chemical Engineers, editor. AIChE symp. series, vol. 85 no. 273. New York: AIChE; 1989. 77 pp.

[6] Hsu CS, Robinson PR. Practical advances in petroleum processing. New York: Springer; 2006. 453 + 411 pp.

[7] Le Page JF, Chatila SG, Davidson M. Raffinage et conversion des produits lourds du petrole. Paris: Editions Technip; 1990. 190 pp.

[8] Self F, Ekholm E, Bowers K. Refining overview - petroleum, processes and products. NY, USA: AIChE, Fuels and Petrochemical Divisions; 2000. 212 pp.

[9] Speight JG. The chemistry and technology of petroleum. In: Chem. industries series. 4th ed. Basel, Switzerland: Marcel Dekker AG; 1999. 955 pp.

[10] Speight JG. The refinery of the future. Amsterdam: Elsevier; 2010. 416 pp.

[11] Trambouze P. Petroleum refining. Materials and equipment. Paris, France: Institut Francais du Petrole Publications/Editions Technip; 2000. 471 + 641 pp.

[12] Treccani P. Emerging technologies for the conversion of residues. In: Poli R, editor. Encyclopaedia of hydrocarbons. Roma, Italy: Istituto Della Enciclopedia Italiana; 2008. p. 137−60.

[13] Wauquier J-P. Petroleum refining V.1: crude oil. Petroleum products. Process flowsheets, V.2: separation processes. Paris, France: Editions Technip; 1995. 471 + 641 pp.

[14] Ancheyta J, Rana MS. Future technology in heavy oil processing. In: Filho PDaP, Pires AP, editors. Pet. eng.-downstream. Unesco EOLSS; 2007. 41 pp. (Available from, http://www.eolss.net/Eolss-sampleAll Chapter.aspx.

[15] Clark B, Graves WG, Lopez-De-Cardenas JE, Gurfinkel ME, Peats AW. Heavy oil. In: Facing the hard truths about energy. National Petroleum Council, Heavy Oil Subgroup of the Technology Task Group; 2007. Topic paper #22, 56 pp.

[16] Delbianco A, Montanari R. Emerging technologies for the conversion of residues. In: Treccani P, editor. Encyclopaedia of hydrocarbons. Roma: Istituto Della Enciclopedia Italiana, Fondata Da Giovanni Treccani S.P.A. Marchesi Grafiche Editoriali S.P.A; 2007. p. 137−70.

[17] Dusseault MB. Hydrocarbons from non-conventional and alternative fossil resources. In: Treccani P, editor. Encyclopaedia of hydrocarbons. Roma: Istituto Della Enciclopedia Italiana, Fondata Da Giovanni Treccani S.P.A. Marchesi Grafiche Editoriali S.p.A; 2007. p. 21−55.

[18] Hydrocarbon processing *refining process handbook*. Houston, TX, USA: Gulf Publishing Co.; 2008. 296 pp.

[19] Morel F, Kressmann S, Kasztelan S. Processes and catalysts for hydrocracking of heavy oil and residues. In: Hydrotreatment and hydrocracking of oil fractions. New York: Elsevier Science B. V; 1997. p. 1−16.

[20] Materials program. In: Gibbons JH, editor. An assessment of oil shale technologies, vol. PB80−210115. Washington, DC, USA: U.S. Government, Office of Technology Assessment; 1980. 517 pp.
[21] Office of technology assessment materials program. In: An assessment of oil shale technologies. Washington, DC, USA: U.S. Government Printing Office NTIS; 1996. p. 119−76 [Chapter 5]. Technology.
[22] Orr AS, Van Sant RW. Residual fuel oils. In: Pet. products handbook. New York: McGraw-Hill; 1960. 8.1−8.31.
[23] Rahimi PM, Gentzis T. The chemistry of bitumen and heavy oil processing. In: Hsu CS, Robinson PR, editors. Practical adv. pet. proc. New York: Springer; 2006. p. 597−634.
[24] Robinson PR. Petroleum processing overview. In: Hsu CS, Robinson PR, editors. Practical adv. pet. proc. New York: Springer; 2006. p. 1−78.
[25] Solari RB. Asphaltene hydroconversion. In: Teh Fu Y, George VC, editors. Asphaltenes asphalts. 2. Developments in pet. science. Elsevier; 2000. p. 149−71.
[26] Biasca FE, Chang E, Bailey RT, Dickenson RL, Johnson HE, Simbeck DR. Upgrading heavy crude oils and residues to transportation fuels: technology, economics and outlook. Report No. Phase 7. Multiclient Review Report. Mountain View, California: SFA Pacific, Inc.; May 2003.
[27] Chem System. Refinery of the future as shaped by environmental regulations. Report No. PERP 02/03S11. White Plains, NY: Nexant; December 2003. 129 pp.
[28] De Bruijn T. Critical review of existing and emerging approaches to moderate upgrading of extraheavy crudes. Report No. 99-35 (CF). Devon, Alberta, Canada: National Centre For Upgrading Technology (Ncut); April 2002. 705 pp.
[29] Flint L. Bitumen & very heavy crude upgrading technology: a review of long term R&D opportunities. LENEF Consulting Limited; March 31, 2004. Report No. Available at: http://www.ptac.org/links/dl/osdfnlreport.pdf.
[30] Frost, Sullivan. Oil refining technologies—new developments and growth opportunities. Report No. D274, San Antonio, TX. May 2004. 240 pp.
[31] Hydrocarbon Publishing Co. Opportunity crudes: technical challenges and economic benefits. Report No. Multi-Client Strategic Report, Southeastern, PA. May 2006.
[32] Ingram C. Heavy oil upgrading. Report No. DOE/MT/95014-2. Atlanta, GA, USA: Clark Atlanta University; September 30, 1996. 17 pp.
[33] Chaffin JM, Liu M, Davison RR, Glover CJ, Bullin JA. Supercritical fractions as asphalt recycling agents and preliminary aging studies on recycled asphalts. Ind Eng Chem Res 1997;36:656−66.
[34] Bousquet J, Laboural T. Deasphalted oils acceptable FCC feed in Europe. Oil Gas J April 20, 1986;62:1.
[35] Subramanian M, Moretta J, Rick B, Martin M. An advanced asphaltene pelletisation process. PTQ 2000;3Q:17−23.
[36] Bonfadini P, Waintraub S, Fonseca CN, Soares GMG. The integration of solvent deasphalting and FCC process: a highly competitive choice for residue conversion. In: Proc. NPRA annual meeting. San Antonio, TX; March 21−23, 1999. Paper AM-99-11, 15 pp.
[37] Energy Information Administration. Refinery capacity: independent satistic & analysis. Report No. Unnumbered. 2013. 62 pp.
[38] Davis TA. Integrated process for deasphalting heavy oils using a gaseous antisolvent. Patent No. US4536283. Exxon Research and Engineering Co.; August 20, 1985.
[39] Davis TA. Method for deasphalting heavy oils using a miscible solvent at a low treat ratio and a carbon dioxide antisolvent. Patent No. US4565623. Exxon Research and Engineering Co.; January 21, 1986.
[40] Pickell MW. Asphaltene and drying oil compositions. Patent No. US2752264. Kerr-McGee Refining Corp.; June 26, 1956.
[41] Garwin L. Hydrocarbon solvent treatment of bituminous materials. Patent No. US4125459. Kerr-McGee Refining Corp.; November 14, 1978.
[42] Nelson SR. Fractionation of heavy hydrocarbon process material. Patent No. US4502944. Kerr-McGee Refining Corp.; March 05, 1985.

[43] Hood RL. Solvent extraction process. Patent No. US5009772. Kerr-McGee Refining Corp.; April 23, 1991.
[44] Gearhart JA, Garwin L. Rose process improves resid feed. Hydrocarb Proc May 1976;55:125–8.
[45] Gearhart JA, Garwin L. Resid-extraction process offers flexibility. Oil Gas J 1976;74(24):63–6.
[46] Nelson S, Corbett RW. Kerr-mcgee's ROSE process expands on proven extraction technology. In: Proc. 3rd UNITAR intern. conf. 'heavy crude tar sands'. Long Beach, California, US, vol. 3; July 22–31, 1985. p. 1822–36.
[47] Nelson S, Roodman RG. ROSE the energy efficient bottom of the barrel alternative. Chem Eng Prog 1985;81(5):63–8.
[48] Nelson SR, Roodman RG. ROSE: the energy efficient bottom of the barrel alternative. Chem Eng Prog 1985;81(5):63–8.
[49] Abdei-Halim T, Floyd R. The ROSE process. In: Meyers RE, editor. Handbook of refining processes. The McGraw-Hill Companies; 2004. p. 15–28.
[50] Residuum Processing Halliburton-KBR ROSE technology and rosemax internals. 2006. 2 pp. Available from: http://www.theyesmen.org/agribusiness/halliburton/kbr/hydroChem/petroRefine/residuumProcessing.html.
[51] Anonymous. Refining processes handbook: deasphalting. Hydrocarb Proc 2008;87(9):67–8.
[52] Patel V, Iqbal R, Eng O. To vacuum or not to vacuum. Hydrocarb Eng June 2007:83–7.
[53] Hedrick BW, Seibert KD, Crewe C. A new approach to heavy oil and bitumen upgrading. In: Proc. NPRA annual meeting. Salt Lake City, UT; March 19–21, 2006. Paper AM-06-29, 14 pp.
[54] Gillis D. Breaking through the bitumen upgrading barriers with the UOP uniflex[tm] process. In: Proc. 5th NCUT upgrading ond relining conlerence. Edmonton, Alberta, Canada; September 14–16, 2009. 24 pp.
[55] Hedrick BW. Process for separating a resin phase from a solution containing a resin, a solvent, and demetallized oil. Patent No. US5145574. UOP Inc.; September 8, 1992.
[56] Hedrick BW. Process for separating a resin phase from a solvent solution containing a solvent, demetallized oil and a resin. Patent No. US5098994. UOP Inc.; March 24, 1992.
[57] Mcgrath MJ, Houde EJ. Upgrading options for processing heavy crudes. In: Proc. AIChE spring meeting. Houston, TX, USA; March 14–18, 1999. 6 pp.
[58] Hedrick BW, Micklich FT, Johnson BH, Meier DG, Whitsura FR, Engelman BR. Phase separator apparatus. Patent No. US5340470. UOP Inc.; August 23, 1994.
[59] Hedrick BW, Micklich FT, Johnson BH, Meier DG, Whitsura FR, Engelman BR. Phase separator apparatus. Patent No. US5474688. UOP Inc.; December 12, 1995.
[60] Hedrick BW. Phase separator apparatus and its use to separate a mixture of liquids and finely divided suspended particles. Patent No. US5599463. UOP Inc.; February 04, 1997.
[61] Hedrick BW. Phase separator apparatus to separate a mixture of liquids and finely divided suspended particles. Patent No. US5770078. UOP Inc.; June 23, 1998.
[62] Houde EJ, Crawley A. Processing of heavy oil utilizing the aurabon process. Report No. DOE/BC/10453-6. Des Plaines, IL, USA: UOP LLC; January 01, 1982. 75 pp.
[63] Sprague SB. How solvent selection affects extraction performance. In: Proc. NPRA annual meeting. Los Angeles, CA, USA; March 23–25, 1986. AM-86-36, 15 pp.
[64] Bonilla JA. Ascot process. Energy Prog 1985;5(4):239–44.
[65] Billon A, Colyar JJ, Kressmann S, Morel F. Deep conversion of heavy oils via combination of catalytic hydrocracking and deasphalting. In: Proc. 16th world pet. congress. Calgary, Canada, vol. 3; June 11–15, 2000. p. 113–6.
[66] Axens Solvahl. 2015. Available from: http://www.axens.net/product/technology-licensing/10094/solvahl.html.
[67] Waintraub S, Fonseca CN, Soares GMG, Campagnolo EA. Conversion of a deasphalting unit for use in the process of supercritical solvent recovery. Braz J Chem Eng 2000;117(3):7.
[68] Kashiwara H. Design consideration of sda process. Kagaku Kogaku 1980;44(7):421–5.
[69] Research association for residual oil processing (Rarop). In: Heavy oil processing handbook. Tokyo, Japan: Agency of Natural Resources and Energy, Ministry of International Trade and Industry; 1991. 183 pp.

[70] Li R, Wang Z, Long J. Combined process of low degree solvent deasphalting and delayed coking. Patent No. US6673234 (Also published as US2002112986, FR2814467, ITMI20011978, DE10147093, CN1344782, CN1142259). China Petroleum & Chemical, Research Institute of Petroleum Processing; January 06, 2002.
[71] Chung KH, Xu Z, Sun X, Zhao S, Xu C. Selective asphaltene removal from heavy oil. PTQ 2006;11(5):99–105.
[72] Chung KH, Xu C, Hu Y, Wang R. Supercritical fluid extraction reveals resid properties. Oil Gas J 1997;95(3):66–9.
[73] Zhao S, Sparks BD, Kotlyar LS, Chung KH. Correlation of processability and reactivity data for residua from bitumen, heavy oils and conventional crudes: characterization of fractions from super-critical pentane separation as a guide to process selection. Catal Today 2007;125(3–4):122–36.
[74] Chung KH, Xu C. Narrow-cut characterization reveals resid process chemistry. Fuel 2001;80(8):1165–77.
[75] Zhao S, Xu C, Wang RA, Xu Z, Sun X, Chung KH. Deep separation method and processing system for the separation of heavy oil through granulation of coupled post-extraction asphalt residue. Patent No. US7597794 (Also published as US20070007168, FR2888245, CN1891784, CA2524995). China University of Petroleum; October 06, 2009.
[76] Zhao S, Xu C, Sun X, Xu Z, Chung KH. China refinery tests new asphaltenes extraction process. Oil Gas J 2010;108(12):52–8.
[77] Rao F, Liu Q. Froth treatment in Athabasca oil sands bitumen recovery process: a review. Energy Fuels 2013;27(12):7199–207.
[78] Tipman RN, Shelfantook WE, Long YC. Solvent process for bitumen separation from oil sands froth. Patent No. US6214213 (Also published as CA2217300, US5876592). Aec Oil Sands LP.; Mocal Energy Ltd.; Athabasca Oil Sands Investment; Imp Oil Resources; Occidental Petroleum Canada; Canadian Oil Sands Investments; Gulf Canada Resources; Murphy Oil Ltd.; Aec Oil Sands LP.; PetroCanada Inc.; April 10, 2001.
[79] Ng SH, Dabros T, Humphries A. Fluid catalytic cracking quality improvement of bitumen after paraffinic froth treatment. Energy Fuels 2007;21(3):1432–41.
[80] Long Y, Dabros T, Hamza H. Analysis of solvent-diluted bitumen from oil sands froth treatment using nir spectroscopy. Can J Chem Eng 2004;82(4):776–81.
[81] Astech. Syncrude/astech innovation in oil sands research. 2003. Available from: http://www.astech.ab.ca/users/folder.asp?FolderID=1889.
[82] Kotlyar LS, Sparks BD, Woods JR, Chung KH. Solids associated with the asphaltene fraction of oil sands bitumen. Energy Fuels 1999;13(2):346–50.
[83] Long Y, Beetge JH, Wang S, Dabros T, Hamza H, Xu Y, et al. Zone settling aid and method for producing dry diluted bitumen with reduced losses of asphaltenes. Patent No. CA2538464 (Also published as US2006196812). Champion Technology Inc.; September 02, 2006.
[84] Romanova UG, Valinasab M, Stasiuk EN, Yarranton HW, Schramm LL, Shelfantook WE. The effect of oil sands bitumen extraction conditions on froth treatment performance. J Can Pet Technol 2006;45(9):36–45.
[85] Romanova UG, Yarranton HW, Masliyah JH, Schramm LL, Shelfantook WE. Investigation of oil sands froth treatment. Can J Chem Eng 2004;82(4):710–21.
[86] Wu XA. Selective creaming: a novel technique to isolate interfacial organics-and mineral solids from water-in-oil emulsions. Energy Fuels 2008;22(4):2346–52.
[87] Birkholz RKO, Tipman NR, Power WJ. Method for processing a diluted oil sand froth. Patent No. CA2232929. Shell Canada Ltd.; September 25, 1989.
[88] Feng X, Wang S. Understanding asphaltene dispersants for paraffinic solvent-based bitumen froth treatment. Energy Fuels 2011;25(6):2601–7.
[89] Nagaishi H, Ikeda K, Kitano K, Chan EW, Gray MR. Observation of heavy oil vaporization under rapid heating. Energy Fuels 1998;12(6):1174–80.

[90] Payne B, Tipman NR, Foulds GA, Valgardson BE, Wamboldt BC. Process for removing solvent from an underflow stream from the last separation step in an oil sands froth treatment process. Patent No. CA2353109. Chevron Canada Resources Ltd.; Shell Canada Ltd.; Western Oil Sands LP.; January 16, 2003.
[91] Rahimi PM, Ellenwood RE, Parker RJ. Partial upgrading of Athabasca bitumen froth by asphaltene removal. In: Proc. 7th UNITAR intern. conf. 'heavy crude tar sands'. Beijing, China, vol. 2; October 27–30, 1998. 8 pp.
[92] Tipman RN, Sankey BM. Process for separation of hydrocarbon from tar sands froth. Patent No. CA2021185. Oslo Alberta Ltd.; January 14, 1992.
[93] Yang X, Hamza H, Czarnecki J. Investigation of subfractions of Athabasca asphaltenes and their role in emulsion stability. Energy Fuels 2004;18(3):770–7.
[94] Chirinos ML, Grosso JL, Layrisse I, Stockwell A. Treatment of viscous crude oils. Patent No. US4781819, US4915819 (Also published as EP0134088, CA1253825). British Petroleum Co., Intevep S.A.; November 01, 1988.
[95] Johnson GR, Rutter PR. Precipitation of asphaltene. Patent No. US4846957 (Also published as EP0242129, JPS63503075, WO8706255). British Petroleum Co, Intevep S.A.; July 11, 1989.
[96] Krasuk J, Solari RB, Aquino LG, Rodriguez JV, Granados A. Solvent deasphalting in solid phase. Patent No. US4572781. Intevep SA; February 25, 1986.
[97] Paraco MA, Chirinos M, Marquina M, Silva F, Morfes G, Siachoque G, et al. Process for improving and recuperating waste, heavy and extra heavy hydrocarbons-mecs. Patent No. CA2584003 (Also published as US2007295644, US7854836, RU2007123776, RU2362794, MX2007005851, CU23770, CO5840270, BRPI0702561, AR061706). Intevep SA; December 17, 2007.
[98] Willauer EE. Process for deasphalting petroleum oils. Patent No. US2538220 (Also published as CA484759). Standard Oil Dev Co.; January 16, 1951.
[99] Long JD. Deasphalting petroleum oil with propane. Patent No. CA499059. Standard Oil Dev Co.; January 05, 1954.
[100] Ditman JG. Propane deasphalting of crude residua of hydrocarbon oil. Patent No. CA512043. Foster Wheeler Corp.; April 19, 1955.
[101] Mertens FT, Ditman JG. Light hydrocarbon deasphalting of crude residue. Patent No. CA531130. Foster Wheeler Corp.; October 02, 1956.
[102] Ditman JG, Dunmyer JC. Solvent deasphalting. Patent No. GB1286176 (Also published as US3627675, JPS4943262, FR2064372, DE2050940). Foster Wheeler Corp.; August 23, 1972.
[103] Godino RL, Morfit O. Recovery and handling of asphalt containing a high concentration of asphaltene. Patent No. CA1036525. Foster Wheeler Corp.; August 15, 1978.
[104] Waghorne RH, Learner CF. Deasphalting with propane and butane. Patent No. US2729589 (Also published as CA552265). Exxon Research and Engineering Co.; January 03, 1956.
[105] Weikart J. Deasphalting process. Patent No. CA548354. Exxon Research and Engineering Co.; November 05, 1957.
[106] Strickland BR. Deasphalting operation. Patent No. CA549003. Exxon Research and Engineering Co.; November 19, 1957.
[107] Waghorne RH, Learner CF. Deasphalting with propane and butane. Patent No. CA552265. Exxon Research and Engineering Co.; January 28, 1958.
[108] Codet HG, Herrmann JW. Deasphalting and demetalizing crude hydrocarbon feeds with hydrofluoric acid. Patent No. US3281350. Esso Research and Engineering Co.; October 26, 1966.
[109] Bushnell JD, Ryan DG. Deasphalting process. Patent No. US3975396 (Also published as JPS5197602, DE2559105). Exxon Research and Engineering Co.; August 17, 1976.
[110] Hemminger CE. Solvent treating before catalytic cracking. Patent No. US3507777. Esso Research and Engineering Co.; April 21, 1970.
[111] Myers HC, Robb AT. Hydrocarbon modified propane deasphalting. Patent No. CA542270. Socony Mobil Oil Co.; June 11, 1957.

[112] Gross HH, Murphey RL, Mcclure MR. Petroleum treating process involving solvent deasphalting and propane fractionation. Patent No. CA623803. Texaco Development Corp.; July 11, 1961.
[113] Beavon DK. Solvent deasphalting. Patent No. CA639922. Texaco Development Corp.; April 17, 1962.
[114] Smith R. Solvent deasphalting. Patent No. CA639896. Texaco Development Corp.; April 17, 1962.
[115] Wilson R, Cole E. Treatment of heavy petroleum oils. Patent No. US3779895 (Also published as CA980283). Texaco Inc.; December 18, 1973.
[116] Doberenz EH, Sullivan JN. Solvent deasphalting process. Patent No. US3278415. Chevron Research Co.; October 11, 1966.
[117] Mills KL. Method for rendering an asphalt or asphaltene product collected in the separation zone of a solvent extraction apparatus free flowing by dispersing an immiscible liquid therewith. Patent No. US3321394. Phillips Petroleum Co.; May 23, 1967.
[118] Vanpool J. Production of low-metal-content gas oil from topped crude oil. Patent No. US3658695. Phillips Petroleum Co.; April 25, 1972.
[119] Crowley RP. Solvent deasphalting. Patent No. US4101415. Phillips Petroleum Co.; July 18, 1978.
[120] Gleim WKT. Removal of heavy asphaltenes from hydrogenation feed stocks. Patent No. US3294678. Universal Oil Products Co.; December 27, 1966.
[121] Gatsis JG. Solvent deasphalting. Patent No. US3830732. Universal Oil Products Co.; August 20, 1974.
[122] Bunas BI. Solvent deasphalting apparatus. Patent No. US4088540 (Also published as CA1100080). UOP Inc.; May 09, 1978.
[123] Uitti KD, Bunas BI. Hydrocarbon deasphalting via solvent extraction. Patent No. US3972807. UOP Inc.; August 03, 1976.
[124] Kellar JS, Carson DB, Boyd DM. Control of hydrocarbon deasphalting process. Patent No. US3981797. UOP Inc.; November 21, 1976.
[125] Bunas BI, Kellar JS. Hydrocarbon deasphalting process and solvent extractor therefor. Patent No. US3998726. UOP Inc.; December 21, 1976.
[126] Murphy JR. Solvent decarbonizing process. Patent No. US3423308. Gulf Research and Development Co.; January 21, 1969.
[127] Van Den Berg GJ, Mooney DG. Process for the preparation of lower boiling hydrocarbons or fractions containing them from residual oils. Patent No. GB1210120 (Also published as DE1816828, NL6717666, NL159135, FR1597013). Shell International Research; October 28, 1970.
[128] Bott TR, Ellis SRM, Panzner F, Bott TR, Ellis SR, Panzner F. Extraction process for asphaltic materials. Patent No. GB2001670. February 07, 1979.
[129] Carey RA. Synthetic lubricating oil stock from metal-contaminated heavy oil. Patent No. US3637483. Ghenron Research Co.; January 25, 1972.
[130] Vallino Jr B. Residual deasphalting and cracking with catalyst demetallization. Patent No. US3379639. Sinclair Research, Inc.; April 23, 1968.
[131] Gearhart JA. Solvent deasphalting. Patent No. US4239616. Kerr-Mcgee Refining Corp.; December 16, 1980.
[132] Achia BU, Shaw DH, Bushnell JD. Method for increasing deasphalted oil production. Patent No. US4522710 (Also published as EP0147113, JPS60192791, JPH055000, CA1249543, AR244780). Exxon Research and Engineering Co.; June 11, 1985.
[133] Bristow DJ, Achia BU, Evers LJ. Process for increasing deasphalted oil production from upgraded residua. Patent No. US4673485 (Also published as EP0160410, JPS60229985, JPH055278, CA1243979). Exxon Research and Engineering Co.; June 06, 1987.
[134] Long Y, Walker A, Dabros T, Kan J. Process for treating heavy oil emulsions using a light aliphatic solvent-naphta mixture. Patent No. CA2435113. Canada Natural Resources; Suncor Energy Inc.; January 11, 2005.
[135] Khusnutdinov IS, Kozin VG, Kopylov AJ. Method of fractionating natural bitumens and high-viscosity petroleums. Patent No. RU2138537. Kz G T Univer; September 27, 1999.
[136] Brons GB, Siskin M. Upgrading of bitumen by hot water treatment. Patent No. CA2118983. US5316659. Wrzeszczynski Kazimierz O; October 3, 1994.

[137] Brons GB, Siskin M. Upgrading of bitumen asphaltenes by hot water treatment containing carbonate. Patent No. US5326456 (Also published as CA2118984). Wrzeszczynski Kazimierz O; October 3, 1994.
[138] Bushnell JD, Glivicky AP, Leighton MD, Sankey BM. Simultaneous deasphalting-extraction process. Patent No. US4125458 (Also published as GB2012809). Exxon Research and Engineering Co.; November 14, 1978.
[139] Savastano C, Cimino R. Process for deasphalting and demetallizing petroleum residues. Patent No. EP612829 (Also published as US5466365, US5346615, RU2119525, MX9401362 JPH06299167 JP3484580 ITMI930347 IT1263961 ES2107736 DK0612829 DE69405123 CA2115488 AU5512394 AU662672 AT157390). Eniricerche S.p.A.; August 31, 1994.
[140] Lerda D, Colombani G, Prevedello A. Process for the removal of heavy metals and asphaltenes from crude petroleum or heavy fractions thereof. Patent No. IT1244487. Eniricerche S.p.A.; July 15, 1994.
[141] Varadaraj R, Wales WE. Additive enhanced solvent deasphalting process. Patent No. US6245222 (Also published as WO0024845, NO20011996, JP2002528595, ID29228, EP1153104, CA2345617, AU6415499). Exxon Research and Engineering Co.; June 12, 2001.
[142] Mann DP, Kukes SG, Coombs DM. Metals removal from oils with a light hydrocarbon and an organophosphorous compound. Patent No. US4518484. Phillips Petroleum Co.; May 21, 1985.
[143] Hood RL, Rettger PL, Bronicki LY. Method of and apparatus for separating solvent in a feed of solvent and deasphalted oil. Patent No. US5814286 (Also published as WO9807492, IL128586, AU4068597). Ormat Process Technologies Inc.; September 29, 1998.
[144] Khajrudinov IR, Sajfullin NR, Gantsev VA, Nigmatullin RG, Mingaraev SS, Gajsin IK, et al. Method of deasphalting of petroleum residue. Patent No. RU2167186. Inst Neftekhimpererabotki An R; May 20, 2001.
[145] Knjaz KAL, Blokhinov VF, Lavrinenko AM, Tjurin AM, Zotkin VA, Boldinov VA, et al. Method of solvent regeneration. Patent No. RU2217475. November 27, 2003.
[146] Tsegel SVG, Malashkevich AV, Mingaraev SS, Reutov AN. Method of deasphalting oil feedstock and oil feedstock deasphalting installation for implementation of method. Patent No. RU2168534. Inventors; June 10, 2001.
[147] Tsegel SVG, Malashkevich AV, Mingaraev SS, Reutov AN. Method of deasphalting oil feedstock and oil feedstock deasphalting installation for implementation of method. Patent No. RU2202594. Inventors; April 20, 2003.
[148] Khajrudinov IR, Sultanov FM, Sajfullin NR, Nigmatullin RG, Moroshkin JG, Timofeev AA, et al. Method for deasphalting of petroleum residues. Patent No. RU2136720. Inst Neftekhimpererabotki An R; September 10, 1999.
[149] Sultanov FM, Khajrudinov IR, Kuznetsov VJ, Teljashev EG. Petroleum residue deasphalting process. Patent No. RU2232792. Inst Neftekhimpererabotki An R; September 10, 2004.
[150] Zajashnikov EN, Knjaz Kov AL, Blokhinov VF, Boldinov VA, Lavrinenko AM, Vishnevskij AV, et al. Oil residue deasphalting process. Patent No. RU2218379. Inventors; December 10, 2003.
[151] Jacobson M, Serrand W, Sweed NH, Weiss H, Dreher I, Zentner U, et al. Improved process for deasphalting residua by reactive recycle of high boiling material. Patent No. EP1194498 (Also published as WO0063320, US2003121828, US6652739, JP2002542375, JP4590108, CN1347442, CA2370591, AU4454200, AU771590, AR023503). ExxonMobil Research and Engineering Co.; April 16, 1999.
[152] Serrand W, Hammond DG, Jacobson M, Pagel JF, Poole MC. Process for deasphalting of residua. Patent No. US5714056 (Also published as WO9704041, JPH11509260, EP0842240, JPH11509260, CA2223649). Exxon Research and Engineering Co.; February 03, 1998.
[153] Hood RL, Rettger P. Process and apparatus for solvent-deasphalting residual oil containing asphaltenes. Patent No. US5914010 (Also published as WO9811971, IL129006, CA2265596, AU4342697). Ormat Process Technologies Inc.; June 22, 1998.
[154] Starshov MI, Mingazetdinov FA, Abdulkhairov RM, Rakutin JV, Zakiev FA. Method for production of pure asphaltenes during deasphalting of heavy crude oils and naturally occurring bitumens. Patent No. RU2155791. Regional Nyj N T Ts Uralo Povo; September 10, 2000.

[155] Du S, Zheng X, Li X. Method of raising processing capacity in mixed C4 deasphalting extraction tower. Patent No. CN1357603. Oil Refinery Jilin Chemical In.; July 10, 2002.
[156] Du S, Zheng X, Li X. Method of resisting solvent recovering rate in supercritical mixed C4 deasphalting tower. Patent No. CN1357602. Oil Refinery Jilin Chemical In.; July 10, 2002.
[157] Li R, Shen H, Wang Y. Combined technologic process for manufacturing petroleum residue. Patent No. CN1485412. China Petroleum & Chemical; March 31, 2004.
[158] Hood RL, Rettger PB, Goldstein RS, Bronicki LY, Doron B, Sinai J. Method of and means for upgrading hydrocarbons containing metals and asphaltenes. Patent No. US5976361 (Also published as US2001002654). Ormat Industries Ltd.; November 02, 1999.
[159] Hood RL, Rettger PB, Goldstein RS, Bronicki LY. Method of and means for upgrading hydrocarbons containing metals and asphaltenes. Patent No. US6274032 (Also published as US2001002654, CA2244598 GR1003831 GR98100301). Ormat Industries Ltd.; August 14, 2001.
[160] Wallace PS, Johnson KA, Penrose CF, Niccum JG. Asphalt and resin production to integration of solvent deasphalting and gasification. Patent No. US6533925 (Also published as WO0216530, TW539742, EP1313824, CA2420138, AU5538401, AR028751). Texaco Development Corp.; March 18, 2003.
[161] Wallace PS, Johnson KA, Kasbaum JL, Niccum JG. Vapor phase neutralization in integrated solvent deasphalting and gasification. Patent No. TW226365B. Texaco Development Corp.; January 11, 2005.
[162] Ingram B, William L, Taylor JJ, Kolstad JS, David KB, Robert H. Multistage solvent extraction process. Patent No. FR2689900 (Also published as DK189187, IT1254320, MX163743, NL194801). Amoco Corp.; May 31, 1995.
[163] Nigmatullin VR, Nigmatullin IR, Nigmatullin RG, Zolotarev PA. Two-step process for propane deasphalting of vacuum residues. Patent No. RU2235110. August 27, 2004.
[164] Oschmann H-J, Schorling P-C. Apparatus for completely deasphalting crude oils. Patent No. DE19644600. Inst Erdoel Und Erdgasforschun; May 7, 1998.
[165] Fiocco RJ, Niessen E. Tray apparatus for deasphalting and extraction. Patent No. US4528068. Exxon Research and Engineering Co.; July 09, 1985.
[166] Fiocco RJ. Cascade sieve tray for extraction and deasphalting. Patent No. US4588563. Exxon Research and Engineering Co.; May 13, 1986.
[167] Savastano C, Cimino R, Meli S. Continuous process for deasphalting and demetallating a residue from crude oil distillation. Patent No. US5354454 (Also published as EP0504982, RU2014345, MX9201247, JPH05105883, ITMI910776, IT1245394, ES2052403. DK0504982. DE69200125. CA2063101. AU1287392. AU641309. AT105324). Enichem Sintesis; October 11, 1994.
[168] Hart PR. Deasphalting process. Patent No. US6106701. Betzdearborn Inc.; August 22, 2000.
[169] Starshov MI, Kajumova NR, Polovnjak VK, Akhunov RM, Abdulkhairov RM. High-viscosity petroleum and natural bitumen deasphalting process. Patent No. RU2174532, RU2175341. Vserossijskij Neftegazovyj Nii; October 10, 2001.
[170] Surong W, Congrong C. Deasphalting method. Patent No. CN1131178. Nanchong Oil Refinery Sichuan; September 18, 1996.
[171] Brecher L, Mones C, Guffey F. Development of the WRITE process for pipeline-ready heavy oil. Report No. WRI 09-R007. Laramie, Wyoming, USA: Western Research Institute; April 2009. 97 pp.
[172] Brecher LE, Guffey F. An update on the development of the wri thermal enhancement (WRITE) process. In: Proc. 4th upgrading refining of heavy oil, bitumen, synthetic crude oil conf. Edmonton, Canada; September 26–27, 2006. p. 178–204.
[173] Hettinger WP, Beck WH. Immobilization of vanadia deposited on sorbent materials during visbreaking treatment of carbo-metallic oils. Patent No. US4469588 (Also published as CA1175000, BR8201773). Ashland Oil, Inc.; September 04, 1984.
[174] Hettinger WP, Beck WH. Sorbent useful in a visbreaking treatment of carbo-metallic oils. Patent No. US4515900 (Also published as EP0175799). Ashland Oil, Inc.; May 7, 1985.
[175] Yan T-Y. Recovery of metal values from petroleum residua and other fractions. Patent No. US4544479. Mobil Oil Corp.; October 1, 1985.

[176] Busch LE, Henderson GO. Demetallizing and decarbonizing heavy residual oil feeds. Patent No. US4427539. Ashland Oil, Inc.; January 24, 1984.
[177] Barger DF. Combination process for upgrading residual oils. Patent No. US4525268. Ashland Oil Inc.; June 25, 1985.
[178] Baird WC, Bearden R. Combined desulfurization and hydroconversion with alkali metal oxides. Patent No. US4007109. Exxon Research and Engineering Co.; February 08, 1977.
[179] Baird WC, Bearden R. Combined desulfurization and hydroconversion with alkali metal hydroxides. Patent No. US4003823. Exxon Research and Engineering Co.; January 18, 1977.
[180] Audeh CA. Co-processing of residual oil and coal. Patent No. US4379045. Mobil Oil Corp.; Abr 05 1983.
[181] Sankey BM, Maa PS, Bearden R. Conversion of the organic component from tar sands to lower boiling products. Patent No. US5795464 (Also published as CA2160834). Exxon Research and Engineering Co.; August 18, 1998.
[182] Yan TY. Demetalation of heavy hydrocarbon oils. Patent No. US4379747. Mobil Oil Corp.; April 12, 1983.
[183] Stratiev D, Shishkova I, Dinkov R, Nikolova R, Mitkova M, Stanulov K, et al. Reactivity and stability of vacuum residual oils in their thermal conversion. Fuel 2014;123:133–42.
[184] Wiehe IA. A phase-separation kinetic model for coke formation. Ind Eng Chem Res 1993;32:2447–54.
[185] Wiehe IA. Petroleum fouling: causes, tools, and mitigation methods. In: Proc. AIChE spring nat. meeting. Orlando, FL; 2006. 1 pp.
[186] Peries JP, Quignard A, Farjon C, Laborde M. Thermal and catalytic asvahl processes under hydrogen pressure for converting heavy crudes and conventional residues. Oil Gas Sci Technol - Rev IFP 1987;43(6):847–53.
[187] Gearhart JA, Washimi K. High severity soaker cracking. In: Proc. NPRA annual meeting. New Orleans, LA; March 22–24, 1992. Paper #AM-92-37, 18 pp.
[188] Washimi K, Ishizuka M. Process for treating heavy petroleum oil resids. Patent No. EP0202099 (Also published as EP0202099, MX169003, KR900000861, JPS61261391, JPH0426360, DD251781, CN86102643). Toyo Engineering Corp., Mitsui Coke Co., Kerr-McGee Corp.; November 20, 1986.
[189] Washimi K, Wada T, Nelson S, Limmer H, Schutter H. High conversion upgrading of extra-heavy oils by a new process combination. In: Proc. 3rd UNITAR intern. conf. 'heavy crude tar sands'. Long Beach, California, US, vol. 3; July 22–31, 1985. p. 1495–509.
[190] Washimi K, Limmer H. New unit to thermal crack resid. Hydrocarb Proc, Intern Ed 1989;68(9):69–70.
[191] Ueda H. Cherry-p process. J Fuel Soc Jpn 1978;57:963–71.
[192] Ueda K. Comprehensive heavy ends reforming refinery process (cherry-p). J Jpn Pet Inst 1976;19(5):417–25.
[193] Ueda K. Method for producing pitch and coke. Patent No. US4014781. Ozaka Gas Co.; March 29, 1977.
[194] Tominaga H, Itoh S, Yashiro M. Hydrovisbreaking of Athabasca bitumen and its reaction mechanism. Bull Jpn Pet Inst 1977;19(1):50–5.
[195] Billon A, Jacquin Y, Peries JP, Toulhoat H. Heavy oil process with hydrovisbreaking hydrodemetallation and hydrodesulfuration. Patent No. US4510042. Institut Francais du Petrole; April 09, 1985.
[196] Billon A, Jacquin Y, Peries J-P, Toulhoat H. Heavy oil process with hydrodemetallation, hydrovisbreaking and hydrodesulfuration. Patent No. US4511458. Institut Francais du Petrole; April 16, 1985.
[197] Billon A, Le Page JF, Peries JP. Process for converting a highly viscous hydrocarbon charge to a less viscous more easily transportable and more easily refinable hydrocarbon fraction. Patent No. US4498974. Institut Francais Du Petrole; February 12, 1985.
[198] Billon A, Peries J-P, Quignard A. Process for the treatment of highly viscous heavy oils at the oil field to effect desalting and transportability thereof. Patent No. US4533462. Institut Francais Du Petrole; August 06, 1985.
[199] Gregoli AA, Rimmer DP. Production of heavy hydrocarbons by in-situ hydrovisbreaking. Patent No. US6016868 (Also published as WO9967504, CA2335771). World Energy Systems, Inc.; January 25, 2000.

[200] Deschamps A, Franckowiak S, Le Page J-F. Process for converting heavy oils or petroleum residues to gaseous and distillable hydrocarbons. Patent No. US4405442. INST FRANCAIS DU PETROLE; September 20, 1983.

[201] Menoufy MF, Ahmed HS, Betiha MA, Sayed MA. A comparative study on hydrocracking and hydrovisbreaking combination for heavy vacuum residue conversion. Fuel 2014;119:106–10.

[202] Aegerter Jr PA, Howell JA, Sughrue EL, Knopp KG. Refining of heavy slurry oil fractions. Patent No. US5080777 (Also published as EP0456058). Phillips Petroleum Co.; November 13, 1991.

[203] Howell JA, Kukes SG. Hydrovisbreaking process for hydrocarbon containing feed streams. Patent No. US4608152 (Also published as EP0183269, ZA8508866, MX167163, JPS61181539, JPH0475782, ES8703510, DD242633, CN85108807, CN1008113, CA1259300, AU4999485, AU571358). Phillips Petroleum Co.; August 26, 1986.

[204] Howell JA, Wilson ME. Multi-stage hydrofining process. Patent No. US4659452. Phillips Petroleum Co.; April 21, 1987.

[205] Carrazza J, Pereira P, Martinez N. Process and catalyst for upgrading heavy hydrocarbons. Patent No. EP814145. Intevep SA; December 19, 1997.

[206] Marzin R, Pereira P, Zacarias L, Rivas L. Resid conversion through the aquaconversion(tm) technology - an economical and environmental solution. In: Proc. 7th UNITAR intern. conf. 'heavy crude tar sands'. Beijing, China, vol. 2; October 27–30, 1998. 10 pp.

[207] Pereira-Almao P, Marzin R, Zacarias L, Cordova J, Carrazza J, Marino M. Steam conversion process and catalyst. Patent No. US5885441 (Also published as EP0870815, SG64488, RU2142497, PT870815, NL1008843, KR100250114, JPH1129778, JP3087116, ES2206779, DE69817653, CN1196379, CN1102952, CN1446887, CN1209445, CA2233699, AT248902). Intevep SA; March 23, 1999.

[208] Pereira P, Flores C, Zbinden H, Guitián J, Solari RB, Feintuch H, et al. Aquaconversion technology offers added value to E. Venezuela synthetic crude oil production. Oil Gas J 2001;99(20):79–85.

[209] Marzin R, Pereira P, Mcgrath MJ, Feintuch HM, Thompson G. New residue process increases conversion, produces stable residue in curacao refinery. Oil Gas J 1998;97(44):79–86.

[210] Elliott JD, Stewart MD. Cost effective residue upgrading: delayed coking, visbreaking & solvent deasphalting. In: Proc. 4th Russian refining technol. conf. Moscow, Russia; September 23–24, 2004. 26 pp.

[211] Maitra P. The jamnagar gasification project. In: Proc. gasification technologies conf. San Francisco, CA, USA; October 9–12, 2012. 16 pp.

[212] Aiba T, Kaji H, Suzuki T, Wakamatsu T. Residue thermal cracking by eureka process. Chem Eng Prog February 1981;77:37–45.

[213] Shigeta M, Maekawa H. Process for thermally cracking heavy petroleum oil. Patent No. CA1107672 (Also published as FR2397450, JPS5420006, JPS5745277). Kureha Chemical Ind. Co. Ltd.; Eureka Ind. Co. Ltd.; August 25, 1981.

[214] Watari R, Shoji Y, Ishikawa T, Hirotani H, Takeuchi T. The development of the new eureka process. In: Proc. NPRA annual meeting. San Antonio, TX; March 29–31, 1987. Paper AM-87-43 10 pp.

[215] Ohba T, Shibutani I, Inomata J, Nagata H, Watari R. The advanced eureka process: environment friendly thermal cracking process. In: Proc. 19th world pet. congress. Madrid, Spain, vol. 2; June 29–July 3, 2008. p. 623–40.

[216] Sugioka R, Kuwahara T, Takao M, Kaji K, Taniuchi M, Tomisaka Y. Characteristics of inclined fluidized bed in liquid-solid system. Driving force for particle circulation between two fluidized beds. Kagaku Kogaku RONBUNSHU 1987;13(4):387–91.

[217] Elliott JD. Optimize coker operations. Hydrocarb Proc 2003;82(9):85–6. 88, 90.

[218] Kovac M, Movik G, Elliott JD. Upgrade refinery residuals into value-added products. Hydrocarb Proc 2006;85(6):57–8. 60, 62.

[219] Craik K, Hortardis A. Ivanhoe energy: set to lead the way in heavy oil upgrading. 2011. 14 pp. Available from: http://www.ivanhoeenergy.com/i/pdf/articles/brca_ivanhoe-energy_energy-october2011-video-bro-spreads.pdf.

[220] Freel B, Graham RG. Method and apparatus for a circulating bed transpot fast pyrolysis reactor system. Patent No. US5792340. Ensyn Technologies Inc.; August 11, 1998.
[221] Freel B, Graham RG. Apparatus for a circulating bed transport fast pyrolysis reactor system. Patent No. US5961786. Ensyn Technologies Inc.; October 16, 1999.
[222] Freel B, Graham RG. Rapid thermal processing of heavy hydrocarbon feedstocks. Patent No. WO0061705 (Also published as US2012279825 (Abandoned), PT2336274, PT1169412, NO20014868, MXPA01010120, ES2429816, ES2429508, EP2336274, EP1169412, DK2336274, DK1169412, CA2369288, BR0009652, AU3798300). Ensyn Group; October 19, 2000.
[223] Freel B, Kriz JF, Clarke D. Modified thermal processing of heavy hydrocarbon feedstocks. Patent No. US7572362 (Also published as US20040069682, CA2444832, DE60327148, EP1420058, CN100347274, RU2004109519, RU2323246, PE10322004, ECSP044976, CO5540064). Ivanhoe Energy; August 11, 2009.
[224] Freel B, Kriz JF, Clarke D. Modified thermal processing of heavy feedstocks. Patent No. US7572365 (Also published as US20040069686). Ivanhoe Energy, Inc.; August 11, 2009.
[225] Freel B, Graham RG. Products produced from rapid thermal processing of heavy hydrocarbon feedstocks. Patent No. EP2275513 (Also published as WO0224835, US2002100711, US7270743, PT1332199, NO20031230, NO330786, MXPA03002341, ES2395116, EP1332199, EP1332199, DK1332199, CA2422534, BR0113937, AU9156301, AT532842, AR033838). Ensyn Petroleum International Ltd.; January 19, 2011.
[226] Ivanhoe Energy Inc. Heavy oil — its time has come. 2005. 2 pp. Available from: www.ivanhoeenergy.com.
[227] Ivanhoe Energy. Ivanhoe energy successfully commissions its new feedstock test facility. 2009. Available from: http://www.ivanhoe-energy.com/s/htl-process.asp?ReportID=682595.
[228] Silverman M. The htl process. 2015. Available from: http://www.ivanhoe-energy.com/s/htl-process.asp.
[229] Graham R, Freel B. Methods and systems for producing reduced resid and bottomless products from heavy hydrocarbon feedstocks. Patent No. EP1970427 (Also published as CA2624746, CN101265415B, EP1970427, US8377287, US8808632, US20080230440, US20130171039, US20140353210). Ivanhoe Energy Inc.; May 09, 2012.
[230] Kalota SA, Pavel SK, Silverman MA. Method, system, and apparatus for lift gas distribution. Patent No. US20120167989 (Also published as CA2823332, CN103442796, EP2658949, WO2012092522.). Ivanhoe Energy Inc.; July 05, 2012.
[231] Silverman MA, Pavel SK, Kalota SA. Improved reactor feed nozzles. Patent No. CA2823341 (Also published as CN103442794, EP2658640, US20120168537, WO2012092520). Ivanhoe Energy Inc.; July 05, 2012.
[232] Pavel SK, Silverman MA, Kalota SA. Method, system, and apparatus for separation in processing of feedstocks. Patent No. CA2823209 (Also published as CN103429311, EP2658631, US20120167767, WO2012092613). Ivanhoe Energy Inc., Ivanhoe Htl Petroleum Ltd.; July 05, 2012.
[233] Kuhach J. Performance of ivanhoe energy's htl crude oil upgrading process using RTP. In: Proc. 4th upgrading refining of heavy oil, bitumen, synthetic crude oil conf. Edmonton, Canada; September 26–27, 2006. p. 123–45.
[234] Koshka E. Maximizing heavy oil value while minimizing environmental impact with htl upgrading of heavy to light oil. In: Proc. 5th NCUT upgrading refining conf. Edmonton, Canada; September 14–16, 2009. 24 pp.
[235] Silverman MA, Pavel SK, Hillerman MD. Htl heavy oil upgrading: a key solution for heavy oil upstream and midstream operations. In: World heavy oil congress; 2011. 7 pp.
[236] Cabrera CA, Hillerman MD, Silverman MA. Htl heavy oil upgrading: Monetizing stranded heavy oil assets in remote locations. In: Proc. world heavy oil congress. Aberdeen, Scotland; September 10–13, 2012. 7 pp.
[237] Cabrera AC, Silverman AM. Upgrade heavy oil more cost-efficiently. Hydrocarb Proc 2012;91(9):55–62.

[238] Silverman MA. Bringing heavy crudes to market. In: Proc. hydrocarbon processing's intern. refining petrochemical conf. Milan, Italy; June 12–14, 2012. 29 pp.
[239] Friedland R. Ivanhoe to buy oilsands assets to utilize htl technology. 2008. Available from: http://www.prnewswire.com/news-releases/ivanhoe-energy-to-acquire-oil-sands-assets-from-talisman-energy-57325502.html.
[240] Sun YD, Yang CH, Zhao H, Shan HH, Shen BX. Influence of asphaltene on the residue hydrotreating reaction. Energy Fuels 2010;24(9):5008–11.
[241] Plain C, Benazzi E, Guillaume D. Residue desulphurisation and conversion. PTQ 2006;11(3):57–63.
[242] Bridge AG, Reed EM, Tamm PW, Cash DR. Chevron hydrotreating processes desulfurize Arabian heavy residua. AIChE Symp Ser 1975;71(148):225–33.
[243] Spieler S, Dahlbert A, Mukherjee U. Upgrading residuum to finished products in integrated hydroprocessing platforms: solutions and challenges. In: Proc. NPRA annual meeting. Salt Lake City, UT, USA; March 19–21, 2006. Paper AM-06-64, 19 pp.
[244] Brossard DN. Chevron RDS/vrds hydrotreating: transportation fuels from the bottom of the barrel. In: Meyers RA, editor. Handbook of pet. refining processes. 2nd ed. New York, USA: McGraw-Hill; 1997. p. 158–70.
[245] Wade R, Maesen T, Vislocky J, Torchia D. Hydrocracking catalyst developments and innovative processing scheme. In: Proc. NPRA annual meeting. San Antonio, TX, USA; March 22–24, 2009. Paper #AM-09-12, 15 pp.
[246] Bachtel RW, Shinmura K, Reynolds BE, Pardoe R, Earls DE, Chabot J. Chevron ocr [(onstream catalyst replacement)] process. In: Proc. AIChE spring nat. meeting. Houston, TX, USA; March 19–23, 1995. Preprint 46f, 17 pp.
[247] Reynolds BE, Yagi K, Bachtel RW. Chevron corp's onstream catalyst replacement (ocr) provides enhanced flexibility to residue hydrotreaters. In: Proc. NPRA annual meeting. New Orleans, LA, USA; March 22–24, 1992. AM-92-61, 12 pp.
[248] Dickenson RL, Schulman BL, Biasca FE, Johnson HE. The bottom-of-the-barrel. Real options to avoid fuel oil. In: Proc. NPRA annual meeting. San Antonio, TX; March 17–19, 1996. Paper No. AM-96-57, 14 pp.
[249] Mart CJ, Ellis EG, Mccaffrey Jr DS. ExxonMobil resid upgrading technologies. In: Proc. intern. pet. conf. exhibition, Petrotech. New Delhi, India; January 17–20, 2005. 6 pp.
[250] Scheffer B, Van Koten MA, Röbschläger KW, De Boks FC. The shell residue hydroconversion process: development and achievements. Catal Today 1998;43(3–4):217–24.
[251] Oelderik JM, Sie ST, Bode D. Progress in the catalysis of the upgrading of petroleum residue: a review of 25 years of R&D on shell's residue hydroconversion technology. Appl Catal 1989;47(1):1–24.
[252] Van Zijll Langhout WC, Ouwerkerk C, Pronk KMA. Development of and experiences with the Shell residue hydroprocesses. AIChE Nat Meeting, Philadelphia, PA, USA. June 8–12). 1980.
[253] Bode D, Van Dongen RH, Van Klinken J. Process for demetallizing hydrocarbon oils. Patent No. CA1094490. Shell Canada Ltd.; January 27, 1981.
[254] Van Dongen RH, Kouwenhoven HW. Process for recovering vanadium and nickel from deactivated catalysts. Patent No. CA1071413 (Also published as FR2288154, ZA7506443, SU741780, SE7511450, SE410474, NO753452, NO143521, NL7511993, NL186587, MX3049, JPS5164412, JPS6012908, IT1043322, GB1526927, DE2545787, BE834257, AU498495, AR224854). Shell Canada Ltd.; February 12, 1980.
[255] Sikonia JG. New data for rcd unibon. Hydrocarb Proc, Intern Ed 1980;60(6, Sect. 1):73–9.
[256] Thompson GJ. UOP rcd unionfining process. In: Meyers RA, editor. Handbook of pet. refining processes. 2nd ed. New York, USA: McGraw-Hill; 1997. p. 176–81.
[257] UOP. Residue hydrotreating. 2015. Available from: http://www.uop.com/processing-solutions/refining/residue-upgrading/%20-%20residue-hydrotreating/.
[258] Zeuthen P. Haldor topsoe: resid hydroprocessing. 2015. Available from: http://www.topsoe.com/processes/resid.

[259] Benito AM, Martínez MT, Fernández I, Miranda JL. Upgrading of an asphaltenic coal residue: thermal hydroprocessing. Energy Fuels 1996;10(2):401–8.
[260] Kressmann S, Morel F, Harle V, Kasztelan S. Recent developments in fixed bed catalytic residue upgrading. Catal Today 1998;43:203–15.
[261] CBI. Chevron lummus global: RDS/vrds/ocr/ufr. 2012. Available from: www.CBI.com/lummus-technology.
[262] Toshima H, Sedlacek Z, Backhouse D, Mayo SL, Plantenga FL. Greater bottoms upgrading with Albemarle's e-bed catalysts. In: Proc. 4th upgrading refining of heavy oil, bitumen, synthetic crude oil conf. Edmonton, Canada; September 26–27, 2006. p. 230–48.
[263] Sayles S, Bailor J. Catalyst addition in ebullated-bed units. PTQ 2005;(Q2):7.
[264] Kressmann S, Boyer C, Colyar JJ, Schweitzer JM, Viguié JC. Improvements of ebullated-bed technology for upgrading heavy oils. Oil Gas Sci Technol - Rev IFP 2000;55(4):397–406.
[265] Nowlan VJ, Srinivasan NS. Control of coke formation from hydrocracked athabasca bitumen. Fuel Sci Technol Intern 1996;14(1–2):41–54.
[266] Sundaram KM, Mukherjee U, Baldassari M. Thermodynamic model of sediment deposition in the LC-fining process. Energy Fuels 2008;22(5):3226–36.
[267] Bannayan MA. Fouling mechanisms and effect of process conditions on deposit formation in H-oil equipment. Stud Surf Sci Catal 1996;100:273–81.
[268] Morel F, Duddy JE, Wisdom L. Consider new technologies to raise diesel yield in bottom-of-barrel products. Hydrocarb Proc 2012:15. Available from: http://www.hydrocarbonprocessing.com/Article/3109412/Consider-new-technologies-to-raise-diesel-yield-in-bottom-of-barrel-products.html.
[269] Arora A, Mukherjee U. Refinery configurations for maximising middle distillates. PTQ 2011;16(3):9.
[270] Wadsworth D. LC-fining options for heavy oil upgrading. In: Proc. NPRA annual meeting. San Diesgo, CA, USA; March 9–11, 2008. Paper AM-08-33, 29 pp.
[271] Arora A, Mukherjee U. Refinery configurations for maximum conversion to middle distillates. In: Proc. NPRA annual meeting. San Antonio, TX, USA; March 20–22, 2011. Paper AM-11-19, 18 pp.
[272] Baldassari M, Mukherjee U. LC-max and other LC-fining process enhancements to extend conversion and onstream factor. In: Proc. AFPM annual meeting. San Diego, CA; March 11–13, 2012. Paper AM-12-73, 504-519.
[273] Colyar JJ, Peer E. H-oil process based heavy crudes refining schemes. In: Proc. 7th UNITAR intern. conf. 'heavy crude tar sands'. Beijing, China, vol. 2; October 27–30, 1998. 10 pp.
[274] Axens IFP Group Technologies. Axens and headwaters technology innovation (HTI) join efforts on ebullated bed technology for heavy oil upgrading. In: Axens news - HW news bull; 2013. USA. 1 pp.
[275] Gauthier T, Heraud JP, Kressmann S, Leplat S, Verstraete J. Residue hydroconversion for Athabasca upgrading – impact of vapourization. In: Proc. 4th upgrading refining of heavy oil, bitumen, synthetic crude oil conf. Edmonton, Canada; September 26–27, 2006. p. 205–29.
[276] Wisdom L, Duddy JE. First commercial residue hydrocracker integrated with gasification to be self-sufficient for its hydrogen consumption requirements. In: Proc. 4th upgrading refining of heavy oil, bitumen, synthetic crude oil conf. Edmonton, Canada; September 26–27, 2006. p. 308–34.
[277] Belinko K, Khulbe CP, Jain AK. Hydrocracking of heavy oil in presence of ultrafine iron sulphate. Patent No. US4963247. Petro-Canada Inc.; October 16, 1990.
[278] Gatsis JG. Hydrocarbon hydrorefining process utilizing a non-stoichiometric vanadium sulfide catalyst. Patent No. US4194967. UOP Inc.; March 25, 1980.
[279] Graeser U, Niemann K. Proven hydrogenation processes for upgrading residua being revived in Germany. Oil Gas J 1982;80(12):121–7.
[280] Graeser U, Eschet G, Holighnaus R. Veba-combi-cracking, a proven technology for high conversion of heavy bottoms. In: Proceedings - refining department, American pet. institute. San Diego, CA, USA; 1986. p. 169–99.
[281] Aldridge CL, Bearden R, Lewis WE. Hydroconversion of heavy feeds by use of both supported and unsupported catalysts. Patent No. CA1334840. Exxon Research and Engineering Co.; March 21, 1995.

[282] Baird WC, Bearden R. Hydroconversion with group ia, iia metal compounds. Patent No. US4127470. Exxon Research and Engineering Co.; November 28, 1978.
[283] Baird WC, Bearden R. Hydroconversion of residua with potassium sulfide. Patent No. US4119528. Exxon Research and Engineering Co.; October 10, 1978.
[284] Aldridge CL, Bearden R, Metrailer WJ. Combination hydroconversion, coking and gasification. Patent No. US4204943. Exxon Research and Engineering Co.; May 27, 1980.
[285] Silva F, Guitian J, Galiasso R, Krasuk J. The HDH process for upgrading heavy Venezuelan crudes. In: Proc. 11th world pet. congress. London, UK, vol. 4. RTD; August 28–September 2, 1983. 10 pp.
[286] Ramírez De Agudelo MM, Galarraga C. Catalyst for the simultaneous hydrodemetallization and hydroconversion of heavy hydrocarbon feestocks. Patent No. US4729980 (Also published as GB2204251). Intevep SA; March 08, 1988.
[287] Ramírez De Agudelo MM, Galarraga C. Catalyst for the simultaneous hydrodemetallization and hydroconversion of heavy hydrocarbon feestocks. Patent No. US4832829. Intevep SA; May 23, 1989.
[288] Ramírez De Agudelo MM, Galarraga C. Catalytic system for the hydroconversion of heavy oils. Patent No. US4888104. Intevep SA; December 19, 1989.
[289] Ramírez De Agudelo MM, Galarraga C. Catalytic system for the hydroconversion of heavy oils. Patent No. US4937218. Intevep SA; June 26, 1990.
[290] Pereira P, Rivas G, Jose C, Granadillo F, Marzin R, Solari B, et al. Dispersed metal sulfide-based catalysts. Patent No. US8551907 (Also published as US2011068045, US2009023965, EP2113304, CA2639777, BRPI0804269). Intevep, S.A; October 08, 2013.
[291] Marzin R, Solari B, Zacarias L. Hydroconversion process for heavy and extra heavy oils and residuals. Patent No. US8679322 (Also published as US2011120908, EP2325285, RU2010120474, MX2010006452, KR20110058639, JP2014122365, JP2011111620, JP5539074, ECSP10010180, CU20100097, CN104109554, CN102071053, CA2703382, BRPI1001712). Intevep, S.A; March 25, 2014.
[292] Reynolds JG, Yu SG, Beret S. Metals-impregnated red mud as a first-stage catalyst in a two-stage, close-coupled thermal catalytic hydroconversion process. Patent No. US4559130. Chevron Research Co.; December 17, 1985.
[293] Reynolds JG, Yu SG, Lewis RT. Red mud as a first-stage catalyst in a two-stage, close-coupled thermal catalytic hydroconversion process. Patent No. US4559129. Chevron Research Co.; December 17, 1985.
[294] Yu SG, Reynolds JG. Presulfided red mud as a first-stage catalyst in a two-stage, close-coupled thermal catalytic hydroconversion process. Patent No. US4560465. Chevron Research Co.; 1985.
[295] Van Nordstrand RA. Hydrocarbons hydroprocessing with imogolite catalyst. Patent No. US4394253. Chevron Research Co.; July 19, 1983.
[296] Van Nordstrand RA. Hydrocarbons hydroprocessing with imogolite catalyst. Patent No. US4446244. Chevron Research Co.; May 01, 1984.
[297] Khadzhiev SN, Kadiev KM, Mezhidov VK, Zarkesh J, Hashemi R, Targhi SKM. Process for hydroconverting of a heavy hydrocarbonaceous feedstock. Patent No. US7585406 (Also published as US2007045156, EP1754770, MXPA06009247, JP2007051284, JP4808101, CU23363, CN103396831, CN1940027, CA2541104). RIPI, NTI Co.; September 08, 2009.
[298] Sue H, Fujita M. Commercial usage for new japanese conversion catalyst. Oil Gas J 1986;84(21):51–6.
[299] Ohtake N, Kuri K. Process for converting heavy hydrocarbon into more valuable product. Patent No. US4770764. Asahi Chemical Ind; September 13, 1988.
[300] Ohtake N, Kaneshima T. An additive for the hydroconversion of a heavy hydrocarbon oil. Patent No. US4863887 (Also published as EP271337, CA1305467, CN1013683, CN87107363, DE3773862). Asahi Chemical Industry Co., Ltd.; September 05, 1989.
[301] Montanari R, Marchionna M, Panariti N, Delbianco A, Collia D. A challenge to sweep-out contaminants from heavy residues. In: Proc. 16th world pet. congress. Calgary, Alberta, vol. 3; June 11–15, 2000. p. 104–10.
[302] Bellussi G, Rispoli G, Molinari D, Landoni A, Pollesel P, Panariti N, et al. The role of MoS2 nano-slabs in the protection of solid cracking catalysts for the total conversion of heavy oils to good quality distillates. Catal Sci Technol 2013;3(1):176–82.

[303] Del Bianco A, Panariti N, Marchionna M. Development of a more efficient use of dispersed Mo-based catalysts to upgrade heavy feedstocks. In: Proc. 210th ACS nat. meeting. Chicago, IL, USA, vol. 40. ACS; August 20–24, 1995. p. 743–6.
[304] Del Bianco A, Panariti N, Di Carlo S, Beltrame PL, Carniti P. New developments in deep hydroconversion of heavy oil residues with dispersed catalysts. 2. Kinetic aspects of reaction. Energy Fuels 1994;8(3):593–7.
[305] Del Bianco A, Panariti N, Di Carlo S, Elmouchnino J, Fixari B, Le Perchec P. Thermocatalytic hydroconversion of heavy petroleum cuts with dispersed catalyst. Appl Catal A General 1993;94(1):1–16.
[306] Del Bianco A, Panariti N, Marchionna M. Upgrading heavy oil using slurry processes. Chem Tech 1995;25(11):35–43.
[307] Panariti N, Del Bianco A, Del Piero G, Marchionna M. Petroleum residue upgrading with dispersed catalysts: Part 1. Catalysts activity and selectivity. Appl Catal A General 2000;204(2):203–13.
[308] Panariti N, Del Bianco A, Del Piero G, Marchionna M, Carniti P. Petroleum residue upgrading with dispersed catalysts: Part 2. Effect of operating conditions. Appl Catal A General 2000;204(2):215–22.
[309] Bearden R, Aldridge CL. Hydroconversion of heavy hydrocarbons. Patent No. US4134825. Exxon Research and Engineering Co.; January 16, 1979.
[310] Bearden R, Aldridge CL. Catalyst for the hydroconversion of heavy hydrocarbons. Patent No. US4226742. Exxon Research and Engineering Co.; October 07, 1980.
[311] Bearden R, Aldridge CL. Hydrocracking with chromic acid and an aliphatic polyhydroxy compound. Patent No. US4604190. Exxon Research and Engineering Co.; August 05, 1986.
[312] Bearden R, Aldridge CL. Catalysts and hydroconversion processes utilizing the same. Patent No. US4578182, US4579838 (Also published as CA1263369). Exxon Research and Engineering Co.; March 25, April 01, 1986.
[313] Bearden R, Aldridge CL. Oil soluble chromium catalyst. Patent No. US4567156. Exxon Research and Engineering Co.; January 28, 1986.
[314] Aldridge CL, Bearden R. Hydrocracking of hydrocarbons. Patent No. US4192735. Exxon Research and Engineering Co.; March 11, 1980.
[315] Aldridge CL, Bearden R. Hydroconversion process. Patent No. US4196072. Exxon Research and Engineering Co.; April 04, 1980.
[316] Cyr T, Lewkowicz L, Ozum B, Lott RK, Lee L-K. Hydrocracking process involving colloidal catalyst formed in situ. Patent No. US5578197 (Also published as CA2088402). Alberta Oil Sands Tech; November 26, 1996.
[317] Strausz OP, Mojelsky TW, Payzant JD, Olah GA, Prakash SGK. Superacid catalyzed hydrocracking of heavy oils and bitumens. Patent No. US5290428 (Also published as JPH0617058). Alberta Oil Sands Tech; March 01, 1994.
[318] Lutz IH. Process for upgrading a heavy viscous hydrocarbon. Patent No. US4454023 (Also published as CA1210358, DE3476419, EP0121376). Alberta Oil Sands Technology & Research Authority; June 12, 1986.
[319] Lott RK, Cyr T, Ozum B, Lee LK, Lewkowicz L. Hydrocracking process involving colloidal catalyst formed in situ. Patent No. CA2088402. Alberta Oil Sands Technology And Research Authority; July 08, 1997.
[320] Mcfarlane RA, Cyr T, Hawkins RWT. Process for dispersing transition metal catalytic particles in heavy oil. Patent No. US5916432 (Also published as CA2216671). Alberta Oil Sands Technology And Research Authority; June 29, 1999.
[321] Mcfarlane RA. A process and system for reducing acidity of hydrocarbon feeds. Patent No. CA2677004 (Also published as CN102625824, US20120152807, WO2011022832). Alberta Research Council Inc., Suncor Energy Inc.; June 17, 2014.
[322] Lott RK, Lee LK. Methods for hydrocracking a heavy oil feedstock using an in situ colloidal or molecular catalyst and recycling the colloidal or molecular catalyst. Patent No. US8431016. Headwaters Heavy Oil, LLC; July 19, 2013.

[323] Wu Z, Zhou Z, Zhou B. Hydrocarbon-soluble molybdenum catalyst precursors and methods for making same. Patent No. US7670984 (Also published as US2007158238, WO2007082118, US2010051507, MY146937, KR20080083059, JP2009522103, JP5112334, EP1973993, CN101370913, CA2636282). Headwaters Tech Innovation, LLC; March 02, 2010.
[324] Wu Z, Zhou Z, Zhou B. Hydrocarbon-soluble molybdenum catalyst precursors and methods for making same. Patent No. US8445399. Headwaters Technology Innovation, LLC; May 21, 2013.
[325] Herbstman S. Hydrotreating heavy residual oils. Patent No. US4125455. Texaco Inc.; November 14, 1978.
[326] Lopez J, Pasek EA. Process for preparing heavy oil hydroprocessing slurry catalyst. Patent No. US4710486. Chevron Research Co.; December 01, 1987.
[327] Lopez J, Pasek EA. Slurry catalyst for hydroprocessing heavy and refractory oils. Patent No. US5094991. Chevron Research Co.; March 10, 1992.
[328] Lopez J. Process for preparing a dispersed group vib metal sulfide catalyst. Patent No. US5484755. Chevron Research Co.; January 16, 1996.
[329] Lopez J. High activity slurry catalyst. Patent No. US5164075. Chevron Research Co.; 1992.
[330] Lopez J, Pasek EA. Dispersed group vib metal sulfide catalyst promoted with group viii metal. Patent No. US4824821. Chevron Research Co.; April 25, 1989.
[331] Aldridge C, Bearden R. High surface area catalysts. Patent No. US4244839. Exxon Research and Engineering Co.; January 13, 1981.
[332] Bearden R, Aldridge CL. Hydrocracking with aqueous phosphomolybdic acid. Patent No. US4637870, US4637871. Exxon Research and Engineering Co.; January 20, 1987.
[333] Gatsis JG. Method of preparing a catalyst for the hydroconversion of asphaltene-containing hydrocarbonaceous charge stocks. Patent No. US4943548. UOP Inc.; July 24, 1990.
[334] Gatsis JG. Method of preparing a catalyst for the hydroconversion of asphaltene-containing hydrocarbonaceous charge stocks. Patent No. US4954473. UOP Inc.; September 04, 1990.
[335] Gatsis JG. Method of preparing a catalyst for the hydroconversion of asphaltene-containing hydrocarbonaceous charge stocks. Patent No. US5171727. UOP Inc.; December 15, 1992.
[336] Que G, Men C, Meng C, Ma A, Zhou J, Deng W, et al. A slurry-bed hydrocracking reactor charged with a multimetallic liquid catalyst and an online fixed-bed hydrocarcking reactor to enhance lightweight oil yield from heavy oil under normal pressure. Patent No. US6660157 (Also published as CN1295112, US20020125172). Petrochina Company Limited, University Of Petroleum; December 09, 2003.
[337] Zhou JS, Deng WA, Liu D, Liang SC, Que GH. Effect of vacuum bottom residue recycling on slurry bed hydrocracking of residue. Acta Pet Sin Pet Proc Sect 2001;17(4):82–5.
[338] Liu D, Li M, Deng W, Que G. Reactivity and composition of dispersed Ni catalyst for slurry-phase residue hydrocracking. Energy Fuels 2010;24(3):1958–62.
[339] Stencel JM, Eklund PC, Bi XX, Derbyshire FJ. Iron ultrafine particle catalysts formed by laser pyrolysis: synthesis, characterization and coal liquefactionactivity. Catal Today 1992;15(2):285–306.
[340] Pesa M. Veba combi-cracking (VCC™) technology. 2015. 11 pp. Available from: http://www.bp.com/content/dam/bp/pdf/Technology/Fischer-Tropsch-Technology-BP.pdf.
[341] KBR Technology. Veba combi-cracking technology. 2013. 3 pp. Available from: http://www.kbr.com/Newsroom/Publications/Brochures/Veba-Combi-Cracking-Technology.pdf.
[342] Pruden BB. Hydrocracking of bitumen and heavy oils at CANMET. Can J Chem Eng 1978;56:277–80.
[343] Belinko K, Patmore DJ, Packwood RH, Ranganathan R. Simultaneous demetalization and hydrocracking of heavy hydrocarbon oils. Patent No. US4376695. Her Majesty the Queen in right of Canada, as represented by the Minister; 1983.
[344] Debruijn TJW, Kriz JF, Silva AE. Hydrocracking of heavy oils in presence of dry mixed additive. Patent No. CA1202588 (Also published as GB2135691, NL8400422, JPS6023482, IT1196021, FR2540883, DE3403979). Canada Majesty in Right of; Petro Canada Inc.; April 01, 1986.
[345] Jain AK, Pruden BB. Hydrocracking of heavy oils in presence of petroleum coke derived from heavy oil coking operations. Patent No. US4999328 (Also published as CA1322746). Petro-Canada Inc.; March 12, 1991.

[346] Ternan M, Parsons BI. Process for catalytically hydrocracking a heavy hydrocarbon oil. Patent No. US4176051 (Also published as CA1094004). Her Majesty The Queen in Right of Canada, As Represented by the Minister of Energy, Mines and Resources; November 27, 1979.
[347] Huber DA, Lee Q, Thomas RL, Frye K, Rudins G. An assessment of the potential for coal/residual oil coprocessing. In: Proc. ACS nat. meeting. Anaheim, CA, USA; 1986. 6 pp.
[348] Lunin L, Chambers LW, Parsons BI. The commercialization of the CANMET hydrocracking process. In: Proc. pet. conf. of the South Saskatchewan section. Regina, Canada; September 15–17, 1985. 15 pp.
[349] Patmore DJ, Khulbe CP, Belinko K. Residuum and heavy oil upgrading with the CANMET hydrocracking process. Proc ACS Div Pet Chem 1981;26:431–9. Preprint.
[350] Muir G, Pruden BB, Skripek M. Residue upgrading by the CANMET hydrocracking process. In: Proc. intern. symp. on heavy oil residue upgrading utilization. Beijing, China; May 5–8, 1991. p. 251–7.
[351] Gatsis J, Gleim W. Conversion of asphaltene-containing hydrocarbonaceous charge stocks. Patent No. US3785297. Universal Oil Prod Co.; March 27, 1973.
[352] Gatsis J, Gleim W. Desulfurization and conversion of black oils. Patent No. US3785958. Universal Oil Prod Co.; January 15, 1974.
[353] Gleim WKT. Role of asphaltene in refining. In: Chilingarian GV, Yen TF, editors. Asphaltenes asphalts. 1. developments in pet. science. Elsevier; 1978. p. 243–58.
[354] Davis H, Adams DG, Fox JV. Evaluation of heavy oil upgrading/refining process designs for hri dynacracking and UOP aurabon process. Report No. Final report. Houston, TX, USA: Texas Consultants, Inc.; April 1984. 53 pp.
[355] Dautzenberg FM, De Deken JC. Reactor developments in hydrotreating and conversion of residues. Catal Rev Sci Eng 1984;26(3–4):421–44.
[356] Gleim WKT. Slurry process for hydrocarbonaceous black oil conversion. Patent No. US3657111. Universal Oil Products Co.; April 18, 1972.
[357] Gleim WKT, Gatsis JG. Slurry process for hydrorefining petroleum crude oil and catalyst for use therein. Patent No. CA947270. Universal Oil Products Co.; May 14, 1974.
[358] Arrigo JT, Gatsis JG, Gleim WKT. Hydrorefining heavy oils. Patent No. GB1047698. Universal Oil Products Co.; November 09, 1966.
[359] O'hara MJ, Gatsis JG. Hydrorefining crudes and heavy oils. Patent No. GB1163008. Universal Oil Products Co.; September 4, 1969.
[360] Gatsis JG. Regenerative hydrorefining of hydrocarbon oils. Patent No. US3249556. Universal Oil Products Co.; May 03, 1966.
[361] Gleim WKT, Gatsis JG. Hydrofining of asphaltic crude oils with a colloidal dispersion of a lead salt as catalyst. Patent No. US3317421. Universal Oil Products Co.; May 02, 1967.
[362] Gatsis JG, Gleim WKT. Catalytic slurry process for effecting the conversion of a hydrocarbonaceous charge stock containing asphaltenes, metallic contaminants, and excessive quantities of sulfurous compounds. Patent No. US3622495. Universal Oil Products Co.; November 23, 1971.
[363] Gillis D, Vanwees M, Zimmerman P, Houde E. Upgrading residues to maximize distillate yields with UOP uniflex™ process. J Jpn Pet Inst 2010;53(1):33–41.
[364] Gillis D, Vanwees M, Zimmerman P. Upgrading residues for high levels of distillate production. PTQ 2010;(Q4):1–10.
[365] Ondrey G. Chementator: improved heavy-crude conversion process boosts diesel yields. Chem Eng 2012;119(3):12 (1 pp.).
[366] Lefebvre H. Maximize vacuum residue conversion and processing flexibility with the UOP uniflex™ process. In: Proc. XVIII Foro de Avances de la Industria de la Refinación. Mexico City. DF, Mexico; July 11–12, 2012. 20 pp.
[367] Haizmann R. Maximize conversion and flexibility: the UOP uniflex process. In: Proc. 6th Russia & CIS BBTC conf; April 13–14, 2011. 20 pp.
[368] Gillis D, Yokomizo G, Van Wees M, Rossi R, Houde E. Residue conversion options to meet marine fuel regulations. PTQ 2010;(Q3):1–12.

[369] KBR. The veba combi-cracking (VCC) process. 2015. 2 pp. Available from: http://www.kbr.com/Technologies/Refining/Veba-Combi-Cracking/Process/Default.aspx.

[370] Doehler W, Kretschmar DIK, Merz L, Niemann K. Veba-combi-cracking - a technology for upgrading of heavy oils and bitumen. In: Proc. ACS div. pet. chem. symp. on adv. resid upgrading. Denver, CO, USA, vol. 32. ACS; April 5−10, 1987. p. 484−9.

[371] Niemann K, Wenzel F. The veba-combi-cracking-technology: an update. Fuel Proc Technol 1993;35(1−2):1−20.

[372] Wenzel F, Kretschmar K. Veba-combi-cracking process-status and future trends. In: Proc. oil sands-our pet. future; Alberta res. council. Edmonton, Alberta, Canada; 1993. p. 238−53.

[373] Lischer H, Kretschmar K, Niemann K, Cotte E, Guitan J, Marruffo F. Deep conversion of heavy oil and bitumen via veba-combi-cracking: an update. In: Proc. 3rd UNITAR intern. conf. on 'heavy crude and tar sands'. Long Beach, CA, USA; July 22−31, 1988. p. 1586−603.

[374] Kretschmar K, Merz L, Niemann K, Guitian J, Krasuk J, Marruffo F. Process for the hydrogenation of heavy and residual oils. Patent No. US4851107 (Also published as CA1302332, DE3634275, EP0263522, EP0516187). Veba Oel Entwicklungs-Gesellschaft Mbh, Intevep S. A; July 25, 1989.

[375] Mayo SL, Motaghi M, Ravi R. Slurry phase hydrocracking: bottoms upgrading for today's market. Chem Eng World March 2014:52−9.

[376] Delbianco A, Meli S, Tagliabue L, Panariti N. ENI slurry technology: a new process for heavy oil upgrading. In: Proc. 19th world pet. congress. Madrid, Spain; Jun 29-Jul 3, 2008. Forum 07, Paper #19−2293, 9 pp.

[377] Panariti N, Rispoli G. Deep conversion of black oils with ENI slurry technology. In: Proc. world energy congress. Montreal, Quebec, Canada, vol. 3; September 12−16, 2010. p. 2646−58.

[378] Bellussi G, Rispoli G, Landoni A, Millini R, Molinari D, Montanari E, et al. Hydroconversion of heavy residues in slurry reactors: developments and perspectives. J Catal 2013;308:189−200.

[379] Bellussi G, Rispoli G, Millini R, Pollesel P. The turning point of the refining industry in europe. Chem Ing Tech 2014;86(12):2150−9.

[380] Montanari R, Rosi S, Marchionna M, Panariti N, Delbianco A. Production of uls fuels from heavy feedstocks. PTQ 2003;8(1):61−6.

[381] Marchionna M, Delbianco A, Panariti N, Montanari R, Rosi S, Correra S. Process for conversion of heavy charges such as heavy crude oils and distillation residues. Patent No. US2003089636 (abandoned). ENI S.p.A.; May 15, 2003.

[382] Marchionna M, Delbianco A, Panariti N, Montanari R, Rosi S, Correra S. Process for conversion of heavy charges such as heavy crude oils and distillation residues. Patent No. US2014197070 (Also published as DE10230403, US2003089636 (abandoned), US2012261309 (abandoned), US2011253594 (abandoned), US2010300934 (abandoned), US2006186021 (abandoned), US2009101540 (abandoned), RU2002117936, RU2298579, MXPA02006667, ITMI20011438, EG23196, CN1398953 CN100467575 CA2392669). ENI S.p.A.; July 17, 2014.

[383] Montanari R, Marchionna M, Rosi S, Panariti N, Delbianco A. Process for the conversion of heavy charges such as heavy crude oils and distillation residues. Patent No. US7691256 (Also published as WO2006066911, US2006157385, RU2007119430, RU2380397, PL382651, PL206827, NO20071892, MX2007006166, JP2008524413, JP4891259, ITMI20042445, EG24829, CN101068908, CA2530894, BRPI0518717, AU2005318406). ENI S.p.A., Snamprogetti S.p.A.; Enitecnologie S.p.A.; April 06, 2010.

[384] Montanari R, Marchionna M, Panariti N, Delbianco A, Rosi S. Process for conversion of heavy feedstocks such as heavy crude oils and distillation residues. Patent No. US8017000 (Also published as WO2004056946, WO2004056947, US2006163115, RU2005117791, RU2352616, PL375815, PL205245, NO20052930, MXPA05006708, JP2006511681, JP4639290, EP1572840, EG23696, ECSP055873, CA2510357, BR0317367, AU2003300217). ENI S.p.A.; Snamprogetti S.p.A.; Enitecnologie S.p.A.; September 13, 2011.

[385] Delbianco A, Panariti N, Correra S, Montanari R, Rosi S, Marchionna M. Process for the conversion of heavy charges such as heavy crude oils and distillation residues. Patent No. US2012261309 (abandoned) (Also published as US2003089636, DE10230403, US2011253594, US2010300934, US2006186021, US2009101540, WO2004058922, RU2002117936, RU2298579, MXPA02006667, ITMI20011438, EG23196, CN1398953, CN100467575, CA2392669). ENI S.p.A.; Snam Progetti; Enitecnologie S.p.A.; July 15, 2012.

[386] Marchionna M, Lami M, Ancillotti F. Catalyst precursors relevant active catalysts and their use in hydrotreating reactions. Patent No. EP0519556 (Also published as IT1248092). SnamProgetti, Enirecherche S.p.A.; December 23, 1992.

[387] Delbianco A, Panariti N, Correra S, Montanari R, Marchionna M, Rosi S. Petroleum residues upgrading: towards 100 percent conversion. In: Proc. AIChE spring meeting, 3rd intern. pet. phase behavior fouling. New Orleans, LA, USA; 2002. p. 520–5.

[388] Montanari R. Snamprogetti's highlights EST technology, RICE Global Engineering & Construction Forum. 2005. 51 pp. (Available from: http://www.forum.rice.edu/wp-content/uploads/roundtables/RT%20031105%20Montanari.pdf.

[389] ENI Press Office. Eni launches the EST project at the sannazzaro refinery. 2011. Available from: http://www.eni.com/en_IT/media/press-releases/2011/05/2011-05-16-Eni-project-EST-Sannazzaro-refinery.shtml.

[390] ENI Press Office. Sannazzaro refinery. 2014. Available from: http://www.eni.com/en_IT/company/operations-strategies/refining-marketing/refining-rf/refinery-sannazzaro.shtml.

[391] Poole MC, Bauman RF, Bearden Jr R, Effron E, Wu FS. Microcat-RC upgrading option for heavy crudes and residua. In: Proc. AIChE nat. meeting. New York, NY, USA. American Institute of Chemical Engineers; April 17–24, 1994. 19 pp.

[392] Bearden R. Microcat-RC™: technology for the hydroconversion upgrading of petroleum residua. In: Proc. 213th ACS nat. meeting. San Francisco, CA, USA, vol. 42; April 13–17, 1997. 20 pp.

[393] Aldridge C, Bearden R. Staged hydroconversion of an oil-coal mixture. Patent No. US4111787 (Also published as FR2356716, ZA7703293, JPS535210, JPS621559, GB1576039, DE2729533, CA1079665, BR7704251, AU2577177, AU504034). Exxon Research and Engineering Co.; September 05, 1978.

[394] Bearden R, Aldridge CL. Hydroconversion process. Patent No. US4067799. Exxon Research and Engineering Co.; January 10, 1978.

[395] Bearden R, Aldridge CL. Hydroconversion of heavy hydrocarbons. Patent No. US4066530 (Also published as BE856347). Exxon Research and Engineering Co.; January 3, 1978.

[396] Aldridge C, Bearden R. Hydroconversion of an oil-coal mixture. Patent No. US4298454. Exxon Research and Engineering Co.; November 03, 1981.

[397] Bearden R, Baird WC, Aldridge CL. Hydroconversion process. Patent No. US4330392. Exxon Research and Engineering Co.; May 18, 1982.

[398] Bearden R, Baird WC. Hydroconversion of heavy hydrocarbons using copper chloride catalyst. Patent No. US4051015. Exxon Research and Engineering Co.; September 27, 1977.

[399] Mertzweiller JK, Bearden R. Hydrorefining of heavy oil with hydrogen and aluminum alkyl compound. Patent No. US4055483. Exxon Research and Engineering Co.; October 25, 1977.

[400] Bearden R. Combined disulfurization and conversion with alkali metals. Patent No. US4076613. Exxon Research and Engineering Co.; February 28, 1978.

[401] Derouane EG, Varghese P. Hydrothermal conversion of heavy oils and residua with highly dispersed catalysts. Patent No. US4578181. Mobil Oil Corporation; March 25, 1986.

[402] Aldridge CL, Bearden R, Eidt CM. A process for coking and hydroconversion of hydrocarbons. Patent No. EP0308195 (Also published as JPH01101393, CA1312033). Exxon Research and Engineering Co.; March 22, 1989.

[403] Sahu R, Song BJ, Im JS, Jeon Y-P, Lee CW. A review of recent advances in catalytic hydrocracking of heavy residues. J Ind Eng Chem 2016;27:12–24.

[404] Zhang S, Liu D, Deng W, Que G. A review of slurry-phase hydrocracking heavy oil technology. Energy Fuels 2007;21(6):3057–62.

[405] Solari B, Guitian J, Krasuk J, Marzin R. HDH, hydrocracking as an alternative for high conversion of the bottom of the barrel. In: Proc. NPRA annual meeting. San Antonio, TX, USA; March 25–27, 1990. Paper AM-90-28, 29 pp.
[406] De Drago G, Guitian J, Krasuk J, Larrauri J, Marzin R, Silva F, et al. Development of HDH process, a refiner's tool for residual upgrading. In: Proc. ACS div. pet. chem. symp. on resid upgrading. Washington, DC, USA, vol. 35. ACS; August 26–31, 1990. p. 584–92.
[407] De Drago G, Guitian J, Krasuk J, Larrauri J, Marzin R, Silva F, et al. Development of the HDH process. Rev Tec INTEVEP 1990;10(1):15–36.
[408] Marzin R, Guitian J, Krasuk J, Silva F, Souto A, Solari B. HDH, a process for residue conversion. In: Proc. 4th UNITAR intern. conf. 'heavy crude tar sands'. Edmonton, Alberta. Canada, vol. 4; August 7–12, 1988. p. 255–70.
[409] Cavicchioli I, De Drago G, Gonzalez G, Krasuk J, Larrauri J, Marzin R. HDH process for heavy crude upgrading. In: Proc. 3rd UNITAR intern. conf. 'heavy crude tar sands'. Long Beach, CA, USA, vol. 3; July 22–31, 1985. p. 1747–79.
[410] Ramirez de Agudelo MM, Larrauri Derteano JM, Krasuk J. Regeneration of an iron based natural catalyst used in the hydroconversion of heavy crudes and residues. Patent No. US4751210 (Also published as GB2204809, IT1219305, CA1313651, DE3816846, FR2615408, JP1293137). Intevep SA.; June 06, 1988.
[411] Ramirez de Agudelo MM, Seaton C, Specht MI. Process for the production of vanadium-containing steel alloys. Patent No. US5242483 (Also published as DE4326259. FR2694573, GB2270908, SE9302560, CA2100442, GB2274280, JP6192785, DE4326259, IT1260942, JP2050998C, JP7091617B, RU2094491). Intevep SA.; September 07, 1993.
[412] Ramirez de Agudelo MM, Manrique M, Seaton C, Hurtado J. Process for the recovery of nickel and magnesium from natural occurring materials. Patent No. US5229088 (Also published as DE4216540, GB2264937, CA2062350, FR2689140, JP5287404, JP2031209C, JP7062182B). Intevep SA.; July 20, 1993.
[413] Castillo E, Morel F. HDHPLUS®: integrated scheme with sequential hydroprocessing. In: Proc. ERTC 11th annual meeting. Paris, France; November 2006. 18 pp.
[414] Pernalete C, Torres B, Moreno A. Reactores de la tecnología HDHPLUS® como componentes de simulación interoperables. Cienc Ing November 2012:1–8. Available from: http://erevistas.saber.ula.ve/index.php/cienciaeingenieria/article/view/3952.
[415] Castillo E. Proceso HDHPLUS®/SHP, la opción para la conversión de residuos de crudos pesados y extrapesados. In: Proc. Jornadas Latinoamericanas de Refinación. Mendoza, Argentina; October 31, 2006. 25 pp.
[416] Martinis J, Castillo E, Morel F. HDHPLUS®/SHP: getting the most out of the bottom of the barrel. In: Proc. 6th intern. bottom of barrel technol. conf. Barcelona, Spain; October 9–10, 2008. 24 pp.
[417] Morel F. HDHPLUS®/SHP: heavy residue hydroconversion technology. In: Proc. 5th NCUT upgrading refining conf. Edmonton, Canada; September 14–16, 2009. 27 pp.
[418] PDVSA Hdhplus® Venezuelan technology. The New PDVSA Contact; 2005 (2), 11 pp. Available from: http://www.latinpetroleum.com/pdf/gabriel_oliveros.pdf.
[419] PDVSA Puerto La cruz refinery expansion project. 2014. 2 pp. Available from: http://abarrelfull.wikidot.com/puerto-la-cruz-refinery-expansion-project.
[420] Kozlowski RH, Spars BG. Conversion of residual oil to middle distillate oils and gasoline. Patent No. US3293169. Chevron Research Co.; December 20, 1966.
[421] Paterson NJ. Converting asphaltenic oils and olefinic gasolines to higher quality gasoline. Patent No. US3472760. Chevron Research Co.; October 14, 1969.
[422] Roselius RR. Heavy oil conversion. Patent No. US3684688. Chevron Research Co.; August 15, 1972.
[423] Lopez J, Pasek EA. Heavy oil hydroprocessing with group VI metal slurry catalyst. Patent No. US4857496. Chevron Research Co.; August 15, 1989.

[424] Lopez J, Pasek EA. Heavy oil hydroprocessing with group VI metal slurry catalyst. Patent No. US4970190. Chevron Research Co.; November 13, 1990.
[425] Lopez J, Pasek EA. Heavy oil hydroprocessing with group VI metal slurry catalyst. Patent No. US5162282. Chevron Research Co.; November 10, 1992.
[426] Kramer DC. Process to prevent catalyst deactivation in activated slurry hydroprocessing. Patent No. US5294329, US5298152. Chevron Research Co.; March 15 and 29, 1994.
[427] Threlkel RS. Low macropore petroleum residuals conversion catalyst. Patent No. WO9604073 (Also published as AT269159, AU3098095, CA2194974, FI970356, CZ9700044, SK10797, EP0804288, CN1154668, CN1078822, JP2001520567, JP3838660). Chevron USA; February 15, 1996.
[428] Cash DR, Armstrong MJ. Residuum hydrotreating/hydrocracking with common hydrogen supply. Patent No. US6179995. Chevron USA; January 30, 2001.
[429] Chen K, Leung PC, Reynolds BE. Process for upgrading heavy oil using a highly active slurry catalyst composition. Patent No. US7238273 (Also published as US2006054535, WO2006031542, NO20071786, MX2007002673, MX2007002654, KR20070057825, KR100876663, JP2008512555, EP1824947, EA200700604, EA016502, CN101084292, CA2579837, BRPI0515087). Chevron USA; March 23, 2006.
[430] Chen K, Leung PC, Reynolds BE, Chabot J. Process for upgrading heavy oil using a highly active slurry catalyst composition. Patent No. US7214309 (Also published as US2006054534, WO2006031543, NO20071784, KR20070051890, KR100902807, JP2008512556, EP1828348, EA200700603, EA012332, CN101044227, CA2579838, BRPI0515094). Chevron USA; March 23, 2006.
[431] Dahlberg AJ, Mukherjee UK, Mayer JF, Louie WSW. High conversion hydroprocessing. Patent No. US7238277 (Also published as WO2006065643, US2006131212, CA2590868, AR053103). Chevron USA; July 03, 2007.
[432] Farshid D, Reynolds B. Process for upgrading heavy oil using a highly active slurry catalyst composition. Patent No. US7390398 (Also published as US2007138055, WO2007078620, KR20080077406, JP2009520061, EP1973995, EA200870066, EA013731, CN101360808, CA2632823, BRPI0619921). Chevron USA; June 21, 2007.
[433] Farshid D, Reynolds B. Integrated in-line pretreatment and heavy oil upgrading process. Patent No. US7431831 (Also published as CN101356252, US20070138058). Chevron USA; October 07, 2008.
[434] Farshid D, Murphy J, Reynolds B. Process for upgrading heavy oil using a reactor with a novel reactor separation system. Patent No. US7431822 (Also published as US2007138056, WO2007078621, KR20080077395, JP2009520062, EP1960498, EA200870067, EA013065, CN101336282, CA2633902, BRPI0619988). Chevron USA; October 7, 2008.
[435] Chen K, Chabot J. Process for recycling an active slurry catalyst composition in heavy oil upgrading. Patent No. US7431824 (Also published as US2006054533, WO2006031571, NO20071781, MX2007002672, KR20070051892, KR100930991, JP2008512557, JP4874977, EP1794265, EA200700610, EA011976, CN101300325, CA2579139, BRPI0515100). Chevron USA; October 07, 2008.
[436] Chen K, Leung PC, Reynolds BE. Highly active slurry catalyst composition. Patent No. US7410928 (Also published as US2007265157, US2006058175, WO2006031575, NO20071780, MX2007002670, KR20070051316, KR100886184, JP2008512242, JP5089390, EP1814662, EA200700609, EA011933, CN101014411, CA2579999, CA2579999, BRPI0515091). Chevron USA; August 12, 2008.
[437] Chen K, Reynolds BE. Highly active slurry catalyst composition. Patent No. US7396799, US7754645 (Also published as US2006058174, WO2006031570, US2009107891, US2007179055, US2008305947, NO20071782, MX2007002674, KR20070051317, KR100886188, JP2008512241, JP5101285, EP1799343, EA200700591, EA011428, CN101027127, CN101027127, CA2579998, BRPI0515071). Chevron USA; July 08, 2008.
[438] Farshid D, Reynolds B. Process for upgrading heavy oil using a highly active slurry catalyst composition. Patent No. US7431823 (Also published as US2007138057, US2009057193). Chevron USA; October 07, 2008.

[439] Dahlberg AJ, Mukherjee UK. High conversion hydroprocessing using multiple pressure and reaction zones. Patent No. US7531082 (Also published as WO2006096368, US2006196809, ZA200707894, TWI338039, JP2008531816, EP1859010, CA2599717, AU2006220972, AR053822). Chevron USA; May 12, 2009.
[440] Mukherjee UK, Dahlberg AJ. High conversion hydroprocessing using multiple pressure and reaction zones. Patent No. US7531082 (Also published as US2006196809, WO2006096368, ZA200707894, TWI338039, JP2008531816, EP1859010, CA2599717. AU2006220972, AR053822). Chevron USA; May 12, 2009.
[441] Farshid D, Reynolds B. Integrated heavy oil upgrading process and in-line hydrofinishing process. Patent No. US7708877 (Also published as US2007138059, KR20080080618 JP2012255158 JP2009520063 JP5081160 EP1960499, EA200870068 EA016773 CA2631855 BRPI0619931). Chevron USA; June 21, 2010.
[442] Farshid D, Murphy J, Reynolds B. Process for upgrading heavy oil using a reactor with a novel reactor separation system. Patent No. US7901569 (Also published as US2009057194). Chevron USA; March 08, 2011.
[443] Chen K, Chabot J. Process for recycling an active slurry catalyst composition in heavy oil upgrading. Patent No. US7972499 (Also published as US20090050526). Chevron USA; July 05, 2011.
[444] Biswas G, Farshid D. Systems and methods for producing a crude product. Patent No. US8372266 (Also published as US2009008290). Chevron USA; February 12, 2013.
[445] Nguyen JV, Brait A, Mironov O, Kuperman AE. Hydroprocessing catalysts and methods for making thereof. Patent No. US8697594 (Also published as CA2817986, CN103260749, EP2658647, US8703637, US8778828, US8802586, US8802587, US8809222, US8809223, US20120168350, US20120168351, US20120172197, US20120172198, US20120172199, US20120172200, US20120172201, US20120172202, US20120172203, US20120172204, US20120172205, US20120172206, US20120172207, WO2012092006). Chevron USA; April 15, 2014.
[446] Nguyen JV, Brait A, Mironov O, Kuperman AE. Hydroprocessing catalysts and methods for making thereof. Patent No. US8802586 (Also published as US20120172198). Chevron USA; August 12, 2014.
[447] Nguyen JV, Brait A, Mironov O, Kuperman AE. Hydroprocessing catalysts and methods for making thereof. Patent No. US8802587 (Also published as US20120172199). Chevron USA; August 12, 2014.
[448] Nguyen JV, Brait A, Mironov O, Kuperman AE. Hydroprocessing catalysts and methods for making thereof. Patent No. US8703637 (Also published as US20120168351). Chevron USA; April 22, 2014.
[449] Nguyen JV, Brait A, Mironov O, Kuperman AE. Hydroprocessing catalysts and methods for making thereof. Patent No. US8809222 (Also published as US20120172200). Chevron USA; August 19, 2014.
[450] Nguyen JV, Brait A, Mironov O, Kuperman AE. Hydroprocessing catalysts and methods for making thereof. Patent No. US8778828 (Also published as US2012172197). Chevron USA; July 15, 2014.
[451] Nguyen JV, Brait A, Mironov O, Kuperman AE. Hydroprocessing catalysts and methods for making thereof. Patent No. US8809223 (Also published as US20120172201). Chevron USA; August 19, 2014.
[452] Yang S, Chabot J, Reynolds BE. Hydroprocessing catalysts and methods for making thereof. Patent No. US8759242 (Also published as US20120004097). Chevron USA; June 24, 2014.
[453] Anonymous. Hpinnovations: demonstration unit to be built for heavy oil upgrader. Hydrocarb Proc 2008;87(5):26 (1 pp.).
[454] Anonymous. Refining and marketing: Chevron trials new upgrading process. Pet Econ 2008;75(5).
[455] Anonymous. News downstream: Chevron plans heavy oil plant. Pet Rev 2008;62(735):8 (1 pp.).
[456] Anonymous. Processing - quick takes: Chevron tests heavy oil hydrocracking technology. Oil Gas J 2008;106(11):10 (1 pp.).
[457] Anonymous. Places'n'faces: base oil from pascagoula? Lubes-n-Greases 2008;14(4):1.
[458] Verstraete J, Guillaume D, Auberger MR. Catalytic hydrotreatment and hydroconversion. In: Hue A-Y, editor. Heavy crude oils - from geology to upgrading. Paris, France: Editions Technip; 2011. p. 299−328 [chapter 18].
[459] Mukherjee U. Reliable platform for high conversion of residue. In: Proc Refinery Technology Meet. Kochi, India; November 10−13, 2013. 16 pp. Available from: https://www.academia.edu/7113736/Paper-Reliable_Platform_for_High_Conversion_of_Residue_final.

[460] Lott RK, Bailey RT. (hc)3™ process - a slurry hydrocracking technology designed to convert bottoms and heavy oils. In: Proc. 7th UNITAR intern. conf. 'heavy crude tar sands'. Beijing, China, vol. 2; October 27–30, 1998. 8 pp.

[461] Lott RK, Lee TLK, Quinn J. (hc)3™ hydrocracking technology. In: Proc. SPE/PS-CIM/CHOA intern. thermal operations heavy oil symp. Calgary, Alberta, Canada; November 1–3, 2005. Paper #98058, 9 pp.

[462] Headwaters Inc. Headwaters incorporated announces results of first commercial scale test of its proprietary (hc)3™ heavy oil hydrocracking technology. In: News Bull.: financial relations board; 2006. p. 1–3. South Jordan, UT, USA.

[463] Lott R. Application of HCAT technology in commercial hydrocracker to upgrade difficult feedstocks. In: Proc. opportunity crudes conf.: challenges, benefits. Houston, TX, USA; May 1–2, 2008. p. 26–7.

[464] Smith KL, Kunnas J. HCAT test run at neste oil porvoo refinery. In: Proc. bottom of the barrel technol. Conf. London, UK; October 2010. 17 pp.

[465] Headwaters Hydrocarbon Technologies Inc. and Alberta Science and Research Authority (hc)3™ Technology. 2003. Available from: http://hcat-hti.com/technology-platforms/hcat/.

[466] Lott R. Methods and mixing systems for introducing catalyst precursor into heavy oil feedstock. Patent No. US2006201854 (Also published as CA2646492, CN101405370, EP1999233, EP2735601, US20090310435, WO2007106783). Headwaters Heavy Oil, LLC; September 14, 2006.

[467] Lott R. Methods and mixing systems for introducing catalyst precursor into heavy oil feedstock. Patent No. US2009310435. Headwaters Heavy Oil, LLC; December 17, 2009.

[468] Lott RK, Chang YH. Systems for increasing catalyst concentration in heavy oil and/or coal resid hydrocracker. Patent No. US8309041. Headwaters Heavy Oil, LLC; November 13, 2012.

[469] Lott RK, Chang YH. Methods for increasing catalyst concentration in heavy oil and/or coal resid hydrocracker. Patent No. US8557105 (Also published as CA2704202, EP2215194, US8034232, US8309041, US20090107881, US20120009094, US20130068658, WO2009058785). Headwaters Heavy Oil, LLC; October 15, 2013.

[470] Pelrine BP, Comolli AG, Lee LK. Ionic liquid catalyst composed of iron promoted with metal including cobalt, molybdenum, palladium, platinum, nickel and/or tungsten in water having acidic pH. Patent No. US6139723. Hydrocarbon Technologies, Inc.; October 31, 2000.

[471] Lott R, Lee LK. Methods for hydrocracking a heavy oil feedstock using an in situ colloidal or molecular catalyst and recycling the colloidal or molecular catalyst. Patent No. US8303802 (Also published as CA2564346, EP2811006, US7449103, US7815870, US8431016, US8440071, US8673130, US20050241991, US20080193345, US20100294701, US20110220553, US20110226667, US20130228494, US20130233765, US20140093433, WO2005104752). Headwaters Heavy Oil, LLC; November 06, 2012.

[472] Lott RK, Chang YH. Methods for increasing catalyst concentration in heavy oil and/or coal resid hydrocracker. Patent No. US8034232. Headwaters Heavy Oil, LLC; October 11, 2011.

[473] Lott RK, Lee LK. Hydroprocessing method and system for upgrading heavy oil using a colloidal or molecular catalyst. Patent No. US7578928 (Also published as CA2564342, CN1950484, EP1753844, EP2272938, US20050241993, WO2005104749). Headwaters Heavy Oil, LLC; August 25, 2009.

[474] Lott RK, Lee LK. Systems for hydroprocessing heavy oil. Patent No. US2014093433. Headwaters Heavy Oil, LLC; April 03, 2014.

[475] Lott RK, Lee LK, Quinn PC. Fixed bed hydroprocessing methods and systems and methods for upgrading an existing fixed bed system. Patent No. US7517446 (Also published as CA2564359, CN1950483, EP1753845, US20050241992, WO2005104786). Headwaters Heavy Oil, LLC; April 14, 2009.

[476] Smith L, Mountainland D, Gendler J. HCAT® residue hydrocracking technology. In: Proc. bottom of the barrel technol. conf. Dubrovnik, Croatia; June 15–16, 2011. 27 pp.

[477] Silverman BM. HCAT concentration effect. In: Proc. bottom of the barrel technol. conf. Dubrovnik, Croatia; May 17, 2013. 29 pp.

[478] Smith KL, Rueter M, Gendler J. Improving resid upgrading economics with HCAT. In: Proc. bottom of the barrel technol. conf. Rome, Italy; May 24–25, 2012. 26 pp.

[479] Smith KL, Gendler J. Improving resid upgrading economics with HCAT. In: Proc. Russia & CIS bottom of the barrel technol. conf. Moscow, Russia; May 2014. 28 pp.

[480] Zarkesh J, Hashemi R. Heavy residue hydroconversion (hrh), a new method for upgrading heavy oil and residue. In: Proc. 4th upgrading refining of heavy oil, bitumen, synthetic crude oil conf. Edmonton, Canada; September 26–27, 2006. p. 88–100.

[481] Rhodey G. A new concept to maximize heavy hydrocarbon upgrading. In: Proc. 4th upgrading refining of heavy oil, bitumen, synthetic crude oil conf. Edmonton, Canada; September 26–27, 2006. p. 102–22.

[482] Romocki SM, Rhodey G. A novel process for heavy residue hydroconversion using a recoverable pseudohomogeneous catalyst (phc) system. In: Proc. intern. thermal operations heavy oil symp. Calgary, Alberta, Canada. Society of Petroleum Engineers; October 20–23, 2008. 7 pp.

[483] Romocki S, Rhodey G. Mobis hrh process: residue hydroconversion using a recoverable nano-catalyst. In: Proc. 5th NCUT upgrading refining conf. Edmonton, Alberta, Canada; September 14–16, 2009. 26 pp.

[484] Zarkesh J, Hashemi R, Ghaedian M, Khakdaman HR, Ahmadpanah SJ, Khadzhiev S, et al. Hrh: nano catalytic process to upgrade extra heavy crude/residual oils. In: Proc. 19th world pet. congress. Madrid, Spain, vol. 2; June 29–July 3, 2008. p. 641–53.

[485] Bahmani M, Farshi A, Masoudian TSK, Rashidzadeh M, Rashtchi M, Sadighi S, et al. An integrated process for the conversion of heavy hydrocarbons to a light distillate and/or mid-distillate. Patent No. EP2154225. RIPI, NTI Co.; February 17, 2010.

[486] Sadighi S, Mohaddecy RS, Masoudian K. New process arrangements for upgrading heavy oils and residua. PTQ, Digit Refin 2014;(Q3):7.

[487] Sue H. Idemitsu hydrocracking process produces distillate from residual oil. In: Proc. 3rd UNITAR intern. conf. 'heavy crude tar sands'. Long Beach, CA, USA, vol. 3; August 8–11, 1985. p. 1780–90.

[488] Armstrong RB, Sadhukhan P, Sue H. Mild resid hydrocracking for middle distillates: a new heavy oil upgrading alternative. In: Proc. 3rd UNITAR intern. conf. 'heavy crude tar sands'. Long Beach, CA, USA, vol. 4; July 22–31, 1985. p. 1837–48.

[489] Itoh T, Sue H. Mild resid hydrocracking for heavy oil upgrading. In: Proc. 5th UNITAR intern. conf. 'heavy crude tar sands' Caracas, Venezuela, vol. 4; August 4–9, 1991. p. 43–8.

[490] Ozaki H, Yamane M, Satomi Y, Suzuka T, Miyake T, Hirato M. Catalyst for treating heavy oils. Patent No. US4115324. Nippon Mining Co.; September 19, 1978.

[491] Kiuchi N, Suzuka T, Miyake T, Uchida S, Hirato M, Souma T, et al. Regeneration of nickel oxide ore catalyst. Patent No. JP52123396. Nippon Mining Co., Hitachi Ltd.; October 17, 1977.

[492] Inooka M. Process for hydrocracking hydrocarbons with hydrotreatment-regeneration of spent catalyst. Patent No. US4525267 (Also published as GB2127320, JPS6239634, WO8204441, DE3152896). Chiyoda Chem Eng Construct Co.; June 25, 1985.

[493] Ando M, Fukui Y, Homma Y, Shiroto Y. Process for hydrotreating heavy hydrocarbon oil. Patent No. US4191636 (Also published as DE2824765, DE2824765, NL7806212, NL180019, JPS541306, JPS5740879, GB1602640, CA1126192). Chiyoda Chem Eng Contruct Co.; March 04, 1980.

[494] Johnson AR, Brown EC. Combination process for the pretreatment and hydroconversion of heavy residual oils. Patent No. US5320741 (Also published as CA2093561, CN1078487, EP0565205). Stone & Webster Engineering Corporation; June 14, 1994.

[495] Silverman MA. SOC technology: a flexible approach to residual oil upgrading. In: Proc. AIChE spring meeting. Houston, TX, USA; March 19–23, 1995. Paper #45a, 9 pp.

[496] Seko M, Ohtake N, Kuri K, Kato K, Asakura N. Super oil cracking (SOC) process for resid upgrading. AIChE Symp Ser 1989;85(273, Fundam. Resid Upgrading):43–8.

[497] Selucky M, Kim SS, Skinner F, Strausz OP. Structure-related properties of Athabasca asphaltene and resins as indicated by chromatographic separation. In: Bunger JW, Li NC, editors. Chemistry of asphaltenes. Washington, DC: ACS; 1981. p. 83–118.

[498] Okui T, Yasumuro M, Tamura M, Shigehisa T, Yui S. Convert oil sands into distillate cost-effectively. Hydrocarb Proc 2006;85(1):79–85.
[499] Yasumuro M, Okui T, Shigehisa T, Tamura M, Okuyama N. Method for hydrocracking petroleum heavy oil. Patent No. CA2426374 (Also published as JP2003327971, BR0301421). Kobe Steel Ltd.; November 15, 2003.
[500] Okui T, Tamura M, Eikawa K, Ferraz De Souza W. Process of hydrocracking heavy oil. Patent No. CA2613806 (Also published as US20080156693, JP2008163097, BRPI0704463). Kobe Steel Ltd.; Petrobras; July 27, 2008.
[501] Okui T, Komatsu N, Okuyama N, Tamura M. Process of hydrocracking petroleum heavy oil. Patent No. CA2536557. Kobe Steel Ltd.; May 24, 2011.
[502] Okui T, Okuyama N, Shimasaki K, Tamura M, Komatsu N, Yasumuro M. Process for hydrocracking of petroleum heavy oil. Patent No. CA2314033. Kobe Steel Ltd.; January 21, 2001.
[503] Liu C, Zhou J, Que G, Liang W, Zhu Y. Hydrocracking of gudao residue with dispersed-phase Mo catalyst. Fuel 1994;73(9):1544–50.
[504] Wang Z-X, Zhang H-Y, Guo A-J, Que G-H, Olah GA, Prakash GKS. Study on hydrogen donation ability of residue hydrocracking system catalyzed by oil-soluble and water-soluble catalysts. ACS Div Fuel Chem 1998;43(3):530–3. Preprints.
[505] Cheng J, Liu GX, Pan P, Luo YH, Que GH. Catalytic hydroconversion of gudao residue using CO - water as hydrogen source. Petrochem Technol 2000;29(6):420–4.
[506] Liu D, Cui W, Zhang S, Que G. Role of dispersed Ni catalyst sulfurization in hydrocracking of residue from karamay. Energy Fuels 2008;22(6):4165–9.
[507] Zhang S, Deng W, Luo H, Lui D, Que G. Slurry-phase residue hydrocracking with dispersed nickel catalyst. Energy Fuels 2008;22(6):3583–6.
[508] Liu D, Kong X, Li M, Que G. Study on a water-soluble catalyst for slurry-phase hydrocracking of an atmospheric residue. Energy Fuels 2009;23(2):958–61.
[509] Luo H, Deng W, Gao J, Fan W, Que G. Dispersion of water-soluble catalyst and its influence on the slurry-phase hydrocracking of residue. Energy Fuels 2011;25(3):1161–7.
[510] Guan CS, Wang ZX, Guo AJ, Que GH. Sulfurization of water-soluble catalyst for suspended bed hydrocracking of residue. Shiyou Xuebao, Shiyou Jiagong/Acta Pet Sin Pet Proc Sect 2004;20(2):75–80.
[511] Ren R, Wang ZX, Guan CS, Guo AJ, Que GH. Study on the sulfurization of water-soluble catalysts for slurry-bed hydroprocessing of residue II XPS analysis of sulfurized product of molybdate. Ranliao Huaxue Xuebao/J Fuel Chem Technol 2005;33(3):299–303.
[512] Ren R, Wang ZX, Guan CS, Guo AJ, Que GH. Study on the sulfurization of water-soluble catalysts for slurry-bed hydroprocessing of residue I effect of sulfurization conditions on the species and dispersity of the sulfurized catalysts. Ranliao Huaxue Xuebao/J Fuel Chem Technol 2005;33(3):293–8.
[513] Bi WD, Que GH. Simulation research of gas holdup in the medium-sized continuous residuum-hydrocracking slurry-bed reactor under environmental conditions. Shiyou Daxue Xuebao/J Univ Pet China 2001;25(3):19–22.
[514] Cheng J, Li J, Liu YH, luo YH, Liu GX, Que GH. Gudao residual hydrocracking with dispersed catalysts using supercritical water-syngas as hydrogen source II. The comparison of residue hydrocracking using different hydrogen sources. J Fuel Chem Technol 2004;32(2):180–4.
[515] Cheng J, Liu Y, Luo Y, Que G. Hydrocracking of gudao residual oil with dispersed catalysts using supercritical water-syngas as a hydrogen source. Part II: the comparison of residue hydrocracking in different hydrogen sources. Pet Sci Technol 2006;24(11):1339–46.
[516] Cheng J, Liu Y, Luo Y, Que G. Hydrocracking gudao residual oil using supercritical water-syngas as hydrogen source in suspension bed. Part III: water-gas shift reaction. Pet Sci Technol 2008;26(17):2088–94.
[517] Guan CS, Wang ZX, Guo AJ, Que GH. Study on slurry catalyst for kelamayi residue hydrocracking. Ranliao Huaxue Xuebao/J Fuel Chem Technol 2003;31(5):434–8.

[518] Rivero R, Rendon C, Gallegos S. Exergy and exergoeconomic analysis of a crude oil combined distillation unit. Energy 2004;29(12–15):1909–27.
[519] Cao X. Optimum selection of vacuum residue processing scheme in oil price boosting area. Pet Proc Petrochem 2009;40(1):1–9.
[520] Corredor LM, Carrillo JA, Parra MJ. Yields and operational conditions for deasphalting of vacuum bottoms using n-C4 + i-C4, n-C5 + i-C5, n-C6 and n-C7 solvents. In: Proc. ACS nat. meeting, div. fuel chem. Boston, MA, USA, vol. 52. ACS; 2007. p. 809–11.
[521] Li R, Huang W, Yang J. Residual oil processing method with high yield of liquid products. Patent No. CN1297984 (Also published as CN1152120). China Petrochemical Corp.; June 06, 2001.
[522] Patel AVK, Moretta JC, Northup AH. Report...Upgrading schemes: heavy oil upgrading options. Hydrocarb Asia 2000;10(1):18–21.
[523] Iqbal R, Khan A, Eng O, Floyd R. Unlocking current refinery constraints. PTQ 2008;13(2):31–3.
[524] Iqbal R. Resid optimization: are vacuum units the most effective starting point?. In: Proc. JPI pet. refining conf. Tokio, Japan; October 20–21, 2004. p. 67 (1 pp.).
[525] Hood RL. Solvent deasphalting unit capable of generating power. Patent No. US5843302 (Also published as CA2224037). Ormat Process Technologies Inc.; June 12, 1998.
[526] Hood RL. Method of and apparatus for processing heavy hydrocarbons. Patent No. US5919355 (Also published as CA2273899). Ormat Industries Ltd.; July 06, 1999.
[527] Hood RL, Rettger PB, Goldstein RS, Bronicki LY. Method of and means for upgrading hydrocarbons containing metals and asphaltenes. Patent No. CA2244598. Ormat Process Technologies Inc.; February 13, 1999.
[528] Friday RJ, Goldstein RS, Rettger PB. Process and apparatus for upgrading hydrocarbon feeds containing sulfur, metals, and asphaltenes. Patent No. EP0984054 (Also published as CA2281058, CN1313577, CN1323339, EP0984054, WO2000014178). Ormat Industries Ltd.; March 08, 2000.
[529] Friday JR, Rettger PB, Goldstein RS. Apparatus for upgrading hydrocarbon feeds containing sulfur, metals, and asphaltenes. Patent No. US6274003. Ormat Ind Ltd.; August 14, 2001.
[530] Doron B. Method of producing combustible products from heavy fuel residue. Patent No. US6365038. Ormat Ind Ltd.; April 02, 2002.
[531] Doron B. Method of and means for producing combustible products from heavy fuel residue. Patent No. CA2182653. Ormat Industries, Ltd.; February 16, 2010.
[532] Bronicki Y. Method of and apparatus for processing heavy hydrocarbon feeds. Patent No. US7297250 (Also published as EP1096002, US2006032789, US2003129109, WO0132807, TR200003193, MXPA02004289, ID27905, GT200000189, EP1096002, EG22312, EA200001012, EA002795, DE60011978, CO5200801, CN1399671, CA2324557, BR0005211, AU1246601, AT270703, AR026308). Ormat Ind Ltd.; November 20, 2007.
[533] Rettger P, Goldstein R, Arnold J, Bronicki Y, Friday JR. Method of and apparatus for upgrading and gasifying heavy hydrocarbon feeds. Patent No. US6702936 (Also published as CA2439038, EP1465967, US20030116470, WO2003060042). Ormat Industries Ltd.; March 09, 2008.
[534] Rettger P, Goldstein R, Arnold J, Bronicki Y, Friday RJ. Method of and apparatus for upgrading and gasifying heavy hydrocarbon feeds. Patent No. US7407571 (Also published as US20040118745). Ormat Industries Ltd.; August 05, 2008.
[535] Bronicki Y, Rettger P, Goldstein R, Friday RJ, Arnold J. Method of and apparatus for upgrading and gasifying heavy hydrocarbon feeds. Patent No. CA2576950. Ormat Industries Ltd.; January 11, 2011.
[536] Rettger PB. Process and apparatus for upgrading hydrocarbon feeds containing sulfur, metals, and asphaltenes. Patent No. US6183627 (Also published as MXPA01002304). Ormat Ind Ltd.; February 06, 2001.
[537] OPTI Canada, news release. 2002. Calgary, Canada. 1 pp. Available from: www.opticanada.com.
[538] Birdgeneau J. The Canadian Long Lake project integrating SAGD, cogeneration and upgrading technologies to develop Athabasca bitumen. In: Proc. 19th world pet. congress. Madrid, Spain, vol. 1; June 29–July 3, 2008. p. 436–43.

[539] Anonymous. OPTI sells 15% stake in joint venture to Nexen. In: Financial post; December 17, 2008. Toronto. 1 pp.
[540] Anonymous. Opti-Nexen Long Lake upgrader begins operation. Edmont J January 23, 2009. Edmonton, Canada. 3 pp. Available from: http://oilsandstruth.org/optinexen-long-lake-upgrader-begins-operation.
[541] Carroll J. Nexen, OPTI Canada may be targets as price drop spurs takeovers. Bloomberg.com; 2009. 4 pp. Available from: http://www.cambridgestrategies.com/Download-document/17-Opti-Canada-May-Be-Targeted-in-Oil-Sands.html.
[542] Iqbal R, Anshumali, Eng O, Niccum P. Heavy oil and bitumen upgrading. Patent No. US7381320 (Also published as US2006042999, WO2006025873, EP1784477, CA2563922). Kellogg Brown and Root, Inc.; June 13, 2008.
[543] Kikuchi K. Upgrading refinery residue by the HSC process and the HSC-ROSE combination. In: Proc. pet. refining. Tokyo, Japan; 1988. 25 pp.
[544] Bakshi AS, Lutz IH. Combination of visbreaking and solvent deasphalting boosts gas-oil yields for fccu feed. Oil Gas J 1986;84(2). 52−54, 57−58.
[545] Furimsky E. Gasification in petroleum refinery of 21st century. Oil Gas Sci Technol - Rev IFP 1999;54(5):597−618.
[546] Netzer D. Alberta bitumen processing integration study. Report No. Final. The Province of Alberta, Economic Development Department. Investment and Industrial Development Branch. Department of Energy. The Alberta Energy Research Institute; March 2006. 164 pp.
[547] Chem Systems. Unconventional heavy oils. Report No. PERP 04/05S9. White Plains, NY, USA: Nexant; November 2005. 112 pp.
[548] Garwin L, Baldwin R. Process for vis-breaking asphaltenes. Patent No. US4389302 (Also published as CA1198390). Kerr-McGee Refining Corp.; June 21, 1983.
[549] Visser D. Process for the conversion of a residual hydrocarbon oil. Patent No. EP0673989. Shell International Research; September 27, 1995.
[550] Visser D. Process for the conversion of a heavy hydrocarbonaceous feedstock. Patent No. EP0372652. Shell International Research; June 13, 1990.
[551] Gorbaty ML, Jacobon M, Siskin M. Residuum conversion process. Patent No. US7033486 (Also published as CA2479779, EP1504076, US2003221992, WO2003085069). ExxonMobil Research and Engineering Co.; April 25, 2006.
[552] Bartilucci M, Jacob SW, Karsner GG. Low efficiency deasphalting and catalytic cracking. Patent No. US5000838. Mobil Oil Corp.; March 19, 1991.
[553] Audeh CA, Rankel LA. Integrated process for whole crude deasphalting and asphaltene upgrading. Patent No. US5192421. Mobil Oil Corp.; March 9, 1993.
[554] Miller CB, Moore HF, Wesley DP, Wombles RH, Jewitt CH, Hayner RE, et al. Demetallation - high carbon conversion process, apparatus and asphalt products. Patent No. US5601697. Ashland Oil Inc.; February 11, 1997.
[555] Miller CB, Moore HF, Wesley DP, Wombles RH, Jewitt CH, Hayner RE, et al. Demetallation - high carbon conversion process, apparatus and asphalt products. Patent No. US5728291. Ashland Oil Inc.; March 17, 1998.
[556] Payton EM, Slover JW. Catalytic cracking of residual oils containing metal contaminants in several stages. Patent No. US3303123. Phillips Petroleum Co.; February 2, 1967.
[557] Ferrughelli DT, Gorbati ML. Heavy oil upgrading process. Patent No. US6171471. Exxon Research and Engineering Co.; January 09, 2001.
[558] Gleim WKT. Regenerative hydrorefining of petroleum. Patent No. US3293172. Universal Oil Products Co.; December 20, 1966.
[559] Bearden R, Hou Z, Gorbaty ML, Ferrughelli DT, Myers RD. Combination slurry hydroconversion plus solvent deasphalting process for heavy oil upgrading wherein slurry catalyst is derived from solvent deasphalted rock. Patent No. US6511937 (Also published as WO0127225, NO20021693, JP2003529629,

EP1228169, CA2386575, AU7870200, AR026009). Exxon Research and Engineering Co.; January 28, 2003.

[560] Bearden R, Hou Z, Gorbaty ML, Ferrughelli DT, Olmstead WN, Meyers RD. Heavy feed upgrading based on solvent deasphalting followed by slurry hydroprocessing of asphalt from solvent deasphalting. Patent No. WO0160952 (Also published as JP2003523451, EP1268713, CN1400997, CA2398252, AU3823501). ExxonMobil Research and Engineering Co.; August 23, 2001.

[561] Aldridge CL, Bearden R, Eidt CM. Combination coking and hydroconversion process. Patent No. US4569751, US4569752, US4750985. Exxon Research and Engineering Co.; February 11, June 14, 1986, 1988.

[562] Voorhies A, Hamner GP. Combination deasphalting-coking-hydrotreating process. Patent No. US3617481. Exxon Research and Engineering Co.; November 02, 1971.

[563] Morel F, Chapus T, Kressman S, Duplan J-L, Billon A, Heinrich G. Process for converting a heavy hydrocarbon fraction using an ebullated bed hydrodemetallization catalyst. Patent No. US5980730. Institut Francais du Petrole; November 9, 1999.

[564] Morel F, Heinrich G, Kressmann S, Billon A, Duplan J-L, Chapus T. Multi-step catalytic process for conversion of a heavy hydrocarbon fraction. Patent No. US6017441. Institut Francais du Petrole; January 25, 2000.

[565] Morel F, Kressmann S, Duplan J-L, Billon A, Chapus T, Heinrich G. Multi-step process for conversion of a petroleum residue. Patent No. US6007703. Institut Francais du Petrole; December 28, 1999.

[566] Morel F, Chapus T, Kressman S, Duplan J-L, Billon A, Heinrich G. Catalytic process for conversion of a petroleum residue using a fixed bed hydrodemetallization catalyst. Patent No. US6117306. Institut Francais du Petrole; September 12, 2000.

[567] Billon A, Morel F, Kressmann S, Kim SD, Ha SK, Heor H. Process for hydrotreatment of a heavy hydrocarbon fraction using permutable reactors and introduction of a middle distillate. Patent No. US6306287. Institut Francais du Petrole; October 23, 2001.

[568] Gueret C, Kressmann S, Verstraete J. Method for the valorization of heavy charges by bubbling-bed deasphalting and hydrocracking. Patent No. US8636896 (Also published as FR2854163, US2005006279, US2011062055, RU2004112558, RU2344160, KR20040093010, KR101088267, CA2464796, EP1505142). Institut Francais du Petrole; January 28, 2014.

[569] Colyar JJ. Effective integration of solvent deasphalting and ebullated-bed processing. Patent No. US7214308 (Also published as US2004163996, RU2005129321, RU2337939, MXPA05008076, WO2004074408, CA2516562, BRPI0407704). Institut Francais du Petrole; May 08, 2007.

[570] Colyar JJ, Kressmann S, Gueret C. Integrated sda and ebullated-bed process. Patent No. US7279090 (Also published as US20060118463, WO2006061120). Institut Francais du Petrole; October 9, 2007.

[571] Colyar JJ, Wisdom L, Billion A, Kressmann S, Mor F. Deep conversion of heavy oils via combination of the h-oil process and solvent deasphalting. In: Proc. 16th world pet. congress. Calgary, Alberta, vol. 3; June 11–15, 2000. p. 111–4.

[572] Sughrue IEL, Tooley PA, Bertus BJ, Grayson BS. Process for converting heavy hydrocarbon oil. Patent No. US5024750 (Also published as EP0435242, NO905576, ES2070991, DK0435242, DE69018599, CA2023860, AT121121). Phillips Petroleum Co.; June 18, 1991.

[573] Beaton WI, Taylor JL, Peck LB, Mosby JF. Catalytic cracking with deasphalted oil. Patent No. US4940529. Amoco Corp.; July 10, 1990.

[574] Mosby JF, Peck LB, Taylor JL, Beaton WI. Resid hydrotreating with resins. Patent No. US5013427 (Also published as WO9101360, JPH04501879, JP2804369, EP0434799, CA2064338). Amoco Corp.; May 07, 1991.

[575] Kolstad JJ, Beaton WI, Taylor JL. Process for deasphalting resid, recovering oils, removing fines from decanted oil and apparatus therefor. Patent No. US5124025. Amoco Corp.; June 23, 1992.

[576] Taylor JL, Beaton WI, Kolstad JJ. Three-stage process for deasphalting resid, removing fines from decanted oil and apparatus therefor. Patent No. US5124026. Amoco Corp.; June 23, 1992.

[577] Beaton WI, Taylor JL, Kolstad JJ, Kennedy BS, Hughes RD. Multi-stage process for deasphalting resid, removing catalyst fines from decanted oil and apparatus therefor. Patent No. US5124027. Amoco Corp.; June 23, 1992.

[578] Taylor JL, Beaton WI, Kolstad JJ. Means for and methods of low sulfur and hydrotreated resids as input feedstreams. Patent No. US5228978. Amoco Corp.; July 20, 1993.

[579] Taylor JL, Kolstad JJ, Beaton WI. Means for and methods of deasphalting low sulfur and hydrotreated resids. Patent No. US5242578. Amoco Corp.; September 7, 1993.

[580] Kolstad JJ, Beaton WI, Taylor JL. Means for and methods of removing heavy bottoms from an effluent of a high temperature flash drum. Patent No. US5258117. Amoco Corp.; November 2, 1993.

[581] Taylor JL, Beaton WI, So BYC, Kolstad JJ. Resid hydrotreating using solvent extraction and deep vacuum reduction. Patent No. US5312543. Amoco Corp.; May 17, 1994.

[582] Inomata M, Fujimura Y, Okada T, Imura K, Sasaki H. Method of refining petroleum. Patent No. US2004069685 (Abandoned) (Also published as EP1350832, RU2270230, PL362971, PL194242, NZ526112, KR20030053062, KR100800286, WO0244307, CN1487989, CN100387690, AU2002222568, AU2256802). JGC Corp.; April 15, 2004.

[583] Fujimura Y, Nagamatsu S, Inomata M, Hagio M. Gefinery- new refining concept to utilize residual oil as gas turbine fuel. ACS Div Pet Chem 2005;50(4):399–400. Preprints.

[584] Dorawala TG, Lafferty WL, Kuhlmann EJ. Conversion increase of vacuum residuums. Patent No. EP0403087 (Also published as JPH0374491, CA2010774). Texaco Development Corp.; December 19, 1990.

[585] Gosselink JW, Van Der HA, Huizinga T, Schaper HPA. Process for the conversion of a residual hydrocarbon oil. Patent No. EP0683218. Shell International Research; November 22, 1995.

[586] Nielsen RH. ENI slurry hydroprocessing technology for diesel fuel. Report No. PEP Review 99-2. Menlo Park, CA: SRI. Consulting; June 2003. 42 pp.

[587] Delbianco A, Montanari R, Panariti N, Rosi S. Integrated process for the conversion of feedstocks containing coal into liquid products. Patent No. US2010219104 (Abandoned) (Also published as WO2005047425, ZA200603668, US2007144944, RU2006115300, RU2360944, ITMI20032207, EP1730252, CN1890350, CN100489061, AU2004289810). ENI S.p.A.; Snam Progetti; Enitecnologie S.p.A.; September 02, 2010.

[588] Heck RH, Diguiseppi FT. Resid hydrocracking using dispersed metal catalysts. Patent No. US5362382. Mobil Oil Corp.; November 8, 1994.

[589] Weinberg HN, Rebick C, Mcelrath KO. Fluid coking with solvent separation of recycle oil. Patent No. US4497705. Exxon Research and Engineering Co.; February 05, 1985.

[590] Rankel LA, Shu P. Process for upgrading heavy crude oils. Patent No. US4530757. Mobil Oil Corp.; July 23, 1985.

[591] Liu SK. Method of producing hydrogen from heavy crude oil using solvent deasphalting and partial oxidation methods. Patent No. US5958365. Atlantic Richfield Co.; September 28, 1999.

[592] Johnson KA, Wallace PS. Filtration of feed to integration of solvent deasphalting and gasification. Patent No. US6274030 (Also published as ZA200106017, WO0039251). Texaco Development Corp.; February 4, 2002.

[593] Barkley RW, Johnson KA, Kasbaum JL, Thacker PS, Wallace PS, Niccum JG. Integration of solvent deasphalting and gasification. Patent No. US6241874 (Also published as ZA200101417). Texaco Development Corp.; June 5, 2001.

[594] Crane SD, Beer GL, Blacker HF. Method for converting a heavy crude oil produced via a wellbore from a subterranean formation into a distillate product stream using a solvent deasphalting process. Patent No. US6054496 (Also published as WO9913024). Atlantic Richfield Co., Arco British Ltd.; April 25, 2000.

[595] Iqbal R, Patel V, Eng O, Motaghi M. Canadian crudes: how much can I process?. In: Proc. NPRA annual meeting. San Antonio, TX, vol. 1; March 22–24, 2009. p. 171–84.

[596] Subramanian M, Hanson FV. Design of supercritical fluid extraction process to upgrade oil sand bitumen by continuous thermodynamics. In: Proc. 7th UNITAR intern. conf. 'heavy crude tar sands'. Beijing, China, vol. 2; October 27–30, 1998. 17 pp.

[597] Angelov G, Shibley PWM. Deasphalting heavy crude oils and heavy crude oil-water emulsions. Patent No. US4634520. Bitumen Development Corp. Ltd.; January 06, 1987.

[598] Marruffo F, Chirinos ML, Sarmiento W, Hernandez - Carstens E. Orimulsion® a clean and abundant energy source. In: Proc. world energy congress. Houston, USA; September 13–18, 1998. p. 73–88.

[599] Wilcox B. Coleson cove Orimulsion conversion project (br-1740). In: Proc. ePRI-DOE-EPA-AWMA. Combined power plant air pollutant control mega symp. Washington, DC, USA; May 19–22, 2003. 5 pp.

[600] Benguigui I, Hood RL, Rettger P. Method of and apparatus for producing power in solvent deasphalting units. Patent No. US5804060 (Also published as WO9721786, AU1147997, CA2238531). Ormat Process Technologies Inc.; September 08, 1998.

[601] Hervish PV, Wescott KR, Briesch MS, Brown SW. Method and system for generating power from residual fuel oil. Patent No. WO9845387 (Also published as US5911875, JP2001527615, DE69815331, CN1254363, CA2286160, EP0973845, AR012342). Siemens Westinghouse Power; October 15, 1998.

[602] Morel F, Peries J. Residue hydroconversion. In: Leprince P, editor. Pet. refining: conv. processes. France: TECHNIP; 2001. p. 411–50.

[603] Bridjanian H, Khadem Samimi A. Bottom of the barrel, an important challenge of the petroleum refining industry. Pet Coal 2011;53(1):13–21.

[604] Sayles S, Romero S. Comparison of thermal cracking and hydrocracking yield distributions. PTQ 2011;(3):9. Paper #1000070.

[605] Rana MS, Ancheyta J, Maity SK, Marroquin G. Comparison between refinery processes for heavy oil upgrading: a future fuel demand. Intern J Oil, Gas Coal Technol 2008;1(3):250–82.

[606] Liu Y, Gao L, Wen L, Zong B. Recent advances in heavy oil hydroprocessing technologies. Recent Pat Chem Eng 2009;2(1):22–36.

[607] Kasztelan S, Harle V, Kressmann S, Morel F. Recent developments in fixed bed catalytic residue upgrading. ACS Div Pet Chem 1997;42(2):340–2. Preprints.

[608] Rana MS, Sámano V, Ancheyta J, Diaz JAI. A review of recent advances on process technologies for upgrading of heavy oils and residua. Fuel 2007;86(9 SPEC. ISS.):1216–31.

[609] Angeles MJ, Leyva C, Ancheyta J, Ramírez S. A review of experimental procedures for heavy oil hydrocracking with dispersed catalyst. Catal Today 2014;220–222:274–94.

CHAPTER 6

Emerging Technologies and Ideas with Potential

1. Introduction

In Chapters 2−4, the characteristics of the troublesome species found in unconventional-opportunity crude oils have been presented. The impact of these species on refining operations has been pointed out as well. This problematic scenario has been addressed by modifications, improvements, and adaptations of commercially available technologies, briefly described in Chapter 5. Regardless of all the advances achieved, some other issues exist, and new or improved technologies are still needed. Most of the existing commercially available technologies abate those species in a comprehensive way by treating all of them as whole heavy and extra-heavy crude oil, bitumen, or their residues. Only asphaltenes have been focused specifically through deasphalting (DA) processes. These technological approaches do not represent cost-effective solutions and therefore do not allow for monetizing the discounts that opportunity crudes offered. In fact, one could think that the real opportunities for R&D have been left aside.

The troublesome compounds are characterized for a high proportion of heteroatoms (S, N, O, and metals) that have to be removed for the manufacturing of transportation fuels and other petroleum products. The chemistry involved in conventional technologies falls into either the C-rejection type or the H-addition type. Meanwhile, some other unconventional treatments, reactions, and chemistries have been proposed for the abatement of these species.

R&D activities in these areas, though subjected to the oil pricing waves, have never stopped, and numerous ideas have been tested and proposed through the years. A vast number of publications have appeared either in the open literature or in patent applications that for one reason or another never made it to the commercial scale. Some have already progressed more than others, but this does not imply a free path to the commercial scale. Whether any of these concepts would eventually progress and emerge into a commercial process remains uncertain. Thus, this chapter will discuss these emerging ideas, first those that address the concerning compounds individually and, later, those dealing comprehensively with the whole BotB fraction.

The Science and Technology of Unconventional Oils. http://dx.doi.org/10.1016/B978-0-12-801225-3.00006-1

2. Asphaltenes

Several ideas have been tested at the lab or bench scale at most, for the abatement of asphaltenes; these are included in this section. Ideas targeting only or mainly asphaltenes are described, though other heteroatomic moieties present in the residue or asphaltenic fraction would be affected. These ideas have been grouped in four categories: (1) conversion induced thermally, (2) chemical reactions, (3) solutions for asphaltenes-derived problems within a production facility, and (4) novel ideas in their infancy stage.

2.1 Thermally Induced Conversion

Thermolysis has been one of the preferred conversion routes, first as a characterization methodology [1–4] and then as a processing means, in conventional technologies. Temperatures range from moderate like those employed in visbreaking to increasingly higher as for pyrolysis, coking, and gasification.

Hot solids in a fluidized bed could induce conversion of vacuum residue, when the residue is solvated with a hydrocarbon under supercritical conditions of the mixture. The process is carried out at a weight ratio of solvent to residue above 2:1. The feed is heated to a temperature below that of the hot solids and fed all together to a reaction zone, where an equilibrium temperature corresponding to the reaction temperature is established. This process is claimed to achieve high conversion rates for asphaltenes, promoting the removal of sulfur, nitrogen, and metals. In addition, high selectivity to naphtha and distillate, and low coke formation, were also claimed under nearly complete conversion [5].

An approach to use the metal contaminants as catalyst proposed to create a metals-enriched fraction by subjecting a portion of the residue to a mixed synthesis gas (syngas)–steam plasma. This fraction containing the metals is used to promote hydrogenation reactions of the residue under simultaneous partial pyrolysis. The catalyst is finely suspended, forming a slurry with the feed, and hydrogen is provided from the reaction of the heavy residue with steam, using plasma [6].

2.2 Chemical Conversion of Asphaltenes

There are chemical reactions that can be carried out with high selectivity for asphaltene conversion that have not been examined, developed, or evaluated for commercial application. Some examples are collected here.

An unusual alkaline metal (Na in particular) catalyst was suggested for promoting desulfurization and asphaltenes hydroconversion (HDC). The crude oil (or fraction) is first hydrotreated under mild conditions to avoid asphaltene conversion or deposition.

The sodium catalysts are used in a second step, after purging the produced H_2S, and under high hydrogen pressure and temperatures [7].

A reaction widely studied in the literature is oxidation catalyzed by ruthenium ions (ruthenium-ion catalyzed oxidation (RICO)) [8–33]. Strausz and coworkers [8–10] introduced RICO reactions into petroleum chemistry in 1985. The basis of the RICO method relies on selectivity for oxidizing aromatic carbon and removing it as CO, while leaving saturated carbon essentially unaffected. Ruthenium (VIII) catalyzes the oxidation of the aromatic asphaltene core. These reactions were proposed for the structural characterization of asphaltenes, since the selective conversion into carboxylic acid provides insights of the original moiety from which they derive [9]. Alkyl-substituted aromatics are oxidized to CO, except at the - site of the alkyl attachment, which is converted to a carboxylic group anchored to the alkyl chain [11]. The major oxidation products were n-alkanoic acids (C_1–C_{35}), α,ω-di-n-alkanoic acids (C_4–C_{35}), smaller amounts of benzene di- through hexacarboxylic acids, and a nondistillable oxidized residue. Mild pyrolysis of the latter, after methylation with diazomethane, produces series of n-alkanes/n-alkenes, n-alkanoic/n-alkenoic acid methyl esters, and free n-alkanoic acids [9,12,13].

As already discussed in Chapter 5, iron ores and minerals have been proposed as additives or catalysts in HDC reactions of the BotB, for processes commercially available, such as CANMET [34–39] (now Uniflex [40–42]) and HDH [43–48]. Synthetic pyrite or pentacarbonyl-iron has been tested for the co-processing of coals and vacuum residues, in syngas-water, observing high rates of conversion to THF-solubles. Conversion in the presence of syngas-water was compared to that in pressurized CO-water and H_2-water. The THF-soluble yield in syngas-water was almost comparable to that obtained with hydrogen-water, indicating certain synergistic effects between the two components of syngas [49]. A point to emphasize is the way that data were handled, which only allows relative comparison among the test results, since the two components of the feed were not individually characterized and conversion was indirectly determined from the THF-insolubles. THF-insolubles were considered to contain the ashes from the minerals present in the feed, the spent catalyst, and the unconverted feed. Finally, changes in the aromaticity of the product light component during propane–asphaltene reaction were observed and taken as an indication of asphaltene reactivity under the different studied atmospheres. However, the differences among the asphaltene yields were relatively insignificant.

Nickel boride (Ni_2B) and boron bromide (BBr_3) were tested in the cleavage of the C–S bonds and of the esters' C–O bond, respectively [23]. The Athabasca n-C_5-asphaltenes were fractionated into occluded maltene and low-molecular-weight and high-molecular-weight (LMA and HMA, respectively) asphaltene. Ni_2B reduction of the HMA fraction yielded 5–18% n-pentane soluble compounds. The unreduced asphaltenes underwent 40% desulfurization and a greater than fourfold drop in the molecular weight (MW). No MW reduction was observed with the LMA fraction. The cleavage of the core sulfidic bonds was attributed as responsible for the decrease in MW. This feature of Athabasca asphaltenes might facilitate upgradability, minimizing coke formation. The common products for both fractions were n-alkanes, cheilanthanes, regular steranes, hopanes, and gammacerane, while the LMA also contained dicyclic terpanes and $C_{21}-C_{25}$ steranes.

A high level of asphaltenes conversion can be achieved via hydrotreatment (HDT) of atmospheric residue (AR) [50]. However, catalyst life cycle and regenerability are negatively impacted by such operation; see Chapters 2 and 5. Furthermore, refractory (difficult-to-remove) N and S compounds that are concentrated in the asphaltenes fraction would require more active catalysts or more severe HDT conditions, or will worsen the quality and effectiveness of the HDT. The presence of metal compounds, some more reactive than those of S- and N-, jeopardize catalyst activity and, thus, product quality [51].

2.3 Upstream Solutions

In order to alleviate the problems caused by asphaltenes in upstream operations, several ideas have been proposed. Unfortunately, most of these have been tested at the lab scale only, and some others have been announced and/or have been the subject of R&D projects, but very little or nothing has been published. Crude oil treatments either at surface level in the production area or in the well have been considered. Treatments in the well merits a separate section and that will be found in Section 6. These merits are the magnitude of the involved challenges, of the R&D efforts and of the impact that a positive achievement would have.

A moderate treatment uses a combination of available technologies but also employs a noncommercial catalyst. The treatment includes catalytic hydrovisbreaking followed by SDA, and the objective was a viscosity reduction of Doba crude oil at production sites. The reduction in viscosity target is that required by pipeline specifications, and the conversion target is such that it would not compromise the crude oil stability. Two catalysts were tested: iron sulfide and molybdenum sulfide. The results showed that the performance of the first overtook that of the latter [52].

An acid treatment carried out in the well causes de-waxing, desulfurization, and DA. An aqueous solution of dithiocarbamate (sodium-methyl or dimethyl) and acetic acid is used to treat the oil or oil-bearing formations [53].

Conditions for performing vacuum pyrolysis of Alberta tar sands have been defined in order to achieve DA and avoid excessive decomposition reactions that would increase the gas yield. These vacuum pyrolysis conditions (total pressure of 10^{-2} bar and temperature of about 500°C) result in high yield of transportable oil. The gas yield was about 11 wt%, and the main components were CO_2 (the most abundant), CO, H_2S, CH_4, and C_2-C_8 hydrocarbons. The pyrolysis oil was de-asphalted and contained maltene in about 98 wt% [54].

2.4 Other Novel Ideas

Ultrafiltration of heavy hydrocarbons has been claimed as a means for removing detrimental qualities of metals and CCR-promoter compounds. The ultrafiltering media are membranes, and the process consists of at least two membrane ultrafiltration stages. Retentates from any stage can be recycled to combine with the feed or can be combined with the retentate streams from the various ultrafiltration stages. The second permeate can be sent to subsequent ultrafiltration stages utilizing the same membrane, with permeate from each such stage being used as feed to the next subsequent downstream stage [55]. Since viscosity and density of heavy oils make ultrafiltration a slow operation, dilution with a suitable solvent that solvates the whole oil may facilitate ultrafiltration [56]. The reported application seems to be more effective for demetallization (DM) than it is for DA. An example will be given in the section describing DM physical methods.

A Sn−Sb intermetallic filter that was specifically designed for the removal of sulfur compounds resulted in only 50% reduction in S. An additional 20% removal of asphaltenes was also observed. This intermetallic filter works under electrokinetic effects (i.e., an applied or induced potential difference). Although the removing capabilities need improvement, the versatility of the device for serial or parallel integration to other units provides more options for examination [57].

Asphatenes adsorption on silica surfaces has been proven as a fractionation and characterization simplification tactic. Further studies and applications of this concept have been explored recently. In fact, it was initially observed during GPC separation that the clay present in the residue was concentrated in the high MW (HMW) fractions together with the asphaltenes. This affinity between the clay and the asphaltenes is thought to be physical, and not chemical. It was also observed that the clay exerted a catalytic effect on asphaltene polymerization, which was most pronounced for the highest MW fraction and gradually decreased with decreasing MW [58].

Fundamental studies of asphaltenes adsorption on clays (particularly kaolin) and on iron oxide nanoparticles [59−68] confirmed the potentiality of an adsorption-based removal method. Adsorption was proven to undergo very fast and to work with both virgin and thermally converted asphaltenes [63,64,67]. The asphaltenes adsorbed on clays could be easily gasified [63], while the nanoparticles were catalytically active on their oxidation [68].

A sequential and rather irreversible adsorption has been proposed [69], in which the first adsorbed layer is formed by the smallest aggregates (three to six molecules), followed sequentially according to solubility [70] (see Section 4.5 in Chapter 2). The three identified fractions were found to be similar in terms of their chemical composition, but structurally different. This structure difference was assigned as responsible for the differences in solubility. Further studies are needed to understand whether changes in reactivity also exist. Studies of the asphaltene—silica—water system indicated that the asphaltene—silica interactions exhibited a time-dependent characteristic, and KCl addition was capable of suppressing a repulsive force component. Meanwhile, a pH increase eliminated the attractive force [71]. In other words, additives promoting the interaction could be used to induce removal, while a change in pH can be employed for a desorption step.

Other siliceous and nonsiliceous materials have been considered as adsorbents, in a process that keeps asphaltenes in solution by the action of a polar solvent. The adsorbent materials might be silica, silica—alumina, acid-treated clays, and activated carbons [72].

Adsorption on metallic surfaces does not appear as interesting as that on silica surfaces. The kinetic results of the interactions between asphaltenes and metal surfaces showed to be too slow to have any potential for a removal application [73]. Solid acids based on transition metal oxides can selectively remove basic asphaltenes. Potential transition metal candidates are tungsten, niobium, their mixtures, and/or their mixtures with tantalum, hafnium, chromium, titanium, or zirconium. The transition metal oxide can be supported on an inorganic refractory oxide support such as alumina. Basic asphaltenes are removed in an adsorption zone. The asphaltene-saturated adsorbent is separated from the feed and subsequently subjected to cracking conditions in the presence of steam. The converted organics are separated, and the regenerated catalyst—adsorbent is recycled back to the adsorption zone [74].

3. Metal Compounds

Total metal content of crude oils and bitumen hardly reaches the percentage level. Regardless of the parts per million (ppm) level of contamination, these compounds cause irreversible damage to refining catalysts, and their removal is highly recommended. Nonetheless, specific technologies for metals removal are not commercially available and are far from being developed, as will become evident in this section.

Metal contaminants concentrate in the BotB fraction and more particularly in the asphaltenic residue. For this reason, DA has been considered and used as an indirect DM method, as for example in the UOP process patented in Ref. [75]. According to this process, a DAO containing 47 ppm metals can be obtained from a VR containing 186 ppm metals (DM = 75%). Additionally, improvements on CCR (from 15.8% to 10.7%), S (from 3.0% to 2.63%), and API gravity (from 10.1 to 13.3) were attained.

Originally, SDA was thought as the production of a valuable stream, which could be fed into other refining processes (e.g., FCC feed preparation [76], gas oil [77], etc.). Rather than feeding the DAO in a conventional FCC, Ashland recommended an RCC unit, while the produced pitch concentrating the metal contaminants is either pretreated or used directly in the manufacture of novel formulations [78,79]. Certain enhancements and modifications of SDA have been oriented toward improving DM capabilities and have been discussed in the previous chapter. In this section, applications of DM under SDA conditions that have not been discussed in Chapter 5 are exemplified.

A tar containing gas oil was used as wash oil and added to the feed of an SDA unit, above the point of introduction. The feed was a residue, and the wash oil was composed of a mixture of DAO (8.5°API gravity, 17.5% CCR, 56 ppm Ni, and 215 ppm V) and a full-range decanted oil, DCO (0.3 ppm V and 0.1 ppm Ni). This wash oil was added at an oil-to-asphalt ratio of 3:1, at 35 bar, and 75–95°C. The obtained product had API gravity of 16.4° and contained 2.5% CCR, 0.7 ppm Ni, and 1.3 ppm V [80]. Since both the residue and the wash oil contained significant amounts of contaminating metals, it seems that the catalyst fines present in the DCO might have DM capabilities.

Combinations of solvents and stages were considered to optimize DM under DA conditions, particularly the use of polar solvents and the recovery of the resins from the DAO. These improvements and intensified applications were discussed in Chapter 2; some additional examples follow. For instance, a first extraction with propane, followed by a second with pentane, yielded a DAO containing 0.3% asphaltenes and 7.0 ppm metals. This process was carried out using an overall solvent–feed volumetric ratio (VR) of 3.8:1.0. The VR feed (8.0°API, 3.1% S, 186 ppm metals, and 14.8% asphaltene) was extracted with 2.5 volumes of propane (40 bar and 95°C). The propane-lean phase was extracted with 4.0 volumes of n-pentane (40 bar and 30°C). The overall DAO yield was better than 94% [81]. This example demonstrates that a combination of solvents and stages can improve simultaneously yield and quality.

In order to increase DM efficiency under SDA conditions, a light acrylic hydrocarbon solvent was suggested as extractant for high metals containing heavy crude or residue using at least a two-stage supercritical method. The recovered low metals containing light fraction from the first stage is combined with a moderate- or low-metals residue or crude, and passes to a second supercritical extraction zone. The heavy fraction from the second extraction zone is recycled to the first zone. The light extracted fraction is stripped off solvent, and the low-metals-containing product is recovered. Then, the solvent is reused [82].

In summary, DA has served to the oil refinery industry with a dual purpose, and most refiner users see it as a way to "kill two birds with one stone." Nevertheless, the larger the metal contamination, the lesser the benefits of DA as a simultaneous DM method are. In the long term, a dedicated technology for removing the metals from the oil will favor downstream processing of both DAO and pitch. Consequently, a comprehensive and more effective value would be added to all oil products.

3.1 Physical Methods

Newer SDA-based patents specifically claim the DM application [83,84]. In any case, conventional SDA products need to be subjected to DM for two different reasons:

1. DAO, even though it contains a largely reduced amount of metal contaminants, has to be DM-treated for meeting feed specifications of the downstream converting processes; and
2. Pitch (asphaltic fraction), by concentrating the metals from the feed, makes DM or any other metal recovery method a potential reclamation method for the higher contaminated crudes.

For this latter purpose, a countercurrent extraction between two immiscible solvents (a heavy and a light), at 1 bar and 70°C in a centrifugal contactor, can be used for leaching the pitch of its constituents. The heavy solvent, consisting of 3–10% H_2O in phenol, dissolves the condensed ring aromatics, the asphaltenes, and most of the metal compounds. The light solvent is a C_8 fraction of a butene–isobutene alkylation, and it dissolves the oil plus the remaining metal compounds. Preferred ratios are two to four volumes of heavy solvent and 2–10 volumes of light solvent to one volume residuum. The use of aqueous phenol as the heavy solvent was necessary to obtain two liquid phases of different density and composition. As the water content of the aqueous phenol solvent is increased, the heavy solvent rejects first the oily constituents, then the resins or maltenes and organometallic compounds, and finally the asphaltenes. The light alkylate solvent rejects these compound classes in the reverse order. Metal recovery was as high as more than 70% [85].

The use of a polar solvent at 0–10°C above its critical temperature and at pressures in the range of 300–1000 psig above its critical pressure for SDA would facilitate DM of the asphaltic residue from a propane-SDA unit. An exemplified feed contained 1–8 wt% S, 5–25% CCR, and 100–2000 ppm metals. Extraction takes place in counterflow with one to five volumes of Me_2O or Me_2NH. The critical temperature of the solvent must be $\geq$ 100°C below the initial boiling point of the feed. The treatment process of Tia Juana VR with a combination of propane SDA produced a DAO (yield = 20–35 wt% based on crude oil) with less than 100 ppm of metal, a pitch extract (yield = 15–20%) with 300–500 ppm metal, and a residue asphaltic pitch (5–10% of the crude) with 1500–7000 ppm metal [86].

The term dendritic separation was coined to mean the simultaneous (all-at-once) separation and isolation of various constituents from a feed, as opposed to the sequential, consecutive separation and isolation of the organic and inorganic constituents. A process of this type was intended for the simultaneous separation of a soluble maltene fraction and the confinement/concentration of the vanadium, nickel, and iron constituents from insoluble asphaltenes (pitches and tars) and from oxidized hydrocarbons (naphthenates) [87].

The phases to be separated are (1) hydrocarbons; (2) derivatives of hydrocarbons; (3) clays and salts (particle sizes of less than 80 pm); and (4) sand, gravel, and added minerals (particle sizes of greater than 80 pm). A mixture of solvents composed of at least two solvents, A and B, is used to play specific roles. Solvent A can act (1) as a surface-active wetting agent for both the hydrocarbon and the mineral substrate by lowering interfacial tension; (2) as demulsifying agent by absorbing water present; (3) as a mineral clay precipitant consolidating agent; and/or (4) as a solvent for the maltene fraction. Solvent B can be made more or less acidic or basic. The mixture of solvents is chosen so that it results in:

1. a biphasic solvent system A + B formed when the solvent mixture temperature is raised above its lower critical solution temperature (LCST; e.g., 40°C in an example case in which A is ethanol in the 10–57% concentration range and B is water),
2. an altered LCST biphasic solvent system containing an aromatic propylene glycol ether (e.g., addition of 1–2 vol% of phenyl propylene glycol ether decreases to 22–26°C the LCST of mixture describe in 1),
3. a biphasic solvent system form by a salt addition (e.g., 10:90 and 60:40 concentrations of N-cyclohexyl pyrrolidone in water are miscible at 50°C but form a biphasic solvent system when a 2% sodium chloride is included in the mixture), and
4. a pH-altered LCST biphasic solvent system (e.g., LCST decreases from 50°C to 5°C when the concentration of NaOH is raised from 0 to 0.5 N and increases from 50°C to 80°C when the concentration of H_2SO_4 is raised from 0 to 1 N, for a solvent system composed of 70% N-cyclohexyl pyrrolidone and 30% water).

This solvent mixture is used to render two or three distinct hydrocarbon layers with specifically partitioned fractions of heavy metals. The top layer is almost devoid of iron and nickel and has a substantially reduced concentration of vanadium. The concentration ratio of the second to the third layer is slightly less than one for iron and slightly less than two for nickel and vanadium. In total numbers, the greater portion of the iron, nickel, and vanadium is found in the third layer. The products in a two-phase system differ because the top is more lipophilic and the bottom is hydrophilic. Partition is defined by the pH of these phases, as each can be more acid or more basic than the other.

Other disclosed extracting solvents include a pyridine–H_2O mixture containing 10–40% water [88] and an aqueous solution of CO_2 [89,90]. The latter is specifically defined for extracting organically bound +2 ionic metal species.

The exploration of mixtures of polar/nonpolar and miscible/immiscible solvent systems is far to be completed. The generation of more than one-phase systems for segregating metal compounds and/or asphaltenes in different phases seems to have great potential. Solvents could be tuned for DM or DA purposes, even to neutralize acid compounds, as it is discussed further in other sections of this book [85–88,91–97].

Other physical means for separation of metal compounds from crude oils include also filtration, ion exchange, sorption, electrophoresis, precipitation, and electroflocculation, and examples of application will be given in the next paragraphs.

Metal contaminants may be separated by filtration through a polymeric membrane. Dilution of the crude oil into a C_5−C_8 cyclohydrocarbon solvent (e.g., cyclopentane, cyclohexane, methylcyclopentane, or cyclooctane) was recommended. Ni content was decreased from 66.0 to 14.0 ppm, and V content from 733.0 to 115.0 ppm [98]. Ultrafiltration with selected membranes, as mentioned above, was found to reduce simultaneously metal content and CCR (crude is filtered in diluted form). The solvent employed has to solvate completely the oil. As an example, an Arab VR was dissolved in cresol at a 1:1 oil−cresol ratio and ultrafiltered using a Nuclepore type C membrane; all of the permeates were combined, and the solvent removed. The permeate yield was 44.7 wt%, and the permeate had CCR 8.6%; the retentate yield was 56−58 wt%, and the retentate had CCR 32.2%, with DM better than 85% [56]. As discussed in Chapter 2, viscosity of heavy oils makes ultrafiltration a difficult method, and dilution with highly solvating solvent is required. These ultrafiltration methods appear to have more drawbacks than benefits, such as expensive membranes and solvents, flexibility and adaptability of the solvent to various crudes, operability, and so on.

Whether ion exchange or sorption, the hydrogen form of Nafion (a fluorosulfonic acid resin: copolymer of tetrafluoroethylene and perfluoro 3,6-dioxa-4-methyl-7-octenesulfonic acid) has been shown to be effective in metals removal. This separation method involves contacting the crude oil diluted in an aromatic hydrocarbon with that strong solid-phase Bronsted acid (Nafion) at 220°C [99].

Sorption methods have been the basis of several developmental attempts. Probably the one that has reached the higher scale level is the asphalt residual treating (ART) process. Although the ART process was commercially demonstrated, both its current status for licensing and whether it is in use are unknown. At the beginning of the 1980s, Engelhard Corp. (now BASF) developed an adsorption process for removal of metal compounds from crude oils. The ART process was described as a selective vaporization of a reduced crude (AR or VR), in the presence of a fluidized proprietary sorbent (ARTCAT), on which most of the metals (92−97%) and part of the asphaltenes (<50%) deposit [100−104]. The ART process claimed to allow increases in the production of transportation fuels while reducing heavy FO yield, without having to hydrocrack. Hence, it reduces not only metal content but also CCR [104]. The process flow scheme is presented in Fig. 6.1 [104−106].

In the ART process, the whole crude or a residue is preheated and subjected to selective vaporization in a high-temperature riser, at a short contact time with an inert solid material of low surface area (ARTCAT with a surface area of 5−20 m^2/g). The asphaltenes and HMW compounds of the feed remain on the contact−sorbent material (basically, all the

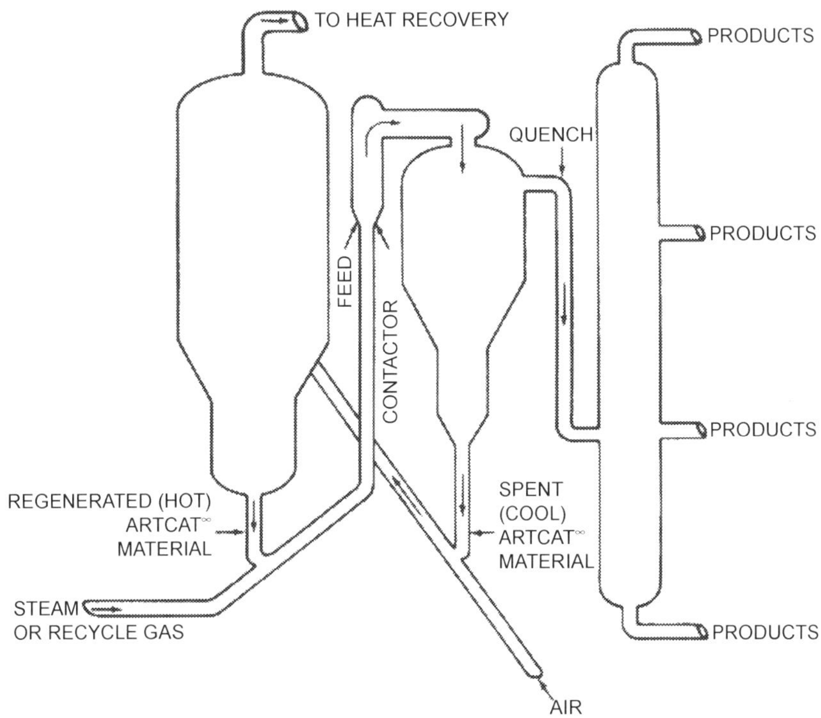

Figure 6.1

The ART process. *Reproduced from Bartholic DB, Haseltine RP. New crude/resid treating process offers savings. Oil Gas J 1981;79(45):242–52, with permission from Oil & Gas J.*

asphaltenes and 30–50% of the S and N, but keeping the hydrogen in the feed). According to this description, molecular conversion is observed only for the high-boiling-point components. The preservation of the H content of the liquid product is explained, assuming that only large molecules (but not the lights) are thermally cracked to lower MW compounds and coke. At the top of the riser, solids and vapors are disengaged, and vapors are quenched rapidly to minimize overcracking of the products. The sorbent particles with the carbonaceous deposits are then transferred to a regenerator, where the organics are burned off in a combustion zone and the sorbent gets reactivated. The recovered regenerated sorbent is mixed with the feed at the base of the fluidizing riser system and recycled to the main reactor by means of steam as lifting vapor. The process resembles much of an FCC operated in a continuous heat-balanced manner with an inventory of the recirculating sorbent material. Combustion of the deposited carbonaceous compounds heats up the sorbent particles in the regenerator, and these hot particles serve as vaporizing agent for the feed in the contactor during the short residence time. Though the unit is heat balanced, surplus heat is produced and could be used for steam and/or electricity generation. In this contactor (reactor), the lighter components of the feed are vaporized and subsequently recovered.

Five patents could be directly or indirectly associated with this Engelhard process [107–109], differing in the contacting and fluidization conditions and having Bartholic as one of the inventors. The sorbent material is described as a clay material, more specifically as kaolin in some of the patents. Spent equilibrium FCC catalyst can also be used as sorbent material [110], particularly if the process is integrated to an FCC, as a feed preparation step [111]. In fact, the ART process is considered to facilitate the subsequent processing in conventional downstream units.

Commercial performance of an integrated ART–RCC unit was reported by Ashland Petroleum [112]. The economics of the ART process for a 100 kbbd unit (CAPEX US$480 MM) was estimated to render net profits (@90%) of US$171.5 MM, before federal taxes and depreciation. This experience also showed a stable operation for a sorbent containing up to 15,000 ppm Ni + V. This high-metal-loading level is controlled by withdrawing spent material and adding fresh material, in order to maintain performance standards (i.e., without detriment of yields and product quality). Comparatively to FCC, the ART process could operate with accumulated metals on the sorbent material at higher levels than those typically tolerable on the catalyst during FCC operations. Although the process could be operated effectively when total metal (Ni + V) content was above 2 wt%, coke production increased to levels considered higher than expected from CCR calculations. Under these conditions, hydrogen production was also observed to increase, consistently with the effect of metals (particularly Ni) on dehydrogenation and coke making. Once again, in FCC a similar result would have been observed at very much lower metal content of the catalyst. This type of increase in coke making was also observed when chloride was contaminating the feed. Moore suggested incorporating a basic nitrogen compound (e.g., ammonia or ammonium polysulfide) mixed with the feed, prior to contacting it to the sorbent material [113]. This suggestion seem to be indicative of acidity being responsible for an increase tendency in coking, probably based on the observation of chloride ions contaminating the feed and the possibility of having alumina as part of the sorbent material.

Ashland Oil made its own contributions to the ART process and its integration within the context of upgrading heavy oils with conventional refinery units [114,115]. One of the first discoveries was the formation of a liquid vanadia coating on the sorbent particles, under the high regeneration temperatures, as vanadium pentoxide level built up. This coating changes the fluid dynamic regime in the sorption vessel, resulting in particles coalescence. Coalescence creates plugging, and fluidization becomes jeopardized. The unit turns into ineffective operation and eventually has to be shut down. Some alternatives for vanadium immobilization were examined [116–120]. The sorbent was promoted with additives capable of forming more stable compounds than V_2O_5, such as oxides of Ti, Zr, Ta, Ba, and In [116]. The sorbent regeneration conditions were kept as mild as possible to prevent vanadium to oxidize to an oxidation state equal to +5 [117]. The sorbent material was reformulated as a high-pore-volume solid with low cracking activity and comprising metal additives (as those already mentioned in Ref. [116]) to immobilize the accumulated vanadium compounds [118]. The preferred method of application for this last described sorbent material was hydrothermal visbreaking [119]. The passivation of the sorbent material was suggested by using rare earth (e.g., lanthanum), as is typically done for catalytic crystalline zeolites [120]. In summary, the ART process could reach commercial demonstration, in its original Engelhard conditions and features, for the purpose of preparing a catalytic cracking (RCC) feed. The fact that an RCC was used indicates that metals removal and CCR decrease were not effective enough for feeding an FCC unit. The process underwent modifications mainly by Ashland, into a visbreaking mode. A number of process patents were derived from these learnings that were applied not only for the BotB upgrading [114,115,121,122], but also for the benefits of the FCC processing of high-metal-containing feeds [120,123–136].

Some other materials and conditions have been tested and examined for DM purposes. Inorganic and carbonaceous acid solids, which have surface areas greater than 100 m^2/g (e.g., silica, silica–alumina, K-10, and similar acid-treated clays), and activated carbons, with surface areas <1000 m^2/g, were recommended as sorbent materials to remove asphaltenes (as mentioned above); these materials can also remove a major portion of the metal contaminants. The preferred feed stream to adsorbent (wt/wt) ratio was reported to be between 0.1 and 10. The steps of the process include contacting the feed stream with an adsorbent, recovering the liquid product, and removing the metals and coke precursors from the adsorbent [72]. The sorbent is regenerated by washing with solvents such as toluene, toluene–methanol, or other appropriate refinery stream. The use of an aromatic solvent helps in removing coke precursor molecules, while a polar component in the solvent system washes off most of the metal-containing moieties. The spent solvent solution is stripped for recovery of a cleaner solvent that is recycled to the process. The stream containing the removed impurities is sent to a coker, to a partial oxidation unit, or to any other disposal technique.

High temperatures, in the range of 250–350°C, are used for the removal of metals using an adsorbent material composed of 50–97 wt% Al_2O_3, and 3–50 wt% alkaline earth metal oxides (CaO and MgO) having a BET surface area of at least 100 m^2/g [137]. Even higher temperatures of about 480°C (at short contact times of <5 s) could improve vanadium removal, with additional benefits of converting the CCR precursors [117].

It seems that adsorption may not require an active surface, as those mentioned in the previous paragraph. An inert material such as fused Al_2O_3, or fixed diatomaceous earths such as Celite 408 or 410, can do the job in the presence of an inert atmosphere (e.g., nitrogen). Macroporosity (average pore diameter of 5000–7000 Å) and very low surface area (2–5 m^2/g) were the required properties for the adsorbent. The incorporation of a hydrogenating component was an optional DM improvement [138].

Other suitable sorbents for metal removal include organoclay compounds, which are formed by contacting the clay mineral (attapulgite, sepiolite, etc.) or a zeolite with an alkylamine base. Contact may take place within an inline filter application or any other adsorption–absorption unit [139]. Attapulgite promoted with antimony pentoxide was reported to remove nearly all the metals (99.5%) [140].

Polarized surfaces of acid or basic nature have been evaluated for metal removal by sorption. The attention on solid parameters has concentrated on surface area, rather than pore structure [141–143]. The success of these polarized surfaces has not been proven commercially, though it appears as an option with potential. Nevertheless, attention should be paid to optimizing the pore structure of the solid material as well.

Although sorption processes seem to be highly effective for DM operations, there is very little detail on sorbent capacity, sorbent saturation levels, and/or contaminated sorbent management (regeneration, reclamation, or disposal). One may infer that combustion might be conveniently applied for regeneration and recycling below the metal saturation limits. A separation reagent for recovering hydrocarbon liquids from carbonaceous materials and applied for the removal and recovery of hydrocarbons from the hosting clays in tar sands [144] could be used for the regeneration step in adsorption processes. This reagent is a surfactant formed in situ, from the reaction of the polar resin components in tar sands bitumen with an inorganic base such as sodium silicate, in sonicated aqueous solution, and in the absence of any organic solvent. The separation reagent must be substantially diluted after being recycled and reused to reduce solvation properties. Since the phenomenon involved in that recovery process is quite different from the requirements for the regeneration of a metal-saturated adsorbent, the investigation of such physical chemistry deem worthy.

The interaction between metal ions and asphaltene molecules (H-bonding) plays a stabilizing role in water-in-oil emulsions [145]; then, it may be expected that creating, reversing, and breaking emulsions may affect the metal-bearing molecules. This is another physicochemical phenomenology worth investigating that has not been considered for a metal removal application.

Electrophoretic separation (currently used for inorganic metal compounds in desalting units) was intended for the DM treatment of thermally cracked residues. The process consists of passing the residual hydrocarbon oil through one or more vessels, each comprising at least one electrode and generating electric field strength of at least 0.4 kV/cm. The process would remove suspended inorganic solid particles from a thermally cracked residue, in which metals were transformed into the corresponding sulfides. A very low effectiveness was observed: metals removal was shown to be just over 10 wt%, with the iron species being better removed [146].

Metal precipitation under supercritical water conditions was claimed as a DM method. Supercritical water processing is carried out at 200–500°C. Under these conditions, certain impurities that might be present or added (e.g., Fe oxides, S^{2-}, amines, NH_3, and CO_3^{2-}) could act as catalysts for further reaction with the original metals (e.g., V, Ni, and Fe), leading to precipitation [147].

Electroflocculation is a combined concept between flocculation and electroflotation or electroprecipitation. It has been applied for water purification and more particularly for heavy metals removal from aqueous systems and/or sludge [148,149]. The design of an electrolytic cell for an electroflocculation process was awarded to Cleaner Technology AS [149]. Flocculation refers to a process where fine particulates dispersed in a liquid medium are caused to clump together into flocs or flakes. The floc may then float to the top of the liquid or settle to the bottom of the liquid, and it can be readily filtered from the liquid. An electric field can be used in different ways to force the flocculation of dispersed small particles [150,151]. In flotation, a gas is bubbled through the suspension or emulsion, forming foam, which retains the particles. The flotation rate depends on the surface tension, the gas bubble diameter, the particle size (and distribution), the gas bubble and particle zeta potentials, temperature, pH, and the residence time. The conventional method uses nozzles to bubble air in the bottom of the flotation tank. Other methods used include dissolved-air flotation, vacuum flotation, and electroflotation. The latter is a continuous method in which water is electrolyzed to produce hydrogen at the cathode and oxygen at the anode, yielding small bubbles of very similar size. In electroprecipitation, as indicated by its name, metal ions are precipitated from the anode material (typically Fe and Al). In a given set of conditions, it is possible to form the powerful oxidizing system Fe^{2+}–H_2O_2 (Fenton's reagent). In emulsified systems, that reagent will break some emulsions by a double effect of oxidative destruction of the emulsifier and by neutralization of the droplet charge. The application of electroflocculation in the presence of the Fenton's reagent, under a given set of conditions, may offer a potential DM option.

3.2 Chemical Treatments and Conversion

The chemical transformation of the metal containing molecules may involve the use of additives or catalysts. The case of unconventional catalysts will be included here, regardless of whether in the absence or in the presence of hydrogen atmosphere. Meanwhile, the use of conventional HDT catalysts was already considered in Chapter 5, together with other commercially available technologies.

The chemical reactivity of the metal moieties is based mainly on the properties derived from the embedding organic core, which can be typified for its donor character conferred by a π-system of surrounding rings. Nevertheless, the steric inhibitory role of these rings has to be kept in mind. This π-electron system will interact with acid (Bronsted or Lewis) functions. An initial attack on the mesocarbons would provoke a destabilization that may lead to decomposition.

In the treatment of crude oil, DM reactions fall in two different broad categories, those occurring in ordinary hydroprocessing (hydrogenation/hydrogenolysis) and those for other chemical transformations of the metal-containing molecules. For instance, promotion of the electrophilic substitution of the surrounding aromatic rings to the metal will weaken the metal bonds to the molecule. For this type of substitution, Friedel—Crafts catalysts could be used. Then, further treatment would be needed for the actual DM to occur.

3.2.1 Chemical Additives

Chemical reagents are typically used for the removal of (water-soluble) light metals, mainly via precipitation. However, light metals in crude oils, particularly Ca, are present in the form of water-insoluble organic salts. In general, metal compounds are rather oil-soluble, and the required additives for their effective removal become very specific. A wide range of chemical agents has been screened for their effectiveness on the DM of crude oils. For instance, most of the decalcifying agents cannot remove other metals. However, one effective class of compounds is Fe-containing agents [152]. The deferrization (Fe-removal) agent TE-4 (reported by the Research Institute of Petroleum Processing, China) not only can remove most of the iron from crude oil but also removes Ca, Na, and Mg, as well as some of the Ni and V [153].

The chelating ability and redox properties (attributed to either the presence of certain functional groups and/or acid—base strength) have been considered in the selection of various chemical agent formulations. These formulations include tartaric acid, triethyl citrate (with and without diethanol amine), succinic acid, mixtures of malic acid and quinoline, oxalic acid, 2-ethylethanoic acid in air, fumaric acid, benzoic acid (alone or in combination with HMTA or diethanolamine), maleic acid, maleic anhydride, malonic acid, and superacid systems (fluorosulfonic and trifluoromethane sulfonic acid) [154]. The amino-carboxilate compounds, carboxylic acids in 1-methylnapthalene, and superacids in dimethylformamide all worked well. The reported example was carried out at 70°C and 600 mole equivalents of the agent [154]. Inorganic acids have been also tested, for example a small amount (i.e. to be considered an additive) of diluted sulfuric acid removed 61% Ni and 55% V from gas oils [155]. An oxidative DM method with sodium hypochlorite and peroxyacetic acid exhibited a high activity, together with the ability for decomposing petroporphyrins [156]. However, the method was found to be rather unselective and stoichiometrically controlled. Other oxidative reactions catalyzed by acids and in the presence of a Friedel—Crafts catalyst were recommended, among others, for DM via metal complexing and chelating [157,158]. This DM catalyst has been formulated based on sulfuric acid and one or more of certain chalcogen-containing compounds with the empirical formula:

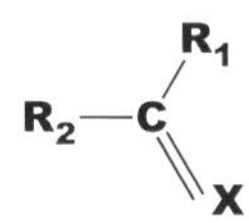

where X is a chalcogen; R_1 and R_2 are selected independently from hydrogen, NR_3R_4, and NR_5; at least one of R_1 and R_2 is, other than hydrogen, a monovalent organic radical; each of R_3 and R_4 is hydrogen or a monovalent organic radical; and R_5 is a divalent organic radical. The chalcogen compound–H_2SO_4 molar ratio is <2. The disclosed novel catalysts involve combinations of the chalcogen compound–sulfuric compositions with one or more transition metal halides and/or with one or more surfactants [157,158]. All these techniques have been criticized as economically unjustified.

It has been discovered that some other oxidants, such as chlorine, sulfuryl chloride, and dinitrogen tetraoxide, can react rapidly with vanadyl porphyrins under mild conditions, and this reaction could be accelerated photochemically. Benzoyl peroxide and butyl hydroperoxide also reacted, but to a lesser extent. Unfortunately, nonporphyrinic vanadium could not be converted under the studied conditions. Similarly, under these conditions, neither strong acids nor any other reagent capable of forming carbonium ions will react with the porphyrins [159].

Regarding photooxygenation of metalloporphyrins, an alkaline medium promotes complexing by breaking the vinyl bridge of the core, which facilitates metal exchange [160]. Since the metal moiety becomes more accessible, it would undergo further reactions, such as the formation of salts when reacting with acids. Examples have been reported involving Mg, Zn, Co, and Cu. In this case, the formed (demetallized) compound can be identified by a 630 nm band of an oxo-violin or purpurin spectrum.

Vanadium and sulfur were simultaneously removed from Boscan asphaltenes by electrolysis in lithium chloride–ethylendiamine, using lithium–ethylendiamine and Raney nickel. The highest V removal achieved was below 85% of the porphyrinic-V [161]. The drawbacks of this reductive desulfurization reaction were the large proportions of Li metal required (Raney Ni was less powerful) and the fact that only metalloporphyrins were preferentially attacked.

3.2.2 Carbonaceous Additives

The use of coke as an additive to perform the selective coking of asphaltenes was suggested as a way of improving DM, but the results were not that positive. The coking reaction, being exothermic and proceeding mainly by a free radical mechanism, did not seem to provide enough control to achieve a good level of selectivity [162].

The *particle capacitance effect* concept defined by the University of Tokyo [163] was taken as the basis for the FTC process, which integrates a thermally produced coke as an additive into the metals removal step. Initially, a porous coke was considered by Fuji Standard Research for process development [164]. The pores of this material are thought to act as a (micro) *pyrolyzer*. Thus, the liquid feed would fill the pores by capillary effect, and since the solid is hot, conversion would take place. The particles remain dry, and the metals accumulate in the pores. Hence, the main requirement for the additive seems to be its pore structure that has to be stable under operating conditions. The FTC Process was developed by Fuji Standard Research Inc. as a similar version of Flexicoking [163–165]. The process units resemble that of an FCC unit (Fig. 6.2), where the equivalent riser becomes the cracker in FTC, the equivalent regenerator acts as gasifier, and the recirculating solid is a material catalytically inert. Fluidization and transport of the particles and gasification of the coke are reported to be easy and highly efficient.

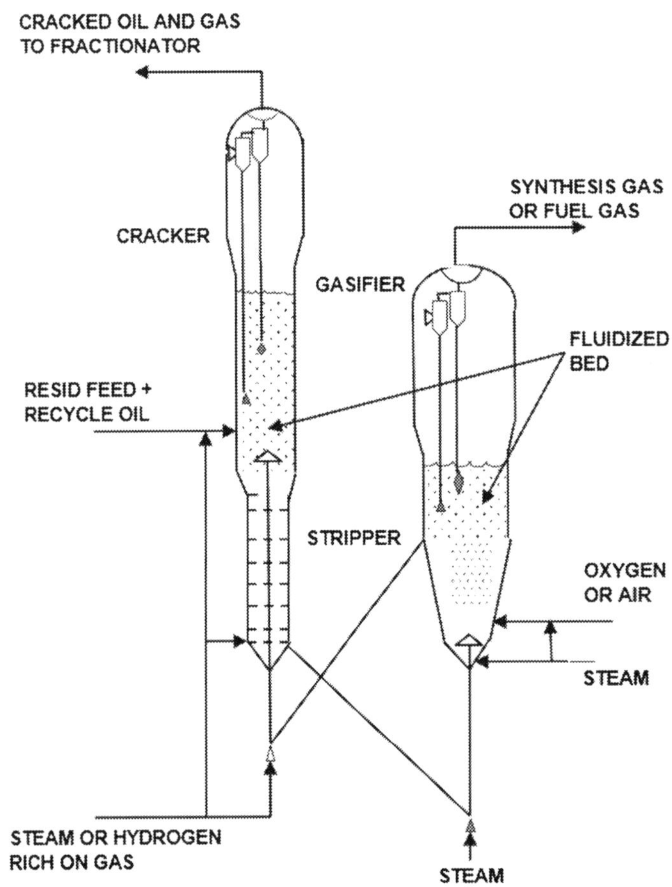

Figure 6.2
FTC process sketch [164].

The process is carried out at temperatures of 430–550°C. The recirculated-solid–oil ratio is 4–10, and the fluidization gas is a steam-containing gas supplied in a concentration higher than 5% of the feedstock [165]. The additive particles are spherical with pore volume of 0.10–1.0 cm^3/g, weight-mean diameter of 0.025–0.25 mm, and specific surface area of 50 to 1500 m^2/g [163]. The lift gas was later changed for hydrogen as an improvement to avoid excessive coke formation caused by the catalytic effect of deposited metals. The hydrogen partial pressure is set between 0.5 to about 5 bar [164]. The process was proven at pilot plant scale, but there is no report on any further developments.

3.2.3 Basic Treatments

Both organic and inorganic bases have been evaluated for the upgrading of crude oils. Besides metals, this type of harsh treatment will also affect sourness and acidity (i.e., sulfur and naphthenic acids), respectively. In fact, treatment with bases either for extraction or neutralization purposes has been considered for acidity abatement and will be mostly discussed in the following section. For instance, an aqueous solution of ethoxylated amine extracts not only heavy metals but also naphthenic acids and sulfur from water-in-oil emulsions [166]. This alkoxylated amine has an $R{-}NH{-}(CH_2CH_2O)_mH$ formula, in which m might be 1 to 10 and R is a hydrocarbon substituent. The processing can be conducted utilizing existing desalter units. The ranges of operating conditions are: ethoxylated amine 0.15–3 molar equivalents of naphthenic acids; water 5–10 wt%; temperature 25–80°C; pressure 4–70 bar; and contact time 3–30 min. During the treatment of a Venezuelan crude with tertbutyl–diethanol amine, a 13% reduction in vanadium was observed for which the total acid number (TAN) was reduced from 2.2 to 1.1 and 17%S was removed. In terms of DM, the reduction in metals (and sulfur) is just marginal, indicating a very low effectiveness of the treatment and anticipating increasing metals contamination of the amine upon reuse.

Adding alkali metals or alkaline earth metals to amines or NH_3 and carrying out the metal extraction from VR and/or asphaltenes, at about 100°C, increased the removal to 77% for Fe, 49% for Ni, and 85% for V [167].

A treatment of crude oils with molten alkaline metal hydroxides (KOH) would remove metal compounds, together with sulfur and nitrogen compounds. In this process, 5–30 wt% of the hydroxide is mixed with the oil. The spent extractant is regenerated by reaction with H_2S, which could be taken from the HDT units. The regenerated extractant is composed by K_2CO_3, K_2S, and unreacted KOH, and an excess of H_2S is needed to totally convert these to KSH. Then, KSH is converted to KOH by reaction with water, and any remaining sulfur is removed by absorption/reaction with Cu or Fe. The reported devanadation was in the range of 90% [168]. Two other processes were described for the treatment of heavy hydrocarbons to remove metals and sulfur with either alkaline metals hydroxides or oxides [169,170]. In this case, high temperatures and hydrogen are also used.

Simultaneously, API gravity increased and CCR is reduced as an indication of a more extensive upgrading. The treatment with alkali metal hydroxides in the presence of H_2 is carried out at temperatures between 400 and 540°C and pressures of 100–210 bar. Treatment under this process of AR Safaniya (3.9 wt% S, Conradson C 12.1 wt%, 20 ppm Ni, 77 ppm V, and 4 ppm Fe), with 13.9 wt% KOH at 435°C and about 120 bar of H_2 resulted in DS and DM of 69% and 94%, respectively, with a 62% reduction in CCR [169]. Meanwhile, with 20 wt% of Na_2O at 445°C and 115 bar H_2, the treatment ensued 86% DS and 95% DM [170]. A more simple treatment at 360°C, Na_2O and H_2 showed less than 30% DM [171].

3.2.4 Acid Treatments

Similarly to the treatment with bases, both organic and inorganic acids have been tested for metals removal from crude oils. Organic peroxy-acids (e.g., peroxyacetic acid) are capable of oxidizing Ni and V at mild conditions, such as 45–50°C. Then, metal products can be washed off with a 5% $NaHCO_3$ aqueous solution. Any acid excess was destroyed by Raney Ni. The oil layer (demetallized VR) is separated, filtered, and dried. The observed DM levels for V and Ni were, respectively, 80% and 74% [172]. Hydroxocarboxylic acids or salts have been recommended for the specific removal of nonporphyrin organically bound iron. Strong mixing is required, but separation and recovery of the demetallized feed are facilitated by using an aqueous solution of the acid [173].

Separation is also facilitated when a solid acid (e.g., acidified active carbon) is employed.[1] In this process, elevated temperatures are required (375–450°C). A solid acid was prepared using a commercially available active C, which was contacted with a mixed solution of 3.3 weight parts of concentrated H_2SO_4 and one weight part of concentrated HNO_3 in a ratio of 4 cm^3 solution/g active C. This contact caused the temperature to rise to 130°C, then the solid was washed free of acid and dried at 140–150°C. This treatment yields two phases, separable by decanting: a thick, dense, lower phase (~20 vol%), and a more volatile, less dense, less viscous, upper demetallized phase (~80 vol%) [141,142].

Extraction with anhydrous liquid polyphosphoric acid at elevated temperatures was carried out using acids containing P_2O_5 equivalents between 72.5 and 86.5 wt%. The metals in the oil are extracted into the acid phase in the form of metal chelates of the linear polyphosphoric acids. Then, the extracted metals are recovered as sulfides by precipitation with H_2S, in a cyclic process, in order to regenerate the polyphosphoric acid [174].

In situ generation and regulation of the oxidizing power of HClO were favorably tuned to achieve DM of nearly 90% for V, Ni, and Fe removal from oil, coal, or coke and their derivatives [175]. The crude oil is mixed with an economically convenient solvent in a

[1] This type of materials was already mentioned above, when discussing adsorption on polarized surfaces.

proportion of 2:1 sample–solvent ratio. The solvent is added to diminish viscosity and improve stirring. A solution of HClO (freshly prepared) is added in a proportion of 20 vol % referred to feed. The formed two phases are separated after 20 min of reaction. Results are presented in Table 6.1.

An acid treatment with aqueous or anhydrous HF was disclosed for the simultaneous DM/DA treatment of crude oil, residues, coke, or coal. The reported example described the treatment of 30 kbb of crude oil that started with a preflash to separate 2100 bb of 95°C⁻ hydrocarbons. Then, the stripped crude was treated in an extraction tower with 12 kbb of anhydrous HF, at 150°C for 10 min. Metal contaminants were converted to insoluble inorganic fluorides, which remained mainly in the extracted phase. However, the raffinate (~25.9 kbb) contained a small amount of HF, and about 10% of the solid metals, and it has to be stripped of HF and sand-filtered. The extracted phase amounted for 4100 bb passed to a stripper tower for recovery of unreacted HF, and then to a sand filter for removal of the solid metal complexes. The filtered extracted phase was passed to a fluidized bed-type thermal cracking (TC) unit to yield 600 tons of metal-free premium coke and 29,700,000 scf H_2. The raffinate bottoms were subjected to HCK [176]. The complete integrated scheme is shown in Fig. 6.3.

Acidified aqueous solutions of ferric salts and stannic chloride reduced V from 1100 to 643 ppm and Ni from 94 to 69 ppm, when treating a Boscan crude oil [177].

Almost complete DM (99%) was achieved by using biphenyl phosphate [178] or hypophosphoric acid, H_3PO_3 [179]. The conditions for (4.5–5.2 wt%) biphenyl phosphate treatment were up to 7 bar and 420°C for 1 h in air or argon, followed by filtration of the cooled product [178].

Process temperature and pressure conditions could be conveniently tuned for inducing DM, using dimethylcarbonate, in the presence of an overpressure of CO_2. Conditions are adjusted to maintain the dimethylcarbonate in its liquid state and form an homogeneous solution with a DAO feed. Then, the entire system is cool to a temperature such as to form three phases. The CO_2 gas is then vented to recover a partly demetallized DAO and primary oil from the light phase. Another DAO and partly demetallized secondary oil are recovered from the intermediate phase. Finally, the recovered dimethylcarbonate is rerouted for its possible reuse [94].

Table 6.1: HClO Treatment Results on a Venezuelan Crude

Composition	V (ppm)	Ni (ppm)	Fe (ppm)	S (%)
Before treatment	1900	455	355	1.70
After treatment	19	4.5	5.5	0.05

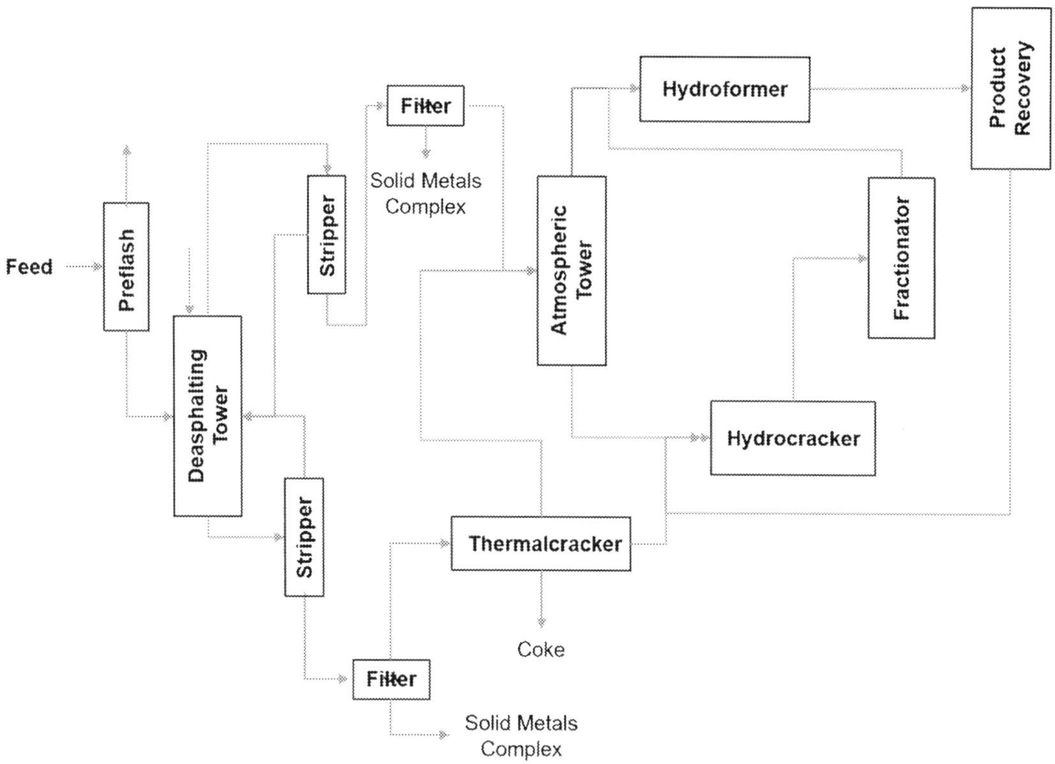

Figure 6.3
Integration of HF extraction to other refining units [176].

Severe extraction conditions (elevated temperature and pressure) were applied with either acids or alkalis. Although these conditions would emulsify the crude oil and facilitate the contact between extractant and feed, neither of the extractants performed satisfactorily. DM was lower than 10%, and DS was not even 30% [180]. Another issue with this process is the limited selection of emulsifying agent since it has to be stable at the leaching pH and should not react or damage in any way the quality of the crude under processing conditions.

Once again, acid treatments of crude oils impose severe changes to the oil chemistry, might be environmentally hostile, would require unsafe conditions, and create hazardous byproducts, resulting as a whole in an uneconomical and ineffective solution.

3.2.5 Thermal Conversion

For the purpose of metals monetization, a conventional visbreaking process might be devoted to DM. The thermal conversion of the metal compounds in the crude to ashes in the presence of 15 to 25 wt% pulverized coal produces metals-enriched solids suitable as a feed for metal-bearing ore in extraction processes [181].

Decomposable additives or catalytic materials are introduced in a hydrovisbreaking process, for reducing the concentration of metals, sulfur, nitrogen, and CCR. A slurry reactor is employed for the processing of feeds, including FCC decanted oil and metal-containing straight run (SR) residues (AR or VR). Three different types of additive or catalytic materials were claimed: (1) a decomposable molybdenum compound (e.g., Molyvan L, a molybdenum dithiophosphate catalyst from R.T. Vanderbilt Co.), (2) a dispersed Mo on alumina or silica, and (3) catalytic cracking catalyst fines (e.g., contained in the DCO). In the first option, the additive accompanies the hydrovisbroken product to downstream processes, while in the latter no externally supplied additive or catalyst is needed [182].

An additive with DM functionality was incorporated to the FCC reactor, which physically differs from the FCC catalyst (e.g., it differs in particle size). The demetallizing additive is fed at the base of the riser. The residue is cracked by contact with a stream of hot regenerated catalyst and demetallized by the additive. The reacting mixture or product, including the metal-containing additive, deactivated cracking catalyst, and cracked products, is discharged from the riser, stripped from oil product, and charged to a first-stage regenerator. The first stage of the regenerator partially regenerates the cracking catalyst and separates it by elutriation from the demetallizing additive. The DM additive accumulates as a dense-phase fluidized bed in a lower portion of the first-stage regenerator. The partially regenerated cracking catalyst is withdrawn from the first-stage regenerator, and regeneration is completed in a second-stage regenerator. The DM additive is recycled from the first-stage regenerator to the base of the riser. A hot regenerated cracking catalyst is withdrawn from the second-stage regenerator and recycled to the riser. A highly active additive is considered a destabilizing factor, since metals and coke would rapidly deposit and deactivate it. Therefore, this coarse catalyst has to be optimized in terms of metals removal and coking. Although metal scavengers are mentioned as possible additives, the preferred additives are spent FCC catalysts with a larger particle size than usual ("coarse size": 100–250 μm, preferably RFCC spent catalysts) or materials like natural occurring clays, minerals, spent catalyst fines from other refinery processes, antimony-based materials, silica, alumina, MgO, or CaO [183].

Tar sand has been tested as a DM additive for crude oil thermal conversion at 425°C. The removed metals remained in the tar sand [184].

3.2.6 Chemical Reactions

Although reactivity of the metal compounds is in general low, certain chemical reactions have been tested for their conversion into more easily removable compounds, for instance a reaction that transforms the organometallic compound into an inorganic salt. For this reaction, an intimate blend of an organic solvent (boiling in the range of about 65 to 425°C) is heated with the residue, at temperatures of 345 to 480°C, in the absence of oxygen for up to 6 h. An insoluble concentrate would contain the removed vanadium. The concentrate is then reacted with a halogen-containing compound to form a vanadium halide, which is then separated. In the case of Fe and Ni, the metal-concentrated fraction is roasted with oxygen at elevated temperatures (up to 1650°C), yielding the corresponding oxides. The recovered metal compounds can then be processed to their reduced state [185]. This treatment is motivated by the metals value and not by the energy value of the oil. Adaptation of such method to refining operations might be seen as a diversification of the business.

A reaction carried out in the presence of an asphaltene extractant (e.g., heptane) containing molten sulfur and a Group VB metal salt was tested for a DM process of oil fractions. Extraction and reaction of an AR of 16.9°API gravity containing 290 ppm V, 37 ppm Ni, 2.23 wt% S, and 10.7% CCR with an extractant composed of 1% V_2S_3 and 1% S at 200°C produced a treated oil with 30 ppm V, <5 ppm Ni, and 1.8% S [186]. Reactions with other sulfur compounds have been also examined. Examples include thiocyanate at temperatures in the range of 250−500°C, such as NH_4SCN in toluene at 385°C, 27 bar nitrogen, for 4 h achieved 70% DM [187]; phosphorus sulfides, such as addition of P_2S_5 and/or P_4S_3 at 300−450°C under high pressure (7−175 bar) for 0.1−2 h [188]; SO_2(l) at 30°C, 2 bar, and 1.2:1 (vol) of SO_2−oil ratio [189]; bisulfite or bisulfide of NH_4, alkaline, and alkaline−earth metals convert asphaltenes to water-soluble sulfonic acid salts, such as sulfonates at 100−300°C under 10−105 bar partial pressure of SO_2 for about 4 h (DA was lower than 20%) [190]; and H_2S and H_2 at 200−500°C and 35−345 bar [191].

Other tested reactants include a combination of Lewis acid and molybdenum chloride [192] and Fe_2O_3 in the presence of a metal phthalocyanine as catalyst [193].

Although chemical reactions can be very selective, various issues have to be considered when seeking for a DM application. First is the low reactivity of the metal- bearing molecules, and then stoichiometry, reactant cost, product recovery/separation, severity required, selectivity, safety, and an environmentally friendly management of exhaust streams and byproducts.

3.2.7 New Catalysts and Catalytic Processes

Catalysts for processing heavy residues have been developed in both supported and unsupported forms. In this section, emphasis will be placed on those systems that did not seem to have reached the commercialization stage.

Several metal-based catalysts have been found in our search. The fast deactivation caused by metal deposition during DM reactions led to the idea of using one-through cheap disposable catalysts but still achieving both high activity and moderate life. Minerals or mineralogical byproducts seem to have been preferred as either cheap or disposable (e.g., manganese nodules) [194−196]. Manganese dioxide was selectively leached with respect to the iron oxide component of the nodule to improve DM performance [195]. The catalyst had a comparatively low aging rate, as the H_2 pressure was reduced below 140 bar, probably due to its unusual macropore structure [194]. A kinetic comparison between Mn nodules and a commercial HDT catalyst formulation (CoMo) indicated a strong H_2 pressure dependence that favors the commercial catalysts at high pressure. Nevertheless, at pressures below 70 bar, the reaction rates were comparable between the two considered catalysts [196]. Regardless of these kinetic results, the use of the Mn nodules was not claimed in the DM process disclosed by the authors [197], whose catalyst was based on conventional HDT active-phase formulations. Mn by itself as a DM active phase does not seem to be very active, since an alumina-supported catalyst, characterized by having a surface area of 175 m^2/g and at least 60% of its pore volume in pores with diameters between 50 and 200 Å, showed a DM activity lower than 50% [198]. However, DM activity was observed to increase by changing the support to SiO_2 and/or increasing the Mn content [199].

Iron-containing natural clays introduced the idea of using this metal as active phase, and it has been demonstrated to be an interestingly active material for the conversion of high-asphaltene-containing residues. In fact, within the nonconventional HDP catalysts, iron-based phases are one of the most studied (and developed) active phases for DM reactions in the absence of or under H_2 pressure (see e.g. Refs. [34–40,42,43,45–48,200]). Bauxite used either as catalyst or as support exhibited a high initial activity, but its microporous structure resulted in rapid deactivation [201–203]. The role of iron in DM reactions has been explained in terms of its affinity for Ni and/or V, more particularly in HDP conditions. During HDP, a partial pressure of H_2S is generated that promotes the formation of iron sulfides. An iron-deficient iron sulfide phase, pyrrhotite, offers specific features for storing Ni and V cations in existing Fe vacancies [204]. However, complete sulfurization of the iron phase will saturate this vacancy, forming either stoichiometric monosulfide (FeS) or disulfide (FeS_2, pyrite) that lacks the desired features. Another iron-based catalyst (MoFeAlSi) was used for the DM of porphyrin and nonporphyrin feeds obtained from Jobo crude oil [205]. In this work, the porphyrin content was measured using the Soret band that is not present in the hydrogenated-metallized compound. Hence, DM cannot be evaluated by the decrease in the Soret band and/or the conversion or disappearance of porphyrins. Although a difference in DM reactivity of those molecules might be expected, its evaluation via Soret band will be mistaken. Mechanistic studies on the hydrodemetallation (HDM) of nickel tetraphenylporphyrin (Ni-TPP) indicated that DM reaction occurs in a series of steps that starts with the hydrogenation of the TPP rings into tetraphenylchlorin (TPC) and tetraphenylisobacteriochlorin (TPiB). Many more steps are needed prior to the actual metal elimination step [206], confirming the fact that porphyrin disappearance or an initial hydrogenation step cannot be taken as evidence for DM. This is the case for some iron–vanadium catalysts evaluated for desporphyrinization [207]. A synthetic catalyst prepared by acid core-impregnation of an iron compound (Fe $\geq$ 1.15%)

onto a porous refractory support has been reported [208]. The active sites are concentrated in the interior of the porous support. VR DM with the presulfided catalyst at 410°C, 140 bar, and space velocity 0.35 h^{-1} was better than 97%. Under ebullating bed conditions and 90% VR conversion, an iron supported on carbon catalyst removed 96% nickel and 90% vanadium [209]. The interesting results found with iron and with the pair FeMo were the basis for the development of catalysts for simultaneous HDM—hydrodesulfurization (HDS) reactions, in which Mo provides HDS activity and Fe stores the removed metals, providing stable performance (minimizing deactivation) [210—215]. These catalysts were prepared in a way that Fe would form a pyrrhotite phase upon sulfiding [211,212]. Since the life cycle would depend on HDM activity and metal content of the feed, the total metal content and the Fe—Mo ratio could be optimized to meet desired performance parameters (i.e., HDS and catalyst stability) [213—215].

Consistently, a Ni nonstoichiometric compound ($NiAs_x$) catalyst exhibiting a pyrrhotite structure was capable of almost 100% DM, even though the tested crude had high metal concentration. Catalyst was supported on Al_2O_3 (surface area 20—300 m^2/g and pore vol 0.5—1.5 cc/g) [216]. Another Ni catalyst was a mineral-derived case developed by Nippon Mining Co., based on a Ni-ore, which could be used in a DM process integrated to metallurgy schemes. The catalyst would crack the heavy oil feed at a temperature between about 350 and 550°C, atmospheric (or slightly above) pressure, seconds to several minutes of contact time, and a catalyst-to-feed ratio of about 1—15. Under these conditions, both vanadium removal and coke formation increased with reaction temperature, indicating that vanadium removal might be adjusted by controlling coke deposition (TC). Advantageously, since the catalyst is Ni-based, the deposited coke can be catalytically gasified, in the presence of air or steam; thus, a gasifier could be employed for catalyst regeneration, and heat and energy integration might be beneficially incorporated [217]. The catalyst is obtained by calcining a nickel-containing ore at temperatures between 850 and 1200°C [218]. Ni in spent catalysts of this type could be reduced and extracted from its ore by the use of an ammoniac ammonium carbonate solution [219]. Commercial units for this reduction process were in operation by 1975 in Nicaro (Cuba), Sered (Czechoslovakia), Greenvale (Australia), and Marinduque (Philippines), though hazardous emissions of Ni dust were thought to be responsible for an abnormally high incidence of respiratory cancer in the regions [220].

The well-known capabilities of metal sulfides for heteroatoms removal (HDS, HDN) led to their consideration for an inexpensive version of HDP catalysts, particularly for the processing of high-metal-containing oils. A systematic comparison of the first-row transition metal sulfides, at low (4%) and high (15%) metal content, and supported on SiO_2 and Al_2O_3 was carried out using a DAO of Morichal crude oil. It was found that group VIII-A metal sulfides showed the highest intrinsic activity for both HDM and HDS reactions [199].

Vanadium active phases seem to be more selective toward devanadation than they are for Ni-removal, as seen from the performance of a catalyst prepared by the impregnation of vanadium oxalate on activated alumina. This V_2O_5 catalyst removed 90–95% of V and 70% Ni, in the presence of hydrogen (125 bar) and 440°C [221]. Potassium was found to promote asphaltenes conversion, keeping DM unaltered when the support was a silica–alumina promoted with chloride [222].

Other materials include activated carbon catalyst, which is used in an unconventional upgrading process [223]. This application would render a spent material that may retain enough metals to represent an option for metallurgical reclamation.

Other proprietary catalysts have been disclosed for particular upgrading processes. One example is the asphaltenic bottom-cracking (ABC) process sketched in Fig. 6.4 [224], which is mentioned here because it has been observed to exhibit a reasonable DM performance. These new noncommercialized ideas on upgrading processes will be discussed in the corresponding section below.

3.2.8 Biocatalytic Treatments

Catalytic biodemetallization (BDM) is the least studied of the bioprocesses for removal of contaminating heteroatoms from crude oils and oil products [225]. Reducing enzymes, such as heme oxygenase and cytochrome c, are able to degrade porphyrin molecules and release the metals, leaving the carbon skeleton intact. A study reported DM of nickel and vanadium petroporphyrins via enzymatic action of peroxidases [226].

The removal of 53% Ni and 27% V from crude oil using chloroperoxidase has been reported [227]. Removal of the metals from the nickel octaethylporphine and vanadium

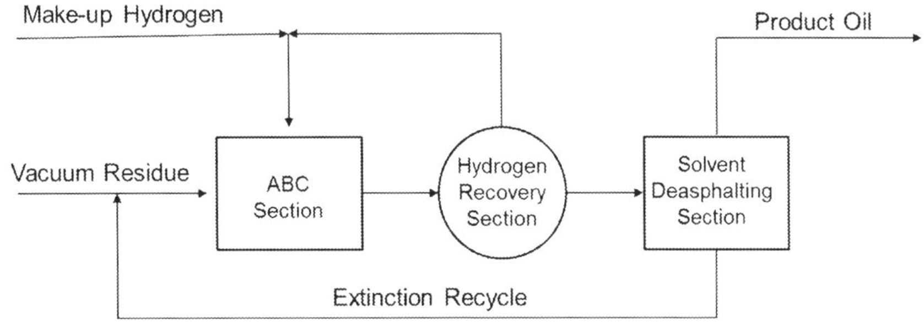

Figure 6.4
Schematics of the asphaltenic bottom cracking (ABC) process. *Reproduced from Takeuchi C, Fukui Y, Nakamura M, Shiroto Y. Asphaltene cracking in catalytic hydrotreating of heavy oils. 1. Processing of heavy oils by catalytic hydroprocessing and solvent deasphalting. Ind Eng Chem Proc Des Dev 1983;22(2):236–42, with permission from ACS Publications.*

octaethylporphine was reported, using this enzyme as the biocatalyst in the presence of hydrogen peroxide and using a ternary solvent system containing toluene, isopropanol, and aqueous buffer (pH 3.0). The ternary system was required to dissolve the petroporphyrin molecules. The enzyme also required chloride and hydrogen peroxide for activation. In addition to the model compounds, an asphaltene fraction from Cold Lake heavy oil was also studied. Removal of 93%, 53%, and 20% metal was reported for the nickel octaethylporphine, vanadium octaethylporphyrin, and asphaltene fractions, respectively. One drawback of this enzymatic system is the chloride requirement, which can result in production of chlorinated molecules. In fact, besides its high peroxidase activity and its versatility, chloroperoxidase is able to halogenate aromatic molecules, as in the case of polycyclic aromatic hydrocarbons (PAHs) [228,229], and unfortunately chloroperoxidase is a feeble protein.

Cytochrome c is a hemeprotein, which catalyzes oxidation reactions on a number of hydrophobic compounds such as substituted phenols, organic sulfides, thiophenes, and PAHs. Cytochrome c has some advantages as a biocatalyst when compared with peroxidases for its application. The catalytically interesting features of cytochrome c [230], whose activity is considered low, are: its heme catalytic group is stable in organic solvents, it functions in water-immiscible organic solvents over a wide range of pH (2−11), and it is economical. Genetic engineering tools have been used for activity and stability improvements [231,232]. Still, limitations are imposed by the partition of reactants between the enzyme active site and the solvent [231]. In order to overcome this problem, a double chemical modification has been performed on cytochrome c [233,234] that, though promising, is not enough for now. García-Arellano et al. [235] using a polyethylene glycol−modified cytochrome c, converted a series of synthetic porphyrins and a petroporphyrin-rich fraction in mild conditions. A very small amount of water (5%v/v) was needed for carrying out the reaction. The DM of the synthetic porphyrins was as high as 95% for V and 74% for Ni. They also succeeded in converting asphaltenes with the same biocatalyst.

A patent on metals removal from a fossil fuel has been registered [236]; it employs heme oxygenase and cytochrome c reductase, such as cytochrome c reductase from *Bacillus megaterium*, *Catharanthus roseuse*, *Escherichia coli*, animal cells, plant cells, or yeast cells. The cytochrome c reductase is used as a substantially cell-free preparation or cell preparation.

Microbial biocatalysts have been also considered. *Pseudomonas azelaica* YA1 strain (FERM P-15,356) has been proved to decompose porphyrin metal complexes [237]; the method employing such a biocatalyst for decomposing the porphyrin-based compounds was also patented [238]. The reaction of vanadium octaethylporphyrin was carried out in a medium containing a meat extract and whole cell spawns. The reaction progress was followed by the disappearance of the UV-vis absorption band of the porphyrin, so metal removal cannot be directly assessed.

The use of amphipathic proteins as flocculating agents in water treatments has been reported. Current advances in protein engineering might be adequate to pursue the design and development of suitable agents that work for porphyrin and nonporphyrin metal compounds [239].

Genetic engineering tools must be applied on two ends of the biotreatment process: on one side for biocatalyst modification to improve the partition limitations, and, on the other, on the whole cell to enhance enzymatic overexpression. Biocatalysts in general are monofunctional and extremely selective to a single molecule or only a few. Therefore, efforts to engineer the bacteria cultures to broaden their functionality and selectivity are required. In the short term, probably a microbial consortium or a biocatalytic system, including more than one biocatalyst, may be the option. So far, the enzyme market favors a preference for the use of whole cells. However, in this case, the molecular size of the reactants and their inherent limitation on membrane transport instead, point toward a cell-free application.

3.3 Combined Methods

A variety of process options has been suggested that include the combination of different physical treatments, chemical treatments, and mixed physical and/or chemical treatments; the promotion of chemical conversion under physical treatments; and so on. These options are discussed in this section.

The use of promoters (e.g., zeolites) or DM catalysts (e.g., [197,240,241]) at the top of an extraction column of an SDA unit was recommended by Mobil for the treatment of asphalt-containing oils to improve metal removal under DA conditions. Employed solvents in the process include C_4-C_{10} refinery streams (naphthas and gasoline streams, LCO, HCO, etc.), benzene, toluene, and glycols. The extraction column is operated at conditions slightly above the supercritical conditions of the employed solvent, which are thought of as means to facilitate the decomposition of metal compounds. Either the added promoter or catalyst would facilitate this decomposition. This *catalytic DA process* enhances DM during the extraction processes. The metal complexes, as well as the asphaltenes, are decomposed. FCC LCO is a particularly preferred solvent inasmuch as it could be recovered including small quantities of catalyst fines, acidic clays, alkali carbonates, and so on. Similarly, the presence of catalyst fines in DCO and its inclusion as a part of the feed comprise an interesting contribution to the employed DM catalyst. The design of the extractor equipment is adapted to improve the desired effect of the promoters and/or catalyst; for example the size or length of the bottom section, where the residue accumulates, can be increased to increase the residence time of the promoters [242,243].

Another suggested promoter is aluminum carbonate or titanium (IV) oxide, which would improve DM at DA extraction temperatures of about 150°C [244].

Adsorption on active carbon of a dissolved SDA pitch was suggested for metals removal. SDA is used for concentrating the metal contaminants in the pitch that is then re-dissolved in an aliphatic solvent, under elevated pressure and temperature conditions. A two-series connected-columns system is used for the adsorption process. Adsorbed asphaltenes were removed from the columns with an aromatic solvent at 360°C and 55 bar. Different combinations of solvents and adsorbents were described for various applications, such as asphaltenes recovery, metals removal, and resins recovery [143].

In summary, not only the combination of emerging concepts and processes but also a sort of hybridization that tries to incorporate new ideas to existing technologies have been claimed in the patent literature. In the case of SDA, both the pitch and the DAO have been treated by adsorption processes [116–118,124–126,140,143]. Similarly, the combinations of precipitation with re-dissolving and extraction [85,86], or uses of DM additives or catalysts in SDA [95], are just examples of many other considered options. These examples are far from exhaustive, and many more possible opportunities for combining concepts of DA with selective DM, or vice versa (selective DA with DM), still remain unexplored. Furthermore, concepts like the dendritic process (separation and isolation all at once [87]) could be applied for upgrading treatments and optimize it toward cost-effective technologies, instead of the conventional sequential combination of units, such as those discussed in Chapter 5.

4. Naphthenic Acids (NAs)

As discussed in Chapter 4 and similarly with the metal compounds, no technology for the specific removal or conversion of NAs has become commercially available, regardless of enormous and long-lasting R&D efforts. These efforts are reflected in the great deal of papers and patents that have been published describing innovative methods, techniques, and processes for these purposes. Both academia and industry have been very productive in this regard. The proposed and considered treatments to reduce the TAN in crude oils and its products are discussed in this section. Some of these practices and concepts have been patented. Part of the associated knowledge has been used for NA manufacturing, commercially. However, most remains at laboratory or bench level, and only a few have reached pilot level without any further development.

4.1 Physical Treatments

A variety of physical treatments has been proposed for the removal of naphthenic acids, probably because these are commonly regarded as the most economical.

4.1.1 Extraction

Extraction is probably the most popular of the physical methods and most used for the manufacture of oil-derived NAs. Most of the patents related to extraction are old enough to have expired, leaving a fertile ground for technology development.

An extraction–reaction protocol was based on the reactivity of NAs with ammonia in ethylene glycol [245,246]. The method protocol has been summarized in Fig. 6.5. Laboratory tests of reaction–extraction processes for removing NAs from crude oils, using glycol–ammonia solution, produced ammonia naphthenate, which remains dissolved in the glycol phase. The dissolved ammonia naphthenate–glycol is separated from the oil and then heated to obtain the NAs. At 50–60°C and a reaction time of over 2 min, NAs were effectively removed from Suizhong crude, by using 3%–6 wt% of ammonia in glycol at a reagent–oil ratio of about 0.3 (wt/wt). The purity of separated NAs was >85% [246]. Petroleum ether can be used to purify the NAs by extracting the neutral oils from the acid removal reagent with a purity >90% [245]. Estimation of the yield gave about 80%, but selectivity was not reported. Nevertheless, it can be observed that the extracted product showed the presence of other compounds [247].

A similar process was used to treat a Korean crude oil, with a TAN value of 8.32 mg_{KOH}/g. In this case, the NAs extractant was 4% ammonia solution in polyethylene glycol, and examined catalysts were CeO, Cu–CeO, ZnO, and SnO, at a reaction temperature of 35–40°C. Although the Cu–CeO catalysts performed the best, none of these catalysts could reduce TAN below 1 mg_{KOH}/g; and only in the presence of 1% KOH (10 M) in 4% NH_3–polyethylene glycol a Cu–Ce(10:90)/Al_2O_3 catalyst, a TAN of 0.56, was attained [248,249]. However, it would also be possible that TAN reduction was a consequence of KOH neutralization, and this questions the need for the other components of the reacting system.

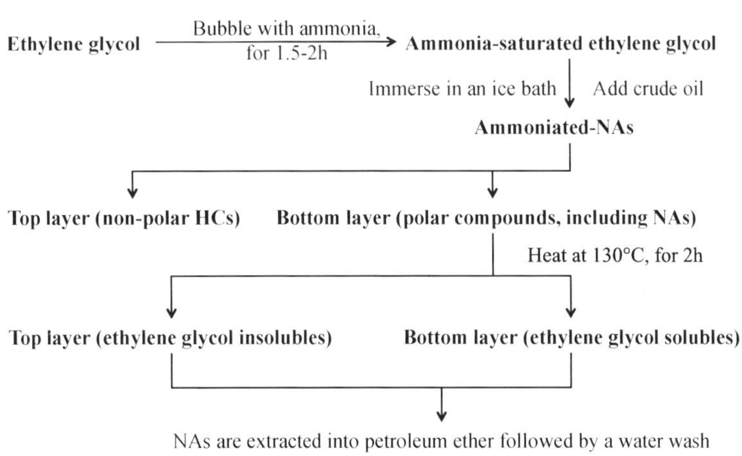

Figure 6.5
Extraction–reaction removal of NAs.

A commonly used extracting solvent is an aqueous-ethanolic sodium hydroxide solution (1 M NaOH in 50% ethanol and 50% water). This extractant is typically applied to sample preparation for analytical purposes [250]. Another analytical preparative technique is liquid chromatography (LC). Preparative LC using a strong anion exchange (SAX) quaternary amine (10 g) was carried out in a solid phase extraction (SPE) system. The ion exchange column (International Sorbent Technology) was conditioned with 30–40 mL of *n*-hexane (Fisher, high-performance LC (HPLC) grade) prior to NAs extraction [251]. An additional preparative extraction, similar to the one described above, includes liquid–liquid extraction with alcoholic solutions. Because this method did not allow the quantitative recovery of the acids, a separation process based on an ion exchange resin (QAE Sephadex A25) was carried out. The isolated acid fraction was then derivatized as methyl esters and quantified by gas chromatography experiments, using the internal standard method. Removal effectiveness ($<82\%$) was checked via the TAN measurement, using the standard ASTM D664-95 (IP 177/96). Characterization results of the recovered acidic fraction showed an average (number) MW of 653 g/mol that, considering the nuclear magnetic resonance (NMR) results of a double bound equivalent of eight and a low aromaticity ($f_a = 1.8\%$), indicated an average chemical formula of $C_{43}H_{73}O_4$ [252].

A KOH-treated silica column was used for the separation of neutral and acidic fractions of Cold Lake bitumen asphaltenes. After adsorption on the column, the neutral fraction was eluted with $CHCl_3$, while a $CHCl_3 + HCOOH$ mixture was employed for the acidic fraction. The metal distribution showed a homogeneous distribution of VO^{2+} ion throughout the different fractions. The concentration of the ESR-determined radicals was significantly higher for the acidic fraction. From this work, two additional relevant observations can be pointed out: (1) the acid fraction decreased the interfacial tension between asphaltenes (in toluene) and diluted NaOH solution; and (2) KOH on silica surface reacted with $CHCl_3$ when accompanied by HCOOH [253].

Extraction can be enhanced by microwave irradiation, as would become evident by the work of several Chinese groups. A Daqing crude VGO with a TAN of 0.63 mg_{KOH}/g was treated with NaOH, at a 0.23:1 solvent-to-oil volume ratio. TAN was reduced to 0.05 mg_{KOH}/g when the mixture was microwave irradiated for 5 min (375 W) under constant atmospheric pressure, and then rested for 25 min. The yield of treated product was 99.3% [254,255]. Similarly, Liaohe AHGO was treated with caustic soda, at a solvent–oil ratio of 0.25:1, under 450 W microwave radiation at atmospheric pressure for 8 min. TAN was reduced from 227.2 to 4.54 $mg_{KOH}/100$ mL with oil recovery 99.0% [256]. Below 375 W, removal is not effective because the energy was not sufficient to provide a required temperature increase and to make the dispersed phase coalesce [257]. Extraction of NAs was affected by the dose of alkali compound in the solvent, pressure, irradiation time, irradiation power, settling time, and oil—solvent volume ratio (O:S). The NAs removal rate maximized at 0.5 bar, 6 min irradiation, 375 W, 25 min settling, and O:S = 10 [258,259]. Microwaves in combination with a nonspecified deacidifying agent were proposed for NAs removal in a Chinese patent. Very little information could be reviewed (only the patent abstract), in which a TAN decrease of 98% was reported. It seems that probably a mixture of NaOH, potassium acid phthalate, and 95% ethanol might have been used as solvent [260]. When extraction is carried out with soda, one may assume that an alcohol solution may be preferred over an aqueous one to facilitate miscibility. Microwave irradiation may be broadly contributing to the process by heating and decreasing viscosity and electronic polarization of the feed.

Ionic liquids have been also studied as extracting solvent for NAs. As an example, a series of 1-n-butyl-imidazolium ionic liquids with three different anions (thiocyanate, octylsulfate and trifluoromethanesulfonate) have been considered for a theoretical and experimental study. All were capable for totally extracting the NAs from a model feed. Capacity and selectivity were theoretically estimated. The thiocyanate ionic liquids exhibited the highest capacity for benzoic acids and the highest selectivity for both benzoic and hexanoic acid. The extraction capabilities of 20% 2-methylimidazole in ethanol (<70%) [261] were lower than those observed with model feed and ionic liquids reported in Ref. [262].

A complex extractant composed of ethanolamine, ethanol, Na_2SO_4, and a proprietary additive (UD10) mixed in specific critical proportions has been proposed. The optimum content of these constituents in the solvent system was reported to be 10 wt% ethanolamine, 35 wt% ethanol, 2 wt% Na_2SO_4, and 100 ppm UD10. The optimum ratio of solvent to oil was 0.3—0.5. Other extraction factors, including extraction temperature, blender rotational speed, phase-splitting time, and extraction—reaction time, were also examined. The method could efficiently prevent the problems of emulsification and high oil loss, typically found during alkaline treatments. NAs extraction was >70% [263,264].

Suggestions for extraction enhancements by incorporating porous and nonporous membranes in the aqueous caustic soda process have been reported. Hydrodynamic factors, concentration of NAs and caustic soda, and temperature have been considered, and a model was proposed [265]. Nevertheless, the advantages of using a membrane might be overwhelmed by the issues and troubles encountered by handling heavy and viscous materials.

Regarding the patents landscape, ExxonMobil is the most prolific owner of intellectual property in the extraction area. Some of the patented processes are briefly exemplified below and grouped by their owner company.

4.1.1.1 ExxonMobil

The following two patents from this company describe a process comprising the steps of:

1. treating the starting crude oil containing NAs with an aqueous solution of an ethoxylated amine [266] or alkolxylated amine [166], forming an emulsion;
2. separating that emulsion into layers, wherein one layer contains the treated lower TAN crude oil; and
3. recovering the layers of treated crude oil and the aqueous layer(s) containing the amine salt.

The important characteristics of the amines are their solubility into both oil and water, allowing dispersability into the oil phase for reaction with NAs and easy removal of the formed salts facilitated for their solubility in water. The amine is added in about 2.5 times the amount of NAs present in the crude. Optionally, amines can be added alone instead of in their aqueous solution, in which case about five to 10 wt% water would be present. Process conditions include temperatures between 25 and 80°C, contact times of three to about 30 min, and pressures from 4 to 65 bar. For heavier crudes, higher temperatures and pressures would be required. Treatment with organic amines combines their neutralization effect with the solubility needed for extraction. For these capabilities, the process can take place in existing desalter units, shown as item (3) in Fig. 6.6. The aqueous layer can be treated to recover the amine and the NAs. Around 70% reduction in TAN was observed for crude oil.

An inversed extraction process can be defined as one in which the NAs are first retained by an adsorbent and then extracted during the adsorbent regeneration step. Cross-linked polymeric amines can be used as the adsorbent in such a process, for treating HACs or acid fractions. The cross-linked polymeric amine is contacted with the feed, and the acid

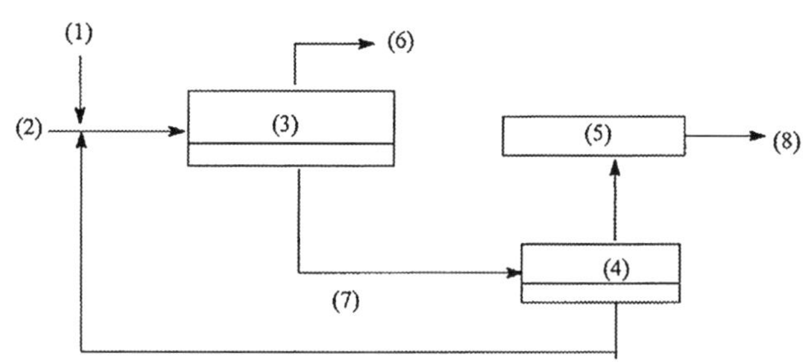

(1) water and ethoxylated amine,
(2) starting crude oil,
(3) the desalter,
(4) the regeneration unit,
(5) the organic acid conversion unit,
(6) treated crude free of organic acids,
(7) lower phase emulsion, and
(8) products.

Figure 6.6
Process for extracting NAs using amines [166,266].

molecules will be attached to it. These resins being insoluble in the crude can be separated or recovered from the crude, for instance by filtration or centrifugation. Once isolated, regeneration is performed by displacing the acids. The contacting is typically carried out at preferably 30–150°C (depending on the thermal stability of the polymeric amine), with molar ratios of amine to acid groups between 1 and 5. Addition of water facilitates the reaction, particularly if the cross-linked polymeric amine is dry. An 83% reduction in TAN was reported [267]. Strong basic polymeric amines with pKa > 9, such as polyvinyl amine or anionic exchange resins, were demonstrated to reduce TAN in only 55% [268].

4.1.1.2 Elf

Once again, an amine (preferably ammonia) is used for the treatment of desalted crude; subsequently, an alcohol (preferably methanol) is used for extracting the formed naphthenate. Experimental results show TAN reduction of 50–70% [269].

4.1.1.3 BP

An extraction process using an immiscible polar solvent was described. The crude and solvent were separated, and the solvent phase was treated with a basic material to neutralize the organic acids, which subsequently were separated from the solvent to be recycled. The polar solvent was selected from alcohols, alcohol derivatives, and ethers. Suitable alcohols include methanol, ethanol, and propanol, with methanol being preferred. Glycols such as polyethylene glycol may also be suitable. Appropriate ethers include glycol ethers. Alcohol derivatives such as alcohol polyalkoxylate may also be employed [270]. An improved version of this process included an additional treatment of the crude oil with a base, in the case the TAN value remains above the desired value [271].

4.1.1.4 Queens University Belfast

A basic ionic liquid was proven to remove either part or all of the NAs from the crude oil. The general structure of suitable ionic liquids; the synthesis of a specific (preferred) ionic liquid, [6-(dimethylamino) hexyl] (ethyl) dimethylammonium bis[(trifluoromethyl) sulfonyl] amide; and the extraction results for a model feed and a crude oil were described in the patent. The reported TAN reductions were of about 80% with the model feed but merely 33% with the crude [272].

4.1.1.5 Chevron

An extracting solvent composed of methanol, water, and ammonia in certain critical ratios was disclosed for NAs extraction from distillates fractions. Selectivity to NAs extraction is controlled by the ratio of ammonia to feed, which must be about 0.25–0.3 parts by weight of ammonia per 100 parts of feed. Extraction of nearly 95% of NAs could be achieved in a diesel feed, but application to crude oil was not feasible since the system became emulsified [273].

4.1.2 Adsorption

A possible way to remove the carboxylic acids from petroleum-distilled fractions is the adsorption in porous materials. Some results obtained until now indicate that ion exchange resins would be costly for this purpose. Nevertheless, adsorption of NAs has been studied on weak and strong anion exchange resins that are commercially available, as well as on zeolites and clays. The capacity and equilibrium constant of adsorption on these resins were compared with those obtained with zeolites, bentonite, and polyvinyl pyridine. The reported adsorption capacity is reproduced in Table 6.2. Resins exhibited higher adsorption capacity than the tested inorganic adsorbents. Best performance was obtained by the macroporous weakly anionic ion exchange resin (A-2X MP) with a weak tertiary amine group, which had the highest adsorption capacity for the removal of NAs from crude oil. Additionally, the weak resin with isoporous structure compared to the strong resins (Indion FFIP and Indion 850) showed weaker interactions with NAs that would facilitate regeneration. The possible mechanism of a quaternary group on such resin interacting with the acid seems to be an ion—dipole interaction unlike hydrogen-bonded structure in the case of the weakly basic ion exchange resins. The rate of uptake of NA is governed by the internal diffusional resistance, which also depends on adsorbed phase concentration [274].

Two other commercial adsorbents (an acid bentonite clay, TCO626 G, from Sud Chemie TONSIL series; and an activated alumina, Selexsorb CD, from UOP) were characterized, and their capacity for removing NAs was evaluated. The adsorbents were pretreated before adsorption, by heating in an oven for 4 h at 150°C, and then cooled to room temperature (RT). Both adsorbents had effective results, indicating that either a negatively charged surface (clay) or positively charged surface (alumina) was active for the adsorption of the naphthenic acids and could reduce TAN values of the oil [275]. There was competition for the adsorption sites among all the compounds present in the examined light oil fraction, indicating probably lack of selectivity, decreased yields, and fast overloading.

Table 6.2: Comparison of Adsorption Capacity of Different Materials [274]

Adsorbent	Loading Capacity (moles/kg)
Na-X	0.155
Polyvinylpyridine	0.60
Bentonite	0.442
Indion FFIP	0.401
Tulsion A-2X MP	0.92
Indion 850	0.884
Tulsion A-2X MP[a]	0.89
Tulsion A-2X MP[b]	0.89

[a]In presence of 5.1% toluene.
[b]In presence of 10.3% toluene.

A NaOH-activated muscovite clay was used for adsorbing NAs from Sudanese HACs Fula and Nile Blend, which had TANs of 8.51 and 1.06 mg_{KOH}/g, respectively. The studied conditions did not render a significant TAN reduction. In fact, no adsorption was recorded for the Nile Blend, and only 26% reduction of the TAN in Fula crude [276].

The enthalpies of adsorption and the isotherms for adsorption of NA onto Na-montmorillonite, Na-kaolinite, and Na-illite were studied by means of calorimetry and static method, at 25°C. The enthalpies of adsorption and adsorption capacity for NA on different clays change in the following order: Na-montmorillonite > Na-illite > Na-kaolinite. Adsorption was proposed to take place in a two-step process: (1) hydrogen bond interaction between the hydroxyl of the carboxylic group and the oxygen in the atom of the Si—O tetrahedron; and (2) van der Waals forces between the large hydrophobic groups of naphthenic acid adsorbed on the clay surface and those in solution. The first adsorption step was considered to be the main contributor to the magnitude of the enthalpy of adsorption [277].

Acid montmorillonite clay commercially available from Sud Chemie was considered for a study that compared the adsorption of NAs from model compounds with that of a diesel sample. The evaluated adsorption capacities for the model compounds ranged from 19.4 g/kg for cyclohexane butyric acid to 25.3 g/kg for cyclopentane carboxylic acid [278]. TAN reduction was always higher for the model compounds than that observed for the real feed, indicating competitive adsorption with other compounds present in the real feed.

In a process that uses a slurry containing solids such as silica, alumina, coke, or montmorillonite clays to generate a separated treated crude and an aqueous layer, the NAs were reduced by up to 86% [279].

One of the causes for environmental concerns about NAs is the fact that soils from pond water of oil sands tailings strongly adsorb them. Adsorption tests conducted at 4°C and a pH of 8.0 ± 0.4 indicated a preferential sorption of NAs with carbon number between 13 and 17 [280].

Heavy NAs removal (200—~750 amu) might be enhanced by the combined effect of emulsification, emulsion breaking, and adsorption. These operations could be performed in the field and thus might be viewed as a way for upstream integration, as well as an application in refining, as one of the initial treatments of the crude oil. The solid adsorbent (0.1—5 wt%, depending on the amount of acids present in the oil) and water (5—10 wt%) are added to the HAC to form an emulsion, unless the oil—water emulsion already exists, in which case this process also serves as an oil—water separation process. The existing separation equipment could be used. In order to attain system emulsification, temperature and pressure are adjusted to between 25 and 80°C and 1—70 bar. Residence times depend on the acidity of the crude, adsorption capacity of the sorbent, and kinetics of the sorption process. It typically ranges between 3 and 30 min. Once emulsification and contact have taken place, the emulsion is broken and separated into layers by using any demulsifying and separation technique, such as demulsification agents, centrifugation, gravity settling, sonication, hydrocyclones, microwaves, and/or electrostatic separation. Powder adsorbents (e.g., silica, alumina, coke, montmorillonite clays such as bentonite, and kaolinite) with particle areas of less than or equal to about 100 to about 1500 μm^2 are suitable for the process. Clays, divided or exfoliated, are preferably used in gel form, with sheets of less than 10 nm thick broken into plates of 50–200 nm. Solids with amphiphilic surfaces are preferred for their high removal capabilities (~85%) [281].

Adsorbents of hydrotalcite type were developed by UOP for the removal of NAs from hydrocarbon streams. Both Mg and Ni mixed oxides with Al were prepared and tested; additionally, another sorbent was formulated as containing disulfonated cobalt phthalocyanine on a mixed-solid oxide of Ni–Mg–Al. Best performance of about 98% removal was accomplished with the latter, while the Ni–Al and Mg–Al mixed oxides could achieve up to 90% removal [282]. However, it is important to point out that the tested feed was kerosene, with a TAN < 0.07 mg_{KOH}/g.

Active phases used in solid phase extraction (SPE) for the quantification of NAs were tested for the development of an analytical method. ENV+, a cross-linked polystyrene-based polymer from Biotage; C18, an octadecyl carbon chain-bonded silica (available from most HPLC columns vendors); and Oasis, a silica-based material loaded with poly-hydrophilic or poly-lipophilic monomers depending on the application (available from Waters), were evaluated. ENV+ exhibited the highest capacity [273].

4.1.3 Blending

Refiners found in blending a *dilution* of the problems brought by NAs, particularly corrosion. Dilution is a guaranteed decrease in TAN values when the diluent is a high-quality crude. Since most refineries are equipped to process crude oils with TAN below 0.5, HACs have to be scheduled for processing when and only when a proper diluent crude is available.

For instance, Fluor Corporation differentiated NAs into two classes, α- and β-naphthenic acids that control the corrosion rate. The characteristic features of the α- and β-fractions are given in Table 6.3. The rate of corrosion was found to decrease by increasing the β/α ratio, by blending appropriate crudes within the CDU/VDU (see Chapter 4) [283].

According to this patent, not all NAs are corrosive; in fact, there seems to be "good" (β) and "bad" NAs. Therefore, crudes having a high concentration of "good" acids would reduce the corrosive action of "bad" acids in another. The question is whether "bad" and "good" NAs can be real. If this is the case, then, are the definitions and differentiating

Table 6.3: Characteristics of α- and β-Acid Fractions

α-NA Fraction	β-NA Fraction
Low molecular weight (125–425)	High molecular weight (325–900)
Moderate to high solubility in aqueous solutions of pH 6–9; typically 0.1–2.5 mg/L; moderate to low solubility in oil, typically <0.1	Low solubility in aqueous solutions of pH 6–9, typically 0 to 0.3 depending on molecular weight; high solubility in oil, typically >0.08
Carboxyl group readily ionizes in aqueous solutions with pKa typically between 4 and 6.5.	Carboxyl group poorly ionizes in aqueous solutions with pKa typically between 5.5 and 7.5.
Neutralizes to form salts	Difficult to neutralize
Iron naphthenate – highly soluble in oil	Iron naphthenate – difficult to form
True boiling point up to ~385°C	True boiling point up to ~360–815°C
No protective surface film formation	Formation of protective and inhibitive surface film
Decomposes at elevated temperatures above 340°C	Fails to readily decompose at elevated temperatures above 340°C
Follows classical NA corrosivity model (i.e., TAN)	Does not follow classical NA corrosivity model

features that categorize the corrosive actions valid? Further NAs molecular characterization is needed for the understanding of surface properties and structure–reactivity relationships. More particularly, reactivity in terms of corrosion should be addressed in these relationships. In this regard, the corrosion ability of 26 organic acids was evaluated, and aliphatic (fatty) acids were compared to NAs. Results indicated an increase in corrosion capabilities of NAs compared to fatty acids. Additionally, in NAs the increase in the number of rings in the molecules decreased the corrosion ability [284], in agreement with Messer disclosure [283]. Apparently, another "good" behavior of NAs has been discovered. Heavy tetra-acid NAs have been recommended as additives for fouling inhibition. This type of tetracarboxylic acids, known as ARN acids, was described in Chapter 4 and is well known for their capabilities in creating Ca^{2+} sludges and deposits. Briefly, ARN acids are HMW molecules with four carboxylic acid groups, each at the end of a long aliphatic chain, forming a four-fingered molecule with polar tips. Their MW ranges from ~1228 to 1236 amu; the most found ARN has an MW of 1232 amu corresponding to the formula of $C_{80}H_{142}O_8$. ARNs are also characterized by the presence of four to eight sites of unsaturation (or four to eight cyclopentyl rings) and for the absence of quaternary carbons, double C-bonds, and aromatic rings. The addition of an effective amount of about 50–1000 ppm of ARNs to a base crude might reduce fouling by nearly 30% [285]. Most likely, this additive might require a limiting Ca-content specification of the base crude oil.

4.2 Chemical Treatments and Processes

Various reactions of acids have been considered for fighting against acidity. As will become clear, the efficiency in TAN reduction depends on the reactivity of the NA compounds present in the feed. In Chapter 4, NAs reactivity was discussed, and it is well understood that reactivity decreases with increases in the MW of the considered NA. Thus, a method is needed that works for the wide range of MWs of NAs; or, in other words, a method that is effective enough regardless of the reactivity of the considered NA compound.

Preliminary results of a catalytic process for TAN reduction have been reported, but very little details were given. Apparently, the whole crude was treated, but if that was the case, that crude cannot be categorized as an HAC (TAN $< 0.9\ mg_{KOH}/g$). The effects of operating conditions were evaluated. At this stage, the process under the most severe conditions reduced TAN in less than 80% [286]. Clearly, it does not seem to be competitive enough, particularly compared to the NAs removal capabilities of some of the physical processes discussed above.

4.2.1 Decomposition

As described in Chapter 4, NAs decompose at temperatures above 300°C; but the larger the complexity of the molecular array, the higher the decomposition temperature. Thermal treatment was identified as an effective approach to decrease acidity, and the acid conversion rate could reach levels above 80%, when an HAC was thermally soaked at 420–440°C. The acid disappearance rate considerably increased with increasing reaction temperature, up to 440°C. At temperatures below 440°C, increases in reaction time favored the acid removal. Obviously, cracking and polymerization reactions occurred under these conditions and would be detrimental to the yield and final properties of the product [287].

Three patents with variations on the conditions of the thermal treatments have been awarded to ExxonMobil. In a first treatment, the feed is heated to 370–425°C for a period of 30–90 min in the presence of a sweep gas such as a mixture of CO and CO_2, keeping CO_2 partial pressure below 0.03 bar and that of H_2O below 0.02 bar. TAN reduction was moderate [288]. Surprisingly, a modification that employs lower temperatures (200–315°C) was claimed to reduce TAN in 85% [289]. At this temperature range, only light NAs could be treated, but heavier NAs molecules will not decompose. For this reason probably, higher severity was disclosed in the patent of Ref. [290]. This process involves temperatures in the range of 260–400°C in a thermal reactor that is swept by an inert gas. Treatment time varies with temperature, at 370–400°C; reaction time ranges from 0.5 to 5 h. The vaporized liquid is recombined with the oil to minimize losses. Experimental data showed significant TAN reduction (>90%) on the recombined oil.

A thermal process includes a first step for water removal. This process comprises a preflash to remove feed water, a mild thermal treatment in a purged low-pressure reactor of two or more stages, and a final step wherein light hydrocarbons that are recovered from either thermal treating or from the preflash are recombined with the reactor effluent to obtain a low-TAN upgraded crude oil. Suitable purge gases include nonoxidizing gases, such as nitrogen, methane, carbon monoxide, well-head gas, fuel gas, and hydrogen. The process operates at 370–400°C, less than 7 bar, and less than 2 h to achieve TAN reduction of up to 90% [291]. From this process description, a possible detrimental effect of water might be implied due to the emphasis for a complete dewatering in the first step.

NAs present in HAC can be substantially converted to CO, to CO_2, and to low amounts of smaller acids (e.g., formic, acetic, propionic, and butyric acids) in steam crackers [292]. Comparing to the previous process [291], it seems that water originally present in the feed may have a negative effect, while a positive effect is observed if added in the form of steam.

Ivanhoe Energy suggested the processing of HAC through their Rapid Thermal Processing (RTP, described in Chapter 5), in the presence of a calcium compound. Calcium is used for the dual function of reducing TAN and SO_x content [293,294].

The effectiveness of applying UV-vis radiation for photolysis of NA mixtures and individual compounds did not result in any significant reduction of the NAs content, nor of the most toxic aryl hydrocarbons. Among the tested wavelength, the UV_{254} radiation source was the most effective, although TAN change was insignificant [295].

A combined process of adsorption followed by thermal decomposition uses a modified spent or coked FCC catalyst as adsorbent for NAs. In this Petrobras patent, NAs are converted to CO, CO_2, and H_2O through the described process. An added benefit is viscosity reduction. The adsorbent FCC catalyst has a 50–70% reduced catalytic activity, and it is obtained by treating spent FCC catalysts with a solution of ammonium hydroxide (10% by weight), in a mass ratio of NH_4OH–mass of spent catalyst of 0.1. The process takes place in a reactor where the ratio of adsorbent to oil is in the range of 0.1–3, at temperatures between 200°C and 500°C, and under pressures from 0.1 to 3 bar and residence times between 1 s and 2 h. TAN was reduced in the range of 74–96% [296]. The high temperatures needed for decomposition may be indicative of a total absence of catalytic activity. Another application, in which heating was provided by microwaves, was limited to streams containing low NA content, with TAN in the range of 0.01 up to 0.8 mg_{KOH}/g. A slurry is formed between the feed and a suitable microwave-absorbent material. Useful microwave-absorbent materials for the process include carbon fines, spent FCC catalysts, spent HDT catalysts, and hydrorefining catalysts were preferred as these absorb radiation on localized sites. Microwaves were thought to be absorbed by the active sites of the adsorbent material or catalyst, leading to a very localized overheating, which induces reaction between the organic acids and the surface or the active sites of the absorbent. Hence, the process can be operated at lower temperatures than those used in a conventional thermal process [297]. Petrobras also disclosed another process for reducing TAN that only applied microwaves on a feed containing emulsified or dispersed water. They claimed that NAs at the oil–water interface pick up the heat and decompose. This proposed process operates at temperatures between 50°C and 350°C, and the microwave radiation is applied at 1 mm to 30 cm to the feed, under pressures between 7 and 5 bar [298].

Finally, a particular case of decomposition is ozonolysis, which in some cases involves ozonation as a first step, followed by thermolysis or any other decomposition means. It has been studied for the degradation of tailing waters [299,300] and more particularly for the degradation of NAs [301,302].

4.2.2 Decarboxylation

Decarboxylation is a particular case of partial decomposition that has received enormous attention and would have minimum impact on yield. This reaction has been even more widely considered for the treatment of fatty acids, for the manufacture of renewable fuels (see, e.g., Refs. [303–306]).

Thermal decarboxylation was found to be the main process responsible for TAN reduction during thermal treatment. The observed increases in TAN reduction with temperature could not be explained as only due to the evaporation of lighter NAs compounds. Moreover, Fourier transform ion cyclotron resonance–mass spectrometry (FT-ICR-MS) did not show any change in the naphthenic acid (O_2 class) composition of the liquid products, indicating that all compounds were decarboxylated roughly, in the same extent, regardless of MW [307].

Catalytic decarboxylation has been widely studied; the most popular catalysts are alkaline and alkaline earth metal oxides. Probably, MgO is the most studied catalyst. As will become evident through the next paragraphs, some of the reported experimental results do not discriminate among decarboxylation, decomposition, adsorption, and/or other reactions, such as naphthenate formation.

A kinetic study was carried out using a mixture of NAs extracted from a Colombian crude oil, using MgO, K–MgO, and Cs–MgO, in the temperature range of 150–350°C. At these temperatures, thermal decarboxylation may be negligible, although kinetic parameters were evaluated [308]. The catalysts were not very active, either, with NAs concentration decreases of less than 16%. Furthermore, changes in CO_2 formation were not reported, indicating that probably evaporation of light NAs and adsorption on the solids were at least partially responsible for the observed decrease in NAs concentration.

A commercial mixture of NAs dissolved in Paraflex HT oil, a pure saturated hydrocarbon mixture, was catalytically converted at 385°C in a batch reactor on alkaline and alkaline earth metal oxides and carbonates. The catalytic activity was evaluated as the acid concentration decrease. The activity of the studied catalysts decreased in the order of $Li_2O > CaO > BaO > MgO > CaCO_3$. However, naphthenate of the catalyst metals was detected in the spent catalysts. The removal rate of TAN and the production rate of CO_2 were independent of the concentration of NA, and the reaction temperature had a great influence on the production rate of CO_2 but a minor effect on the removal rate of NAs [309].

Alkaline earth metal oxide catalysts were compared with ZnO, for conversion of NAs in HVGO. During reaction, alkaline earth metal oxides were converted to hydroxides and/or carbonates; observation of naphthenate formation was not reported. However, TAN reduction was explained as due to multiple effects. Catalytic decarboxylation was the main removal path with MgO and ZnO. Meanwhile, NAs on or with CaO underwent multiple transformation pathways (catalytic decarboxylation, neutralization, and TC). In the case of BaO, neutralization was the governing reaction [310].

A theoretical and experimental study to develop a catalytic decarboxylation process to remove petroleum acids from high-acid crude oils was conducted. Two testing rigs were employed, fixed and fluid beds, Fig. 6.7 shows the respective schemes in (A) and in (B), respectively. Model NA compounds were spiked into the feed to cover a wider range of acidity. These model compounds were extracted from diesel cuts. The experimental results on the fluid bed were better than those on the fixed bed, reaching decarboxylation of 97% for a HAC of 12.52 TAN, over a (MLC500) solid acid catalyst at a volume space velocity of 8 h^{-1}, catalyst-to-oil ratio of 7.5, and temperature of 460°C [311]. No details were given of the catalyst in the published article, but a web search indicated it is a commercial residue-cracking catalyst produced by SCC, the largest manufacturer and supplier of petroleum and chemical catalysts in China [312]. The theoretical results did not contribute any additional insights into the process, since the main conclusion was that the energy barrier was lower in the presence of a catalyst than that of thermal decarboxylation.

Another combined effort of theoretical and experimental study was carried out for understanding of the NA catalytic decarboxylation process [313]. Theoretical discrete Fourier transform calculations predicted the attack of the hydroxyl group on the ortho-position of the aromatic ring as a key step. This step had a transition barrier of >30 kcal/mol, which was consistent with the experimental observations [314]. Among the various solid catalysts tested, MgO exhibited the highest activity toward the decarboxylation reaction of both saturated and aromatic model NA compounds at a temperature range of 150–250°C, when the reaction test was run for 4 h. The benefits of

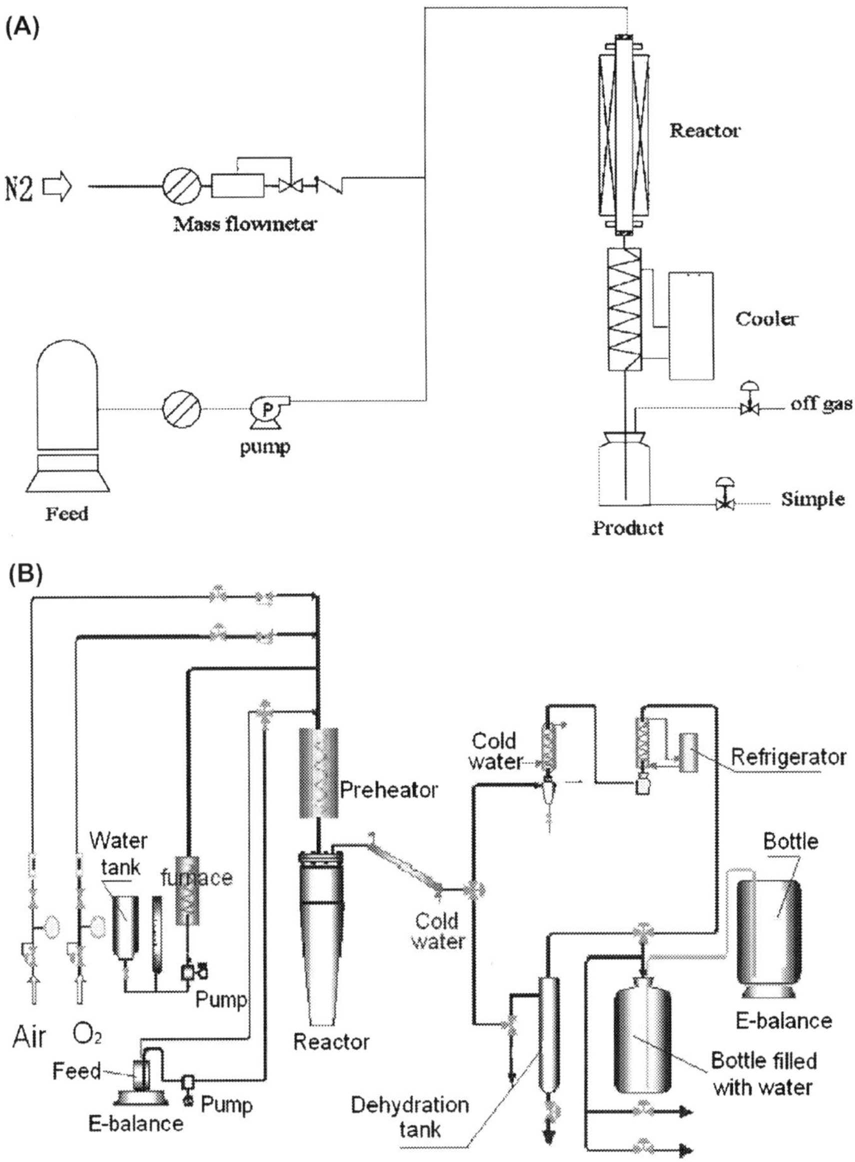

Figure 6.7

Experimental rigs for catalytic decarboxylation: (A) fixed bed, and (B) fixed fluid bed. *Reproduced from Fu X, Dai Z, Tian S, Long J, Hou S, Wang X. Catalytic decarboxylation of petroleum acids from high acid crude oils over solid acid catalysts. Energy Fuels 2008;22(3):1923–29, with permission from ACS Publications.*

MgO were determined from the combined results on the formation of CO_2 and the conversion of acid. A multiple role of MgO explained its effectiveness, namely its ability to adsorb acidic compounds via acid–base interactions and its catalytic capabilities for promoting decarboxylation and cracking reactions at moderate temperatures. Thus, the direct application of MgO to a HAC would result in significant NA removal and lower TAN [315,316]. Nevertheless, among all the catalysts tested, it was only in the case of Ag_2O as catalyst that the CO_2 yield was found to match well with the acid conversion and with the concentration of the decarboxylated product. The presence of naphthalene in the reaction product was interpreted as the occurrence of a direct decarboxylation pathway [317].

Simple model compounds of light NAs were catalytic decarboxylated using alkaline earth metal oxides (Mg, Ca, Sr, and Ba) as catalysts. Commercially available oxides as well as laboratory-prepared samples were included in the study. These simple NAs were decarboxylated completely or to a very high extent. Activity evaluation was based on the change in concentration of NAs in the feed to the product. The highest activity was found with MgO and CaO. However, a white solid suspension eluted from the reactor, and upon analyses Mg—naphthenates were identified. Since CO_2 evolution was not reported, there remain doubts regarding whether the reaction that took place was decarboxylation [318].

A very generic patent from the California Institute of Technology describes a treatment where oil is contacted with a metal oxide to upgrade quality. The process is carried out in the temperature range of 200—450°C. The metal oxide can be one of the alkaline earth metal oxides (e.g., Mg, Ca, Sr, and Ba), an oxidative transition metal oxide (e.g., Cu, Mn, Pb, Ni, Ce, La, Y, Zr, and Ag), or basic clay sorbents in combination with the metal oxide to adsorb NAs or catalyze their conversion [319].

UniPure Corporation describes a chemical treatment process whereby the crude oil is mixed with an alkaline earth metal oxide and is subjected to reaction at 250—350°C and 25—35 bar. Calcium oxide was the preferred reactant material, and barium or magnesium oxide comprised the other alternatives. These oxide materials reacted with the NAs and sulfur compounds to form alkaline earth metal carbonates and alkaline earth sulfides, respectively, at the conditions employed [320].

4.2.3 Hydrotreatment

HDT and, in general, hydroprocessing are not practical; neither economical for processing a whole crude. Typically, they are carried out on a per-need basis on selected refinery streams. In recent times, residues HDT has become more popular. As a rule of thumb, the heavier the processed cut (or the original crude oil), the shorter the catalyst life cycle is. The HDT available technologies for residues of heavy oils and bitumen were described in Chapter 5, including the issues faced when processing this type of opportunity crudes. Troublesome species in heavy oils and bitumen are not necessarily present in HACs. Hence, refineries with existing HDT units pursued small adaptations recommended by the licensors and have tried the processing of NAs-containing streams.

HDT catalysts are formulated to contain Group VIB metals and nonnoble Group VIII metals, supported on alumina. These metals or mixtures of metals are present as oxides on the fresh catalyst and sulfides under operating conditions. Preferred catalysts include cobalt–molybdenum (1–5% Co as oxide, 10–25% Mo as oxide), nickel–molybdenum (1–5% Ni as oxide, 10–25% Mo as oxide), and nickel–tungsten (1–5% Ni as oxide, 10–30% W as oxide) on alumina. Especially preferred are nickel–molybdenum and cobalt–molybdenum catalysts. Under HDT conditions NAs may undergo three different reactions: hydrodeoxygenation (HDO; the elimination of OH group as H_2O), decarboxylation (CO_2 elimination), and decarbonylation (CO elimination). HDO is expected to have the highest yields in the presence of hydrogen for a sulfur-containing feed. However, C-elimination reactions would be favored in the absence of either H_2 or S. The formation of water in an acid environment accelerates corrosion and also would have a negative effect on alumina-supported catalysts. Ordinarily, alumina would hydrolyze under such conditions, threatening catalyst integrity. Additionally, CO is considered a metal poison. Furthermore, the simultaneous presence of CO_2, NH_3, and H_2S in the product gas may bring complications to its handling.

ExxonMobil disclosed a series of patents claiming the use of different HDT catalyst formulations for the treatment of NAs containing hydrocarbon streams in the absence of hydrogen. The limited capabilities of HDT catalysts for converting NAs appeared to be evident. In a first process, reaction conditions were 285–345°C, and LHSV from 1 to 8 h^{-1}. Experimental results show 20–50% TAN reduction [321]. In a second set of three patents, the process was operated at 200–425°C and at a pressure between atmospheric and 70 bar. The reactor was swept with a hydrogen-containing gas [322] or an inert gas [322,323] to maintain the water–CO_2 partial pressure below 4 bar. A dispersed catalyst of Group VB/VIB/VIIB/VIII metals with preferred metals, including Mo, V, Fe, Ni, Co, and Cr, was used. Under the experimental conditions used, TAN reduction was about 80% in the absence of catalyst and H-sweep; it was 85–95% with catalyst and H-sweep. The effect of water partial pressure was detrimental for TAN reduction, but there was not a reported effect of the absence of H-sweep, though others had shown sweeping to be beneficial [289]. A small pore hydrotreating catalyst was suggested for selectively converting lower MW NAs, at temperatures of 200–370°C. Preferred catalysts included nickel–molybdenum and cobalt–molybdenum catalysts such as KF-840 and KF-756, which are commercially available from Albemarle. Up to 91% reduction in TAN was observed [324].

In the next two patents, ExxonMobil described processes operated at HDT conditions, at a partial pressure of hydrogen sulfide. A HAC was treated at 200–370°C, 2–140 bar pressure of a hydrogenating gas containing 0.05–25 mol% H_2S. Conventional catalysts cobalt–molybdenum (1–5% Co as oxide, 5–25% Mo as oxide), nickel–molybdenum (1–5% Ni as oxide, 5–25% Mo as oxide), and nickel–tungsten (1–5% Ni as oxide,

5–30% W as oxide) on alumina were preactivated by sulfiding. Low-acidity metal oxide supports were preferred in order to minimize hydrocracking and/or hydroisomerization reactions. The experimental results showed the impact of H_2S in the treat gas on the first-order kinetic rate constants. Regardless, the TAN reduction was not significant, <57% [325,326].

A slurry-phase low-pressure process was disclosed by Texaco. The catalyst included one or more metals, nonnoble Group VIII and/or Group VIB of the periodic table. HDT was carried out at 250–500°C, 13–105 bar, and 0.1–5 h^{-1}. Total TAN suppression was reported as possible [327]. In a second patent, similar conditions were employed for a conventional hydroprocessing catalyst, but the results show lower TAN reduction of about 83–97% [328].

Statoil-Hydro's approach has been to try to understand the HAC quality issues that affect processing and reduce their attractiveness to customers. Then, particular solutions may be offered for their safe and smooth processing. As discussed in Chapter 4, not only do these crude oils cause production and transportation problems, but also their corrosive capabilities cause severe damage to refinery equipment. Therefore, precautions and mitigation actions have to be taken. Since naphthenic corrosion is difficult to predict, some efforts were devoted to construct an experimental rig for assessing corrosion potential of different fluids under varying operating conditions. Thanks to a breakthrough in the understanding of the fundamental mechanism of calcium naphthenate formation, the group concentrated on developing an ultimate solution for removing the acids prior to refining [329,330]. NA removal (NAR) is Statoil's proprietary process for removing NAs selectively, under mild conditions using catalytic HDT. The hydrogenation process of crude oil is carried out at 1–50 bar and 100–300°C, over a conventional NiMo or CoMo residue catalyst. Only 5–6% of TAN reduction was attained. There was no reduction in the content of any one of the other contaminants (sulfur, nitrogen, or metals) [331]. Hydrogen consumption is claimed to be low, and supposedly no yield loss was reported to occur [332].

The integration of an HDT to a coker, for the HDO of distillates derived from the upgraded bitumen, was disclosed by Suncor Energy. No data were reported [333].

In summary, considering that HDT of whole crude oil is not economical and even less if the oil is an opportunity crude, together with the marginal results on TAN reduction, it seems HDT is not the solution and there is still need for an effective technology for NAs abatement.

4.2.4 Esterification

Another process option for treating crude uses alcohols to form esters with the organic acids contained within the HACs. A kinetic study of the NAs esterification with methanol indicated that reaction occurred at temperatures between 150 and 250°C. TAN reductions above 90% could be attained at 250°C when a Colombian HAC (TAN = 7.3 mg_{KOH}/g) was let to react for 6 h [334].

Operating conditions for the esterification of NAs with ethanol and methanol are the subject matter in the work of Sartori et al. [335]. The process is carried out above 250°C, at pressures of 10–30 bar and contact time of up to 30 min. Experimental results show TAN reduction of up to nearly 90%.

The conversion of a fraction of NAs isolated from the lubricating oil portion of a Gulf Coast crude oil was investigated [336]. The whole scheme, separation and reaction, is shown in Fig. 6.8. The separation method was already described and presented in Chapter 4 and is included here for completion purposes, as part of the whole scheme. The isolated NAs that were utilized for this investigation had a narrow MW range of about 309–338 amu (average value was 317 as calculated from an acid number of 177). These NAs were practically free of acids of MW < 292 (C_{19}^{-}) and MW > 348 (C_{23}^{+}). The average number of carbon atoms per molecule was 20.7, with an average molecular formula of $C_{20.7}HS_{5.2}0_{2.1}$. The collected evidence indicates that the acids were monobasic compounds containing on average about 2.6 rings per molecule. The corresponding naphthenic hydrocarbons derived from the conversion of the NAs were prepared following the reaction scheme included in Fig. 6.8: esterification, reduction, halogenation (with iodine), reduction, and hydrogenation. The esterification step was carried out by passing ethyl alcohol vapor through the mixture while at 115–120°C.

The first reduction step used sodium in anhydrous ethyl alcohol as reductant, while gaseous hydrogen chloride plus zinc dust, in glacial acetic acid, were used in the second reduction step. An intermediate step converted the esters in iodides with resublimed iodine in the presence of red phosphorus, at 70°C. The obtained naphthenic hydrocarbons had higher viscosity indices than those of typical naphthene-based oils [336]. It seems that these additional four steps to convert the ethyl naphthenates into hydrocarbons are rather complex and impractical for refinery operations. However, the increase in viscosity index might look attractive for the manufacture of specialty lube oils.

Tin oxide was found to be an effective component, in a mixed-solid catalyst for esterification reactions of HACs with methanol. Experimental results indicated that conversion was improved by properly extending the pore size of the catalyst. Process conditions for the treatment of Shuizhong 36–1 crude oil were an LHSV of 1.0 h^{-1}, methanol–oil mass ratio of 0.02, and temperature of 300°C. TAN of the crude oil was reduced from 2.8 mg_{KOH}/g to 0.34 mg_{KOH}/g [337]. Similar results were obtained when HVGO was treated with this catalyst [338]. TAN reduction went only to 0.5 mg_{KOH}/g, when the treatment of this crude took place in a fixed bed reactor, on an Al_2O_3-supported SnO [339]. Results were better when this catalyst was used for treating a diesel fuel, reducing TAN from 1.7 mg_{KOH}/g to <0.1 mg_{KOH}/g, at a methanol–oil ratio of 0.010.

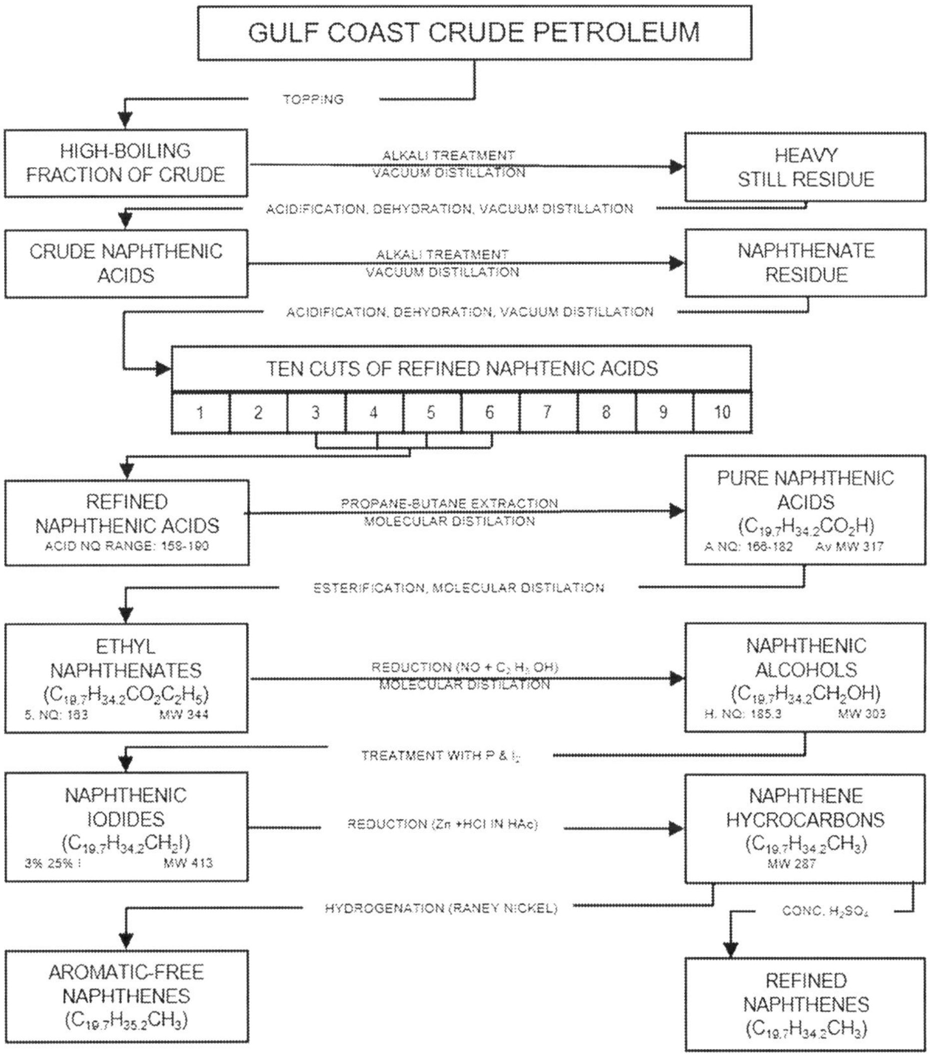

Figure 6.8

Conversion of NAs into naphthenic hydrocarbons. *Reproduced from Goheen GE. Conversion of naphthenic acids to naphthene hydrocarbons. Chem Const Ind Eng Chem 1940;32(4):503–08, with permission from ACS Publications.*

The optimum reaction temperature was 280°C. Apparently, an improved color is an additional benefit of esterification [340]. Kinetic data followed a second-order reaction equation with or without the SnO catalyst [341].

An alumina-supported ZnO catalyst was examined for the esterification of NAs in VGO. Conversion was nearly 80% at 260°C after 1 h of reaction. The esterified VGO had higher kinematic viscosity and a higher flash point [342].

ExxonMobil is the most prolific company in esterification, as well as in many of the other approaches for TAN reduction. Two of its patents will be reviewed next. The first process is a treatment conducted in the presence of a metal carboxylate catalyst, either added or created in situ. Process conditions were temperatures below 450°C and pressures of about 10–30 bar. Any excess methanol may be recovered and reused, but there was no need for removing the esters from the treated crude oil. Experimental results show significant TAN reduction, especially with Ca, Na, and Mg salts [343]. Similar conditions were applied for basic catalysts formed by trace amounts of Group IA and IIA metal carbonates, hydroxides, phosphates, and mixtures of a hydroxide and phosphate [344].

Besides catalysis, microwave dielectric heating represents another means for lowering severity as per residence time is concerned, since it can be significantly reduced from 6–10 h to 5 min. The esterification of natural NAs in the original HAC was assisted with microwave irradiation using acid catalysis, including sulfuric and p-toluenesulfonic acid. Various alcohols (methanol, ethanol, n-butanol, and tert-butyl alcohol) were tested. Depending on the catalyst and the steric and nucleophilic properties of the alcohols, the yield of naphthenic esters ranged from 31% to 89% [345].

Catalytic esterification has been a study option for making a less demanding process. However, catalysts have to address the complexity of the NAs mixture, which derive in broad distribution of molecular sizes, reactivity, and properties. Whether more active catalysts or more reactive intermediates would be discovered is still a matter of investment on research.

4.2.5 Neutralization

Neutralization with organic bases can be seen as a particular case of esterification, while the use of inorganic bases will form water- or oil-soluble naphthenates as discussed in Section 4.2.6.

A tetraalkylammonium hydroxide was proposed for a neutralization reaction carried out at 50–350°C. The preferred base was tetramethylammonium hydroxide [346]. The tetraalkylammonium hydroxide needs to be regenerated with an alcohol. Experimental data showed TAN reductions in the range of 93–95%.

Polymeric bases were proved to be effective for the neutralization of NAs; examples are cross-linked polymeric amines such as polypropylene amine [347] and anionic exchange resins having amino groups. The HAC is mixed with the base, and the reaction proceeds over <3 to ~ 15 h. The reaction duration depends on the temperature, the nature of the crude oil, and its acid content. More than 80% of the acids could be neutralized, and the process is thought to be environmentally amicable [348]. Polymeric resins have been proven useful for NAs abatement by physical methods (e.g., adsorption and extraction), as described above.

In a comparison of organic amines with aqueous ammonia, diethylamine and diethylenetriamine have been found to effectively reduce TAN. Increasing the alkalinity of the media and decreasing steric hindrance of the base improved TAN reduction. The treatment is carried out under the effects of an electric field. In the case of aqueous ammonia, a 3:1 ratio of ammonia–NA and 40 mg/L dosage of demulsifiers were optimal. Commercial demulsifiers LA42 (nonionic) LGA (cationic) were tested. Best results were observed with a 1:1 ratio of LA42–LGC and an electric field of 1500 V/cm [349]. The reactivity of inorganic bases for NAs neutralization could be higher than that of organic bases. However, other factors affect the effectiveness of the process. If sodium hydroxide and sodium carbonate solutions are used as neutralizers, the crude oil gets emulsified upon TAN reduction. This effect was already mentioned in a previous section. A model for explaining the effect of the basic-solution–oil volume ratio and the demulsifier dose was developed to predict the interfacial behavior. Neutralization experiments were carried out under desalting conditions in a DP-2 instrument (electrostatic precipitator). TAN reduction was maximum with a volume ratio of 4 wt% sodium hydroxide solution to crude oil of 4:1, a demulsifier dose of 20 mg/L (LA42–LGA ratio of 1:1), and temperature of 40°C. An unstable oil-in-water (o/w) emulsion was formed that quickly separated [350]. Stable water-in-oil (w/o) and o/w emulsions coexist when the volume ratio of sodium hydroxide to oil approaches 1. The o/w emulsion was categorized as an electriferous colloid type of solution, in which oil drops are tiny colloid particles dispersed in the water phase. The lipophilic groups of the surfactant adsorb on the surface of the colloid particle, while the hydrophilic groups point outward. The surface of an oil drop is surrounded by negative ion groups, forming steady colloid particles. The colloid particles exclude one another and stay well-distributed and steady due to the interfacial electrokinetic potential of negative charges. Therefore, the small size and dispersion level of the oil drops make separation very difficult to achieve. Upon increases of the sodium hydroxide dose, provided electrolytes serve to neutralize the negative ion–surrounded colloid particles. Therefore, repulsive forces decrease accordingly and coalescence might occur, by oil drops colliding. Consequent agglomeration would break the emulsion. On the other hand, naphthenate ions formed by alkali action would adsorb on the surface of emulsified drops. Asphaltene and resin molecules adsorb on the water–oil interface, forming a firm interfacial membrane that causes a barrier to assemble. When a nonionic demulsifier (e.g., LA42) is added, it accumulates on the interface, mitigating the adsorption capacity for asphaltene and resin molecules. The hydrophilic group of demulsifiers stretches into the water phase, and the lipophilic group lies on the oil–alkali interface, thus forming a loose interfacial film that is easy to break. The unreacted sodium hydroxide solution continues to react with NA, and the TAN reduction can proceed thoroughly. Addition of cationic LGA demulsifier serves to neutralize negative charges on the surface of oil drops, facilitating separation of the oil from the basic solution. The combination of the two types of demulsifiers was recommended for a comprehensive treatment involving neutralization and separation [349,350].

The pH of formation of ferric hydroxides and naphthenates, and the consumption of neutralizer required for their formation, depend on the nature of the neutralizer and the concentration of the initial reagents. A solid phase may appear during neutralization with a soda solution, due to the fact that the precipitation pHs of ferric hydroxide and ferric naphthenate are very close. Formation of ferric naphthenate takes place during titration and extraction of iron with a mixture of a neutral salt and a basic salt. Meanwhile, if a fourfold excess of acid is added, only a neutral salt would be formed. The highest yield of ferric naphthenate was obtained when ammonia was used as the neutralizer, since it prevents the formation of solid phases [351].

A multifunctional complex agent was defined for a combined treatment that comprises desalting and deacidification [352–354]. The benefits of this technology compared with the conventional alkali treatment include feed flexibility for adapting to various acidity levels, less alkali consumption (reduced in 67–75%), no discharge of polluted waste alkali, better oil yield, no need for intensive mixing, and it can be applied in existing desalters [352]. The treatment of SR diesel reduced yield losses by 50–60% and product TAN to less than 0.017 mg_{KOH}/g [353]. In order to use existing facilities, NaOH neutralization has to minimize wastage and emulsification, while maximizing conversion and reaction rate [354].

Caustic soda washing is the current method for NAs manufacturing from crude oil, based on sodium naphthenate formation. However, this method brings about numerous issues, such as the low percentage of NAs even in HACs, the broad range of NAs reactivity, the low water solubility of heavy naphthenate soaps, the emulsifying character of sodium naphthenate, the large volumes of (hazardous) fluids required, and the presence of other caustic-extractable compounds (e.g., S-compounds). Originally (in the 1930s), a neutralization–separation method for NAs manufacturing involved crude oil distillation in the presence of caustic soda or the prior caustic–crude reaction followed by distillation [355].

The neutralization product is a mixture of salts, contaminated with large quantities of asphalt or residual pitch that would make the acids probably unusable. Nevertheless, Napfining, a NAs-removing technology for jet fuels–diesel streams offered commercially by Merichem, is based on caustic washing. Since Merichem is the world's largest producer of NAs (in a Tuscaloosa, Alabama, plant), one may assume the production route might be neutralization–extraction. If that is the case, the patent of Ref. [356] may be associated with that technology. Merichem announced they have developed proprietary technology for the treatment of HACs [357]. Due to the issues associated with caustic washing and disposal of spent materials, Merichem also offers to the refiners a completed package that includes caustic management services.[2]

[2] See, regarding information in the company's website, http://www.merichem.com/caustic-mgmt/naphthenic-acids.

In general, the particular case of caustic soda washing is a mature technology, currently applied in refineries (and production fields) to sweeten operations for the removal of S-compounds from light ends and streams. Supporting technologies involved in caustic handling, regeneration, and reuse are well known in the industry, as well as those associated with the recovered S-compounds.

In the patents arena, ExxonMobil has been very active in the neutralization area using both liquid and solid bases. Other companies have also disclosed processes using either type of bases, to a lesser extent.

4.2.5.1 Liquid Bases

The following patents describe processes for using bases in liquid form, either liquid base per se or dissolved solid bases. Liquid bases are employed at temperatures below 40°C, under mild stirring, to minimize formation of stable emulsions. Treatment time is 2 min to 2 h.

An aqueous base solution was used to treat crude oil at the well head. As discussed above, an unstable emulsion results that can easily break into a treated oil layer and an acidic aqueous layer. Formation of a stable emulsion is prevented either by adding the oil to the aqueous solution and mixing with mild agitation, or by adding a nonionic surfactant. Up to 100% reduction in TAN was proven to be possible [358]. Similarly, another well-head process treats an acidic calcium-containing crude oil with an aqueous NaOH or KOH solution, and the unstable emulsion formed is processed similarly as before. The presence of calcium helps the acids to accumulate in the interfacial zone between the two layers. Experimental results showed 100% TAN removal and 96% Ca removal [359].

An aqueous solution of hydroxides of alkaline and alkaline earth metals and ammonium hydroxide has been disclosed as neutralizing bases at ambient temperature [360,361]. In addition to the aqueous solution of alkaline earth metal hydroxide, an emulsifier was included for the NAs neutralization in HACs, and for the separation of the oil and aqueous phases, respectively. A phase transfer agent was recommended for aiding hydrogen ions transfer from the acid in the oil to the base in the aqueous phase [362].

4.2.5.2 Solid Bases

Most of the basic solids can perform on multiple functions, such as adsorbents and catalysts (already described above). For neutralization purposes, solid bases are typically mixed with the crude and heated to temperatures above 30–50°C.

Alkali metal trialkylsilanolates, for example trimethylsilanolates of sodium, potassium, or lithium, used as a solid (or dissolved as a liquid) showed TAN reduction between 78% and 92%, when the treatment is performed at 60–170°C. The neutralization products were the mixture of salts and trialkylsilanol [363].

Alkaline earth metal carbonates, preferably magnesium carbonate, are another disclosed example of a solid base that has been applied in water or organic-medium slurry. The neutralization reaction is carried out at temperatures between 30 and 300°C and takes 1−20 h. Magnesium carbonate reduced TAN to a minimum of 1.0 mg_{KOH}/g [364].

Oxides, hydroxides, and hydrates of Group IA (alkaline metals) or Group IIA (alkaline earth metals) have been tested and showed TAN reduction varying between 25% and 100%. Treatment with CaO and MgO showed the best results. The process runs at around 100−150°C and atmospheric pressure, and residence time is from 2 to 8 h [365].

Overbased detergents are a very fine suspension of solid reversed micelles in oil, with particle size of 2−100 nm. In overbased detergents, the particle core is typically composed by a carbonate or hydroxide of alkaline earth metal (e.g., $CaCO_3$ or $Ca(OH)_2$). This core is surrounded by surfactant molecules acting in the kinetic stability of the dispersion in oil. Overhead detergents, besides their antiwear uses for diesel and lube oils, are also known by their acid neutralization capabilities. In this case, sulfonate and phenate detergents prepared from calcium and containing at least 3 wt% Ca have been recommended for neutralizing NAs. A 35% TAN reduction was observed for neutralization carried out below 200°C and with contact time from 1 to 20 h [366].

The addition of Group IIA metal naphthenate salts can reduce TAN 100%. An additional reported benefit was the mitigation of materials-handling problems associated with the use of liquids for neutralization and the consequent emulsion formation. The same process could also be performed with CaO and MgO. These latter materials showed the best results when contacted with the crude oil at around 100−150°C and atmospheric pressure, for times in the range of 2−8 h [367]. Meanwhile, higher temperatures (200−300°C) and probably longer residence times of up to 20 h are required when manganese oxide is used for neutralization. Evaluation experiments were based on the disappearance or reduction of the carboxyl group peak in infrared (IR) spectroscopy [368].

A hybrid neutralization−extraction process is described in a BP patent, in which a Group IA metal salt neutralizes the NAs and a polar solvent extracts the resulting NAs−salts. The solvent and crude oil are separated, and the solvent phase is treated to remove the neutralization salts and thus can be recycled. The suggested solvent treatment was a reaction with a Bronsted acid, which is capable of reacting with the Group IA metal neutralization salts to reconvert the organic acids and the Group IA metal salt of the Bronsted acid. Alternatively, the solvent phase may be contacted with an acid ion exchange resin to reconvert the organic acids and remove the Group IA metal from the polar solvent [369]. In another patent, the suggested treating medium was composed of an aliphatic alcohol and a Group IA/IIA metal salt. The neutralization salts were extracted into the alcohol. The alcohol and crude oil were separated, and the alcohol phase was treated through an acid ion exchange column to convert a portion of the neutralization salts into the corresponding esters of the organic acids and the alcohol [370].

The use of dolomitic lime in the presence of a dibasic acid catalyst in a slurry-phase reactor was disclosed in a patent assigned to Interglobal Desulphurisations Systems. The primary intention was to transfer sulfur compounds from the hydrocarbon phase to the aqueous phase. The treatment seems to be particularly selective to light NAs present in the naphtha fraction. Lamentably, changes in TAN were not reported [371].

4.2.6 Removal

The use of a chemical reaction to facilitate the removal of NAs from HACs follows. Formation of inorganic naphthenates appears as the most studied and applied reaction. The search for an easy-to-separate (solid) phase is a common aspiration.

In the course of an ongoing investigation into the phenomenon of NA corrosion, a new test method evolved [372,373], and it was envisioned as a substitute to the TAN or NAN, as an indicator for NAs corrosion potential [374]. This method involves the formation of soluble iron naphthenates and could serve as the basis for inspiration of a NAs removal process. The method protocol consists of reacting an oil sample with pure iron powder within an autoclave heated to the testing temperature. The result is based on the amount of dissolved iron found in the oil sample. The oil sample can dissolve an amount of iron for a given time at a given temperature, depending on the NA corrosion. The test has demonstrated that 1 m^2/g iron powder reacts with NAs at temperatures of 230–270°C. Indirectly, it was also demonstrated that reaction rate increases with surface area [372–374]. Some other factors were identified as affecting the effectiveness of the method: iron powder amount, reaction temperature, and contact time [375,376].

Similarly, Ca^{2+} also reacts with NAs in aqueous solution, as has been investigated in Ref. [377]. In this study, the nucleation period and particle growth of calcium naphthenate were monitored upon addition of a Ca^{2+} solution to an aqueous high-pH (11.2–11.5) NAs or fatty acids solution. The formation and growth of calcium naphthenate particles shifted the baseline of the near-IR spectra. Formation of calcium soaps was taken as indication of solubility products, defined as the ion concentration products where no particle growth was detected. The effect of particle size distribution on the optical properties (optical density and scattering) introduced some quantitative limitations due to the changing particle size throughout the reaction time. Nevertheless, the study served to confirm the occurrence of this reaction between NAs present in oil and Ca^{2+} present in the formation and/or the production water [329,378]. Production experience of the Statoil-operated Heidrun field in the Norwegian Sea showed that crude NAs react to form calcium naphthenate when exposed to calcium ions in produced formation water. Since the start of production in 1995, the crude oil has consistently shown an unusually high content of NAs (compared with other Norwegian fields) of about 2 wt% [329]. In crude oil, the partitioning of NAs between oil and water and the dissociation equilibria in water are factors to consider when dealing with HACs. These have been studied in connection with their interfacial tension and micellization behavior, by means of near-IR spectroscopy. It was shown that the method gives information about the induction period and about the rate of reaction for particle growth. Solubility products can also be estimated [378].

The solubilizing effect of NAs on asphaltenes would be lost upon removing NAs from a crude oil and may cause the asphaltene to aggregate more easily at the water–oil interface [379]. Furthermore, increases in the interfacial tension and in the w/o emulsion stability were observed upon NAs removal, indicating that NAs play an important role in the stability of w/o emulsions. The effect of NAs interaction with divalent cations on the interfacial tension and the role of oil-soluble nonionic surfactants were studied. The continuous lowering in the interfacial tension upon NAs–cations interaction was explained as due to the lesser penetration of the hydrophobic acid structures into the interfacial layer than that of the unsaturated acids, forming more well-defined oil–water interfaces. Longer interfacial distances between the acid and the divalent cation would hinder 2:1 stoichiometric complexes from forming [380].

4.2.7 Chemical Additives

Chemical additives are employed in the prevention of production and processing problems; more specifically, the formation of dense emulsion in upstream operations and desalter upsets in downstream processing. Very little is known on a technical basis about the action and functioning of these additives, as well as of their chemical composition. However, there is clear evidence of the mutual interference among certain additives, and sometimes commercial vendors advise their customers regarding the inconvenience, but only after problems become evident.

The complex composition of opportunity crudes (heavy and acid) affects differently each of the refining units. The combined effects of asphaltenes and NAs create severe complications in the operation and operability of the desalter unit. Testing crude samples is recommended for identifying potential problems [381,382]. However, variability in the composition of a given traded crude oil makes the problems unpredictable. A combined approach of chemical analyses and additives injection points has been suggested by GE. They also indicated the need for usage of multiple additives to cope with the many problems brought by those compounds present in opportunity crudes. These suggestions, including the type of additive at each injection point, are collectively shown in Fig. 6.9 [381].

A combined naphthenate-inhibitor and demulsifier reduce or eliminate desalter problems and also result in better water quality. Laboratory studies reported the effect of this type of new chemicals on coalescence, sedimentation, and dense emulsion formation [383].

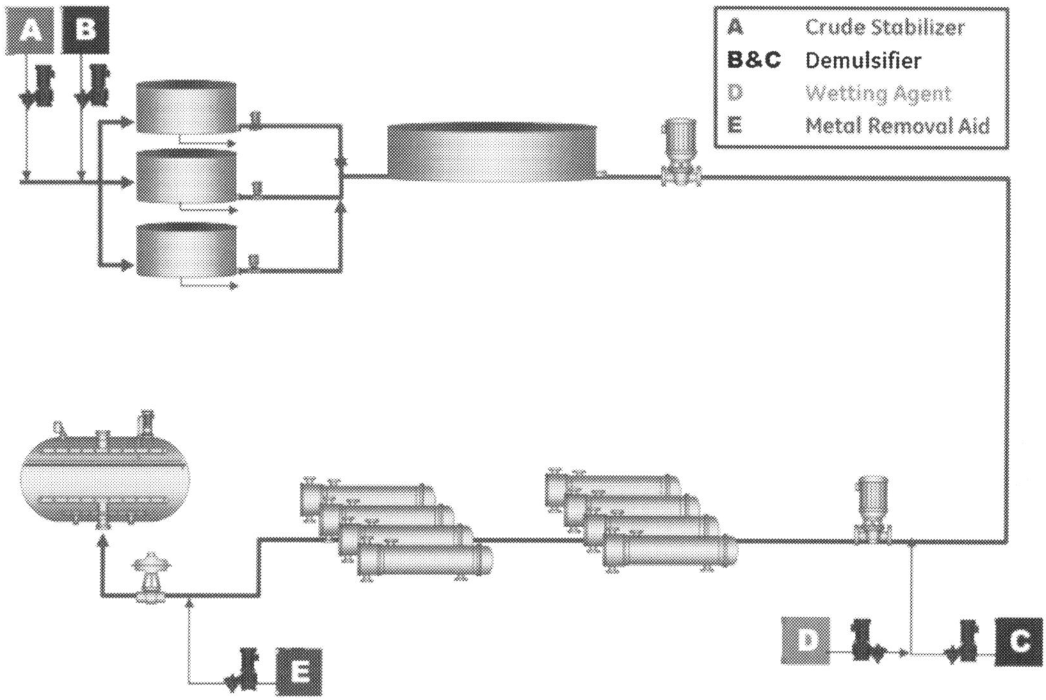

Figure 6.9
Suggested injection points for chemical additives. *Reproduced from Dion M. Challenges and solutions for processing opportunity crudes. In: Proc. AFPM annual meeting. Orlando, FL, USA. March 23–25, 2014. 13 pp., with permission from GE.*

Emulsion stability is increased by the presence of Ca^{2+} in HACs, negatively impacting desalting operations. A chemical metal-complexing agent has been recommended by Baker Petrolite for the removal of Ca (and Fe) naphthenate. This complexing agent converts metallic species into water-soluble compounds. In this process, the metals complexing agent is injected into the desalter wash water, under high mixing energy; then the formed soluble metal compounds are removed with desalter effluent water, but NAs remain in the oil phase [384,385].

Although the chemistry and physics involved in this type of process operate in a sort of inverse manner of what is intended for TAN reduction, the lessons learned can serve as inspiration for new approaches and solutions.

4.3 Biotechnologies

NAs account for most of the acute aquatic toxicity of oil sands process-affected water (OSPW). Reduction of the toxicity of OSPW has been targeted by ways that decrease the concentration of NAs below the toxic levels. Biodegradation of NAs has been studied, particularly for pursuing interests in remediation and reclamation of the tailings ponds derived from the exploitation of tar sands. Studies have shown that NAs are susceptible to biodegradation, which decreases their concentration and reduces environmental impact (i.e., toxicity). Although these studies have been carried out in the aqueous phase, the relevance of the involved biochemistry will become obvious.

Early research looked at aerobic degradation of simple model compounds and showed that mineralization of these simple compounds was possible, starting with the carboxylic side chain but also including the ring structures. Alkylated compounds, representative of true NAs, were more resistant to degradation. The biodegradation of NAs extracted from tailings water was demonstrated under aerobic conditions through both CO_2 production and a reduction in acute toxicity [386]. Changes in NAs concentration during incubation of aerobic cultures in tailing pond waters, from Alberta, Canada, were followed by high-performance liquid chromatography (HPLC) and gas chromatography−mass spectrometry (GC−MS). The elimination of acute toxicity was associated with a 90% decreased NAs concentration (from 100 to 10 mg/L) [387].

The rhizosphere cultures of five plant species native to Alberta were enriched by incubating for 4 months, with the addition of 0.5% bitumen as the sole carbon source. Biodegradation of a NAs commercial preparation from Merichem revealed a high potential for aerobic degradation. Changes in the concentration and composition of the NAs mixtures during incubation were followed using HPLC and gas chromatography−electron impact mass spectrometry. Concentrations did not significantly change in the sterile control, but they decreased by up to 90% after 10 days of incubation in the viable cultures. Lower MW NAs were preferentially degraded, while the proportion of HMW acids increased during incubation [388].

As mentioned above, NAs are adsorbed on pond soils, and the study of their effect on exposed sediment microbial communities has been considered in a report where the extent of NAs degradation by these communities exposed to various amounts of OSPW was determined. ^{14}C-labeled NA model compounds were used as surrogates: S1 was a monocyclic NA [^{14}C-cyclohexane carboxylic acid], and S2 was a bicyclic NA surrogate (^{14}C-decahydronaphthoic acid (DHNA)). The extent of about 30% degradation of the S1 NA was relatively uniform in all considered sediments after 14 days. In contrast, degradation of S2 was significantly lower in non-process-affected sediments. Enrichment *Pseudomonas putida* and *Pseudomonas fluorescens* co-culture was obtained from an active tailings settling pond and using commercially available NAs as the sole carbon source. This co-culture removed >95% NAs from waters contaminated with the commercial NAs but only partially degraded the NAs in OSPW. Selected microbial cultures could be induced for the degradation of bicyclic NAs, indicating that probably some other could be engineered to process the more recalcitrant NAs found in process-affected wetlands [389].

However, it seems that native sediment cultures already have degraded some NAs, leaving the more recalcitrant compounds to accumulate.

A fundamental study examined the microbial degradation of commercially derived and oil sands–derived NAs by oil sands–derived microbial cultures. Utilization of NAs by oil sands–derived microbial cultures was minimal during the incubation period. In static and semicontinuous tests, uptake of commercial NAs in laboratory microcosms resulted in microbial biomass that was similarly ^{13}C enriched (from 0.3 to 2.9‰) depending on the length of incubation [390].

NAs susceptibility to biodegradation was found to be affected by differences in alicyclic features associated with the carboxylic group, such as carbon chain length, chain branching, the oddness or evenness of the carbon chain containing the carboxylic group, positions where alkyl groups are substituted on the cyclic ring, geometrical isomerism, and the number of cyclic rings. The possibility of indigenous species degrading some NAs and leaving the recalcitrant NAs behind was suggested [391]. Bioaugmentation with selected bacteria was recommended as a plausible option to further degrade the more refractory classes of NAs in the tailings pond waters. Two routes for bioaugmentation employed clay materials to concentrate (1) NAs or (2) NAs-degrading bacteria in their surfaces and/or pores. Alternatively, bioaugmentation may take advantage of the synergistic association between algae and bacteria consortia to promote efficient aerobic degradation. Finally, another option is biostimulation with nutrients to promote the growth and activity of the microorganisms [392].

NAs found in oils differ from those in tailings, primarily due to a lack of compounds with carbon numbers of 22 or greater in the latter. Clemente has reviewed the biodegradation of aqueous solutions of NAs [393]. He found that aerobic biodegradation of NAs preparations has been sufficiently demonstrated in laboratory studies, resulting in the decrease of the NAs concentrations and in changes to the distribution of the residual acids. The changes indicated that the LMW acids are more susceptible to biodegradation than the higher MW acids. These laboratory findings are consistent with the observations from the analyses of samples from oil sands tailings waters. LMW NAs were observed to be biodegradable and at the dilute concentrations found in tailing waters. A fairer comparison could come from a treatment involving a concentrating step, which then will require a larger amount of microbial biomass [394]. Still, biodegradation shown to be a potential solution for a range of NAs originally present in crude oil and leached into the tailings represents an alternative subject worth investigating. Although bioremediation objectives differ from those pursued in biological upgrading of HAC NAs, research results, lessons learned, and derived knowledge could be applied on NAs abatement. Bioremediation requires the reduction of the concentration of these compounds to avoid reaching toxic levels. Meanwhile, biological upgrading would ideally involve biotransformation of NAs to compounds that are easier to handle and more valuable. In other words, eliminating toxicity is quite different from removal.

Ultra-high-resolution FT-ICR-MS was used for the identification of the molecular structural changes caused by anaerobic biodegradation on six genetically related oils. The level of biodegradation was established by observing the changes undergone by selected biomarkers. Increasing degradation changed the relative abundances of heteroatomic N-, S-, and O-containing compounds, indicating a certain degree of selective biodegradation. At moderate biodegradation, (1) NSO compounds associated with long alkyl side chains were removed, regardless of the NSO core; (2) O_1, O_3, and SO_{3-9} compound series were mineralized under conditions of mild biodegradation (n-alkanes, isoprenoids altered but still present); and (3) simultaneous microbial degradation and generation changed the relative concentration of O_2 species Z-series (hydrogen deficiency) and their alkyl distributions: (a) acyclic fatty acids decreased; (b) five-ring NAs (hopanoic acids, $Z = -10O_2$ species) increased; (c) monocyclic NAs ($Z = -2$) were enriched initially and then decreased; (d) di-, tri-, and tetracyclic NAs ($Z = -4$, -6, and -8) were preferentially produced; and (e) the acyclic to 2–4-ring-cyclic ratio of NAs can be associated with the biodegradation extent [395].

Anaerobic biodegradation was studied using methanogenic microcosms. Methane dominates the product, and mineralization becomes a secondary reaction. Nonetheless, acetate or H_2 is needed for the methanogens metabolism and should be provided by some of the anaerobic microorganisms present in the consortium [396]. Could these microorganisms be isolated and engineered to metabolize NAs to an intermediate *mineralizable* compound?

The reactions observed in the biodegradation of NAs (side chain and ring oxidation and mineralization to CO_2) are therefore not directly applicable to oil upgrading. The work performed clearly shows that despite their toxicity, NAs are ready targets for microbial attack. Future upgrading research needs to look for more suitable reactions and for systems capable of catalyzing those reactions. Microbes have the potential to aid in the processing or transportation of heavy crude oils if interesting reactions are harnessed. Another approach could focus on the surface properties of the microbes or of microbial-derived products as aids or additives for flocculating emulsions or dispersing components. The natural environment of bacteria is either at the oil–water interface or in the bulk water phase. The most attractive opportunities for biological upgrading may therefore be in dealing with surface active components such as NAs or in flocculating oil-in-water emulsions [397].

A commercially applied biotechnology is phytodegradation. In the case of NAs, algae have been considered an option for aqueous systems. A variety of algae species have been tested in phytodegradation, including blue-green algae (*Oscillatoria* sp., *Aphanizomenon* sp., and *Anbaena* sp.), green algae (*Selenastrum* sp., *Nannochloris* sp., *Ankistrodesmus* sp., *Scenedesmus* sp., *Haematococcus* sp., and *Chlorella* sp.), and diatoms (*Naviculla* (1), *Naviculla* (2), and *Nitzschia* spp.). A model NA (4-methylcyclohexaneacetic acid, *cis*- and *trans*-forms) and a NAs—oil sand mixture were evaluated. A complete uptake of the model NA, both *cis*- and *trans*-isomers, was presumably phytodegraded by the diatom algae *Naviculla* (2) sp. at a concentration of ~5.5 mg/L within a period of 14 days. However, there was no evidence for the phytodegradation of the petroleum oil sands—NAs mixtures, except for experiments utilizing the green algae, possibly *Selenastrum* sp. The differences in the phytodegradation of the model NA by the diatoms appear to be associated with differences in transport mechanisms by the algae [398].

4.4 Process Improvements

Since the most troublesome effects of NAs are observed in the refinery front-end units, mainly the desalter and the CDUs, the desalter has been the focus of incremental enhancements. New Ca-removal agents and demulsifiers are said to promote different chemistries, but the technical grounds of these improvements are not reported. Announcements on the benefits of desalting operations are associated with the trade names and commercial codes of the products rather than with any meaningful chemical description (see Section 4.2.7) [399—403].

A new micromixer and a membrane dispersion extractor were developed for NAs removal from HACs. Performance of the system was tested for gasoline (1.84 mg_{KOH}/g) and crude oil (3.2 mg_{KOH}/g). The de-acidification efficiency was 90%, with a residence time in the micromixer below 0.5 s, at a water—oil phase ratio of 1:50. There was no emulsification phenomenon during the process, and the two phases spontaneously separated during settling, in the absence of electric field [404].

The effect of different force fields on removing NAs from diesel fuel has been studied, including gravitational field, high-voltage electric field, and centrifugal field. The time needed for phase separation decreases in the following order: gravitational > high-voltage electric > centrifugal. The centrifugal field additionally contributes to demulsification, thus minimizing entrainment between phases. Phase separation and acid removal were reported to be satisfactory when the centrifuge was operated at an angular velocity of 3000 rpm and residence time of 60 s [405].

5. Crude and Residue Upgrading

Certain improvements, variations, modifications, and new concepts for catalytic processes and catalysts have been proposed; some of them are carried out under hydrogen pressure, though differing from typical hydroprocessing.

5.1 Emerging Processes

5.1.1 Heavy Oil Treating (HOT) Process

The use of water as a source of hydrogen (or oxygen) represents a cheaper alternative for residue upgrading. The requirement is either a catalyst that dissociatively activates the water molecule or a reducing reactive material that, upon its own oxidation, incorporates hydrogen into the oil.

In the HOT process, steam is used in the presence of a ferrous oxide (wüstite) containing catalyst (probably iron ores). The process consists of three fluidized bed reactors (cracker, regenerator, and desulfurizer). In the cracker, the Fe^{2+} is oxidized to Fe^{3+} (magnetite) and produces hydrogen. The so-produced H_2 is used for the hydrogenolysis of heavier molecules to produce light fractions. Produced coke deposited on the magnetite is partially oxidized to regenerate the catalyst [406–410]. Previously, Nippon Mining Co.'s experience with Ni ores-based catalysts [217,218] was mentioned (see the discussion of the SOC process[3] in Chapter 5). In the case of the Fe-ore catalyst, the involved chemical reactions are summarized below [411]:

$$3FeO + H_2O \rightarrow Fe_3O_4 + H_2 \text{ reactor} \tag{6.i}$$

$$Fe_3O_4 + \text{asphaltenes} \rightarrow \text{hydrocarbons} + \text{coke}-Fe_3O_4 \text{ reactor} \tag{6.ii}$$

$$\text{coke}-Fe_3O_4 + O_2 \rightarrow 3FeO + CO + CO_2 \text{ regenerator} \tag{6.iii}$$

$$FeO + SO_2 + 3CO \rightarrow FeS + 3CO_2 \text{ regenerator} \tag{6.iv}$$

$$3FeO + 5O_2 \rightarrow Fe_3O_4 + 3SO_2 \text{ desulfurizer} \tag{6.v}$$

A scrubber is used to separate the effluent from the reactor system into hydrogen gas, liquefied petroleum gas (LPG), liquid products, and bottoms. The bottoms containing unconverted feed and regenerated catalyst are recycled to the cracker.

Although a research study indicated that the catalyst did not deactivate upon redox cycles [412], in practice it did under the HOT process conditions. During process, an accumulation of wüstite was observed. The catalyst became denser, and porosity decreased. It seems that both the kinetics of the redox reactions and the stoichiometry involved have to be balanced accordingly at process conditions, in order to prevent deactivation [411].

5.1.2 Genoil Hydroconversion Upgrader (GHU)

The upgrading technology currently being developed by Genoil, Inc., was originally conceptualized by Canadian Environmental Equipment and Engineering Technologies Inc. (CE3). The Genoil Hydroconversion Upgrader (GHU) is a catalytic process technology for reducing viscosity or upgrading sour (high-sulfur), acidic, heavy crudes, bitumen, and

[3] Nippon Mining catalysts did not seem to have been incorporated in the development of the SOC process, although it is not clear whether the Ni-ores can be used as the ultrafine powder additive.

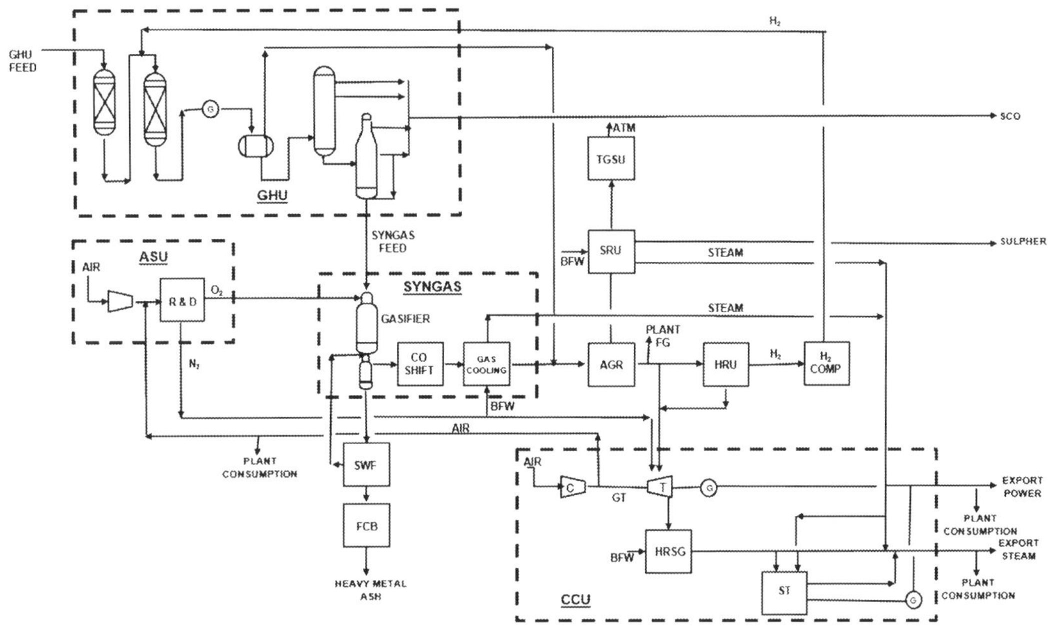

Figure 6.10

Genoil hydroconversion upgrader (GHU). *Reproduced from Genoil Inc. The genoil hydroconversion upgrading system (ghu®) for heavy and extra heavy crude. (Unpublished Descriptive Brochure). 2009; 20 pp., with permission from Gulf Pub.*

refinery residues in the presence of hydrogen. The original reducing viscosity concept was meant for field and surface installations, while GHU is seen as a refinery unit.

Genoil, Inc., has a number of publications containing a high-level description of the technology.[4] A simplified scheme is presented in Fig. 6.10. However, a more complete description can be found in their issued patent [413], which, according to Ref. [414] and p. 13 in Ref. [415], corresponds to GHU. Thus, the process consists of three steps:

1. heating and mixing feed and hydrogen in a primary vessel, followed by (catalytic or not) hydrogenation and separation of volatile light ends from heavier ends. A portion of the hydrotreated heavy ends stream is used for quenching the separation vessel;
2. removing light volatiles from the primary vessel and directing them to a secondary vessel for further separation; and
3. directing the heavy nonvolatiles from the primary vessel into a sequence of a (primary) hydrogen treatment loop. The loop includes a noncatalytic hydrogen treatment followed

---

4 Visit http://www.genoil.ca/main-publications.html and http://www.genoil.ca/main-technologies/ghu-pilot-test/ghu-process.html.

by another hydrogenation (catalytic or not) step, with intermediate heating and mixing to improve dispersion and reaction. A portion of the reaction product is recycled to the loop to adjust quality, and another portion is used for the quenching of the primary vessel. A blender vessel collects the entire product and disengages from the gas product to recover and recycle H_2-rich gas stream to the primary vessel.

For Genoil [413], hydrogenation reactions are defined as nondestructive and destructive, which probably can be more accurately defined as hydrogenation and hydrogenolysis reactions. These reactions are carried out in mixers specially designed for homogeneous (noncatalytic) hydrogenation, and fixed bed reactors for catalytic hydrogenation and HDC. In those mixers, preheated hydrogen is thoroughly mixed with the liquid feed to "above-normal" saturation levels, allowing a less severe operation. The operating conditions described are very similar to those of conventional HDP, and no comparison has been published vis-à-vis the different process options. The so-called hydrogenation loop operates at high temperatures ($\sim 520°C$) and would achieve high conversions, which may have plugging and fouling consequences. Fixed-bed operation will become jeopardized. In fact, the discussion on the HDT section (see Section 3.2.1 in Chapter 5) clearly points out the inconveniences of considering fixed-bed reactors to upgrade heavy crudes, particularly for high conversion conditions of high CCR and high metals crudes.

The treatment of an oil with 8.5°API resulted in DS of over 99.5% and a pitch conversion level of 95%. Most of the reported data concern the quality of the synthetic crude oil (SCO) produced; data on yields and the effect of other contaminants on the performance have not been reported. Nevertheless, a prevision for using the first reactors as switching guard reactors has been mentioned for cases with higher metal content feeds that require the use, removal, and reloading of HDM catalysts while in service.

Various feeds ranging from 6.5° to 17.5°API gravity have been tested in their 10 bpd pilot plant. GHU is advertised as having 75% lower CapEx and OpEx in comparison with competitive processes [416,417], and as being a self-sustained technology producing "bottomless barrels." *Self-sustainability* is defined as the capacity to run without needing external hydrogen or natural gas [418,419]. However, this latter statement can only be possible if the unconverted residue can be sent to a gasifier, and the syngas then used for hydrogen recovery to supply the Genoil GHU, and the remaining syngas used as fuel gas or to generate power and steam by adding an Integrated Gasification Configuration Cycle (IGCC) unit into the overall plant configuration. These units have not been considered in the costs estimates [415,420].

Genoil and Haiyitong Inc. (HYT) established a US$700 million contract for building a ~20 kbpd GHU at a Nampaihe Town, Huanghua City, Hebei, refinery (with current capacity of 2.5 MT/Y) in northeastern China in 2006. The GHU unit for HYT was designed to process heavy oil and refinery residues. The project would improve feedstock flexibility in the refinery and hopefully margins [421]. Genoil and HYT established a joint company, Dora Energy Technology (DET), for executing the project in early January 2016. A new feasibility study announced by DET indicated a 32% ROI and 7-year payback [422].

5.1.3 Chattanooga Process

Chattanooga Corporation has developed a process based on a pressurized fluid bed reactor, in which heated hydrogen is used as lift gas, heat carrier, and reactant [423]. Reactions are carried out at high temperatures (~540°C) and low pressure (30–45 bar). Feeder is adapted to process a variety of feedstocks, including bitumen, tar sands, oil shale, and heavy oil. In the case of tar sands, the dry oil sands are fed through lock hoppers, using a flow of cold hydrogen gas, directly into the reactor. Part of the technology is the design of the hydrogen heater, which is fueled by process off-gases and supplemented by natural gas or product oil, depending upon economic conditions. A hot gas filter is used to remove particulate solids from the reactor overhead gases. Under process conditions, asphaltenes might deposit on the solids and would be removed from the SCO. For tar sands also, a very large solid flow has to be heated up to reactor temperature and then cooled down as low as possible for removal, making heat recovery one of the most important challenges of the technology economics. The hydrocarbon vapors are condensed and separated from the gas stream. An amine scrubbing is used to recovered hydrogen and separates it from the H_2S that is sent to the S-plant; H_2 is recycled to the reactor [424,425]. The process flow chart is presented in Fig. 6.11 [423–426].

One of the first patents claimed hydrocracking and hydrogenation to occur [427]. However, the relatively low hydrogen pressure, the very short vapors residence time, as well as the absence of active catalytic materials made us presume it was very unlikely for hydrogenation or hydrogenolysis to occur. Probably, low H pressure would inhibit the formation of diolefins and may result in a more stable (toward oxidation and gum formation) product. The undergoing reactions can be described more accurately as TC and slight hydrogenation. Nevertheless, this level of conversion allows for an easy removal of waste solids.

The conversion reactor has been tested at pilot plant level with Colorado oil shale, Kentucky oil shale, and bitumen–sand at the National Center for Upgrading Technology (NCUT) in Devon, Alberta, Canada. Tests demonstrated effective hydrogen fluidization and high extraction efficiency, and produced 32–36°API SCO from bitumen [424]. The process has been covered by a set of patents in the United States and in Canada [428–431], though some more recent patent applications have been abandoned [429,432].

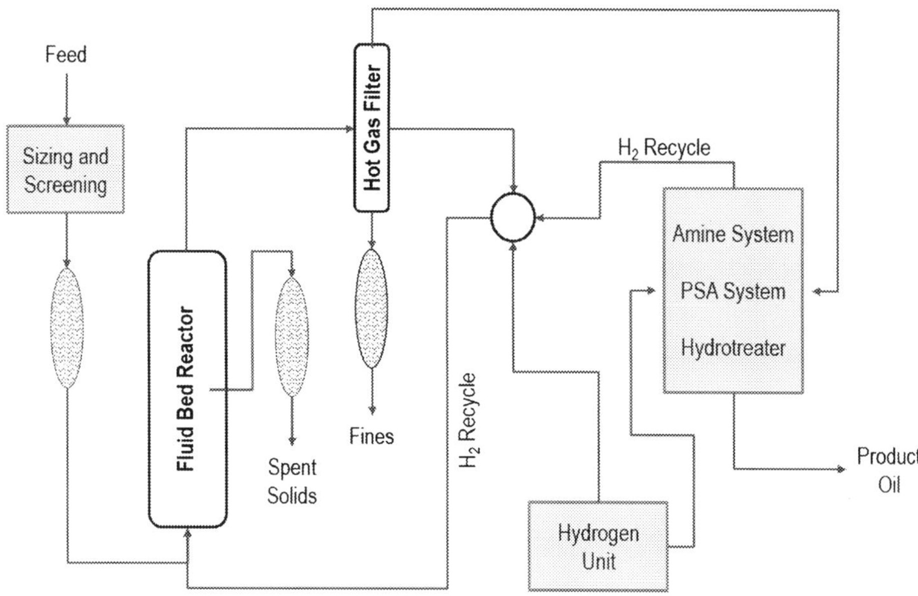

Figure 6.11
Flow diagram of the Chattanooga process, based on description from Ref. [426].

5.1.4 Standalone and Other Combined Processes

In the late 1980s, a process was conceptualized by Jorgensen et al. and applies an instantaneous thermal shock to the bitumen molecules using a superheated hydrogen-containing gas stream or steam (as a hydrogen source) [433–435]. The initial concepts were assigned to British Petroleum (now BP). The energetic shock was supposedly provided by an electric arc specifically designed for that purpose [433]. The feed is preheated to temperatures between 360 and 460°C, and the reactive gas to 600–1200°C, prior to mixing. Energy of 62–250 kJ/mol is assumed to be released by the reactive gas to the heavy hydrocarbon feed. Further development took place after 2000 and was funded by French and Australian companies. This includes jet injectors for improving the feeder, addition of an extraction stage, demulsifying steps, and so on. The produced SCO is said to possess a higher H–C ratio than the original oil and to be stable [436–439].

A selenium compound (e.g., SeO_2) can be used for the conversion of a heavy oil or residue under suitable reaction conditions ($T = 250–550°C$ and $p = 70–650$ bar), specifically in a $H_2–H_2S$ atmosphere ($H_2–H_2S$ ratio = 2–10). This process obtained interesting levels of conversion of a Hondo VR (e.g., API from 6.7° to 21.4°, CCR from 11.8% to 6.3%, S from 6.1 to 2.7 wt%, Ni + V = from 135 + 289 ppm to 30 + 55 ppm, and liquid yield of 78%) [440].

Nowadays, most FCC feeds are pretreated by HDT and more particularly residue feeds for RFCC; the novelty of the integrated process of Ref. [441] is that the spent FCC catalyst is recycled to the pre-treatment unit to remove metals, CCR, and S in the presence of H_2. Although the pretreating unit uses high-pressure H_2 (55–70 bar) at high temperatures (400–455°C), it is not an HDT unit. The catalyst is regenerated by coke burning exclusively. DM was in the range from 90% to 98%, even with feedstocks containing 700–800 ppm total Ni and V. Comparatively, this process technology results similar to the FTC process discussed in Section 3.2.2 and shown in Fig. 6.2.

Besides SDA, where supercritical conditions are used to facilitate solvent recovery, conversion in the presence of a supercritical solvent opens new opportunities. Supercritical conditions improve solvency capabilities, enabling the formation of a *substantially* single-phase system. This single-phase status could last for a period of time long enough to achieve the desired conversion. In Table 6.4, the critical conditions of certain compounds typically used as process solvents are shown.

In the case of water, supercritical processing offers additional advantages, such as better solubility of hydrogen, CO, and organic compounds; the cage created by the solvation layer inhibits recombination reactions of dissolved molecules and increased reactivity of water itself. Standard Oil proposed conversion processes of heavy hydrocarbons in

Table 6.4: Supercritical Conditions of Certain Compounds Typically Used as Solvents

Compound	Critical Temperature (°C)	Critical Pressure (bar)
H_2O	374.1	221
CH_4	−82.6	46
C_2H_4	9.4	50
C_2H_6	32.3	49
C_3H_6	91.9	46
C_3H_8	96.8	42.5
CH_3OH	239.6	81
C_2H_5OH	240.9	61
Hexan-1-ol	337	40
2-Methoxyethanol	302	52
1,4-Dioxane	314	51
Tetrahydrofuran	267	51
Acetone (C_3H_6O)	235.1	47
Dichloromethane	237	60
Chloroform	263	54
Acetonitrile	275	48
CO_2	31.1	74
NH_3	133	114
N_2O	37	72
Xe	17	58

supercritical water. One of them uses an olefin of C_5^- [442], and the other uses olefin in addition to acidic water, a halogen, and a hydrogen halide [443]. Although reported process conditions were temperatures of at least 300°C and pressures above 140 bar, the description referred to those of supercritical water.

In another supercritical solvent (n-alkanes solvents, tetralin, and decalin), conversion to liquid distillates was better than 82% and DM was almost 100%, under conditions that yielded HDN and HDS better than 80% when using an activated carbon catalyst under hydrogen partial pressure [223]. The process economy is adjusted, controlling the amount of solvent used. Fig. 6.12 shows the reported scheme [444]. Several solvents have been tested, and though the role of the solvent is not completely understood, one could believe it to be associated with the effect of dilution. Diffusion limitations of the large molecules would decrease with the decrease in density and viscosity of the reacting system. Moreover, the solubility of asphaltenes in the supercritical solvent would be enhanced, facilitating their accessibility to hydrogen. Considering that deactivation of the carbon catalyst was insignificant in a fixed bed reactor, a slurry reactor may minimize any other drawbacks of the reactor system.

Some other hydrocarbon solvents (aromatic and aliphatic) under supercritical conditions were tested for VR HCK, using acid-treated activated carbon as catalyst, at 400°C and 35 or 70 bar. At these conditions, the highest conversion was lower than 70%. While the

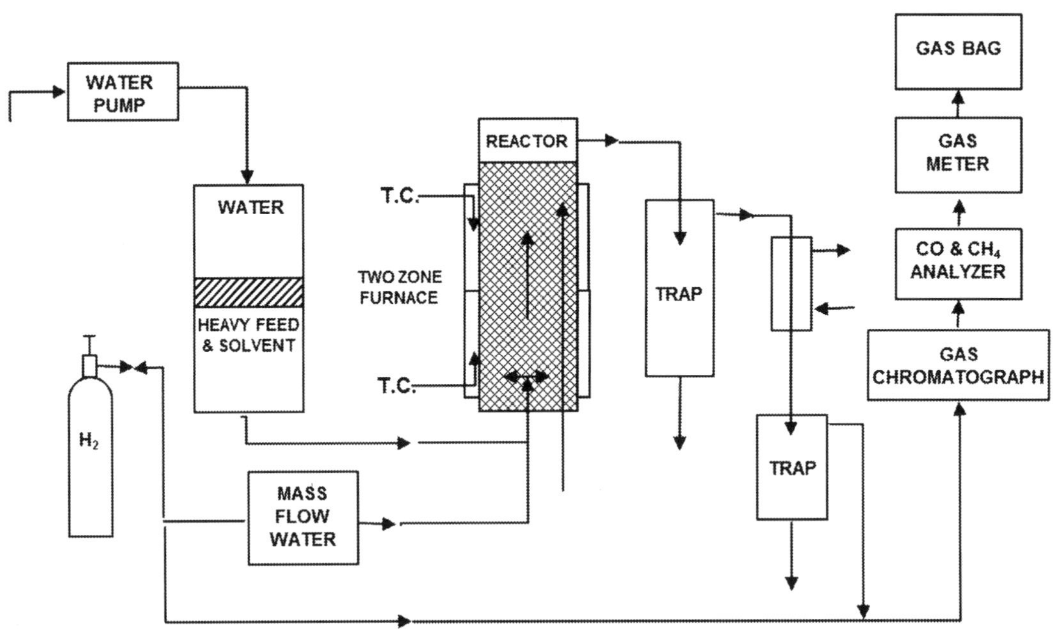

Figure 6.12
Catalytic supercritical upgrading process.

effect of the solvents was minuscule, the surface acidity of the catalyst showed a greater effect on conversion [445].

The asphaltenic bottom-cracking (ABC) process [224], which scheme was shown in Fig. 6.4, was mentioned in this chapter as an example of catalytic conversion of asphaltenes, with important capabilities in DM. The process is a high-pressure, fixed-bed reactor using a proprietary catalyst [446]. The asphaltenes are selectively converted with a low hydrogen consumption and decrease significantly the metal content in the product. Catalyst was tested for 6 months, processing various heavy feedstocks showing a stable performance. The yields and product quality reported indicated almost complete conversion of the feed and total metal removal. Various HDT catalysts [447–449] have been patented by Chiyoda, who developed this and some other HDT processes [450,451]. The net effect of the ABC pretreatment was found to be an increase in life of the downstream catalysts and an increase in distillate yields, obvious consequences of the significant decrease in metal content of the product oil.

5.2 Other Slurry and Dispersed Catalysts

Besides the proprietary catalytic systems previously reported in Section 3.2.3 of Chapter 5, a large number of other formulations have been proposed, studied, or tested for slurry-phase processes. Similarly to the commercially available technologies, most studied metals are those typically used on supported HDP catalysts (i.e., group IV or/and group VIII nonnoble transition metals) (Mo, W, Ni, Co, and, because cheaper cost, Fe), and also to a lesser extent other transition metals such as V (because originally present in the feeds), Cr, Cu, Zn, and so on. Process conditions fall within the range of HDC, that is, temperatures between 400 and 480°C and pressures in the range of 30–200 bar. As could be expected in catalytic HDC, higher temperatures increase not just conversion and heteroatoms removal but also coke yield. Meanwhile, higher hydrogen pressures would reduce coke formation but by increasing operating costs.

In a slurry-phase reactor, the catalyst is homogeneously distributed in the reactor containing the VR or whole heavy oil. Dispersed systems improve this distribution thoroughly and easier than any fine powder. On the other hand, while fine powders may suffer from attrition, dispersed systems are more difficult to recover and recycle.

The soluble systems are typically used at a concentration from 300 to 2000 wppm of metal based on feed. On the high-concentration side, product quality is best; however, depending on the metal used and recovery capabilities, this side may become prohibitively expensive. Oil-soluble (organometallic salts and complexes) and water-soluble compounds are considered precursors of the active phases that disperse in the reacting product. The oil-soluble precursor is typically dissolved in a lighter oil fraction, to ensure a high dispersion into the more dense and viscous heavy fraction or heavy oil. The water-soluble precursors are better used in the form of water—oil emulsion or a colloidal system. Colloidal catalysts might react with organic compounds in the oil and form dispersed micelles, by ligand exchange and molecular complexation processes. Some of the sulfur moieties in the crude could form these colloids and would delay coke growth at higher temperature [452].

These soluble catalysts are converted to metal sulfides under process conditions, but also they can be formed from original powder sulfides, from oxides, or from salts of the metal by mixing into organic acids. Hence, two topics have been widely studied, proposed, and tested. On one hand is sulfurization of the precursors (conditions, S-source, etc.) [453—460], and on the other is dispersion of the active materials during HDC [461—473].

Although metals are converted to sulfides, existing morphological differences might affect the performance of the dispersed system. For instance, Cold Lake bitumen has been upgraded on exfoliated MoS_2 as a dispersed catalyst, and the results were compared with MoS_2 prepared in situ by the decomposition of molybdenum naphthenate. Liquid yield and reduction in coke deposition were similar between the catalysts, while the exfoliated MoS_2 showed better HDN and removal of asphaltene and CCR. These features of the exfoliated MoS_2 were explained as a consequence of its increased rim-edge sites [474].

The number of compounds and materials that have been formulated for use in slurry-phase reactors is quite large. A brief description of reported examples follows.

Description	Conditions (°C/bar)	Remarks	Refs.
V and Mn acetylacetonates; ($V/Mn = 2.0{:}1.0$)	225/35 (H_2 + 1—20 mol% H_2S)	V and Mn sulfides produced in situ. Asphaltene conv > 96%; DM > 94%;	[475]
Pb compounds		DN and DS > 50%.	[476]
Metal naphthenates + catalytic support	400—450/100—150	Acid support material for enhancing conversion. One or more metals.	[477—480]
Me-carbonyl	420—480/70—180	H_2 flow adjusted for stripping volatiles.	[481]
Mo-dithiocarboxylate, $(NH_4)_2MoS_4$, $Mo(S_2PO_2)$, $NH_4(MoO_4)_7$, and $(NH_4)_3PMo_{12}O_{40}$		Promotion of hydrogenation and desulfurization, and inhibition of poly-condensation reactions.	[482]
MoO_2-acetylacetonate	410/110	CH_4 as an H-source.	[483]
Mo-naphthenate + S_2 + Co-octylate + di-2-ethylhexyl phosphate	—	S promotes precursors decomposition at T < 240°C and activates faster metals in the oil.	[484]
Mo-dithiocarbamate, Mo-dithiophosphate	—	Difficulties in forming MoS_2 affect catalytic performance.	[467]
Fe + Mo-dithiocarbamate (Molyvan-A)	—	95% conversion.	[485]

Description	Conditions (°C/bar)	Remarks	Refs.
Mo-dithiocarboxylate + Ni-naphthenate or Fe-naphthenate	430/30	Slight metals synergism; conversion decreased when coke decreased.	[452]
Layered (NH_4-Co)-molybdate + oleic acid	–	Metal synergy observed when metals are bound in the same compound.	[486]
Mo, W, Ni, or Co compounds	–	Ni–Mo combination best HDS.	[487]
Multimetallic water-soluble compounds	–	Best results with Ni–Mo combination.	[468,488,489]
Me-nitrates precursors	–	Used as water-soluble catalysts.	[457,490]
Microcrystalline MoS_2	–	Decreased coke formation.	[491]
Mo, V, and Re sulfide	315–370/170–350	Aqueous-phase sulfiding.	[492]
Alkali metals, alkaline earth metals, and transition metals	–	Catalyst preparation.	[493]
Fe-carbonyl	420/110	CH_4 as an H-source.	[494]
Amphipathic Ni compound	420–435/50–80	Batch reaction; various Ni-sulfide crystalline phases after reaction.	[459]
$Zr_3(PO_4)_4$ + $Cu_3(PO_4)_2$ + $Mo(CO)_6$	350–450/170 H_2 < 6000 ft.^{3}/bbl	Catalyst particularities: 80–150 m^2/g surface area, 0.2–1.0 cm^3/g pore volume.	[495]
$Zr_3(PO_4)_4$, + $Co_3(PO_4)_2$ + $FePO_4$; 1–3:1 Zr-(Co + Fe) and 1:2 Fe–Co molar ratios	–		[496] [497]
Red mud and Me/red mud	470–500/150	Self-activation with S from feed. No significant benefit.	[498–501]
Fe oxides or Fe sulfides + Me-phthalocyanines	380/140	Coke yield suppressed to 0.4 wt%.	[193]
Fe compound + petroleum coke	450/150	CanMet alternative cat-system.	[502]
Fe/activated-C	450/100	Coke formation inhibitor.	[503]
Coal-coated Fe, Co, and Mo		Coal + pitch co-processing.	[504]
Fly ash	480/250	Coke precursors reduction.	[505]
Supported HDT catalyst	390–460/80–160	Alumina support; Co/Mo ratio = 0.5; NiMo; NiMo + Fe-ore; NiMo + additives (Si, Al, and Ti oxide).	[506–510]
	–	NiMo on mesoporous alumina or silica-alumina: conversion <70%.	[511]
	350–450/50	NiMo on carbon; dispersion increased activity.	[512]
Metal-promoted (Co, Mo, Pd, Pt, Ni, or W) iron-ionic liquid	–	Preferred catalyst: 8–16 wt% Fe + 0.1–1.5 wt% Mo.	[513]

6. In Situ Upgrading and Downhole Refining

As indicated in Chapter 1, more than half of the world oil reserves are heavy oil and bitumen, with vast deposits in Canada and Venezuela. These, unconventional-opportunity crudes are more difficult to produce and to refine than conventional oils. The need for moving this very viscous, heavy oil or bitumen from the production field to the refining center or shipping terminal introduced the concept of *primary upgrading*, which at the time was used indistinguishably from *field upgrading*. As such, the upgrading target was to improve viscosity to an acceptable pipeline requirement [514]. The extension of upgrading depends on the initial properties of the oil. In Fig. 6.13, the ranges of these properties for the referred types of oil are exemplified.

Dilution with a less dense, less viscous fluid looked to be a suitable short-term alternative, and consequently *diluted bitumen* (*dilbit*) started to reach the marketplace. However, the diluting solvent should be either very light oil or a light fraction, and neither of them was readily available at the field area. Production of dilbit furnished the producer with two new roles: to be a trader and to become a blender.

Besides the dilbit option, bitumen could also be upgraded to SCO, using, for instance, technologies such as those described in the previous chapter. So far, C-rejection technologies have been preferred. One of these technologies, not mentioned previously, is the WRITE process. The WRITE technology was originally developed for the treatment of tank sludge and roughly consists of water separation, followed by a topping treatment to recover valuable distillates, and then the residue is sent to a coker [515]. Recently, it was rebranded as a candidate process for field upgrading, for the production of pipeline-ready heavy oil [516].

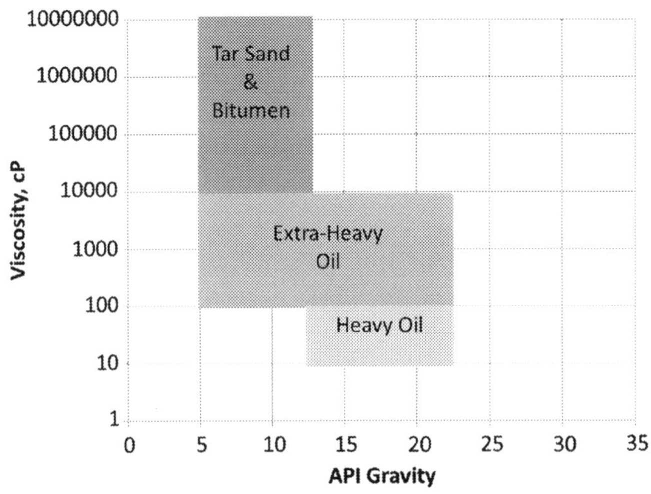

Figure 6.13
Viscosity and API gravity for different types of opportunity crudes.

Initially, all those technological approaches available in the short term were implemented, allowing the Canadian authorities in 2009 to predict an oil production of 2.8 Mbpd by 2015 [517]. Actual production by 2015 was about 3.9 Mbpd, indicating that development took place more accelerated than forecasted, which could be seen as the effect of a period of high oil prices. Volatility of the oil market dropped the oil prices to levels seen at the end of the 1990s. The need returned for suitable field-upgrading technologies to meet pipeline specifications and ensure cost-effective processing. The situation is particularly complicated for oil sands, which are a mixture of sand, bitumen, and water, containing roughly 10% bitumen, 5% water, and 85% inorganic matter (75% sand and 10% silt and clay). Two categories of production methods are mining and recovery/extraction. Since oil is trapped in the sand, it has to be recovered and/or extracted. The first method, surface mining, removes the top fertile soil layers to expose the underlying oil sand. The second type of method, recovery or extraction, involves dissolving the bitumen or heating at surface or subsurface levels, respectively, to render fluid oil in situ in the well or the field. In situ production methods advantage mining and surface processing, due to the minimal land surface disturbance, and give access to deeper resources that are inaccessible to mining. Regardless of the method, extraction systems are typically required to separate the bitumen from the clay, sand, and water. Furthermore, the recovered bitumen needs upgrading to produce higher grade oil with at least minimum pumpability (i.e., an SCO). Depending on the level of upgrading, SCO can be sold at a higher crude oil price with lesser or no discount [518].

The results of a 1998 market-driven study indicated that centralizing upgrading in a refinery has a lower capital cost than field upgrading, only because gained benefits came from utilizing existing infrastructure [519]. This study missed the fact that extensive modifications and revamping were needed to adapt existing refineries to the processing of heavy oils and bitumen. Furthermore, the crude or bitumen has to be brought to the refinery grounds, which will require field upgrading anyway. Still, 10 years later, there was no refinery processing any bitumen [520]. The market study found some economical merits in partial upgrading, but these benefits depended on a synergistic fit between a given refinery and the upgrader operation. The integration of bitumen production with downstream refineries was recommended, and taking advantage of possible synergies gave this option considerable merit [519]. First, in order to define the integration opportunities, the basis of the available production methods has to be understood. By 2007, 14 production methods and 29 technologies were identified as being practiced, on demonstration, or under development. These methods were reviewed and compared by Clark et al. [521]. Definitively, production alternatives are numerous; a summarizing chart [522] is shown in Fig. 6.14.

Some of those recovery methods shown in Fig. 6.14 that in the author's opinion may offer opportunities for integration with in situ upgrading (ISU) on the surface, near the well bore, or downhole will be further discussed in the next subsections. In the search for

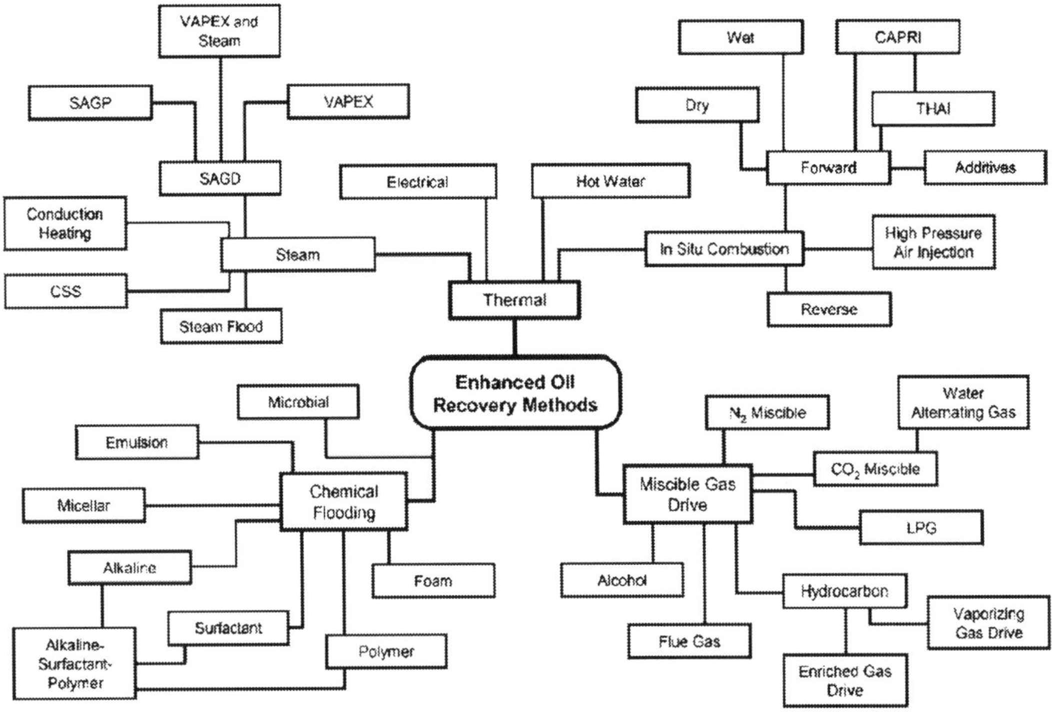

Figure 6.14

Enhanced oil recovery methods. *Reproduced from Shah A, Fishwick R, Wood J, Leeke G, Rigby S, Greaves M. A review of novel techniques for heavy oil and bitumen extraction and upgrading. Energy & Environ Sci 2010;3(6):700–14, with permission from RSC.*

integration of upstream and downstream processes, the idea of upgrading (or partial upgrading) within or at the oil well has received certain attention, particularly from the downstream sector more than that on the upper side. The term for this new field of R&D was coined *downhole refining*, which is also known as *in situ refining*. Upstream, this idea was taken as value destruction since it would increase production costs, regardless of the potential advantages that improvements in the oil quality may create.

Nevertheless, the definition of a long-term vision and of strategic objectives was the only guarantee of a sustainable development. Thus, two Canadian organizations, Natural Resources of Canada (NRCan) and the Alberta Energy Research Institute (AERI), funded a study for defining opportunities for stable long-term R&D. Seven technology development opportunities were identified as focused areas [523]:

- integration to bitumen recovery processes,
- moderate upgrading processes,
- optimization of hydrogen utilization,
- advanced catalysis,

- synthetic hydrocarbons
- physical separation technologies, and
- biotechnology.

An aggressive R&D plan was defined, funded, and executed, and most of those achievements will be discussed in this section. Downhole refining has tried to incorporate much of the knowledge derived from different areas, such as production techniques, C-rejection processes, slurry-phase reactions, dispersed catalysts, and so on. So far, some processes have been demonstrated at pilot plant level and during field tests, but none has reached commercial demonstration or is available for licensing.

6.1 The Resource Technology Association (RTA) Geotreater Process

In the 1980s, RTA started to develop a treatment process to recover oil from oil sands froths. Experimental results indicated a significant reduction in viscosity. Thus, a concept for a surface and subsurface partial upgrading process was defined for integration upstream with the objective of meeting pipeline specifications and solving transportation issues of heavy oils and bitumen. Geotreatment is a mild thermal treatment applied post production. At the surface level, the crude oils are cleaned and/or stabilized using conventional surface equipment, before or after the geotreatment. This surface treatment is integrated to the underground conversion.

The targeted conversion reactions aiming toward viscosity reduction can be achieved by a moderate thermal treatment under elevated pressure [524,525] or oxidative heating [526]. Asphaltenes present in the feeds would tend to form coke upon heating. More particularly, if heating is provided through the reactor walls, coke will deposit on the walls and accumulate thereon in the reactor. In direct oxidative heating, coke formation is minimized since heat is provided by the exothermic oxidation reaction of part of the oil components. Notwithstanding, the core of the technology is the innovative subsurface reactor that supposedly overcomes these problems [527,528].

The reactor is a vertical tube composed of three zones: (1) a heat exchange section, (2) a combustion zone, and (3) a reaction zone. In the heat exchanger, heat is transferred from the effluent product stream to the incoming feed stream. The oxidizing agent is introduced in the combustion zone into the core portion of the feed. The reaction zone is the volume where feed reaches reaction temperature. Since heat is provided by combustion, there can be substantial overlap between the combustion zone and the reaction zone. The schematic of such a reactor is shown in Fig. 6.15.

A relatively narrow and very long reactor minimizes back-mixing, and thus phase separation and equipment fouling. Typical process conditions include temperatures in the range of 375−435°C, and pressure greater than 70 bar up to about 280 bar. The moderate temperatures would lower coke yield, and the high pressures would reduce gas formation

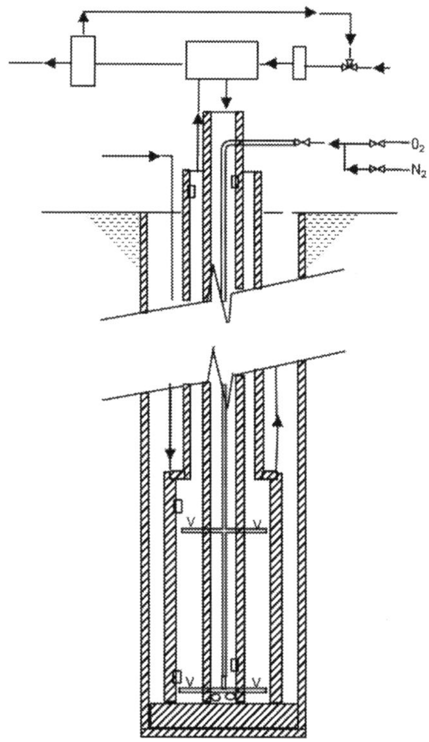

Figure 6.15
Diagram of geotreater reactor, based on description from Refs. [527-528].

and would shift reactions toward liquids. The higher proportion of the liquid phase would dilute the asphaltenes and their trend to coking.

In the well bore option, the reactor can be placed in a way such that the outer reactor wall serves as the well casing. In this arrangement, the inner conduit is suspended within the outer shaft in order to allow for the thermal expansion expected during warmup.

Numerous experimental runs and a variety of feeds were tested at pilot plant level (50 bpd) wherein the direct and indirect heating modes were tested. The process was demonstrated during a 31-day run, in early 1988, at 610 m depth, with direct heating for 10 days and indirect electric heating for 21 days [529,530]. While direct heating has technical advantages, it is riskier and costlier. Instead, indirect heating would be limited to feeds with the lowest coking tendencies.

6.2 Thermal Processes

Besides the in situ method of using a solvent to reduce heavy oil viscosity at surface level, other production technologies have been developed for reducing the energy requirements of recovery. In situ thermal processing technologies have been progressing with great acceleration, and even more in recent times when the drop in oil prices created urgency for improvements in economy. These methods include cyclic steam injection (CSI), steam-assisted gravity drainage (SAGD), in situ combustion (ISC), underground gasification, in situ pyrolysis (ISP), and so on. Comparatively, each of these technologies offers different advantages and disadvantages, in terms of economic, environmental, geological, and physical characteristics. Some of these technologies can synergize with each other; thus, benefits may be found in combining processes toward improvements in efficiency and effectiveness. SAGD recovery is still the method of preference to enhance production. However, in situations in which the availability of steam is difficult or otherwise limited because the resource is too deep, in arctic regions, or offshore, solvent-assisted becomes the option. Other options are the direct heating of the formation and injecting steam or generating it downhole. Various sources of heating have been examined, such as chemical, electricity, induction, microwaves, radiofrequency (RF) heating, and so on.

In summary, integrated approaches for on-site and ISU, including field upgrading as well as downhole upgrading, would involve an increased level of operations at production sites. Examples are [531]:

- steam-assisted or vapor-assisted production, which allows the development of a confined liquid and gas chamber along the length of the production wells;
- thermal treatments in the reservoir;
- incorporation of processing elements in the reservoir;
- development of lower cost hydrogen generation; and
- sustainable ways of upgrading to higher quality levels of bitumen and extra-heavy oils.

Some of the process technologies for shale oil production from oil shale, which fall within the thermal processes category are adaptations of those commercially applied on tar sands (and vice versa), and some others may serve as the basis for new technologies for heavy oils and bitumen. In oil shale, the R&D efforts devoted to this area are already paying back, with processing technologies that have enabled the accelerated exploitation seen in recent times. A review of these methods can be found in Ref. [532]. The results and derived knowledge might be useful for scientists and engineers interested in the science and technology associated with opportunity crudes.

6.2.1 Direct Heating

Direct heating as a methodology for enhanced recovery and heavy oil and bitumen production has received significant attention. Shell Oil Corporation is probably the most prolific owner of intellectual property on the subject. These methodologies for oil production fall out of the scope of this book, and only those dealing directly with upgrading will be considered in this section.

Heat sources were suggested to be positioned in a hydrocarbon-containing formation within a relatively thin layer of hydrocarbon material, in a staggered pattern near the edges of the layer. Thus, the closed proximity of the sources and the hydrocarbons allows reaching a desired temperature by a large proportion of hydrocarbons in that layer. A mixture of hydrocarbons, H_2, and/or other formation fluids may be produced from the formation. Heat may be applied to the formation to raise the temperature of a portion of the formation to a pyrolysis temperature [533].

A thermally induced hydrogenation, including direct heating of the reservoir to hydrocarbons conversion temperatures, and the direct injection of hot hydrogen have been disclosed by UOP [534,535]. A similar process by the Institute of Gas Technology injects H_2 at about 150°C and is left in the well until viscosity and sulfur are reduced to the desired levels [536]. Another version of direct hydrogenation converts the well into a reactor, by fracturing and then sealing it. Hot hydrogen is then introduced and allowed to react for a period of time [537].

One way of generating the heat for the thermal treatment is by burning part of the oil in the well, known as the *in situ combustion* (ISC) technique. The technology has been used in the United States for nearly 90 years. Initially, many field trails failed and created a negative reputation that has precluded a widespread acceptance among operators. Although failures were adjudicated to poor oil combustion characteristics and operational problems, now it is known that the geological characteristics of the reservoir bore part of the responsibility. It is believed that the status of the technology will support safe and productive operations [538].

The *toe-to-heel ISC* (THAI) process was described in Ref. [539] as comprising an injection well and a production well, the former being completed relatively higher in the reservoir and the latter completed relatively low. The production well has a horizontal leg perpendicularly oriented to linear and a laterally extending, upright combustion front propagated from the injection well. The leg is positioned in the path of the advancing front. The front advances along the leg, induced by keeping open the production well.

Precision combustion initially intended to patent a broad idea of ISC, by injecting oxygen into a chamber within an oil deposit and thereby combusting oil at the chamber walls [540]. Then, they disclosed a more defined method oriented to enhanced oil recovery from underground reservoirs, consisting of injecting air at the reservoir floor level, removing the combustion products from the reservoir top, and collecting oil from a horizontal production well near the reservoir floor. A bleed well for combustion products collection is constructed above the air injection; these combustion products can be passed to an expansion turbine for energy recovery [541,542]. In order to improve combustion efficiency and recovery effectiveness, this company proposed to incorporate a combustion catalyst, with the expectation of concurrently provoking crude upgrading [543]. More suggestions were given for better control and improved efficiency for ISC, namely incorporation of a catalyst, an ignition fuel, and a diluent that could be either water or CO_2 [544].

ISC is considered to be the most energy efficient of thermal-enhanced oil recovery methods. However, formation of deposits complicates operation and might cause severe damage to the formation and to the well. Deposition increases with the decrease of API gravity of the oil in the reservoir. Asphaltenes deposition is a common problem found in production operations that sometimes calls for repairs. Asphaltene-compatible workover fluids have to be used in such methodologies. A proper balance of saturates and aromatic composition is recommended for such fluids [545]. However, the experience gained in inhibiting or avoiding asphaltene precipitation has to be incorporated into any ISU method.

The use of dispersed water-soluble additives and catalysts was thought of as an attempt to inhibit deposition reactions. The kinetic study of ISC of Huntington Beach, United States (18.5°API) [546], and Hamaca, Venezuela (10.5°API), oils in the presence of aqueous solutions of metallic (Cu, Ni, Fe, Sn, Zn, Al, and Cd) salts and ketals (ketyl acetates) indicated a negligible or negative effect of the additives. Actually, Cu, Ni, and Cd salts had little or no effect, while Fe, Sn, Zn, and Al increased fuel laydown. Further studies including two California oils, Cymric light (36°API) and Cymric heavy (12°API), as well as Saudi "tar" (19.7°API) showed increases in the combustion frontal velocities for Fe, Sn, and Zn salts, which led to changes in the H—C ratios of the fuel and in heats of combustion, air requirements, and upgraded crude density. Fuel deposition differed among the tested crudes, and with the exception of Zn in Huntington Beach oil, the presence of the additives increased deposition [547]. Nevertheless, by controlling the combustion front and within a specific range of temperatures, significant upgrading occurs, which was explained due to the selective burning of the immobile heavy ends, while most of the light ends flow toward the producing well [548].

In situ pyrolysis processes are typically used for the processing of oil shale, for the conversion of kerogen into oil, gas, and coke. A fuel cell—based process for electricity production with in situ carbon capture reduces CO_2 emissions by balancing heat, increases retorting temperature to produce a portion of char, and generates electricity using product gas with a subsurface fuel cells [549]. Chinese Liushuhe oil shale was pyrolyzed at laboratory scale in an autoclave, at underground temperatures and pressure conditions. The effects of the presence of water and of metallic ions were investigated. While there was a clear effect of the presence of water, metals did not affect pyrolysis. Water generates thermal bitumen as an intermediate product, which is then converted to shale oil and gas. With increasing temperature, the yield of bitumen maximized at 300°C, while that of shale oil and gas increased gradually [550]. Since pyrolysis is energy intensive, a combination with in situ coke combustion has been proposed as a way to reduce energy requirements. Adjustments on starting time and duration of each process would define the total energy reduction [551]. Attention should be paid to heat value recovery, to any potential ISU of the oil, and to reusing or recycling the combustion products as means for decreasing green house gas emissions.

6.2.2 Steam

The inefficient and expensive surface mining techniques with higher environmental impact could be replaced by technologies based on steam injection, which also represents a possibility to open resources that other methods cannot. SAGD and cyclic steam stimulation (CSS) are currently in commercial use. SAGD, originally developed by the Alberta, Canada, Department of Energy (formerly AOSTRA), is cheaper than CSS methods [518]; however, it is still important to remember that the oil produced from these technologies would still need upgrading. In SAGD, two horizontal wells are drilled one on top of the other, and separated by a constant vertical distance, typically 5 m; steam is injected in the upper one as a heat carrier; and steam flows by gravity to the oil well (the lower one) through the porous media, reducing the viscosity of bitumen. A SAGD sketch [552] is presented in Fig. 6.16.

AOSTRA disclosed a process for ISU in which steam is injected in combination with (H_2-free) CO at 260°C into the reservoir. It was claimed that under these conditions, steam and CO undergo a water–gas shift (WGS) reaction in the reservoir, producing H_2 and CO_2, which were supposed to upgrade the oil [553].

A burner was defined for downhole steam generation that includes an injector and a cooling liner. Steam enters the burner through holes in the cooling liner. Combustion occurring within the cooling liner heats the steam, increases its quality, and may superheat it. The heated, high-quality steam and combustion products exit the burner and enter an oil-bearing formation. This burner was described in detail in Ref. [554].

A study for comparing the effect of steam under ISC conditions was performed by forward combustion experiments of dry and wet Turkish heavy oil reservoirs. Combustion was carried out in a vertical tube packed with crushed limestone and saturated with crude oil and water. Temperature increased up to when combustion stabilizes, then decreased as the front approached the tube outlet end. The presence of moisture decreased the combustion rate, and consequently, the rates of heat generation and maximum temperature were lower. In both cases, carbonate minerals were observed to decompose with the concomitant formation of CO_2. The effects on the crude oil parallel those caused by the increases in temperature, such as decreased H–C ratio, fuel consumption rate, and increased °API gravity [555].

Precision Combustion holds two Canadian patents on steam-based methods for oil recovery and upgrading, in an oil-bearing formation. The first one consists of injecting a

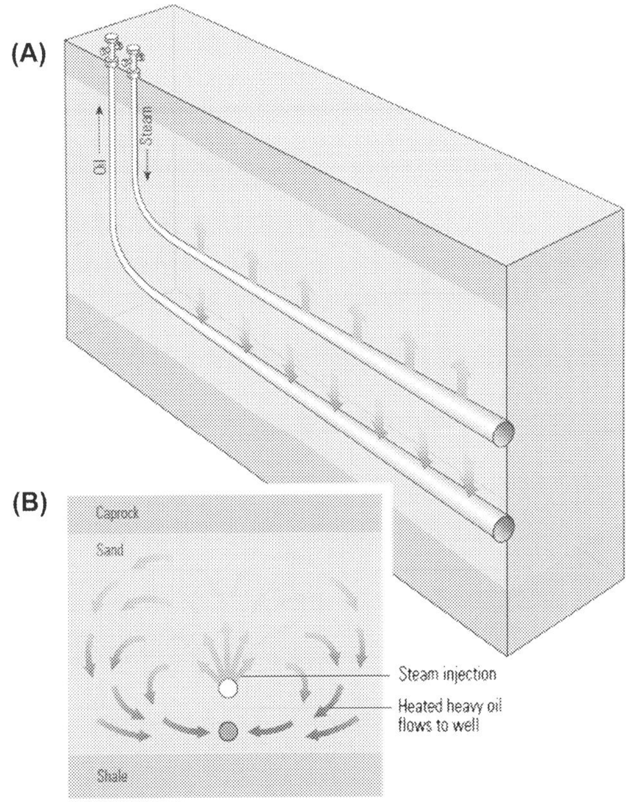

Figure 6.16

Sketch of SAGD technology: (A) injection/recovery system; (B) flow diagram of the complete process. *Reproduced from Total E&P. Extra heavy oil and bitumen: extra reserves for the future. Paris, France: Total Group: 2007. 28 pp. Available from: http://www.total.com/sites/default/files/atoms/file/total-extra-heavy-oils-and-bitumen-reserves-for-the-future, with permission from Schlumberger.*

steam-containing hot gas phase fluid into direct contact with the heavy oil in the reservoir. The temperature of the fluid is set high enough to induce steam cracking [556]. The second invention provides a SAGD-based method consisting of the injection of a fixed gas into an interior region of a steam chamber, displacing steam toward the chamber walls. The advantageously heated oil drains toward the reservoir floor and is recovered via a production well [557].

In situ gasification has been thought of as a way of generating hydrogen to be used by bitumen upgrading. Oil sands in situ gasification might potentially result in more energy efficiency and less contamination than the combination of recovery and upgrading together, since it would reduce acid gases emission, which might be sequestered by reservoir minerals. Additionally, water consumption may drop, and bitumen heavy metals and sulfur compounds may remain downhole. Based on pilot observations of Marguerite

Lake ISC, a new in situ bitumen gasification process was designed using SAGD well configuration. The combined in situ methods were achieved by a cyclic injection of steam and oxygen, resulting in greater energy produced per unit of energy invested, with less than half the water usage [558].

A method of enhanced recovery was based on the direct hydrogenation of the crude oil by the cyclic injection of superheated steam and hot/high-pressure hydrogen into a preheated formation zone next to the well. Hydrogen is left in the shut-well to soak the oil for a period of time to allow reaction. Fluid under pressure is injected into the formation through a separated well to drive the crude to the production well for recovery [559,560]. A similar process was previously awarded to Pan American Petroleum Corp., though this old process did not include the cyclic repetition of the steam stimulation and hydrogenation step [561].

A series of patents and patent applications have been disclosed by Yeda R&D, on variations of the recovery and ISU in the presence of steam. These variations include the incorporation of CO into the reservoir [562,563], inclusion of an Fe-based catalyst [562], conditions for partial upgrading [564], supercritical (water) conditions [565,566], and combinations of the above [564,565,567]. The process conditions for these processes varied within broad ranges: 250–450°C, 35–210 bar, and 25 s–10 min.

A combined DA and cracking result was observed when bitumen was treated with supercritical water at 440°C and 300 bar. Under these conditions, liquid product yield was maximized, while coke yield was minimized [568].

6.2.3 CO_2 and Other Small Molecules

Similarly to SAGD, other molecules have been proposed to assist production of heavy oils and bitumen. For instance, in the VAPor EXtraction process (VAPEX) hydrocarbon solvents are injected into the upper well to reduce viscosity by dilution of the bitumen. Compared to SAGD, this method is more energy efficient but also more expensive. Advantageously, partial upgrading occurs, converting the bitumen to crude oil [518].

A combined process including in situ combustion and SAGD, VAPEX, or gaseous solvent, employing horizontal production wells, has been disclosed. The co-enhancing agent might be water, steam, and/or a nonoxidizing gas, CO_2 being the preferred agent [569].

In the case of CO_2, a simplified model was developed to assess the density of the upgraded crude. The model parameters included the mass ratio of solvent to crude, as well as temperature, pressure, and the properties of the crude oil. In order to simulate the dispersion and concentration of injected CO_2, the model considered the interaction dynamics in the reservoir and the greatly changing CO_2 density when close to its critical point. The generation of a new phase within the crude and the differences in mobility between the new crude matrix and the precipitate readily enable removal of asphaltenes. Thus, an upgraded crude low in heavy metal, sulfur, and nitrogen would be recovered [570]. In terms of recovery, changes in viscosity are more important than those in density. However, in terms of upgrading and under subcritical or critical conditions, the changes in CO_2 density would have a larger effect on hydrocarbon solubility and direct consequences on asphaltenes precipitation.

A process for ISU of a heavy hydrocarbon with an API gravity of less than or equal to about 8° includes the injection of a light solvent into the well at reservoir conditions. Upgraded oil (sort of DilBit) will be produced from the well [571]. Depending of the choice of the light hydrocarbon, asphaltenes deposition might occur, causing damage to the production infrastructure, to the well, and/or to the formation.

In situ hydrovisbreaking renders production fluids composed of virgin heavy hydrocarbons, heavy hydrocarbons, lighter liquid hydrocarbons, residual reducing gases, hydrocarbon gases, and other components. At surface level, these can be separated into an SCO ($525°C^-$) and a residue ($525°C^+$). The latter can be partially oxidized to produce clean reducing gas and fuel gas for steam generation. The reducing gas and steam can be injected back to the formation for the in situ hydrovisbreaking process [572–574]. The commercial basis for one production unit was estimated for a 20-year project life to be a 52-ha production pattern, replaced annually and producing 32 kbpd heavy oil or bitumen and 18 kbpd SCO (0–22°API). A typical bitumen formation would support the operation of multiple production units simultaneously [575]. Petroleum Equities Inc. described this combined process as consisting of four steps (see Fig. 6.17):

1. Preheating a portion of a heavy oil reservoir with saturated steam
2. Establishing hydrovisbreaking conditions in situ by injecting superheated steam and hot hydrogen into the formation
3. On-surface separation of the raw production into SCO and residue
4. Partial oxidation of the residue

In this process, a specifically designed combustor could be operated for producing either superheated steam or both hot hydrogen and superheated steam. H_2 is burnt with a deficient O_2 stream, in order to leave excess H_2 for the hydrovisbreaking reaction. The superheated steam would aid in reaching visbreaking temperatures [576]. A computer simulation predicted a significant increase in recovery as well as an important improvement in oil quality [577]. An experimental verification of the hydrovisbreaking step was carried out in an autoclave mixing 500 g of crude oil, 500 g of simulated reservoir sand, and a given amount of brine. The autoclave was then pressurized to 30 bar, with N_2, H_2, or a $N_2{-}H_2$ mixture stirred, and heated to the desired temperature (275–370°C) for the desired residence time. API gravity increases were in the range of 4.2–11.1, and viscosity reductions were better than 98% [578].

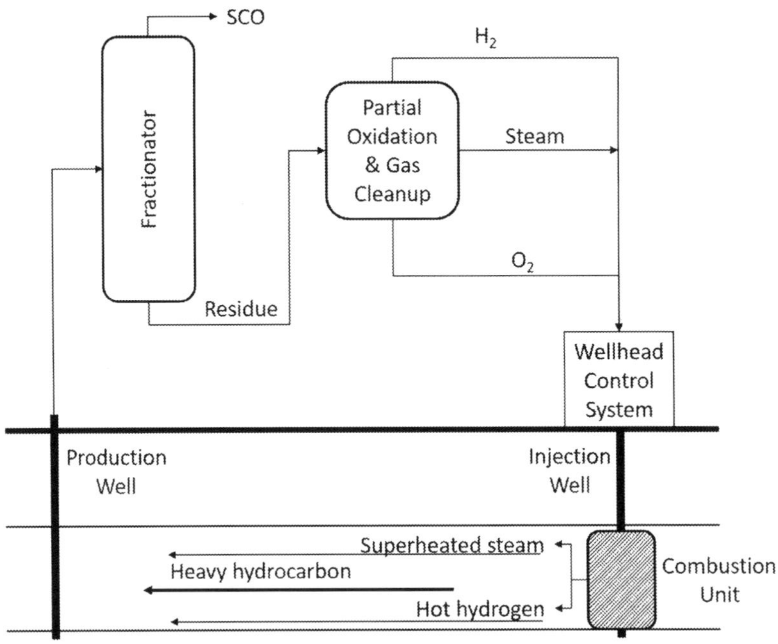

Figure 6.17
Schematic of ISU process proposed by Petroleum Equity, Inc., based on description from Ref. [576].

A controlled energy-efficient visbreaking process has been described as based on the selective absorption of electromagnetic energy by specific molecules in the heavy crude oil. A precise temperature control throughout the treatment volume is assured by providing heat via electromagnetic energy transfer at the molecular level [579].

6.3 Catalytic Processes

In the case of (extra-)heavy oils and bitumen that are too viscous to be transported through the pipelines, integration of upstream operations to downstream types of processes may provide potential economic benefits. If thermal processes have been thought of favorably, catalytic upgrading may be even better. This option needs, first, to demonstrate its feasibility. Upgrading at locations such as in situ, near-well-bore, or downhole brings enormous challenges. A successful in situ catalytic upgrading (ISCU) technology involves placing a catalyst in or near a production well, contacting it with the oil and co-reactants (H_2, water, or CO), providing pressure and heat to reaction temperatures, and recovering the upgraded oil. Additionally, another complicating feature is the presence of brine in the downhole environment.

Similar to conventional upgrading technologies and based on refining processing catalysts, two options appear possible: a solid pelleted catalyst or a dispersed system. Both options have been considered and tested: solid pelleted catalysts have been examined up to the field scale, while dispersed systems have been explored up to the pilot plant level, mimicking reservoir conditions and materials. The results prior to 1997 have been reviewed by Weissman [580]; some of these results are cited here, but we focused our discussion on more recent achievements.

6.3.1 Heterogeneous and Nanocatalysts of Conventional Formulations

One way to provide the needed heat for catalytic reaction was addressed by modifications of the combustion tube apparatus. The concept of downhole catalytic upgrading of heavy oil using ISC was proposed first by Moore et al. [581] and Weissman et al. [582,583]. Middle Eastern heavy crude oil and the corresponding brine were tested at reservoir conditions. Besides heating the feed or the catalyst bed, ISC also generated reactive gases. ISC has been demonstrated to cause partial upgrading purely due to thermal effects (see above). A hydroprocessing catalyst was located in the heated zone and provided additional conversion. Catalytic upgrading resulted in about 50% DS and an increase of 8°API gravity. Additionally, the heated catalyst was found to be efficient at converting CO to additional H_2. This test was taken as a demonstration that ISCU could be performed on modified ISC systems [582].

The catalytic upgrading process in situ (CAPRI) represents an improvement over the THAI process described above (see Ref. [539]), to which it was interdependently integrated. In this case, the leg that is located in the path of the advancing front contains an upgrading catalyst. The hot combustion gases react with the commingled oil over the catalyst to provide ISU [584]. A sketch of the THAI–CAPRI operating array [585] is shown in Fig. 6.18A.

The first results corresponded to a laboratory-scale CAPRI test of a THAI partially upgraded pilot oil from Petrobank's White Sands project. The CAPRI reactor was packed with a mixture of gravel and catalyst, for mimicking the core showed in Fig. 6.18B. Viscosity was improved from 433 mPa s to 87 mPa s, and API gravity from 13° to 16° [586]. An increase of 5.2°API was obtained when Lloydminster heavy oil was upgraded under a THAI–CAPRI arrangement [587]. The THAI–CAPRI process claims to achieve recovery of up to 87% under which asphaltenes hydrous pyrolysis and WGS reactions take place. Hydrogenation routes of the thermally or catalytically cracked oil may be responsible for the production of light hydrocarbons, increases in saturates, and reduction of heavy ends in the catalytically upgraded crude. The environmental benefits include a large reduction in heavy metals and sulfur, with potential reductions in refinery loadings [588]. Various analytical methods have been used for the evaluation of the level of upgrading observed during THAI-driven ISCU on heavy Wolf Lake oil. In addition to

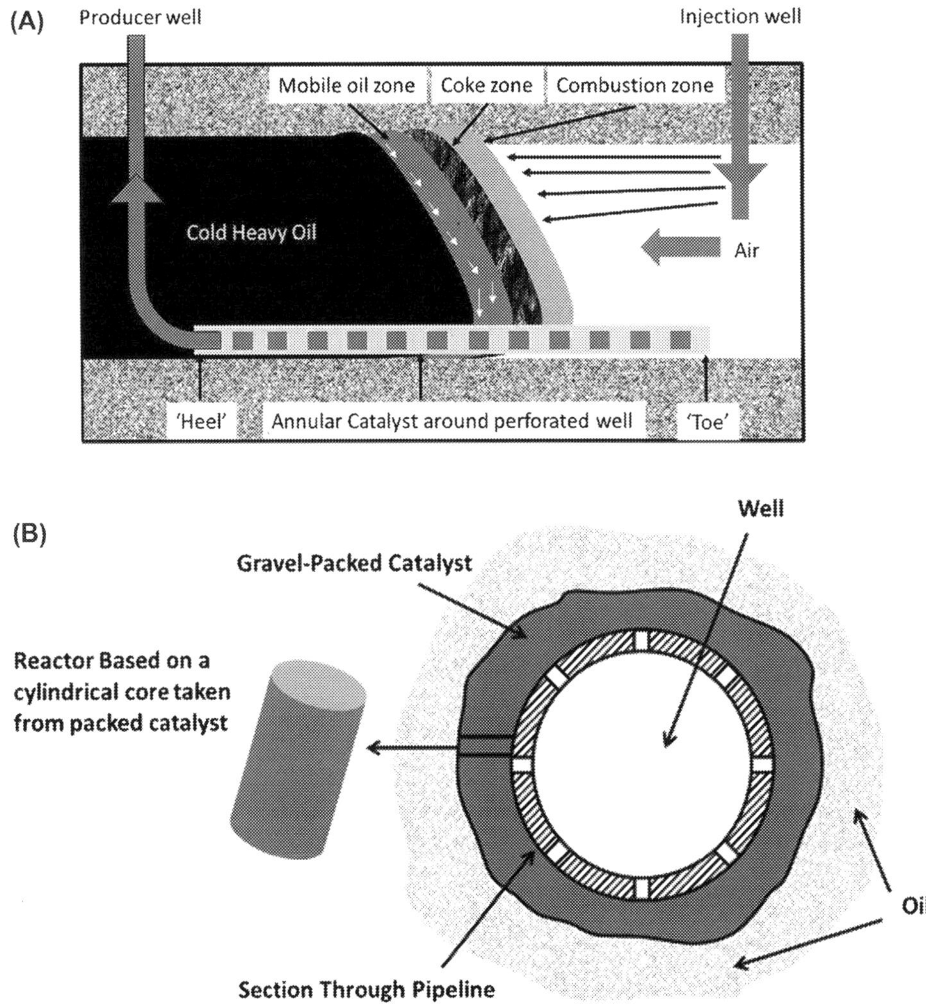

Figure 6.18

The THAI/CAPRI: (A) operating array; (B) details of the CAPRI reactor. *(A) Reproduced from Hart A, Shah A, Leeke G, Greaves M, Wood J. Optimization of the CAPRI process for heavy oil upgrading: effect of hydrogen and guard bed. Ind Eng Chem Res 2013;52(44):15394–406, with permission from ACS Publications; (B) Reproduced from Kendall R. Using timelapse seismic to monitor the THAI™ heavy oil production process. CSEG Rec 2009;34(7). Available from: http://csegrecorder.com/articles/view/using-timelapse-seismic-to-monitor-the-thai-heavy-oil-production-process, with permission from CSEG)*

analyzing the upgraded oil's density and viscosity, the oil was also analyzed by gas chromatography, elemental analysis, simulated distillation (SimDis), and microactivity test (MAT). Experiments were carried out in semiscaled 3D combustion cells, with a conventional HDT catalyst that was gravel packed around the horizontal producer well. Upgrading produced a 60 mPa s light oil that could be converted under MAT conditions into gasoline and diesel. The removed heavy metals and sulfur were retained by the packing materials [589]. The effects of using hydrogen, methane, nitrogen, and a blended gas mixture in a THAI–CAPRI arrangement were investigated, with a Co–Mo/γ-Al_2O_3 catalyst at a reaction temperature of 425°C, pressure 10 bar, and gas-to-oil vol-ratio of 50. The spent catalyst was regenerated ex situ by thermal oxidation of the carbonaceous deposits. The API gravity improvements, on a 14°API crude oil, were 4°API with H_2, 3°API with CH_4, 2.9°API with THAI gas, and 2.7°API with N_2. The reduction in specific surface area of the catalyst was: 57.2% for H_2, 68% for CH_4, and 96% for N_2 from the original surface area of 214 m^2/g. After 6 h operation in N_2, the spent catalyst contained 23.5 wt% coke, and 6 wt% and 3 wt% less coke when using H_2 and CH_4, respectively. The regeneration conditions employed only recovered 48.5% of the catalyst specific surface area [590]. Catalyst deactivation in the CAPRI process was also studied in the laboratory, under oil upgrading conditions with a 0.49 Pa s/13°API crude oil at 20 bar, temperatures in the range of 350–425°C, and residence time of 9.2 min, in the presence or absence of H_2. The increase in pressure from 10 to 20 bar (compared to the test reported in Ref. [590]) not only reduced coke deposition and extended catalyst life but also improved further the quality of the upgraded oil, increasing API in 7° and reducing viscosity in 87.2% [591]. In all cases, catalyst deactivation was exclusively associated with coke deposition. The poisoning effect of metals originally present in the feed was not reported, though the irreversible deactivation caused by these elements might be the cause of less than 50% recovery of surface area upon regeneration. In an attempt to draw the potential of a THAI–CAPRI operation under steam-addition production processes, a similar experimental study was carried out at the laboratory scale on a crude oil with a viscosity of 1.091 Pa s, at a reaction temperature of 425°C, pressure of 20 bar, 500 gas-to-oil vol-ratio, and 0.02–0.1 steam-to-oil vol-ratio. The tested catalyst was a γ-Al_2O_3-supported CoMo. Blank experiments, in the absence of steam and after 15 h time-on-stream (t.o.s.), showed: 85.5% viscosity decrease, 27.5 wt% coke deposition on the catalyst, 3.4% DS, and 17% DM. At the same time on stream, the presence of steam diminished coke deposition to 17 wt%, with further reduction of up to 11.3 wt% upon increases of the steam-to-oil vol-ratio. Over the same range of conditions, viscosity was reduced in 88–92%, DS = 16–25%, and DM = 43–70.5%, as the steam-to-oil ratio increased [592]. The set of reactions occurring in THAI–CAPRI have been proposed [522] to be:

$$\text{Thermal cracking (pyrolysis)} \rightarrow \text{heavy residue / light oil} + \text{coke} \tag{6.vi}$$

$$\text{High-temperature oxidation: coke and VR} \rightarrow CO + CO_2 + H_2O \tag{6.vii}$$

$$\text{Carbon rejection} \rightarrow CH_x/CH_y + C\ (y > x) \tag{6.viii}$$

$$\text{Hydrogen addition: } CH_x + H_2 \rightarrow CH_y\ (y > x) \tag{6.ix}$$

The hydrogen responsible for the hydrogenation and hydrogenolysis reactions was thought to be produced via WGS and/or gasification reactions [593]. A potential design for the

CAPRI *reactor* has been disclosed as a well liner, consisting of an elongate cylindrical outer liner member concentrically placed with an elongate inner liner member [594].

A combined ISC–ISU test was carried out at the laboratory scale, using materials from the Llancanelo field in Argentina. In a first set of experiments, ISC was tested, and then the combined process, using a commercial NiMo-type catalyst. Two laboratory combustion tube tests were completed to validate this concept. For the first test, a tube was packed with ISC packing based on the Llancanelo core. For the second test, a heated catalyst bed was positioned in the downstream region of the combustion tube. The reactive upgrading gases formed at the ISC stage, such as CO and possibly H_2, would move the oil onto the heated bed of catalyst. Although the first test showed a smooth ISC operation, during the second test, excessive pressure drops developed in the system that required a change in the air injection rate. This pressure drop was associated with coke deposition on the catalyst, indicating a need for periodic regeneration. Nevertheless, the second test demonstrated the ISU stage produced upgraded oil, with reduced density and viscosity [595]. In previous sections, we discussed the inconvenience of fixed-bed HDT for the upgrading of heavy oils and their residues. Even for refinery operations, where regeneration is an almost routine operation, fixed-bed HDT has been proven uneconomical, for the particular case of opportunity crudes. At this point, whether fixed-bed HDP (CAPRI) is an option for ISU remains uncertain.

Notwithstanding, the devoted efforts have shifted gears to dispersed and ultradispersed nanocatalysts for ISU. As pointed out in previous sections, H-addition upgrading of heavy oils and bitumen has been mostly studied on conventional HDP-type catalyst formulations (see Chapter 5). Furthermore, the best performing catalysts in the unsupported or dispersed particle mode are also composed by the same metals (Mo, Ni, and Co) present in active phases on conventional HDT catalysts. In the case of dispersed catalysts, the performance of oil-soluble or lipophilic precursors, under high H_2 pressure, was always better than emulsions of water-soluble precursors and of dispersed powders (see Section 5.2).

Since some production processes introduce water or steam to the oil well, the study of catalytic aquathermolysis of heavy oils with water-soluble and oil-soluble metallic compounds, amphiphilic and ionic liquids, and dispersed particles of different metals has been reported to produce important viscosity reductions. Downhole aquathermolysis might produce SCOs in situ with a reduced viscosity and increased API gravity, and thus reduced energy demands and decreased emissions. A nano-nickel catalyst can be used for the aquathermolysis reactions of extra-heavy oil at 280°C. When the catalyst is used in the colloidal form, the present surfactant contributes to the production of a W/O emulsion, after upgrading [596].

Scott and Pereira [597] have reviewed the different process configurations and catalytic systems tested, including THAI–CAPRI or, in general, ISC–ISU with NiMo–Al_2O_3,

ISCU with an amphiphilic metal compound in aquathermolysis conditions, and dispersed nanocatalysts in a hydrocarbon carrier containing dissolved hydrogen. Methods for nanocatalyst preparation were also examined by these authors experimentally. More particularly, the possibility of using transient w/o emulsions had positive results for the preparation of nanosized Ni, Mo, and NiMo catalysts [598].

A high-heat-capacity fluid such as VR has been considered for steam replacement as a source of heat to the well. Additionally, a dispersed catalyst and dissolved H_2 would be the essential complements for ISCU. A test was conducted in a sand-packed pilot plant reactor for evaluating the injection of dispersed catalyst and H_2 in Athabasca VR. This experiment showed Athabasca VR conversion of only 23 wt%. However, after several runs with Athabasca VR + catalyst + H_2, it was noticed that metals were being deposited on the packing material. The effect of locating the catalyst particles on the surface of the porous medium was studied as well as the quantification of metal accumulation in the reactor packing. A mixture of a commercial VR and bitumen was flowed through the reactor at reservoir conditions, without further catalyst addition. Under these conditions, conversion of the residue fraction was 32 wt%, API gravity increased from 6° to 12.1°, and a significant reduction in viscosity was observed [599].

Pereira's R&D group from Calgary University has been considering nanosized ultradispersed (UD) catalysts and adsorbents for approaching an "underground refinery," and as a surface upgrading alternative [600,601]. Some of the preliminary adsorption work was already described in this chapter. Preparation and testing of multimetalllic nanocatalysts provided initial understanding of their potential application as dispersed systems for downhole upgrading. The direct introduction of nanocatalysts into the formation was tested experimentally at the laboratory scale, by evaluating the propagation of the dispersed-emulsified catalyst particles through a sand-packed bed. Particles in the micron-range size were trapped and created pressure drop through the system. Similarly, the presence of a surfactant in the catalyst suspension promoted emulsion formation with the connate water of the sand bed. Finally, although nanosized catalysts were capable of propagating through the sand bed, there was a certain level of retention. While the micron-sized catalyst particles were held responsible for the pressure drop and for irreversible permeability damage, the retained nanoparticles were considered neither to affect the pressure nor the permeability [602].

Various ultradispersed Ni–W–Mo nanocatalysts were prepared in a VGO matrix, in situ the upgrading reactor. Athabasca bitumen HDP tests performed at 320–340°C, 35 bar, and a H_2 flow rate of 1 cm^3/min resulted in significant inhibition of coke formation (51.3%) while moderately improving quality (e.g., at 340°C: DS ~50%, DN ~35%, MCR reduction ~25%, viscosity decrease ~20%, and API improvement <10%) [603]. Similar results were found when a water-soluble colloidal version of the multimetallic catalyst was tested at similar conditions [604]. Under these conditions, the considered multimetallic catalytic systems increased gas yield considerably, though the proportion of CO_2 in the gas mixture decreased by 50%. The accumulative volume of coke precursor gases, such as ethylene and propylene, increased with reaction severity [605]. This increase in gas yield was considered to be the driver for the improved bitumen recovery evaluated from the material balance (Fig. 8 in Ref. [606]), which could not be seen as absolute recovery (Fig. 7 in Ref. [606]). Actually, this latter drawing showed a negative effect on recovery when using the nanocatalyst (cf. Figs. 6 and 7 in Ref. [606]).

Processes combining the findings discussed in the last two paragraphs have been disclosed, one for downhole HDP [607] and another including a hot fluid (e.g., VR or DAO), H_2 addition, and a catalyst [608]. The University of Calgary has also disclosed the ultradispersed catalysts described above [609–612].

6.3.2 Emerging Reacting Systems

An effective recovery of heavy oil or bitumen is one that produces all components present within the oil and does not leave behind any residue. In practical terms, asphaltenic components would flow together with the rest of the component and would not precipitate or otherwise deposit in the flow path. As discussed previously, neither thermal techniques nor solvents have shown to be that effective. Novel ISU methods are in their infancy, and still there is space for new ideas. Under HDP conditions, dispersed systems and nanoparticle catalysts have demonstrated capabilities for asphaltene conversion. Under less severe conditions, breaking of C–S bonds within asphaltenic components is feasible on nanosized metal particles, resulting in saturates and aromatics increases while simultaneously reducing asphaltene content and contributing to a reduced viscosity and density.

Although it has been reported that the formation minerals can act as catalyst for downhole HDC, using a hydrogen donor, methane, or steam for improving viscosity, API gravity, and distillate fractions of heavy crude oils [613], robust evidence does not exist to prove that would be the case.

As mentioned in this chapter, the presence of steam in the reservoir gives an opportunity for aquathermolysis if the proper conditions are met. Hence, nonconventional catalyst formulations have been proposed for promoting aquathermolysis. For instance, at laboratory scale, an important upgrading (API gravity improvement of up to 6°) was obtained by aquathermolysis of Midway Sunset crude oil and well-head oil, at 425°C and in the presence of Co(II) 2-ethylhexanoate or Fe(III) 2-ethylhexanoate [614,615]. In this work, the action of Co- or Fe- was not explained as catalytic, though Co was reported to have a better effect than Fe. Catalytic aquathermolysis was studied by various authors [616–619] and by Fan et al. [620–624] on heavy crude oils from the Liaohe fields in China.

Prior to studying catalytic effects, blank experiments were carried out to understand the aquathermolysis results. Although steam is known to have an effect on reducing viscosity, the attained reduction is not permanent and increases back to nearly the original value [621], which was attributed to observed increases in the MW of the asphaltenes and resins in the crude oil; see Yi et al. [625]. The more permanent effect on viscosity reduction has been commonly explained as due to hydrogenation and hydrogenolysis reactions, which involve H-transfer between an H-donor molecule and an H-receptor molecule. Confirmation of this hypothesis was given by the fact that the viscosity reduction effect of steam under aquathermolysis conditions and in the presence of hydrogen donor compounds was found to be larger and permanent. Studies on the effect of the concentration of the hydrogen donor indicated that 0.8% tetralin was found to reduce viscosity in 80% using 20% water, at 240°C [624].

For a downhole upgrading operation, the key problem was identified as being the selection of a more effective, commercially available, and economically feasible hydrogen donor [619]. A first catalytic material to be considered comes from the minerals present in the formation. Laboratory-scale tests indicated that under the conditions studied and for the particular reacting system considered, the minerals not only showed catalytic properties [622] but also synergized with the catalyst employed [623]. This catalyst, $VOSO_4-NiSO_4-FeSO_4$, tested in the laboratory was capable of reducing viscosity up to 42% and produced relevant data that could be used for a field test [621]. The results of the field test [620] were explained assuming that $VOSO_4$ and $NiSO_4$ were active for hydrogenation and hydrogenolysis, while $FeSO_4$ was responsible for WGS reactions that provided the reacting hydrogen. The combined action of the formation minerals and of the catalyst was explained through the main chemical reactions occurring [620,621]:

$$\text{Reforming reaction: } CH_4 + H_2O \rightarrow H_2 + CO + CO_2 \text{ catalyst} \tag{6.x}$$

$$\text{WGS reaction: } CO + H_2O \rightarrow H_2 + CO_2 \text{ catalyst} \tag{6.xi}$$

$$\text{Heavy oil} \rightarrow \text{(lighter, hydrogenated) hydrocarbons catalyst} \tag{6.xii}$$

$$H_2S + CO \rightarrow COS + H_2 \text{ minerals} \tag{6.xiii}$$

$$COS + H_2O \rightarrow CO_2 + H_2S \text{ minerals} \tag{6.xiv}$$

$$\begin{gathered} 6Ca_{0.167}Al_{2.23}Si_{3.67}O_{10}(OH)_2 + 6H_2O + 12OH^- \\ \rightarrow Ca^{2+} + 14Al(OH)_2 + 2H_4SiO_4^- \text{Minerals} \end{gathered} \tag{6.xv}$$

$$KAlSi_3O_8 + 8H_2O \rightarrow K^+ + Al(OH)_4^- + 3H_4SiO_4^- \tag{6.xvi}$$

Both, electromagnetic (microwaves) heating [626,627] and low-frequency vibration [617,618] were found to enhance the catalytic effect. The low-frequency vibration-assisted catalytic aquathermolysis was observed to decrease the average MW of heavy oil, increase the saturate and aromatic contents, decrease the resin and asphaltene contents, and decrease the heteroatoms content of heavy oil. In this case, the role of the electromagnetic vibration was seen as keeping the catalyst dispersed and favoring oil–catalyst contact [617,618]. Greff et al. explained the role of the electromagnetic heating as the creation of hydrogen-donating species in situ, promoted by the presence of a Ni nanocatalyst [626,627].

Under catalytic aquathermolysis conditions, a nano-keggin-$K_3PMo_{12}O_{40}$ mainly affected O-containing functional groups in the heavy oils, though details were not given on the functional groups bearing the O [616]. It would be interesting to know whether the catalyst has the same effect on carboxylic groups and could be used for TAN reduction.

Methane as a source of hydrogen was examined using an iron carbonyl catalyst. A viscosity reduction of two orders of magnitude (from 500 to 1.3 Pa s), DS of 14%, and 41% conversion of VR was observed upon the treatment of Hamaca crude oil, with 250 ppm Fe, at 410–420°C and 110 bar [494]. A molybdenum acetylacetonate catalyst gave slightly better results in a comparative study [483].

A mixed oxide catalyst has been disclosed to promote the decomposition and cracking of hydrocarbons, and its use was suggested for enhanced recovery, in situ DS, and hydrocarbons removal from contaminated soils. The catalyst is composed of 15–35% SiO_2, 1–6% Al_2O_3, 5–20% Fe_2O_3, 10–30% CaO, 2–4% TiO_2 or B_2O_3, and 8–12% transition metal salts. The main feature of the catalyst is having a ratio of particle surface area (in cm^2) to weight (in g) of 3000 or finer. The transition metal salt might be a halide of Fe, Co, or Cu. The SiO_2, Al_2O_3, Fe_2O_3, and CaO components can be substituted by a mixture of 30–50% Portland cement and 30–50% volcanic ash. The catalyst can be incorporated into a water or steam phase in the corresponding recovery processes [628].

A study of the kinetics and the thermodynamic equilibrium of asphaltene sorption onto nanoparticles of nickel and/or palladium oxides supported on fumed silica was carried out using asphaltenes from a Colombian crude oil. After adsorption, asphaltenes were thermally cracked and catalytically oxidized. The presence of the metal oxides increased asphaltenes adsorption compared to the bare fumed silica support. A pseudo-second-order kinetic model fits the obtained data. The nanoparticles promoted decomposition by decreasing activation energy and reaction temperature [629]. Similarly, nanoparticles of Fe, Co, and Ni oxides were tested for adsorption followed by either cracking or steam gasification, and this was effective for conversion of asphaltenes under conditions that could be applied for ISU [630]. A study of the adsorption followed by oxidation of asphaltenes from Athabasca VR was carried out using Fe_4O_3 and Al_2O_3 nanoparticles. Although lighter asphaltene molecules adsorbed on Fe_4O_3 faster and with a greater loading than the heavier ones, catalytic oxidation proceeded equally among all asphaltenes regardless of their MW [68]. Particle size showed to be more important than any other textural property of alumina, proving than microsized alumina performed better than nanosized alumina [631]. In addition to iron oxide and alumina, Co_3O_4, TiO_2, MgO, CaO, and NiO were also studied for the same application. Adsorption capacities for asphaltenes were found to follow the order of: CaO > Co_3O_4 > Fe_3O_4 > MgO > NiO > TiO_2. The presence of the nanoparticles caused a decrease of about 140°C in the reaction temperature of oxidation [632]. A controlled oxidation of asphaltenes has been patented without limitation to the extent of the reaction, pointing out only that the level of oxidation is adapted to a desired application or use [633].

In situ adsorption systems have been recommended for coping with several of the issues found with heavy oils, bitumen, and tar sands exploitation. One of the cases is the emission of major gaseous pollutants, such as H_2S, for which in situ capture would contribute toward an environmentally neutral final product. Ultradispersed colloidal particles of NiO and iron oxide and hydroxide have shown to effectively remove H_2S from the oil phase [634–636]. Aqua-thermolysis reactions taking place during SAGD for heavy oil recovery might produce gaseous H_2S. Sequestration, scavenging, capturing, or converting it into a chemically inactive species is deemed necessary for sustainable recovery and upgrading. The uptake rate and H_2S capacity of colloidal FeOOH particles in heavy oil matrices increased as the concentration of FeOOH increased [635]. In situ prepared colloidal particles showed better performance than commercial α-Fe_2O_3 nanoparticles. In situ prepared ZnO colloidal particles completely removed H_2S, even at high temperatures [637]. In situ-prepared colloidal metal oxides in an oil sand matrix under recovery conditions (e.g., ZnO, CuO, NiO, and Al_2O_3) have been tested. All the considered metal oxides reacted stoichiometrically with $H_2S_{(g)}$ at the selected temperature and pressure, except Al_2O_3. However, there was an advantageous increase in reactivity by preparing in situ ultradispersed particles of the oxide, as demonstrated for the case of ZnO [638]. The concomitant removal of asphaltenes contributes to a simultaneous improvement of the oil quality [634,639–641]. Surface coverage calculations revealed multilayer adsorption when model solutions were adsorbed onto nanoparticles [639]. Core materials were also proven to have adsorption capacity, not only for asphaltenes but also for cracked products [67]. It would be expected that core materials would saturate on these molecules during thermal treatments, ISC, and/or ISU.

Strong hydrogen donors may enable hydrogenation and hydrogenolysis reactions. It is believed that microwave radiation may create hydrogen-donating species from molecules present in the oil. Tests for this idea were performed using a 2.45 GHz emitter in the presence of a nanosized nickel catalyst. Increased cracking and vaporization were observed upon the combined actions of the microwaves and the catalyst [626,627]. Microwaves have been used as an energy source for upgrading of opportunity crudes, and specific applications have been discussed in this chapter. Microwave-induced in situ/ downhole upgrading is probably one application that is difficult to realize. The possibility of injecting microwave-absorbing materials into the reservoir probably would be an easy part of the technology. However, locating these absorbers in close proximity to the oil or at best in direct contact with it would be the challenge. A review article discusses the numerous applications that have been suggested, mentioning also that none have been reduced to practice and are far from being considered for an implementation [642].

An experimental study aiming for the effective formation of emulsions at reservoir conditions and the modeling of the whole process was undertaken as a necessary step for the understanding of ISU processes. It is known that heavy compounds, such as asphaltenes, form rigid film structures at the water—oil interface, limiting coalescence phenomena and separation efficiency either by gravity or by conventional electro-coalescence. A laboratory procedure based on electrical stability was defined, and the amount of demulsifier required for effective water separation was determined. Additionally, coalescence efficiency was measured by following the growth of dispersed water droplets inside the emulsion [643].

Upgrading of emulsions by converting or reacting the water molecules is thought to be a means for emulsion breaking in a single step. In this case, the WGS reaction would break apart the water molecule in the emulsion and generate hydrogen for ISU. A test was carried out to verify whether the generated hydrogen was capable of catalytically desulfurizing emulsified oil using dispersed catalysts. A proof of the validity of the approach was taken from the effectiveness of syngas in the process [644].

An ISU process for the extraction of bitumen from tar sands uses sodium silicate with the aid of sonication at low temperature and ambient atmosphere. Surfactants were formed in situ during the processing. The upgraded bitumen was virtually free of asphaltenes and preasphaltenes (thermal asphaltenes). Kinetic studies revealed that surfactants generation is autocatalytic. The reaction rate (sonication time requirement) can be reduced from hours to minutes by the addition of a free radical initiator (e.g., H_2O_2) [645].

Oil sands and derived fractions are known to be thermally unstable and sensitive to air; reaction with O_2 starts even at low temperatures. An ISU via two-stage low-temperature oxidation has been proposed. Oxidation at low temperature, in the first step, will create labile C—O bonds that will break via free radicals. In the second step, at higher temperatures, bond cleavage will be promoted by these free radicals. In practice, for an "in the field" application, this process would be performed by first injecting air into a formation at low temperature, then starting a steam-soak or steam flood. Experimentally, lab-scale runs examined the effects of oxygen partial pressure, temperature, reaction time, and the presence of rock and brine. Viscosity reduction was observed when using lower O_2 partial pressures, higher second-stage temperatures, and longer run times [646]. Although viscosity reduction was occasionally observed, the expectation that free radicals created in a first step would promote bond breaking in the second might be wrong. Free radicals may well promote bond-making reactions such as addition, oligomerization, and polymerization. The effect of the introduction of oxygen functions on the properties of the O-containing bitumen has been reviewed by Speight [647].

In situ oxidation and sulfonation are other reactions that were proposed in the early stage of bitumen discovery, as a way for exploitation. The reactions were also proposed to be followed by aqueous leaching. Oxidation conditions and type of oxidant (air, oxygen, and ozone) were investigated, and extensive experimental data have been accumulated. These reactions create derivative compounds with greater surface tension-reducing powers and superior ability for bitumen extraction [648]. Various methods and conditions for sulfonation into water-soluble bituminous derivatives were also studied using alkali solutions of sulfites and/or bisulfites, under oxidation conditions. The emulsifying and dispersing powers of the water-soluble sulfonated compounds were suggested for enhanced recovery of tar sands [649]. The sulfonation of Athabasca bitumen with SO_2 did not proceed to the emulsifiers derivatives, to a high extent. Furthermore, bitumen oxidation with O_2 and with SO_2 may render esters that have to be hydrolyzed to become emulsifiers [650]. A deeper study of these reactions not only would provide evidence for their real potential, if any, but also might help with the understanding of other ISU methods that flood the well with oxidizing agents. Under these oxidizing conditions, either emulsifying derivatives or SO_2 might be formed, changing the chemical environment, promoting other reactions, and/or inhibiting the desired ones. The possibility of the aerobic oxidation of asphaltenes might provide a pathway for bio-ISU. In the 1970s, it was studied for weathering and storage considerations, but its effects on bitumen quality might serve as means for in situ recovery processes [651].

A combination of process units has been proposed for the integrated production and upgrading of heavy oils. The thermally produced heavy oil is combined with a hydrogen donor diluent, and the mixture is subjected to TC. The cracked products are fractionated into four fractions: (1) light end vapor, (2) intermediate liquid, (2) gas oil, and (4) pitch. The gas oil fraction is hydrogenated with a hydrogen-containing gas stream to produce the hydrogen donor. The pitch is subjected to partial oxidation to produce the hydrogen-containing gas stream and a byproduct gas stream containing steam. The steam is injected into the reservoir to enhance the oil recovery [652].

6.4 Biocatalytic Alternatives

Biological treatment (BDT) of crude oil and intermediate refining steams represents an attractive processing alternative that operates at mild conditions, with negligible energy requirements. However, catalyst and process development have faced enormous challenges. In the case of catalysts, the dilemma started with the decision of using living cells or molecular biocatalysts. The latter would have facilitated the introduction of the idea into the refining world, which will not be pleased having to handle bacteria cultures, in chemical reactors. The refiner already faces the problems caused by microbial-induced corrosion (MIC) and associates the presence of any microorganism with that phenomenon.

Furthermore, the pathology and toxicity associated with many of the considered bacterial strains do not help either.

Enzymes are considered the most active and most selective catalysts one can find, exhibiting very high turnover frequencies at very low severity (i.e., RT and atmospheric pressure). These features, for other sectors (food, pharma, etc.), are very precious, since they are probably capable of converting a specific compound or a particular functional group into a precise product. Hence, a single enzyme could not effectively work on the thousands of compounds present in crude oil. Thus, a multifunctional or multienzymatic catalyst would be needed, which made this approach economically prohibitive. Consequently, living cells as biocatalysts were left as the economically feasible option.

A living microorganism offers the advantage of being self-regenerative, since they probably will reproduce faster than they would die. However, they live in an aqueous phase that is immiscible or partially miscible with crude oil. Furthermore, the actual catalyst resides inside the cell, and reactants would need to diffuse through the cell membrane and look for it. Alternatively (or luckily), the desired active enzyme resides in the cell membrane, making diffusion needs not so severe. These issues add up to complicate process development. The process concept has to address mass transport between two liquid phases and between the gas and liquid phases, design for continuous operation and high throughput, minimizing CapEx and OpEx, product separation, and biocatalyst recovery. Initially, impeller-based stirred reactors were thought as a mixing device, but mechanical stirring of a refinery-size bioreactor made the process more energy intense. An electrically driven emulsion phase contactor has been suggested for efficiently contacting immiscible phases, in the absence of surfactants [653]. Oak Ridge National Laboratory developed one of these reactor types for dispersing the biocatalyst aqueous phase into the hydrocarbonaceous phase [654].

Biocatalytic conversion of crude oils was a popular study topic in the 1990s, lasting for almost 30 years. By 2004, a book was published collecting research results [655]; in 2007, another book reported the science and technology advances [656]. During that period, particular attention was given to biodesulfurization (BDS), though biodenitrogenation (BDN), BDM, and bioconversion (BDC) were also considered in lesser proportions.

The fact that certain microorganisms coexist with the oil in the reservoir and catalyze or metabolize it to yield biodegradation products constitutes clear evidence of bioconversion [386−388,391,393,395,657−669]. Microorganism activity has been the basis for the development of enhanced recovery technologies (microbial enhanced oil recovery, or MEOR) specifically designed for heavy oils and bitumen (see, e.g., Refs. [670−672]). The search for microorganisms with appealing catalytic activity identified potential possibilities for BDS, BDN, BDM, and BDC, as will be discussed in this section.

BDS was the most active R&D area, and the learnings and findings have been reviewed [225,673–684]. Some other technologies have been incorporated in a complementary way, such as sonochemistry [685,686], emulsification [687], supporting/immobilization [688], and electrokinetics [689].

A decline in BDS R&D activities was the consequence of two concurring factors: (1) the high level of DS required for the production of ultra-low-sulfur fuels (ULSFs), and (2) the production of sulfoxides compounds rather than S-free compounds. Recent R&D activities have been trying to address these issues and mostly searching for improvements in effectiveness. However, we have already pointed out the relevance of oxidation–sulfonation reactions for ISU, and probably the learnings from BDS could be incorporated in the design and definition of new technologies. Since the desired level of upgrading at the field application is the reduction in viscosity, these reactions could provide derived surfactants and render a lower viscosity and easier to recover emulsified oil.

BDN and BDM have been investigated to a lesser extent. Most of the studies on conversion of N-containing compounds concerned degradation, and very little is known about the actual removal of nitrogen. Metals removal has been evaluated using whole cells as well as molecular biocatalysts. This work was considered above. Besides, the changes in metal-bearing compounds have been considered in connection with asphaltene conversion, as well.

These three reactions BDS, BDN, and BDM, together with asphaltene conversion, have been discussed in upgrading and BDC works [397,670,690–705]. Microbial conversion of asphaltenes [706,707] represents a potential solution for reducing viscosity in a downhole method. The activity of (chloro)peroxidase [708–711] and of Cytochrome c [231,233,234,712] have been proven to provide significant levels of conversion.

7. Radiation-Based Methods

Radiation in different forms (neutrons, electrons, X-rays, γ-rays, etc.) can be delivered to petroleum crude oil with energy that is many orders of magnitude in excess of that required to break large hydrocarbon molecules. Studies on radiation-induced cracking (RTC) of hydrocarbons were carried out in the late 1950s and early 1960s. Reviews on the radiation chemistry of organic molecules [713] and the effect of high temperature (carried out at Exxon laboratories) [714] were published in 1956 and 1958, respectively. The investigations on radiation chemistry on different materials carried out in Kazakhstan found that γ- and β-irradiations resulted in increasing adsorbents capacity, improving weak links between neighboring grains in HTSC ceramic, facilitating technological processes on raw material utilization, oil refining, sintering of ceramic materials, decreasing the electric resistance temperature coefficient of nichromium, and synthesizing new types of water-soluble and water-swelling polymers [715].

7.1 Description and Terminology

The typical terminology used in radiation chemistry, or radiochemistry, is rather specific. In order to facilitate the understanding of this section for a reader not familiarized with radiochemistry, the definition of some common terms follows. The energy lost by a particle per unit length of its path has been termed its *linear energy transfer* and should not be confused with the term *energy transfer*, which is applied to the transfer of ionization or excitation energy from one molecule to another, or from one part of a molecule to another part of the same molecule. Two types of intermolecular energy transfer have been proposed: charge transfer and excitation transfer. The former must involve transfer of an electron, and this may occur either by a resonance process between two molecules of the same kind or by a nonresonant process between two molecules of a different kind, with or without chemical decomposition. The mechanism of excitation transfer is not clear to date. It has been shown that energy absorbed by a major component may be transported with high efficiency to a minor component over distances on the order of 50 Å or more, before finally being liberated as the fluorescence radiation of the minor component. New concepts were defined to explain this transfer: sensitized fluorescence, exciton transfer, photon transport, and photon cascade. It is accepted that linear energy losses for excitation energy transfer are proportional to substance density (ρ), then volume energy losses should be proportional to $\rho^{3/2}$.

The amount of energy absorbed, also known as the *dose*, is measured in units of Grays (Gy), where 1 Gy is equal to 1 J per kilogram, or megarads (MR or Mrad), where 1 MR is equal to 1,000,000 ergs per gram. The beam energy required depends directly on the application being used. Low-energy accelerators range from 150 keV to 2.0 MeV. Medium-energy accelerators have energies between 2.5 and 8.0 MeV. High-energy accelerators have beam energies above 9.0 MeV. Accelerator power is a product of electron energy and beam current. Available beam powers range from 5 kW to about 300 kW. This definition of (absorbed) dose can be expressed by Eq. (6.1):

$$D_a = \frac{P \times t}{M} F_p \tag{6.1}$$

where D_a is the average dose in kilograys (kGy), P is the emitted power of the radiation source in kilowatts (kW), t is the treatment time in seconds (s), and M is the mass of the material in kilograms (kg). F_p is a dimensionless factor that accounts for the fraction of emitted power absorbed by the material in a typical industrial irradiation process. F_p may be in the range from 0.25 to 0.50. This power fraction depends on the size and shape of the object and on the way it is oriented in the radiation field. The rearrangement of Eq. (6.1) gives the mass processing rate shown in Eq. (6.2):

$$\frac{M}{t} = \frac{PF_p}{D_a} \tag{6.2}$$

This relationship has been called a *unity rule* because the processing rate is 1 kg/s with an emitted power of 1 kW, an average dose of 1 kGy, and a power absorption fraction of 1.0. Although these values do not correspond to most industrial processes, the unity rule is easy to remember, and it provides a basis for quickly estimating actual processing rates by scaling with the appropriate values of F_p, P, and D_a.

The absorption of ionizing energy in any material causes a temperature rise according to Eq. (6.3):

$$\Delta T = \frac{D_a}{c} \tag{6.3}$$

where ΔT is the temperature rise in °C, and c is the thermal capacity of the material in joules per gram per °C.

Yields are usually expressed in terms of the energy yield, G. This is the number of molecules produced in a given reaction for each 100 eV of energy absorbed. The material produced is written in brackets, and the radiation is written as a subscript. Thus, $G_\alpha(\mathrm{A}) = 2$ means that, in a certain reaction induced by α-particle irradiation, two molecules of A are produced for each 100 eV of energy absorbed. *Indirect action* arises from the nonspecificity of absorption of the energy of ionizing radiations, from which it follows that if material A is present in very much greater quantity than material B, then A will absorb most of the radiation energy, and chemical effects on B are more likely to arise via the primary entities formed from A than from a direct effect of the radiation on B.

7.2 Radiation-Induced Chemistry

Reactions induced by plasma, X- and γ-rays, electrons produced in machines, β-particles from nuclear disintegration, protons, deuterons, α-particles, and neutrons, giving rise to extranuclear changes, were initially believed to occur via ionic mechanisms, due to the ionizing nature of the radiation used. All ionizing radiations ultimately transfer energy to an irradiated system via particles. In the case of X-rays or γ-rays, the effective particles are high-energy electrons ejected by the interaction of photons with atoms. After the formation of the initiating molecular pairs, a variety of secondary processes may occur that contrast from gaseous systems to condensed matter. The increase in collision rate will favor breakdown to molecular products rather than to radicals: the former by increasing the probability of energy removal before decomposition, and the latter by tending to cause recombination of radicals once formed [713].

In contrast to other examples in radiation chemistry, there seems to be little or no difference between the effects produced on hydrocarbons by the various kinds of radiation. A direct comparison has been made for cyclohexane: 2 meV electrons, 14 meV deuterons, and 35 meV α-particles were all shown to produce the same energy yield as hydrogen.

Both C—C and C—H bonds can be broken when hydrocarbons are irradiated, and the highly reactive fragments formed are capable of a variety of reactions, including reaction with each other. The sensitivity of unsaturated hydrocarbons toward radiation decreases with unsaturation, that is, (1) multiple bonds are not broken, and they inhibit the rupture of neighboring single bonds; (2) triple bonds are more effective in this respect than double bonds, and two double bonds are more effective than one; (3) the effect of a multiple bond is least marked when at the end of a chain; and (4) the *cis* form of an ethylenic compound is broken down more readily than the *trans* form. Hence, aromatic compounds are highly radiation resistant and very strong absorbers for the excess energy of the radiation-generated radicals. Furthermore, the comparison of unsaturated compounds with their corresponding saturated hydrocarbon showed that, for the former, radiation yields of gaseous products were lower, while the total yields and the yields of polymeric products were greater. Polymerization yields are higher for molecules that have a multiple bond at the end of the chain than for those that have that type of bond in other positions. In other words, the principal reaction of unsaturated hydrocarbons on irradiation is polymerization. As in the case of unsaturated aliphatic hydrocarbons, the predominant effect of the irradiation of aromatic hydrocarbons is the production of polymers [713].

It was found early that product distributions of RTC of hydrocarbons are similar to those obtained with TC. The radiation yields (10^4—10^5) at 320—510°C increase with increasing temperature. Yields are lower in the liquid than the vapor phase, and increase at lower intensity. Higher cracking conversions were consistently obtained with radiation than with thermal blanks, for both pure and mixed feeds, at different levels of conversion, and at different temperatures. RTC results were qualitatively explained by assuming a free-radical mechanism. A Rice—Herzfeld radical mechanism was proposed, confirming that atmospheric pressure RTC is an ordinary radical chain reaction. The mechanism of radiation decomposition of hydrocarbons is believed to be a chain process of low chain length [714].

The conventional TC of hydrocarbons is usually the result of C—C bond cleavage to give two radicals; similarly, RTC would follow a free-radical mechanism (Rxn. 6.xvii):

$$RC_nH_{2n}R' \rightarrow [RH_{2m}C_m]^\bullet + [C_{n-m}H_{2(n-m)}R']^\bullet \qquad (6.xvii)$$

followed by radical reactions such as the hydrogen abstraction and β-cleavage:

$$[C_mH_{2m}R]^\bullet + RC_nH_{2n}R \rightarrow RC_mH_{2m+1} + [R_iC_nH_{2n-1}R']^\bullet \qquad (6.xviii)$$

$$[RC_nH_{2n-1}R']^\bullet \rightarrow [RC_kH_{2k}]^\bullet + R'C_{n-k}H_{2(n-k)-1} \qquad (6.xix)$$

Reactions, Rxns. (6.xviii) and (6.xix) represent possible propagation steps of a chain process in which small organic radicals abstract hydrogen from heavy hydrocarbon, and large radicals dissociate into an alkene and a smaller radical [716]. The chain process is not observed at temperatures lower than ~300°C, but thermal initiation (similar to

Rxn. 6.xvii) becomes significant only at temperatures above ~500°C. The termination of the chain process takes place by the dimerization or disproportionation of radicals (Rxns. 6.xx and 6.xxi).

$$[C_kH_{2k}R]^\bullet + [C_bH_{2b}R']^\bullet \rightarrow RH_{2k}C_k - C_bH_{2b}R' \quad (6.xx)$$

$$\rightarrow RH_{2k+1}C_k + C_bH_{2b-1}R' \quad (6.xxi)$$

Early works reported that at low temperatures, under conditions where nuclear radiation is the driving force for the cracking of hydrocarbons, low radiation yields are obtained and the cracked product is rich in hydrogen [714]. It means that the olefins formed could be the result of both dehydrogenation and cracking. The radical yields are extremely high and increase with increasing temperature, and there is a tendency for yields to decrease with increasing intensity of radiation, and to be lower in the condensed phase.

Two reactions are important in the radiation chemistry of pure carboxylic acids: one is decarboxylation (Rxn. 6.xxii) to yield hydrocarbons; the other is dehydrogenation (Rxn. 6.xxiii) to give an unsaturated acid.

$$CH_3(CH_2)_xCH_2COOH \rightarrow CH_3(CH_2)CH_3 + CO_2 \quad (6.xxii)$$

$$CH_3(CH_2)_XCH_2(CH_2)_YCOOH \rightarrow CH_3(CH_2)_XCH{=}CH(CH_2)_YCOOH + H_2 \quad (6.xxiii)$$

The effect of chain length on the efficiency of the two processes is not a simple one and is poorly understood. The reactivity toward decarboxylation decreases down to C_{22} and rises again at C_{30}. Dehydrogenation reaches a maximum at C_{16} and a minimum near C_{24} [717].

Thiols are readily affected by radiation due to the ease with which they are oxidized. A higher oxidation yield is observed with oxygen present and with an increase in pH. Reduction of disulfides to thiols is not generally observed, but it has been claimed to occur with molecular hydrogen present [713].

The advantage of RTC lies in generating the chain-propagating radicals at a lower temperature than that required for TC. Another difference between heating and radiation processing (RP) is that radiation-induced energy has the capability of breaking any bonds, while in the heating case primarily "weak" bonds are broken. High-energy electron treatments allow one to create conditions for the desirable order of bond breaking and, thus, for potential designing of the output.

7.3 Electron Beam Source

In general, the components of an electron beam system are vacuum pumping equipment, large high-voltage power supply and complex shielding, as well as the requirement for in-plant engineering and maintenance expertise. There are more than 1500 high-current electron beam accelerators being used industrially throughout the world.

An electron beam is produced while the electron accelerator is on, requiring adequate shielding and proper procedures to protect workers in the facility and the environment. The machine can be turned off for maintenance and when not in use. The power requirements for an electron accelerator are directly related to dose and hourly throughput. However, as the processing capacity of the accelerator rises, its cost increases in a lower proportion than the power increases, making the economy of scale more attractive.

Electron accelerators are generally described by how deeply they can penetrate a material and how much material at what dose they can treat during a set period of time. Penetrating ability is determined by the energy level of the electrons or voltage of the accelerator and expressed in million electron volts (MeV). The useful electron energy range for material processing is 0.1–10 MeV. At energy levels below 0.1 MeV, electrons will not penetrate solid materials. A useful length of penetration of electrons in water [718] has been reported and can be seen in Table 6.5.

Accelerators can generally be classified according to exactly how they generate accelerated electrons. The four main types of accelerators are: electrostatic direct-current (DC), electrodynamics DC, RF linear accelerators (LINACs), magnetic-induction LINACs, and continuous-wave (CW) machines. In general, DC accelerators are characterized by high power output and high efficiency, while LINAC systems are typically much more compact and can generate higher beam energies. However, they are also considerably less efficient. Similarly, CW machines can be fairly compact and can achieve high beam energies. Currently, industrial electron accelerators are available commercially in various ranges of energy–power combinations, namely with low to medium energy levels (less than 5 MeV) and relatively high beam powers (150–500 kW), or with higher energy levels (7–10 MeV) and relatively low beam powers (10–20 kW). These available energy–power combinations are shown in Fig. 6.19. Economics and other practical considerations have limited the voltage rating for large DC industrial accelerators to no more than 5 MeV.

Table 6.5: Useful Penetration Depth of Electrons of Various Energies (MeV) in Water (in)

Energy	Treatment on One Side	Treatment on Both Sides
1	0.11	0.28
2	0.22	0.57
3	0.35	0.83
4	0.48	1.17
5	0.61	1.51
6	0.76	1.84
7	0.88	2.17
8	1.01	2.54
9	1.14	2.84
10	1.26	3.17

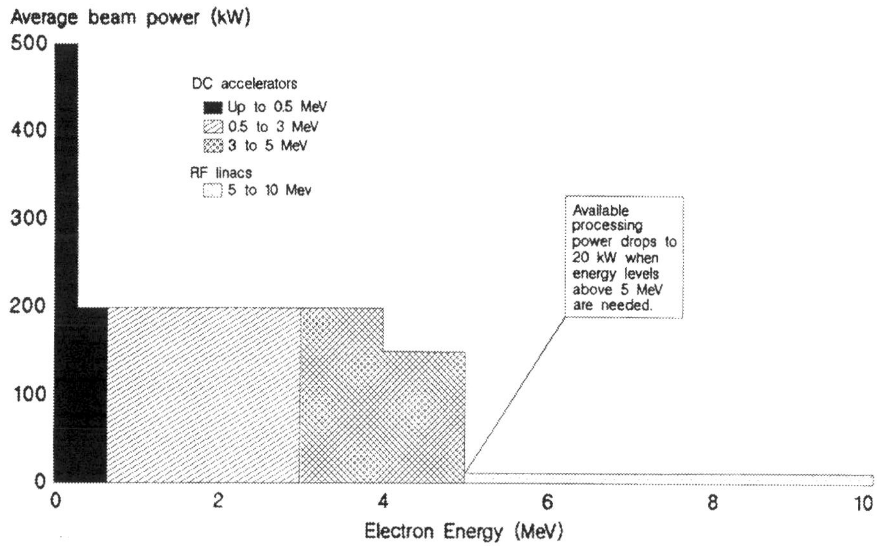

Figure 6.19
Energy-level/beam power combinations for existing commercial electron accelerators.

Worldwide, there are only a few companies that manufacture large electron beam systems, and each of them markets a range of products with various energy delivery levels. For laboratory work and for industrial (low-throughput) treatment processes, compact and moderate-cost low-energy e-beam equipment is commercially available. For larger requirements and minimum footprint, self-shielded equipment dispensing a high-current medium-energy (500 keV) e-beam is available. Currently, industrial applications of electron accelerators include polymerization reactions, curing, sterilization, other medical materials and applications, wires, ceramics, and so on. For R&D activities, an electron–positron accelerator (LINAC) has been normally used. As an example, IBA Industrial, one of the few manufacturers of large electron beam systems, markets the Rhodotron TT100, a potential candidate for application in cold cracking of petroleum. The TT100 can provide about 100 mA of electrons in the 5–7 MeV (penetration) range (hence 500–700 kW). When operating at full beam power (700 kW), the electron beam system will directly consume about 1400 kW, and may require another 25–30% in terms of kW for cooling. Examples of Rhodotron currently offered by IBA and their characteristics are collected in Table 6.6.

A Rhodotron schematic is shown in Fig. 6.20. The electrical field in the single coaxial-shaped cavity of the Rhodotron is radial and oscillates at a frequency of either 107.5 or 215 MHz, depending on the Rhodotron model. In reference to Fig. 6.20, (1) electrons are fired by the heated filament of the electron source located at the outer wall of the cavity. (2) The electrons are introduced into the cavity when the electric field is such that it will accelerate the electrons inward, toward the hollow coaxial cylinder in the center. (3) The

Table 6.6: Commercially Available Options of Rhodotron

	TT100	TT200	TT300	TX400	TX1000
Energy (MeV)	10	10	10	7	5–7.5
Beam power range (kW)	0.5–35	0.5–80	0.5–190	0.5–280	0.5–700
Power consumption at full energy (kW)	210	310	452	596	1270
Number of passes	12	10	10	7	6
Diameter (m)	1.6	3.0	3.0	3.0	3.0
Height (m)	1.75	3.0	3.0	3.0	3.3

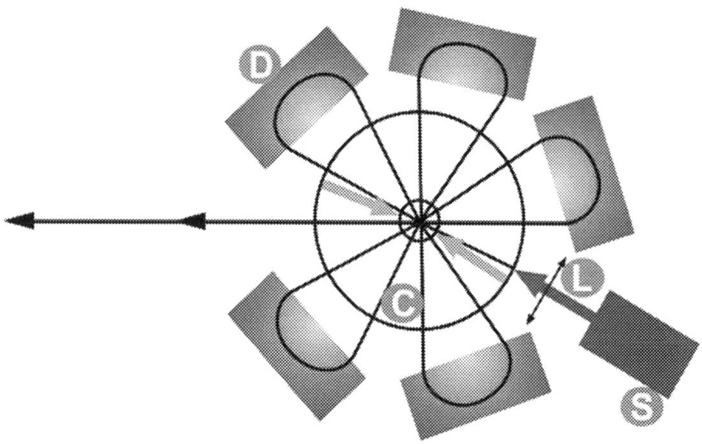

Figure 6.20
Rhodotron schematic. *Adapted from Abs M, Capdevila JM, Delvigne T, Genin F, Jongen Y, Nguyen A. Rhodotron accelerators for industrial electron-beam processing: a progress report. In: Proc. 5th European particle accelerator conf. Sitges, Barcelona, Spain. June 10–14, 1996. 2687–89.*

electrons pass through openings in the inner cylinder while the electric field is reversing. (4) On emerging from the inner cylinder, the electrons are further accelerated (toward the outer cavity wall) under the influence of the new reversed field. (5) Using beam deflection magnets, the electrons are reintroduced into the main body of the accelerator (6) for additional crossings of the cavity in order to reach the required energy level and leave the cavity through a beam line.

7.4 Radiation-Induced Processes

A method based on a combination of microwaves, plasma, hydrogen donors, hydrogen partial pressure, and metals has been proposed for the upgrading of low-value hydrocarbon. In this process, the hydrogen donor and molecular hydrogen are introduced in the reacting vessel that contains a plasma initiator made of tungsten, iron, or their mixtures, or made of a nonmetal other than silica, such as calcium aluminate, carbon, and iron oxide. Microwave radiation at a frequency of at least 0.3 GHzm is applied to the reaction zone once the feed is introduced. Hydrogen is kept at less than a 6:1 $C_{in\text{-}the\text{-}feed}$: hydrogen ratio, during upgrading. A typical range of conditions includes microwave frequency around 0.915, 2.45, or 5.8 GHz, and a pulsed microwave with pulses that last for 1−10 ms and are off during 3 ms up to 1 min [719,720].

Plasma, being a hot gas composed of neutral atoms, ions, and electrons, is very reactive and could have additional effects on the reactants. For instance, hydrogen plasma may have capabilities for hydrogenation and/or hydrogenolysis of hydrocarbons and methane plasma for methylation. For this reason, probably hydrogen plasmas results are more interesting, for the conversion of heavy oil and large hydrocarbons [721,722]. Nevertheless, the high reactivity of plasmas would preclude selectivity, and gas and coke yield might be high.

A huge number of radicals are formed when electrons (from a linear accelerator) collide with organic molecules. Thus, the irradiation of hydrocarbon mixtures or of crude oil changes their composition, which is reflected in gross changes to the sample viscosity, density, evolution of gases, and/or formation of precipitates. Consequently, the yield of individual distillation cuts (gasoline, diesel, petroleum coke, etc.) is affected.

The pioneering methods of radiolysis conversion were covered in a patent awarded to Exxon [723]. The hydrocarbons were cracked by exposure to high-energy ionizing radiations having an energy content equivalent to at least 30 eV, until reaching a total dosage of $2.6 \times 10^{15}-0.2 \times 10^{21}$ eV/g, at incipient cracking temperature in the range of 400−510°C. Process was meant to be applied to shale and tar fuel oils, asphalts, and petroleum. The products comprise a gaseous mixture of hydrogen and hydrocarbons containing 1−3 carbon atoms, together with higher molecular weight hydrocarbons. Gaseous product contains between 1% and 20% hydrogen. The process could take place in the vapor or liquid phase at pressures of 1−16 bar and be conducted continuously with conventional separations of unreacted and reacted products and recycling of the former. Suitable radiations are α-particles, β-rays, accelerated electrons, γ-rays, X-rays, and neutrons from the usual sources. When using a nuclear reactor, the hydrocarbons may serve as a moderator. When employing a γ-ray source such as Co-60, the reactor may comprise a stainless-steel pipe coil cast in an electrically heated aluminum block having a hole in the center for insertion of the source; when using high-velocity electrons, a similar apparatus having a thin metal window for admission of the electrons may be employed. When using nuclear reactors, the apparatus may consist of a 20 ft. long, 1 1/4 in. aluminum pipe of which only the last 30 in. are inserted into the reactor while the remainder is encased in a stainless-steel, insulated, tubular furnace [723].

The radiolysis of lubricating oils and residues was carried out using γ-rays (from Co-60) and 19 MeV electrons at 25−60°C and a dose range of 1−80 megarads by Reyes Lujan [724].

The G values for H and CH_4, and their variation with dose rate, ranged between 4×10^{-4} and 25 Mrads. Polymerization of the paraffinic compounds and inertia of aromatics were observed over a wide range of intensity.

Mustafaev and Guileva converted a heavy petroleum fraction at 430°C. Conversion was about 70%, and the main products were naphtha and diesel. Their economical evaluation resulted in 10% improvement compared to conventional catalytic refining [725].

The refining treatment performed by Lykhterova et al. to heavy oils (heavy oil, bitumen, pitch, and coke), using electron beams, included ozone addition. They reported a sulfur content reduction as due to formation of water-soluble sulfones [726].

More recently, researchers from the Science Research Institute of Experimental and Theoretical Physics in Almaty, Kazakhstan, pursued experimental studies of high-energy electron irradiation treatment of crude oils, fractions, and pure-compound mixtures. Different radiation and thermal processing conditions have been covered: mostly batch experiments, energy of 2–5 MeV, electrons generated by an electron accelerator, current densities from 1 to 6 $\mu A/cm^2$, a total dose of 1–4 kG, rates of exposure of 1–4 kG/s, and temperatures from RT to 450°C. An advantageous presence of ozone was confirmed [727]. Synergetic effects on yields and the hydrocarbon content of oil products were experimentally observed to occur between ionized ozone-containing air and ionizing radiation. Two variants of combined action of ionizing radiation (electron and γ-rays) and ozone-containing air were considered:

1. bubbling pretreatment of feedstock by ionized air at RT and subsequent RTC at higher temperatures; and
2. simultaneous feedstock bubbling by ozone-containing air and γ-irradiation at low temperatures (25–40°C).

Radiation-induced processing (RP) was performed in the mode of "distillation under the beam" using a special facility. Cracking was observed to start in the ranges of 1.2–1.5 kGy/s and 380–400°C. The general effects were a decrease in the required dose, decrease in the temperature of cracking (an average starting temperature of about 40°C), 8% to 10% decrease in the liquid yield, 10% increase gasoline proportion in the overall liquid product, and isomers and aromatics enrichment in the gasoline fraction. It was shown that preliminary bubbling of heavy oil feedstock by ozonized air allows RTC reaction temperature reductions and it improves characteristics of the light fractions. In conditions of continuous feedstock bubbling by ozone-containing air, RTC at RT was characterized by high yields of light fractions. An additional (unproved) statement indicated that radiation ozonolysis reduces concentration of high-molecular aromatic compounds, by disintegrating the condensed polyarene nuclei in molecules of pitches and asphaltenes. A considerable decrease of their average molecular mass, partial elimination of sulfur and nitrogen, and enrichment of the residue by oxygen were supposed to be achieved. At the same time, the degree of hydrogen deficiency in molecules became lower as a result of opening of a part of aromatic rings [727]. Although higher yields of light fractions were obtained, no stability of the ozonized product fractions was provided. No comparison was made with conventional thermal processing of the same crude oil sample.

The ozone-containing air is said to promote conversion of practically all the crude oil components and/or oil products. Electrophilic O_3 addition reactions or ozone-induced radical oxidation by molecular oxygen seems to take place. Sulfide oxidation to sulfoxides at 20°C was noticeable, proceeding according to the scheme shown in Rxn. (6.xxiv).

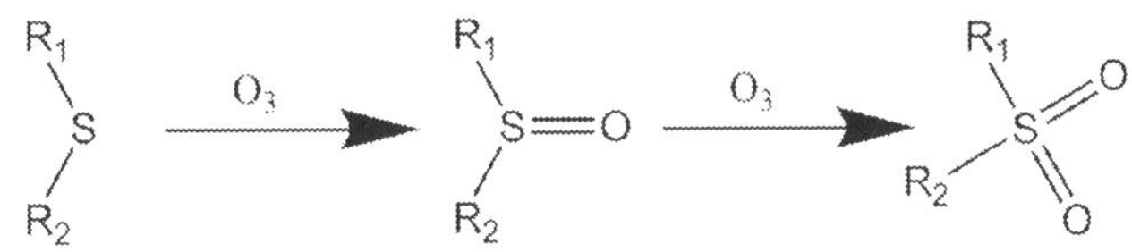

(6.xxiv)

The S-profile in the product showed 60% being present in the gas phase, probably part of the removed sulfur from the light molecular compounds. Although S is believed to be concentrated in the heavier molecules, the profile in Fig. 6.21 seems to indicate the opposite. These findings on ozonolysis were applied in a two-stage method: for oil DS consisting of a first stage in which the S-containing compounds are strongly oxidized; and the actual S-removal takes place at the second stage by means of conventional methods.

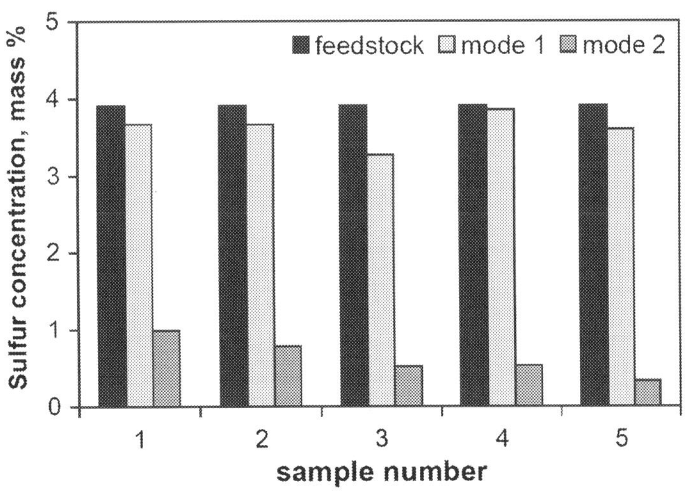

Figure 6.21

Concentration of total sulfur in the liquid product (LP), Mode 1: bubbling by ionized air near the room temperature; Mode 2: RTC at 400°C, LP fractional contents: (1) C_4-C_{21}; (2–3) C_4-C_{30}; (4) C_4-C_{34}; (5) C_4-C_{37}. *Reproduced from Zaykina RF, Zaykin YA, Yagudin SG, Fahruddinov IM. Specific approaches to radiation processing of high-sulfuric oil. Radiat Phys Chem 2004;71(1–2):467–70, with permission from Elsevier.*

Since sulfoxides and sulfones could be washed off from the treated products, that conventional method could easily be a simple prewash of the product, which can be applied even as an intermediate step between RTC stages. In irradiation technologies, an excited inexpensive air mixture is a byproduct of the accelerator action. It comprises monatomic oxygen, excited oxygen molecules, and various oxygen-containing complexes, together with ozone. The ozone-containing air is thought to excite the S-compounds by energetic electron bombardment and is responsible for their oxidation. Two different dosage levels were examined, but no quantitative analyses were provided. The preferred conditions (mild mode) involved $P = 2$ kGy/s, $D = 70$ kGy [728]. The general statements made on utility were not supported, such as the formation of heavier S-compounds yielding a partial desulfurization of the lighter fractions.

A study of the cracking of an AR (370°C$^+$) from Mordovo–Karmalskoe bitumen (Tatarstan, Russia) induced by a high dose e-beam (average dose rate: 8 kGy/s) was used to explain the reduction in sulfur content of the produced distillates. The AR was characterized as having a density of 0.967 g/cm^3, a total sulfur content of 3.7 wt%, and an asphaltenes content of 15.5 wt%. The source of the electron radiation (energy of 8 MeV, pulse duration of 6 μs, pulse generation frequency of 300 Hz, average beam current of ≤800 μA, initial width of scanned beam of 245 mm, and scanning frequency of 1 Hz) was an ELU-6E linear accelerator [716]. The experimental rig is sketched in Fig. 6.22.

The effect of various gases was examined in two different modes: (A) bubbling mode and (B) gas lift mode. The volatile radiolytic products were removed together with the bubbling gas. In agreement with previously discussed work [727,728], the end product of TC contains more low-boiling components than the end product of the RTC. At the same time, TC was three times slower; it yielded the lowest conversion of feedstock to end product (about 37 wt%) and the higher content of unsaturated hydrocarbons (about 16 wt%). The use of a propane–butane mixture produces an end product with a lower average boiling point than the use of helium or methane [716]. The formation of the S-containing free radicals might take place by electron transfer from another free radical (propagation reaction, e.g., Rxn. 6.xxv) and/or might be initiated by electron addition to sulfide compounds (Rxn. 6.xxxi). Regarding S-profile, a termination reaction produces S-containing heavier molecules from free radicals (see Rxn. 6.xxvi).

$$\bullet R + RSH \rightarrow RH + RS\bullet \qquad (6.xxv)$$

$$2RS\bullet \rightarrow RSSR \qquad (6.xxvi)$$

$$RS\bullet + RHC{=}CHR' \rightarrow RS{-}CHR{-}C\bullet HR' \qquad (6.xxvii)$$

$$RS{-}CHR{-}C\bullet HR' + RSH \rightarrow RS{-}CHR{-}CH_2R' + RS\bullet \qquad (6.xxviii)$$

$$e^- + R_2S \rightarrow R_2S^- \qquad (6.xxix)$$

$$R_2S^- + R_2S \rightarrow RSSR^- + R_2 \qquad (6.xxx)$$

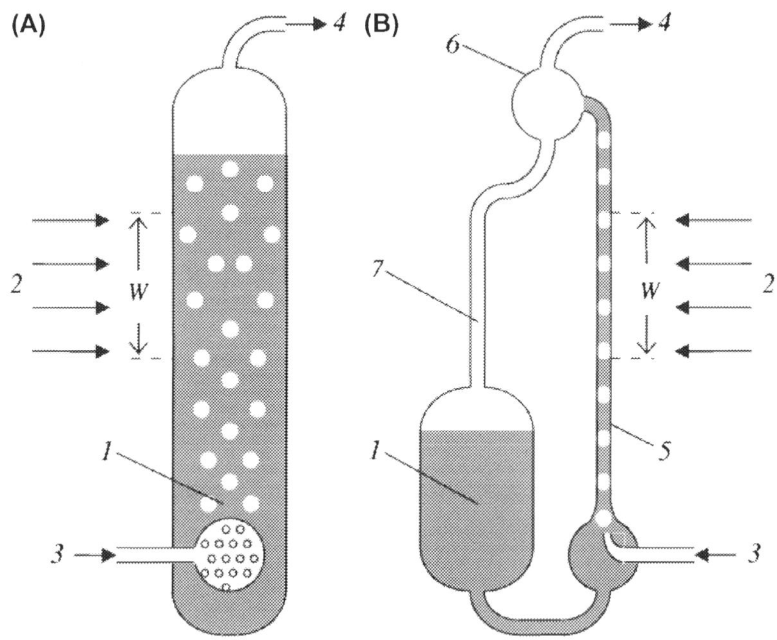

Figure 6.22

Irradiation modes: (A) bubbling mode and (B) gas lift mode; (1) bitumen in a receiver, (2) electron beam, (3) carrier gas, (4) vapor–gas mixture, (5) gas lift tube, (6) separation vessel, and (7) return tube. *Reproduced from Bludenko AV, Ponomarev AV, Chulkov VN, Yakushev IA, Yarullin RS. Electron-beam decomposition of bitumen–gas mixtures at high dose rates. Mendeleev Commun 2007;17:227–29, with permission from Elsevier.*

The formation of hydrogen sulfide is accompanied by the formation of free sulfur. Since sulfur is an excellent radical scavenger, it reduces the total yield of bitumen decomposition (Rxn. 6.xxvi through Rxn. 6.xxxix), and hydrogen sulfide affects it indirectly by its role in the formation of free sulfur.

$$e^- + HSR \rightarrow HS^- + \bullet R \quad (6.xxxi)$$

$$HS^- + H^+ \rightarrow H_2S \quad (6.xxxii)$$

$$\bullet H + HSR \rightarrow H_2 + \bullet SR \quad (6.xxxiii)$$

$$\bullet H + HSR \rightarrow H_2S + \bullet R \quad (6.xxxiv)$$

$$RH^+ + H_2S \rightarrow RH + H_2S^+ \quad (6.xxxv)$$

$$\rightarrow RH + H^+ + \bullet SH \quad (6.xxxvi)$$

$$H + H_2S \rightarrow H_2 + HS\bullet \quad (6.xxxvii)$$

$$H + HS\bullet \rightarrow H_2 + S \quad (6.xxxviii)$$

$$2HS\bullet \rightarrow H_2S + S\bullet \quad (6.xxxix)$$

Irradiation at 350–450°C of a viscous, highly aromatic, low-paraffin (1.5%), high-sulfur (2%), high-metal-content crude (Karazhanbas field, Kazakhstan) with high-energy electrons (2 MeV) confirmed the refractory nature of the aromatic compounds. The high polycyclic aromatic rings proved more radiation resistant than the monocyclic aromatics. As expected, the paraffinic components were converted to produce increases in both light products and branched alkanes (increasing the octane number in the gasoline fraction). Large branches on the aromatic rings tend to de-alkylate to form toluene, xylene, and styrene. The condensation reactions create products with increasing MW and higher aromatization. Reactivity follows the order of alkanes, resins, asphaltenes, and coke [729].

Three different feeds from the Kazakhstan fields were considered for irradiation processing: (1) the fuel oil produced in Kazakhstan by the Atyrau refinery, (2) heavy crude from the Karazhanbas field, and (3) paraffinic oil from the Kumkol field. The oil feedstocks were irradiated by 4 MeV electrons at a temperature of 400°C and dose rate of 1 kGy/s. The effect of increasing dose was also studied. The dependence of the yields on irradiation doses is nonlinear, due to saturation for doses >10 kGy, mainly due to alkylation reactions of the free radicals to the olefinic double bonds. An interesting feature of the radiation-induced conversion was the observation of isomerization reactions during the treatment of Karazhanbas oil under mild conditions, though these were not observed in Kumkol oil processed similarly. This reactivity change was attributed to the effects of energy transfer from paraffin to an aromatic component of the hydrocarbon mixture. Therefore, characteristics of the initial hydrocarbon contents will define the heightened rate of its radiation destruction and the proportion of iso-alkane content in the gasoline fraction of RTC products. Aspects of radiation-thermal processing are shown in the experimental results for different crude oils that are characterized by essentially different proportions of hydrocarbon types [730]. The G-values of 7000 seem high because normal G-values are 1–2 for many chemical reactions. Chain reactions can reach much bigger numbers than one or two, but 7000 seems too high for the molecules present in crude oils. Although comparisons are made with TC, the chemical distillation analysis and mass balances were not rigorous, and therefore the comparison with conventional refining might not be accurate.

Another attempt for desulfurization was followed with crude oil and oil products having a sulfur concentration higher than 3 wt%. The combination of preliminary oil processing by ionized ozone-containing air at room temperature and subsequent high-temperature radiation processing provided desulfurization of light fractions and reduced total S-amount in the liquid product by concentrating it in the high-molecular compounds [731].

Earlier approaches tested oil residua, heavy crude oil, and bitumen. Mortuk and Shilikty feeds (density = 0.998 g/cm^3) were tested to evaluate the efficiency of RTC in breaking heavy molecules of pitches and asphaltenes. The obtained distillates yields under different processing treatments are shown in Fig. 6.23. In this figure, TC is thermal cracking, TCC

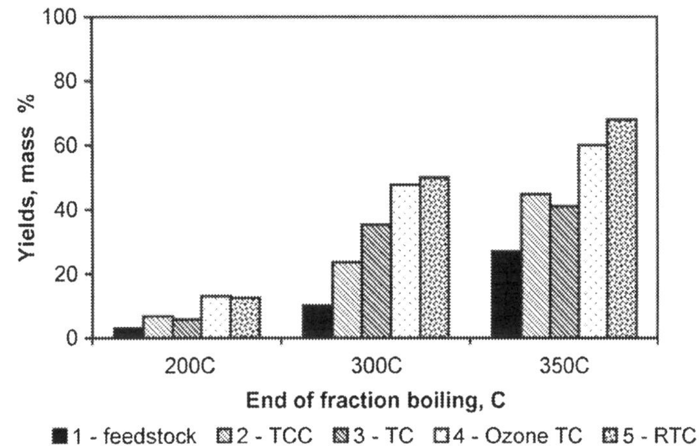

Figure 6.23

Distillate yields from Mortuk bitumen under different treatments. *Reproduced from Zaykin YA, Zaykina RF. Bitumen radiation processing. Radiat Phys Chem 2004;71(1–2):471–74, with permission from Elsevier.*

is thermocatalytic cracking, Ozone TC is TC in the presence of O_3, and RTC is radiation-induced TC. As can be seen, the distillates yields were always better for the RTC treatment [732].

Gasoline analysis confirmed the effects of energy transfer between the paraffins and the aromatics, previously observed with Karazhanbas crude oils [729]. The composition and quality of the gasoline obtained by RTC are compared with that of TCC in Table 6.7. The main noticeable difference of the gasoline produced by RTC is the higher proportion of isoparaffins, which is clearly reflected in the octane number improvements [732].

An empiric equation (Eq. 6.4) predicting an iso/normal paraffin ratio of the gasoline yield was developed as a function of the dose and treatment temperature (T in K); ρ and ρ_o are the density of the feedstock and of the gasoline fraction, respectively. This equation (Eq. 6.4) was derived from considering an energy transfer mechanism for a hypothetical two-component hydrocarbon mixture (an iso- and a normal paraffin) [733]. The good

Table 6.7: Quality Comparison of Gasoline Obtained From TCC and RTC Processing

Field	Type of Processing	Iso- to n-Paraffin Ratio	Aromatics–Naphthenes Ratio	Unsaturated	Octane Number
Mortuk	TCC	0.59	1.69	18	68
	RTC	1.68	1.50	10.4	75
Shilikty	TCC	0.61	1.77	19.3	66
	RTC	1.75	1.19	14.8	72

agreement found between the predicted iso-/n-ratio and the experimental results seems to verify the validity of the energy transfer assumption. Hence, the degree of paraffin isomerization during RTC seems to be determined by the transfer of excess excitation energy to the denser and radiation-resistant medium [732].

$$Y_{iso-}^{normal} = \frac{1040}{TD_a} \frac{[(1 + 190/T)\rho - \rho_o]}{\rho_o} \tag{6.4}$$

The factor (1 + 190/T) was considered as a correction for the difference in thermal expansion of the heavy fraction compared to that of the light fraction; the 1040/D_aT equals 1 for the characteristic irradiation mode (D_a = 1.5 kGy, T = 420°C), which favored paraffin isomerization of the gasoline fraction.

An attempt to compare the application of electron beam irradiation for transformation of heavy and/or viscous hydrocarbons into light and well-flowing substances and enhancement of composition of hydrocarbon mixtures with conventional technologies of TC was published in Refs. [734,735]. The comparison of product yield from TC and RTC was reported in Ref. [735] and is included here in Table 6.8.

This RTC process was named hydrocarbon enhancement electron-beam technology (HEET), and its proposed scheme is shown in Fig. 6.24.

Typical processing conditions for HEET include temperatures of 350–420°C and atmospheric pressure. The preprocessing block is used for feed preheating (400 ± 50°C). Then, a pumping system would deliver the feed into the reaction chamber where it is irradiated by high-energy electrons produced by an electron accelerator. All parameters of the reactor are automatically controlled.

The patent application filed by Petrobeam [736] discloses a method based on the learnings from Refs. [727–735,737,738], for the deep destructive processing of hydrocarbon chains utilizing self-sustaining radiation cracking of hydrocarbon chains under a wide variety of irradiation conditions and temperature ranges (from RT to 400°C). Several modes of

Table 6.8: Refined Fractions From Fuel Oil Products

Boiling Temperature (°C)	Feed Fractions (wt%)	TC (wt%)	HEET (wt%)
<200	None	10	15
200–300	2	15	27
300–360	8	15	18
360–450	38	25	20
>450	52	30	10
Gas	None	5	10

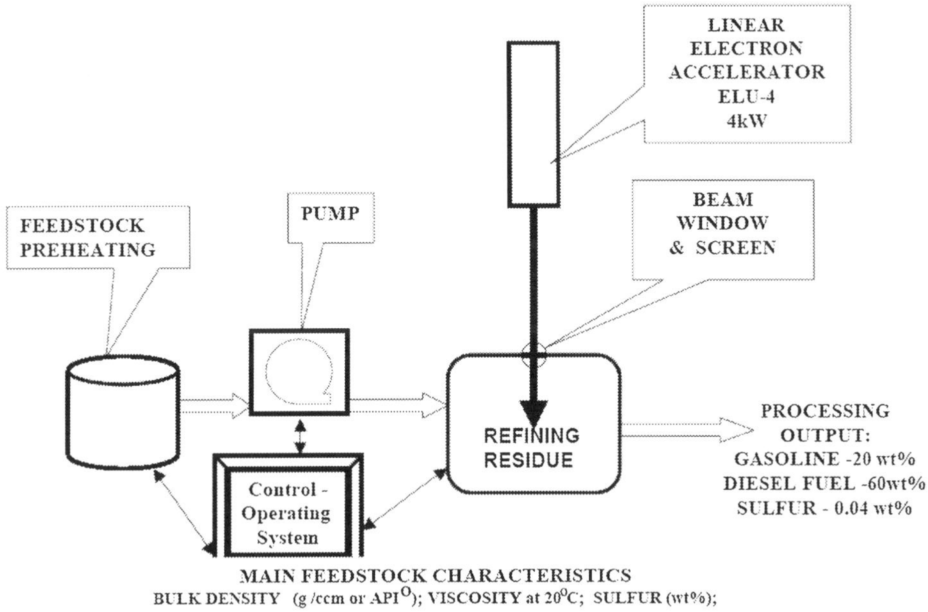

Figure 6.24

Schematic of hydrocarbon enhancement electron-beam technology (HEET) refinery. *Reproduced from Mirkin G, Zaykina RF, Zaykin YA. Radiation methods for upgrading and refining of feedstock for oil chemistry. Radiat Phys Chem 2003;67(3–4):311–14, with permission from Elsevier.*

application were disclosed, including (1) a special case of RTC referred to as high-temperature radiation cracking (HTRC), (2) low-temperature radiation cracking (LTRC), and (3) cold radiation cracking (CRC). As an example, a petroleum feedstock was subjected to irradiation to initiate and/or at least partially propagate a chain reaction between components of the petroleum feedstock.

More recent work focused on trying to understand the low yields obtained. Chemical as well as physical changes of the oil upon radiation affect the subsequent continuation of the process. These changes include:

- The thixotropic properties of heavy oil and bitumen structures and the conversion of oil at low temperature under shear stress lead to viscosity drop and cracking rate increases.
- Dose dependence of conversion differs between the light fraction and the heavy fraction; at high doses, products saturate fast.
- Cracking and polymerization are competing reactions.
  - Dose rate increases at lower dose values promote light fraction yields; cracking products from heavy factions are unstable, and their polymerization may continue after irradiation.
  - Excess irradiation increases polymerization more than it does on cracking.

- Under storage, RTC products aged to more radiation-refractory compounds (via polymerization); dependence on dose rate becomes more uncertain for aged RP products.

A model considering all these findings was proposed and was proved to provide satisfactory predictions for the low-temperature RTC process [739]. This model predicted an initial cracking rate of 0.113 s^{-1}, at 120°C and under a dose rate of 80 kGy/s. This model was then tested to relate the radiation-generated unstable species with the isomerization reactions. Those unstable molecular states are thought to be responsible for propagation of the chain-cracking reactions. Good agreement was found between experimental results and those predicted by the model [740].

Three different locations for the HEET upgrading application were envisioned: (1) in-field for transportation purposes, (2) at the refinery gate to improve the ability to process heavy crude oils, and (3) as a unit in a future generation refinery. However, there is not quantitative or economical comparison to support any of these alternatives. HEET is supposed to be simple and compact (smaller footprint), while conventional configurations are very complex (multistage and multiline) and vast. Comparatively, while the footprint of HEET facilities can be measured in hectares, a refinery of the same capacity might comprise square kilometers. However, HEET might be just a treatment step for unconventional oils, while a refinery includes many more processing units for a larger variety of gas, liquid, and solid products. The schematic of operation for HEET is overly simplified for refining, as it only includes a high-energy electron beam focused on a flowing hydrocarbon stream. Chain lengths of 1000–2000 C-atoms are not reasonable in crude oil systems for the G-values reported; for example, a very heavy asphaltene of MW of about 5000 (does it exist?) would contain less than 400 C-atoms. Typically, for all the described processes, the energy of the radiation employed exceeds that of the promoted chemical reactions in the range of five orders of magnitude [727–732,734,735,737,738].

Currently, electron beam technology (RTC HEET) is not commercially available. The following section will focus on the reported economics and energetics involved in this type of processes. However, at this point, it is worth summarizing the technical drawbacks derived from the preceding discussion.

- The HEET refinery schematic as presented in Fig. 6.24 is overly simplified, even as a pretreatment step. Quality of the radiation-upgraded crude seems to be poorer than any conventional syncrude, and clearly a stabilizing additional unit will be required. However, the kind of needed reprocessing or postprocessing is not yet understood.
- So far, technology has not been adequately tested; more accurate definitions of yield, quality, and stability of products are needed.
- A heavy cut rather than the whole crude should be treated, since light molecules might be masking some radiation effects.

- A valid comparison between TC and RTC is needed.
- Continuous rather than batch experiments are needed and separation stages included, to fully characterize the produced fractions.
- Processing of crude oil or refined products at STP is uncertain.
- The cost-effectivity and environmental impact of the use of radiation for treating crude oil need to be assessed.
- Sulfur removal has to be demonstrated through a robust S-mass balance, since S-compounds seem to be moving around rather than actually being removed.
- Presently, refinery acceptance looks uncertain.

7.5 Energy and Economics

Energy consumption for conversion reactions of hydrocarbons has been the subject of several articles from Zaykin's team [715,734,735]. A technical comparison of the processing capabilities of HEET, TC, and TCC methods was given in Ref. [83]. The involved reactions in HEET are similar to those in classic TC and TCC technologies. In the case of HEET, irradiation replaces heating in the most energy-consuming stage of the chain reaction, initiation. Apparently, HEET may be more energy efficient, because chemical conversion proceeds at a minimal processing temperature and pressure. These features have been taken as HEET being environmentally and operationally safer than TC processes.

Basically, the energy consumed for the initiation of cracking, in the form of either heat or electron energy, is only required to create a generous concentration of active radicals, since it does not change the product enthalpy. All this energy is transformed into heat. Additional enthalpy is supplied to the product only at the second stage of thermally activated chain development, and it is equal to the corresponding reaction activation enthalpy (if all the matter has reacted). In case of electron irradiation, this initial energy is not excessive, since the energy is transferred directly to molecules of the material being converted. In contrast, the much larger TC energy consumption is caused by the fact that feedstock has to be heated to higher temperatures, in order to transfer a relatively small energy amount into the molecules. This temperature is ruled by an Arrhenius-type equation, according to which the probability that a molecule will consume energy E is proportional to $\exp(-E/kT)$.

Energy requirements of a standard TC process compared with those of HEET are presented in Table 6.9 and illustrated in Fig. 6.25, where E_{Total} is the total energy consumption; E_{th} is the energy consumed for the feedstock heating up from 20°C to reaction temperature (650°C for TC and 400°C for HEET); E_{init} is the energy spent for reaction initiation; E_r is the energy spent for endothermic reaction of chain development; $E_d = E_{Total} - E_r$ is the dissipated energy after the treated product is cooled back to 20°C; and $H_p - H_f$ is the treated product total enthalpy change.

Table 6.9: Comparative Energy Balances for Thermal Cracking of Oil Residue by Heating (HTC) at 650°C and for Thermal Cracking of Oil Residue by Radiation (HEET) at 400°C [735]

Energy Type	Energy Expense (kJ/kg)	
	TC	HEET
E_{th}	1078 (up 400°C)	1078 (up 400°C)
E_{init}	567	5
E_r	214 (up 650°C)	214 (up 400°C)
$E_d = E_{Total} - E_r$	−1645 (down to 20°C)	−1083 (down to 20°C)
$H_p - H_f$	214	214

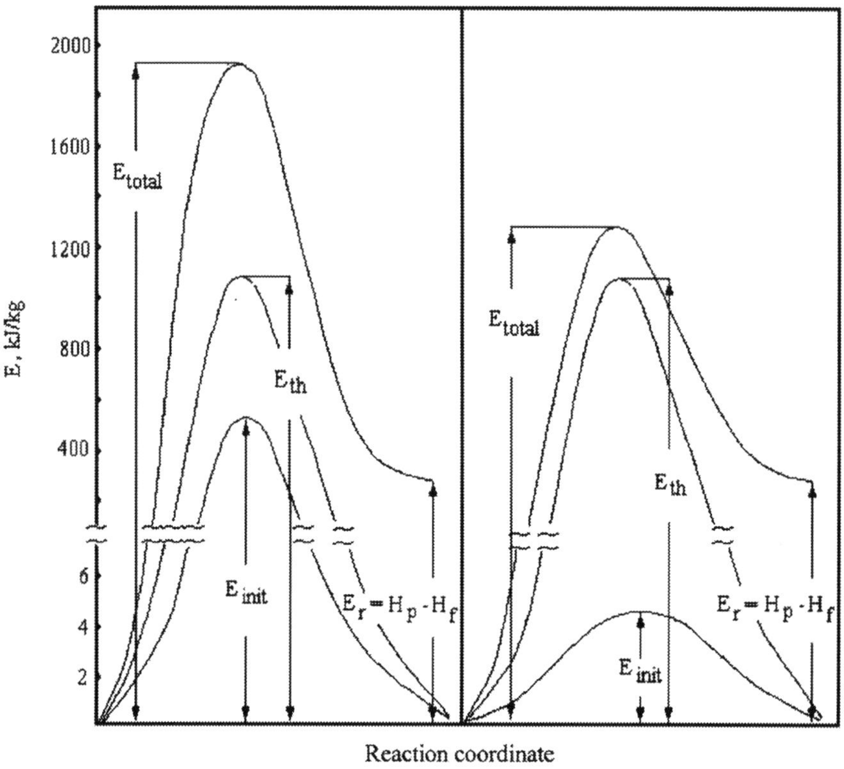

Figure 6.25

Heat balance diagram for conventional thermal cracking of fuel oil at 650°C and radiation-thermal cracking at 400°C. *Reproduced from Zaykin YA, Zaykina RF, Mirkin G. On energetics of hydrocarbon chemical reactions by ionizing irradiation. Radiat Phys Chem 2003;67(3–4):305–09, with permission from Elsevier.*

Considering the activation energy for the initiation reaction as derived from Fig. 6.25, one can infer a larger rate increase with temperature for TC than for HEET (higher activation energy for TC than for RTC). There is no explanation in the paper on how the value of E_{init} was calculated for the TC process [735].

From the data in Table 6.9, E_{Total} was 1431 kJ/kg for TC and 869 kJ/kg for HEET, indicating much greater energy savings for HEET than for TC. However, in terms of primary energy (fuel usage), the efficiency of the energy-generating system has to be considered. For TC, the energy is delivered via a charge furnace that may have a minimum of 80% of thermal efficiency (poor for modern standards) requiring a primary energy supply of about 1790 kJ/kg. Meanwhile, for HEET the energy is delivered via electricity that in turn is generated from fuel. Energy efficiency is of 20% to 30% for RF LINACs and 60% to 80% for DC accelerators [718]. However, energy efficiency for a large-scale power generation plant might average about 60%. Therefore, assuming (the best case scenario) an energy efficiency of about 45% would translate the requirement of primary energy for HEET into 1940 kJ/kg. Hence, the HEET energy advantages vanished when primary energy data are compared. In general, the energy consumption for hydrocarbon processing was not adequately addressed in the publication of Ref. [735].

Considering a dose of about 50 kGy (50 kJ/kg), which has been typically used in the more recent experiments, about 0.014 kW per kilogram of crude oil would be spent during processing. Translating this power at \$0.06/kW, the radiation exposure cost (processing cost) becomes \$0.0008/kg or \$0.13/barrel for a heavy crude oil sample of 10°API gravity. This rough economic estimation resulted in RP costs significantly lower, about \$4/barrel for conventional refining. However, the quality of a refined product is certainly better than that of the RP product.

In terms of throughput, each kilowatt of beam output can treat (at 70% capture efficiency) about 2500 kg/h to a dose of 1 kGy (3600 × 0.7). Hence, for a treatment dose of 15 kGy, using a 700 kW beam, one can treat 2500 × 700/15 = 16,666 kg/h or about 3355 bpd of a 10°API crude. Energy integration could be improved by heating the feed while cooling the electron source, leading to 25–30% savings of power.

Total pricing of an e-beam facility does not scale down proportionately, especially at lower powers. Modular applications of RP for small-volume refineries would introduce larger inefficiencies. Furthermore, radiation energy of a given source varies widely, so trying to compare heat with beam energy is quite specious. Calculations of the efficiency of power generation from fossil fuels, transmission losses in electricity, accelerator efficiency, and energy losses and dissipation, among others, introduce unlimited sources of inaccuracy. A typical refinery processing crude oil to produce high-quality products estimates not only its energy efficiency but the economics as well, in a very rigorous and accurate manner. A full economics of an RP is still pending; the rough estimations made to date cannot be compared to any known public assessment of refining. A more detailed techno-economic analysis is needed, based on a more accurate and representative determination of performance and product yields, quality, and stability.

The Department of Energy (DoE) concluded that the (published) attempts at HEET economic evaluations (CapEx, OpEx, IRR, NPV, Payback, etc.) have been based on inconsistent data [741]. This conclusion was justified due to the following facts:

- The value added and cost-effectiveness of e-beam processing are highly biased by the poorness of the electrical efficiency of linear accelerators, of the processing costs, and of product yields, stability, and quality.
- A rigorous life cycle analysis (well-to-wheel) is needed, but speculatively, one may anticipate that the results of that LCA might indicate a need for CapEx reduction.
- In order to improve the OpEx estimation, a valid total energy evaluation and proper energy balance are required. Most likely, a conclusion of such estimation would point toward costs reduction needs.

8. Sonochemistry

The chemical effects of acoustic waves, and more particularly ultrasound (frequency in the range of 20 kHz to 10 MHz), on the matter have been a research topic for more than 100 years. Substantial efforts started to be seen in synthetic chemistry by 1980 and in petroleum chemistry by the 1990s. This section has no intention of reviewing the subject. Instead, a brief description of the principles will be given: references will direct the reader to examples and more detailed explanations, and the relevant upgrading applications to (heavy and acid) oils will be discussed.

8.1 Description

The sonochemical phenomenon is the effect that acoustic waves have on chemical reactions and processes. While this effect is minimal in solids, it is of significant impact in liquids. The speed of sound in liquids is about 1.5 km/s, and frequencies are in the range from 15 kHz to 10 MHz, corresponding to wavelengths of 10–0.01 cm, respectively. Acoustic waves in the ultrasonic range are produced by either magnetostrictive or piezoelectric transducers. A piece of iron or nickel surrounded by an electric coil is the core of magnetostrictive transducers. The metal expands and contracts at ultrasonic frequencies under the influence of the magnetic field generated by the electric current that flows through the coil. In piezoelectric transducers, a quartz crystal oscillates at ultrasonic frequencies under alternating current.

Ultrasound inflicts an oscillating pressure on a fluid that will induce motion and mixing (acoustic streaming), at low intensity. At higher intensities, waves propagate by oscillating pressures that alternately stretch and compress the liquid. Thereby, during expansion tiny microbubbles or cavities are created, but then collapse during compression. Cavitation appears when the pressure within the liquid drops sufficiently lower than its vapor pressure.

In summary, exposing a liquid to a strong acoustic field causes the generation of cavitation microbubbles. Gases dissolved in the bulk liquid can enter the microbubbles. The acoustic wave (macro level) transfers focused energy to the vapor inside the bubbles (micro level), resulting in their growth and the collapse of these microbubbles. During the compression stage, a release of significant energy occurs that is both short lived (nanoseconds) and concentrated in microvolumes. Extremely high shear forces are generated that result in micromixing. Furthermore, these localized spots concentrate extremely high pressures and temperatures, capable of producing highly reactive free radicals. Therefore, acoustic cavitation can be defined as the formation, growth, and implosive collapse of bubbles in liquids upon their irradiation with sound waves of high intensity (Fig. 6.26). In this way, the transfer of acoustic energy in liquids is such an extremely focused process that enables highly energy-demanding chemical reactions and processes to occur.

The conditions within the cavitating bubbles can be extreme, with temperatures as high as 5000−7000°C and pressures up to 500−10,000 bar. The exact value of these parameters continues to be debated. The energy is dissipated in microvolumes with little influence on the temperature of the bulk liquid. Rapid cooling, estimated by some individuals to be on the order of 10^6-10^9°C/s, follows the collapse of the bubbles.

Cavitation occurs over a definite threshold of ultrasonic intensity. The bubbles may be stable, resulting in acoustic micromixing plus increased compressibility. Although the direct effect is negligible in solids, when the solids are immersed in a liquid, cavitation induces

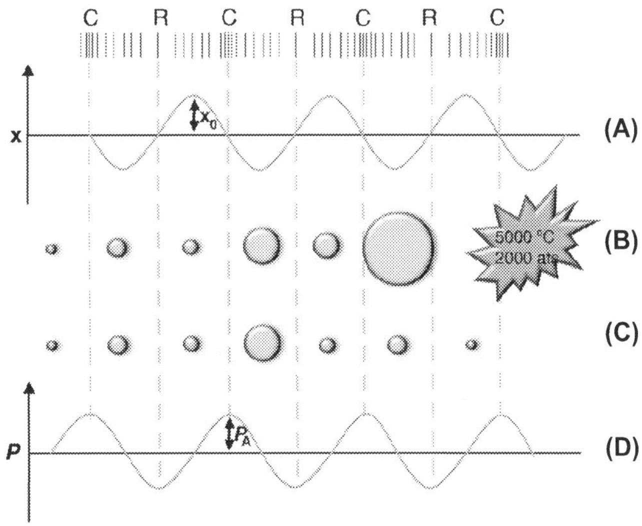

Figure 6.26

Stable and transient cavitation bubbles: (A) displacement, (B) transient cavitation, (C) stable cavitation, and (D) pressure. *Reproduced from Santos HM, Lodeiro C, Capelo-Martínez J-L. The power of ultrasound. In: Ultrasound in chemistry. Wiley-VCH Verlag GmbH & Co. KGaA; 2009. p. 1−16, with permission from Wiley-VCH Verlag.*

interesting phenomena. For heterogeneous systems, bubbles near a solid–liquid or solid–gas interface collapse asymmetrically. If cavitation occurs near an extended solid surface, the bubble collapse generates a nonspherical cavity that pushes high-speed jets of liquid directly onto the surface. Liquid quickly travels from the bulk into the bubble, producing a jet of liquid moving at around 100 m/s. The surface gets heated by those jets and their associated shock waves. Actually, substantial changes are introduced to the surface that, in the case of catalysts, could mean activation. Furthermore, acoustic energy can produce high-velocity interparticle collisions of suspended solid particles in a liquid irradiated by ultrasound. The additional changes caused by those collisions include surface morphology, composition, and reactivity. These high-velocity jets have been used for applications such as cleaning, removing impurities from spent catalyst, and pushing dyes into fabrics.

A theoretical model of the chemical effects of ultrasonic cavitation predicted the accumulation of OH radicals inside cavitation bubbles and the nonlinearity of bubble motion. OH accumulates in the vapor phase of the bubble, increasing in concentration with each oscillation. The formation of water from hydrogen and oxygen results in heat release. Hydrogen peroxide formation follows an unusual fall and then rise due to increase and decreases in the amount of hydrogen–oxygen mixture dissolved into solution. This unusual phenomenon is due to the rapid expansion of the bubble after collapse dropping species concentration and thus slowing reaction rates during the rarefaction phase of the bubble oscillation cycle. The model is recognized to fail in investigating the overall sonochemical reaction rates [742].

8.2 Applications

Applications of sonochemistry take advantage of any or all of the effects that ultrasound causes in materials. As pointed out in the previous description, three different effects can be distinguished:

- Cavitation is the most important of the effects since the most significant chemical consequences result from the formation and collapse of the microbubbles. While the sonic reactor is kept at 1 bar and room temperature, the conditions created by the violent implosion of the microbubbles are very much higher. Theoretical considerations estimate that the pressure at the center of a collapsing microbubble might be around 5200 bar, which is 20–150 times the range of pressure applied in HDC processes. Comparatively, temperature can be one order of magnitude higher (~5000°C) than the highest temperature applied in HDC.
- Compression and rarefaction of the solvent/liquid are caused by the fast fluids movement derived from the changes in sonic pressure.
- Microstreaming is the consequence of placing a large amount of vibrational energy into small volumes with little heating.

Ultrasound might be used either as a pretreatment or as a reaction driver. This will depend on whether the observed ultrasonic effect is the result of (1) a mechanical improvement or (2) a real sonochemical effect. Applications of the first type include cleaning, degreasing, sterilizing, extraction, degasing, defoaming, emulsifying, dissolving, homogenization, dispersing, impregnation, crystallization, disaggregation, precipitation, particle size reduction, etc. As such, it has been proposed for oil recovery from tar sands [743,744].

A sonic reactor is needed (i.e., a reactor capable of providing sonication during the course of the reaction) for sonochemical applications. A variety of reacting systems can be sonochemically promoted, for example synthetic chemistry [745,746], catalysts (biocatalysts, heterogeneous and homogeneous) preparation [747] and activation [748], phase transfer catalysis [749], polymerization [750], electrochemistry, decomposition [751], separation [752,753], and so on.

The potential of sonochemistry has been applied at the laboratory scale in many fields of synthetic chemistry. However, industrial applications have been delayed for the difficulties in both scaling up and the commercial availability of large equipment. Ultrasound can be delivered at frequencies in the range of 25–40 kHz, providing an energy density of 20–600 W/cm^2. Regardless of the application, the first challenge is defining the optimum conditions for sonication. Feed properties and reaction parameters will affect cavitation and would determine the type of sonic system required. Examples to consider regarding the feed properties are vapor pressure, dissolved gases, phase transition, viscosity, and the presence or formation of solid particles; and among reaction parameters, pressure, temperature, and phase changes are probably the most important. The type of sonic system required is defined by frequency, power, geometrical design, and size of the chemical reactor. There are three methods of introducing ultrasound into a reaction medium:

- An ultrasonic source directly immersed in the reacting vessel.
- An ultrasonic tank containing a sonicated liquid for transmitting sonic energy to the reactor walls of an immersed reaction vessel.
- Transducers installed on the outer surface of the reactor walls directly transfer the ultrasonic vibrations.

Proposed and available industrial reactor designs have been reviewed and summarized by Mason [754], namely cleaning baths, submersible transducer assemblies, tube reactors, probe systems, flow systems, near-field acoustic processors, liquid whistle reactors, liquid whistle reactors, and push–pull reactors. According to Mason, two of the major drawbacks of the sonochemical methods have been addressed and mitigated: both reactor erosion and noise hazards can be diminished with adequate reactor design and by conventional sound insulation, respectively.

A whole crude oil was effectively pretreated with ultrasonic irradiation for dewatering and desalting. The pretreatment at 80°C with a standing-wave field of 0.38 W/cm^2 for 5 min increased the rate of dewatering in 93% and that of desalting in 88%. The finished product met refinery specifications [755].

The BotB compounds, particularly asphaltenes, are very complex, and intermolecular interactions lead to association and aggregation. Ultrasound might be capable of disaggregating such associations [756]. Moreover, thermal bond breaking may occur under the cavitation conditions; C–C, C–H, and C–X (X = heteroatom) can be randomly (nonselectively) cracked and form free radicals. The reactive free radicals would rapidly react and create new radicals, or recombine and terminate the reaction [757]. This lack of selectivity of the thermal energy will drive also the cracking of acyclic compounds, increasing gas and coke formation [758].

In the case of asphaltenes, dehydrogenation and cracking were found to be the main reactions occurring under ultrasound [759]. However, this work did not report hydrogen balance or crude oil analyses for supporting those conclusions. As mentioned above, two effects can be provided by ultrasound: high temperature and pressure at localized cavitation centers, and microstreaming. If an ultrasonic treatment is capable of dissociating the asphaltenes, then the presence of a surfactant would prevent any further asphaltenes agglomeration [760].

Similarly, ultrasound was found to promote DM reactions on model metalloporphyrinic compounds. Addition of a surfactant and hydrogen peroxide enhanced the decomposition efficiency of the ultrasonic treatment to about 90% DM after 1 h. Metalloporphyrins underwent DM reaction through an oxidative intermediate. Although metal mass balances were not better than 90%, there was a good agreement between porphyrins decomposition and free-metal recovery [761].

Another study considered the effect of a mild ultrasonic treatment in the presence of an adsorbent, for improving simultaneous DA and DM of VRs of heavy oils. The adsorbent was sulfur-modified polystyrene resin, and the treatment was carried out on petroleum asphaltenes dissolved in THF or 1-methylnaphthalene. A sketch exemplifying the various roles of the process components is shown in Fig. 6.27. The solvent was playing a dispersing role of the macromolecular structures. Meanwhile, the role of the ultrasound was to disaggregate micelle-like structures and the existing associations in asphaltenes. These associations were believed to be either van der Waals interactions between long-chain alkyl groups substituted on polyaromatic rings, or charge transfer interactions, such as hydrogen bonding or aromatic plane stacking. The treatment also dissociated the soluble metalloporphyrin compounds concentrated in the asphaltenic fraction. Finally, the role of the adsorbent was to interact selectively with the metallic compounds and fix these on its surface while accelerating the asphaltene conversion into soluble maltenes [762].

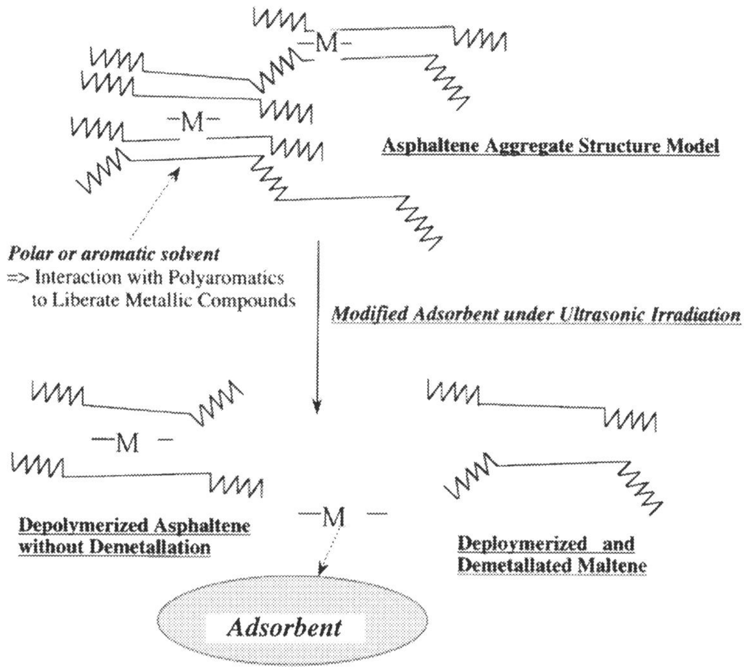

Figure 6.27
Ultrasound and adsorption treatment scheme of asphaltene aggregate. *Reproduced from Sakanishi K, Yamashita N, Whitehurst DD, Mochida I. Depolymerization and demetallation treatments of asphaltene in vacuum residue. Catal Today 1998;43(3–4):241–47, with permission from Elsevier.*

Early work of the group of Professor Teh-Fu Yen at the University of Southern California reported that at lab-scale, a batch reacting system composed of 8.6 g heavy oil, 250 mL toluene, 2 mL H_2O_2, and 250 mL of 10% aqueous sodium silicate sonicated for 8 h converted 47% of the asphaltene. Additionally, the gravity of the crude oil increased from 17.7° to 20°API. Whether the API gravity was measured on the whole product or on the deasphalted oil was omitted [763]. Although the mixture might have undergone some reaction, there was no convincing evidence of a chemical effect of ultrasound on the heavy oil. CanMet unsuccessfully tried to reproduce these results in their own laboratories, using Athabasca bitumen [764]. Thereon, by the end of the 1990s, the R&D activities of Yen's group became focused toward a continuous process for improving quality by asphaltene content reduction. Specifically, the effects of cavitation and added surfactant were reported to be 35% conversion of bitumen asphaltenes into gas oil and resin fractions in 15 min [759].

Equipment limitations lead the way for modular systems arranged according to capacity. In this approach, process units must provide enough flexibility and small footprint to multiply into enough units to cope with large capacities needed at the refinery level. Intermediate products should also have to comply with specification of the downstream units. In order to process 20 kbpd, using a cylindrical pipe type of reactor, with attached transducers (e.g., Cylsonic, a wire cleaner by Branson [765]), 24 units were estimated to be needed at a rate of 583 gal/min, a power density of 100 W/in.^{2}, and a residence time of 2.5 min. A cost of US$0.25/bbl was assessed under those conditions and premises [757]. The volume of each pipe is about 0.2 gal (2.9 in. of internal diameter and 7.5 in. of length). The Cylsonic specifications, with a capacity for handling only 0.2 gal, require a total of nearly 7300 pipes, arranged in 24 modules of about 300 pipes. This arrangement seems to be a nuisance, in terms of footprint. The same company offers a liquid processor (Pentagonal, an ultrasonic inline liquid processing) with a capacity of 4 gal, which may contribute to a smaller footprint.

A study directed toward catalytic improvements by ultrasound was conducted using a Ni–AI catalyst, in water–ethanol solutions, for the DS of benzo- and dibenzothiophene. The roles of the ultrasonic irradiation were found to be in: (1) increasing the reaction rate, (2) changing catalyst morphology (smaller particle size) and cleaning the catalyst surface, and (3) promoting water and ethanol dissociation [766].

Similarly, the DS of 4.6-dimethyldibenzothiophene, a refractory compound present in diesel fractions, showed the improvements of ultrasound for the base-catalyzed oxidative desulfurization (ODS) of this compound using H_2O_2 and Na_2CO_3. An extraction of the formed sulfone was carried out using acetonitrile. More than 90% S-removal at moderate temperature without added hydrogen was achieved, and the effective mixing of ultrasound enabled a better contact within the different reacting liquid phases [745].

A more recent study on the sonochemical effects on the conversion of HGO derived from Athabasca bitumen, in an inert atmosphere and in the absence of any additives, resulted in almost negligible chemical effects, such as DS of 7%, DN of 11%, and a viscosity reduction of 5%. The authors proposed a radical change mechanism to explain the minimal effect of ultrasound [767].

8.3 Technological Achievements

Application of sonochemistry ideas for the upgrading of heavy oils and bitumen started by the end of the 1980s and continued through the first decade of 2000, including attempts for commercial demonstration. Currently, these efforts seem to be fading away.

8.3.1 Eneresource Inc./Tar Sands Energy Ltd.

Preliminary developments of using ultrasound for the treatment of heavy oils and bitumen targeted viscosity reduction based on the emulsifying/dispersing application described above. The combined effect of ultrasound and an aqueous solution of an inorganic base (sodium silicate) on tar sands resulted in the recovery of liquid hydrocarbons. A surfactant is formed under the influence of sonication from the reaction of polar resin components of the feed and the inorganic base. The dispersing effect of sonication and the presence of the self-created surfactant give rise to a microemulsion of polar-external micelles, in the absence of any organic solvent. This surfactant penetrates the bitumen and facilitates the removal of bitumen from the sand particles. The polar-organic asphaltene materials are carried into the aqueous phase by the anionic tail of the surfactant and stabilized within the micelle structure. Meanwhile, the lighter and nonpolar hydrocarbons separate from the emulsion and rise to the top. This lighter oil fraction is recovered by skimming. The heavier asphaltenic fraction forms complexes with the polyvalent metals and associates as charcoal-like agglomerates, which would deposit to the bottom of the treatment tank. A free radical initiator (e.g., benzoyl peroxide) might be optionally added in small quantities in order to increase the rate of bitumen separation [144,768,769].

8.3.2 SulphCo/International Ultrasonic Technology (Sonocracking Technology)

This company, founded in 1999 by Rudolf Gunnerman, based its development on the research findings of Prof. Yen. During the late 1990s and early 2000s, partial oxidation of S-compounds was considered as an economical route for desulfurization. The formed oxisulfides (mainly sulfones) could be easily extracted, and removal of ~90% was reported [770]. An advantageous use of hydrogen peroxide was reported for the ODS of model compounds present in the diesel fraction. Examples of novel catalysts for this application included phosphotungstic acid [771] and Ti-containing molecular sieves [771]. As mentioned in the previous section, Sadeghi et al. [768] also observed an enhancement on ultrasonic treatment from using peroxides. Prof. Yen was a co-inventor in that latter patent.

SulphCo during the early 2000s tried to exploit this idea into the development of a process for the desulfurization of heavy oils, by acquiring a subsequent patent from Prof. Yen. In this patent, later assigned to SulphCo a combination of a hydroperoxide in an aqueous-organic medium and ultrasound were used for converting organic S-compounds into their corresponding sulfones [772].

SulphCo's process development centered on making it continuous and adapted to existing ultrasonic equipment. In a first attempt, a surfactant was included on top of the hydroperoxide or an aliphatic hydrocarbon [773]. The total reacting mixture passed continuously through the sonic reactor. The treated product spontaneously separated into two phases, an aqueous and an organic. At that time, the concept was demonstrated on a diesel fuel. More than 98% conversion was reported. Other reported benefits besides conversion included nitrogen removal and hydrogenation, which resulted in an increased cetane. The treatment of the emulsified feed containing organic peroxides and/or a transition metal catalyst, especially Ni, Ag, W, Co, and Mo, was also proposed, but a patent was not awarded [774]. Another proposed co-reactant was dialkyl ether [775].

A modular unit was designed based on those results, and a 1000 bpd first scalable unit was announced would be built and tested in a Californian refinery by April 2004 [776]. San Joaquin heavy crude oil was tested in a 1000 bpd unit, although it was not clear where that unit was built. API gravity was improved from 15° to 25°API, sulfur content was reduced from 1.5% to 0.8%, and residuals and asphaltenes were removed by ~85%. A collaborating partnership started between SulphCo and the oil giant ChevronTexaco in 2002. They built a 2 kbpd pilot plant in Richmond, California, and later, in July 2005, Chevron quietly dropped the Sonocracking project.

A 25 kbpd unit for processing heavy crude oil based on 2–3-point increases on API gravity and 20–40% of S-removal was designed, and costs were estimated. An investment cost of about 5% to 10% of that of a conventional HDS unit and an operating cost of US$0.15/bbl were estimated [777]. This design was prepared but never built.

Further improvements were approached by incorporating microwaves in combination with ultrasound [778], through improvements in ultrasound generation [779–781], by optimizing process conditions for increasing conversion [782] and by geometrical modifications of the reactor design [783].

All these achievements made SulphCo to quadruplicate into a billion-dollar market value and be awarded the 2006 Excellence in Crude Oil Refining Technology of the Year Award, by Frost & Sullivan [784]. The technology was supposedly offered to turn low-grade crude oil into high-priced, clean-burning fuels, for just 20 cents a barrel. *Sonocracking* was the trade mark given to the ultrasound technology that could turn S-sour crude into sweeter intermediate-grade oil that is worth US$10 more per barrel. However, at the same time, investors, lenders, and corporate partners like Caterpillar, Clark Oil, Sinclair, and ChevronTexaco were disappointedly walking away [785], and new prospects were announced.

OIL-SC in South Korea built a unit and tested 2000 bpd of Arab Medium crude oil, demonstrating an increase in 5°API gravity. A proprietary catalyst that was not described in any previous patent application was employed in that testing run [786]. At that time (beginning June 2006), a plan for installing 210 kbpd of Sonocracking capacity (seven 30-kbpd units) at Fujairah Oil Technology (FOT) in the United Arab Emirates was also announced. Trans Gulf Petroleum, owned by Fujairah government and SulphCo, created FOT as a 50/50 joint venture. The equipment was going to be manufactured by NTG Technology in Germany [786].

SulphCo was expecting that after the demonstration unit had been started up, they would partner with a major oil company for worldwide licensing of the process. That never happened; it seems neither scale nor conversion met the expectations. Furthermore, the results of that commercial-scale run directed SulphCo to limit the scope of the technology and to emphasize the technology application on fuel blendstocks. In 2009, a new view was defined for Sonocracking: limiting the technology to a fuel finishing step for ultra-low fuels production. The best results were obtained for the kerosene and diesel fractions with 80% and 90% DS, respectively [787]. These findings translated into additional limitations of Sonocracking scope: (1) distillate fraction feed treatment, and (2) production of ULSD. Thus, two process configuration options were visualized: (1) a Sonocracking treatment prior to HDS or (2) the two-step ODS process (Fig. 6.28).

In 2011, SulphCo filed for relief under Chapter 7 bankruptcy. They revealed in the Assets Overview published as part of the 363 sale that the first commercial trial of the technology was a 40 h continuous run with SR naphtha on ~150–500 ppm S, and the average S-content of the gasoline product was 20 ppm (87–96% removal) at a flow rate of 3000 bpd [788].

By 2013 and under Chapter 7, the SulphCo ODS technology was defined as a two-step process: (1) the ultrasound-assisted conversion of the sulfur compounds to their oxidized analogs employing its patented and proprietary Sonocracking technology, and (2) the subsequent removal of these oxidized compounds via adsorption, extraction, water wash, or other similar separation techniques.

On March 18–19, 2013, Heritage Global Partners and Pluritas, LLC sustained a global online intellectual property auction of SulphCo Sonocracking technology [789].

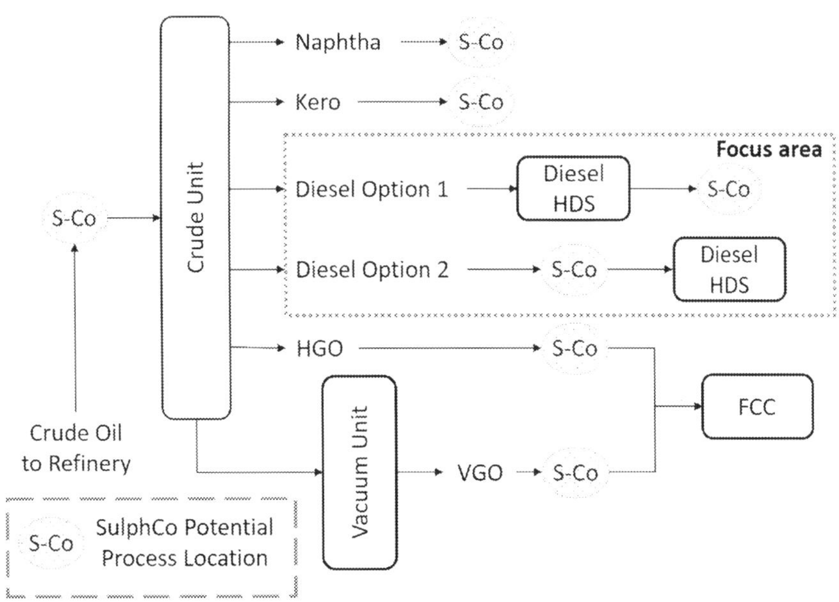

Figure 6.28
Focus areas for diesel treatment using the SulphCo technology as an ODS process.

International Ultrasonic Technology acquired the intellectual property (nine US awarded patents [772,773,775,779−783] and related world patents on the ultrasonic technology) and equipment owned by SulphCo for *ultrasound-assisted oxidative desulfurization*. The new owner has offered to commercialize and to market the technology, though this time with emphasis on transportation fuels desulfurization [790].

8.3.3 Petrosonic Inc.

The president of SulphCo, Mark Cullen, left the company around the time of the bankruptcy filings. He had filed six patent applications, which have priority dates back to 2001. These applications were assigned later to Petrosonic in 2006. Basically, he claimed a sonication treatment for the DS of crude oil, oil fractions, fuels, and/or organic liquids, in the presence or absence of an oxidizing agent [791−796]. Three of these applications were granted, and the others were abandoned. The granted patents include a complaint note from SulphCo.

The three granted patents focus on the ODS−ODN sonication treatment of fossil fuels or petroleum-derived fractions or liquids, at temperatures between 125°C and 500°C, in the presence of hydroperoxides, a surface active agent (or any other emulsion stabilizer), and transition metal (preferably, Fe^{2+}, Fe^{3+}, and Cu^{2+}) catalysts (particularly for crude oil applications). Ultrasound is applied at a frequency of about 10−20 kHz, at a displacement amplitude of 30−60 μm, and at energy density of 1−20 W/cm^3. The effects of the treatment were reported to include: (1) opening and saturation of fused-ring aromatics, (2) olefins hydrogenation, (3) API gravity increase, (4) cetane number improvement, and (5) partial oxidation of S- and N-compounds [794−796]. An additional step for removal of the oxidized S- and N-compounds was needed, and though patents were filed for two extraction ideas, these were never awarded [797,798].

Currently, the status of this company is unknown, and there has not been any further report on the technology. Nevertheless, there seems to be a connection between Petrosonic Ltd., and Petrosonic Energy, Inc., since they appear as common assignees in the patent application of Ref. [799].

8.3.4 Sonoprocess−Sonoro Energy, Ltd. and Petrosonic Energy Inc.

SONIC Technology Solutions Inc. purchased the intellectual property and patent rights of an upgrading process for oil oxidation chemistry from Northern Oil Research Technologies, Inc., in 2008. This oxidation chemistry comprises stages for DA, biological oxidation, chemical oxidation, and solvent recovery. SONIC planned to incorporate that chemistry into its existing DA and chemical oxidation Sonoprocess technology, for the production of low-sulfur fuel oil, from Canadian heavy oils and bitumen. The Sonoprocess technology was developed by SONIC's joint venture, PetroSonic Energy Systems Inc. [800].

In 2010, an incorporated company, Bearing Mineral Exploration (BME), created in 2008 with interests on minerals exploration and production, started to seek new business opportunities. In June 2012, BME changed its name, articles, and business interests to Petrosonic Energy, Inc. At that time, sonication assets were acquired from Sonoro Energy, including intellectual property rights. These rights include the patent applications of Ref. [799], for a DA process carried out on heavy oils in the presence of a solvent and under ultrasonic energy.

Although Sonoro Energy possesses additional intellectual property on related areas, such as a sonic reactor [801] and oil treatment [802], these did not seem to be part of the negotiation with Petrosonic Energy.

Petrosonic Energy defined the Sonoprocess as a simple, cost-effective, commercially available process for producing heavy oil with densities and viscosities to meet 99% of pipeline specifications. Thus, API gravity is increased by 5−10°API, in crudes with API >8°API and performed at DS of 40% and DM > 70%. In the sonicating reactor, ultrasound is applied at low frequency, high energy, and high amplitude; it improves mass transfer efficiency; and SDA becomes the first step of the process. Compared to conventional SDA, the DA required time is reduced from 6−10+ hours to 2 min. In a second step, the solvent is recovered, purified, and recycled to the reactor. A subsidiary company built a 3 kbpd unit for the manufacturing of synthetic fuel oil, in Albania. The FEED of a pilot plant of the Sonoprocess technology was prepared in collaboration with the Western Research Institute of Laramie, Wyoming, USA.

In May 2014, Petrosonic Energy granted to Kuai Le GU, LLC a nonexclusive license to research, develop, practice, perform, make, use, manufacture, assemble, modify, sell, offer for sale, distribute, import, and otherwise exploit Sonoprocess in the People's Republic of China. In June 2014, Gas Liquids Engineering of Calgary completed an independent engineering feasibility analysis and validation of the Sonoprocess sonic DA technology.

More recently, an enhancement that enables the use of propane or LPG for a DA application on heavy oils was announced as a new Sono Hi-P technology. The improvement consists of a modification of the sonication chamber to allow application of higher pressures than before. This improvement is seen as a risk-mitigating option that would allow a wider deployment of the sonic technology [803]. However, by the first quarter of 2015, the company (and, thus, the technology) was struggling for its survival, by seeking new sources of financing.

8.3.5 ExxonMobil

Applications from this company include aqueous and nonaqueous defoaming in cokers and alkylation reactors. A sonication probe is inserted directly into the foam, and sonicating pulses of energy density >25 W/cm^2 are applied until the foam collapses [804].

After realizing that sonicated reduction in viscosity was only a temporary effect, the search for a more permanent effect on heavy oils took some of their efforts. Sonication was combined with low-temperature (20–70°C) acid treatment, using sulfuric acid, hydrochloric acid, perchloric acid, acetic acid, para-toluene sulfonic acid, alkyl toluene sulfonic acids, mono di- and trialkyl phosphoric acids, C_{3-16} organic carboxylic acids, succinic acid, or petroleum naphthenic acid. Sonication is applied at frequencies in the range from 15 kHz to 10 MHz, and energy density in the range of 25–800 W/cm^2 [805].

Another application from this company was already discussed in the naphthenic acids section. In this application, ultrasound was used to enhance adsorption of naphthenic acids on specifically selected solids, for the production of a reduced TAN oil from HACs [281].

8.3.6 Others

Petronetics proposed a method for cracking heavier hydrocarbons by providing ultrasound at a rate sufficient to induce cavitation that creates high enough temperatures and pressures for bond breaking. The upgraded petroleum product exhibits lower distillation curves and decreased contaminants content [806].

A combination of effects has been proposed in pulsed rotor units (PRUs) for the treatment of liquid–liquid, liquid–solids, and gas–liquid systems:

- a mechanical action on the particles of a heterogeneous medium, which consists of percussive, shearing, and pulverizing loads and contacts with the working parts of the PRU;
- a hydrodynamic action, which is expressed in high shearing stresses in the liquid, and pressure and flow-rate pulsations of the liquid; and
- a hydro-acoustic action on the liquid due to small-scale pressure pulsations, vigorous cavitation, shock waves, and nonlinear acoustic effects.

The experimental results showed a shift in the distillation curve of a PRU-treated crude to a lower temperature ($T_{50} = 265°C$) from that of the untreated crude ($T_{50} = 328°C$) [807].

Several groups have considered the ultrasonic upgrading of heavy fractions [808–810]. However, as already mentioned, the sole application of ultrasonic cavitation seems to be a temporary effect. Most likely, cracking may occur via free radicals. The initially formed lighter and reactive radicals would recombine, producing heavier compounds, by recombination, condensation, oligomerization, and polymerization reactions. In order to maintain the effectiveness of the cavitation treatment, aromatic fractions or polarizing solvents were suggested to be blended with the upgraded oil [811,812]. Additionally, more complex reacting mixtures have been proposed even for light fractions (diesel), for example oxidizing agents, solvents, emulsifiers, and catalysts [813].

9. Mechanochemistry

Mechanical action appears to be the most commercially accessible method for nonthermal activation of chemical reactions. However, besides solid management and manufacturing, not very many applications can be found in other industrial sectors.

9.1 Description

The coined term of mechanochemical processing (MCP) refers to chemical reactions and phase transformations that take place due to the application of mechanical energy. In preparation of this section, more than 70 references were considered. This review shows that mechanochemical devices and applications have been the subject of numerous books, articles, and patents; and, although applications are devoted mainly to solid-state and surface chemistry, there are also some applications concerning crude oil refining, including both processes and catalysts.

It seems now accepted that mechanical energy can be transformed into chemical energy and/or transfer to existing chemical bonds. The chemical reactions observed to occur at room temperature under mechanical action have been explained as due to (1) a "quasi-autoclave" regime, which is created for a local increase in pressure and temperature; (2) a mechanically activated surface or catalyst with an extended defect structure; and (3) the generation of newly formed surfaces with a greater amount of unsaturated coordination.

MCP has had a long history, and the produced materials have already found a number of potential technological applications (e.g., in areas such as hydrogen storage materials, heaters, gas absorbers, fertilizers, catalysts, cosmetics, and waste management). The way in which mechanical energy is applied may vary, and different forms of mechanical stress forces have been studied. For instance, a ball mill, which is a powder-processing technique involving deformation, cold welding, fracturing, and rewelding of powder particles, is used for mechanical alloying.

Boldyrev devoted more than 50 years of research on this area. He distinguished two types of effects upon mechanical treatment, namely mechanochemical reactions (MRs) and mechanical activation (MA). These two cases differ on the influence that a stress field (caused by the mechanical deformation) of a solid has on its reactivity. The limiting cases are based on the relaxation times of the stress field. Thus, MR concerns relaxation times shorter than the chemical reaction time; while, in the opposite case, MA takes place. MCP in a high-energy mill induces a dramatic alteration of the structure and surface properties of solids. Mechanical impact on the solid generates stress fields within the particles that can relax via several channels: heat release, development of the surface area as a result of brittle fracture of particles, generation of various sorts of structural defects, and stimulation of solid-phase chemical reactions (dehydration, oxide crystallization, brittle fracture and agglomeration of oxide particles, and accumulation of structural defects resulting in the successive reductive transformation) [814–816].

For mixtures containing two or several components, MCP involves additional effects related to the fragmentation, mixing, and compacting of particles. The main processes occurring at the interface are (1) formation of solid solutions followed by a chemical reaction, and/or (2) formation of mechano-composites when no mixing at the atomic level or chemical interaction takes place.

The main subjects for Boldyrev's research on solids MA were one- and multicomponent systems and the nature of the defects occurring during activation in one-component systems. According to him, MA occurs by changes in the concentration of point defects, dislocations, anomalous distribution of point defects in crystalline lattices, the formation of quasi-crystalline structures, and the influence of these defects on solid-state transformations (polymorphous transitions, thermal decomposition, reduction, and chemical dissolution) [817].

A proposed phenomenological model of MA was based on the concept of a solution of defects of different types in the crystal lattice in which the concentration of the defects and the equilibrium among them depend on the energetic stress (i.e., the dissipation equation). The Gibbs–Folmer treatment of critical dimensions and conditions for growth of a crystal nucleus are extended to the process of dispersion and multiplication of defects during MA [818].

The obtained results on the mechanical treatment of quartz, periclase, and zinc ferrite prove that distribution of the defects depends mainly on the mechanism of treatment. The kind, concentration, and distribution of the defects depend on the specific features of the substance. The formation of surface and volume defects is the reason for mechanical-induced metastability, which in turn might be influencing reaction processes. The increase in surface area accompanying particle size reduction may be of importance for increased reactivity. The investigations reveal that these parameters are affected primarily by structural changes [819].

Examples of mechanochemical reactions comprise those for which evidence of bond breaking or bond forming was substantiated. Zhang and Saito published a review on the fundamental aspects of mechanochemistry of inorganic and organic materials and its engineering applications [820], based on reactions that can be easily carried out by dry grinding (neither aqueous solutions nor heating). Examples of the subsequent applications of these mechanochemical reactions included nonthermal production of $(Sr,Ba)CO_3$ from celestine ($SrSO_4$) and barite ($BaSO_4$), dehydrochlorination of PVC without heat treatment, and the respective extraction of tungsten and rare earths from scheelite ($CaWO_4$) and from bastnaesite ($ReCO_3F$).

Other examples of MRs are the hydrothermal processes, as those observed in autoclaves during solids MCP. When solids contain some amount of free or chemical-bound water, high-energetic activators enable MRs. The estimations of the optimal value of the water content and data on the investigation of the MR between calcium hydroxide and hydrated silica were presented as experimental confirmation of the hydrothermal regime [821].

A complex methodology was developed to quantify the conditions of mechanical treatment in laboratories and commercial ball mills. The experimental detection of the collision frequency and the impact energy has been paralleled by an accurate numerical modeling of the balls and powder dynamics. The obtained results have been further confirmed by a direct inspection of the motion of balls and powder inside a transparent quartz reactor by means of a high-speed video recording. The temperature of mechanochemical reactors has been also continuously monitored during the course of mechanical treatments. The methodology has been systematically applied to characterize on an absolute basis the kinetic behavior of a large variety of mechanical-induced structural and chemical transformations. In particular, systematic investigations have been performed on the amorphization of intermetallic and binary systems, on the formation of extended solid solutions in immiscible systems, on the ignition of combustion-like reactions in metal—metalloid mixtures, and on the hydrogen storage in nanocrystals and amorphous multicomponent metallic alloys. Mechanochemical transformations appeared to be basically ruled by two different mechanisms characterized by simple asymptotic trends and sigmoidal curves, respectively. An attempt to rationalize the observed mechanisms on a phenomenological basis has been made [822].

In a recent review, the successful industrial penetration of MCP in the engineering of new materials and of heterogeneous catalysis, as well as in extractive metallurgy, was pointed out. The capabilities of MCP for driving processes take place under nonequilibrium conditions, creating a well-crystallized core of nanoparticles with disordered near-surface shell regions and performing simple dry time-convenient one-step syntheses. Furthermore, MCP underpins technological applications that, like preparing new nanomaterials with the desired properties, can be achieved in a reproducible way with high yield and under simple and easy operating conditions [823].

9.2 Mechano-Upgrading

MCP has been applied for the treatment, upgrading, and conversion of crude oil and some of its fractions and products. From the review showed in this section, it will become apparent that upon MCP treatment of crude oil or hydrocarbon fractions, mechanochemical reactions occur and have been evidenced by the presence of new compounds in the products that were originally absent. The spectroscopic characterization of the products has proved to be useful in the determination of the chemical changes induced by MCP. The conversion, the product yield, and selectivity depend on the chemical nature of the reacting molecules, the energy involved, and the presence and nature of any catalyst. The refining applications include upgrading of heavy oils, bitumen, residuess, as well as distillates (gasoline and diesel). Studied reactions include hydrogenation, cracking, isomerization, and partial oxidation. Biomass also has been tested, and the cracking of lignin-related materials was also observed upon MCP treatments, for example [824–827].

Several examples of heavy oil and bitumen upgrading by MCP means have been reported. Changes in distillation yields of natural bitumen and of Baltic shale oil were obtained by grinding in a vibrating mill at 20–40°C for about 30 min in nitrogen. Distillation yielded nearly 50% distillate fuels [828]. Another increase in distillate yields from crude oils, residues, and bitumen can be attained by mechanical ozonization. Nearly 50% of the atmospheric bottom fraction was converted into light hydrocarbons (gasoline and diesel). The product contains up to 16 wt% olefins and decreased sulfur content [829]. The HMW heteroatomic compounds (resins and asphaltenes) in ozonolysis products can be used for surfactants manufacturing. The specific properties of the polyfunctional carboxylic acids or their salts obtained by alkaline hydrolysis form the basis for the production of high-effective demulsifiers that could be used in petroleum dehydration processes. A similar ozonolysis treatment was applied on biomass as well (e.g., seeds, bulbs, and grass) [830].

Production of carbon fibers was achieved by a treatment with MCP of three petroleum-cracking residues and a pitch-like material, followed by another treatment with petroleum ether and $CHCl_3$. The C-fibers derived from the pitch-like material contained symmetric aromatic clusters of five to six layers with side alkyl chains, a high C content, and fused aromatic rings, which were not present in the other residues [831].

Spectroscopic studies showed the possibility of the occurrence of a mechanocracking process of residual components of the Astrakhan gas condensate by the effect of an electromagnetic field in a swirling bed, at 280–300°C. Phase redistribution occurred with a new product of higher resin and asphaltene content [832].

Conversion of asphaltenes from Mantanzas (Cuba) crude oil to liquid fuels was carried out by grinding under Ar. The average MW of these asphaltenes dropped by a factor of three during grinding, from which the authors determined the presence of two labile bonds in the asphaltene molecule [833].

Related applications, such as the treatment of lighter petroleum fractions, are included here as indicative of the potentiality of MCP processing. The performance properties of liquid fuels were found to change upon MCP. The MA of SR distillates from Talakansk oil (360–400°C) leads to a change of its hydrocarbon composition, decreases in its paraffin content and its pour point, and increases in density and viscosity values [834].

Lighter hydrocarbons, such as gasoline and individual n-alkanes C_6, C_7, and C_8, were subjected to MCP in the presence of a high-silica zeolite in H-form (H-ZSM) and its metal modified version (Zn, Fe, Ni, Ag, and Pt). Under MR conditions, besides gaseous hydrocarbons, heavier hydrocarbons (branched and linear) were formed. Mainly, iso-structures were detected. Selectivity depends on molecular size of the feed and on catalyst type. Under MA conditions, hydrocarbon conversion was associated with the presence of gaseous phase bubbles in the mixture. The bubbles heating upon compression provides a sufficient local temperature rise for alkane isomerization [835].

The possibility to produce hydrogen and methane from hydrocarbon gases by means of mechanochemically initiated cracking was demonstrated. The presence of a solid phase increased conversion. Crystal quartz was exemplified since, during mechanical treatment, it efficiently generated active centers of radical nature [836]. This application of MA and MR has been filed in Russia as a method involving the addition of solid-phase particles into a reactor–activator for hydrocarbons. The solid particles (e.g., quartz, quartz–glass, silica gel, Al_2O_3, and iron chips) are filled up to one-third of the reactor space. Mechanical activation is carried out for 5–30 min, followed by holding for 60 min without activation. The resulting fuels were claimed to have improved thermophysical properties [837].

The composition of propane–butane gas mixtures was changed significantly when subjected to MCP, in a centrifugal ball mill in the absence and/or presence of some mineral particles. Product composition depended on duration of the mechanical coercion and solid or porous nature of mineral additives. The presence of quartz particles at temperatures not exceeding 100°C resulted in complete conversion (to methane, hydrogen, and small amounts of carbon). In the absence of minerals or in the presence of porous silica gel or aluminum, oxide conversion decreased. Direct transformation of mechanical energy to hydrocarbon intramolecular energy was considered responsible for their conversion. The catalytic properties of the minerals were neglected as effective conversion agents. The process was recommended for the removal of heavier hydrocarbons from natural gas and for complex mixtures, at the casing head of the well [838].

Specific MCP devices have been tested and evaluated. Care should be exercised when extrapolated to other MCP devices. The mechanic device described in Ref. [839] responds to a well-differentiated design from all of those described in other reviewed patents, particularly those designed for grinding or comminution applications, which are basically mills of all types. It is described as an ejector with a cylindrical receiving chamber and pipes for gas and liquid supply, mounted along the chamber axis. The tube outlet is formed by a multinozzle attachment. The ejector contains a contraction, a mixing chamber, and a diffuser with optimal geometry and gas–fluid dynamics. The device consists of an inlet piping with a multinozzle block hermetically sealed in the cross-section and connected with a cylindrical channel and eventually with an expanding diffuser, with the one-half angle not in excess of 4°. The ratio of the cylindrical channel cross-section area to the sum of the nozzle hole areas, at the outlet from the multinozzle block, ranged from 2.1 to 5.9. This device has been claimed to be useful for the upgrading of unconventional oils. The treatment method involves a physicomechanical effect on a moving flow of oil and/or oil products, created by changing pressure along the flow. This treatment aims to increase distillate yields, with simple equipment and low-energy consumption. A pressure of at least 3.5 bar is applied at the flow inlet, and subsequently, it is reduced to a value not in excess of 0.5 bar. Formation of a vapor–liquid mixture is promoted, and then a pressure increase of at least 1 bar is applied, at the flow outlet. A twofold cycle (feed passing through the device, and the product recycled once again through the device) was believed to produce better results. The pressure drop is thought to be responsible for causing *fractional distillation* of the hydrocarbon feed. Further processing included the separation of the gas–vapor and liquid phases and the chemical stabilization of the produced liquid fraction [840].

The effect of MCP on the emulsification of high-viscosity petroleum products was reported. Thermally stable water-in-oil emulsions, having a fivefold decreased viscosity compared to the original oil product, were prepared for the first time by mechanochemical treatment in the presence of a surfactant. The critical fluxing temperature, thermal properties, and viscosity of asphalt emulsions prepared by mechanochemical treatment are identical to those of residual oil [841]. Road asphalt was produced from Mongolian high-paraffin crude oils. The paraffin presence is a serious obstacle for the production of quality asphalt from this type of crude. The rationale for explaining this behavior was based on reactivity enhancement of n-alkanes by MCP. In fact, their transformation into more reactive compounds, under mild MCP conditions (and the effect of water–acid solutions), enabled interactions in the hydrocarbonaceous media, which are not observed in the absence of MCP [842].

The manufacturing of pavement asphalt from bituminous materials was carried out by mixing the bitumen at 130–155°C with a comminuted additive. The additive was prepared by comminution of S-compound, water ~40 wt%, $Na_2S_2O_3$, Na_2CO_3, NaSCH, hydroquinone, Fe_2O_3, SiO_2, NH_3, C_6H_5OH, naphthalene, tar, and cyanic compounds. The method eliminates emission of SO_2 and H_2S if town gases are used as the S-source [843]. Another related application is the separation of bitumen, lignite, and humic acids by grinding in a vibration mill. This treatment also caused chemical changes, such as increase in the content of quinoid, CO-group, CH_2–C, O– moieties, H–C and O–C atomic ratios, and phenolic OH groups, and decrease in the content of COOH groups [844].

As mentioned above, the application of MCP to solids is more broadly known. Among the range of considered materials, tar sands and coal have received a great deal of attention. An oil-producing method from tar sands was based in MCP separation. The detailed explanation of the process and its scheme can be found in Ref. [845].

The method and apparatus for treating ores (mineral particles surrounded by hydrocarbon compounds) in a continuous process involve crushing or otherwise comminuting in a slurry phase and heating to about 80°C. The slurry is intensively sheared to cause separation, by shear fracture of the hydrocarbon layers surrounding the particles in the grains. The conditioned slurry is blended with a peroxide aqueous solution. The peroxide enters the grains and is decomposed therein, creating bubbles of free oxygen within the grains that disrupt the hydrocarbon envelope. In decomposing, the peroxide increases the hydrophilicity of the particle surfaces. The resulting hydrocarbonaceous phase is separated from the substrate particles by flotation, accelerated by the attached oxygen bubbles [845]. This mixer for treating the ore is described in Ref. [846], and it is characterized by having a low axial flow rate and a high radial flow rate.

MCP treatment of coal in its different forms is an application very close and related to oil refining. Some relevant findings will be described in the following paragraphs. A review on coals MCP [847] discusses the effect of milling on properties and structure. The grinding of the sludge remaining after coal hydrogenation yielded benzene-soluble organic matter containing aromatic hydrocarbons. These hydrocarbons were produced by mechanical degradation of the (benzene-insoluble) organic compounds originally present in the sludge. Upon grinding, C—C and C—O bonds of these compounds seemed to have been broken [848]. The combination of stepwise comminution and extraction (with alkane—benzene solvent) of gas coal caused mechanical-oxidative degradation of the original coal, as indicated by an increase of phenolic OH content in extracted organic products [849]. When coal was ground for 20 min under Ar, the analysis of solid residues and products showed mechanical degradation. There was a decrease of C content, a certain degree of de-aromatization, and an increase of O concentration [850].

Upon grinding, the dispensability, porosity, and structural—chemical parameters of brown coal, coal, and anthracite changed, and the sorption activity and binding properties were enhanced. The mechanical destruction resulted in a reduction of the (C—H) atomic ratio and an increase in the volatiles yield. Simultaneously, the specific surface area and specific electric resistivity increased, while the density was reduced [851]. The media in which the coals are ground and the presence of catalysts (e.g., NaOH, Mo, and Fe^{3+}) significantly affect the mechanochemical activation. Moreover, MA had a positive effect on H-content and liquid fuel production from brown coal and anthracite, and increased the potential utilization of brown coal as sorbent or binder [852]. In fact, the use of a catalyst (e.g., Fe^{3+} and Mo^{6+}) during brown coal grinding resulted in changes in hydrogenation behavior, indicated by a decrease in aromaticity of the organic structure of the coal. The MCP of coal in the presence of a paste-forming solvent activates its donor capacity. Thus, when a suitable catalyst is present, MA enables hydrogen transfer from the paste-forming solvent to coal [853].

Coking coal and lignite were ground in a vibrating mill in air and at a low temperature in nitrogen, and gas coal was milled in a vacuum ball mill (initial grain size 1−3 mm). The coal powders had increased electrical resistivity and lower density than the initial coals. During milling, the C−H ratio decreased both in air and in vacuum. The amounts of benzene- and benzene−ethanol-soluble matter increased during milling. The IR spectra revealed changes in the substituents groups of the aromatic rings, increases in the intensity of bands corresponding to alcohol and CO groups, and decreases in those of phenolic and ether groups. Milling of lignite caused destruction of ester groups [854].

The MCP of vacuum bottoms from brown-coal hydroliquefaction (softening point 134°C and solid content 50 wt%) gave rise to a drop in the CH_2−CH_3 ratios in the distillates, asphaltenes, and preasphaltenes, which is indicative of the mechanical degradation of alkyl chains and CH_2 bridges in these components. An additional decrease in the softening point (to 106°C) could be taken as further evidence of the mechano-upgrading [855].

X-ray diffraction of ground brown coals showed the weakening of molecular interactions with a subsequent increase in the distance between the C layers and on changes in the structure of elemental structural units, which might be responsible for the increase in reactivity [856]. Their mechanochemical treatment in a centrifugal−planetary mill caused a partial destruction of covalent cross-links in organic media, predominantly oxygen-containing groups. As a result, a considerable change in the supermolecular structure occurs, the content of the graphite-like phase decreases, and there is a sharp increase in the flexibility of macromolecular chains, the rate of permeation of THF in the organic matter, its swelling propensity, and the yield of soluble materials [857]. The main chemical change upon MCP concerns the methylene structure and its reflection on an MW decrease [858]. The IR and EPR spectroscopy analyses of the products from grinding high-volatile bituminous B and coking coals showed a decrease in the amount of C in aliphatic chains and in the length of aliphatic chains. The ground coals contain more aromatic C in −CH− aromatic groups. For sub-bituminous and high-volatile bituminous coals, this correlation is reversed. The amount of C in condensed aromatic rings decreases, and the number of aromatic rings decreases from 3−4 to 1−2 [859]. Further characterization of the ground products of high-volatile bituminous B coal indicated a decrease in the C and the H content of the coal and an increase in the O content. The C−O and C−C bonds were broken, and structural deformation was also observed [860].

The effect of the atmosphere present during MCP of a hydrogenation slurry of brown coal was studied, in a vibratory mill, using air, H_2, Ar, and hydrogen-containing gas for 30 min. The composition and structure of both crystal and amorphous phases and of distillates boiling below 360°C changed to an extent determined by the gas used for the treatment. The treatment in hydrogen brought about the most profound changes [861]. The analysis of the liquids recovered from that treatment in a vibratory mill showed that the yield of

$360^{\circ}C^{-}$ distillates increased more than twofold when the grinding was conducted in hydrogen; meanwhile, the asphaltene content increased, especially when the grinding was conducted in Ar or air [862]. The IR and X-ray spectroscopic analysis indicated that this mechanical treatment increased the MW of the asphaltenes, decreased the size of their aromatic nuclei, and decreased their paraffinic component content. In contrast, the structure of the resins was only slightly affected [863]. The changes caused by vibrational grinding in air and Ar consisted of an increase in the number of alcoholic and of phenolic OH groups, decreased contents of carboxylic groups and aromatic esters, and a decreased degree of substitution of aromatic rings. These changes increased coals' reactivity with alkaline $KMnO_4$. The yield of humic acids from brown coals increased after grinding, but the MW of these acids decreased from 5000–50,000 to 5000–10,000 [864]. A deeper analysis (by IR and NMR spectroscopy) of the grinding in hydrogen products showed an increased in aliphatic CH_2 and CH_3 groups, phenolic OH, and aromatic CH groups. Meanwhile, grinding in a neutral medium increased the contents of phenolic and alcoholic OH and ester groups [865].

Hydrogenation of liptobiolitic coal in a petroleum residue, as solvent, was investigated in the presence of an iron-containing ore catalyst at 400–430°C and pressure of 70 bar. Conversion into gaseous and liquid products, in the range of 94% to 97%, was observed. The catalyst increased conversion by 21–23 wt%. Under these conditions, the liquid yield ($200^{\circ}C^{-}$ light hydrocarbons) was increased up to 24–28 wt%. The distillate products consisted mainly of paraffins, and most of the aromatic hydrocarbons were alkylbenzenes [866].

Hydrogenation of solid-C (graphite) was observed when hydride-forming metals (e.g., Zr) were present, and a high-energy impact milling action was applied. The MR was explained as due to activation of H_2 through the formation and decomposition of metal hydrides. This graphite hydrogenation MR was carried out under flow conditions [867].

9.3 MCP of Catalysts and Adsorbents

In addition to the preceding referenced examples, MCP (both MA and MR) has been broadly applied in catalysis, including reaction and catalyst manufacturing and activation, particularly in the preparation of nanocatalysts. The effect of the MCP of catalysts and materials represents a special case of solid mechanochemical treatments. Both activation and reactions have been proven to occur during and upon MCP. The MCP of inorganic oxides and mixtures falls into a simple scheme involving the concurrent manifestation of grinding, particle aggregation, and primary crystallite coalescence. The properties of the resulting powders are determined by the position of dynamic equilibrium among the processes taking place. Affecting the course of one or several processes makes it possible to shift the position of equilibrium and thus to obtain powders with different aggregation extents and different particle sizes and size distribution curves, or to carry out the treatment under conditions favorable for mechanochemical synthesis. Some examples are given in the work of Erofeev et al., who prepared and activated catalysts for hydrocarbon conversions [868,869].

MoS_2 fine particles were activated by comminution and tested in an autoclave for HDS activity. In contrast to conventional HDS catalysts, the comminuted MoS_2 fine particles catalyzed more HDS of 4,6-dimethyldibenzothiophene than that of dibenzothiophene (DBT). Furthermore, dry comminution gave better results than those obtained with wet comminution [870]. The active surface area of the comminuted MoS_2 increased with the volume fraction of grinding media, whereas the BET surface area obtained first increased and then reached a plateau. The catalytic activities for hydrogenation of 1-methylnaphthalene (1-MN) and HDS of DBT were evaluated on the comminuted MoS_2. Activity increased upon comminution, and the activity values correlated with the changes in active surface area [871]. MCP of unsupported mixtures of $CoS-MoS_2$ and $NiS-MoS_2$ were tested on the hydrogenation of 1-MN, DBT HDS, and coal liquefaction. The reaction rates using the mixed catalysts of individually comminuted metal sulfides increased proportionally to the weight fraction of the constituents. On the other hand, those rates using the comminuted mixture of binary metal sulfides were higher. New catalytic sites were supposed to be created on the surface of the comminuted binary mixture of the metal sulfides [872].

The effect of MCP on the properties of a high-silica zeolite resulted in a decrease of its acidity, crystallinity, and surface area. During testing of the MCP-treated zeolite, for upgrading low-octane SR gasoline, the concentration of aromatics decreased while that of alkenes, naphthenes, and isoalkanes increased as well as the product yield. The treated catalyst was also used for the aromatization of propane—butane mixtures. The aromatic yield was increased by about 10%, and the life cycle of the treated catalyst was nearly duplicated, in comparison to that of the untreated catalyst [873].

Kaolinite is one of the most studied materials. Its surface was modified by grinding kaolinite—quartz mixtures and resulted in structural bulk changes (delamination). MCP treatment of kaolinite results in a new surface structure [874]. Significant modifications were observed on the kaolinite surface, such as surface hydroxyls replaced with H_2O molecules and changes in the surface structure of the O—Si—O [875]. H_2 generation during kaolinite dry grinding increased with time up to a maximum concentration. This H_2 generation is considered to occur as a result of three processes: (1) structural destruction (delamination) and loss of hydroxyl groups, (2) transformation of liberated hydroxyls into H_2O molecules, and (3) H_2 generation through reaction between surface H_2O molecules and mechano-radicals created by the rupture of Si—O or Al—O—Si bonds. Although the surface area reached a plateau after 240 min grinding, the H_2 concentration continued to increase, indicating that surface mechano-radicals were created during this latter grinding stage [876].

Other catalyst MCP preparation examples included a weakly agglomerated 15 nm α-Al_2O_3 powder, and either yttrium- or calcium-stabilized zirconia [877].

The surface activation and chemical changes induced by MCP have been used as the basis for improving solid-based mixing and separation techniques. For instance, the structural modification of fumed silica to develop adsorption–desorption characteristics was studied. Treatment time was determinant for changes in the pore structure of the resulting material; both micropores and mesopores were affected. Pretreatments change the mesopore structure of final adsorbents to a greater extent than that observed on micropores [878]. Mesopores are important for asphaltene adsorption. The potentiality of using silica and siliceous materials for asphaltenes removal has been mentioned earlier in this chapter.

10. General Remarks

Emerging technologies for the processing of unconventional-opportunity crude oils fall into two categories: those that have reached the maximum development level and are available for commercial licensing, though no plant has been built yet; and those that have not yet climbed to that level and remain like new ideas or ideas tested at the most up to pilot plant scale.

10.1 Abatement of Families of Individual Compounds

In this chapter, ideas that have not made it to commercial scale and some emerging concepts for the abatement of the families of contaminant compounds of the BotB were presented and discussed.

10.1.1 Asphaltenes

As pointed out in Chapter 2, the development of a cost-effective alternative for current SDA is still needed. Increasing refinery flexibility toward the processing of complex diets would represent a competitive advantage for the possessor of that technological capability. The pressure imposed by current and future fuels specifications, the volatility of oil prices, and the constant decline in refining margins call for the incorporation of new ideas for the removal of contaminants and troublesome compounds.

The limitations of current SDA technologies have precluded their extensive deployment. Only its combination with other process units seems to have better commercialization possibilities. Old noncommercialized ideas and new emerging concepts may become more competitive in the future, particularly if the developmental work incorporates synergies to broaden the scope of the application.

10.1.2 Metals

Since the work on DM started, a large number of patents have been awarded that were never commercialized. Therefore, knowledge is now available for exploitation. Our search retrieved more than 1000 references, directly or indirectly related to DM (excluding HDM). After a first revision, 560 were directly concerned with DM, from which 225 were patents.

The fact that metals are concentrated in the heavy bottom of the crude oil, together with the asphaltenes, leads to the idea that a simultaneous removal of both components would make the process more efficient. Thus, DA was the first approach followed for removing metals. Combinations of solvents and post-DA treatments of DAO and pitch have been intended to improve the DM yields. However, results indicate that DA alone cannot achieve the required high DM yields. New metal-targeting features have to be brought in.

Physical and chemical treatments have been explored for DM, and they include combining and integrating DA with some other processes, adsorption–absorption, separation methods, acid and basic treatments, chemical reactions, catalytic (reductive and oxidative) removal, hydrotreating, HDC, and bioprocesses. Acid and basic treatments involve highly reactive reagents, which have strong hazards associated with their use and may not represent economical and environmentally friendly solutions. Furthermore, so far the achieved DM levels are comparable to or lower than those offered by SDA. Among the explored chemical reagents, CO_2, SO_2, and H_2S are typically available within a refinery. Although so far the results do not look very attractive, the new chemistry appears promising. Nonconventional catalysts based on Fe, Mn, and V; phosphates of Zr, Co, Fe, and Cu; and $NiAs_x$ have shown high DM activity (95%$^+$). Some of them required high hydrogen pressure, but some others did not. The highest activity was shown by $NiAs_x$.

Emerging concepts and knowledge were identified and discussed. These can be the basis for future developmental work. Polarized surfaces, solvent improvements for DA, physical methods, supercritical conditions, basic solids, nonconventional catalysts, oxidation reactions, particle capacitance effect, and process combinations and integration are part of these concepts and knowledge.

Although the history of DM is long, it seems that opportunities still exist for developing new processes and technologies, which might be economically more effective and efficient than those previously considered.

10.1.3 Acids

Although very sophisticated techniques have been used to separate, identify, characterize, and study the molecular characteristics and chemical behavior of NAs, very little is known and much further research is needed. A general molecular formula and very accurate MWs have been determined. Characterization studies of HACs have contributed to establish MS as a technique to map the fingerprint of a given crude oil.

A broad range of process areas have been examined for acidity abatement, including physical and chemical treatments and processes, as well as improvement to existing processes to diversify their use into the considered application. Blending, adsorption, extraction, neutralization, catalytic and thermal conversion (decarboxylation, hydrotreating, and decomposition), and microwave treatment have been reported.

The lack of a commercial application for any of them is an indication that a good balance of costs versus benefits has not been achieved yet. Hence, the opportunity for a cost-effective technology for acidity abatement remains open.

10.2 Whole Crude and Residue Upgrading

10.2.1 Downhole and In Situ Processing

The development level reached by thermal methods for recovery of heavy oil and oil sand deposits opens opportunities for upgrading in the field, near the well bore, or downhole that could not be possible before. The opportunities vary and depend on the thermal method and/or the reactivity of the employed heat carrier (steam, CO_2, organic vapors, etc.). Regardless of the appealing of the ideas, these upgrading processes entail not only great technological challenges but also economic considerations.

During some of these recovery processes, pyrolysis, aquathermolysis, and oxidation reactions may coexist because of the coexistence of bitumen, water, and air—oxygen, in an environment of increased temperature and pressure [879]. These observations can be taken as a confirmation of the existence of chemical reactor(s) in the heated well and a certain level of upgrading occurring during thermal recovery. The improvement of the upgrading induced by thermal in situ recovery methods could have the initial target of preparing pipeline-ready oil or an ultimate target of recovery a higher quality crude oil, which would not have the discount penalty. Notwithstanding, the final objective for ISU is to reduce the total energy and costs involved in production and surface upgrading.

Some of the proposed ideas were reviewed in this chapter. The THAI—CAPRI process take advantage of the heat produced by THAI—ISC, but it suffers the drawbacks associated with catalyst deactivation in fixed-bed HDT. Emerging alternatives have been found in dispersed ultrafine catalyst. As per slurry-phase systems, dispersed catalyst performed better than pelleted ones, even in the case of ultra-finely divided conventional catalysts, for ISU. Injecting nanosized catalysts into the mobile oil zone during the ISC process showed potential but needs further investigation.

Other downhole conditions have been tested at laboratory scale. All the variables known to affect performance have been considered, though some of them would not be technically feasible downhole. Economically feasible pathways for ISU using dispersed catalysts have been proposed for additional investigation [880]. Hence, challenges still exist and mainly concern providing heat, high temperatures and pressures, and sources of co-reacting upgrading atmosphere (H_2, water, CO, CO_2, etc.).

Bioprocessing of heavy oils and bitumen may find more opportunities downhole or in situ in the future, than it found in refining in the past. The level of upgrading required upstream is lower than that demanded for fuels production. Although R&D efforts are not as intense as they were in the 1990s, significant advances have been achieved. Bioprocessing research may need to focus now on reservoir conditions and make it happen. Some other reactions promoted by biocatalysts of no interest to refining would probably have better potential for upstream integration. As it has been pointed out, microorganisms promote oxidation of HMW aliphatics, aromatics, NAs, and S- and N-compounds, as well as hydrogenation and dehydrogenation of lower MW compounds. These reactions and their extension on microbial catalysts might produce the level of upgrading needed at field locations.

Although in all the areas reviewed extraordinary achievements and advances have been attained, there is still space for innovation. New out-of-the-box ideas are needed for optimizing energy balances, diminishing environmental impact, and processing costs. Conventional thinking does not seem to have any opportunity for solving the problems found with unconventional-opportunity crudes. While refining research during the last five or six decades has been devoted to removing heteroatomic contaminants, integrating refining processes upstream might necessitate the incorporation of slight amounts of these moieties in the crude oil. Catalysis research also has to focus on new forms and formulations of catalysts, to promote maybe the same reactions but through different mechanisms.

One thing is totally clear: the need for an R&D team constituted by multiple skills, from upstream and downstream, with a broad range of experiences in both time and type. For instance, R&D researches usually have minimal or no operational experience, and vice versa operators typically have no knowledge of the R&D environment and methodologies.

10.2.2 Radiation-Induced Conversion

Application of radiation cracking to hydrocarbons is not new; however, its consideration for the upgrading of heavy oils is more recent. Nevertheless, any of the different forms of radiation (neutrons, electrons, X-rays, γ-rays, etc.) has not succeeded in adding value to the processing of crude oil. In general, tested samples need to be irradiated with energy that is many orders of magnitude in excess of that required to break hydrocarbon bonds. Different degrees of conversion have been assessed, though polymerization and recombination reactions compete with cracking (in some instances in a more favorable way). In fact, since both C—C and C—H bonds can be broken upon irradiation, the highly reactive formed fragments can undergo a variety of reactions.

Most of the reported results in the literature correspond to the e-beam upgrading of crude oils (from the Kazakhstan and Caspian regions) containing high concentrations of water, rarely found in refinery feedstocks. The presented comparisons indicated that conversion strongly depended on feedstock composition and/or reacting atmosphere. Ozone (and ozone-enriched air) was used to create a reactive atmosphere (oxidant) to convert S-compounds into sulfones and sulfoxides. On one hand, the effect of water needs to be assessed, and in the other, rigorous mass and energy balances are still pending. For instance, conclusive remarks were made on the desulfurization capabilities based on lower S-content of the (lighter) liquid products (compared to the original present in the feedstock). However, sulfide compounds were only oxidized and remained in the heavier fractions.

The reported attempts on comparing economics between conventional refining and RP were unsubstantiated. The concluding advantages of e-beam treatments over refining processes were based in an oversimplified process scheme and the assumption of an e-beam product of similar quality than that of a refinery.

Linear accelerators seem to be the preferred source for radiation-induced chemistry, probably because residual radiation when electrical energy is turned off will be minimum or none. Although technology advancements have lowered both capital and operating costs, further costs and footprint decreases are needed. Some other improvements are required as well, such as increasing beam guns efficiency and energy capture by the target molecules.

In terms of IP, there seems to be a very unpopulated field with huge opportunities for R&D. As mentioned above, high-energy electron beams have been the most deeply studied radiation source so far. Consequently, e-beam-derived technology is also the most advanced, and Petrobeam Ltd was the company that was championing it for years. Nonetheless, a pilot plant has not been built yet.

10.2.3 Sonochemistry

Application of sonochemistry at the level capacity of a commercial refinery faced scaling-up challenges, mainly due to the lack of availability of large equipment. Modular alternatives were considered. Still, it was demonstrated that the pure effect of ultrasound was temporal, and efforts were taken to make it more permanent. These efforts included the addition of surface active agents and solvents, for instance.

Conversion and more particularly oxidative desulfurization were tested. Initially, research results reported C—C and C—H bonds breaking under cavitation conditions. These reactions were never proven to occur at a larger scale. Furthermore, ultrasound did not sufficiently enhance the action of peroxides to partially oxidize the S-compounds. In fact, the efficiency of ultrasound was found to be lower than that of high-pressure hydrodynamic cavitation [881]. Proposed sonochemical ODS in the presence of peroxides is based on the formation of OH radicals. According to Gogate et al., the yield of OH radicals can only be about 20 μmol/L, when applying 1 h ultrasound with the most efficient equipment. Since heavy oil, residues, and bitumen contain sulfur in the percentage range, the required (OH) radicals yield increase is three to four orders of magnitude greater than that achievable as assessed by Gogate et al. [881]. The S concentration in fuels in the ppm range may give better opportunities for this method. Nevertheless, the additional step required to extract or remove the formed oxi-sulfide compounds subtracts competitiveness from the system.

10.2.4 Mechanochemistry

While sonochemical processes are localized in space and time, mechanochemical processes are characterized by the localization in space. In mechanochemistry, a specific particle region and not the whole particle would undergo chemical transformation at any instant of time. The sequential accumulation of such sites in the particle created at different times by the mechanical energy would determine the process kinetics.

The successful applications of MCP in coal conversion have not been resembled in oil refining. Since MCP-treated or -manufactured catalysts have shown performance advantages and mechano-upgrading of hydrocarbons has been proven to occur, a potential exists for a mechanochemical process for upgrading opportunity crudes.

10.3 Summary

Most of the R&D activities has centered on catalysis and process design. The lack of cost-effective solutions has not resulted from the lack of R&D efforts. In fact, the devoted efforts started more than six decades ago and continued through the twenty-first century. The long-lasting history of these efforts can be taken as indicative of the need for a holistic interdisciplinary approach, incorporating new knowledge and contributions from diverse areas. A deeper knowledge on the molecular structures of the BotB compounds is being acquired, and soon a complete, extremely detailed picture of the whole residue will enable the creation of novel options.

Some of the most recent upgrading concepts were based on unconventionally exotic methods, like those using other types of energy sources such as mechanical, electron beam, γ-rays, microwaves, plasma, and ultrasound. They failed in providing robust and competitive data consistently. Compared with commercially available upgrading technologies, these unconventional methods have failed in demonstrating any potential benefit so far.

The conventional existing technologies, commercially available were presented and discussed in the previous chapter. Although some of these technologies might represent a better option than others and they pass the technoeconomic feasibility test, none of these can be robustly categorized as cost-effective. Their opportunity is attached to concurrent high oil prices and large differentials.

References

[1] Ignasiak T, Kemp-Jones AV, Strausz OP. The molecular structure of Athabasca asphaltene. Cleavage of the carbon-sulfur bonds by radical ion electron transfer reactions. J Org Chem 1977;42(2):312–20.

[2] Mcintyre DD, Montgomery DS, Strausz OP. Carboxylic acids bound to the mineral matter in heavy cretaceous oils of Alberta. AOSTRA J Res 1986;2:251–65.

[3] Payzant JD, Mojelsky TW, Strausz OP. Improved methods for the selective isolation of the sulfide and thiophenic classes of compounds from petroleum. Energy Fuels 1989;3(4):449–54.

[4] Payzant JD, Montgomery DS, Strausz OP. The identification of homologous series of benzo[b] thiophenes, thiophenes, thiolanes, and thianes assessing linear carbon framework in the pyrolysis oil of Athabasca asphaltene. AOSTRA J Res 1988;4:117–31.

[5] Mccall TF, Ramamurthy P, Shah KV, Silverman MA, Van Sickels MJ. Supercritical hydrocarbon conversion process. Patent No. US7144498 (Also published as WO2005074440, US2005167333, US2007090021). Kellogg Brown and Root, Inc.; December 05, 2006.

[6] Garifzjanova GG, Garifzjanov GG, Tukhvatullin AM, Jarullin MR. Method of processing high-viscosity high-sulfur petroleum. Patent No. RU2187536. Ooo Plazmokhim Fa.; August 20, 2002.

[7] Bearden R, Hamner GP. Combined catalytic and alkali metal hydrodesulfurization and conversion. Patent No. US3976559. Exxon Research and Engineering Co.; August 24, 1976.

[8] Mojelsky TW, Ignasiak TM, Frakman Z, Mcintyre DD, Lown EM, Montgomery DS, et al. Structural features of Alberta oil sand bitumen and heavy oil asphaltenes. Energy Fuels 1992;6(1):83–96.

[9] Strausz OP, Mojelsky TW, Faraji F, Lown EM, Peng P. Additional structural details on Athabasca asphaltene and their ramifications. Energy Fuels 1999;13(2):207–27.

[10] Strausz OP, Mojelsky TW, Lown EM, Kowalewski I, Behar F. Structural features of boscan and Duri asphaltenes. Energy Fuels 1999;13(2):228–47.

[11] Wang Z, Que G, Liang W, Qian J. Investigation on chemical structure of resins and pentane asphaltenes in vacuum residua. Acta Pet Sin (Pet Proc Sect) 1999;15(6):39–46.

[12] Mojelsky TW, Montgomery DS, Strausz OP. The side chains associated with the. Undistillable aromatic and resin components of Athabasca bitumen. AOSTRA J Res 1986;2:177–84.

[13] Strausz OP. Structural investigations of Alberta oil sand bitumens. In: Proc. ACS nat. meeting, div. pet. chem. preprints. Toronto, ON, Canada, vol. 33. ACS; 1988. p. 264–8.

[14] Strausz OP. Recent advances in the chemistry of alberta bitumens and heavy oils. In: Proc. 4th UNITAR intern. conf. 'heavy crude tar sands'. Edmonton, Alberta, Canada; August 7–12, 1989. p. 605–28.

[15] Strausz OP. Bitumen and heavy oil chemistry. AOSTRA-technical handbook on oilsands, Bitumens heavy oils. 1989. p. 34–73.

[16] Strausz OP. Structural features of Athabasca bitumen related to upgrading performance. In: Proc. ACS nat. meeting, div. pet. chem. preprints. Dallas, TX, USA, vol. 34. ACS; 1989. p. 395–400.

[17] Payzant JD, Mcintyre DD, Mojelsky TW, Torres M, Montgomery DS, Strausz OP. The identification of homologous series of thiolanes and thianes possessing a linear carbon framework from petroleums and their interconversion under simulated geological conditions. Org Geochem 1989;14(4):461–73.

[18] Chen HH, Mojelsky TW, Payzant JD, Lown EM, Henry D, Strausz OP. Chemical changes in Alberta oil sands during steam treatment. AOSTRA J Res 1991;7(1):17–35.

[19] Strausz OP, Lown EM. Structural features of Athabasca bitumen related to upgrading performance. Fuel Sci Technol Intern 1991;9(3):269–81.

[20] Strausz OP, Mojelsky TW, Lown EM. The molecular structure of asphaltene: an unfolding story. Fuel 1992;71:1355–63.

[21] Strausz OP, Lown EM, Mojelsky TW. Molecular structure of asphaltene. In: Proc. ACS nat. meeting, div. pet. chem. preprints. San Diego, CA, USA, vol. 39. ACS; March 13–18, 1994. p. 199–207.
[22] Strausz OP, Lown EM, Mojelsky TW. Oil-soluble catalysts for the hydrocracking of Athabasca bitumen. In: Proc. 210th ACS nat. meeting, div. pet. chem. preprints. Chicago, IL, USA, vol. 40. ACS; August 20–24, 1995. p. 741–2.
[23] Peng P, Morales-Izquierdo A, Hogg A, Strausz OP. Molecular structure of Athabasca asphaltene: sulfide, ether, and ester linkages. Energy Fuels 1997;11(6):1171–87.
[24] Strausz OP, Lown EM, Mojelsky TW, Peng P. About the structure of petroleum asphaltenes. ACS Div Fuel Chem 1998;43(4):917–23. Preprints.
[25] Peng P, Morales-Izquierdo A, Lown EM, Strausz OP. Chemical structure and biomarker content of jinghan asphaltenes and kerogens. Energy Fuels 1999;13(2):248–65.
[26] Peng P, Fu J, Sheng G, Morales-Izquierdo A, Lown EM, Strausz OP. Ruthenium-ions-catalyzed oxidation of an immature asphaltene: structural features and biomarker distribution. Energy Fuels 1999;13(2):266–77.
[27] Peng P, Morales-Izquierdo A, Lown EM, Strausz OP. Oxidation of an immature asphaltene: structural features and biomarker distribution. Energy Fuels 1999;13(2):266–77.
[28] Strausz OP, Peng P, Murgich J. Covalent structure and molecular architecture of Athabasca asphaltene. ACS Div Fuel Chem 1999;44(4):743–5. Preprints.
[29] Artok L, Su Y, Hirose Y, Hosokawa M, Murata S, Nomura M. Structure and reactivity of petroleum-derived asphaltene. Energy Fuels 1999;13(2):287–96.
[30] Strausz OP, Peng P, Murgich J. About the colloidal nature of asphaltenes and the MW of covalent monomeric units. Energy Fuels 2002;16(4):809–22.
[31] Ali MF, Siddiqui MN, Al-Hajji AA. Structural studies on residual fuel oil asphaltenes by rico method. Pet Sci Technol 2004;22(5–6):631–45.
[32] Zhang H, Yan Y, Cheng Z, Sun W. Structural changes of sub-fractions in residue hydrotreating products by ruthenium catalyzed oxidation. Pet Sci Technol 2008;26(16):1945–62.
[33] Zhan GZ, Shaohui G, Suoqi Z, Guangxu Y, Lanqi S, Libo C. Alkyl side chains connected to aromatic units in dagang vacuum residue and its supercritical fluid extraction and fractions (sfefs). Energy Fuels 2009;23(1):374–85.
[34] Reynolds BE, Rogers JL, Broussard RA. Evaluation of residue conversion options. In: Proc. NPRA annual meeting. San Antonio, TX; March 16–18, 1997. Paper #AM-97-51, 14 pp.
[35] Waugh RJ, Chambers LW, Menzies MA, Patmore DJ. CANMET hydrocracking handles wide range of heavy crudes. In: Proc. NPRA annual meeting. San Francisco, CA. USA; March 20–22, 1983. 18 pp.
[36] Pruden BB, Denis JM, Muir DCG. Upgrading of Cold Lake heavy oil in the CANMET hydrocracking demonstrations plant. In: Proc. 4th UNITAR intern. conf. 'heavy crude tar sands'. Edmonton, Canada; August 7–12, 1988. p. 249–54.
[37] Waugh WB, Miur G, Skripek M. Advanced hydroprocessing of heavy oil and refinery residues. In: Proc. 13th world pet. congress. Buenos Aires, Argentina; October 20–25, 1991. Topic 18, Paper #4. 6 pp.
[38] Pruden BB. Hydrocracking of bitumen and heavy oils at CANMET. Can J Chem Eng 1978;56:277–80.
[39] Kriz JF, Ternan M. Hydrocracking of heavy asphaltenic oil in presence of an additive to prevent coke formation. Patent No. US5296130 (Also published as CA2111665). Canadian Minister of Energy; March 22, 1994.
[40] Gillis D, Vanwees M, Zimmerman P, Houde E. Upgrading residues to maximize distillate yields with UOP uniflex™ process. J Jpn Pet Inst 2010;53(1):33–41.
[41] Gillis D, Zimmerman P, Houde E, Haizmann R, Van Wees M. Upgrading residues to maximize distillate yields. In: Proc. NPRA annual meeting. San Antonio, TX, vol. 2; March 22–24, 2009. p. 780–98.
[42] Ondrey G. Chementator: improved heavy-crude conversion process boosts diesel yields. Chem Eng 2012;119(3). 12 (1 p.).

[43] Munoz TA, Pachano JR, Pinzon P, Krasuk JH. HDH, a process for short residue conversion. Proceedings65. Refining Department, American Pet. Institute; 1986. p. 249–78.
[44] De Drago G, Guitian J, Krasuk J, Larrauri J, Marzin R, Silva F, et al. Desarrollo del proceso HDH (development of the HDH process). Rev Tec INTEVEP 1989;9(2):111–33.
[45] De Drago G, Guitian J, Krasuk J, Larrauri J, Marzin R, Silva F, et al. Development of HDH process, a refiner's tool for residual upgrading. In: Proc. ACS div. pet. chem. symp. on resid upgrading. Washington, DC, USA, vol. 35. ACS; August 26–31, 1990. p. 584–92.
[46] De Drago G, Guitian J, Krasuk J, Larrauri J, Marzin R, Silva F, et al. Development of the HDH process. Rev Tec INTEVEP 1990;10(1):15–36.
[47] Silva F, Guitian J, Galiasso R, Krasuk J. The HDH process for upgrading heavy Venezuelan crudes. In: Proc. 11th world pet. congress. London, UK, vol. 4. RTD; August 28–September 2, 1983. 10 pp.
[48] Krasuk JH, Silva FJ, Galiasso RE, Souto A. Hydroconversion and upgrading of heavy crudes of high metal and asphaltene content. Patent No. US4591426. Intevep S.A.; May 27, 1986.
[49] Hata K-A, Wada K, Mitsudo T-A. Iron-catalyzed coprocessing of coals and vacuum residues using syngas-water as a hydrogen source. Energy Fuels 1998;12:1181–90.
[50] Callejas MA, Martınez MT. Hydrocracking of a Maya residue. Kinetics and product yield distributions. Ind Eng Chem Res 1999;38:3285–9.
[51] Callejas MA, Martınez MT. Hydroprocessing of a Maya residue. 1. Intrinsic kinetics of asphaltene removal reactions. Energy Fuels 2000;14:1304–8.
[52] Dehkissia S, Larachi F, Rodrigue D, Chornet E. Lowering the viscosity of Doba-Chad heavy crude oil for pipeline transportationsthe hydrovisbreaking approach. Energy Fuels 2004;18:1156–68.
[53] Kissel CL. Well or oil treating method to effect dewaxing. Patent No. GB2304731. Amvac Chemical Corp.; March 26, 1997.
[54] Pakdel H, Roy C. Recovery of bitumen by vacuum pyrolysis of Alberta tar sands. Energy Fuels 2003;17:1145–52.
[55] Feimer JL, Desjardine LT. Multi-stage ultrafiltration process (OP-3711). Patent No. US5256297 (Also published as WO9413756). Exxon Research and Engineering Co.; October 26, 1993.
[56] Osterhuber EJ. Upgrading of heavy oils by solvent dissolution and ultrafiltration. Patent No. US4797200. Exxon Research and Engineering Co.; January 10, 1989.
[57] Iacocca R, Yen TF, Randall G, Duffield R. Treatment of fluids. Patent No. CA2297094 (Also published as EP1024894, WO1999004898). Klinair Environmental Technologies (Ireland) Limited; February 4, 1999.
[58] Ignasiak TM, Kotlyar L, Samman N, Montgomery DS, Strausz OP. Preparative gel permeation chromatography of Athabasca asphaltene and the relative polymer-forming propensity of the fractions. Fuel 1983;62(3):363–9.
[59] López-Linares F, Carbognani L, González MF, Sosa-Stull C, Figueras M, Pereira-Almao P. Quinolin-65 and violanthrone-79 as model molecules for the kinetics of the adsorption of C7 Athabasca asphaltene on macroporous solid surfaces. Energy Fuels 2006;20(6):2748–50.
[60] Gonzalez MF, Stull CS, Lopez-Linares F, Pereira-Almao P. Comparing asphaltene adsorption with model heavy molecules over macroporous solid surfaces. Energy Fuels 2007;21(1):234–41.
[61] Carbognani L, Gonzalez MF, Lopez-Linares F, Sosa Stull C, Pereira-Almao P. Selective adsorption of thermal cracked heavy molecules. Energy Fuels 2008;22(3):1739–46.
[62] López-Linares F, Carbognani L, Sosa-Stull C, Hassan A, Pereira-Almao P, Spencer RJ. Adsorption studies of heavy molecules on solid sorbents. In: Proc. AIChE spring nat. meeting. New Orleans, LA; 2008. 1 pp.
[63] Sosa-Stull C, Lopez-Linares F, Carbognani L, Pereira-Almao P. Adsorption of virgin and thermal cracked Athabasca asphaltenes for further gasification. In: Proc. AIChE spring nat. meeting. New Orleans, LA; 2008. 1 pp.
[64] López-Linares F, Carbognani L, Sosa-Stull C, Pereira-Almao P, Spencer RJ. Adsorption of virgin and visbroken residue asphaltenes over solid surfaces. 1. Kaolin, smectite clay minerals, and Athabasca siltstone. Energy Fuels 2009;23(4):1901–8.

[65] Lopez-Linares F, Carbognani L, Hassan A, Pereira-Almao P, Rogel E, Ovalles C, et al. Adsorption of Athabasca vacuum residues and their visbroken products over macroporous solids: influence of their molecular characteristics. Energy Fuels 2011;25(9):4049–54.
[66] Lopez-Linares F, Carbognani L, Pereira-Almao P, Rogel E, Ovalles C, Pradhan A, et al. Molecular properties of virgin and thermal cracked Athabasca vacuum residue influencing their adsorption over macroporous surfaces. In: Proc. 241st ACS nat. meeting. Anaheim, CA; March 27–31, 2011. 1 p.
[67] López-Linares F, Carbognani L, Spencer RJ, Pereira-Almao P. Adsorption studies in Athabasca core sample: virgin and mild thermal cracked residua. Energy Fuels 2011;25(8):3657–62.
[68] Nassar NN, Hassan A, Carbognani L, Lopez-Linares F, Pereira-Almao P. Iron oxide nanoparticles for rapid adsorption and enhanced catalytic oxidation of thermally cracked asphaltenes. Fuel 2012;95:257–62.
[69] Castillo J, Fernandez A, Ranaudo MA, Acevedo S. New techniques and methods for the study of aggregation, adsorption, and solubility of asphaltenes. Impact of these properties on colloidal structure and flocculation. Pet Sci Technol 2001;19(1–2):75–106.
[70] Acevedo S, Escobar O, Echevarria L, Gutierrez LB, Mendez B. Structural analysis of soluble and insoluble fractions of asphaltenes isolated using the PNP method. Relation between asphaltene structure and solubility. Energy Fuels 2004;18(2):305–11.
[71] Abraham T, Christendat D, Karan K, Xu Z, Masliyah J. Asphaltene-silica interactions in aqueous solutions: direct force measurements combined with electrokinetic studies. Ind Eng Chem Res 2002;41:2170–7.
[72] Gorbaty ML, Ferrughelli DT, Olmstead WN, Miseo S, Soled SL, Robbins WK. Selective adsorption process for resid upgrading (law815). Patent No. US6245223 (Also published as US2001036564, WO9931199). Exxon Research and Engineering Co.; June 12, 2001.
[73] Xie K, Karan K. Kinetics and thermodynamics of asphaltene adsorption on metal surfaces: a preliminary study. Energy Fuels 2005;19(4):1252–60.
[74] Murrell LL, Grenoble DC, Long RB. Separating basic asphaltenes using transition metal oxide acid catalysts. Patent No. US4424114. Exxon Research and Engineering Co.; January 03, 1984.
[75] Stine LO. Petroleum crude oil conversion process. Patent No. US3696024. Universal Oil Products Co.; October 3, 1972.
[76] Hemminger CE. Solvent treating before catalytic cracking. Patent No. US3507777. Esso Research and Engineering Co.; April 21, 1970.
[77] Vanpool J. Production of low-metal-content gas oil from topped crude oil. Patent No. US3658695. Phillips Petroleum Co.; April 25, 1972.
[78] Miller CB, Moore HF, Wesley DP, Wombles RH, Jewitt CH, Hayner RE, et al. Demetallation - high carbon conversion process, apparatus and asphalt products. Patent No. US5601697. Ashland Oil Inc.; February 11, 1997.
[79] Miller CB, Moore HF, Wesley DP, Wombles RH, Jewitt CH, Hayner RE, et al. Demetallation - high carbon conversion process, apparatus and asphalt products. Patent No. US5728291. Ashland Oil Inc.; March 17, 1998.
[80] Murphy JR. Solvent decarbonizing process. Patent No. US3423308. Gulf Research and Development Co.; January 21, 1969.
[81] Gatsis JG. Solvent deasphalting. Patent No. US3830732. Universal Oil Products Co.; August 20, 1974.
[82] Coombs DM. Two-feed oil extraction and demetalization process. Patent No. CA1232859. Phillips Petroleum Co.; February 16, 1988.
[83] Hedrick BW. Process for separating a resin phase from a solution containing a resin, a solvent, and demetallized oil. Patent No. US5145574. UOP Inc.; September 8, 1992.
[84] Hedrick BW. Process for separating a resin phase from a solvent solution containing a solvent, demetallized oil and a resin. Patent No. US5098994. UOP Inc.; March 24, 1992.
[85] Doberenz EH, Sullivan JN. Solvent deasphalting process. Patent No. US3278415. Chevron Research Co.; October 11, 1966.

[86] Long RB, Solomon HJ. Removal of organometallic compounds from asphaltic residuals. Patent No. US3511774. Esso Research and Engineering Co.; May 12, 1970.
[87] Di Tullio V. A process for low temperature separation and isolation of crude heavy oil. Patent No. CA2304972. Deinking/Solvent Extraction Technologies Ltd.; October 12, 2001.
[88] Beach LK, Shewmaker JE. Extraction of metals from heavy petroleum oils. Patent No. US2922760. Esso Research and Engineering Co.; January 26, 1960.
[89] Blum SC, Sartori G, Gorbaty ML, Savage DW, Dalrymple DC, Wales WE. CO2 treatment to remove organically bound metal ions from crude. Patent No. US6093311 (Also published as CA226652C, CN115123C, CN123248A, EP0931122, WO1998014534). ExxonMobil Research and Engineering, Co.; July 25, 2000.
[90] Blum SC, Sartori G, Gorbaty ML, Savage DW, Dalrymple DC, Wales WE. CO2 treatment to remove organically bound metal ions from crude. Patent No. US6187175. ExxonMobil Research and Engineering, Co.; February 13, 2001.
[91] Khusnutdinov IS, Kozin VG, Kopylov AJ. Method of fractionating natural bitumens and high-viscosity petroleums. Patent No. RU2138537. Kz G T Univer; September 27, 1999.
[92] Lerda D, Colombani G, Prevedello A. Process for the removal of heavy metals and asphaltenes from crude petroleum or heavy fractions thereof. Patent No. IT1244487. Eniricerche SPA.; July 15, 1994.
[93] Mann DP, Kukes SG, Coombs DM. Metals removal from oils with a light hydrocarbon and an organophosphorous compound. Patent No. US4518484. Phillips Petroleum Co.; May 21, 1985.
[94] Savastano C, Cimino R. Process for deasphalting and demetallizing petroleum residues. Patent No. EP612829 (Also published as US5466365, US5346615, RU2119525, MX9401362 JPH06299167 JP3484580 ITMI930347 IT1263961 ES2107736 DK0612829 DE69405123 CA2115488 AU5512394 AU662672 AT157390). Eniricerche S.p.A.; August 31, 1994.
[95] Varadaraj R, Wales WE. Additive enhanced solvent deasphalting process. Patent No. US6245222 (Also published as WO0024845, NO20011996, JP2002528595, ID29228, EP1153104, CA2345617, AU6415499). Exxon Research and Engineering Co.; June 12, 2001.
[96] Bushnell JD, Glivicky AP, Leighton MD, Sankey BM. Simultaneous deasphalting-extraction process. Patent No. US4125458 (Also published as GB2012809). Exxon Research and Engineering Co.; November 14, 1978.
[97] Brons GB, Siskin M. Upgrading of bitumen asphaltenes by hot water treatment containing carbonate. Patent No. US5326456 (Also published as CA2118984). Wrzeszczynski Kazimierz O; October 3, 1994.
[98] Kulkarni SS, Chang YA, Gatsis JG, Funk EW. Membrane separation of hydrocarbons using cycloparaffinic solvents. Patent No. US4750990. UOP Inc.; June 14, 1988.
[99] Farcasiu M. Process for demetalizing petroleum utilizing strong solid-phase bronsted acids. Patent No. US4460458. Mobil Oil Corp.; July 17, 1984.
[100] Bartholic DB, Haseltine RP. Development of engelhard's ART process. In: Proc. 46th midyear meeting of the American Pet. Institute. Chicago, IL, USA; March 11, 1981. p. 335–9.
[101] Suchanek AJ, Christian BR. New diversity shown for the ART process. In: Proc. heavy oil oil sands technical symp. Calgary, Canada; March 16, 1988. AM-88–74, 22 pp.
[102] Bartholic DB, Barger DF, Center AM. The applicability of the ART process to whole Maya crude upgrading. In: Proc. NPRA annual meeting. San Antonio, TX, USA; March 24–26, 1985. 85-AM-34, 16 pp.
[103] Haseltine RP, Caldwell DL, Logwinuk AK. The ART process offers increased refinery flexibility. In: Proc. NPRA annual meeting. San Francisco, CA, USA; March 20–22, 1983. AM-83-43, 25 pp.
[104] Bartholic DB, Haseltine RP. New crude/resid treating process offers savings. Oil Gas J 1981;79(45):242–52.
[105] Bartholic DB, Haseltine RP. New process cleans up FCC feed. Oil Gas J 1981;79(22):89–92.
[106] Bartholic DB, Center AM, Christian BR, Suchanek AJ. Petroleum refinery of the future. In: Mcketta JJ, editor. Petroleum processing handbook. New York, NY, USA: Marcel Dekker, Inc. CRC Press; 1992. p. 108–29.

[107] Bartholic DB. Treatment of petroleum stocks containing metals. Patent No. US4256567. Engelhard Minerals and Chemicals Corp.; March 17, 1981.

[108] Bartholic DB, Flanders RL. Upgrading crude oil and its residual fractions by vaporizing the charge in a falling curtain of contact particles. Patent No. US4435272. Engelhard Corp.; March 06, 1984.

[109] Bartholic DB, Reagan WJ. Process for upgrading tar sand bitumen. Patent No. US4804459. Engelhard Corp.; February 14, 1989.

[110] Reagan W, Byrne JW, Himpsl FL. Pretreating hydrocarbon feed stocks using deactivated FCC catalyst. Patent No. US4384949. Engelhard Min & Chem; May 24, 1983.

[111] Bartholic DB. Preparation of FCC charge from residual fractions. Patent No. US4243514, US4309274. Engelhard Minerals and Chemicals Corp.; 1981, 1982, January 06, January 05.

[112] Busch LE, Hettinger Jr WP, Krock RP. Reduced crude conversion - 2. Demetallization unit broadens RCC feed slate. Oil Gas J 1984;82(52):54−6.

[113] Moore AS. Selective vaporization process. Patent No. US4569754 (Also published as EP0186403, CN85109037, CA1250243). Engelhard Corp.; February 11, 1986.

[114] Busch LE, Walters PW, Zandona OJ. Combination process for upgrading residual oils. Patent No. EP0066387 (Also published as US4894141). Ashland Oil, Inc.; December 08, 1982.

[115] Busch LE, Henderson GO. Demetallizing and decarbonizing heavy residual oil feeds. Patent No. US4427539. Ashland Oil, Inc.; January 24, 1984.

[116] Beck HW, Carruthers JD, Cornelius EB, Hettinger Jr WP, Kovach SM, Palmer JL, et al. Immobilization of vanadium deposited on sorbent materials during the treatment of oils containing heavy metals and coke precursors. Patent No. EP0063683 (Also published as US4549958, WO8203226). Ashland Oil, Inc.; November 03, 1982.

[117] Hettinger WP, Carruthers JD, Watkins WD. Immobilization of vanadium deposited on sorbent materials during the treatment of oils containing heavy metals and coke precursors. Patent No. EP0065626 (Also published as JPS57182387, JPS6253038, CA1175375, BR8202229, AU8131182, AU534630). Ashland Oil, Inc.; December 1, 1982.

[118] Hettinger WP, Beck WH. Immobilization of vanadia deposited on sorbent materials during visbreaking treatment of carbo-metallic oils. Patent No. US4469588 (Also published as CA1175000, BR8201773). Ashland Oil, Inc.; September 04, 1984.

[119] Hettinger WP, Beck WH. Sorbent useful in a visbreaking treatment of carbo-metallic oils. Patent No. US4515900 (Also published as EP0175799). Ashland Oil, Inc.; May 7, 1985.

[120] Beck HW, Lochow J, Charles F, Nibert CW. Passivation of vanadium accumulated on catalytic solid fluidizable particles. Patent No. US4515683 (Also published as EP0140007, NO843643, NO167586, JPS6096687, JPS6335197, CA1212367). Ashland Oil Inc.; May 07, 1985.

[121] Barger DF. Combination process for upgrading residual oils. Patent No. US4525268. Ashland Oil Inc.; June 25, 1985.

[122] Busch LE, Walters PW, Zandona OJ. Oil upgrading by thermal and catalytic cracking. Patent No. US4569753. Ashland Oil Inc.; February 11, 1986.

[123] Hettinger Jr WP, Kovach SM, Beck HW. Process for cracking high-boiling hydrocarbons using continuous addition of acidity enhancing additives. Patent No. US4432890, US4496665, US4552647 (Also published as EP0123014, MX163298, JPS59170188, PS6049676, CA1217441, AU2463084). Ashland Oil Inc.; 1981, 1985, March 30, January 29, November 12.

[124] Beck HW, Carruthers JD, Cornelius EB, Hettinger WP, Kovach SM, Palmer JL, et al. Immobilisation of vanadium deposited on catalytic materials during conversion of oil containing coke precursors and heavy metals. Patent No. EP0063712 (Also published as US4485184, MX162472, JPS57182386, JPS62194, CA1190879). Ashland Oil Inc.; November 03, 1982.

[125] Beck WH, Carruthers JD, Cornelius EB, Hettinger WP, Kovach SM. Immobilisation of vanadia deposited on catalytic materials during the conversion of oil that contains coke precursors. Patent No. EP0073874 (Also published as US4432890). Ashland Oil Inc.; March 16, 1983.

[126] Beck WH, Carruthers JD, Cornelius EB, Hettinger Jr WP, Kovach SM, Palmer JL, et al. Immobilization of vanadia deposited on catalytic materials during carbo-metallic oil conversion. Patent No. EP0074945

(Also published as JPS5832691, JPS5949275, WO8203225, AU7323481, US4750987, CA1183792). Ashland Oil, Inc.; March 30, 1983.
[127] Hettinger Jr WP, Kovach SM. Cracking high-boiling hydrocarbons using high ratio of catalyst residence time to vapor residence time. Patent No. US4374019. Ashland Oil, Inc.; February 15, 1983.
[128] Kovach SM, Palmer JL. Method to passivate metals deposited on a cracking catalyst during reduced crude processing. Patent No. EP0077044 (Also published as US4551231). Ashland Oil Inc.; April 20, 1983.
[129] Myers GD. Use of naphtha as riser diluent in carbo-metallic oil conversion. Patent No. US4376038 (Also published as EP0061137, ZA8201681, MX163043, JPS57190088, JPH0216354, CA1168613, BR8201577, AU8131382, AU3043984, AU570253). Ashland Oil Inc.; March 08, 1983.
[130] Walters P, Benslay RM, Barger DF. Carbometallic oil conversion with ballistic separation. Patent No. US4390503 (Also published as EP0065274, CA1181573). Ashland Oil Inc.; June 28, 1983.
[131] Hettinger WP. Passivating heavy metals in carbo-metallic oil conversion. Patent No. US4454025. Ashland Oil Inc.; June 12, 1984.
[132] Hettinger Jr WP, Beck HW. Process and catalyst for the conversion of oils that contain carbon precursors and heavy metals. Patent No. EP0112956. Ashland Oil, Inc.; July 11, 1984.
[133] Hettinger WP, Beck HW, Kovach SM. Carbo-metallic oil conversion catalysts. Patent No. US4440868 (Also published as EP0083163, WO8302070, MX163139, JPS58112052, CA1183828). Ashland Oil Inc.; April 03, 1984.
[134] Kovach SM. Immobilization and neutralization of contaminants in crude oil. Patent No. US4437981. Ashland Oil, Inc.; March 20, 1984.
[135] Zandona OJ, Hettinger Jr WP, Kovach SM, Beck HW. Process and catalyst for the conversion of carbo-metallic containing oils. Patent No. US4457833. Ashland Oil Inc.; July 03, 1984.
[136] Hettinger WP, Beck HW. Carbo-metallic oil conversion process and catalysts. Patent No. US4877514. Ashland Oil Inc.; October 31, 1989.
[137] Reid JS, Szymanski T, Mesher SA. Removal of metal impurities from hydrocarbon streams. Patent No. US6372124 (Also published as US2002014439, CA2437769, CN1735677, DE10296268, WO2002064703). Saint-Gobain Norpro Corp.; April 16, 2002.
[138] Mitchell DS. Removing metalliferous contaminants from hydrocarbons. Patent No. US3947347. Chevron Research Co.; March 30, 1976.
[139] Murphy CB, Fabri JM, Mahoney RP. Organoclay compositions for purifying contaminated liquids and methods for making and using them. Patent No. US6627084 (Also published as US2003015473). Polymer Ventures Inc.; September 30, 2003.
[140] Farha FE, Fendley EC. Promoted absorbents for the removal of contaminants. Patent No. US6752919 (Also published as US2004023018). Chemical Products Industries, Inc.; June 22, 2004.
[141] Blytas GC. Removal of metal compounds from hydrocarbons. Patent No. US4048061. Shell Oil Co.; September 13, 1977.
[142] Blytas GC. Process for demetallizing heavy hydrocarbons. Patent No. US4116820. Shell Oil Co.; Septmber 26, 1978.
[143] Bott TR, Ellis SRM, Panzner F, Bott TR, Ellis SR, Panzner F. Extraction process for asphaltic materials. Patent No. GB2001670. February 07, 1979.
[144] Sadeghi M-A, Sadeghi K, Kuo J-F, Jang L-K, Yen TF. Sonication method and reagent for treatment of carbonaceous materials. Patent No. US4891131. Tar Sands Energy Ltd.; 1990.
[145] Singh S, Mclean JD, Kilpatrick PK. Fused ring aromatic solvency in destabilizing water-in-asphaltene-heptane-toluene emulsions. J Dispers Sci Technol 1999;20(1 & 2):279–93.
[146] Kruka VR, Wielers AFH. Method to remove metals from residuals. Patent No. US5746907 (Also published as WO9531517, EP759963, CN1148404, CA2190369, DE69501891, JPH10505364, AU2614095). Shell Internationale Research Maatschappij B. V.; Shell Canada Ltd.; May 05, 1998.
[147] Di Tullio V, Herrbach P. Supercritical water refining of petroleum residues and simultaneous hydrolysis-decarboxylation of waste glycerides, fats, and proteins. Patent No. US2002173682 (Abandoned) (Also

published as CA2002074881, CA2348947, WO02074881). Deinking/Solvent Extraction Technologies Ltd.; November 21, 2002.

[148] Daigle EO, Tenhundfeld GJ, Gupta P. Treatment of fluids and/or sludge with electro-plasma. Patent No. WO2005102019 (Also published as US2008190770 (Abandoned)). Cap Technologies, LLC; November 21, 2005.

[149] Syversen U, Sundell J, Saur I, Rubach S, Kjaerheim G. Water purification plant. Patent No. US6099703 (Also published as WO9720777, EP0880471, NO303009). CLEANER TECHNOLOGY AS; August 08, 2000.

[150] Koren JPF, Syversen U. State-of-the-art electroflocculation. Filtr Sep 1995;32(2):153−6.

[151] Saur IF, Rubachs S, Forde JS, Kjaerheim G, Syversen U. Electroflocculation: removal of oil, heavy metals and organic compounds from oil-in-water emulsions. Filtr Sep 1996;33(4):295−303.

[152] Weers J. New chemical process removes crude oil contaminants. In: Proc. NPRA annual meeting. Salt lake city, UT; March 19−21, 2006. AM-06-31, 10 pp.

[153] Liu G, Xu X, Gao J. Study on the deferrization and desalting for crude oils. Energy Fuels 2004;18:918−23.

[154] Reynolds JG. Removal of nickel and vanadium from heavy crude oils by exchange reactions. In: Proc. 227th ACS nat. meeting. Anaheim, CA, USA, vol. 49; March 28−April 1, 2004. p. 79−80.

[155] Mukhopadhyay M, Mukhopadhyay AK. Nonhydrocarbon constituents in petroleum. Chem Age India 1974;25(2):103−9.

[156] Gould KA. Oxidative demetallization of petroleum asphaltenes and residue. Fuel 1980;59(10):733−6.

[157] Young DC. Demetallizing organometallic compounds using sulphuric acid-urea adducts. Patent No. US5099014. Union Oil Co.; March 24, 1992.

[158] Young DC. Acid catalyzed oxidative reactions. Patent No. US5105040. Union Oil Co.; April 14, 1992.

[159] Sugihara JM, Branthaver JF, Willcox KW. Oxidative demetallation of oxovanadium(IV) porphyrins. In: Yen TF, editor. The role of trace metals in pet; 1975. p. 183−93.

[160] Wasser PKW, Fuhrhop JH. Photooxygenation of metalloporphyrins and metallochlorins. Ann NY Acad Sci 1973;206:533−48.

[161] Sugihara JM, Okada T, Branthaver JF. Reductive desulfuration on vanadium and metalloporphyrin contents of fractions from boscan asphaltenes. J Chem Eng Data 1965;10:190−4.

[162] Salusinszky AL, Snow RJ. Partial coking of hydrocarbon residues. In: Chemeca 81: des. change, proc. 9th Australas. chem. eng. conf.; 1981. p. 561−8.

[163] Miyauchi T, Ikeda Y, Kikuchi T. Process for thermal cracking of heavy petroleum oils. Patent No. US4490243 (Also published as GB2079777, JPS5718783, JPH022917, CA1178230). Terukatsu Miyauchi; December 25, 1984.

[164] Miyauchi T, Ikeda Y, Kikuchi T. Process for thermal cracking of heavy petroleum oils. Patent No. US4772378 (Also published as CA1280710, CN1014153, CN86101895, JP61200196, GB2172610). Fuji Standard Res Inc.; September 20, 1988.

[165] Miyauchi T, Ikeda Y, Kikuchi T, Kubota O, Tsutsui T. A new fluid thermal cracking of residual oil with high-quality fluidization and metal effect elimination. In: Proc. 12th world pet. congress. Houston, USA, vol. 4; April 26−May 1, 1987. p. 335−41.

[166] Varadaraj R, Pugel TM, Savage DW. Removal of naphthenic acids in crude oils and distillates. Patent No. US6096196 (Also published as WO9950375, RU2205857, NO20004808, NO325474, JP2002509979, ES2172983T, EP1066360, DK1066360, DE69900888, CN1295608, CA2322223, BR9909182, AU745351). Exxon Research and Engineering Co.; August 1, 1999.

[167] Rieve RW, Shalit H, Rothrock JJ. Removal of metal contaminants from heavy hydrocarbon fractions. Patent No. US3617530. Atlantic Richfield Co.; November 2, 1971.

[168] Aldridge CL, Hamner GP, Mason RB. Removal of sulfur, nitrogen, and metal compounds from residual oils. Patent No. US3367861. Esso Research and Engineering Co.; February 06, 1968.

[169] Baird WC, Bearden R. Combined desulfurization and hydroconversion with alkali metal hydroxides. Patent No. US4003823. Exxon Research and Engineering Co.; January 18, 1977.

[170] Baird WC, Bearden R. Combined desulfurization and hydroconversion with alkali metal oxides. Patent No. US4007109. Exxon Research and Engineering Co.; February 08, 1977.
[171] Bearden R. Residua desulfurization with sodium oxide and hydrogen. Patent No. US4007110. Exxon Research and Engineering Co.; February 08, 1977.
[172] Gould KA. Vanadium and nickel removal from petroleum utilizing organic peroxyacid. Patent No. US4197192 (Also published as US2001036564). Exxon Research and Engineering Co.; April 10, 1980.
[173] Reynolds JG. Demetalation of hydrocarbonaceous feedstocks using hydroxo-carboxylic acids and salts thereof. Patent No. US4789463. Chevron Research Co.; December 6, 1986.
[174] Tilley GL, Young DC. Demetallization of residual oils with polyphosphoric acids. Patent No. US3622505. Union Oil Co.; November 23, 1971.
[175] Paris-Marcano LC. Process for recovering metals and for removing sulfur from materials containing them by means of an oxidative extraction. Patent No. US5017280. Laboratorios Paris C. A.; May 21, 1991.
[176] Codet HG, Herrmann JW. Deasphalting and demetalizing crude hydrocarbon feeds with hydrofluoric acid. Patent No. US3281350. Esso Research and Engineering Co.; October 26, 1966.
[177] Michlmayr MJ. Upgrading metal-contaminated petroleum oils containing vanadium and/or nickel. Patent No. US4039432. Chevron Research Co.; August 2, 1977.
[178] Kukes SG. Demetallization of hydrocarbon-containing feed streams with phosphorus compounds. Patent No. US4529503. Phillips Petroleum Co.; July 7, 1985.
[179] Kukes SG, Battiste DR. Demetalization of heavy oils with phosphorous acid. Patent No. US4522702. Phillips Petroleum Co.; June 11, 1985.
[180] Gomez RAM. Treatment of crude oils. Patent No. US6955753 (Also published as WO0136564, US2001036564, MXPA02004947, EP1238044, GB2371558, AU1390). RMG Services PTY Ltd.; October 18, 2005.
[181] Yan T-Y. Recovery of metal values from petroleum residua and other fractions. Patent No. US4544479. Mobil Oil Corp.; October 1, 1985.
[182] Aegerter Jr PA, Howell JA, Sughrue EL, Knopp KG. Refining of heavy slurry oil fractions. Patent No. US5080777 (Also published as EP0456058). Phillips Petroleum Co.; November 13, 1991.
[183] Haddad JH, Owen H, Schatz KW. Process and apparatus for catalytic cracking of residual oils. Patent No. US4875994. Mobil Oil Corp.; October 24, 1989.
[184] Cha CY, Boysen JE, Branthaver JF. Process for removing heavy metal compounds from heavy crude oil. Patent No. US5041209. Western Research Institute; August 20, 1991.
[185] Mason RB, Hamner GP. Recovering vanadium from residues obtained in the refining of crude oil. Patent No. US3384448. Esso Research and Engineering Co.; May 21, 1968.
[186] Long RB. Process for removing metals from petroleum residua. Patent No. US3849297. Exxon Research and Engineering Co.; November 19, 1974.
[187] Delaney DD, Fenton DM. Removal of metals from hydrocarbon oils. Patent No. US4414104. Union Oil Co.; November 08, 1983.
[188] Kukes SG, Johnson MM, Nowack GP. Demetallization of heavy oils. Patent No. US4421638. Phillips Petroleum Co.; December 20, 1983.
[189] Espenscheid WF, Yan T-Y. Desulfurization of residual oil. Patent No. US4035287. Mobil Oil Corp.; July 12, 1977.
[190] Gleim WKT. Removal of heavy asphaltenes from hydrogenation feed stocks. Patent No. US3294678. Universal Oil Products Co.; December 27, 1966.
[191] Gleim WKT. Noncatalytic crude oil hydrofining process. Patent No. US3303126. Universal Oil Products Co.; February 07, 1967.
[192] Gleim WKT. Hydrorefining metal-containing sour crudes treated with a lewis acid and a molybdenum chloride. Patent No. US3483117. Universal Oil Products Co.; December 09, 1969.
[193] Bearden R, Aldridge CL. Hydroconversion process. Patent No. US4067799. Exxon Research and Engineering Co.; January 10, 1978.

[194] Chang CD, Silvestri AJ. Demetalation of petroleum residua using manganese nodules. Ind Eng Chem Proc Des Dev 1976;15(1):161–5.
[195] Chang CD, Naro A, Silvestri AJ. Catalyst for the demetalation of a hydrocarbon charge stock. Patent No. US3772185. Mobil Oil Corp.; 1973.
[196] Oleck SM, Sherry HS. Fresh water manganese nodules as a catalyst for demetalizing and desulfurizing petroleum residua. Ind Eng Chem Proc Des Dev 1977;16(4):525–8.
[197] Oleck SM, Sherry HS, Fischer RH, Milstein D. Process for demetalation and desulfurization of petroleum oils. Patent No. US4089774 (Also published as DE2638498, ZA7605023, NZ181809, JPS5230282, JPS5731457, GB1550684). Mobil Oil Corp.; May 16, 1978.
[198] Ward JW. Demetallization and desulfurization of petroleum feedstocks with manganese on alumina catalysts. Patent No. US4045331. Union Oil Co.; August 30, 1977.
[199] Ramirez de Agudelo MM, Gonzalez Reyes L, Gajardo P. The role of transition metal sulfides in hydrotreatment - first row transition metal sulfides. Rev Tec INTEVEP 1984;4(1):19–26.
[200] Gillis D. Breaking through the bitumen upgrading barriers with the UOP uniflex^{tm} process. In: Proc. 5th NCUT upgrading ond relining conlerence. Edmonton, Alberta, Canada; September 14–16, 2009. 24 pp.
[201] Johnson CA, Maruhnic P, Nongbri G, Wolk RH. Catalyst development for demetallization of petroleum residua. In: Proc. AIChE nat. meeting. Houston, TX, USAA; March 16–20, 1975. 347 (1 p.).
[202] Wolk RH, Rovesti WC. Low sulfur fuel oil from high metals-containing petroleum residuum. Patent No. US3819509. Hydrocarbon Research, Inc.; June 26, 1974.
[203] Iannibello A, Marengo S, Girelli A. Bauxite-based catalysts in heavy crude oil hydrotreating. Appl Catal 1982;3(3):261–72.
[204] Ramirez de Agudelo MM, Gajardo P, Reyes LG. The role of transition metal sulfides in hydrotreatment part I - iron sulfide as catalyst. In: Proc. ACS div. pet. chem. preprints. Kansas City, USA, vol. 27; 1982. p. 712–20.
[205] Galiasso R, Garcia J, Caprioli L, Pazos JM, Soto A. Reactions of porphyrinic and non porphyrinic molecules during hydrodemetalation of heavy crude oils. ACS Div Pet Chem 1985;30(1):50–61. Preprints.
[206] García-López AJ, Cuevas R, Ramírez J, Ancheyta J, Vargas-Tah AA, Nares R, et al. Hydrodemetallation (HDM) kinetics of Ni-tpp over Mo/Al2O3-tio2 catalyst. Catal Today 2005;107–108:545–50.
[207] Scott CE, Embaid BP, Gonzalez-Jimenez F, Hubaut R, Grimblot J. Behavior of iron-vanadium sulfide catalysts for hydrotreating reactions. J Catal 1997;166(2):333–9.
[208] Elkes GJ, Pazos JM. Demetallization catalyst for petroleum. Patent No. GB2032795. British Petroleum Co. Ltd.; September 26, 1980.
[209] Fukuyama H. Development of carbon catalyst for heavy oil hydrocracking. In: Proc. NPRA annual meeting. San Antonio, TX; March 17–19, 2002. Paper #AM-02-14, 10 pp.
[210] Ramirez de Agudelo MM, Galarraga C. A stable catalyst for heavy oil processing. I. Methodology and initial concepts. Rev Soc Venez Catal 1990;5:117–21.
[211] Ramirez de Agudelo MM, Galarraga C. A stable catalyst for heavy oil processing. II. Preparation and characterization. J Catal 1992;134:98–106.
[212] Ramírez De Agudelo MM, Galarraga C. Catalyst for the simultaneous hydrodemetallization and hydroconversion of heavy hydrocarbon feestocks. Patent No. US4729980 (Also published as GB2204251). Intevep S.A.; March 08, 1988.
[213] Ramirez de Agudelo MM, Galarraga C. A stable catalyst for heavy oil processing. III. Activity and selectivity. Chem Eng J 1991;46(2):61–8.
[214] Ramirez de Agudelo MM, Galarraga C. A stable catalyst for heavy oil processing: IV. Active phases and functionality. Rev Tec INTEVEP 1990;10(2):173–82.
[215] Ramírez De Agudelo MM, Galarraga C. Catalyst for the simultaneous hydrodemetallization and hydroconversion of heavy hydrocarbon feestocks. Patent No. US4832829. Intevep S.A.; May 23, 1989.

[216] Kukes SG, Nowack GP. Demetallization of hydrocarbon feed streams with nickel arsenide. Patent No. US4551230. Phillips Petroleum Co.; November 5, 1985.
[217] Kiuchi N, Suzuka T, Miyake T, Uchida S, Hirato M, Souma T, et al. Regeneration of nickel oxide ore catalyst. Patent No. JP52123396. Nippon Mining Co, Hitachi Ltd.; October 17, 1977.
[218] Ozaki H, Yamane M, Satomi Y, Suzuka T, Miyake T, Hirato M. Catalyst for treating heavy oils. Patent No. US4115324. Nippon Mining Co.; September 19, 1978.
[219] Ishigaki S. Demetalization of heavy residual oil by cracking, in relation to metal smelting. CEER 1975;7(9):7–13.
[220] Dewally D. Influence of physicochemical properties, methods or preparation and purity of nickel compounds on their biological effects. In: Proc. nickel in the human environment. Lyon, France; March 8–11, 1983. p. 243–54.
[221] Hamilton HA, Mcilvried HG, Sebulsky RT. Catalytic removal of vanadium and nickel from oils. Patent No. US3553106. Gulf Research and Development Co.; January 5, 1971.
[222] Gleim WKT. Conversion of heavy hydrocarbon charge stocks. Patent No. US3849292. Universal Oil Products Co.; November 19, 1974.
[223] Scott DS, Radlein D, Piskorz J, Majerski P, Debruijn TJW. Upgrading of bitumen in supercritical fluids. Fuel 2001;80(8):1087–99.
[224] Takeuchi C, Fukui Y, Nakamura M, Shiroto Y. Asphaltene cracking in catalytic hydrotreating of heavy oils. 1. Processing of heavy oils by catalytic hydroprocessing and solvent deasphalting. Ind Eng Chem Proc Des Dev 1983;22(2):236–42.
[225] Le Borgne S, Quintero R. Biotechnological processes for the refining of petroleum. Fuel Proc Technol 2003;81(2):155–69.
[226] Fedorak PM, Semple KM, Vazquez-Duhalt R, Westlake DWS. Chloroperoxidase-mediated modifications of petroporphyrins and asphaltenes. Enzyme Microb Technol 1993;15:429–37.
[227] Mogollon L, Rodriguez R, Larrota W, Ortiz C, Torres R. Biocatalytic removal of nickel and vanadium from petroporphyrins and asphaltenes. Appl Biochem Biotechn 1998;70–72:765–77.
[228] Vázquez-Duhalt R, Ayala M, Márquez-Rocha FJ. Biocatalytic chlorination of aromatic hydrocarbons by chloroperoxidase of caldariomyces fumago. Phytochemistry 2001;58(6):929–33.
[229] Ayala M, Robledo NR, Lopez-Munguia A, Vazquez-Duhalt R. Substrate specificity and ionization potential in chloroperoxidase- catalyzed oxidation of diesel fuel. Environ Sci Technol 2000;34(13):2804–9.
[230] Ryu K, Heo J, Yoo IK. Removal of dibenzothiophene and its oxidized product in anhydrous water-immiscible organic solvents by immobilized cytochrome c. Biotechn Lett 2002;24(2):143–6.
[231] Torres E, Sandoval JV, Rosell FI, Mauk AG, Vazquezduhalt R. Site-directed mutagenesis improves the biocatalytic activity of iso-1-cytochrome-C in polycyclic-hydrocarbon oxidation. Enzyme Microb Technol 1995;17(11):1014–20.
[232] Villegas JA, Mauk AG, Vazquez-Duhalt R. A cytochrome C variant resistant to heme degradation by hydrogen peroxide. Chem Biol 2000;7:237–44.
[233] Garcia-Arellano H, Valderrama B, Saab-Rincon G, Vazquez-Duhalt R. High temperature biocatalysis by chemically modified cytochrome c. Bioconjugate Chem 2002;13:1336–44.
[234] Tinoco R, Vazquezduhalt R. Chemical modification of cytochrome C improves their catalytic properties in oxidation of polycyclic aromatic hydrocarbons. Enzyme Microb Technol 1998;22(1):8–12.
[235] García-Arellano H, Buenrostro-Gonzalez E, Vazquez-Duhalt R. Biocatalytic transformation of petroporphyrins by chemical modified cytochrome C. Biotechn Bioeng 2004;85(7):790–8.
[236] Xu G-W, Mitchell KW, Monticello DJ. Process for demetalizing a fossil fuel. Patent No. US5726056. Energy Biosystems Corp.; March 10, 1998.
[237] Tomita F, Yokota A, Saito K, Yamamoto C, Abe A. Decomposition of porphyrin. Patent No. JP9176168. Petroleum Industry Development Center (Sekiyu Sangyo Kasseika Center); Mitsubishi Oil Co Ltd.; July 08, 1997.

[238] Tomita F, Yokota A, Saito K, Abe A. Method for decomposing porphyrin-based compound. Patent No. JP11187890. Petroleum Energy Center; July 13, 1999.

[239] Marison IW, Doerries SC. Flocculating agents. Patent No. US2003008343 (Abandoned). Optima Environnement S.A.; January 30, 2003.

[240] Rosinski EJ, Stein TR, Fischer RH. Catalyst for residual demetalation and desulfurization. Patent No. US3876523. Mobil Oil Corp.; April 8, 1975.

[241] Milstein D, Fischer RH. Demetalation and desulfurization of residual oil utilizing hydrogen and trickle beds of catalysts in three zones. Patent No. US4054508. Mobil Oil Corp.; October 18, 1977.

[242] Audeh CA, Yan TY. Supercritical selective extraction of hydrocarbons from asphaltic petroleum oils. Patent No. US4354928. Mobil Oil Corp.; October 19, 1982.

[243] Yan TY, Audeh CA. Selective extraction of hydrocarbons from asphaltic petroleum oils. Patent No. CA1198996. Mobil Oil Corp.; January 7, 1986.

[244] Ikematsu M, Honjiyou I, Sakai K. Solvent deasphalting of asphaltene-containing hydrocarbons. Patent No. JP58127785, JP58138787, JP58171473, JP58171474, JP58171475, JP58171476, JP58171477, JP59027985, JP59027986 (Also published as US4525269, US4514287, US4502950, FR2519646, DE3300413). Nippon Oil Co., Ltd.; July 29, 1983.

[245] Wang Y, Chu Z, Qiu B, Liu C, Zhang Y. Removal of naphthenic acids from a vacuum fraction oil with an ammonia solution of ethylene glycol. Fuel 2006;85(17–18):2489–93.

[246] Wang Y, Chu Z, Qiu B, Liu C, Zhang Y. Separating naphthenic acid from high-acidity crudes by reaction-extraction process. Pet Refin Eng 2006;36(1):36–8.

[247] Rikka P. Spectrometric identification of naphthenic acids isolated from crude oil [MSc thesis from Texas State University-San Marcos, Department of Chem. & Biochemistry]. December 2007. 68 pp.

[248] Shohaimi MN, Jaafar J, Wan Abu Bakar W. Catalytic deacidification optimization of korean crude oil based on response surface methodology. Clean Technol Environ Policy 2014;17(6):1513–22.

[249] Shukri NM, Bakar WAWA, Jaafar J, Majid ZA. Removal of naphthenic acids from high acidity korean crude oil utilizing catalytic deacidification method. J Ind Eng Chem 2015;28:110–6.

[250] Barth T, Høiland S, Fotland P, Askvik KM, Pedersen BS, Borgund AE. Acidic compounds in biodegraded petroleum. Org Geochem 2004;35(11–12 SPEC. ISS.):1513–25.

[251] Jones DM, Watson JS, Meredith W, Chen M, Bennett B. Determination of naphthenic acids in crude oils using nonaqueous ion exchange solid-phase. Anal Chem 2001;73(3):703–7.

[252] Saab J, Mokbel I, Razzouk AC, Ainous N, Zydowicz N, Jose J. Quantitative extraction procedure of naphthenic acids contained in crude oils. Characterization with different spectroscopic methods. Energy Fuels 2005;19(2):525–31.

[253] Khulbe KC, Mann RS, Macphee JA. Separation of acidic fraction from the Cold Lake bitumen asphaltenes and its relationship to enhanced oil recovery. Fuel Proc Technol 1996;46(1):63–9.

[254] Huang M, Zhao S, Li P, Huisingh D. Removal of naphthenic acid by microwave. J Clean Prod 2006;14(8):736–9.

[255] Huang MF, Zhao SL, Yang K, Li P. Removal of acid from diesel oil by using microwave irradiation. Acta Pet Sin (Pet Proc Sect) 2003;19(6):81–4.

[256] Kong L, Wu J, Li P, Zhao S. Laboratory study on acid removal from heavy atmospheric distillate by microwave radiation. Pet Proc Petrochem 2004;35(10):22–4.

[257] Kong L, Zhang Z, Li P, Zhao S. Study on separating naphthenic acids from diesel fuel by microwave irradiation. China Pet Proc Petrochem Technol 2004;(3):45–50.

[258] Kong L, Zhao S, Zhang Q, Li P, Zhuang L. Removing naphthenic acids from diesel oil by microwave irradiation. Pet Sci Technol 2006;24(2):157–66.

[259] Zhao S, Li P, Kong L. Microwave demulsification in removing naphthenic acids from diesel oil. Pet Sci Technol 2007;25(12):1583–92.

[260] Zhao SL. Technological process of eliminating naphthenic acid from diesel oil. Patent No. CN1740276. Liaoning Petrochemical University; March 01, 2006.

[261] Shi LJ, Shen BX, Wang GQ. Removal of naphthenic acids from beijiang crude oil by forming ionic liquids. Energy Fuels 2008;22(6):4177−81.

[262] Kamarudin H, Mutalib MIA, Man Z, Azmi M. Extraction of naphthenic acids from liquid hydrocarbon using imidazolium ionic liquids. In: Proc. intern. conf. on environment sci. eng. Rome, Italy, Sep 26−27. IPCBEE, vol. 32; 2012. p. 17−23.

[263] Qi L, Ren XG, Song YJ, Ren SM. Separation and recovery of naphthenic acids from crude oil with high TAN. Beijing Huagong Daxue Xuebao (Ziran Kexueban)/J Beijing Univ Chem Technol Nat Sci Ed 2004;31(1):11.

[264] Ren XG, Qi L, Song YJ, Ren SM. Process for separation of naphthenic acids from high TAN heavy crude oil of Sudan. Guocheng Gongcheng Xuebao/The Chin J Proc Eng 2004;4(5):401−5.

[265] Netke SA, Pangarkar VG. Extraction of naphthenic acid from kerosene using porous and nonporous polymeric membranes. Sep Sci Technol 1996;31(1):63−76.

[266] Varadaraj R, Savage DW, Wales WE. Removal of naphthenic acids in crude oils and distillates. Patent No. US5961821 (Also published as EP1068280, WO9950376, CA2323051, RU2208622, ES2172318T, AU745496, NO20004806). Exxon Research and Engineering Co.; October 5, 1999.

[267] Sartori G, Savage DW, Ballinger BH, Dalrymple DC. Process for decreased the acidity of crudes using crosslinked polymeric amines (law871). Patent No. US6121411. Exxon Research and Engineering Co.; September 19, 2000.

[268] Sartori G, Savage DW, Ballinger BH, Dalrymple DC. Process for extraction of naphthenic acids from crudes. Patent No. US6281328. ExxonMobil Research and Engineering Co.; August 28, 2001.

[269] Rousseau G, Bayle G, Crolet JL. Process for deacidifying crude oils and a device therefore. Patent No. GB2356404. ELF Exploration and Production; May 23, 2001.

[270] Duncum SN, Osborne CG. A process for deacidifying crude oil. Patent No. WO0234864 (Also published as AU9405701). BP Exploration Operating Co. Limited; May 02, 2002.

[271] Duncum SN, Osborne CG. Process for deacidifying a crude oil system. Patent No. US6464859 (Also published as CN1334857, EP1155101, WO2000046322). BP Exploration Operating Co., Ltd.; October 15, 2002.

[272] Collins IR, Plechkova NV. Process for deacidifying crude oil. Patent No. EP1911829 (Also published as WO2007138307). University of Belfast; April 16, 2008.

[273] Headley J, Du J-L, Peru K, Janfada A. Solid phase extraction (SPE) and quantification of naphthenic acid in milli-q, simulated groundwater and river waters (Unpublished Presentation). 2007. Available from:, http://www.traceorganic.com/2007/presentations/Jing-LongDu_Solid%20Phase%20Extraction%20(SPE)%20and%20Quantification%20of%20%20Naphthenic%20Acid.pdf. 30 pp.

[274] Gaikar VG, Maiti D. Adsorptive recovery of naphthenic acids using ion-exchange resins. React Funct Polym 1996;31(2):155−64.

[275] Silva JP, De Senna LF, Do Lago DCB, Da Silva Jr PF, Dias EG, De Figueiredo MAG, et al. Characterization of commercial ceramic adsorbents and its application on naphthenic acids removal of petroleum distillates. Mat Res 2007;10(2):219−25.

[276] Saad OM, Gasmelseed GA, Hamid AHM. Separation of naphthenic acid from sudanese crude oil using local activated clays. J Appl Ind Sci 2014;2(1):14−8.

[277] Zou L, Han B, Yan H, Kasperski KL, Xu Y, Hepler LG. Enthalpy of adsorption and isotherms for adsorption of naphthenic acid onto clays. J Colloid Interface Sci 1997;190(2):472−5.

[278] Silva JP, Costa ALH, Chiaro SSX, Delgado BEPC, De Figueiredo MAG, Senna LF. Carboxylic acid removal from model petroleum fractions by a commercial clay adsorbent. Fuel Proc Technol 2013;112:57−63.

[279] Varadaraj R. Removal of acids from oils. Patent No. EP1390447. Exxon Research and Engineering Co.; February 15, 2002.

[280] Janfada A, Headley J, Peru K, Barbour S. A laboratory evaluation of the sorption of oil sands naphthenic acids on organic rich soils. J Environ Sci Health - Part A Toxic/Hazardous Subst Environ Eng 2006;41(6):985−97.

[281] Varadaraj R. Removal of acids from oils. Patent No. US6454936 (Also published as CA2438462, DE60214537T2, EP1390447, US20020139711, WO2002072735). Exxonmobil Research And Engineering Co.; September 24, 2002.
[282] Gillespie RD, Arena BJ. Naphthenic acid removal as an adjunct to liquid hydrocarbon sweetening. Patent No. US5389240. UOP LLC; February 14, 1995.
[283] Messer B, Tarleton B, Beaton M. Compositions, configurations, and methods of reducing naphthenic acid corrosivity. Patent No. US8118994 (Also published as WO2005040313, US2008164137, JP2007508437, EP1678278, CN102382681, CN1894389, CA2541997). Fluor Corp.; February 21, 2012.
[284] Zhang Q, Tian S, Huang S, Wang J. Relationship between naphthenic acid structure and its corrosion performance. Shiyou Xuebao, Shiyou Jiagong/Acta Pet Sin (Pet Proc Sect) 2012;28(4):652–6.
[285] Levine SW, Brons GB, Lutz GA, Leta DP, Walters CC. Addition of high molecular weight naphthenic tetra-acids to crude oils to reduce whole crude oil fouling. Patent No. US8663455 (Also published as CA2746631, CN102245743, CN102245743, EP2373764, US20100147739, WO2010068261). ExxonMobil Research And Engineering Company; March 04 2014.
[286] Kumar R, Thorat TS, Chithra V, Rathore V, Rao PVC, Choudary NV. Processing opportunity crude oils - catalytic process for high-acid crudes. Hydrocarb World 2009;4(2):64–8.
[287] Shen H, Wang Y, Li R. Investigation on thermal treatment for decreasing the acidity of sour crude. Pet Proc Petrochem 2004;35(2):32–5.
[288] Blum SC, Olmstead WN. Viscosity reduction by heat soak-induced naphthenic acid decomposition in hydrocarbon oils. Patent No. US5976360 (Also published as WO9714766, EP0948581, CN1200139). Exxon Research and Engineering Co.; November 02, 1997.
[289] Blum SC, Olmstead WN, Bearden R. Thermal decomposition of naphthenic acids. Patent No. US5820750 (Also published as WO9625471, SG68588, MX9603335, KR100451614, EP0809683, CN1175972, CA2212775, AU708575). Exxon Research and Engineering Co.; October 13, 1998.
[290] Bearden R, Blum SC, Olmstead WN, Robbins WK. Process for reducing total acid number of crude oils. Patent No. US5891325 (Also published as WO0009631, EP1109879, CN1312849, CA2338623). Exxon Research and Engineering Co.; April 6, 1999.
[291] Bienstock MG, Matragrano JG, Patel RD, Bearden R. Thermal process for reducing total acid number of crude oil. Patent No. US6086751 (Also published as WO9910452, ES2172912, EP1062298, DK1062298, DE69804025, CA2294952, AU9040698, AU735810). Exxon Research and Engineering Co.; July 11, 2000.
[292] Buchanan JS, Mccoy JN, Brown SH, Wolfenbarger JA, Graham JE, Keusenkothen PF. Steam cracking of high tan crudes. Patent No. US8277639 (Also published as US2007066860). Exxonmobil Chem Patents Inc.; October 02, 2007.
[293] Freel B, Kriz J,F, Clarke D. Modified thermal processing of heavy feedstocks. Patent No. US7572365 (Also published as US20040069686). Ivanhoe Energy, Inc.; August 11, 2009.
[294] Freel B, Kriz J,F, Clarke D. Modified thermal processing of heavy hydrocarbon feedstocks. Patent No. US7572362 (Also published as US20040069682, CA2444832, DE60327148, EP1420058, CN100347274, RU2004109519, RU2323246, PE10322004, ECSP044976, CO5540064). Ivanhoe Energy; August 11, 2009.
[295] Mcmartin DW, Headley JV, Friesen DA, Peru KM, Gillies JA. Photolysis of naphthenic acids in natural surface water. J Environ Sci Health - Part A Toxic/Hazardous Subst Environ Eng 2004;39(6):1361–83.
[296] Pravia ORC, De L. Alvarenga Baptista CM, Gomes JR, Peixoto Bugueta PC, Cerqueira HS, Moreira EM. Process for reducing the naphthenic acidity of petroleum oils. Patent No. US7504023 (Also published as US2004026299 (Abandoned), US2006283781, WO2004005434, BR0202552, AU2003244845). Petroleo Brasileiro S.A.; March 17, 2009.
[297] Cardoso MJB, Moreira EM. Process for reducing the acidity of hydrocarbon mixtures. Patent No. US7514657 (Also published as BRPI0503793, US2007056880). Petroleo Brasileiro S.A.; April 07, 2009.
[298] Naves GPAP, Pinho ADR, Cardoso MJB, Da Cunha PSMD, Coutinho RCC, Moreira EM. Process for reducing the naphthenic acidity of petroleum oils or their fractions. Patent No. US7507329

(Also published as US2006201855, AR049786, MXPA05007911, FR2883571, BR0306059). Petroleo Brasileiro S.A.; March 24, 2006.

[299] Lunin VV, Frantsuzov VK, Likhterova NM. Deepening of crude oil processing using ozonation and high-energy impacts. In: Proc. 17th world pet. congress. Rio de Janeiro, Brazil, vol. 3; September 1–5, 2002. p. 141–50.

[300] Pereira AS, Islam MS, Gamal El-Din M, Martin JW. Ozonation degrades all detectable organic compound classes in oil sands process-affected water; an application of high-performance liquid chromatography/obitrap mass spectrometry. Rapid Commun Mass Spectrom 2013;27(21):2317–26.

[301] Scott AC, Zubot W, Mackinnon MD, Smith DW, Fedorak PM. Ozonation of oil sands process water removes naphthenic acids and toxicity. Chemosphere 2008;71(1):156–60.

[302] Perez-Estrada LA, El-Din MG, Martin JW. Degradation of naphthenic acids by ozonation in various conditions: uncovering the ospw ozonation. In: Proc. intern. conf. on unconventional oils the environment. Calgary, AB; July 4–6, 2011. p. 214–6.

[303] Maher KD, Kirkwood KM, Gray MR, Bressler DC. Pyrolytic decarboxylation and cracking of stearic acid. Ind Eng Chem Res 2008;47(15):5328–36.

[304] Mäki-Arvela P, Kubickova I, Snåre M, Eränen K, Murzin DY. Catalytic deoxygenation of fatty acids and their derivatives. Energy Fuels 2007;21(1):30–41.

[305] Santillan-Jimenez E, Crocker M. Catalytic deoxygenation of fatty acids and their derivatives to hydrocarbon fuels via decarboxylation/decarbonylation. J Chem Technol Biotechn 2012;87(8):1041–50.

[306] Wilson III GR, Edwards T, Corporan E, Freerks RL. Certification of alternative aviation fuels and blend components. Energy Fuels 2013;27(2):962–6.

[307] Smith DF, Rodgers RP, Rahimi P, Teclemariam A, Marshall AG. Effect of thermal treatment on acidic organic species from Athabasca bitumen heavy vacuum gas oil, analyzed by negative-ion electrospray fourier transform ion cyclotron resonance (FT-ICR) mass spectrometry. Energy Fuels 2009;23(1):314–9.

[308] Mejía C, Quiroga H, Vásquez C, Peña D. Influence of basic catalysts on the decarboxylation kinetics of naphthenic acids. Rev Tec la Fac Ing Univ del Zulia 2010;33(3):245–53.

[309] Yang L, Nelson AR, Gray MR. Liquid-phase catalytic decarboxylation of naphthenic acids. In: Proc. AIChE spring nat. meeting. New Orleans, LA; 2008. 1 p.

[310] Ding L, Rahimi P, Hawkins R, Bhatt S, Shi Y. Naphthenic acid removal from heavy oils on alkaline earth-metal oxides and ZnO catalysts. Appl Catal A General 2009;371(1–2):121–30.

[311] Fu X, Dai Z, Tian S, Long J, Hou S, Wang X. Catalytic decarboxylation of petroleum acids from high acid crude oils over solid acid catalysts. Energy Fuels 2008;22(3):1923–9.

[312] Sinopec Corp. Oil refining catalyst. 2014. Available from: http://english.sinopec.com/products_service/catalyst/.

[313] Goddard W,A, Tang Y. Project overview: improved processes to remove naphthenic acids (Unpublished Presentation). 2002. p. 23.

[314] Zhang A, Ma Q, Tang Y. Catalytic decarboxylation for naphthenic acid removal from crude oils - a theoretical and experimental study. ACS Div Pet Chem 2004;49(2):218–21. Preprints.

[315] Zhang A, Ma Q, Wang K, Liu X, Shuler P, Tang Y. Naphthenic acid removal from crude oil through catalytic decarboxylation on magnesium oxide. Appl Catal A General 2006;303(1):103–9.

[316] Zhang A, Ma Q, Wang K, Tang Y, Goddard WA. Improved processes to remove naphthenic acids. Report No. Final Technical Report Grant #DE-FC26-02NT15383. Pasadena, CA, USA: California Institute of Technology; December 09, 2005. 96 pp.

[317] Zhang A, Ma Q, Tang Y, Goddard W,A. Improved processes to remove naphthenic acids. Report No. Annual Technical Progress Report (DOE Award number: DE-FC26-02NT15383). Pasadena, CA, USA: California Institute of Technology; April 28, 2004. 18 pp.

[318] Oh HY, Park JH, Rhee YW, Kim JN. Decarboxylation of naphthenic acid using alkaline earth metal oxide. J Ind Eng Chem 2011;17(4):788–93.

[319] Tang Y, Zhang A. Process to upgrade oil using metal oxides. Patent No. US2006016723 (Abandoned) (Also published as WO2006014486, JP2008504409, EP1781760). California Institute of Technology; January 26, 2006.
[320] Ohsol EO, Gillespie TE, Pinkerton JW. Process to upgrade crude oils by destruction of naphthenic acids, removal of sulfur and removal of salts. Patent No. US5985137 (Also published as CA2235691, EP1060230, WO1999043766). Unipure Corp.; November 16, 1999.
[321] Trachte KL, Angelo JB, Aldous KK. Catalytic reduction of acidity of crude oils in the absence of hydrogen. Patent No. US5871636. Exxon Research and Engineering Co.; February 16, 1999.
[322] Bearden RJ, Blum SC, Olmstead WN. Process for reducing total acid number of crude oil. Patent No. US5914030 (Also published as WO9910451, ZA9807449, ZA9807446, MY117164, MY116422, EP1034236, CN1272869, CA2300173, AR013450, AR013449). Exxon Research and Engineering Co.; June 22, 1999.
[323] Bearden RJ, Blum SC, Olmstead WN. Process for reducing total acid number of crude oil. Patent No. US5928502. Exxon Research and Engineering Co.; July 27, 1999.
[324] Trachte KL, Robbins WK. Process for selectively removing lower molecular weight naphthenic acids from acidic crudes. Patent No. US5897769 (Also published as NO324649B). Exxon Research and Engineering Co.; April 27, 1999.
[325] Halbert TR, Riley KL, Trachte KL, Vannauker DL. Process for reduction of total acid number in crude oil. Patent No. US5910242 (Also published as EP0899319, JP11140463, CA2242394). Exxon Research and Engineering Co.; June 8, 1999.
[326] Halbert TR, Riley KL, Trachte KL, Vannauker DL. Reduction of total acid value of crude oil. Patent No. JP11140463. Exxon Research and Engineering Co.; May 25, 1999.
[327] Chakka S, Caspary MT, Decanio SJ. Process for upgrading a hydrocarbon oil. Patent No. US5928501. Texaco Inc.; July 27, 1999.
[328] Chakka S, Caspary MT, Decanio SJ. Process for upgrading a hydrocarbon oil. Patent No. US6547957. Texaco, Inc.; April 15, 2003.
[329] Statoil-Hydro. Calcium naphthenate deposition (Unpublished Web site technology news). 2000. Available from:, http://www.statoil.cz/STATOILCOM/technology/SVG03268.nsf?opendatabase&lang=en&artid=A95EDC38D4E147DFC1256FBD0044CD8A.
[330] Statoil-Hydro. Naphthenate chemistry breakthrough. 2007. Available from: http://www.statoilhydro.com/en/technologyinnovation/RefiningAndProcessing/OilRefining/NaphthenateChemistryBreakthrough/Pages/default.aspx.
[331] Grande K, Sorlie C. Process for removing essentially naphthenic acids from a hydrocarbon oil. Patent No. US6063266. Den Norske Stats Oljeseskap A/S; May 16, 2000.
[332] Statoil-Hydro. Refining naphthenic crudes. 2008. Available from: http://www.statoilhydro.com/en/technologyinnovation/RefiningAndProcessing/OilRefining/RefiningNaphthenicCrudes/Pages/default.aspx.
[333] Marr HG, Winsor JF. In-line hydrotreatment process for low TAN synthetic crude oil production from oil sand. Patent No. US2005161371 (Abandoned) (Also published as CA2455149). Suncor Energy Inc.; July 28, 2005.
[334] Quiroga-Becerra H, Mejía-Miranda C, Laverde-Cataño D, Hernández-López M, Gómez-Sánchez M. A kinetic study of esterification of naphthenic acids from a colombian heavy crude oil. CTyF - Cienc Tecnol Futuro 2012;4(5):21−31.
[335] Sartori G, Savage DW, Dalrymple DC, Ballinger BH, Blum SC, Wales WE. Esterification of acidic crudes. Patent No. US6251305. Exxon Research and Engineering Co.; June 26, 2001.
[336] Goheen GE. Conversion of naphthenic acids to naphthene hydrocarbons. Chem Const Ind Eng Chem 1940;32(4):503−8.
[337] Sun XY, Wang YZ, Liu CG. Catalyst for esterification process to decrease the acidity of crude oil. Shiyou Xuebao, Shiyou Jiagong/Acta Pet Sin (Pet Proc Sect) 2007;23(3):92−5.
[338] Wang YZ, Liu YP, Liu CG. Removal of naphthenic acids of a second vacuum fraction by catalytic esterification. Pet Sci Technol 2008;26(12):1424−32.

[339] Wang YZ, Li JY, Sun XY, Duan HL, Song CM, Zhang MM, et al. Removal of naphthenic acids from crude oils by fixed-bed catalytic esterification. Fuel 2014;116:723–8.
[340] Wang Y, Sun X, Liu Y, Liu C. Removal of naphthenic acids from a diesel fuel by esterification. Energy Fuels 2007;21(2):941–3.
[341] Wang Y-Z, Liu Y-P, Liu C-G. Kinetics of the esterification of low-concentration naphthenic acids and methanol in oils with or without SnO as a catalyst. Energy Fuels 2008;22(4):2203–6.
[342] He B, Wang J, You X, Qiu Z, Mao R, Hu Z. Deacidification of vacuum gas oil by catalytic esterification. Shiyou Huagong/Petrochem Technol 2013;42(11):1282–5.
[343] Sartori G, Savage DW, Blum SC, Dalrymple DC, Wales WE. Metal compounds as accelerators for petroleum acid esterification. Patent No. US5948238 (Also published as WO0020532, PT1119596, NO20011653, JP2002526634, ID28914, ES2179680, EP1119596, DK1119596, DE69902399, CA2345467, AU6262799, AU755479, AT221566). Exxon Research and Engineering Co.; September 07, 1999.
[344] Siskin M, Manalastas PV, Sartori G. Process for treatment of petroleum acids (law824). Patent No. US6190541, US6767452 (Also published as WO0068341, EP1200538, CA2372819, AU768439). Exxon Research and Engineering Co.; 2001, 2004, February 20, July 27.
[345] Cirin-Novta V, Kuhajda K, Kevresan S, Kandrac J, Grbovic L, Vujic D. Microwave-assisted preparation of naphthenic acid esters. J Serbian Chem Soc 2006;71(12):1263–8.
[346] Sartori G, Savage DW, Ballinger BH. Process for neutralization of petroleum acids. Patent No. GB2304729 (Also published as US5683626). Exxon Research and Engineering Co.; March 26, 1997.
[347] Sartori G, Savage DW, Ballinger BH, Dalrymple DC. Method of decreasing acidity of crude oils and fractions. Patent No. EP0924286 (Also published as US6121411, NO985879, NO318135, ES2267161, DK0924286, DE69834896, CA2252040, AU9714198, AU743069). Exxon Research and Engineering Co.; June 23, 1999.
[348] Zhuang L, Lu Z, Tian Y, Zhai Y. Process for separation of petroleum acids from crude oil. China Pet Proc Petrochem Technol 2005;(2):37–9.
[349] Lu R, Xu X, Yang J, Gao J. Reduction of total acid number of crude oil and distillate. Energy Sources, Part A Recovery, Util Environ Eff 2007;29(1):47–57.
[350] Lu RJ, Xu XR, Yang JY, Gao JS. Reduction of the total acid number of crude oil with basic solution. Pet Sci Technol 2007;25(11):1415–27.
[351] Grishchenko AE, Berestovoi AM, Piskunov IN. Effect of neutralizers and the process parameters on the composition of compounds of iron (III) with naphthenic acid. Izv Vysshikh Uchebnykh Zaved Tsvetnaya Metall 1983;(4):31–5.
[352] Tang X, An R, Zhang H, Zhang F. Process for removing acid from distillates with an acid removal agent. Pet Proc Petrochem 2005;36(5):8–12.
[353] Tang XD, An R, Cui YX, Yang J. Study of application on the technology of refining distillate oil by deacidification-solvent. Xinan Shiyou Xueyuan Xuebao/J Southwest Pet Inst 2006;28(4):78–81.
[354] Tang XD, Xiao HF, Cui YX. Lab research on the technology for deacidification solvent of the crude oil with high acid number. J Southwest Pet Univ 2007;29(3):104–6.
[355] Sowers WH. Process for reducing organic acidity in mineral oils. Patent No. US2112313. Pure Oil Co.; 1938.
[[356] Davis GE. Neutral oil recovery process for the production of naphthenic acids. Patent No. US5011579. MERICHEM Co.; April 30, 1991.
[357] Kittrell N. Removing acid from crude oil. In: Proc. crude oil quality association meeting. New Orleans, LO; February 2006. 19 pp.
[358] Varadaraj R, Savage DW. Process for decreasing acidity of a crude oil. Patent No. EP0881274 (Also published as US6030523 NO316075B AU740689 ID20373). Exxon Research and Engineering Co.; December 02, 1998.
[359] Varadaraj R, Savage DW. Process for neutralization of petroleum acids (law810). Patent No. US6030523 (Also published as WO0050539, NO20014153, MY135914, BR0008535, AU2742700, AU758916). Exxon Research and Engineering Co.; February 29, 2000.

[360] Greaney MA. Method for reducing the naphthenic acid content of crude oil and fractions. Patent No. US6531055 (Also published as WO0179386, PT1285044, JP2004501219, ES2220760, EP1285044, DK1285044, DE60103270, CA2404267, AU2001249542, AU4954201, AT266711). ExxonMobil Research and Engineering Co.; March 11, 2003.
[361] Greaney MA. Method for reducing the naphthenic acid content of crude oil and its fractions. Patent No. US6627069. ExxonMobil Research and Engineering Co.; September 30, 2003.
[362] Collins IR, Duncum SN, Osborne CG. Process for reducing the acidity of oil. Patent No. WO0075262 (Also published as US2002125175 (Abandoned), EP1194505, CN1353746, AU4940600). BP Exploration Operating Co. Limited; December 14, 2000.
[363] Sartori G, Savage DW, Gorbaty ML, Ballinger BH. Process for neutralization of petroleum acids using alkali metal trialkylsilanolate. Patent No. GB2304730 (Also published as US5643439). Exxon Research and Engineering Co.; March 26, 1997.
[364] Sartori G, Savage DW, Gorbaty ML, Ballinger BH, Dalrymple DC. Method of decreasing acidity of crude oils and fractions. Patent No. EP0924284 (Also published as NO985878, ES2241095, DE69829832, CA2252033, AU9714098, AU745948). Exxon Research and Engineering Co.; June 23, 1999.
[365] Sartori G, Savage DW, Gorbaty ML, Ballinger BH, Blum SC, Anderson MP, et al. Process for decreasing the acid content and corrosivity of crudes. Patent No. US6022494 (Also published as WO9708271 EP0935644). Exxon Research and Engineering Co.; February 08, 2000.
[366] Gorbaty ML, Martella DJ, Sartori G, Savage DW, Ballinger BH, Blum SC, et al. Process for neutralization of petroleum acids using overbased detergents. Patent No. US6054042 (Also published as WO9708275 EP0856040 NO316026B CA2226750). Exxon Research and Engineering Co.; April 25, 2000.
[367] Blum SC, Sartori G, Savage DW, Gorbaty ML, Ballinger BH, Anderson MP, et al. Process for decreasing the acid content and corrosivity of crudes. Patent No. US6679987. ExxonMobil Research and Engineering Co.; January 20, 2004.
[368] Sartori G, Savage DW, Ballinger BH, Dalrymple DC. Method of decreasing acidity of crude oils and fractions. Patent No. EP0924285 (Also published as NO985881, ES2209044, DK0924285, DE69818619, CA2252928, AU9713998, AU745956). ExxonMobil Research and Engineering Co.; June 23, 1999.
[369] Duncum SN, Osborne CG. Process for the deacidification of crude oil. Patent No. WO0218519 (Also published as AU8229501). BP Exploration Operating Co. Limited; March 07, 2002.
[370] Duncum SN, Osborne CG. Process for deacidifying crude oil. Patent No. WO02099013 (Also published as AU2002256827). BP Exploration Operating; March 20, 2003.
[371] Darcy JJ. Desulfurization and hydrocarbon quality enhancement process. Patent No. US5858212 (Also published as WO9800478). InterGlobal Desulfuruzations Systems, Inc.; January 12, 1999.
[372] Hau JL, Yepez O, Specht MI, Lorenzo R. The iron powder test for naphthenic acid corrosion studies. In: Proc. corrosion 99. San Antonio, TX. USA. NACE International; April 25–30, 1999. Paper #379, 16 pp.
[373] Hau JL, Yepez OJ, Torres LH, Vera JR. Measuring naphthenic acid corrosion potential with the Fe powder test. Rev Metal (Madrid) 2003;(Spec.Volume):116–23.
[374] Yepez O, Vera J. Method of determining the corrosiveness of naphthenic acid in crude oil refinery streams. Patent No. US6294387 (Also published as EP1039290, CA2300608). Intevep S.A.; 2001.
[375] Zhang L, Tian S, Zhao S, Yan J. A method to predict corrosivity of petroleum acid by iron powder test. Pet Proc Petrochem 2006;37(11):58–61.
[376] Yepez O. On the chemical reaction between carboxylic acids and iron, including the special case of naphthenic acid. Fuel 2007;86(7–8):1162–8.
[377] Havre TE. Near-IR spectroscopy as a method for studying the formation of calcium naphthenate. Colloid Polym Sci 2004;282(3):270–9.
[378] Havre TE. Formation of calcium naphthenate in water/oil systems, naphthenic acid chemistry and emulsion stability [Doktor Ingeniør thesis from Norwegian University of Science and Technology, Department of Chem. Eng.]. October 2002. 168 pp.

[379] Hemmingsen PV, Kim S, Pettersen HE, Rodgers RP, Sjoblom J, Marshall AG. Structural characterization and interfacial behavior of acidic compounds extracted from a North Sea oil. Energy Fuels 2006;20(5):1980–7.

[380] Brandal Ø. Interfacial (o/w) properties of naphthenic acids and metal naphthenates, naphthenic acid characterization and metal naphthenate [Doctor Ingenier thesis from Norwegian University of Science and Technology, Department of Chem. Eng.]. February 2005. 151 pp.

[381] Dion M. Challenges and solutions for processing opportunity crudes. In: Proc. AFPM annual meeting. Orlando, FL, USA; March 23–25, 2014. 13 pp.

[382] Srinivasan V, Subramaniyam M, Shah P. Processing strategies for metallic and high-acid crudes: refiners can minimise the risk of processing crudes that are acidic with high metals content through innovative treatment chemistries. PTQ 2013;18(5):51–7.

[383] Hurtevent C, Ubbels S. Preventing naphthenate stabilised emulsions and naphthenate deposits on fields producing acidic crude oils. In: Proc. SPE 8th intern. symp. oilfield scale. Aberdeen, UK; May 31–June 1, 2006. p. 69–75.

[384] Bieber S. A new process for removing calcium from crude oils containing calcium naphthenates. In: Proc. crude oil quality association meeting. Chicago, IL; May 26, 2005. 14 pp. Available from: http://www.coqa-inc.org/docs/default-source/meeting-presentations/20050526calcium-removal.pdf?sfvrsn=2.

[385] Weers JJ, Bieber S. Calcium removal from high TAN crudes. PTQ 2005;Q3:5.

[386] Herman DC, Fedorak PM, Mackinnon MD, Costerton JW. Biodegradation of naphthenic acids by microbial populations indigenous to oil sands tailings. Can J Microbiol 1994;40(6):467–77.

[387] Clemente JS, Mackinnon MD, Fedorak PM. Aerobic biodegradation of two commercial naphthenic acids preparations. Environ Sci Technol 2004;38(4):1009–16.

[388] Biryukova OV, Fedorak PM, Quideau SA. Biodegradation of naphthenic acids by rhizosphere microorganisms. Chemosphere 2007;67(10):2058–64.

[389] Del Rio LF, Hadwin AKM, Pinto LJ, Mackinnon MD, Moore MM. Degradation of naphthenic acids by sediment micro-organisms. J Appl Microbiol 2006;101(5):1049–61.

[390] Videla PP. Examining oil sands dissolved carbon and microbial degradation using stable isotope analysis [MSc biology thesis from University of Waterloo]. 2007. 124 pp.

[391] Quagraine EK, Headley JV, Peterson HG. Is biodegradation of bitumen a source of recalcitrant naphthenic acid mixtures in oil sands tailing pond waters? J Environ Sci Health - Part A Toxic/Hazardous Subst Environ Eng 2005;40(3):671–84.

[392] Quagraine EK, Peterson HG, Headley JV. In situ bioremediation of naphthenic acids contaminated tailing pond waters in the Athabasca oil sands region, demonstrated field studies and plausible options: a review. J Environ Sci Health - Part A Toxic/Hazardous Subst Environ Eng 2005;40(3):685–722.

[393] Clemente JS, Fedorak PM. A review of the occurrence, analyses, toxicity, and biodegradation of naphthenic acids. Chemosphere 2005;60(5):585–600.

[394] Sobolewski A, Van Poppelen P. The design approach to achieving the impossible in biological treatment (Unpublished Presentation). 1995. p. 24. Available from:, www.microbialtech.com.

[395] Kim S, Stanford LA, Rodgers RP, Marshall AG, Walters CC, Qian K, et al. Microbial alteration of the acidic and neutral polar NSO compounds revealed by fourier transform ion cyclotron resonance mass spectrometry. Org Geochem 2005;36(8):1117–34.

[396] Holowenko FM, Mackinnon MD, Fedorak PM. Naphthenic acids and surrogate naphthenic acids in methanogenic microcosms. Water Res 2001;35(11):2595–606.

[397] Kirkwood KM, Foght JM, Gray MR. Prospects for biological upgrading of heavy oils and asphaltenes. In: Vazquez-Duhalt R, Quintero-Ramirez R, editors. Studies surf. sci. catal. Amsterdam: Elsevier B.V.; 2004. p. 113–43.

[398] Headley JV, Du JL, Peru KM, Gurprasad N, Mcmartin DW. Evaluation of algal phytodegradation of petroleum naphthenic acids. J Environ Sci Health - Part A Toxic/Hazardous Subst Environ Eng 2008;43(3):227–32.

[399] Xu X, Yang J, Gao J. Effects of demulsifier structure on desalting efficiency of crude oils. Pet Sci Technol 2006;24(6):673–88.
[400] Xu X, Zhang Y, Zhan M, Gu B. Effect of demulsifer composition on desalting efficiency of crude oils. Huadong Huagong Xueyuan Xuebao 1992;18(2):186–92.
[401] Xu Z. The effects of decalcifying agents on crude oil decalcification and its action mechanism. Pet Proc Petrochem 2004;35(10):29–33.
[402] Xu Z, Tan L, Yu L, Wang X. Commercial test of crude oil de-calcification agent. Pet Refin Eng 2004;34(10):46–8.
[403] Zhan M, Ma F, Xu X, Yang J, Zhang Y. Study on deep desalting of Shengli huangdao crude oil I. Effect of demulsifier composition. Hua Dong Li Gong Da Xue/J East China Univ Sci Technol 1995;21(1):37–41.
[404] Xu J, Luo G, Chen G. Study on oil deacidification with a novel micro-mixer. Pet Proc Petrochem 2004;35(5):47–9.
[405] Zhu K, Tu Y, Huang Y, Shen X. The effect of different force fields on removing naphthenic acids from diesel fuel. Pet Proc Petrochem 2004;35(3):15–7.
[406] Ozaki H, Murakami T, Suzuka T, Yoshioka S, Ohse H. New upgrading process for residual oil. In: Proc. 32nd conf. of Canadian soc. for chem. eng. Vancouver, BC, Canada, vol. 2; October 3–6, 1982. p. 704–12.
[407] Ozaki H, Murakami T, Suzuka T, Yoshida S, Ohse H. Cracking of residual oils with simultaneous generation of hydrogen. In: Proc. ACS div. pet. chem. Seattle, WA, USA, vol. 28. ACS; 1983. p. 717–41.
[408] Ozaki H, Suzuka T. Recent development of upgrading of heavy oils. Nenryo Kyokaishi 1986;65(6):370–7.
[409] Ozaki H, Hisamitsu T, Komori K. Desulfurization and upgrading of heavy residual oils over hydrotreating catalysts. Stud Surf Sci Catal 1989;44(Success. Des. Catal.):123–30.
[410] Ozaki H, Sakamoto M. Catalytic hydrotreating of heavy residual oils in hds units. Stud Surf Sci Catal 1989;53:191–200.
[411] Fukase S, Suzuka T. Residual oil cracking with generation of hydrogen: deactivation of iron oxide catalyst in the steam-iron reaction. Appl Catal A, General 1993;100(1):1–17.
[412] Murakami Y, Miyamoto A, Kamitomai S, Kurahashi K, Kosaki Y. Production of concentrated hydrogen from dilute, lean reducing gas by use of the steam-oxide process (part 1). Behavior of iron oxide medium in the reduction-oxidation cycles. J Jpn Pet Inst 1982;25:380–8.
[413] Satchwell R, Lourenco J, Cochrane PJ, Facey RM, Rodrigues J, Bannard K, et al. Process for treating crude oil using hydrogen in a special unit. Patent No. US7001502 (Also published as CA2249051, CA2306069, EP1121400, WO2000018854). Canadian Enviromnental Equipment & Engineering Technologies, Inc.; CE^3 Technologies Inc.; Genoil Inc.; Lifschultz Enterprises Co., LLC; February 21, 2006.
[414] Castañeda LC, Muñoz JAD, Ancheyta J. Current situation of emerging technologies for upgrading of heavy oils. Catal Today 2014;220–222:248–73.
[415] Chung P, Bugg T. Genoil hydroconversion upgrader (ghu®). 2012. Unpublished Electronic presentation. Available from: http://www.genoil.ca/images/stories/genoil/PDF/genoil_technical_2008%20.pdf. 56 pp.
[416] Anonymous. Catalytic hydrocracking process converts heavy materials. Hydrocarb Proc 2004;83(9). 27 (1 p.).
[417] Lifschultz DK. Upgrading heavy oil products. PTQ 2006;11(1):83–6.
[418] Anonymous. The bottomless barrel. In: Oilsands rev.; 2007. p. 16–7. Calgary, Alberta. Canada.
[419] Runyan J. Is bottomless-barrel refining possible? Future energy demand will involve innovative methods that can upgrade residual products. Hydrocarbon proc. September 2007. p. 81–90.
[420] Genoil Inc. The genoil hydroconversion upgrading system (ghu®) for heavy and extra heavy crude (Unpublished Descriptive Brochure). 2009. 20 pp.
[421] Hebei Zhongjie Petrochemical. Haiyitong upgrader project in China. 2015. New York, NY, USA. 2 pp.

[422] Press Release. Hebei zhongjie petrochemical group Co. Ltd. & genoil sign comprehensive contract to build one of the world's most advanced residue upgrading refineries with significantly increased profitability. PR Newswire; January 19, 2016. 5 pp. Available from: http://www.prnewswire.com/news-releases/hebei-zhongjie-petrochemical-group-co-ltd—genoil-sign-comprehensive-contract-to-build-one-of-the-worlds-most-advanced-residue-upgrading-refineries-with-significantly-increased-profitability-565742571.html.

[423] Chattanooga Corp. Chattanooga process: synthetic crude oil. 2006. 14 pp. Available from: http://www.chattanooga-corp.com/documents/ChattanoogaCorp%20-%20Changing%20the%20Technology%202006%20Oct17.pdf.

[424] Chattanooga Corp. Chattanooga process - brochure. 2006. 2 pp. Available from: http://www.chattanooga-corp.com/documents/Chattanooga_Process.pdf.

[425] Chattanooga Corp. Chatanooga process: poster. 2008. Available from: http://www.chattanooga-corp.com/news.html.

[426] Karpenski M. A process using heated pressurized hydrogen fluidized bedproducing high grade sco from oil sand, heavy oil, bitumen, oil shale. In: Proc. oil sands heavy oil technol. conf. Calgary, Alberta. Canada; July 16, 2008. 19 pp.

[427] Kirkbride CG, Doyle JA, Hildebrandt F. Process and apparatus for converting oil shale or tar sands to oil. Patent No. US6139722. Chatanooga Corp.; October 31, 2000.

[428] Kirkbride CG, Doyle JA, Hildebrandt F. Process and apparatus for converting oil shale or tar sands to oil. Patent No. CA2316416. Chatanooga Corp.; September 09, 2000.

[429] Kirkbride C, Doyle J, Hildebrandt F. Process and apparatus for converting oil shale of tar sands to oil. Patent No. US6319395 (Also published as US20020054836 (Abandoned)). Chatanooga Corp.; November 20, 2001.

[430] Kirkbride CG, Doyle JA, Hildebrandt F. Process and apparatus for converting oil shale or tar sands to oil. Patent No. CA2236183. Chatanooga Corp.; August 25, 2001.

[431] Doyle JA, Hildebrandt F. Process and apparatus for converting oil shale or oil sand (tar sand) to oil. Patent No. CA2467372. Chatanooga Corp.; November 14, 2005.

[432] Doyle J, Hildebrandt F. Process and apparatus for converting oil shale or oil sand (tar sand) to oil. Patent No. US2005252832 (Abandoned) (Also published as WO2006124048). Chatanooga Corp.; November 17, 2005.

[433] Diaz H, Jorgensen P, Vernet P. Electric arc conversion process. Patent No. US4566961 (Also published as DE3461102, EP0120625, FR2542004, JPS60500625, NO844314, WO198400351). British Petroleum Company, plc; January 26, 1986.

[434] Blouri B, Jorgensen P. Conversion of heavy oils and residues into light liquid hydrocarbons under pressure of hydrogen activated by thermal shock. Patent No. FR2588878. British Petroleum Co.; April 24, 1987.

[435] Jorgensen P, Jorgensen E. Process for conversion of heavy or not readily fusible hydrocarbons and residues into branched, light liquid hydrocarbons of low melting point with the aid of CO + xh2 gas mixtures made reactive by energy shock. Patent No. FR2622893. BP France; May 12, 1989.

[436] Jorgensen P. System for the conversion of hydrocarbons. Patent No. AU2004201428. Carbon Resources Ltd.; April 29, 2004.

[437] Jorgensen P. Deep conversion combining the demetallization and the conversion of crudes, residues or heavy oils into light liquids with pure or impure oxygenated compounds. Patent No. US6989091 (Also published as WO0023540, US2012055847, US2010260649, US2005276735, US2005211602, US2004065589, PT1129153, PL347248, PL191375, OA12680, NZ510706, NO20011828, JP2003525310, EP1129153, EP1342773, EA003082, DK1129153, CU23154, DE69931064, CN1323338, CN100489066, CA2346181, BR9915551, AU6418599, AT324422, AP1341). Air Liquide, World Energy Systems Corp, Carbon Resources Limited, Lanisco Holdings Ltd.; January 24, 2006.

[438] Jorgensen P. Deep conversion combining the demetallization and the conversion of crudes, residues or heavy oils into light liquids with pure or impure oxygenated compounds. Patent No. US7967954.

Air Liquide, World Energy Systems Corp, Carbon Resources Limited, Lanisco Holdings Ltd.; June 28, 2011.

[439] Jorgensen P. Deep conversion combining the demetallization and the conversion of crudes, residues or heavy oils into light liquids with pure or impure oxygenated compounds. Patent No. US2012055847 (Abandoned). World Energy Systems Corporation; March 08, 2012.

[440] Stapp PR. Upgrading of residual oils using a selenium catalyst where sulfur and metallic impurities are reduced. Patent No. US4597855. Phillips Petroleum Co.; July 1, 1986.

[441] Murphy JR. Integrated heavy oil cracking process utilizing catalyst separated from cracking in pretreating zone. Patent No. US4048057. Pullman Inc., USA; September 13, 1977.

[442] Paspek SC, Coker CD. Conversion of high boiling organic materials to low boiling materials. Patent No. US4594141. The Standard Oil Company; June 10, 1986.

[443] Paspek SC. Upgrading heavy hydrocarbons with supercritical water and olefins. Patent No. US4483761. The Standard Oil Company; November 20, 1984.

[444] Piskorz J, Radlein DSAG, Majerski P, Scott DS. Hydrotreating of heavy hydrocarbon oils in supercritical fluids. Patent No. US5496464. Canada Natural Resources; March 05, 1996.

[445] Viet TT, Lee J-H, Ryu JW, Ahn I-S, Lee C-H. Hydrocracking of vacuum residue with activated carbon in supercritical hydrocarbon solvents. Fuel 2012;94:556–62.

[446] Takeuchi C, Nakamura M, Shiroto Y. Upgrading of residual oils by asphaltene conversion. ACS Div Pet Chem 1979;24(2):666–77.

[447] Shiroto Y, Ono T, Asaoka S, Nakamura M. Catalysts for hydrotreatment of heavy hydrocarbon oils containing asphaltenes. Patent No. US4422960, US4444655 (Also published as GB2069363, NL8005913, NL189446, MX162469, MX171865, JPS56115638, JPS601056, IT1134471, FR2475931, DE3041676, CA1153750). Chiyoda Chem Eng Construct Co.; December 27, 1983.

[448] Inooka M, Wakabayashi M, Matsuda M, Kasuya M, Ohguchi Y, Nakamura M, et al. Demetallation of hydrocarbons with catalysts supported on sepiolite. Patent No. US4152250. Chiyoda Chem Eng Construct Co.; May 01, 1979.

[449] Inooka M. Process for hydrocracking hydrocarbons with hydrotreatment-regeneration of spent catalyst. Patent No. US4525267 (Also published as GB2127320, JPS6239634, WO8204441, DE3152896). Chiyoda Chem Eng Construct Co.; June 25, 1985.

[450] Ando M, Fukui Y, Homma Y, Shiroto Y. Process for hydrotreating heavy hydrocarbon oil. Patent No. US4191636 (Also published as DE2824765, DE2824765, NL7806212, NL180019, JPS541306, JPS5740879, GB1602640, CA1126192). Chiyoda Chem Eng Contruct Co.; March 04, 1980.

[451] Inooka M. Processing heavy hydrocarbon oils. Patent No. US4486295 (Also published as GB2031011, NL7907428, DE2940015, CA1132077). Chiyoda Chem Eng Construct Co.; December 04, 1984.

[452] Shen R, Zhao H, Liu C, Que G. Hydrocracking of Liaohe vacuum residue on bimetallic oil-soluble catalysts. Pet Proc Petrochem 1998;29(11):9–12.

[453] Angeles MJ, Leyva C, Ancheyta J, Ramírez S. A review of experimental procedures for heavy oil hydrocracking with dispersed catalyst. Catal Today 2014;220–222:274–94.

[454] Liu D, Cui W, Zhang S, Que G. Role of dispersed Ni catalyst sulfurization in hydrocracking of residue from karamay. Energy Fuels 2008;22(6):4165–9.

[455] Ren R, Wang ZX, Guan CS, Guo AJ, Que GH. Study on the sulfurization of water-soluble catalysts for slurry-bed hydroprocessing of residue I effect of sulfurization conditions on the species and dispersity of the sulfurized catalysts. Ranliao Huaxue Xuebao/J Fuel Chem Technol 2005;33(3):293–8.

[456] Ren R, Wang ZX, Guan CS, Guo AJ, Que GH. Study on the sulfurization of water-soluble catalysts for slurry-bed hydroprocessing of residue II XPS analysis of sulfurized product of molybdate. Ranliao Huaxue Xuebao/J Fuel Chem Technol 2005;33(3):299–303.

[457] Guan CS, Wang ZX, Guo AJ, Que GH. Sulfurization of water-soluble catalyst for suspended bed hydrocracking of residue. Shiyou Xuebao, Shiyou Jiagong/Acta Pet Sin (Pet Proc Sect) 2004;20(2):75–80.

[458] Liu D, Zhang SY, Cui WL, Deng WA, Que GH. Effect of sulfuration conditions on the coke-restraining activity of water solubility Ni catalyst. Shiyou Huagong Gaodeng Xuexiao Xuebao/J Petrochem Univ 2009;22(3):42–50.

[459] Zhang S, Deng W, Luo H, Lui D, Que G. Slurry-phase residue hydrocracking with dispersed nickel catalyst. Energy Fuels 2008;22(6):3583–6.

[460] Jin HN, Liu D, Zhang SY, Zhang H, Que G. Synthesis and characterization of dispersed catalyst used in heavy oil hydrocracking. In: Proc. 232nd ACS nat. meeting exposition. San Francisco, CA, vol. 232. PETR 21; September 10–14, 2006. 8 pp.

[461] Luo H, Deng WA, Zhang SY, Que GH. Interaction mechanism of adjuvant in residue slurry-bed hydrocracking I. Effects of adjuvant on colloid character of the reaction system. Shiyou Xuebao, Shiyou Jiagong/Acta Pet Sin (Pet Proc Sect) 2009;25(2):241–5.

[462] Luo H, Deng WA, Gao JJ, Zhang SY, Que GH. The effect of adjuvant on residue asphaltene in slurry-bed hydrocracking. Shiyou Xuebao, Shiyou Jiagong/Acta Pet Sin (Pet Proc Sect) 2009;25(5):646–50.

[463] Zhang SY, Luo H, Wang JQ, Deng WA, Que GH. Synergic effect between surfactants and catalysts in residue hydrocracking. Shiyou Xuebao, Shiyou Jiagong/Acta Pet Sin (Pet Proc Sect) 2008;24(5):515–9.

[464] Zhang S, Deng W, Tian G, Liu D, Que G. Effect of slurry oil-soluble catalyst on hydrocracking of asphaltene. Shiyou Huagong/Petrochem Technol 2008;37(11):1143–6.

[465] Zhang S, Deng W, Liu D, Que G. New development of slurry-bed heavy oil hydrocracking process. Pet Refin Eng 2007;37(2):1–6.

[466] Liu D, Du H, Zhang J, Que G. Reverse microemulsion synthesis and characterization of nano nickel sulfide catalyst for residue slurry-phase hydrocracking. Energy Fuels 2015;29(5):3353–8.

[467] Watanabe I, Otake M, Yoshimoto M, Sakanishi K, Korai Y, Mochida I. Behaviors of oil-soluble molybdenum complexes to form very fine MoS_2 particles in vacuum residue. Fuel 2002;81(11–12):1515–20.

[468] Zhou JS, Deng WA, Liu D, Liang SC, Que GH. Effect of vacuum bottom residue recycling on slurry bed hydrocracking of residue. Acta Pet Sin (Pet Proc Sect) 2001;17(4):82–5.

[469] Panariti N, Del Bianco A, Del Piero G, Marchionna M. Petroleum residue upgrading with dispersed catalysts: Part 1. Catalysts activity and selectivity. Appl Catal A General 2000;204(2):203–13.

[470] Yuansheng L, Jun W, Lijing J, Zhongqing Z, Jiduan L, Shuyan R, et al. Hydrocracking of heavy oil and residuum with a dispersing-type catalyst. Patent No. US6004454. China Petrochemical Corp., Fushun Research Inst. of Petrol.; December 21, 1999.

[471] Yuansheng L, Jun W, Lijing J, Zhongqing Z, Jiduan L, Shuyan R, et al. Dispersing-type catalyst for hydrocracking of heavy oil and residuum, and a method for the preparation thereof. Patent No. US5948721 (Also published as CA2190404). China Petrochemical Co., Fushun Research Inst. of Petrol.; September 07, 1999.

[472] Mcfarlane RA, Cyr T, Hawkins RWT. Process for dispersing transition metal catalytic particles in heavy oil. Patent No. US5916432 (Also published as CA2216671). Alberta Oil Sands Technology And Research Authority; June 29, 1999.

[473] Luo H, Deng W, Gao J, Fan W, Que G. Dispersion of water-soluble catalyst and its influence on the slurry-phase hydrocracking of residue. Energy Fuels 2011;25(3):1161–7.

[474] Tye CT, Smith KJ. Cold Lake bitumen upgrading using exfoliated MoS_2. Catal Lett 2004;95(3–4):203–9.

[475] Gleim WT. Slurry hydrorefining of black oils with mixed vanadium and manganese sulfides as catalysts. Patent No. US3694352. Universal Oil Products Co.; September 26, 1972.

[476] Gleim WKT, Gatsis JG. Hydrofining of asphaltic crude oils with a colloidal dispersion of a lead salt as catalyst. Patent No. US3317421. Universal Oil Products Co.; May 02, 1967.

[477] Wang Z-X, Zhang H-Y, Guo A-J, Que G-H, Olah GA, Prakash GKS. Study on hydrogen donation ability of residue hydrocracking system catalyzed by oil-soluble and water-soluble catalysts. ACS Div Fuel Chem 1998;43(3):530–3. Preprints.

[478] Bearden R. Microcat-RC™: technology for the hydroconversion upgrading of petroleum residua. In: Proc. 213th ACS nat. meeting. San Francisco, CA, USA, vol. 42; April 13–17, 1997. 20 pp.
[479] Ramírez De Agudelo MM, Galarraga C. Catalytic system for the hydroconversion of heavy oils. Patent No. US4888104. Intevep S.A.; December 19, 1989.
[480] Ramírez De Agudelo MM, Galarraga C. Catalytic system for the hydroconversion of heavy oils. Patent No. US4937218. Intevep S.A.; June 26, 1990.
[481] Lott RK, Cyr T, Lee LK. Process for reducing coke formation during hydroconversion of heavy hydrocarbons. Patent No. CA2004882. Alberta Oil Sands Technology and Research Authority; July 07, 1991.
[482] Liu C, Zhou J, Que G, Liang W, Zhu Y. Hydrocracking kinetics of gudao residue in the presence of dispersed-phase Mo catalyst. In: Proc. 205th ACS nat. meeting, symp. on resid upgrading. Denver, CO, USA, vol. 38. ACS; March 28–April 02, 1993. p. 446–52.
[483] Ovalles C, Filgueiras E, Morales A, Rojas I, De Jesus JC, Berrios I. Use of a dispersed molybdenum catalyst and mechanistic studies for upgrading extra-heavy crude oil using methane as source of hydrogen. Energy Fuels 1998;12(2):379–85.
[484] Kushiyama S, Aizawa R, Kobayashi S, Koinuma Y, Uemasu I, Ohuchi H. Effect of addition of sulphur and phosphorus on heavy oil hydrotreatment with dispersed molybdenum-based catalysts. Appl Catal 1990;63(1):279–92.
[485] Lee LK, Pradhan VR, Popper G, Comolli AG. Dispersed slurry catalysts for hydroconversion of carbonaceous materials. In: Proc. 212th, ACS; div. fuel chem. Orlando, FL, vol. 41; August 25–29, 1996. p. 946–50.
[486] Jeon SG, Na JG, Ko CH, Yi KB, Rho NS, Park SB. Preparation and application of an oil-soluble CoMo bimetallic catalyst for the hydrocracking of oil sands bitumen. Energy Fuels 2011;25(10):4256–60.
[487] Lee DK, Koon PS, Yoon WL, Lee IC, Woo SI. Residual oil hydrodesulfurization using dispersed catalysts in a carbon-packed trickle bed flow reactor. Energy Fuels 1995;9(1):2–9.
[488] Li Y, Wang J, Jia L, Wang X. Water soluble catalyst used for residue suspension hydroconversion. Pet Proc Petrochem 1999;30(5):21–4.
[489] Wang X, Li Y, Zhang Z, Jia L. Study on homogeneous catalyst and process for residue suspended bed hydrocracking. Pet Refin Eng 2000;30(4):50–2.
[490] Guan CS, Wang ZX, Guo AJ, Que GH. Study on slurry catalyst for kelamayi residue hydrocracking. Ranliao Huaxue Xuebao/J Fuel Chem Technol 2003;31(5):434–8.
[491] Liu C, Zhou J, Que G, Liang W, Zhu Y. Hydrocracking of gudao residue with dispersed-phase Mo catalyst. Fuel 1994;73(9):1544–50.
[492] Wagner RE. Two-phase hydrocarbon-converting system. Patent No. US3663431. Union Oil Co.; May 16, 1972.
[493] Cordova J, Pereira Almao P, Guitian J, Andriollo A, Cirilo A, Granadillo F. Production of oil soluble catalytic precursors. Patent No. US6043182. Intevep S.A.; March 28, 2000.
[494] Ovalles C, Filgueiras E, Morales A, Scott CE, Gonzalez-Gimenez F, Pierre Embaid B. Use of a dispersed iron catalyst for upgrading extra-heavy crude oil using methane as source of hydrogen. Fuel 2003;82(8):887–92.
[495] Kukes SG. Demetallization of hydrocarbon containing feed streams. Patent No. US4457835. Phillips Petroleum Co.; July 3, 1984.
[496] Kukes SG, Parrott SL. Catalyst for hydrofining of oil and process for preparing the catalyst. Patent No. US4681869. Phillips Petroleum Co.; July 21, 1987.
[497] Kukes SG, Hawley GR. Demetallization of hydrocarbon containing feed streams. Patent No. US4557824. Phillips Petroleum Co.; December 10, 1985.
[498] Reynolds JG, Yu SG, Beret S. Metals-impregnated red mud as a first-stage catalyst in a two-stage, close-coupled thermal catalytic hydroconversion process. Patent No. US4559130. Chevron Research Co.; December 17, 1985.
[499] Reynolds JG, Yu SG, Lewis RT. Red mud as a first-stage catalyst in a two-stage, close-coupled thermal catalytic hydroconversion process. Patent No. US4559129. Chevron Research Co.; December 17, 1985.

[500] Nguyen-Huy C, Kweon H, Kim H, Kim DK, Kim D-W, Oh SH, et al. Slurry-phase hydrocracking of vacuum residue with a disposable red mud catalyst. Appl Catal A General 2012;447–448:186–92.
[501] Liu C, Shen R, Liang W. Hydro-conversion of gudao vacuum residue in the presence of industrial wastes. Pet Proc Petrochem 1996;27(8):60–3.
[502] Jain AK, Pruden BB. Hydrocracking of heavy oils in presence of petroleum coke derived from heavy oil coking operations. Patent No. US4999328 (Also published as CA1322746). Petro-Canada Inc.; March 12, 1991.
[503] Terai S, Fukuyama H. Upgrading heavy oil using iron/active carbon catalyst. PTQ 2000;5(3):31–7.
[504] Fouda SA, Kelly JF. Simultaneous hydrocracking or coprocessing of +525 petroleum pitch/coal slurries. Patent No. CA1276902. Her Majesty In Right Of Canada As Represented By The Minister Of Energy, Mines And Resources Canada; November 27, 1990.
[505] Khulbe CP, Ranganathan R, Pruden BB, Khulbe CP, Ranganathan R, Pruden BB. Hydrocracking of heavy oils/fly ash slurries. Patent No. US4299685 (Also published as CA1124194). November 10, 1981.
[506] Li C, Shi B, Cui M, Shang H-Y, Que G-H. Application of Co-Mo/cnt catalyst in hydro-cracking of gudao vacuum residue. J Fuel Chem Technol 2007;35(4):407–11.
[507] Matsumura A, Sato S, Kondo T, Saito I, De Souza WF. Hydrocracking marlim vacuum residue with natural limonite. Part 2: experimental cracking in a slurry-type continuous reactor. Fuel 2005;84(4):417–21.
[508] Mochida I, Zhao X-Z, Sakanishi K. Suppression of sludge formation by two-stage hydrocracking of vacuum residue at high conversion. Ind Eng Chem Res 1990;29(12):2324–7.
[509] Mochida I, Zhao X-Z, Sakanishi K. Catalytic two-stage hydrocracking of Arabian vacuum residue at a high conversion level without sludge formation. Ind Eng Chem Res 1990;29(3):334–7.
[510] Tugsuu T, Yoshikazu S, Enkhsaruul B, Monkhoober D. A comparative study on the hydrocracking for atmospheric residue of mongolian tamsagbulag crude oil and other crude oils. Adv Chem Eng Sci 2012;2(3):402–7.
[511] Purón H, Pinilla JL, Berrueco C, Montoya De La Fuente JA, Millán M. Hydrocracking of Maya vacuum residue with NiMo catalysts supported on mesoporous alumina and silica–alumina. Energy Fuels 2013;27(7):3952–60.
[512] Kouzu M, Kuriki Y, Uchida K, Sakanishi K, Sugimoto Y, Saito I, et al. Catalytic hydrocracking of petroleum residue over carbon-supported nickel-molybdenum sulfides. Energy Fuels 2005;19(3):725–30.
[513] Pelrine BP, Comolli AG, Lee LK. Ionic liquid catalyst composed of iron promoted with metal including cobalt, molybdenum, palladium, platinum, nickel and/or tungsten in water having acidic pH. Patent No. US6139723. Hydrocarbon Technologies, Inc.; October 31, 2000.
[514] Colyar J. Has the time for partial upgrading of heavy oil and bitumen arrived? PTQ 2009;14(5):43–56.
[515] Johnson LA, Satchwell RM, Glaser RR, Brecher LE. Process for recovery of tank bottom wastes. Patent No. US5259945 (Also published as EP0678116). Western Research Institute; November 09, 1993.
[516] Brecher L, Mones C, Guffey F. Development of the WRITE process for pipeline-ready heavy oil. Report No. WRI 09-R007. Laramie, Wyoming, USA: Western Research Institute; April 2009. 97 pp.
[517] Humphries M. North american oil sands: history of development, prospects for the future. Intern J Energy, Environ Econ 2009;16(2–3):291–314.
[518] Patel S. Canadian oil sands: opportunities, technologies and challenges. Hydrocarb Proc 2007;86(2). 65–68+70+72+74.
[519] Crandall GR. Canadian heavy crude/bitumen markets: drivers and challenges. In: Proc. 7th UNITAR intern. conf. 'heavy crude tar sands'. Beijing, China; October 27–30, 1998. 10 pp.
[520] Golden S. Canadian crude processing challenges. PTQ 2008;13(1):53–61.
[521] Clark B, Graves WG, Lopez-De-Cardenas JE, Gurfinkel ME, Peats AW. Heavy oil. In: Facing the hard truths about energy. National Petroleum Council, Heavy Oil Subgroup of the Technology Task Group; 2007. Topic paper #22, 56 pp.

[522] Shah A, Fishwick R, Wood J, Leeke G, Rigby S, Greaves M. A review of novel techniques for heavy oil and bitumen extraction and upgrading. Energy & Environ Sci 2010;3(6):700–14.
[523] Flint L. Bitumen & very heavy crude upgrading technology: a review of long term R&D opportunities. LENEF Consulting Limited; March 31, 2004. Report No. Available at: http://www.ptac.org/links/dl/osdfnlreport.pdf.
[524] Bain RL, Larson JR, Gertenbach DD, Gillespie DW, Leto JJ. Viscosity reduction processing at elevated pressure. Patent No. US4778586. Resource Technology Associates; October 18, 1988.
[525] Bain RL, Gertenbach DD, Larson JR, Leto JJ, Gillespie DW. Viscosity reduction processing at elevated pressure. Patent No. CA1292716. Resource Technology Associates; December 03, 1991.
[526] Bain RL, Larson JR. Viscosity reduction by direct oxidative heating. Patent No. US4818371. Resource Technology Associates; April 04, 1989.
[527] Bain RL, Larson JR. Apparatus for thermal treatment of a hydrocarbon stream. Patent No. US5008085 (Also published as CA1298803). Resource Technology Associates; April 16, 1991.
[528] Bain RL, Larson JR. Apparatus for thermal treatment of a hydrocarbon stream. Patent No. CA1327952. Resource Technology Associates; March 22, 1994.
[529] Luhning R. Removing constraints on oil sands potential: pipeline transportation of emerging partially upgraded bitumen. J Can Pet Technol 2003;42(8):7–8.
[530] Luhning RW, Anand A, Blackmore T, Lawson DS. Pipeline transportation of emerging partially upgraded bitumen. In: Proc. 3rd annual Canadian intern. pet. conf. the 53th annual technical meeting of the pet. soc. of CIM Calgary. Canada, vol. 33; June 11–13, 2002. 20 pp.
[531] Pereira-Almao P, Larter S. An integrated approach to on site/in situ upgrading. In: Proc. 19th world pet. congress. Madrid, Spain, vol. 1; June 29–July 3 2008. p. 456–61.
[532] Speight JG. Shale oil production processes. Elsevier Inc; 2012.
[533] De Rouffignac EP, Vinegar HJ, Wellington SL, Shahin GT, Berchenko IE, Stegemeier GL, et al. In situ thermal processing of a hydrocarbon containing formation with heat sources located at an edge of a formation layer. Patent No. US6702016. Shell Oil Co.; March 09, 2004.
[534] Stine LO. Method for in situ conversion of hydrocarbonaceous oil. Patent No. US4444257. UOP LLC; April 24, 1984.
[535] Stine LO. In situ conversion of hydrocarbonaceous oil. Patent No. US4448251. UOP LLC; May 15, 1984.
[536] Schora FC, Janka JC. Enhanced oil recovery. Patent No. US4141417. Institute of Gas Technology; February 27, 1979.
[537] Gregoli AA. Method of in-situ hydrogenation of carbonaceous material. Patent No. US4501445. Cities Service Company; February 20, 1985.
[538] Sarathi PS. Nine decades of combustion oil recovery-a review of in situ combustion history and assessment of geologic environments on project outcome. In: Proc. 7th UNITAR intern. conf. 'heavy crude tar sands'. Beijing, China, vol. 2; October 27–30, 1998. p. 1189–200.
[539] Greaves M, Turta A. Oilfield in situ combustion process. Patent No. US5626191. Petroleum Recovery Inst.; Alberta Research Council Inc.; May 06, 1997.
[540] Pfefferle WC. Method for cagd recovery of heavy oil. Patent No. US2007187094 (Abandoned) (Also published as WO2008060311, CA2642750). Precision Combustion Inc.; August 16, 2007.
[541] Pfefferle WC. Method for in situ combustion of in-place oils. Patent No. CA2616572. Precision Combustion Inc.; June 27, 2008.
[542] Pfefferle WC. Method for in-situ combustion of in-place oils. Patent No. US7581587 (Also published as US2007256833). Precision Combustion Inc.; September 01, 2009.
[543] Pfefferle WC. Downhole catalytic combustion for hydrogen generation and heavy oil mobility enhancement. Patent No. US2005239661 (Abandoned). Precision Combustion Inc.; October 27, 2005.
[544] Pfefferle WC. Reducing the energy requirements for the production of heavy oil. Patent No. US7665525 (Also published as US2006260814). Precision Combustion Inc.; February 23, 2010.

[545] Jamaluddln AKM, Nazarko TW, Sills S, Fuhr B. Asphaltene-compatible fluid design for workover operations. In: Proc. 6th UNITAR intern. conf. 'heavy crude tar sands'. Houston, TX; February 12–17, 1995. p. 579–85.
[546] Shallcross DC. Modifying in-situ combustion performance by the use of water-soluble additives. In: Proc. Asia-Pacific conf. Sydney, Australia; September 13–15, 1989. 12 pp.
[547] Castanier LM, Brigham WE. Modifying in-situ combustion with metallic additives. In Situ 1997;21(1):27–45.
[548] Castanier LM, Brigham WE. Upgrading of crude oil via in situ combustion. J Pet Sci Eng 2003;39(1–2):125–36.
[549] Mulchandani H, Brandt AR. Oil shale as an energy resource in a CO_2 constrained world: the concept of electricity production with in situ carbon capture. Energy Fuels 2011;25(4):1633–41.
[550] Zheng D, Li S, Ma G, Wang H. Autoclave pyrolysis experiments of chinese liushuhe oil shale to simulate in-situ underground thermal conversion. Oil Shale 2012;29(2):103–14.
[551] Bauman JH, Deo M. Simulation of a conceptualized combined pyrolysis, in situ combustion, and CO_2 storage strategy for fuel production from green river oil shale. Energy Fuels 2012;26(3):1731–9.
[552] Curtis C, Kopper R, Decoster E, Guzman-Garcia A, Huggins C, Knawer L, et al. Heavy-oil reservoirs. Oilfield Rev 2002;14(3):30–52.
[553] Hyne JB, Tyrer JD. Use of hydrogen-free carbon monoxide with steam in recovery of heavy oil at low temperatures. Patent No. US4487264. Alberta Oil Sands Technology and Research Authority; December 11, 1984.
[554] Klassen M, Gokulakrishan P, Fuller C, Hamer A, Langdon JE, Ware CH. System, method and apparatus for hydrogen-oxygen burner in downhole steam generator. Patent No. US7770646 (Also published as WO2008045408, US2008083537). World Energy Systems Inc.; August 10 2010.
[555] Bagci S, Kok MV. In-situ combustion laboratory studies of Turkish heavy oil reservoirs. Fuel Proc Technol 2001;74(2):65–79.
[556] Pfefferle WC. Method for high temperature steam. Patent No. CA2621172 (Also published as WO2007050189). Precision Combustion Inc.; May 3, 2007.
[557] Pfefferle WC. Method for SAGD recovery of heavy oil. Patent No. CA2577680. Precision Combustion Inc.; August 09, 2007.
[558] Kapadia PR, Wang J, Kallos MS, Gates ID. Practical process design for in situ gasification of bitumen. Appl Energy 2013;107:281–96.
[559] Rose LC, Allen JC, Ware CH. Recovery of oil by in situ hydrogenation. Patent No. CA1240262. Worldenergy Systems Inc.; August 09, 1988.
[560] Ware CH, Rose LC, Allen JC. Recovery of oil by in situ hydrogenation. Patent No. US4597441. World Energy Systems, Inc.; July 01, 1986.
[561] Hujsak KL. Underground hydrogenation of oil. Patent No. US3327782. Pan American Petroleum Corp.; June 27, 1967.
[562] Berkowitz N, Dunn SR. Method for upgrading heavy and semi-heavy oils. Patent No. CA2242774. Yeda Res and Dev Co. Ltd.; February 18, 2000.
[563] Berkowitz B, Dror I, Dunn SR. Method for extracting and upgrading of heavy and semi-heavy oils and bitumens. Patent No. CA2316084 (Also published as CA2519736). Yeda Res and Dev Co. Ltd.; February 02, 2002.
[564] Berkowitz N. Method for recovering and partially upgrading bitumen from oil sands, tar sands, oil shales, h eavy oil reservoirs. Patent No. CA2000251. Yeda Res and Dev Co. Ltd.; April 06, 1991.
[565] Berkowitz N, Dunn SR. Method for recovering and concurrently partly upgrading heavy oils, bitumens and oil shale organic matter, and for extracting certain wastes. Patent No. CA2208046. Yeda Res and Dev Co. Ltd.; January 03, 1999.
[566] Berkowitz N, Dror I, Dunn SR. Method for extracting and upgrading heavy and semi-heavy oils, bitumen, and cleaning of toxic wastes. Patent No. US7947165, US8372347 (Also published as CA2519736, US2007056881). Yeda Res and Dev Co. Ltd.; May 24, 2000.

[567] Berkowitz N, Dunn SR. Method for extracting and upgrading heavy and semi-heavy oils, bitumen, and cleaning of toxic wastes. Patent No. CA2252218. Yeda Res and Dev Co. Ltd.; May 13, 2000.
[568] Morimoto M, Sugimoto Y, Sato S, Takanohashi T. Bitumen cracking in supercritical water upflow. Energy Fuels 2014;28(2):858–61.
[569] Ayasse C. Oilfield enhanced in situ combustion process. Patent No. US7493952 (Also published as US2006207762). Archon Technologies Ltd.; February 24, 2006.
[570] Munroe N. Solvent based enhanced oil recovery for in-situ upgrading of heavy oil sands. Report No. Final Technical Report Grant #DE-FG26-06NT42745. Miami, FL, USA: Applied Research Center, FIU Engineering Center; June 22, 2009. 35 pp.
[571] Vallejos C, Vasquez T, Siachoque G, Layrisse I. Process for in-situ upgrading of heavy hydrocarbon. Patent No. US6405799 (Also published as CA2312621). Intevep S.A.; June 18, 2002.
[572] Gregoli AA, Rimmer DP, Graue DJ. Upgrading and recovery of heavy crude oils and natural bitumen by in situ hydrovisbreaking. Patent No. US6016867. World Energy Systems, Inc.; January 25, 2000.
[573] Gregoli AA, Rimmer DP. Production of heavy hydrocarbons by in-situ hydrovisbreaking. Patent No. US6016868 (Also published as WO9967504, CA2335771). World Energy Systems, Inc.; January 25, 2000.
[574] Graue DJ. Upgrading and recovery of heavy crude oils and natural bitumen by in situ hydrovisbreaking. Patent No. US6328104 (Also published as US6016867, WO1999067505). World Energy Systems, Inc.; December 11, 2001.
[575] Herron EH. In situ hydrovisbreaking (Unpublished Presentation). 2003. 16 pp.
[576] Herron EH. Commercial application of in situ upgrading of heavy oil. 2003. 4 pp. Available from: http://www.petroleumequities.com/commercialapplication.html.
[577] Herron EH. Underground refining converts heavy hydrocarbons to synthetic crude oil. 2002. 4 pp. Available from: http://www.petroleumequities.com/converting.html.
[578] Herron EH. Experimental verification of in situ upgrading of heavy oil. 2003. 3 pp. Available from: http://www.petroleumequities.com/experimentalverification.html.
[579] Kasevich R. Down hole physical upgrading of heavy crude oils by selective energy absorption. Patent No. US2006180304 (Also published as CA2595293, CN101142372, EP1871984, WO2006078946). KSN; August 17, 2006.
[580] Weissman JG. Review of processes for downhole catalytic upgrading of heavy crude oil. Fuel Proc Technol 1997;50(2–3):199–213.
[581] Moore RG, Laureshen CJ, Mehta SA, Ursenbach MG, Belgrave JDM, Weissman JG, et al. Downhole catalytic upgrading process for heavy oil using in situ combustion. J Can Pet Technol 1999;38:76–92.
[582] Weissman JG, Kessler RV, Sawicki RA, Belgrave JDM, Laureshen CJ, Mehta SA, et al. Down-hole catalytic upgrading of heavy crude oil. Energy Fuels 1996;10(4):883–9.
[583] Weissman JG, Kessler RV. Downhole heavy crude oil hydroprocessing. Appl Catal A General 1996;140(1):1–16.
[584] Ayasse C, Greaves M, Turta A. Oilfield in situ hydrocarbon upgrading process. Patent No. US6412557 (Also published as EP1060326, CA2255071, WO9930002). Petroleum Recovery Inst; Alberta Research Council Inc.; July 02, 2002.
[585] Kendall R. Using timelapse seismic to monitor the THAI™ heavy oil production process. CSEG Rec 2009;34(7). Available from: http://csegrecorder.com/articles/view/using-timelapse-seismic-to-monitor-the-thai-heavy-oil-production-process.
[586] Shah A, Wood J, Fishwick R, Greaves M, Rigby S. In-situ up-grading of heavy oil/natural bitumen: CAPRI process optimisation. In: Proc. 8th world congress of chem. eng. Montreal, Canada; August 23–27, 2009. 6 pp.
[587] Xia TX, Greaves M, Werfilli WS, Rathbone RR. Downhole conversion of Lloydminster heavy oil using THAI-CAPRI process. In: Proc. intern. thermal operations heavy oil symp. intern. horizontal well technol. conf. (SPE/PS-CIM/CHOA) Calgary, Alberta. Canada; November 4–7, 2002. Paper #78998, 16 pp.

[588] Greaves M, Xia TX. Downhole upgrading of Wolf Lake oil using thai/capri processes - tracer tests. In: Proc. ACS nat. meeting, div. fuel chem. Anaheim, CA, USA, vol. 49; March 4–8, 2004. p. 69–72.

[589] Greaves M, Xia TX. Downhole catalytic process for upgrading heavy oil: produced oil properties and composition. J Can Pet Technol 2004;43(9):25–30.

[590] Hart A, Leeke G, Greaves M, Wood J. Down-hole heavy crude oil upgrading by CAPRI: effect of hydrogen and methane gases upon upgrading and coke formation. Fuel 2014;119:226–35.

[591] Hart A, Shah A, Leeke G, Greaves M, Wood J. Optimization of the CAPRI process for heavy oil upgrading: effect of hydrogen and guard bed. Ind Eng Chem Res 2013;52(44):15394–406.

[592] Hart A, Leeke G, Greaves M, Wood J. Downhole heavy crude oil upgrading using CAPRI: effect of steam upon upgrading and coke formation. Energy Fuels 2014;28(3):1811–9.

[593] Xia TX, Greaves M. Downhole upgrading Athabasca tar sand bitumen using THAI - SARA analysis. In: Proc. SPE intern. thermal operations heavy oil symp. Margarita Island; March 12–14, 2001. 15 pp.

[594] Ayasse C. Well liner segments for in situ petroleum upgrading and recovery, and method of in situ upgrading and recovery. Patent No. US7909097 (Also published as US20100096120, CA2641281, MX2011004043, RU2475629). Archon Technologies Ltd.; March 22, 2011.

[595] Cavallaro AN, Galliano GR, Moore RG, Mehta SA, Ursenbach MG, Zalewski E, et al. In situ upgrading of Llancanelo heavy oil using in situ combustion and a downhole catalyst bed. J Can Pet Technol 2008;47(9):23–31.

[596] Li W, Zhu JH, Qi JH. Application of nano-nickel catalyst in the viscosity reduction of Liaohe extra-heavy oil by aqua-thermolysis. J Fuel Chem Technol 2007;35(2):176–80.

[597] Scott CE, Pereira-Almao P. Catalysis for heavy oils and bitumen upgrading. Curr Top Catal 2014;11:1–24.

[598] Contreras C, Scott CE, Pereira-Almao PR. Preparation of Ni, Mo and NiMo nanoparticles from transient w/o emulsions. In: Proc. 238th ACS nat. meeting, div. pet. chem. preprints, vol. 54; August 16–10, 2009. p. 56–8.

[599] Coy Plazas LA. Experimental simulation of a hot fluid injection process for in-reservoir upgrading [MSc thesis from University of Calgary, Chem. Pet. Eng.]. January 2013. 176 pp.

[600] Hashemi R, Nassar NN, Pereira Almao P. Nanoparticle technology for heavy oil in-situ upgrading and recovery enhancement: opportunities and challenges. Appl Energy 2014;133:374–87.

[601] Hashemi R. In-situ upgrading and recovery enhancement of Athabasca bitumen by ultra-dispersed nanocatalysts [Ph.D. thesis from University of Calgary, Department of Chem. Pet. Eng.]. December 2013. 223 pp.

[602] Zamani A, Maini B, Pereira-Almao P. Experimental study on transport of ultra-dispersed catalyst particles in porous media. Energy Fuels 2010;24(9):4980–8.

[603] Hashemi R, Nassar NN, Pereira-Almao P. In situ upgrading of Athabasca bitumen using multimetallic ultradispersed nanocatalysts in an oil sands packed-bed column: Part 1. Produced liquid quality enhancement. Energy Fuels 2014;28(2):1338–50.

[604] Galarraga CE, Pereira-Almao P. Hydrocracking of Athabasca bitumen using submicronic multimetallic catalysts at near in-reservoir conditions. Energy Fuels 2010;24(4):2383–9.

[605] Hashemi R, Nassar NN, Pereira-Almao P. In situ upgrading of Athabasca bitumen using multimetallic ultradispersed nanocatalysts in an oil sands packed-bed column: Part 2. Solid analysis and gaseous product distribution. Energy Fuels 2014;28(2):1351–61.

[606] Hashemi R, Nassar NN, Pereira-Almao P. Enhanced heavy oil recovery by in situ prepared ultradispersed multimetallic nanoparticles: a study of hot fluid flooding for Athabasca bitumen recovery. Energy Fuels 2013;27(4):2194–201.

[607] Moini AB, Kantzas A, Pereira-Almao P, Larter S. Catalytic down-hole upgrading of heavy oil and oil sand bitumens. Patent No. US2010212893 (Abandoned) (Also published as WO2008058400). University of Calgary; August 26, 2010.

[608] Pereira Almao P, Chen Z, Maini B, Scott-Algara C. In situ upgrading via hot fluid injection. Patent No. CA2810022 (Also published as CA2864788, WO2013177683). In Situ Upgrading Technologies Inc.; November 30, 2013.
[609] Pereira-Almao P, Ali-Marcano VA, Lopez-Linares F, Vasquez A. Ultradispersed catalyst compositions and methods of preparation. Patent No. US7897537. University of Calgary; March 01, 2011.
[610] Pereira-Almao P, Ali-Marcano VA, Lopez-Linares F, Vasquez A. Ultradispersed catalyst compositions and methods of preparation. Patent No. US8283279. University of Calgary; October 09, 2012.
[611] Pereira-Almao P, Ali-Marcano VA, Lopez-Linares F, Vasquez A. Ultradispersed catalyst compositions and methods of preparation. Patent No. US8298982. University of Calgary; October 30, 2012.
[612] Pereira-Almao P, Ali-Marcano VA, Lopez-Linares F, Vasquez A. Ultradispersed catalyst compositions and methods of preparation. Patent No. US8304363 (Also published as US2011105307 US2011105308 US2011105315 US2007161505 WO2007059621 CA2630365). University of Calgary; November 06, 2012.
[613] Vallejos C, Vasquez T, Ovalles C. Process for the downhole upgrading of extra heavy crude oil. Patent No. US5891829. Intevep S.A.; April 06, 1999.
[614] Reynolds JG, Murray AM, Nuxoll EV, Fox GA, Thorsness CB, Khan MR. Upgrading of heavy oil from the san joaquin valley of California by aqueous pyrolysis. Mat Res Soc Symp 1997;432:333–8.
[615] Reynolds JG, Thorsness CB. Mild upgrading of midway sunset crude oil from the San Joaquin Valley of California by aqueous pyrolysis...Catalysis and modeling. ACS Div Pet Chem 1997;42(2):476–8. Preprints.
[616] Chen Y, Wang Y, Lu J, Wu C. The viscosity reduction of nano-keggin-k3pmo12o40 in catalytic aquathermolysis of heavy oil. Fuel 2009;88(8):1426–34.
[617] Xu H, Pu C, Wu F. Low frequency vibration assisted catalytic aquathermolysis of heavy crude oil. Energy Fuels 2012;26(9):5655–62.
[618] Xu HX, Pu CS, Wu FP. Mechanism of underground heavy oil catalytic aquathermolysis. Ranliao Huaxue Xuebao/J Fuel Chem Technol 2012;40(10):1206–11.
[619] Zhao FJ, Liu YJ, Zhao TH, Wen SB, Zhao G. Advance in catalytically upgrading heavy oil by aquathermolysis using hydrogen donor. Oilfield Chem 2006;23(4):379–84.
[620] Fan H, Liu Y. Downhole catalyst upgrades heavy oil. Oil Gas J 2002;100(11):60–2.
[621] Fan H, Liu Y, Zhang L, Zhao X. The study on composition changes of heavy oils during steam stimulation processes. Fuel 2002;81(13):1733–8.
[622] Fan H, Zhang Y, Lin Y. The catalytic effects of minerals on aquathermolysis of heavy oils. Fuel 2004;83(14–15):2035–9.
[623] Fan H-F, Liu Y-J, Zhong L-G. Studies on the synergetic effects of mineral and steam on the composition changes of heavy oils. Energy Fuels 2001;15(6):1475–9.
[624] Liu Y, Fan H. The effect of hydrogen donor additive on the viscosity of heavy oil during steam stimulation. Energy Fuels 2002;16(4):842–6.
[625] Yi Y, Li S, Ding F, Yu H. Change of asphaltene and resin properties after catalytic aquathermolysis. Pet Sci 2009;6(2):194–200.
[626] Greff JH, Babadagli T. Use of nano-metal particles as catalyst under electromagnetic heating for viscosity reduction of heavy oil. In: Proc. intern. pet. technol. conf. Bangkok, Thailand, vol. 2; February 7–9, 2012. p. 1641–53.
[627] Greff J, Babadagli T. Use of nano-metal particles as catalyst under electromagnetic heating for in-situ heavy oil recovery. J Pet Sci Eng 2013;112:258–65.
[628] Newton JP. Metal oxide-based catalyst composition for in-situ recovery of heavy petroleums, tar sands, and contaminated soils. Patent No. WO2004009510 (Also published as EP1597213, MXPA05000949, CN1684920, CA2493260, US2008116111). January 29, 2004.
[629] Franco CA, Montoya T, Nassar NN, Pereira-Almao P, Cortés FB. Adsorption and subsequent oxidation of Colombian asphaltenes onto nickel and/or palladium oxide supported on fumed silica nanoparticles. Energy Fuels 2013;27(12):7336–47.

[630] Nassar NN, Hassan A, Pereira-Almao P. Application of nanotechnology for heavy oil upgrading: catalytic steam gasification/cracking of asphaltenes. Energy Fuels 2011;25(4):1566−70.

[631] Nassar NN, Hassan A, Pereira-Almao P. Effect of the particle size on asphaltene adsorption and catalytic oxidation onto alumina particles. Energy Fuels 2011;25(9):3961−5.

[632] Nassar NN, Hassan A, Pereira-Almao P. Metal oxide nanoparticles for asphaltene adsorption and oxidation. Energy Fuels 2011;25(3):1017−23.

[633] Duyvesteyn WPC, Morley RL. Oxidation of asphaltenes. Patent No. US7811444, US8529687 (Also published as US20070284283, US20100320118). Marathon Oil Canada Corporation; October 12, 2013.

[634] Husein MM, Patruyo L, Pereira-Almao P, Nassar NN. Scavenging $H_2S(g)$ from oil phases by means of ultradispersed sorbents. J Colloid Interface Sci 2010;342(2):253−60.

[635] Nassar NN, Husein MM. Ultradispersed particles in heavy oil: Part I, preparation and stabilization of iron oxide/hydroxide. Fuel Proc Technol 2010;91(2):164−8.

[636] Husein MM, Patruyo L, Pereira P. In-situ removal of $H_2S(g)$ using ultradispersed iron oxide nanoparticles. In: Proc. AIChE annual meeting. Salt Lake City, UT; November 4−9, 2007. 328 t, 1 p.

[637] Nassar NN, Husein MM, Pereira-Almao P. Ultradispersed particles in heavy oil: Part II, sorption of $H_2S(g)$. Fuel Proc Technol 2010;91(2):169−74.

[638] Nassar NN, Pereira-Almao P. Capturing $H_2S(g)$ by in situ-prepared ultradispersed metal oxide particles in an oilsand-packed bed column. Energy Fuels 2010;24(11):5903−6.

[639] Abu Tarboush BJ, Husein MM. Adsorption of asphaltenes from heavy oil onto in situ prepared NiO nanoparticles. J Colloid Interface Sci 2012;378(1):64−9.

[640] Abu Tarboush BJ, Husein MM. Dispersed Fe_2O_3 nanoparticles preparation in heavy oil and their uptake of asphaltenes. Fuel Proc Technol 2015;133:120−7.

[641] Nassar NN, Husein MM, Al-Jabari ME. Investigation on asphaltene deposition onto different surfaces of ultradispersed catalysts. In: Proc. 8th world congress of chem. eng. Montreal, QC; August 23−27, 2009. 5 pp.

[642] Mutyala S, Fairbridge C, Paré JRJ, Bélanger JMR, Ng S, Hawkins R. Microwave applications to oil sands and petroleum: a review. Fuel Proc Technol 2010;91(2):127−35.

[643] Dalmazzone C, Noïk C, Glénat P, Dang F. Development of a methodology for the optimization of dehydration of extraheavy-oil emulsions. SPE J 2010;15(3):726−36.

[644] Ng FTT. Upgrading heavy oil/bitumen emulsions via in situ hydrogen generation. ACS Div Fuel Chem 1998;43(3):456−9. Preprints.

[645] Sadeghi KM, Sadeghi M-A, Yen TF. Novel extraction of tar sands by sonication with the aid of in situ surfactants. Energy Fuels 1990;4(5):604−8.

[646] Xu HH, Okazawa NE, Moore RG, Mehta SA, Laureshen CJ, Ursenbach MG, et al. In situ upgrading of heavy oil. J Can Pet Technol 2001;40(8):45−53.

[647] Speight JG, Moschopedis SE. Effect of oxygen functions on the properties of bitumen fractions. J Can Pet Technol 1978;17(3):73−5.

[648] Moschopedis SE, Speight JG. Oxidation of a bitumen in relation to its recovery from tar and sand formations. In: Proc. ACS nat. meeting, div fuel chem. Los Angeles, CA, USA, vol. 19; April 1−5, 1974. p. 192−201.

[649] Moschopedis SE, Speight JG. Oxidation of bitumen in relation to its recovery from tar-sand formations. Fuel 1974;53(1):21−5.

[650] Moschopedis SE, Speight JG. Sulphoxidation of athabasca bitumen. Fuel 1978;57(10):647−9.

[651] Jha KN, Rao PM, Strausz OP. Oxidation of the athabasca oil sand and its fractions. In: Proc. ACS div fuel chem preprints. Miami Beach, FL, USA, vol. 23; September 10−15, 1978. p. 91−7.

[652] Sweany GA. Process for stimulating and upgrading the oil production from a heavy oil reservoir. Patent No. US4284139. Conoco, Inc.; August 18, 1981.

[653] Scott TC, Depaoli DW, Sisson WG. Further development of the electrically driven emulsion-phase contactor. Ind Eng Chem Res 1994;33(5):1237−44.

[654] Borole AP, Kaufman EN. Process considerations in crude oil biodesulfurization. In: Proc. ACS spring meeting. Anaheim, CA, USA, vol. 44; 1999. p. 23–7.
[655] Vazquez-Duhalt R. Petroleum biotechnology: developments and perspectives. Studies surf. sci. catal., vol. 151; 2004. 3 pp.
[656] Ramirez-Corredores MM, Borole AP. Biocatalysis in oils refining. Studies surf. sci. catal., vol. 164. Amsterdam Elsevier; 2007. 406 pp.
[657] Pan Y, Liao Y, Shi Q, Hsu CS. Acidic and neutral polar NSO compounds in heavily biodegraded oils characterized by negative-ion ESI FT-ICR MS. Energy Fuels 2013;27(6):2960–73.
[658] Huang J, Nemati M, Hill G, Headley J. Batch and continuous biodegradation of three model naphthenic acids in a circulating packed-bed bioreactor. J Hazard Mat 2012;201–202:132–40.
[659] Paslawski JC, Headley JV, Hill GA, Nemati M. Biodegradation kinetics of trans-4-methyl-1-cyclohexane carboxylic acid. Biodegradation 2009;20(1):125–33.
[660] Paslawski J, Nemati M, Hill G, Headley J. Biodegradation kinetics of trans-4-methyl-1-cyclohexane carboxylic acid in continuously stirred tank and immobilized cell bioreactors. J Chem Technol Biotechn 2009;84(7):992–1000.
[661] Gunawan Y, Nemati M, Dalai A. Biodegradation of a surrogate naphthenic acid under denitrifying conditions. Water Res 2014;51:11–24.
[662] Johnson RJ, Smith BE, Rowland SJ, Whitby C. Biodegradation of alkyl branched aromatic alkanoic naphthenic acids by pseudomonas putida kt2440. Intern Biodeterior Biodegrad 2013;81:3–8.
[663] Mckenzie N, Yue S, Liu X, Ramsay BA, Ramsay JA. Biodegradation of naphthenic acids in oils sands process waters in an immobilized soil/sediment bioreactor. Chemosphere 2014;109:164–72.
[664] Tahhan RA, Abu-Ateih RY. Biodegradation of petroleum industry oily-sludge using jordanian oil refinery contaminated soil. Intern Biodeterior Biodegrad 2009;63(8):1054–60.
[665] Meredith W, Kelland SJ, Jones DM. Influence of biodegradation on crude oil acidity and carboxylic acid composition. Org Geochem 2000;31(11):1059–73.
[666] Quesnel DM, Bhaskar IM, Gieg LM, Chua G. Naphthenic acid biodegradation by the unicellular alga dunaliella tertiolecta. Chemosphere 2011;84(4):504–11.
[667] Scott AC, Mackinnon MD, Fedorak PM. Naphthenic acids in Athabasca oil sands tailings waters are less biodegradable than commercial naphthenic acids. Environ Sci Technol 2005;39(21):8388–94.
[668] Altunina LK, Svarovskaya LI, Rozhenkova ZA, Golovko AK. Oil degradation by reservoir microorganisms. In: Adv. prod. appl. lube base stocks, proc. int. symp. field full j. title; 1994. p. 354–61.
[669] Toor NS, Han X, Franz E, Mackinnon MD, Martin JW, Liber K. Selective biodegradation of naphthenic acids and a probable link between mixture profiles and aquatic toxicity. Environ Toxicol Chem 2013;32(10):2207–16.
[670] Premusic ETL, Mow S. Biochemically enhanced oil recovery and oil treatment. Patent No. US5297625 1994. ASS Universities Inc./Brookhaven Science; March 29, 1999.
[671] Lazar I, Petrisor IG, Yen TF. Microbial enhanced oil recovery (meor). Pet Sci Technol 2007;25(11):1353–66.
[672] Bryant RS, Bailey SA, Stepp AK, Evans DB, Parli JA, Kolhatkar AR. Biotechnology for heavy oil recovery. In: Proc. 7th UNITAR intern. conf. 'heavy crude tar sands'. Beijing, China, vol. 2; October 27–30, 1998. 7 pp.
[673] Mohebali G, Ball AS. Biocatalytic desulfurization (BDS) of petrodiesel fuels. Microbiology 2008;154(8):2169–83.
[674] Mjalli FS, Ahmed OU, Al-Wahaibi T, Al-Wahaibi Y, Alnashef IM. Deep oxidative desulfurization of liquid fuels. Rev Chem Eng 2014;30(4):337–78.
[675] Boniek D, Figueiredo D, Dos Santos AFB, De Resende Stoianoff MA. Biodesulfurization: a mini review about the immediate search for the future technology. Clean Technol Environ Policy 2014;17(1):29–37.
[676] Srinivas Raju K, Sridevi V, Chandana Lakshmi MVV, Thomas KK. Biodesulfurization of fossil fuels- an overview. Intern J Pharma Bio Sci 2013;4(3):B719–27.

[677] Mužic M, Sertić-Bionda K. Alternative processes for removing organic sulfur compounds from petroleum fractions. Chem Biochem Eng Q 2013;27(1):101−8.

[678] Wei Z, Ning X. Development and application of the bio-desulfurization technology. Adv Mater Res 2012;518−523:178−82.

[679] Sohrabi M, Kamyab H, Janalizadeh N, Huyop FZ. Bacterial desulfurization of organic sulfur compounds exist in fossil fuels. J Pure Appl Microbiol 2012;6(2):717−29.

[680] Chandra Srivastava V. An evaluation of desulfurization technologies for sulfur removal from liquid fuels. RSC Adv 2012;2(3):759−83.

[681] Liu L, Lu H, Qian J, Xing J. Progress in the technology for desulfurization of crude oil. China Pet Proc Petrochem Technol 2010;(4):1−6.

[682] Debabov VG. Microbial desulfurization of motor fuel. Appl Biochem Microbiol 2010;46(8):733−8.

[683] Xu P, Feng J, Yu B, Li F, Ma C. Recent developments in biodesulfurization of fossil fuels. Adv Biochem Eng Biotech 2009;113:255−74.

[684] Tuli D, Kumar A. Non-conventional technologies for fuel desulphurization. In: Proc. 19th world pet. congress. Madrid, Spain, vol. 2; June 29−July 3, 2008. p. 751−9.

[685] Bhasarkar JB, Dikshit PK, Moholkar VS. Ultrasound assisted biodesulfurization of liquid fuel using free and immobilized cells of rhodococcus rhodochrous mtcc 3552: a mechanistic investigation. Bioresour Technol 2015;187:369−78.

[686] Tang Q, Lin S, Cheng Y, Liu SJ, Xiong JR. Enhanced biodesulfurization of bunker oil by ultrasound pre-treatment with native microbial seeds. Biochem Eng J 2013;77:58−65.

[687] Dinamarca MA, Rojas A, Baeza P, Espinoza G, Ibacache-Quiroga C, Ojeda J. Optimizing the biodesulfurization of gas oil by adding surfactants to immobilized cell systems. Fuel 2014;116:237−41.

[688] Xu XH, Qiu XL, Mu TZ, Zhou JM, Bai B, Song ZY, et al. Biodesulfurization of immobilized sulfur oxidation bacteria thioalkalivibrio versutus d301 with Fe_3O_4 magnetic nanoparticles. Guocheng Gongcheng Xuebao/The Chin J Proc Eng 2015;15(2):301−6.

[689] Boshagh F, Mokhtarani B, Mortaheb HR. Effect of electrokinetics on biodesulfurization of the model oil by rhodococcus erythropolis ptcc1767 and bacillus subtilis dsmz 3256. J Hazard Mat 2014;280:781−7.

[690] Borole AP. Biological upgrading of petroleum - recent developments and research needs. In: Proc. AIChE spring nat. meeting. Orlando, FL; April 23−27, 2006. 1 p.

[691] Bachmann RT, Johnson AC, Edyvean RGJ. Biotechnology in the petroleum industry: an overview. Intern Biodeterior Biodegrad 2014;86:225−37.

[692] Strubinger A, Ehrmann U, León V, Desisto A, González M. Changes in Venezuelan Orinoco belt crude after different biotechnological approaches. J Pet Sci Eng 2015;127:421−32.

[693] Steinbakk S. Bioconversion of heavy oil [MSc thesis from Norwegian University of Science and Technology]. Biotechn; May 2011. 141 pp.

[694] Shong RG. Bioprocessing of crude oils. In: Proc. ACS spring meeting. Anaheim, CA, USA, vol. 44; 1999. p. 1−4.

[695] Ginosar DM, Bala GA, Anderson RP, Fox SL, Stanescue MA. Bioprocessing of high-sulfur crudes via appliaction of critical fluid biocatalysis. Report No. INEEL/EXT-02-00711. Idaho, US: Idaho National Laboratory; May 01, 2002. 22 pp.

[696] Engineering Resources, I. Biological upgrading of coal liquids. Quarterly report. Report No. DOE/PC/91059 T4. December 1992. 5 pp.

[697] Yuste L, Corbella ME, Turiegano MJ, Karlson U, Puyet A, Rojo F. Characterization of bacterial strains able to grow on high molecular mass residues from crude oil processing. FEMS Microbiol Ecol 2000;32(1):69−75.

[698] Premuzic ET, Lin MS. Induced biochemical conversions of heavy crude oils. J Pet Sci Eng 1999;22(1−3):171−80.

[699] Premuzic ET, Bohenek MSLM, Zhou WM. Bioconversion reactions in asphaltenes and heavy crude oils. Energy Fuels 1999;13(2):297−304.

[700] Nash LJ, Palmer SR. Biocatalytic upgrading of petroleum. In: Proc. 217th ACS nat. meeting, Anaheim, CA, USA, vol. 44; March 21–25, 1999. p. 18–22.
[701] Leon V, Kumar M. Biological upgrading of heavy crude oil. Biotechn Bioproc Eng 2005;10(6):471–81.
[702] Kilbane II JJ, Le Borgne S. Petroleum biorefining: the selective removal of sulfur, nitrogen, and metals. In: Studies surf. sci. catal.; 2004. p. 29–65.
[703] Hao R, Lu A. Conversion and degradation of crude oil by bacillus sp3. Chin J Geochem 2007;26(2):201–6.
[704] Ferrari MD, Neirotti E, Albornoz C, Mostazo MR, Cozzo M. Biotreatment of hydrocarbons from petroleum tank bottom sludges in soil slurries. Biotechn Lett 1996;18(11):1241–6.
[705] Bosecker K. Microbial deterioration of hydrocarbons. In: Proc. 6th intern. conf. on stability handling of liquid fuels. Vancouver, BC, Canada, vol. 2; October 13–17, 1998. p. 763–73.
[706] Uribe-Alvarez C, Ayala M, Perezgasga L, Naranjo L, Urbina H, Vazquez-Duhalt R. First evidence of mineralization of petroleum asphaltenes by a strain of neosartorya fischeri. Microb Biotechn 2011;4(5):663–72.
[707] Ramirez M. Biorefining of crude oil. In: Schwartz S, Masciangioli T, Boonyaratanakornkit B, editors. Bioinspired chemistry for energy: a workshop. National Academies Press; 2008. 3 pp.
[708] Zhi L, Jiang Y, Wang Y, Hu M, Li S, Ma Y. Effects of additives on the thermostability of chloroperoxidase. Biotechnol Prog 2007;23(3):729–33.
[709] Ayala M, Hernandez-Lopez EL, Perezgasga L, Vazquez-Duhalt R. Reduced coke formation and aromaticity due to chloroperoxidase-catalyzed transformation of asphaltenes from Maya crude oil. Fuel 2012;92(1):245–9.
[710] Juarez-Moreno K, Díaz De León JN, Zepeda TA, Vazquez-Duhalt R, Fuentes S. Oxidative transformation of dibenzothiophene by chloroperoxidase enzyme immobilized on (1d)-γ-Al_2O_3 nanorods. J Mol Catal B Enzym 2015;115:90–5.
[711] Ayala M, Verdin J, Vazquez-Duhalt R. The prospects for peroxidase-based biorefining of petroleum fuels. Biocatal Biotransformation 2007;25(2):114–29.
[712] Borole AP, Cheng CL, Davison BH. Substrate desolvation as a governing factor in enzymatic transformations of pahs in aqueous-acetonitrile mixtures. Biotechnol Prog 2004;20(4):1251–4.
[713] Collinson E, Swallow AJ. The radiation chemistry of organic substances. Chem Rev 1956;56(3):471–568.
[714] Lucchesi PJ, Tarmy BL, Long RB, Baeder DL, Longwell JP. High temperature radiation chemistry of hydrocarbons. Ind Eng Chem 1958;50:879–84.
[715] Ahsenova TI, Daukeev DK, Iskakov BM, Zaykin YA, Mazhrenova NR, Nurkeeva AS. Investigations on radiation processing in Kazakhstan. Radiat Phys Chem 1995;46(4–6):1401–4.
[716] Bludenko AV, Ponomarev AV, Chulkov VN, Yakushev IA, Yarullin RS. Electron-beam decomposition of bitumen–gas mixtures at high dose rates. Mendeleev Commun 2007;17:227–9.
[717] Whitehead WL, Goodman C, Breger IA. The decomposition of fatty acids by alpha radiation. J Chim Phys 1951;48:184–9.
[718] Mentzer Morrison R. An economic analysis of electron accelerators and cobalt-60 for irradiating food. USDA, Economic Res. Service; 1989. Technical Bulletin Number 1762, 47 pp. Available from: http://www.ers.usda.gov/publications/TB1762/tb1762.pdf.
[719] Murphy WJ. Upgrading of low value hydrocarbons using a hydrogen donor and microwave radiation. Patent No. US5328577 (Also published as EP601798, MX9307331, CA2103123). Exxon Research and Engineering Co.; July 12, 1994.
[720] Murphy WJ, Shaw DH. Upgrading of low value hydrocarbons using a hydrogen donor and microwave radiation. Patent No. US5181998 (Also published as EP0435590, JPH03291244, CA2031958). Exxon Research and Engineering Co.; July 03, 1991.
[721] Kubanek G, Munz RJ, Gauvin WH. Heavy oil processing in steam and hydrogen plasmas. Can J Chem Eng 1986;64(5):803–7.

[722] Kong PC. Method for cracking hydrocarbon compositions using a submerged reactive plasma system. Patent No. US5626726. Lockheed Idaho Technologies Company; May 06, 1997.
[723] Long RB, Longwell JP, Hibshman HJ. High temperature hydrocarbon radiolysis. Patent No. GB866752. Exxon Research and Engineering Co.; April 26, 1961.
[724] Reyes Lujan J. Radiolysis of petroleum products. Rev Mex Fis 1968;17(1):1–17.
[725] Mustafaev I, Guileva N. The principles of radiation-chemical technology of refining petroleum residues. Radiat Phys Chem 1994;46(4–6):1313–6.
[726] Lykhterova NM, Lunin VV, Lomonosova MV. Unconventional methods for processing of heavy petroleum feedstocks. Netf I Gaz 1998;6(3–5):2.
[727] Zaykin YA, Zaykina RF. Stimulation of radiation-thermal cracking of oil products by reactive ozone-containing mixtures. Radiat Phys Chem 2004;71(1–2):475–8.
[728] Zaykina RF, Zaykin YA, Mamonova TB, Nadirov NK. Radiation methods for demercaptanization and desulfurization of oil products. Radiat Phys Chem 2002;63(3–6):621–4.
[729] Zaykina RF, Zaykin YA, Mamonova TB, Nadirov NK. Radiation-thermal processing of high-viscous oil from karazhanbas field. Radiat Phys Chem 2001;60(3):211–21.
[730] Zaykina RF, Zaykin YA, Mirkin G, Nadirov NK. Prospects for irradiation processing in the petroleum industry. Radiat Phys Chem 2002;63(3–6):617–20.
[731] Zaykina RF, Zaykin YA, Yagudin SG, Fahruddinov IM. Specific approaches to radiation processing of high-sulfuric oil. Radiat Phys Chem 2004;71(1–2):467–70.
[732] Zaykin YA, Zaykina RF. Bitumen radiation processing. Radiat Phys Chem 2004;71(1–2):471–4.
[733] Zaykin YA, Zaykina RF. Criteria of synergetic effects in radiation-induced conversion of complex hydrocarbon mixtures. Oil Gas Kazakhstan 2004;2:64–73.
[734] Mirkin G, Zaykina RF, Zaykin YA. Radiation methods for upgrading and refining of feedstock for oil chemistry. Radiat Phys Chem 2003;67(3–4):311–4.
[735] Zaykin YA, Zaykina RF, Mirkin G. On energetics of hydrocarbon chemical reactions by ionizing irradiation. Radiat Phys Chem 2003;67(3–4):305–9.
[736] Zaykin YA, Zaykina RF. Self-sustaining cracking of hydrocarbons. Patent No. US8192591 (Also published as AR058345, BRPI0619990, CA2633885, CN101336283, EA200801452, EA016698, EP1973994, MX2008007791, US2010051444, US2012305384, WO2007070698). Petrobeam Inc.; June 05, 2012.
[737] Zaykin YA, Zaykina RF, Silverman J. Radiation-thermal conversion of paraffinic oil. Radiat Phys Chem 2004;69(3):229–38.
[738] Zaykina RF, Zaykin YA. Radiation technologies for production and regeneration of motor fuel and lubricants. Radiat Phys Chem 2002;65(2):169–72.
[739] Zaikin YA, Zaikina RF. Polymerization as a limiting factor for light product yields in radiation cracking of heavy oil and bitumen. Radiat Phys Chem 2013;84:6–9.
[740] Zaikin YA. On the nature of radiation-excited unstable states of hydrocarbon molecules in heavy oil and bitumen. Radiat Phys Chem 2013;84:2–5.
[741] National Energy Technology Laboratory. A literature review on cold cracking of petroleum crude oil. U.S. Department of Energy: Office of Oil and Natural Gas Office of Fossil Energy; 2006 [Energy Policy Act of 2005 Section 1406]. 30 pp.
[742] Gong C, Hart DP. Ultrasound induced cavitation and sonochemical yields. J Acoustical Soc Am 1998;104(5):2675–82.
[743] Abramov OV, Abramov VO, Myasnikov SK, Mullakaev MS. Oil sand processing by ultrasonic technique. Chem Pet Eng 2008;44(1–2):102–7.
[744] Abramov OV, Abramov VO, Myasnikov SK, Mullakaev MS. Extraction of bitumen, crude oil and its products from tar sand and contaminated sandy soil under effect of ultrasound. Ultrason Sonochemistry 2009;16(3):408–16.
[745] Deshpande A, Bassi A, Prakash A. Ultrasound-assisted, base-catalyzed oxidation of 4,6-dimethyldibenzothiophene in a biphasic diesel–acetonitrile system. Energy Fuels 2005;19(1):28–34.

[746] Suslick KS. Ultrasound in synthesis. In: Scheffold R, editor. Modern synthetic methods. Berlin: Germany; 1986. 60 pp.

[747] Zheng X, Zhu L, Yan A, Bai C, Xie Y. Ultrasound-assisted cracking process to prepare MoS_2 nanorods. Ultrason Sonochemistry 2004;11(2):83–8.

[748] Cataldo F. Ultrasound-induced cracking and pyrolysis of some aromatic and naphthenic hydrocarbons. Ultrason Sonochemistry 2000;7(1):35–43.

[749] Tang Q, Lin S, Cheng Y, Liu S, Xiong J-R. Ultrasound-assisted oxidative desulfurization of bunker-C oil using tert-butyl hydroperoxide. Ultrason Sonochemistry 2013;20(5):1168–75.

[750] Suslick KS, Price GJ. Applications of ultrasound to materials chemistry. Annu Rev Mat Sci 1999;29(1):295–326.

[751] Lovett S, Berruti F, Behie LA. Ultrapyrolytic upgrading of plastic wastes and plastics/heavy oil mixtures to valuable light gas products. Ind Eng Chem Res 1997;36(11):4436–44.

[752] Fontana A, Braekman-Danheux C, Labani A, Jung CG. Ultrasonic removal of heavy metals and pyrolytic valorization of waste oils. Chim Nouv 1994;12(47):1383–5.

[753] Meegoda JN, Perera R. Ultrasound to decontaminate heavy metals in dredged sediments. J Hazard Mat 2001;85(1–2):73–89.

[754] Mason TJ. Industrial sonochemistry: potential and practicality. Ultrasonics 1992;30(3):192–6.

[755] Ye G, Lu X, Han P, Peng F, Wang Y, Shen X. Application of ultrasound on crude oil pretreatment. Chem Eng Proc: Proc Intensif 2008;47(12):2346–50.

[756] Shedid SA. An ultrasonic irradiation technique for treatment of asphaltene deposition. J Pet Sci Eng 2004;42(1):57–70.

[757] Yen TF. Correlation between heavy crude sources and types and their refining and upgrading methods. In: Proc. 7th UNITAR intern. conf. heavy crude tar sands. Beijing, China, vol. 2; October 27–30, 1998. 7 pp.

[758] Yen TF. Future sources of heavy crude and their production and upgrading technology. In: Proc. ACS div. fuel chem. preprints. Anaheim, CA, USA, vol. 44; 1999. p. 76–9.

[759] Dunn K, Yen TF. A plausible reaction pathway of asphaltene under ultrasound. Fuel Proc Technol 2001;73(1):59–71.

[760] Lin JR, Yen TF. An upgrading process through cavitation and surfactant. Energy Fuels 1993;7(1):111–8.

[761] Tu SP, Yen TF. The feasibility studies for radical-induced decomposition and demetalation of metalloporphyrins by ultrasonication. Energy Fuels 2000;14(6):1168–75.

[762] Sakanishi K, Yamashita N, Whitehurst DD, Mochida I. Depolymerization and demetallation treatments of asphaltene in vacuum residue. Catal Today 1998;43(3–4):241–7.

[763] Lee AS, Xu XW, Yen TF. Upgrading of heavy crude oils at low temperature and ambient atmosphere. In: Proc. 4th UNITAR intern. conf. 'heavy crude tar sands'. Edmonton, Alberta, Canada, vol. 5; August 7–12, 1989. p. 109–16.

[764] Patmore DJ. Ultrasound treatment of bitumen/water emulsions. Report No. ERL 90–66(CF), CANMET. Energy Research Laboratories. NRCan; 1990.

[765] Branson. Industrial cleaning literature. 2005. Available from: http://mastersonics.com/documents/mmm_basics/general_info/ultrasonics_faq/branson-components-products.pdf.

[766] Scott CE, Altafulla JR, Bolivar C, De Navarro CU. Sonocatalytic desulfurization. In: Studies surf. sci. catal.; 2002. p. 343–6.

[767] Gopinath R, Dalai AK, Adjaye J. Effects of ultrasound treatment on the upgradation of heavy gas oil. Energy Fuels 2006;20(1):271–7.

[768] Sadeghi M-A, Sadeghi K, Kuo J-F, Jang L-K, Yen TF. Treatment of carbonaceous materials. Patent No. US4765885. Eneresource Inc.; August 23, 1988.

[769] Sadeghi M-A, Sadeghi K, Kuo J-F, Jang L-K, Yen TF. Treatment of carbonaceous materials. Patent No. US5017281. Tar Sands Energy Ltd.; May 21, 1991.

[770] Zannikos F, Lois E, Stournas S. Desulfurization of petroleum fractions by oxidation and solvent extraction. Fuel Proc Technol 1995;42(1):35–45.
[771] Collins FM, Lucy AR, Sharp C. Oxidative desulphurisation of oils via hydrogen peroxide and heteropolyanion catalysis. J Mol Catal A Chem 1997;117(1–3):397–403.
[772] Yen TF, Mei H, Lu SH-M. Oxidative desulfurization of fossil fuels with ultrasound. Patent No. US6402939 (Also published as WO0226916, RU2235754, NO20031403, NO331096, MXPA03002702, EP1335962, CA2422866, AU9304301, AR030820). SulphCo, Inc.; June 11, 2002.
[773] Gunnerman RW. Continuous process for oxidative desulfurization of fossil fuels with ultrasound and products thereof. Patent No. US6500219 (Also published as US2003014911, WO02074884, RU2233862, NO20034161, MXPA03008432, EP1409614, CA2440668). SulphCo, Inc.; December 31, 2002.
[774] Gunnerman RW, Moote PS, Cullen MT. Ultrasound-based petroleum oxidation-desulfurization-denitrogenation of petroleum fractions. Patent No. US2003051988 (Abandoned). SulphCo, Inc.; March 20, 2003.
[775] Gunnerman RW. Continuous process for oxidative desulfurization of fossil fuels with ultrasound and products thereof. Patent No. US6827844 (Also published as US2004079680). SulphCo, Inc.; December 07, 2004.
[776] Denney D. Technology applications: improved heavy-oil reactor. J Pet Technol 2004;56(11):26–7.
[777] Frost, Sullivan. Oil refining technologies—new developments and growth opportunities. Report No. D274, San Antonio, TX. May 2004. 240 pp.
[778] Gunnerman R. Upgrading of petroleum by combined ultrasound and microwave treatments. Patent No. US2006180500 (Abandoned) (Also published as CA2601578, CN101155900A, EP1848788, WO2006088567). SulphCo, Inc.; August 17, 2006.
[779] Gunnerman R, Richman C. Loop-shaped ultrasound generator and use in reaction systems. Patent No. US7275440 (Also published as CA2584312, CN101057137, EP1812790, US2006101919, US2006260405, WO2006055368). SulphCo, Inc.; October 02, 2006.
[780] Gunnerman RW. Corrosion resistant ultrasonic horn. Patent No. US6652992 (Also published as CA2488134, DE60316472, EP1573909, WO2004062101). SulphCo, Inc.; November 25, 2003.
[781] May J, Richman CI, Gunnerman RW. Power driving circuit for controlling a variable load ultrasonic transducer. Patent No. US7408290 (Also published as CN101548402, EP1854281, US20060238068, WO2006093602). SulphCo, Inc.; August 05 2008.
[782] Gunnerman RW. Conversion of petroleum resid to usable oils with ultrasound. Patent No. US7300566 (Also published as CA2559490, CN101124303, EP1733007, US20050205463). SulphCo, Inc.; November 27, 2007.
[783] Gunnerman RW. High-throughput continuous-flow ultrasound reactor. Patent No. US7559241 (Also published as CA2567649, CN100466997, CN101018509, EP1748882, US20050274600, US20080067055 (Abandoned), WO2005118277). SulphCo, Inc.; July 14, 2009.
[784] Frost, Sullivan. Frost & sullivan lauds sulphco's superiority in crude oil refining technologies. 2006. 2 pp. Available from: http://www.ti.frost.com/prod/servlet/press-release.pag?docid=70923230.
[785] Alpert B. A crank case?. In: Barron's; January 23, 2006. US Edition. 3 pp. Avauilable from: http://online.barrons.com/articles/SB113780731535052643.
[786] Genck W. Ultrasonics makes sound progress. Chem Proc 2006;69(3):35–42.
[787] SulphCo. Oxidative desulfurization. In: Proc. IAEE. Houston, TX. USA; June 11, 2009. 18 pp.
[788] Pluritas. SulphCo: section 363 sale. 2013. Available from: http://www.pluritas.com/_public/_public_sulphco/SulphCo_Asset_Overview.pdf.
[789] Heritage Global Partners and Pluritas. Sulphco's sonocracking technology intellectual pproperty auction. 2013.
[790] Iut. Sonocracking technology to desulfurize hydrocarbon. 2014. Available from: http://www.internationalultrasonics.com/sonocracking-desulfurization.html.
[791] Cullen M. Treatment of crude oil fractions, fossil fuels, and products thereof with sonic energy. Patent No. US2005167336 (Abandoned). Petrosonics, LLC; April 04, 2005.

[792] Cullen M. Treatment of crude oil fractions, fossil fuels, and products thereof with sonic energy. Patent No. US2005182285 (Abandoned). Petrosonics, LLC; August 18, 2005.
[793] Cullen M. Treatment of crude oil fractions, fossil fuels, and products thereof with sonic energy. Patent No. US2006157339 (Abandoned). Petrosonics, LLC; July 20, 2006.
[794] Cullen M. Treatment of crude oil fractions, fossil fuels, and products thereof. Patent No. US7871512. Petrosonics, LLC; January 18, 2011.
[795] Cullen M. Treatment of crude oil fractions, fossil fuels, and products thereof. Patent No. US8409426 (Also published as CA2534450, CN1839194, CN101565633, DE602004020082, EP1668095, EP2025736, US7871512, US20040074812, US20110108465, WO2005019383). Petrosonics, LLC; April 02, 2013.
[796] Cullen M. Treatment of crude oil fractions, fossil fuels, and products thereof with sonic energy. Patent No. US7081196 (Also published as US2004035753, WO2004101100, US2006157339, US2005167336, US2005182285, SI1844831, SI1635924, PT1844831, PT1635924, HK1088269, ES2428153, ES2298783, EP1844831, EP1635924, DK1844831, DK1635924, DE602004010906, CY1107363, CA2528802, AU2011200452, AU2004238288, AT381962). Petrosonic, LLC; July 25, 2006.
[797] Cullen M. Sulfone removal process. Patent No. US2004200759 (Abandoned). October 14, 2004.
[798] Cullen M. Process for generating and removing sulfoxides from fossil fuel. Patent No. US2004222131 (Abandoned). November 11, 2004.
[799] Hill J, Arato C. Method for treating heavy crude oil. Patent No. WO2009111871 (Also published as EP2260089, US2013277275, EA201071060). Sonic Technology Solutions; Sonoro Energy, Ltd.; Petrosonic Ltd.; Petrosonic Energy Inc.; September 17, 2009.
[800] Sumel AR. Sonic completes successful testing and heavy oil upgrading acquistion. Sonic Technology Solutions; 2008. 2 pp. Available from: www.SONICTSI.com.
[801] Arato C, Janke T. Sonic reactor. Patent No. WO2014134724. Sonoro Energy Ltd.; September 12, 2014.
[802] Arato C, Janke T. Method for treatment of oil and sand cuttings. Patent No. WO2014134726. Sonoro Energy Ltd.; September 12, 2014.
[803] Petrosonic Energy. Petrosonic energy, Inc. Invents and files patent on a new breakthrough solution, sono Hi-P. 2014. Calgary, Canada. 1 p. Available from: http://www.petrosonic.net/s/NewsReleases.asp?ReportID=669385&_Type=News-Releases&_Title=Petrosonic-Energy-Inc.-Invents-and-Files-Patent-on-a-New-Breakthrough-Solution.
[804] Varadaraj R. Defoaming of foams utilizing sonication. Patent No. US6590000 (Also published as CA2438410, DE60220391, EP1377352, US2002128328, WO2002072226). Exxonmobil Research And Engineering Co.; July 08, 2003.
[805] Varadaraj R. Viscosity reduction of oils by sonic treatment. Patent No. US6544411 (Also published as CA2438319, EP1379610, US20020125174, US20030132139 (abandoned), WO2002072734). Exxonmobile Research And Engineering Co.; April 08, 2003.
[806] Austin D. Method to upgrade hydrocarbon mixtures. Patent No. US2003019791 (abandoned) (Also published as WO2002102746). Petronetics, LLC; January 30, 2003.
[807] Promtov M. Cavitation technology for quality improvement of hydrocarbon fuels. Chem Pet Eng 2008;44(1):63–6.
[808] Chakma A, Berruti F. Ultrasonic visbreaking of Athabasca bitumen. Heavy crude and tar sands. In: Proc. 5th UNITAR intern. conf. 'heavy crude tar sands'. Caracas, Venezuela, vol. 4; August 4–9, 1991. p. 101–4.
[809] Chakma A, Berruti F. The effects of ultrasonic treatment on the viscosity of Athabasca bitumen and bitumen-solvent mixtures. J Can Pet Technol 1983;32(5):48–51.
[810] Andreev VG, Bryukhovetskii OS, Ponomareva OI, Rodionov VN. Use of the ultrasonic cavitation effect ot reduce oil viscosity and increase its rate of transport. Pet Abstr 1996;37(40). 109 (1 p.).
[811] Basin MB, Vainora B, Gimbutas A, Tugusi SO, Barilciuk MV, Bednov BV, et al. Method for processing of residual oil. Patent No. LT1967. Akcine Bendrove "Mazeikiu Nafta"; January 25, 1996.
[812] Basin MB, Vainora B, Gimbutas A, Tugusi SO, Barilciuk MV, Bednov BV, et al. Cavitation method for processing petroleum refining residues. Patent No. LT3883. Akcine Bendrove "Mazeikiu Nafta"; April 25, 1996.

[813] Yu G, Zhong Q, Zhou X, Zhang N, Li H, Hongguan W. Ultrasonic oxidation-extraction and deep desulfurization method for diesel oil. Patent No. CN103450931. December 18, 2013.
[814] Boldyrev VV. Mechanochemistry and mechanical activation of solids. Bull Acad Sci USSR, Div Chem Sci 1990;39(10):2029–44.
[815] Boldyrev VV. Mechanochemical synthesis in inorganic chemistry. In: G AY, editor. Nauka sb of ussr acad sci; 1991. p. 5–32. Novosibirsk, Russia.
[816] Boldyrev VV. Mechanochemistry and mechanical activation of solids. Russ Chem Rev 2006;75(3):177–89.
[817] Boldyrev VV. Mechanochemistry and mechanical activation. Mat Sci Forum 1996;225–227(PART 1):511–20.
[818] Buyanov RA, Molchanov VV. Phenomenological model of mechanochemical activation in catalyst technology and catalysis. Khimiya Interesakh Ustoichivogo Razvitiya (Chem Sustain Dev) 2001;9(3):369–77.
[819] Steinike U, Tkacova K. Mechanochemistry of solids - real structure and reactivity. J Mat Synth Proc 2000;8(3/4):197–203.
[820] Zhang Q, Saito F. Fundamental research works on mechanochemistry of inorganic and organic materials and its engineering applications. Recent Res Dev Chem Eng 1999;3:69–78.
[821] Kosova NV, Khabilbullin AK, Boldyrev VV. Hydrothermal reactions under mechanochemical treating. Solid State Ionics 1997;101–103(Pt. 1):53–8.
[822] Delogu F, Deidda C, Mulas G, Schiffini L, Cocco G. A quantitative approach to mechanochemical processes. J Mat Sci 2004;39(16–17):5121–4.
[823] Baláž P, Achimovičová M, Baláž M, Billik P, Cherkezova-Zheleva Z, Criado JM, et al. Hallmarks of mechanochemistry: from nanoparticles to technology. Chem Soc Rev 2013;42(18):7571–637.
[824] Abduazimov KA, Kristallovich EL, Aripov KN. An investigation of the change in nitrolignin during mechanical treatment. Chem Nat Compd (Translation of Khimiya Prirodnykh Soedinenii) 1999;34(6). 743 (1 p.).
[825] Lee DY, Matsuoka M, Sumimoto M. Mechanochemistry of lignin. IV. Mechanochemical reactions of phenylcoumaran models. Holzforschung 1990;44(6):415–8.
[826] Lee DY, Sumimoto M. Mechanochemistry of lignin. III. Mechanochemical reactions of b-o-4 lignin model compounds. Holzforschung 1990;44(5):347–50.
[827] Mikushina IV, Troitskaya IB, Dushkin AV, Bazarnova NG. Change of the chemical composition of wood during mechanochemical treatment. Khimiya Interesakh Ustoichivogo Razvitiya (Chem Sustain Dev) 2002;10(4):443–7.
[828] Vol-Epshtein AB, Platonov VV, Khrenkova TM, Valdez Martinez JC, Chizhevskii AA, Shpil'berg MB. Manufacture of liquid distillate products by mechanical-chemical treatment of natural bitumen and oil shale. Khimiya Tverd Topl (Solid Fuel Chem) 1987;(5):37–40.
[829] Golovko AK, Kamyanov VF, Shabotkin IG. Initiated low-temperature cracking of ozonized petroleum and heavy petroleum ends. Eurasian Chemico-Technological J 2004;6(2):99–105.
[830] Golovko AK, Kamyanov VF, Filimonova TA. Soil amendments by ozonolysis of petroleum resins and asphaltenes. Eurasian Chemico-Technological J 2005;7(2):89–97.
[831] Zamanova LV, Khrenkova TM, Kirda VS, Nikolaeva LV, Fedoseev SD. Structural features of petroleum-based raw materials for production of carbon fibers. Khimiya Tverd Topl (Solid Fuel Chem) 1991;(2):128–32.
[832] Strakhova NA, Rosental DA, Kortovenko LP. Mechanochemical treatment of residual components of the astrakhan gas condensate. Neft Pet Chem 2003;43(1):55–9.
[833] Platonov VV, Valdez JC, Klyavina OA, Vol-Epshtein AB, Shpil'berg MB, Chizhevskii AA, et al. Mechanical degradation of petroleum asphaltenes from the mantansas field, Cuba. Khimiya Tverd Topl (Solid Fuel Chem) 1989;(4):102–8.
[834] Pikula PA, Golovko AK, Kamyanov VF. The composition and properties of straight-run and mechano-activated 360-400 deg C distillate from yakutsk petroleum. Eurasian Chemico-Technological J 2001;3(3):205–10.

[835] Dneprovsky KS, Golovko AK, Lomovsky OI, Vosmerikov AV. The mechanochemical treatment effect on the composition of gasoline fractions. Pet Coal 1999;41(3–4):166–8.

[836] Gamolin OE, Golovko AK, Lomovskiy OI, Kamyanov VF, Mamylov SG. Mechanochemical conversion of gaseous oil hydrocarbons in the systems gas-solid. Khimiya Interesakh Ustoichivogo Razvitiya (Chem Sustain Dev) 2005;13(2):165–72.

[837] Gamolin OE, Golovko AK, Kam'yanov VS, Lomovskii OI. Processing of hydrocarbon gases and liquid hydrocarbons. Patent No. RU2256690. Institut Khimii Nefti SO RAN IKhN SO RAN; July 20, 2005.

[838] Gamolin OY, Golovko AK, Lomovsky OI, Kam'yanov VF. Mechanically activated chemical conversion of gaseous hydrocarbons. Eurasian Chemico-Technological J 2003;5(4):305–10.

[839] Alferov MJ, Koss AV, Penzin RA. Ejector and method of its operation. Patent No. RU2209350. July 27, 2003.

[840] Koss AV, Penzin RA. Method of preparation of oil and/or petroleum products for refining and plant for realization of this method. Patent No. RU2287355 (Also published as WO2007075111, US2009000940 (Abandoned)). November 20, 2006.

[841] Red'kina NI, Khodakov GS. Mechanochemistry and processing characteristics of aqueous emulsions of high-viscosity petroleum products. Theor Found Chem Eng 2002;36(4):395–9.

[842] Bayasgalan K, Bembel VM, Golovko AK, Bat-Ochir S. Reactions of n-alkanes in the presence of activation-catalysts. ACS Div Pet Chem 2007;52(1):8. Preprints.

[843] Rutkowski M, Gawel I, Kalisiewicz B. Method of modifying bituminous substances. Patent No. PL138643. Politechnika Wroclawska, Pol.; January 28, 1989.

[844] Khrenkova TM, Antonova VM, Nikanorova LP, Chizhevskii AA. Composition and structure of humic acids obtained by ultrasonic treatment of lignite and their physiological activity. Khimiya Tverd Topl (Solid Fuel Chem) 1988;(5):17–22.

[845] Conaway LM, Keller MR, Noble RK. Method and apparatus for separating bitumen from particulate substrates. Patent No. US2004129646 (Abandoned). Continuum Environmental LLC, USA; July 08, 2004.

[846] Conaway LM, Keller MR, Noble RK, Caffey D. Flow-through mixing apparatus. Patent No. US2006104157 (Abandoned). Continuum Environmental LLC, USA; May 18, 2006.

[847] Khrenkova TM, Kirda VS. Mechanical activation of coals. Khimiya Tverd Topl (Solid Fuel Chem) 1994;(6):36–42.

[848] Khrenkova TM, Kirda VS, Chizhevskii AA. Effect of dispersion on the solid residue of a hydrogenation slurry. Khimiya Tverd Topl (Solid Fuel Chem) 1985;(6):77–8.

[849] Khrenkova TM, Kirda VS, Dubrova SN. Study of the products of conversion of gas coal during its stepwise comminution. Kolloidn Zhurnal 1983;45(3):605–7.

[850] Khrenkova TM, Kirda VS, Koshevnik AY, Andreeva AI. Study of products of the mechanochemical transformations of coking coal. Khimiya Tverd Topl (Solid Fuel Chem) 1984;(2):30–6.

[851] Lebedev VV, Khrenkova TM, Goldenko NL, Golovina GS. Effect of mechanical grinding on coal properties. Tr Instituta Goryuchikh Iskop Mosc 1973;29(1):84–9.

[852] Lebedev VV, Nikanorova LP, Khrenkova TM, Golovina GS, Cheredkova KI, Goldenko NL. Mechanochemical modification of coals. Khimiya Tverd Topl (Solid Fuel Chem) 1985;(2):35–40.

[853] Khrenkova TM, Rubinchik VB, Kirda VS. Change of hydrogenation behavior of brown coals during grinding in the presence of a catalyst. Khimiya Tverd Topl (Solid Fuel Chem) 1991;(4):50–4.

[854] Khrenkova TM, Lebedev VV, Goldenko NL, Golovina GS. Nature of a change in the properties of coals during crushing. Khimiya Tverd Topl (Solid Fuel Chem) 1975;(1):11–7.

[855] Chizhevskii AA, Khrenkova TM, Zotova OV, Gorlov EG. Composition and structure of components of coal hydrogenation sludge and their change in mechanical treatment. Khimiya Tverd Topl (Solid Fuel Chem) 1987;(3):105–9.

[856] Kirda VS, Khrenkova TM. Effect of mechanical treatment on the structural characteristics of brown coals. Khimiya Tverd Topl (Solid Fuel Chem) 1990;(2):15–8.

[857] Kuznetsov PN, Kuznetsova LI, Borisevich AN, Pavlenko NI. Effect of mechanochemical treatment on the supermolecular structure of brown coal. Khimiya Interesakh Ustoichivogo Razvitiya (Chem Sustain Dev) 2003;11(5):743–9.
[858] Khrenkova TM, Kirda VS, Antonova VM. Destructive transformations in mechanical action on residual coal. Khimiya Tverd Topl (Solid Fuel Chem) 1991;(5):17–20.
[859] Khrenkova TM, Goldenko NL, Kirda VS. Structure of chloroform-soluble products of mechanochemical transformations of coal. Khimiya Tverd Topl (Solid Fuel Chem) 1984;(3):27–30.
[860] Khrenkova TM, Kirda VS. Changes in the chemical composition of coal under mechanical action in solvents. Khimiya Tverd Topl (Solid Fuel Chem) 1988;(3):13–6.
[861] Kirda VS, Chizhevskii AA, Khrenkova TM. Physicochemical changes of substances in heavy distillates from mechanically treated hydrogenation slurries. Khimiya Tverd Topl (Solid Fuel Chem) 1989;(5):63–7.
[862] Chizhevskii AA, Khrenkova TM. Mechanochemical activation of a coal hydrogenation sludge in different gas media. Khimiya Tverd Topl (Solid Fuel Chem) 1988;(4):92–7.
[863] Kirda VS, Chizhevskii AA, Khrenkova TM. Chemical structure of asphaltenes from a mechanically activated brown-coal hydrogenation sludge. Khimiya Tverd Topl (Solid Fuel Chem) 1990;(5):25–31.
[864] Ekaterinina LN, Khrenkova TM, Motovilova LV. Chemical changes in coals on grinding. Khimiya Tverd Topl (Solid Fuel Chem) 1977;11(4):36–7.
[865] Khrenkova TM, Goldenko NL. Study of products of the mechanical breakdown of high-volatile coal used in hydrogenation. Khimiya Tverd Topl (Solid Fuel Chem) 1978;(5):44–5.
[866] Sharypov VI, Kuznetsov BN, Beregovtsova NG, Startsev AN, Parmon VN. Catalytic hydroliquefaction of barzass liptobiolitic coal in a petroleum residue as a solvent. Fuel 2006;85(7):918–22.
[867] Morozova OS, Leonov AV, Khomenko TI, Korchak VN. Mechanism of h2 activation on metals under mechanochemical treatment. Mat Sci Forum 2001;360–362:415–20 [Metastable, Mechanically Alloyed and Nanocrystalline Materials].
[868] Erofeev VI, Gornostaev VV, Koval LM, Tikhonova NV. Catalyst for converting aliphatic C_2–C_{12}-hydrocarbons, method for preparation thereof, and a method for converting aliphatic C_2–C_{12}-hydrocarbons into high antiknock gasoline and/or aromatic hydrocarbons. Patent No. RU2235590, RU2236289. Russia: Obshchestvo s Ogranichennoi Otvetstvennost'yu "Tomskneftekhim"; September 10 and 20, 2004.
[869] Erofeev VI, Gornostaev VV, Ermizin KV, Kuznetsov NN, Kritonov VD, Maskaev GP, et al. Hydrocarbon feedstock pyrolysis catalyst, method for preparation thereof, and a method of hydrocarbon feedstock pyrolysis into lower olefins. Patent No. RU2247599, RU2223144, RU2238142 (Also published as RU2003132496). Russia: Obshchestvo s Ogranichennoi Otvetstvennost'yu "Tomskneftekhim"; March 10, 2005.
[870] Kouzu M, Hiramatsu T, Uchida K, Ikazaki F, Kuriki Y. Characterization of HDS activity of mechanically comminuted MoS_2 fine particle. Sekitan Kagaku Kaigi Happyo Ronbunshu 2001;38:411–4.
[871] Uchida K, Kuriki Y, Kouzu M, Orita H, Toda K, Itoh N, et al. Preparation of MoS_2 catalyst for ultra-deep desulfurization of diesel oil by a media agitation mill. Effect of media filling ratio. Funtai Kogaku Kaishi 2002;39(9):679–84.
[872] Kuriki Y, Uchida K, Shimada K, Ohshima S, Yumura M, Ikazaki F. Preparations of binary metal sulfides for hydrotreating catalyst by using the mechanical comminution. Trans Mat Res Soc Jpn 2000;25(1):143–6.
[873] Vosmerikov AV, Velichkina LM, Vosmerikova LN, Korobitsyna LL, Ivanov GV. Use of mechanochemical technology in zeolite catalysis. Khimiya Interesakh Ustoichivogo Razvitiya (Chem Sustain Dev) 2002;10(1–2):45–51.
[874] Frost RL, Mako E, Kristof J, Horvath E, Kloprogge JT. Mechanochemical treatment of kaolinite. J Colloid Interface Sci 2001;239(2):458–66.

[875] Frost RL, Mako E, Kristof J, Horvath E, Kloprogge JT. Modification of kaolinite surfaces by mechanochemical treatment. Langmuir 2001;17(16):4731–8.

[876] Kameda J, Saruwatari K, Tanaka H. H_2 generation during dry grinding of kaolinite. J Colloid Interface Sci 2004;275(1):225–8.

[877] Karagedov GR, Lyakhov NZ. Mechanochemical grinding of inorganic oxides. KONA Powder Part J 2003;21:76–87.

[878] Gun'ko VM, Voronin EF, Mironyuk IF, Leboda R, Skubiszewska-Zieba J, Pakhlov EM, et al. The effect of heat, adsorption and mechanochemical treatments on stuck structure and adsorption properties of fumed silicas. Colloids Surf A Physicochem Eng Asp 2003;218(1–3):125–35.

[879] Kapadia PR, Kallos MS, Gates ID. A review of pyrolysis, aquathermolysis, and oxidation of Athabasca bitumen. Fuel Proc Technol 2015;131:270–89.

[880] Pereira-Almao P. In situ upgrading of bitumen and heavy oils via nanocatalysis. Can J Chem Eng 2012;90(2):320–9.

[881] Gogate PR, Shirgaonkar IZ, Sivakumar M, Senthilkumar P, Vichare NP, Pandit AB. Cavitation reactors: efficiency assessment using a model reaction. AIChE J 2001;47(11):2526–38.

[882] Total E&P. Extra heavy oil and bitumen: extra reserves for the future. Paris, France: Total Group; 2007. 28 pp. Available from: http://www.total.com/sites/default/files/atoms/file/total-extra-heavy-oils-and-bitumen-reserves-for-the-future.

[883] Abs M, Capdevila JM, Delvigne T, Genin F, Jongen Y, Nguyen A. Rhodotron accelerators for industrial electron-beam processing: a progress report. In: Proc. 5th European particle accelerator conf. Sitges, Barcelona, Spain; June 10–14, 1996. p. 2687–9.

[884] Santos HM, Lodeiro C, Capelo-Martínez J-L. The power of ultrasound. In: Ultrasound in chemistry. Wiley-VCH Verlag GmbH & Co. KGaA; 2009. p. 1–16.

CHAPTER 7

Shale Oils

Lucia M. Petkovic
Idaho National Laboratory, Idaho Falls, ID, United States

1. Introduction

Rocks from which fossil hydrocarbons have been or can be recovered are called source rocks. These are sedimentary rocks rich in organic matter formed from the decomposition of buried biomass that underwent degradation through geological times. As time passed, the decaying biomass was buried more and more deeply in the Earth's crust and submitted to higher temperatures and pressures. Organic matter derived primarily from planktonic organisms buried and submitted to moderate temperatures produced and expelled a liquid known as petroleum or crude oil. At higher temperatures it produced and expelled gases known as natural gas [1]. Organic matter originated from higher plants, which contained significant amounts of lignin, generated condensed structures and became what is currently known as coal. Oil shales can be regarded as a type of source rock that is rich in organics but that has not expelled its organic matter. By applying heat treatments to oil shales, an organic liquid, namely, shale oil, may be produced [2].

The organic matter found in oil shales is known as kerogen. However, different definitions of the word kerogen have appeared over the years [3]. One popular definition is that kerogen is organic matter found in source rocks that is insoluble in relatively nonpolar solvents while the word bitumen is reserved for the organic matter that is soluble. Oil shales contain organic matter mainly as kerogen and slightly as bitumen [4].

Kerogen is a complex material that includes large hydrocarbon molecules containing nitrogen, oxygen, and sulfur. Kerogen average molecular weight may be around several thousands. For example, Torrente and Galan [5] estimated that an oil shale from Spain contained kerogen having an average molecular weight of 3000 and proposed a formula like $C_{200}H_{300}SN_5O_{11}$. Comprehensive reviews on origin, evolution, and structure of kerogen have been published. According to Reinsalu and Aarna [6] and Urov and Sumberg [7], the terms oil shale, shale oil, and shale gas are sometimes used inconsistently and may cause confusion. For example, some deposits classified as oil shales may not be shales at all. In this chapter, oil shale will refer to any rock (shale or not) from which an oil is obtained by transforming its

The Science and Technology of Unconventional Oils. http://dx.doi.org/10.1016/B978-0-12-801225-3.00007-3

kerogen through thermal treatment. This treatment is commonly known as retorting and produces shale oil. Retorting may also produce a gas, known as shale gas, that contains methane, hydrogen, carbon monoxide, carbon dioxide, nitrogen, different hydrocarbons like ethylene, and also hydrogen sulfide and other impurities.

The term oil shale has also been applied to source rocks that contain raw petroleum (not just kerogen) in their porous structure. For example, the US. Energy Information Administration uses the term shale oil to refer to raw petroleum trapped in porous structures and the term kerogen to refer to the oil obtained from retorting [8].

Raw petroleum trapped in porous structures is typically extracted applying well stimulation technologies such as hydraulic fracturing or fracking. Similarly, gas can be extracted applying these technologies. The terms tight oil and tight gas, instead of shale oil and shale gas, would be a better terminology to refer to oil and gas obtained in any of those cases.

This chapter deals with the production and upgrading of shale oils recovered from retorting technologies. Technologies for upgrading tight oil and tight gas will not be discussed in this chapter.

2. Oil Shale Resources

There are massive amounts of oil shale around the world in more than 600 deposits located in at least 40 countries. In general, those deposits are mostly unexploited. However, under favorable economic conditions and improved extraction technologies they might become the main resource for future energy needs.

Qualities and quantities of shale oil that could be produced from those deposits is difficult to estimate because of a number of factors including determining how different countries have estimated either their resources or the potential to actually mine them. Anyway, a rough estimate cites total world resources of shale oil as of year 2011 in the order of 4.8 trillion of barrels, with the United States having the largest known resource estimated at around 3.7 trillion barrels [9]. China, Russia, Congo, and Brazil also possess significant oil shale resources [9,10].

Assessment of oil shale resources is often based on the Fischer assay method (ASTM 3904). This method measures the amount of oil, water, and spent shale obtained upon prescribed laboratory-scale heating conditions. An oil shale that would produce 25+ gallons of oil per ton (GPT) is considered of high grade and 15+ GPT is considered potentially viable. Considering just the 25+ GPT of the Green River Formation in the United States, this resource would represent a 50-year supply of oil for the United States. If the 15+ GPT is also included, a 165-year supply of oil is estimated [11].

3. Production of Shale Oil

Oil shales may be burned directly to produce heat. This has been practiced by humankind since ancient times because oil shales burn directly without the need of any treatment. Nowadays, Estonia produces more than 90% of its electricity by burning oil shales [12].

A rich oil shale that produces about 25+ GPT contains enough energy to cover the energy required for its processing and to offer excess energy for useful products. Although the general thermal efficiency depends mainly on the amount of oil that can be harvested from a given shale, it can also depend on the particular technology applied to extract the oil [13].

Production of shale oil has already been applied commercially in China, Estonia, and Brazil. One of the world largest commercial shale oil producers is the Fushun Mining Group in China. As of 2007 it reported producing 300,000 tons per year of shale oil from 6.6 million tons of oil shale [14]. Commercial producers are also VKG Oil in Estonia and Petrobras in Brasil.

As of 2012, commercial technologies producing shale oil are the Kiviter lump shale retorting and the Galoter particulate oil shale retorting in Estonia, the Petrosix lump shale retorting in Brazil, the Fushun retorting system in China, the Alberta Oil Sands Technology and Research Authority, and Taciuk Processing in Australia [15]. In the United States important technology developments have been accomplished but long-time continuous commercial production has not materialized yet [15].

Any technology to produce shale oil includes usually three general steps: mining, retorting to extract the oil, and oil upgrading. However, additional steps are also involved to mitigate the important environmental impact of land and water usage and disposal of spent shale. These issues and concerns have been reported elsewhere [16–18].

Mining oil shale can be of two general types: surface mining and in situ retorting. In surface mining, the oil shale is mined by traditional methods such as either open pit or underground mining [15]. Once mined, the shale is crushed and loaded into a reactor known as a retort, where the temperature is increased to about 400–500°C to decompose the kerogen and release a mixture of liquid and gaseous products known as shale oil and shale gas [15]. Other oil extraction methods such as ultrasonic [19], solvent extraction [20], and supercritical carbon dioxide are also found in the literature [21]. Higher temperature processing by heating up oil shales to 1000–1200°C for production of town gas was used in Estonia in the 1950–60s [22].

The residue material after retorting known as spent shale or char is formed by most of the inorganic material along with trapped carbonaceous heavy compounds. In some cases, this char may be used to produce additional heat for the retorting process. In the 1980s the US

Bureau of Mines had a research program to study the caustic extraction of alumina and soda from retorted dawsonite-bearing oil shale as a domestic source of alumina for production of aluminum metal [23].

Numerous retorting technologies have been patented and several have been developed at different scales [24]. They differ in retort design and in the manner by which heat is applied to the oil shale particle. In aboveground retorting, heat may be applied by direct combustion of a portion of oil shale or another fuel within the same retort. Heat may also be applied indirectly by contacting hot gases or solids with the oil shale [25]. Examples of aboveground retorting technologies are the Nevada–Texas–Utah retort process invented in 1923, the Kiviter technology used in Estonia since 1921, the Fushun technology, and the Chevron's Staged Turbulent Bed Retorting process [26,27].

Neither mining nor handling of raw shale or disposing of spent shale is required for in situ retorting. Retorting in this case is achieved by slowly heating the underground shale formation directly and retrieving the oil in a manner similar to pumping petroleum [28]. Heating times are in the order of years and can be accomplished through different technologies. Considering the heating mechanism, three main technology types are available: conduction heating, convection heating, and radiation heating technologies. Examples of conduction heating technologies include the Shell Oil In situ Conversion Process, the Exxon Mobil Electrofrac, and the Independent Energy Partners Geothermic Fuels Cells technology. Examples of convection heating technologies are the Chevron's Technology for the Recovery and Upgrading of Oil from Shale and the General Synfuels International's Omnishale process. Examples of radiation heating technologies include the Raytheon Radio Frequency/Critical Fluids and the Lawrence Livermore National Laboratory's Radio Frequency technology. Detailed analysis of the retorting technologies can be found in several published review articles; see, for instance, Refs. [14,19,29].

Interest in oil shales has cycled over the years because of the volatility of the oil market. The interest of companies and governments in oil shales' exploitation becomes extremely serious within certain circumstances, such as evidences that a \$50+ oil barrel price will persist long enough, under threats on energy security or from the perception that a low inventory market situation would arise (limited or deficient supply, declining production or peak oil, etc.) [30].

4. Properties

Shale oil is usually a viscous material of high molecular weight [31]. Decades ago, Rovere and coworkers [32] and Guo and Ruan [33] analyzed shale oil samples and identified several hundreds of compounds including hydrocarbons (i.e., alkanes, alkenes, and aromatics), phenols, ketones, nitriles, pyridines, dihydroindoles, quinolines, and tetrahydroquinolines. Further analyses of shale oil samples have identified about 30,000 compounds [34].

Using 2D gas chromatography Dijkmans and coworkers [35] detected 20 different classes of compounds in the shale oil: paraffins, isoparaffins, olefins/mononaphthenes, dinaphthenes, monoaromatics, naphthenoaromatics, diaromatics, naphthenodiaromatics, triaromatics, thiols/sulfides, benzothiophenes, naphthenobenzothiophenes, dibenzothiophenes, pyridines, anilines, quinolines, indoles, acridines, carbazoles, and phenols. Elemental sulfur may also be found in some shale oils [36]. Alfonso and coworkers found alkylphenols and carboxylic acids in shale oil [37]. Geng and coworkers found phenols, indanols, naphthols, phenylphenols, fluorenols and phenanthrenols, aliphatic ketones, esters, and aromatic ketones [38].

Shale oil properties and composition depend on many factors including type of oil shale and recovery method. Oil shales may show differences in structure and reactivity even when they were deposited at the same geological time and in the same region [39]. Retorting temperature, residence time, particle size and heating rate have a significant effect on oil yield and composition [40,41]. The slow heating applied to in situ retorting produces an oil that is different from the one that would be obtained by traditional mining and retorting.

Laboratory-scale analyses have confirmed that nitrogen, oxygen, and sulfur content in the shale oil will depend not only upon the oil shale origin but also upon the processing applied to produce the oil [42,43].

Rapid pyrolysis of shale oils was compared to conventional retorting by Shadle and coworkers [44]. Rapid pyrolysis produced a wide range of high molecular weight components that contained more heteroaromatic functionality than in the case of conventional retorting. The fuel obtained was of low quality and would also require upgrading.

Nazzal and coworkers [45,46] reported that particle grain size of the oil shale influences both yield and composition of the shale oil produced by pyrolysis. Oil yield increased and hydrocarbon gases decreased with increasing particle grain size. Concentration of polycyclic aromatics (PCAs) increased with decreasing particle grain size because of an increment in secondary reactions leading to aromatization. When compared to pyrolysis under an inert gas, pyrolysis under steam increased significantly the yield of oil and also the concentration of PCAs. Yanik and coworkers [47] also reported significant differences in the yields and compositions of the shale oils produced under four different conditions, i.e., slow pyrolysis, flash pyrolysis, and extraction with super- and subcritical water.

Microwave-assisted solvent extraction using a variety of solvents was studied by Al-Gharabli and coworkers [48]. They found that methanol exhibited the highest extractive capacity.

Jin et al. [49] produced shale oils by applying three different pyrolysis systems and found important variations in the nature and distribution of nitrogen-containing compound depending on the processing conditions. Tong et al. [50,51] produced shale oil samples by applying different heating rates ranging from 5 to 20°C/min to the same final temperature of 520°C. They analyzed the oil produced by positive-ion electrospray ionization Fourier transform ion cyclotron resonance mass spectrometry and found that increasing the heating rate resulted in an increase of total and basic nitrogen contents and the number of basic nitrogen compounds.

Wang et al. [43,52,53] studied the effects of residence time, temperature, and heating rates during pyrolysis of oil shale for production of shale oil. Samples retorted for different residence times between 6 and 40 min showed that the liquid yield increased and then leveled off as the residence time increased. The gases produced also increased while the spent shale amount decreased as residence time increased. The atomic H/C ratio of the oil decreased from 2.007 to 1.768, the oxygen and sulfur content decreased, and the nitrogen content slightly increased with increasing residence time. The initial stages of the retorting process produced higher boiling point oils than the later stages. Longer residence times decreased the aliphatic content and increased the aromatic content in the oil. The alkene/alkane gas ratio increased from 0.27 to 0.65 with increasing residence time from 6 to 60 min. This was an indication that secondary cracking reactions occur at longer residence times.

Although increasing retorting temperature from 430 to 520°C improved the production of both oil and gas, the oil/gas yield ratio decreased. Higher temperatures increased the nitrogen content in the derived oil and decreased the atomic H/C ratio and oxygen content. The sulfur content was not affected by the temperature.

Heating rates were varied from 5 to 20°C/min to a final temperature of 520°C. It was found that the production of noncondensable gases increased with increasing heating rate. The carbon and nitrogen contents of the oil also increased, while the content of hydrogen and oxygen decreased as the heating rate increased. The sulfur content was almost not affected. Higher heating rates decreased the amount of saturates and aromatics and increased the amounts of light oil, asphaltenes, and non-hydrocarbons in the oil. High amount of methane and minor amounts of C_2-C_4 hydrocarbons were found in the gas fraction.

Higher heating rates shifted the maximum concentration of C_1-C_4 hydrocarbons to higher temperature and increased the total content of C_1-C_4 hydrocarbons. At the same time, the alkene/alkane ratio in the gas phase decreased from 0.45 to 0.29 with increase in heating rate from 5 to 20°C/min. This lowering of the alkene/alkane gas ratio was explained by the occurrence of coking reactions at higher heating rates. Furthermore, while shale oil is being produced by heat treatment of oil shale, some of the minerals present within the shale may play a catalytic role and promote chemical reactions that affect the composition of the product [54].

If compared to petroleum crudes, shale oils are more olefinic and contain higher concentrations of nitrogen, oxygen, and sulfur compounds. Nitrogen in shale oils may be as high as 4.2 wt% while in petroleum crudes it is less than 1.0 wt% [31,35,55]. Petroleum usually contains between 0.1 wt% and 1.0 wt% of oxygen while shale oils may contain more than 5 wt%. Sulfur contents as high as 8.3 wt% in oil shale have been reported [56].

Shale oils also contain high amounts of ash and toxic minerals [56] and have higher specific gravity than petroleum because of the presence of heavy nitrogen, sulfur, and oxygen compounds [36]. Significant amounts of polycyclic aromatic compounds (PACs), which are known health hazards, are found in shale oils. These PACs include significant concentrations of PAC hydrocarbons, sulfur-PAC, and nitrogen-PAC [57].

The higher amounts of heteroatom-containing compounds and numerous impurities make shale oil quality very low. However, the enormous amounts of oil shale available position them as a significant resource for future production of fuels and chemicals.

5. Upgrading of Shale Oil

Shale oils can be used directly as fuel in industrial furnaces or can be processed to obtain fuels and chemicals similar to products obtained from petroleum-refining operations. However, in both cases the shale oil has to be treated to eliminate some or most of its undesirable components.

Shale oils may have a boiling range within the boiling range of gas oil fractions obtained from petroleum but it is not recommended to use them directly as fuel or heating oil because of their high content of aromatics, heavy metals, nitrogen, and sulfur components that are not only carcinogenic but also form soot and pollutant emissions [56]. Furthermore, nitrogen compounds in shale oil undergo or promote polymerization processes that increase the oil viscosity during storage and also produce undesired color and odor problems [58]. Nitrogen compounds are also catalyst poisons. Sulfur is also detrimental as residues of high-sulfur shale oils have been shown to have a tendency to condensate and polymerize [59]. A review of shale oil fuel stability and mechanisms for their control and prevention was published in 1992 [60].

The high concentrations of heteroatom-containing compounds strongly influence the type of treatment that would be required to obtain marketable fuels from shale oils [61]. To develop methods for shale oil upgrading, a good understanding of the amounts and type of heteroatom-containing compounds found in shale oils is required. This goal has been the focus of numerous research articles for the past 70 years [50,62–64].

Nitrogen-containing compounds in shale oil are often classified as basic, weakly basic, and neutral. The basic compounds are usually derivatives of pyridine, quinolone, acridine, and amines. The weakly basic ones are derivatives of pyrrole, indole, and carbazole, while the neutral compounds are derivatives of nitrile and amides [58]. Usually, the basic nitrogen compounds, which are the most poisonous for refining catalysts are in higher amounts than the rest. If shale oil is submitted to distillation, nitrogen compounds are found in higher amounts in the heavier fractions, as it is the case with opportunity crudes. Tang and coworkers [65] found that about 33% of total nitrogen in shale oil is in molecules with a high boiling point.

Upgrading of shale oils may include several steps that are dictated by the initial composition of the oil, the particular product(s) that are desired, and the availability of electric power and process water particularly in remote locations [36]. As crude shale oil exits the retorting step, it generally contains emulsified water and suspended solids. These are eliminated through dewatering and desalting [36]. Arsenic and iron present in shale oil must be removed because they are catalyst poisons in downstream processes. Arsenic and iron are water soluble and their elimination is accomplished by washing [66]. Highly dispersed transition metal sulfides formed in situ from the decomposition of dithiocarbamate complexes of transition metals have also been studied for dearsenation of shale oils [67]. Dearsenation can also be accomplished through hydrotreating (HDT). In this case, arsenic is retained as thermally stable metal arsenides and metal arsenous sulfides on the catalyst [68]. Once the catalyst capacity to retain arsenic is reached, the catalyst must be replaced.

To convert shale oils into products that are comparable to petroleum or its fractions, the usual upgrading strategy is hydroprocessing (HDP). During the HDP of shale oil, typical HDT reactions of hydrodesulfurization (HDS), hydrodenitrogenation (HDN), and hydrodeoxygenation, saturation of aromatics and olefins, and sometimes cracking and hydrocracking (HCK) take place converting shale oil or its distillates to lighter hydrocarbon fractions such as gasoline, diesel oil, and jet fuel [36].

Numerous articles on upgrading of Estonian shale oil and its distillate fractions were published by Luik and coworkers [69–79]. They studied the effect of hydrogenation on the properties of kukersite retort oil and its distillate fractions using several catalysts. They concluded that it is indeed possible to upgrade Estonian shale oil into diesel fuel or diesel fuel additive. With respect to upgrading distillate fractions, shale oil fractions boiling within 180–240°C, 240–320°C, and above 320°C were considered for diesel fraction, fuel oil fraction, and heavy fuel oil fraction, respectively. Each of those fractions was hydrogenated under various reaction conditions to establish the effects of hydrogenation on the yield, composition, and properties of the products obtained. They concluded that the type and amount of hydrogenated products depended on complex effects of the pyrolysis conditions, temperature, time, and catalysts. So, there was no unique solution for all fractions and neither the best conditions nor catalysts could be recommended for an optimum total result. The HDT of Estonian shale oil and its fractions, using Co–Mo catalysts reported in Refs. [71,74,76,77], indicated that the chemical composition of the shale oil and fractions after HDT resembled those of petroleum products [73]. Because hydrogenation reactions occur more easily than the desired hydrogenolysis reactions that break the C–N bond in the N-compounds, the consumption of hydrogen can be in the order of 3–7 moles of hydrogen per mole of nitrogen compound [66]. This high hydrogen consumption is the main disadvantage of the application of HDN to shale oils.

HDN applied to shale oil was studied by Saban et al. [66] and by Jeong et al. [80]. Both research groups used commercial Co–Mo/Al_2O_3 and Ni–Mo/Al_2O_3 presulfided catalysts. They found that nitriles decomposed faster than five- and six-membered nitrogen heterocyclic compounds [66] and that larger molecules were more difficult to convert [80].

Yoshida and coworkers also did research on the chemical structure changes in shale oil under HDT using red-mud/sulfur and Ni–Mo catalysts [81]. Their results showed that Ni–Mo catalysts were more effective for the removal of heteroatoms than red-mud/sulfur.

Using a conventional sulfided NiO–MoO_3/γ-Al_2O_3 catalyst to hydrotreat shale oil distillate fractions, Benyamna and coworkers [82] reported that heteroatoms were removed more easily from the light fractions containing paraffins than from heavier fractions containing aromatics.

Chishti and Williams [83] found that HDT of shale oil using a Ni–Mo catalyst reduced the content not only of nitrogen and sulfur compounds but also of three and four ring polyaromatic hydrocarbons and at the same time increased the content of single and two ring PACs.

Wang and coworkers studied the upgrading of shale oil by coupling the reactions of hydrogenation and ring opening of aromatics using W–Ni/γ-Al_2O_3 catalysts [84].

Catalytic HDT of shale oil over Ni–Mo and Co–Mo catalysts was also studied by Williams and Chishti [85]. Aromatic nitrogen and sulfur compounds, and alkene content decreased with the treatment. The treatment was favored by higher hydrogen pressure. Both Ni–Mo and Co–Mo displayed similar efficiency for HDN and HDS reactions [86].

Tang and coworkers [65] used a commercial Ni–W/Al_2O_3 catalyst to study the HDN of shale oil. Their results showed that basic nitrogen compounds have higher HDN reactivity than nonbasic nitrogen. The same research group developed kinetic models applicable to HDS [87,88].

Various commercial HDT catalysts including Co–Mo/Al_2O_3, Ni–W/Al_2O_3, and Ni–MoW/Al_2O_3 were tested by Yu and coworkers [89] to improve the quality of a diesel fraction (200−360°C) obtained from Fushun (China) shale oil. The experiments showed that the Ni−MoW catalyst was the most active for heteroatom removal and that relatively mild conditions were sufficient to produce clean diesel from Fushun shale oil distillate.

Yu and coworkers [90] also studied the hydrogenation of shale oil distillates using sulfided catalysts Ni−W/Al_2O_3 and Co−Mo/Al_2O_3. The results showed that higher temperature, higher hydrogen pressure, and longer residence time improved significantly the removal of nitrogen and to a lesser extent sulfur. The Ni−W/Al_2O_3 catalyst was much more active than the Co−Mo/Al_2O_3.

Beside HDT, other methods have also been studied to improve the quality of shale oils and shale oil fractions. For example, solvent extraction combined with aqueous alkali washing was successful applied by Chi and coworkers [91] to obtain a product that satisfied the specifications of diesel fuel.

Han and coworkers [31] applied different acidic reagents such as phosphoric acid and sulfuric acid to remove the basic portion of nitrogen compounds and obtained a significant improvement in the shale oil quality. Adsorption using zeolites instead of severe HDT has been proposed to remove residual nitrogen compounds from hydrotreated shale oil [92].

Sometimes reduction of molecular weight of shale oils is desired. In these cases, conventional catalytic cracking or HCK may be applied [36]. Because both shale oil molecular weight distribution and heteroatom content are different from petroleum, numerous articles have been published on the catalysts and the conditions that optimize shale oil cracking [93−103]. Because of the large concentration of heteroatomic contaminants in shale oil, a HDT stage prior to HCK is required to reduce heteroatom content and thus avoid poisoning of the (more expensive) HCK catalyst, by the reduction of heteroatom content. For example, Landau and coworkers [96−98] studied a three-stage process. The first stage catalysts were sulfided Ni−Mo or Co−Mo supported on wide-pore alumina. This first stage produced 97% and 79% removal of sulfur and nitrogen, respectively. The second stage catalysts were sulfided Ni−Mo, Ni−Mo−Cr, or Co−Mo−Cr supported on combined HY zeolite and alumina. The reduction in sulfur was to 100 −200 ppm and nitrogen to 7−30 ppm. The third stage catalysts were sulfided Ni−Mo supported on combined HY zeolite, ZSM-5 zeolite, and alumina [96]. The third stage produced 80% of the product in the naphtha boiling range. They also studied [98] HCK of hydrotreated shale oil and its atmospheric residue in a fixed bed reactor pilot plant with two Ni−Mo-zeolite catalysts based on mono- (HY + Al_2O_3) and bi- (HY + H-ZSM-5 + Al_2O_3) zeolite supports. Their results showed that light naphtha (<100°C), heavy naphtha (<200°C), and jet fuel (160−280°C) in the liquid product were 10−15% higher with the bi-zeolite catalyst.

Significant efforts in shale oil HDP using conventional petroleum HDP catalysts have been done by petroleum-refining companies. In the 1990s Chevron, UOP Inc., and Sunoco proposed different schemes to treat shale oil for gasoline production [96]. These technologies included HDT, HCK, and catalytic cracking. The high content of nitrogen in the feedstock was always a challenge to achieve desired nitrogen reduction levels.

6. Alternative Uses of Shale Oil

In addition to the interest in producing fuels from shale oil, alternative applications of shale oil and its fractions have been reported. For example, Nikitin and co-workers [104] studied shale oil and its high boiling point 230–320 and 320–360°C fractions as raw material for the manufacturing of metal corrosion inhibitors.

Although compounds containing heteroatoms are undesirable for applications of shale oil as fuel, they have been suggested as the source of valuable chemicals [55,58]. Furthermore, nitrogen compounds in raw shale oils and residues can promote binding with inorganic materials. The application of shale oil for asphalt paving applications appeared interesting and thus it has been studied [105]. For this purpose, the shale oil was separated into two fractions of different viscosity. The harder fraction was observed to deteriorate with time. However, the softer fraction showed properties desirable in asphalt recycling applications. Then, a shale oil residue was evaluated for its application as an asphalt additive. The result was unsuccessful because of higher embrittlement and fatigue cracking when compared to petroleum-derived asphalts [106].

Preparation of pharmaceutical products such as sulfonated shale oils has been an old application of shale oils that has been around for more than a century [107,108]. Those compounds have antiinflammatory, antibacterial, antimycotic, and ulcer-healing effects in both animals and humans [109–111]. Production of ammonium bituminosulfonate (ichthyol), an antifungal medicine, from shale oil was also studied [112,113]. The process included purification of the oil, followed by sulfonation, and neutralization of the sulfonic acids by ammonium hydroxide.

Grigoryeva and Kekisheva have reported on the use of water-soluble phenols obtained from shale oil in bactericidal detergent compositions [114]. Shale oil may contain up to 26 wt% of phenols and research work on whether the phenols should be separated from the oil before or after HDT was performed by Luik and coworkers [72]. Phenols obtained from shale oils can also find applications as feedstock for the production of alkyl resorcinol epoxy resins [24,115].

Production of plasticizers for rubber from high-sulfur shale oil has been studied by Fainberg and coworkers [116].

Laboratory-scale studies of the delayed coking process of vacuum distillate bottoms of a retorted shale oil as a source of carbon for manufactured graphite was undertaken by Sadeghi and coworkers [117]. In addition, production of carbon black from shale oil has also been reported [118].

Copyrolysis of heavy shale oil with polyethylene waste was studied, having as a goal the possibility to transform the wastes into liquid fuels [119]. The oil obtained was highly unsaturated and would require hydrogenation. In addition, the heavy shale oil produced much coke and gas at the pyrolysis temperatures required to decompose polyethyelene. To overcome this a two-stage process was proposed.

7. General Remarks

Fuel production from oil shales requires heating to convert the organic substance in oil shale, known as kerogen, into a liquid product known as shale oil. The oil shale can be mined using conventional methods and heated above ground using reactors known as retorts. Mining can be avoided by applying what is known as in situ retorting. In in situ-retorting heat is applied directly to the rocks underground and the shale oil obtained is extracted in a manner similar to petroleum extraction. Once the shale oil is obtained, regardless of its intended application, it must be upgraded because its high levels of nitrogen, sulfur, and oxygen compounds and other impurities make it unstable and prone to polymerization. Furthermore, if shale oil is used directly it produces high levels of pollutants. Upgrading of shale oil usually involves dewatering and desalting to eliminate arsenic and iron, HDT to eliminate heteroatom-containing compounds, and catalytic cracking and HCK to convert shale oil to desired gasoline, jet fuel, or diesel fractions.

Production of shale oil has already been applied commercially in China, Estonia, and Brazil. In the United States important technology developments have been accomplished but long-time continuous commercial production has not materialized yet.

Interest in oil shales has cycled over the years and has been influenced by petroleum price. Currently, in most countries hydrocarbons recovered from shale oils are not competitive against petroleum-derived hydrocarbons.

References

[1] Radovic L. Energy and fuels in society: analysis of bills and media reports. McGraw-Hill Custom Publishing; 1997. p. 484.

[2] Hepbasli A. Oil shale as an alternative energy source. Energy Sources 2004;26(2):107–18.

[3] Durand B, Kerogen. In: Durand B, editor. Insoluble organic matter from sedimentary rocks. Institut Francais du Petrole; 1980. p. 525.

[4] Palayangoda SS, Nguyen QP. Thermal behavior of raw oil shale and its components. Oil Shale 2015;32(2):160–71.

[5] Torrente MC, Galan MA. Kinetics of the thermal decomposition of oil shale from puertollano (Spain). Fuel 2001;80(3):327–34.
[6] Reinsalu E, Aarna I. About technical terms of oil shale and shale oil. Oil Shale 2015;32(4):291–2.
[7] Urov K, Sumberg A. Characteristics of oil shales and shale-like rocks of known deposits and outcrops. Oil Shale 1999;16(3):1–64.
[8] Energy Information Administration International Energy Outlook 2014: world petroleum and other liquid fuels. 2014. p. 62. Available from: http://www.eia.gov/forecasts/ieo/.
[9] World energy council world energy resources: Survey. 2013. Available from: https://www.worldenergy.org/wp-content/uploads/2013/10/WER_2013_2_Oil.pdf. Report 468 p.
[10] Speight JG. Chapter 2-oil shale resources. In: Shale oil production processes. Boston: Gulf Professional Publishing; 2012. p. 35–73.
[11] USGS In-place oil shale resources examined by grade in the major basins of the green river formation, Colorado, Utah, and Wyoming. 2016. Available from: https://pubs.usgs.gov/fs/2012/3145/FS12-3145.pdf.
[12] Metsaots K, Printsmann A, Sepp K. Public opinions on oil shale mining heritage and its tourism potential. Scand J Hosp Tour 2015;15(4):380–99.
[13] Bunger JW, Russell CP. Thermodynamics of shale oil production. In: Ogunsola OI, Hartstein AM, Ogunsola O, editors. Oil shale: a solution to the liquid fuel dilemma; 2009. p. 89–101.
[14] Qian J, Wang J, Li S. World's oil shale available retorting technologies and the forecast of shale oil production. In: Chung JS, editor. Proceedings of the Eighteenth International Offshore and Polar Engineering Conference; 2008. p. 19–20.
[15] Speight JG. Chapter 4-mining and retorting. In: Shale oil production processes. Boston: Gulf Professional Publishing; 2012. p. 93–122.
[16] Gallo P. Comparison of environmental impacts of shale oil production, coal-gasification, coal liquefaction and coal-fired electric-power generation. In: Proc. The Tenth oil shale Symposium. Golden, co. USA, April 21–22; 1977. p. 192–213.
[17] Parks N. Shale oil: alternative energy or environmental degradation? Bioscience 2008;58(6):490.
[18] Chappell WR. Environmental and health consequences of shale oil production in the USA. J Geol Soc 1980;137(SEP):571–4.
[19] Zhao P, Zhao Y, Zou C, Gu T. Study on ultrasonic extraction of kerogen from Huadian oil shale by solvents. Oil Shale 2013;30(4):491–500.
[20] Li GX, Han DY, Cao ZB, Yuan MM, Zai XY. Studies on Fushun shale oil furfural refining. Oil Shale 2011;28(3):372–9.
[21] Wu T, Xue Q, Li X, Tao Y, Jin Y, Ling C, et al. Extraction of kerogen from oil shale with supercritical carbon dioxide: molecular dynamics simulations. J Supercrit Fluids 2016;107:499–506.
[22] Francu J, Harvie BA, Laenen B, Siirde A, Veiderma M. A study on the Eu oil shale industry – viewed in the light of the Estonian experience. Report No. Belgium (Europe): European Academies Science Advisory Council; March 2007. p. 63.
[23] Asai G, Mauser JE, Lincaln RL, Henry JL. Caustic extraction of alumina and soda from dawsonite·bearing oil shale. Report No. Report of Investigations 8796. Washington (DC, USA): Bureau of Mines; 1983. p. 25.
[24] Raidma E. Diglycidyl ethers from shale oil alkyl resorcinols .2. Methods for production and characteristics of uncured alkyl resorcinol epoxy-resins. Oil Shale 1995;12(3):259–69.
[25] Speight JG. Shale oil production processes. Elsevier Inc.; 2012.
[26] Gusca J, Siirde A, Eldermann M. Energy related sustainability analysis of shale oil retorting technologies. In: Valtere S, editor. International scientific conference environmental and climate technologies, CONECT 2014; 2015. p. 216–21.
[27] Pan Y, Zhang X, Liu S, Yang S, Ren N. A review on technologies for oil shale surface retort. J Chem Soc Pak 2012;34(6):1331–8.
[28] Sohns HW, Carpenter HC. In situ oil shale retorting. Chem Eng Prog 1966;62(8):75–8.
[29] Pan Y, Jipeng Ning JM, Shuangchun Y. Research on in-situ oil shale mining technology. IOSR J Pharm Sci Invent 2012;1(1):1–7.

[30] Fischer PA. Hopes for shale oil are revived. World Oil 2005;226(8):69–74.
[31] Han DY, Li GX, Cao ZB, Zhai XY, Yuan MM. A study on the denitrogenation of Fushun shale oil. Energy Sources, Part A: recovery, Utilization Environ Effects 2013;35(7):622–8.
[32] Rovere CE, Crisp PT, Ellis J, Bolton PD. Chemical characterization of shale oil from Condor, Australia. Fuel 1983;62(11):1274–82.
[33] Guo SH, Ruan Z. The composition of Fushun and Maoming shale oils. Fuel 1995;74(11):1719–21.
[34] Bae E, Na J-G, Chung SH, Kim HS, Kim S. Identification of about 30 000 chemical components in shale oils by electrospray ionization (ESI) and atmospheric pressure photoionization (APPI) coupled with 15 T fourier transform ion cyclotron resonance mass spectrometry (FT-ICR MS) and a comparison to conventional oil. Energy & Fuels 2010;24:2563–9.
[35] Dijkmans T, Djokic MR, Van Geem KM, Marin GB. Comprehensive compositional analysis of sulfur and nitrogen containing compounds in shale oil using GC × GC–FID/SCD/NCD/TOF-MS. Fuel 2015;140:398–406.
[36] Speight JG. Chapter 6-refining shale oil. In: Shale oil production processes. Boston: Gulf Professional Publishing; 2012. p. 139–63.
[37] Afonso JC, Schmal M, Cardoso JN. Acidic oxygen compounds in the Irati shale oil. Industrial Eng Chem Res 1992;31(4):1045–50.
[38] Geng C, Li S, Ma Y, Yue C, He J, Shang W. Analysis and identification of oxygen compounds in Longkou shale oil and Shenmu coal tar. Oil Shale 2012;29(4):322–33.
[39] Amer MW, Marshall M, Fei Y, Jackson WR, Gorbaty ML, Cassidy PJ, et al. A comparison of the structure and reactivity of five Jordanian oil shales from different locations. Fuel 2014;119:313–22.
[40] Han XX, Jiang XM, Cui ZG. Studies of the effect of retorting factors on the yield of shale oil for a new comprehensive utilization technology of oil shale. Appl Energy 2009;86(11):2381–5.
[41] Na JG, Im CH, Chung SH, Lee KB. Effect of oil shale retorting temperature on shale oil yield and properties. Fuel 2012;95:131–5.
[42] Al-Ayed OS, Matouq M. Factors affecting sulfur reactions in high sulfur oil shale pyrolysis. J Energy Resour ASME 2009;131(1):4.
[43] Wang S, Jiang X, Han X, Tong J. Effect of retorting temperature on product yield and characteristics of non-condensable gases and shale oil obtained by retorting Huadian oil shales. Fuel Process Technol 2014;121:9–15.
[44] Shadle LJ, Seshadri KS, Wang YHC. Characterization of shale oils. 2. Analysis of the flash pyrolysis products of oil-shale in the green river formation. Fuel Process Technol 1994;37(2):121–42.
[45] Williams PT, Nazzal JM. Pyrolysis of oil shales: influence of particle grain size on polycyclic aromatic compounds in the derived shale oils. J Inst Energy 1999;72(491):48–55.
[46] Nazzal JM. The influence of grain size on the products yield and shale oil composition from the pyrolysis of Sultani oil shale. Energy Convers Manag 2008;49(11):3278–86.
[47] Yanik J, Yuksel M, Saglam M, Olukcu N, Bartle K, Frere B. Characterization of the oil fractions of shale oil obtained by pyrolysis and supercritical water extraction. Fuel 1995;74(1):46–50.
[48] Al-Gharabli SI, Azzam MOJ, Al-Addous M. Microwave-assisted solvent extraction of shale oil from Jordanian oil shale. Oil Shale 2015;32(3):240–51.
[49] Jin JM, Kim S, Birdwell JE. Molecular characterization and comparison of shale oils generated by different pyrolysis methods. Energy Fuels 2012;26(2):1054–62.
[50] Tong J, Liu J, Han X, Shi Q, Jiang X. Influence of heating rate on basic nitrogen-containing species content in Dachengzi shale oil studied by positive-ion electrospray ionization FT-ICR mass spectrometry. Oil Shale 2013;30(1):76–89.
[51] Tong J, Liu J, Han X, Wang S, Jiang X. Characterization of nitrogen-containing species in Huadian shale oil by electrospray ionization Fourier transform ion cyclotron resonance mass spectrometry. Fuel 2013;104:365–71.
[52] Wang S, Jiang X, Han X, Tong J. Effect of residence time on products yield and characteristics of shale oil and gases produced by low-temperature retorting of Dachengzi oil shale. Oil Shale 2013;30(4):501–16.

[53] Wang S, Liu J, Jiang X, Han X, Tong J. Effect of heating rate on products yield and characteristics of non-condensable gases and shale oil obtained by retorting Dachengzi oil shale. Oil Shale 2013;30(1):27–47.
[54] Carter SD, Citiroglu M, Gallacher J, Snape CE, Mitchell S, Lafferty CJ. Secondary coking and cracking of shale oil vapors from pyrolysis of hydropyrolysis of a Kentucky Cleveland oil-shale in a 2-stage reactor. Fuel 1994;73(9):1455–8.
[55] Hsu-Chou RSY, Mobashery S, Yen TF. Denitrogenation of shale oil by oxime formation from pyrroles. Energy Sources 1998;20(9):857–66.
[56] Akash BA, Jaber JO. Characterization of shale oil as compared to crude oil and some refined petroleum products. Energy Sources 2003;25(12):1171–82.
[57] Williams PT, Nazzal JM. Polycyclic aromatic compounds in shale oils: influence of process conditions. Environ Technol 1998;19(8):775–87.
[58] Guo S, Qin K. Nitro-containing compounds in the Chinese light shale oils. Oil Shale 1993;10(2–3):165–77.
[59] Fainberg V, Ishai I, Hetsroni G, Leichter S. High-sulphur shale oil as a prime matter for bitumen production. Oil Shale 1995;12(4):305–16.
[60] Fathoni AZ, Batts BD. A literature-review of fuel stability studies with a particular emphasis on shale oil. Energy & Fuels 1992;6(6):682–93.
[61] Bunger JW, Russell CP, Cogswell DE. Molecular composition of shale oil. In: Ogunsola OI, Hartstein AM, Ogunsola O, editors. Oil shale: a solution to the liquid fuel dilemma; 2009. p. 103–13.
[62] Ball JS, Vanmeter R. Determination of nitrogen in shale oil and petroleum – application of established methods. Anal Chem 1951;23(11):1632–4.
[63] Shue FF, Yen TF. Concentration and selective identification of nitrogen-containing and oxygen-containing-compounds in shale oil. Anal Chem 1981;53(13):2081–4.
[64] Fainberg V, Podorozhansky M, Hetsroni G, Laichter S. Oxygen and nitrogen-compounds of Israeli shale oil. Fuel Sci Technol Int 1994;12(4):581–91.
[65] Tang X, Li S, Yue C, He J, Hou J. Hydrodenitrogenation of basic and non-basic nitrogen compounds in shale oil over NiW/Al_2O_3 catalyst. In: Xu P, Wang Y, Su Y, Hao L, editors. Advances in applied science and industrial technology, pts 1 and 2; 2013. p. 45–8.
[66] Saban MD, Skala DU, Jovanovic JA, Mayn VW, Rahimian IGH. Hydrodenitrogenation of Aleksinac shale oil distillates in a pilot trickle-bed reactor. Fuel Process Technol 1992;30(1):15–31.
[67] Singhal GH, Brown LD, Ryan DF. Synthesis of transition-metal arsenides and arsenic metal sulfides and their role in the dearsenation of shale oils. J Solid State Chem 1994;109(2):219–22.
[68] Sikonia JG. Arsenic management in shale oil upgrading. J Environ Geochem Health 1985;7(2):64–8.
[69] Luik H, Vink N, Lindaru E, Maripuu L. Upgrading of Estonian shale oil distillation fractions. 5. Hydrogenation of heavy mazute. Oil Shale 2000;17(1):25–30.
[70] Luik H. Hydrogenation of Estonian oil-shale and shale oil. Oil Shale 1994;11(2):151–60.
[71] Luik H, Lindaru E, Vink N. Upgrading of Estonian shale oil. 2. Effect of hydrogenation on the properties of kukersite retort oil. Oil Shale 1996;13(3):193–6.
[72] Luik H, Muurisepp AM, Vink N, Liiv M. Upgrading of Estonian shale oil. 3. Hydrogenation of phenols. Oil Shale 1996;13(3):197–203.
[73] Luik H, Vink N, Lindaru E. Upgrading of Estonian shale oil. 1. Effect of hydrogenation on the chemical composition of kukersite retort oil. Oil Shale 1996;13(1):13–9.
[74] Luik H, Lindaru E, Vink N, Maripuu L. Upgrading of Estonian shale oil distillation fractions. 1. Hydrogenation of the "diesel fraction". Oil Shale 1999;16(2):141–8.
[75] Luik H, Maripuu L, Vink N, Lindaru E. Upgrading of Estonian shale oil distillation fractions – 3. Hydrogenation of light mazute. Oil Shale 1999;16(4):331–6.
[76] Luik H, Vink N, Lindaru E, Maripuu L. Upgrading of Estonian shale oil distillation fractions. 2. The effect of time and hydrogen pressure on the yield and composition of "diesel fraction" hydrogenation products. Oil Shale 1999;16(3):249–56.

[77] Luik H, Vink N, Lindaru E, Maripuu L. Upgrading of Estonian shale oil distillation fractions – 4. The effect of time and hydrogen pressure on the yield and composition of light mazute hydrogenation products. Oil Shale 1999;16(4):337–42.

[78] Luik H, Vink N, Maripuu L, Lindaru E. Upgrading of Estonian shale oil distillation fractions. 6. The effect of time and temperature on the yield and composition of heavy mazute hydrogenation products. Oil Shale 2000;17(1):31–6.

[79] Luik H, Luik L, Johannes I, Tiikma L, Vink N, Palu V, et al. Upgrading of Estonian shale oil heavy residuum bituminous fraction by catalytic hydroconversion. Fuel Process Technol 2014;124:115–22.

[80] Jeong SY, Bunger JW, Russell CP. Effect of molecular-size on intrinsic reaction-kinetics in catalytic hydroprocessing of shale oil. Energy & Fuels 1994;8(5):1143–8.

[81] Yoshida R, Miyazawa M, Yoshida T, Narita H, Maekawa Y. Chemical structure changes in Condor shale oil and catalytic activities during catalytic hydrotreatment. Fuel 1996;75(1):99–102.

[82] Benyamna A, Bennouna C, Moreau C, Geneste P. Upgrading of distillate fractions of Timahdit Moroccan shale oil over a sulfided $NiO-MoO_3/\gamma-Al_2O_3$ catalyst. Fuel 1991;70(7):845–8.

[83] Chishti HM, Williams PT. Aromatic and hetero-aromatic compositional changes during catalytic hydrotreatment of shale oil. Fuel 1999;78(15):1805–15.

[84] Wang H, Dai F, Li Z, Li C. Upgrading shale oil distillation to clean fuel by coupled hydrogenation and ring opening reaction of aromatics on $W-Ni/\gamma-Al_2O_3$ catalysts. Energy & Fuels 2015;29(8):4902–10.

[85] Williams PT, Chishti HM. Upgraded oils from the catalytic hydrotreatment of shale oil. J Inst Energy 2000;73(497):196–203.

[86] Williams PT, Chishti HM. Reaction of nitrogen and sulphur compounds during catalytic hydrotreatment of shale oil. Fuel 2001;80(7):957–63.

[87] Tang X, Li S, Yue C, He J, Gong Y. Lumping kinetics of hydrodesulfurization for crude Longkou shale oil. In: Xu P, Wang Y, Su Y, Hao L, editors. Advances in applied science and industrial technology, pts 1 and 2; 2013. p. 12–6.

[88] Tang X, Li S, Yue C, He J, Hou J. Lumping kinetics of hydrodesulfurization and hydrodenitrogenation of the middle distillate from Chinese shale oil. Oil Shale 2013;30(4):517–35.

[89] Yu H, Li S, Jin G. Catalytic hydrotreating of the diesel distillate from Fushun shale oil for the production of clean fuel. Energy Fuels 2010;24(8):4419–24.

[90] Yu H, Li S, Jin G. Hydrodesulfurization and hydrodenitrogenation of diesel distillate from Fushun shale oil. Oil Shale 2010;27(2):126–34.

[91] Chi YL, Li SY, Li XB. Study on shale oil diesel refining. Oil Shale 2005;22(3):359–66.

[92] Ellis J, Korth J. Removal of nitrogen-compounds from hydrotreated shale oil by adsorption on zeolite. Fuel 1994;73(10):1569–73.

[93] Cottingh PL, Carpente HC. Hydrocracking prehydrogenated shale oil. Industrial Eng Chem Process Des Dev 1967;6(2):212–7.

[94] Cottingham PL, Nickerson LG. Hydrocracking of Insitu shale oil. Abstr Pap Am Chem Soc 1976;172(SEP3):60.

[95] Cottingham PL, Nickerson LG. Diesel fuel and burner fuels via shale oil hydrocracking. Erdol Kohle Erdgas Petrochem 1976;29(5):214–23.

[96] Landau MV, Herskowitz M, Givoni D, Laichter S, Yitzhaki D. Medium-severity hydrotreating and hydrocracking of Israeli shale oil. 1. Novel catalyst systems. Fuel 1996;75(7):858–66.

[97] Landau MV, Herskowitz M, Givoni D, Laichter S, Yitzhaki D. Medium severity hydrotreating and hydrocracking of Israeli shale oil – II. Testing of novel catalyst systems in a trickle bed reactor. Fuel 1998;77(1–2):3–13.

[98] Landau MV, Herskowitz M, Givoni D, Laichter S, Yitzhaki D. Medium severity hydrotreating and hydrocracking of Israeli shale oil III. Hydrocracking of hydrotreated shale oil and its atmospheric residue for full conversion to motor fuels. Fuel 1998;77(14):1589–97.

[99] Levy JH, Mallon RG, Wall GC. Vapor-phase cracking and coking of 3 Australian shale oils – kinetics in the presence and absence of shale ash. Fuel 1987;66(3):358–64.

[100] Shultz EB, Guyer JJ, Linden HR. High temperature, vapor-phase cracking of hydrocarbons – pyrolysis of crude shale oil. Industrial Eng Chem 1955;47(12):2479–82.
[101] Tait AM, Hensley AL. Catalysts for single-stage hydrocracking of whole shale oil. Abstr Pap Am Chem Soc 1981;182(AUG):37–PETR.
[102] Tait AM, Hensley AL. Jet fuels from shale oil by single-stage hydrocracking. Abstr Pap Am Chem Soc 1982;183(MAR):51–FUEL.
[103] Tait AM, Hensley AL. Catalysts for the one-stage hydrocracking of shale oils. Erdol Kohle Erdgas Petrochem 1982;35(1):38.
[104] Nikitin EE, Tishina EA, Englin AB, Fainberg VS, Shimonaev GS, Churshukov ES, et al. Corrosion protection by shale oil. Prot Metals 1990;26(5):627–32.
[105] Mahboub K, Simpson AL, Oduroh PK, Robertson RE, Harnsberger PM, Taulbee DN, et al. A study of the suitability of the KENTORT-II eastern shale oil for asphalt paving applications. Fuel 1992;71(12):1527–37.
[106] Thomas KP, Harnsberger PM. Evaluation of an eastern shale oil residue as an asphalt additive. Fuel Sci Technol Int 1996;14(9):1253–80.
[107] Cholcha W, Leuschner J, Leuschner F. Experimental studies on the tolerability of pale sulfonated shale oil following local and systemic application. Arzneimittel-Forschung/Drug Res 1994;44-1(2):170–7.
[108] Cholcha W, Leuschner J, Leuschner F. Safety studies of dark sulfonated shale oil following local and systemic application. Arzneimittel-Forschung/Drug Res 1994;44-2(7):844–9.
[109] Martens J, Gayko G. Application of sulfonated shale oils in horses under special consideration of a new pale sulfonated shale oil – clinical report. Prakt Tierarzt 2002;83(6):532–7.
[110] Gayko G, Cholcha W, Kietzmann M. Antiinflammatory, antibacterial and antimycotic effects of dark sulfonated shale oil (ichthammol). Berl Und Munchener Tierarztliche Wochenschr 2000;113(10):368–73.
[111] Beckert S, Warnecke J, Zelenkova H, Kovnerystyy O, Stege H, Cholcha W, et al. Efficacy of topical pale sulfonated shale oil in the treatment of venous leg ulcers: a randomized, controlled, multicenter study. J Vasc Surg 2006;43(1):94–100.
[112] Fainberg V, Hetsroni G, Leichter S. Ichthyol production from Israeli shale oil. Oil Shale 1995;12(1):39–45.
[113] Listemann H, Scholermann A, Meigel W. Antifungal activity of sulfonated shale oils. Arzneimittel-Forschung/Drug Res 1993;43-2(7):784–8.
[114] Grigoryeva L, Kekisheva L. The use of shale oil derived alkylresorcinols in bactericidal detergent compositions. Oil Shale 1998;15(1):37–43.
[115] Raidma E. Diglycidyl ethers from shale oil alkyl resorcinols. 1. Initial raw phenols and the structure of reaction-products. Oil Shale 1994;11(3):241–9.
[116] Fainberg V, Hetsroni G, Feiglin T, Schwartz A. High-sulfur shale oil as a component of rubber stocks. Oil Shale 1997;14(2):133–42.
[117] Sadeghi MA, Sadeghi KM, Yen TF. Formation of mesophase using precursors from shale oil. Energy Sources 1993;15(3):391–401.
[118] Zhou CL. Production of carbon black from shale oil in China. Oil Shale 1996;13(2):83–6.
[119] Tiikma L, Tamvehus H, Luik L. Coprocessing of heavy shale oil with polyethylene waste. J Anal Appl Pyrolysis 2007;79(1–2):191–5.

CHAPTER 8

Final Remarks and Future Aspirations

1. Introduction

The decay of the conventional sources of crude oils by the end of the 20th century opened the door to a transition toward less conventional oils. The largest remaining reserves of crude oil are those of unconventional oils (UOs) and among these, opportunity crudes (OCs) and more particularly heavy, extraheavy, bitumen, and tar sands represent about 70% of the proven reserves [1]. The production of UOs is required to devote great effort toward technology development, up to the point of achieving economic feasibility. An increasing trend in oil prices at the end of the 1990s enabled the startup of production of some of these resources (OCs) in an accelerated way. By 2005, oil producers successfully managed to control overproduction and to balance the crude oil supply and demand, removing the excess inventory from the system and contributing to the highest increase in crude oil price ever seen in history [2]. The new sources were more difficult to produce and/or to recover, and furthermore the resulting crude oil was of a poorer quality, generating difficulties for its processing. The OCs with increasing exploitation costs have contributed to a trend of increasing prices since the end of the last century and through the first 14 years of this century. However, technology advances make "unconventional" a dynamic concept; someday unconventional may shift to conventional, as new technologies improve the economy.

However, oil quality is also a determinant component for oil price. Conventional crude oils were originally valued by their hydrocarbon composition, particularly by their potential in producing fuels. In this regard, probably a distillation curve[1] would have been more than enough information for the refiner to make a purchasing decision. Later on, the appearance of more contaminated crude oils, the changes in density and viscosity of the sources, and the Clean Air Act introduced new factors into the value equation for crude oils. Then, American Petroleum Institute (API) specific gravity and sulfur were considered as quality markers, and the lower the API gravity and the higher the S-content, the lower the price of the crude oil. The appearance of even poorer-quality crude oils from unconventional resources increased the importance of quality in the price equation. Some crude oils within the UOs market exhibited a very low API gravity and/or contained acid compounds that could cause damage

[1] A distillation curve is also known as the true boiling point curve (TBP), which shows the volume percentage that distillates at each temperature.

The Science and Technology of Unconventional Oils. http://dx.doi.org/10.1016/B978-0-12-801225-3.00008-5

to the facilities of the oil industry. Thus bitumen, extraheavy, heavy, and high-acid crude (HAC) oils fall into this group. To be able to sell these crude oils, the producer had to offer a significant discount on the price of a conventional oil maker to the oil purchaser. These crude oils were coined opportunity crudes since they were offered under an opportunity price. Besides OCs, shale oils with the largest production costs are also considered UOs and regardless of their higher API and lower S-content they have to be discounted.

Geographically, OCs introduced new players in the oil supply market. Regarding OCs, the largest reserves of bitumen, extraheavy and heavy oils are located in Venezuela and Canada. Meanwhile, HACs are more widely spread around the globe; new players include China, West Africa, and Brazil. Large reserves of shale oils are located in the United States, Estonia, Russia, Zaire, Brazil, Italy, and Canada.

Currently, the oil industry faces a scenario that has passed through record high oil prices into a period of decline and volatility. World reserves of light crude oils are in accelerated decline, while the proportion of UOs, and more particularly of OCs, increases continuously. Consequently, the refinery diet is continuously changing with the increases in the supply of OCs. However, regardless of discounted prices, refiners are struggling to keep margins in good shape.

The economic benefits for refiners behind the discounted prices of UOs could be significant, if and only if the refinery implements facilities that provide feedstock flexibility. The conventional technologies currently in operation keep refiners struggling to optimize margins. The incorporation of UOs is motivated only during high oil prices and/or large discounts. In this situation, refining typically overcomes the many technical challenges by mitigating the derived effects of trying to maintain unit integrity and reliability, monitoring, controlling, and adjusting product quality, and making sure environmental compliance is in place. Clearly, there is room for improvement.

In summary, the current market scenario could bring to the supply side an increased proportion of more sour, heavier, more acidic, and/or harder to produce or recover oil, while on the demand side more stringent regulations and specifications impose higher-quality products. Therefore refiners need to address the situation with increased flexibility and more cost-effective technologies.

The R&D activities and efforts, though biased by price volatility, have continuously enriched both the knowledge base and the technology portfolio. Almost a century has passed since the discovery of the crude oil components and compounds (the "bad" actors) considered in this book. Thus a vast number of documents (journal articles, conference papers, books, theses, patents, etc.) have been published over time. Nearly 6000 records constitute the database of references supporting this book. Although in many instances the full text document could not be accessed, most of the references have been reviewed and more than 3300 have been described, discussed, and cited. The most prolific area is that of asphaltenes, probably because of the complexity and challenges posed by these compounds; almost 40% of the effort has been devoted to this field.

We have described and discussed the scientific knowledge of the troublesome species (the actors) responsible for the refining issues of unconventional oil in three separated chapters. The commercially available technologies, the emerging technologies, and the proposed concepts and ideas under development were also presented in two other chapters. As the whole, clear needs for R&D still exist.

One of the areas that have provided a good push toward knowledge advances is the analytical and characterization techniques and methodologies. On one side, simplification, analysis time, and precision of the quantitation of compounds and properties have facilitated the determination of the different compounds and provided capabilities for their online monitoring. An example is the H-cell for the simultaneous determination of asphaltenes and total acid number (TAN). This method utilizes microscale diffusion that differentiates it from distillation and chemical-based methods and affords better accuracy [3]. On the other side, improvements in resolution have been leading the way to a better molecular identification and characterization of oils and oil products. These advances in resolution have driven improvements in isolation methods and subfractionation techniques that may contribute to the simplification of the samples and, indirectly, to the simplification of the interpretation. More than 20,000 compounds [4] could be detected with the use of high-resolution Fourier transform ion cyclotron resonance mass spectrometry in the heavy fractions, including ~3000 acid compounds [5] and another ~3000 basic compounds [6]. Ultrahigh resolution is needed when the metal compounds are targeted as well. In this case, ~85,000 compounds could be detected [7]. Obviously, the complexity is enormous and new challenges emerge any time a new technique becomes available or an improvement is incorporated into an old one.

The advances in modeling have been even more significant than those of characterization. Both hardware and software capabilities have improved enormously allowing the modeling of more complex molecules and systems, their properties, and behavior. Nonetheless, there is still room for improvement. Incorporation of phenomenology and of the knowledge derived from the more sophisticated experimental techniques could provide new advances in the future.

The fuel manufacturing processes are of enormous complexity. The increase in UOs' proportion in the refinery diet, together with the increasingly stringent environmental requirements, implies that the refiner has to do more with the poorer-quality crude slate and still make a profit out of it. Therefore more difficult challenges call for the development and implementation of cost-effective technologies that allow the realization of the potential economic benefits.

2. The Actors

In general, the bottom of the barrel (BotB) term denotes the heavier molecules present in crude oils, and therefore the residue from distillation. However, it is more typically used for OCs to refer to the compounds, whose conversion or removal becomes mandatory when a smooth refining operation is desired. The most detrimental context of the BotB term concerns heavy/extraheavy oils and bitumen. The contaminant levels in commercial bitumen grades look similar on average to those measured in heavy oil samples. Therefore processing issues and challenges would be similar as well [8].

Asphaltenes, metal-bearing compounds, and naphthenic acids constitute "bad" actors, whose behavior precludes or obstructs the smooth refining operation. Their physical properties and reactivity have severe consequences not only in refining, but also in production and transportation. A lack of comprehensive knowledge of the heavier molecules may be one of the reasons for the inexistence of cost-effective technologies. Regarding refining, from their entrance into the refining processes, the desalting process and the dewatering unit, to the final product blending systems, the impact of these actors is significantly negative. OCs processing creates difficulties along almost all the refinery units since these crude oils are extremely hard to desalt, difficult to vaporize, thermally unstable, corrosive, incompatible, produce high multiunsaturated products, and carry fouling[2] [9] precursors or materials. The phenomenology derived from the properties and behavior of these bad actors has been discussed in previous chapters; however, a brief description follows.

2.1 Asphaltenes

Asphaltenes are macromolecules, exhibiting the highest molecular weight (MW) of all the molecules in crude oil. Heteroatomic moieties are concentrated in this BotB subfraction. They are a solubility class consisting of a mixture of compounds that are soluble in toluene and insoluble in light *n*-paraffins, typically either C_5 or C_7. Compounds differing in MW, BP, molecular size, and many more properties are present in this fraction. The complexity of the mixture and of the constituting molecules has precluded the structural identification of these compounds. Consequently, the proposed molecular models try to respond to characterization results and/or behavioral features.

The macrostructural organization of asphaltenes is another controversial subject in which the defendants of each molecular structural model have set their own view based in that

[2] Fouling is defined as the formation of an unexpected phase that interferes with processing. While the fouling phase is often a solid, it could be a liquid when a gas is expected or it could be an emulsion when separate liquid phases are expected.

model. The current consensus is that the macrostructural arrangement is hierarchical and starts with molecules and ends up in aggregates, the differences being the considered interactions and the intermediate organization. The supramolecular assembly of molecules proposed by Gray [10] incorporates all possible interactions and detected experimental arrangements, namely, aromatic $\pi-\pi$ stacking, H-bonding, acid−base interactions, metal coordination complexes, interactions between cycloalkyl and alkyl groups, and the formation of hydrophobic pockets. Additionally, it includes a range of architectures and molecular structural types creating porous networks and host−guest complexes [10].

The solubility of asphaltenes is their defining characteristic. Additionally, other troublesome properties comprise their trend for self-association and aggregation, surface activity, stability (phase transitions), and reactivity. Asphaltenes are also the main responsible compounds for the high density and viscosity of bitumen and heavy OCs. Hence dilution or conversion is needed for the transportation and processing of asphaltenic crude oils. If dilution is chosen, the first consideration is compatibility. Any given OC may show incompatibility with other crude oils and diluents, leading to the formation of sludge, precipitates, or deposits that will upset operation of the refinery unit involved. Furthermore, some OCs are self-incompatible under certain process conditions.

Moreover, precipitation, aggregation, stability, and adsorption on metallic surfaces lead to fouling and deposition. Depending on which unit is upset by the fouling or deposits formation, the problem could affect product specifications or generate further upsets in other units, e.g., the wastewater treatment plant. The surface activity of asphaltenes also provokes the formation of water/oil emulsions and rigid films at the water/oil interface [11]. These behaviors lead not only to fouling and deposition, but also to the poor performance of refining units, more particularly the desalter and dewatering units.

Finally, the low reactivity of the polycondensed aromatic rings present in the asphaltenes' molecules leads to coke formation upon thermal conversion. This coke formation propensity, together with the low reactivity of many other molecular features of asphaltenes, cause catalyst deactivation by fouling and pore plugging. The coke-forming potential of asphaltenes categorizes them as coke precursors and therefore their association with carbon Conradson residue (CCR) and microcarbon residue (MCR).

The lack of a precise definition of asphaltenes and of the understanding of their behavior leads to a poor definition of the problems these compounds cause. Consequently, the worse the problem definition, the higher the probability of not finding an appropriate solution [12].

2.2 Metal Compounds

The metal compounds are the least characterized of the considered bad actors. Metals are present in crude oils in the form of inorganic as well as organic compounds. Heavy and light metals are present in crude oils, e.g., nickel, vanadium, iron, silicon, cadmium, mercury, aluminum, sodium, calcium, and magnesium, among others. Abundance of these metals ranges from less than a part per billion (ppb) to more than several thousand parts per million (ppm). The most abundant transition metals tend to concentrate in the BotB fraction; the less abundant transition metals are present in lighter fractions, particularly Hg and Cd. Transition metals are present typically in the form of organometallic complexes. In heavy oils and bitumen, Ni and V are usually the most abundant metals and in concentration ranges of up to 200 ppm Ni and 2800 ppm V.

Within the organometallic complexes, metal porphyrins have been identified as the preferential form of V and to a lesser extent of Ni, but almost nothing has been said about Fe (indigenous compounds). Iron content is typically lower than V and Ni in crude oils; however, it is higher in shale oils, even in cases where V and Ni are below detectability limits. This element is also present as nonindigenous compounds (iron oxides and sulfides) resulting from the corrosion of pipelines, heat exchangers, and other vessels. Regarding these transition metal compounds, the knowledge on their chemical nature is moderate for the porphyrins but very limited for the rest, which have been grouped simply as nonporphyrinic. Similarly, separation and isolation methodologies have not advanced enough and the understanding of the reactivity is very limited, precluding new ideas from emerging. Furthermore, the metals–asphaltenes interactions complicate the picture. At this time it is not clear whether metal compounds are trapped, shielded, or occluded by asphaltenes nor if they are a structural part of the asphaltene molecules.

Knowledge of the association between the most abundant metals and other heteroatoms within the organic matrix is very limited as well. Vanadium, for instance, is present as vanadyl in porphyrins, thus associated with an inorganic oxygen atom. The S-content and V-content of a large number of vacuum resid (VR) samples from different crude oils have been plotted in Fig. 8.1. This plot seems to indicate that S-content tends to increase with V-content. However, the rate of increase changes across the plot. The gaps created among the curves connecting the data points may be indicative of the existence of crude oil families. This plot raises more questions and challenges for the understanding of the molecular features of the heteroatomic compounds.

Light metals, alkaline, alkaline earth, and some of the heavy metals are mainly present in inorganic form as salt that could be removed during the desalting process, except in HACs, in which organic salts (mainly naphthenates) are found. The most abundant naphthenates are those of Ca and Fe.

The chemistry of these two groups of organometallic compounds present in OCs (porphyrins and naphthenates) is quite different and generates dissimilar problems that affects distinct units. One of these problems is caused by the poisoning effect of the

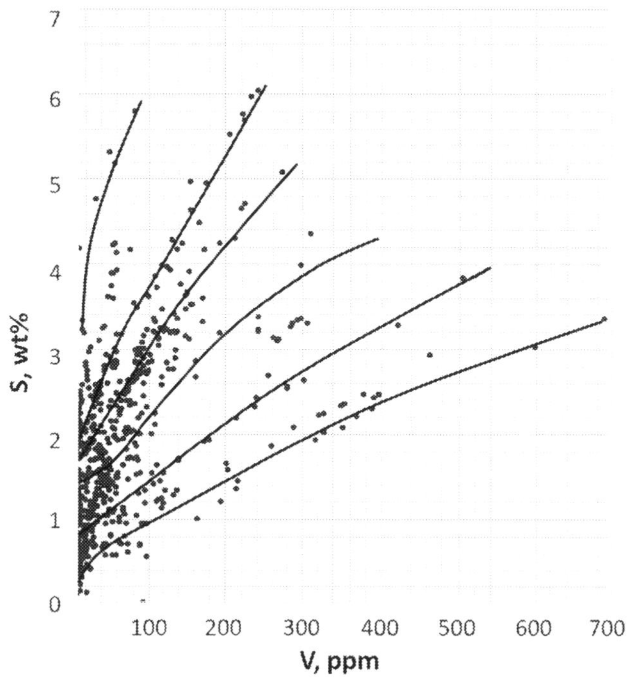

Figure 8.1
Content of S and V in vacuum resids (VRs) from different crude oils.

transition metals on catalysts and another by the surfactant properties of naphthenates. Consequently, transition metals will affect catalysts used for processing BotB feeds [vacuum gas oil (VGO) and VR], fluidized catalytic cracking (FCC), and hydrotreatment (HDT). On the other hand, naphthenates as surface active materials promote the formation of emulsions and foams, and are adsorbed and deposited on metal surfaces. The most affected units are the desalter and heat exchangers.

Iron and calcium are also the major metallic contaminants in shale oil, while Ni and V are typically low. The X-ray diffraction results of a Bakken shale oil sediment allowed the identification of the crystalline phases of anhydrite (Ca_2SO_4), magnetite (Fe_3O_4), and pyrrhotite (nonstoichiometric FeS). It appears that much of the iron in the Bakken crude comes from very small particles of iron oxide and pyrrhotite [13]. Meanwhile, calcium carbonates are the main forms for calcium (e.g., dolomite, calcite, etc.).

2.3 Naphthenic Acids

Naphthenic acids (NAs), cycloaliphatic acids, represent the major proportion of acid compounds in crude oil. A large proportion of acidity-bearing compounds is present in HACs. The MW distribution of NAs in HACs serves as a fingerprint for the identification of such crude oil. Some of the mineral elements present in the oil well react with NAs, leading to naphthenate formation, which will remain in the oil and impact the refining processes as indicated next.

Several methods have been defined for the analytical determination of acidity. Some acidity assessment methods have been standardized and others are proprietary methods. Historically, the value of the acidity resulting from these methods has been associated with NAs content and with the effects NAs have on refining processes. The discrepancies among the acidity values obtained by the different methodologies have raised controversial discussions. It is clear that the reasons behind the controversy are that besides NAs there are other acid compounds in crude oil and therefore not only NAs are responsible for the problems caused by acidity.

The increasing content of NAs in crude oils increases acidity to levels where processing becomes jeopardized. Besides acidity, NAs exhibit surface activity and acid—base type of reactivity. All these properties give rise to the problems these compounds cause in refining processing units. NAs acidity is responsible for part of the corrosion observed in refining units. The main product of naphthenic corrosion is Fe—naphthenate, which is oil soluble and may contaminate the refining products, but also may poison catalysts. NAs are also known to accelerate sulfidic corrosion by reacting and solubilizing Fe ions from the passivating layer. The unit most affected by corrosion is the crude (atmospheric) distillation unit (CDU), particularly in areas of high temperature. Additionally, these compounds negatively affect other refining units, causing poor desalter performance, fouling, catalyst poisoning, product degradation, and environmentally troubled discharges. The environmental concerns derived from NAs cascade into social problems, promoting and underpinning R&D activities with a broad scope. All the issues need to be addressed holistically.

The incorporation of shale oils to the diet not only increased the corrosion potential of NAs, but also gave rise to the accumulation of other highly corrosive species (e.g., amine salts).

Compared to the previous bad actors, NAs are more reactive toward various types of reactions. All these reactions constitute potential opportunities for the abatement of these compounds and a great deal of new ideas have been proposed.

2.4 Other Contaminants

Other indigenous and nonindigenous compounds can be found contaminating crude oils [14]. Organic heteroatomic compounds (besides metals and acids, already discussed) are present in crude oil. S- and N-compounds are particularly relevant in UOs, and more particularly for the BotB fractions in which they concentrate. The reactivity of these BotB compounds is lower than that of those present in lighter fractions. In Chapter 4, the reactions in which S-compounds and NAs synergize to accelerate corrosion were presented. A variety of additives for wax and asphaltene control are based on S-compounds and would have environmental impact. Besides the N-compounds originally present in a crude oil, also basic N-constituents may be added from blending cracked/thermally converted products into the crude system. Other N-compounds are used as additives for corrosion control as biocides or as H_2S scavengers. These compounds in combination with phenols (or other heterocycles) form color bodies and promote gum formation in the naphtha and kero fractions. Some of these compounds may have negative environmental impact.

The biological origin and the length of the genesis make shale oil composition intermediate between that of OCs and oils of biological origin (biomass derived). The absolute content, the relative proportion, and the nature of the present heteroatoms, metals, and heavy compounds also set differences among shale oils and conventional oils. One may think that these differences to conventional oils would make them similar to OCs. This is not the case; the oxygen content in shale oils is typically higher than it would be in conventional oils, while the most abundant metal is iron.

Characterization results of shale oils are scarce and show large variability [15—23]. In some instances, O-content is higher than that of N- or S-content, but in others N-content is the greatest while in others still S-content is the highest. The content variability of heteroatomic moieties in shale oil is the consequence of the several methods employed in its recovery. In general, S-content in pyrolysis-recovered shale oils is comparable to that present in conventional crude oils, but N-content is higher. The O-content typically low in crude oils is much more significant in shale oils. The high polarity of O-compounds may give rise to atypical incompatibility between crude oil and shale oil, which differentiates clearly from that observed with crude oils. During HDT, hydrodeoxygenation may take place with the concomitant formation of water that will shift specifications off the products. In low-S shale oils, the lubricity of the distillates will also be negatively affected.

Besides heteroatomic species, other specifications of the fuels will also be upset by the higher proportion of paraffins in shale oils, such as octane number in gasoline and pour point, cloud point, and cold properties in distillates.

Estonian shale oils have been studied the most. However, since these oils are extracted by retorting and sometimes at supercritical conditions, they end up clean enough with low levels of contamination. Shale oil produced by fracking is expected to be more contaminated and may require a pretreatment step prior to refining. A current challenge for R&D is the development of methods to improve the recovery of the organic phase from the source rock.

Organic chlorides are some of the nonindigenous compounds, which are artificially introduced into the crude. Solvents containing organic chlorides may be used at production sites as a paraffin solvent. Disposal of organic chlorides used as degreasing agents and by some reclaiming operations is another source of this contaminant. Chlorides end up mostly in the light fractions and would form hydrochloric acid by reaction with hydrogen. Thus during reformer feed (naphtha) pretreatment the presence of organic chloride contamination can cause extreme and rapid corrosion. The effects on pipelines and vessels may lead to leaks and explosions. Furthermore, scaling of ammonium chloride (during HDT of N-compounds ammonia is formed) and iron chloride causes plugging of pumps and exchangers. The presence of chlorides also affects the specifications of final products and it is particularly dangerous in aviation gasoline, where engine damage may occur.

Arsenic is a contaminant that may be present in crude oils in the ppb range, while its presence in shale oils can be higher. It will irreversibly poison the catalysts by chemically reacting with active sites, as, e.g., forming NiAs or CoAs.

Methanol (as well as glycols) may also be found in crude oils from deep water wells, where used seasonally, to prevent the formation of hydrates; this is more common in late summer and early fall (peak of the hurricane season in the Gulf of Mexico). Its miscibility with water will cause significant impact on the wastewater treatment systems. Methanol will remain in the aqueous phase in the desalter and could drastically upset the balance of the system leading to Environmental Protection Agency permit excursions, because of poor performance of the biotreating microorganisms on other hydrocarbons and toxins that may even be killed if toxicity increases beyond tolerable levels.

Other additives causing different problems are incorporated in the crude oil either at upstream facilities or in the refinery. These contain, for instance, phosphorus for well stimulation (causes fouling in refining); silicone used in defoamers introduced during oil transport and tertiary oil recovery, as well as in coker units (catalyst poison); surfactants used for emulsion breaking, dispersants, stabilizers, deposit control, etc. (environmental impact); and mineral acids used for well stimulation (causes corrosion and will have environmental impact) [14].

2.5 Summary

In Table 8.1, a summary of the actors, crude oil properties (associated with the presence of those actors), and their impact in refining is presented.

These areas of the refinery and every other area must be aware of the problems that these actors may cause. Some of the problems in a refinery unit propagate downstream and cascade into other problems. The quality of the feed has to be carefully assessed to anticipate any correcting action that may be needed. Furthermore, characterizing,

Table 8.1: Actors, Properties, and Impact in Refining

Actor or Property	Comment	Impact
API	UOs may have SG values closed to that of water.	Water–oil separation
Asphaltenes	Solubility/miscibility, stability, association-aggregation, surface activity, low reactivity	Compatibility, fouling, deposition, catalyst deactivation, unit performance upsets, product specifications
CCR	Coke formation	Catalyst deactivation and yields
Mineral elements	Na, Mg, Ca, As, Se, Hg, Cd, halides. Formation of naphthenates with alkali metals	Corrosion, fouling, scaling, catalyst deactivation, unit upsets, emulsion formation
NAs	Acidity, surface activity, high and broad reactivity	and stability, compliance of wastewater and environment, soaps, sludge, product specifications
N and S	High H_2 consumption	Corrosion
Transition metals	Ni/V/Fe catalyst poisons	Irreversible catalyst deactivation, shorten life cycle or run time
Viscosity	UOs may need viscosity reduction for pumping purposes	High transportation costs

API, American Petroleum Institute; *CCR*, carbon Conradson residue; *NAs*, naphthenic acids; *SG*, specific gravity; *UOs*, unconventional oils.

monitoring, and tracking of specific properties and contaminants are critical in modern refineries. These actors and their properties determine not only the criteria for selection of crude slates for existing facilities but also decisions on revamp or new capital investments.

3. Approaches

The increasing proportion of UOs in the refinery diet and the increasing proportion of the considered actors in the processing feed would eventually force the refiner to move from mitigating strategies to taking direct care of either removing the troublesome species or converting them into value-added products. The growth of the market for asphalt and fuel oil (FO) no longer parallel the increasing availability of OCs. Instead, the increasing demand for high-quality products imposes changes in refining approaches that pass through the proper conversion of the BotB actors and the appropriate consideration of the issues that these actors cause. The challenges faced for realizing the opportunity benefits that UOs represent are multiple [24–29] and have been visually presented in Fig. 8.2. Every single factor outlined in this figure is affected by processing UOs, and sooner or later the refiner would need to address them.

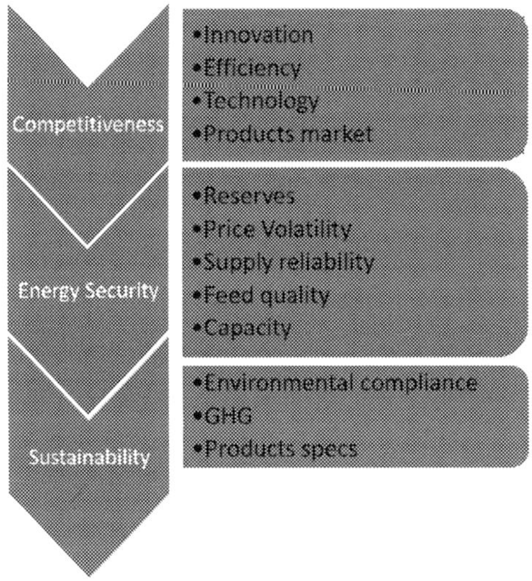

Figure 8.2
Refining challenges.

Historically, refining has grown by timely developed and adopted suitable technologies. However, for the UOs market volatility has not helped either for the needed sustained investment in technology development, or in the upgrading of refineries in both cases, e.g., improvements of existing units or implementing new process technologies. Efforts for the development of new technologies are isolated, small, and local, where instead they should be global and collaborative [28].

Volatility of the market moves conditions from favorable to unfavorable, in short periods of time. The key for competitiveness includes a large-scale, flexible, and complex configuration, operated with high efficiency index and incorporating innovation, not only of technology, but also in crude selection and procurement that can realize any emerging opportunity in the sector. Furthermore, any required measures for energy conservation and efficiency would be implemented to enable potential profits.

The market (demand, prices, etc.), quality, distribution, and yields of products directly determine refinery margins. Thus increasing the yield of high-value products from cheaper OCs would favor margins. Feedstock flexibility is advantageous to the refinery for selecting and purchasing feeds at cheaper prices [26].

The search for cost-effective technological solutions represents enormous challenges that could be only overcome by the contribution of multidisciplinary teams. Achievements can be reached more easily through a better understanding of the physics and chemistry of the troublesome compounds. Moreover, consolidating the accumulated fundamental knowledge on molecular characterization; reaction pathways, mechanisms, and kinetics; catalytic reactions and catalyst features; process thermodynamics; fluid phase behavior at process conditions; stability and reactivity of intermediate and final products; property–structure relationships, etc. is a titanic laborious task. However, this knowledge is out there and can be reached. Assembling the right "intellectually skillful" R&D team is the challenge, but most important is recognizing that the real opportunity offered by "opportunity crudes" is still for R&D.

3.1 Refining Diet

The constant changes in crude slate entering the refinery are the result of worldwide geopolitical uncertainties, of the turns in demand from pushes to deceleration, and in supply from pulls to oversupply. This volatility is perhaps the greatest challenge for refining sustainability. In this turbulent scenario, only refineries with installed capacity for flexible feedstock uptake could benefit the most.

Heavy crude oils and bitumen are commercially available in different trading blends. These blends are needed for transportation purposes from the field to the market. The types of blends are graphically defined in Fig. 8.3.

A synthetic crude oil (SCO) results from the blend of the liquid product of the primary upgrader (coker). It is a blend of naphtha, distillate, and gas oil range of materials, with significantly reduced residue or no residue at all. The proportion of naphtha is lower than that of diesel and gas oil compared to conventional crude oils. The S-content is lower than that of the original crude oil or bitumen, but it is relatively high in aromatics. Therefore the cetane number of the diesel and the smoke point of the jet fuels can be far beyond specifications

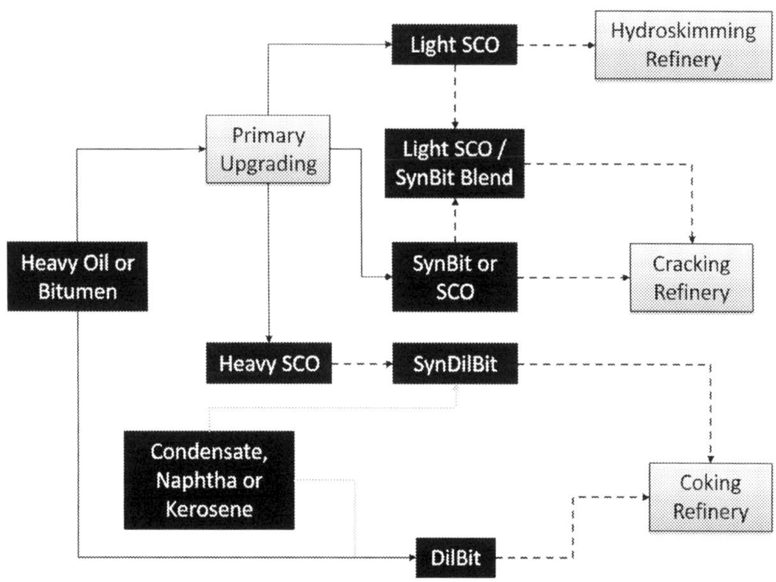

Figure 8.3
Definitions of commercially available unconventional oil blends. *SCO*, synthetic crude oil.

[30]. In Canada, some companies hydrotreat the SCO and produced a sweet light SCO, e.g., Suncor Oil Sands Blend A, Syncrude Sweet Blend, Husky Sweet Blend, Premium Albian Synthetic, etc. Even though these blends contain lower S and N, the retained species are more refractory compounds that will impact S- and N-content in the final gasoline and distillates. Sour blends of SCO containing different proportions of straight run heavy gas oils and hydrotreated diluent are also available, e.g., Suncor's trade blends Suncor Oil Sand-V, Suncor Oil Sand-H, or Suncor Custom Cracked, Albian Heavy Synthetic, etc.

Raw bitumen can also be diluted (dilbit) for transportation purposes to reach the market from the field, e.g., Cold Lake Blend, Bow River, Lloyd trade blends (Lloydminster Kerrobert; Lloydminster Hardisty; and Western Canadian Blend), etc. It is nominally composed of 30% diluent and 70% bitumen. The type of diluent (condensate, naphtha, kerosene, etc.) varies among regions and oil source, but it is neither cheap nor plentiful. The density of the diluent varies in the range of 650–750 kg/m^3 [31].

Another grade is that in which the diluent is a light SCO, thus this grade becomes a synbit, e.g., Christina Lake Blend, MacKay Heavy, etc. In this case, the SCO density would be in the range of 840–870 kg/m^3, and nominally the blend would contain 50% SCO. In theory, synbit would represent a higher refinery value than dilbit (improved yield value). There are many more marketed dilbit blends than synbit blends, because of the higher cost of SCO compared to lighter diluents [31].

Finally, syndilbits are blends of diluent, hydrotreated light SCO, and bitumen, e.g., Wabasca Heavy, Western Canadian Select, etc. An exemplification of quality[3] and composition for the different types of commercial blends is shown in Fig. 8.4, where the features described earlier can be graphically observed.

Fracking technology has enabled the production of shale oil beyond expectations, contributing significantly to US energy independence and balance of its export/import volumes. Globally, it has also added new volumes to the underinventory, rebalancing the supply/demand equilibrium and lowering the price. In turn, low oil prices do not favor the high operating costs of fracking and are preventing a massive exploitation.

Several refineries have incorporated shale oils to their diet and certain problems have appeared in their way. In fact, the value chain of the industry has been affected to different extents along the various segments. The most affected segments have been transportation and refining. The reported troubles include:

- Incompatibility
- Fouling and deposition

[3] Quality data for crude oils and blends can be found at http://www.crudemonitor.ca/home.php. Additional tools are also available for blending, comparison, etc.

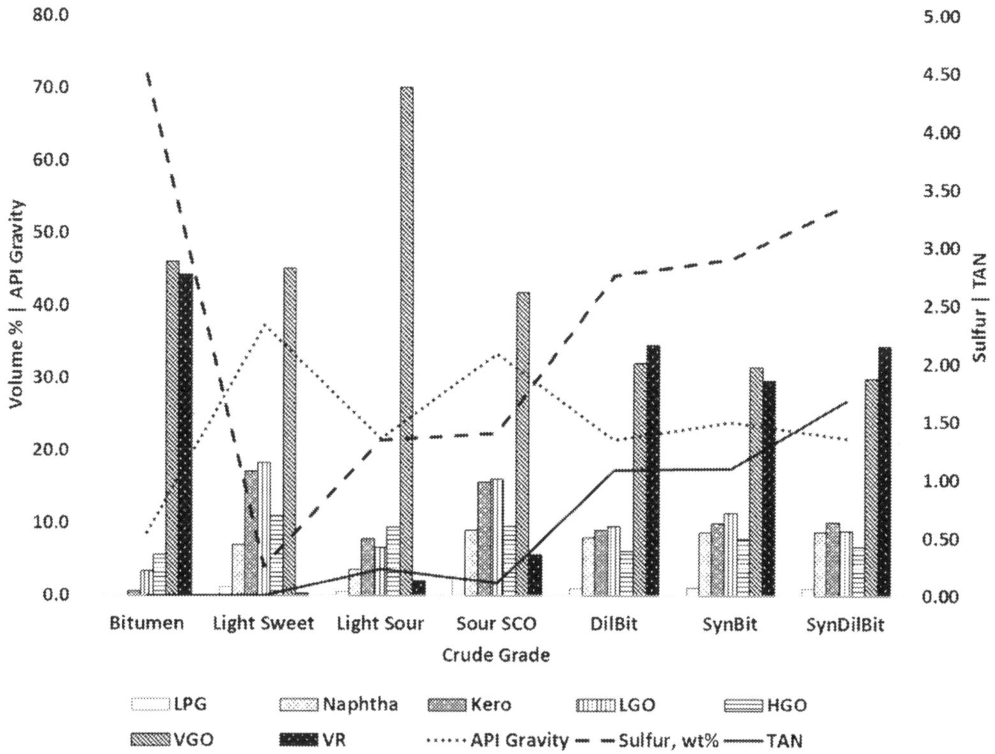

Figure 8.4
Exemplification of quality: API gravity, S-content (wt%), total acid number (TAN) (mgKOH/g), and composition: liquid petroleum gas (LPG), naphtha, kero, light gas oil (LGO), heavy gas oil (HGO), vacuum gas oil (VGO), and vacuum resid (VR) (volume % distilled fraction) of commercial grades of bitumen-derived crude oils. *SCO*, synthetic crude oil.

- Corrosion and accumulation of corrosive salts
- Pipelines plugging by scaling and wax deposition
- H_2S presence
- Water contamination of distillates (water and sediments off-specifications of fuels)
- Pour point and cloud point off-specifications of distillates

A typical assay of Bakken crude shows this crude as light and sweet with an API gravity of 42 degree and S-content of 0.19 wt%. A Bakken sample could be fractionated mainly into a gasoline cut and a light cycle oil (LCO) cut. The gasoline fraction was highly paraffinic, with low octane number (RON of 61 and MON of 58). Meanwhile, the LCO fraction had an aniline point of 69°C and an API gravity of 37.6 degree, resulting in a diesel index of 59. The crude also contained a heavy end [32]. While a jet that meets specifications could be obtained from the hydrotreated shale oil-derived fraction, the resulting diesel will have a low cetane number, high pour point, and would be unstable for storing (high water and sediments) [33].

Mitigating solutions have been practiced, including dilution and additives. A larger incorporation of shale oils in the refinery crude slate would impose additional operating costs and negatively affect the margin. Anticipation is the best option, therefore it is recommended to include [34]:

- Technical assessments, both in the lab and the field, to evaluate the impact of the shale oil on transportation, storage, and processing;
- Analytical tests to identify crude blends that would be affected by asphaltene precipitation, or any other incompatibility range;
- Tools to characterize feedstocks and analyze deposits;
- Risk assessment models, advanced monitoring technology, and analytical tools to predict and eliminate fouling and corrosion threats;
- Evaluations to determine the most appropriate treatments and processing strategy.

Nevertheless, fundamental solutions are needed and R&D has to step in immediately before massive exploitation returns.

A KBR evaluation study from 2005 anticipated two main aspects, which would most likely affect operations and profitability in refining. The first factor was the shift toward heavier crude oils because of higher crude prices and differentials. The second aspect concerned the effect on product slate from that shift on the diet that would require a reduction in the FO and coke yield by converting them to more profitable products such as transportation fuels [35]. Another point to consider regards the crude selection and planning required ahead of any intended purchase. Thus the screening of UOs should include not only the gathering of quality data but also the identification of the impact on oil movements, desalting/dewatering units, catalyst consumption, product specifications, wastewater treatment plant, corrosion, fouling, tankage, etc. Furthermore, prior to purchasing, the mitigation strategy should be developed and implemented long before the crude oil is received in the refinery. Then, monitoring and controlling measures should be executed during the running period [36].

One of the factors considered during crude oil purchases is the discount offered by traders based on oil quality [API, S-content, and more recently total acid number (TAN)]. We have discussed how the lower the quality, the larger the discount gets, but also the greater the impact on refinery margins may become if technical considerations were not in place. Furthermore, those properties that set the discount are not enough and are far from accurate when making refining business decisions.

Feedstock valuation, planning, and scheduling are concurrent activities in any modern refinery with the objectives of optimizing crude blends in terms of maximizing the use of existing assets and profits. The products that can be produced from a given crude within a refinery configuration determine its value. Thus not only will crude oil composition affect its value but also the configuration of the refinery where it will be processed. Several methodologies exist for measuring and generating data, but the challenge is converting these raw data into useful information, for which complex simulation models have to be utilized for optimizing profitability [37]. Decisions that were previously economically biased currently have a stronger technical component. The balance between economics and reliability, environmental and safe operating targets would constitute the "fit for purpose" criterion [38]. In summary, for UOs the incorporation of technical evaluation criteria in the decision-making process was highly encouraged [39].

The refinery planning systems provide coordination of crude oil purchases over a period of several months. Planning precedes scheduling and uses forecasted product demands. Then, crudes are purchased based on the monthly refinery plan. Meanwhile, the scheduling systems [40,41] plan the activities over days to weeks and correct or adjust for deviations from the forecast, accounting for changes in demand or plant capabilities or limitations. Incorporation of UOs to the diet has complicated the job of the planner and scheduler.

Planning and scheduling decisions require a full crude assay to optimize the diet for a given refinery configuration that would maximize refinery margins or otherwise minimize processing costs. This laboratory-based process is a costly and time-consuming process that also involves simulation/linear programming (LP) and refining economy tools. Simpler and faster tools based on spreadsheet calculations have been proposed. Although facilitating prescreening of crude oils under different scenarios, the rigorous methods that reveal the true value of crude oils for the refiner will not be replaced. An example is given in Ref. [42], a simple tool considering the differential in price and operating costs between the candidate crude and Brent. The operating costs differential was based mainly on hydroprocessing (HDP) (hydrogen consumption and management) and evacuation of residuals.

Upon crude arrivals and based on crude assay results, crude schedulers need to assign destination tank/s for each crude, blend crudes as needed to meet the targets for yields and qualities of the fractionated products from the CDU, and determine the charging rate to each crude unit. The scheduler relies on different algorithms to sort out such complicated tasks; continuous development of algorithms of LP, nonlinear programming, mixed-integer LP, etc. are in progress (see, for instance, Refs. [43,44]).

Furthermore, the incorporation of UOs in the crude slate also require modifications in the crude management strategies. Two relevant practices are typically carried out: crude oil pretreatment in the tank farm and an appropriate slop handling. The former involves an improved solids control, asphaltenes stabilization, reductions in water draw in tank, and oil undercarry; while the latter includes segregation, injection practice, and treatment. Additional pretreatments may also include new techniques for dehydration and solids removal [45].

3.2 Refinery Units

The previous section anticipated the problems caused by the presence of bad actors and of other contaminating elements, and by the UOs' properties directly associated with their presence. In this section, we briefly identify the location of these problems within each of the affected refining units.

3.2.1 Desalting

During desalting, the oil is washed with water and let to settle in a tank or decanter. The separation of oil and water is based on their relative density differences; the similarity in density between heavy oils and water makes this separation problematic. Besides the limitations introduced by the similarity of this physical property, other issues are brought about by the presence of natural indigenous surface-active compounds and acidity-bearing compounds, leading to incompatibility, emulsions, corrosion, etc. [46–48]. A more detail discussion on the surface activity of asphaltenes, NAs, and naphthenates were given, respectively, in Chapters 2 and 4, including the problems caused by the formation of emulsions and films in the desalter.

The formation of emulsions and the large presence of solids by, for instance, Western Canadian feedstocks put a high demand on refinery crude oil desalting capabilities. To optimize these units for the processing of these crudes, a combination of chemical additives, including oil-soluble emulsion-breaking chemicals, asphaltene stabilizers, solids-wetting agents, and polymers is required [49–54]. Thus emulsion formation turns into oil undercarry (wastewater treatment plant issue) and water carryover. Formation or presence of naphthenates promote stability of the resulting emulsions and are carried over into the desalted crude, causing additional trouble downstream the desalter. The increase in the rag layer would potentially cause high oil entrainment in the brine water and create difficulties in maintaining the desired basic sediment, water, and salt removal at the desalter outlet. Furthermore, difficulties in the detection and monitoring the emulsion/rag layer often result in relatively higher injection rates of chemical additives [55].

In normal operations, overhead water is recycled to optimize use of the wastewater treatment facilities. However, with HACs, if overhead water is recycled, organic acids will accumulate in the wash water and potentially repartition into the crude oil. Although lower pH generally destabilizes emulsions and higher pH results in saponification of acids, the low pHs may drive organic acids into the oil phase [56].

The presence of large concentrations of Ca in shale oils creates a potential synergy with HACs. The problem associated with naphthenic acids and (Ca) naphthenates would be aggravated if HACs and shale oils are combined in the refinery diet. Other mineral contaminants, like the siliceous minerals present in shale oil, would remain solid under desalting conditions. In the presence of asphaltenes, these minerals would adsorb them and concentrate in the rag layer, upsetting the desalting performance.

The desalter upsets leading to contamination of the effluent with oil undercarry and with an increased proportion of oil-coated solids generate a series of problems downstream, including: increased corrosion, increased fouling and scaling on the heat exchanger network, downstream units operability (particularly those units dealing with the BotB fractions, i.e., VGO, VR), and increased energy consumption.

3.2.2 The CDU and the Vacuum Distillation Unit (VDU)

The incorporation of HACs in the crude slate increases corrosion in the crude units. Devices for the control and monitoring of NAs corrosion (NAC) need to be installed. Some Canadian heavy vacuum gas oils (HVGOs) exhibit a TAN of 5–6 mgKOH/g. This level of acidity certainly required appropriate metallurgy such as 317L stainless steel or better alloys with ~3% molybdenum, in areas of high corrosion sensitivity [21]. If the HAC incorporation is a long-term situation, then capital expenditures in metallurgical upgrades should be considered.

Similarly, the CDU overhead condensing system would also be subjected to NAC and corrective actions should also be implemented. In both cases, appropriate modeling and chemical program technologies can mitigate increased corrosion risk.

The lowering of the API gravity of the refinery crude slate by 5–15 degree brings a set of common problems that were described earlier. The design of the VDU is critical in defining volume yields, product yields and quality, and run length when processing heavy oils and bitumen blends without considerable losses [57]. Bitumen-derived crude oils and heavy oils are known to produce cracked gas at more than double, and sometimes triple, the rate of common conventional crudes, which requires improved vacuum systems with either new ejector condenser systems or new liquid ring vacuum pumps. In tanks and pretreatment units, incompatibility will cause storage tank sludging or crude preheat exchanger fouling. For the latter preheater, a double-fired heater design was recommended [55]. High-solid content and unstable asphaltenes will also lead to fouling in the furnaces and heat exchangers [49]. In a coking refinery processing commercial grades of heavy oil or bitumen blends, the high vacuum column operating pressure is set to reduce overall liquid volume yields, and if the design of the coker fractionator was based on the processing of conventional heavy feedstocks, then the life cycle would be as short as 6 months [58]. Major modifications to the crude units and their ancillary equipment are needed [59].

The associative behavior and interactions of NAs, asphaltenes, naphthenates, and among all these species lead to scaling and polymeric fouling [60]. Typically, fouling control technologies are employed as mitigation options. Fouling in refineries has been reviewed by Deshannavar [61].

Although in theory a higher vacuum column cutpoint appears to be advantageous, it may not be the most economic option in the long run, even if technically feasible. A higher cutpoint will minimize the yield of the lower value VR, but will maximize the yield of higher-contaminated VGO and HVGO feeds, for the FCC, and the hydrocracker (HCK) units, respectively. These feeds need to be hydrotreated prior to FCC and HCK causing the problems described in the next subsections.

3.2.3 Fluidized Catalytic Cracking

The FCC process converts heavier molecules into lighter ones by using an acid zeolite catalyst. The feedstock is a low-N, low-CCR VGO. Basic nitrogen compounds and the high content of coke precursors lead to a poor performance caused by catalyst deactivation. Furthermore, any excess heat evolution during the regeneration cycle has to be balanced by the use of a catalyst cooling system.

Other compounds present in UOs have detrimental effects in FCC. The incorporation of OCs in the diet impose the need for feed pretreatment. A VGO hydrotreater is required; not all refineries have one in their configuration. Asphaltene, resin, and metal constituents are responsible for the deposition of N-species, metal poisons, and coke on the catalyst, in FCC as well as in HDP. Metal poisons and some N-compounds ultimately cause the irreversible deactivation of the catalysts. Consequently, an increase in catalysts consumption for both units negatively affects the economics of the processes. Although metal passivators mitigate the action of the deposited metals on the zeolite FCC catalyst, the presence and exposure of Ni to the reacting hydrocarbons may induce dehydrogenation reactions with the concomitant formation of additional coke.

The loss in conversion activity affects FCC yields, while the change in quality of the VGO affects product quality. In comparison to the yields obtained with Brent, the incorporation of a proportion of a light sweet SCO will decrease liquid petroleum gas and gasoline yield, while increasing LCO and clarified slurry oil. These changes are even more pronounced if instead the feed is composed by VGO derived from a sour SCO, in addition to Brent VGO [62].

The deposits of unreacted asphaltenes and coke can be considered reversible deactivation in FCC since regeneration will burn these species in the regenerator, restoring activity and providing the heat of reaction. Operation of the FCC unit involves the careful management of the heat balance. Hence an excessive carbonaceous deposition would create stress to the unit. Then, coke precursors such as asphaltenes and Ni accumulation and exposure threaten operability by disturbing the heat balance.

While shale oils contain low amounts of FCC feed, the feed for isomerization and reforming can be much larger than that present in conventional oils. Bakken crude, for instance, contains less than 28% of the 360°C$^+$ fraction. While in Arabian light the proportion between the 360°C$^+$ fraction and the 360°C$^-$ fraction is 1:1, in Bakken shale oil it can be 1:2 [32]. Furthermore, the presence of Fe and Ca have negative effects on catalyst performance. Finely dispersed Fe particles in the feed (either as organic compounds or as colloidal inorganic particles) can deposit on the catalyst surface, reducing its effectiveness [63]. In FCC, the combination of Fe, Si, Ca, Na, and other contaminants in the deposits on the catalyst surface could lead to the formation of phases with low melting temperature. In the regenerator, where the highest temperatures are achieved, the pore structure collapses irreversibly. Back into the reactor (riser), the catalyst would exhibit limited access of feed molecules and conversion will be reduced [63].

Fe, in combination with Ca and/or Na, has a greater negative effect on catalyst performance than does Fe alone. The symptoms of Fe and Ca poisoning include a loss of bottoms cracking as these feed metal moieties block the pore structure of the catalyst particles preventing reacting molecules from accessing the active sites [64]. New catalyst formulations have been proposed with optimized matrix and mesoporosity to provide resistance to Fe and Ca poisoning [65].

3.2.4 HDP

The units' capacity most affected by a decrease in API of the crude slate in the refinery are the hydrotreaters. An increase in the volume of feed streams and intermediate products that need HDT is caused by the higher S- and N-content typically present in heavy oils and bitumen-derived crude grades. Larger feed volumes and higher contamination require larger volumes of reactor for the incremental volume of catalyst needed.

The increase in S and N will also increase the formation of ammonium bisulfide that will lead to increased demand for wash water to minimize corrosion potential of the sour water stream. Additionally, the potential for higher amine loading on the recycle gas scrubber caused by the increased H_2S yield, if the amine rate remains unchanged, has to be assessed [55].

Sour heavy crude oils will cause sulfidic corrosion. The combination of NAs and S in sour-acid heavy crude oils will synergistically accelerate corrosion. In HDP, the deposition of corrosion products is a type of fouling on fixed bed reactors. In the absence of corrosion mitigating actions, and of any other corrosion control aid, the deposition of metal ores formed upstream of the unit in question would increase. The tendency of deposition at the top of the bed leads to pressure drops that may shorten the life cycle of the unit run, leading to unexpected shutdowns. Part of the catalyst bed would need to be removed and fresh (or regenerated) catalyst loaded. Typically, guard beds are used to prevent pressure drops and increase time of the run (e.g., [66–68]).

Metal compounds as mentioned earlier are responsible for the irreversible deactivation of HDP catalyst. The hydrodemetalization reactions may occur at micropores' mouth because of the large volume of the metal-bearing molecules. Metals will then accumulate in agglomerated structures, with time on stream, restricting accessibility to other reacting molecules. Hydrogenation activity would suffer the most, increasing coke formation and building up pressure drops. Furthermore, besides the physical effect of fouling, interaction with the active phase leads to the formation of mixed sulfides and sintering, causing deleterious poisoning.

Consequently, the incremental metal accumulation with time on stream accelerates coke deposition, leading to a significant shortening of the run cycle when processing OCs' derived fractions [69]. While in FCC, metal passivators mitigate the action of the metals, in HDP the life cycle reduction or bed plugins would require more frequent shutdown for unloading spent catalysts and loading new ones. For all these reasons, moving bed HDP are preferred to mitigate deactivation and extend the life cycle.

HDP may be a better option for the processing of shale oil, in comparison with FCC. Most hydrotreaters would have a guard bed that may be enough to protect the catalytic bed from some of the contaminant inorganics present in the oil. Oxygen, nitrogen, and sulfur will be removed, while olefins will be hydrogenated [33]. However, in anticipation of trouble, a complete characterization, beyond crude assay, of the considered oil should be carried out prior to attempting its refining.

3.2.5 Visbreaking and Coking

High-temperature corrosion of NAs affects not only CDU and VDU, but also the coker and visbreaker if installed in the refinery [70]. The decomposition of NAs forming light carboxylic acids during visbreaking may explain the corrosion observed at the top trays of the fractionation column of the visbreaker unit at ~200°C [71]. While in the drums of the coker unit, coke forms a protective layer, other components of the unit can be subjected to high-temperature corrosion (above 475°C). Examples are the furnace, soaking drums, lower part of the tower, high-temperature exchangers, and fractionation tower and its side pipes. Metallurgy revision is required when TAN is thought to increase in the feed.

Coke formation cannot be tolerated in visbreaking. In fact, it is coking that limits feedstock conversion, rather than FO specifications. Asphaltenes as coke precursors enforce a careful control of operating variables [49]. For this reason, visbreaking is considered an obsolete technology for the processing of heavy oils and bitumen-derived blends. Therefore revamping of the units to hydrovisbreaking or substitution for a coker has been recommended [72].

3.2.6 Wastewater Treatment

During desalting of HACs, naphthenate salts that are much less water soluble are therefore more difficult to remove in the desalter. Consequently, HACs require a larger amount of water, a better control of the pH and conductivity, a larger use of chemical additives, and an increased amount of caustic soda. Then, a larger volume of wastewater could be produced, containing a major concentration of contaminants and environmental pollutants. Asphaltenes, NAs, and naphthenates not only stabilize the water and solids in the crude oil continuous phase of the rag layer resulting in a complex fluid rag layer, they also concentrate in this layer and are carried over into the wastewater [73,74].

Treatment chemicals and application methods are needed to minimize the impact of high solids content of the desalter effluent water on wastewater treatment plants. Reducing wastewater plant upsets and improving performance are key factors in this regime.

3.2.7 Other Units and Equipment

Foaming is one of the problems associated with the presence of NAs. The plants devoted to gas–oil separation may experience formation of foams, which causes oil losses in the separated gas stream, revenue losses, and possible damage to compressors [75].

Asphaltenes' tendency to agglomerate and solubility characteristics provoke fouling, deposits, and precipitation in heat exchangers and tanks. These types of equipment are also subject to mineral deposits and scaling when processing OCs.

Incorporation of shale oils in the crude slate could impact the tank farm through the buildup of waxes and solids, the presence of H_2S, and the potential incompatibility with other crudes. The latter requires continuous monitoring, while the others could be mitigated by the use of additives (dispersants, scavengers, etc.). The heat exchangers, both the cold and the hot trains, would also be affected by fouling.

3.3 Configurations

The configuration of a given refinery determines the type of crudes in the diet that can be processed and the type, yield, and quality of the product slate that can be made. Meanwhile, the number and type of installed processes and the level of integration among them defines the complexity of such configuration. Typically, refinery complexity is reflected by the Nelson Complexity Index, which is a cost-based index that provides insight into refinery construction costs, replacement value, conversion capability, and product slate [76]. The simplest configuration is a refinery with just a CDU, which could only process very high-quality crude oils. Then, the hydroskimming refinery will additionally include naphtha and/or diesel upgrading. The incorporation of conversion units, such as the inclusion of an FCC unit increases complexity from hydroskimming to the level of a cracking refinery. Finally, a coking refinery with an installed coker will have higher complexity and will be capable of converting the least reactive BotB fraction of crude oils. For a given crude oil a more complex refinery will make more value-added products than a less complex one. Therefore increasing complexity will increase the value of the crude oil since a complex refinery could render the same product slate than a lower complex one from a crude oil of lower quality (lower price).

3.3.1 Description

At this point, after having described the effect of the relevant compounds present in UOs, the changes in the refinery diet, and their impact in the refining units, it is clear that changes in the configuration are required for incorporating UOs reliably, safely, and economically. More than five decades ago, the search and evaluation of possible configurations for the processing of UOs started. The adaptation of existing technologies and the identification of required incremental improvements was driven by the market changes caused by the growing OCs supply and the decline of conventional oils [77,78].

The commercially available technologies were described in detail in Chapter 5. These processes are based on either C-rejection or H-addition to modify the low H/C ratio of OCs to the fuels' level. The C-rejection technologies are usually better suited to upgrading relatively heavy, high metal residues rather than H-addition technologies that generally require a low CCR and low metals feed. Respectively, feed containing more than about 150−200 ppm of metals and CCR above ~12−15% are better processed via C-rejection; above ~20 ppm metals and CCR of ~7% HDP can be applied, below these levels FCC can be applied [79], as indicated in the example shown in Fig. 8.5.

C-rejection is also economically preferred, particularly coking as can be deduced from the curve plotted in Fig. 8.6. The comparison indicates that moving from visbreaking to solvent deasphalting (SDA) and then to coking, the relative increase in CapEx is always smaller than the relative increase in margin. More comparative details were given in Chapter 5; nevertheless it is important to mention that the observed trend is derived from the increase in lights yields, product quality, and decrease in low-quality FO yield [80]. Combinations of coking with either SDA or visbreaking have also been proposed to decrease the load to the coker and produced intermediate streams to blend in the feed for FCC or HCK [81].

However, the quality and yields of the higher value-added products cannot satisfy market specifications and additional H-addition processes are needed not only to meet specifications but also to stabilize the products. Configurations using C-rejection (predominantly coking)−HDP combinations [82] are currently employed, respectively, as primary and secondary upgrading preferred technologies for the production of blends of light sweet SCO [83]. The conversion of the coker unit is accordingly adjusted to maximize the yield of value-added products as well as to minimize the hydrogen requirements [84]. One of the proposed combinations also suggested the inclusion of a

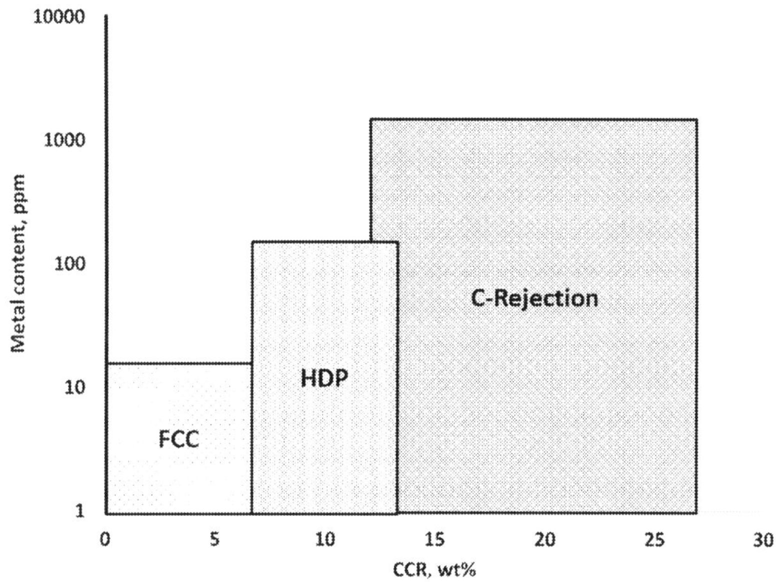

Figure 8.5
Example of choice of technology type in terms on feed quality parameters. *CCR*, carbon Conradson residue; *FCC*, fluidized catalytic cracking; *HDP*, hydroprocessing.

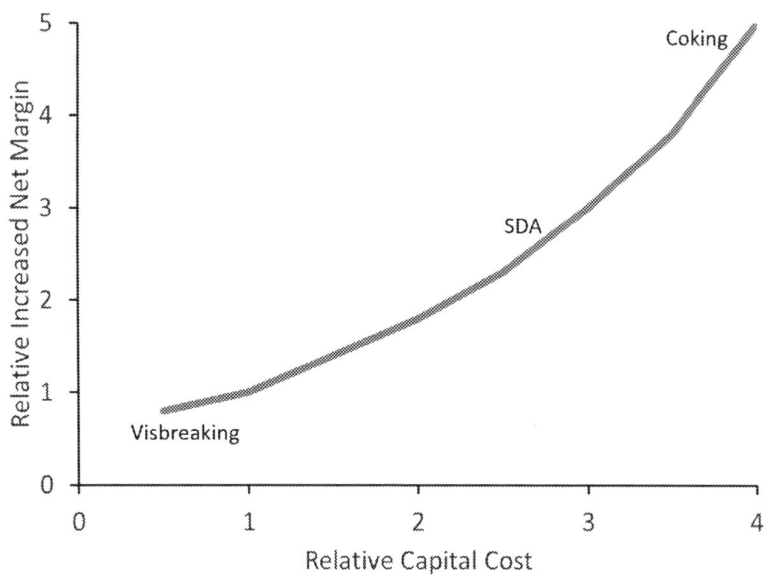

Figure 8.6
Relative margins and capital cost for C-rejection technologies. *SDA*, solvent deasphalting.

gasifier (if economically feasible) [85] for conversion of residuals and production of syngas that may be used for rendering the additional needed hydrogen [86,87], the synthesis of chemicals, or power generation [88].

Technology selection and costs still pose challenges to the refiner. Nevertheless and regardless of the economic feasibility of these configurations, none of them would really take advantage of the discounted price opportunity, nor get the best value from the crude oil. Not long ago, Runyan remarked that refineries would need to invest in finding new methods to convert the BotB into market-demanded lighter products to minimize unwanted residues and meet the specifications of the wanted products [89].

3.3.2 Evolution

The history of refining shows that its evolution was initially driven by the market, particularly the demand for transportation fuels. Later, refinery growth responded to economic drivers while later still refining technologies advances had to respond to the pressure from environmental regulations (Table 8.2). It seems that a new driver is already in place: the drastic shift in crude quality. Therefore this industry has never stopped changing; it has been evolving according to the external pressures imposed.

In this regard, market, economy, and environmental pressures all point toward an increasing need for larger volumes of higher-quality products, under more stringent environmental regulations in refinery emissions and requirements for more energy efficient operations. In a scenario of poorer-quality crude slate, refining costs can only increase. The increases in CapEx have already been mentioned; suffice to say that environmental compliance is another source of cost upsurges.

The number of operating refineries worldwide has changed from 674 in 2004 [90] to 662 in 2009 [91] and to 643 in 2014 [92], while the processing capacity (in mbpd) has moved from 82.4 to 88.3 and to 88.0, respectively. Technip reported in 2014 that only 30 grassroots refineries had been built since 1958, which had added 400 kbpd capacity [93]. Therefore these changes evidence the execution of revamping projects, addition of units, and/or changes in configurations. The share of the United States in these changes was high, since the number of operating refineries had dropped from 324 in 1981 to 153 by 2002 [94], 132 by the end of 2004 [90], 129 by 2009 [91], and 123 by the end of 2014 [92]. In the United States, the last major refinery was built in 1976. However, expansion projects have been executed to offset the effect of the closures since then [95]. The circumstances are no different in Western Europe [94].

The existing refining capacity per geographical region and for four grades of crude oils, namely, light sweet, light sour, medium, and heavy, corresponding to increasing level of refinery complexity is shown in Fig. 8.7 (data have been taken from Ref. [91]). As can be seen, it is North America that is better prepared for processing heavy oils.

Table 8.2: Evolution and Growth of Refining

Historical Event	Year	Process Name	Purpose	Products, By-Products, etc.
Oil well drilling	1850			
Kero lamp	1862	Atmospheric distillation	Produce kerosene	Naphtha, tar, etc.
	1870	Vacuum distillation	Lubricants (original)	Asphalt, residual coker feedstocks
Light bulb and car	1905			
WW I	1913	Thermal cracking	Increase gasoline	Residual, bunker fuel
Car massification	1916	Sweetening	Reduce sulfur and odor	Sulfur
	1930	Cracking feedstocks	Produce fuels blendstocks	Gasoline, kero, diesel
		Thermal reforming	Improve octane number	Residual
	1932	Hydrogenation	Remove sulfur	Sulfur
	1932	Coking	Produce gasoline basestocks	Coke
	1933	Solvent extraction	Improve lubricant viscosity index	Aromatics
	1935	Solvent dewaxing	Improve pour point	Waxes
	1935	Catalytic polymerization	Improve gasoline yield and octane	Petrochemical feedstocks
	1937	Catalytic cracking	Higher octane gasoline	Petrochemical feedstocks
WW II	1939	Visbreaking	Reduce viscosity	Increased distillate, tar
	1940	Alkylation	Increase gasoline octane and yield	High-octane aviation gasoline
	1940	Isomerization	Produce alkylation feedstock	Naphtha
Car globalization	1942	Fluidized catalytic cracking	Increase gasoline yield and octane	Petrochemical feedstocks
	1950	Deasphalting	Increase cracking feedstock	Asphalt
	1952	Catalytic reforming	Convert low-quality naphtha	Aromatics
Clean Air Act	1954	Hydrodesulfurization	Remove sulfur	Sulfur
	1956	Inhibitor sweetening	Remove mercaptan	Disulfides
	1957	Catalytic isomerization	Molecules with high octane	Gasoline blend, alkylation feed
Baby boomers	1960	Hydrocracking	Improve quality and reduce sulfur	Fuels blend, alkylation feed
Oil embargo	1974	Catalytic dewaxing	Improve pour point	Wax
	1975	Residue hydrocracking	Increase gasoline yield	Heavy residuals
Environmental care	1990	Etherification	Lead ban, octane booster	Reformulated gasoline

During this time, an increasing trend of HACs in the refinery diet was faced by the industry. An extension of the economic life of some existing refineries may be possible if advantage is taken of the potential increase in margin when processing this type of OC. More recently, the shale oil boom in the United States led to a greater incorporation of this

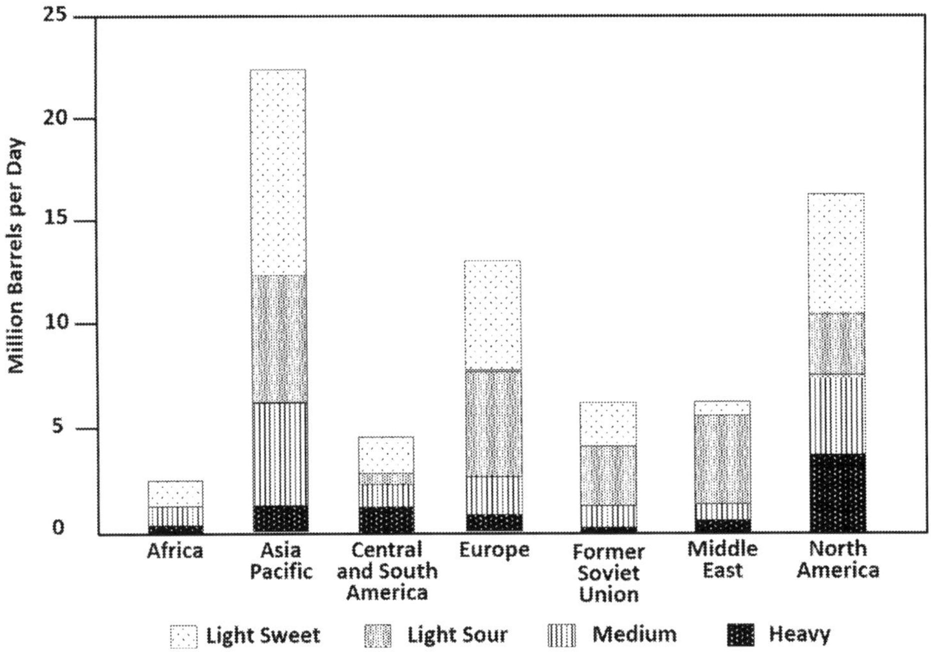

Figure 8.7

Installed refinery capacity per geographical region and for four crude grades.

type of crude in the refinery diet. As mentioned earlier, a trend for lower oil prices was triggered as well. Compared to upstream operations, refining operations and margins have benefited significantly more from this recent fall in oil prices. Regrettably, refining R&D activities is still in a declining mode, even though actors' abatement remains technically and economically unsolved.

3.3.3 Drivers and Options

In this status quo, refiners are often constrained to convert a given crude slate to a specific product set, with no consideration of the required changes at molecular level and the penalty cost of such conversion. In doing so, refiners may be expending substantial capital while ignoring economic synergistic opportunities that could be found if a deeper look at crude slate is taken closer to the molecules. These ideas constitute the principles of molecular management, which tries to capture the best value of every molecule naturally occurring in crude oils. A suggestion for such considerations when defining refinery configurations has been reported [96]. The molecular view, for instance, of the VR fraction from a heavy OC would provide a better refining approach than the mere BP and specific gravity. In this case, a combination of C-rejection (SDA and coking) and H-addition would significantly increase the higher value-added products and minimize the low value and lower demanded FO, asphalt, and coke. In this configuration, the deasphalted oil (DAO) will be hydrocracked while the bottoms are sent to a coker [96].

The capital expenditures needed for the processing of OCs can only be justified if enough economic benefits could be achieved. The increase in conversion capacity from a simple hydroskimming refinery to a cracking refinery and further to a coking refinery (refer to Section 3.3 for a detailed explanation of complexity increases with conversion) will broad the range of diet quality grades that could be processed. The benefits should be reflected in the margins. Nevertheless, one would need to be aware that margins depend not only on refinery configuration, but also on location, oil and products market, oil prices, product prices, and so many other factors. Fig. 8.8 tries to illustrate these features with data reported by the International Energy Agency [97]. Notice the higher margins for processing Brent in a cracking refinery versus a hydroskimming refinery in Northwest Europe. Meanwhile, the difference in margins was insignificant when comparing the processing of West Texas Intermediate (WTI) in a cracking or in a coking refinery in the United States (Fig. 8.8A). Regarding locations, it has been shown that for a cracking refinery, margins have been higher in the United States > Singapore > Northwest Europe, along the period of low oil prices (2014–2016), when processing conventional oils [98]. The impact on margins when UOs are incorporated in the diet [Heavy Louisiana Sweet (HLS) and Maya] or directly processed (Bakken) can be appreciated in Fig. 8.8B.

Two relevant aspects can be distinguished: (1) the margin of Bakken, a UO behaves comparatively similar to those of conventional oils (both magnitude and processing, in either a cracking or a coking refinery), and (2) addition of deeper conversion units (coking) consistently outperform when incorporating OCs (heavy oils) in comparison with cracking refineries handling light sweet Brent crude. Examples of refineries that historically have benefited from the processing of UOs by anticipating suitable revamps and capital investments included US Sunoco's 150 kbpd Eagle Point refinery in New Jersey and 85 kbpd Tulsa refinery in Oklahoma; in Germany, ConocoPhillips' 260 kbpd Wilhelmshaven and Shell's 107 kbpd Harburg complex; the 80 kbpd Gothenburg facility in Sweden; and Total's 137 kbpd Dunkirk refinery in France [99].

The refining strategies for changing capabilities differ from one company to another. An example was given in the case of Shell that changed its capability by divesting in Europe, selling around 700 kbpd of simpler refining capabilities, and investing USD 7.5 billion in China in a complex refinery [100]. Certain refineries have been subjected to upgrades and revamps to prepare themselves for the processing of OCs. In this regard, a small proportion of HACs could be taken by revamping the CDU and installing corrosion control and monitoring devices in refineries with a high-quality blend diet. Some conventional refineries with an original specification of a maximum TAN of 0.5 mgKOH/g have been upgraded to process feed blends with higher TAN (~1.5–2.0 mgKOH/g).

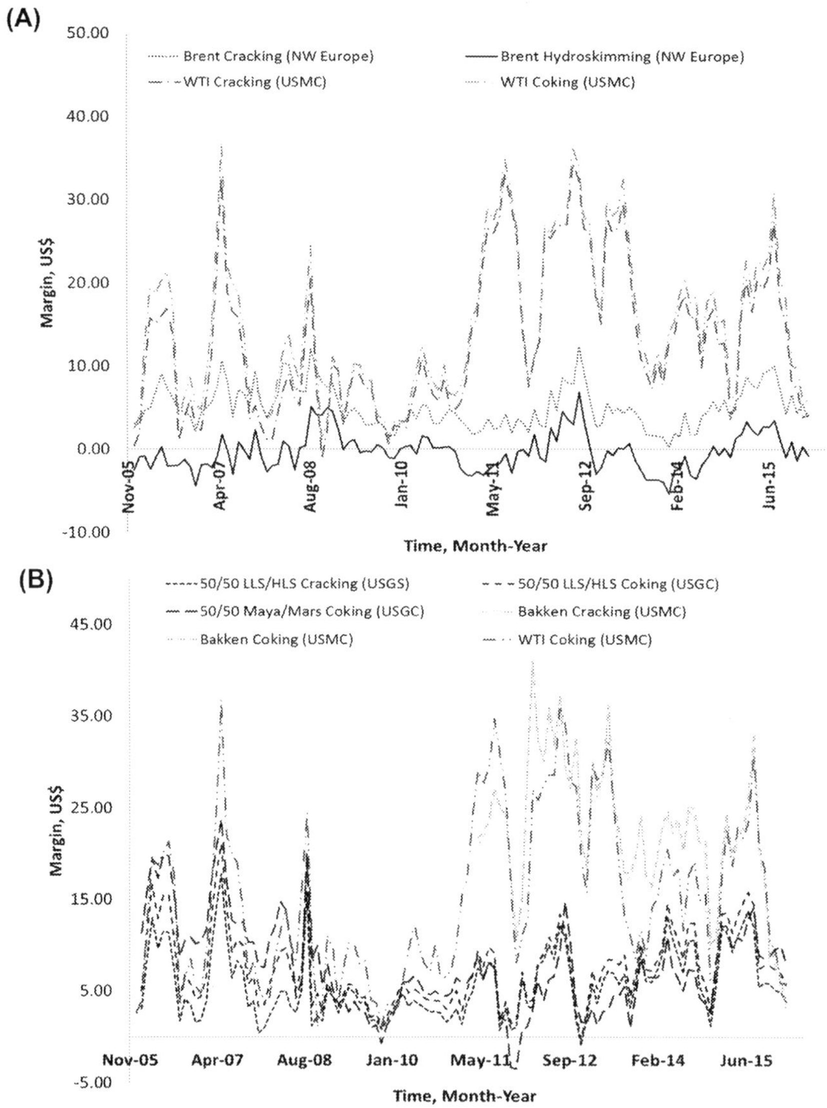

Figure 8.8

Refining margins comparison: (A) conventional oils in different refinery complexities; (B) effect of the incorporation of unconventional oils. (See text for explanation on crude oils trade names, types, and quality grades.)

Nevertheless, these refineries have to complement these revamps with additional devices for controlling dosing of chemical additives to the affected units.

The best practices for the safe refining of HACs include four key management elements:

1. Preestimation or assessment of risks;
2. Constant monitoring of feed properties and vulnerabilities;

3. Careful design of a mitigation strategy (blending, passivation, chemical additives, or metallurgy upgrade); and
4. Periodic revision of technical and economic value.

Asia Pacific and in particular China have been dominating HAC production, consumption, and imports for the last 15 years. A significant increase in HAC has taken place in South Korea, Taiwan, Malaysia, Singapore, and India. Chinese exports to the US West Coast have also been rising. On the other hand, the HAC exporting countries in Latin America and Africa have not improved their refining capacity for this type of crude oil, keeping it obsolete and inadequate. Therefore increased production is placed in the open market. Sudan and China represent a higher effort for building refining capacity capable of handling its surge of incremental HAC output; nevertheless, the surplus production outpaces the specially built new refining capacity [101].

As shown in Fig. 8.3, the different grades of commercially available blends of bitumens and heavy oils clearly required different refinery configurations. Additionally, modifications to the existing units are also required [102]. The light SCO grade is a bottomless SCO that gives rise to other refinery constraints. The processing of the poor-quality VGO and distillates, which also constitute larger volume proportions of the crude, imposes stress on the HDT units, both the feed pretreating for FCC and the kero/diesel units. The yields and final quality of the blendstocks are affected negatively, such as S- and N-content, diesel cetane, and kerosene smoke point [103].

Dilbit crudes, often called "dumbbell" crudes, exhibit abnormal high-volume proportions of naphtha/light-kero and residue fractions while those of the kerosene and diesel range are low. A proportional increase in capacity for processing the excess volume of the larger fractions will be required, such as the CDU, including the atmospheric column and the atmospheric column overhead cooling and reflux system; and in the capacity of the VDU, including column, pump, quench, and heat recovery system. The high content of high-S residue and in some instances a high TAN of the dilbit have to be considered carefully according to previously described problems. The consequent impact on FO and asphalt quality may justify additional deep conversion units. The capacity for processing the additional naphtha needs to be considered as well. This additional capacity should account for any coker naphtha if a coker is installed in the refinery, and in this case the coker would also need revamping for increasing capacity. These naphtha streams will need HDT and reforming for upgrading the naphtha octane and meeting gasoline specifications. Some refineries would need incremental capacity for these two units, and also the ancillary units (e.g., S-recovery).

In the case of synbit, volume proportion of the fractions does not fit a "dumbbell" distribution, thus the CDU and VDU are not typically affected; however, the presence of bitumen can be felt in other units. This crude grade tends to be high in aromatic content and refractory S-compounds, creating stresses in the HDP units for meeting intermediate stream and product specifications. The difficulties faced by the VGO hydrotreater during the pretreatment of the FCC feed affect product yields and quality from the FCC. Furthermore, the life cycle of the hydrotreater is shortened. Although increasing the cutpoint in the VDU would favor an increased yield of higher value-added products, the shortening of the life cycle in catalytic processes has to be considered in the economic evaluation [55]. The increased capacity of all these units and ancillary units has to be considered as well [102].

The processing of syndilbits is the least fortunate refining case since the constraints created for dilbits and synbits add up together and even synergize. One of the synergistic effects may be asphaltene stability, for instance.

Therefore refiners planning to process heavy oils or bitumen blends in general will have to install additional residue conversion capacity. Moreover, to maintain acceptable FCC yields and quality, FCC feed pretreating capacity is also required. Combinations and configurations of the hydrotreating units can be seen in Ref. [104]. Probably, hydrocracking capacity would add flexibility to the processing of the increased volume of VGO-range barrels of these crudes and meet market needs for distillate fuels. The lower H/C ratio of the diet would also need to be compensated for with additional hydrogen consumption, which in turn would require improvements in hydrogen management [62].

Up to now, modifications have been introduced to address the issues found when processing OCs. Several refinery configurations have been proposed, and some have been built and are currently in operation. Iqbal et al. have described the basis of some of these configurations [35], making clear that for the processing of OC a simple refinery would need to face the disadvantages of having to produce a large amount of low-quality FO, needing to purchase very expensive light crude oils to compensate the quality of the OC, and having very limited capability for adjusting product slate and specifications. A visbreaking-based refinery would be capable of processing OCs, but will produce a large proportion of unmarketable high-S FO by having to blend the unconverted thermal tars. Similarly, a coking refinery or an SDA-based refinery will not be able to capture all the benefits because of the production of large quantities of low value-added coke in the former and not having conversion capabilities in the latter. An HDP-based refinery will be able to produce high value-added products in a larger proportion, but these are high CapEx and OpEx units that need to address all the problems described in the previous sections and also in previous chapters. The recommendation from these authors [35] was combinations of SDA—coking—HDP according to the original complexity of the refinery and the geographical location.

SDA is adept at extracting good-quality FCC feed from petroleum residue, because the more aromatic molecules in the residue and most of the metals and carbon residue are rejected with the asphaltene product, while the more saturated, less contaminated components are concentrated in the deasphalted oil for cracking in the FCC unit (FCCU). At the same time, such units can be relatively inexpensive, easy to operate, and do not consume hydrogen or catalyst [105]. Increasing the utilization of asphaltenes from a ROSE SDA unit (see Chapter 5; Section 2.1.1 for a description of the ROSE process technology) as feed to delayed cokers as well as to gasification units and pelletization for shipment as a solid fuel have accelerated the installation of new SDA units over the past few years. The arrangement of dense-phase FCC catalyst coolers to provide advantages in catalyst cooler reliability and turndown flexibility compensate for the excess of coke formed. Plus, countercurrent regeneration enables the FCC operation to alternate between complete CO combustion and partial CO combustion as needed to accommodate changing feedstocks and changing operating objectives. Together or separately, these technologies can provide refiners with the ability to economically produce more refined products from an increasingly supply of heavy crude oils [106]. In Chapter 5, we presented six different configurations for SDA-based refineries (the reader is invited to revisit that chapter).

At this point, it might be clear that conventional process technologies would exhibit strong limitations for the processing of UOs. Most of the time these limitations are overcome or mitigated by their blending with crude oils of higher quality grade or their fractions. Therefore, the market conditions play an important role (technical and economical) and represent a high risk for the refiner. Thus, the success of a given process depends on the marketplace and within a limited crude slate. It is clear that a combination of more than one process technology would be advantageous. The combination of C-rejection and H-addition may bring the best of two worlds. In a scenario of low-grade pet-coke surplus (or decreasing coke demand), high crude oil prices, and low natural gas prices, H-addition processes would become incentivized. Under these circumstances, both CCR and metal content would determine the technology of choice for HDP. Nevertheless, it seems that these combinations of the many commercially available technologies are still not enough for capturing the whole value from the crude, or all the economic benefits. As mentioned previously, for the HDT of the heavier fractions, moving bed reactors are preferred. The problems found with the ebullated bed have redirected attention toward the slurry bed. In both cases, high CapEx and OpEx have limited the number of units that are currently operating commercially.

A comparative evaluation of eight configurations for the processing of heavy oils was carried out considering the following line-outs of crude distillation units (CDU: atmospheric, ADU, and vacuum, VDU) delayed coking unit (DCU), FCC. Reformer, hydrodesulfurization (HDS), HDT, HCK, and SDA [107]:

conf. 1. CDU−DCU−FCC−Reformer−HDT
conf. 2. ADU−FCC−HDS−Distillate−HDT (H_2 plant and C_3-recovery)
conf. 3. ADU−DCU−FCC−HDS−HDT (H_2 plant and C_3-recovery)
conf. 4. CDU−DCU−HCK−HDT (H_2 plant)
conf. 5. CDU−DCU−HCK−FCC−HDT−Reformer (H_2 plant)
conf. 6. CDU−SDA−FCC−Reformer−HDT

conf. 7. CDU−SDA−FCC−HDT (H_2 plant)
conf. 8. CDU−SDA−FCC−HDT−Reformer (H_2 plant)

The results of the technoeconomic evaluation pointed out that conf. 2 provided the highest gross profit and best payout period. Meanwhile, the maximum gasoline yield was obtained with conf. 1, which placed it in fifth place in terms of gross profit and payback period. Instead, conf. 4, which ranked second in profit and payback, provided the maximum distillate [107]. Therefore a given configuration may better suit a specific scenario, depending on the economic drivers considered. The refining of shale oils should start by removing the solids. A three-step method has been proposed, involving centrifugation, settling, and filtration. Following, arsenic could be removed by adsorption [33]. Shale oils with composition similar to that of Bakken oil would give rise to overloading of units for light cuts, such as reformer and isomerization, and at the same time would grow short on feed for the FCCU and the coker. A diet predominantly constituted by shale oil would allow the VDU and coker units to shut down. The refiner could send the entire atmospheric residue to the FCCU but still the refinery could be short on FCC feed. Bypassing a portion of the whole crude around the CDU to fill up the FCCU capacity may be considered, in which case excess gas yield would impact the gas processing units and tank farm [32]. As mentioned previously, compared to FCC, HDP might be a better approach for shale oils. Nevertheless, the hydrotreated (retorted) shale oil may still exhibit high pour points and viscosity (even higher than the original shale oil) and will not meet pipeline specifications. One option would be to subject the hydrotreated oil to thermal conversion or vice versa, i.e., to subject the raw shale oil to thermal conversion followed by HDT. Another alternative is the use of drag reducing agents and/or pour point improving additives to the hydrotreated oils or fractionated blendstocks [33]. The multiple recovery methods for shale oil result in a wide variation of composition and levels of contamination.

Therefore refineries' configuration may be adjusted to optimize the performance and get the best value from the different compositions. The choice of configuration and operating model is crucial and varies among the different geographic regions where the many drivers will affect differently the value creation along the chain. The value drivers have been categorized by Lewe et al. [108] who also discussed in detail the impact on the refining value chain. A summary of these value drivers is collected in Table 8.3.

Refining companies run business models that are considered integrated. Four principal operating models can be currently identified in the industry. There is no hierarchy for these models:

1. Upstream integration: 50% of the diet is provided by a single source of crude oil, either as equity crude or as a long-term supply contract.

Table 8.3: Refinery Value Drivers

Input	Asset Related	Output
Crude Fungibility	**Scale and Technology**	**Fuel and Energy**
Local or regional balances	World scale or subscale	Merchant only versus retail
Pipeline or imports by ship	Distillation and conversion	Export versus local sales
Multiple asset optimization	Technology	
Trading and Hedging	**Fiscal and Regulatory Regime**	**Specialties**
Feedstock	Tax	Specialist markets (marine,
Products	Regulation	aviation, asphalt) with
Currency	Environment	dedicated assets
		Brand quality
Energy Imports	**Supply Chain**	**Petrochemicals**
Electricity	Location	Which value chain
Steam	Logistics infrastructure	Joint venture or sole ownership
	Working capital optimization	Export versus local sales
Blendstocks	**Feed Slate Flexibility**	**Lubricants**
Gasoline	Dedication of technology	Base oil plant
Biofuel	Ability to change baskets	Blending plant and storage
Gas to liquids	Operational flexibility	

2. Merchant–refiner: Neither upstream nor downstream integration is in place, allowing the refiner to react quickly and flexibly to both crude and downstream supply opportunities; using this model the refiner easily adjusts to operations or integrates into a larger logistics hub.
3. Downstream integration: More than 50% of the refining products are marketed through secured channels either through equity or long-term contractual arrangements.
4. Vertical integration: The requirements for both upstream and downstream integrated refiners are fulfilled at the same time, thus capturing value in taking the most advantages across the value chain [108].

Refiners create and maximize value; in doing so, value drivers (Table 8.3) are considered in connection with the integrated business model in operation. These four models are business integrated models, which do not involve either technology integration or integrating technologies. The conventional existing technologies commercially available and installed in operating refineries have been presented and discussed (see Chapter 5). Although some of these technologies might represent a better option than others and pass the technoeconomic feasibility lenses, none of these can be robustly categorized as cost-effective, and their market niche is limited to a scenario of high oil prices and/or large differentials. Therefore none of the configurations based on conventional technologies would capture the opportunities and full benefit of OCs. Meanwhile, licensors keep offering either stand-alone process units, combinations of these, or integrated versions of such combinations. In spite of the current scenario, refiners seem to be reluctant to invest in those technologies. As a consequence, commercialization has moved enormously slow.

4. The Future

At this point, it may be evident that there are many reasons for processing UOs. First, the conventional oils supply is rapidly declining. Furthermore, the largest existing reserves of conventional oils are located in regions characterized by political upheaval. It seems logic that operators seeking a secure supply may require a dose of UOs. Second, refining is losing competitiveness because most of the installed capacity is not very sophisticated, nor feedstock flexible. Instead, most of the installed configurations are suitable for conventional oils processing. Refinery profitability can be improved by switching to UOs processing. However, by proceeding with minimum revamps, mitigation strategies or by keeping the constraints of existing assets, the improvement in profitability is only marginal. It is easy to understand that by living with the limitations of existing facilities that were originally designed for lighter crude oils, refiners will limit the amount of UOs to be added to the diet. A careful evaluation of the options of new process technologies would minimize negative surprises.

In the last few years more than ever, facilities have been shut down, particularly small and less efficient ones. Plants have been turned into terminals, sold, or placed for sale. Thus opportunities call for upgrading existing facilities to increase flexibility and/or conversion capacity to return to competitiveness. This trend has given rise to building fewer but larger and more complex refineries as well as expansion of simpler ones into more complex configurations. The average refinery capacity of about 50 kbpd by the 1970s had risen to 260 kbpd by 2007 [109]. Operators who made such adjustments prior to price increases caught the benefits earlier than expected; examples include Greek Hellenic Petroleum, Italian Saras, and Austrian OMV [99].

Most of the time refiners will not consider any changes in the refinery configuration, nor the construction of new process technology without securing feedstock supply. However, securing UOs supply is not the end-game and the refiner has to consider other drivers for supporting profit margins improvement. In the short term, maximizing propylene and distillate yield and quality, minimizing high-sulfur FO yield and carbon footprint, mitigating fouling and corrosion, and implementing the latest BotB technologies are being considered [99].

The sudden shale oil boom has resulted in a new opportunity for the refining industry. The apparent high quality based on API and S (light sweet) and their high reactivity could be misleading in terms of processability, because of large quality variability and high levels of sediments, iron, and alkali metals. It would be expected that the incorporation of shale oil in a refinery processing conventional oil could be managed with proper planning and scheduling, and the needed mitigating methods for the contaminant present. However, refineries including OCs should anticipate deeper problems associated with corrosion, incompatibility, fouling, and deposits. Furthermore, refiners would need to look to the future to bring products to meet specifications.

The mitigating methods currently available and practiced for coping with the problems caused by UOs actors are extremely limited and highly specific to the crude source. Each problem has to be defined on a case-by-case basis, and finding a possible effective solution requires its testing and evaluation of suitability and economical acceptance. Therefore processing UOs might require significant modifications of the refinery configuration. In the short term, probably the inclusion of new units dedicated to changing the properties of the new crude blend diet into the specifications of the current operating units might have to be considered instead of the mitigating actions in place.

Although the advance in scientific knowledge is evident, no technological breakthrough has been realized. The reactivation of the funding flow to R&D activities should be considered in the very short term. In 2007, Plotkin [110] listed the areas he considered further R&D efforts should focus on (our comments in brackets):

- New concepts such as integrated processing and management of hydrogen content (which can be considered a way of molecular management described earlier);
- Integration of energy consumption between process segments (Instead, we consider taking integration one step further to technology integration as a whole.);
- Online sensors for monitoring and correcting process operations. New techniques based on nanotechnology are expected to emerge with applications in this area (Instead, we consider taking this approach one step forward as well, by functionalizing nanotechnologies. Thus the design of hybrid multifunctional applications to include treatment, monitoring, and control in a single technology would provide an integral solution.);
- Catalysis and biocatalysis for low-temperature conversion of bitumen and extraheavy oil (we have considered and discussed in previous chapters, this type of development and its adaptation, including other biotreatments for a downhole crude pretreatment application);
- Better understanding of the molecular structure and properties of bitumen and extra-heavy oil components;
- Production of value-added products through integration with petrochemicals.

Multiple physical processes and chemical reactions have been examined and tested for the abatement of the bad actors present in UOs. However, the economic benefits of these alternatives do not seem to break the commercialization barrier and they worsen in periods of low oil prices.

The lack of knowledge of the chemical details of these compounds and of the understanding of the reactivity may have precluded new ideas from emerging. Furthermore, the complicating effect of the various interactions among them, and the possibility of other compounds being trapped, shielded, occluded by asphaltenes, or even being a structural part of the asphaltene molecules make the study of these compounds a very difficult one to approach. Further research is needed. The immediate focus may be on the elucidation of the molecular structures and the physical form of these compounds. Clarification of the interactions, aggregation, adsorption, and chemically bound molecular entities in the crude oil matrix is desirable. Once these are achieved then the understanding of their reactivity would become easier, and perhaps new ideas for cost-effective processes would flourish.

In the median term, however, emerging ideas could be implemented. Slurry conversion processes have begun operations (in Italy and China), at moderate scales (<25 kbpd), indicating that technological rollout has started already.

Processes specifically for demetalization or deacidification are not commercially available. Instead, metals are removed by either deasphalting, conversion (thermal or catalytic), or HDP, if the refinery has such units. Meanwhile, in the case of acidity, mitigation practices are in place where HACs are being processed. The limitations and problems encountered in doing so were briefly described earlier. In the long term, the implementation of brand new process technologies has to be considered.

Although in all the areas reviewed extraordinary achievements and advances have been attained, there is still plenty space for innovation and invention. For example, new ideas that take crude pretreatment far beyond just desalting and dewatering to the production of a cleaner, less troublesome crude oil. New out-of-the-box ideas are needed for optimizing energy balances, diminishing environmental impact and processing costs. Conventional thinking does not seem to have any opportunity for solving the problems found with UOs.

Refining research during the last five or six decades has been devoted to removing heteroatomic contaminants. Consequently, the refining technology scenario has not changed significantly in the last decades. Among the emerging technologies for the processing of OCs, some have reached the maximum development level and are available for commercial licensing, while others have not reached that level and either have passed the proof-of-concept scale or remain simply ideas.

Most of the R&D activities have been centered on catalysis and process design. In the future, catalysis research may focus on new forms and formulations of catalysts, to promote the same reactions but through different mechanisms.

Some of the most recent upgrading concepts were based on unconventionally exotic methods, like those using other types of energy sources such as mechanical, electron-beam, γ-rays, microwaves, plasma, and ultrasound. But they failed in providing robust and competitive data consistently. Compared with commercially available upgrading technologies, these unconventional methods have failed to demonstrate any potential benefit, so far.

The long-lasting history of the R&D efforts and the reduced number of emerging technological solutions may be indicative of the lack of holistic interdisciplinary approaches, incorporating new knowledge and contributions from all relevant areas. New ideas are still needed to feed the technology development pipeline. These ideas would probably derive from the careful and holistic analysis of the cumulative knowledge of the characteristics, features, and behavior of the molecules that have to be treated. The many ways in which undesirable compounds behave and react could be the basis or inspiration for new processes that contribute to their abatement and render the most value for the refiner.

The technological integration of refining processes to upstream operations might necessitate the incorporation of slight amounts of those heteroatomic moieties in the crude oil. Furthermore, the current thoughts on how far refining has to go, when carried out down the hole, need to be revisited and probably adjusted to the industry needs. These adjustments require not only multidisciplinary views but also consideration of a case-by-case basis. Historically, the R&D efforts for downhole (in situ) refining have been sporadic, at a frequency worse than what has happened with the rest of R&D activities.

Integration in the oil industry may be seen as a way of managing risks, particularly those associated with the market such as buffering negative earnings, securing supply, or securing demand. Thus integration is defined as the degree to which the company owns and controls its upstream suppliers and downstream buyers along the value chain. This vertical integration of the value chain is more intense in the national oil companies than it is for the international oil companies, especially after nationalization of many of the upstream assets in the 1970s [100]. A study proposed and evaluated an integrated complex for the Canada oil sands that would include oil sands processing, upgrading, refining, the extraction of petrochemical basic compounds from by-products present in oil sands process streams, and petrochemical plants producing both high-quality fuels and new world-scale quantities of petrochemicals [111]. A CapEx of about USD 10 bn was estimated in 2006 for building an integrated facility in the Edmonton area [112]. In this type of integrated cluster, the role of FCC and of the cokers is seen as key in the making of more light olefins. All of the refining processes designed to make light olefins enable a bridging between refiners and the petrochemical markets [2]. The value of propylene to refiners allowed for an advantageous positioning of this material in the chemical market worldwide. Unfortunately, so far integrated projects of this type have not been realized. Instead, the present wave of low oil prices caught the growth plans of Canadian operators. The market-driven integration mode that created upstream–primary upgrading- and upstream–primary and secondary upgrading-integrated projects will no longer be able to fulfill the one-time aspirations of the Canadian Association of Petroleum Producers for producing 8.0 mbpd by 2035. In fact, this expected growth was reduced in 2014 to

5.2 mbpd, to 4.8 mbpd in 2015, and to 4.0 mbpd in 2016.[4] This old volume growth strategy has been now replaced by a new strategy based on cost reductions. Obviously, cost reductions could be enabled by technology advances. Therefore integrated technology packages, which provide integral technological solutions, are needed more than ever. However, concepts for technology integration are scarce.

A similar situation may arise with the evolution and implementation of environmental regulations. The European legislation for Euro VI has set standards that have been copied or followed by other countries. Speciation took a leading role in the definition of these new standards. Among the changes introduced by Euro VI are (1) the inclusion of ammonia and formaldehyde in the emissions pollutants; (2) the substitution of the term "miles per gallon" with "grams of carbon dioxide emitted per kilometer travelled"; and (3) the split of total hydrocarbons into methane and nonmethane total hydrocarbons [113,114]. These new standard parameters will affect not only the specifications of transportation fuels but also the market of the refining products. New and cleaner alternative fuels will enter the market in competition with fossil fuels. Once again, refining technologies should respond with cost-effective technologies for producing cleaner products from dirtier crude oils. Furthermore, these technologies should not have a negative environmental impact in terms of emissions and effluents.

Regarding integration, there do not seem to be economic incentives for upstream, while for downstream, a sensation of cost increments seems to exist. The lack of mutual knowledge of each other's operations and businesses, particularly in the R&D teams, does not help the situation. Whereas the production technologies have been evolving rapidly and changes in the quality of the produced crude oil and bitumen typically accompany the implementation of such technologies, the refining technologies have not evolved in response to those advances.

Probably, a view of a pretreatment that can be applied downhole or at the well surface would be more adequate than the original in situ refining, and may find the economic incentives needed for upstream. Downhole the applications of biotreatments or nanotreatments (see Chapter 6) might be adaptable or reengineered to well conditions. Currently, at the well surface few Canadian companies are practicing the Paraffinic Froth Treatment technology [115–120] for treating the mined bitumen produced and so removing water, solids, and part of the asphaltenes (see details of this upstream technology in Chapter 5). The implementation of this technology has questioned the need for additional (primary) upgraders in the region [121]. Similarly, a defined pretreatment considering downstream operational issues would save the refiners from some of the problems they are now facing.

[4] See news report at http://www.albertaoilmagazine.com/2015/09/mining-project-oil-sands/.

In summary, the future seems to be offering to R&D the real opportunities for taking the best advantages in the development of unconventional resources and most particularly those from the so-called, opportunity crudes.

5. Conclusions

Under the current scenario, refiners have been reluctant to invest either in conventional technologies or in the commercially available improved versions. The evolution of existing technologies has been limited mainly to mitigating measures for coping with the problems caused by the bad actors present in UOs, instead of facing these actors' abatement. R&D interests peak and decline as refining investments come and go, paralleling the volatile behavior of oil prices, regardless of the conscious perception that without the basic and fundamental knowledge, technological advances would be just dreams.

Meanwhile, the market changes, the social demands are harder to meet, and the tougher environmental regulations are driving the R&D trends into improving process economics. For existing facilities, the incorporation of UOs into the refinery diet still needs cost-effective improvements or cost-reducing modifications. However, in the long run more drastic changes may need to take place, and integral technologies to be implemented along the value chain of the oil industry are needed. These achievements can only be gained by a better understanding of the properties, chemistry, and behavior of these oils, and their impact on the processes and products. It is clear that further research is needed toward the development of new process technologies.

Unconventional oils have challenged creativity and R&D. Fundamental understanding is specifically required in many more aspects, such as the molecular characterization of the reacting species; pathways, mechanisms, and kinetics of the chemical reactions; factors affecting product and reaction selectivity (coke and coke precursors formation, fouling species, poisons, etc.); process thermodynamics; fluid-phase behavior at process conditions; intermediate and final products stability and reactivity; property—structure relationships, etc. A deeper knowledge of the molecular structures of the BotB compounds is being acquired, thus there is hope that this knowledge would enable the creation of novel alternatives. Nevertheless, we have not been able to realize the opportunities and fulfill their demonstration at industrial-scale operations by taking full advantage of the accumulated scientific and technical experience, or incorporating the results from the more advanced methods of process modeling, or creating the multidisciplinary R&D teams. Therefore immense opportunities still exist in a broad range of areas. These limitless opportunities may be just incremental improvements to existing technologies, but they can go further toward a breakthrough with the reinvention of an "out-of-the-box" new way of refining. Eventually, refiners would be forced to incorporate the treatment and conversion technologies required for processing UOs. We have shown how the means used by upstream and midstream to sort their problems complicate the issues faced by the refiner. New ideas that tackle refining and the problems with UOs at the oil well are still needed. Integral solutions that deal with the initial problems and do not propagate into other problems elsewhere are ideal and urgently required. An integrated industry should examine the situation holistically and probably would consider managing the molecules throughout the complete value chain, because at the end of the chain the removal of contaminants paves the pathway to the market. In the extremes of production, market volatilities (supply/demand imbalances) could be better managed in a more integrated value chain that in turn may mitigate instabilities in margins as well.

Finally, a complete package of proprietary technologies for the treatment and processing of UOs represent a competitive advantage for the possessor, and more particularly for the integrated companies. The original concept of economic (market-driven) integration is no longer valid and needs to evolve into a technology-integrated one.

References

[1] Alboudwarej H, Felix JJ, Taylor S, Badry R, Bremner C, Brough B, et al. Highlighting heavy oil. Oilfield Rev 2006;18(2):34–53.

[2] Swaty TE. Global refining industry trends: the present and future. Hydrocarbon Process 2005;84(9):35–46.

[3] Alabi OO, Bowden SA, Parnell J. Simultaneous and rapid asphaltene and TAN determination for heavy petroleum using an H-cell. Anal Methods 2014;6(11):3651–60.

[4] Mckenna AM, Purcell JM, Rodgers RP, Marshall AG. Heavy petroleum composition. 1. Exhaustive compositional analysis of Athabasca bitumen HVGO distillates by Fourier transform ion cyclotron resonance mass spectrometry: a definitive test of the Boduszynski model. Energy Fuels 2010;24(5):2929–38.

[5] Qian K, Robbins WK, Hughey CA, Cooper HJ, Rodgers RP, Marshall AG. Resolution and identification of elemental compositions for more than 3000 crude acids in heavy petroleum by negative-ion microelectrospray high-field fourier transform ion cyclotron resonance mass spectrometry. Energy Fuels 2001;15(6):1505–11.

[6] Klein GC. Petroleomics: applications in the fingerprinting of the acidic and basic crude oil components detected by electrospray ionization Fourier transform ion cyclotron resonance mass spectrometry. Doctor of Philosophy Thesis from Florida State University, Department of Chemistry and Biochemistry; October 12, 2005. 145 pp.

[7] Mckenna AM, Williams JT, Putman JC, Aeppli C, Reddy CM, Valentine DL, et al. Unprecedented ultrahigh resolution FT-ICR mass spectrometry and parts-per-billion mass accuracy enable direct characterization of nickel and vanadyl porphyrins in petroleum from natural seeps. Energy Fuels 2014;28(4):2454–64.

[8] Vetters E, Lemieux A. Oil sands bitumen processability. In: Proc. Crude Oil Quality Association Meeting. New Orleans, LA, USA, February 10–11; 2010. 12 pp. Available from: http://www.coqa-inc.org/docs/default-source/meeting-presentations/20100211_Vetters_Oil_Sands_Processability.pdf.

[9] Wiehe IA. Petroleum fouling: causes, tools, and mitigation methods. In: Proc. AIChE Spring Nat. Meeting, Orlando, FL; 2006. 1 pp.

[10] Gray MR, Tykwinski RR, Stryker JM, Tan X. Supramolecular assembly model for aggregation of petroleum asphaltenes. Energy Fuels 2011;25(7):3125–34.

[11] Chrisman E, Lima V, Menechini P. Asphaltenes – problems and solutions in E&P of Brazilian crude oils. In: Abdul-Raouf ME-S, editor. Crude oil emulsions – composition stability and characterization. InTech: On line; 2012. p. 1–26. Available from: http://www.intechopen.com/books/crude-oil-emulsions-composition-stability-and-characterization/asphaltenes-problems-and-solutions-in-e-p-of-brazilian-crude-oils.
[12] Becker HL. Asphaltene: to treat or not. In: Proc. SPE Permian basin oil and gas recovery, Midland, TX, March 21–23; 2000. 8 pp.
[13] Holubnyak YI, Bremer JM, Mibeck Ba F, Hamling JA, Huffman BW, Klapperich RJ, et al. Understanding the souring at Bakken oil reservoirs. In: Proc. International Symposium on Oilfield Chemistry 2011, The Woodlands, TX, vol. 2; 2011. p. 803–8.
[14] Kremer LN. Crude oil and quality variations. PTQ 2004;Autumm:87–93.
[15] Chen X, Shen B, Sun J, Wang C, Shan H, Yang C, et al. Characterization and comparison of nitrogen compounds in hydrotreated and untreated shale oil by electrospray ionization (ESI) Fourier transform ion cyclotron resonance mass spectrometry (FT-ICR MS). Energy Fuels 2012;26(3):1707–14.
[16] Kumar R, Bansal V, Badhe RM, Madhira ISS, Sugumaran V, Ahmed S, et al. Characterization of Indian origin oil shale using advanced analytical techniques. Fuel 2013;113:610–6.
[17] Tong J, Liu J, Han X, Wang S, Jiang X. Characterization of nitrogen-containing species in Huadian shale oil by electrospray ionization Fourier transform ion cyclotron resonance mass spectrometry. Fuel 2013;104:365–71.
[18] Savest N, Hruljova J, Oja V. Characterization of thermally pretreated Kukersite oil shale using the solvent-swelling technique. Energy Fuels 2009;23(12):5972–7.
[19] Rheams KF, Neathery TL. Characterization and geochemistry of Devonian oil shale north Alabama-south central Tennessee. In: Proc. ACS National Meeting, Symposium on Characterization and Chemistry of Oil Shales Petroleum Chemistry, St. Louis, Missouri; 1984. p. 113–25.
[20] Dalai NS, Suryan MS, Shen MS. Electron Spin Resonance and thermogravimetric investigations of the pyrolysis of eastern shales. Preprints of the ACS symposium on characterization and chemistry of oil shales. In: Proc. ACS National Meeting, Symposium on Characterization and Chemistry of oil Shales Petroleum Chemistry, St. Louis, Missouri; 1984. p. 34–40.
[21] Jong KM, Petrakis L, Takayasu M, Friedlaender FJ. Physichochemical properties of magnetically separated oil shale solids. In: Proc. ACS National Meeting, Symposium on Characterization and Chemistry of oil Shales Petroleum Chemistry, St. Louis, Missouri; 1985. p. 159–65.
[22] Qin KZ, Wang RA, Jia SS, Lao YX, Guo SH. Characterization and geochemistry of devonian oil shale north Alabama-south central Tennessee. In: Proc. ACS National Meeting, Symposium on Characterization and Chemistry of Oil Shales Petroleum Chemistry, St. Louis, Missouri; 1984. p. 104–12.
[23] Levy JH, Stuart WI. Thermal properties of Australian oil shales: characterization by thermal analysis and infrared spectrophotometry. In: Proc. ACS National Meeting, Symposium on Characterization and Chemistry of Oil Shales Petroleum Chemistry, St. Louis, Missouri; 1984. p. 61–70.
[24] Dion M. Challenges and solutions for processing opportunity crudes. In: Proc. AFPM Annual Meeting, Orlando, FL, USA, March 23–25; 2014. 13 pp.
[25] Lu T. Challenges faced in high TAN crude processing. In: Proc. Intern. Conf. Petrofed. New Delhi, India, April 16; 2012. 36 pp.
[26] Bankapur BN. Challenges for margin improvement in refinery. J Petrotech Soc 2008;5(1):8–11.
[27] Lu T. Challenges in opportunity crude processing. 2012 (Unpublished Presentation). 17 pp. Available from: http://www.world-petroleum.org/docs/docs/petrofed2012/Presentation/Session%201/1_Thomas%20Ting%20King%20Lu.pdf.
[28] Basu AS. Challenges in refining sector. In: Unpublished Presentation at 8th Summer School of Petrotech Society, Gurgaon, India, June 3–7; 2013. 57 pp. Available from: http://www.petrotechsociety.org/Presentations/8th_Summer_School/Session%204%20Vartika%20Shukla.pdf.
[29] Konen J, Goodman M. Challenges with synthetic crude processing – enabled with increased hydrotreating performance. In: Proc. AFPM Annual Meeting, San Diego, CA; 2012. p. 328–42.

[30] Yui S. Producing quality synthetic crude oil from Canadian oil sands bitumen. J Jpn Pet Inst 2008;51(1):1–13.
[31] Canadian Association of Petroleum Producers (CAPP). Canada's oil sands overview and bitumen blending primer. 2012 (Unpublished Presentation. US National Academy of Science, October 23).
[32] Bryden K, Habib Jr ET, Topete OA. Processing shale oils in FCC: challenges and opportunities. Hydrocarb Proc 2013;92(9):10.
[33] Fenton DM, Hennig H, Richardson RL. The chemistry of shale oil and its refined products. In: Proc. ACS National Meeting, Symposium on Oil Shale Tar Sands, and Related Materials, San Francisco, CA, USA, March 25; 1981. p. 315–25.
[34] Baker Hughes. Technology solutions for shale oil. Brochure; 2013. 2 pp. Available from: www.bakerhughes.com/shaleoilprocessing.com.
[35] Iqbal R, Eng O, Anshumali. Economic routes to process challenged and opportunistic crudes. In: Proc. NPRA Annual Meeting, San Francisco, CA, March 13–15; 2005. Paper #AM-05-42, 15 pp.
[36] Kremer LN. Crude oil management: reduce operating problems while processing opportunity crudes. In: Proc. Intern. Conf. on Refinery Proc. Orlando, FL, USA, April 23–27; 2006. p. 1. 188b.
[37] Stommel J, Snell B. Consider better practices for refining operations. Hydrocarb Proc 2007;86(10):105–9.
[38] Ohmes R. Characterising and tracking contaminants in opportunity crudes. PTQ 2014;1:1–12.
[39] Hartmann JCM. Crude valuation for crude selection. PTQ 2003;Winter:123–7.
[40] Kelly JD, Mann JL. Crude oil blend scheduling optimization: an application with multimillion dollar benefits - part 1. Hydrocarb Proc 2003;82(6):47–54.
[41] Kelly JD, Mann JL. Crude oil blend scheduling optimization: an application with multimillion dollar benefits – part 2. Hydrocarb Proc 2003;82(7):72–9.
[42] Voolapalli RK, Parihar P, Kumar R. Improve feedstock selection for your refinery. Hydrocarb Proc 2012;6:65–71.
[43] Reddy PCP, Karimi IA, Srinivasan R. Novel solution approach for optimizing crude oil operations. AIChE J 2004;50(6):1177–97.
[44] Pan M, Li X, Qian Y. New approach for scheduling crude oil operations. Chem Eng Sci 2009;64(5):965–83.
[45] Price X. Addressing the challenges associated with Canadian crudes. In: Proc. Crude Oil Quality Association Meeting. New Orleans, LA, USA, February 10–11; 2010. 38 pp. Available from: http://www.coqa-inc.org/docs/default-source/meeting-presentations/20100211_Price_Challenges.pdf.
[46] Anonymous. Crude incompability problems at heavy crude unit desalter. Hydrocarb Asia 2007;17(4):68–70.
[47] Kremer LN. Handling desalter and corrosion issues for opportunity crude oils. In: Proc. Opportunity Crudes Conf.: Challenges, Benefits, Houston, TX, April 26–May 2. Hydrocarbon Publishing Co.; 2008. Presentation 22 pp.
[48] Garcia-Swofford SN, Lordo SA. The benefits of optimizing your desalter operation. In: Proc. AIChE Spring Nat. Meeting, Houston, TX; 2007. 1 p.
[49] Baker Hughes. Planning ahead for effective Canadian crude processing. Report No. Unnumbered, White Paper. June 28, 2010. 4 pp.
[50] Lordo S, Garcia JM, Garcia-Swofford S. Desalter acidification additives and their potential impacts on crude units. In: Proc. Corrosion, New Orleans, LO; 2008. p. 085561–0855612.
[51] Choi DW. How to buy and operate desalters. Hydrocarb Proc 2005;84(3):75–9.
[52] Lindemuth PM, Lessard RB, Lozynski M, Kremer M. Improve desalter operations. Hydrocarb Process 2001;80(9):67–70.
[53] Stark JL, Nguyen J, Kremer LN. New method prevents desalter upsets from blending incompatible crudes. Oil Gas J 2002;100(11):89–91.
[54] Green K. Improve processing of 'opportunity crudes'. Hydrocarb Proc 2004;83(6 Section 1):53–5.

[55] Turini K, Turner J, Chu A, Vaidyanathan S. Processing heavy crudes in existing refineries. In: Proc. AIChE Spring Meeting, 7th Global Congress on Proc. Safety, Chicago, IL, March 13−17; 2011. 31 pp.
[56] Shafizadeh A, Mcateer G, Sigmon J. High acid crudes. In: Proc. Crude oil quality association meeting. New Orleans. ChevronTexaco; January 30, 2003. 50 pp. Available from: http://www.coqg.org/20030130High%20Acid%20Crudes.pdf.
[57] Golden S. Canadian crude processing challenges. PTQ 2008;13(1):53−61.
[58] Process Notes. Oil sands crude − profits and problems?. PTQ; 2012. 1 p.
[59] Barletta T, Kurzym KJ. Consider retrofits to handle high-viscosity crudes. Hydrocarb Proc 2004;83(9). 11 pp.
[60] Nalco and Exxon Energy Chemicals. Inhibition of naphthenic acid corrosion and other related issues (Unpublished Presentation at Refining High Acid Crudes II). 2002. 72 pp.
[61] Deshannavar UB, Rafeen MS, Ramasamy M, Subbarao D. Crude oil fouling: a review. J Appl Sci 2010;10:3167−74.
[62] Brierley G, Gembicki V, Cowan T. Changing refinery configuration for heavy and synthetic crude processing. In: Proc. NPRA Annual Meeting, Salt Lake City, UT, March 19−21; 2006. Paper AM-06-16, 23 pp.
[63] Yaluris G, Cheng WC, Peters M, Hunt LJ, Boock LT. The effects of Fe poisoning on FCC catalysts. In: Proc. NPRA Annual Meeting, New Orleans, LA, vol. 6; 2001. 26 pp.
[64] Yaluris G. The effects of fe poisoning on FCC catalysts: an update. Catalagram 2002;91:8.
[65] De Graaf B, Radcliffe C, Evans M, Diddams P. Processing shale oil in an FCC unit: catalyst and profit optimisation. Digit Refin August 2015;6. Available from: www.digitalrefining.com/article/1001071.
[66] Myers GA, (Atlantic Richfield Co.). Guard bed system for removing contaminant from synthetic oil. Patent No. US3876533. April 8, 1975.
[67] Paraskos JA, Yanik SJ, (Gulf Research and Development Co.). Hydrodesulfurization process employing a guard chamber. Patent No. US4053391. October 10, 1977.
[68] Nelson WL. Status of residual desulfurization. Oil Gas J 1972;70(31):98−100.
[69] Hauser A, Marafi A, Almutairi A, Stanislaus A. Comparative study of hydrodemetallization (HDM) catalyst aging by Boscan feed and Kuwait atmospheric residue. Energy Fuels 2008;22(5):2925−32.
[70] Haynes D. Naphthenic acid bearing refinery feedstocks and corrosion abatement. In: Proc. AIChE − Symp. Chicago, IL. Nalco Co. Proprietary Knowledge; October 2006. 30 pp.
[71] O'Kane JM, Rudd TF, Cooke D, Dean FWH, Powell SW. Detection and monitoring of naphthenic acid corrosion in a Visbreaker unit using hydrogen flux measurements. In: Proc. Corrosion, San Antonio, TX, USA, March 14−18. NACE International; 2010. Paper No. 10351. 15 pp.
[72] Bridjanian H, Khadem Samimi A. Bottom of the barrel, an important challenge of the petroleum refining industry. Pet Coal 2011;53(1):13−21.
[73] Varadaraj R, Brons C. Molecular origins of crude oil interfacial activity part 3: characterization of the complex fluid rag layer formed at crude oil − water interfaces. Energy Fuels 2007;21(3):1617−21.
[74] Varadaraj R, Brons C. Molecular origins of heavy crude oil interfacial activity part 2: fundamental interfacial properties of model naphthenic acids and naphthenic acids separated from heavy crude oils. Energy Fuels 2007;21(1):199−204.
[75] Callaghan IC, Mckechnie AL, Ray JE, Wainwright JC. Investigation of the foaming properties of crude oil. Part 1. Identification of the crude oil components responsible for foaming. Soc Pet Eng AIME 1983:5. SPE-12342
[76] Burn J. Global refining capacity. Oil Drum March 30, 2011:13. Available from: http://www.theoildrum.com/node/7641.
[77] Haensel V. New petroleum refining technology. In: St-Pierre LE, Brown GR, editors. Future sources of org. raw mat.: CHEMRAWN I. Pergamon; 1980. p. 435−48.
[78] Schulman BL, Biasca FE, Johnson HE, Dickenson R. New routes—processes and uses—for minimizing heavy fuel oil production. In: Proc. NPRA Anual Meeting, San Antonio, TX, March 22−24; 1994. Paper No. AM-94-27, 11 pp.

[79] Phillips G, Liu F. Advances in residue upgrading technologies offer refiners cost effective options for zero fuel production. In: Proc. European Refining Technol. Conf. Paris, France; 2002. 31 pp.

[80] Foster Wheeler. Maximising the benefits of residue upgrading – the challenges (Unpublished Presentation). 2009. 59 pp.

[81] Weber C, Yeung S. Opportunity crudes: to process or not to process?. In: Proc. Crude Oil Quality Association Meeting. Tulsa, OK, USA, October 26–27; 2011. 25 pp.

[82] Runyan J. Integration of hydroconversion upgrader into refinery for bottoms and crude oil quality upgrading - case study. In: Proc. World Heavy Oil Congress, Edmonton, Alberta, Canada, March 10–12, vol. 1; 2008. p. 255–61.

[83] Foster Wheeler. Maximising the benefits of residue upgrading – how do I choose what's best for me? (Unpublished Presentation). 2009. 13 pp.

[84] Brown W, Pinchuk R, Monaghan G. Benefits from coking of heavy feedstocks. PTQ 2005;10(4):85–91.

[85] Zuideveld PL, Chen Q, Van Den Bosch PJWM. Integration of gasification with thermal residue conversion in refineries. In: Proc. Gasification Technologies Conf., San Francisco, California, October 8–11; 2000. 15 pp.

[86] Phillips G, Liu F. Invest in the future. Hydrocarb Eng 2003;8(9):63–8.

[87] Zuideveld P, Wolff J. New methods upgrade refinery residuals into lighter products, special report on clean fuels. Hydrocarb Proc 2006;85(2):73–4. 77–79.

[88] Sieli G, Gupta N. A winning combination. Hydrocarb Eng September 2009:42–6.

[89] Runyan J. Is bottomless-barrel refining possible? Future energy demand will involve innovative methods that can upgrade residual products. Hydrocarb Proc September, 2007:81–90.

[90] Worldwide refining survey: worldwide refineries-capacities. Oil Gas J 2004;12:2.

[91] Worldwide refining survey. Oil Gas J 2009;12:67.

[92] Worldwide refining survey: worldwide refineries-capacities. Oil Gas J 2014;12:3.

[93] Technip oil refining. March 1, 2014. 16 pp. Available from: http://www.technip.com/en/publications/brochures/oil-refining.

[94] Robinson PR. Petroleum processing overview. In: Practical advances in petroleum processing; 2006. p. 1–78.

[95] Chem System. Refinery of the future as shaped by environmental regulations. Report No. PERP 02/03S11. White Plains (NY): Nexant; December 2003. 129 pp.

[96] Motaghi M, Shree K, Krishnamurthy S. Consider new methods for bottom of the barrel processing—part 1. Advanced methods use molecule management to upgrade heavy ends. Hydrocarb Proc 2010;89(2):35–40.

[97] International Energy Agency. IEA/KBC monthly global indicator refining margins. In: Oil market report, monthly report; 2016. Available from: https://www.iea.org/media/omrreports/MHM.xls.

[98] Fitzgibbon T, Kloskowska A, Martin A. Impact of low crude prices on refining. Energy Insights February 2015.

[99] Goldhammer B, Weber C, Christensen P, Yeung S, Garrett T, Yeung T. Future of opportunity crudes processing. PTQ 2011;16(4):7.

[100] Meijknecht J, Correljé AF, Van Holk B. A Cinderella story?: Restructuring of the European refining sector. Report No. 291264. The Hague (Netherlands): Clingendael International Energy Programme; April 2012.

[101] Processing News. Refiner demand for discounted feeds will increase trade of high-acid crudes. Oil Gas J 2008;106(11):52–5.

[102] Bieber S. Operational issues processing western Canadian crude oil. In: Proc. Crude Oil Quality Association Meeting, Philadelphia, PA, USA, June 18; 2009 (Unpublished).

[103] Brierley G. Impacts of processing synthetic crude blends. In: Proc. UOP Western U.S. Conf., July 20; 2005 (Unpublished Presentation).

[104] Plumail JC. Adapting refineries to enhance conversion of residua into distillates. In: Proc. Refining Round Table, Moscow, Russia, September 28; 2004. 36 pp.

[105] Niccum PK, Northup AH. Processing heavy ends: part I. PTQ 2008;13(4):75–86.
[106] Niccum PK, Northup AH. Processing heavy ends: part II. PTQ 2008;13(4):81–5.
[107] Garg MO, Kumar S, Nanoti SM, Sharma YK. Refinery configurations: designs for heavy oil. Hydrocarb Process 2011;90(10).
[108] Lewe T, Kaushal V, Alberich J, Gaspar D, Forrest R, Walters N, et al. Refining 2021: who will be in the game?. 2012. Available from: http://www.atkearney.com/paper/-/asset_publisher/dVxv4Hz2h8bS/content/refining-2021-who-will-be-in-the-game-/10192.
[109] Veasna B, Kojima M. Refining crude oil – part 2. Pet Sect Brief Note 2008;11:4.
[110] Plotkin JS. Exploiting unconventional heavy oils. PTQ 2007;12(1):65–7.
[111] Laureshen CJ, Clark PD, Du Plessis MP. Adding value to Alberta's oil sands. J Can Pet Technol 2006;45(8):54–9.
[112] Netzer D. Alberta bitumen processing integration study. Report No. Final. The Province of Alberta, Economic Development Department. Investment and Industrial Development Branch. Department of Energy. The Alberta Energy Research Institute; March 2006. 164 pp.
[113] Harrison S. Driving refining change. Part 1. PTQ Digit Refin 2014:8 (Q3).
[114] Harrison S. Driving refining change: Part 2. PTQ Digit Refin 2014:3 (Q3).
[115] Wu XA. Selective creaming: a novel technique to isolate interfacial organics-and mineral solids from water-in-oil emulsions. Energy Fuels 2008;22(4):2346–52.
[116] Feng X, Wang S. Understanding asphaltene dispersants for paraffinic solvent-based bitumen froth treatment. Energy Fuels 2011;25(6):2601–7.
[117] Rao F, Liu Q. Froth treatment in Athabasca oil sands bitumen recovery process: a review. Energy Fuels 2013;27(12):7199–207.
[118] Nhieu P, Liu Q, Gray MR. Role of water and fine solids in onset of coke formation during bitumen cracking. Fuel 2016;166:152–6.
[119] Froth treatment explained. Oil Sands Magazine; July 20, 2015. 7 pp. Available from: http://www.oilsandsmagazine.com/bitumen-froth-treatment-process-overview-secondary-extraction-oilsands-bitumen-production/.
[120] Parffinic froth treatment. Oil Sands Magazine; July 20, 2015. 13 pp. Available from: http://www.oilsandsmagazine.com/oilsands-mining-solvent-paraffinic-froth-treatment-pft/.
[121] Why Alberta doesn't need any new upgraders. Oil Sands Magazine; November 5, 2015. 6 pp. Available from: http://www.oilsandsmagazine.com/why-upgrading-bitumen-in-alberta-no-longer-makes-sense/.

Abbreviations and Acronyms

Acid, base, and neutral	ABN
Activation energy	E_a
Aquaconversion	AQC
Atmospheric equivalent boiling point	AEBP
Atmospheric pressure chemical ionization	APCI
Atmospheric pressure photoionization	APPI
Atmospheric residue	AR
Antisolvent	AS
Barrel	bbl
Boiling point	BP
Barrel per day	bpd
Canadian Crude Quality Technical Association	CCQTA
Crude Oil Quality Group	COQG
Cyclic steam injection	CSI
Conradson carbon residue	CCR
Deasphalting	DA
Decanted oil (slurry oil)	DCO
Demetallization	DM
Desulfurization	DS
Equation of state	EOS
Electron paramagnetic resonance	EPR
Electrospray ionization	ESI
Extended X-ray absorption fine structure	EXAFS
Fluidized catalytic cracking	FCC
Fourier transform ion cyclotron resonance	FT-ICR
Fragmentation	FRAG
Gas chromatography	GC
Gel permeation chromatography	GPC
Heavy cycle oil	HCO
Heavy to light process	HTL
Heavy vacuum gas oil	HVGO
High-performance liquid chromatography	HPLC
Hydrocarbon	HC
Hydrocarbon enhancement electron-beam technology	HEET
Hydroconversion	HDC
Hydrocracking	HCK
Hydrodeasphalting	HDAf
Hydrodemetallization	HDM
Hydrodenitrogenation	HDN
Hydrodesulfurization	HDS
Hydroprocessing	HDP
Hydrotreatment	HDT

In-situ combustion	ISC
In-situ catalytic upgrading	ISCU
In-situ pyrolysis	ISP
In-situ upgrading	ISU
Light cycle oil	LCO
Mass spectrometry	MS
Microbial-induced corrosion	MIC
Microcarbon residue	MCR
Molecular weight	MW
Naphthenic acid	NA
Oil sands process-affected water	OSPW
Polyaromatic hydrocarbon	PAH
Polycyclic aromatic	PAC
Polynuclear aromatic	PNA
Radiation-induced processing	RP
Radiation-induced thermal cracking	RTC
Rapid thermal pyrolysis process	RTP
Refining processing costs	RPC
Refractive index	RI
Residue catalytic cracking	RCC
Room temperature	RT
Ruthenium-ion catalyzed oxidation	RICO
Saturates, aromatics, resins, and asphaltenes	SARA
Scanning electron microscopy	SEM
Sequential elution fractionation	SEF
Specific gravity	SG
Silica/alumina ratio	SAR
Size exclusion chromatography	SEC
Small-angle neutron scattering	SANS
Small-angle X-ray scattering	SAXS
Solid-phase extraction	SPE
Solubility parameter	SP
Solvent deasphalting	SDA
Steam-assisted gravity drainage	SAGD
Straight run gas oil	SRGO
Synthetic crude oil	SCO
Tailings pond water	TPW
Thermal cracking	TC
Time-of-flight	TOF
Time-on-stream	tos
Total acid number	TAN
Vacuum gas oil	VGO
Vacuum residue	VR
Visbreaking	VB
X-ray absorption near edge spectroscopy	XANES
X-ray photoelectron spectroscopy	XPS

Index

'*Note*: Page numbers followed by "f" indicate figures and "t" indicate tables'.

D

E

N

O

T

W

X

Y

Printed in the United States
By Bookmasters